D0845357

Quality Control and Industrial Statistics

Quality Control and Industrial Statistics

Acheson J. Duncan, Ph.D.
Professor Emeritus of Statistics
Department of Mathematical Sciences
The Johns Hopkins University

1986 Fifth Edition

Homewood, Illinois 60430

ISBN 0-256-03535-0

Library of Congress Catalog Card No. 85–81998

Printed in the United States of America

1 2 3 4 5 6 7 8 9 0 K 3 2 1 0 9 8 7 6

To
my wife Helen

PREFACE

This book discusses statistical principles and procedures pertinent to the control of the quality of output of a given production process and to research in general on quality problems.

The revision undertaken in this fifth edition has had three major goals. One goal has been to bring the book up to date with respect to sampling standards. The discussion of standards for acceptance sampling by attributes still centers on Mil. Std. 105D, but note is made of ANSI/ASQC Standard Z1.4–1981 and how this differs from Mil. Std. 105D. The revision that is currently (1986) underway of ISO Standard 2859–1974, which is essentially the same as Mil. Std. 105D, is also discussed. Similarly, Mil. Std. 414 is retained as the center of the discussion of standards for acceptance sampling by variables for percent nonconforming, but note is made of ANSI/ASQC Standard Z1.9–1980 and ISO Standard 3951–1981 and how these newer variables standards match the attribute standards better. The use of acceptance charts by the ISO standard is also noted. The latest department of defense standard for continuous sampling, designated as Mil. Std. 1235B, is discussed.

A second major goal of the current revision has been to make the text more cost oriented. A section has been added that explains the use of prior information about variations in the process average in conjunction with knowledge of the unit cost of inspection and the cost of passing a nonconforming item to derive Bayesian minimum cost sampling plans. This section also presents the Deming argument that with prior knowledge as to process variations in quality and availability of information on costs that permit the computation of a lot break-even quality level, the use of a standard acceptance sampling plan, such as a Mil. Std. 105D plan, may be an economically nonoptimum procedure and that acceptance without inspection or 100 percent inspection would be better. Note is made of Joyce Orsini's rules for inspection and correction of product in a "state of chaos." In the discussion of control charts examples of average run length curves are presented for the various types of charts.

A third major goal has been to bring the book more in harmony with the growing employment of computers in statistical quality control. A new section on exploratory data analysis introduces procedures that are not only colorful and meaningful, but also readily lend themselves to the use of a computer. The chapter on cusum charts has been rewritten to improve the presentation of the subject matter, but it has also been enlarged to cover new material. Again cusum procedures readily lend themselves to the use of a computer. An Appendix VI has been added to the book in which are listed computer programs selected from those that have been reviewed over the years in the *Journal of Quality Technology.*

Other additions of material have been (1) an enlargement of the historical section to cover early developments in England more fully, (2) a description of the 1984 addition to British Std. 6001 (the British counterpart of U.S. Mil. Std. 105D) of a supplement on sampling plans indexed by nominal limiting quality that in 1985 became the basis for a new Part II of ISO-2859 and the presentation in this text of a table of sampling plans indexed by nominal limiting quality levels, the plans being compatible with those of Mil. Std. 105D with respect to lot-size ranges and sample sizes, (3) an account of Schilling's lot-sensitive sampling plans for high quality product, (4) an introduction to Ellis R. Ott's analysis of means, (5) an account of G. J. Hahn's prediction limits, and (6) a brief discussion of Hoadley's quality measurement plan (QMP). An example of a QMP box-plot printout is given.

Chapters 25–36 of the text contain fairly standard material and revision of this part of the book has taken the form of updating the references. Those that have been added include papers and other sources that pertain directly to the material covered in the text, but in some cases the added references pertain to related material outside the immediate scope of the text that the author believes the reader might wish to pursue.

No attempt has been made to discuss "Taguchi Methods," about which there has recently been much talk, since they are beyond the scope of the text. Those who wish to investigate these methods could well start with the paper on "Quality Engineering by Design—The Taguchi Approach" by Thomas B. Barker of the Xerox Corporation that appears in the 39th [1985] *Annual Quality Congress Transactions,* pp. 675–88, and continue with the paper by Haghu N. Kackar on "Off-Line Quality Control, Parameter Design, and the Taguchi Method" that was published with discussion in Vol. 17 (1985) of the *Journal of Quality Technology,* pp. 176–209.

Also, no attempt has been made to present a general discussion of "robust" methods of estimation, since again such a discussion is judged to be beyond the scope of the text. For this the reader is referred to the book by Peter J. Huber on *Robust Statistics.*

With reference to acceptance sampling of lots, the text refrains from using the term "compliance sampling" since it is believed that it can lead to confusion and misunderstanding. The quality of a unit of product can properly be said to be in compliance or not in compliance with given product specifications. In acceptance sampling, lots are simply accepted or rejected. Even if a sampling plan has a zero acceptance number, it should not be stated that an accepted lot is in compliance with product specifications. Only if every item in a lot has been tested and found to comply with product specifications, can it be said that the lot as a whole is in compliance. When product safety and other considerations of that kind are of primary interest, attention will focus on the limiting quality level of a sampling plan. Hence, in this text instead of talking about "compliance sampling," a section is devoted to tables of sampling plans that are indexed by limiting quality values, as was noted above.

The fifth edition follows the current trend in using the term "nonconformity" in place of the term "defect" and the term "nonconforming" in place of the term "defective," wherever it seems appropriate. The symbols used in the fourth edition have continued to be used in the fifth. In this connection it is to be especially noted that for the most part a prime is still used to denote a universe value.

The order of presentation of subject matter in the fourth edition has been generally preserved in the fifth. A significant amount of new material has been added to Chapter 14 and the title of the chapter has been changed. Other new material has been inserted at various points as seemed appropriate.

As noted above, this text is a presentation of statistical principles and procedures related to statistical quality control. In the "new quality philosophy" that is currently evolving in the United States, much attention is being given to principles of management. Except for the discussion of minimum cost statistical procedures, this book has little to say about management problems per se. For the discussion of such problems and their solution the reader is referred to W. Edwards Deming's 1986 book *Out of the Crisis* and to special issues of *The Juran Report* which consist of proceedings of annual conferences on quality improvement sponsored by the Juran Institute, Inc., 88 Danbury Road, Wilton, Conn. 06897. For broader aspects of the application of statistics to quality problems, the reader would do well to keep in touch with the current activities of the recently formed American Statistical Association Committee on Quality and Productivity.

In preparation of this fifth edition I am greatly indebted to W. Edwards Deming and his former student, Joyce Orsini, for provision of a copy of her doctor's thesis. I am also indebted to Dr. Deming for a copy of the page proof of chapter 15 of *Out of the Crisis* and for his review of a preliminary draft of the part of this text that relates to his work. In addition, I am much indebted to Richard M. Brugger of the U.S. Army Armament, Munitions, and Chemical Command for provision of a copy of Mil. Std. 1235B on continuous sampling and for his review of my write-up of the standard. Of significance has been the review of the initial draft of the fifth edition arranged for by the publisher. I found this very helpful. I was glad to have the reviewer's corrections of errors noted in the fourth edition as well as a similar set of corrections provided the publisher by Glenn F. Lindsay of the Naval Postgraduate School. In all cases, however, responsibility for what appears in this text is solely mine.

Churchill Eisenhart of the National Bureau of Standards kindly provided material related to the early history of statistical quality control in Britain. H. A. Croasdale of the British Standards Institution kindly sent me a copy of Supplement 1 to British Std. 6001 and a copy of the newly adopted Part II of ISO Std. 2859. Conversations with August B. Mundel have been helpful with respect to current activities of ISO TC-69 and I am indebted to Kathy Al-Meer of the ASQC staff for keeping me posted regarding last minute developments relating to the issuance of new or revised ASQC standards.

I wish to thank the various publishers and authors who have kindly given permission to use copyrighted material. I also wish to thank Kay Lutz of the departmental secretarial staff for all the typing and other work she did for me.

Acheson J. Duncan

CONTENTS

Part 5. Some Statistics Useful in Industrial Research

1

Introduction

1. HISTORICAL REVIEW

Quality control is as old as industry itself. From the time man began to manufacture, there has been an interest in quality of output. As far back as the Middle Ages the medieval guilds insisted on a long period of training for apprentices and required that those seeking to become mastercraftsmen offer evidence of their ability. Such rules were, in part, aimed at the maintenance of quality. In more modern times, factory inspection and research, pure food and drug acts, activities of professional societies, all have sought for years to assure the quality of output. Quality control has thus had a long history.

On the other hand, statistical quality control is new. The science of statistics itself goes back only two to three centuries, and its greatest development has been in the last 80 years. Early applications were made in astronomy and physics and in the biological and social sciences, but it was not until the 1920s that statistical theory began to be applied effectively to quality control. A factor in the birth of statistical quality control in the 20s was the development, in immediately preceding years, of an exact theory of sampling.

The first to apply the new statistical methods to the problem of quality control was Walter A. Shewhart of the Bell Telephone Laboratories. In a memorandum prepared on May 16, 1924, Shewhart made the first sketch of a modern "control chart."[1] The new technique was subsequently developed in various memoranda and articles;[2] and in 1931 Shewhart published a book

---

[1] *Industrial Quality Control,* July 1947, p. 23.

[2] See, e.g., W. A. Shewhart, "Finding Causes of Quality Variations," *Manufacturing Industries,* February 1926, pp. 126–28; "Quality Control Charts," *Bell System Technical Journal,* October 1926, pp. 593–603; and "Quality Control," ibid., October 1927, pp. 722–35. Also note W. L. Robertson, "Quality Control by Sampling," *Factory and Industrial Management,* September 1928, p. 503; and ibid., October 1928, p. 725.

on statistical quality control which bore the title *Economic Control of Quality of Manufactured Product.* [3] This book set the pattern for subsequent applications of statistical methods to process control. Two other Bell System men, H. F. Dodge and H. G. Romig, took the leadership in developing the application of statistical theory to sampling inspection,[4] the culmination of their work being the now well-known Dodge-Romig *Sampling Inspection Tables.* The work of Shewhart, Dodge, and Romig constitutes much of what today comprises the theory of statistical quality and control.

In the early 30s, these Bell System men, in collaboration with the American Society for Testing and Materials, the American Standards Association, and the American Society for Mechanical Engineers, undertook to popularize the newer statistical methods in the United States. Shewhart also visited London, where he met with prominent British statisticians and engineers.[5]

In the United States, despite the publicity given to the newer methods, the rate of adoption was at first slow. Professor H. A. Freeman, who was promoting statistical quality control at Massachusetts Institute of Technology, ascribed this sluggish response of the early years to (1) "a deep-seated conviction of American production engineers that their principal function is to so improve technical methods that no important quality variations remain, and that in any case, the laws of chance have no proper place among 'scientific' production methods," and (2) "the difficulty of obtaining industrial statisticians who are adequately trained in this fairly complicated field."[6] In 1937, probably not more than a dozen single enterprises in American mass-production industries had introduced the new technique into their ordinary operations.[7]

This initial coolness of American industry toward statistical quality control was rapidly overcome during World War II. The outbreak of the conflict in 1939 set the United States to thinking of national defense. This meant enlargement of military personnel and matériel. The armed services began to enter the market as large consumers of American output and, as such, had an increasing influence of quality standards.

The influence of the military on the adoption of statistical quality control was of two kinds. First, the armed services themselves adopted scientifically designed sampling inspection procedures. The initial step in this development

---

[3] New York: D. Van Nostrand Co., Inc., 1931.

[4] See, e.g., H. F. Dodge, "Using Inspection Data to Control Quality," *Manufacturing Industries,* 16 (1928), pp. 517–19, 613–15; and H. F. Dodge and H. G. Romig, "A Method of Sampling Inspection," *Bell System Technical Journal,* October 1929, pp. 613–31; H. F. Dodge, "Statistical Aspects of Sampling Inspection Plans," *Mechanical Engineering,* October 1935, pp. 645–46; H. F. Dodge and H. G. Romig, "Single Sampling and Double Sampling Inspection Tables," *Bell System Technical Journal,* January 1941, pp. 1–61.

[5] See p. 5 below.

[6] H. A. Freeman, "Statistical Methods for Quality Control," *Mechanical Engineering,* April 1937, p. 261.

[7] Ibid.

of military sampling inspection procedures was undertaken shortly after America's entry into the war when at the invitation of the government[8] a group of distinguished engineers from the Bell Telephone Laboratories were brought to Washington to develop a sampling inspection program for Army Ordnance.[9] The Army Ordnance and Army Service Forces sampling inspection tables that came out in 1942 and 1943 were largely the work of these men. The same men also undertook an extensive training program to acquaint government personnel in the use of the new procedures and tables.

The second line of influence of the military was the establishment of a widespread educational program for industrial and other personnel.[10] As early as December 1940, the American Standards Association initiated at the request of the War Department a project that eventuated in the *American War Standards Z1.1–1941* and *Z1.2–1941,* "Guide for Quality Control and Control Chart Method of Analyzing Data," and *American War Standard Z1.3–1942,* "Control Chart Method of Controlling Quality During Production."[11] These gave concise statements of American control chart practice and served as text material for subsequent training courses. In July 1942, an intensive 10-day course in statistical quality control was given at Stanford University to representatives of war industries and procurement agencies of the armed services.[12] Later the course was shortened to eight days and given at Los Angeles. These courses were made possible by the financial backing of the Engineering, Science, and Management War Training Program of the United States Office of Education. The success of this early educational program led the Office of Production Research and Development of the War Production Board to establish similar courses throughout the country.[13] Ac-

[8] The person responsible for this invitation was General Leslie E. Simon who with the advice and guidance of Shewhart had started work on statistical quality control at the Picatinny Arsenal as early as 1934. His book, *An Engineers' Manual of Statistical Methods* (New York: John Wiley & Sons, 1941), came out in 1941. (See Littauer [P '50] and Simon and Cohen [P '71]. Professor S. B. Littauer was an instructor in the wartime training program in quality control and Professor A. C. Cohen served at the Picatinny Arsenal from 1941 to 1944.)

[9] The group consisted principally of George W. Edwards, director of Quality Assurance of the Bell Labs; H. F. Dodge of the Bell Labs; and G. R. Gause, then of Army Ordnance. H. G. Romig of the Bell Labs assisted.

[10] The material of this paragraph is based on a letter from H. F. Dodge and on Holbrook Working and Edwin G. Olds, *Organizations Concerned with Statistical Quality Control* (OPRD, Quality Control Reports, No. 2).

[11] Members of the ASA Emergency Technical Committee Z1 on Quality Control were A. G. Ashcroft; W. E. Deming; H. F. Dodge, chairman; J. Goullard, secretary; L. E. Simon; and R. E. Wareham. H. F. Dodge of the Bell Telephone Laboratories did most of the organizing and writing of the technical material with the heavy active cooperation of W. Edwards Deming of the U.S. Census Bureau. These standards are currently (1985) published as ANSI Z1.1–1985, ANSI Z1.2–1985, and ANSI Z1.3–1985 bound as one document.

[12] These courses were planned by Professors Eugene L. Grant and Holbrook Working of Stanford University and Dr. W. Edwards Deming.

[13] Professor Holbrook Working was borrowed from Stanford University to develop this program. The work was subsequently made the principal activity of a special branch of the OPRD, and in March 1945, its headquarters was transferred to the Carnegie Institute of Technology.

cordingly, from 1943 to 1945, 810 organizations from 35 different states sent representatives to attend one or more of the 33 intensive courses on statistical quality control that were offered by the OPRD. Among these 810 organizations were 43 different educational institutions, most of whose representatives were faculty members who attended to prepare themselves to give instruction in quality control.

Mention should be made at this point of the wartime research in statistical quality control and especially of the Statistical Research Group, Columbia University, Applied Mathematics Panel, Office of Scientific Research and Development. This research group was organized in July 1942, and continued in existence until September 1945. Members of the staff were drawn from universities and research organizations throughout the country.[14] It advised and assisted the Army, the Navy, and the Office of Scientific Research and Development on statistical aspects of problems arising in their activities. It investigated certain problems of a predominantly statistical or probabilistic character and developed, reviewed, and expounded statistical techniques. Included in its contributions to statistical quality control was the preparation of the Navy manual on sampling inspection by attributes, which was later expanded and its underlying principles explained in *Sampling Inspection,* published by McGraw-Hill Book Company (1948). Another contribution to industrial statistics was its volume on *Selected Techniques of Statistical Analysis for Scientific and Industrial Research and Production and Management Engineering,* published by McGraw-Hill (1947). One of the most important contributions to statistical quality control was the development by Professor A. Wald of sequential sampling. Indeed, so important was sequential analysis deemed by the U.S. government that it withheld publication of Wald's original paper until June 1945.

The training courses and research programs sponsored by the federal government produced in various industrial centers nuclei of interested and trained personnel. Subsequent to the completion of introductory courses, quality control societies were formed in various localities, the meetings of which offered continuing opportunities for the exchange of ideas and the education of new members. In Buffalo, the Society of Quality Control Engineers in cooperation with the University of Buffalo began in July 1944 the

---

[14] This group was organized by Professor Harold Hotelling of Columbia University under the auspices of Warren Weaver, chief of the Applied Mathematics Panel. Members of the senior staff included Kenneth J. Arnold, University of Wisconsin; Rolling F. Bennet, Columbia University; Julian H. Bigelow, Institute for Advanced Study; Albert H. Bowker, Stanford University; Churchill Eisenhart, National Bureau of Standards (on leave from University of Wisconsin); H. A. Freeman, Massachusetts Institute of Technology; Milton Friedman, University of Chicago; M. A. Girschick, U.S. Bureau of the Census; Millard W. Hastay, National Bureau of Economic Research; Harold Hotelling, Columbia University; Edward Paulson, University of North Carolina; L. J. Savage, University of Chicago; Herbert Solomon, Stanford University; George J. Stigler, Brown University; A. Wald, Columbia University; W. Allen Wallis, University of Chicago; and J. Wolfowitz, Columbia University. See Statistical Research Group, Columbia University, *Techniques of Statistical Analysis* (New York: McGraw-Hill, 1947), p. ix.

publication of *Industrial Quality Control* under the editorship of Martin Brumbaugh. The new publication quickly attained a national circulation. All of this led to a growing use of statistical quality control in the later years of the war. Shortly after the war, sufficient interest developed to form a national organization known as the American Society for Quality Control. George D. Edwards, director of Quality Assurance of the Bell Telephone Laboratories, was the first president of the new organization, and Walter A. Shewhart was made its first honorary member. Born on February 16, 1946, the American Society for Quality Control had by 1985, 198 local sections and 13 divisions[15] with a membership in excess of 39,000. It took over the publication of *Industrial Quality Control* and became the leading force in promoting the use of statistical quality control on the American continent. A section was also set up in Japan.

In Britain the development of statistical quality control paralleled and became entwined with that in the United States. Statistical analyses relating to the variation in quality of output were undertaken as early as the 1920s by Bernard Dudding of the General Electric Company's research laboratories at Wembley. The need to study small-sample variations in means against a probability background was emphasized in his 1929 company paper. Since reference is made in this paper to articles by Shewhart and others in *The Bell System Technical Journal,*[16] it is not clear whether the application of sampling theory as a control for mean values occurred to Dudding before he came across these earlier papers. In 1925 L. H. C. Tippett, a statistician for the British cotton industry, published his paper on the distribution of sample ranges from a normal universe which was made use of by Shewhart in his book on the *Economic Control of Quality of Manufactured Product.*

In 1931 Egon S. Pearson of University College, London, visited the United States where he spent some days with Shewhart. As a result, Shewhart was invited to come to England in May 1932 to give three lectures in University College on "The role of statistical method in industrial standardization." Following this, things moved rapidly in England. Pearson gave a paper on "A survey of the uses of statistical method in the control and standardization of the quality of manufactured products" at the December 1932 meeting of the Royal Statistical Society. Shortly afterward the statistical society set up an industrial and agricultural research section and authorized a supplement to the regular journal for publication of papers pertaining to this new field of application. It is interesting to note that the provisional committee to arrange for this new section had representatives from Imperial Chemical

---

[15] The 13 ASQC divisions are the following: Administrative Applications, Aerospace and Defense, Automotive, Biomedical, Chemical and Process Industries, Electronics, Energy, Food, Drug and Cosmetics, Human Resources, Inspection, Reliability, Statistics, and Textile and Needle Trades.

[16] See, for example, W. A. Shewhart, "Quality Control Charts," *Bell System Technical Journal* 5 (1926), pp. 593–606; and "Quality Control," *Bell System Technical Journal* 6 (1927), pp. 722–35.

Industries, the London School of Hygiene, the Boot Trade Research Association, the Department of Applied Statistics, University of College, London, and the School of Agriculture, Cambridge;[17] and at its first meeting the opening paper was read by R. H. Pickard, director of the British Cotton Industry Research Association. Following Shewhart's lectures the British Standards Institution set up a small committee on Statistical Methods in Standardization and Specification, including Egon Pearson, Dudding, and representatives from various British industries. An outcome was BS 600–1935 *The Application of Statistical Methods to Industrial Standardization and Quality Control* which bore Egon Pearson's name.[18] Another "guide" subsequently issued was BS 2564 (1955) *Control chart technique when manufacturing to a specification* by B. P. Dudding and W. J. Jennett (originally prepared for the General Electric Company in 1944). An excellent review of these early beginnings of statistical quality control in Britain is contained in a paper by E. G. Pearson in *The Statistician* 22 (1973), pp. 165–78.

The response of British industry to the new statistical methods was quick and widespread. By 1937, they were being applied to such products as coal, coke, cotton yarns, cotton textiles, woolen textiles, spectacle glass, lamps, building materials, and manufactured chemicals.[19] Application was further stimulated by World War II, and in 1952, the Royal Statistical Society added the journal *Applied Statistics* to its series of publications.

From the United States and Great Britain quality control techniques have spread to other countries. Under the guidance of Dr. W. Edwards Deming, statistical quality control in Japan has developed into one of the finest systems in the world.[20] In Europe there was formed the European Organization for Quality Control. Technical Committee 69 on *Applications of Statistical Methods* of the International Organization for Standardization (ISO)[21] has undertaken to issue standards on sampling plans and control charts and today almost all industrialized nations use statistical methods to control quality.

After control charts and sampling plans, other techniques, such as correlation, analysis of variance, and the design of experiments, found their way into the industrial laboratories and research departments. The use of these tools was urged as early as the better known control charts and sampling plans.[22] Since the first two were already established techniques, their initiation

---

[17] *Supplement to the Journal of the Royal Statistical Society* 1, no. 1 (1934), p. 2.

[18] Unfortunately, this is out of print, since the plates were destroyed during World War II in the bombing of London.

[19] H. A. Freeman "Statistical Methods for Quality Control," *Mechanical Engineering,* April 1937, p. 261.

[20] This development is reviewed in a paper by J. M. Juran in *Reports of Statistical Application Research* 22 (1975), pp. 66–72, issued by the Union of Japanese Scientists and Enginners.

[21] The American National Standards Institute (ANSI) is the U.S. member body of ISO. ASQC serves as the secretariat for the Technical Advisory Group (TAG) to ANSI with respect to the work of ISO/TC69.

[22] For example, see A. E. R. Westman, "Statistical Methods in Ceramic Research," *Journal of the American Ceramics Society,* March 1927, p. 133.

into industrial usage was not accompanied by the fanfare and publicity that went with the introduction of the techniques that peculiarly characterize quality control per se. The fact that they wee primarily concerned with research rather than with process control or inspection also may have kept them out of the limelight. Nevertheless, it is safe to say that these powerful statistical tools are today doing as much for industrial research as they have done for biological, agricultural, and social research.[23]

In recent years there have been several new lines of development. One of these has been the response surface analysis associated largely with the name of Professor George E. P. Box. This systematized exploratory technique has found a ready welcome in the chemical industry together with the closely associated procedure known as *evolutionary operation.*

In probability and statistical theory there has been a growing interest in stochastic processes, and in the industrial field this has led to the development of cumulative sum control charts and a procedure known as adaptive quality control. The former was primarily a British development associated largely with the names of E. S. Page and G. A. Barnard, but great interest has more recently been taken in it in the United States. (See in particular papers by N. L. Johnson and F. S. Leone, and also James Lucas and D. W. Marquardt.) "Adaptive quality control" has been promoted by G. E. P. Box and G. M. Jenkins.

A third major development has been reliability engineering. Because of its broader scope this has been developed not only by the American Society for Quality Control but also by such organizations as the Institute of Electrical and Electronics Engineers (IEEE). A very important branch of reliability engineering is "life testing." An important standard in this area is Mil. Std. 781C *Reliability Design Qualification and Production Acceptance Tests: Exponential Distribution* issued by the Naval Electronic Systems Command. Mil. Std. 781D *Reliability Testing for Engineering Development, Qualification and Production,* which will supersede 781C, is expected to be issued in 1986 together with Military Handbook 781 *Reliability Test Methods, Plans and Environments for Engineering Development, Qualification and Production.* The extremely high standards generally required by reliability engineering, especially in the development of missile systems, have created a whole new dimension of the quality problem. In the attempt to motivate workers to turn out near perfect quality, special "zero defects" programs were developed. In 1966 those active in this area organized the American Society for Zero Defects. The name was subsequently changed and in 1985 the organization was known as the American Society for Performance Improvement.

---

[23] See such publications as L. H. C. Tippett, *Technological Applications of Statistics* (New York: John Wiley & Sons, 1950); K. A. Brownlee, *Industrial Experimentation* (2d rev.; Brooklyn, N.Y.: Chemical Publishing Co., Inc., 1948); Owen L. Davies (ed.), *Statistical Methods in Research and Production* (3d ed.; London: Oliver & Boyd, Ltd., 1957), and *The Design and Analysis of Industrial Experiments* (London: Oliver & Boyd, Ltd., 1954); Carl A. Bennett and Norman L. Franklin, *Statistical Analysis in Chemistry and the Chemistry Industry* (New York: John Wiley & Sons, 1954); and Norman L. Johnson and Fred C. Leone, *Statistics and Experimental Design in Engineering and the Physical Sciences* (New York: John Wiley & Sons, 1964).

In general, there has been growing recognition that worker participation in the direction of a quality control program is highly desirable. Not only does the worker often have valuable suggestions for improvement but participation gives him a personal interest in better quality. In Japan, worker participation has been effected through quality control circles; in the United States it is being developed through what is called participation problem solving.

In the United States recent years have seen the rise of consumerism which has led to federal government mandatory standards pertaining primarily to product safety. An especially significant development was the creation (1972) of a Consumer Product Safety Commission with the power to promulgate safety standards for all types of consumer products. A parallel development has been increased interest in standardization and the promotion by the American National Standards Institute of a product certification program, the latter to be carried out through third-party testing laboratories at the cost of the manufacturer. The greatly enhanced concern about environmental conditions has also led to programs for control of quality of air and water.

In the face of the development of federal mandatory safety standards, the American Society for Testing and Materials (ASTM) established a new committee (F–15) on Consumer Product Safety to work with the U.S. Consumer Product Safety Commission on the development of standards. The aim has been to keep industry posted as to areas in need of control and to develop on a voluntary, consensus basis standards that can be ultimately adopted by the federal government as mandatory standards. The Consumer Product Safety Commission is not allowed to include sampling plans in its product standards, but it can include these in its certification rules, so that sampling plans recommended by an ASTM committee for use with a given product specification may still become part of CPSC procedures.

An important development with respect to quality control standards was the organization in 1974 of a Quality Assurance Committee Z1 by the American National Standards Institute. Its scope is described as follows.

> In the field of quality assurance, to develop and review generic standards having general application, including the development of appropriate guidelines of general nature and to provide advice to other standards activities which cover quality assurance for specific areas, and to interface with appropriate international activities. The preparation of individual product or industry standards is not included within the scope of this committee.

The American Society for Quality Control has been designated the secretariat of this ANSI Committee Z1.

A very recent development in the United States has been the burst of interest in the quality of goods and services in general, in the role to be played by top management in the improvement of quality and in the increase in productivity that can be attained through the reduction in scrap or rework of nonconforming product. This development was fostered by the pressure of competition with foreign, especially Japanese, products and was spearheaded by the publication in 1982 by the Massachusetts Institute of Technology Center for Advanced Engineering Study of W. Edwards Deming's book *Quality, Productivity and Competitive Position* which is currently (1986) being superseded by his newer book *Out of the Crisis.*

In response to this increased interest in quality the American Statistical Association established early in 1984 an Ad Hoc Committee on Quality and Productivity to study the contributions that might be made by statisticians.[24] In turn, at its meeting in January 1985, the National Society of Professional Engineers approved the formation of a "not-for-profit" American Quality and Productivity Institute and an invitation to other organizations such as the American Statistical Association to be cosponsors. As this book goes to press, the full effect of this recent burst of interest in quality is yet to materialize.

In conclusion, note may be made of the cooperation of the American Soceity for Quality Control and the American Statistical Association in the publication of *Technometrics*—a journal of statistics for the physical, chemical, and engineering sciences. The journal *Industrial Quality Control,* published by the American Society for Quality, was discontinued in 1968 and replaced by two journals *Quality Progress* and the *Journal of Quality Technology.* Note may also be made of the establishment of an International Academy of Quality and the formation of a section of the International Statistical Institute known as the International Association for Statistics in Physical Sciences.[25]

2. WHAT STATISTICAL QUALITY CONTROL CAN DO

In the discussion in the previous section of the recent burst of American interest in quality, it was noted that a key element was the increase in productivity and reduction in cost that can accompany an improvement in the quality of goods and services. Examples of this are given in two newspaper articles published as far back as 1949 when the post-World War II development of quality control in the United States was underway.

On April 10, 1949, the *New York Times* ran the following headline on its financial page.

QUALITY CONTROLS TO REDUCE LOSSES
Industrial Production Savings Up to 25 percent
Promised under System to End Waste

---

[24] For the views of Gerald J. Hahn and Thomas J. Boardman, the Chairman and Vice Chairman respectively of this committee, on "The Statisticians Role in Quality Improvement," the reader is referred to the Forum of the March 1985 issue of the American Statistical Association *Amstat News.*

[25] In a brief review of this kind it is impossible to mention the work of all those who have contributed to the development of statistical quality control. For a fuller account the reader is referred to the articles listed at the end of this chapter. A shorter historical review is given in the author's article "Quality Control, Statistical" in the *Encyclopedia of Statistical Sciences.*

The supporting article was an account of a talk by C. W. Kennedy, quality control engineer of the Federal Products Corporation. Highlights of the article were as follows:

> Startling records in improvement of employee relations, lower break-even points, elimination of waste and lower costs are being obtained. . . .
>
> Quality control has become the first point of attack in methods improvement, because in many plants waste and losses from rejections, scrap, junk, salvage and reworked production were as high as 25 percent of total output. A range between 5 percent and 25 percent is not unusual, while a properly designed system can reduce these losses to a maximum of 1 percent and hold that figure.
>
> One of the most spectacular examples of large cost reductions made possible by scientific quality control is the reduction in price of penicillin . . . drug manufacturers . . . concede that quality control made new low price levels practicable.
>
> Case examples . . . show increases in output of as much as 10 percent, in addition to reduction of waste, salvage, scrap and reworked goods. . . .
>
> A further benefit of adequate quality control is improved employe morale. . . . 90 percent of the employes engaged in production wants to "do their work the right way, and subconsciously resent doing it any way which causes rejections."
>
> The new methods are unique because they require no capital outlays and only a minimum of re-education of supervisory personnel.

Further examples of what statistical quality control can do were reported in *The Wall Street Journal,* May 26, 1949. The *Journal* wrote:

> Gillette Safety Razor, for instance, discovered through a quality control system that its screw machine department was producing a lot more scrap than charts indicated was necessary. For every $45,000 worth of parts turned out each month, there was $4,000 of scrap, $1,000 of parts that needed reworking, $2,500 spent on inspection of finished products and $1,500 for preventive inspection during processing.
>
> Investigation showed 20 percent of the trouble was caused by improper operation. In one instance, a piece of material wasn't being located properly in the machine. In another, a slight alteration in the tool was necessary. In still a third case, tool-sharpening habits had to be changed.
>
> Altogether, corrections cut the output of scrap and reworks from $5,000 to $1,500 a month. But that wasn't the total saving. Since control charts kept a running check on operations, a final inspection wasn't needed and another $2,500 in monthly costs was eliminated.
>
> Partly as a result of a similar analysis of its carding department, Bigelow-Sanford Carpet Co. expects to save $1,800,000 during the first year's operation of its quality control program ending in July. Carding machines lay wool fibers parallel before they are twisted into yarn. Control charts showed the carding machines were capable of preparing the wool so as to produce much less variation in the thickness of the yarn.
>
> Variations in thickness mean variations in tensile strength which can cause costly breaks in the yarn in later operations or defects in finished carpets. Bigelow-Sanford was losing $2 million a year in discounts granted on carpets with abnormal variations in thickness and other defects.

"It took the combined efforts of operators, supervisors and engineers to work out changes in operating methods and equipment which resulted in a 25 percent reduction in thickness variations," says R. F. Hurst, Bigelow-Sanford's quality control director.

The foregoing examples illustrate what can be done with control charts. Other examples could be found to illustrate the advantage of having an inspection procedure based on sound statistical theory. This is particularly true when sampling inspection is employed. For with a scientifically designed sampling plan, a high percentage of acceptance can be obtained for product of standard quality, while at the same time an upper limit can be placed on the fraction nonconforming of outgoing material.

Still other examples could be given of the gains to be obtained by employing statistical theory in industrial research. In general, statistical quality control can and has saved money.

3. THIS TEXT

This text undertakes to present the basic principles and procedures of statistical quality control. It is not a manual of rules of thumb but a discourse on the assumptions and principles of theory that underlie modern quality control practice. It aims to promote understanding of the fundamental bases of statistical quality control and should therefore be useful in the comprehension not only of current procedure but also of future developments.

The body of the text begins with a section of fundamentals: probability, frequency distributions, and sampling. These are the concepts and tools that are applied in subsequent sections. A supplementary section of Part 1 describes "exploratory data analysis." Parts 2 and 3 pertain to inspection: the first deals mostly with acceptance sampling, the second with rectifying inspection. Part 2 also contains some discussion of economic procedures when information is available on process quality and on break-even quality levels. Part 4 takes up control charts. Part 5 covers that part of statistical theory that is especially pertinent to industrial research. It includes the estimation of lot and process characteristics, the theory of testing hypotheses, tests of differences, analysis of variance, multiple comparisons, analysis of components of variance, regression, analysis of covariance, and design of experiments.

The symbols in the text conform, for the most part,[26] with the set of symbols for control charts and sampling plans proposed by the Standards Committee of the American Society for Quality Control.[27] As in previous

---

[26] An important exception is that in this text a double prime, e.g., $\bar{X}''$, is used to designate a "standard value" (See p. 484.) whereas in ASQC Std. A–1, 1978, a subscript zero is used, e.g., $\bar{X}_0$.

[27] See ASQC Standard A1–1978–*Definitions, Symbols, Formulas, and Tables for Control Charts,* and ASQC Standard A2–1978–*Definitions and Symbols for Acceptance Sampling by Attributes.*

editions, a prime is used to distinguish universe values from sample values. Since this is an important distinction for clear thinking, it has been carefully made throughout the text. A primed quantity is thus a quantity that refers to either a universe of individual values or a universe of samples derived from the universe of individual values.[28] A glossary of symbols is given in Appendix III.

It will be noted that the text now uses the terms *nonconformity* and *nonconforming unit* where previous editions have used the terms *defect* and *defective unit.* This change was made to bring the text in line with the latest ASQC recommended practice. In ASQC/ANSI Standard Z1.4 a nonconformity is distinguished from a defect as follows:

> Nonconformity: A departure of a quality characteristic from its intended level or state that occurs with severity sufficient to cause an associated product or service not to meet a specification requirement.

> Defect: A departure of a quality characteristic from its intended level or state that occurs with a severity sufficient to cause an associated product or service not to satisfy intended normal, or foreseeable, usage requirements.

In Chapters 10 and 13 dealing with Mil. Std. 105D and Mil. Std. 414 the terms *defect* and *defective unit* are used in the sense of a *nonconformity* and *nonconforming unit* since these are the meanings given to these terms in these standards.

4. SELECTED REFERENCES*

Acceptance Sampling (B '50), Dudding (P '52), Freeman (P '37), Grant (P '53), Hahn and Boardman (P '85), Littauer (P '50), Pearson, E. S. (P '33 and P '73), Simon and Cohen (P '71), Tippett (P '62), Van Rest (P '53), Weaver (P '51), and Working and Olds (B).

---

[28] For discussion of universes of samples, see beginning of Chapter 6.

* B and P refer to the Book and Periodical sections, respectively, of the Cumulative List of References in Appendix V.

Part 1

Fundamentals

2

Probability

1. AN OLD GAMBLING GAME

Most of us have gambled at one time or another. We have matched pennies or rolled dice or played roulette. In every case we came immediately face to face with the theory of probability. We may not have known it by that name, but we did speak of *the odds* and were keenly interested in our *chances* of winning.

If we were matching pennies, we probably reasoned something as follows: A penny has two sides. If I shake it in my hand, a head should turn up as often as a tail. If my opponent does the same thing, he should also get a head as often as he gets a tail. In both our cases a head and a tail have an equal chance of occurring. Under these conditions, it is a fair game for us to match on an odd-and-even basis. Why? Because four possible results can happen, as indicated in the following diagram.

		My penny	
		H	T
My opponent's penny	H	HH	HT
	T	TH	TT

I can get a head and so can my opponent; I can get a head and he can get a tail; I can get a tail and he can get a head; and, finally, I can get a tail and he can get a tail. Since heads and tails are equally likely on each coin, each of these four combinations are also equally likely. The first and the last yield like faces ("evens") and the middle two yield different faces ("odds"). "Evens" and "odds" have equal probabilities; therefore, the matching game is a fair one.

The above reasoning is typical of what is called the probability calculus, somewhat simplified but essentially sound. Since the probability calculus is used extensively in the theory of statistical quality control, it is well to secure at the start an understanding of fundamentals. It is the purpose of this chapter to explain the meaning of probability and the basic principles of the probability calculus. These will be the tools we shall use in subsequent chapters.

2. PROBABILITY

2.1. The Axioms of Probability

Probability is formally discussed with reference to a set of elementary events. Denote this set by the symbol E.

Consider now a given collection of subsets of the elementary events of E which we will designate as the subsets $A_1, A_2, \ldots, A_n$. Some of these may overlap. Let there be assigned to each A_i a nonnegative real number $P(A_i)$ which we will call the probability that an elementary event belongs to A_i. Postulate further that the probability that an elementary event belongs to *some* one of these subsets has the value 1 (= certainty), i.e., that $P(E) = 1$. Also postulate that if the subsets $A_1, A_2, \ldots, A_k$ have no elementary events in common, then

$$P(A_1 + A_2 + \ldots + A_k) = P(A_1) + P(A_2) + \ldots + P(A_k) \tag{2.1}$$

The above are the probability axioms appropriate for a finite set E. To illustrate, consider the elementary events that can result from rolling a die. These are the occurrences of the numbers 1, 2, 3, 4, 5, or 6. This is the set of events E. Consider the subsets $A_1 = 1, 2$, $A_2, = 3, 4, 5$ and $A_3 = 1, 4, 5, 6$. Let P_1 be the probability of an event belonging to the subset A_1, P_2 the probability of its belonging to the subset A_2, and P_3 the probability of its belonging to the subset A_3. Then the axioms say that the P_i must all be nonnegative real numbers and that the probability of an event belonging to the entire set 1, 2, 3, 4, 5, 6 must be one. Since subsets A_1 and A_2 above are mutually exclusive, the axioms also say that

$$P(A_1 + A_2) = P(A_1) + P(A_2)$$

or that

Prob of a 1 or 2 or 3 or 4 or 5 must equal the
Prob of a 1 or 2 plus the Prob of a 3, 4, or 5

Since A_3 overlaps with A_1 and A_2, it is not true that $P(A_1 + A_3) = P(A_1) + P(A_3)$. Nor is it true that $P_1 + P_2 + P_3 = 1$.

If the collection of events E is infinite, an additional axiom is needed to prevent bizarre results. This axiom reads:

If we have a sequence of infinite subsets each one containing the one that follows and if the "area" common to all these subsets approaches zero as the number of such subsets is indefinitely increased, then the probability of a point belonging to the nth subset also approaches zero as n gets larger and larger.

For example, if the E consists of the infinite number of points on the real axis between 0 and 1, and if we consider the successive infinite subsets of points 0—0.5, 0—0.25, 0—0.125, 0—0.0625, . . . , 0—$1/n$, then the probability of a point falling in the subset 0—$1/n$ approaches zero as n approaches infinity. In simple language this axiom requires that the probability of a single point in an infinite number of points be zero. For example, if E consists of the points in the interval 0 to 100, the probability of an outcome equal to exactly 20, say, is 0.

2.2. A Word of Caution

In the formation of subsets of events the reader is cautioned to note that a consistent basis of classification must be adhered to. For example, in referring to a box of beads with varying weights and colors, it would not be proper to talk about the probability of drawing a *red* bead *or* a *heavy* bead. For the first subset (red) is based on a classification by *color* and the second subset (heavy) is based on a classification by *weight.* Where more than one bais of classification is involved, the proper procedure is to form subsets on the basis of a multiple system of classification. If there are three color classes, say, red, white, and blue, and two weight classes, heavy and light, we could consistently form the six subsets: heavy, red; heavy, white; heavy, blue; light, red; light, white; and light, blue. The subsets are thus formed by cross-classification and we have subsets of compound events.

2.3. The Frequency Approach to Probability

In statistical quality control, the above model for probability is applied mostly to random mass phenomena. In such a context it becomes what is known as the frequency approach to probability and is based on our intuitive notions of randomness when related to a large number of events. Consider again the rolling of a die. If this die is unbiased and is rolled in a random manner, our intuition suggests that equal probabilities of $1/6$ shoud be assigned to the occurrence of a 1, a 2, a 3, a 4, a 5, or a 6. (For the previous subsets A_1, A_2, and A_3, this would be the assignment of probabilities of $2/6$, $3/6$, and $4/6$.) If this die is rolled a large number of times, the actual relative frequencies that will result will be expected to be close to these assigned probabilities, at least sufficiently to use the assigned probabilities in making practical decisions.

2.4. The Subjective Approach to Probability

Although the frequency approach to probability has wide application in statistical quality control, there are some, possibly many, occasions when the concept of relative frequency in a large number of trials seems inappropriate. We may be interested, for example, in the probability that the workers in a given department will go out on strike next week, or the probability that a competitor will decide to build a new plant nearby. Such probabilities are *rational degrees of belief.* They are subjective, personal. They are not related to frequencies of events, actual or conceptual.

Subjective probabilities need further axioms for their definition and ordering. For discussion of these, the reader is referred to the works of Fishburn, Lindley, and Savage referenced at the end of this chapter. It will be sufficient here to note that we can talk about probabilities as rational degrees of belief and can develop a probability calculus for them in the same manner as we can for probabilities that are viewed as relative frequencies. In this book, however, it will be presumed that we are using the frequency approach unless otherwise stated.

3. RANDOMNESS

In applying the frequency approach to probability, the key concept is that of randomness. Since in the frequency approach the concept of randomness is an intuitive concept and cannot be defined, we need to look at it closely.

Although a positive statement as to the existence of randomness must be cautiously made, it is usually not difficult to decide when a procedure is nonrandom. Consider a penny that we shake in our hands and record whether it comes up heads or tails. If the penny is well shaken, we intuitively feel that a head is as likely to occur as a tail, so we would assign a probability of one-half to each result. But what do we mean by "well shaken"? This is hard to answer. On the other hand, if we always lay a penny in our left hand with the tail down and if our "shaking" simply consists of turning the coin over from the left to the right hand so that it now sits with the tail side up, we have not shaken the coin in a random manner. Indeed, if this method were used, the coin would always turn up tails.

Let us take another illustration. Suppose that as a manufacturer we continuously buy certain parts from a given supplier. The supplier claims that only 5 percent of the material he sells is not in conformity with specifications. We are not willing to accept his word, so we have a "sampling plan" to check on each lot. Our plan is to take one item from a lot and inspect it. If the item meets specifications, we accept the whole lot; if it does not, we reject the whole lot. The question to which we want an answer is this: If the lots contain only 5 percent nonconforming parts, as claimed, how many lots would we reject under our sampling plan? That is, how hard would we be on the supplier?

The reader will note that this problem is essentially the same as the previous problem, in which we shook a penny in our hand. The answer is therefore similar. We argue that the probability of a nonconforming part in each lot is, by assumption, 0.05. (Each lot is 5 percent nonconforming.)

If the part we take for inspection is drawn *at random,* then 5 percent of the time we shall get a nonconforming part and 95 percent of the time a conforming part. Hence our answer is that we shall accept 95 percent of the lots and reject 5 percent.

Again, it will be noted that the argument hinges on the assumption of randomness. How can we tell when we have drawn a part *at random?* As we have indicated, we must give a rather vague reply. The best we can do is to say that we have used a method that the "collective intuition" of statisticians strongly suggests is random and when tested behaves the way we would expect a random procedure to behave. One such method, for example, might be to use a table of random numbers. (See the following section.)

Although a positive statement thus presents difficulties, we can, as noted above, often make a negative statement. Suppose that in our example the parts come in boxes, and that in the course of transportation the defective parts tend to be shaken toward the bottom of the box. Suppose, further, that in drawing our part for inspection we always reach to the bottom of the box. Under these conditions our selection will not be a random one, and we shall reject many more lots than 5 out of 100.

As a third illustration, consider the output of a machine. Suppose that during a given span of time the machine turns out 3 percent nonconforming units; that is, the probability of a nonconforming unit in the output of this period is 0.03. If we take the output for a small part of the period, numbering, say, 1,000 units, how many nonconforming units can we expect to get? The answer is 30, if the machine is operating in a *random manner* during the whole span of time. As before, it will be easier to say when the machine is not operating in a random manner than to say when it is operating in a random manner. This will be studied in more detail in a later section. For the moment, we may note that, if the output ran in cycles, say, 3 nonconforming units, then 97 conforming units, then 3 nonconforming units followed by 97 conforming units, and so on, the machine would not be operating in a random manner. Again, if there was an increase in the percentage of nonconforming units as the machine grew older, it could not be said to be operating in a random manner. Say the first 1,000 items had no nonconforming items, the second 1,000 items had 1 nonconforming item, the tenth 1,000 items had 9 nonconforming items, the hundredth 1,000 had 99 nonconforming items, and so forth. In this second case, to know how many defective items we would get in a sample of 1,000 items from the output of the machine, we would have to know how old the machine was at the time we took the sample.

4. HOW TO DRAW A RANDOM SAMPLE

In industry, the theory of probability is frequently used in connection with sampling. For example, a decision regarding the acceptance or rejection of a lot may be based on the inspection of a sample from this lot. If the theory of probability is to be properly used to calculate the risk of a wrong decision resulting from the analysis of such sample data, it is required that the sample be random. We shall discuss in this section preferred methods for getting random samples. As noted in Section 3, the procedures outlined are in the last analysis based upon our intuitive notions of randomness.

The group of items from which a sample is taken will be called the universe. In reality the universe may be a lot or batch of items submitted for sampling inspection. In theory it will be the set of elementary events E.

A method of selecting a random sample that is very popular among satisticians is to use tables of random sampling numbers. These tables are collections of the digits 0 to 9 that have been put together in a way that our intuition strongly suggests will yield a random arrangement. One of the earlier tables, for example, was derived essentially by taking the 19th and 20th digits in a 20-place log table. A later table used an electronic device that was equivalent in principle to selecting a number from a 32-place roulette wheel after it had made about 3,000 revolutions. The initial tables derived in such ways as these are tested to see if the digits generated do have the property of being random. Sometimes such tests indicate that the desired results can easily be obtained by some simple adjustment. For example, in one case it was found that the initial generating process had yielded many more 6s than would be expected in a random series. The table was thus corrected by replacing some of the 6s by other digits selected at random.

When an electronic computer is employed to take a random sample, the computer itself is frequently given the task of deriving the "random" numbers. A popular procedure for computer derivation of random numbers is known as the multiplicative congruential method.[1] This generates a series of "pseudo random" numbers by using residuals obtained from division. Let X (modulo 2^6) be taken to mean the residual remaining after X is divided by $2^6(= 64)$. For example, if $X = 130$, then $X(\text{mod } 2^6) = 2$. If we set

$$X_0 = 1 \text{ and } X_{n+1} = 5X_n \,(\text{mod } 2^6)$$

we will get the residuals mod 2^6 of powers of 5. The first 16 residuals are 1, 5, 25, 61, 49, 53, 9, 45, 33, 37, 57, 29, 17, 21, 41, and 13, and after this the series repeats. (Hence the tag "pseudo.") In general, if we set

$$X_0 = 1 \text{ and } X_{n+1} = 5^k X_n \,(\text{mod } 2^{s+2})$$

where k is an odd integer, we will generate a series of numbers with period 2^s. The taking of residuals in the above procedure satisfies our intuitive feeling that the results will be random, and statistical tests of numbers so generated have indicated that they apparently do appear in random order within any one cycle. Their repetition from cycle to cycle, of course, guarantees a uniform distribution over many cycles.

---

[1] Herbert A. Meyer (ed.), *Symposium on Monte Carlo Methods* (New York: John Wiley & Sons, 1956), pp. 17–21; and Operations Research Center, Massachusetts Institute of Technology, *Notes on Operations Research, 1959* (Cambridge: Technology Press), pp. 233–34.

Since electronic computers are not always readily accessible and in many cases their use may not be feasible, published tables are probably the most readily available source of random numbers. Some of the better known of such tables are the following: Table XXXIII of R. A Fisher and F. Yates, *Statistical Tables for Biological, Agricultural and Medical Research;* Interstate Commerce Commission, Bureau of Transport Economics and Statistics, *Table of 105,000 Random Decimal Digits;* and the Rand Corporation's *A Million Random Digits,* published by The Free Press (Glencoe, Ill.) and reproduced in part in the *Journal of the American Statistical Association,* 1952–1954. Table DD of Appendix II contains two pages of random numbers taken from the Fisher and Yates tables.

The use of tables of random numbers varies. Generally, to draw a random sample from a given universe, the members of the universe are associated with the set of random numbers or some subset thereof. Then a sample is taken from the set or subset of random numbers and the corresponding items of the universe are selected. This gives a random sample of the size desired.

The example, suppose that there are 100 bottles on a rack and we wish to draw a random sample of 5. We note that the numbers in Table DD are grouped in clusters of two digits each. If we number the bottles from 00 to 99, we can select a random sample of five bottles by simply picking any five numbers from Table DD. Say we open to Table DD, Random Numbers III, and let us pick the first five two-figured numbers in the third column of the first block on that page. The random numbers will thus be 68, 27, 23, 76, and 28. The bottles with these numbers will constitute our sample of bottles.

Possibly a better way of proceeding is to put our pencil down at random on a digit in the table. If it is even, we use Random Numbers IV; if odd, we use Random Numbers III. The first two-figured number to the right of this that is less than 26 may be used to indicate what column to start in, and the next two-figured number to the right may be used to indicate the row. If some such method is employed, we will have greater assurance that our starting point will be random. When the starting point is once picked, movement in any direction will give a random sample of numbers. If more numbers are needed than available in Table DD, reference should be made to one of the larger sets named above. Even these may not be large enough and resort may have to be made to an electronic computer.

If a universe is finite and if there are no physical difficulties in selecting items from the universe, random numbers provide an excellent way for picking a random sample. If the members of the universe cannot be equally well selected (the bottles may be stacked so that interior bottles are completely surrounded) or if the members themselves cannot be given numbers (such as the screws in a large box), then random numbers cannot be applied. In these cases, other methods will have to be used.

Sometimes if random numbers cannot be applied in a given situation, they may be applied to a situation derived from the first. Thus, if bottles in the inside of a stack cannot be reached, it might be that subsequently, when the bottles are being taken off the rack, they could be numbered in the order of their withdrawal, and a random sample taken as the bottles are removed from the stack. Possibly a better procedure would be to take the sample as the bottles are initially placed on the shelf. These bottles could then be laid aside and stacked on the outside so they would be readily available for sample inspection.

If the members of the universe cannot be numbered, it may be that they can be thoroughly mixed in some way so that their position in the container is made as random as possible. Theoretically, if a universe is "thoroughly mixed" any "chunk" will be a random sample. The trouble in practice is to get agreement on when a universe is "thoroughly mixed."

Sometimes universes are not homogeneous but are made up of strata that are homogeneous. It might happen, for example, that a given lot consists of the output of three machines. If the ouputs are kept separate, we really have three distinct universes each of which may be sampled separately. If the outputs are thoroughly mixed, we can no longer sample the output of each machine separately but can take only a single random sample. The latter is not so efficient a method of sampling, however, as that of sampling each stratum before mixing.[2]

If a universe tends to become stratified because of standing, shaking, transportation, or the like, it is better to take random samples from each stratum separately and weight the results by the size of the stratum than to sample the universe as a whole. For example, suppose that the railroad transport of a given item causes the lighter items to move to the top of the car while the heavier items find their way to the bottom; then it is better to sample the top and bottom of the car separately, if that is feasible. When a universe is not thoroughly mixed but the stratification tendencies are not known, it is better to sample all sections of the universe then to look upon any one section as a random sample of the whole, something that could be done if the universe were thoroughly mixed.

In summary, if we can assign numbers to the members of a universe and can withdraw the items easily, random sampling numbers can be used to make the method of selection a random one. If we cannot assign numbers to the members of a universe, it should be thoroughly mixed, if possible, before sampling. This is most likely to be successful with liquids. With powders and collections of solid units, attempts at mixing through shaking, rotation, and so on, may lead to stratification instead of thorough mixing. In general, it is best to use random numbers if at all possible. These will hold up in court.

The general rules to be borne in mind in drawing a random sample are:

---

[2] See any statistics text that discusses stratified sampling.

1. *Adopt a method of selection that will give every member of the universe an equal chance of being drawn.*
2. *Avoid any method that associates the selection of an item with the classification of the item being selected.*

5. WHEN IS A PROCESS OPERATING IN A RANDOM MANNER?

If a process is operating in a random manner, then any part of the output may be viewed as a random sample from the output as a whole. Hence it is important to know whether a process is operating in a random manner.

As noted above, we can describe a random process only indirectly by indicating when a process is not random. A process is obviously not operating in a random manner when it is running in cycles or shows bunchiness or has a trend up or down. Sometimes these nonrandom characteristics are so distinct as to be obvious. In other cases it may require some statistical test to decide whether a process is operating in a random manner.

Tests used to determine whether a series of items is random take various forms. One is to study runs in the data. If there is a long run above or below average, it is an indication that nonrandom forces are at work. If there is a long run up or down, it indicates nonrandomness. If the runs in the data are too few in number, it indicates nonrandomness; and if the runs form cycles, it indicates nonrandomness. The theory of runs as applied to detecting nonrandomness will be studied in detail in Section 7 of Chapter 18.

Another type of test is to study the variation in samples from the process. In Chapters 5 and 6 we shall see that, when a process is operating in a random manner, samples from such a process conform to certain statistical patterns. When samples from a given process do not appear to conform to the appropriate pattern, we conclude that it is not operating in a random manner.

The principal device that is used to study process variability is the control chart. This presents the variation in samples of output over time or according to some assigned order; and, by showing up runs or revealing extreme variations, it may bring to light nonrandom fluctuations. A process that over a considerable range of output and time shows no nonrandom variations within the limits of a control chart, nor any points outside the control limits, is said to be "in control" with respect to the type of variation depicted by the chart. It would be very unsafe to assume that a process is operating in a random manner if it is not "in control." How to get a process "in control" is discussed in Part 4.

The reader should be warned at this point that very few, if any, industrial processes ever operate in a purely random manner for a significant length of time. The concept of a process operating in a random manner that is developed here is an "ideal" or abstraction to which our theoretical discussion

FIGURE 2.1
Graph of a Series of Random Numbers

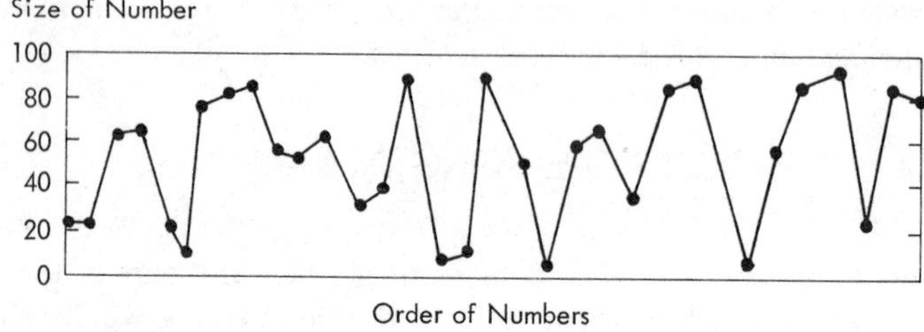

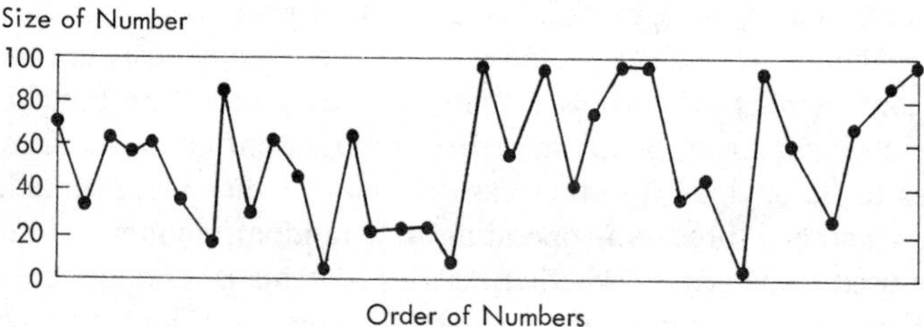

will apply. When the theory is actually applied to a real situation, minor deviations from the theoretical ideal may have to be overlooked. When theory calls for randomness, we should make every reasonable effort to be assured that nonrandom influences are not present. From an engineering point of view, however, it would seem safe to assume, for purposes of theory, that a process is operating in a random manner, if the process does not show lack of control on a standard control chart, although more detailed analysis might turn up minor instances of nonrandom variation. Like all theory, the application of statistical theory becomes an art the success of which depends on the judgment as well as the knowledge of the one applying it.

Figure 2.1 shows what a purely random series looks like. This is the graph of a set of random numbers plotted as a time series. Figure 18.9 shows variations in an industrial series, and Figure 18.3 illustrates a control chart.[3]

6. EMPIRICAL PROBABILITIES

In industry we rarely have prior knowledge of probabilities of the quality of output. Most of these must therefore be obtained empirically. A machine, for example, is turning out parts. How can we tell from an examination of the machine how many nonconforming parts it will produce? The answer is obvious: we cannot. All we can do in a given situation is to let the machine produce a large number of parts and count the number that do not conform

---

[3] See Chapter 19 for a discussion of how this chart was constructed.

to specifications. Then we can use this fraction nonconforming of past production as an estimate of the probability of a nonconforming part being produced by the machine. This empirically determined probability is what we have to use to predict the fraction nonconforming of future output. Of course, the machine must be operating in a random manner if the fraction nonconforming of past production is to be used as an estimate of the probability of a nonconforming unit.[4] It must continue to operate in a random manner if the probability of a nonconforming item is to be used successfully to predict the fraction nonconforming of future output. In some cases, empirical knowledge of one machine can be used to make predictions about a similar machine which has not been tested. Basically, however, our knowledge of industrial probabilities is empirical.

7. PREDICTIONS OF RELATIVE FREQUENCIES

In using probability theory to predict relative frequencies of random events, note should be made of two aspects. First, the phenomena to which probability theory is to be applied should be *mass* phenomena. If probability is to be used to predict the relative frequency of a given type of event in a sequence of events selected at random from the set of all events, then *the sequence must be a long one.* If we shook a penny only 10 times, it would hardly be safe to predict that 5 times it would come up heads. If we inspected only 20 lots, we could not safely predict that just 1 of them would be rejected. If we took only 100 items from the output of a machine, we would not have much assurance that just 1 of them would not conform to specifications. On the other hand, if we shook a penny 2,000 times or if we inspected 10,000 lots or examined 20,000 units of output, we might, with considerable assurance, predict that about 1,000 times the penny would come up heads or about 500 of the lots would be rejected or about 600 of the units would not conform. This is why gambling houses rarely go broke. They fix the odds in their favor and then allow the "law of large numbers" to iron out temporary runs against them. This should be remembered in quality control. In Parts 2 and 3 we shall take up various sampling inspection plans and shall talk about the probability of various results. It must be remembered that these probabilities are predictions of what would happen "in the long run." They tell what may be expected to happen in many inspections, not in a small number.

The second point to be made is that our *predictions are approximate, not exact.* If we shake a penny a large number of times, we can predict that *about* half of the time it will turn up heads. If we inspect a large number of lots, we can predict that *about* 5 percent of them will be rejected. If we examine a large number of items produced by our machine, we can predict that *about* 3 percent will not conform to specifications. How far off our

[4] See Section 3 above.

predictions will be we never know beforehand. Experience indicates, however, that if the phenomena are random and if there are a large number of them, our prediction of relative frequency will be sufficiently accurate for most practical purposes.

8. TWO IMPORTANT THEOREMS

There are two important theorems that are used time and time again in the calculation of probabilities. These are the addition theorem and multiplication theorem.

8.1. The Addition Theorem

It was postulated in Section 2 that if the sets of events $A_1, A_2, \ldots, A_n$ were all disjoint, i.e., there was no overlapping, then the probability of an event belonging to any one of the sets $A_1, A_2, \ldots, A_n$ would equal the sum of the probabilities of an event belonging to the individual sets. Symbolically, it was postulated that

$$P(A_1 + A_2 + \ldots + A_n) = P(A_1) + P(A_2) + \ldots + P(A_n) \tag{2.1}$$

Suppose, however, that there is some overlapping. Suppose, for example, the sets A_1 and A_2 contain some events in common, i.e., they are not mutually exclusive. Then it is no longer true that $P(A_1 + A_2) = P(A_1) + P(A_2)$. In the case of the overlapping of the two sets the correct relationship is given by

THEOREM (2.1). *If two sets are not mutually exclusive, the probability of belonging to either one or the other is the sum of the individual probabilities minus the probability of belonging to both simultaneously.* In symbols this is

$$P(A_1 + A_2) = P(A_1) + P(A_2) - P(A_1A_2) \tag{2.2}$$

where $P(A_1A_2)$ refers to the probability of an event belonging to both A_1 and A_2 (i.e., the intersection of A_1 and A_2). This is the addition theorem for two sets that are not mutually exclusive. Its validity is readily seen from Figure 2.2. If we call A_1' the set of events that belong to A_1 and not A_2, if we call A_2' the set of events that belong to A_2 and not A_1, and if we call A_1A_2 the set of events that belong to both A_1 and A_2, then it follows from (2.1) that

$$P(A_1) = P(A_1') + P(A_1A_2)$$
$$P(A_2) = P(A_2') + P(A_1A_2)$$
$$P(A_1 + A_2) = P(A_1') + P(A_2') + P(A_1A_2)$$

Hence

$$P(A_1 + A_2) = P(A_1) + P(A_2) - P(A_1A_2)$$

FIGURE 2.2
Illustration of the Addition Theorem for Two Intersecting Sets

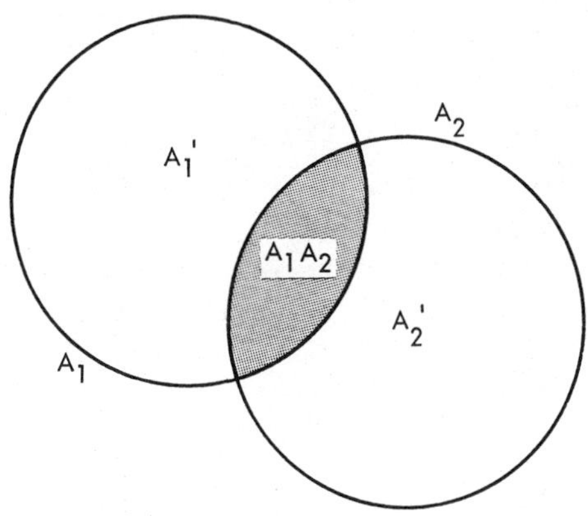

since we must eliminate one of the $P(A_1A_2)$'s that are counted twice when we add $P(A_1)$ *and* $P(A_2)$. For three sets we would have (see Figure 2.3) the following:

$$P(A_1+A_2+A_3)=P(A_1)+P(A_2)+P(A_3)-P(A_1A_2)-P(A_1A_3)-P(A_2A_3)+P(A_1A_2A_3) \quad (2.3)$$

It will be left to the reader to prove this.

To illustrate the addition theorem for sets that are not mutually exclusive,

FIGURE 2.3
Illustration of the Addition Theorem for Three Intersecting Sets

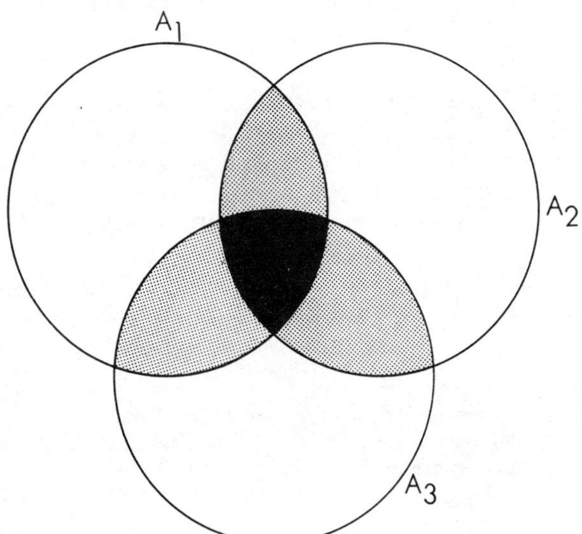

TABLE 2.1
Probabilities Pertaining to the Joint Operation of a Turbine and Boiler

		Turbine		
		In Order	*Out of Order*	
Boiler	In Order	0.9506	0.0294	0.9800
	Out of Order	0.0194	0.0006	0.0200
		0.9700	0.0300	1.0000

consider the following problem. A power company has a steam turbine run by a coal-burning boiler. On any day, any one of four possible situations might exist. The turbine might be in good working condition, but the boiler might not; the boiler might be in good working condition, but the turbine might not; they both might be out of order; or they both might be in order. Let the probabilities of these various events be those given in Table 2.1.

It is readily seen that the probability of the turbine *or* the boiler being out of order is $0.03 + 0.02 - 0.0006 = 0.0494$. The reader should be careful to note that it is *not* $0.03 + 0.02 = 0.05$; for in this way of computing the result, the probability of both the turbine and boiler being out of order would be incorrectly counted twice.

8.2. The Multiplication Theorem

The multiplication theorem relates to the case in which events are cross-classified with respect to two or more bases of classification, say classifications A and B. Events that belong both to class A_i of classification A and simultaneously to class B_j of classification B can be said to form the A_iB_j set. Each such "two-dimensional" set will thus be distinguished by two subscripts, an i and a j. It might be easier in this case if we start with an illustration and develop the multiplication theorem with reference to this.

Suppose that we have a lot of 10,000 parts distributed as follows with respect to noncomformity to specifications:

	Supplied by Company X	*Supplied by Company Y*	*Supplied by Company Z*	*Total*
Having a major nonconformity	100	50	50	200
Having a minor nonconformity	400	150	250	800
Having an incidental nonconformity	600	400	300	1,300
Having no nonconformities	3,900	2,400	1,400	7,700
Total	5,000	3,000	2,000	10,000

If we look at the marginal totals of this table, we see immediately that with respect to a part drawn at random from the lot, the probability of its having a major nonconformity is 0.02, the probability of its having a minor defect is 0.08, the probability of its having an incidental nonconformity is 0.13, and the probability of it having no nonconformities is 0.77. Likewise, the probability of a part's being supplied by Company X is 0.50, the probability of its being supplied by Company Y is 0.30, and the probability of its being supplied by Company Z is 0.20. These are called marginal probabilities. They all follow directly from our definition of probability. It also follows from our definition that the probability of a part's having the *joint* property of having a major nonconformity *and* being supplied by Company X equals 100/10,000 = 0.01. Similar calculations yield the following table of joint probabilities.[5]

	Supplied by Company X	*Supplied by Company Y*	*Supplied by Company Z*	*Total*
Having a major nonconformity	0.010	0.005	0.005	0.020
Having a minor nonconformity	0.040	0.015	0.025	0.080
Having an incidental nonconformity	0.060	0.040	0.030	0.130
Having no nonconfomities	0.390	0.240	0.140	0.770
Total	0.500	0.300	0.200	1.000

Consider, now, some further probabilities. Consider the following question: If we know that a part was supplied by Company X, what is the probability that it has a major nonconformity? This is what is called a *conditional probability.* The symbolic way of representing a conditional probability is $P(A_i|B_j)$, which is to be read "probability of A_i, given B_j." In the present instance, the probability of a part's having a major nonconformity, given that it is supplied by Company X, is computed by dividing 100 (the number of parts supplied by Company X having a major nonconformity) by 5,000 (the total number supplied by Company X). This yields 100/5,000, or 0.02. The conditional probabilities for the various classes of nonconformance for each supplier are thus as follows:

	Given that X Was Supplier	*Given that Y Was Supplier*	*Given that Z Was Supplier*
Having a major nonconformity	100/5,000	50/3,000	50/2,000
Having a minor nonconformity	400/5,000	150/3,000	250/2,000
Having an incidental nonconformity	600/5,000	400/3,000	300/2,000
Having no nonconformities	3,900/5,000	2,400/3,000	1,400/2,000
Total	1,000	1,000	1,000

[5] These are simply the items in the previous table divided by 10,000.

Now it will be noticed that the joint probability of a part's having a major nonconformity *and* having been supplied by Company X is equal to the probability of a part's being supplied by Company X (5,000/10,000) times the conditional probability of a part's having a major nonconformity, *given* that it was supplied by Company X (100/5,000). For

$$\frac{5{,}000}{10{,}000} \times \frac{100}{5{,}000} = \frac{100}{10{,}000}$$

The above illustrates what is known as the *multiplication theorem.* In general terms the multiplication theorem runs:

THEOREM (2.2). *The joint probability of* A_i *and* B_j *is equal to the probability of* A_i *times the probability of* B_j *given* A_i.

It can also be stated as follows:

THEOREM (2.2). *The joint probability of* A_i *and* B_j *is equal to the probability of* B_j *times the probability of* A_i *given* B_j.

In symbols we have

(2.4) $$P(A_iB_j) = P(A_i)P(B_j|A_i) \quad \text{or} \quad P(A_iB_j) = P(B_j)P(A_i|B_j)$$

An important corollary of the multiplication theorem pertains to the special case when the probabilities of the two classifications are independent. Independence means that the conditional probabilities of Classification I are the same for each of the subsets making up Classification II. For example, with reference to the lot of 10,000 parts, if the classification by type of nonconformity is independent of the classification by supplier, then the probability of a part with a major or minor nonconformity would be the same for parts supplied by Company X as for parts supplied by Company Y or for parts supplied by Company Z. Thus, for the case of independence, the joint probabilities of the distribution of 10,000 parts might be as follows:

	Supplied by Company X	*Supplied by Company Y*	*Supplied by Company Z*	*Total*
Having a major nonconformity	0.010	0.006	0.004	0.020
Having a minor nonconformity	0.040	0.024	0.016	0.080
Having an incidental nonconformity	0.065	0.039	0.026	0.130
Having no nonconformities	0.385	0.231	0.154	0.770
Total	0.500	0.300	0.200	1.000

It will be noticed that in the case of independence the joint probabilities of any row are in proportion to those of any other row and likewise the joint probabilities of any column are in proportion to those of any other column.

For the case of independence, the multiplication theorem reads as follows:

COROLLARY (1) OF THEOREM (2.2). *If the probabilities of two classifications are independent, the joint probability of* A_i *and* B_j *equals the probability of* A_i *times the probability of* B_j. *Thus,*

(2.5) $$P(A_iB_j) = P(A_i)P(B_j)$$

if the probabilities are independent. For more than the two classifications we have the following statement of the multiplication theorem for independent probabilities:

COROLLARY (2) OF THEOREM (2.2). *If the probabilities of* N *classifications are independent, the joint probability of* A_i, B_j, . . . , N_s *equals the product of all the individual probabilities. Thus,*

(2.6) $$P(A_i, B_j, \ldots, N_s) = P(A_i)P(B_j) \ldots P(N_s)$$

if the probabilities are independent. These are very important forms of the multiplication theorem and are used frequently in the probability calculus.

To illustrate the multiplication theorem, let us return to the boiler-turbine data of the previous section and view them as being cross-classified. The probability of the boiler's going out of order on any given day in the year is 0.02 and the probability of the turbine's going out of order on any given day is 0.03. Assume that these two occurrences are independent of each other. Then, by the multiplication theorem for independent probabilities, we have

Probability that the boiler will be out of order but the turbine will not equals (0.02)(0.97) = 0.0194.

Probability that the turbine will be out of order but the boiler will not equals (0.03)(0.98) = 0.0294.

Probability that both the boiler and the turbine will be out of order on the same day equals (0.02)(0.03) = 0.0006.

Hence by the addition theorem the probability that the boiler-turbine combination will be out of order on a given day in the year equals 0.0194 + 0.0294 + 0.0006 = 0.0494. A quicker way of solving the problem would have been to calculate the probability of the combination's *not* being out of order on any given day and subtract this from 1 to get the probability of its being out of order. Thus the probability of the boiler-turbine combination's being out of order equals 1 − (0.98)(0.97) = 1 − 0.9506 = 0.0494.

9. BAYES' THEOREM

In the previous section, we discussed the conditional probabilities of the form: Given the supplier, what is the probability of getting a part of specified quality, i.e., having a major nonconformity, having a minor nonconformity, and so on? In many instances the reverse type of problem is the one we are interested in. We know the quality of the part, since it has been inspected. What we want is the probability that it was supplied by a specified company. Suppose, for example, that the part in hand had a minor nonconformity but not a major nonconformity. Then, referring back to the original distribution of 10,000 parts, we could compute the following conditional probabilities:

Given that the part has a minor nonconformity:

$$\text{The probability it was supplied by X} = \frac{400}{800} = 0.5000$$

$$\text{The probability it was supplied by Y} = \frac{150}{800} = 0.1875$$

$$\text{The probability it was supplied by Z} = \frac{250}{800} = 0.3125$$

Frequently, the probabilities do not come in a form convenient for carrying out calculations similar to the above. For example, what we may have are the probability of getting a part with a minor nonconformity if it came from Supplier X (= 400/5,000), the probability of getting such a part if it came from Supplier Y (= 150/3,000), and the probability of such a part if it came from Supplier Z (= 250/2,000). We also may have the probability that a part came from Supplier X (= 5,000/10,000), the probability that a part came from Supplier Y (= 3,000/10,000), and the probability that a part came from Supplier Z (= 2,000/10,000). With these probabilities we would calculate the conditional probability that a part came from Supplier X given that it had a minor nonconformity, by carrying out the following set of computations. First, we would compute the joint probability of a part with a minor nonconformity and the fact that it came from Supplier X. This would be given by the product of

(I) The probability of a part with a minor nonconformity given that it came from Supplier X

times

(II) The probability the part came from Supplier X

which for our data would be $\left(\frac{400}{5{,}000}\right)\left(\frac{5{,}000}{10{,}000}\right)$. Then we could do the same calculations for Supplier Y and Supplier Z which would yield $\left(\frac{150}{3{,}000}\right)\left(\frac{3{,}000}{10{,}000}\right)$ and $\left(\frac{250}{2{,}000}\right)\left(\frac{2{,}000}{10{,}000}\right)$, respectively. Since the sum of these three joint probabilities would be the total probability of getting a part with a minor nonconformity, the conditional probability that a part came from Supplier X given that it had a minor nonconformity would be simply the ratio of the first of these joint probabilities to their sum, i.e., Prob (Company X supplied the part with a minor nonconformity)

$$= \frac{\left(\frac{400}{5{,}000}\right)\left(\frac{5{,}000}{10{,}000}\right)}{\left(\frac{400}{5{,}000}\right)\left(\frac{5{,}000}{10{,}000}\right) + \left(\frac{150}{3{,}000}\right)\left(\frac{3{,}000}{10{,}000}\right) + \left(\frac{250}{2{,}000}\right)\left(\frac{2{,}000}{10{,}000}\right)}$$

$$= \frac{\frac{400}{10{,}000}}{\frac{400}{10{,}000} + \frac{150}{10{,}000} + \frac{250}{10{,}000}} = \frac{400}{800} = 0.5000$$

This way of calculating the probability of a source given a specific result is known as Bayes' theorem, the name being that of an English clergyman who developed these notions many years ago. In general, Bayes' theorem may be stated as follows:

If $P(A_i|B_j)$ is the probability of $A = A_i$ given $B = B_j$ and $P(B_j)$ is the probability of $B = B_j$, then with the knowledge that $A = A_\alpha$, the probability that $B = B_\beta$ is given by

$$P(B = B_\beta|A = A_\alpha) = \frac{P(A = A_\alpha|B = B_\beta)P(B = B_\beta)}{\sum_j P(A = A_\alpha|B = B_j)P(B = B_j)} \tag{2.7}$$

This conditional probability that $B = B_\beta$ given that $A = A_\alpha$ is called a *posterior probability,* since it gives the probability of B after the event A has happened. These posterior probabilities stand in contrast to the unconditional probabilities $P(B_j)$, which are called *prior probabilities.* They are the probabilities of B prior to the occurrence of A. For further discussion the reader is referred to Section 1 of Chapter 24.

10. PERMUTATIONS AND COMBINATIONS

In calculations of probability it is often necessary to determine the number of ways of doing something or the number of combinations that can be made of certain things. It is therefore fitting to close this chapter with a brief discussion of *permutations* and *combinations.*

A *permutation* is simply an arrangement. The word *boy* is a permutation of the letters *b, o,* and *y.* Others would be *byo, oby, yob, oyb,* and *ybo.* A *combination* is a set or collection. The word *boy* is a combination of the three letters *b, o,* and *y.* The groups *byo, oby,* and so on, are the same combination as *boy,* since a combination takes account merely of the elements it contains and has nothing to do with the order or arrangement of these elements, as does a permutation.

It will be noted that six permutations were made of the three letters *b o y.* In general, $n!$, i.e., $n(n-1)(n-2) \ldots 1$ different permutations can be made from n elements. The reason for this is as follows: If we can do something in s ways and something else in t ways, then we can do them both successively in s times t ways. Thus, when we are arranging the letters *b o y,* we can consider that there are three places to be filled by the three letters. When we come to fill the first place, we have three choices, we can pick either b or o or y. Having filled the first place with one of these three, we have left two choices for filling the second place. Finally, having filled the first and second places, we have only one choice left to fill the last place. Hence we can pick (i.e., arrange) the three letters in $3 \times 2 \times 1 = 3! = 6$ different ways.

The same argument can be used to show that if we have n things and wish to permute or make arrangements of r of these n things, the number of ways we can do this is $n(n-1)(n-2)\ldots$ to r factors or $n(n-1)\ldots(n-r+1)$. If we both multiply and divide this by $(n-r)(n-r-1)\ldots 1$ we get $n!/(n-r)!$ Hence, the number of permutations of n things taken r at a time is

$$P_r^n = \frac{n!}{(n-r)!} \tag{2.8}$$

The following will illustrate the use of this permutation formula. The ordinary telephone dial contains 10 different numbers that can be dialed. If repetitions are not allowed, how many different seven-digit telephone numbers can be constructed for use with this dial? The answer is $P_r^n = 10!/3! = (10)(9)(8)(7)(6)(5)(4) = 604{,}800$. If repetitions were permitted, it would be simply $(10)(10)(10)(10)(10)(10)(10) = 10^7 = 10{,}000{,}000$.

The number of different *combinations* that may be made of n things taken r at a time is, of course, less than the number of permutations of n things taken r at a time. For here we are concerned only with the constituency of the group and not with the order or arrangement. In general, for each combination of r things there are $r!$ different permutations. Hence, if C_r^n is the number of combinations of r things that may be made from n things, then the total number of permutations of r things that may be made from n things is $C_r^n \times r!$ But it was seen above that the number of permutations of r things that may be made from n things is $P_r^n = n!/(n-r)!$. Hence $C_r^n \times r! = n!/(n-r)!$ and therefore

$$C_r^n = \frac{n!}{r!(n-r)!} \tag{2.9}$$

is the number of combinations of r things that can be made from n things.

Again an example will illustrate the use of this formula. A publisher makes a practice of using two different colors on the covers of his books. If there are 10 different colors from which it is possible for him to choose, how many different color combinations can he use for book covers? The answer is $10!/2!8! = 45$.

11. PROBLEMS

2.1. A sorting machine separates the output of a given process into four grades. The sorting of 100,000 items yielded the following results:

Grade	*No. in Grade*
I	21,293
II	58,177
III	16,465
IV	4,065
Total	100,000

a. What is the probability of the process turning out an item of Grade I? Of Grade II?

b. What is the probability of the process turning out an item of Grade I *or* Grade II?

c. In the next 10,000 items produced, how many would you expect to have of Grade III?

d. Suppose that you took two items from the process, what is the probability that they will *both* be of Grade II? Of Grade II or better? What is the probability that a sample of 10 items will all be of Grade II *or better?*

e. State explicitly the assumptions that underlie your answers to *a, b, c,* and *d.* Are your probabilities a priori or empirical probabilities?

2.2. In a lot of 10,000 parts,

20 percent have incidental nonconformities,
10 percent have minor nonconformities,
3 percent have major nonconformities,
6 percent have both incidental and minor nonconformities but no major nonconformities,
1 percent have both incidental and major nonconformities, but no minor nonconformities,
1 percent have both minor and major nonconformities, but no incidental nonconformities, and
0.5 percent have all three types of nonconformities.

If a part is selected at random from such a lot, what is the probability—

a. It will be nonconforming?

b. It will have only an incidental or minor nonconformity?

2.3. Among a work force of 1,000 employees,

30 percent are women and 70 percent are men,
80 percent had some form of accident insurance during the past year,
8 percent had accidents during the past year,
10 percent are women who had accident insurance but did not have an accident during the past year,
1 percent are women who had an accident during the past year but did not have accident insurance,
1 percent are men who had an accident during the past year but did not have accident insurance, and
2 percent are women who had an accident during the past year and did have accident insurance.

If an employee is selected at random from the rolls of the company, what is the probability it will be either a man or a woman who had an accident during the year but was covered by accident insurance?

2.4. During a week a process turns out 20,000 units of product. Of these, 15,000 were made from raw material supplied by Company A and 5,000 from raw material supplied by Company B. The 15,000 units made from A's raw material were graded as follows:

Grade	*No. in Grade*
I	3,218
II	8,543
III	2,782
IV	457

a. About how many of the 5,000 units made from B's raw material would you expect in each of these four grades if you believed the probabilities of the items falling in the various grades were independent of the company supplying the raw material? A unit selected at random from the output is found to be of Grade III. What is the probability it was made from raw material B?

b. What is the answer to the second part of *a* if the number of units in each of the grades made from raw material B were 362, 1,777, 2,310, and 551, respectively?

2.5. The following are the results of a week's work by two plant crews:

	No. Nonconforming	*No. Conforming*
Crew A	151	563
Crew B	85	307

Would you infer from these data that the probability of a nonconforming unit is independent of the crew that produced it? Justify your answer.

2.6. A product is made up of two parts, Part A and Part B. The manufacturing process for each is such that the probability of a defective Part A is 0.08 and the probability of a nonconforming Part B is 0.05. What is the probability that the assembled product will not contain at least one nonconforming part, i.e., the probability the assembled product will not be nonconforming?

2.7. During three years' operations under relatively constant conditions, accidents occurred as follows:

	Day of the Week					
	Mon.	*Tues.*	*Wed.*	*Thurs.*	*Fri.*	*Total*
Dept. A	32	21	23	29	35	140
B	14	10	11	10	15	60
C	12	12	13	11	11	59
D	24	22	23	21	22	112
Total	82	65	70	71	83	371

a. An accident is known to have occurred, but the day and department are unknown. Estimate the probability that it occurred in Department A on a Monday. In Department B on a Friday.

b. Estimate the probability that the accident occurred in Department C. In Department B.

c. Estimate the probability that the accident occurred in Department A, given that it happened on Monday. In Department A, given that it occurred on Wednesday.

d. Which of the probabilities you calculated in *a, b,* and *c* are joint probabilities? Which are marginal probabilities? Which are conditional probabilities?

2.8. Births of infants in four hospitals of a large city were as follows during a given year:

	Hospital				
Race of Baby	*A*	*B*	*C*	*D*	*Total*
White	3,575	2,979	1,192	2,384	10,130
Black	463	386	154	308	1,311
Other	168	140	56	112	476
Total . . .	4,206	3,505	1,402	2,804	11,917

a. A baby is known to have been born yesterday but the hospital and race are unknown. Estimate the probability that it was a black baby born in Hospital B. A black baby born in Hospital D.

b. Estimate the probability that the baby (whatever its race) was born in Hospital A.

c. Whatever the hospital, estimate the probability that the baby was a black baby.

d. Given that the baby was born in Hospital C, estimate the probability of its being a white baby. Given that the baby was a white baby, estimate the probability it was born in Hospital C.

e. Which of the probabilities you calculated in *a, b, c,* and *d* are joint probabilities? Which are marginal probabilities? Which are conditional probabilities?

f. In a subsequent year it is known that 4,108 babies were born in Hospital A; 3,612 were born in Hospital B; 1,337 in Hospital C; and 2,815 in Hospital D. It was also known that of the 11,872 babies 10,231 were white; 1,162 were black; and 479 were of other races. Assume that the distribution of babies by race within hospitals had the same general stochastic characteristics in this later year as in the earlier year and estimate the number of white babies born in Hospital B. The number of black babies born in Hospital A. What probability theorem did you use?

2.9. *a.* Suppose that the probability of a nonconforming door lock in an auotomobile is 1 out of 10,000. What is the probability that both of the doors on a two-door sedan will have nonconforming locks?

b. Suppose that the probability of a nonconforming tire casing is 1/8,000 and the probability of a nonconforming valve is 1/4,000. What is the probability that the tire of a car will be nonconforming either because the casing is nonconforming or because the valve is nonconforming or both?

2.10. Five hundred radios are stacked on shelves. Describe in detail how you would pick a random sample of 10 radios.

2.11. Bottles of a certain soft drink are shipped in cases that hold 24 bottles apiece. Describe how you would select a random sample of 50 bottles from an incoming truckload of 200 cases.

2.12. *a.* Plot the data of Table 3.2 on a chart, selecting the figures from left to right on each line. Study the chart to see if you can spot any obvious evidence of nonrandom variation.

b. Do the same for the data of Section (i) of Problem 3.3*a*, running down each column, starting with the left-hand column.

c. Repeat *b*, using the data of Section (ii) Problem 3.3*a*.

2.13. Suppose that a wheel has 0.5 of its perimeter painted red, 0.2 yellow, and 0.3 blue. When the wheel is spun, it may come to rest with a fixed marker opposite the red section of the wheel, the blue section, or the yellow section. If the wheel comes to rest with the marker opposite the red section, a draw is made from a red bag containing 93 white balls and 7 black balls. If it comes to rest with the marker opposite the yellow section, a draw is made from a yellow bag containing 94 white and 6 black balls. Finally, if it comes to rest with the marker opposite the blue section, a draw is made from a blue bag containing 95 white and 5 black balls.

The wheel is spun and a ball is drawn from one of the bags. The ball is *black*. What is the probability it was drawn from the blue bag?

2.14. Of a given item in a storeroom, 40 percent were produced by the factory's morning shift, 35 percent were produced by the afternoon shift, and 25 percent by the evening shift. The percent nonconforming in the output of the morning shift is 4 percent, in the output of the afternoon shift it is 6 percent, and in the output of the evening shift 10 percent.

An item is selected from this storeroom and on inspection is found to be nonconforming. What is the probability it was produced by the evening shift?

2.15. A product is assembled in four stages. At the first stage there are 6 assembly lines, at the second stage 5, at the third stage 4, and at the last stage 2. In how many different ways may the product be routed through the assembly process?

2.16. An inspector visits five different machines during the course of a day. To prevent the operators from knowing just when to expect him, he varies the order of his visits. In how many different orders can he visit the five machines?

2.17. Twenty nuts are to be put on 20 bolts. How many different combinations of nuts and bolts may be made?

2.18. An inspector takes 1 item from each of 15 trays. In how many ways may he pick a sample containing two nonconforming items? (Ignore here the order of selection of the items.)

2.19. A product has 20 points at which it may be nonconforming. In how many different ways may the product have just one nonconformity? Just two nonconformities? Just three nonconformities?

2.20. A product has 5 ways in which it might be critically nonconforming and 10 ways in which it might have a minor nonconformity. In how many different ways may it have just one critical nonconformity? One critical nonconformity and one minor nonconformity? Three minor nonconformities? Two critical nonconfomrities and two minor nonconformities?

2.21. A shop employs 2 foremen, 20 union men, and 5 nonunion men. In how many ways may a shop committee of one foreman, three union men, and one nonunion man be constituted?

2.22. A lot contains 100 items. How many different samples of 10 each may be made from these 100 items? If the lot contains nine nonconforming units, how many different samples containing two nonconforming units and eight conforming units may be selected? What, then, is the probability of drawing a random sample of 10 with 2 nonconforming units from a lot of 100 containing 9 nonconforming units?

12. SELECTED REFERENCES*

References on Probability. Fishburn (P '67 and P '69), Kolomogorov (B '50), Lindley (P '58), Munroe (B '51), Nagel (B '39), Neyman (B '38 and B '50), Savage, L. J. (B '54), Smith, J. G. and Duncan (B '44), Weaver, W. (P '48), and Von Mises (B '39)

References on Random Numbers, etc. Kogure (P '58), Page (P '59), Rand Corp. (B '55)

---

* B and P refer to the Book and Periodical sections, respectively, of the Cumulative List of References in Appendix V.

3

Frequency Distributions of Sample Data

It is an essential nature of industrial processes that their output varies from item to item. A product may be said to be uniform, but actually no two units will be alike. All we need are fine enough instruments to detect the differences, and it will be seen that the output is not uniform but essentially variable in nature. Our immediate purpose is to examine the tools that the statistician has devised for studying this interproduct variation.

In this chapter we shall discuss methods of describing the variability of sample data. In doing this we shall bear in mind that we are primarily interested in learning about the characteristics of the "universe" from which the given sample data have come. This universe might, for example, be a specified lot of items on hand or it might be the infinite set of items that could theoretically be produced by a given process if that process were able to continue indefinitely. Since we shall be dealing with sample data, the symbols used will generally be symbols for samples. In the next chapter we shall be concerned with methods of describing the variability of universes.

1. TYPES OF DATA

1.1. Discrete Data

In analyzing variability it is essential to distinguish two types of data: discrete data and continuous data. Discrete data necessarily vary in jumps. For example, data that arise from counting are discrete. Table 3.1 shows the number of noncomformities that were counted in each of 29 pieces of woolen cloth. Each nonconformity is an integral unit. It either exists or it does not. We cannot speak of "half a nonconformity" or "1¼ nonconformities" unless we are speaking of averages. For each piece the number of nonconformities

TABLE 3.1
Number of Nonconformities Found in 29 100-Yard Pieces of Woolen Cloth

No. of Piece	*No. of Nonconformities*	*No. of Piece*	*No. of Nonconformities*	*No. of Piece*	*No. of Nonconformities*
1	2	11	1	21	1
2	0	12	0	22	2
3	0	13	2	23	2
4	1	14	1	24	1
5	1	15	1	25	0
6	2	16	5	26	0
7	2	17	0	27	0
8	1	18	3	28	1
9	0	19	1	29	4
10	1	20	1		

SOURCE: Data provided by Bendix Radio Division, Bendix Aviation Corp.

is a whole number, and the variation in this number from piece to piece must proceed in multiples of 1. Table 3.1 thus illustrates discrete data.

1.2. Continuous Data

Continuous data refer to data that can potentially take on any value within a given range. This means that within that range there is no inherent restriction on the values that may be taken. Because of the limitations of measurement, all data vary in jumps. It is important to keep in mind, however, that some data have the potentiality of being measured as accurately as the instruments allow. Such data are inherently continuous, even though, for any level of measurement, they may vary in jumps.

Table 3.2 (p. 42), illustrates continuous data. It contains measurements of 145 bomb bases. The highest is 0.850 inches and the lowest 0.813 inches. Other measurements have the potentiality of any value between 0.813 and 0.850 inches, and possibly outside these limits. Measurements are made to the nearest thousandth of an inch, so that in reality the data vary in multiples of this unit. Each measurement ends on an integral 1,000th of an inch, and there are no measurements to a 10,000th of an inch. This limitation on the values appearing in the table is a limitation of measurement, however, and not an inherent characteristic of the data themselves. Essentially the data are continuous, since measurements can be made as fine as instruments allow.

The distinction between discrete and continuous data is important in subsequent analysis.

2. CONSTRUCTION OF A FREQUENCY DISTRIBUTION

In analyzing data, one of the first steps is to count the number of times a given variation occurs. If we list the various values of a variable in order

TABLE 3.2
Overall Heights of Fragmentation Bomb Bases (samples taken at random from production of April 20, 1944: specification 0.830 ± 0.010 inches)

Group No.	*a*	*b*	*c*	*d*	*e*
1	0.831	0.829	0.836	0.840	0.826
2	0.834	0.826	0.831	0.831	0.831
3	0.836	0.826	0.831	0.822	0.816
4	0.833	0.831	0.835	0.831	0.833
5	0.830	0.831	0.831	0.833	0.820
6	0.829	0.828	0.828	0.832	0.841
7	0.835	0.833	0.829	0.830	0.841
8	0.818	0.838	0.835	0.834	0.830
9	0.841	0.831	0.831	0.833	0.832
10	0.832	0.828	0.836	0.832	0.825
11	0.831	0.838	0.844	0.827	0.826
12	0.831	0.826	0.828	0.832	0.827
13	0.838	0.822	0.835	0.830	0.830
14	0.815	0.832	0.831	0.831	0.838
15	0.831	0.833	0.831	0.834	0.832
16	0.830	0.819	0.819	0.844	0.832
17	0.826	0.839	0.842	0.835	0.830
18	0.813	0.833	0.819	0.834	0.836
19	0.832	0.831	0.825	0.831	0.850
20	0.831	0.838	0.833	0.831	0.833
21	0.823	0.830	0.832	0.835	0.835
22	0.835	0.829	0.834	0.826	0.828
23	0.833	0.836	0.831	0.832	0.832
24	0.826	0.835	0.842	0.832	0.831
25	0.833	0.823	0.816	0.831	0.838
26	0.829	0.830	0.830	0.833	0.831
27	0.850	0.834	0.827	0.831	0.835
28	0.835	0.846	0.829	0.833	0.822
29	0.831	0.832	0.834	0.826	0.833

SOURCE: Lester A. Kauffman, *Statistical Quality Control at the St. Louis Division of American Stove Company* (OPRD, Quality Control Reports, No. 3, August 1945), p. 11.

of size and give the frequencies with which each size occurs, we have what is called a frequency distribution.

2.1. Discrete Distributions

Consider the data of Table 3.1. These have been rearranged in Table 3.3 in a form that shows how often each number of nonconformities has occurred. It tells us that there were 8 pieces of cloth in which there were 0 nonconformities, 12 pieces in which there was 1 nonconformity, 6 pieces in which there were 2 nonconformities, and so forth. Since the number of nonconformities

TABLE 3.3
Distribution of Number of Nonconformities in 29 Pieces of Woolen Cloth

Number of Nonconformities	Pieces with Designated Number of Nonconformities	
	Number	*Fraction*
0	8	0.2759
1	12	0.4137
2	6	0.2069
3	1	0.0345
4	1	0.0345
5	1	0.0345
Total	29	1.0000

FIGURE 3.1
Graph of a Discrete Frequency Distribution (data in Table 3.3)

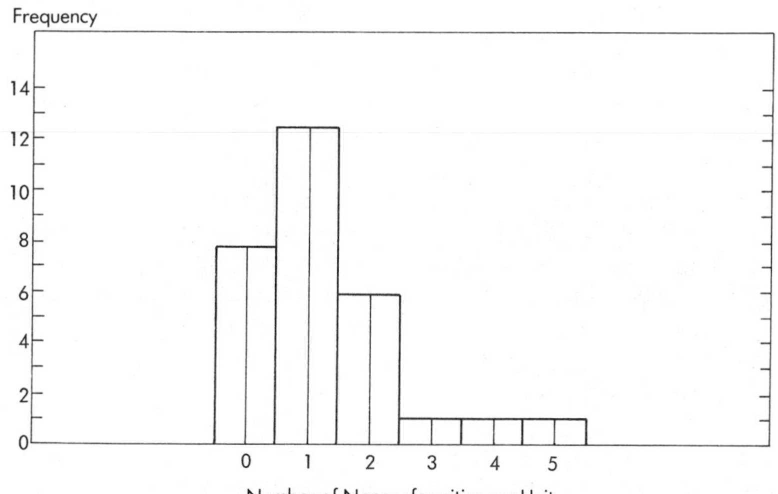

is a discrete variable, the distribution of Table 3.3 is a discrete frequency distribution.

The frequencies of Table 3.3 are represented in Figure 3.1 by vertical bars. Such a diagram is called a histogram. Although the bars touch,[1] it should be remembered that in the case of discrete data the frequencies apply only to points on the X-axis and not to a range of values. Figure 3.1 shows

[1] Sometimes the frequencies are represented by vertical lines at the various abscissa values, instead of by bars. Such a procedure clearly distinguishes the discrete case from the continuous case, in which contiguous bars must be used.

the distribution of the nonconformities of 29 given pieces of cloth but it has its greatest significance in suggesting how the nonconformities of this kind of cloth in general might be distributed.

2.2. Continuous Distributions

In making a frequency distribution of continuous data, it is almost always necessary to group the data into class intervals. Very few of the actual measurements will occur more than once unless we have a very large number of cases. Hence, in order to get an overall picture for purposes of generalization, grouping is usually imperative.

In grouping the data into classes or intervals, it is simplest to keep the size of the interval constant. The two problems are: (1) What size interval shall be chosen? and (2) How shall the intervals be located? Consider each of these questions with reference to the data of Table 3.2. These data run from 0.813 to 0.850 inches, a range of 0.037 inches. If this whole range is to be broken down into subintervals of constant size, the question of what size interval becomes the question of how many intervals. If the answer is, say, five intervals, then each interval should be 0.007 or 0.008 inches ($0.037/5 = 0.0074$). If it is 10 intervals, then each should be 0.003 or 0.004 inches. The answer to be given in any concrete case depends on the judgment of the statistician. If we take too many intervals, we will not get the general picture we are trying to obtain. If we take too small a number of intervals, the general picture will be blurred; it will lack sufficient detail to give it proper characterization. The best compromise appears to be atained in most cases by using 10 to 15 intervals. With reference to the data of Table 3.2, it would appear best to use intervals of 0.003 inches.

The second question is how to locate the intervals on the abscissa scale. For example, the intervals could run 0.811 to 0.813 inclusive, 0.814 to 0.816 inclusive, and so on, or they could run 0.812 to 0.814 inclusive, 0.815 to 0.817 inclusive, and so forth, or 0.81250 to 0.81549 inclusive, 0.81550 to 0.81849 inclusive, and so on. In practice, it will probably be best to try several locations to see which gives the best looking distribution. Generally, a distribution which is regular and symmetrical is to be preferred to one that is not. Since the data are presumably measured to the nearest 0.001 inch, it is also probably best to have the intervals start at a point like 0.8125. For this will enable each case to be assigned to an interval without ambiguity and will put each case in its proper interval. For the data of Table 3.2, the intervals of 0.8125 up to but not including 0.8155, 0.8155 up to but not including 0.8185, 0.8185 up to but not including 0.8215, and so forth, were selected. This yields a total of 13 intervals.

When the data of Table 3.2 are assigned to the various intervals and the number in each interval counted, the result is the frequency distribution presented in Table 3.4. It probably gives us a good picture of how the heights of the bomb bases in general vary in size. A graphic representation of the distribution of Table 3.4 is shown in Figure 3.2.

TABLE 3.4
Frequency Distribution of Overall Heights of Fragmentation Bomb Bases

*Height** *(Inches)*	*Absolute Frequency*	*Relative Frequency*
0.8125–	2	0.0138
0.8155–	3	0.0207
0.8185–	4	0.0276
0.8215–	5	0.0345
0.8245–	14	0.0965
0.8275–	21	0.1448
0.8305–	55	0.3793
0.8335–	23	0.1586
0.8365–	7	0.0483
0.8395–	6	0.0414
0.8425–	2	0.0138
0.8455–	1	0.0069
0.8485–	2	0.0138
Total	145	1.0000

* Lower limits.

2.3. Cumulative Distributions

If the frequencies of a frequency distribution are cumulated from one end or the other, usually from the lower to the higher end of the scale, we have a cumulative distribution. The cumulative distribution constructed from Table 3.1 is shown in Table 3.5 and the cumulative distribution constructed from

FIGURE 3.2
Graph of a Continuous Frequency Distribution (data in Table 3.4)

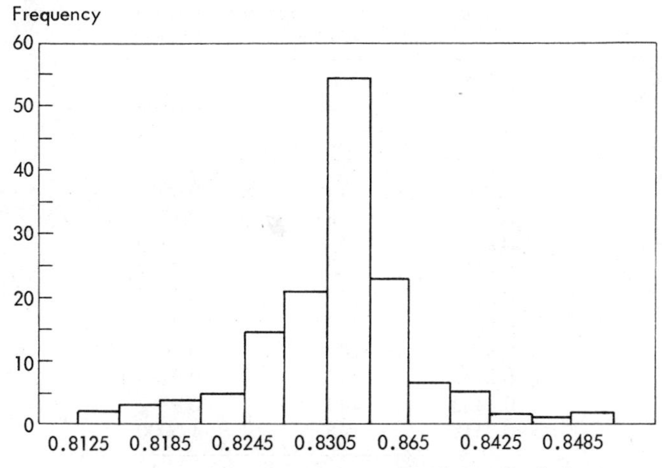

TABLE 3.5
Cumulative Distribution of the Number of Nonconformities in 29 Pieces of Woolen Cloth

Number of Nonconformities	Pieces with Number of Nonconformities Equal to or Less Than that Designated	
	Number	*Fraction*
0	8	0.2759
1	20	0.6896
2	26	0.8965
3	27	0.9310
4	28	0.9655
5	29	1.0000

Table 3.2 is shown in Table 3.6. Graphs of the two cumulative distributions are shown in Figures 3.3 and 3.4.

It will be noted that a distribution cumulated from the lower end has the following properties:

For $X = -\infty$ or lower limit, cumulative relative frequency = 0.

For $X = +\infty$ or upper limit, cumulative relative frequency = 1.00

TABLE 3.6
Cumulative Distribution of the Overall Heights of Fragmentation Bomb Bases

Upper Class Limit of Height	Bomb Bases with Heights Less Than that Designated	
	Number	*Fraction*
0.8125	0	0.0000
0.8155	2	0.0138
0.8185	5	0.0345
0.8215	9	0.0621
0.8245	14	0.0965
0.8275	28	0.1931
0.8305	49	0.3379
0.8335	104	0.7172
0.8365	127	0.8759
0.8395	134	0.9241
0.8425	140	0.9655
0.8455	142	0.9793
0.8485	143	0.9862
0.8515	145	1.0000

FIGURE 3.3
Graph of a Discrete Cumulative Frequency Distribution (data in Table 3.5)

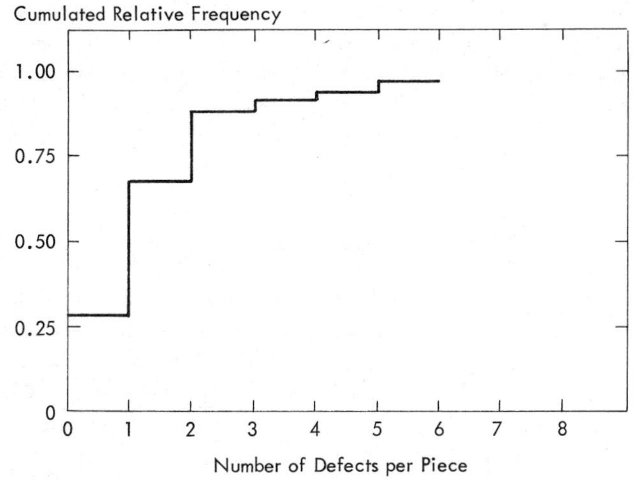

FIGURE 3.4
Graph of a Continuous Cumulative Frequency Distribution (data in Table 3.6)

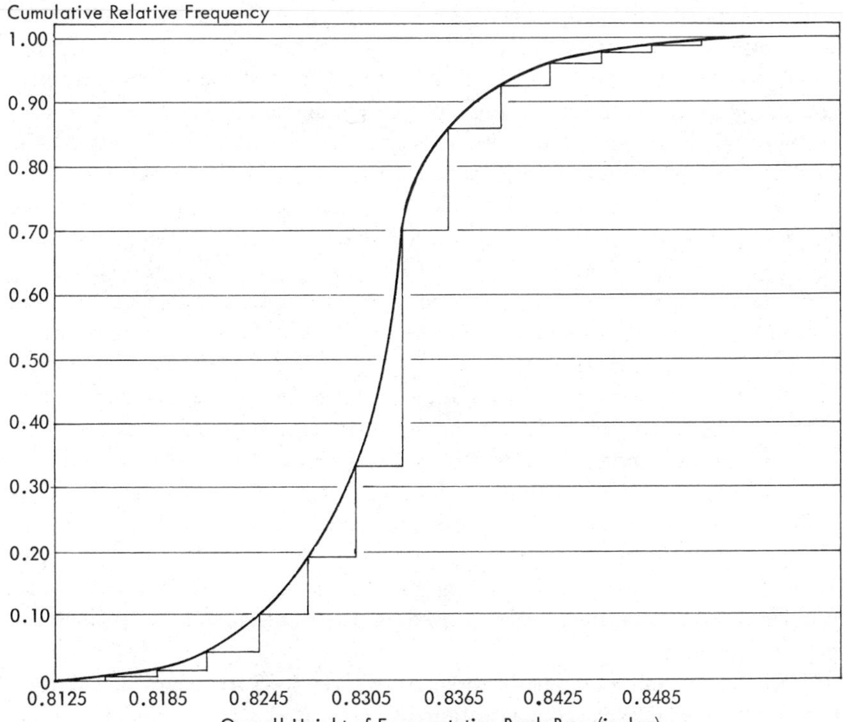

Cumulative relative frequency for X_1 is less than or equal to cumulated relative frequency for X_2 if $X_1 < X_2$.

Since the bomb data are continuous, the cumulative distribution of Figure 3.3 can be smoothed by a curve. (See Figure 3.4.) This shows approximately how the cumulative distribution function would look if the class intervals were made infinitesimally small and the number of cases indefinitely increased. It is probably a better representation of the cumulative distribution for the whole output (the universe) of bomb bases than the broken line. A smooth curve of this kind is called an ogive.

3. CHARACTERISTICS OF A FREQUENCY DISTRIBUTION

A frequency distribution gives a "bird's-eye view" of the variation in a set of data. First of all, it tells where the data are on the scale of variation and whether the data have any central tendency. It also tells where this concentration, if any, is located. The frequency distribution of Table 3.1 shows an obvious concentration at the smaller number of nonconformities, while the distribution of Table 3.2 shows a marked concentration in the center of the range of variation.

Some distributions have more than one point of concentration. These are said to be multimodal distributions. A distribution with a single point of concentration is called unimodal. When the number of cases is large, distributions of production data are usually unimodal. When multimodal distributions occur, it is likely to be an indication that the data are not homogeneous. Such distributions would arise if part of the output were produced under one set of conditions and part under another.

A frequency distribution also tells about the general variability of the data. It answers the question: How much variability is there? If one distribution is narrow and another is broad, both being constructed on the same horizontal scale, it means that the first has less variablity than the second. A graph of a distribution can be varied at will by modifying the horizontal scale. Consequently, whether a distribution shows little or great variation should be judged with reference to the scale itself and not by the appearance of the graph of the distribution.

The third characteristic of the data revealed by a frequency distribution is the symmetry of its variation. Is the distribution symmetrical about its point of central tendency, or is it lopsided? If the latter, the distribution is said to be skewed. A distribution that has most of its cases close to the point of central tendency on the left but many cases spread out for some distance on the right is said to be positively skewed. The distribution of Table 3.1 is positively skewed. If the tail of the distribution runs the other way, it is said to be negatively skewed. A distribution that is symmetrical has no skewness.

The fourth and final part of the picture given by a frequency distribution pertains to the relative concentration of cases at the center and along the tails of the distribution. This characteristic of a distribution is called its kurtosis. If a distribution has relatively high concentration in the middle and out on the tails, but relatively little in between, it has large kurtosis. If it is relatively flat in the middle and has thin tails, it has little kurtosis. The distribution of Table 3.4 has more than average, but not extreme kurtosis.

The various characteristics of a frequency distribution are illustrated in Figures 3.5–3.8.

4. MEASURING THE CHARACTERISTICS OF A DISTRIBUTION

In some problems a general picture of a distribution is all that is needed. For this a frequency table or histogram is sufficient. In many problems,

FIGURE 3.5
Illustrations of Central Tendency

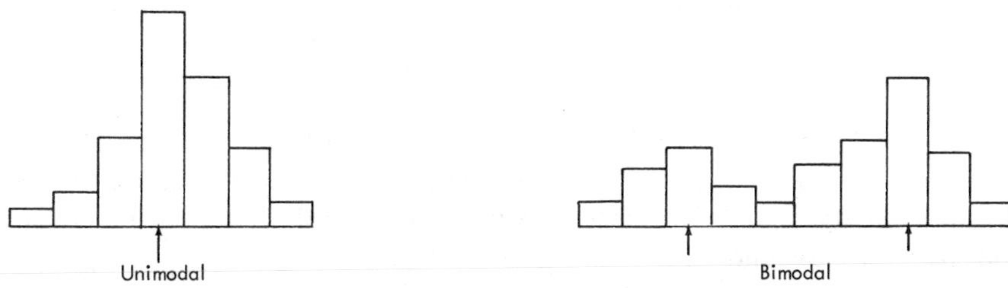

FIGURE 3.6
Illustrations of Variability

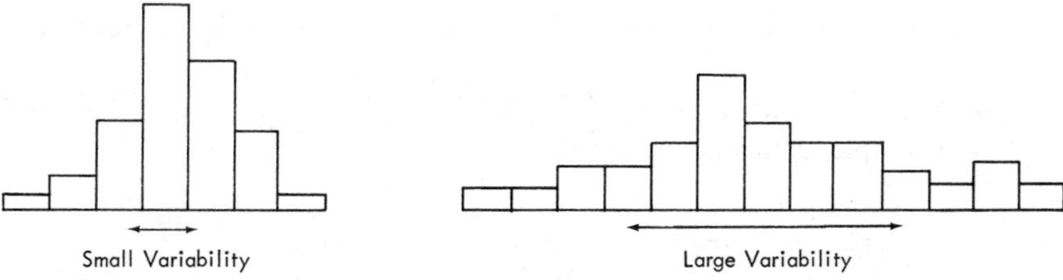

FIGURE 3.7
Illustrations of Skewness

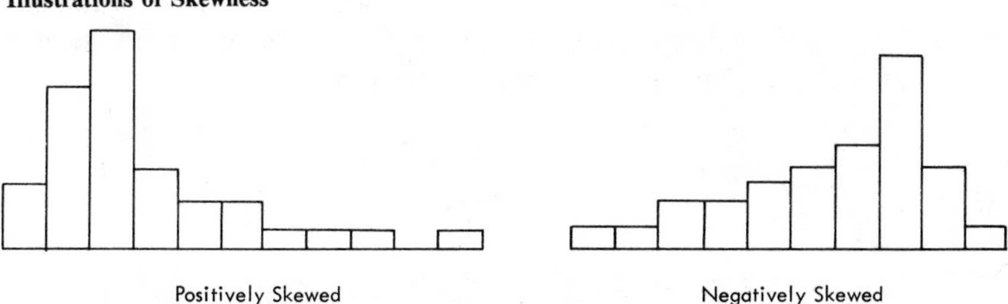

FIGURE 3.8
Illustrations of Kurtosis

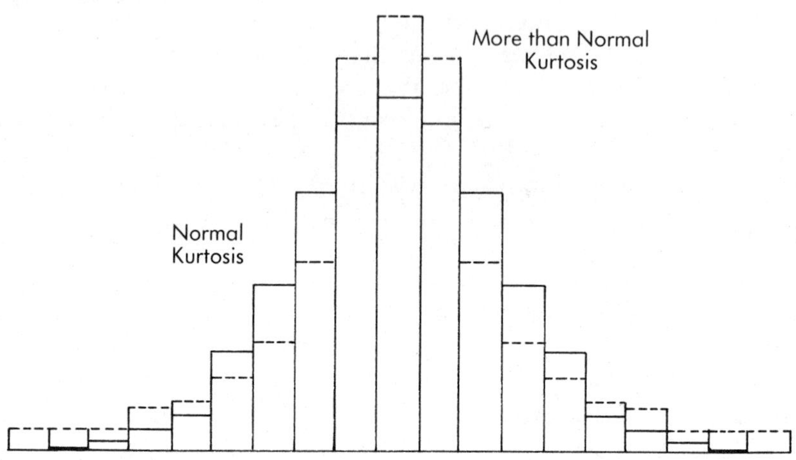

however, it is desirable to have numerical measurements of the characteristics of a distribution. The advantages accruing from the derivation of numerical measures are as follows:

1. Four or five figures occupy much less space than a table or chart, and in many cases they provide all the information desired.
2. Numerical measures force the statistician to sharpen his ideas pertaining to a characteristic and thus improve his understanding of it.
3. Numerical measures permit comparisons between one distribution and another.
4. Numerical measures help in making inferences about a universe from a sample from that universe.

4.1. Measures of Location and Central Tendency

We shall describe briefly three different measures of location and central tendency. These are the arithmetic mean, the median, and the mode.

4.1.1. The Arithmetic Mean. *4.1.1.1. Definition.* The arithmetic mean is a commonly used measure of location. It is represented by the symbol $\bar{X}$. If the data are ungrouped, the mean is given by the formula

$$\bar{X} = \frac{\sum_j X_j}{n} \tag{3.1}$$

If the data are classified or grouped, the mean is given by the formula

(3.1a)

$$\bar{X} = \frac{\sum_j F_j X_j}{n}$$

In these formulas F_j is the frequency of the *jth* value or class (X_j), n is the total number of cases, and Σ means to sum.[2]

Thus if the data are discrete and are arranged in a frequency distribution, the mean may be computed by multiplying each value by the frequency with which it occurs, adding the products, and dividing by the sum of the frequencies (= total number of cases, n). If the data are continuous and are arranged in a frequency distribution with class intervals, the mean may be computed by multiplying the midpoint of each interval by the frequency of the interval, adding, and dividing by the sum of the frequencies. A mean computed from grouped data in this way may be slightly different from the mean calculated from the same data in ungrouped form, because of the assumption that all cases in any class interval have the value of the midpoint of the interval. The difference, however, will not be great if the grouping has been done wisely.[3]

4.1.1.2. Short Method for Computing a Mean of Grouped Data. When the frequencies are large and the measurements are given to several significant figures, the calculation of the mean of grouped data can be greatly facilitated by using a short method. The essential features of this short method are (1) to carry out the calculation with reference to a conveniently selected origin and (2) to measure the variable in multiples of the class interval. The answer so obtained is then adjusted to refer the mean to the initial origin and express it in original units.

Figure 3.9 depicts graphically the various quantities that are involved in the short method. In this diagram, X is the midpoint of any interval; A is the arbitrarily selected origin (usually the midpoint of some interval); x is the deviation of X from the mean $\bar{X}$; and $\xi = (X - A)/i$. The short method consists of computing

(3.2)

$$\mu_{a\,1} = \frac{\sum_j F_j \xi_j}{n} i$$

where i is the size of the class interval, and then finding $\bar{X}$ from the relationship

(3.3)

$$\bar{X} = A + \mu_{a\,1}$$

a formula that is derived in Appendix I (2).

The computations involved in the short method may be illustrated with reference to the data of Table 3.4. The work may be summarized as follows:

---

[2] See Appendix I (1).

[3] It can be surprisingly great in certain instances, however, if the grouping has not been done wisely. The maximum possible error when the class interval is of constant size i is $i/2$.

Lower Limit of Class Interval X_j	*Midpoint*	F_j	ξ_j	$F_j\xi_j$
0.8125–	0.8140	2	−6	−12
0.8155–	0.8170	3	−5	−15
0.8185–	0.8200	4	−4	−16
0.8215–	0.8230	5	−3	−15
0.8245–	0.8260	14	−2	−28
0.8275–	0.8290	21	−1	−21
0.8305–	0.8320 (= *A*)	55	0	0
0.8335–	0.8350	23	1	23
0.8365–	0.8380	7	2	14
0.8395–	0.8410	6	3	18
0.8425–	0.8440	2	4	8
0.8455–	0.8470	1	5	5
0.8485–	0.8500	2	6	12
		$n = 145$		$-27 = \sum_j F_j \xi_j$

Here A is taken as the midpoint of the interval 0.8305–0.8335 and has the value 0.8320. The ξ_j's are marked off from this point. By definition,

$$\mu_{a\,1} = \frac{\sum_j F_j \xi_j}{n} i$$

Hence, for these data $\mu_{a\,1} = \frac{-27}{145}(0.003) = -0.0005586$. Consequently, $\bar{X} = 0.8320 - 0.0006 = 0.8314$.

In employing the short method, it is usually desirable that A be picked as the midpoint of some interval that has a large frequency and is near the middle of the distribution. No hard-and-fast rule has to be followed. It need only be remembered that A should be selected so as to save labor, and the more the better. Putting A way at one end will make for large ξ_j's at the other end; hence this is usually to be avoided. Putting A in an interval with a large frequency will eliminate multiplying by this frequency (since $\xi_j = 0$ for the A interval), and hence the larger the frequency of the A interval, the more the labor saved. In practice, we often have to compromise between these two tendencies.

4.1.1.3. The Mean as a Measure of Location. The mean is a mathematical measure of location that is based on all the data. It is the "center of gravity." Sample means from bell-shaped or "normal" universes (see Chapter 4) vary less from sample to sample or at least do not vary more than other measures of location. The sample mean is thus said to be an efficient estimate[4] of the universe mean.

---

[4] See Chapter 24, Section 1.2.

FIGURE 3.9
Illustration of Quantities Used in Short Method for Calculating $\bar{X}$ and Other Statistics

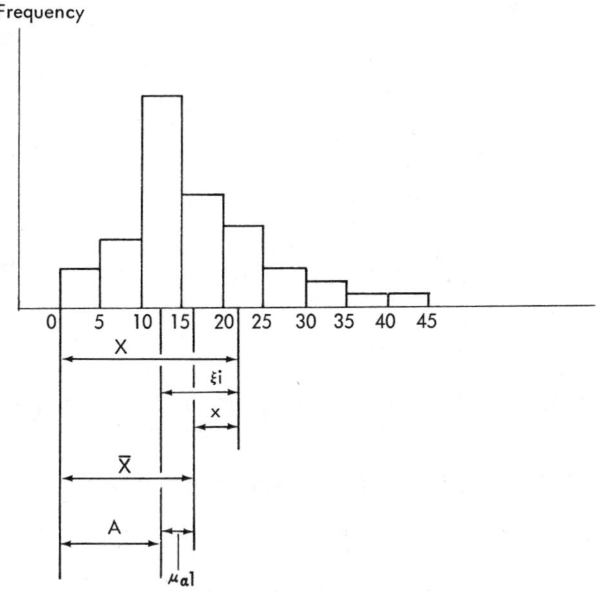

If a distribution is unimodal and symmetrical, the mean is also a measure of central tendency. If it is skewed, the mean will be pulled away from the point of central tendency in the direction of the tail of the distribution. In such a case, the mean is not a "typical" value. These relations are illustrated in Figures 3.10 and 3.11.

4.1.2. The Median. *4.1.2.1. Definition.* The "median" of a set of data is the value above which there are as many cases as there are below it.[5] It is usually represented by the symbol *Mi.*

4.1.2.2. Computation of Median of Ungrouped Data. To compute the median for ungrouped data, arrange the values in order of size. If the number of cases is odd, the median is the value of the middle case. If the number of cases is even, the median is taken as the arithmetic average of the two middle cases. For example, the median of 6, 8, 4, 9, 7, 3, 5, is 6; the median of 4, 9, 5, 7, 8, 11, is 7.5; and the median of 7, 9, 3, 10, 5, 2, 5, is 5.

4.1.2.3. Computation of Median of Grouped Data. To compute the median of grouped data, it is necessary to interpolate for the $n/2$th case. The procedure is to sum the frequencies by intervals beginning at either end and to stop at the interval which contains the middle case. Suppose that the number of cases down to the limit of this interval is n' (lower limit if

[5] In instances in which there is more than one case having exactly the same value as the median, these extra cases must be considered as belonging to the group "above" or "below" the median.

FIGURE 3.10
Illustration of the Mean of a Symmetrical Frequency Distribution

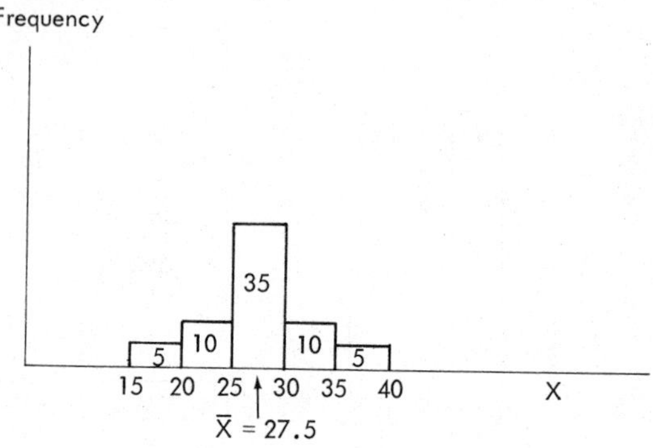

the counting was begun at the bottom; upper limit if the counting was begun at the top). Then the median is found by adding to the lower limit or subtracting from the upper limit (whichever is used) the quantity $\frac{|n/2 - n'|}{F_o}$ *(i)*, where F_o is the frequency of the interval in question and i is the size of the class interval.

To illustrate, let us compute the median of the distribution of Table 3.4. For these data $n/2 = 145/2 = 72.5$. Adding from the bottom, we find that there are 49 cases up to, but not including, the interval 0.8305 to 0.8335. In this interval there are 55 cases, so that up to the next interval there are 104 cases. The 72.5th case therefore falls in the interval 0.8305 to 0.8335. As was seen, up to 0.8305 there are 49 cases. Hence, there are $72.5 - 49 = 23.5$ cases yet to go to reach the median. By interpolation, the median is located at $0.8305 + \frac{23.5}{55}(0.003) = 0.8317$.

FIGURE 3.11
Illustration of the Mean of an Asymmetrical Frequency Distribution

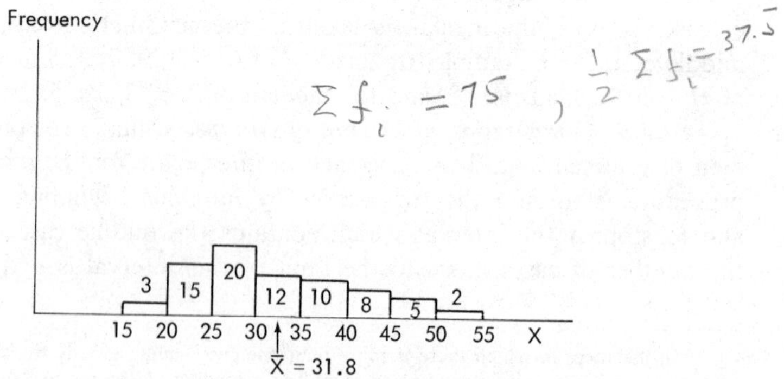

It may be noted that if a graph has been made of the cumulative frequency distribution of the data, the median may be interpolated directly on this graph. For example, turn to Figure 3.4 and note that a horizontal line drawn through the 50 percent point on the vertical axis will intersect the smooth curve at a point whose abscissa is 0.8322, which is approximately the same as that obtained above. This graphic interpolation is better than the "straight-line" interpolation used above, since it probably gives results closer to the median of the universe.

4.1.2.4. Computation of Median of Discrete Data. The computation of the median of discrete data may be illustrated with reference to the data of Table 3.1. For these data $n/2 = 29/2 = 14.5$. There are 8 cases with 0 defects and 12 with 1 defect. Both the 14th and 15th cases have 1 defect. Hence, in this instance, the median is 1. If the 14th case had been 1 and the 15th case had been 2, the median would have been taken as 1.5, although there is no case with that value.

4.1.2.5. The Median as a Measure of Location. The median is a position average that is unaffected by the values of the cases on either side of it. Unlike the mean, therefore, it is not affected by extreme values. If a distribution is skewed, the median is not pulled away from the point of central tendency as much as is the mean (see Figures 3.12 and 3.13). If a universe has excessive kurtosis, medians of samples from such a universe may vary less from sample to sample than the means will.

4.1.3. The Mode. The mode of a set of data is the value at which the greatest number of cases is concentrated. It is represented by the symbol *Mo.* If the data are grouped, the interval with the largest number of cases is the modal interval. If the frequency distribution is smoothed out by a frequency curve, the mode is the abscissa value at which the peak of the

FIGURE 3.12
Comparison of the Mean and Median of a Symmetrical Frequency Distribution

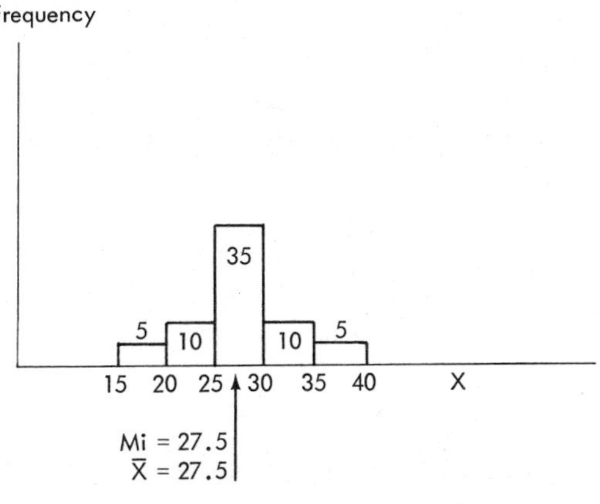

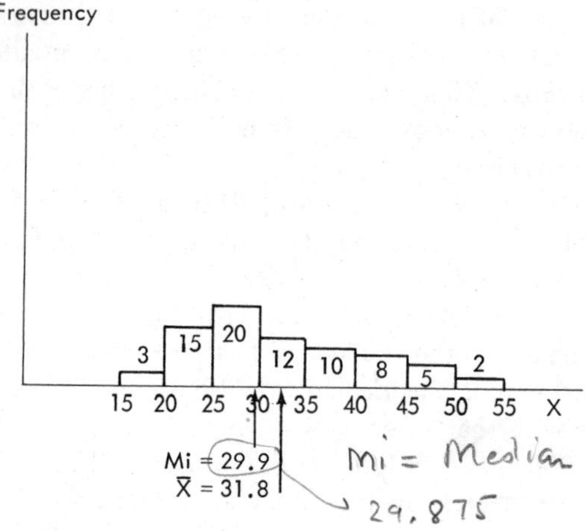

FIGURE 3.13
Comparison of the Mean and Median of a Positively Skewed Frequency Distribution

curve occurs. In Table 3.3 the mode is 1, and in Table 3.4 the modal interval is 0.8305 to 0.8335.

By definition the mode directly indicates the point of central tendency. However, except for discrete data, it is hard to give the mode an exact value. For grouped data, a modal interval takes the place of an exact mode. If a curve is fitted to the grouped data, many problems arise that make a precise determination of its peak a difficult, or at least a lengthy process. For this reason and because it varies from sample to sample more than the mean and median do, the mode is not used much.

4.2. Measures of General Variability

We shall consider five measures of general variability: the range, the average deviation, the root-mean-square deviation, the standard deviation, and the semiquartile range.

4.2.1. The Range. The simplest measure of general variability is the range. This is the difference between the highest and the lowest value of a set of data. It applies only to ungrouped data[6] and is represented by the symbol R. Thus

(3.4) $$R = X_H - X_L$$

The range of the data of Table 3.2, for example, is $0.850 - 0.813 = 0.037$ inches. In small samples the range is a relatively sensitive measure of general

[6] Of course, for grouped data we could take the difference between the upper limit of the highest interval and the lower limit of the lowest interval as a rough measure of the range.

variability, but in large samples the range is not so good as the measures discussed in the following paragraphs.

4.2.2. The Average Deviation *4.2.2.1. Definition.* If we take a central point such as the mean of a set of data and compute the mean deviation of the cases from this central point, we will get a measure of general variability. This is called the average deviation and is represented by "AD."

4.2.2.2. Computation of AD for Ungrouped Data. Consider the cases 4, 7, 9, 11, 14. Their mean is 9, and the average deviation from the mean is

$$\frac{|4-9|+|7-9|+|9-9|+|11-9|+|14-9|}{5}=\frac{5+2+0+2+5}{5}=\frac{14}{5}=2.8$$

Note that the absolute values of the deviations are taken, not their signed values, for the sum of the latter would be zero.[7]

The average deviation does not have to be measured from the mean. In fact, it is considered best to measure the average deviation from the median, since it has its least value when so measured.[8] In the present instance this makes no difference, since the distribution is symmetrical and the mean and the median are the same.

4.2.2.3. Computation of AD for a Discrete Distribution. The computation of the average deviation of a discrete distribution may be illustrated with reference to the distribution:

X_j	F_j
1	5
2	10
3	15
4	10
5	5
	45

This has a mean of 3. The 5 cases which have an X value of 1 deviate by 2 points from the mean of 3. The 10 cases with an X value of 2 deviate by 1 point from the mean of 3. The 15 cases with an X value of 3 deviate by 0 points from the mean of 3, and so on. The average of the absolute deviations from the mean is computed as follows:

---

[7] For $\sum_j (X_j - \bar{X}) = \sum_j X_j - \sum \bar{X} = n\bar{X} - n\bar{X} = 0$, since $\bar{X} = \sum_j X_j/n$ by definition.

[8] The reason for this is that the median, by definition, comes at the middle of the distribution. If some other point is taken to measure the average deviation, the deviations of some cases will be increased and those of other cases will be decreased. The deviations of the points lying between the median and the new point, however, will not be decreased as much as the deviations of an equal number of points on the other side of the median will be increased. Hence the sum of the absolute deviations from the new point will be greater than from the median.

Deviation $x_j = X_j - \bar{X}$	Absolute Value of Deviation $\lvert x_j \rvert$	Frequency F_j	$F_j \lvert x_j \rvert$
−2	2	5	10
−1	1	10	10
0	0	15	0
+1	1	10	10
+2	2	5	10
		$n = 45$	$\sum_j F_j \lvert x_j \rvert = 40$

$$AD_{\bar{X}} = \sum_j \frac{F_j \lvert x_j \rvert}{n} = \frac{40}{45} = 0.89$$

4.2.2.4. Computation of AD for Grouped Data. We can compute the average deviation for grouped data in exactly the same way we computed it for a discrete distribution. In this we take the midpoint of each interval as the value of all cases in the interval.

For a special method of computing the average deviation from grouped data, see J. G. Smith and A. J. Duncan, *Elementary Statistics and Applications,* pp. 222–23.

4.2.2.5. The Average Deviation as a Measure of General Variability. The average deviation is easy to compute for small samples, but it is hard to handle in mathematical analysis. owing to the use of absolute values. It is seriously affected by extreme variations.

4.2.3. The Root-Mean-Square Deviation. *4.2.3.1. Definition.* Another measure of general variability is the root-mean-square deviation (RMSD). This is the square root of the average of the squared deviations from the mean. The formula for the root-mean-square deviation is

(3.5)

$$\text{RMSD} = \sqrt{\frac{\sum_j F_j (X_j - \bar{X})^2}{n}}$$

By squaring and then later taking the square root, the difficulty created by the signs of the deviations is eliminated.

4.2.3.2. The Root-Mean-Square Deviation as a Measure of General Variability. The square of the RMSD is a quantity that is later identified as the sample second *moment* around the mean and given the symbol μ_2. (See Section 4.3 below.) When μ_2 is used as an estimator of the second moment around the mean of the universe[9] (μ_2'), it is found to have a downward bias. As noted in the next section, this can be corrected by multiplication by the factor $n/(n - 1)$.

[9] For the precise definition of μ_2, see Section 4.3 of Chapter 4.

4.2.4. The Variance and the Standard Deviation. *4.2.4.1. Definitions.* Because of the bias inherent in the use of $(\text{RMSD})^2$ to estimate μ_2', it has become customary to work with an unbiased statistic. This is the quantity

(3.6)

$$s^2 = \frac{n}{n-1}(\text{RMSD})^2 = \frac{\sum_j F_j(X_j - \bar{X})^2}{n-1}$$

in which the bias is removed by substituting $n - 1$ for n.

The quantity s^2 is called the *sample variance.* The *universe variance* is defined as the second moment about the mean for the universe (μ_2') and is here given the alternate symbol σ'^2.

The square root of the sample variance is called the sample *standard deviation (s).* Likewise the square root of the universe variance is called the universe standard deviation (σ').

4.2.4.2. Computation of the Standard Deviation of Ungrouped Data. For ungrouped data the computation of the standard deviation may be illustrated by the following example:

$$\text{Let } X = 5, 3, 9, 7. \text{ Then } \bar{X} = \frac{5+3+9+7}{4} = \frac{24}{4} = 6$$

$$\text{and } s^2 = \frac{(5-6)^2 + (3-6)^2 + (9-6)^2 + (7-6)^2}{4-1} = \frac{1+9+9+1}{3}$$

$$= \frac{20}{3} = 6.67, \text{ Hence } s = \sqrt{6.67} = 2.58.$$

4.2.4.3. Computation of the Standard Deviation of Grouped Data. For grouped data the computation of the standard deviation may be illustrated as follows:

X_j	*Midpoint*	F_j	$x_j = X_j - \bar{X}$	x_j^2	$F_j x_j^2$
0.5—	1	5	−2	4	20
1.5—	2	10	−2	1	10
2.5—	3	15	0	0	0
3.5—	4	10	+1	1	10
4.5—	5	5	+2	4	20
		$n = 45$	[Note: $\bar{X} = 3.0$]		$\sum_j Fx_j^2 = 60$

$$s = \sqrt{\frac{\sum_j F_j x_j^2}{n-1}} = \sqrt{\frac{60}{44}} = 1.17$$

4.2.4.4. Short Method for Computing the Standard Deviation of Grouped Data. The short method for computing the mean of grouped data may be extended to yield a short method for computing the standard deviation or variance. This is done by simply adding a column for $F_j\,(\xi_j)^2$. Thus we have for the data of Table 3.4:

X_j	*Midpoint*	F_j	ξ_j	$F_j(\xi_j)$	$F_j(\xi_j)^2$
0.8125–		2	–6	–12	72
0.8155–		3	–5	–15	75
0.8185–		4	–4	–16	64
0.8215–		5	–3	–15	45
0.8245		14	–2	–28	56
0.8275–		21	–1	–21	21
0.8305–	0.8320	55	0	0	0
0.8335–		23	1	23	23
0.8365–		7	2	14	28
0.8395–		6	3	18	54
0.8425–		2	4	8	32
0.8455–		1	5	5	25
0.8485–		2	6	12	72
		145		–27	567

Let

$$\mu_{a\,1} = \frac{\sum_j F_j\,(\xi_j)}{n}\,i \text{ and } \mu_{a\,2} = \frac{\sum_j F_j\,(\xi_j)^2}{n}\,i^2$$

We then have the following "short formulas":[10]

$$\mu_2 = \mu_{a\,2} - \mu_{a\,1}^2$$

$$s^2 = \frac{n}{n-1}\,\mu_2$$

$$s = \sqrt{\left(\frac{n}{n-1}\right)\mu_2}$$

On applying these formulas to the above data, we have

$$s^2 = \frac{145}{144}\left[\frac{567}{145} - \left(\frac{-27}{145}\right)^2\right](0.003)^2 = 0.00003512$$

and

$$s = 0.00593 \text{ inches.}$$

4.2.4.5. The Variance and Standard Deviation as Measures of General Variability. The variance (s^2) is a mathematically defined quantity that is based on all the data. It can be handled analytically with relative ease

[10] See Appendix I (3).

since it does not involve such troublesome procedures as taking absolute values. It has the desirable property that s^2 is an unbiased estimate of the universe variance σ'^2 and sample values of s^2 from bell-shaped or normal[11] universes vary less or at least do not vary more than other unbiased estimates of the universe variance.

Whereas s^2 is an unbiased estimate of the universe variance σ'^2, it should be noted that s is *not* an unbiased estimate of the universe standard deviation σ'. To correct the bias in s we can in the case of a normal universe divide by the c_4 factor. For values of n up to 25, this is given in Table M of Appendix II. For larger values of n we use the formula

$$c_4 = \sqrt{\frac{2}{n-1}}\left[\frac{\Gamma(n/2)}{\Gamma((n-1)/2)}\right] \tag{3.7}$$

where $\Gamma(\cdot)$ is the gamma function, values of which may be found in the U.S. Department of Commerce, *Handbook of Mathematical Functions.*

For our numerical example we have:

$$\begin{aligned} &\text{An unbiased estimate of } \sigma'^2 = s^2 &&= 0.00003512 \\ &\text{An unbiased estimate of } \sigma' \ = s/c_4 &&= (0.00593)/0.9987 \\ & &&= 0.00594 \end{aligned}$$

The value for σ' assumes that the sample comes from a normal universe which is probably not true in this instance. (See Section 3.6 of Chapter 27.) In any case, the correction for bias will be very small for a sample of size 145.

4.2.5. Quartiles and the Semiquartile Range. Just as a median divides a distribution in half, so the quartiles divide it into quarters. These are all calculated like the median. For grouped data, for example, the lower quartile is interpolated as the $n/4$th case and the upper quartile as the $3n/4th$ case. The median is, of course, the second quartile. The lower quartile (Q_1) is thus a value such that 25 percent of the cases lie below it and 75 percent of the cases above it. The upper quartile (Q_3) is a value such that 25 percent of the cases lie above it and 75 percent of the cases below it. Half the difference between Q_3 and Q_1 is known as the semi-quartile range (Q) (see Figure 3.14).

The computation of the quartiles and the semiquartile range for grouped data[12] may be illustrated with reference to the data of Table 3.4. The first quartile (Q_1) is the value of the $\frac{145}{4}$th or 36¼th case. To interpolate, we count 28 cases up to, but not including, the class limit 0.8275. This leaves 8¼ cases to go. In the next interval there are 21 cases. Therefore, Q_1 falls at $0.8275 + \frac{8.25}{21}(0.003) = 0.8287$. In the same manner, by working from

[11] See Section 6.1 of Chapter 4.

[12] For ungrouped data, see A. L. Bowley, *Elements of Statistics* (5th ed.; London: P. S. King & Son, Ltd., 1926), pp. 106–7.

FIGURE 3.14
Illustration of the Semiquartile Range

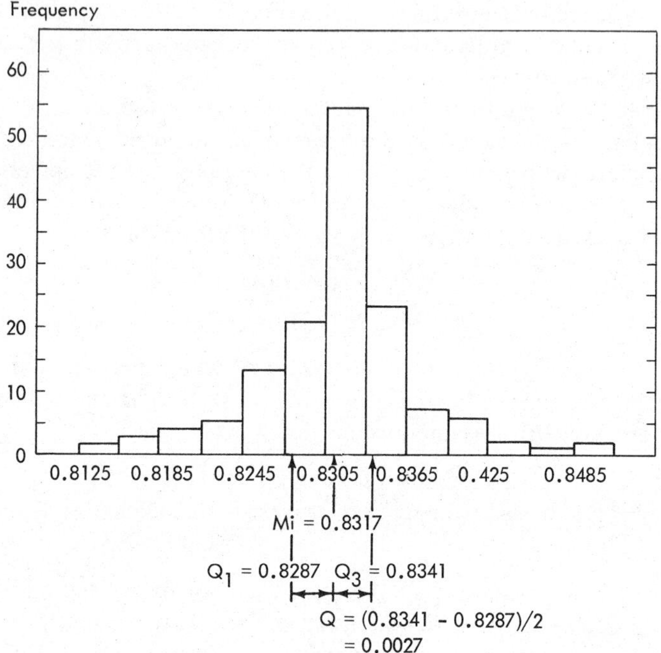

the other end, it can be shown that $Q_3 = 0.8365 - \frac{18.25}{23}(0.003) = 0.8341$. Hence the semiquartile range is $Q = \frac{0.8341 - 0.8287}{2} = 0.0027$ inches. Like the median, the quartiles and hence the semiquartile range can be computed directly from the cumulative distribution. Thus from Figure 3.4 we find that the 75 percent point comes at 0.8340 and the 25 percent point comes at 0.8288, results which are very close to those obtained by linear interpolation.

It is to be noted that the semiquartile range is usually less than both the average deviation and the RMSD. If a distribution is extremely U-shaped, however, it is possible for the semiquartile range to be the largest of the three measures.

4.3. Measures of Skewness and Kurtosis

The statisticians have also developed various measures of skewness and kurtosis. These will be discussed more fully in Chapter 27. It need only be noted here that there are certain quantities called moments about the arithmetic mean, defined by the formula

(3.8)
$$\mu_k = \frac{\Sigma Fx^k}{n}$$

where $x = X - \bar{X}$. The first moment about the mean is always zero[13] because $\Sigma Fx = 0$, and the second moment about the mean adjusted for bias is the variance, i.e.,

$$\mu_1 = 0$$

$$\mu_2 \frac{n}{n-1} = s^2$$

A measure of skewness often used in statistics is

(3.9)
$$\gamma_1 = \left(\frac{\mu_3^2}{\mu_2^3}\right)^{1/2}$$

and a measure of kurtosis is[14]

(3.10)
$$\gamma_2 = \frac{\mu_4}{\mu_2^2} - 3$$

Bell-shaped frequency distributions[15] will have approximately $\gamma_1 = 0$ and $\gamma_2 = 0$.

4.4. Summary of Measurements of the Distribution of Bomb Bases

In previous sections we have computed various measurements of the distribution of the heights of bomb bases presented in Table 3.4. These measurements together with measurements of skewness and kurtosis may be summarized as follows. All measurements except the range pertain to the data after they have been grouped.

Measurements of central tendency:
- Mean = 0.8314 inches
- Median = 0.8317 inches
- Midpoint of modal group = 0.8320 inches

Measurements of general variability:
- Range of original ungrouped data = 0.0370 inches
- Standard deviation of grouped data = 0.00593 inches
- Semiquartile range = 0.0027 inches

Measurement of relative skewness:[16]
- $\gamma_1 = -0.08$

---

[13] See above, footnote 7.

[14] $\beta_2 = \mu_4/\mu_2^2$ is also frequently used to measure kurtosis.

[15] See Chapter 4, Section 3.2.

[16] Computed in Chapter 27, Section 3.5

Measurement of kurtosis:[16]

$\gamma_2 = 1.73$

These figures indicate a concentration of cases around 0.8310 to 0.8320 inches. They show that the general variability of the data is slight relative to the mean value. The above figures also show practically no skewness, but a significantly large kurtosis, i.e., an exceptionally large concentration of data around the point of central tendency coupled with a relatively large number of cases far out on the tails.

4.5. Summary of Measures of a Frequency Distribution

Measure	*Symbol*	*Definition*	*Remarks*
I. Measures of central tendency:			
a. Mean*	$\bar{X}$	$\sum FX/n$	The center of gravity. Based on all the cases. Easy to compute, can be handled mathematically. Has relatively small sampling error. Affected by extreme cases.
b. Median	*Mi*	50 percent cases $< Mi$ 50 percent cases $> Mi$	Divides area in half. Not seriously affected by extreme cases. Hard to handle mathematically. Sampling error larger than that of $\bar{X}$, if universe is normal.
c. Mode	*Mo*	The *X* with greatest frequency	The typical value. Hard to locate in the universe.
II. Measures of general variability:			
a. Range*	*R*	$X_H - X_L$	Very easy to compute. A sensitive measure of universe variability only if sample is small. Affected by extreme variations.
b. Average deviation	A.D.	$\frac{\Sigma F\lvert x\rvert}{n}$	Based on all the cases. Easy to compute for small samples. Hard to handle mathematically. Affected by extreme variations.
c. Root-mean-square deviation	RMSD	$= \sqrt{\frac{\Sigma Fx^2}{n}}$	Based on all the cases. Easy to handle mathematically. Has relatively small

Measure	Symbol	Definition	Remarks
			sampling error. Affected by extreme variations. $(\text{RMSD})^2$ is a biased estimate of σ'^2.
d. Standard deviation*	$s =$	$\sqrt{\frac{\Sigma Fx^2}{n-1}} = \sqrt{\frac{n\mu_2}{n-1}}$	Based on all the cases. Easy to handle mathematically. Has relatively small sampling error. Affected by extreme variations. s^2 is an unbiased estimate of σ'^2.
e. Semiquartile range	Q	$\frac{Q_3 - Q_1}{2}$	Not seriously affected by extreme cases. Measures the variability of the central portion of the distribution. The quartile range, $Q_3 - Q_1$, subtends 50 percent of the cases.
III. Measure of skewness:			
a. Coefficient of skewness*	γ_1	$\left(\frac{\mu_3^2}{\mu_2^3}\right)^{1/2}$	Based on all the data. Can be handled mathematically. Affected by extreme variations. Zero for a symmetrical distribution.
IV. Measure of kurtosis:			
a. Coefficient of kurtosis*	γ_2	$\frac{\mu_4}{\mu_2^2} - 3$	Approximately 0 for a bell-shaped distribution.

* The most commonly used measures.

5. EXPLORATORY DATA ANALYSIS

It is well to conclude this chapter with discussion of a procedure for preliminary study of a batch of data that John W. Tukey has called *Exploratory Data Analysis.*[17] There are two advantages of this procedure. Minimization of prior assumptions allows the data to guide the choice of appropriate models and the procedure lends itself readily to computer programming and use of computer graphics. We shall restrict ourselves here to describing the language and concepts related to exploratory analysis of a single batch of data. Our discussion is based largely on *Applications, Basics and Computing of Exploratory Data Analysis* by Paul F. Velleman and David C. Hoaglin and the reader is referred to this for related computer programs.

[17] See his 1977 book with that title.

5.1. Stem-and-Leaf Displays

A stem-and-leaf display of a batch of data is a simple way of sorting the data values in numerical order and displaying them. The procedure is based primarily on the decision as to what digits in a data value will be taken as "leading digits" and, in consequence of that decision, which will be "trailing digits." If a data value is 23.461, for example, 23 might be selected as its leading digits, which would make 461 its trailing digits. In a stem-and-leaf display of a batch of such data, sorting would be done on the basis of the leading digits. Of the trailing digits, only the leftmost digit (in our example, the digit *4*) would be retained and shown in the display. The leading digits of a data value will thus indicate the location of its stem and the leftmost digit of its trailing digits will be the value of the leaf for that data value.

To illustrate the procedure let us make a stem-and-leaf display of the yarn-strength data in the first column of Problem 3.3a(1) below. An examination of the data suggests that the digits in the hundreds and tens columns can be appropriately taken as leading digits and accordingly serve as locations for stems. The digit in the units column will in each case be taken as a leaf. If measurements of yarn-strength had been made to one or more decimals, the digits to the right of the decimal point (being trailing digits to the right of the leftmost trailing digit) would have been disregarded. The initial stem-and-leaf display of these data will thus be that pictured in Figure 3.15.

The stem-and-leaf display of Figure 3.15 has one or more leaves for every stem, but there may be displays in which some stems do not have any leaves. For example, if in the given set of data the yarn for which a strength of 122 pounds is recorded had had a strength of 119 pounds, the stem at location 12 would have been shown without a leaf. Sometimes it may be desirable to show more than one line per stem. With respect to the given data, for example, it could be that 10 tests were made by one man and 15 by another. In such a case, measurements by man A could be shown on the first line for each stem and measurements by man B on a second line.

A full stem-and-leaf display will show "depths." These are the figures

FIGURE 3.15
Initial Stem-and-Leaf Display of 25 Measurements of Yarn Strength

	Stems	Leaves	
	6	6	
	7	98	
	8	5968	
Stems	9	176356	Leaves
	10	7312	
	11	71105	
	12	2	
	13	28	

FIGURE 3.16
Full Stem-and-Leaf Display for 25 Measurements of Yarn Strength

	Unit = 1	
Depths	*10*	*5 Represents 105*
1	6	6
3	7	98
7	8	5968
(7)	9	176356
12	10	7312
8	11	71105
3	12	2
2	13	28

shown in Figure 3.16 to the left of the stem column. A figure on a line in the depth column shows the cumulative number of leaves on that line and lines closer to the nearer end of the batch. The figure 7 in the third line of the depth column of Figure 3.16, for example, indicates that the cumulated number of leaves on the first, second and third lines is seven. The figure 3 in the second line from the bottom of the depth column indicates that the cumulated number of leaves in the next-to-last and last lines is three. The line containing the middle value of the data shows the number of leaves on *that* line in parentheses. If the batch contains an even number of data values, no single value will be exactly in the middle and in that case we talk about the two "middle values." If one of these is on one stem and the other on the stem next to it, depths in the stem-and-leaf display are shown for all stems. Evaluation of the middle of the batch when there is an even number of cases is discussed in the next section.

5.2. "Letter-Value" Displays

If the 25 measurements of yarn strength in our batch are arranged in order of size, we can assign to each data value its depth from the nearer end. Such an arrangement can easily be obtained from Figure 3.15 by arranging the leaves on each stem in order of size. If there is an odd number of data values in the batch, as there is in this instance, the data value with the greatest depth will be the middle value or the *Median* (represented in a "Letter-Value" display by the letter M). If there is an even number of cases, there will be two "middle values" of equal depth and the depth of the median will be taken as lying half way between. In general the depth of the median is defined as

$$\text{Depth of median} = d(M) = \frac{(n+1)}{2}$$

which is a fractional value if n is even. In our example, there are 25 data values so the depth of the median will be 13 and Figure 3.16 gives the median value as equal to 97. If there had been 26 data values, the depth of the median would have been 13½. When there are an even number of data values, the median is conventionally taken as the mean of the two middle data values. Thus if a yarn strength measurement of 131 pounds is added to the data of Figure 3.16, the median would be taken as (97 + 101)/2 = 99.

If we apply the above procedure to the upper and lower halves of the batch of data, we can derive what are referred to as its upper and lower "hinges." Thus the depth of a hinge is defined by the relationship

$$d(H) = ([d(M)] + 1)/2$$

where the symbol [ ] means "integer part of." Hinges are similar to quartiles where the latter are defined such that one quarter of the data lies below the lower quartile and one quarter of the data lies above the upper quartile.[18] The depth of the hinges, however, is calculated from the depth of the median with the result that the hinges may be slightly closer to the median than the quartiles. Nevertheless, Velleman and Hoaglin suggest that eventually hinges may come to be called "quarters." For our example, the depth of a hinge is (13 + 1)/2 = 7, which from Figure 3.16 is shown to yield a lower hinge of 89 and an upper hinge of 115, being respectively the seventh data value from the lower end of the ordered batch and the seventh data value from the upper end of the ordered batch.

Proceeding further we can obtain "eights" of the batch by defining the depth of an "eight" as equal to

$$d(E) = ([d(H)] + 1)/2$$

For our example this yields $d(E) = (7 + 1)/2 = 4$ and the eights themselves equal 85 and 117 respectively, since these are the fourth data values from the lower and upper ends respectively of the ordered data batch.

The median, it will be noted, is represented by the capital letter *M*, a hinge by the capital letter *H* and an eighth by the capital letter *E*. Further subdividing points are generally not given names but are represented by the capital letters *D, C, B, A, Z, Y, X, W,* and so forth. The definition of depth remains the same, however. Thus we have

$$d(D) = ([d(E)] + 1)/2$$

which for our example yields $d(D) = (4 + 1)/2 = 2½$ and *D* values of (78 + 79)/2 = 78.5 and (122 + 132)/2 = 128.

Further analysis consists of computing midpoints of the capital letter values and the spreads between the upper and lower values for the same letter. All this is shown in a capital letter display. For our example, such a display is presented in Figure 3.17.

It will be noted in Figure 3.17 that the mid-values for *H, E* and *D*

---

[18] See p. 61.

FIGURE 3.17
Letter-Value Display for Measurements of Yarn Strength

Letter	*Depth*	*Lower Value*		*Upper Value*	*Mid-Value**	*Spread*
M	13		97			
H	7	89		115	102	26
E	4	85		117	101	32
D	2.5	78.5		128	103.25	49.5
	1	66		138	102	72

* These are actually the midpoints of the ranges of the capital letter values.

and the average of the extreme values of the entire set of data, called the midrange (here equal to 102), are all about the same. They are larger than the median, however, which suggests a slight upward skewness in the middle part of the data. If there had been an upward trend in the *H, E* and *D* mid-values, it would have indicated an overall upward skewness in the data. If there had been a downward trend in these mid-values, it would have indicated a downward skewness.

The spreads between the letter-values are measures of general variability. The *H*-spread indicates the spread of the middle half of the data and is similar to an interquartile range.[19] The *E*-spread indicates the spread of the middle three quarters of the data and the *D*-spread indicates the spread of the middle seven eights of the data.

5.3. Boxplots

A skeletal boxplot is a graphic display of some of the information provided by a capital-letter display. It shows graphically the relative positions of the median, the upper and lower hinges and the extremes. The middle portion of the data that lies between the hinges is shown by a box. (The length of the box along the data-value scale measures the *H*-spread. The width of the box has no significance.) Lines are drawn from each end of the box to the extreme value on that side. These are called "whiskers" and the whole chart may be referred to as a "box-and-whisker plot." A skeletal box-and-whisker plot for the measurements of yarn-strength in our example is presented in Figure 3.18.

In exploratory data analyses an important concept is that of a "fence." Inner fences for a batch of data are defined as

$$\text{Lower hinge} - (1.5 \times H\text{-spread})$$

$$\text{Upper hinge} + (1.5 \times H\text{-spread})$$

[19] See pp. 61 and 62 for discussion of the semiquartile range.

FIGURE 3.18
A Skeletal Boxplot of Yarn Strength Data

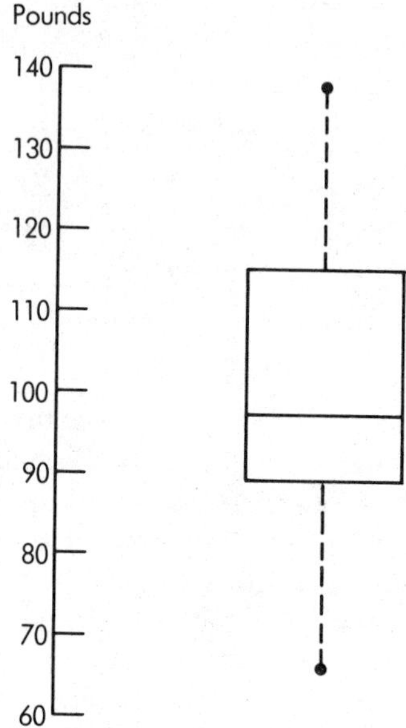

and outer fences are defined as

$$\text{Lower hinge} - (3 \times \text{H-spread})$$
$$\text{Upper hinge} + (3 \times \text{H-spread})$$

A data value that lies beyond an inner fence is termed "outside" and a data value that lies beyond an outer fence is termed "far outside." The outermost data value at each end of a batch that still lies within the inner fence at that end is referred to as an "adjacent value." Figure 3.18 shows a skeletal boxplot. A full boxplot would run the whiskers only to adjacent values and show outside and far outside values individually, labeling each for identification. The fences are not shown and serve only to define the data values that are outside and far outside values. The full boxplot thus clearly calls attention to outlying observations that may need special investigation.

In our example of yarn-strength measurements inner fences are at 50 and 154 and outer fences are at 11 and 193 and there are thus no outliers. If our example, however, had contained additional measurements of 45 pounds and 198 pounds, these values would have shown up as outliers. Our stem-leaf display in this case would have had additional stems at 4 and 19 with leaf values of 5 and 8, respectively. The median depth would have been 14, since the number of measurements would have been 27, and the hinges would have had the depths of 8. The median or M value would have continued to be the same, however, and the H values would also have remained unchanged. Thus the H-spread would have been unchanged and the inner fences would still have been at 50 and 154 and the outer fences at 11 and 193. The data value 45, however, would have fallen below the inner fence of 50 and would have shown up as an outside value. The data value 198 would have fallen beyond the outer fence at 193 and would have shown up as a far outside value. The full boxplot for the augmented data would thus have been that shown in Figure 3.19. This is the same as Figure 3.18 except

FIGURE 3.19
A Boxplot with Outliers

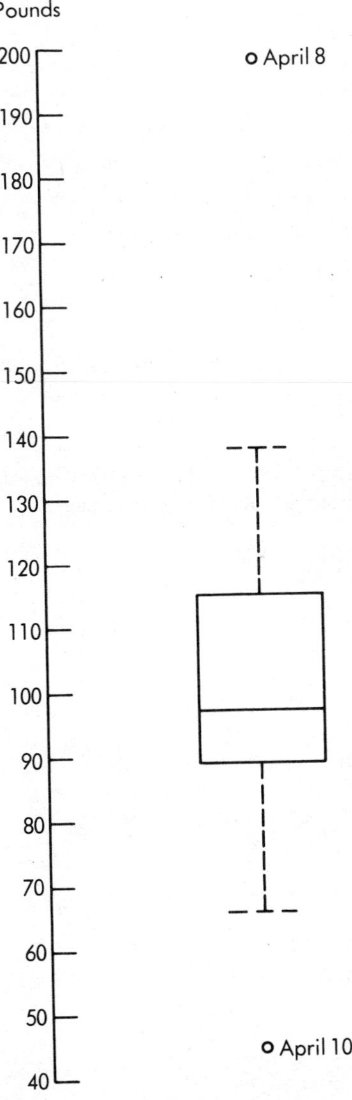

that adjacent values are indicated by cross lines, whiskers are run only to the adjacent values and not the extremes, and the data values 45 and 198 are shown as outlying dots, labeled in this case by the date of measurement to aid in finding the cause of extreme variation.

Boxplots are thus a simple graphical procedure that provides much information about the distribution of a set of data and full boxplots may highlight situations that need special investigation. Comparison between different sets of data can be readily made by presenting a series of boxplots. A major advantage is that all this can be done on a computer.

6. PROBLEMS

3.1. Classify the following data as to whether they are discrete or continuous:

a. Weekly number of accidents in a given plant.
b. Tensile strength of cotton yarn in pounds per square inch.
c. Percentage of tanks with front discharge silver solder leaks.
d. Monthly number of machines rejected.
e. Daily employment.
f. Diameters of metal knobs.

3.2. A gauge can be read accurately to the nearest 10,000th of an inch. Are the measurements made from such a gauge discrete or continuous? Justify your answer.

3.3. *a.* Make frequency distributions of the following sets of data and construct a histogram for each as shown on top of page 73.

(i) Yarn-Strength Tests of 22s Yarns (Pounds).[20]

66	92	99	94
117	137	85	105
132	91	95	103
111	84	89	96
107	96	102	100
85	97	100	101
89	100	98	98
79	105	97	97
91	104	104	97
97	137	114	101
138	80	111	102
103	104	98	98
111	104	99	94
86	106	102	100
78	84	91	98
96	92	95	99
93	86	111	92
101	104	104	102
102	132	97	87
110	94	98	99
95	99	102	62
96	102	109	92
88	101	88	100
122	104	91	96
115	107	103	98

(ii) Shear Strength of Welds of Stainless Steel in Pounds per Weld; Gauge 0.044, Condition 1/4 H:[21]

2385	2305	2230	2340	2280
2280	2310	2210	2440	2230
2330	2340	2220	2370	2290
2360	2330	2190	2340	2270
2350	2340	2230	2360	2290
2350	2350	2160	2340	2270
2370	2360	2270	2330	2270
2310	2360	2400	2380	
2280	2340	2350	2350	
2310	2280	2360	2360	
2310	2290	2360	2390	
2330	2350	2300	2360	
2280	2330	2350	2400	
2290	2280	2340	2320	
2190	2285	2290	2360	
2280	2250	2250	2350	
2270	2340	2270	2340	
2260	2330	2340	2320	
2250	2350	2310	2350	
2260	2275	2360	2330	
2270	2190	2300	2320	
2270	2240	2430	2300	

[20] United States Department of Agriculture, War Food Administration, Office of Marketing Services, *Results of Fiber and Spinning Tests of Some Varieties of Upland Cotton Grown in the United States, Crop of 1944* (Processed, April 1945).

[21] Supplied by a manufacturing company.

(iii) Results of Daily Inspection of Fuel Tanks: Percent with Leaks:[22]

57.6	75.5	72.4	72.7
55.7	61.4	73.3	80.5
56.4	46.8	80.5	86.8
58.7	52.8	70.1	73.8
50.5	59.5	80.0	55.1
50.5	45.9	57.8	53.6
44.0	40.0		
43.3	45.9		
40.8	59.6		
32.6	55.9		
59.3	54.6		
49.4	67.3		
45.5	60.5		
50.0	53.3		
35.5	46.4		
53.8	49.5		
62.7	72.0		
75.0	82.3		

(iv) Ranges of 150 Samples of Four Items Each Drawn at Random from a Normal Universe:[23]

1.5	1.2	3.1	1.3	0.7	1.3
0.1	2.9	1.0	1.3	2.6	1.7
0.3	0.7	2.4	1.5	0.7	2.1
3.5	1.1	0.7	0.5	1.6	1.4
1.7	4.2	3.0	1.7	2.8	2.2
1.8	2.3	3.3	3.1	3.3	2.9
2.2	1.2	1.3	1.4	2.3	2.5
3.1	2.1	3.5	1.4	2.8	2.8
1.5	1.9	2.0	3.0	0.9	3.1
1.9	1.7	1.5	3.0	2.6	1.0
2.9	1.8	1.4	1.4	3.3	2.4
1.8	4.1	1.6	0.9	2.1	1.5
0.9	2.9	2.5	1.6	1.2	2.4
3.4	1.3	1.7	2.6	1.1	0.8
1.0	1.5	2.2	3.0	2.0	1.8
2.9	2.5	2.0	3.0	1.5	1.3
2.2	1.0	1.7	4.1	2.7	2.3
0.6	2.0	1.4	3.3	2.2	2.9
1.6	2.3	3.3	2.0	1.6	2.7
1.9	2.1	3.4	1.5	0.8	2.2
1.8	2.4	1.2	4.7	1.3	2.1
2.9	3.0	2.1	1.8	1.1	1.4
2.8	1.8	1.8	2.4	2.3	2.2
2.1	1.2	1.4	1.6	2.4	2.1
2.0	1.1	3.8	1.3	1.3	1.0

[22] Assume that number inspected each day is the same. Actually, it varied around 180. Data from Lester A. Kauffman, *Statistical Quality Control at the St. Louis Division of the American Stove Company* (OPRD, Quality Control Reports, No. 3, August 1945), p. 17.

[23] Computed from W. A. Shewhart, *Economic Control of Quality of Manufactured Product* (New York: D. Van Nostrand Co., Inc., 1931), p. 442.

(v) Shear Strength of Welds of "Material 56s" in Pounds per Weld; Gauge 0.016, Condition 1/4 *H*:[24]

146	162	150	148	160
148	160	144	154	162
146	146	152	152	164
140	156	154	152	150
150	150	138	154	150
146	146	154	154	152
146	154	160	158	144
142	150	106	158	150
152	148	152	156	152
148	148	150	160	154
158	148	150	162	152
152	150	150	154	
146	152	152	148	
156	150	158	152	
152	152	152	146	
150	152	156	158	
150	152	156	162	
146	148	152	160	
148	152	154	160	
152	146	154	158	
156	152	144	154	

b. In what respects do the five distributions differ?

c. Do the data of *a* (i) appear to be homogeneous? Discuss.

d. Comment on the data of *a* (v).

3.4. Make cumulative distributions of the data of Problem 3.3.

3.5. The following are four sets of measurements:

(i) 106.2, 105.9, 105.8, 106.1, 105.9.

(ii) 107.1, 106.4, 105.9, 106.5, 106.2.

(iii) 106.5, 106.4, 106.5, 106.3, 105.8.

(iv) 106.6, 106.7, 106.3, 106.9, 106.4.

All data are measured in millimeters.

a. Compute the mean of each set.

b. Compute the mean of the whole 20 items.

c. Compute the mean of the four group means. How does this compare with what you computed in *b?*

3.6. *a.* Compute the means of each of the four columns in Problem 3.3*a* (i).

b. Compute the mean of the four means. Is your answer the same as the mean of all the 100 items? Would this be true if the last column contained 20 items only? What would you have to do in this case?

---

[24] Supplied by a manufacturing company.

3.7. Calculate the arithmetic means of—

a. The data of Problem 3.3*a*(i).

b. The data of Problem 3.3*a*(ii).

c. The data of Problem 3.3*a*(iii).

d. The data of Problem 3.3*a*(iv).

e. The data of Problem 3.3*a*(v). Repeat this, leaving out the value 106. Compare your answers.

3.8. The following is the distribution of the shear strength of 94 spot welds:[25]

Pounds per Weld	*Frequency*
137.5–	2
141.5–	4
145.5–	19
149.5–	35
153.5–	17
157.5–	12
161.5–	5
	94

Compute the arithmetic mean of this distribution.

3.9. The following is the distribution of the dimension of a part:[26]

Dimension	*Frequency*
0.2490–	1
0.2495–	7
0.2500–	11
0.2505–	27
0.2510–	20
0.2515–	9
0.2520–	5
0.2525–	2
	82

Compute the mean of this distribution.

3.10. Compute the median of—

a. The data of Problem 3.3*a*(i). (Use grouped form.)

b. The data of Problem 3.3*a*(ii). (Use grouped form.)

c. The data of Problem 3.3*a*(iii). (Use grouped and ungrouped forms.)

d. The data of Problem 3.3*a*(iv). (Use grouped form.)

e. The data of Problem 3.3*a*(v). (Use grouped and ungrouped forms.) Recompute, leaving out the value of 106. How much is *Mi* affected by the 106? Compare with Problem 3.7*e*.

---

[25] Supplied by a manufacturing company.

[26] See G. R. Armstrong and P. C. Clarke, "Frequency Distribution vs. Acceptance Table," *Industrial Quality Control,* September 1946, p. 23. The nature of the dimension of the part was not revealed.

3.11. Compute the median of—

a. The data of Problem 3.8.

b.. The data of Problem 3.9.

3.12. What is the modal interval of the distribution of Problem 3.3*a* (i)? Problem 3.3*a* (iii)? Problem 3.3*a* (v)?

3.13. If a distribution has two peaks, is the highest one "the mode"? Explain. What would you do if you had a distribution with two peaks?

3.14. What is the range of the data of Problem 3.3*a* (i)? 3.3*a* (ii)? 3.3*a* (iii)? 3.3*a* (iv)? 3.3*a* (v)?

3.15. Compute the ranges of each group of data in Problem 3.5.

3.16. Compute the average deviation of each group of data in Problem 3.5.

3.17. Compute the average deviation from the median of the data of Table 3.1.

3.18. Compute the average deviation from the mean of the distribution of Problem 3.8.

3.19. Compute the average deviation from the median of the distribution of Problem 3.9.

3.20. Compute the standard deviation of—

a. 5, 6, 4, 2, 7, 9.

b. 105, 106, 104, 102, 107, 109.

c. 1050, 1060, 1040, 1020, 1070, 1090.

d. 0.05, 0.06, 0.04, 0.02, 0.07, 0.09.

Compare your results. If the standard deviation of x is s, what is the standard deviation of $x + c$, where c is a constant? The standard deviation of cx? (This is an important conclusion and should be remembered.)

3.21. Compute the standard deviation of the distribution of Problem 3.8. Compare the value found for the standard deviation with the value of the average deviation computed in Problem 3.18.

3.22. Using the "short method," compute the standard deviation of the distribution given in Problem 3.9. Problem 3.3*a* (i). Problem 3.3*a* (iii).

3.23. Using the "short method," compute the standard deviation of the distribution given in Problem 3.3*a* (v). Recompute the standard deviation of the distribution after the extreme value 106 has been omitted. Compare your results.

3.24. Compute the semiquartile range for the distribution of Problem 3.8. Compare with AD and s.

3.25. Compute the semiquartile range for the distribution of Problem 3.9. Compare with AD and s.

3.26. Compute the semiquartile range for the distribution of the data of Problem 3.3*a* (i). Compare with s.

3.27. Compute the semiquartile range for the distribution of the data of Problem 3.3*a* (iii). Compare with s.

3.28. What is the nature of the skewness of the distribution of Problem 3.8? Problem 3.9? The distribution of the data of Problem 3.3*a* (ii)? Figure 3.11?

3.29. Make an exploratory data analysis of the data in the third column of Problem 3.3*a* above and compare with the analysis of the data in the first column that is presented in Section 4.6 above.

3.30. Make an exploratory data analysis of the data in column *a* of Table 3.2 and compare with the results of the frequency distribution analysis presented in this chapter.

3.31. Make an exploratory data analysis of the data in the second column of Problem 3.3*a* (ii) above.

7. SELECTED REFERENCES*

American Society for Testing and Materials (B '76), Armstrong and Clarke (P '46), Bidlack, Close, and Warren (P '45), Brumbaugh (P '45), Kennedy (B '48), Rutherford (B '48), Schrock (B '50), Shewhart (B '31), Smith, E. S. (B '47), Tippett (B '50), Tukey (B '77), and Velleman and Hoaglin (B '81).

---

* B and P refer to the Book and Periodical sections, respectively, of the Cumulative List of References in Appendix V.

4

Universe Frequency Distributions

1. DISTRIBUTIONS OF FINITE UNIVERSES

The last chapter was concerned primarily with methods of describing the variability of a given finite set of data. These were viewed as a sample from a larger set or universe, and the analysis of the given sample data was aimed primarily at acquiring knowledge of this universe. The methods used to analyze a sample set of data, however, can also be used to describe a finite universe itself. The only difference is that universe symbols would be used instead of sample symbols. There is no need, therefore, for further discussion of methods of analysis of finite universes, and we shall devote our full attention to the description and analysis of infinite universes.

2. DISTRIBUTIONS OF INFINITE UNIVERSES

The concept of an infinite universe is an abstraction of great importance in statistical theory. Being infinite we cannot physically construct the distribution of such a universe. We can, however, often conceive of a procedure that would approach this distribution as a limit as the number of cases is indefinitely increased.

Consider, for example, the rolling of an unbiased die. In the first *60 rolls* we might get 8 (or 13.3 percent) ones, 11 (or 15.6 percent) twos, 5 (or 8.3 percent) threes, 15 (or 25 percent) fours, 12 (or 20 percent) fives, and 9 (or 15 percent) sixes. In the first *600 rolls* we might get 111 (or 18.5 percent) ones, 95 (or 15.8 percent) twos, 97 (or 16.1 percent) threes, 105 (or 17.5 percent) fours, 93 (or 15.95 percent) fives, and 99 (or 16.5 percent) sixes. If we continue to roll indefinitely the proportion of the total number of rolls yielding a one, a two, a three, a four, a five, or a six will in each case come closer to 16⅔ percent. This uniform distribution of relative frequency that is thus approached can be viewed as the distribution of results to be obtained in an infinite number of rolls.

Again suppose that we indefinitely measure a continuous variable such as the length of a rod cut by a given lathe. After the first 1,000 rods let us construct a histogram of rod lengths in the manner described in Chapter 3 with the exception that we show the *relative frequencies* rather than the absolute frequencies and we represent these by the *areas* of the rectangles instead of their heights. Suppose further that after another 3,000 rods we construct an area histogram of relative frequency for the whole 4,000 rods, but we now use twice as many class intervals one half the former size. Let us continue in this way. As we quadruple the number of rods, let us double the number and halve the size of the class intervals. Then as we proceed, the tops of the rectangles of our histogram will ultimately trace out a smooth curve. This will be a graphic description of the distribution of the hypothetically infinite universe of rod lengths that would be turned out by the given lathe. It is assumed, of course, that throughout the buildup of the distribution, the operation of the lathe remains constant, the variation in the lengths of the rods turned out being entirely due to chance forces. It will be noted that the area under the curve and not its ordinate measures the relative frequency. A frequency curve of the type just described will later be identified as a density function.

3. FORMULAS FOR DISTRIBUTIONS OF INFINITE UNIVERSES

In many cases, if we know the kind of chance forces that are at work, we can write down a mathematical formula for the distribution of an infinite universe. To give the reader some understanding of how this is done we shall present two examples, one pertaining to a discrete variable, the other to a continuous variable.

3.1. Derivation of a Formula for the Distribution of a Discrete Variable: An Example

To illustrate the derivation of a mathematical formula for a discrete distribution, let us consider the rolling of a *pair* of dice. We shall assume that the dice are unbiased and are rolled in a random manner. The list of possible results is shown in Figure 4.1. This indicates, for example, that a sum of 4 may be obtained (1) by a 1 on Die I and a 3 on Die II, (2) by a 2 on each of the dice, and (3) by a 3 on Die I and a 1 on Die II. For unbiased dice thrown in a random manner, any one of the 36 combinations shown in Figure 4.1 would be equally likely. In an infinite number of rolls the relative frequency of getting a sum of 4 would be the relative frequency with which a 4 appears in these 36 combinations. Since this appears 3 times out of 36 possibilities, the relative frequency of a 4 would be $3/36$.

FIGURE 4.1
Sums Yielded by Different Combinations in the Rolling of a Pair of Dice

		Die I					
		1	2	3	4	5	6
Die II	1	2	3	4	5	6	7
	2	3	4	5	6	7	8
	3	4	5	6	7	8	9
	4	5	6	7	8	9	10
	5	6	7	8	9	10	11
	6	7	8	9	10	11	12

A summary of the relative frequencies with which various sums will appear is given in Table 4.1. It will be noted that for a sum *(S)* up to and including 7, the relative frequency will be given by the formula

$$\text{(4.1a)} \qquad \text{Relative frequency of } S = \frac{S-1}{36} \qquad 2 \leq S \leq 7$$

For values of S equal to or greater than 7, we have

$$\text{(4.1b)} \qquad \text{Relative frequency of } S = \frac{13-S}{36} \qquad 7 \leq S \leq 12$$

The distribution of the infinity of sums obtainable by the rolling of the pair of dice is thus mathematically described by formulas (4.1a) and (4.1b). A picture is shown in Figure 4.2 on p. 82. It will be noted that we have a triangular-shaped distribution.

TABLE 4.1
Relative Frequency of Sums Yielded by a Pair of Dice

Sum	*Relative Frequency*
2	1/36
3	2/36
4	3/36
5	4/36
6	5/36
7	6/36
8	5/36
9	4/36
10	3/36
11	2/36
12	1/36

FIGURE 4.2
Distribution of the Sums Yielded by a Pair of Dice

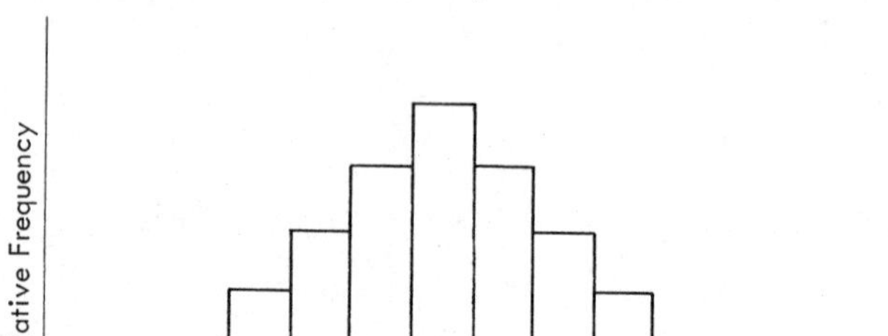

3.2. Derivation of a Formula for the Distribution of a Continuous Variable: An Example

We shall now illustrate the derivation of a formula for the distribution of a continuous variable. Suppose that we have a variable X which fluctuates around a given level μ, this variation being the sum of the variations in a large number (n) of elementary variables V. Thus

$$X = \mu + V_1 + V_2 + \ldots + V_n \tag{4.2}$$

Further suppose that each V independently takes on only two values, $+\epsilon$ and $-\epsilon$, the ϵ's being the same for all V's and generally small with respect to μ. The variable X will thus take on the values $\mu - n\epsilon$, $\mu - (n - 2)\epsilon$, $\mu - (n - 4)\epsilon$, . . . μ . . ., $\mu + (n - 4)\epsilon$, $\mu + (n - 2)\epsilon$, and $\mu + n\epsilon$. For simplicity let n be an even number.

Since each V can independently take on two different values, there will altogether be 2^n different combinations of values for a set of n V's. (Recall that if we can do something in s ways followed by something else that can be done in t ways, we can do the two in succession in $s \times t$ ways.) Finally let us assume that the variables V behave so that each of these 2^n combinations of values is equally likely to occur.

An example of the above situation would be the following: Suppose that we are on the beach with a large bucket half full of sand. We also have a group of n children. We play the following game. Each child in turn tosses a penny. If it comes up heads, he adds (from the beach) a teaspoonful of sand to the bucket. If it comes up tails, he takes a spoonful out of the bucket. After each round of n tosses, we weigh the bucket and then restore the weight to its original amount. The weight of the bucket after each round will be the X of equation (4.2). The μ will be the weight of the bucket just half full, and ϵ will be the weight of a teaspoonful of sand (assumed constant).

There will be altogether 2^n different ways in which the children may combine to add or subtract teaspoonfuls of sand in a round of n tosses, and on the assumption of independent action by each child, each of these 2^n combinations will be equally likely.

Now under the assumed conditions the value of X depends on how many of the V's have $+\epsilon$ values. Let this number be Y. Then the relative frequency with which X will take the different values $\mu - n\epsilon$, $\mu - (n - 2)\epsilon, \ldots \mu \ldots \mu + (n - 2)\epsilon$, $\mu + n\epsilon$ will be as follows:

Relative frequency with which $X = \mu + 2\left(Y - \frac{n}{2}\right)\epsilon$ is

(4.3)
$$\frac{n!}{Y!(n-Y)!}\bigg/ 2^n$$

Since out of the 2^n different combinations this will be the proportion that will have Y values of $+\epsilon$ and $n - Y$ values of $-\epsilon$. It will be noted that when $Y = n/2$, $X = \mu$. If we set $y = Y - n/2$, then formula (4.3) becomes

Relative frequency with which $X = \mu + 2y\epsilon$ is

(4.4)
$$\frac{n!}{(n/2 + y)!(n/2 - y)!}\left(\frac{1}{2}\right)^n$$

Now it can be shown that if we set $h = 4/n$ and if we let n get indefinitely large, then formula (4.4) can be approximated[1] by

Relative frequency with which $X = \mu + 2y\epsilon$ is

(4.5)
$$\sqrt{\frac{h}{2\pi}}\,e^{-hy^2/2} = \sqrt{\frac{h}{2\pi}}\,e^{-z^2/2}$$

where $z = y\sqrt{h}$ and $e = 2.718+$. It will be noted from Figure 4.3 on p. 84 that as y goes from $-n/2$ to 0 to $+n/2$, X goes from $\mu - n\epsilon$ to μ to $\mu + n\epsilon$, and z goes from $-\frac{n\sqrt{h}}{2}$ to 0 to $\frac{n\sqrt{h}}{2}$. Intervals between points on the X scale will be of width 2ϵ, but intervals between points on the z scale will be of width $\sqrt{h}$. If we construct rectangles on these z points (see Figure 4.4, p. 84) with heights $e^{-z^2/2}/\sqrt{2\pi}$, then the areas of these rectangles will measure approximately the relative frequencies with which z takes on the values $-\frac{n\sqrt{h}}{2}$, $-\frac{(n-2)\sqrt{h}}{2}$, $-\frac{(n-4)\sqrt{h}}{2}$, and so on [i.e., the relative frequencies with which X takes on the values $\mu - n\epsilon$, $\mu - (n - 2)\epsilon$, $\mu - (n - 4)\epsilon$, and so forth]. The relative frequency with which

[1] For an outline of the proof, see Appendix I (15). In Appendix I (15), $1/h = \sigma_x'^2$, since $\sigma_x'^2 = n/4$ if $p' = 1/2$, as it does in our example.

z takes on the values z_1 to z_2 will thus be the sum of the areas of the rectangles from z_1 to z_2 which is symbolically

$$\sum_{z=z_1}^{z=z_2} \left(\frac{e^{-z^2/2}}{\sqrt{2\pi}}\right)\sqrt{h}$$

FIGURE 4.3
Relationships between Values of X, Y, y, and z, Assuming n Even

X	Y	$y = Y - n/2$	$z = y\sqrt{h}$
$\mu - n\epsilon$	0	$-n/2$	$-n\sqrt{h}/2$
$\mu - (n-2)\epsilon$	1	$-(n-2)/2$	$-(n-2)\sqrt{h}/2$
$\mu - (n-4)\epsilon$	2	$-(n-4)/2$	$-(n-4)\sqrt{h}/2$
.			
.			
.			
$\mu - 2\epsilon$	$(n-2)/2$	-1	$-\sqrt{h}$
μ	$n/2$	0	0
$\mu + 2\epsilon$	$(n+2)/2$	1	$\sqrt{h}$
.			
.			
.			
$\mu + (n-4)\epsilon$	$n-2$	$(n-4)/2$	$(n-4)\sqrt{h}/2$
$\mu + (n-2)\epsilon$	$n-1$	$(n-2)/2$	$(n-2)\sqrt{h}/2$
$\mu + n\epsilon$	n	$n/2$	$n\sqrt{h}/2$

FIGURE 4.4
A Histogram of z Values

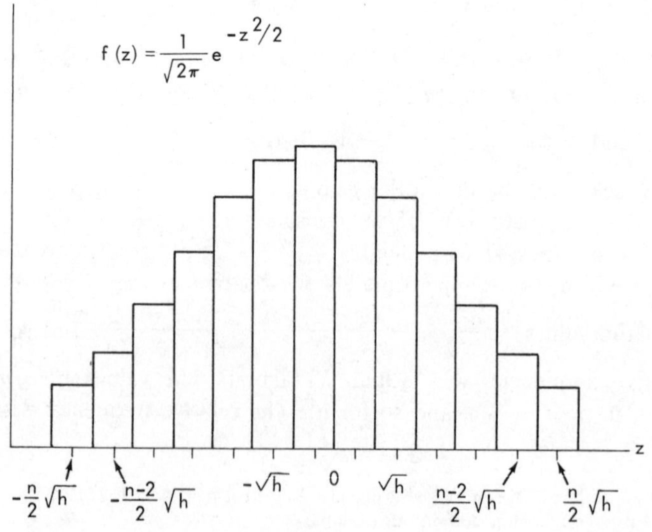

FIGURE 4.5
Graph of the Cumulative Form of the Normal Distribution

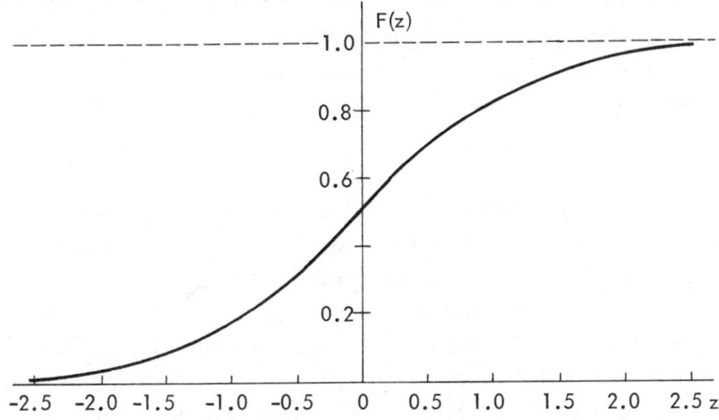

As n gets larger and larger, however $h(=4/n)$ gets smaller and smaller, the points on the z-axis get closer and closer, and the sum of the areas of the given rectangles represented by the above formula approaches the integral

$$\int_{z_1}^{z_2} \frac{e^{-z^2/2}}{\sqrt{2\pi}}\,dz, \text{ where } dz = \sqrt{h}$$

In the limit, then, the variable z may be treated as if it were continuous and the cumulative distribution of the universe of z values would be given by equation (4.6), which reads as follows:

(4.6) Cumulative relative frequency of values less than or equal to z is

$$F(z) = \int_{-\infty}^{z} \frac{e^{-t^2/2}}{\sqrt{2\pi}}\,dt$$

where t is the variable of integration.

Now it turns out that (4.6) is a very important distribution formula in statistical theory. It has thus been given a special name. In general, a continuous variable with a cumulative distribution of the form of formula (4.6) is said to have a *standard normal distribution* or to be *normally distributed.* A graph of 4.6 is shown in Figure 4.5.

The normal formula crops up in many theoretical statistical problems. In general if a continuous variable X is the *sum* of many independent contributing variables V, then X will tend to have a normal distribution.[2] The distribution is described in more detail in Section 6.1 below.

---

[2] See, for example, E. T. Whitaker and G. Robinson, *The Calculus of Observations* (London: Blackie & Son Ltd., 1932).

3.3. General Comments on Formulas

In general a formula for the relative frequencies with which a discrete variable takes on a specified set of values is an explicit function of these values. In the first example above, the relative frequencies with which S took on the values 0 to 7 were given simply by the relationship: Relative frequency = $\frac{S-1}{36}$.

Formulas for cumulative relative frequencies for a continuous variable generally take the form of an integral such as equation (4.6). In most cases in statistical theory the indicated integration cannot be accomplished analytically and we have to be content with its symbolic statement. When actual values are desired, the integration is carried out numerically. For important distributions with widespread usage, tables of the cumulative distribution have been computed. Cumulative distributions generally have the **S**-shaped form shown in Figure 4.5. It will be noted that the curve never decreases as the variable increases and that $F(-\infty) = 0$ and $F(\infty) = 1$.

The rate at which the cumulative relative frequency changes will be given by the derivative of the cumulative distribution. In (4.6) the derivative of $F(z)$ with respect to z is

(4.7)
$$f(z) = \frac{dF(z)}{dz} = \frac{e^{-z^2/2}}{\sqrt{2\pi}}$$

This is called the *density function* of z. It will be noted that the density function measures the slope of the cumulative curve. A graph of (4.7) is shown in Figure 4.6; it is the derivative of Figure 4.5.

The density function is the "frequency curve" that is the limit approached by an area histogram of relative frequency such as discussed in Section 2. It is to be remembered that a relative frequency or probability is measured by an *area* under the curve of the density function and not by its ordinate.

FIGURE 4.6
Graph of the Density Function of the Normal Distribution

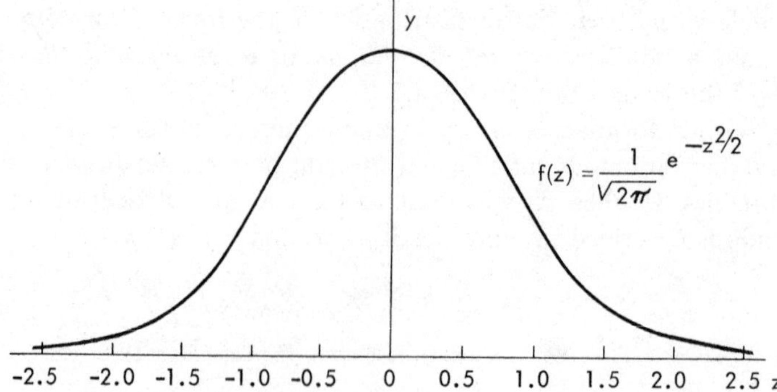

4. DISTRIBUTIONS OF RELATIVE FREQUENCY AS PROBABILITY DISTRIBUTIONS

It is desirable to note at this point that the variables for which we will give distribution formulas will be random variables, and distributions of relative frequency will therefore be distributions of probability, the relative frequencies being the probabilities. We will consequently use the terms *relative frequency* and *probability* interchangeably. For example, we will use *the normal frequency distribution* and *the normal probability distribution* as synonymous terms. Generally, probability will be used rather than relative frequency.

4.1. Expected Values

When the distribution of a universe is viewed as a probability distribution, its mean is often referred to as the *expected value.*

If a random variable X is discrete and takes on the values X_1, X_2, . . ., X_k with probabilities or relative frequencies P_1, P_2, . . ., P_k, then the expected value of X is defined as

(4.8)
$$E(X) = \sum_{i=1}^{k} P_i X_i = \text{mean } X \text{ by definition.}$$

The symbol E is to be read as an operator calling for the expected value or mean value of the quantity in parentheses. Thus in the dice example discussed in Section 3.1, the expected value or mean of X is

$$E(X) = \tfrac{1}{36}(2) + \tfrac{2}{36}(3) + \tfrac{3}{36}(4) + \tfrac{4}{36}(5) + \tfrac{5}{36}(6) + \tfrac{6}{36}(7) + \tfrac{7}{36}(8) + \tfrac{4}{36}(9) + \tfrac{3}{36}(10) + \tfrac{2}{36}(11) + \tfrac{1}{36}(12) = 7$$

If X is continuous with density function $f(X)$, the expected value of X is defined as

(4.8a)
$$E(X) = \int_{-\infty}^{\infty} Xf(X)dX$$

For example, if the density function of z is $f(z) = \dfrac{1}{\sqrt{2\pi}} e^{-z^2/2}$, the expected value of z is

$$E(z) = \int_{-\infty}^{\infty} z\frac{e^{-z^2/2}}{\sqrt{2\pi}} dz$$

Since the integrand is an odd function, being negative to the left of zero and positive to the right of zero, and since the function is otherwise symmetrical about zero, the integral will obviously be zero. Hence the expected value or mean of z is zero. Of course, in general, $E(X)$ can have any value.

4.2. Variances

The variance of a probability distribution can be viewed as the expected value of the square of the deviations from the mean. Thus if X is a discrete random variable that takes on the values $X_1, X_2, \ldots, X_k$ with probabilities or relative frequencies $P_1, P_2, \ldots, P_k$ and has an expected value or mean value $= \bar{X}'$, then the variance of X is defined as

$$\sigma'^2_X = \text{Var}(X) = E(X - \bar{X}')^2$$

(4.9)
$$= \sum_{i=1}^{k} P_i(X_i - \bar{X}')^2$$

For the dice example we would have

$$\begin{aligned}\sigma'^2_X &= \tfrac{1}{36}(2-7)^2 + \tfrac{2}{36}(3-7)^2 + \tfrac{3}{36}(4-7)^2 \\ &+ \tfrac{4}{36}(5-7)^2 + \tfrac{5}{36}(6-7)^2 + \tfrac{6}{36}(7-7)^2 \\ &+ \tfrac{5}{36}(8-7)^2 + \tfrac{4}{36}(9-7)^2 + \tfrac{3}{36}(10-7)^2 \\ &+ \tfrac{2}{36}(11-7)^2 + \tfrac{1}{36}(12-7)^2 = 5.83\end{aligned}$$

If X is continuous, with a density function $f(X)$ and mean $\bar{X}'$, the variance is defined as

(4.9a)
$$\sigma'^2_X = \text{Var}(X) = \int_{-\infty}^{\infty} (X - \bar{X}')^2 f(X)dX$$

Thus for the density function $f(z) = \dfrac{1}{\sqrt{2\pi}} e^{-z^2/2}$, we have

$$\sigma'^2_z = \text{Var}(z) = \int_{-\infty}^{\infty} (z-0)^2 \frac{e^{-z^2/2}}{\sqrt{2\pi}}\, dz$$

which when we integrate by parts, yields

$$\sigma'^2_z = \text{Var}(z) = 1$$

The standard deviation is the square root of the variance.

4.3. Moments

The kth moment of a probability distribution around the zero origin is the expected value of the kth power of X. Thus

(4.10)
$$\mu'_{0k} = E(X)^k$$

For a discrete variable this takes the form

$$\mu'_{0k} = \sum P_i X_i^k$$

For a continuous variable it takes the form

$$\mu'_{0k} = \int_{-\infty}^{\infty} X_k f(X) dX$$

The kth moment of a probability distribution *around the mean* of the distribution is the expected value of the kth power of $(X - \bar{X}')$. Thus

$$\mu'_k = E(X - \bar{X}')^k$$

(4.10a) $$= \sum P_i(X_i - \bar{X}')^k \qquad \text{[If } X \text{ is discrete]}$$

$$= \int_{-\infty}^{\infty} (X - \bar{X}')^k f(X) dX \qquad \text{[If } X \text{ is continuous]}$$

The skewness of a universe distribution is often measured by the quantity

$$\gamma'_1 = (\mu'^2_3 / \mu'^3_2)^{1/2}$$

and its kurtosis by

$$\gamma'_2 = \mu'_4 / \mu'^2_2 - 3$$

Karl Pearson has built a system of frequency curves[3] based on values of γ'_1 and γ'_2.

4.4 Independence

In closing this section a word should be said about independence in relation to formulas for probability distributions.

Independence is related to joint probability distributions. For discrete variables, for example, the joint probability of X and Y may be written $P(X, Y)$; it gives the probability that $X = X_1$, *say, and simultaneously* that $Y = Y_1$. Now if the formula for this joint distribution factors into two parts, one of which contains only X and the other only Y, the variables X and Y will be said to be independent. For example, if $P(X, Y) = a^2 e^{-a(X+Y)}$, this factors into $(ae^{-aX})\,(ae^{-aY})$ and X and Y are independent. Compare the discussion of independence in Chapter 2.

For continuous variables the condition for independence is that the joint *density function* factor into two parts, each dependent only on one of the variables, i.e., $f(X, Y) = f_1(X) f_2(Y)$. For if $f(X, Y) = f_1(X) f_2(Y)$, then

$$\int\int f(X, Y) dX dY = \int\int f_1(X) f_2(Y) dX dY$$
$$= \int f_1(X) dX \cdot \int f_2(Y) dY$$

and the joint cumulative probability factors into two parts, each dependent only on one variable.

For n discrete variables, independence exists if

[3] Actually Pearson uses $\beta_1 = \gamma'^2_1$ and $\beta_2 = \gamma'_2 + 3$. See W. P. Elderton, *Frequency Curves and Correlation* (2d ed.; London: Charles and Edwin Layton, 1927).

$$P(X_1, X_2, \ldots, X_n) = P(X_1)P(X_2) \cdots P(X_n)$$

and for n continuous variables, if

$$f(X_1, X_2, \ldots, X_n) = f_1(X_1)f_2(X_2) \ldots f_n(X_n)$$

5. SOME IMPORTANT DISTRIBUTIONS OF DISCRETE VARIABLES

There are several distributions of discrete variables that occur very frequently in statistical applications. We shall note the *binomial distribution,* the *Poisson distribution,* and the *hypergeometric distribution.*

5.1. The Binomial Distribution

The formula for the binomial distribution is

(4.11) $$P\left(\frac{X}{n}\right) = \frac{n!}{X!(n-X)!}p'^X(1-p')^{n-X}$$

in which $P(X/n)$ stands for the probability of X out of n, p' and n are special constants or parameters, and X/n is a variable that takes on the

FIGURE 4.7
Graph of the Binomial Distribution for $n = 10$ and $p' = 0.20$

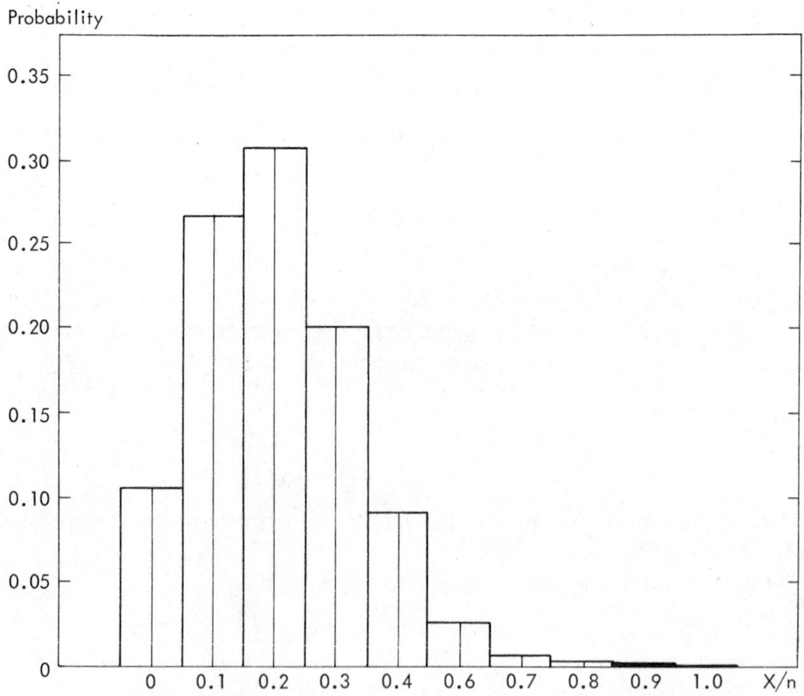

values $0/n$, $1/n$, $2/n$, . . ., n/n. By way of illustration $P(X/n)$ might be the probability of X nonconforming items in a sample of size n when sampling from an infinite universe which is fraction p' nonconforming.

Figures 4.7 (on p. 90) and 4.13 (on p. 101) show graphs of two binomial distributions, one for which $p' = 0.2$, $n = 10$; the other for which $p' = 0.5.$, $n = 10$. If we write $p = X/n$, then the mean of any binomial distribution when p is taken as the variable, is[4]

(4.12)
$$E(p) = p'$$

and the standard deviation of any binomial distribution, when p is taken as the variable, is[5]

(4.13)
$$\sigma'_p = \sqrt{\frac{p'(1-p')}{n}}$$

If X is the variable, then we have

(4.12a)
$$E(X) = p'n$$

and

(4.13a)
$$\sigma'_X = \sqrt{np'(1-p')}$$

These are important formulas that are used in acceptance sampling and control chart analysis.

If we experiment with various values of p', we find the following:

1. If $p' = 0.5$, the binomial distribution is symmetrical around its mean (cf. Figure 4.13, p. 101).
2. If $p' < 0.5$, the right-hand tail of the distribution is longer than the left, and the distribution is skewed to the right or positively (cf. Figure 4.7).
3. If $p' > 0.5$, the left-hand tail of the distribution is longer than the right, and the distribution is skewed to the left or negatively.
4. The standard deviation of the binomial distribution has its maximum value when $p' = 0.5$.

Tables of the binomial distribution have been published for various values of n. We shall note three such tables. For values of n from 0 to 49, the most readily available table is likely to be the National Bureau of Standards, *Tables of the Binomial Probability Distribution* (No. 6 of its Applied Mathematics Series). This gives to 7 places (1) the ordinates of the binomial distribution and (2) the sum of the ordinates from X to n for values of p' from 0.01 to 0.50 in steps of 0.01 and values of n from 2 to 49. Harry G. Romig's

---

[4] See Appendix I (6).

[5] See Appendix I (7).

50–100 Binomial Tables gives similar probabilities (to 6 places) for values of n from 50 to 100 in steps of 5. The Harvard University Computation Laboratory, *Tables of the Cumulative Binomial Probability Distribution,* gives to 5 places the cumulative probabilities from X to n for values of n in steps of 1 from 1 to 50, in steps of 2 from 50 to 100, in steps of 10 from 100 to 200, in steps of 20 from 200 to 500, and in steps of 50 from 500 to 1,000. In addition to probabilities for the decimal values of p' from 0.01 to 0.50 in steps of (0.01), the Harvard tables also give probabilities for such fractional values as $\frac{1}{16}$, $\frac{1}{12}$, $\frac{1}{3}$, and the like. For approximate results, use may be made of the nomograph given in Figure 7.5.

5.2. The Poisson Distribution

The formula for the *Poisson distribution* is

(4.14)
$$P(X) = \frac{u'^X e^{-u'}}{X!}$$

in which e is the limit as y approaches ∞ of $(1 + 1/y)^y$ and equals 2.718 +, u' is a special constant or parameter, and X is a variable that takes on the values 0, 1, 2, . . ., n. Both the binomial and the Poisson distributions are of capital importance in the theory of sampling. As we shall see later, the Poisson distribution may be used either to approximate the binomial distribution or as a distribution in its own right. Figure 4.8 shows a graph

FIGURE 4.8
Graph of the Poisson Distribution for $u' = 1.0$

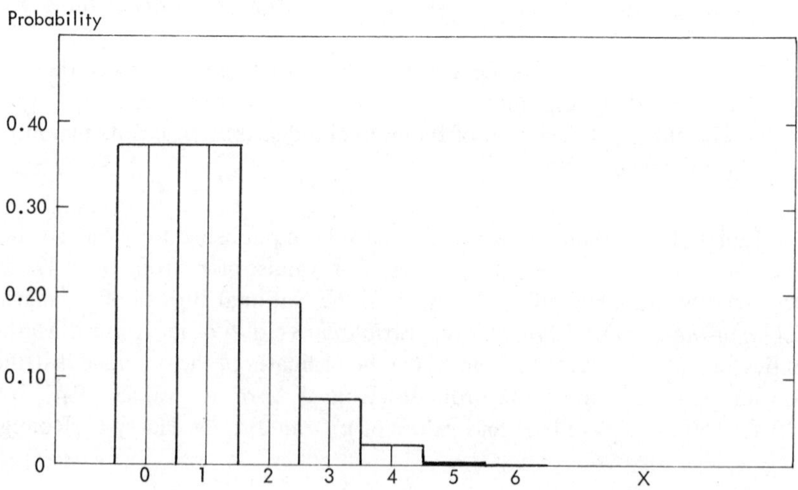

of the Poisson distribution for $u' = 1.0$. The mean of the Poisson distribution is[6] u', and its standard deviation is[7] $\sqrt{u'}$.

Tables of probabilities for the Poisson distribution have been prepared by E. C. Molina and are entitled *Poisson's Exponential Binomial Limit.* These give both individual and cumulative probabilities for values of u' from 0.001 to 100. (Molina's tables use a for u'.). For example, for u' (or a) = 1.0 they give the following:[8]

Values of X	*Individual Probabilities or Ordinates*	*Cumulative Probabilities or Sum of Ordinates Beginning at* X
0	0.367879	1.000000
1	0.367879	0.632121
2	0.183940	0.264241
3	0.061313	0.080301
4	0.015328	0.018988
5	0.003066	0.003660
6	0.000511	0.000594
7	0.000073	0.000083
8	0.000009	0.000010
9	0.000001	0.000001

Figure 4.8 was constructed from these data. A table that is especially good for small values of u' has been published by the General Electric Company. Molina's Table II is abridged in Table E of Appendix II of this text.

Figure 4.9 on p. 94 is a chart for determining graphically the cumulative probabilities of the Poisson distribution. In many cases it is a ready substitute for Molina's Table II. It has been taken from the Dodge-Romig *Sampling Inspection Tables* and will be referred to as the Dodge-Romig Chart.

To find the probability of X_1 or less, we enter the Dodge-Romig chart with the abscissa value equal to u', proceed vertically until we hit the line $X = X_1$, and then read off the answer on the vertical scale. For example, if $u' = 1.0$, the probability of 2 or less is read from the chart as 0.92. This agrees with the result given in Molina's tables, as indicated above.

It will be noticed that the Dodge-Romig chart gives the probability of X or less, whereas Molina's tables give the probability of X or more. The chart also goes only as high as $u' = 30$ and has a corresponding limitation on values of X.

---

[6] See Appendix I (8).

[7] See Appendix I (9).

[8] Reproduced with permission from Table I (p. 3) and Table II (p.3) of *Poisson's Exponential Binomial Limit* by E. C. Molina, of the Bell Telephone Laboratories, Inc. (New York: D. Van Nostrand Co., Inc., 1949).

FIGURE 4.9
Cumulative Probability Curves for the Poisson Distribution (the Dodge-Romig Chart)

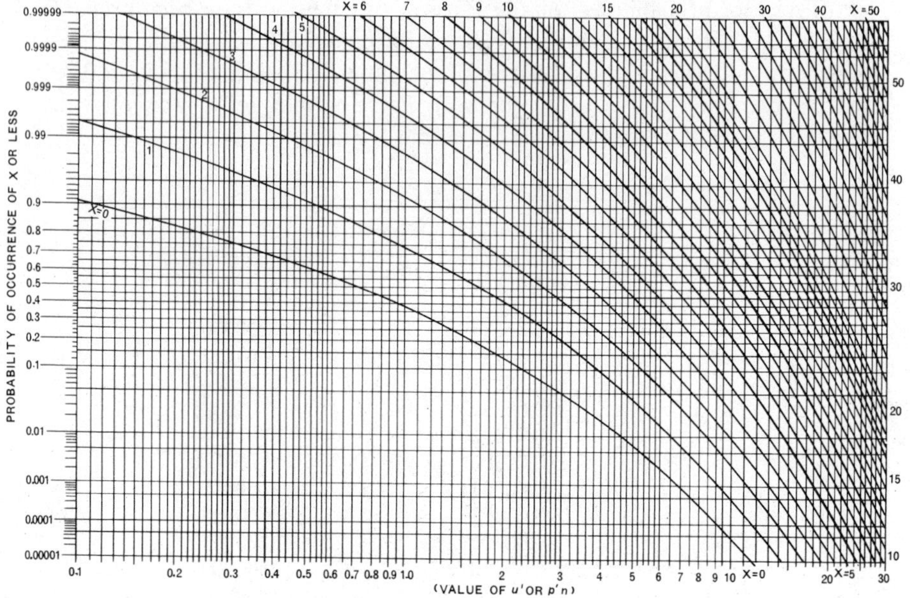

SOURCE: Reproduced, with permission, from H. F. Dodge and H. G. Romig, *Sampling Inspection Tables* (New York: John Wiley & Sons, 1944).

5.3. The Hypergeometric Distribution

The formula for the hypergeometric distribution is

$$\textbf{(4.15)} \quad P(X/n) = \frac{C^{N-m}_{n-X} C^{m}_{X}}{C^{N}_{n}} = \frac{\dfrac{(N-m)!}{(n-X)!(N-m-n+X)!} \cdot \dfrac{m!}{X!(m-X)!}}{\dfrac{N!}{n!(N-n)!}}$$

in which $P(X/n)$ again stands for the probability of X out of n, and n, N, and m are special parameters. The variable X/n takes on the values of $0/n$, $1/n$, up to m/n if $m \leq n$, otherwise up to n/n. The symbol C^{m}_{X}, for example, is the combinational formula and stands for the number of combinations of X out of m. By way of illustration, $P(X/n)$ might be the probability of getting X nonconforming items in a random sample of n taken without replacement from a lot of N items of which m are nonconforming.

A hypergeometric distribution looks very much like a binomial distribution. If the variable is taken as $p = X/n$ and if we set $p' = m/N$, then the mean and standard deviation of the hypergeometric distribution are[9]

[9] See A. M. Mood and F. A. Graybill, *Introduction to the Theory of Statistics* (2d ed.; New York: McGraw-Hill, 1963), p. 113.

$$E(p) = p' \tag{4.16}$$

$$\sigma'_p = \sqrt{\frac{p'(1-p')}{n}\left(\frac{N-m}{N-1}\right)} \tag{4.17}$$

The hypergeometric distribution has been tabulated by G. J. Lieberman and D. B. Owen in *Tables of the Hypergeometric Probability Distribution.* Results are given for $N = 2$ to 100 and $n = 1$ to 50 and for a few other special values of n and N.

If the Lieberman-Owen tables are not available or results are wanted for values outside the range of the tables, we can evaluate formula (4.15) with the help of tables of factorials (see Table CC of Appendix II). To illustrate, let $N = 50$, $n = 10$, and $m = 5$. Then by (4.15)

$$P(0/n) = C^{45}_{10}/C^{50}_{10} = \frac{45!}{10!35!}\frac{10!40!}{50!}$$

Table CC gives

$$\log(45!) = 56.0778$$
$$\log(40!) = 47.9116$$

yielding

$$\log(45!)(40!) = 103.9894$$

and

$$\log(35!) = 40.0142$$
$$\log(50!) = 64.4831$$

yielding

$$\log(35!)(50!) = 104.4973$$

Hence

$$\log\left[\frac{(45!)(40!)}{(35!)(50!)}\right] = 103.9894 - 104.4973 = -0.5079$$

On taking antilogs, we have

$$P(0/n) = 0.3105$$

6. SOME IMPORTANT DISTRIBUTIONS OF CONTINUOUS VARIABLES

6.1. The Normal Distribution

The most important of all distribution formulas for continuous variables is that for the normal distribution. The density function for the normal distribution in its most general form is

(4.18)
$$f(X) = \frac{1}{\sigma'\sqrt{2\pi}} e^{-(X-\bar{X}')^2/2\sigma'^2}$$

where $\bar{X}'$ is the mean of the distribution, σ' is its standard deviation, and $e = 2.718 +$. If we set $z = (X - \bar{X}')/\sigma'$, the density function for z is

(4.18a)
$$f(z) = \frac{1}{\sqrt{2\pi}} e^{-z^2/2}$$

which is the formula for the standard normal density function. An example of conditions that might produce the normal distribution was given above in Section 3.2.

The graph of the normal density function in its general form is given in Figure 4.10, and a graph of the standardized form is given in Figure 4.11. In either form this curve is called the normal frequency curve, or the normal probability curve, or simply the normal curve. It will be noted that the curve is symmetrical and bell-shaped. The points of inflection of the density function fall at $X = \bar{X}' + \sigma'$ and $X = \bar{X}' - \sigma'$ or $z = \pm 1$, and the range $\bar{X}' - 3\sigma'$ to $\bar{X} + 3\sigma'$ or $z = -3$ to $z = +3$ subtends 99.7 percent of the area under the curve. For this distribution $\gamma_1' = (\mu_3'^2/\mu_2'^3)^{1/2} = 0$ and $\gamma_2' = \mu_4'/\mu_2'^2 - 3 = 0$.

No formula exists for the cumulative form for the normal distribution since the density function cannot be integrated. Extensive tables of the cumulative function have been computed, however. A graph of the cumulative function for the standardized normal variable $z = (X - \bar{X}')/\sigma'$ is shown above in Figure 4.5. The area under the curve of Figure 4.11 from $-\infty$ to z_1 is the ordinate of the cumulative curve at z_1. Tables of the cumulative function are thus often termed Tables of Areas under the Normal Curve. Such a table is reproduced in Table A2 of Appendix II. This gives the area under the standard normal curve (i.e., under the normal density function) from

FIGURE 4.10
The Normal Density Function: General Form

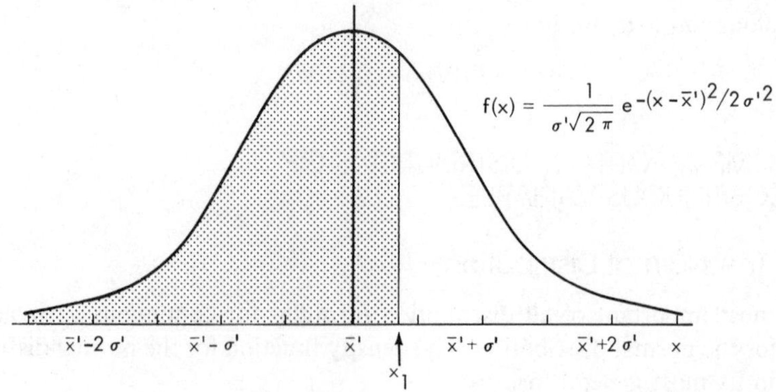

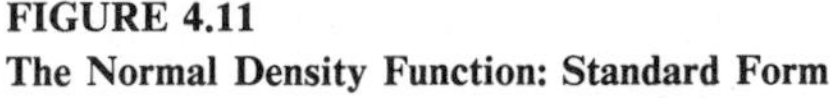

FIGURE 4.11
The Normal Density Function: Standard Form

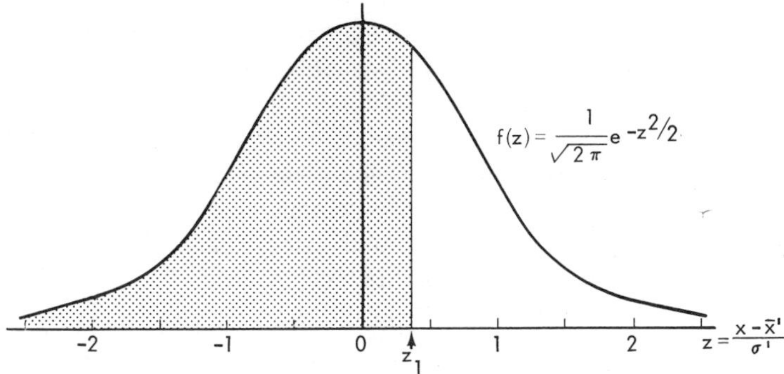

$-\infty$ to z, which is the relative frequency or probability of values equal to or less than z.

It will be noted that to use the normal area table, it is first necessary to express the variable in standardized form, i.e., as a deviation from its mean value measured in standard deviation units. For example, suppose that it is known that the distribution of X is normal in form and that the mean of the distribution is 100 and its standard deviation is 10. Suppose that the problem is to find the probability that X will be less than 120. To solve this problem we proceed as follows:

1. Compute $x = X - \bar{X}'$. Thus $x = 120 - 100 = 20$.
2. Compute $z = x/\sigma'$. Thus $z = 20/10 = 2$.
3. Locate z in the normal area table and read off the probability that a standardized normal variable will be equal to or less than this value of z. Thus for $z = 2$, we find the probability of $z \leq 2$ is 0.9772. This is the answer to our problem.

Table A2 gives probabilities for only positive values of z. Since the normal curve is symmetrical, however, the table will readily yield probabilities for negative values of z. Suppose, for example, that in the previous problem we wished to find the probability of an X less than or equal to 80. Our computations would then be as follows:

1. $x = X - \bar{X}' = 80 - 100 = -20$.
2. $z = -20/10 = -2$.
3. Our table shows that the probability of a z less than or equal to $+2$ is 0.9772. Since $-z$ is as far from the mean in a negative direction as $+z$ is in a positive direction and since the normal curve is symmetrical about its mean, the probability of a z being less than or equal to -2 is simply 1 minus the probability of z being less than or equal to $+2$. Hence the probability that z is less than or equal to -2 is

$1 - 0.9772 = 0.0228$. This is therefore the probability that the X in our problem is less than or equal to 80.

Although Table A2 gives cumulative probabilities, we can find the probability that X lies between certain limits by subtracting the cumulative probabilities for the two limits. For example, suppose that in the previous problem we wanted to find the probability that X lies between 80 and 120. First, we compute the probability that X is less than or equal to 120. This was found above to be 0.9772. Second, we compute the probability that X is less than or equal to 80. This was found to be 0.0228. Then we take the difference between the two probabilities, which yields $0.9772 - 0.0228 = 0.9544$. Hence the probability that X lies between 80 and 120 is 0.9544.

Finally, to get the probability that X exceeds a certain value, we subtract from 1 the probability that it is equal to or less than that value. For example, to find the probability that the X of the previous problem exceeds 120, we subtract from 1 the probability that X is less than or equal to 120. This yields $1 - 0.9772 = 0.0228$. Hence the probability that X is greater than 120 equals 0.0228. Since the curve is symmetrical about 100, that is, of course, the same as the probability that X is less than 80.[10]

Table A1 of Appendix II gives the ordinates of the standard normal curve. These are the values of the standard density function for specified values of z. Their most common use is in plotting the normal curve.

6.2. The Negative Exponential Distribution

A distribution that is of special interest in reliability and life testing is the negative exponential distribution. The density function is

$$f(X) = \frac{1}{\theta} e^{-X/\theta} \qquad 0 < X < \infty \tag{4.19}$$

where θ is the mean and the variance is θ^2. The cumulative form of the distribution is given by

$$F(X) = \int_0^X f(t)dt = 1 - e^{-X/\theta}$$

A graph of the density function is shown in Figure 4.12. Values of the cumulative distribution can be readily derived from tables of the negative exponential function which are generally available.[11]

---

[10] Since the variable is assumed to be continuous, the inclusion or omission of the expression "or equal to" does not affect the numerical results.

[11] See, for example, Charles D. Hodgman, *Mathematical Tables from Handbook of Chemistry and Physics* (Cleveland, Ohio: Chemical Rubber Publishing Co., 1939).

FIGURE 4.12
The Density Function of the Negative Exponential Distribution

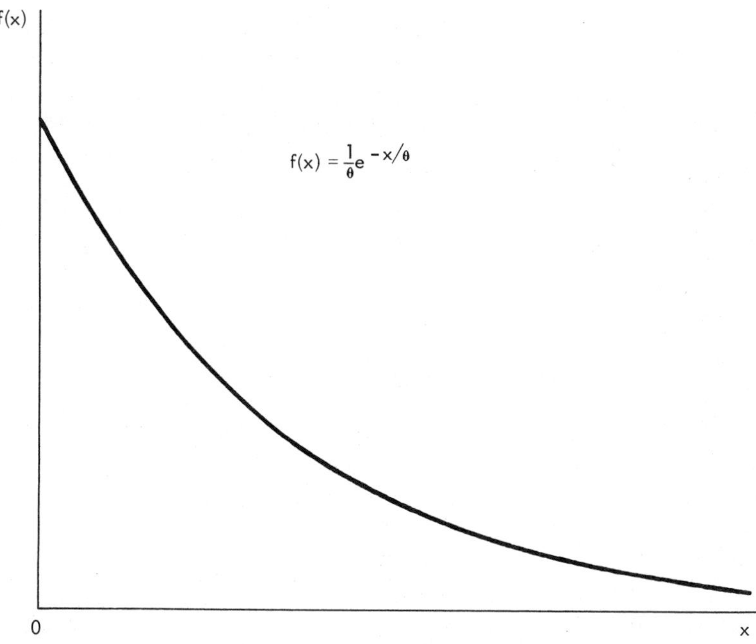

7. SOME INTERRELATIONSHIPS AMONG THE VARIOUS DISTRIBUTIONS

7.1. The Normal Distribution as an Approximation to the Binomial Distribution

The binomial distribution is a discrete distribution. To find the probability of getting $X = X_1$ or less, it is necessary to add up the probability of getting $X = X_1$, the probability of getting $X = X_1 - 1$, the probability of getting $X = X_1 - 2$, and so on, down to the probability of getting $X = 0$. This summing of individual probabilities may be easy enough if X_1 is close to 0 — say, 1 or 2 — but may become very time consuming if X_1 takes on such values as 10 or 15 or 100. Since tables of the binomial distribution may not generally be available or may not cover the desired value of n, various short-cut methods have been devised for computing the sum of the individual probabilities of a binomial distribution.

One of the most common of the short-cut methods for computing the sum of the ordinates of the binomial distribution is to use the normal distribution as an approximation to the binomial distribution. It can be shown by mathematical analysis that the ordinates (probabilities) of a binomial distribution can under certain circumstances be approximated by the corresponding ordinates of the normal curve whose mean and standard deviation are the same as that of the binomial distribution.[12] The sum of the ordinates of a normal distribution over any given range of values can, in turn, be approximated by the area under the normal curve for that range. (Compare Section 3.2 above). Since the latter can be computed with relative ease, the normal distribution affords, in certain circumstances, a convenient instrument for approximating a sum of binomial probabilities.

The normal curve gives an excellent approximation to the binomial distribution when p' is close to 0.5. In fact, for $p' = 0.5$, the approximation is good for n as small as 10 (see Figure 4.13). As p' deviates from 0.5, the approximation gets worse and worse (see Figure 4.14). On the other hand, for values of p' significantly different from 0.5, the approximation of the normal distribution to the binomial distribution gets better, the larger the value of n. Even if p' is as low as 0.10 or as high as 0.90, if n runs above 50 the normal approximation does not give bad results. Below 0.10 or above 0.90, the Poisson distribution is commonly used to approximate the binomial distribution, although the normal distribution still does fairly well[13] so long as $p'n \geq 5$. Approximation by the Poisson distribution is taken up below.

The use of the normal approximation to the binomial distribution may be illustrated by an example. Suppose that the probability of a nonconforming unit in the output of a given machine is 0.25. A random sample of 100 units is taken from the output of the machine and is inspected for nonconformities. What is the probability that it will contain 15 or less nonconforming units? It is shown in the next chapter that this probability is given by a binomial distribution with a mean of 0.25 and a standard deviation of $\sqrt{\dfrac{0.25(1-0.25)}{100}} = 0.043$. This binomial distribution can be approximated by a normal curve, the mean of which is 0.25 and the standard deviation of which is 0.043. Hence the answer to the question asked is given approximately by the area of this normal distribution below $z = \dfrac{0.155-0.25}{0.043} = -2.21$. From Table A2 of Appendix II it is seen that the probability of getting a normal variable which lies below $z = -2.21$ is 0.0135. Hence the probability

---

[12] See Appendix (15).

[13] If $p' > 0.10$, the normal curve may do well, even though $p'n < 5$. For example, if $p' = 0.20$ and n is 10, i.e., $p'n = 2$, the probability that $X = 4$ or more is 0.1209. This is the exact answer given by the binomial distribution. If p is treated as a normal variable, with a mean of $p' = 0.2$ and a standard deviation of $\sqrt{\dfrac{p'(1-p')}{n}} = 0.126$, the probability of p's being greater than 0.35 is found to be 0.01170, an error of 0.0039. Likewise, at the other end, the probability that $X = 0$ is given exactly by the binomial as 0.1074. By using the normal approximation, the probability of p's being less than 0.05 is 0.1170, an error of 0.0096. (Note that the area under the normal curve is measured from 0.35 and not 0.4 and from 0.05 and not 0. See Figure 4.14.)

FIGURE 4.13
Comparison of the Normal Distribution with the Binomial Distribution for which $n = 10$ and $p' = 0.50$

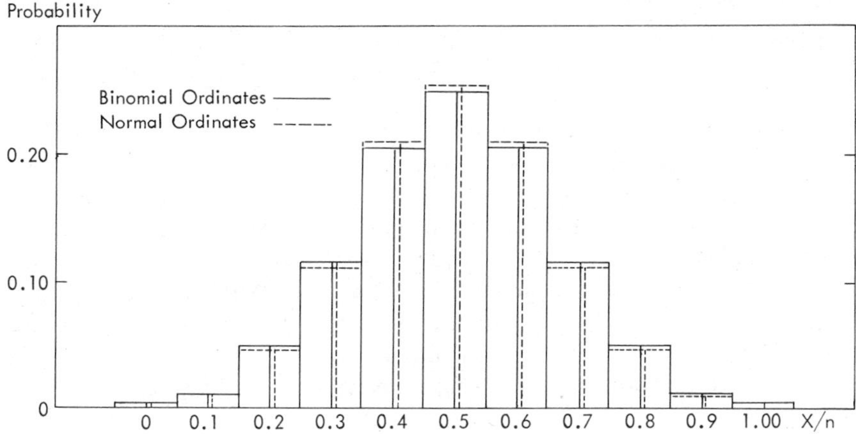

of getting 15 or fewer nonconforming units in the present instance is roughly 1 out of 100.

It will be noticed that in measuring the area under the normal curve we started at 0.155 and not at 0.15. This gives a better approximation than if 0.15 had been used. A better approximation to a sum of binomial tail probabilities is always obtained by measuring the normal area from one half a place beyond the point from which the summation starts. In other words, to find the probability of X or *less* we set

FIGURE 4.14
Comparison of the Normal Distribution with the Binomial Distribution for which $n = 10$ and $p' = 0.20$

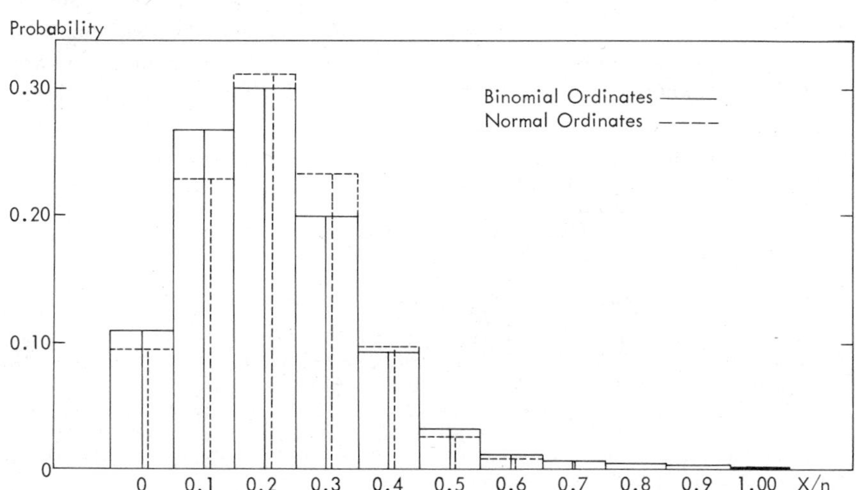

(4.20)
$$z = \left(\frac{X+0.5}{n} - p'\right) \Big/ \sigma'_p$$

To find the probability of X or *more* we set

(4.20a)
$$z = \left(\frac{X-0.5}{n} - p'\right) \Big/ \sigma'_p$$

The reason for this use of a "correction factor" of 0.5 can readily be seen from a study of Figure 4.13. The rectangles representing the binomial probabilities all have bases of unity. The sum of a given set of binomial probabilities is thus graphically represented by the sum of the areas of the corresponding rectangles. It is the sum of the areas of these rectangles that is approximated by the area under the normal curve. With reference to Figure 4.13, it is obvious that the curve area for approximating, for example, the probability of $X = 7$ or more should start at $X = 6.5$ not $X = 7$ since that is where the area of the rectangle erected at $X = 7$ begins.

7.2. The Poisson Distribution as an Approximation to the Binomial Distribution

For small values of p' and large values of n—say, for $p' < 0.10$ and certainly for $p'n < 5$—the binomial distribution is better approximated by the Poisson distribution than by the normal distribution.[14] The Poisson distribution is a discrete distribution like the binomial, but its formula permits ready computation and summation. As noted in Section 5.2 extensive tables exist and rough answers can quickly be obtained from the Dodge-Romig chart, Figure 4.9.

To illustrate the use of the Poisson approximation to the binomial distribution, consider the following problem: A process turns out material with a probability of a nonconforming unit equal to 0.02. A sample of 400 units is taken. What is the probability of 4 or less nonconforming units? As shown in Chapter 5 the exact answer is given by the binomial distribution. The Harvard binomial tables give it as 0.09733. To find an approximate answer from the Poisson distribution, set $\mu' = p'n = (0.02)(400) = 8$. Go to Figure

---

[14] See Appendix I (16). On the long tail of the binomial distribution the Poisson probabilities are greater than, and the normal probabilities are less than, the binomial probabilities. Some comparisons may be of interest.

If $p' = 0.02$ and $n = 100$, i.e., $p'n = 2$, the probability that $X = 4$ or more is given by the Poisson as 0.1429, and the probability that $X = 0$ is given as 0.1353. If p is treated as a normal variable with a mean of $p' = 0.02$ and a standard deviation of $\sqrt{\frac{p'(1-p')}{n}} = 0.0014$, it is found that the probability of p's exceeding 0.035 is 0.1423 and the probability of its being less than 0.005 is also 0.1423. These results differ from the Poisson probabilities by 0.0006 and 0.0070 and will be even closer than this to the binomial probabilities.

If $p' = 0.005$ and $n = 2000$, i.e., $p'n = 10$, we find the probability of 4 or less is given by the Poisson distribution as 0.0293 and the probability of 16 or more as 0.0487. The normal approximation yields tail areas below 0.00225 and above 0.00775 of 0.0409, which differ from the Poisson results by 0.0116 and 0.0078.

FIGURE 4.15
Comparison of Binomial and Poisson Ordinates when $p' = 0.02$ and $n = 25$

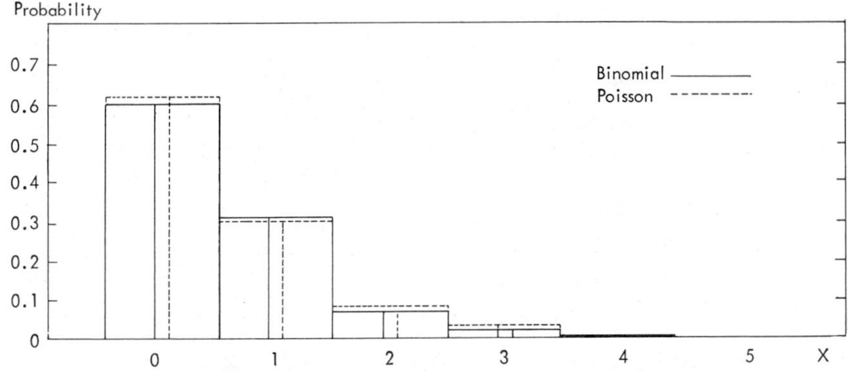

4.9, locate the point 8 on the abscissa scale, proceed vertically to the curve $X = 4$ and read off on the ordinate scale the probability of 4 or less unconforming units. The result is 0.10. The approximation is seen to be very good.

When n is large but p' is small, we sometimes have a "gray area" in which the normal approximation and Poisson approximation do about equally well. In such cases we use whichever is most convenient. Generally this is the Poisson distribution.

7.3. Approximation to the Hypergeometric Distribution

If *both* N *and* n *are large and if neither* m *nor* N − m *is very small,* the hypergeometric distribution can be approximated by a normal distribution. Thus if the variable is $p = X/n$, then p is approximately normally distributed with a mean $= p'$ and a standard deviation equal to

(4.21)
$$\sigma'_p = \sqrt{\frac{p'(1-p')}{n}\left(\frac{N-n}{N-1}\right)}$$

when $p' = m/N$. By this approximation the sum of hypergeometric probabilities from 0 to X can be taken as approximately equal to the area under a standard normal curve below

(4.22)
$$z = \left(\frac{X+0.5}{n} - p'\right) \Big/ \sqrt{\frac{p'(1-p')}{n}\left(\frac{N-n}{N-1}\right)}$$

It will be noted that $\sqrt{\frac{N-n}{N-1}} = \sqrt{\frac{1-n/N}{1-1/N}}$ and that this is greater than 0.95 or almost 1 when $n/N \leq 0.10$. In such cases there is little to distinguish between the normal approximation to the hypergeometric distribution and the normal approximation to a binomial with $p' = m/N$. The use of the "correction factor" of 0.5 will again be noted.

If N, m, *and* N − m *are large relative to* n *and* X, the hypergeometric can be approximated by a binomial distribution with $p' = m/N$ and n (binomial) = n (hypergeometric). In the more useful cumulative form we have

$$\textbf{(4.23)} \qquad \sum_{X=0}^{c} C_{n-X}^{N-m} C_X^m / C_n^N \doteq \sum_{X=0}^{c} \frac{n!}{X!(n-X)!} \left(\frac{m}{N}\right)^X \left(1 - \frac{m}{N}\right)^{n-X}$$

The proof of this is given in Appendix I (17). We shall call this approximation Binomial I.

If N *and* n *are large, but* m *or* N − m *is relatively small,* the hypergeometric distribution can be approximated by a binomial distribution with $p' = n/N$ and n (binomial) = m (hypergeometric). In the cumulative form we have, because of the symmetry of m and n in equation (4.15),

$$\textbf{(4.24)} \qquad \sum_{X=0}^{c} \frac{C_{n-X}^{N-m} C_X^m}{C_n^N} \doteq \sum_{X=0}^{c} \frac{m!}{X!(m-X)!} \left(\frac{n}{N}\right)^X \left(1 - \frac{n}{N}\right)^{m-X}$$

We shall call this approximation Binomial II.

In cases in which the p' for the binomial approximation is small and n or m are not too small, it is possible to use the Poisson distribution as an approximation to the hypergeometric distribution.

A comparison of the binomial and Poisson approximations for $p'n = 0.5$ is shown in Table 4.2. The normal approximation is not shown in this case since it does rather badly for $p'n$ as small as 0.5.

8. TCHEBYCHEV'S INEQUALITY

When we do not have a formula for a distribution but know something about its characteristics, then Tchebychev's inequality is of some use. This inequality reads as follows:[15]

The probability that a variable X *should deviate from its mean by more than* k *times its standard deviation is equal to or less than* 1/k². *In symbols,*

$$\textbf{(4.25)} \qquad P(X < \bar{X}' - k\sigma' \text{ or } X > \bar{X} + k\sigma') \leq 1/k^2$$

Camp and Meidel have made an extension of Tchebychev's inequality that increases its usefulness without seriously restricting its application. The Camp-Meidel extension runs as follows:[16]

If the distribution of X *is unimodal, the probability that* X *should deviate from its mean by more than* k *times its standard deviation is equal to or less than* 1/2.25k².

These inequalities enable us to place limits on the variation of a variable if we have knowledge of its mean and standard deviation. For example, if

---

[15] For proof, see Appendix I (10).

[16] See H. L. Rietz, *Mathematical Statistics* (Carus Mathematical Monographs, No. 3) (Chicago: Open Court Publishing Co., 1927), Sec. 49. *Unimodal* is here meant to imply that the distribution is monotonically decreasing on both sides of its one mode.

TABLE 4.2

Comparison of Hypergeometric, Binomial I and II, and Poisson Probabilities for Selected Values of *n*, *N*, and *p′*

X	$p'n = 0.5$											Poisson	X
	$p' = 0.25$, $n = 2$			$p' = 0.1$, $n = 5$				$p' = 0.02$, $n = 25$					
	Hypergeometric		Binomial I	Hyper-geometric	Binomial II	Hyper-geometric	Binomial I	Hyper-geometric	Binomial II	Hyper-geometric	Binomial I		
	$N = 8$, $m = 2$	$N = 40$, $m = 10$	$N \to \infty$, $m \to \infty$	$N = 20$, $m = 2$	See Text	$N = 100$, $m = 10$	$N \to \infty$, $m \to \infty$	$N = 250$, $m = 5$	See Text	$N = 500$, $m = 10$	$N \to \infty$, $m \to \infty$	$n \to \infty$, $p' \to 0$	
0......	0.53572	0.55769	0.56250	0.55263	0.56250	0.58375	0.59049	0.58783	0.59049	0.59586	0.60359	0.60653	0
1......	0.42857	0.38462	0.37500	0.39474	0.37500	0.33940	0.32805	0.33249	0.32805	0.31968	0.30795	0.30326	1
2......	0.03571	0.05769	0.06250	0.05263	0.06250	0.07021	0.07290	0.07789	0.07290	0.07393	0.07542	0.07582	2
3......						0.00637	0.00810	0.00742	0.00800	0.00969	0.01180	0.01264	3
4......						0.00025	0.00045	0.00036	0.00045	0.00080	0.00132	0.00158	4
5......						0.00001	0.00001	0.00001	0.00001	0.00004	0.00011	0.00016	5
6......										0.00000	0.00001	0.00001	6

the distribution of X is unimodal and if[17] $\bar{X}' = 100$ and $\sigma' = 10$, then the chance that X will be above 130 or below 70 equals $1/2.25(3)^2$ (= 0.049) or less.

9. FORECASTING WHAT A PROCESS WILL DO

If a process is operating in a random manner, the distribution of the output will be stable. This is just one aspect of the general stability of the results of chance forces. If a process is operating in a random manner, therefore, the distribution of a large sample of the output may be taken as a good picture of the distribution of the output in general. The relative frequencies of the sample distribution can then be taken as probabilities for forecasting future output.

For example, Figure 4.16 suggests that the variation in the heights of bomb bases is a random one. Consequently, the sample of 145 cases that make up Table 3.2, although not large enough to yield a precise representation of the distribution of the universe, is probably large enough to give us a rough picture of what to expect in the future. For example, from Figure 3.4 it may be predicted that approximately 7 percent of the output will fall above 0.8405 inches and another 5 percent below 0.8195 inches, or that 12 percent of the product will lie outside the limits 0.8195 and 0.8405.[18] Since these are the specification limits for this product,[19] it may be predicted that the proportion nonconforming will be about 12 percent.

Consider, also, the following: Suppose that a process is operating in a random manner and the output is distributed in the form of a normal distribution. Suppose, further, that we can make a good estimate of the mean and standard deviation of the process. Then, under these conditions, we can lay off a distance of $3\sigma'$ on either side of the mean and expect 99.7 percent of the items to fall within these limits. Hence, if we plot the output of a process on a chart which shows the mean and $3\sigma'$ limits (cf. Figure 4.16) and if we find several points falling outside these limits, we can conclude either that the process is not normally distributed or that it is not operating in a random manner, temporarily at least. If a process is operating in a random manner and we know or have good estimates of the mean and standard deviation of the process, but if the form of the distribution is unknown, then we can generally use the Camp-Meidel version of Tchebychev's inequality and say that not more than $100/2.25(3)^2$ percent of the cases will fall outside $\bar{X}' \pm 3\sigma'$. This is slightly under 5 percent.

For example, Figure 4.16 shows that 3 of the 145 bomb base heights fell outside the $3\sigma'$ limits. Since we saw in Chapter 3 that the distribution

---

[17] It will be noted that $\bar{X}'$ and σ' are universe values. They may be estimated, however, from the mean and standard deviation of a large random sample.

[18] See Figure 4.16 for a study of the randomness of the data.

[19] See Table 3.2. The limits are 0.830 ± 0.01.

FIGURE 4.16
Variation in Heights of Fragmentation Bomb Bases (see Table 3.2)

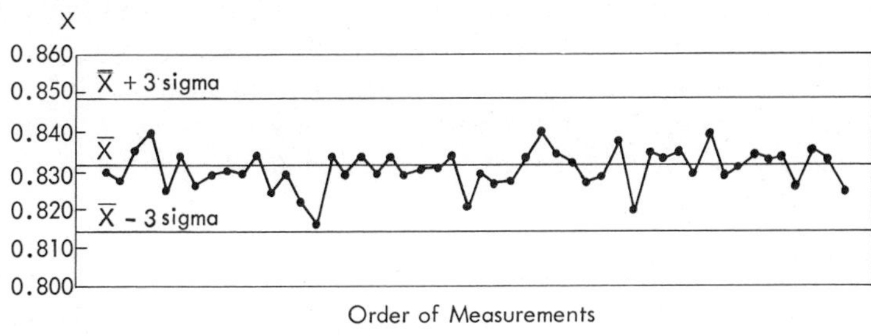

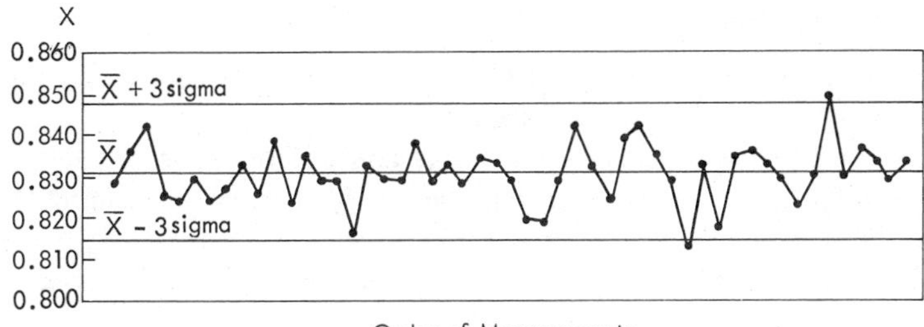

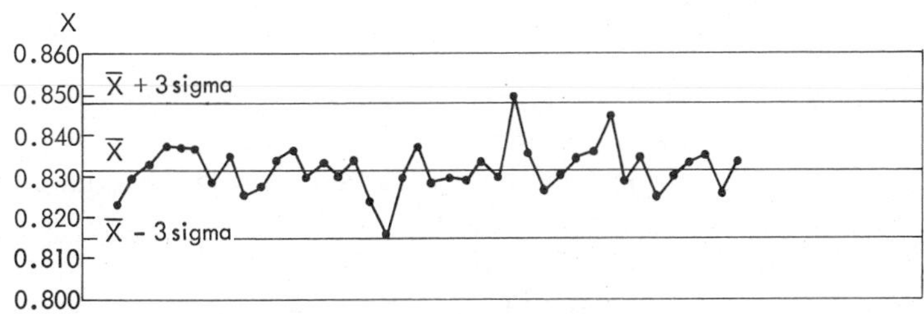

of bomb base heights has more than normal kurtosis, this number of cases beyond $3\sigma'$ was not unexpected. For the future we can predict that the number beyond the 3σ limits of Figure 4.16 will probably not exceed 5 out of 100. If we should find more than this, we can conclude that something has happened to the process.

Knowledge of what a process will do if operating in a random manner is the foundation of the theory of control charts. This is more fully discussed in Part 4.

10. DISTRIBUTION OF SUMS AND DIFFERENCES

An important branch of distribution theory consists of certain theorems about the distribution of sums, differences, products, and ratios. This section will discuss the distributions of sums and differences.

10.1. Three Important Theorems

There are three important theorems about the distribution of a sum or difference. The first theorem deals with the mean. We have

THEOREM (4.1). *Given the distribution of a random variable* X *and the distribution of another random variable* Y, *the mean value of the distribution of the variable* X ± Y *equals the mean value of* X ± *the mean value of* Y. In brief,

(4.26) $$E(X \pm Y) = E(X) \pm E(Y) = \bar{X}' \pm \bar{Y}'$$

This theorem is proved in Appendix I (11). A corollary to the above is

(4.26a) $$E(aX \pm bY) = aE(X) \pm bE(Y) = a\bar{X}' \pm b\bar{Y}'$$

since the expected value of a constant times a variable equals the constant times the expected value of the variable.

A second theorem pertaining to sums and differences is

THEOREM (4.2). *Given the distribution of a random variable* X *and the distribution of another random variable* Y, *the variance of the variable* X ± Y *equals the variance of* X *plus the variance of* Y, *if* X *and* Y *are independent,* i.e.,

(4.27) $$\sigma'^2_{X\pm Y} = \sigma'^2_X + \sigma'^2_Y \text{ if } P(XY) = P(X)P(Y)$$

This is proved in Appendix I (13). A corollary to this is

COROLLARY OF THEOREM (4.2). *The variance of* $aX \pm bY$ *equals* $a^2\sigma'^2_X + b^2\sigma'^2_Y$ *if* X *and* Y *are independent*
since the variance of a constant times a variable equals the square of the constant times the variance of the variable.

The following theorem will be stated without proof.[20]

THEOREM (4.3). *If the distributions of* X *and* Y *are both normal, the distribution of the sum or difference,* X ± Y, *is also normal.*

The foregoing theorems are used continuously in statistical theory and will be applied frequently in this text. They also have direct applications.

10.2. Statistical Tolerances

Suppose, for example, that a rod is put together in sections. From a study of the manufacturing process, we know the means and standard deviations of the lengths of the individual sections and wish to find the mean and standard deviation of the lengths of the assembled rod. The answer is given

---

[20] See P. G. Hoel, *Introduction to Mathematical Statistics* (4th ed.; New York: John Wiley & Sons, 1971), or any other book on mathematical statistics. Hoel proves that the *mean* of a sample from a normal universe is normally distributed. But a mean is simply a sum divided by a constant. Also see Section 3.2 of this chapter.

immediately by application of Theorems 4.1 and 4.2. From Theorem 4.1 we know that the mean of the lengths of the assembled rods is the sum of the means of the individual sections. From Theorem 4.2 we know that the standard deviation of the lengths of the assembled rods is the square root of the sum of the variances of the individual lengths. If it happens that the lengths of the individual sections are all normally distributed, we also know that the length of the assembled rod will be normally distributed.

The relationships just discussed are the foundation of what are called statistical tolerances. Suppose that the engineering tolerance in the length of each section of the rod is ±0.001 inches. If there are 4 sections to the rod, then the sum of the engineering tolerances will be ±0.004 inches. However, if rods are assembled by picking the sections at random, then the sum of the engineering tolerances will have little practical significance. This may be demonstrated as follows:

Suppose that the standard deviation of the length of each section is 0.0005 inches and suppose that the mean length falls at the center of the tolerance range, i.e., at the standard value. In other words, the engineering tolerances are $\pm 2\sigma'$ from the mean. Under these conditions, the standard deviation of the length of the assembled rod will be

$$\begin{aligned}\sigma' &= \sqrt{(0.0005)^2 + (0.0005)^2 + (0.0005)^2 + (0.0005)^2}\\ &= 0.0005\sqrt{4} = 0.0010 \text{ inches}\end{aligned}$$

and

$$3\sigma' = 0.0030 \text{ inches}$$

If the individual lengths are normally distributed, the length of the assembled rod will be normally distributed. Hence, if we use random assembly, the probability that a rod will deviate from its standard length by more than 0.003 inches is very small—say, less than 3 out of 1,000. These limits are 0.001 inches less than the engineering tolerances.

The practical conclusion is that, if the tolerances of the assembled rod are the important factors, we may lighten up on the tolerances of the individual sections without much fear that the overall tolerance will be exceeded. For example, under the given conditions, about 5 percent of the individual sections will exceed their tolerances. This 5 percent need not be rejected, however, but can be allowed to pass with the others. For, with random assembly, extremely few cases will occur where all the individual sections exceed their tolerances and thus give a total rod length in excess of its tolerance. Even cases in which two of the individual sections exceed their tolerances will be very rare.[21]

[21] For an interesting article on special phases of statistical tolerances, see F. S. Acton and E. G. Olds, "Tolerances—Additive or Pythagorean?" *Industrial Quality Control,* November 1948, pp. 6–12.

11. DISTRIBUTIONS OF PRODUCTS AND QUOTIENTS

11.1 Important Theorems

An important theorem regarding the distribution of a product is the following:

THEOREM (4.4). *Given the distribution of a random variable* X *and the distribution of a random variable* Y, *the mean value of the product* XY *equals the product of the mean values, if* X *and* Y *are independent,* i.e.,

(4.28) $$E(XY) = E(X)E(Y) \quad \text{if} \quad P(XY) = P(X)P(Y)$$

COROLLARY OF THEOREM (4.4).

(4.28a) $$E(X/Y) = E(X)E\left(\frac{1}{Y}\right) \quad \text{if} \quad P(X/Y) = P(X)P\left(\frac{1}{Y}\right)$$

Another important theorem pertains to the variance of a product of two independent random variables. Let $X = \bar{X}' + x$ and $Y = \bar{Y}' + y$ where $\bar{X}'$ and $\bar{Y}'$ are the means of X and Y, respectively. Then $XY = (\bar{X}' + x)(\bar{Y}' + y) = \bar{X}'\bar{Y}' + x\bar{Y}' + y\bar{X}' + xy$. If x and y are relatively small compared with $\bar{X}'$ and $\bar{Y}'$, we can neglect the term xy. If we do so, the corollary to Theorem (4.2) immediately yields

(4.29) $$\sigma'^2_{XY} = \bar{Y}'^2\sigma'^2_X + \bar{X}'^2\sigma'^2_Y$$

If the xy term is kept, we get[22] the exact formula for the variance of xy, viz,

(4.29a) $$\sigma'^2_{XY} = \bar{Y}'^2\sigma'^2_X + \bar{X}'^2\sigma'^2_Y + \sigma'^2_X\sigma'^2_Y$$

A third important theorem pertains to the variance of a ratio. We have $X/Y = (\bar{X}' + x)/(\bar{Y}' + y)$ which will approximately equal[23] $\bar{X}'/\bar{Y}' + \dfrac{x\bar{Y}' - y\bar{X}'}{\bar{Y}'^2}$ if x and y are small relative to $\bar{X}'$ and $\bar{Y}'$. Hence, on applying the Corollary to Theorem 4.2, we have

THEOREM (4.5). *If the variations in the random variables* X *and* Y *are small compared with* $\bar{X}'$ *and* $\bar{Y}'$ *and if* X *and* Y *are independent, then*

(4.30) $$\sigma'^2_{X/Y} \doteq \frac{\bar{Y}'^2\sigma'^2_X + \bar{X}'^2\sigma'^2_Y}{\bar{Y}'^4}$$

Since the distribution of a sum or difference is normal if the distribution of individual variables is normal, the distribution of a product or quotient will, in practical applications, be approximately normal if the distribution of the individual variables are both normal and if the division variable does not in practice become zero, or take on values very close to zero.

---

[22] See Appendix I (14).

[23] This is obtained by carrying out the division algebraically and noting that the remainder contains terms in xy and y^2 and hence is of higher order of approximation than the first three terms given above.

11.2 Distributions of Areas, Volumes, and Ratios

Often in industrial research we have to find the area or volume of a given product, or we have to calculate a ratio such as pounds per cubic foot. In many of these instances the above theorems will be found useful.

For example, to find an area, we measure the length and width of our object. Any variation in the length *(L)* and width *(W)* will produce a variation in area. If the variations are independent, it follows immediately (from Theorems 4.5 and 4.6) that the mean area equals the mean width times the mean length and the variance of the area equals

$$\sigma_A^2 = \bar{W}'^2\sigma_L'^2 + \bar{L}'^2\sigma_W'^2 + \sigma_L'^2\sigma_W'^2$$

Similarly for a volume. This would be obtained by multiplying width *(W)* by length *(L)* by depth *(D)*. If the variations are independent, the mean volume *(V)* would be the product of the mean width by the mean length by the mean depth and its variance would be

$$\sigma_V'^2 = \bar{W}'^2\bar{D}'^2\sigma_L'^2 + \bar{L}'^2\bar{D}'^2\sigma_W'^2 + \bar{W}'^2\bar{L}'^2\sigma_D'^2 + \bar{W}'^2\sigma_L'^2\sigma_D'^2 + \bar{L}'^2\sigma_W'^2\sigma_D'^2 + \bar{D}'^2\sigma_W'^2\sigma_L'^2 + \sigma_L'^2\sigma_W'^2\sigma_D'^2$$

The area and volume would in each case be approximately normally distributed.

The distribution of a ratio can also be found approximately by the above theorems. Suppose, for example, we wish to determine a tensile strength in terms of pounds per square inch. To do this we measure the total weight and the total area and then take their ratio. If $\bar{W}'$ is the average or standard weight and w the variation from this average and if $\bar{A}'$ is the average or standard area and a the variation from this average, then any tensile strength is give by

$$\bar{S}' + s = \frac{\bar{W}' + w}{\bar{A}' + a}$$

where $\bar{S}'$ is the average or standard tensile strength. But if w and a are relatively small, the right-hand member is equal approximately to

$$\frac{W'}{\bar{A}'} + \frac{\bar{A}'w - \bar{W}'a}{\bar{A}'^2}$$

Since the average or standard tensile strength will equal the average or standard weight divided by the average or standard area, we have

$$\bar{S}' = \frac{\bar{W}'}{\bar{A}'}$$

and

$$s = \frac{\bar{A}'w - \bar{W}'a}{\bar{A}'^2}$$

Thus, if w and a are small relative to W and A, and if they are independent, the tensile strength will have a mean of approximately $\frac{\bar{W}'}{\bar{A}'}$ and a variance approximately equal to

$$\sigma_S'^2 = \frac{\bar{A}'^2\sigma_w'^2 + \bar{W}'^2\sigma_a'^2}{\bar{A}'^4}$$

If in practical applications w and a are approximately normally distributed and are relatively small, then s will also be approximately normally distributed.

The above results will often enable us to compute probabilities for areas, volumes, and ratios if we have knowledge of the distributions of the individual measurements that go to make up those areas, volumes, and ratios.

12. PROBLEMS

4.1. For $n = 5$ and $p' = 0.1$, compute the ordinates of the binomial distribution for $X/n = 0, \frac{1}{5}, \frac{2}{5}, \frac{3}{5}, \frac{4}{5}$, and $\frac{5}{5}$. Graph your results. Compute the mean and standard deviation of this distribution.

4.2. For $n = 5$ and $p' = 0.5$, compute the ordinates of the binomial distribution for $X = 0, \frac{1}{5}, \frac{2}{5}, \frac{3}{5}, \frac{4}{5}$, and $\frac{5}{5}$. Graph the results. Compute the mean and standard deviation of this distribution.

4.3. For $n = 20$ and $p' = 0.4$, compute the ordinates of the binomial distribution for $X/n = 0, \frac{1}{20}, \frac{2}{20}, \frac{3}{20}, \frac{4}{20}$, and $\frac{5}{20}$. Compute the mean and standard deviation of this distribution.

4.4. For $n = 7$ and $p' = 0.2$, compute the ordinates of the binomial distribution for $X/n = 0, \frac{1}{7}, \frac{2}{7}, \frac{3}{7}, \frac{4}{7}, \frac{5}{7}, \frac{6}{7}$, and $\frac{7}{7}$. Graph your results.

4.5. For $n = 50$ and $p' = 0.2$, compute the ordinates of the binomial distribution for $X/n = 0, \frac{1}{50}, \frac{2}{50}$, and $\frac{3}{50}$. Graph your results. Compute the mean and standard deviation of this distribution.

4.6. For $n = 100$ and $p' = 0.01$, compute the ordinates of the binomial distribution for $X/n = 0, \frac{1}{100}, \frac{2}{100}$, and $\frac{3}{100}$. Graph your results. Compute the mean and standard deviation of this distribution.

4.7. Check your answers to Problems 4.1–4.4 with the National Bureau of Standards, *Tables of the Binomial Probability Distribution.*

4.8. Take u' (or a) $= 0.5$ and find from Molina's tables the ordinates of the Poisson distribution for $X = 0, 1, 2, 3, 4$, and 5. Compare with the binomial ordinates of Problem 4.1.

4.9. Take u' (or a) $= 1.4$ and find from Molina's tables the ordinates of the Poisson distribution for $X = 0, 1, 2, 3, 4, 5, 6$, and 7. Compare your results with the binomial ordinates computed in Problem 4.4.

4.10. Take u' (or a) $= 1$ and find from Molina's tables the ordinates of the Poisson distribution for $X = 0, 2$, and 3. Compare the results with the binomial ordinates computed in Problems 4.5 and 4.6. To which of these do the Poisson ordinates come the closer?

4.11. Take $u' = 5$ and find from Molina's tables the sum of the ordinates of the Poisson distribution for $X \geq 5$; $X \geq 10$; $X < 5$; $X < 10$; $X > 5$; $X \leq 5$; $X \leq 3$.

4.12. Take $u' = 2$ and find from Table E the sum of the ordinates of the Poisson distribution for $X \geq 2$; $X \geq 4$; $X < 2$; $X > 4$; $X > 2$; $X < 4$; $X \leq 2$; $X \leq 4$.

4.13. Take $u' = 0.5$ and find from Table E the sum of the ordinates of the Poisson distribution of $X \geq 4$. Compare with the sum of the binomial ordinates for $X/n = 4/5$ and $5/5$ computed in Problem 4.1.

4.14. Take $u' = 1.4$ and find from Table E the sum of the ordinates of the Poisson distribution for $X \geq 5$. Compare with the sum of the binomial ordinates for $X/n = 5/7$, $6/7$, and $7/7$ computed in Problem 4.4.

4.15. Take $u' = 1$ and find from Table E the sum of the ordinates for $X \leq 3$. Compare with results of Problem 4.5. Problem 4.6. Which gives the closer results?

4.16. Find the answers to Problem 4.11 by using Figure 4.9.

4.17. Find the answers to Problem 4.12 by using Figure 4.9.

4.18. Find the answer to Problem 4.13 by using Figure 4.9.

4.19. In 1939, a public utility company reported the following data pertaining to the disability of its male workers:

No. of Absences of One Day or Longer as a Result of Disability	*Percent of Male Workers*
0	44.8
1	31.5
2	13.0
3	6.7
4 or more	4.0

Find the Poisson distribution that comes the closest to "fitting" these data.

4.20 *a.* For $N = 50$, $n = 5$, $m = 5$, find in the Lieberman-Owen tables of the hypergeometric distribution the ordinates of the distribution for $X/n = 0$, $1/5$, and $2/5$. Compare with problems 4.1 and 4.8. Check the table value for the ordinate at $X/n = 1/5$ using formula (4.15).

b. Repeat the first part of *a* for $N = 20$, $n = 5$, $m = 2$ and compare with results obtained in *a*. Take the sum of the ordinates for 0, $1/5$, and $2/5$ and compare with results given by formula (4.24) and the binomial tables.

4.21. For $N = 500$, $n = 50$, $m = 10$, compute from formula (4.15) the ordinates of the hypergeometric distribution for $X/n = 0$, $1/50$, $2/50$, and $3/50$. Compare with results of Problems 4.5 and 4.15.

4.22. *a.* For $N = 100$, $n = 50$, $m = 50$, find in the Lieberman-Owen tables of the hypergeometric distribution the probability of $X/n \leq 22/50$.

b. For the data of *a* use the normal approximation to the hypergeometric distribution to compute the probability of $X/n \leq 22/50$ and compare with the exact results given in *a.*

4.23. A normal curve has a mean of 0 and a $\sigma' = 1$.

a. Compute the area under the curve.

(i) Above $z = 1.62$.

(ii) Below $z = 1.62$.

(iii) Above $z = 2.48$.

(iv) Between $z = 1.00$ and $z = 2.00$.

(v) Between $z = -1.62$ and $z = 0.33$.

4.24. A normal curve has a mean of 120 and a standard deviation of 10. What is the—

a. Value of X corresponding to $z = 0.5$?

b. Value of X corresponding to $z = 2$?

c. Value of z corresponding to $X = 125$?

d. Area under curve between $X = 120$ and $X = 125$?

e. Area under curve between $X = 100$ and $X = 125$?

4.25. What proportion of the area under a normal curve lies—

a. Between $z = 1.20$ and $z = 1.35$?

b. Between $z = 1.25$ and $z = 1.30$?

c. Between $z = -1.282$ and $z = +1.282$?

d. Above $z = 1.960$?

e. Below $z = -1.282$?

f. Below $z = 1.645$?

g. Above $z = 3.08$?

4.26. Between what values of z does the middle 8.26 percent of the area of a normal curve lie?

4.27. Below what value of z does 75 percent of the area of a normal curve lie?

4.28. Above what value of z does 75 percent of the area of a normal curve lie?

4.29. The maximum temperatures reached by a given heating process show an average of 113.30° Centigrade and a standard deviation of 5.6°. Assume that these maximum temperature variations are random and normally distributed and determine—

a. What percent of the maximum temperatures are less than 116.10°?

b. What value is exceeded by 57.78 percent of the maximum temperature readings?

c. What limits include the middle 50 percent of the maximum temperature readings?

d. What percent of maximum temperatures lie between 114.98° and 116.66°?

4.30. The outside diameters of certain bushings are normally distributed. The mean of the distribution is 2.000 inches, and its standard deviation is 0.003 inches.

Determine the probability that the outside diameter of a bushing will equal or exceed 2.009 inches. Be less than 1.994 inches. Lie between 1.997 and 2.003 inches. Between 1.994 and 2.006 inches.

4.31. The skein strength of a certain type of cotton yarn varies from piece to piece in accordance with a normal distribution. The mean of the distribution is 90 pounds and its standard deviation is 11 pounds. Determine the probability that the skein strength of a piece of cotton yarn of the given type will exceed 110 pounds. 115 pounds. Be less than 80 pounds. 70 pounds. Lie between 75 and 105 pounds.

4.32. Suppose that the mean of a normally distributed variable y is 0.4 and its standard deviation is 0.1095. Find the probability that y should be less than 0.275. Compare this result with the sum of the binomial ordinates called for in Problem 4.3. Would the comparison have been better or worse if we had found the probability that y would be less than 0.25?

4.33. The life of a given part is known to have a negative exponential distribution with a mean life of 1,000 hours. What is the probability that a part will last less than 500 hours? Less than 200 hours?

4.34. A variable X is given as a *product* of a large number of other variables, i.e.,

$$X = AV_1V_2V_3 \ldots V_n, \text{ where } A \text{ is a constant.}$$

How would you expect $\log X$ to be distributed? Justify your answer.

4.35. The form of the distribution of a given quality characteristic (X) is not well known, but there are good estimates of its mean and standard deviation. The $\bar{X}' = 90$ and the $\sigma' = 11$. From these figures set limits on the variation of X, assuming that the process operates in a random manner. What is the maximum probability that these limits will be exceeded in a single case?

4.36. A rod is made up of five sections. A study of the individual sections shows that the end sections have mean lengths of 1.001 inches and the three middle sections mean lengths of 1.999 inches. The standard deviation of the length of each section is 0.0004 inches.

If random assembly is employed, what will be the average length of the assembled rods? What will be the standard deviation of assembled lengths?

If the lengths of the individual sections are normally distributed, what is the probability that the assembled rod will have a length in excess of 8.002 inches?

4.37. In a certain chemical process three bottles of a standard fluid are emptied into a larger container. If the standard deviation of the contents of the individual bottles is 0.07 ounces, what is the standard deviation of the volume of liquid emptied into the larger container?

4.38. Seven wires are twisted together to make a wire of larger dimension. If the standard deviation of the diameter of each component wire is 0.005 inches, what is the standard deviation in the diameter of the completed wire?

4.39. Sheets of tin are first cut into strips approximately 4 inches wide, and these, in turn, are cut into sections which are rolled up and soldered together to make cylinders. The sections are rolled so as to make the 4-inch side the height of the cylinder. Numerous measurements show an actual average height of 4.007 inches, with a standard deviation of 0.008 inches. The average diameter of the cylinders is 2.489 inches, with a standard deviation of 0.020 inches.

Suppose that it is not practical to measure the volume of the cylinder directly and estimate from the measurement of height and diameter the mean value of the volume contained in the cylinder. What is the standard deviation of this volume? If these cylinders are capped to make cans, what proportion of the cans would you estimate to have volumes in excess of 22 cubic inches? [The volume of a cylinder equals the area of the base (πr^2) times the height (h), i.e., $V = \pi r^2 h$. Treat this as a product of two factors, πr^2 and h. Note that the σ' of the base area equals $2\pi r \sigma'_r$ and that σ'_r equals one half of the standard deviation of the diameter.]

4.40. The length of a rectangular metal sheet has an average dimension of 6.100 feet, and its width has an average dimension of 3.950 feet. The standard deviation of the length is 0.008 feet, and the standard deviation of the width is also 0.008 feet. What is the standard deviation of the area of the sheet? Suppose that the sheets are to be finished and that the amount of work involved in the finishing depends directly on the area. The unit finishing job consists of 100 sheets. What is the standard deviation in the total area of these 100 sheets?

4.41.* The quantity $m_X(t)$ is defined as equal to $E(e^{tX})$ and is called the moment generating function of X. Show that if e^{tX} is expanded in a power series and the operator E applied term by term, then

$$E(e^{tX}) = 1 + t\mu'_{01} + \frac{t^2}{2!}\mu'_{02} + \frac{t^3}{3!}\mu'_{03} + \ldots$$

Then show that if we take the kth derivative of $E(e^{tX})$ with respect to t and set $t = 0$ in the result, we will get μ'_{0k}.

4.42.* Show that if c is a constant, then the moment generating function of cX is the same as the moment generating function of X with t replaced by ct, i.e., $m_{cX}(t) = m_X(ct)$.

4.43.* Show that if c is a constant, then $m_{X+c}(t) = e^{ct} \cdot m_X(t)$. [*Note:* $e^{t(X+c)} = e^{ct}e^{tX}$.]

4.44.* Show that if X and Y are independent, then $m_{X+Y}(t) = m_X(t) \cdot m_Y(t)$. [*Hint:* If X and Y are independent, $f(X, Y) = g(X)h(Y)$.]

4.45.* Show that if X has a Poisson distribution with mean u', the moment generating function for X is $e^{-u'} e^{u'e^t}$. Verify that the variance equals u'.

4.46.* Show that if x is normally distributed with mean zero and variance σ'^2, the moment generating function for x is $e^{t^2\sigma'^2/2}$. Verify that $\gamma'_1 = 0$ and $\gamma'_2 = 0$. [*Hint:* By definition $m_x(t) = \int_{-\infty}^{\infty} (e^{tx}) \frac{e^{-x^2/2\sigma'^2}}{\sigma'\sqrt{2\pi}} dx$. Complete the square and get an integral the value of which is 1 leaving $e^{t^2\sigma'^2/2}$.]

4.47.* Show that if X has a negative exponential distribution, viz, $f(X) = \frac{1}{\theta} e^{-X/\theta}$, the moment generating function for X is $(1 - \theta t)^{-1}$. Verify that the mean equals θ and the variance equals θ^2.

---

* Students with training in the calculus will find starred problems will give them knowledge of statistical theory that will be very useful in reading the text.

4.48.* Show that if X and Y are independent and normally distributed with mean 0 and variances $\sigma_X'^2$ and $\sigma_Y'^2$, respectively, then $W = aX + bY$ is distributed normally with mean 0 and variance $a^2\sigma_X'^2 + b^2\sigma_Y'^2$. [*Hint:* Use results of Problems 4.42, 4.46, and 4.44.]

4.49.* If the joint density function for X and Y is

$$\frac{1}{\sigma_X'\sigma_Y' 2\pi} e^{-\{[(X-\bar{X}')^2/2\sigma_x'^2]+[(Y-\bar{Y}')^2/2\sigma_y'^2]\}}$$

show that X and Y are independent.

4.50.* A complex piece of equipment contains many components. After the equipment has been put in smooth running order, failures in the equipment, due to failures in a component, occur at random over time. When a component fails it is immediately replaced by another, and the equipment continues to run as before. For a long period of time during the life of the equipment, say from time T_1 to T_2, these random failures occur at a constant rate λ. In the "debugging period" T_0 to T_1, the failure rate may be much higher than λ, and after T_2, when the equipment begins to wear out, the failure rate may again move to higher levels. Between T_1 and T_2, however, the failure rate is steady.

Under the above conditions, the number of failures that will occur on the average in the period T_1 to T_2 will be $\lambda(T_2 - T_1)$. Let this period be subdivided into many small periods of size Δt, these being so small that the chance of more than one failure occurring in the same Δt period can be neglected. Then we say that the probability of a failure occuring in any period Δt equals $\lambda(T_2 - T_1)\Delta T/(T_2 - T_1) = \lambda\Delta t$, for this is the relative frequency with which a failure would occur on the average in a specified interval Δt.

The probability that the equipment will not fail in n successive Δt periods but fail in the $(n + 1)$th period is

$$(1 - \lambda\Delta t)^n \lambda\Delta t$$

Let n become larger and larger and Δt smaller and smaller, the product $n\Delta t$ being kept equal to t. Show then that

$$\lim_{\Delta t \to 0} (1 - \lambda\Delta t)^n \lambda\Delta t \to \lambda e^{-\lambda t}\Delta t$$

thus proving that when there is a constant failure rate λ, the distribution of "life" or "time between failures" (t) has the form of a negative exponential distribution with mean $1/\lambda$ (cf. Problem 4.47 above).

4.51.* Consider again a piece of complex equipment that is operating with a constant failure rate λ. The number of failures that will occur on the average in a fixed period of time of length S will be λS. Show that if we observe many separate periods of operation of lengths S with a constant failure rate λ, the distribution of the number of failures per period will follow a Poisson distribution with mean λS.

4.52.* The *gamma function* is defined as

* Students with training in the calculus will find starred problems will give them knowledge of statistical theory that will be very useful in reading the text.

$$\Gamma(m) = \int_0^\infty X^{m-1}e^{-X}dX$$

Show by repeated integration by parts that when m is an integer $\Gamma(m) = (m - 1)!$ Note that $\Gamma(m)$ is taken as the definition of $(m - 1)!$ when m is not an integer.

4.53.* If the life of an item has the density function

$$f(X) = \frac{\beta(X - \gamma)^{\beta-1}e^{-(X - \gamma)^\beta/\alpha}}{\alpha} \text{ for } X \geq \gamma$$

$$= 0 \text{ elsewhere}$$

with $\alpha > 0$ and $\beta > 0$, then X is said to have a Weibull distribution. Show that the mean of the Weibull distribution is

$$E(X) = \gamma + \alpha^b\Gamma(b + 1)$$

where $b = 1/\beta$ and the variance of the Weibull distribution is

$$\text{Var }(X) = \alpha^{2b}\{\Gamma(2b + 1) - [\Gamma(b + 1)]^2\}$$

Show that if $\gamma = 0$ and $\beta = 1$, the Weibull distribution reduces to the negative exponential distribution.

4.54.* Suppose that the cumulative distribution of the life of a given item is $Q(t)$. In other words let $Q(t)$ be the probability that an item will fail by time t. The probability that an item will have a life of t or longer is therefore $1 - Q(t)$. Call this survival probability the "reliability" and designate it by $R(t)$, i.e., $R(t) = 1 - Q(t)$. If an item has survived for a time t, the probability it will fail in the next Δt time units is $Q(t + \Delta t) - Q(t) = [(1 - R(t + \Delta t)) - (1 - R(t))] = -[R(t + \Delta t) - R(t)]$. If we divide by Δt and let Δt approach zero, we have a measure of the rate at which probability of failure (or the probability of survival) changes with t. Set $q(t) = \frac{dQ(t)}{dt}$. Then it will be recognized that $q(t)$ is the "density function" of the distribution of life. Hence we have

$$\frac{dR(t)}{dt} = -q(t)$$

If we divide $-q(t)$ by $R(t)$ we get what is called the instaneous failure rate or hazard rate. If we started, say, with a number of items N, all operating under identical conditions, then by time t we would on the average have $NR(t)$ left. During the next small interval Δt, there would on the average be $\frac{dQ(t)}{dt}\Delta t\left(= -\frac{dR(t)}{dt}\Delta t\right)$ failures and the failure rate or number of failures during Δt per number of survivals at time t would be

$$-\frac{dR(t)}{dt}\Delta t/R(t)$$

* Students with training in the calculus will find starred problems will give them knowledge of statistical theory that will be very useful in reading the text.

Call this instantaneous failure rate or hazard rate $\lambda(t)$. Then we have

$$\lambda(t) = -\frac{1}{R(t)}\frac{dR(t)}{dt} = q(t)/R(t)$$

a. Show that the reliability function $R(t)$ is given by

$$R(t) = \exp[-\int_0^t \gamma(t)dt]$$

b. Prove that if $\lambda(t)$ = a constant k, then

$$R(t) = e^{-kt}$$

What is the density function of life in this case?

c. Prove that if the density function of life is given by the Weibull distribution, i.e., if

$$q(t) = \frac{\beta(t-\gamma)^{\beta-1}e^{-(t-\gamma)^{\beta}/\alpha}}{\alpha}$$

then

$$R(t) = e^{-(t-\gamma^{\beta}/\alpha}$$

d. For the Weibull case plot $R(t)$ for $\gamma = 10$, $\alpha = 1$, and $\beta = 0.5$; for $\gamma = 10$, $\alpha = 1$, and $\beta = 1$; and $\gamma = 10$, $\alpha = 1$, and $\beta = 2$. What kind of density function do we have when $\beta = 1$?

e. Show that for a Weibull density function, the hazard rate is

$$\lambda(t) = \frac{\beta(t-\gamma)^{\beta-1}}{\alpha}$$

f. For the Weibull case plot $\lambda(t)$ for $\gamma = 10$, $\alpha = 1$, $\beta = 0.5$; for $\gamma = 10$, $\alpha = 1$, and $\beta = 1$; and for $\gamma = 10$, $\alpha = 1$, and $\beta = 2$.

13. SELECTED REFERENCES*

Acton and Olds (P '48), Bache (P '79), Burr (P '76), Eaton (P '47), Evans (P '74 and Ps '75), Factory Management (P '53), Feller (B '68), Fry (B '28), General Electric (B '62), Goodman (P '62), Hall and Sampson (P '73), Hinkley (P '69), Harvard University Computing Laboratory (B '55), Hoel (B '71), Lieberman and Owen (B '61), Molina (B '49), Mood and Graybill (B '63), Pearson and Hartley (B '58), Reitz (B '27), Romig (B '53), Shewhart (B '31), Tukey, *Propagation of Errors,* Wilks (B '48), and Working (B '43).

* B and P refer to the Book and Periodical sections, respectively, of the Cumulative List of References in Appendix V.

5

The Sampling Distribution of a Proportion or Fraction

In subsequent sections of this text considerable use is made of what are technically known as sampling distributions. To explain what is meant by a sampling distribution, let us consider a process that is operating in a random manner, and let us suppose that a large number of samples, all of the same size, are taken from the process. Furthermore, let us suppose that each sample is inspected and the fraction of the sample that does not conform to specifications is determined. Then, if a frequency distribution is made of these many sample fractions nonconforming, we would have an approximation to the "sampling distribution of the fraction nonconforming" for the given size sample and the given process.

The above example may be varied somewhat by supposing that we are sampling from a finite lot instead of from the indefinitely large output of a process. In such a case it will be presumed that sampling is made with replacement. In other words, let us suppose that, after each *item* is inspected and its quality noted, the item is returned to the lot and the latter thoroughly mixed before the next item is drawn. If we carry out our sampling this way, always taking the same size sample, the distribution of the resulting sample fractions nonconforming will be "the sampling distribution of the fraction nonconforming" for the given size sample and the given lot.

Of course, sampling with replacement is rarely, if ever, undertaken as a practical sampling procedure in industry. It is merely noted here as a concrete means of getting a "sampling distribution." Actually, the forms of most sampling distributions are derived by mathematical analysis, although occasionally resort is had to experiments with sampling to find out something about sampling distributions that cannot be derived by straightforward mathematical analysis.

It should also be noted that in strict theory the number of samples taken from the process or lot is supposed to be infinite. Thus the concept of a sampling distribution is that of a limit which is approached as more and more samples are taken. In other words, a sampling distribution is the distribution of a certain characteristic among an infinity of samples. It is the distribution of a universe—the universe of all possible samples.

In this chapter we shall study the sampling distribution of a proportion or fraction. In the next chapter we will consider the sampling distributions of means, variances, ranges, and other sample statistics.

1. THE SAMPLING DISTRIBUTION OF A PROPORTION OR FRACTION

1.1. When the Universe Is Relatively Large

1.1.1. Derivation. In this section we shall consider the case of a universe that is "infinite" or at least large relative to (say, no less than 10 times) the size of the sample. When the universe is relatively small, it will be presumed that each item is returned to the universe and the latter thoroughly mixed before the next item is drawn. It will also be assumed that many samples will be drawn (theoretically, an "infinite number") before the distribution of sample proportions is made. All samples, of course, will be presumed to be random samples. With these assumptions, the theory will be developed with reference to the following example.

Suppose that variations in the quality of output of a machine are purely random. Suppose, further, that if the machine were allowed to operate for an indefinite, or at least a very long, time, the fraction of nonconforming units it would turn out would be equal to p'. It may be said then that the probability of a nonconforming unit from this machine is p'.

Let us take a random sample of 10 units from the output of the machine. Since the machine is hypothetically operating in a random manner, any 10 units would constitute a random sample. The sample being random and the universe being "infinitely" large, we can assume that the probability of any one unit's being nonconforming is independent of the character of any other unit. When the universe is relatively small and sampling is without replacement, this assumption of independence is no longer valid, and we must use another line of attack. This is done in Section 1.2.

Consider, first, the probability of getting no nonconforming units in our sample. This is given immediately by the multiplication theorem for independent probabilities. Since the probability of a nonconforming unit is p', the probability of a conforming unit is $1 - p'$. Hence the probability of getting no nonconforming units in a sample of 10 is $(1 - p')^{10}$. In brief, $P(0) = (1 - p')^{10}$. For example, if $p' = 0.20$, then $1 - p' = 0.80$, and the probability of no nonconforming units in a sample of 10 is $(0.80)^{10} = 0.1074$.

Consider, next, the probability of getting just 1 nonconforming unit out of 10. The probability that the first will be nonconforming and the rest all conforming is $p'(1-p')^9$. Likewise, the probability that the first will be conforming, the second nonconforming, and the last 8 conforming is $(1-p')p'(1-p')^8$, which is again $p'(1-p')^9$. In general, there are 10 possibilities for the appearance of the 1 nonconforming unit. Since the probability of each possibility is $p'(1-p')^9$, the probability of any one of these results is $10p'(1-p')^9$. This follows from the addition theorem. Hence the probability that a sample of 10 will have just 1 nonconforming unit is

$$P(0.1) = 10p'(1-p')^9$$

If $p' = 0.2$ and $(1-p') = 0.8$, then $P(0.1) = 0.2684$.

Consider, further, the probability of getting just 2 nonconforming units out of 10. The probability that any given 2 units (say, the first 2 units) will be nonconforming and the rest all conforming is $p'^2(1-p')^8$. But there are many different ways in which the 2 nonconforming units can occur. The total number of such ways is the same as the number of different combinations of 2 things that can be made from 10 things, i.e., $C_2^{10} = 10!/2!8! = 45$. Hence, by the addition theorem, the probability of 2 nonconforming units out of 10 is

$$P(0.2) = 45p'^2(1-p)^8$$

If $p' = 0.2$ and $(1-p') = 0.8$, then $P(0.2) = 0.3020$.

If the same reasoning is continued, it will be seen that the probability of X nonconforming units out of 10 is

$$P\left(\frac{X}{10}\right) = C_X^{10}p'^X(1-p')^{10-X}$$

Repeated application of this formula gives the following results:

Fraction of Nonconforming Units in Sample = X/10	*Probability of* X *Nonconforming Units Out of* 10		*Evaluation for* p′ = 0.20
0	$1p'^0(1-p')^{10}$	=	0.1074
0.1	$10p'^1(1-p')^9$	=	0.2684
0.2	$45p'^2(1-p')^8$	=	0.3020
0.3	$120p'^3(1-p')^7$	=	0.2013
0.4	$210p'^4(1-p')^6$	=	0.0881
0.5	$252p'^5(1-p')^5$	=	0.0264
0.6	$210p'^6(1-p')^4$	=	0.0055
0.7	$120p'^7(1-p')^3$	=	0.0008
0.8	$45p'^8(1-p')^2$	=	0.0001
0.9	$10p'^9(1-p')^1$	=	0.0000
1.0	$1p'^{10}(1-p')^0$	=	0.0000

The foregoing tabulation gives us the probabilities of the various possible results of sampling. It is the "sampling distribution" of the fraction nonconforming in a sample of 10 and tells us how the fractions nonconforming of many samples of that size would be distributed. A picture of the distribution is shown[1] in Figure 4.7.

If a sample of n is taken, the sampling distribution of the fraction nonconforming is given by the formula

$$P\left(\frac{X}{n}\right)=\frac{n!}{X!(n-X)!}p'^{X}(1-p')^{n-X} \qquad X=0,\ldots,n \tag{5.1}$$

This will be recognized as the formula for the binomial distribution. *Thus, whatever the universe is "infinitely" large, the sampling distribution of a proportion or fraction is the binomial distribution. This is still approximately true for a finite universe that is large relative to the size of the sample drawn.*

1.1.2. Illustration of Use. We have just seen that when we inspect a random sample of items from a relatively large universe and determine whether they are nonconforming or conforming, the variation in the fraction nonconforming from sample to sample will be given, at least approximately, by the binomial distribution. For example, if we take a random sample of 20 items from a lot of 1,000, 10 percent of which are nonconforming, the probability of getting 0 nonconforming units in the sample is approximately

$$\frac{20!}{0!20!}(0.10)^0(0.9)^{20}=0.121$$

and the probability of getting 1 nonconforming unit is approximately

$$\frac{20!}{1!19!}(0.10)^1(0.9)^{19}=0.268$$

By the addition theorem, the probability of getting 0 *or* 1 nonconforming units is approximately

$$0.121+0.268=0.389$$

Hence, if we make it a rule to accept all lots which have 1 or less nonconforming units in a sample of 20, we would, on the average, accept lots that are 10 percent nonconforming 389 times out of 1,000. This example illustrates the use to which the binomial distribution is put in sampling inspection.

1.2. When the Universe Is Relatively Small

If a sample comprises a significant proportion of the universe and if each *item* is not replaced before the next item is drawn, the distribution of sample results is not adequately represented by the binomial distribution; for in this case, the probability of a nonconforming unit may be materially changed as the sample is drawn. It is the purpose here to derive the exact distribution of a sample proportion or sample fraction in such circumstances.

---

[1] See above p. 90.

To keep the argument concrete, consider a lot of 50 units from which we intend drawing a sample of 10. The "sampling ratio" is thus 20 percent. Suppose that this lot actually contains 5 nonconforming units and 45 conforming units; i.e., the lot fraction nonconforming is 0.10. Then the probabilities of drawing 0, 1, 2, . . ., 5 nonconforming units in a sample of 10 may be derived as follows:

First, note that the total number of combinations of 10 units each that may be made from 50 units[2] is C_{10}^{50}. Second, note that the total number of combinations of 10 conforming units each that may be made from 45 conforming units is C_{10}^{45}. Hence the probability of 10 conforming units (i.e., 0 nonconforming units) in a sample of 10 units will, in this case, be

$$P\left(\frac{0}{10}\right) = \frac{C_{10}^{45}}{C_{10}^{50}} = 0.3105$$

For among all possible combinations of 10 units each that may be drawn from the lot of 50 units, it is this proportion that will have no nonconforming units. Hence this is the probability of 0 nonconforming units in a sample of 10. Accordingly, if many samples of 10 were taken at random from the lot, with replacement of each *sample* after it had been inspected and a thorough mixing of the lot, then the relative frequency of samples having 0 nonconforming units would approximate 0.3105.

Next note that the total number of combinations of 9 conforming units each that may be made from 45 conforming units is C_9^{45}. Also the number of "combinations" of 1 nonconforming unit each that may be made from 5 nonconforming units is C_1^5. Furthermore, each one of the C_9^{45} combinations of 9 conforming units can be combined with[3] the C_1^5 "combinations" of 1 nonconforming unit, to yield a total of $C_9^{45} \times C_1^5$ combinations of 10 units, 9 of which are conforming and 1 nonconforming. Hence the probability that a random sample of 10 will contain 1 nonconforming unit and 9 conforming units is

$$P\left(\frac{1}{10}\right) = \frac{C_9^{45} C_1^5}{C_{10}^{50}} = 0.4313$$

In general, if a lot contains N units, m of which are nonconforming, the probability that a random sample of n units will contain X nonconforming units ($X \leq n$ or $X \leq m$ if $m \leq n$) is

$$P\left(\frac{X}{n}\right) = \frac{C_{n-X}^{N-m} C_X^m}{C_n^N} \qquad (5.2)$$

We recognize this as the formula for the hypergeometric distribution that was described in the previous chapter.[4]

[2] See above, Chapter 2, Section 10.

[3] See above, Chapter 2, Section 10.

[4] See above, Chapter 4, Section 5.3. Also see J. G. Smith and A. J. Duncan, *Sampling Statistics and Applications* (New York: McGraw-Hill, 1945), p. 54.

For the problem to which the above discussion refers, this hypergeometric formula yields the following results:

Fraction of Nonconforming Units in Sample = X/n	*Probability in a Sample of 10*
0	$C_{10}^{45}/C_{10}^{50}=0.3105$
0.1	$C_{9}^{45}C_{1}^{5}/C_{10}^{50}=0.4313$
0.2	$C_{8}^{45}C_{2}^{5}/C_{10}^{50}=0.2099$
0.3	$C_{7}^{45}C_{3}^{5}/C_{10}^{50}=0.0442$
0.4	$C_{6}^{45}C_{4}^{5}/C_{10}^{50}=0.0040$
0.5	$C_{5}^{45}C_{5}^{5}/C_{10}^{50}=0.0001$
	1.0000

Whether to use the binomial distribution or the hypergeometric distribution in a particular case will depend on the "sampling ratio." Theoretically, if the universe is finite, the exact answer is given by the hypergeometric distribution. The approximation given by the binomial distribution is very good, however, provided that the sample is not a relatively large part of the universe. Just where to draw the line is somewhat arbitrary. As noted previously, usually the hypergeometric distribution is not used until the sample becomes greater than 10 percent of the universe; some prefer not to use it until the sample exceeds 20 percent. For several numerical comparisons of the binomial and hypergeometric distributions, the reader is referred to Table 4.2 of Chapter 4.

2. PROBLEMS

5.1. A lot contains 30 items, 6 of which are nonconforming. What is the probability that a random sample of 5 items from the lot will contain no nonconforming items?

5.2. A sampling plan calls for the taking of a random sample of 50 from a lot. If the sample contains 0 or 1 nonconforming items, the lot is accepted without further inspection. If it contains 2 or more nonconforming items, the lot is rejected.

a. What is the probability of accepting a lot that contains 10,000 items, 500 of which are nonconforming?

b. What is the probability of accepting a lot that contains 1,000 items, 50 of which are nonconforming?

c. What is the probability of accepting a lot that contains 100 items, 5 of which are nonconforming?

5.3. An inspector draws a sample of 500 items from a large lot in which 25 percent of the items are nonconforming. What is the probability that the sample will contain 100 or less nonconforming units?

5.4. A random sample of 200 items is taken from a lot of 5,000 items that contains 600 nonconforming items. What is the probability of getting 20 or less nonconforming items in the sample?

5.5. A random sample of 100 items is taken from a lot of 10,000 items that is 10 percent nonconforming. What is the probability of getting 5 or less nonconforming units in the sample? Use, first, the normal approximation to the binomial distribution and then the Poisson approximation. Compare your results.

5.6. A sample of 1,000 is taken from a large lot containing 1 percent nonconforming units. What is the probability that the sample will contain 14 or more nonconforming units?

5.7. A sampling plan calls for the drawing of 100 units from a large lot. If 3 or less units are found to be nonconforming, the lot is accepted. If 4 or more are found to be nonconforming, the lot is rejected. If the lot is 5 percent nonconforming, what is the chance of its being accepted?

5.8. A sampling plan calls for taking a random sample of 200 units from a large lot. If 5 or less nonconforming units are found in the sample, the lot is accepted. If 6 or more nonconforming units are found in the sample, the lot is rejected. What is the chance of accepting a lot that is 8 percent nonconforming?

5.9. A manufacturer claims that his product is only 2 percent nonconforming. You take a sample of 900 items and find 31 nonconforming. What would you conclude about the accuracy of the manufacturer's claim? Justify your answer. What would you conclude if you found 23 nonconforming items? 5 nonconforming items?

5.10. *a.* A sample of 40 is taken from a large lot containing 6 percent nonconforming units. Using the *Tables of the Binomial Probability Distribution,* compute the probability of getting 2 or less nonconforming units; the probability of 2 or more nonconforming units.

b. Compute the second of the above probabilities by means of—

(i) The Poisson approximation to the binomial distribution.

(ii) The normal approximation to the binomial distribution.

(iii) The binomial distribution itself.

5.11. A sample of 300 items is taken from a lot of 700 items containing 14 percent nonconforming items. Compute the probability of getting 25 or less nonconforming units.

5.12. A sampling plan calls for taking a random sample of 100 items from a lot. If 3 or less are nonconforming, the lot is accepted. If 4 or more are nonconforming, the lot is rejected. What is the chance of accepting a lot of 400 items of which 20 are nonconforming? Compare your answer with that obtained in Problem 5.7.

5.13. A sample of 30 items is taken from a lot of 50 containing 4 nonconforming items. Compute the probability of getting 1 or less nonconforming items in this sample.

5.14. Past records indicate that lots bought from a given manufacturer have varied in quality as follows:

10% were 1% nonconforming
50% were 2% nonconforming
35% were 3% nonconforming
5% were 4% nonconforming

Lots are large and are inspected by testing a sample of 100 items. You find in a given case that the sample contains 4 nonconforming items. What is the probability the lot is one of the 4 percent nonconforming lots? One of the 3 percent nonconforming lots? One of the 1 percent nonconforming lots? [*Hint:* Cf. Chapter 2, Section 9.]

3. SELECTED REFERENCES*

Bowley (B '26), Cowden (B '57), Feller (B '68), Fry (B '28), Molennar (P '73), and Working (B '43).

---

* B and P refer to the Book and Periodical sections, respectively, of the Cumulative List of References in Appendix V.

6

Other Important Sampling Distributions

In the previous chapter we were concerned with the distribution of sample proportions among many random samples from the same universe—the sampling distribution of a proportion or fraction, as it was called. Here we shall be concerned with distributions of sample means, variances, standard deviations, ranges, and other sample statistics among many random samples from the same universe. These will be called the sampling distribution of the mean, the sampling distribution of the variance, the sampling distribution of the standard deviation, the sampling distribution of the range, and so forth. They will give us the probability that the mean or variance or standard deviation or range of a sample will exceed any specified value, given certain characteristics of the universe from which the sample was drawn. They are, in fact, distributions of universes of sample means, variances, standard deviations, ranges, and so on, derived by computing and making distributions of these statistics for all possible samples of the given size from the designated universe of individual values.[1]

In general, the distribution of a universe of sample means, variances, and so forth, obtained by sampling from a given universe of individual values will depend upon the character of the universe of individual values. In the case of sampling distributions of a fraction or proportion discussed in the previous chapter, this aspect of sampling distributions did not arise since the distribution of individuals was always a simple dichotomous one. As he reads this chapter the reader should carefully note what assumptions are being made about the universe of individual values, for some conclusions will be general, others will not be. *It will be generally assumed that sampling*

[1] The word *universe* when used alone usually means the universe of individual values from which a sample is taken. Technically speaking, however, the set of all possible samples of a given size that might be taken from the universe of individual values is itself a universe of samples. For this reason means, variances, and other parameters of sampling distributions are throughout the text marked with primes to show they are universe values.

is from an "infinite" universe (say a process) or from a universe that is so large compared with the sample (say at least 10 times as large) that we need not be concerned with the finiteness of the universe. For sampling from relatively small finite universes the reader is referred to W. G. Cochran, *Sampling Techniques,* and the references contained therein.

1. SAMPLING DISTRIBUTION OF AN ARITHMETIC MEAN

1.1. Sampling Distribution of the Mean Number of Spots on a Pair of Dice

In rolling a pair of dice we are in fact taking a sample of two from a distribution of individuals that has a probability of $1/6$ for each number of spots. (See Table 6.1.) The distribution of the sum of the spots on a pair of dice was derived in Section 3.1 of Chapter 4, and to get the distribution of the mean number of spots we simply divide the sum by 2. (See Table 6.2) If the mean

TABLE 6.1
Distribution of Spots on a Die

X = *No. of Spots*	*Probability*
1	$1/6$
2	$1/6$
3	$1/6$
4	$1/6$
5	$1/6$
6	$1/6$

TABLE 6.2
Probability Distribution of the Mean of the Spots on a Pair of Dice

Mean $\bar{X} = \frac{X_1 + X_2}{2}$	*Probability*
1.0	$1/36$
1.5	$2/36$
2.0	$3/36$
2.5	$4/36$
3.0	$5/36$
3.5	$6/36$
4.0	$5/36$
4.5	$4/36$
5.0	$3/36$
5.5	$2/36$
6.0	$1/36$

and not the total number of spots was used in the game of dice, Table 6.2 would guide us in placing our bets.

1.2. Distribution of Means of Samples from Any Universe

Some of the procedures followed in statistical quality control resemble very closely a game of dice in which we concern ourselves with means instead of sums. For example, we may wish to control a boring machine. If at regular intervals we take samples of four holes bored by the machine and compute the mean diameter of the holes, we are doing much the same thing as rolling four dice and taking the average of the spots that turn up. In such a case, we would be interested in the distribution of the mean diameter of a sample of four borings.

In the dice problem we know how the spots on a single die were distributed (Table 6.1), and from this we were able to find out how the sum and mean number of spots on a pair of dice were distributed (Tables 4.1 and 6.2). We accomplished the latter by simply making all possible combinations of the spots on a pair of dice. If, for some reason or other, a die were not available for us to look at and the only evidence available were the results of rolling a die, we could have proceeded empirically. Thus we could have rolled a single die a large number of times and then counted the number of times a 1 turned up, a 2, a 3, and so on. This would have given us the empirical equivalent of Table 6.1, or what we might call the distribution of the universe of individual values. Then, as before, we could form all possible combinations of the spots on a pair of dice[2] and with the help of the empirically derived distribution of individual values, find the probabilities of various mean values.

The latter procedure is what would be open to us in the case of the boring machine. Assume that the boring machine is operating in a random manner around a constant level, i.e., is under control at a fixed level.[3] Then we could, if we wished, take measurements of individual borings and construct a distribution of these measurements. We could next form all possible combinations of four measurements and from the empirically derived distribution of individual measurements find the expected distribution of means of samples of 4. According to the theory of probability, this would tell us how the means of many samples of 4 from the process would tend to be distributed.

As an oversimplified example, suppose that we found the distribution of individual measurements to be as shown[4] in Table 6.3. Furthermore, for the sake of simplification, suppose that we take means of two instead of four borings. Then the various possible combinations of two measurements,

---

[2] Assuming, of course, that the dice were both exactly alike.

[3] See above, Chapter 2, Section 5.

[4] The distribution is purposely made discrete to keep the illustration simple. In reality, of course, it would be continuous.

TABLE 6.3
Hypothetical Distribution of Hole Diameters

X	P
1.9 inches	0.1
2.0 "	0.4
2.1 "	0.3
2.2 "	0.2

the means of the combinations, and their joint probabilities would be as follows:

Combination	$\bar{X}$	*Probability*
1.9, 1.9	1.90	(0.1)(0.1) = 0.01
1,9, 2.0	1.95	(0.1)(0.4) = 0.04
1.9, 2.1	2.00	(0.1)(0.3) = 0.03
1.9, 2.2	2.05	(0.1)(0.2) = 0.02
2.0, 1.9	1.95	(0.4)(0.1) = 0.04
2.0, 2.0	2.00	(0.4)(0.4) = 0.16
2.0, 2.1	2.05	(0.4)(0.3) = 0.12
2.0, 2.2	2.10	(0.4)(0.2) = 0.08
2.1, 1.9	2.00	(0.3)(0.1) = 0.03
2.1, 2.0	2.05	(0.3)(0.4) = 0.12
2.1, 2.1	2.10	(0.3)(0.3) = 0.09
2.1, 2.2	2.15	(0.3)(0.2) = 0.06
2.2, 1.9	2.05	(0.2)(0.1) = 0.02
2.2, 2.0	2.10	(0.2)(0.4) = 0.08
2.2, 2.1	2.15	(0.2)(0.3) = 0.06
2.2, 2.2	2.20	(0.2)(0.2) = 0.04
		1.00

From this we could derive the expected distribution of means of samples of two. The result would be that shown in Table 6.4 on the following page.

Fortunately, we do not have to rely upon such calculations every time we wish to acquire some knowledge about the expected distribution of sample means; for it can be shown by mathematical analysis that all distributions of sample means have certain properties in common. These may be presented in the form of three theorems. All assume random sampling.

The first theorem about the distribution of sample means is the following:[5]

THEOREM (6.1). *The mean of the distribution of sample means is the*

[5] See Appendix I (20). It will be noted that we are comparing two universes. One is the universe of individual values; the other is the universe made up of the means of all possible samples of the given size from the universe of individual values. The theorem is valid for relatively small finite universes as well as for infinite or relatively large universes.

TABLE 6.4
Distribution of Means of All Possible Samples of Two from the Universe of Individual Values Given in Table 6.3

$\bar{X}$	P
1.90	0.01
1.95	0.08
2.00	0.22
2.05	0.28
2.10	0.25
2.15	0.12
2.20	0.04
	1.00

mean of the universe of individual values from which the samples are taken.

This theorem is illustrated by Tables 6.1 to 6.4. The first two compare the distribution of the spots on a single die with the distribution of the mean number of spots on a pair of dice. It will be noticed from Figures 6.1 and 6.2 that the central values of the two distributions are the same. This is demonstrated numerically as follows:

The mean value of the spots on a single die (Table 6.1) is

$$E(X) = \sum PX = \tfrac{1}{6}(1+2+3+4+5+6)$$
$$= \tfrac{21}{6} = 3.5$$

The mean value of the distribution of sample means (Table 6.2) is

$$E(\bar{X}) = \sum P\bar{X} = \tfrac{1}{36}(1.0) + \tfrac{2}{36}(1.5) + \tfrac{3}{36}(2) + \tfrac{4}{36}(2.5)$$
$$+ \tfrac{5}{36}(3) + \tfrac{6}{36}(3.5) + \tfrac{5}{36}(4) + \tfrac{4}{36}(4.5)$$
$$+ \tfrac{3}{36}(5) + \tfrac{2}{36}(5.5) + \tfrac{1}{36}(6) = 3.5$$

Hence the mean of the distribution of sample means is the sme as the mean of the universe of individual values.

It is not so obvious that the means of Tables 6.3 and 6.4 are the same. Nevertheless, we have, for Table 6.3,

$$E(X) = \sum PX = (0.1)(1.9) + (0.4)(2.0) + (0.3)(2.1) + (0.2)(2.2) = 2.06$$

And for Table 6.4 we have

$$E(\bar{X}) = \sum P\bar{X} = (0.01)(1.9) + (0.08)(1.95) + (0.22)(2.0) + (0.28)(2.05)$$
$$+ (0.25)(2.10) + (0.12)(2.15) + (0.04)(2.2) = 2.06$$

Again the mean of the distribution of sample means is the same as the mean of the universe of individual values.

The second general theorem regarding the distribution of sample means is the following:[6]

THEOREM (6.2). *The variance of the distribution of sample means equals the variance of the universe of individual values divided by* n, *the size of the sample.*

This theorem may also be illustrated with reference to Tables 6.1 to 6.4. Thus the variance of the individual spots on a single die (Table 6.1) is

$$\sigma'^2_X = \sum P(X - \bar{X})^2 = \tfrac{1}{6}(1 - 3.5)^2 + \tfrac{1}{6}(2 - 3.5)^2 + \tfrac{1}{6}(3 - 3.5)^2$$
$$+ \tfrac{1}{6}(4 - 3.5)^2 + \tfrac{1}{6}(5 - 3.5)^2 + \tfrac{1}{6}(6 - 3.5)^2 = 2.916$$

The variance of the mean number of spots on a pair of dice (Table 6.2) is

$$\sigma'^2_{\bar{X}} = \sum P(\bar{X} - \bar{\bar{X}})^2 = \tfrac{1}{36}(1 - 3.5)^2 + \tfrac{2}{36}(1.5 - 3.5)^2 + \tfrac{3}{36}(2 - 3.5)^2$$
$$+ \tfrac{4}{36}(2.5 - 3.5)^2 + \tfrac{5}{36}(3 - 3.5)^2 + \tfrac{6}{36}(3.5 - 3.5)^2$$
$$+ \tfrac{5}{36}(4 - 3.5)^2 + \tfrac{4}{36}(4.5 - 3.5)^2 + \tfrac{3}{36}(5 - 3.5)^2$$
$$+ \tfrac{2}{36}(5.5 - 3.5)^2 + \tfrac{1}{36}(6 - 3.5)^2 = 1.458$$

If we take the ratio $\sigma'^2_{\bar{X}}$ to σ'^2_X, we get $1.458/2.916 = \tfrac{1}{2}$ or $1/n$, which illustrates the theorem.

With reference to Table 6.3, we have

$$\sigma'^2_X = \sum P(X - \bar{X})^2 = \sum PX^2 - 2\bar{X} \sum PX + \bar{X}^2 \sum P$$
$$= \sum PX^2 - \bar{X}^2$$
$$= 0.1(1.9)^2 + 0.4(2.0)^2 + 0.3(2.1)^2 + 0.2(2.2)^2 - (2.06)^2$$
$$= 0.0084$$

Likewise, with reference to Table 6.4, we have

$$\sigma'^2_{\bar{X}} = \sum P(\bar{X} - \bar{\bar{X}})^2 = \sum P\bar{X}^2 - \bar{\bar{X}}^2$$
$$= 0.01(1.9)^2 + 0.08(1.95)^2 + 0.22(2.00)^2 + 0.28(2.05)^2 + 0.25(2.10)^2$$
$$+ 0.12(2.15)^2 + 0.04(2.20)^2 - (2.06)^2$$
$$= 0.0042$$

Thus, as in the dice example, we have

$$\sigma'^2_{\bar{X}} / \sigma'^2_X = \frac{0.0042}{0.0084} = \frac{1}{2} = \frac{1}{n}$$

[6] See Appendix I (21).

An important corollary of Theorem 6.2 is as follows:

COROLLARY OF THEOREM (6.2). *The standard deviation of the distribution of sample means equals* $1/\sqrt{n}$ *times the standard deviation of the universe of individual values.*

The standard deviation of the distribution of sample means is commonly called the standard error of the mean.

A third theorem pertaining to the distribution of sample means will be stated without proof.[7]

THEOREM (6.3). *The form of the distribution of sample means approaches the form of a normal probability distribution as the size of the sample is increased.*

Theorem 6.3 is illustrated in a crude way by Figures 6.1 to 6.5. Figure 6.1 shows that the distribution of the spots on a single die is "rectangular"

FIGURE 6.1
Graph of the Distribution of Spots on a Single Die

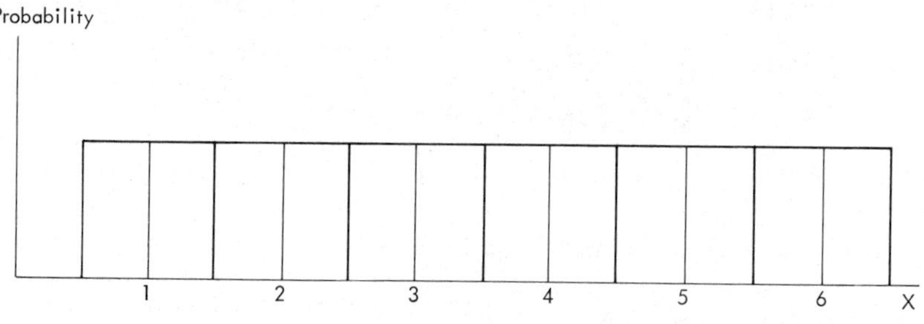

FIGURE 6.2
Graph of the Distribution of the Mean Number of Spots on a Pair of Dice

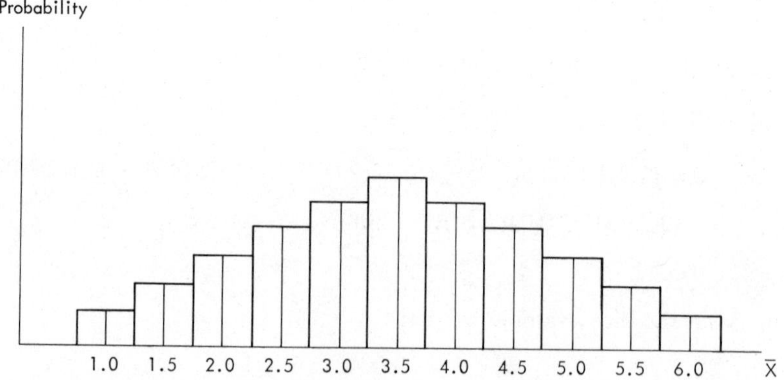

[7] In textbooks on mathematical statistics this is called the Central Limit Theorem. See, e.g., S. S. Wilks, *Mathematical Statistics* (New York: John Wiley & Sons, 1962) pp. 257–58. For normal universes, see Problem 6.24. Also see Section 3.2 of Chapter 4.

FIGURE 6.3
Graph of the Distribution of the Mean Number of Spots on a Set of Three Dice

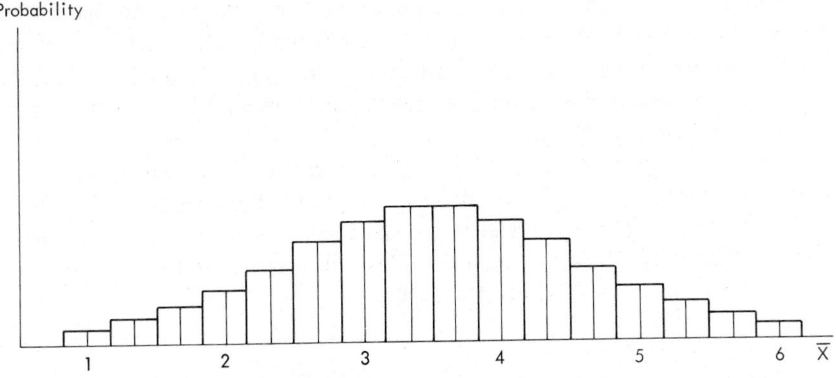

FIGURE 6.4
Graph of a Hypothetical Distribution of Individual Hole Diameters

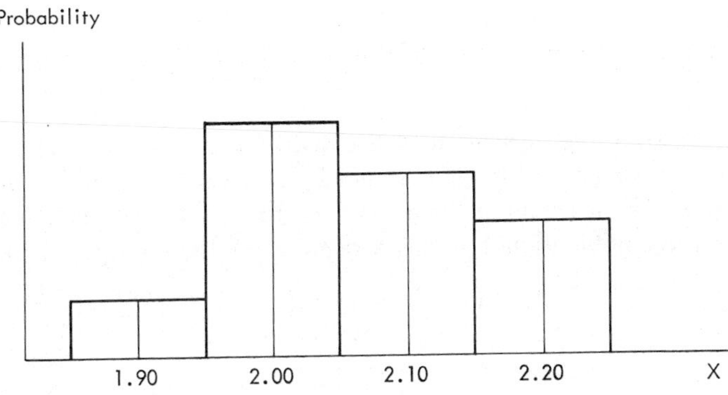

FIGURE 6.5
Graph of the Distribution of Mean Diameters of Samples of Two Holes from the Universe of Figure 6.4

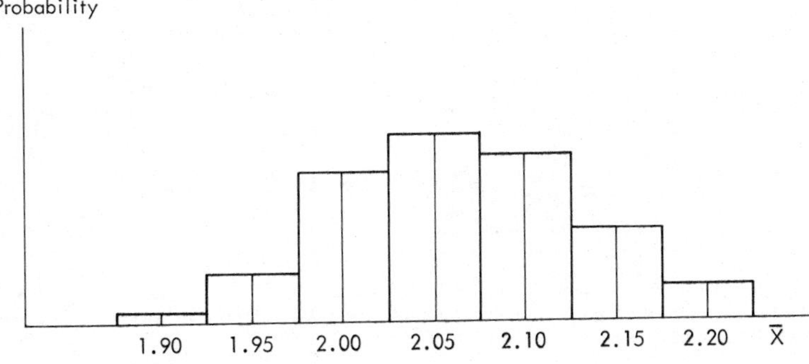

in shape, whereas Figure 6.2 shows that the distribution of the mean number of spots on a pair of dice is triangular. If three dice are averaged, we get the results shown in Figure 6.3, which resembles very closely a normal distribution. Again Figure 6.5 is more normal than Figure 6.4. Unless, therefore, the distribution of the universe of individual values is very bizarre, satisfactory results can be obtained by assuming that the distribution of means of samples is a normal distribution.

Theorem 6.3 is of capital importance because if we can estimate the mean and standard deviation of the universe of individual values, we immediately have most of the knowledge we need about sampling fluctuations in means of samples from this universe. This will be of considerable use when we come to the study of control charts.

2. SAMPLING DISTRIBUTION OF A VARIANCE AND STANDARD DEVIATION

In this section we shall talk about the sampling distribution of the sample variance $s^2 = \Sigma(X - \bar{X})^2/(n - 1)$.

2.1. Sampling Distribution of the Variance of the Spots on a Pair of Dice

We can illustrate the derivation of the distribution of sample variances for a specified universe of individual values by referring again to the rolling of dice. The spots on a pair of dice have a variance as well as a mean. For example, the combination 1, 3 has a mean of 2. Its variance is, therefore,[8]

$$s^2 = (1 - 2)^2 + (3 - 2)^2 = 2.00$$

TABLE 6.5
Distribution of the Variances of Spots on a Pair of Dice

s^2	*Probability*
0	6/36
0.50	10/36
2.00	8/36
4.50	6/36
8.00	4/36
12.50	2/36

[8] Note that in this case $n - 1 = 2 - 1 = 1$.

If we consider all the various combinations to be obtained from a pair of dice, we get the following results:

Combination	s^2	*Number of Ways in which Listed Combination Can Occur*
1, 1	0	1
1, 2	0.50	2
1, 3	2.00	2
1, 4	4.50	2
1, 5	8.00	2
1, 6	12.50	2
2, 2	0	1
2, 3	0.50	2
2, 4	2.00	2
2, 5	4.50	2
2, 6	8.00	2
3, 3	0	1
3, 4	0.50	2
3, 5	2.00	2
3, 6	4.50	2
4, 4	0	1
4, 5	0.50	2
4, 6	2.00	2
5, 5	0	1
5, 6	0.50	2
6, 6	0	1

From this we may derive directly the distribution of the variance of spots on a pair of dice. The results are presented in Table 6.5 and Figure 6.6.

If we roll a pair of dice and each time compute the variance of the spots that appear, Table 6.5 tells us that $6/36$ of the time s^2 will equal 0; $10/36$ of the time s^2 will equal 0.50; $8/36$ of the time it will equal 2.00; and so forth.

2.2. Distribution of Variances of Samples from Any Universe

If we were interested in the variability of a process rather than, or in addition to, its mean value, we might seek to control that variability by studying the variability of samples taken from the process. In such a case, we might be interested in how sample variances would be expected to fluctuate from sample to sample. If we knew or derived empirically the distribution of individual values, we could, as we did in the case of the mean, derive the expected distribution of sample variances.

Consider again the distribution of hole diameters of Table 6.3. This we may suppose represents the distribution that would result if the boring machine operated indefinitely in a random manner at the level set. Given this

FIGURE 6.6
Graph of the Distribution of the Variances of Spots on a Pair of Dice

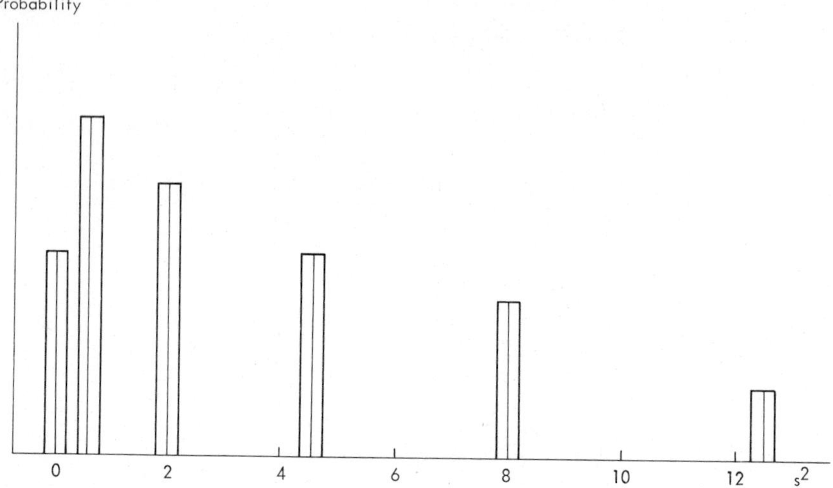

distribution of individual holes, we can derive the distribution of variances of all possible samples of 2 from this universe. As before, we form all possible combinations of results, compute the variance of each, and note its probability. Thus we have the following tabulation:

Combination	s^2	*Probability*
1.9, 1.9	0	(0.1)(0.1) = 0.01
1.9, 2.0	0.0050	(0.1)(0.4) = 0.04
1.9, 2.1	0.0200	(0.1)(0.3) = 0.03
1.9, 2.2	0.0450	(0.1)(0.2) = 0.02
2.0, 1.9	0.0050	(0.4)(0.1) = 0.04
2.0, 2.0	0	(0.4)(0.4) = 0.16
2.0, 2.1	0.0050	(0.4)(0.3) = 0.12
2.0, 2.2	0.0200	(0.4)(0.2) = 0.08
2.1, 1.9	0.0200	(0.3)(0.1) = 0.03
2.1, 2.0	0.0050	(0.3)(0.4) = 0.12
2.1, 2.1	0	(0.3)(0.3) = 0.09
2.1, 2.2	0.0050	(0.3)(0.2) = 0.06
2.2, 1.9	0.0450	(0.2)(0.1) = 0.02
2.2, 2.0	0.0200	(0.2)(0.4) = 0.08
2.2, 2.1	0.0050	(0.2)(0.3) = 0.06
2.2, 2.2	0	(0.2)(0.2) = 0.04
		1.00

From this we get the distribution of sample variances shown in Table 6.6.

Fortunately, we do not have to carry out such computations when we

TABLE 6.6
Distribution of Variances of All Possible Samples of 2 from the Universe of Individual Values Given in Table 6.3

s^2	P
0	0.30
0.0050	0.44
0.0200	0.22
0.0450	0.04
	100

wish to find out something about this distribution of sample variances. For, as in the case of the mean, certain general properties of the distribution of sample variances can be derived by mathematical analysis. These will be stated in the form of two theorems. The first theorem regarding the distribution of sample variances is as follows:[9]

THEOREM (6.4). *The mean of the distribution of sample variances equals the variance of the universe of individual values.*

This theorem is illustrated in Tables 6.5 and 6.6. The variance of the spots on a single die, it will be recalled, is 2.916 (Table 6.1). The mean of the variances of two dice is found from Table 6.5 to be 2.916. In other words, the mean of variances of samples of 2 is the variance of the universe of individual values, as required by the theorem. Likewise, the variance of individual hole diameters is 0.0084 (Table 6.3), and the mean value of the variances of samples of 2 from this universe of individual values (Table 6.6) is 0.0084. Again they are equal.

The second theorem pertains to the variance of the distribution of sample variances. We have the following:[10]

THEOREM (6.5). *The variance of the distribution of sample variances is equal to* $\frac{1}{n}\left[\mu_4' - \frac{n-3}{n-1}\mu_2'^2\right]$. (The μ''s are moments of the universe of individual values. See Section 4.3 of Chapter 4.)

Theorem 6.5 can be illustrated with reference to the distribution of Table 6.5. This is the distribution of the variances of the spots on two dice and is thus an example of the special case $n = 2$. The second and fourth moments of the universe of individual values (i.e., of the distribution of Table 6.1) are $\mu_2' = 2.916$ and $\mu_4' = 14.729$. Hence, according to our formula, the variance of the distribution of Table 6.5 should be

[9] See Appendix I (22). Again it will be noted that we are comparing two universes. One is the universe of individual values; the other is the universe made up of the variances of all possible samples of the given size from the universe of individual values.

[10] See Appendix I (23).

$$\sigma'^2_{s^2} = \frac{1}{2}\left[14.729 - \frac{(-1)}{1}(2.916)^2\right] = 11.62$$

The actual value computed directly from the table is also 11.62.

In many applications of Theorem 6.5 an error arises because μ'_4 and μ'_2 have to be estimated from the sample data and, unless n is very large, the sampling errors in the values of μ'_4 and μ'_2 are likely to be considerable. If μ'_4 and μ'_2 are known or can be adequately estimated from supplementary data, then Theorem 6.5 can be applied with reasonable accuracy.

Unfortunately, there is no general theorem regarding the form of the distribution of sample variances. Hence, if we can estimate the moments μ'_2 and μ'_4 of the universe of individual values but otherwise know nothing about its form or are unwilling to make any assumptions regarding it (other than its being unimodal), the best we can do[11] is to apply the Camp-Meidel extension of Tchebychev's inequality.[12] This will give us rough limits on the probable variation in s^2, which in some cases may be satisfactory.

2.3. Distributions of Variances of Samples from a Normal Universe

It often happens in industrial processes that the distribution of quality characteristics is normal in form or approximately normal. In these instances, we can do much better than when the form of the universe is unknown; for it can be shown that, if the universe is normal, the distribution of sample variances has the form (except for a constant) of a special distribution known as the χ^2 distribution.[13]

Fortunately, tables have been constructed giving abscissa values for selected ordinates of the cumulative χ^2 distribution. The abscissa values of χ^2 for selected cumulative probabilities will be found in Table C of Appendix II. The cumulation of χ^2 probabilities is from the upper and not the lower end, so the probabilities are those of a χ^2 equal to or *greater than* that listed in the table. It will be noted that the χ^2 distribution varies with a quantity designated as ν, which in the present instance equals the size of the sample minus 1. For each ν there is a different χ^2 distribution. A graph of the χ^2 density function for $\nu = 4$ is shown in Figure 6.7.

To find the probability that the variance of a sample of n items from

---

[11] We could try some rough calculations such as those undertaken with reference to the hypothetical data of Table 6.3, but the amount of calculation rapidly becomes prohibitive as the sample size is increased.

[12] See Chapter 4, Section 3.

[13] The formula for the χ^2 density function is

$$y = \frac{e^{-x^2/2}(\chi^2)^{(\nu-2)/2}}{2^{\nu/2}\left(\frac{\nu-2}{2}\right)!}$$

See A. M. Mood and F. A. Graybill, *Introduction to the Theory of Statistics* (2d. ed.; New York: McGraw-Hill, 1963), p. 227.

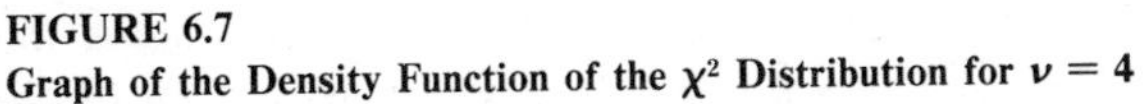

FIGURE 6.7
Graph of the Density Function of the χ^2 Distribution for $\nu = 4$

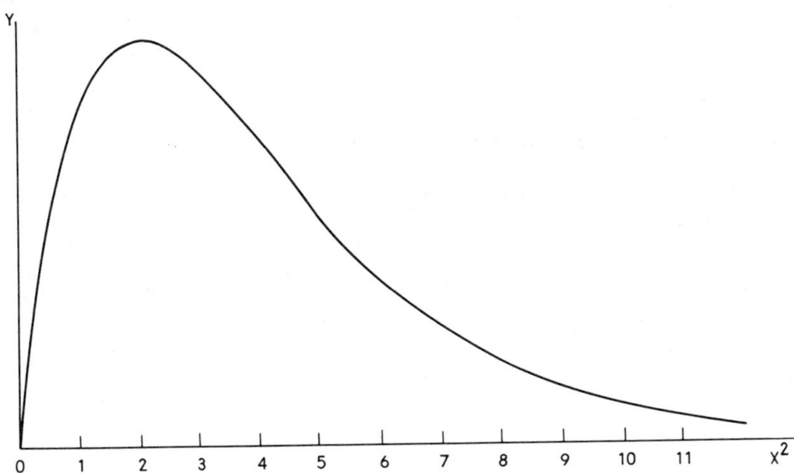

a specified normal universe will equal or exceed a given value, say s_1^2, we compute the quantity $(n - 1)s_1^2/\sigma'^2$, where σ'^2 is the variance of the universe, and treat this as a value of χ^2. For example, suppose that we are to draw a sample of 10 items from a normal universe whose variance is 25 and suppose that we wish to find the probability that the sample variance will exceed 49. To find the desired probability, we compute

$$\frac{(n-1)s_1^2}{\sigma'^2} = \frac{9(49)}{25} = 17.64$$

We then enter the χ^2 table at the line $\nu = n - 1 = 10 - 1 = 9$ and note that the value 17.46 falls approximately at the point cumulative probability $= 0.04$. In other words, the probability that the sample variance will equal or exceed 49 is about 4 out of 100.

We can also work backward. Suppose we know, as before, that the universe is normal and that the variance of the universe is 25 and suppose that we want to find the value of the sample variance that in samples of 10 would be exceeded, say, only once in 1,000 times. To solve this problem we set $(n - 1)s_1^2/\sigma'^2$ equal to the 0.001 point of the χ^2 distribution for $\nu = n - 1$ and solve for s_1^2. Specifically, we have

$$\frac{9(s_1^2)}{25} = 27.877$$

which yields $s_1^2 = 77.4$. In other words, if we draw random samples of 10 from a normal universe whose variance is 25, we will get sample variances that, on the average, exceed 77.4 not more than once in 1,000 times.

The table of the χ^2 distribution goes only as high as $\nu = 30$. This is because, for higher values of ν, the χ^2 distribution approaches very closely the normal distribution. For values of $\nu > 30$, we can safely assume that the quantity $\sqrt{2\chi^2} - \sqrt{2\nu - 1}$ has a normal distribution with a mean of 0 and a standard deviation of 1.

In concluding this section, it may be noted that the mean value of the χ^2 distribution is ν. This means that the mean value of $\frac{(n-1)s^2}{\sigma'^2}$, which is distributed like χ^2, is ν. But in this case, $\nu = n - 1$, so the mean value of $\frac{(n-1)s^2}{\sigma'^2}$ is $n - 1$ or the mean value of s^2 is σ'^2, which agrees with Theorem 6.4. It may also be noted that the variance of the χ^2 distribution is 2ν. Consequently, the quantity $\frac{(n-1)s^2}{\sigma'^2}$, which is distributed like χ^2 with $\nu = n - 1$, has a variance equal to $2(n - 1)$. This means that s^2 has a variance equal to $\frac{2}{n-1}\sigma'^4$ and that the standard deviation of s^2 is[14] $\sqrt{\frac{2}{n-1}}\sigma'^2$. It will be noted that we are assuming here that the universe of individual values is normal.

2.4. Distribution of Standard Deviations of Samples from Any Universe

No exact formulas exist for the mean and standard deviation of the distribution of sample standard deviations from any universe. The following theorems, however, are useful for large samples.

THEOREM (6.6). *If* n *is large, the mean of the distribution of sample standard deviations from any universe is approximately the standard deviation of the universe,* i.e., E(s) $\doteq \sigma'$.

The argument for this theorem is very rough and runs as follows: From our general notions of probability, it can be inferred that, if n is large, the standard deviation of any sample will not differ much from the standard deviation of the universe. Certainly, therefore, the mean of sample standard deviations will be close to the standard deviation of the universe. Hence, for large samples, $E(s) \doteq \sigma'$.

---

[14] In a normal universe $\gamma_2' = \frac{\mu_4'}{\mu_2'^2} - 3 = 0$, so that $\mu_4' = 3\mu_2'^2$. (See p. 65 above.) Substituting this in the general formula for the variance of the variance (p. 139) gives

$$\sigma'^2_{s^2} = \frac{1}{n}\left[\mu_4' - \frac{n-3}{n-1}\mu_2'^2\right] = \frac{1}{n}\left[3\mu_2'^2 - \frac{n-3}{n-1}\mu_2'^2\right]$$

$$= \frac{2}{n-1}\mu_2'^2 \text{ and } \sigma'_{s^2} = \mu_2'\sqrt{\frac{2}{n-1}}$$

which agree with the formulas derived from the χ^2 distribution.

A second theorem is as follows:[15]

THEOREM (6.7). *If* n *is large, the standard deviation of the distribution of sample standard deviations from any universe equals approximately*

$$\frac{\sigma'}{\sqrt{2n}}\sqrt{\frac{\gamma_2'}{2}+1}$$

It is to be emphasized that Theorems 6.6 and 6.7 should be used only for large samples; n should at least equal 30, and much larger values would be preferable.

2.5. Distribution of Standard Deviations of Samples from a Normal Universe

When we wish to compute probabilities, it makes no difference whether we work with the variance or the standard deviation, which is the square root of the variance; for the probability that s^2 will exceed a specified value s_1^2 is the same as the probability that s will exceed s_1. *Hence the discussion of Section 2.3 above pertaining to variances of samples from a normal universe can be carried over with little modification to the analysis of sampling fluctuations in standard deviations of samples from a normal universe.*

To illustrate this, consider the following. Suppose that we are to take a sample of 10 from a normal universe whose standard deviation is 5 and suppose we want to find the probability that the standard deviation of the sample will equal or exceed 7. We compute $(n - 1)s^2/\sigma'^2$ as before and look the result up in a χ^2 table with $\nu = n - 1 = 9$. Since in our example the standard deviations are merely the square roots of the values chosen in our previous example pertaining to the variance, the χ^2 and the probability derived from it are also the same. Thus

$$\frac{(n-1)s^2}{\sigma'^2}=\frac{9(49)}{25}=17.64$$

and the probability that s will equal or exceed 7 is about 0.04. In computing probabilities of sample results, therefore, it makes no difference whether we use s or s^2; the answers are the same.

Although the probabilities yielded by the use of the standard deviation are the same as those yielded by the use of the variance, it is to be noted that the mean value of the distribution of sample standard deviations from a normal universe is not, in small samples, the square root of the mean value of the distribution of sample variances, nor, if n is small, is the standard deviation of the distribution of sample standard deviations equal to the square root of the standard deviation of sample variances. Thus the mean value of the distribution of sample variances from any universe[16] is σ'^2, but the mean

[15] See Appendix I (24).

[16] See Appendix I (22).

value of the distribution of sample standard deviations from a normal universe is[17]

$$\sqrt{\frac{2}{n-1}}\frac{\left(\frac{n-2}{2}\right)!}{\left(\frac{n-3}{2}\right)!}\sigma'$$

The quantity

$$\sqrt{\frac{2}{n-1}}\frac{\left(\frac{n-2}{2}\right)!}{\left(\frac{n-3}{2}\right)!}$$

has been designated as c_4, *and its values for* $n = 2$ to $n = 25$ are recorded in Table M in Appendix II. Similarly, the standard deviation of the distribution of sample variances from a normal universe is[18] $\sqrt{\frac{2}{n-1}}\sigma'^2$, but the standard deviation of the distribution of sample standard deviations from a normal universe is[19] approximately $\sigma'/\sqrt{2(n-1)}$. Although statisticians like to work with s^2 in control chart analysis, quality control engineers have preferred to use the mean and standard deviation of s instead of s^2.

3. SAMPLING DISTRIBUTION OF THE RANGE OF A SMALL SAMPLE FROM A NORMAL UNIVERSE

In small samples the range and standard deviation are likely to fluctuate together. If the standard deviation is large, the range is also likely to be large. If the standard deviation is small, the range is likely to be small. In large samples, however, the occurrence of one extreme value will cause the range to be large, but it may have much less effect on the standard deviation. If we are interested, therefore, in analyzing or controlling variability and use small samples, the range may often be employed as a substitute for the standard deviation with little loss in inefficiency.[20] In addition to being almost

[17] W. A. Shewhart, *Economic Control of Quality of Manufactured Product* (New York: D. Van Nostrand Co., Inc., 1931), p. 184.

[18] See Appendix I (23) and footnote 14, on p. 142.

[19] See Appendix I (24) and note that, for a normal universe, $\gamma_2 = 0$. For a normal universe the exact formula for the standard deviation of the distribution of sample standard deviations is $\sigma'\sqrt{1-c_4^2}$. The proof is as follows: For a normal universe $E(s) = c_4\sigma'$ and $E(s^2) = \sigma'^2$. Hence the standard deviation of s, which is defined as $\sqrt{E(s-E(s))^2}$, has the value

$$\sigma'_s = \sqrt{E(s-E(s))^2} = \sqrt{E(s^2) - 2E(s)E(s) + [E(s)]^2}$$
$$= \sqrt{E(s^2) - [E(s)]^2} = \sqrt{\sigma'^2 - c_4^2\sigma'^2} = \sigma'\sqrt{1-c_4^2}$$

[20] See Chapter 24, Section 1.2, for a technical discussion of what is meant by the "efficiency" of a statistic.

as efficient as the standard deviation in small samples, the range is easy to calculate. For that reason where electronic computer facilities programmed for the calculation of the standard deviation are not readily available, the use of the range is usually preferred in quality control analysis.

Fortunately, tables of the distribution of the relative range $w = R/\sigma'$ (see Figure 6.8) have been worked out for a normal universe. Table D1 of Appendix II gives values of w for various levels of cumulated probability for $n = 2$ to $n = 12$. In other words, it gives values of w for which the probability of an equal or lower value is that indicated in the table. The table also gives values for the mean of the distribution of w and its standard deviation.

To illustrate some of the uses of this table, consider the following: A universe is known to have a standard deviation of 20. A random sample of 5 is taken from this universe. What is the probability that the range of the sample will exceed 80? What value has a probability of 0.001 of being exceeded by the sample range? To answer the first question, find $w = R/\sigma' = 80/20 = 4$ and note from Table D1 that for $n = 5$ the probability that w will exceed 4 is between 0.05 and 0.025. Linear interpolation yields 0.040. This is the probability that the sample range will exceed 80. To answer the second question, note that, for $n = 5$, the upper 0.001 points for w is 5.48. For $\sigma' = 20$, this yields $R = \sigma'w = 20(5.48) = 109.6$. Hence 109.6 is the value for which the chances are 0.001 that it will be exceeded by the sample range.

Consider another problem. Suppose that we had 10 samples of 5 each,

FIGURE 6.8
Graph of the Density Function of the Distribution of the Relative Range $w = R/\sigma'$ for $n = 5$

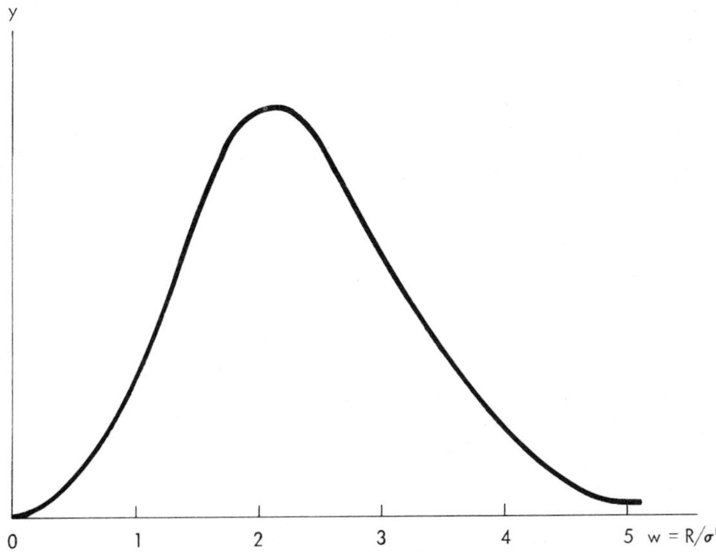

the ranges of which are 31, 30, 29, 28, 37, 34, 32, 35, 39, and 30. What estimate can we make of the standard deviation of the universe from which the samples came? The following is a procedure that is frequently used in situations of this kind. Calculate the average range of the 10 samples. This is 33.5. Table D1 shows that for samples of 5 the mean value of w, which equals R/σ', is 2.326. This means that the average value of R is $2.326\sigma'$. Hence, if 33.5 is taken as an estimate of the mean value of R, then σ' may be estimated at 33.5/2.326 = 14.4. In other words, to take 33.5 as an estimate of the average of all sample ranges from the given universe (with $n = 5$) is the equivalent of taking the standard deviation of the universe as 14.4. In quality control the mean value of w is commonly represented by d_2. *The estimate of* σ' derived above is the equivalent of taking $\sigma' = \bar{R}/d_2$.

4. THE LARGEST AND SMALLEST OF A SAMPLE AS "TOLERANCE LIMITS" FOR THE INDIVIDUALS OF A UNIVERSE

If a sample is small, there is little assurance that the range of the sample will include a high percentage of the members of the universe. For example, if we take a single sample of five washers from a manufacturing process and find that these five washers vary in thickness from 0.121 to 0.129 inches—i.e., the range equals 0.008 inches—we do not have a high degree of assurance that a large percentage of the washers produced by the process in the future will lie within these same limits, 0.121 to 0.129 inches. This is true even though the process is operating and continues to operate in a random manner. If we take a sample of 1,000 washers, however, then there is a high degree of assurance that the range of this sample of 1,000 will include a large percentage of washers produced in the future, assuming that the process is operating and continues to operate in a random manner.

The above is merely common sense. S. S. Wilks, however, has provided a formula that makes our reasoning somewhat more precise. If we wish an assurance of ϵ that at least 100 β percent of the contents of a lot or the output of a random process will be included between the largest and smallest values in a random sample from the lot or process, then the size of the sample must be such that[21]

$$n\beta^{n-1} - (n-1)\beta^n = 1 - \epsilon \tag{6.1}$$

[21] S. S. Wilks, *Mathematical Statistics* (1946 ed.), p. 93. H. Scheffé and J. W. Tukey in "A Formula for Sample Sizes for Population Tolerance Limits," *Annals of Mathematical Statistics* 15 (1944), p. 217, give an approximate formula for n. viz,

$$n \doteq \tfrac{1}{4}\chi^2_{1-\epsilon}(1+\beta)/(1-\beta)$$

where $\chi^2_{1-\epsilon}$ is the entry in Table C of Appendix II for $\nu = 4$ and $P = 1 - \epsilon$. Also see Z. W. Birnbaum and H. S. Zuckerman, "A Graphical Determination of Sample Size for Wilks' Tolerance Limits," *Annals of Mathematical Statistics* 20 (1949), pp. 313–16.

This equation is valid regardless of the form of the distribution of the items in the lot or the form of the distribution of the items turned out by the process, provided that the variable is continuous and sampling is random.

To illustrate the use of equation (6.1), suppose we wish to determine "tolerance limits" for our lot or process that will include at least 99 percent of the items and suppose that we wish an assurance of 0.95 that the limits we set will do this. Then the size of our sample *(n)* would have to be such that

$$n(0.99)^{n-1} - (n-1)(0.99)^n = 0.05$$

This yields $n = 473$. If we wish to have an assurance of 0.99 that our tolerance limits will include at least 99 percent of the items of the universe, then n must be 660. Figure 6.9 presents graphic representations of equation (6.1) for $\beta = 0.99$ and 0.9973.

5. A PREDICTION INTERVAL FOR A FUTURE SAMPLE RESULT GIVEN A CURRENT SAMPLE

Quite different from the tolerance limits described in the previous section are the "prediction intervals" that have been developed by Gerald J. Hahn. Whereas the tolerance limits pertain to the percentage of the universe that may be expected with a given probability to lie within the range of a given sample from that universe, the Hahn prediction intervals pertain to the range within which a single future observation, or more generally some function of a future sample of m observations, may be expected to lie with a stated probability, given a current sample of n observations. More precisely a prediction interval has been defined in general as follows.[22]

> Suppose that $Y_1, \ldots, Y_n$ are the observations in a random sample of size n from a distribution with some unknown parameters. Aslo, suppose that $x_1, \ldots, x_m$ are the observations in an independent (future) random sample of size m from the same distribution. Let $g(X_1, \ldots, X_m)$ be some statistic (function of the observations) for the second sample; for example, this statistic may be the mean of the second sample. A two-sided 100γ percent prediction interval to contain the future statistic $g(X_1, \ldots, X_m)$ with probability γ is an interval with lower and upper endpoints $g_L(Y_1, \ldots, Y_n)$ and $g_U(Y_1, \ldots, Y_n)$, which are functions of the observations in the first sample, such that the interval encloses the future statistic with probability γ; that is,
>
> $$\Pr\{g_L(Y_1, \ldots, Y_n) \le g(X_1, \ldots, X_m) \le g_U(Y_1, \ldots, Y_n)\} = \gamma$$
>
> for any possible values of the unknown parameters of the underlying distribution.

When the universe is normal, a 0.95 prediction interval for a single future observation, given a current sample of n observations, can be derived

---

[22] The quotation is from the article by Gerald J. Hahn and Wayne Nelson, "A Survey of Prediction Intervals and their Applications," that originally appeared in the *Journal of Quality Technology* 5 (1973), p. 185. American Society for Quality Control. Reprinted by permission.

FIGURE 6.9
Graph of the Relationship between the Size of the Sample and the Degree of Assurance that the Range of the Sample Will Include at Least 100β Percent of the Items of the Universe for β = 0.99 and 0.9973

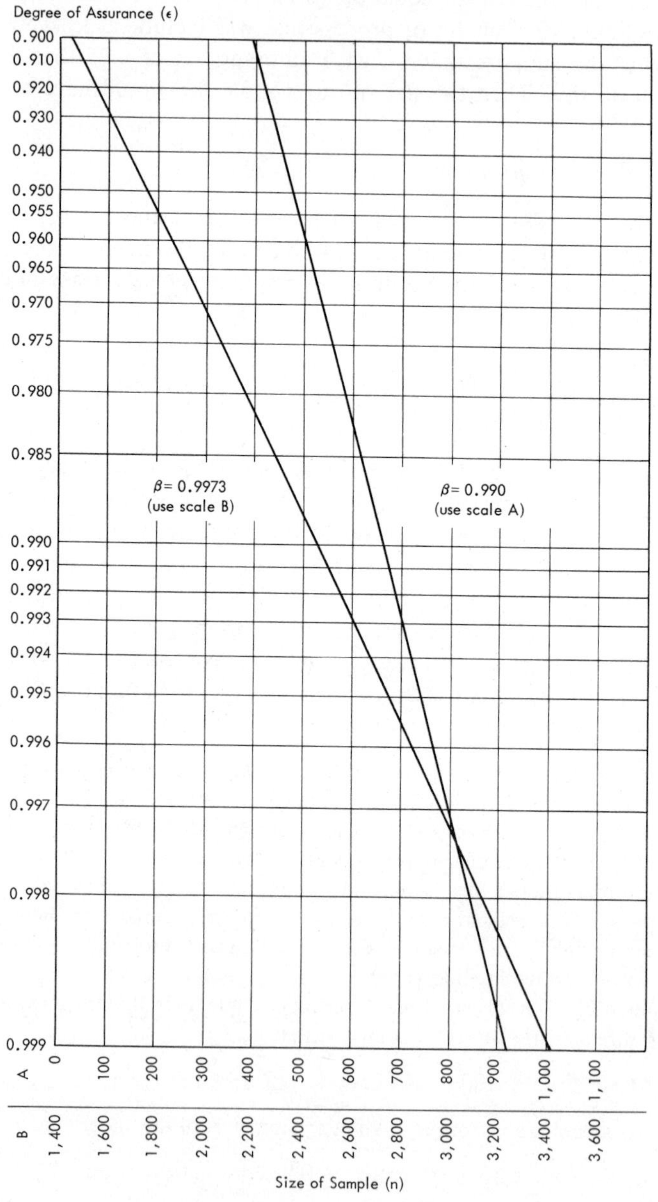

from Table 6.7 taken from Hahn's paper in the *Journal of Quality Technology* 1 (1969), p. 169. To use this table, we calculate the mean $(\bar{Y})$ and standard deviation (s) of the current sample and set the 0.95 prediction limits for a future single observation from the same universe equal to $\bar{Y} \pm ks$. The derivation of the table is given in Hahn's article in the *Journal of the American Statistical Association* 64 (1969), pp. 878–88.

TABLE 6.7
Factors k for Obtaining Two-Sided 0.95 Prediction Limits for a Single Future Observation Given Previous Sample of Size n from a Normal Universe*

Previous Sample Size, n	k	*Previous Sample Size,* n	k	*Previous Sample Size,* n	k
2	15.562	14	2.236	26	2.099
3	4.969	15	2.215	27	2.094
4	3.558	16	2.197	28	2.088
5	3.041	17	2.181	29	2.083
6	2.777	18	2.168	30	2.079
7	2.616	19	2.156	31	2.075
8	2.508	20	2.145	41	2.046
9	2.431	21	2.135	51	2.029
10	2.372	22	2.127	61	2.016
11	2.327	23	2.119	81	2.002
12	2.291	24	2.112	101	1.994
13	2.261	25	2.105	201	1.977
				501	1.967
				∞	1.960

* This table originally appeared in the *Journal of Quality Technology* 1 (1969), p. 169. American Society for Quality Control. Reprinted by permission.

For discussion of prediction intervals for other quantities than a single future observation and the derivation of prediction intervals for other than a normal universe, the reader is referred to the Hahn-Nelson article quoted above. Care must be taken not to confuse "prediction intervals" with "confidence intervals" discussed in Chapter 24.

6. SAMPLING DISTRIBUTION OF THE STATISTIC

$$t = \frac{\bar{X} - \bar{X}'}{s/\sqrt{n}}$$

In Section 1 we found that sample means tend to be normally distributed. To make use of this conclusion, however, we must have some knowledge of the standard deviation of the sampled universe. This may be estimated from previously collected data or from the sample itself, if it is large enough.

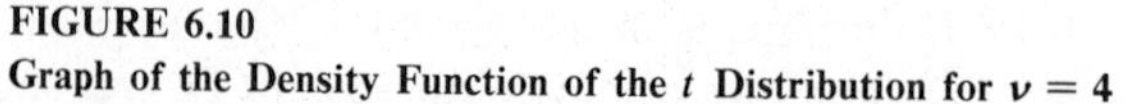

FIGURE 6.10
Graph of the Density Function of the t Distribution for $\nu = 4$

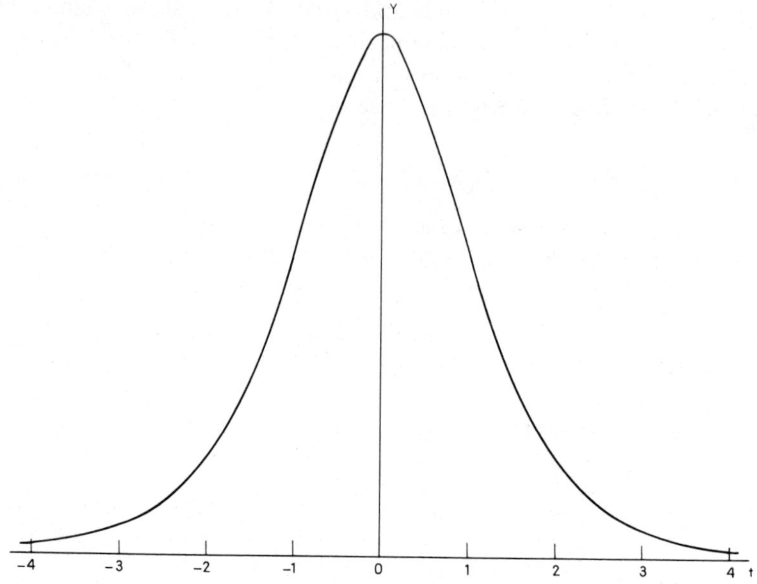

When a sample is small, however—say, less than 30—estimation of the standard deviation of the universe from the sample may not give satisfactory results. For small samples, more exact methods must be employed.

To solve the difficulty raised by estimating the universe standard deviation from a small sample, the exact distribution for a normal universe has been found[23] for sample values of the statistic $t = \dfrac{\bar{X} - \bar{X}'}{s/\sqrt{n}}$, where s is the sample standard deviation. This distribution is called the t distribution (also frequently Student's distribution).[24] The distribution shows how the statistic t, computed for many samples from a normal universe, would tend to be distributed. A graph of the density function is shown in Figure 6.10.

Table B of Appendix II gives values of t corresponding to selected cumulated probabilities. In this case, it is to be noted that the probabilities are cumulated from both ends simultaneously. In other words, the probability given in the table is the probability that a sample t should exceed in *absolute value* the value of t listed in the table. Since the t distribution is symmetrical around its mean, if we want the cumulated probability from one end only, we halve the probability listed in the table. It will also be noted that the t

---

[23] See J. G. Smith and A. J. Duncan, *Sampling Statistics and Applications* (New York: McGraw-Hill, 1945), pp. 236–41, and other references at end of chapter.

[24] "Student" was the pen name used by W. S. Gosset, who developed this distribution.

distribution, like the χ^2 distribution, depends on a parameter ν, which is related to the size of the sample. In the problems discussed here, ν will always be taken equal to $n - 1$. There are different t distributions, therefore, for different values of ν.

To illustrate the use of the t distribution, consider the following problem. Suppose that we have a random sample of 9 items for which the mean is 119 and the value of s is 21. We want to know whether this sample might reasonably have come from a normal universe whose mean is 100. To reach a conclusion, we compute[25]

$$t = \frac{(\bar{X} - \bar{X}')}{s/\sqrt{n}} = \frac{(119 - 100)}{(21)/\sqrt{9}} = \frac{119 - 100}{21/3} = \frac{19}{7} = 2.72$$

Table B tells us that, for $\nu = 9 - 1 = 8$, the probability of an absolute value of t greater than 2.72 is about 3 out of 100. This is a small probability. We would therefore be led to the conclusion that the sample did not come from the supposed universe.[26]

7. SPECIAL STATISTICS BASED ON THE RANGE

In 1950, P. B. Patnaik[27] showed that the square of the average range has a distribution that is approximately of the form of a χ^2 distribution. For the case in which $\bar{R}$ is the average of ranges of g random samples from normal universes with a common variance, each sample containing m cases, Patnaik has worked out[28] conversion factors d_2^* and degrees of freedom (ν) for various values of g and m. An extension of Patnaik's table is presented in Table D3.

The significance of this work by Patnaik is that in any analysis using s we can replace s by the more readily computed $\bar{R}/d_2^*$. Thus the statistic $\frac{(\bar{X} - \bar{X}')}{\bar{R}/d_2^*\sqrt{n}}$ will have approximately a t distribution and the statistic $\nu\bar{R}^2/(d_2^*)^2\sigma'^2$ will have approximately a χ^2 distribution, the degrees of freedom (ν) in each case being given by Table D3. When g exceeds 1, the use of the Patnaik approximation involves a loss in degrees of freedom that may run as high as 25 percent. For the special case $g = 1$ exact tables of the distribution of the statistic $(\bar{X} - \bar{X}')/R$ have been worked out by E. Lord.[29] In this case, there is not much loss in efficiency if m is small, say 8 or less.

---

[25] It will be recalled that s is defined as equal to $\sqrt{\frac{\Sigma (X - \bar{X})^2}{n - 1}}$.

[26] For further discussion of the theory of testing hypotheses, see Chapter 25.

[27] "The Use of Mean Range as an Estimator of Variance in Statistical Tests," *Biometrika* 37 (1950), pp. 78–87.

[28] In this test d_2^* is used in place of Patnaik's c, m in place of Patnaik's n, and g in place of Patnaik's m.

[29] E. Lord, "The Use of Range in Place of Standard Deviation in the t-Test," *Biometrika* 34 (1947), pp. 41–67.

Another important statistic that has been developed is the "studentized" range.[30] In Section 3 above, we discussed the relative range $w = R/\sigma'$. This might have also been called the standardized range since the sample range is expressed in terms of the universe standard deviation. The "studentized" range is R divided by the square root of an independent, unbiased estimate of σ'^2. Examples would be $q = (X_H - X_L)/s$ or $(X_H - X_L)d_2^*/\bar{R}$ if $\bar{R}/d_2^*$ is used in place of s. Here s^2 is an independent unbiased estimate of the variance of the X universe. The upper 0.05 point of the distribution of the studentized range is given in Table D2. The degrees of freedom to be used are those for s or $\bar{R}/d_2^*$, the latter being given by Table D3.

The use of Patnaik's approximation will be illustrated in Chapter 26 and that of the studentized range in Chapter 29.

8. PROBLEMS

6.1. *a.* Derive the distribution of the mean number of spots on a set of three dice.

b. Compute the mean and standard deviation of this distribution directly from the distribution itself. Compare your results with the mean and standard deviation computed from Theorem 6.1 and the corollary of Theorem 6.2.

6.2. Past data suggest that the mean diameter of bushings turned out by a manufacturing process is 2.257 inches and the standard deviation is 0.008 inches. Estimate the probability that a sample of 4 bushings will have a mean diameter equal to or greater than 2.263 inches. Make a similar estimate for a sample of 16. Upon what assumptions is your analysis based?

6.3. The number of nonconformities per radio averages 3.24, and the standard deviation of the number of nonconformities is 1.8. What is the probability that a set of 4 radios will have an average number of nonconformities greater than 5? Less than 1? What are these same probabilities for a set of 9 radios? State the assumptions on which your analysis rests.

6.4. A process is estimated to have an average value of 10.264 and a standard deviation of 2.51. Samples of 5 are taken from the process at stated intervals and their means plotted on a chart. Find an upper limit which will be exceeded by the sample averages not more than once out of 1,000 times on the average, provided that the process operates in a random manner.

6.5. A process is estimated to have an average value of 101.7 and a standard deviation of 11.5. Samples of 4 are taken from the process at stated intervals and their means plotted on a chart. Find a lower limit below which the sample averages will fall not more than 2 times out of 1,000 on the average, provided that the process operates in a random manner.

6.6. A process is estimated to have an average value of 75.4 and a standard deviation of 8.4. Samples of 9 are taken from the process at stated intervals, and their means plotted on a chart. If the process operates in a random manner, what is the probability that a sample mean will fall outside the limits 67.0 and 83.8?

---

[30] See E. S. Pearson and H. O. Hartley (eds.), *Biometrika Tables for Statisticians* (2d ed.; Cambridge, England; The University Press, 1958), Vol. 1, pp. 51–52.

6.7. A process is estimated to have an average value of 36.7 and a standard deviation of 3.8. Samples of 5 are taken from the process at stated intervals and their means plotted on a chart. What is the probability that a sample mean will fall within the limits 32 and 40?

6.8. *a.* With the aid of the computations carried out in Problem 6.1, derive the distribution of the variances of the number of spots on a set of three dice.

b. Compute directly the mean and standard deviation of this distribution and compare the results with those obtained by application of Theorems 6.4 and 6.5.

6.9. A process is estimated to have the following distribution:

Nonconformities per Unit	*Probability*
0	0.3
1	0.4
2	0.2
3	0.1

a. Derive the distribution of the variances of the number of nonconformities per unit in samples of 3 from this process.

b. Compute the variance of the process, and from Theorem 6.4 compare the mean of the distribution of variances of samples of 3.

c. Compute the mean of the distribution derived in *a* and compare the results with those obtained in *b.*

d. Compute the standard deviation of the distribution derived in *a* and compare with the result obtained by applying Theorem 6.5.

6.10. A process is estimated to have the following distribution in which quality measurements (X) are grouped in class intervals of size 1 with lower limits 0, 1, 2, etc.

X	P
0 –	0.05
1 –	0.15
2 –	0.20
3 –	0.35
4 –	0.15
5 –	0.10
	1.00

a. Compute the mean and standard deviation of this distribution.

b. Estimate the mean and standard deviation of variances of samples of 5 from this distribution.

c. Compute the limits $\bar{X}_{s^2} \pm 3$ standard deviations of s^2 and place a limit on the probability of exceeding these limits.

6.11. The standard deviation of the weights of cans of tomatoes in a lot of 10,000 equals 1.4 ounces. Assume that the weights are normally distributed, and find the probability that a sample of 4 cans will have a standard deviation that exceeds 2.0 ounces.

6.12. The standard deviation of a process is estimated to be 4.3. If the output is normally distributed, what is the probability that in a sample of 16 units the standard deviation will exceed 6? What is the probability that it will be less than 1?

6.13. If the standard deviation for the 10,000 cans in Problem 6.11 equals 1.4 ounces, what is the probability that the range of a sample of 5 cans will exceed 7.5 ounces?

6.14. The standard deviation of a process the output of which is normally distributed is estimated at 5.5. What is the probability that in a sample of 9 units of output the range will exceed 11?

6.15. *a.* Work out the distribution of the ranges of the number of spots on a pair of dice and also the distribution of relative ranges R/σ' (σ' can be computed from Table 6.1.). Compute the average and standard deviation of the distribution of relative ranges. Compare your results with those of Table D1 in Appendix II. Should they agree?

b. Do the same for three dice.

6.16. You wish to be able to say, with a 0.95 chance of being correct, that the range of a sample you take will include at least 99 percent of the future output of a process that is operating in a random manner. How large should your sample be?

6.17. You wish to be able to say, with a 0.99 chance of being correct, that the range of a sample you take will include at least 99.73 percent of the future output of a process that is operating in a random manner. How large should your sample be?

6.18. You take a sample of 500 items from a process that is operating in a random manner and claim that at least 99 percent of the future output will be within the range of your sample? What degree of assurance do you have that your claim will be correct?

6.19. You take a sample of 800 items from a process that is operating in a random manner and claim that at least 99.73 percent of the future output will lie within the range of your sample. What degree of assurance do you have that your claim will be correct?

6.20. The mean and standard deviation of a normally distributed process are unknown. A sample of 5 is taken from the process and the mean of the sample $\bar{X}$ is found to be 80 and the sample standard deviation s is found to be 20. Determine an interval that you can claim with 0.95 assurance will include the value of the next item taken from the process.

6.21. What would the answer to Problem 6.20 be if the sample taken from the process had consisted of 15 items and their mean and standard deviation had been $\bar{X} = 52$ and $s = 3.7$?

6.22. The average weight of the 10,000 cans of tomatoes in Problem 6.11 is 21.7 ounces. The standard deviation is 1.4 ounces. If a sample of 4 cans is tested, what is the probability that the average weight of the 4 cans will exceed 22.5 ounces?

6.23. Suppose that the standard deviation of the weights of the 10,000 cans in Problem 6.22 is not known, but the standard deviation of the sample of 4 cans is found to be 1.4 ounces. What is the probability that the mean of the sample exceeds 22.5 ounces?

6.24. The mean and standard deviation of a normally distributed process are unknown. A sample of 5 is taken from the process and the mean of the sample is found to be 80 and the value of s is 20. If the mean of the process is 100, what is the probability of an absolute value of t greater than $\frac{|80-100|}{20/\sqrt{5}}$?

6.25. Twenty samples of 5 each from a normal universe have an average range $(\bar{R})$ equal to 0.006. If the mean of the universe is 0.830, is the probability greater or less than 0.025 that the mean of another independent sample of 4 items from the same universe should be equal to or greater than 0.833? (*Hint:* Let $t = \frac{\bar{X} - \bar{X}'}{\bar{R}/d_2^*\sqrt{4}}$ and use Table D3 of Appendix II to compute the degrees of freedom.) Is it likely that the range of the means of another independent set of 10 samples of 5 items each from the same universe should be in excess of 0.004? $\left(\textit{Hint: } \text{Take } q = \frac{\bar{X}_H - \bar{X}_L}{\bar{R}/d_2^*\sqrt{5}}.\right)$

6.26. Ten samples of 8 each from a normal universe have an average range $\bar{R}$ equal to 3.4. If the mean of the universe is 20, estimate the probability that the mean of another independent sample of 9 items should fall short of 19. Is it likely that the range of another independent set of 6 sample means of 8 items should be in excess of 1.3?

6.27.* Using the results of Problem 4.44, show that the mean of a sample of n items from an "infinite" normal universe with mean zero and variance σ'^2 is normally distributed with zero mean and variance σ'^2/n. (*Hint:* Show that the sample mean has a moment generating function for a normal distribution with the given means and variance.)

6.28.* As noted in Section 2.3 above, if X has a χ^2 distribution, its density function is

$$f(X) = \frac{e^{-X/2}X^{(\nu-2)/2}}{2^{\nu/2}\left(\frac{\nu-2}{2}\right)!}$$

The parameter ν is called the degrees of freedom. Show that the moment generating function for a χ^2 variable is

$$m_x(t) = \frac{1}{(1-2t)^{\nu/2}}$$

[*Hint:* By definition

$$m_x(t) = \int_0^\infty \frac{e^{tX}e^{-X/2}X^{(\nu-2)/2}}{2^{\nu/2}\left(\frac{\nu-2}{2}\right)!}\,dX$$

* Students with training in the calculus will find starred problems will give them knowledge of statistical theory that will be very useful in reading the text.

By setting $Y = (1 - 2t)X$ we can get an integral that equals 1, leaving $1/(1 - 2t)^{\nu/2}$.]

6.29* From the results of the previous problem show that the mean of a χ^2 variable is ν and its variance is 2ν. [*Note:* Var $(X) = E(X - \bar{X}')^2 = E(X - E(X))^2 = E(X^2) - [E(X)]^2$.]

6.30* Show that if X has a χ^2 distribution with ν_1 degrees of freedom and Y has a χ^2 distribution with ν_2 degrees of freedom, and if X and Y are independent of each other, then $X + Y$ has a χ^2 distribution with $\nu_1 + \nu_2$ degrees of freedom. (*Hint:* Note Problem 4.44.)

6.31* If z has a standard normal distribution, then we have

$$\text{Prob. } (-z_1 \leq z \leq z_1) = \int_{-z_1}^{z_1} \frac{e^{-z^2/2}}{\sqrt{2\pi}}\,dz$$

$$= 2\int_{0}^{z_1} \frac{e^{-z^2/2}}{\sqrt{2\pi}}\,dz$$

owing to the symmetry of the normal distribution. Because of this same symmetry, Prob. $(-z_1 \leq z \leq z_1) =$ Prob. $(z^2 \leq z_1^2)$. Hence

$$P(z^2 \leq z_1^2) = 2\int_{0}^{z_1} \frac{e^{-z^2/2}}{\sqrt{2\pi}}\,dz = \int \frac{e^{-z^2/2}}{2^{1/2}\sqrt{\pi}}(z^2)^{-1/2}d(z^2)$$

Taking it as given that $(-\frac{1}{2})! = \sqrt{\pi}$ (see Mood and Graybill, p. 128), show that z^2 has a χ^2 distribution with 1 degree of freedom (i.e., $\nu = 1$).

6.6.32* From the results of Problems 6.30 and 6.31 show that if z is a standardized normal variable, then the sum of the squares of n independent z values has a χ^2 distribution with n degrees of freedom.

These results are of major importance in the analysis of variance. They will also give some feeling as to why $(n - 1)s^2/\sigma'^2 \left[= \frac{\Sigma(X - \bar{X})^2}{\sigma'^2}\right]$ has a χ^2 with $n - 1$ degrees of freedom. In this case the deviations are measured from the sample mean and only $n - 1$ of them are independent. For further discussion of this see Chapter 29, Section 1.7.

6.33* A definite multiple integral has the form

$$\int_{Z_1}^{Z_2}\int_{Y_1}^{Y_2}\int_{X_1}^{X_2} f(X, Y, Z)dX\,dY\,dZ$$

If the variables are subjected to a transformation, say $U = u(X, Y, Z)$, $V = v(X, Y, Z)$, and $W = w(X, Y, Z)$, then the differential element $dXdYdZ$ must be modified to get $dUdVdW$. The factor that accomplishes this modification is known as the Jacobian. By definition

---

* Students with training in the calculus will find starred problems will give them knowledge of statistical theory that will be very useful in reading the text.

$$J\left(\frac{X, Y, Z}{U, V, W}\right) = \begin{vmatrix} \frac{\partial X}{\partial U} & \frac{\partial X}{\partial V} & \frac{\partial X}{\partial W} \\ \frac{\partial Y}{\partial U} & \frac{\partial Y}{\partial V} & \frac{\partial Y}{\partial W} \\ \frac{\partial Z}{\partial U} & \frac{\partial Z}{\partial V} & \frac{\partial Z}{\partial W} \end{vmatrix}$$

The transformed integral becomes

$$\int_{W_1}^{W_2}\int_{V_2}^{V_1}\int_{U_1}^{U_2} g(U, V, W)\, J\left(\frac{X, Y, Z}{U, V, W}\right) dU\,dV\,dW$$

a. Given the transformation

$$t = z/\sqrt{W},\ s = \sqrt{W},$$

show that the Jacobian $J\left(\frac{z, W}{t, s}\right)$ has an absolute value equal to $2s^2$.

b. Given the transformation

$$F = \frac{X/\nu_1}{Y/\nu_2} \text{ and } G = Y$$

show that the Jacobian is $\nu_1 G/\nu_2$.

c. Given the linear transformation

$$W_1 = a_{11}X_1 + a_{12}X_2 + a_{13}X_3$$

$$W_2 = a_{21}X_1 + a_{22}X_2 + a_{23}X_3$$

$$W_3 = a_{31}X_1 + a_{32}X_2 + a_{33}X_3$$

where $\sum_j a_{ij}a_{kj} = 0;\ \sum_j a_{ij}{}^2 = 1.$

6.34.* The density function for the t distribution is

$$f(t) = \frac{\left(\frac{\nu-1}{2}\right)!}{\sqrt{\nu\pi}\left(\frac{\nu-2}{2}\right)!}\ \frac{1}{[1 + t^2/\nu]^{(\nu+1)/2}}$$

Let z have a standard normal distribution and let νW have a χ^2 distribution with ν degrees of freedom. Then—

a. Write the joint density function for z and νW, assuming that they are independent.

b. Let $t = z/\sqrt{W}$, $s = \sqrt{W}$, and note that the Jacobian of the transformation is $2s^2$. Rewrite the joint density function in terms of t and s.

c. Let $X = \left(1 + \frac{t^2}{\nu}\right)\nu s^2/2$ and integrate out X leaving $f(t)$ above.

---

* Students with training in the calculus will find starred problems will give them knowledge of statistical theory that will be very useful in reading the text.

This problem proves that if z is a standardized normal variable and νW has a χ^2 distribution and if z and W are independent, then the ratio $z/\sqrt{W}$ has a t distribution.

9. SELECTED REFERENCES*

Cowden (B '57), Feller (B '68), Hahn (P '70), Hahn and Nelson (P '73), Hald (B '52), Hall (P '84), Hoel (B '71), Kendall (B '48), Lord (P '47), Maritz and Jarrett (P '78), Mood and Graybill (B '63), Mosteller and Tukey (P '49), Nelson (P '63), Olmstead (P '46), Owen and Frawley (P '71), Patnaik (P '50), Pearson and Hartley (B '58), Singh (P '70), Smith, J. G. and Duncan (B '45), Stevens (P '39), and Wald and Wolfowitz (P '49).

---

* B and P refer to the Book and Periodical sections, respectively, of the Cumulative List of References in Appendix V.

Part 2

Lot Acceptance Sampling Plans

7

Acceptance Sampling by Attributes: Single-Sampling Plans

1. ACCEPTANCE SAMPLING

One of the major fields of statistical quality control is acceptance sampling. A company receives a shipment of goods. It samples the shipment and either accepts it as conforming to its standards or rejects it. If the company rejects the lot as below standard, it may be returned to the supplier or it may be kept, depending on how badly the goods are needed or on what arrangements have been made with the supplier. Possibly there will be a price concession on rejected lots. Some companies do not return rejected lots to a supplier until they are reasonably assured through further inspection of the lot that it is of low quality. When a government agency has its own inspectors in a supplier's plant, no return of rejected lots is necessary since they are refused before they are shipped. Frequently a company's own output is submitted to acceptance sampling at various stages of production. A given lot of product is sampled and either accepted for further processing or for shipment to customers, or rejected.

It is to be emphasized that the purpose of acceptance sampling is to determine a course of action, not to estimate lot quality. Acceptance sampling prescribes a procedure that, if applied to a series of lots, will give a specified risk of accepting lots of given quality. In other words, acceptance sampling yields quality assurance.

It is also to be emphasized that acceptance sampling is not an attempt to "control" quality. The latter is the purpose of control charts, which will be discussed in Part 4. These guide the engineer in modifying production so as to turn out better products. This is real quality control. An acceptance sampling plan merely accepts and rejects lots. If the lots are all of the same quality, it will accept some and reject others, and the accepted lots will be no better than the rejected lots.

The indirect effects of acceptance sampling on quality are likely to be much more important than the direct effects. When a supplier's product is being rejected at a high rate, one of two things may happen. The supplier may take steps to improve his production methods or the customer may be led to seek other and better sources of supply. In many cases, large companies send out their own experts to help suppliers solve quality problems. Acceptance sampling thus indirectly improves quality of production through its encouragement of good quality by a high rate of acceptance and its discouragement of poor quality by a high rate of rejection.

Furthermore, if acceptance sampling is used by a manufacturer at various stages in production, it may have beneficial effects in general on the quality of production. If a company relies heavily on final 100 percent inspection of the goods shipped to customers, a careless attitude toward quality may be generated among production personnel. Workers and management alike may feel that no bad material will be shipped because all substandard material will be caught in final inspection. Emphasis in the production department may then be on quantity of output rather than quality. Final 100 percent inspection, however, may not be as effective as believed; for because of its monotonous and boring character, a good percentage of nonconforming material may actually get by, the more the poorer the quality submitted for inspection. Even if final inspection is perfect, a poor attitude toward quality by production personnel will lead to large scrap and rework. On the other hand, under a program of acceptance sampling, the cost of screening and reworking rejected lots may be placed on the production department. Production personnel will then become quality conscious and there will be an interest in quality on the part of both inspection and production. The rule will be: "Make it right the first time." These psychological aspects of acceptance sampling are of major importance.

Acceptance sampling is likely to be used under the following conditions:

1. When the cost of inspection is high and the loss arising from the passing of a nonconforming unit is not great. It is possible in some cases that no inspection at all will be the cheapest plan.
2. When 100 percent inspection is fatiguing and a carefully worked-out sampling plan will produce as good or better results. As noted above, 100 percent inspection may not mean 100 percent perfect quality, and the percentage of nonconforming items passed may be higher than under a scientifically designed sampling plan.
3. When inspection is destructive. In this case sampling must be employed.

In deciding whether to adopt acceptance sampling of a continuing series of lots or to discontinue or modify an acceptance sampling procedure already in use, careful consideration should be given to the information that is available with respect to (1) the distribution of the quality of the lots turned out by the given process, (2) the unit cost of inspection and (3) the cost of passing a nonconforming unit. For further discussion of the attainment of an economic inspection procedure the reader is referred to Section 2 of Chapter 14.

The type of acceptance sampling that is discussed in Chapters 7 to 10 is an "attribute" inspection that merely grades the product as nonconforming or conforming. Inspection and acceptance sampling based on quantitative measurements are discussed in Chapters 11 to 15. The present chapter will discuss attribute sampling plans that require only one sample. The next chapter will discuss double and sequential sampling, and Chapter 9 will discuss multiple sampling. Chapter 10 will describe the set of plans issued by the Department of Defense for sampling inspection by attributes, known as Military Standard 105D.

Often to an acceptance sampling plan there is attached a provision for further inspection of lots rejected by the sampling plan. In these cases, acceptance sampling becomes part of a broader program of *rectifying inspection.* These programs of rectifying inspection give a definite assurance regarding the average quality of the items passed by the program that is not inherent in mere acceptance sampling. Programs of rectifying inspection are discussed in Part 3.

2. FRACTION-NONCONFORMING ATTRIBUTE SAMPLING PLANS

2.1. The Operating Characteristic Curve of a Single-Sample Fraction-Nonconforming Sampling Plan

A single-sample fraction-nonconforming sampling plan specifies the sample size that should be taken and the number of nonconforming units that cannot be exceeded without the lot's being rejected. The latter is called the acceptance number and in this text is symbolized by the letter c. For example, under a particular sampling plan, a sample of 100 is taken from a given lot. If two or less nonconforming units are found, the lot is accepted; if three or more nonconforming units are found, the lot is rejected. This sampling plan would be designated as the plan $n = 100$, $c = 2$.

The discriminatory power of a sampling plan is revealed by its operating characteristic curve, or OC curve as it is called. This curve shows how the probability of accepting a lot varies with the quality of the material offered for inspection. The concept is a simple one but needs careful exposition.

2.1.1. Type B OC Curves. let us suppose we have a process that is operating in a random manner so as to turn out $100p'$ percent nonconforming items. The product of this process will be said to be of quality p'. If lots of size N are made up of this product, the fractions nonconforming of the lots will follow a binomial distribution (since the process is assumed to be operating in a random manner[1]). If each lot is submitted to a sampling

[1] Cf. Chapter 2, Section 3.

inspection plan such as described above, then the probability of accepting a lot will be the proportion of the lots from the given process that will in the long run be accepted under the plan. The operating characteristic curve for the sampling plan will be the curve that shows how this probability of acceptance varies with the product quality p'. More precisely this is known as an OC curve of Type B and is to be distinguished from an OC curve of Type A which we shall discuss subsequently.[2]

The computation of the ordinates of a Type B OC curve is a simple matter. To draw lots of N items at random from a (theoretically infinite) process and then to draw random samples of n from these lots is in essence the equivalent of drawing random samples of n directly from the process. Hence the probability under the given sampling plan of accepting a lot from a process of product quality p' is the probability that a random sample of size n from an infinite universe with fraction nonconforming p' will contain c or less nonconforming items. This is a probability given by the binomial distribution and can either be computed from the binomial formula (4.11) or one of its approximations,[3] taken from a table of binomial probabilities or read from Figure 7.5.

The Type B OC curve for the sampling plan $n = 100$, $c = 2$ is shown in Figure 7.1. This shows, for example, that if product quality is 0.05, the probability of lot acceptance is 0.12. Again, if product quality is 0.02, the probability of lot acceptance is 0.68; if it is 0.01, the probability of acceptance is 0.92. Since the p''s were small, the Poisson approximation was used to compute the probabilities. They were in fact found from Figure 4.9 by entering the abscissa scale with $u' = p'n$ and reading off P_a from the vertical scale for the given value of c. For $p' = 0.05$, for example, we entered the abscissa scale at $u' = (0.05)(100) = 5$, went up to the curve marked $X = 2$, and read off on the vertical scale $P_a = 0.12$. Instead of using the Poisson approximation we could have employed the binomial distribution directly, using one of the published tables of this distribution or Figure 7.5, or as a last resort computing the probabilities directly from formula (4.11).

2.1.2. Type A OC Curves. A Type A OC curve gives the probability of accepting an isolated lot. This may be interpreted as the proportion of lots that would be accepted in an infinite series of lots all identical to the one in question.

The Type A OC curve is to be carefully distinguished from the Type B OC curve. The Type B OC curve just discussed gives the probability of accepting a lot from a randomly operating process turning out product of average quality p'. To be more explicit, if lots of 500 are taken from a process that is operating randomly with average fraction nonconforming 0.05, the fractions nonconforming of the lots will vary in accordance with a binomial

---

[2] The distinction between Type A and Type B OC curves is due to Harold F. Dodge and Harry G. Romig. See their *Sampling Inspection Tables* in Cumulative List of References.

[3] Cf. Chapter 4, Section 7.

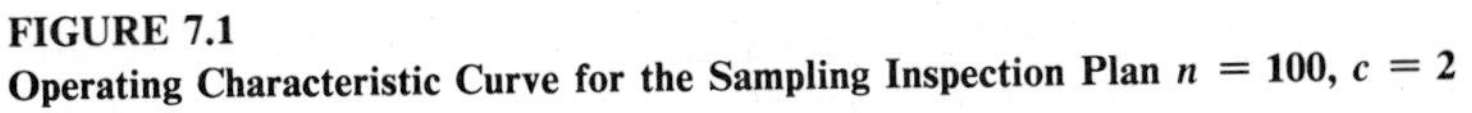

FIGURE 7.1
Operating Characteristic Curve for the Sampling Inspection Plan $n = 100$, $c = 2$

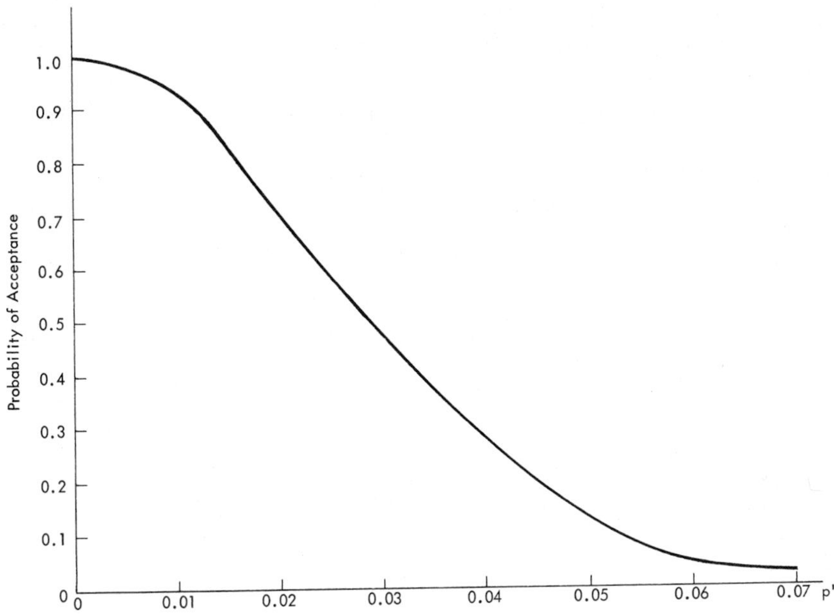

distribution the average of which will be 0.05. Some lots will have only 15 nonconforming items, say; others will have 35 nonconforming items; and in some very few lots the number nonconforming may fall outside this range. In general, however, the number of nonconforming items in a lot will be about 25. The ordinate of a Type B OC curve at $p' = 0.05$ is the proportion of these binomially distributed lots that will on the average be accepted by the given sampling plan. Consider now a second infinite series of lots of 500 items each. Let these lots all be identical in that they have exactly 5 percent nonconforming items. The ordinate of a Type A OC curve at $p' = 0.05$ is then the proportion of such lots that will on the average be accepted by the given sampling plan. The Type B OC curve describes how a consumer is likely to view the operating characteristics of a sampling plan when he is buying a steady stream of material from a given supplier. The Type A OC curve describes how a consumer is likely to view the operating characteristics of a sampling plan when he buys isolated lots of material or thinks about the quality of individual lots rather than the average quality of a stream of lots.

In calculating the probability of acceptance for a Type A OC curve we can no longer adopt the view that we have a sample from a process or "infinite" universe. Instead the appropriate model is that of sampling from a finite universe, the lot. In this case the probabilities of acceptance will be given by the hypergeometric distribution.[4] They will thus either be taken from the Lieberman-Owen *Tables of the Hypergeometric Probability Distribution* or computed from formula (4.15) or one of its approximations.[5] For example, if the sampling plan $n = 50$, $c = 1$ is applied to lots of size 100, the Lieberman-Owen tables give immediately that the probability of acceptance of lots that are 2 percent nonconforming is 0.75. This would be the ordinate of the Type A OC curve for the given sampling plan at point $p' = 0.02$ for lots of size 100. If the sampling plan was $n = 100$, $c = 2$, and lots were of size 200, we have figures that are beyond the range of the Lieberman-Owen tables and have to resort to the special approximation given by formula (4.24). In this case the probability of acceptance of lots of 0.02 quality can be approximated by the probability of getting two or less nonconforming units in a sample of four from an infinite universe in which there are 50 percent nonconforming units, i.e., by the binomial sum:

$$\begin{aligned} P(X \le 2) &= \sum_{X=0}^{2} \frac{p'N!}{X!(p'N - X)!} \left(\frac{n}{N}\right)^{X} \left(1 - \frac{n}{N}\right)^{p'N - X} \\ &= \sum_{X=0}^{2} \frac{4!}{X!(4 - X)!} \left(\frac{100}{200}\right)^{X} \left(1 - \frac{100}{200}\right)^{4-X} \\ &= 0.69 \end{aligned}$$

Generally the probabilities given by formula (4.24) can be obtained directly from tables of the binomial probability distribution.

It will be noted that a Type A OC curve depends on the lot size involved. How the Type A OC curve for a given sampling plan varies with lot size is illustrated in Figure 7.2. It is seen that as the lot size gets large, the Type A OC curve rapidly approaches the OC curve for infinite lots. The Type A OC curve for an infinite lot, however, is mathematically identical with a Type B OC curve. It follows therefore that a Type B OC curve can be taken as a good approximation to a Type A OC curve when lots are large, say more than 10 times the sample, with the exception, of course, that for the Type A curve the abscissa p' is to be read as lot quality and not product quality. Type A OC curves for large lots will be called Type A_L, and Type A OC curves for small lots will be called Type A_S.

To simplify subsequent discussion of OC curves we shall generally talk about OC curves of sampling plans without distinguishing between Type A and Type B. When we refer to "the OC curve" of a sampling plan we shall mean its Type B OC curve or its approximate mathematical counterpart, the Type A_L curve. Graphs will show the abscissa as p'. If the Type B interpretation is to be made, p' will be interpreted as product quality; if the Type A interpretation is to be made, p' will be interpreted as lot quality.

---

[4] Cf. Chapter 4, Section 5.3.

[5] Cf. Chapter 4, Section 7.3.

FIGURE 7.2
Type A Operating Characteristic Curves for Sampling Plan n = 40, c = 1.*

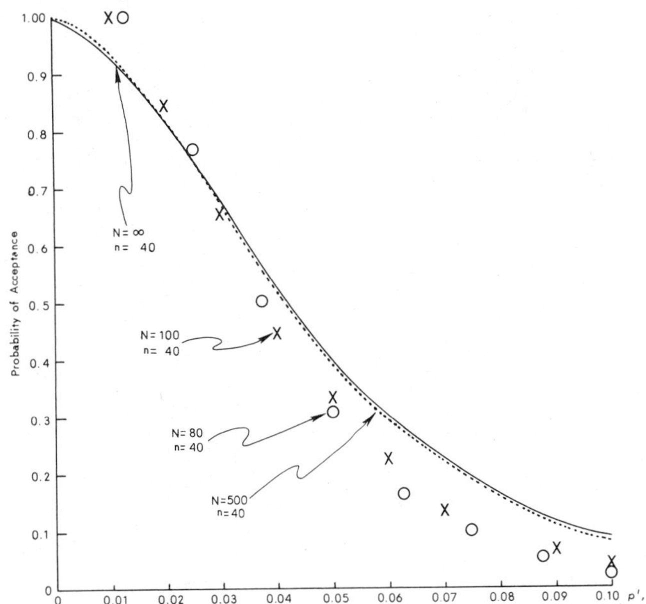

* For N = 80, 100, and 500, isolated points are shown since these are the only ones possible.

When we wish to talk about Type A_S OC curves, we shall specifically note this.

2.1.3. Variation in OC Curves with n and c. The sampling inspection plan that would discriminate perfectly between good and bad lots would have a Type A OC curve that is **z**-shaped. This would run horizontally at a probability of acceptance = 1.0 until p' is such that $p'n = c$, at which point it would drop vertically, and then for higher values of p' would run horizontally again at a probability of acceptance of zero. Under such a program, *all* lots with p' less than or equal to the maximum allowable fraction nonconforming would be accepted, and *all* lots of p' greater than this maximum allowable fraction nonconforming would be rejected. Such a program would give perfect control over the quality of inspected material.

Unfortunately, a **z**-shaped Type A OC curve can only be attained by perfect 100 percent inspection. It can be approached, however, by increasing the sample size. This tendency to become more **z**-shaped as the sample size is increased is true for both Type B and Type A OC curves. For large lots the tendency is illustrated in Figure 7.3 (p. 168). (Note that c is kept proportional to n.) Thus the precision with which a plan separates good and bad lots increases with the size of the sample. The best size to use is always a compromise between the greater precision of larger samples and their greater cost.

FIGURE 7.3
Operating Characteristic Curves for Samples of Different Size

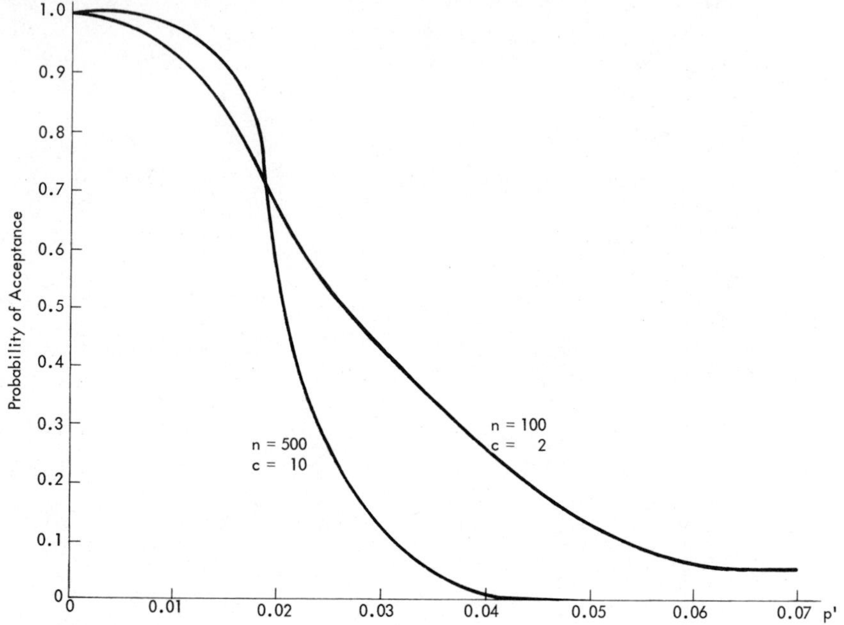

FIGURE 7.4
Operating Characteristic Curves for Different Acceptance Numbers

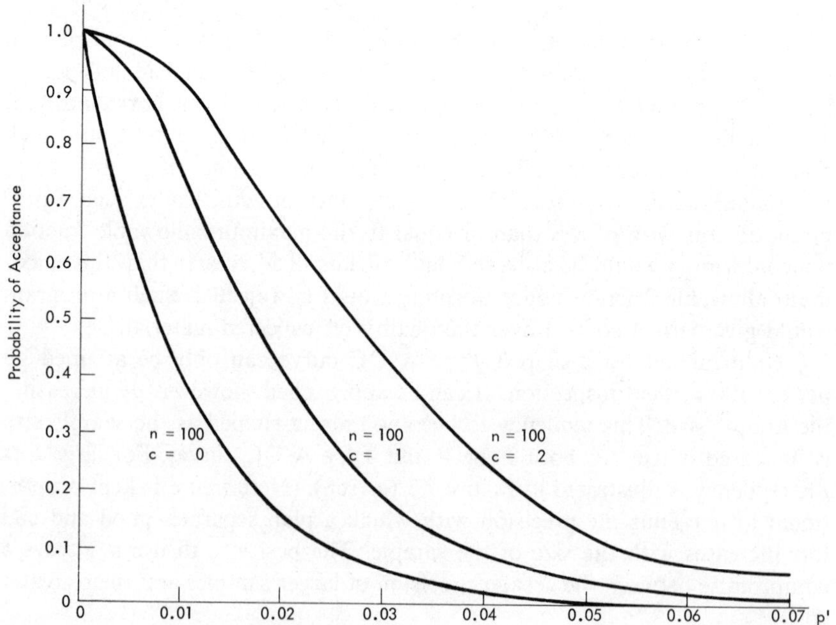

Figure 7.4 shows how the OC curve for a plan varies with the acceptance number alone. As c is diminished, the plan is tightened up, and the effect is to lower the OC curve. As c is increased, the plan becomes more lax, and the effect is to raise the OC curve.

2.1.4. A Comment. Before closing this section the author would like to refer to a comment of a friend to the effect that quality control texts tend to overdramatize OC curves. This is undoubtedly true. Texts tend to be concerned with theory, and quality control theory deals largely with risks. In order that the student may have a clear understanding of theory, these risks must be spelled out at length. Great care must be taken in making precise statements. This is particularly true of OC curves, with the result that OC curves may seem to be more important than they really are.

It must also be remembered that OC curves are purely theoretical concepts. We say they indicate the proportion of lots that would be accepted in a series of lots from a randomly operating process or in a series of identical lots, but in actuality we rarely if ever have a series of lots from a strictly random process or a series of identical lots. In actuality, the series of lots that come in for inspection is from a process that is not operating in a random manner nor are the lots all identical. The proportion of such lots accepted by the sampling plan would not be that given by the OC curve but something different. It might be helpful to the realistic man to view the OC curve as simply giving him some idea of *what to expect.* If a process is operating at the p'_1 level, he is to expect that a certain proportion of lots will on the average be accepted by the sampling plan. If it shifts to another level p'_2, he must expect a different proportion of acceptances. The OC curves thus give a description of what to expect under this condition or that condition. They do not directly give the proportion of acceptances during a given period in which conditions are changing.

2.2. Characterization of an Acceptance Sampling Plan

Although the complete story is only told by the full OC curve for a sampling plan, interest sometimes centers on certain parts of the curve. Sometimes we would like to know what lot or product quality will yield a high probability of acceptance. The producer would be particularly interested in this since it would indicate to him what his target would have to be to get a high rate of acceptance. A sampling plan is thus often characterized by its 0.95 or 0.99 point. These may be designated the $p'_{0.95}$ and $p'_{0.99}$ points. At the other end, a consumer might be particularly interested in what lot or product quality will yield low probabilities of acceptance. He might thus be interested in the $p'_{0.10}$ or $p'_{0.05}$ point. Generally, the two points that are most frequently referred to are the $p'_{0.95}$ point and the $p'_{0.10}$ point. In some theoretical studies, interest has centered on the $p'_{0.50}$ point or "point of indifference." Other aspects of the OC curve that have been of theoretical interest are the point of inflection of the curve, the slope of the curve at the point of indifference, and the slope of the curve at the point of inflection. These characteristics will not be developed in this text.

It should be carefully noted again by the reader that the symbol p'_{P_a} has different meanings depending on the type of OC curve that is referred to. If we are referring to a Type A OC curve, p'_{P_a} means the *lot* quality that has a probability P_a of being accepted under the given sampling plan. If we are referring to a Type B OC curve, p'_{P_a} means the *process* average quality for which the probability of lot acceptance is P_a.

When a consumer establishes a sampling plan for a continuing supply of material, he commonly does so with reference to what is called an *acceptable quality level* or AQL. This is the poorest level of quality or maximum fraction nonconforming for the supplier's process that the consumer would consider to be acceptable as a process average for the purposes of acceptance sampling. It will be noted that the AQL so defined is a characterization of the supplier's process and not of the sampling plan used by the consumer. It is possible, of course, to design a sampling plan such that is $p'_{0.95}$ point or $p'_{0.99}$ point or some other particular point will be the AQL. (The appropriate OC curve here would be of Type B.) If, however, a particular AQL is referenced in the setting up of a given sampling plan, it is not to be inferred that its $p'_{0.95}$ or $p'_{0.99}$ point, say, equals this AQL unless it is so stated. It is to be emphasized that the AQL is a constraint relating to the judicial act of inspection. It is not intended to be a specification on the product nor a target for production. It is simply the standard which the consumer indicates he will use in judging the product. This is what is meant in the definition of the AQL by the phrase "for the purposes of acceptance sampling."

A consumer may also be interested in the quality of individual lots as well as in the quality level of a supplier's process. In such cases he may establish a *lot tolerance fraction nonconforming* or p'_t indicating the poorest quality he is willing to tolerate in an individual lot. This lot tolerance fraction nonconforming is an aspect of the consumer's standards of acquisition and as so defined is not a characterization of a sampling plan. It is possible of course to design a sampling plan such that its $p'_{0.10}$ point or $p'_{0.05}$ point coincides with a specified p'_t. (The appropriate OC curve here would be of Type A.) If it is noted, however, that a particular p'_t was referenced in setting up a sampling plan, it is not to be inferred that this is its $p'_{0.10}$ or $p'_{0.05}$ point unless it is so stated.

Two aspects of a sampling plan that are often discussed in the literature are the "producer's risk" and "consumer's risk." The "producer's risk" generally pertains to a Type B OC curve. If P_a is the probability of acceptance given by such a curve, then the producer's risk is the probability of rejection or $1 - P_a$. It is usually used in reference to rejection of lots from a process the average quality of which equals the AQL. The "consumer's risk" generally pertains to a Type A OC curve. It is the probability of accepting a lot the quality of which is equal to the lot tolerance fraction nonconforming p'_t.

The above are the more important aspects of sampling plans that are referenced in the design of such plans.

2.3. Derivation of a Single-Sample Fraction-Nonconforming Plan with Specified p'_1, p'_2, α and β

2.3.1. Type B and Type A_L OC Curves. We sometimes[6] wish to design a sampling plan such that its OC curve passes through two designated points. (One point is not enough to determine a single-sampling plan, but two points are.) Thus a consumer may wish a plan such that the probability of acceptance is $1 - \alpha$ for material of p'_1 quality and is β for material of p'_2 quality. In this section we shall restrict the discussion to Type B and Type A_L, OC curves. Designs with respect to Type A_S OC curves are discussed in the next section.

To illustrate the problem let us suppose that a consumer wants a plan that will not accept, more than 10 percent of the time, material that is 8 percent nonconforming or worse. He also would like the supplier to aim at producing material that is no worse than 1 percent nonconforming, and if the supplier does so, he wants to have a plan that will accept such material at least 95 percent of the time. In other words, the consumer wants a plan with $p'_1 = 0.01$, $p'_2 = 0.08$, $\alpha = 0.05$, $\beta = 0.10$.

To find a plan that will meet these requirements *exactly* may be impossible, for the sample size and acceptance number of a plan must be integers. A reasonable approximation to the desired result may be obtained, however, by the use of Table 7.1. This gives, for various acceptance numbers c, two values of $p'n$, one that will make the probability of acceptance under the plan equal to 0.95 and another that will make the probability of acceptance equal to 0.10. Since the n for any single-sampling plan remains constant, the difference between the $p'n$ giving a 0.95 chance of acceptance and the $p'n$ giving a 0.10 chance of acceptance arises from the difference in p'. The last column of Table 7.1 gives the ratio of these two p''s; it is the ratio of the p'_2 to the p'_1 for a plan which has the acceptance number c.

With reference to the problem under consideration, the ratio of the selected p'_2 (i.e., 0.08) to the selected p'_1 (i.e., 0.01) is 8. For $c = 1$, the ratio $p'_{0.10}/p'_{0.95}$ is 10.96, and for $c = 2$ it is 6.50. The desired ratio 8 lies between

[6] In this text much of the discussion of the design of acceptance sampling plans concentrates on how to design a plan the OC curve for which will pass through, or come close to passing through, two predetermined points. It is not to be inferred from this, however, that it is common practice to design plans in such a way. In a textbook it serves as a relatively simple way of introducing the reader to basic concepts. It also shows how the design of a sampling plan can be viewed as the equivalent of designing a test of a hypothesis (see Chapter 25), although this similarity may be more misleading than helpful in dealing with real acceptance sampling problems. For procedures employed in the design of sampling plans in common use the reader is referred to Chapter 10 and the discussion of Military Standard 105D and to the rectifying inspection plans of Chapter 16.

TABLE 7.1
Table for Two-Point Design of a Single-Sampling Plan with α Approximately 0.05 and β Approximately 0.10

c	$p'n_{0.95}$	$p'n_{0.10}$	$p'n_{0.10}/p'n_{0.95} = p'_{0.10}/p'_{0.95}$
0	0.051	2.30	45.10
1	0.355	3.89	10.96
2	0.818	5.32	6.50
3	1.366	6.68	4.89
4	1.970	7.99	4.06
5	2.613	9.28	3.55
6	3.285	10.53	3.21
7	3.981	11.77	2.96
8	4.695	12.99	2.77
9	5.425	14.21	2.62
10	6.169	15.41	2.50
11	6.924	16.60	2.40
12	7.690	17.78	2.31
13	8.464	18.96	2.24
14	9.246	20.13	2.18
15	10.04	21.29	2.12

SOURCE: From F. E. Grubbs, "On Designing Single Sampling Inspection Plans," *Annals of Mathematical Statistics* 20 (1949), p. 256. Probabilities based on the Poisson distribution. For a more comprehensive table, see A. H. Bowker and G. J. Lieberman, *Handbook of Industrial Statistics* (Englewood Cliffs, N.J.: Prentice-Hall, 1955), p. 832.

these two figures, so that the statistician may take his pick. If he leans to the conservative side, he will pick $c = 1$; if to the liberal side, he will pick $c = 2$. In either case he must decide whether he is to hold the α or the β. If he decides on the former, he finds the value of n by dividing the $p'n_{0.95}$ for the selected c by 0.01; if he decides on the latter, he divides the $p'n_{0.10}$ for the selected c by 0.08.

c taken as 1		*c taken as 2*	
β Held α Adjusted	*α Held β Adjusted*	*β Held α Adjusted*	*α Held β Adjusted*
$n = \frac{p'n_{0.10}}{0.08}$	$n = \frac{p'n_{0.95}}{0.01}$	$n = \frac{p'n_{0.10}}{0.08}$	$n = \frac{p'n_{0.95}}{0.01}$
$= \frac{3.89}{0.08}$	$= \frac{0.355}{0.01}$	$= \frac{5.32}{0.08}$	$= \frac{0.818}{0.01}$
$\doteq 49$	$\doteq 36$	$\doteq 67$	$\doteq 82$

The results that follow from the various possible choices are summarized in the accompanying tabulation. The various plans that might be selected are thus $n = 49$, $c = 1$; $n = 36$, $c = 1$; $n = 67$, $c = 2$; and $n = 82$, $c = 2$. The actual α's and β's for these plans are as follows[7]:

Plan	α	β
A. $n = 49$, $c = 1$	0.08	0.10
B. $n = 36$, $c = 1$	0.05	0.23
C. $n = 67$, $c = 2$	0.03	0.10
D. $n = 82$, $c = 2$	0.05	0.04

It will be seen that plan A increases slightly the percentage of rejections of p'_1 quality, while giving the desired β. Plan B holds fast on the α but increases greatly the β. Plan C holds to the desired β but reduces α. Finally, plan D holds α and reduces β. In other words, plans A and C give the desired consumer protection, but plan A is tougher than desired on the producer, while plan C is easier on the producer. Plans B and D treat the producer as desired, but plan B gives less protection to the consumer than desired, while plan D gives more protection. Plans A and C come closest to the desired standard. If the ratio of the selected p'_2 to the selected p'_1 had been less than 8—say, between 2 and 4—it would have been possible to find a plan that came closer to the desired criteria than the above plans do.

A graphical procedure[8] for deriving a sampling plan for a given p'_1, p'_2, α, and β is afforded by the nomograph of Figure 7.5. Two lines are drawn, one connecting p'_1 with $1 - \alpha$, the other connecting p'_2 with β. Their intersection gives the region of the desired plan. If it is desired to hold α, the first line is followed either to the c-line just above the intersection point or to the c-line just below it (assuming that the point of intersection does not of course fall exactly on a c-line) and the alternate n's are read off from the chart. The line through p'_2 to β can be used in a similar way. The final result is again four alternate plans that have approximately the desired criteria.

The tabular method described above makes use of the Poisson approximation to the binomial distribution to compute the probabilities involved. Tables have also been prepared that enable the procedure to be based on binomial probabilities. These will be found, together with a description of the method of employing them, in the *Annals of Mathematical Statistics,* Vol. XX, pp. 242–56.

---

[7] These were computed from Figure 4.9. For example, for $c = 1$ and $p'n = (0.01)(49) = 0.49$, $P_a = 0.915$, and $\alpha = 1 - 0.915 = 0.085$.

[8] Harry R. Larson, "A Nomograph of the Cumulative Binomial Distribution," *Industrial Quality Control* 23 (1966), pp. 270–78. Also see Shaul P. Ladany, "Graphical Determination of Single-Sample Attribute Plans for Industrial Small Lots," *Journal of Quality Technology* 3 (July 1971), pp. 115–19.

FIGURE 7.5
Nomograph of the Cumulative Binomial Distribution

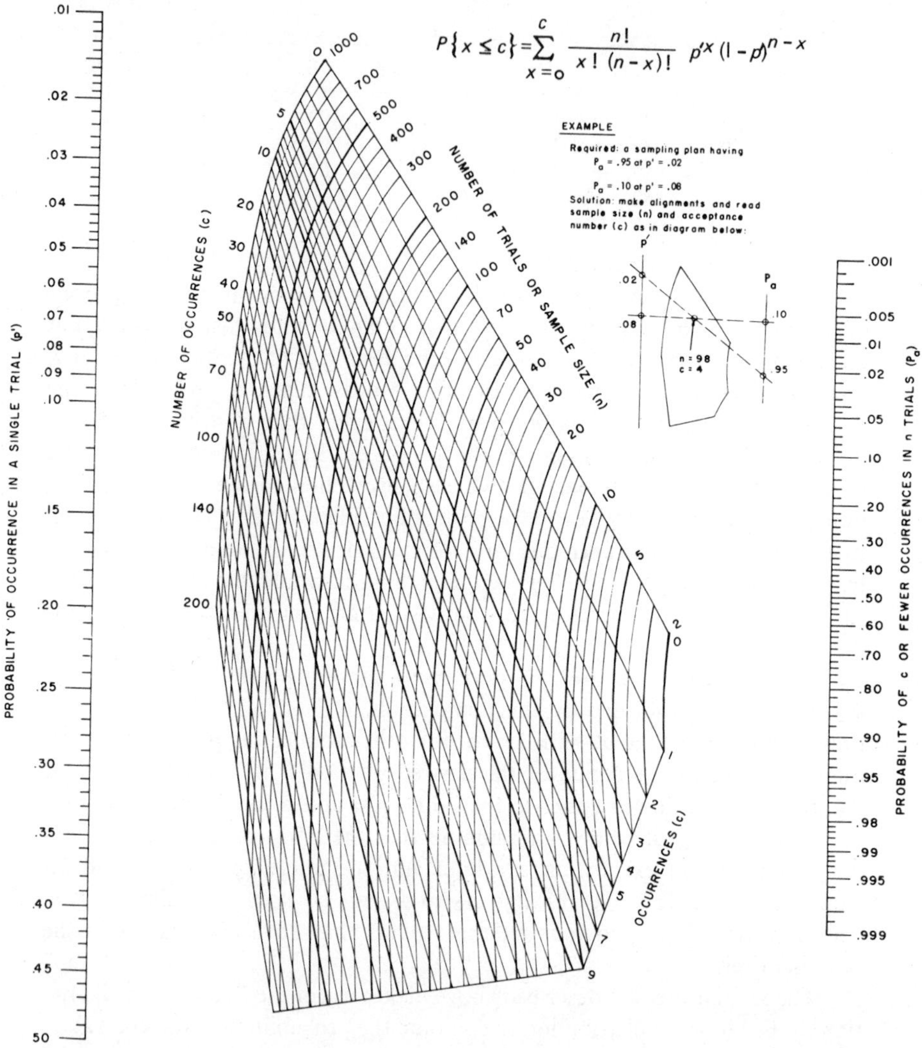

Note: If p' is less than 0.01, set kp' on the p'-scale and n/k on the n-scale, where $k = 0.01/p'$, rounded upward conveniently. Nomograph reproduced with permission from Harry R. Larson. "A Nomograph of the Cumulative Binomial Distribution," from the *Western Electric Engineer,* April 1965.

It should be pointed out in concluding this section that for a specified α and β for Type B or Type A_L OC curves, Table 7.1 is a classification of single-sampling plans according to the ratio of their $p'_{0.10}$ to their $p'_{0.95}$ points. This ratio accordingly serves as a means of indexing that can be readily used in designing plans for specified p'_1 and p'_2. Since the ratio is a basic characteristic of any OC curve, it will be used in matching single-sampling plans with more complicated plans for the sake of comparing their relative efficiencies.

2.3.2. Type A_S OC Curves. When reference is to an OC curve of Type A and lots are relatively small, the derivation of a single-sample fraction-nonconforming sampling plan that will come close to having stated p'_1, p'_2, α, and β criteria can easily be effected by trial-and-error methods using tables of the hypergeometric distribution or a binomial approximation. For example, if the criteria that are $p'_1 = 0.01$, $p'_2 = 0.08$, $\alpha = 0.05$, and $\beta = 0.10$, as in the problem of the previous section, if the OC curve is to be of Type A, and if lots are of size 100, say, then search in the Lieberman-Owen *Tables of the Hypergeometric Probability Distribution* suggests that the closest solution is $n = 40$, $c = 1$. For this the true α will be 0 and the true β will be 0.097, which is better than required.

If the lot size N exceeds 100, we cannot use the Lieberman-Owen tables. We can, however, fall back on the binomial approximation to the hypergeometric distribution given by formula (4.24). We thus use binomial probability tables to find values of n and c that will satisfy the following equations:

$$\sum_{X=0}^{c} \frac{p'_1N!}{X!(p'_1N - X)!}\left(\frac{n}{N}\right)^X\left(1 - \frac{n}{N}\right)^{p'_1N - X} = 0.95$$

$$\sum_{X=0}^{c} \frac{p'_2N!}{X!(p'_2N - X)!}\left(\frac{n}{N}\right)^X\left(1 - \frac{n}{N}\right)^{p'_2N - X} = 0.10$$

In the National Bureau of Standards tables of the binomial distribution we take $n = p'_1N$ in the first equation and $n = p'_2N$ in the second. The p of the tables is taken equal to n/N and r is taken equal to $c + 1$. The probabilities in the table are read as $1 - P_a$. To illustrate, suppose that lots are of size 200 and we have the same criteria as before, viz. $p'_1 = 0.01$, $p'_2 = 0.08$, $\alpha = 0.05$, and $\beta = 0.10$. Search of the National Bureau of Standards tables of the binomial distribution under $n = p'_1N = 2$ and $n = p'_2N = 16$, suggests that $n = 44$, $c = 1$ will be a plan having approximately the desired characteristics. For with $p'_1 = 0.01$ and $p'_1N = 2$, and with $p'_2 = 0.08$ and $p'_2N = 16$, the sampling plan $n = 44$, $c = 1$, yields 0.952 for the first equation and 0.097 for the second, which is a good fit.

2.3.3. Plans Designed with Reference to AQL and p'_t Values. Little has been said so far as to the bases for selecting p'_1 and p'_2. In simplified situations it is likely that the p'_1 will be identified with the AQL. For suppose that p'_1 is set equal to the AQL and suppose that α (the producer's risk) is made small. Then if the supplier turns out an average quality of product equal to the AQL and if the process is operating randomly, a large percentage of the supplier's lots will be accepted. It is also likely that the p'_2 will be identified with the lot tolerance percent defective, for the consumer may want protection on individual lots.

Now it will be noted that setting p'_1 equal to the AQL implies reference to an OC curve of Type B, whereas setting p'_2 equal to the p'_t implies reference to an OC curve of Type A. Hence, we would not mathematically be fitting a sampling plan to two points on the same OC curve unless the inspection lots are large so that the two types of curves are practically identical. This creates no problem, however. If we assume a Type B OC curve in designing our sampling plan and if the plan is in fact applied to small lots, then the true consumer's risk will be less than the β used in the design of the plan. (See Figure 7.2.) Using a Type B OC curve, then, to fit a sampling plan to $p_1' = \text{AQL}$, $p_2' = p_t'$, $\alpha =$ producer's risk for AQL quality, and $\beta =$ consumer's risk, will in fact err on the conservative side and give the consumer more protection than he wishes.

It is possible, however, to be exact and to determine a sampling plan that will satisfy the AQL criterion with reference to a Type B OC curve and also satisfy the p_t' criterion with reference to a Type A OC curve. For example, suppose that in the previous problem it had been stated that the AQL was 0.01 and that the consumer and supplier had agreed to a producer's risk for AQL quality of 0.05. Further it had been stated that the consumer had set $p_t' = 0.08$ and was willing to take a consumer's risk of 0.10. It is known that $N = 200$. To find a single-sampling plan that will meet these criteria we will set up one relationship based on the AQL and another based on the p_t'. Thus, we will have the following equations to be approximately satisfied, viz, the binomial equation

$$\sum_{X=0}^{c} \frac{n!}{X!(n-X)!}(p_1')^X(1-p_1')^{n-X} = 1-\alpha, \text{ where } p_1' = \text{AQL}.$$

and the hypergeometric equation

$$\sum_{X=0}^{c} \frac{C_{n-X}^{N-p_2'N}\, C_X^{p_2'N}}{C_M^N} = \beta, \text{ where } p_2' = p_t'.$$

If conditions are such that we can use the Poisson approximation to the binomial and the binomial approximation to the hypergeometric given by (4.24), the above equations become

$$\sum_{X=0}^{c} (p_1'n)^X e^{-p_1'n}/X! = 1-\alpha.$$

$$\sum_{X=0}^{c} \frac{(p_2'N)!}{X!(p_2'N-X)!}\left(\frac{n}{N}\right)^X\left(1-\frac{n}{N}\right)^{p_2'N-X} = \beta$$

In terms of the problem the above means that we must find an n and a c such that (1) on the Dodge-Romig chart (Figure 4.9) the ordinate will be approximately 0.95 when $u' = 0.01n$ and (2) the entry in a binomial table with sample size equal to $p_2'N = 16$ and probability of a defective unit $= n/200$ will yield a cumulative probability for c or less of approximately 0.10.

The solution of the above problem can be obtained by trial and error. Analyses for Type A OC curves and for Type B OC curves carried out in previous sections suggest that c will be in the neighborhood of 1. We shall thus set up the following calculations:

c	$(p_1'n)_{0.95}$ *Read from Figure 4.9 for Given* c	$(n/N)_{0.10}$ *Obtained from Binomial Tables for* $n = p_2'N$ *and Given* c	n *for First Criterion*	n *for Second Criterion*
0	0.05	0.13	5	26
1	0.36	0.22	36	44
2	0.82	0.30	82	60

where $(p_1'n)_{0.95}$ and $(n/N)_{0.10}$ mean those values of $p_1'n$ and n/N such that the probability of c or less is approximately 0.95 and 0.10, respectively. It is obvious that $c = 1$ gives two values of n that are the closest. If we wish to hold the producer's risk for AQL quality at 0.05, we will take $n = 36$, but this will give a consumer's risk for lots of size 200 equal to 0.19. If we wish to hold the consumer's risk at 0.10, we will take $n = 44$. This gives a producer's risk equal to 0.07. A compromise plan of $n = 40$, $c = 1$, will yield a producer's risk of 0.06, and a consumer's risk of 0.14.

2.4. Chain Sampling Plans

When tests are destructive or very costly, small sample plans may be chosen as a matter of practical expediency. The acceptance numbers for such plans will usually be zero. Plans with zero acceptance numbers are maverick plans, however, in that their OC curves are convex throughout.[9] A result is that the probability of acceptance starts to drop rapidly for the smallest values of the percent nonconforming and this is hard on the producer. Compare, for example, the OC curves of Figure 7.6A with those of 7.6B.

For certain circumstances, Harold F. Dodge has suggested[10] that chain sampling plans might be a desirable substitute for ordinary single sampling plans with zero acceptance numbers in that they have OC curves that are concave at the start. These chain sampling plans are offered for cases "where there is repetitive production under the same essential conditions and where the lots or batches of product to be sampled are offered for acceptance in substantially the order of their production."[11]

A chain sampling plan runs as follows. For each lot, a sample of n units is selected and each unit is tested for conformance to specifications.

---

[9] "Convex" is used here in the mathematical sense that the second derivative is positive throughout. With reference to physical terminology, the curves are sometimes said to be "concave downwards." In the text, "concave" in contrast to "convex" is used to refer to a curve for which the second derivative is negative throughout.

[10] See Harold F. Dodge. "Chain Sampling Inspection Plan," *Industrial Quality Control* No. 4 (1955), pp. 10–13.

[11] Ibid., p. 10.

FIGURE 7.6
General Shape of Operating Characteristic Curves, Single-Sampling Plans

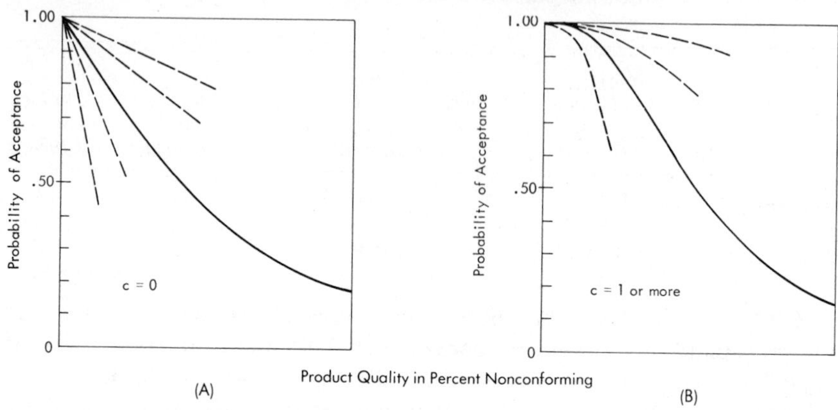

SOURCE: Reproduced with permission from Harold F. Dodge. "Chain Sampling Inspection Plan," *Industrial Quality Control* 19 (1955), p. 10.

If a sample from a given lot contains zero nonconforming units, the lot is accepted. The lot is also accepted if only one of the sample units is nonconforming provided no nonconforming units were found in the previous i samples. If the sample contains two or more nonconforming items, the lot is rejected.

The OC curve for a chain sampling plan is given by the formula

(7.1)
$$P_a = P(0, n) + P(1, n)[P(0, n)]^i$$

where $P(0, n)$ and $P(1, n)$ are the probabilities of getting 0 and 1 nonconforming unit out of n, respectively. To illustrate, consider the chain sampling plan $n = 10$, $c = 0$, $i = 1$. For this we have the following values of p' and P_a in the OC curve:

p'	P_a
0.01	(0.904) + (0.091)(0.904) = 0.987
0.02	(0.817) + (0.167)(0.817) = 0.953
0.03	(0.737) + (0.228)(0.737) = 0.905
0.04	(0.665) + (0.277)(0.665) = 0.849
0.05	(0.599) + (0.315)(0.599) = 0.788
0.06	(0.539) + (0.344)(0.539) = 0.724
0.10	(0.349) + (0.387)(0.349) = 0.484
0.15	(0.197) + (0.347)(0.197) = 0.265
0.20	(0.107) + (0.268)(0.107) = 0.136
0.25	(0.056) + (0.188)(0.056) = 0.067

The curve is shown in Figure 7.7, where with OC curves for other values of i, it is contrasted with the OC curves for the single-sampling plan $n = 10$, $c = 0$.

FIGURE 7.7
Operating Characteristic Curves of Some Chain Sampling Plans

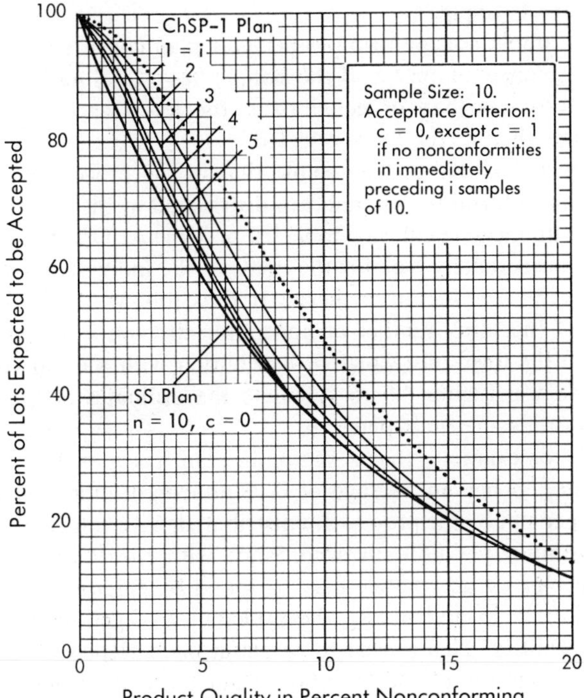

SOURCE: Reproduced with permission from Harold F. Dodge, "Chain Sampling Inspection Plan," *Industrial Quality Control* 19 (1955), p. 11.

Dodge warns that for chain sampling to be properly used the following conditions should be met:

a. The lot should be one of a series in a continuing supply as mentioned above,
b. Lots should normally be expected to be of essentially the same quality,
c. The consumer should have no reason to believe that the lot currently sampled is poorer than the immediately preceding ones, and
d. The consumer must have confidence in the supplier, confidence that the supplier would not take advantage of a good record to slip in a bad lot now and then when it would have the best chance of acceptance.

3. SINGLE-SAMPLING NONCONFORMITIES-PER-UNIT ACCEPTANCE SAMPLING PLANS

Sometimes there is need for a nonconformities-per-unit sampling plan instead of a fraction-nonconforming plan. If the material submitted for inspection is cloth, linoleum, and the like or if it consists of large units, such as television sets and refrigerators, then a nonconformities-per-unit plan may be the most appropriate one to apply.

A single-sample nonconformities-per-unit plan consists of a sample size n and an acceptance number c. If the sample has a total number of nonconformities that is less than or equal to c, the lot is accepted. If it has a number greater than c, the lot is rejected.

For Type B OC curves we assume that in the process as a whole the number of nonconformities per unit are distributed in the form of a Poisson distribution with mean equal, say, to u'.[12] For Type A_L OC curves we assume that this is approximately true for the items in the individual lots. (We shall not discuss Type A_S OC curves here.) It thus follows that[13] the sampling distribution of the total number of nonconformities in a sample of n units will also have the form of a Poisson distribution with a mean equal to nu'. Hence we can use Figure 4.9 to calculate the ordinates of the OC curves for nonconformities-per-unit plans.

Suppose, for example, that a sampling plan has $n = 100$ and $c = 17$. Then, if a process has an average of 0.15 nonconformities per unit and if the items turned out by the process are distributed, with respect to the number of nonconformities per unit, in accordance with the Poisson distribution, the probability of lot acceptance may be found by entering Figure 4.9 with $u' = 100(0.15) = 15$ and noting the probability of getting a lot of 17 or less nonconformities. This is seen to be about 0.75. Hence, for the specified process the probability of lot acceptance under the given sampling plan is 0.75.

From the foregoing it will be noted that the OC curve for a nonconformities-per-unit plan with sample size n and acceptance number c is approximately the same as the OC curve for a fraction-nonconforming plan with the same sample size and same acceptance number. The only difference is that in the former case the abscissa scale will read "number of nonconformities per unit" and in the latter case "fraction nonconforming." (Alternative forms are "number of nonconformities per hundred units" and "percent nonconforming.") Most of what has been written about fraction-nonconforming plans in the earlier part of this chapter can likewise be said about nonconformities-per-unit plans with "number of nonconformities" replacing "number of nonconforming units" and "number of nonconformities per unit" replacing fraction nonconforming."

4. FORMATION OF LOTS FOR INSPECTION

In concluding this chapter a word should be said about the formation of lots for inspection purposes. First, it should be noted that an inspection lot

[12] See Chapter 20, Section 2.

[13] Cf. Maurice G. Kendall, *The Advanced Theory of Statistics,* Vol. I (2d ed.; London, Charles Griffin & Co., Ltd., 1948), p. 243.

is not necessarily a shipment lot. Usually it is a lot especially formed for inspection purposes, generally by or under the supervision of the inspector.

Several principles should be followed in forming lots. First, if a lot is homogeneous, the larger it is the better. For large lots allow large samples at low unit cost, and large samples give good OC curves.

The pooling of nonhomogeneous production or shipment lots to get one large inspection lot is not good, however. For if the original lots are inspected separately, the sampling plan will accept most of the good lots and reject most of the poor ones, tending to average up the quality accepted. If the original lots are pooled before inspection, the quality of lots accepted will be an average figure, poorer than that obtained by separate inspection. For example, let the original lots fall into two groups, those that are 0.01 nonconforming and those that are 0.05 nonconforming, and let the sampling plan be $n = 100$, $c = 2$. Then, if there are an equal number of each kind of lots, the average quality of the lots passed by the plan will be

$$[0.92(0.01) + 0.12(0.05)]/(0.92 + 0.12) = 0.0146$$

On the other hand, if the lots are pooled before inspection so that the lots thus formed all run around 3 percent nonconforming, then lots that are passed will be about 3 percent nonconforming or about double that obtained by using separate lots.

Accordingly, in forming lots, we are torn by two desires: (1) to get the economies of large lots[14] and (2) to get the benefits of discrimination that follow from using homogeneous lots that differ in quality. Any particular solution must be a compromise between these conflicting goals.

* * * * *

(If the reader so desires, it is possible at this point to jump to Chapter 16 and study single-sample rectifying inspection schemes before going on to double and sequential sampling.)

5. PROBLEMS

7.1. Make a graph of the Type B OC curve for each of the following fraction-nonconforming sampling plans:

a. $n = 50$, $c = 1$.

b. $n = 150$, $c = 5$.

c. $n = 200$, $c = 4$.

d. $n = 300$, $c = 10$.

7.2. Make a graph of the Type B OC curve for each of the following fraction-nonconforming sampling plans:

[14] A procedure for forming homogeneous "grand lots" is discussed in Leslie Simon's *An Engineers' Manual of Statistical Methods* and in Edward G. Schilling's *Acceptance Sampling in Quality Control.*

a. $n = 40$, $c = 1$.

b. $n = 80$, $c = 1$.

c. $n = 80$, $c = 2$.

d. $n = 160$, $c = 2$.

7.3. Suppose that lots contain 600 items and make graphs of Type A OC curves for plans *(a)* and *(c)* of Problem 7.1. Compare with the curves derived in Problem 7.1.

7.4. Suppose that lots contain 500 items and make graphs of Type A OC curves for plans *(a)* and *(d)* of Problem 7.2.

7.5. Find a single-sample fraction-nonconforming sampling inspection plan that will come close to having a $p_1' = 0.03$, a $p_2' = 0.10$, $\alpha = 0.05$, and $\beta = 0.10$.

7.6. Find a single-sample fraction-nonconforming sampling inspection plan that will come close to having a $p_1' = 0.03$, a $p_2' = 0.10$, $\alpha = 0.05$, and $\beta = 0.05$.

7.7. Find a single-sample fraction-nonconforming sampling inspection plan that will come close to having a $p_1' = 0.03$, a $p_2' = 0.10$, $\alpha = 0.01$, and $\beta = 0.10$.

7.8. Find a single-sample fraction-nonconforming sampling inspection plan that will come close to having a $p_1' = 0.03$, a $p_2' = 0.10$, $\alpha = 0.01$, and $\beta = 0.05$.

7.9. Find a single-sample fraction-nonconforming sampling inspection plan that will come close to having a $p_1' = 0.005$, a $p_2' = 0.05$, $\alpha = 0.05$, and $\beta = 0.10$.

7.10. Find a single-sample fraction-nonconforming sampling inspection plan that will come close to having a $p_1' = 0.005$, a $p_2' = 0.05$, $\alpha = 0.01$, and $\beta = 0.10$.

7.11. Find a single-sample fraction-nonconforming sampling inspection plan that will come close to having a $p_1' = 0.03$, a $p_2' = 0.08$, $\alpha = 0.05$, and $\beta = 0.10$.

7.12. Find a single-sample fraction-nonconforming sampling inspection plan that will come close to having a $p_1' = 0.015$, a $p_2' = 0.04$, $\alpha = 0.05$, and $\beta = 0.10$.

7.13. A company uses the following sampling plan. It takes a sample equal to 10 percent of the lot being inspected. If 2 percent or less of the sample is nonconforming, the lot is accepted. If more than 2 percent of the sample is nonconforming, the lot is rejected.

a. If a lot numbers 2,000 items of which 80 are nonconforming, what is the probability of its being accepted under the above plan?

b. If a lot numbers 1,000 items of which 40 are nonconforming, what is the probability of its being accepted under the above plan?

c. If a lot numbers 5,000 items of which 200 are nonconforming, what is the probability of its being accepted under the above plan?

d. If 0.04 is the lot tolerance fraction nonconforming, what can you say about the protection given by the above plan?

7.14. A company has the following plan. It takes a sample equal to 10 percent of the lot being inspected. If 1 percent or less of the sample is nonconforming, the lot is accepted. If more than 1 percent of the sample is nonconforming, the lot is rejected.

a. If a lot contains 4,000 items of which 80 are nonconforming, what is the probability of its being accepted under the above plan?

b. If a lot contains 1,000 items of which 20 are nonconforming, what is the probability of its being accepted under the above plan?

c. If a lot contains 10,000 items of which 200 are nonconforming, what is the probability of its being accepted under the above plan?

d. If 0.02 is the lot tolerance fraction nonconforming, what can you say about the protection given by the above plan?

7.15. An AQL is given as 0.02. The p'_t is stated to be 0.07. Find a single-sample fraction-nonconforming sampling plan such that the producer's risk for AQL quality will be 0.03 and the consumer's risk will be 0.10.

7.16. An AQL is given as 0.025. The p'_t is stated to be 0.09. Find a single-sample fraction-nonconforming sampling plan such that the producer's risk for AQL quality will be 0.02 and the consumer's risk 0.08.

7.17. Draw the OC curve for each of the following nonconformities-per-unit sampling plans.

a. $n = 50, c = 4$.

b. $n = 300, c = 4$.

c. $n = 300, c = 24$.

d. $n = 750, c = 24$.

7.18. Find a single-sample nonconformities-per-unit sampling inspection plan that will come close to having a u'_1 of 3 nonconformities per 100 units, a u'_2 of 10 nonconformities per 100 units, $\alpha = 0.05$, and $\beta = 0.10$. (*Hint:* This is worked in exactly the same way as Problem 7.5 and has the same answer except that c represents the maximum total number of allowable nonconformities.)

7.19. Find a single-sample nonconformities-per-unit sampling inspection plan that will come close to having a u'_1 of 1.5 nonconformities per 100 units, a u'_2 of 4 nonconformities per 100 units, $\alpha = 0.05$, and $\beta = 0.10$. (Compare Problem 7.12.)

6. SELECTED REFERENCES*

ASQC Standard A2 (B '78), Barnett (P '74), Bowker and Lieberman (B '55 and B '72), Brumbaugh (P '45), Burgess (P '48), Burr (B '76), Cowden (B '57), Deming (B '50), Dodge (P '55 and P '69), Dodge and Romig (B '59),Godfrey and Mundell (P '84), Grant and Leavenworth (B '80), Grubbs (P '49), Guenther (P '73), Hahn (P '74), Hamaker (P '49 and P '50), Hill, Horsnell, and Warner (P '59), Horsnell (P '54), Jaech (P '80), Johnson and Kotz (P '85), Ladany (P '71), Schilling (B '82), Soundararajan (P '78), Soundararajan and Govindaraju (P '83), Stephens (P '78), and Wetherill and Campling (P '66).

The reader is also referred to Sections 2, 3 and 4 of Chapter 14 below and to the references cited there.

Cost-Oriented References. The interested reader will find excellent discussions in the following references: Davies (P '59), Guenther (P '71), Guthrie and Johns (P '59), Hald (P '60), Hald and Thyregod (B '71), Hill (P '60), Homiling (P '62), Pfanzal (P '63), Van der Waerden (P '60), and Wetherill (P '60).

* B and P refer to the Book and Periodical sections, respectively, of the Cumulative List of References in Appendix V.

8

Acceptance Sampling by Attributes: Double and Sequential Fraction-Nonconforming Sampling Plans

1. DOUBLE-SAMPLING PLANS

A double-sampling plan is designated by five numbers n_1, n_2, c_1, c_2, and c_3, c_1 being less than c_2 and c_2 being less than or equal to c_3. The plan works as follows: A sample of size n_1 is taken from a given lot. If it contains c_1 or less nonconforming units, it is immediately accepted. If it contains more than c_2 nonconforming units, it is immediately rejected. If the number of nonconforming units is greater than c_1 but not more than c_2, a second sample of size n_2 is taken. If in the *combined* samples there are c_3 or less nonconforming units, the lot is accepted. If there ae more than c_3 nonconforming units, the lot is rejected. Frequently c_2 is taken equal to c_3. The loss in flexibility because of this restriction does not severely affect the efficiency of the sampling plan in the region of practical interest,[1] and for this reason, plans of this simpler type only will be discussed in this chapter. When $c_2 = c_3$, a double-sampling plan is described by the four numbers n_1, n_2, c_1, and c_2 and in the remainder of this chapter will be so designated.

A double-sampling plan has two possible advantages over a single-sampling plan. First, it may reduce the total amount of inspection; for the first sample taken is less than that called for under a comparable single-sampling plan, and, consequently, in all cases in which a lot is accepted or rejected on the first sample, there may be a considerable saving in total inspection. It is also possible to reject a lot without completely inspecting the entire

[1] Cf. H. C. Hamaker and R. Van Strik, "The Efficiency of Double Sampling for Attributes," *Journal of the American Statistical Association* 50 (1955), pp. 830–49. Military Standard 105D does not have c_2 generally equal to c_3, however. See Chapter 10.

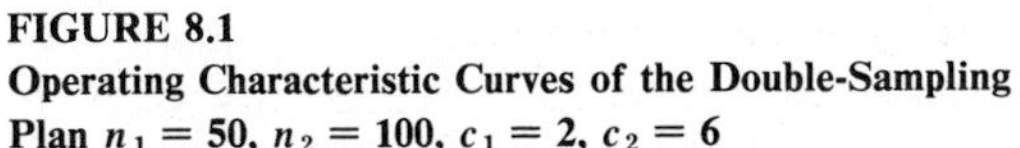

FIGURE 8.1
Operating Characteristic Curves of the Double-Sampling Plan $n_1 = 50$, $n_2 = 100$, $c_1 = 2$, $c_2 = 6$

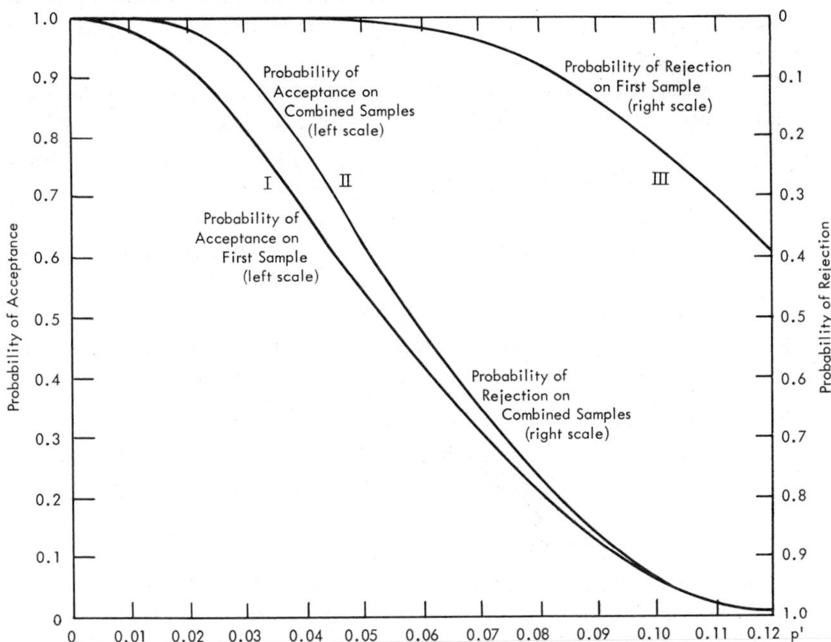

second sample. Second, a double-sampling plan has the psychological advantage of giving a lot a second chance.[2] To some people, especially the producer, it may seem unfair to reject a lot on the basis of a single sample. Double sampling permits the taking of two samples on which to make a decision.

1.1. Operating Characteristic Curves

Like a single-sampling plan, a double-sampling plan has an operating characteristic curve that gives the probability of acceptance as a function of incoming quality. A double-sampling plan also has supplementary OC curves that show the probability of acceptance on the first sample and the probability of rejection on the first sample. Principal and supplementary OC curves for the plan $n_1 = 50$, $n_2 = 100$, $c_1 = 2$, and $c_2 = 6$ are shown in Figure 8.1.

1.1.1. Type B and Type A_L OC Curves. The OC curves for double-sampling can, like the OC curves for single sampling, be of Type A or Type B. In this section we shall describe the computations for Type A_L OC curves. The computations would be the same for Type B OC curves, but in this

[2] This is, of course, no real advantage. If a single- and double-sampling plan have the same OC curve, they will be equally "fair."

case p' would represent the process quality level instead of lot quality. To simplify the computations we shall assume that p' is sufficiently small to use the Poisson distribution for approximating the probabilities involved. Computations appropriate for Type A_S OC curves will be described in the next section.

The computation of the ordinates of the principal and supplementary OC curves of Type A with large lots will be illustrated for the plan of Figure 8.1 for the lot fraction nonconforming of 0.06. For this lot fraction nonconforming, the probability of acceptance on the first sample, i.e., the probability of 2 or less nonconforming units out of 50, is found by the Poisson approximation to be[3] 0.423. This is the ordinate of curve I in Figure 8.1 at the abscissa point $p' = 0.06$.

The probability of acceptance on the combined samples equals the probability of acceptance on the first sample plus the probability of acceptance on the second sample. For the given plan, the latter is the sum of—

1. The probability of 3 nonconforming units in the first sample times the probability of 3 or less nonconforming units in the second. For a lot fraction nonconforming of 0.06, this is equal to $0.22 \times 0.15 = 0.033$.
2. The probability of 4 nonconforming units in the first sample times the probability of 2 or less nonconforming units in the second sample. For a lot fraction nonconforming of 0.06, this is equal to $0.17 \times 0.06 = 0.010$.
3. The probability of 5 nonconforming units in the first sample times the probability of 1 or less nonconforming units in the second sample. For a lot fraction nonconforming of 0.06, this equals $0.10 \times 0.02 = 0.002$.
4. The probability of 6 nonconforming units in the first sample times the probability of 0 nonconforming units in the second sample. For a lot fraction nonconforming of 0.06, this equals $0.05 \times 0.002 = 0.0001$.

The total of these four probabilities is approximately 0.045.[4] This is the probability of acceptance on the second sample. The total probability of acceptance for a lot fraction nonconforming of 0.06 is thus $0.423 + 0.045 = 0.468$, or approximately 0.47. This is the ordinate of curve II at abscissa point 0.06.

Curve II is the principal operating characteristic curve of the plan, since it gives the probability of final acceptance. The difference between curve II

---

[3] Computed from Figure 4.9 or taken from Table E of Appendix II.

[4] It will be noted that this is greater than the probability of getting 6 or less nonconforming units in the combined sample of 150. For the lot is accepted if 2 or less nonconforming units appear in the first sample, whereas, if a second sample were taken in these cases, the total nonconforming units might exceed 6.

and curve I gives the probability of acceptance on the second sample. Curve II, when read with reference to the right-hand scale, also gives the probability of rejection on the combined samples. Curve III, when read with reference to the right-hand scale, gives the probability of rejection on the first sample. This is derived by simply computing the probability of more than 6 nonconforming units in the first sample. For a lot fraction nonconforming of 0.06, the probability of rejection on the first sample is 0.034. The difference between curve II and curve III gives the probability of rejection on the second sample.

In summary, for a lot fraction nonconforming of 0.06, the probability of acceptance on the first sample is 0.423, the probability of acceptance on the second sample is 0.045, and the probability of final acceptance is 0.468. The probability of rejection on the first sample is 0.034, the probability of rejection on the second sample is 0.498,[5] and the probability of final rejection is 0.532.

1.1.2. Type A_S OC Curves. When we have the problem of deriving a Type A OC curve for small lots, it is likely that we will be able to use either the Lieberman-Owen *Tables of the Hypergeometric Probability Distribution* or some table of the binomial probability distribution and approximation formula (4.24) to determine the desired probabilities. As a last resort, we can of course use formula (4.15).

The computations will again be illustrated with reference to the double-sampling plan $n_1 = 50$, $n_2 = 100$, $c_1 = 2$, $c_2 = 6$, and points on the various OC curves will be computed for $p' = 0.06$. We shall assume that lots are of size $N = 500$, which means that we cannot use the Lieberman-Owen tables in this case but will have to use the special binomial approximation given by the formula (4.24).

Using formula (4.24) we find that the probability of acceptance on the first sample, i.e., the probability of getting 2 or less nonconforming items in a sample of 50 from a lot of 500 containing $(0.06)(500) = 30$ nonconforming items, is approximately

$$\sum_{X=0}^{c_1} \frac{(p'N)!}{X!(p'N-X)!}\left(\frac{n}{N}\right)^X\left(1-\frac{n}{N}\right)^{p'N-X}$$

or

$$\sum_{X=0}^{2} \frac{30!}{X!(30-X)!}\left(\frac{50}{500}\right)^X\left(1-\frac{50}{500}\right)^{30-X}$$

which is shown by the binomial tables to be 0.411. This is the probability of acceptance on the first sample for a lot of 0.06 quality.

To find the probability of acceptance on the second sample we proceed as in the previous section. The computations may be summarized as follows:

[5] Computed by subtracting the sum of the other probabilities from 1.

X	Probability of Exactly X in First Sample	Nonconforming Items Left $= p'N - X$	$c_2 - X$	Probability of $c_2 - X$ or Less in Second Sample	Product of Probabilities
3	0.236	27	3	0.125	0.0295
4	0.177	26	2	0.053	0.0094
5	0.102	25	1	0.016	0.0016
6	0.047	24	0	0.003	0.0001
					Sum = 0.0406

The probability 0.236, for example, is given by a single term of the special binomial formula (4.24). Thus

$$P(X) = \frac{(p'N)!}{X!(p'N-X)!}\left(\frac{n}{N}\right)^X\left(1-\frac{n}{N}\right)^{p'N-X}$$

$$= \frac{30!}{3!27!}\left(\frac{50}{500}\right)^3\left(1-\frac{50}{500}\right)^{27}$$

which is shown by the binomal tables to be 0.236. The probabilities in the fifth column are given by the same special binomial formula which in this case takes the form

$$P(y \le c_2 - X) = \sum_{y=0}^{c_2-X} \frac{(p'N-X)!}{y!(p'N-X-y)!}\left(\frac{n_2}{N-n_1}\right)^y\left(1-\frac{n_2}{N-n_1}\right)^{p'N-X-y}$$

Thus for $X = 3$ we have

$$P(y \le 6-3) = \sum_{y=0}^{6-3} \frac{(30-3)!}{y!(30-3-y)!}\left(\frac{100}{450}\right)^y\left(1-\frac{100}{450}\right)^{30-3-y}$$

$$= \sum_{y=0}^{3} \frac{27!}{y!(27-y)!}(0.22)^y(0.78)^{27-y}$$

which is shown by the binomial tables to be 0.125. The sum of the probability products again gives the probability of acceptance on the second sample, viz, 0.041. The total probability of acceptance will be $0.411 + 0.041 = 0.452$.

The probability of rejection on the first sample will be the probability of getting 7 or more nonconforming items in a random sample of 50 from a lot of 500 containing 30 nonconforming items. Using again the special binomial formula (4.24), we have

$$\text{Prob}\,(X \ge 7) = \sum_{X=7}^{30} \frac{30!}{X!(30-X)!}\left(\frac{50}{500}\right)^X\left(1-\frac{50}{500}\right)^{30-X}$$

which is shown by the binomial tables to be 0.026. This then is the probability of rejection on the first sample. The probability of rejection on the second equals $1 - 0.452 - 0.026 = 0.522$ and the total probability of final rejection is 0.548. The results do not deviate very much in this instance from those obtained for the case of large lots.

1.2. The Average Sample Number Curve with Complete Inspection of the Second Sample

Besides principal and supplementary operating characteristic curves, a double-sampling plan also has what may be called an average sample number curve (ASN curve). In single sampling the size of the total sample inspected is always constant; in double sampling it will vary, depending on whether a second sample is necessary or not. The probability of a second sample will vary with p'. Consider once again the double-sampling plan $n_1 = 50$, $n_2 = 100$, $c_1 = 2$, and $c_2 = 6$ and assume a Type B or Type A_L OC curve. If $p' = 0.06$, the chance of acceptance on the first sample (i.e., the chance of getting two or less nonconforming units) is 0.423, and the chance of rejection on the first sample (i.e., the chance of getting more than six nonconforming units) is 0.034. (See Section 1.1.1 above.) The probability of a decision on the first sample is 0.457, equal to the sum of 0.423 and 0.034, and the probability of having to take a second sample is $1 - 0.457 = 0.543$. With complete inspection of the second sample the *average* size sample is equal to the size of the first sample *times* the probability that there will only be one sample *plus* the size of the combined samples *times* the probability that a second sample will be necessary. For the given sampling plan, therefore, the average sample number with complete inspection of the second sample for a p' of 0.06 would be $50(0.457) + 150(0.543) = 104$. This is the ordinate of the average sample number curve for the abscissa point 0.06. Other points may be computed in the same way. The final results for the plan in question are shown in Figure. 8.2. The general formula for an average sample number curve of a double-sampling plan with complete inspection of the second sample is

(8.1) $$\text{ASN} = n_1P_1 + (n_1 + n_2)(1 - P_1) = n_1 + n_2(1 - P_1)$$

where P_1 is the probability of a decision on the first sample.

1.3. The ASN Curve with Curtailed Inspection

In practice, inspection of the second sample is usually terminated and the lot rejected as soon as the number of nonconforming items in the combined samples is found to exceed the acceptance number c_2. This is called curtailed inspection.

It will be noted that for the first sample of double sampling and the one sample used in single sampling, inspection is not generally curtailed. This is because it is desirable to have complete inspection of a fixed number of items in order to secure "for the record" an unbiased estimate of the quality of the material submitted for inspection. Such information will have

FIGURE 8.2
Average Sample Number Curves for Single- and Double-Sampling Inspection Plans

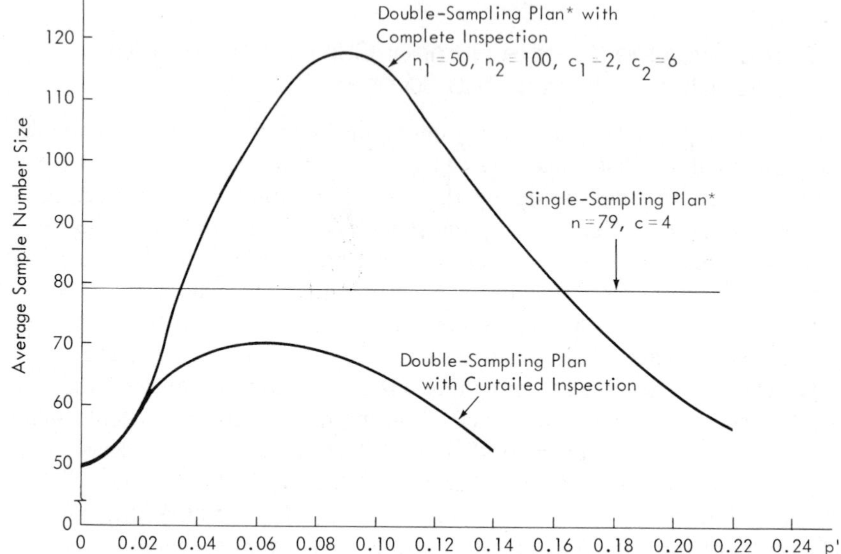

* All plans have approximately the same operating characteristic curve.

nothing to do directly with the operation of the sampling plan but is useful information to have for administrative purposes. If inspection was not complete, but was curtailed, estimates of quality could be badly biased. As an extreme case, suppose that in single sampling the rejection number was 4, and suppose the first 4 items happened to be defective. If inspection were stopped at this point, it could only be reported that 100 percent of the items inspected were nonconforming. Scarcely anyone could be convinced, however, that the whole lot was nonconforming.

The formula for the ASN curve for a double-sampling plan with curtailed inspection for Type B or Type A_L OC curves is as follows:[6]

$$\text{ASN} = n_1 + \sum_{k=c_1+1}^{c_2} P_{n_1:k}\Big[n_2 P''_{n_2:c_2-k} + \frac{c_2-k+1}{p'} P'_{n_2+1:c_2-k+2}\Big] \tag{8.2}$$

where

n_1 = size of first sample,
n_2 = size of second sample,

[6] See Irving W. Burr, "Average Sample Number under Curtailed or Truncated Sampling," *Industrial Quality Control,* February 1957, pp. 5–7.

c_1 = acceptance number for first sample,
c_2 = acceptance number for combined samples,
$P_{n:X}$ = probability of exactly X nonconforming items out of n,
$P'_{n:X}$ = probability of X or more nonconforming items out of n,
$P''_{n:X}$ = probability of X or less nonconforming items out of n,
and k is a summation variable.

The following table is used to compute the ASN curve for a double-sampling plan with curtailed inspection:

(1)	(2)	(3)	(4)	(5)	(6)	(7)	(8)
k	$c_2 - k$	$P_{n_1:k}$	$P''_{n_2:c_2-k}$	$P'_{n_2+1:c_2-k+2}$	$\frac{c_2-k+1}{p'}$	$n_2(4) + (5)(6)$	(3)(7)
c_1+1							
c_1+2							
.							
.							
.							
c_2							

From this table we have ASN = $n_1 + \Sigma(8)$.

To illustrate the use of the above table, the computation of the ordinate of the ASN curve at $p' = 0.08$ is shown in Table 8.1 for the sampling plan $n_1 = 50$, $n_2 = 100$, $c_1 = 2$, and $c_2 = 6$ with curtailed inspection. It will be noticed that this ordinate is much less than that for the same plan with complete inspection.

In Figure 8.2 a comparison is made between the average sample number curves with complete and curtailed inspection for the double-sampling plan $n_1 = 50$, $n_2 = 100$, $c_1 = 2$, and $c_2 = 6$, and the "average sample number curve" for a single-sampling plan with approximately the same operating characteristic curve. The "average sample number curve" for the single-sam-

TABLE 8.1
Computation of Ordinate of ASN Curve at $p' = 0.08$ for the Double-Sampling Plan $n_1 = 50$, $n_2 = 100$, $c_1 = 2$, and $c_2 = 6$ with Curtailed Inspection

(1) k	(2) $c_2 - k$	(3) $p_{n_1:k}$	(4) $P''_{n_2:c_2-k}$	(5) $P'_{n_2+1:c_2-k+2}$	(6) $\frac{c_2-k+1}{p'}$	(7) $n_2(4)$ + (5)(6)	(8) (3)(7)
3	3	0.195	0.04238	0.90483	50.0	49.48	9.6
4	2	0.195	0.01375	0.95984	37.5	37.37	7.3
5	1	0.156	0.00302	0.98737	25.0	24.99	3.9
6	0	0.104	0.00033	0.99719	12.5	12.50	1.4
							22.2

ASN (at $P' = 0.08$) = $n_1 + \Sigma(8) = 50 + 22.2 = 72.2$

pling plan is, of course, a horizontal straight line, since the same number is always taken in each sample. It will be noted that in Figure 8.2 the ASN curve for the double-sampling plan with curtailed inspection is below the ASN for the single-sampling plan over the whole range of p'.

Although reduction in the amount of inspection can thus be obtained by curtailment on the second sample, it should be emphasized that for material of the designated acceptable quality the main saving we expect to receive from double sampling is the ability to get by with samples of n_1, less than the n for single sampling.

1.4. Double-Sampling Plans with Specified p'_1, p'_2, α, and β

It is of interest to be able to derive a double-sampling plan that has a specified p'_1, p'_2, α, and β, just as was done in Chapter 7 for single sampling. With double sampling, however, the problem is more complicated. First of all, specification of the p'_1, p'_2, α, and β is not enough to yield a unique double-sampling plan. This means that we must draw upon other considerations in making a final decision about a plan. The popular procedure has been to assign fixed relationships between n_1 and n_2 that lead to good plans.

If we set n_2 equal to some constant multiple of n_1, it is possible to construct a table for double-sampling plans that is similar to Table 7.1. The reasoning runs as follows. If p' is small and we restrict ourselves to Type B and Type A_L OC curves, the probabilities involved in double sampling can be well approximated by the Poisson distribution. This means that with a fixed relationship between n_2 and n_1, the probabilities of acceptance for a given set of c_1 and c_2 values will be simply a function of $p'n_1$. Hence, plans for which the p'_2 and p'_1 bear a constant ratio can be made to have identical OC curves (within the approximation given by the Poisson distribution) by simply varying the n_1. To see this, suppose for a given plan that P_1 is the probability of acceptance for lots of p'_1 quality and P_2 is the probability of acceptance for lots of p'_2 quality. Then, if the p'_1 and p'_2 are *both* multiplied by a factor *a, their ratio being maintained,* and if at the same time n_1 is divided by a, the values of P_1 and P_2 (and in general the OC curve) will remain unchanged because the $p'n_1$'s are held constant.

Bearing the above in mind, we can derive a table of double-sampling plans for sets of values of c_1 and c_2, say $c_1 = 0$, $c_2 = 1$; $c_1 = 0$, $c_2 = 2$; $c_1 = 1$, $c_2 = 2$, and so on. For each of these plans we can, for an arbitrarily selected n_1 (and $n_2 = kn_1$), construct the OC curve and determine the values of p' for which the probabilities of acceptance are $1 - \alpha$ and β, respectively. Finally we can compute the ratio of these two values of p'. The result will be a table, similar to Table 7.1 for single sampling, from which we can derive a double-sampling plan given the ratio p'_2/p'_1.

The above procedure will lead to an extensive table of sampling plans, however, and not all plans will be equally good. To simplify matters we

TABLE 8.2
Values Useful in Deriving a Double-Sampling Plan with a Specified p'_1 and p'_2 ($n_2 = 2n_1$, $\alpha = 0.05$, and $\beta = 0.10$)

Plan Number	R*	*Acceptance Numbers*		*Approximate Values of* $p'n_1$ *for*			*Approximate* $(ASN)/n_1$ *for 0.95 Points*†
		c_1	c_2	P = 0.95	0.50	0.10	
(1)	*(2)*	*(3)*	*(4)*	*(5)*	*(6)*	*(7)*	*(8)*
1	14.50	0	1	0.16	0.84	2.32	1.273
2	8.07	0	2	0.30	1.07	2.42	1.511
3	6.48	1	3	0.60	1.80	3.89	1.238
4	5.39	0	3	0.49	1.35	2.64	1.771
5	5.09	1	4	0.77	1.97	3.92	1.359
6	4.31	0	4	0.68	1.64	2.93	1.985
7	4.19	1	5	0.96	2.18	4.02	1.498
8	3.60	1	6	1.16	2.44	4.17	1.646
9	3.26	2	8	1.68	3.28	5.47	1.476
10	2.96	3	10	2.27	4.13	6.72	1.388
11	2.77	3	11	2.46	4.36	6.82	1.468
12	2.62	4	13	3.07	5.21	8.05	1.394
13	2.46	4	14	3.29	5.40	8.11	1.472
14	2.21	3	15	3.41	5.40	7.55	1.888
15	1.97	4	20	4.75	7.02	9.35	2.029
16	1.74	6	30	7.45	10.31	12.96	2.230

* $R = p'_2/p'_1$.
† ASN is without curtailment on second sample.
SOURCE: These values are taken in part from the Chemical Corps Engineering Agency, Manual No. 2, *Master Sampling Plans for Single, Duplicate, Double and Multiple Sampling* (Army Chemical Center, Md., 1953), and adapted in part from H. C. Hamaker, "The Theory of Sampling Inspection Plans," *Philips Technical Review* 11 (1950), p. 266.

can eliminate plans that seem to be intuitively bad. For example, if n_2 is taken equal to $2n_1$, then it would seem to be intuitively good to have c_2 at least 3 times c_1. For if c_2 were less than 3 times c_1, it would mean that we would accept a lot on the first sample of n_1 items with a higher percentage of nonconforming items than we would accept on the combined sample of $3n_1$ items.[7]

Considerations such as the above underlie Table 8.2. Here $n_2 = 2n_1$, $\alpha = 0.05$, and $\beta = 0.10$. Plans are offered for which c_2 is at least 3 times c_1. Values of $p'n_1$ are given for three points on each of the OC curves and also a number from which the ASN without curtailed inspection can be computed for product quality for which the probability of acceptance of a lot is 0.95. The ASN with curtailed inspection on the second sample will be only slightly less than the tabular figure for this same product quality.

Table 8.3 is a table similar to Table 8.2 except that for this table n_1 is

[7] See Hamaker and Van Strik, "The Efficiency of Double Sampling for Attributes," *Journal of the American Statistical Association,* 50 (1955), pp. 830–49.

TABLE 8.3
Values Useful in Deriving a Double-Sampling Plan with a Specified p_1' and p_2' ($n_1 = n_2$, $\alpha = 0.05$, and $\beta = 0.10$)

Plan Number	R*	*Acceptance Numbers*		*Approximate Values of* p′n₁ *for*			*Approximate (ASN)/n₁ for 0.95 Point*†
		c_1	c_2	P = 0.95	0.50	0.10	
(1)	*(2)*	*(3)*	*(4)*	*(5)*	*(6)*	*(7)*	*(8)*
1	11.90	0	1	0.21	1.00	2.50	1.170
2	7.54	1	2	0.52	1.82	3.92	1.081
3	6.79	0	2	0.43	1.42	2.96	1.340
4	5.39	1	3	0.76	2.11	4.11	1.169
5	4.65	2	4	1.16	2.90	5.39	1.105
6	4.25	1	4	1.04	2.50	4.42	1.274
7	3.88	2	5	1.43	3.20	5.55	1.170
8	3.63	3	6	1.87	3.98	6.78	1.117
9	3.38	2	6	1.72	3.56	5.82	1.248
10	3.21	3	7	2.15	4.27	6.91	1.173
11	3.09	4	8	2.62	5.02	8.10	1.124
12	2.85	4	9	2.90	5.33	8.26	1.167
13	2.60	5	11	3.68	6.40	9.56	1.166
14	2.44	5	12	4.00	6.73	9.77	1.215
15	2.32	5	13	4.35	7.06	10.08	1.271
16	2.22	5	14	4.70	7.52	10.45	1.331
17	2.12	5	16	5.39	8.40	11.41	1.452

* $R = p_2'/p_1'$.
† ASN is without curtailment on second sample.
SOURCE: Taken from Chemical Corps Engineering Agency, Manual No. 2, *Master Sampling Plans for Single, Duplicate, Double and Multiple Sampling* (Army Chemical Center, Md., 1953).

taken equal to n_2 and c_2 is at least twice c_1. The reader may find it of interest to compare the two tables. In making the comparison it should be noted that the ASN in the tables is computed without allowing for curtailment on the second sample.

An example will illustrate the use of Table 8.2. Suppose that we want a double-sampling plan with the $p_1' = 0.01$, the $p_2' = 0.08$, $\alpha = 0.05$, and $\beta = 0.10$. Then we have $R = p_2'/p_1' = \dfrac{0.08}{0.01} = 8$. Plan 2 in Table 8.2 has an R nearest to this. For this $c_1 = 0$, $c_2 = 2$. Deciding to hold the p_2', β point, we find that n_1 should equal $\dfrac{(p'n_1)_{0.10}}{p_2'} = \dfrac{2.42}{0.08} = 30$. This gives $n_2 = 2(30) = 60$. The desired criteria are therefore those approximated by the double-sampling plan $n_1 = 30$, $n_2 = 60$, $c_1 = 0$, $c_2 = 2$. The 0.95 point will fall at about $p' = 0.010$, the 0.50 point at about $p' = 0.036$ and the 0.10 point at about $p' = 0.081$. The ASN without curtailment on the second sample will be about $(30)(1.511) = 45$ items for lots of 0.01 quality.

It will be noted that Table 8.2 gives only approximate values based on the Poisson approximation. At the p'_2 point in particular, actual probabilities of acceptance may deviate from 0.10 by 1 to 2 points.

1.5. Other Tables of Double-Sampling Plans

Paul Peach in *An Introduction to Industrial Statistics and Quality Control* (2d ed.; Raleigh, N.C.: Edwards & Broughton Co., 1947), gives tables of double-sampling plans for which c_3 does not equal c_2. They are indexed by the p'_2/p'_1 ratio and have $\alpha = 0.05$, $\beta = 0.05$.

Still other double-sampling plans will be found in Statistical Research Group, Columbia University, *Sampling Inspection,* and in its successors, Military Standard 105D, ANSI/ASQC Z1.4 and ISO Std. 2859. The latter are discussed in Chapter 10.

A third set of double-sampling plans will be found in H. F. Dodge and H. G. Romig, *Sampling Inspection Tables.* These are part of rectifying inspection programs and will be discussed in Part 3. They may be used, however, as simple acceptance sampling plans if so desired.

2. ITEM-BY-ITEM SEQUENTIAL-SAMPLING PLANS

In single sampling the number of items sampled is definitely fixed by the plan. In double sampling, it was seen that the number of items sampled is fixed in part by the plan but is also determined in part by the results of the sampling process. If a lot is accepted or rejected on the first sample, there is no need to take a second sample. It was because of this that double sampling was found to offer the possibility of reduced sampling costs.

Success with double sampling suggested that an even greater reduction in cost might be attained if we took a sequence of samples and allowed the number of samples to be determined entirely by the results of the sampling process. This fruitful suggestion led to a whole new approach to sampling, which is known as sequential sampling.

Historically, there seem to have been at least three independent developments of sequential sampling.[8] The one best known in the United States was that developed by A. Wald when he was a member of the Statistical Research Group organized at Columbia University during World War II. Another was that developed by G. A. Barnard in England. Both of these men were concerned with sequential sampling that allows for a decision each time a new sample item is inspected or what is called item-by-item sequential sampling. A third development was by W. Bartky. This was along the lines of multiple sampling or "group" sequential sampling. In this chapter we shall be concerned primarily with the work of A. Wald.

Item-by-Item sequential sampling is based fundamentally upon the notion

[8] See references at end of this chapter.

of a "random walk." Suppose, for example, that two gamblers, A and B, each have a capital of $10. They agree to play the following game. A perfectly unbiased coin is to be tossed in a random manner. If the coin turns up heads, A pays B $1. If the coin turns up tails, B pays A $1. They agree to play until either one has lost all his money to the other. If the coin were biased, the game could still be played, but the outcomes, A's ruin or B's ruin, would not now be equally likely.

In item-by-item sequential sampling, we play the same kind of game. As we sample, we rack up a plus value, say, each time a sample item is good and a minus value each time the item is bad. If at any time the score gets above a certain value, we stop sampling and accept the lot. If at any time the score gets below a certain value, we reject the lot. A special feature of the Wald approach is the manner in which the parameters of the game are determined.

2.1. Wald's Sequential Probability Ratio Plan

Wald starts with a given p'_1, p'_2, α, and β, and determines a scoring method and acceptance and rejection boundaries that will satisfy these requirements at the same time that it yields an especially efficient sampling procedure. Wald adopts the "sequential probability ratio" as a good method of keeping score. This is the ratio of the probability of getting the cumulated sample result if material were of p'_2 quality to the probability of getting this result if material were of p'_1 quality. If this score at any time gets larger than a given figure determined by the assigned α and β, the lot is rejected. If it falls below a second figure, again determined by α and β, the lot is accepted. Note that under this scheme, high score leads to rejection, low score to acceptance.

Actually it is not necessary to compute the sequential probability ratio (SPR) in practice, for Wald showed that the SPR scoring system can be reduced to a much simpler system which consists merely in plotting the cumulated number of nonconforming items on a chart or entering the same on an inspection form. A typical chart is Figure 8.3. If the graphic method is employed, the cumulated sample results are plotted successively on the chart. For each point the abscissa is the total number of units drawn up to that time and the ordinate is the total number of these units that are nonconforming. If the plotted points stay within a zone marked by two successive parallel lines, the sampling is continued without a decision. As soon as a point falls *on or above* the upper line the lot is rejected. As soon as a point falls *on or below* the lower line, the lot is accepted. Figure 8.4 indicates how the same game can be played on an inspection form instead of a chart.

In the following section we shall describe how such a plan is derived so as to yield a specified p'_1, p'_2, α, and β. In other sections we shall give formulas for its operating characteristic curve and its average sample number curve.

FIGURE 8.3
Sequential-Sampling Chart

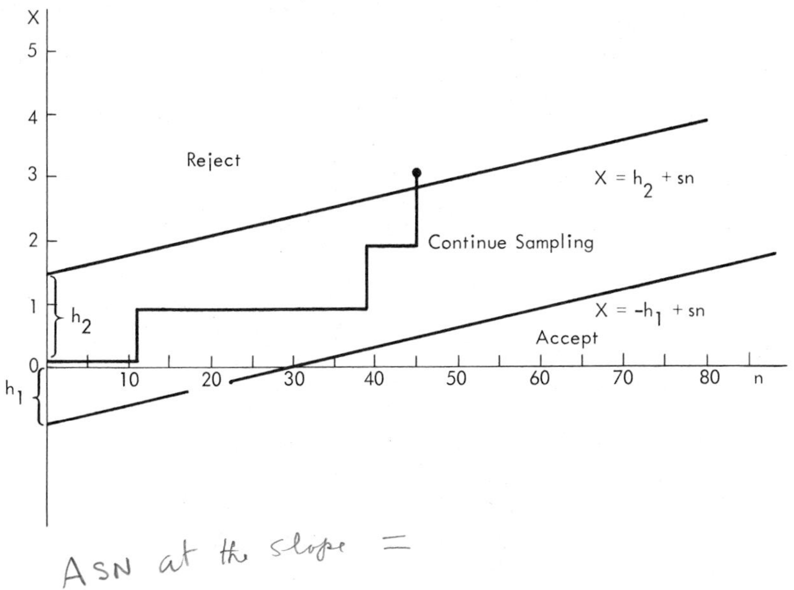

FIGURE 8.4
Illustration of a Table of Acceptance and Rejection Numbers for the Sequential-Sampling Plan $p_1' = 0.01$, $p_2' = 0.08$, $\alpha = 0.05$, $\beta = 0.10$

n	A	R	n	A	R	n	A	R	n	A	R
			19	*	2	37	0	3	55	0	4
2	*	2	20	*	3	38	0	3	56	0	4
3	*	2	21	*	3	39	0	3	57	0	4
4	*	2	22	*	3	40	0	3	58	0	4
5	*	2	23	*	3	41	0	3	59	0	4
6	*	2	24	*	3	42	0	3	60	0	4
7	*	2	25	*	3	43	0	3	61	1	4
8	*	2	26	*	3	44	0	3	62	1	4
9	*	2	27	*	3	45	0	3	63	1	4
10	*	2	28	*	3	46	0	3	64	1	4
11	*	2	29	*	3	47	0	3	65	1	4
12	*	2	30	*	3	48	0	3	66	1	4
13	*	2	31	0	3	49	0	4	67	1	4
14	*	2	32	0	3	50	0	4	68	1	4
15	*	2	33	0	3	51	0	4	69	1	4
16	*	2	34	0	3	52	0	4	70	1	4
17	*	2	35	0	3	53	0	4			
18	*	2	36	0	3	54	0	4			

* No acceptance until 31 items have been inspected.

2.1.1. Derivation of an SPR Plan with a Specified p_1', p_2', α, and β. It is shown in Appendix I (25) that the equations for the two limit lines that will give the desired p_1', p_2', α, and β are

$$X = -h_1 + sn$$
(8.3)
$$X = h_2 + sn$$

where

(8.4)
$$\begin{cases} h_1 = \log \dfrac{1-\alpha}{\beta} \Big/ \log \left[\dfrac{p_2'(1-p_1')}{p_1'(1-p_2')}\right] \\ h_2 = \log \dfrac{1-\beta}{\alpha} \Big/ \log \left[\dfrac{p_2'(1-p_1')}{p_1'(1-p_2')}\right] \\ s = \dfrac{\log \left[\dfrac{1-p_1'}{1-p_2'}\right]}{\log \left[\dfrac{p_2'(1-p_1')}{p_1'(1-p_2')}\right]} \end{cases}$$

In Statistical Research Group, Columbia University, *Sequential Analysis of Data: Applications,* pages 2.39–2.42, tables will be found that give h_1, h_2, and s for $\alpha = 0.05$ and $\beta = 0.10$ and 0.50 for values of $p_1' = 0.0002$ to 0.10 and values of p_2' from 0.002 to 0.35. Selected sections of this table are reproduced in Table R of Appendix II. (It is to be carefully noted that the symbol "s" is used in sequential sampling to represent the slope of the lines of acceptance and is in no way to be taken to represent a standard deviation as in Chapter 2 and elsewhere.)

2.1.2. The OC Curve for an SPR Sampling Plan. Type B and Type A_L OC curves for an SPR plan can be sketched in from three points. The plan is so designed, it will be recalled, that the probability of acceptance for p' equal to p_1' is $1 - \alpha$ and the probability of acceptance for p' equal to p_2' is β. This gives two points on the curve. It can also be proved[9] that s always lies between p_1' and p_2', and the probability of acceptance for a lot fraction nonconforming equal to s is[10] $\dfrac{h_2}{h_1 + h_2}$. This gives a third point on the curve near the middle. Further points can be obtained from the parametric equations:[11]

Product or lot quality
$$p' = \frac{1 - \left(\dfrac{1-p_2'}{1-p_1'}\right)^{\theta}}{\left(\dfrac{p_2'}{p_1'}\right)^{\theta} - \left(\dfrac{1-p_2'}{1-p_1'}\right)^{\theta}}$$

---

[9] See Appendix I (26).

[10] Statistical Research Group, Columbia University, *Sequential Analysis of Statistical Data: Applications* (New York: Columbia University Press, 1945), p. 2.48.

[11] A. Wald, *Sequential Analysis* (New York: John Wiley & Sons, 1947), p. 51.

(8.5)

$$\text{Probability of acceptance} \quad P_a = \frac{\left(\frac{1-\beta}{\alpha}\right)^{\theta} - 1}{\left(\frac{1-\beta}{\alpha}\right)^{\theta} - \left(\frac{\beta}{1-\alpha}\right)^{\theta}}$$

These give, for arbitrarily selected values of θ, the p' and P_a coordinates of points on the OC curve. For $\theta = 1$, we have $p' = p'_1$, and for $\theta = -1$ we have $p' = p'_2$. For $\theta = 0$, we have $p' = s$. Hence, $\theta = \pm 0.5$ will give two good intermediate points.

2.1.3. The ASN Curve. The average sample number curve can be sketched from several points.[12] At $p' = 0$, the ASN $= h_1/s$; at $p' = 1$, ASN $= h_2/(1 - s)$. These are the two extreme points on the curve. They are also minimum points. The curve rises to a maximum at some point between p'_1 and p'_2.

(8.6)

$$\begin{cases} \text{At } p' = p'_1, & \text{the ASN} = \dfrac{(1-\alpha)h_1 - \alpha h_2}{s - p'_1} \\[2ex] \text{At } p' = p'_2, & \text{the ASN} = \dfrac{(1-\beta)h_2 - \beta h_1}{p'_2 - s} \\[2ex] \text{At } p' = s, & \text{the ASN} = \dfrac{h_1 h_2}{s(1-s)} \end{cases}$$

The Statistical Research Group, Columbia University, gives these five points on the ASN curve for the various plans listed in their table (also see Table R of Appendix II). Use of this table will permit the sketching of an ASN curve without further computation. If desired, additional points can be obtained from the formula.[13]

(8.7)

$$\text{ASN} = \frac{P_a \log \dfrac{\beta}{1-\alpha} + (1 - P_a) \log \dfrac{1-\beta}{\alpha}}{p' \log(p'_2/p'_1) + (1-p')\log\left[(1-p'_2)/(1-p'_1)\right]}$$

As indicated above, the ASN curve for an SPR sampling plan is especially low at the extremes and may therefore lead to considerable saving in costs. In fact, Wald and Wolfowitz have shown that no other attributes sampling plan with the same two points on the OC curve can have a smaller ASN at those points.[14] For intermediate points, however, it is possible (but not likely) that SPR sampling will give larger samples on the average than a single- or double-sampling plan. In general, it has been estimated that SPR

---

[12] Statistical Research Group, Columbia University, *Sequential Analysis of Statistical Data: Applications,* pp. 2.64–2.65.

[13] Wald, *Sequential Analysis,* p. 53.

[14] A. Wald and J. Wolfowitz, "Optimum Character of the Sequential Probability Ratio Test," *Annals of Mathematical Statistics* 19 (1948), pp. 326–39.

sampling may reduce the ASN to approximately 50 percent of that required by a comparative single-sampling plan.[15]

2.2. Barnard's Approach to Sequential Sampling

Whereas Wald started with the sequential probability ratio and derived a sequential-sampling chart with boundary lines determined by specified values of p'_1, p'_2, α, and β, Barnard[16] started with the boundary lines and indicated how we might compute the OC curve that followed therefrom. His scoring system is also somewhat different. It consists of scoring 1 for each good item and $-b$ for each bad one. The score starts at 0; when it rises to or exceeds a certain value H, the lot is accepted; when it equals or falls below a second value $-H'$, the lot is rejected. Some feel that this method of scoring is administratively simpler than the Wald method. The "parameters" of the Wald and Barnard procedures are related as follows:

$$b = \frac{1-s}{s} \text{ or } s = \frac{1}{b+1}$$

(8.8)
$$H = h_1/s \text{ or } h_1 = \frac{H}{b+1}$$

$$H' = h_2/s \text{ or } h_2 = \frac{H'}{b+1}$$

Following Barnard's method, F. J. Anscombe has worked out tables of sequential sampling plans that are indexed with respect to[17] simple values of h_1 and h_2. Each pair of values determines a special class of OC curves, a particular member of which is determined by the value assigned to s. These tables give data with which to sketch the OC curves and ASN curves for any given values of s. Likewise the ratios of p'_α to p'_β are given for several sets of values of α and β, so that plans of the Barnard type can be found for specified p'_2/p'_1 ratios and specified α and β.

3. GROUP SEQUENTIAL- AND MULTIPLE-SAMPLING PLANS

Sometimes it is found more practical to take sequences of groups of items instead of sequences of items. The ordinary sequential chart may be set up,

---

[15] A. Wald, *Sequential Analysis,* p. 57.

[16] G. A. Barnard, "Sequential Tests in Industrial Statistics," *Journal of the Royal Statistical Society* Ser. B, 8 (1946), pp. 1–21.

[17] Anscombe uses $R_1 = h_1$ and $R_2 = h_2$. See "Tables of Sequential Inspection Schemes to Control Fraction Defective," *Journal of the Royal Statistical Society* Ser. A. 107, Part 2 (1949), pp. 180–206.

but, instead of plotting points item by item, the results are cumulated by groups. For example, the results of the first 20 items are plotted, then the results of the first 40, then the first 60, and so forth. Decisions are made as before, but at less "frequent" intervals in terms of the inspection process.

Using group methods affects[18] the OC curve and the ASN. The former is not affected very much and under certain conditions may not be affected at all. The average sample number is very likely to be increased, however; for taking groups means that a decision that might be made on the kth item inspected will have to be put off until the whole group is inspected (unless the kth itme is the last in the group). It may also mean that a decision will be postponed to later groups, which will increase the amount of inspection still further. Despite its disadvantages, practical considerations in the plant may make group sequential sampling preferable to item-by-item sequential sampling.

If h_1 and h_2 are integers—say, each equals 1—and the size of the group is taken equal to $1/s$ or to the integer just greater than $1/s$, if this is not an integer, group sequential sampling will give (approximately if $1/s$ is not an integer) the same results as item-by-item sequential sampling; for in this case no acceptance can be made on the basis of any group unless the whole group has been found effective, and no acceptance will be possible with a smaller sized group than $1/s$. (It will be recalled that s is the slope of the limit lines and represents the increase in n corresponding to a rise of 1 in X.) If h_1 and h_2 are close to, but not exactly, integers, good results will still be given by taking the group size equal to $1/s$ or to the integer just greater than $1/s$ if this is not an integer. For this reason, $1/s$ or the integer just greater than $1/s$, if $1/s$ is not an integer, is a preferable size group.

Group sequential plans are often truncated. That is, after a number of samples have been taken, the plan calls for acceptance or rejection. They then become "multiple"-sampling plans. These are discussed in the next chapter.

4. PROBLEMS

8.1. Make a graph of the principal and supplementary OC curves (of Type A_L or B) for the double-sampling plan $n_1 = 50$, $n_2 = 100$, $c_1 = 2$, and $c_2 = 7$. If the process average equals 0.05, what is the probability that a lot will be accepted on the first sample? Rejected on the first sample? Finally accepted?

8.2. Make a graph of the principal and supplementary OC curves (of Type A_L or B) for the double-sampling plan $n_1 = 50$, $n_2 = 50$, $c_1 = 2$, and $c_2 = 5$. If the process average equals 0.05, what is the probability that a lot will be accepted on the first sample? Rejected on the first sample? Finally accepted?

---

[18] Wald, *Sequential Analysis*, pp. 101–3.

8.3. Repeat Problem 8.1 for Type A_s OC curves assuming that the lot size is 500.

8.4. Repeat Problem 8.2 for Type A_s OC curves assuming that the lot size is 250.

8.5. Assume that in the plan of Problem 8.1, inspection of the second sample is stopped as soon as the rejection number is reached, but that the lot is not accepted until the whole of the second sample is inspected. Derive the ASN curve for this curtailed double-sampling plan. Compare the results with the ASN of a single-sampling plan that has approximately the same OC curve as that of the double-sampling plan. (*Hint:* Pick the p'_1, α point and the p'_2, β point from the OC curve for the double-sampling plan and find a single-sampling plan, the OC curve of which runs through these points.)

8.6. Repeat Problem 8.5 for the plan of Problem 8.2.

8.7. Find a double-sampling plan that will come close to having $p'_1 = 0.03$, $p'_2 = 0.10$, $\alpha = 0.05$, and $\beta = 0.10$. Compare with Problem 7.5.

8.8. Find a double-sampling plan that will come close to having $p'_1 = 0.005$, $p'_2 = 0.05$, $\alpha = 0.05$, and $\beta = 0.10$. Compare with Problem 7.9.

8.9. Find a double-sampling plan that will come close to having $p'_1 = 0.03$, $p'_2 = 0.08$, $\alpha = 0.05$, and $\beta = 0.10$. Compare with Problem 7.11.

8.10. Find a double-sampling plan that will come close to having $p'_1 = 0.015$, $p'_2 = 0.04$, $\alpha = 0.05$, and $\beta = 0.10$. Compare with Problem 7.12.

8.11. *a.* Construct an item-by-item sequential-sampling plan that will have $p'_1 = 0.02$, $p'_2 = 0.08$, $\alpha = 0.05$, and $\beta = 0.10$. Do this both graphically and in tabular form.

b. An inspector tests 40 units without finding a single nonconforming one. Under the plan in *a,* would he have decided to accept the lot before reaching the 40th unit, or would he have to continue further before reaching a decision? Would he have come to a decision to reject the lot if he had found the 10th, 18th, and 23d units nonconforming?

c. Sketch the OC curve for this plan.

d. Sketch the ASN curve for this plan. How does this compare with a single-sampling plan with the same p'_1 and p'_2?

8.12. *a.* Repeat Problem 8.11, *a, c, d,* omit *b,* using $p'_1 = 0.01$, $p'_2 = 0.04$, $\alpha = 0.05$, and $\beta = 0.10$.

b. An inspector tests 70 units without finding a single nonconforming one. Under the plan derived in *a,* would he have decided to accept the lot before reaching the 70th unit, or would he have to continue sampling before making a decision? Would he have come to a decision to reject the lot if he had found the 20th, 35th, and 65th units nonconforming?

8.13. *a.* Devise a group sequential-sampling plan that comes close to having the same OC curve as the item-by-item sequential-sampling plan of Problem 8.11.

b. "Truncate" the plan in *a* after the seventh sample; i.e., modify the plan so that some decision is made on the eighth or ninth sample.

8.14. Repeat Problem 8.13, using the item-by-item sequential-sampling plan of Problem 8.12.

5. SELECTED REFERENCES*

Anscombe (P '49), Barnard (P '46), Bartky (P '43), Bowker and Lieberman (B '55 and B '72), Burr (B '76 and P '57), Champernowne (P '53), Cowden (B '57), Deming (B '50), Dodge and Romig (B '59), Grant and Leavenworth (B '80), Hamaker (P '50 and P '53), Hamaker and Van Strik (P '55), Jackson (P '60), Maghsoodloo and Bush (P '85), Mood and Graybill (B '63), Peach (P '45 and B '47), Schilling (B '81), Shafer (P '46), Statistical Research Group, Columbia University (B '45 and B '48), U.S. Dept. of Army, Chemical Corps Engineering Agency (B '53, Manual No. 2), Vogholhar and Wetherill (P '60), Wald (B '47), Wald and Wolfowitz (P '48), Weiss (P '62) and Wetherill (P '61).

---

* B and P refer to the Book and Periodical sections, respectively, of the Cumulative List of References in Appendix V.

9

Acceptance Sampling by Attributes: Multiple Fraction-Nonconforming Sampling Plans

1. METHODS OF DESCRIBING MULTIPLE-SAMPLING PLANS

If we follow the Wald approach to item-by-item sampling and the "group sequential" sampling derived therefrom,[1] a multiple-sampling plan will be described in terms of acceptance and rejection boundaries for accumulated groups of samples. Such a description of a multiple-sampling plan will run like this:

Cumulated Sample Size	*Acceptance Number*	*Rejection Number*
20	0	4
40	1	5
60	3	6
80	5	8
100	8	10
120	9	11
140	10	11

The plan will operate as follows. If, at the completion of any stage, the number of nonconforming items equals or falls below the acceptance number, the lot is accepted. If, during any stage, the number of nonconforming items

[1] See Chapter 8.

equals or exceeds the rejection number, the lot is rejected. Otherwise another sample is taken. This multiple decision procedure continues until the seventh sample is taken when a decision to accept or reject must be made. The first sample is usually inspected 100 percent for the sake of the record, but inspection is often stopped as soon as the rejection number is reached in any stage subsequent to the first.

Enters and Hamaker,[2] following the lead of Barnard, have developed an approach to multiple sampling that is thought by some to be administratively simpler than that following the Wald line of development. The Barnard-Enters-Hamaker way of describing a multiple-sampling plan runs as follows. Start with an initial score I. Add S for each group of n items inspected and subtract the number of nonconforming items found. Accept the lot if at any stage the cumulated score reaches or exceeds A; reject the lot if the cumulated score becomes 0 or less. At the end of m stages, accept the lot if the final score is greater than the initial score; otherwise reject the lot. The multiple-sampling plan

$$M(I = 2,\ S = 2, A = 4, m = 8)$$

will thus have the following scoring system. Start with a score of 2, add 2 for each group inspected and subtract the number of nonconforming units found in the group. The lot is accepted when the total score reaches 4; it is rejected when the total score becomes 0 or less. At the end of 8 groups, if the score is greater than 2, accept; otherwise reject.

It is claimed that inspectors find the Barnard-Enters-Hamaker way of describing a plan easier to follow than one prescribing a series of acceptance and rejection numbers. It should be noted, however, that this simpler method of describing a multiple-sampling plan can readily be converted into the more conventional form. Thus the above multiple-sampling plan M (2,2,4,8) can be rewritten as follows:

Cumulated Sample Size	*Acceptance Number*	*Rejection Number*
n	0	4
$2n$	2	6
$3n$	4	8
$4n$	6	10
$5n$	8	12
$6n$	10	14
$7n$	12	16
$8n$	15	16

[2] J. H. Enters and H. C. Hamaker, "Multiple Sampling in Theory and Practice." (Privately reprinted in 1951 by the Royal Statistical Society from *Statistical Method in Industrial Production, Papers given at a conference held in Sheffield in 1950.*)

2. COMPUTATION OF THE OC CURVE FOR A MULTIPLE-SAMPLING PLAN

The following is an explanation of the computation of a Type B or Type A_L OC curve for a multiple-sampling plan described in the conventional manner. The process is simple enough and can be done readily if an orderly procedure is followed. The steps consist in calculating the probability of accepting, rejecting, and continuing to sample at each stage in the plan.[3] We shall illustrate the method with reference to the sampling plan of Figure 9.1.

Step 1. Note that P_a is to be calculated at $p' = 0.05$. For this value of p' write on a strip of paper the probabilities of exactly 0 out of 20 (= 0.3585), 1 out of 20 (= 0.3774), 2 out of 20 (= 0.1887), and 3 out of 20 (= 0.0596); the probabilities of 1 or more out of 20 (= 0.6415), 2 or more out of 20 (= 0.2642), 3 or more out of 20 (= 0.0755), and 4 or more out of 20 (= 0.0159); and the probabilities of 1 or less out of 20 (= 0.7359) and 2 or less out of 20 (= 0.9246) (all evaluated at $p' = 0.05$). These may be found in tables of the binomial probability distribution.[4]

Step 2. Set up a work sheet such as shown in Figure 9.1 with heavy lines drawn around the acceptance values and rejection values. Be sure there is enough room in each cell to write in as many numbers as the difference between the acceptance and rejection numbers.

Step 3. For the first stage write in the probabilities of the results that lead to acceptance, rejection, and continuance of sampling. Note that the value written above the rejection limit is the probability of equaling or exceeding the rejection number and that written below the acceptance limit is the probability of equaling or falling below the acceptance number.

Step 4. For *each* of the results of the first stage that lead to a second sample, compute the probability of acceptance and rejection at the second stage and also the probability of each of the results that will lead to a third stage. To carry out these computations it was found easiest to set the probability of exactly 1 in the first sample (0.3774) in the computing machine as a multiplicand and to multiply successively by the probability of exactly 0, the probability of exactly 1 (giving a total of 2 at the second stage), the probability of exactly 2 (giving a total of exactly 3), the probability of exactly 3 (giving a total of exactly 4), and the probability of 4 or more. Thus $0.3774 \times 0.3585 = 0.1353$ which is the probability of exactly 1 at the second stage and hence equals the probability of acceptance at the second stage. Again $0.3774 \times 0.3774 = 0.1424$ which is the probability of exactly 2 at the second stage; $0.3774 \times 0.1887 = 0.0712$ which is the probability of exactly 3 at the second stage; $0.3774 \times 0.0596 = 0.0225$ which is the probability of exactly

[3] See Statistical Research Group, Columbia University, *Sampling Inspection* (New York: McGraw-Hill, 1948), pp. 189 ff.; and Dudley J. Cowden, *Statistical Methods in Quality Control* (Englewood Cliffs, N.J.: Prentice-Hall, 1957), Chapter 37.

[4] See, for example, the National Bureau of Standards' Table.

FIGURE 9.1
Work Sheet for Computing a Point on the OC Curve for a Multiple-Sampling Plan (illustration shows the calculation of P_a at $p' = 0.05$ for the multiple-sampling plan given below)

Number of Nonconforming Items							
11						.0019	.0017
10					.0016 .0008 .0024	.0027	.0010
9					.0041 .0031 .0072	.0026	
8				.0056 .0021 .0077	.0160 .0485 .0645		
7				.0139 .0078 .0217			
6			.0213 .0124 .0033 .0370	.0279 .0246 .0525			
5		.0157 .0142 .0060 .0359	.0304 .0309 .0125 .0738	.0265 .0959 .1224			
4	.0159	.0225 .0356 .0225 .0806	.0289 .0618 .0396 .1303				
3	.0596	.0214 .0712 .0712 .1638	.0587 .1545 .2132				
2	.1887	.0676 .1424 .2100					
1	.3774	.1353					
0	.3585						

MULTIPLE-SAMPLING PLAN

Cumulated Sample Size	*Acceptance Nos.*	*Rejection Nos.*
20	0	4
40	1	5
60	3	6
80	5	8
100	8	10
120	9	11
140	10	11

Probability of—

Exactly 0 out of 20 = 0.3585
Exactly 1 out of 20 = 0.3774
Exactly 2 out of 20 = 0.1887
Exactly 3 out of 20 = 0.0596

. . . .

1 or more out of 20 = 0.6415
2 or more out of 20 = 0.2642
3 or more out of 20 = 0.0755
4 or more out of 29 = 0.0159
1 or less out of 20 = 0.7359
2 or less out of 20 = 0.9246

Probability of Acceptance

=0.3585
+0.1353
+0.2132
+0.1224
+0.0645
+0.0026
+0.0010
=0.8975

4 at the second stage; and $0.3774 \times 0.0159 = 0.0060$ which is the probability of 5 or more at the second stage and is thus a component of the probability of rejection at the second stage—all starting with exactly one at the first stage. Next we take 0.1887 and multiply it successively by 0.3585, 0.3774, 0.1887, and 0.0755 to get 0.0676, 0.0712, 0.0356, and 0.0142 which are the probabilities that starting with exactly 2 at the first stage we shall get exactly 2, 3, 4, or 5 or more at the second stage. Finally we multiply 0.0596 by 0.3585, 0.3774, and 0.2642 to get 0.0214, 0.0225, and 0.0157 which are the probabilities that starting with exactly 3 at the first stage we shall get 3, 4, or 5 or more at the second stage. The probabilities are then added up in each cell to yield a probability of exactly 2 at the second stage = 0.2100, a probability of exactly 3 at the second stage = 0.1638, a probability of exactly 4 at the second stage = 0.0806, and a probability of 5 or more = 0.0359, which is the probability of rejection at the second stage.

Subsequent Stages. The process is repeated at each succeeding stage until we have run through all seven stages. Then the total probability of acceptance at $p' = 0.05$ is obtained by adding up all the probabilities of acceptance at the individual stages. This is done in Figure 9.1.

3. COMPUTATION OF THE ASN CURVE FOR A MULTIPLE-SAMPLING PLAN

If inspection is not curtailed when a lot is rejected prior to completion of a given stage of sampling, the ASN of a multiple-sampling plan is given by the formula

(9.1) $$\text{ASN} = P_1 n_1 + P_2(n_1 + n_2) + \ldots + P_k(n_1 + n_2 + \ldots + n_k)$$

where n_i is the size of the ith group in the multiple sampling plan (in this book the n_i's are all equal) and P_i is the probability of a decision at each stage. The ASN with curtailment is slightly less than this but not materially so. The P_i's will be obtained from the probabilities computed in deriving the OC curve and are, of course, functions of the lot or product quality p'.

For the sampling plan for which a point on the OC curve was computed in the previous section, we have the ASN at $p' = 0.05$ equal to

$$\begin{aligned}&(0.3585 + 0.0159)(20) \qquad + (0.1224 + 0.0077)(80)\\ &+(0.1353 + 0.0359)(40) \qquad + (0.0645 + 0.0024)(100)\\ &+(0.2132 + 0.0370)(60) \qquad + (0.0026 + 0.0019)(120)\\ &+(0.0010 + 0.0017)(140) = 47.72\end{aligned}$$

4. DESIGN OF A MULTIPLE-SAMPLING PLAN WITH A SPECIFIED p_1', p_2', α, AND β

Although there are tables of multiple-sampling plans in Military Standard 105D, ANSI/ASQC Z1.4, and ISO Std. 2859 (see Chapter 10) and a similar set in the Statistical Research Group's *Sampling Inspection,* none of these is indexed with reference to the ratio of the p'_2 to the p'_1. There are two tables of multiple-sampling plans, however, that are so indexed. These are the Army Chemical Corps' *Master Sampling Plans for Single, Duplicate, Double and Multiple Sampling*[5] and tables offered by Enters and Hamaker in their article on "Multiple Sampling in Theory and Practice."[6] Selections from the former are presented in Table 9.1 (p. 210) and the latter are shown in Table 9.2 (p. 211). The plans of Table 9.1 are indexed according to the ratio $p'_{0.10}/p'_{0.95}$ and those of Table 9.2 according to $p'_{0.05}/p'_{0.95}$. In Table 9.1 plans are described in conventional form; in Table 9.2, in the Barnard-Enters-Hamaker form. Values in both tables are based on the Poisson approximation.

In column (2) of Table 9.1 is listed the ratio of the p'_2 to the p'_1 of each plan (given $\alpha = 0.05$ and $\beta = 0.10$), in column (3) are given the acceptance and rejection numbers, in columns (4) to (6) are values to be used in computing three points on the OC curve, and in column (7) values to be used in computing the ASN for p' equal to the p'_1. The use of the table may be explained by an example. Suppose, as previously, that we desire a plan with a $p'_1 = 0.01$ and a $p'_2 = 0.08$ with $\alpha = 0.05$ and $\beta = 0.10$. We then have $R = p'_2/p'_1 = 8$. The plan in Table 9.1 with a ratio closest to this is plan No. 5, for which the ratio is $8.06 \doteq 8$. We can then readily find a plan with the 0.95 point approximately at 0.01 and the 0.10 point approximately at 0.08. The values given in the table are "unity values" based on the Poisson approximation. The group size for the plan will be given by dividing the unity value for the 0.95 point by the specified p'_1 (i.e., by 0.01) or by dividing the unity value for the 0.10 point by the specified p'_2 (i.e., by 0.08). It makes little difference in this case since R is close to 8. Suppose that we use the p'_1 point. Then we get $0.093/0.01 = 9.3$. To be conservative we shall use a group size of 10. Hence the plan is as follows:

Cumulated Sample Size	*Acceptance Numbers*	*Rejection Numbers*
10	*	2
20	*	2
30	0	2
40	0	3
50	0	3
60	0	3
70	0	3
80	1	3
90	2	3

* Acceptance not permitted at this sample size.

[5] See Cumulative List of References.

[6] See Cumulative List of References.

TABLE 9.1
Values Useful in Designing a Multiple-Sampling Plan*, $\alpha = 0.05$, $\beta = 0.10$

Number of Plan	R†	*Acceptance and Rejection Numbers*	*Approx. Values of* p′n$_i$ *for* $P_a = 0.95$	0.50	0.10	*(ASN)/*n$_i$ *at Approx.* $P_a = 0.95$
(1)	*(2)*	*(3)*	*(4)*	*(5)*	*(6)*	*(7)*
1	18.46	Ac * * 0 0 1 2 3	.048	.38	.89	3.243
		Re 2 2 2 2 3 4 4				
2	12.15	Ac * * * 0 0 1 2	.065	.31	.79	4.373
		Re 2 2 2 2 2 3 3				
3	9.95	Ac * * 0 0 1 2 4	.10	.43	1.00	3.461
		Re 2 2 2 3 3 4 5				
4	8.91	Ac * * 0 0 0 0 0 2	.088	.34	.78	3.876
		Re 2 2 2 2 2 3 3 3				
5	8.06	Ac * * 0 0 0 0 0 1 2	.093	.36	.75	4.077
		Re 2 2 2 3 3 3 3 3 3				
6	7.04	Ac * 0 0 1 1 1 2 3	.18	.62	1.27	2.828
		Re 2 3 3 3 4 4 4 4				
7	6.20	Ac * 0 1 1 2 3 4	.24	.74	1.48	2.515
		Re 2 3 3 3 4 5 5				
8	4.95	Ac * 0 1 2 4 4 5	.31	.84	1.55	2.606
		Re 2 3 4 5 6 6 6				
9	4.61	Ac * 0 0 1 2 3 4 6	.31	.78	1.43	3.268
		Re 3 3 4 4 5 6 7 7				
10	4.29	Ac 0 2 3 5 7 9 10	.68	1.73	2.93	1.727
		Re 4 5 7 9 10 11 11				
11	4.02	Ac * 1 2 3 4 6 7	.47	1.14	1.89	2.380
		Re 3 4 5 6 6 8 8				
12	3.75	Ac * 1 1 2 3 5 7	.56	1.23	2.11	2.839
		Re 3 4 5 6 6 8 8				
13	3.56	Ac * 1 1 3 4 5 7 9	.59	1.26	2.10	2.872
		Re 3 5 6 7 8 9 10 10				
14	3.23	Ac 0 2 3 4 6 8 11	.96	1.92	3.10	2.218
		Re 4 5 8 9 10 12 12				
15	3.03	Ac 0 3 6 8 10 12 14	1.20	2.34	3.64	1.891
		Re 4 7 9 11 12 14 15				
16	2.69	Ac 1 3 6 9 11 14 17	1.56	2.75	4.20	1.839
		Re 5 7 10 13 15 18 18				
17	2.54	Ac 1 3 6 9 13 16 18	1.60	2.83	4.06	1.911
		Re 5 8 11 13 16 18 19				
18	2.35	Ac 1 5 7 10 13 17 22	2.00	3.34	4.70	1.982
		Re 6 9 12 16 19 21 23				
19	2.16	Ac 1 5 9 13 18 22 25	2.40	3.77	5.19	2.138
		Re 7 10 13 18 22 25 26				
20	1.94	Ac 3 8 13 18 24 30 36	3.74	5.46	7.26	1.967
		Re 8 15 20 25 30 34 37				

* Plans 1–6, 9, 12–14, and 16–20 have the same *R*s and acceptance and rejection numbers as the plans on pages 276, 309, 311, 323, 371, 352, 354, 341, 355, 302, 304, 295, 317, 306, and 320 respectively of SRG's *Sampling Inspection* (New York: McGraw-Hill, 1948). All plans are from Chemical Corps tables.

† $R = p_2'/p_1'$.

TABLE 9.2
Values Useful in Designing an 8-Stage Multiple-Sampling Plan, $\alpha = 0.05$, $\beta = 0.05$ (Bernard-Enters-Hamaker Type Description)

Plan Number	*Ratio* $p'_{0.05}/p'_{0.95}$	*The Plan*			*Approx. Values of* $p'n_i$ *for*		*Inverse Relative Efficiency* at* $P_a = 0.50$
		I	S	A	$P_a = 0.095$	$P_a = 0.50$	
(1)	*(2)*	*(3)*	*(4)*	*(5)*	*(6)*	*(7)*	*(8)*
1	25.40	0	1	2	0.075	0.534	0.905
2	8.94	1	1	2	0.318	1.146	0.774
3	5.50	1	1	3	0.344	0.891	0.771
4	3.92	2	1	4	0.475	1.012	0.721
5	3.10	2	2	4	1.122	2.060	0.707
6	2.68	3	2	5	1.282	2.190	0.720
7	2.47	3	3	5	1.973	3.211	0.735
8	2.24	3	3	6	1.97	3.021	0.74
9	2.11	4	3	7	2.12	3.146	0.739
10	1.99	3	4	7	2.66	3.920	0.715
11	1.92	4	4	8	2.98	4.070	0.751
12	1.83	4	5	8	3.64	5.012	0.720
13	1.69	5	6	10	4.56	5.996	0.746

* The "Inverse Relative Frequency" is the ratio of the ASN to the sample size for a single sampling plan with essentially the same operating characteristics.

SOURCE: From J. H. Enters and H. C. Hamaker, "Multiple Sampling in Theory and Practice." See Cumulative List of References.

The ASN at $p' = 0.01$ is approximately $(10)(4.077) \doteq 41$. The ASN given in the table does not allow for curtailment of inspection in case the lot is rejected before a particular sample is completely inspected, but at the p'_1 point, the ASN with curtailment will be only slightly less than without it.

Table 9.2 is used in a manner similar to Table 9.1. Again we start with the ratio p'_2/p'_1. This time, however, the ratio is for $p'_{0.05}/p'_{0.95}$. Suppose as before that we have the $p'_1 = 0.01$ and the $p'_2 = 0.08$, with $\alpha = 0.05$, $\beta = 0.05$. Then $R = 8$. The plan in Table 9.2 with R nearest to 8 is Plan No. 2. For this plan $np'_{0.95} = 0.318$. The $np'_{0.05}$ point can be obtained by multiplying the $np'_{0.95}$ point by 8.94, which yields $np'_{0.05} = 0.318(8.94) = 2.843$. If we wish to hold the p'_2 and β, we determine the group size by dividing the unity value 2.843 by p'_2, which in this case yields[7] $n = \dfrac{2.843}{0.08} = 35.5 \doteq 36$. The plan will go as follows: Take up to 8 successive samples of 36 each. The starting score is 1. At the end of the inspection of each sample, add 1 to the score and subtract the number of nonconforming units found. If at the end of any stage the total score is 2 or more, stop sampling and accept the lot. If at the end of any stage, the total score is 0 or less, reject the lot. If the sampling runs to 8 samples, accept if the final score is

[7] We could have held the p'_1 and α, and secured $n = 0.318/0.01 \doteq 32$.

greater than 1; otherwise reject. This plan will have a $p'_{0.95}$ of $\frac{0.318}{36} = 0.0088$ and a $p'_{0.50}$ of $\frac{1.146}{36} = 0.032$. At the latter point, the multiple plan will on the average require only 77 percent of the amount of sampling inspection required by a comparable single-sampling plan.

5. PROBLEMS

9.1. *a.* For a Type B OC curve, compute the probability of acceptance P_a at $p' = 0.10$ under the multiple-sampling plan:

Cumulated Sample Size	*Accept*	*Reject*
31	0	4
62	2	5
93	3	8
124	4	9
155	6	10
186	8	12
217	11	12

b. How well does your result agree with Table 9.1?

c. What lot quality will have approximately a 0.95 chance of acceptance under the above plan? Approximately a 0.50 chance of acceptance?

d. If the cumulated sample size ran 310, 620, 930, . . ., 2,170, what would be the 0.95, 0.50, and 0.10 points of the OC curve? How does the ratio of the 0.10 point to the 0.95 poinf for the plan 31, 62, ..., 217 compare with the same ratio for the plan 310, 620, . . ., 2,170?

9.2. *a.* For a Type B OC curve, compute the probability of acceptance P_a at $p' = 0.005$ under the multiple-sampling plan:

Cumulated Sample Size	*Accept*	*Reject*
100	*	3
200	1	4
300	2	5
400	3	6
500	4	6
600	6	8
700	7	8

b. Let $p' = 0.0047$ and determine from Table 9.1 the approximate probability of acceptance. How well does this agree with the result obtained in *(a)*?

c. What lot quality will have approximately a 0.50 chance of acceptance under the above plan? Approximately a 0.10 chance of acceptance?

d. If the cumulated sample size ran 10, 20, 30, . . ., 60, what would be the 0.95, 0.50, and 0.10 points of the OC curve? How does the ratio of the 0.10 point to the 0.95 point compare for the 10, 20, 30, . . ., 60 plan compare with the same ratio for the 100, 200, 300, . . ., 600 plan?

9.3. *a.* According to a Barnard-Enters-Hamaker type description, a multiple-sampling plan is given as $M(I = 3, S = 4, A = 7, m = 8)$. Convert this to conventional form for group size $n = 133$.

b. For this plan compute the probability of acceptance at $p' = 0.02$ and compare with Table 9.2. What would be the approximate 0.05 and 0.50 points for the plan? Compare the 0.95 and 0.50 points of this plan with the same points of plan No. 20 of Table 9.1 for group size $n = 133$.

c. What would the approximate 0.95 and 0.50 points be in each case for group sizes of $n = 100$?

9.4 *a.* According to a Barnard-Enters-Hamaker type description, a multiple-sampling plan is given as $M(I = 3, S = 2, A = 5, m = 8)$. Convert this to conventional form for group size $n = 128$.

b. For this plan compute the probability of acceptance at $p' = 0.01$ and compare with Table 9.2. What would be the approximate 0.05 and 0.50 points for the plan? Compare the 0.95 and 0.50 points of this plan with the same points of plan No. 16 of Table 9.1 for group size $n = 156$.

c. What would the 0.95 and 0.50 points be in each case for sample group sizes of $n = 50$?

9.5. For the multiple-sampling plan of Problem 9.1, compute the ASN without curtailment at $p' = 0.10$. How does this compare with the ASN without curtailment at the 0.95 point? (*Hint:* Use Table 9.1.) How would the latter compare with the ASN with curtailment at the same value of p'?

9.6. For the multiple-sampling plan of Problem 9.2, compute the ASN without curtailment at $p' = 0.10$. How does this compare with the ASN without curtailment at the 0.95 point? (*Hint:* Use Table 9.1.) How would the latter compare with the ASN with curtailment at the same value of p'?

6. SELECTED REFERENCES*

Barnard (P '46), Bartky (P '43), Cowden (B '57), Enters and Hamaker (P '51), Schilling and Johnson (P '80), Schilling (B '82), Statistical Research Group. Columbia University (B '48), and U.S. Dept. of the Army, Chemical Corps Engineering Agency (B '53, Manual No. 2).

* B and P refer to the Book and Periodical sections, respectively, of the Cumulative List of References in Appendix V.

10

Acceptance Sampling by Attributes: The U.S. Department of Defense Mil. Std. 105D and Its Civilian Counterparts, ANSI/ASQC Std. Z1.4 and ISO Std. 2859

Standard military sampling procedures for inspection by attributes were developed during World War II.[1] Army Ordnance tables and procedures were prepared in 1942, and these later became (with modifications) the Army Service Forces tables. Navy tables were issued in 1945 and were adopted in 1949 as a joint Army-Navy standard. This JANSTD was superseded by Mil. Std. 105A in 1950. Since then there have been minor modifications in the 105 standard issued as 105B and 105C. Military Standard 105D that is discussed here is the outcome of a study by an American-British-Canadian Working Group that sought to derive a common standard for the three countries. Mil. Std. 105D was issued by the U.S. government in 1963. In 1971, it was adopted by the American National Standards Institute as ANSI Standard Z1.4. and in 1974, except for minor editorial changes, it was adopted by the International Organization for Standardization as ISO Std. 2859. In 1981 there was some editorial revision of ANSI Z1.4 and addition of some new material by an ASQC writing group and the revised standard was issued as ANSI/ASQC Standard Z1.4. (See Section 4 below.) Revision of ISO Std. 2859 is currently (1986) under consideration. (See Section 5 below.) There are no plans, however, to change the basic tables so that these remain the same for the three standards. Thus the discussion in this chapter of the technical aspects of Mil. Std. 105D also applies to the other two standards.

---

[1] For a discussion of the development of quality control in the United States during World War II, see Chapter 1 and the references listed there.

1. DESCRIPTION OF MILITARY STANDARD 105D

It should be noted at the beginning of our discussion that in Mil. Std. 105D the term *defect* is used in the broad sense of a nonconformity to specifications and the term *defective unit* is used to refer to a nonconforming unit. In order to avoid confusion, the discussion of the standard that follows will be in terms of defects and defective units, but they will have the same broad meaning as in the standard itself. ANSI/ASQC Z1.4 and a proposed revision of ISO Std. 2859 use the terms *nonconformity* and *nonconforming unit* and are thus in accord with current recommended practice.

The sampling plans discussed in previous chapters have been individual plans that were designed to have certain characteristics, say a given p_1', p_2', α, and β. Mil. Std. 105D is a set of individual plans organized in a system of sampling schemes. A sampling scheme consists of a combination of a normal sampling plan, a tightened sampling plan, and a reduced sampling plan with rules for switching from one to the other. There may also be a provision for discontinuance of inspection if a specified number of consecutive lots remain on tightened inspection. The operation of these schemes is explained as the discussion proceeds.

The focal point of Mil. Std. 105D is the acceptable quality level or AQL. In applying the standard it is expected that in a conference (at a high level) between a supplier and a military agency it will be made clear to the supplier what, for the purposes of acceptance sampling, the agency considers to be an acceptable quality level for a given product characteristic. It is expected that the supplier will be submitting for inspection a series of lots of this product, and it is the purpose of the sampling procedures of Mil. Std. 105D so to constrain the supplier that he will produce product of at least AQL quality. This is done not only through the acceptance and rejection of a particular sampling plan but by providing for a shift to another, tighter sampling plan whenever there is evidence that the contractor's product has deteriorated from the agreed upon AQL.

Mil. Std. 105D is thus indexed with respect to a series of AQLs. For fraction-defective plans the AQLs run from 0.01 percent to 10 percent. For defects-per-unit plans there are an additional 10 AQLs running up to 1,000 defects per 100 units. It will be noted that for the lower AQL levels the same sampling plan can be used to control either a fraction defective or number of defects per unit. The AQLs indexed in the standard are so selected that each bears the ratio 1.585 to the AQL immediately below it. The reason for this is noted below.

In addition to an initial decision on an AQL it is also necessary in applying Mil. Std. 105D to decide on the "inspection level." This determines the relationship between the lot size and the sample size. Three general levels of inspection are offered. Level II is designated as normal. Level I may be specified when less discrimination is needed. Level III when more discrimination is needed. There are also four special levels. The decision on the inspection level is based on the type of product involved. For simple inexpensive items the level can be low; for complex expensive items it should be high. If inspection is harmful, a low level may be used. The inspection level is adopted at the initiation of the sampling program and is not generally changed thereafter.

For a specified AQL and inspection level, and a given lot size, Mil. Std. 105D gives a normal sampling plan that is to be used as long as the supplier is apparently producing product of AQL quality or better. It also gives a tightened plan to which a shift is to be made if there is evidence that quality has deteriorated. The rule is that a switch from the normal plan to the tightened plan will be made if two out of five consecutive lots have been rejected on original inspection. Normal inspection is reinstituted if five consecutive lots have been accepted on original inspection. If 10 consecutive lots remain under a tightened plan, inspection is stopped pending action on quality. It will be noted that this shift to and from tightened inspection pertains to a particular supplier and is quite independent of the originally adopted inspection level which relates to the type of product and not a supplier.

Mil. Std. 105D offers three types of sampling plans, the choice generally being in the hands of the government inspectors in charge. The three types are single-, double-, and multiple-sampling plans. The selection of one or another type is made usually on the basis of administrative convenience.

Because of the possibility of choosing from three types of plans, the standard does not immediately give a sample size but initially gives a sample size code letter. This together with the decision on the type of plan gives the specific sampling plan to be used.

The steps in the use of the standard may be summarized as follows:

1. Decide on the AQL.
2. Decide on inspection level.
3. Determine lot size.
4. Enter table to find sample size code letter.
5. Decide on type of sampling plan to be used.
6. Enter proper table to find the plan to be used.
7. Begin with normal inspection and follow the switching rules and the rule for discontinuance of inspection as called for.

The standard contains OC curves for the various single-sampling plans. An example is given in Figure 10.1. These are Type B OC curves. For AQLs up to 10 percent and sample sizes less than 80, the binomial distribution was used in their computation; for larger AQLs and sample sizes, the Poisson distribution was used since this is the appropriate distribution for defects-per-unit plans. The OC curves for the matching double- and multiple-sampling plans are roughly comparable with those for the corresponding single-sampling plans. The standard also gives numerical values for a number of the OC curves. Numerical values for Figure 10.1 are given here in Table 10.1 shown on pp. 218 and 219.

The OC curves presented in the standard are for the initial sampling plan. The standard does not give the "scheme" OC curves that would result from following the switching rules while maintaining the process average at a given p' value. The latter are given, however, in the 1981 revision of ANSI Z1.4 which is discussed in Section 4 below. For examples see Figure 10.4 below. Mil. Std. 105D also gives ASN curves for sampling without curtailment for its various double and multiple plans. See Figure 10.2 (p. 220).

As an adjunct to the sampling procedures described above, Mil. Std. 105D calls for shifts to reduced inspection if the quality is running especially good. For this, production must be running at a steady rate, the last 10 lots must all have been accepted on original normal inspection and the total number of defectives (or defects) must be less than a value set forth in Table 10.10 below. The basic idea of reduced inspection is to save the government money when quality is running good.

The standard has supplementary tables giving the 10 percent and 5 percent points of the OC curves. These are included for the benefit of those who are interested in limiting the risk of acceptance for individual lots worse than a specified lot tolerance fraction defective or lot tolerance number of defects per unit.[2] There is in addition a table of AOQL values (explained in Chapter 16) for normal and tightened inspection plans for those interested in this aspect.

The standard differentiates critical defects, major defects, and minor defects. It is frequent practice to choose an AQL = 1 percent for major defects and an AQL = 2.5 percent for minor defects. No critical defects are acceptable.

The full text of the standard with supporting tables exclusive of those for reduced inspection for double- and multiple-sampling plans is reproduced in the next section.

2. THE TEXT OF MIL. STD. 105D

1. Scope

1.1. Purpose. This publication establishes sampling plans and procedures for inspection by attributes. When specified by the responsible authority, this publication shall be referenced in the specification, contract, inspection instructions, or other documents and the provisions set forth herein shall govern. The "responsible authority" shall be designated in one of the above documents.

1.2. Application. Sampling plans designated in this publication are applicable, but not limited, to inspection of the following:

a. End items.
b. Components and raw materials.
c. Operations.

---

[2] For sets of attribute sampling plans directly indexed by limiting quality (LQ) values, see Section 3 of Chapter 14.

FIGURE 10.1

A Sample of OC Curves from Military Standard 105D (Curves are for single sampling. Curves for double and multiple sampling are matched as closely as practicable. All are for sample size code letter K.)*

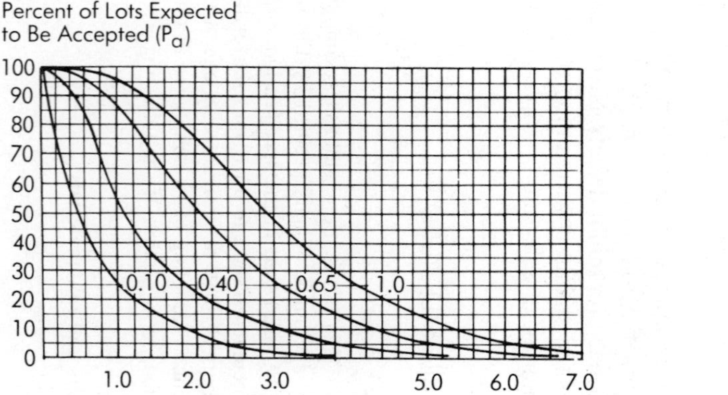

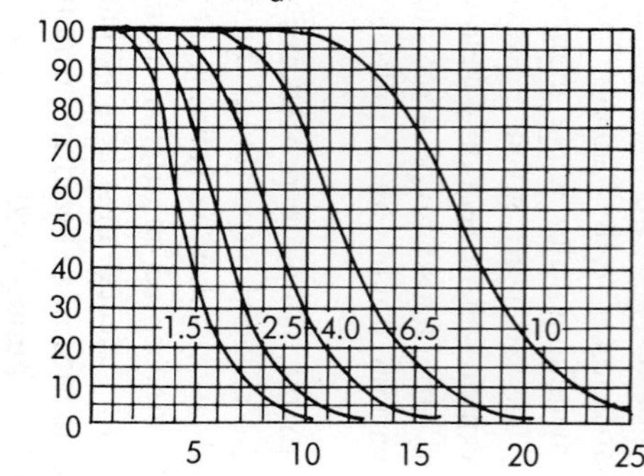

Quality of Submitted Lots (in percent defective for AQLs $\leq$ 10; in defects per hundred units for AQLs $>$ 10).

Note: Figures on curves are Acceptable Quality Levels (AQLs) for normal inspection.

TABLE 10.1
Tabulated Values for Operating Characteristic Curves for Single-Sampling Plans, Code Letter K (Mil. Std. 105D Table X-K-1)

P_a	Acceptable Quality Levels (normal inspection)											
	0.10	0.40	0.65	1.0	1.5	2.5	X	4.0	X	6.5	X	10
	p (in percent defective or defects per hundred units)											
99.0	0.0081	0.119	0.349	0.658	1.43	2.33	2.81	3.82	4.88	5.98	8.28	10.1
95.0	0.0410	0.284	0.654	1.09	2.09	3.19	3.76	4.94	6.15	7.40	9.95	11.9
90.0	0.0840	0.426	0.882	1.40	2.52	3.73	4.35	5.62	6.92	8.24	10.9	13.0
75.0	0.230	0.769	1.382	2.03	3.38	4.77	5.47	6.90	8.34	9.79	12.7	14.9
50.0	0.554	1.34	2.14	2.94	4.54	6.14	6.94	8.53	10.1	11.7	14.9	17.3
25.0	1.11	2.15	3.14	4.09	5.94	7.75	8.64	10.4	12.2	13.9	17.4	20.0
10.0	1.84	3.11	4.26	5.35	7.42	9.42	10.4	12.3	14.2	16.1	19.8	22.5
5.0	2.40	3.80	5.04	6.20	8.41	10.5	11.5	13.6	15.6	17.5	21.4	24.2
1.0	3.68	5.31	6.73	8.04	10.5	12.8	18.3	16.1	18.3	20.4	24.5	27.5
	0.15	0.65	1.0	1.5	2.5	X	4.0	X	6.5	X	10	X
	Acceptable Quality Levels (tightened inspection)											

Note: All values given in above table based on Poisson distribution as an approximation to the binomial.

FIGURE 10.2
Average Sample Number Curves for Double and Multiple Sampling—Normal and Tightened Inspection (Mil. Std. 105D, Table IX)

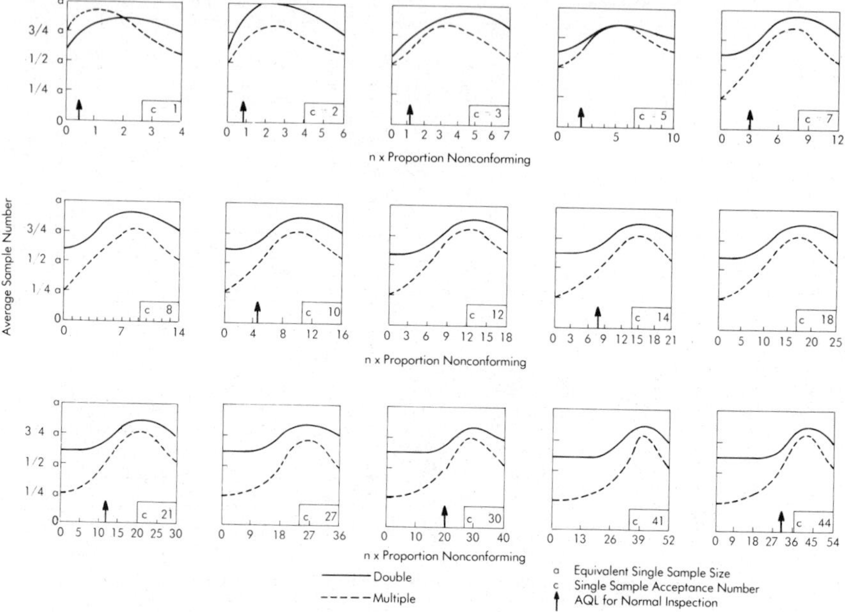

d. Materials in process.
e. Supplies in storage.
f. Maintenance operations.
g. Data or records.
h. Administrative procedures.

These plans are intended primarily to be used for a continuing series of lots or batches. The plans may also be used for the inspection of isolated lots or batches, but, in this latter case, the user is cautioned to consult the operating characteristic curves to find a plan which will yield the desired protection (see 11.6).

1.3. Inspection. Inspection is the process of measuring, examining, testing, or otherwise comparing the unit of product (see 1.5) with the requirements.

1.4. Inspection by Attributes. Inspection by attributes is inspection whereby either the unit of product is classified simply as defective or nondefective, or the number of defects in the unit of product is counted, with respect to a given requirement or set of requirements.

1.5. Unit of Product. The unit of product is the thing inspected in order to determine its classification as defective or nondefective or to count the number of defects. It may be a single article, a pair, a set, a length, an area, an operation, a volume, a component of an end product, or the end product itself. The unit of product may or may not be the same as the unit of purchase, supply, production, or shipment.

2. Classification of Defects and Defectives

2.1. Method of Classifying Defects. A classification of defects is the enumeration of possible defects of the unit of product classified according to their seriousness. A defect is any nonconformance of the unit of product with specified requirements. Defects will normally be grouped into one or more of the following classes; however, defects may be grouped into other classes, or into subclasses within these classes.

2.1.1. Critical Defect. A critical defect is a defect that judgment and experience indicate is likely to result in hazardous or unsafe conditions for individuals using, maintaining, or depending upon the product; or a defect that judgment and experience indicate is likely to prevent performance of the tactical function of a major end item such as a ship, aircraft, tank, missile or space vehicle. NOTE: For a special provision relating to critical defects, see 6.3.

2.1.2. Major Defect. A major defect is a defect, other than critical, that is likely to result in failure, or to reduce materially the usability of the unit of product for its intended purpose.

2.1.3. Minor Defect. A minor defect is a defect that is not likely to reduce materially the usability of the unit of product for its intended purpose, or is a departure from established standards having little bearing on the effective use or operation of the unit.

2.2. Method of Classifying Defectives. A defective is a unit of product which contains one or more defects. Defectives will usually be classified as follows:

2.2.1. Critical Defective. A critical defective contains one or more critical defects and may also contain major and or minor defects. NOTE: For a special provision relating to critical defectives, see 6.3.

2.2.2. Major Defective. A major defective contains one or more major defects, and may also contain minor defects but contains no critical defect.

2.2.3. Minor Defective. A minor defective contains one or more minor defects but contains no critical or major defect.

3. Percent Defective and Defects per Hundred Units

3.1. Expression of Nonconformance. The extent of nonconformance of product shall be expressed either in terms of percent defective or in terms of defects per hundred units.

3.2. Percent Defective. The percent defective of any given quantity of units of product is one hundred times the number of defective units of product contained therein divided by the total number of units of product, i.e.:

$$\text{Percent defective} = \frac{\text{Number of defectives}}{\text{Number of units inspected}} \times 100$$

3.3. Defects per Hundred Units. The number of defects per hundred units of any given quantity of units of product is one hundred times the number of defects contained therein (one or more defects being possible in any unit of product) divided by the total number of units of product, i.e.:

$$\text{Defects per hundred units} = \frac{\text{Number of defects}}{\text{Number of units inspected}} \times 100$$

4. Acceptable Quality Level (AQL)

4.1. Use. The AQL, together with the Sample Size Code Letter, is used for indexing the sampling plans provided herein.

4.2. Definition. The AQL is the maximum percent defective (or the maximum number of defects per hundred units) that, for purposes of sampling inspection, can be considered satisfactory as a process average (see 11.2).

4.3. Note on the Meaning of AQL. When a consumer designates some specific value of AQL for a certain defect or group of defects, he indicates to the supplier that his (the consumer's) acceptance sampling plan will accept the great majority of the lots or batches that the supplier submits, provided the process average level of percent defective (or defects per hundred units) in these lots or batches be no greater than the designated value of AQL. Thus, the AQL is a designated value of percent defective (or defects per hundred units) that the consumer indicates will be accepted most of the time by the acceptance sampling procedure to be used. The sampling plans provided herein are so arranged that the probability of acceptance at the designated AQL value depends upon the sample size, being generally higher for large samples than for small ones, for a given AQL. The AQL alone does not describe the protection to the consumer for individual lots or batches but more directly relates to what might be expected from a series of lots or batches, provided the steps indicated in this publication are taken. It is necessary to refer to the operating characteristic curve of the plan, to determine what protection the consumer will have.

4.4. Limitation. The designation of an AQL shall not imply that the supplier has the right to supply knowingly any defective unit of product.

4.5. Specifying AQLs. The AQL to be used will be designated in the contract or by the responsible authority. Different AQLs may be designated for groups of defects considered collectively, or for individual defects. An AQL for a group of defects may be designated in addition to AQLs for individual defects, or subgroups, within that group. AQL values of 10.0 or less may be expressed either in percent defective or in defects per hundred units; those over 10.0 shall be expressed in defects per hundred units only.

4.6. Preferred AQL. The values of AQLs given in these tables are known as preferred AQLs. If, for any product, an AQL be designated other than a preferred AQL, these tables are not applicable.

5. Submission of Product

5.1. Lot or Batch. The term lot or batch shall mean "inspection lot" or "inspection batch," i.e., a collection of units of product from which a sample is to be drawn and inspected to determine conformance with the acceptability criteria, and may differ from a collection of units designated as a lot or batch for other purposes (e.g., production, shipment, etc.).

5.2. Formation of Lots or Batches. The product shall be assembled into identifiable lots, sublots, batches, or in such other manner as may be prescribed (see 5.4). Each lot or batch shall, as far as is practicable, consist of units of product of a single type, grade, class, size, and composition, manufactured under essentially the same conditions, and at essentially the same time.

5.3. Lot or Batch Size. The lot or batch size is the number of units of product in a lot or batch.

5.4. Presentation of Lots or Batches. The formation of the lots or batches, lot or batch size, and the manner in which each lot or batch is to be presented and identified by the supplier shall be designated or approved by the responsible authority. As necessary, the supplier shall provide adequate and suitable storage space for each lot or batch, equipment needed for proper identification and presentation, and personnel for all handling of product required for drawing of samples.

6. Acceptance and Rejection

6.1. Acceptability of Lots or Batches. Acceptability of a lot or batch will be determined by the use of a sampling plan or plans associated with the designated AQL or AQLs.

6.2. Defective Units. The right is reserved to reject any unit of product found defective during inspection whether that unit of product forms part of a sample or not, and whether the lot or batch as a whole is accepted or rejected. Rejected units may be repaired or corrected and resubmitted for inspection with the approval of, and in the manner specified by, the responsible authority.

6.3. Special Reservation for Critical Defects. The supplier may be required at the discretion of the responsible authority to inspect every unit of the lot or batch for critical defects. The right is reserved to inspect every unit submitted by the supplier for critical defects, and to reject the lot or batch immediately, when a critical defect is found. The right is reserved also to sample, for critical defects, every lot or batch submitted by the supplier and to reject any lot or batch if a sample drawn therefrom is found to contain one or more critical defects.

6.4. Resubmitted Lots or Batches. Lots or batches found unacceptable shall be resubmitted for inspection only after all units are re-examined or retested and all defective units are removed or defects corrected. The responsible authority shall determine whether normal or tightened inspection shall be used, and whether reinspection shall include all types or classes of defefects or for the particular types or classes of defects which caused initial rejection.

7. Drawing of Samples

7.1. Sample. A sample consits of one or more units of product drawn from a lot or batch, the units of the sample being selected at random without regard to their quality. The number of units of product in the sample is the sample size.

7.2. Representative Sampling. When appropriate, the number of units in the sample shall be selected in proportion to the size of sublots or subbatches, or parts of the lot or batch, identified by some rational criterion. When representative sampling is used, the units from each part of the lot or batch shall be selected at random.

7.3. Time of Sampling. Samples may be drawn after all the units comprising the lot or batch have been assembled, or samples may be drawn during assembly of the lot or batch.

7.4. Double or Multiple Sampling. When double or multiple sampling is to be used, each sample shall be selected over the entire lot or batch.

8. Normal, Tightened and Reduced Inspection

8.1. Initiation of Inspection. Normal inspection will be used at the start of inspection unless otherwise directed by the responsible authority.

8.2. Continuation of Inspection. Normal, tightened or reduced inspection shall continue unchanged for each class of defects or defectives or successive lots or batches except where the switching procedures given below require change. The switching procedures shall be applied to each class of defects or defectives independently.

8.3. Switching Procedures.

8.3.1. Normal to Tightened. When normal inspection is in effect, tightened inspection shall be instituted when 2 out of 5 consecutive lots or batches have been rejected on original inspection (i.e., ignoring resubmitted lots or batches for this procedure).

8.3.2. Tightened to Normal. When tightened inspection is in effect, normal inspection shall be instituted when 5 consecutive lots or batches have been considered acceptable on original inspection.

8.3.3. Normal to Reduced. When normal inspection is in effect, reduced inspection shall be instituted providing that all of the following conditions are satisfied:

a. The preceding 10 lots or batches (or more, as indicated by the note to Table VIII) [Table 10.10 of this text] have been on normal inspection and none has been rejected on original inspection; and
b. The total number of defectives (or defects) in the samples from the preceding 10 lots or batches for such other number as was used for condition "*a*" above is equal to or less than the applicable number given in Table VIII. [Table 10.10, p. 238.] If double or multiple sampling is in use, all samples inspected should be included, not "first" samples only; and
c. Production is at a steady rate; and
d. Reduced inspection is considered desirable by the responsible authority.

8.3.4. Reduced to Normal. When reduced inspection is in effect, normal inspection shall be instituted if any of the following occur on original inspection:

a. A lot or batch is rejected; or
b. A lot or batch is considered acceptable under the procedures of 10.1.4; or
c. Production becomes irregular or delayed; or
d. Other conditions warrant that normal inspection shall be instituted.

8.4. Discontinuation of Inspection. In the event that 10 consecutive lots or batches remain on tightened inspection (or such other number as may be designated by the responsible authority), inspection under the provisions of this document should be discontinued pending action to improve the quality of submitted material.

9. Sampling Plans

9.1. Sampling Plan. A sampling plan indicates the number of units of product from each lot or batch which are to be inspected (sample size or series of sample sizes) and the criteria for determining the acceptability of the lot or batch (acceptance and rejection numbers).

9.2. Inspection Level. The inspection level determines the relationship between the lot or batch size and the sample size. The inspection level to be used for any particular requirement will be prescribed by the responsible authority. Three inspection levels: I, II, and III, are given in Table I [Table 10.2, p. 228] for general use. Unless otherwise specified, Inspection Level II will be used. However, Inspection Level I may be specified when less discrimination is needed, or Level III may be specified for greater discrimination. Four additional special levels: S–1, S–2, S–3 and S–4, are given in the same table and may be used where relatively small sample sizes are necessary and large sampling risks can or must be tolerated.

NOTE: In the designation of inspection levels S–1 to S–4, care must be exercised to avoid AQLs inconsistent with these inspection levels.

9.3. Code Letters. Sample sizes are designated by code letters. Table I [Table 10.2, p. 228] shall be used to find the applicable code letter for the particular lot or batch size and the prescribed inspection level.

9.4. Obtaining Sampling Plan. The AQL and the code letter shall be used to obtain the sampling plan from Tables II, III or IV [Tables 10.3 to 10.9, pp. 229–237]. When no sampling plan is available for a given combination of AQL and code letter, the tables direct the user to a different letter. The sample size to be used is given by the new code letter not by the original letter. If this procedure leads to different sample sizes for different classes of defects, the code letter corresponding to the largest sample size derived may be used for all classes of defects when designated or approved by the responsible authority. As an alternative to a single sampling plan with an acceptance number of 0, the plan with an acceptance number of 1 with its correspondingly larger sample size for a designated AQL (where available), may be used when designated or approved by the responsible authority.

9.5. Types of Sampling Plans. Three types of sampling plans: Single, Double and Multiple, are given in Tables II, III and IV, respectively [Tables 10.3 to 10.9, pp. 229–237]. When several types of plans are available for a given AQL and code letter, any one may be used. A decision as to type of plan, either single, double, or multiple, when available for a given AQL and code letter, will usually be based upon the comparison between the administrative difficulty and the average sample sizes of the available plans. The average sample size of multiple plans is less than for double (except in the case corresponding to single acceptance number 1) and both of these are always less than a single sample size. Usually the administrative difficulty for single sampling and the cost per unit of the sample are less than for double or multiple.

10. Determination of Acceptability

10.1. Percent Defective Inspection. To determine acceptability of a lot or batch under percent defective inspection, the applicable sampling plan shall be used in accordance with 10.1.1, 10.1.2, 10.1.3, and 10.1.4.

10.1.1. Single Sampling Plan. The number of sample units inspected shall be equal to the sample size given by the plan. If the number of defectives found in the sample is equal to or less than the acceptance number, the lot or batch shall be considered acceptable. If the number of defectives is equal to or greater than the rejection number, the lot or batch shall be rejected.

10.1.2. Double Sampling Plan. The number of sample units inspected shall be equal to the first sample size given by the plan. If the number of defectives found in the first sample is equal to or less than the first acceptance number, the lot or batch shall be considered acceptable. If the number of defectives found in the first sample is equal to or greater than the first rejection number, the lot or batch shall be rejected. If the number of defectives found in the first sample is between the first acceptance and rejection numbers, a second sample of the size given by the plan shall be inspected. The number of defectives found in the first and second samples shall be accumulated. If the cumulative number of defectives is equal to or less than the second acceptance number, the lot or batch shall be considered acceptable. If the cumulative number of defectives is equal to or greater than the second rejection number, the lot or batch shall be rejected.

10.1.3. Multiple Sample Plan. Under multiple sampling, the procedure shall be similar to that specified in 10.1.2, except that the number of successive samples required to reach a decision may be more than two.

10.1.4. Special Procedure for Reduced Inspection. Under reduced inspection, the sampling procedure may terminate without either acceptance or rejection criteria having been met. In these circumstances, the lot or batch will be considered acceptable, but normal inspection will be reinstated starting with the next lot or batch (see 8.3.4*(b)*).

10.2 Defects per Hundred Units Inspection. To determine the acceptability of a lot or batch under Defects per Hundred Units inspection, the procedure specified for Percent Defective inspection above shall be used, except that the word "defects" shall be substituted for "defectives."

11. Supplementary Information

11.1. Operating Characteristic Curves. The operating characteristic curves for normal inspection, shown in Table X [see Figure 10.2, p. 228], indicate the percentage of lots or batches which may be expected to be accepted under the various sampling plans for a given process quality. The curves shown are for single sampling; curves for double and multiple sampling are matched as closely as practicable. The OC curves shown for AQLs greater than 10.0 are based on the Poisson distribution and are applicable for defects per hundred units inspection; those for AQLs of 10.0 or less and sample sizes of 80 or less are based on the binomial distribution and are applicable for percent defective inspection; those for AQLs of 10.0 or less and sample sizes larger than 80 are based on the Poisson distribution and are applicable either for defects per hundred units inspection, or for percent defective inspection (the Poisson distribution being an adequate approximation to the binomial distribution under these conditions). Tabulated values, corresponding to selected values of probabilities of acceptance (P_a, in percent) are given for each of the curves shown, and, in addition, for tightened inspection, and for defects per hundred units for AQLs of 10.0 or less and sample sizes of 80 or less.

11.2. Process Average. The process average is the average percent defective or average number of defects per hundred units (whichever is applicable) of product submitted by the supplier for original inspection. Original inspection is the first inspection of a particular quantity of product as distinguished from the inspection of product which has been resubmitted after prior rejection.

11.3. Average Outgoing Quality (AOQ). The AOQ is the average quality of outgoing product including all accepted lots or batches, plus all rejected lots or batches after the rejected lots or batches have been effectively 100 percent inspected and all defectives replaced by nondefectives. [See Chapter 16, Section 1 of this text.]

11.4. Average Outgoing Quality Limit (AOQL). The AOQL is the maximum of the AOQs for all possible incoming qualities for a given acceptance sampling plan. AOQL values are given in Table V–A [Table 10.11, p. 239] for each of the single sampling plans for normal inspection and in Table V–B [Table 10.12 of this text] for each of the single sampling plans for tightened inspection. [See Chapter 16, Section 1 of this text.]

11.5. Average Sample Size Curves. Average sample size curves for double and multiple sampling are in Table IX [Figure 10.3, p. 229]. These show the average sample sizes which may be expected to occur under the various sampling plans for a given process quality. The curves assume no curtailment of inspection and are approximate to the extent that they are based upon the Poisson distribution, and that the sample sizes for double and multiple sampling are assumed to be 0.631n and 0.25n respectively, where n is the equivalent single sample size.

11.6. Limiting Quality Protection. The sampling plans and associated procedures given in this publication were designed for use where the units of product are produced in a continuing series of lots or batches over a period of time. However, if the lot or batch is of an isolated nature, it is desirable to limit the selection of sampling plans to those associated with a designated AQL value, that provide not less than a specified limiting quality protection. Sampling plans for this purpose can be selected by choosing a Limiting Quality (LQ) and a consumer's risk to be associated with it. Tables VI–A [Table 10.13, p. 241] and VII* give values of LQ for the commonly used consumer's risks of 10 percent and 5 percent respectively. If a different value of consumer's risk is required, the OC curves and their tabulated values may be used. The concept of LQ may also be useful in specifying the AQL and Inspection Levels for a series of lots or batches, thus fixing minimum sample size where there is some reason for avoiding (with more than a given consumer's risk) more than a limiting proportion of defectives (or defects) in any single lot or batch.

3. DISCUSSION OF MIL. STD. 105D

It is to be emphasized that Mil. Std. 105D is an organized system of sampling *schemes* to be applied to a series of lots from some production process.

* Not reproduced in this text.

TABLE 10.2
Sample Size Code Letters (Mil. Std. 105D, Table I)

Lot or Batch Size			*Special Inspection Levels*				*General Inspection Levels*		
			S-1	S-2	S-3	S-4	I	II	III
2	to	8	A	A	A	A	A	A	B
9	to	15	A	A	A	A	A	B	C
16	to	25	A	A	B	B	B	C	D
26	to	50	A	B	B	C	C	D	E
51	to	90	B	B	C	C	C	E	F
91	to	150	B	B	C	D	D	F	G
151	to	280	B	C	D	E	E	G	H
281	to	500	B	C	D	E	F	H	J
501	to	1,200	C	C	E	F	G	J	K
1,201	to	3,200	C	D	E	G	H	K	L
3,201	to	10,000	C	D	F	G	J	L	M
10,001	to	35,000	C	D	F	H	K	M	N
35,001	to	150,000	D	E	G	J	L	N	P
150,001	to	500,000	D	E	G	J	M	P	Q
500,001	and	over	D	E	H	K	N	Q	R

TABLE 10.3
Master Table for Normal Inspection—Single Sampling (Mil. Std. 105D, Table II–A)

Sample size code letter	Sample size	Acceptable Quality Levels (normal inspection).																									
		0.010	0.015	0.025	0.040	0.065	0.10	0.15	0.25	0.40	0.65	1.0	1.5	2.5	4.0	6.5	10	15	25	40	65	100	150	250	400	650	1000
		Ac Re	Ac Re	Ac Re	Ac Re	Ac Re	Ac Re	Ac Re	Ac Re	Ac Re	Ac Re	Ac Re	Ac Re	Ac Re	Ac Re	Ac Re	Ac Re	Ac Re	Ac Re	Ac Re	Ac Re	Ac Re	Ac Re	Ac Re	Ac Re	Ac Re	Ac Re
A	2	↓	↓	↓	↓	↓	↓	↓	↓	↓	↓	↓	↓	↓	↓	0 1	↓	↓	1 2	2 3	3 4	5 6	7 8	10 11	14 15	21 22	30 31
B	3	↓	↓	↓	↓	↓	↓	↓	↓	↓	↓	↓	↓	↓	0 1	↑	↓	1 2	2 3	3 4	5 6	7 8	10 11	14 15	21 22	30 31	44 45
C	5	↓	↓	↓	↓	↓	↓	↓	↓	↓	↓	↓	↓	0 1	↑	↓	1 2	2 3	3 4	5 6	7 8	10 11	14 15	21 22	30 31	44 45	↑
D	8	↓	↓	↓	↓	↓	↓	↓	↓	↓	↓	↓	0 1	↑	↓	1 2	2 3	3 4	5 6	7 8	10 11	14 15	21 22	30 31	44 45	↑	↑
E	13	↓	↓	↓	↓	↓	↓	↓	↓	↓	↓	0 1	↑	↓	1 2	2 3	3 4	5 6	7 8	10 11	14 15	21 22	30 31	44 45	↑	↑	↑
F	20	↓	↓	↓	↓	↓	↓	↓	↓	↓	0 1	↑	↓	1 2	2 3	3 4	5 6	7 8	10 11	14 15	21 22	↑	↑	↑	↑	↑	↑
G	32	↓	↓	↓	↓	↓	↓	↓	↓	0 1	↑	↓	1 2	2 3	3 4	5 6	7 8	10 11	14 15	21 22	↑	↑	↑	↑	↑	↑	↑
H	50	↓	↓	↓	↓	↓	↓	↓	0 1	↑	↓	1 2	2 3	3 4	5 6	7 8	10 11	14 15	21 22	↑	↑	↑	↑	↑	↑	↑	↑
J	80	↓	↓	↓	↓	↓	↓	0 1	↑	↓	1 2	2 3	3 4	5 6	7 8	10 11	14 15	21 22	↑	↑	↑	↑	↑	↑	↑	↑	↑
K	125	↓	↓	↓	↓	↓	0 1	↑	↓	1 2	2 3	3 4	5 6	7 8	10 11	14 15	21 22	↑	↑	↑	↑	↑	↑	↑	↑	↑	↑
L	200	↓	↓	↓	↓	0 1	↑	↓	1 2	2 3	3 4	5 6	7 8	10 11	14 15	21 22	↑	↑	↑	↑	↑	↑	↑	↑	↑	↑	↑
M	315	↓	↓	↓	0 1	↑	↓	1 2	2 3	3 4	5 6	7 8	10 11	14 15	21 22	↑	↑	↑	↑	↑	↑	↑	↑	↑	↑	↑	↑
N	500	↓	↓	0 1	↑	↓	1 2	2 3	3 4	5 6	7 8	10 11	14 15	21 22	↑	↑	↑	↑	↑	↑	↑	↑	↑	↑	↑	↑	↑
P	800	↓	0 1	↑	↓	1 2	2 3	3 4	5 6	7 8	10 11	14 15	21 22	↑	↑	↑	↑	↑	↑	↑	↑	↑	↑	↑	↑	↑	↑
Q	1250	0 1	↑	↓	1 2	2 3	3 4	5 6	7 8	10 11	14 15	21 22	↑	↑	↑	↑	↑	↑	↑	↑	↑	↑	↑	↑	↑	↑	↑
R	2000	↑	↑	1 2	2 3	3 4	5 6	7 8	10 11	14 15	21 22	↑	↑	↑	↑	↑	↑	↑	↑	↑	↑	↑	↑	↑	↑	↑	↑

↓ = Use first sampling plan below arrow. If sample size equals, or exceeds, lot or batch size, do 100 percent inspection.

↑ = Use first sampling plan above arrow.

Ac = Acceptance number.

Re = Rejection number.

TABLE 10.4
Master Table for Tightened Inspection—Single Sampling (Mil. Std. 105D, Table II–B)

Sample size code letter	Sample size	Acceptable Quality Levels (tightened inspection)																									
		0.010	0.015	0.025	0.040	0.065	0.10	0.15	0.25	0.40	0.65	1.0	1.5	2.5	4.0	6.5	10	15	25	40	65	100	150	250	400	650	1000
		Ac Re	Ac Re	Ac Re	Ac Re	Ac Re	Ac Re	Ac Re	Ac Re	Ac Re	Ac Re	Ac Re	Ac Re	Ac Re	Ac Re	Ac Re	Ac Re	Ac Re	Ac Re	Ac Re	Ac Re	Ac Re	Ac Re	Ac Re	Ac Re	Ac Re	Ac Re
A	2	↓	↓	↓	↓	↓	↓	↓	↓	↓	↓	↓	↓	↓	↓	↓	↓	↓	↓	1 2	2 3	3 4	5 6	8 9	12 13	18 19	27 28
B	3	↓	↓	↓	↓	↓	↓	↓	↓	↓	↓	↓	↓	↓	↓	0 1	↓	↓	1 2	2 3	3 4	5 6	8 9	12 13	18 19	27 28	41 42
C	5	↓	↓	↓	↓	↓	↓	↓	↓	↓	↓	↓	↓	↓	0 1	↓	↓	1 2	2 3	3 4	5 6	8 9	12 13	18 19	27 28	41 42	↑
D	8	↓	↓	↓	↓	↓	↓	↓	↓	↓	↓	↓	↓	0 1	↓	↓	1 2	2 3	3 4	5 6	8 9	12 13	18 19	27 28	41 42	↑	↑
E	13	↓	↓	↓	↓	↓	↓	↓	↓	↓	↓	↓	0 1	↓	↓	1 2	2 3	3 4	5 6	8 9	12 13	18 19	27 28	41 42	↑	↑	↑
F	20	↓	↓	↓	↓	↓	↓	↓	↓	↓	↓	0 1	↓	↓	1 2	2 3	3 4	5 6	8 9	12 13	18 19	↑	↑	↑	↑	↑	↑
G	32	↓	↓	↓	↓	↓	↓	↓	↓	↓	0 1	↓	↓	1 2	2 3	3 4	5 6	8 9	12 13	18 19	↑	↑	↑	↑	↑	↑	↑
H	50	↓	↓	↓	↓	↓	↓	↓	↓	0 1	↓	↓	1 2	2 3	3 4	5 6	8 9	12 13	18 19	↑	↑	↑	↑	↑	↑	↑	↑
J	80	↓	↓	↓	↓	↓	↓	↓	0 1	↓	↓	1 2	2 3	3 4	5 6	8 9	12 13	18 19	↑	↑	↑	↑	↑	↑	↑	↑	↑
K	125	↓	↓	↓	↓	↓	↓	0 1	↓	↓	1 2	2 3	3 4	5 6	8 9	12 13	18 19	↑	↑	↑	↑	↑	↑	↑	↑	↑	↑
L	200	↓	↓	↓	↓	↓	0 1	↓	↓	1 2	2 3	3 4	5 6	8 9	12 13	18 19	↑	↑	↑	↑	↑	↑	↑	↑	↑	↑	↑
M	315	↓	↓	↓	↓	0 1	↓	↓	1 2	2 3	3 4	5 6	8 9	12 13	18 19	↑	↑	↑	↑	↑	↑	↑	↑	↑	↑	↑	↑
N	500	↓	↓	↓	0 1	↓	↓	1 2	2 3	3 4	5 6	8 9	12 13	18 19	↑	↑	↑	↑	↑	↑	↑	↑	↑	↑	↑	↑	↑
P	800	↓	↓	0 1	↓	↓	1 2	2 3	3 4	5 6	8 9	12 13	18 19	↑	↑	↑	↑	↑	↑	↑	↑	↑	↑	↑	↑	↑	↑
Q	1250	↓	0 1	↓	↓	1 2	2 3	3 4	5 6	8 9	12 13	18 19	↑	↑	↑	↑	↑	↑	↑	↑	↑	↑	↑	↑	↑	↑	↑
R	2000	0 1	↑	↓	1 2	2 3	3 4	5 6	8 9	12 13	18 19	↑	↑	↑	↑	↑	↑	↑	↑	↑	↑	↑	↑	↑	↑	↑	↑
S	3150			1 2																							

↓ = Use first sampling plan below arrow. If sample size equals or exceeds lot or batch size, do 100 percent inspection.
↑ = Use first sampling plan above arrow.

Ac = Acceptance number.
Re = Rejection number.

TABLE 10.5

Master Table for Reduced Inspection—Single Sampling (Mil. Std. 105D, Table II–C)

Sample size code letter	Sample size	Acceptable Quality Levels (reduced inspection)†																									
		0.010	0.015	0.025	0.040	0.065	0.10	0.15	0.25	0.40	0.65	1.0	1.5	2.5	4.0	6.5	10	15	25	40	65	100	150	250	400	650	1000
		Ac Re	Ac Re	Ac Re	Ac Re	Ac Re	Ac Re	Ac Re	Ac Re	Ac Re	Ac Re	Ac Re	Ac Re	Ac Re	Ac Re	Ac Re	Ac Re	Ac Re	Ac Re	Ac Re	Ac Re	Ac Re	Ac Re	Ac Re	Ac Re	Ac Re	Ac Re
A	2	↓	↓	↓	↓	↓	↓	↓	↓	↓	↓	↓	↓	↓	↓	0 1	↓	↓	1 2	2 3	3 4	5 6	7 8	10 11	14 15	21 22	30 31
B	2	↓	↓	↓	↓	↓	↓	↓	↓	↓	↓	↓	↓	↓	0 1	↑	↓	0 2	1 3	2 4	3 5	5 6	7 8	10 11	14 15	21 22	30 31
C	2	↓	↓	↓	↓	↓	↓	↓	↓	↓	↓	↓	↓	0 1	↑	↓	0 2	1 3	1 4	2 5	3 6	5 8	7 10	10 13	14 17	21 24	↑
D	3	↓	↓	↓	↓	↓	↓	↓	↓	↓	↓	↓	0 1	↑	↓	0 2	1 3	1 4	2 5	3 6	5 8	7 10	10 13	14 17	21 24	↑	↑
E	5	↓	↓	↓	↓	↓	↓	↓	↓	↓	↓	0 1	↑	↓	0 2	1 3	1 4	2 5	3 6	5 8	7 10	10 13	14 17	21 24	↑	↑	↑
F	8	↓	↓	↓	↓	↓	↓	↓	↓	↓	0 1	↑	↓	0 2	1 3	1 4	2 5	3 6	5 8	7 10	10 13	↑	↑	↑	↑	↑	↑
G	13	↓	↓	↓	↓	↓	↓	↓	↓	0 1	↑	↓	0 2	1 3	1 4	2 5	3 6	5 8	7 10	10 13	↑	↑	↑	↑	↑	↑	↑
H	20	↓	↓	↓	↓	↓	↓	↓	0 1	↑	↓	0 2	1 3	1 4	2 5	3 6	5 8	7 10	10 13	↑	↑	↑	↑	↑	↑	↑	↑
J	32	↓	↓	↓	↓	↓	↓	0 1	↑	↓	0 2	1 3	1 4	2 5	3 6	5 8	7 10	10 13	↑	↑	↑	↑	↑	↑	↑	↑	↑
K	50	↓	↓	↓	↓	↓	0 1	↑	↓	0 2	1 3	1 4	2 5	3 6	5 8	7 10	10 13	↑	↑	↑	↑	↑	↑	↑	↑	↑	↑
L	80	↓	↓	↓	↓	0 1	↑	↓	0 2	1 3	1 4	2 5	3 6	5 8	7 10	10 13	↑	↑	↑	↑	↑	↑	↑	↑	↑	↑	↑
M	125	↓	↓	↓	0 1	↑	↓	0 2	1 3	1 4	2 5	3 6	5 8	7 10	10 13	↑	↑	↑	↑	↑	↑	↑	↑	↑	↑	↑	↑
N	200	↓	↓	0 1	↑	↓	0 2	1 3	1 4	2 5	3 6	5 8	7 10	10 13	↑	↑	↑	↑	↑	↑	↑	↑	↑	↑	↑	↑	↑
P	315	↓	0 1	↑	↓	0 2	1 3	1 4	2 5	3 6	5 8	7 10	10 13	↑	↑	↑	↑	↑	↑	↑	↑	↑	↑	↑	↑	↑	↑
Q	500	0 1	↑	↓	0 2	1 3	1 4	2 5	3 6	5 8	7 10	10 13	↑	↑	↑	↑	↑	↑	↑	↑	↑	↑	↑	↑	↑	↑	↑
R	800	↑	↑	0 2	1 3	1 4	2 5	3 6	5 8	7 10	10 13	↑	↑	↑	↑	↑	↑	↑	↑	↑	↑	↑	↑	↑	↑	↑	↑

↓ = Use first sampling plan below arrow. If sample size equals or exceeds lot or batch size, do 100 percent inspection.

↑ = Use first sampling plan above arrow.

Ac = Acceptance number.

Re = Rejection number.

† = If the acceptance number has been exceeded, but the rejection number has not been reached, accept the lot, but reinstate normal inspection (see 10.1.4).

TABLE 10.6

Master Table for Normal Inspection—Double Sampling (Mil. Std. 105D, Table III–A)

Sample size code letter	Sample	Sample size	Cumulative sample size	Acceptable Quality Levels (normal inspection)																									
				0.010	0.015	0.025	0.040	0.065	0.10	0.15	0.25	0.40	0.65	1.0	1.5	2.5	4.0	6.5	10	15	25	40	65	100	150	250	400	650	1000
				Ac Re	Ac Re	Ac Re	Ac Re	Ac Re	Ac Re	Ac Re	Ac Re	Ac Re	Ac Re	Ac Re	Ac Re	Ac Re	Ac Re	Ac Re	Ac Re	Ac Re	Ac Re	Ac Re	Ac Re	Ac Re	Ac Re	Ac Re	Ac Re	Ac Re	Ac Re
A				↓	↓	↓	↓	↓	↓	↓	↓	↓	↓	↓	↓	↓	↓	•	↓	↓	•	•	•	•	•	•	•	•	•
B	First	2	2	↓	↓	↓	↓	↓	↓	↓	↓	↓	↓	↓	↓	↓	•	↑	↓	0 2	0 3	1 4	2 5	3 7	5 9	7 11	11 16	17 22	25 31
	Second	2	4																	1 2	3 4	4 5	6 7	8 9	12 13	18 19	26 27	37 38	56 57
C	First	3	3	↓	↓	↓	↓	↓	↓	↓	↓	↓	↓	↓	↓	•	↑	↓	0 2	0 3	1 4	2 5	3 7	5 9	7 11	11 16	17 22	25 31	↑
	Second	3	6																1 2	3 4	4 5	6 7	8 9	12 13	18 19	26 27	37 38	56 57	
D	First	5	5	↓	↓	↓	↓	↓	↓	↓	↓	↓	↓	↓	•	↑	↓	0 2	0 3	1 4	2 5	3 7	5 9	7 11	11 16	17 22	25 31	↑	↑
	Second	5	10															1 2	3 4	4 5	6 7	8 9	12 13	18 19	26 27	37 38	56 57		
E	First	8	8	↓	↓	↓	↓	↓	↓	↓	↓	↓	↓	•	↑	↓	0 2	0 3	1 4	2 5	3 7	5 9	7 11	11 16	17 22	25 31	↑	↑	↑
	Second	8	16														1 2	3 4	4 5	6 7	8 9	12 13	18 19	26 27	37 38	56 57			
F	First	13	13	↓	↓	↓	↓	↓	↓	↓	↓	↓	•	↑	↓	0 2	0 3	1 4	2 5	3 7	5 9	7 11	11 16	↑	↑	↑	↑	↑	↑
	Second	13	26													1 2	3 4	4 5	6 7	8 9	12 13	18 19	26 27						
G	First	20	20	↓	↓	↓	↓	↓	↓	↓	↓	•	↑	↓	0 2	0 3	1 4	2 5	3 7	5 9	7 11	11 16	↑	↑	↑	↑	↑	↑	↑
	Second	20	40												1 2	3 4	4 5	6 7	8 9	12 13	18 19	26 27							
H	First	32	32	↓	↓	↓	↓	↓	↓	↓	•	↑	↓	0 2	0 3	1 4	2 5	3 7	5 9	7 11	11 16	↑	↑	↑	↑	↑	↑	↑	↑
	Second	32	64											1 2	3 4	4 5	6 7	8 9	12 13	18 19	26 27								
J	First	50	50	↓	↓	↓	↓	↓	↓	•	↑	↓	0 2	0 3	1 4	2 5	3 7	5 9	7 11	11 16	↑	↑	↑	↑	↑	↑	↑	↑	↑
	Second	50	100										1 2	3 4	4 5	6 7	8 9	12 13	18 19	26 27									
K	First	80	80	↓	↓	↓	↓	↓	•	↑	↓	0 2	0 3	1 4	2 5	3 7	5 9	7 11	11 16	↑	↑	↑	↑	↑	↑	↑	↑	↑	↑
	Second	80	160									1 2	3 4	4 5	6 7	8 9	12 13	18 19	26 27										
L	First	125	125	↓	↓	↓	↓	•	↑	↓	0 2	0 3	1 4	2 5	3 7	5 9	7 11	11 16	↑	↑	↑	↑	↑	↑	↑	↑	↑	↑	↑
	Second	125	250								1 2	3 4	4 5	6 7	8 9	12 13	18 19	26 27											
M	First	200	200	↓	↓	↓	•	↑	↓	0 2	0 3	1 4	2 5	3 7	5 9	7 11	11 16	↑	↑	↑	↑	↑	↑	↑	↑	↑	↑	↑	↑
	Second	200	400							1 2	3 4	4 5	6 7	8 9	12 13	18 19	26 27												
N	First	315	315	↓	↓	•	↑	↓	0 2	0 3	1 4	2 5	3 7	5 9	7 11	11 16	↑	↑	↑	↑	↑	↑	↑	↑	↑	↑	↑	↑	↑
	Second	315	630						1 2	3 4	4 5	6 7	8 9	12 13	18 19	26 27													
P	First	500	500	↓	•	↑	↓	0 2	0 3	1 4	2 5	3 7	5 9	7 11	11 16	↑	↑	↑	↑	↑	↑	↑	↑	↑	↑	↑	↑	↑	↑
	Second	500	1000					1 2	3 4	4 5	6 7	8 9	12 13	18 19	26 27														
Q	First	800	800	•	↑	↓	0 2	0 3	1 4	2 5	3 7	5 9	7 11	11 16	↑	↑	↑	↑	↑	↑	↑	↑	↑	↑	↑	↑	↑	↑	↑
	Second	800	1600				1 2	3 4	4 5	6 7	8 9	12 13	18 19	26 27															
R	First	1250	1250	↑	↑	0 2	0 3	1 4	2 5	3 7	5 9	7 11	11 16	↑	↑	↑	↑	↑	↑	↑	↑	↑	↑	↑	↑	↑	↑	↑	↑
	Second	1250	2500			1 2	3 4	4 5	6 7	8 9	12 13	18 19	26 27																

↓ = Use first sampling plan below arrow. If sample size equals or exceeds lot or batch size, do 100 percent inspection.

↑ = Use first sampling plan above arrow.

Ac = Acceptance number

Re = Rejection number

• = Use corresponding single sampling plan (or alternatively, use double sampling plan below, where available).

TABLE 10.7
Master Table for Tightened Inspection—Double Sampling (Mil. Std. 105D, Table III–B)

Sample size code letter	Sample	Sample size	Cumulative sample size	Acceptable Quality Levels (tightened inspection)																									
				0.010	0.015	0.025	0.040	0.065	0.10	0.15	0.25	0.40	0.65	1.0	1.5	2.5	4.0	6.5	10	15	25	40	65	100	150	250	400	650	1000
				Ac Re	Ac Re	Ac Re	Ac Re	Ac Re	Ac Re	Ac Re	Ac Re	Ac Re	Ac Re	Ac Re	Ac Re	Ac Re	Ac Re	Ac Re	Ac Re	Ac Re	Ac Re	Ac Re	Ac Re	Ac Re	Ac Re	Ac Re	Ac Re	Ac Re	Ac Re
A				↓	↓	↓	↓	↓	↓	↓	↓	↓	↓	↓	↓	↓	↓	↓	↓	↓	↓	•	•	•	•	•	•	•	•
B	First	2	2	↓	↓	↓	↓	↓	↓	↓	↓	↓	↓	↓	↓	↓	↓	•	↓	↓	0 2	0 3	1 4	2 5	3 7	6 10	9 14	15 20	23 29
	Second	2	4																		1 2	3 4	4 5	6 7	11 12	15 16	23 24	34 35	52 53
C	First	3	3	↓	↓	↓	↓	↓	↓	↓	↓	↓	↓	↓	↓	↓	•	↓	↓	0 2	0 3	1 4	2 5	3 7	6 10	9 14	15 20	23 29	↑
	Second	3	6																	1 2	3 4	4 5	6 7	11 12	15 16	23 24	34 35	52 53	
D	First	5	5	↓	↓	↓	↓	↓	↓	↓	↓	↓	↓	↓	↓	•	↓	↓	0 2	0 3	1 4	2 5	3 7	6 10	9 14	15 20	23 29	↑	↑
	Second	5	10																1 2	3 4	4 5	6 7	11 12	15 16	23 24	34 35	52 53		
E	First	8	8	↓	↓	↓	↓	↓	↓	↓	↓	↓	↓	↓	•	↓	↓	0 2	0 3	1 4	2 5	3 7	6 10	9 14	15 20	23 29	↑	↑	↑
	Second	8	16															1 2	3 4	4 5	6 7	11 12	15 16	23 24	34 35	52 53			
F	First	13	13	↓	↓	↓	↓	↓	↓	↓	↓	↓	↓	•	↓	↓	0 2	0 3	1 4	2 5	3 7	6 10	9 14	↑	↑	↑	↑	↑	↑
	Second	13	26														1 2	3 4	4 5	6 7	11 12	15 16	23 24						
G	First	20	20	↓	↓	↓	↓	↓	↓	↓	↓	↓	•	↓	↓	0 2	0 3	1 4	2 5	3 7	6 10	9 14	↑	↑	↑	↑	↑	↑	↑
	Second	20	40													1 2	3 4	4 5	6 7	11 12	15 16	23 24							
H	First	32	32	↓	↓	↓	↓	↓	↓	↓	↓	•	↓	↓	0 2	0 3	1 4	2 5	3 7	6 10	9 14	↑	↑	↑	↑	↑	↑	↑	↑
	Second	32	64												1 2	3 4	4 5	6 7	11 12	15 16	23 24								
J	First	50	50	↓	↓	↓	↓	↓	↓	↓	•	↓	↓	0 2	0 3	1 4	2 5	3 7	6 10	9 14	↑	↑	↑	↑	↑	↑	↑	↑	↑
	Second	50	100											1 2	3 4	4 5	6 7	11 12	15 16	23 24									
K	First	80	80	↓	↓	↓	↓	↓	↓	•	↓	↓	0 2	0 3	1 4	2 5	3 7	6 10	9 14	↑	↑	↑	↑	↑	↑	↑	↑	↑	↑
	Second	80	160										1 2	3 4	4 5	6 7	11 12	15 16	23 24										
L	First	125	125	↓	↓	↓	↓	↓	•	↓	↓	0 2	0 3	1 4	2 5	3 7	6 10	9 14	↑	↑	↑	↑	↑	↑	↑	↑	↑	↑	↑
	Second	125	250									1 2	3 4	4 5	6 7	11 12	15 16	23 24											
M	First	200	200	↓	↓	↓	↓	•	↓	↓	0 2	0 3	1 4	2 5	3 7	6 10	9 14	↑	↑	↑	↑	↑	↑	↑	↑	↑	↑	↑	↑
	Second	200	400								1 2	3 4	4 5	6 7	11 12	15 16	23 24												
N	First	315	315	↓	↓	↓	•	↓	↓	0 2	0 3	1 4	2 5	3 7	6 10	9 14	↑	↑	↑	↑	↑	↑	↑	↑	↑	↑	↑	↑	↑
	Second	315	630							1 2	3 4	4 5	6 7	11 12	15 16	23 24													
P	First	500	500	↓	↓	•	↓	↓	0 2	0 3	1 4	2 5	3 7	6 10	9 14	↑	↑	↑	↑	↑	↑	↑	↑	↑	↑	↑	↑	↑	↑
	Second	500	1000						1 2	3 4	4 5	6 7	11 12	15 16	23 24														
Q	First	800	800	↓	•	↓	↓	0 2	0 3	1 4	2 5	3 7	6 10	9 14	↑	↑	↑	↑	↑	↑	↑	↑	↑	↑	↑	↑	↑	↑	↑
	Second	800	1600					1 2	3 4	4 5	6 7	11 12	15 16	23 24															
R	First	1250	1250	•	↑	↓	0 2	0 3	1 4	2 5	3 7	6 10	9 14	↑	↑	↑	↑	↑	↑	↑	↑	↑	↑	↑	↑	↑	↑	↑	↑
	Second	1250	2500				1 2	3 4	4 5	6 7	11 12	15 16	23 24																
S	First	2000	2000			0 2																							
	Second	2000	4000			1 2																							

↓ = Use first sampling plan below arrow. If sample size equals or exceeds lot or batch size, do 100 percent inspection.
↑ = Use first sampling plan above arrow.
Ac = Acceptance number
Re = Rejection number
• = Use corresponding single sampling plan (or, alternatively, use double sampling plan below, where available).

TABLE 10.8
Master Table for Normal Inspection—Multiple Sampling (Mil. Std. 105D, Table IV–A)

Sample size code letter	Sample	Sample size	Cumulative sample size	Acceptable Quality Levels (normal inspection)																									
				0.010	0.015	0.025	0.040	0.065	0.10	0.15	0.25	0.40	0.65	1.0	1.5	2.5	4.0	6.5	10	15	25	40	65	100	150	250	400	650	1000
				Ac Re	Ac Re	Ac Re	Ac Re	Ac Re	Ac Re	Ac Re	Ac Re	Ac Re	Ac Re	Ac Re	Ac Re	Ac Re	Ac Re	Ac Re	Ac Re	Ac Re	Ac Re	Ac Re	Ac Re	Ac Re	Ac Re	Ac Re	Ac Re	Ac Re	Ac Re
A				↓	↓	↓	↓	↓	↓	↓	↓	↓	↓	↓	↓	↓	↓	•	↓	↓	•	•	•	•	•	•	•	•	•
B				↓	↓	↓	↓	↓	↓	↓	↓	↓	↓	↓	↓	↓	•	↑	↓	‡	‡	‡	‡	‡	‡	‡	‡	‡	‡
C				↓	↓	↓	↓	↓	↓	↓	↓	↓	↓	↓	↓	•	↑	↓	‡	‡	‡	‡	‡	‡	‡	‡	‡	‡	↑
D	First	2	2	↓	↓	↓	↓	↓	↓	↓	↓	↓	↓	↓	•	↑	↓	* 2	* 2	* 3	* 4	0 4	0 5	1 7	2 9	4 12	6 16	↑	↑
	Second	2	4															* 2	0 3	0 3	1 5	1 6	3 8	4 10	7 14	11 19	17 27		
	Third	2	6															0 2	0 3	1 4	2 6	3 8	6 10	8 13	13 19	19 27	29 39		
	Fourth	2	8															0 3	1 4	2 5	3 7	5 10	8 13	12 17	19 25	27 34	40 49		
	Fifth	2	10															1 3	2 4	3 6	5 8	7 11	11 15	17 20	25 29	36 40	53 58		
	Sixth	2	12															1 3	3 5	4 6	7 9	10 12	14 17	21 23	31 33	45 47	65 68		
	Seventh	2	14															2 3	4 5	6 7	9 10	13 14	18 19	25 26	37 38	53 54	77 78		
E	First	3	3	↓	↓	↓	↓	↓	↓	↓	↓	↓	↓	•	↑	↓	* 2	* 2	* 3	* 4	0 4	0 5	1 7	2 9	4 12	6 16	↑	↑	↑
	Second	3	6														* 2	0 3	0 3	1 5	1 6	3 8	4 10	7 14	11 19	17 27			
	Third	3	9														0 2	0 3	1 4	2 6	3 8	6 10	8 13	13 19	19 27	29 39			
	Fourth	3	12														0 3	1 4	2 5	3 7	5 10	8 13	12 17	19 25	27 34	40 49			
	Fifth	3	15														1 3	2 4	3 6	5 8	7 11	11 15	17 20	25 29	36 40	53 58			
	Sixth	3	18														1 3	3 5	4 6	7 9	10 12	14 17	21 23	31 33	45 47	65 68			
	Seventh	3	21														2 3	4 5	6 7	9 10	13 14	18 19	25 26	37 38	53 54	77 78			
F	First	5	5	↓	↓	↓	↓	↓	↓	↓	↓	↓	•	↑	↓	* 2	* 2	* 3	* 4	0 4	0 5	1 7	2 9	↑	↑	↑	↑	↑	↑
	Second	5	10													* 2	0 3	0 3	1 5	1 6	3 8	4 10	7 14						
	Third	5	15													0 2	0 3	1 4	2 6	3 8	6 10	8 13	13 19						
	Fourth	5	20													0 3	1 4	2 5	3 7	5 10	8 13	12 17	19 25						
	Fifth	5	25													1 3	2 4	3 6	5 8	7 11	11 15	17 20	25 29						
	Sixth	5	30													1 3	3 5	4 6	7 9	10 12	14 17	21 23	31 33						
	Seventh	5	35													2 3	4 5	6 7	9 10	13 14	18 19	25 26	37 38						
G	First	8	8	↓	↓	↓	↓	↓	↓	↓	↓	•	↑	↓	* 2	* 2	* 3	* 4	0 4	0 5	1 7	2 9	↑	↑	↑	↑	↑	↑	↑
	Second	8	16												* 2	0 3	0 3	1 5	1 6	3 8	4 10	7 14							
	Third	8	24												0 2	0 3	1 4	2 6	3 8	6 10	8 13	13 19							
	Fourth	8	32												0 3	1 4	2 5	3 7	5 10	8 13	12 17	19 25							
	Fifth	8	40												1 3	2 4	3 6	5 8	7 11	11 15	17 20	25 29							
	Sixth	8	48												1 3	3 5	4 6	7 9	10 12	14 17	21 23	31 33							
	Seventh	8	56												2 3	4 5	6 7	9 10	13 14	18 19	25 26	37 38							
H	First	13	13	↓	↓	↓	↓	↓	↓	↓	•	↑	↓	* 2	* 2	* 3	* 4	0 4	0 5	1 7	2 9	↑	↑	↑	↑	↑	↑	↑	↑
	Second	13	26										* 2	0 3	0 3	1 5	1 6	3 8	4 10	7 14								
	Third	13	39										0 2	0 3	1 4	2 6	3 8	6 10	8 13	13 19								
	Fourth	13	52										0 3	1 4	2 5	3 7	5 10	8 13	12 17	19 25								
	Fifth	13	65										1 3	2 4	3 6	5 8	7 11	11 15	17 20	25 29								
	Sixth	13	78										1 3	3 5	4 6	7 9	10 12	14 17	21 23	31 33								
	Seventh	13	91										2 3	4 5	6 7	9 10	13 14	18 19	25 26	37 38								
J	First	20	20	↓	↓	↓	↓	↓	↓	•	↑	↓	* 2	* 2	* 3	* 4	0 4	0 5	1 7	2 9	↑	↑	↑	↑	↑	↑	↑	↑	↑
	Second	20	40									* 2	0 3	0 3	1 5	1 6	3 8	4 10	7 14									
	Third	20	60									0 2	0 3	1 4	2 6	3 8	6 10	8 13	13 19									
	Fourth	20	80									0 3	1 4	2 5	3 7	5 10	8 13	12 17	19 25									
	Fifth	20	100									1 3	2 4	3 6	5 8	7 11	11 15	17 20	25 29									
	Sixth	20	120									1 3	3 5	4 6	7 9	10 12	14 17	21 23	31 33									
	Seventh	20	140									2 3	4 5	6 7	9 10	13 14	18 19	25 26	37 38									

↓ = Use first sampling plan below arrow (refer to continuation of table on following page, when necessary). If sample size equals or exceeds lot or batch size, do 100 percent inspection.
↑ = Use first sampling plan above arrow.
Ac = Acceptance number.
Re = Rejection number.
• = Use corresponding single sampling plan (or alternatively, use multiple sampling plan below, where available).
‡ = Use corresponding double sampling plan (or alternatively, use multiple sampling plan below, where available).
* = Acceptance not permitted at this sample size.

TABLE 10.8 *(concluded)*

Master Table for Normal Inspection—Multiple Sampling (Mil. Std. 105D, Table IV–A, Continued)

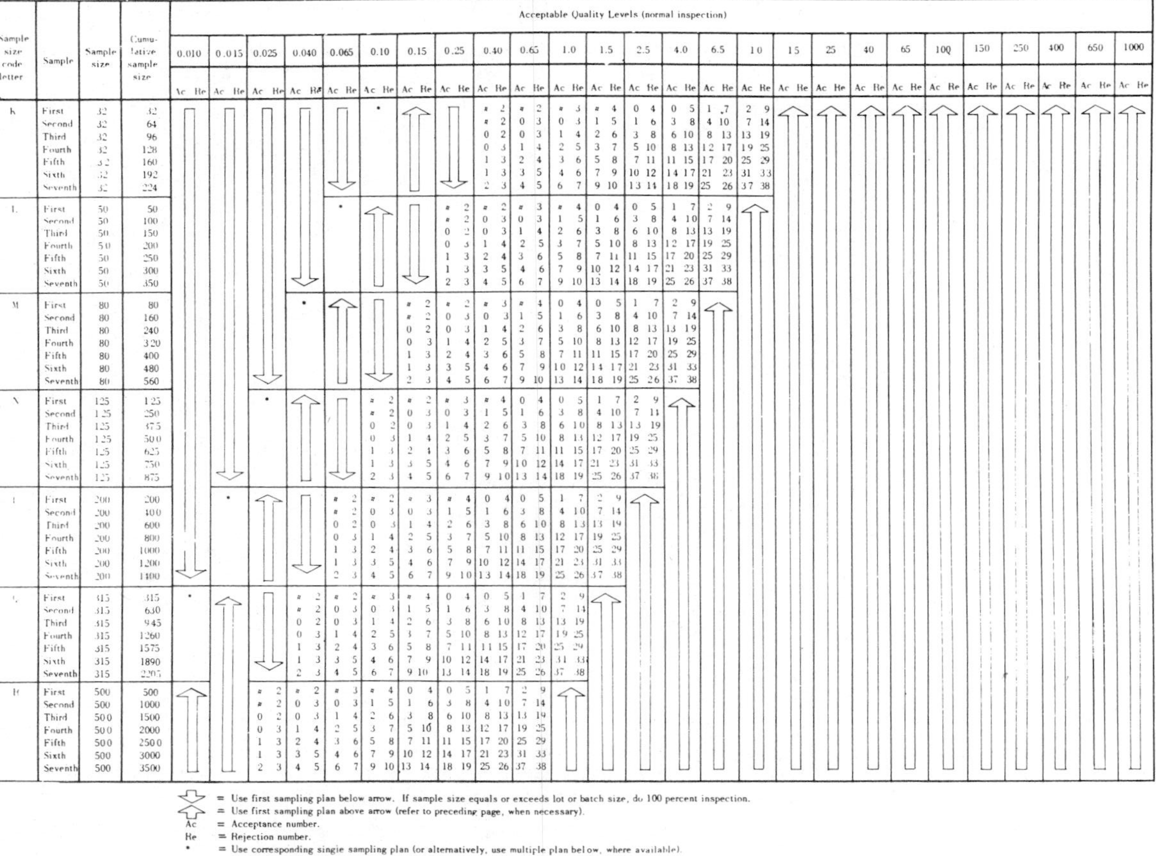

Sample size code letter	Sample	Sample size	Cumulative sample size	0.010	0.015	0.025	0.040	0.065	0.10	0.15	0.25	0.40	0.65	1.0	1.5	2.5	4.0	6.5	10	15	25	40	65	100	150	250	400	650	1000
				Ac Re	Ac Re	Ac Re	Ac Re	Ac Re	Ac Re	Ac Re	Ac Re	Ac Re	Ac Re	Ac Re	Ac Re	Ac Re	Ac Re	Ac Re	Ac Re	Ac Re	Ac Re	Ac Re	Ac Re	Ac Re	Ac Re	Ac Re	Ac Re	Ac Re	Ac Re
K	First	32	32	↓	↓	↓	↓	↓	•	↑	↓	# 2	# 2	# 3	# 4	0 4	0 5	1 7	2 9	↑	↑	↑	↑	↑	↑	↑	↑	↑	↑
	Second	32	64	↓	↓	↓	↓	↓		↑	↓	# 2	0 3	0 3	1 5	1 6	3 8	4 10	7 14	↑	↑	↑	↑	↑	↑	↑	↑	↑	↑
	Third	32	96	↓	↓	↓	↓	↓		↑	↓	0 2	0 3	1 4	2 6	3 8	6 10	8 13	13 19	↑	↑	↑	↑	↑	↑	↑	↑	↑	↑
	Fourth	32	128	↓	↓	↓	↓	↓		↑	↓	0 3	1 4	2 5	3 7	5 10	8 13	12 17	19 25	↑	↑	↑	↑	↑	↑	↑	↑	↑	↑
	Fifth	32	160	↓	↓	↓	↓	↓		↑	↓	1 3	2 4	3 6	5 8	7 11	11 15	17 20	25 29	↑	↑	↑	↑	↑	↑	↑	↑	↑	↑
	Sixth	32	192	↓	↓	↓	↓	↓		↑	↓	1 3	3 5	4 6	7 9	10 12	14 17	21 23	31 33	↑	↑	↑	↑	↑	↑	↑	↑	↑	↑
	Seventh	32	224	↓	↓	↓	↓	↓		↑	↓	2 3	4 5	6 7	9 10	13 14	18 19	25 26	37 38	↑	↑	↑	↑	↑	↑	↑	↑	↑	↑
L	First	50	50	↓	↓	↓	↓	•	↑	↓	# 2	# 2	# 3	# 4	0 4	0 5	1 7	2 9	↑	↑	↑	↑	↑	↑	↑	↑	↑	↑	↑
	Second	50	100	↓	↓	↓	↓		↑	↓	# 2	0 3	0 3	1 5	1 6	3 8	4 10	7 14	↑	↑	↑	↑	↑	↑	↑	↑	↑	↑	↑
	Third	50	150	↓	↓	↓	↓		↑	↓	0 2	0 3	1 4	2 6	3 8	6 10	8 13	13 19	↑	↑	↑	↑	↑	↑	↑	↑	↑	↑	↑
	Fourth	50	200	↓	↓	↓	↓		↑	↓	0 3	1 4	2 5	3 7	5 10	8 13	12 17	19 25	↑	↑	↑	↑	↑	↑	↑	↑	↑	↑	↑
	Fifth	50	250	↓	↓	↓	↓		↑	↓	1 3	2 4	3 6	5 8	7 11	11 15	17 20	25 29	↑	↑	↑	↑	↑	↑	↑	↑	↑	↑	↑
	Sixth	50	300	↓	↓	↓	↓		↑	↓	1 3	3 5	4 6	7 9	10 12	14 17	21 23	31 33	↑	↑	↑	↑	↑	↑	↑	↑	↑	↑	↑
	Seventh	50	350	↓	↓	↓	↓		↑	↓	2 3	4 5	6 7	9 10	13 14	18 19	25 26	37 38	↑	↑	↑	↑	↑	↑	↑	↑	↑	↑	↑
M	First	80	80	↓	↓	↓	•	↑	↓	# 2	# 2	# 3	# 4	0 4	0 5	1 7	2 9	↑	↑	↑	↑	↑	↑	↑	↑	↑	↑	↑	↑
	Second	80	160	↓	↓	↓		↑	↓	# 2	0 3	0 3	1 5	1 6	3 8	4 10	7 14	↑	↑	↑	↑	↑	↑	↑	↑	↑	↑	↑	↑
	Third	80	240	↓	↓	↓		↑	↓	0 2	0 3	1 4	2 6	3 8	6 10	8 13	13 19	↑	↑	↑	↑	↑	↑	↑	↑	↑	↑	↑	↑
	Fourth	80	320	↓	↓	↓		↑	↓	0 3	1 4	2 5	3 7	5 10	8 13	12 17	19 25	↑	↑	↑	↑	↑	↑	↑	↑	↑	↑	↑	↑
	Fifth	80	400	↓	↓	↓		↑	↓	1 3	2 4	3 6	5 8	7 11	11 15	17 20	25 29	↑	↑	↑	↑	↑	↑	↑	↑	↑	↑	↑	↑
	Sixth	80	480	↓	↓	↓		↑	↓	1 3	3 5	4 6	7 9	10 12	14 17	21 23	31 33	↑	↑	↑	↑	↑	↑	↑	↑	↑	↑	↑	↑
	Seventh	80	560	↓	↓	↓		↑	↓	2 3	4 5	6 7	9 10	13 14	18 19	25 26	37 38	↑	↑	↑	↑	↑	↑	↑	↑	↑	↑	↑	↑
N	First	125	125	↓	↓	•	↑	↓	# 2	# 2	# 3	# 4	0 4	0 5	1 7	2 9	↑	↑	↑	↑	↑	↑	↑	↑	↑	↑	↑	↑	↑
	Second	125	250	↓	↓		↑	↓	# 2	0 3	0 3	1 5	1 6	3 8	4 10	7 14	↑	↑	↑	↑	↑	↑	↑	↑	↑	↑	↑	↑	↑
	Third	125	375	↓	↓		↑	↓	0 2	0 3	1 4	2 6	3 8	6 10	8 13	13 19	↑	↑	↑	↑	↑	↑	↑	↑	↑	↑	↑	↑	↑
	Fourth	125	500	↓	↓		↑	↓	0 3	1 4	2 5	3 7	5 10	8 13	12 17	19 25	↑	↑	↑	↑	↑	↑	↑	↑	↑	↑	↑	↑	↑
	Fifth	125	625	↓	↓		↑	↓	1 3	2 4	3 6	5 8	7 11	11 15	17 20	25 29	↑	↑	↑	↑	↑	↑	↑	↑	↑	↑	↑	↑	↑
	Sixth	125	750	↓	↓		↑	↓	1 3	3 5	4 6	7 9	10 12	14 17	21 23	31 33	↑	↑	↑	↑	↑	↑	↑	↑	↑	↑	↑	↑	↑
	Seventh	125	875	↓	↓		↑	↓	2 3	4 5	6 7	9 10	13 14	18 19	25 26	37 38	↑	↑	↑	↑	↑	↑	↑	↑	↑	↑	↑	↑	↑
P	First	200	200	↓	•	↑	↓	# 2	# 2	# 3	# 4	0 4	0 5	1 7	2 9	↑	↑	↑	↑	↑	↑	↑	↑	↑	↑	↑	↑	↑	↑
	Second	200	400	↓		↑	↓	# 2	0 3	0 3	1 5	1 6	3 8	4 10	7 14	↑	↑	↑	↑	↑	↑	↑	↑	↑	↑	↑	↑	↑	↑
	Third	200	600	↓		↑	↓	0 2	0 3	1 4	2 6	3 8	6 10	8 13	13 19	↑	↑	↑	↑	↑	↑	↑	↑	↑	↑	↑	↑	↑	↑
	Fourth	200	800	↓		↑	↓	0 3	1 4	2 5	3 7	5 10	8 13	12 17	19 25	↑	↑	↑	↑	↑	↑	↑	↑	↑	↑	↑	↑	↑	↑
	Fifth	200	1000	↓		↑	↓	1 3	2 4	3 6	5 8	7 11	11 15	17 20	25 29	↑	↑	↑	↑	↑	↑	↑	↑	↑	↑	↑	↑	↑	↑
	Sixth	200	1200	↓		↑	↓	1 3	3 5	4 6	7 9	10 12	14 17	21 23	31 33	↑	↑	↑	↑	↑	↑	↑	↑	↑	↑	↑	↑	↑	↑
	Seventh	200	1400	↓		↑	↓	2 3	4 5	6 7	9 10	13 14	18 19	25 26	37 38	↑	↑	↑	↑	↑	↑	↑	↑	↑	↑	↑	↑	↑	↑
Q	First	315	315	•	↑	↓	# 2	# 2	# 3	# 4	0 4	0 5	1 7	2 9	↑	↑	↑	↑	↑	↑	↑	↑	↑	↑	↑	↑	↑	↑	↑
	Second	315	630		↑	↓	# 2	0 3	0 3	1 5	1 6	3 8	4 10	7 14	↑	↑	↑	↑	↑	↑	↑	↑	↑	↑	↑	↑	↑	↑	↑
	Third	315	945		↑	↓	0 2	0 3	1 4	2 6	3 8	6 10	8 13	13 19	↑	↑	↑	↑	↑	↑	↑	↑	↑	↑	↑	↑	↑	↑	↑
	Fourth	315	1260		↑	↓	0 3	1 4	2 5	3 7	5 10	8 13	12 17	19 25	↑	↑	↑	↑	↑	↑	↑	↑	↑	↑	↑	↑	↑	↑	↑
	Fifth	315	1575		↑	↓	1 3	2 4	3 6	5 8	7 11	11 15	17 20	25 29	↑	↑	↑	↑	↑	↑	↑	↑	↑	↑	↑	↑	↑	↑	↑
	Sixth	315	1890		↑	↓	1 3	3 5	4 6	7 9	10 12	14 17	21 23	31 33	↑	↑	↑	↑	↑	↑	↑	↑	↑	↑	↑	↑	↑	↑	↑
	Seventh	315	2205		↑	↓	2 3	4 5	6 7	9 10	13 14	18 19	25 26	37 38	↑	↑	↑	↑	↑	↑	↑	↑	↑	↑	↑	↑	↑	↑	↑
R	First	500	500	↑	↑	# 2	# 2	# 3	# 4	0 4	0 5	1 7	2 9	↑	↑	↑	↑	↑	↑	↑	↑	↑	↑	↑	↑	↑	↑	↑	↑
	Second	500	1000	↑	↑	# 2	0 3	0 3	1 5	1 6	3 8	4 10	7 14	↑	↑	↑	↑	↑	↑	↑	↑	↑	↑	↑	↑	↑	↑	↑	↑
	Third	500	1500	↑	↑	0 2	0 3	1 4	2 6	3 8	6 10	8 13	13 19	↑	↑	↑	↑	↑	↑	↑	↑	↑	↑	↑	↑	↑	↑	↑	↑
	Fourth	500	2000	↑	↑	0 3	1 4	2 5	3 7	5 10	8 13	12 17	19 25	↑	↑	↑	↑	↑	↑	↑	↑	↑	↑	↑	↑	↑	↑	↑	↑
	Fifth	500	2500	↑	↑	1 3	2 4	3 6	5 8	7 11	11 15	17 20	25 29	↑	↑	↑	↑	↑	↑	↑	↑	↑	↑	↑	↑	↑	↑	↑	↑
	Sixth	500	3000	↑	↑	1 3	3 5	4 6	7 9	10 12	14 17	21 23	31 33	↑	↑	↑	↑	↑	↑	↑	↑	↑	↑	↑	↑	↑	↑	↑	↑
	Seventh	500	3500	↑	↑	2 3	4 5	6 7	9 10	13 14	18 19	25 26	37 38	↑	↑	↑	↑	↑	↑	↑	↑	↑	↑	↑	↑	↑	↑	↑	↑

Acceptable Quality Levels (normal inspection)

↓ = Use first sampling plan below arrow. If sample size equals or exceeds lot or batch size, do 100 percent inspection.
↑ = Use first sampling plan above arrow (refer to preceding page, when necessary).
Ac = Acceptance number.
Re = Rejection number.
• = Use corresponding single sampling plan (or alternatively, use multiple plan below, where available).
= Acceptance not permitted at this sample size.

TABLE 10.9
Master Table for Tightened Inspection—Multiple Sampling (Mil. Std. 105D, Table IV–B)

Sample size code letter	Sample	Sample size	Cumulative sample size	Acceptable Quality Levels (tightened inspection)																									
				0.010	0.015	0.025	0.040	0.065	0.10	0.15	0.25	0.40	0.65	1.0	1.5	2.5	4.0	6.5	10	15	25	40	65	100	150	250	400	650	1000
				Ac Re	Ac Re	Ac Re	Ac Re	Ac Re	Ac Re	Ac Re	Ac Re	Ac Re	Ac Re	Ac Re	Ac Re	Ac Re	Ac Re	Ac Re	Ac Re	Ac Re	Ac Re	Ac Re	Ac Re	Ac Re	Ac Re	Ac Re	Ac Re	Ac Re	Ac Re
A				↓	↓	↓	↓	↓	↓	↓	↓	↓	↓	↓	↓	↓	↓	↓	↓	↓	↓	•	•	•	•	•	•	•	•
B				↓	↓	↓	↓	↓	↓	↓	↓	↓	↓	↓	↓	↓	↓	•	↓	↓	++	++	++	++	++	++	++	++	++
C				↓	↓	↓	↓	↓	↓	↓	↓	↓	↓	↓	↓	↓	•	↓	↓	++	++	++	++	++	++	++	++	++	↑
D	First	2	2	↓	↓	↓	↓	↓	↓	↓	↓	↓	↓	↓	↓	•	↓	↓	# 2	# 2	# 3	# 4	0 4	0 6	1 8	3 10	6 15	↑	↑
	Second	2	4	↓	↓	↓	↓	↓	↓	↓	↓	↓	↓	↓	↓		↓	↓	# 2	0 3	0 3	1 5	2 7	3 9	6 12	10 17	16 25	↑	↑
	Third	2	6	↓	↓	↓	↓	↓	↓	↓	↓	↓	↓	↓	↓		↓	↓	0 2	0 3	1 4	2 6	4 9	7 12	11 17	17 24	26 36	↑	↑
	Fourth	2	8	↓	↓	↓	↓	↓	↓	↓	↓	↓	↓	↓	↓		↓	↓	0 3	1 4	2 5	3 7	6 11	10 15	16 22	24 31	37 46	↑	↑
	Fifth	2	10	↓	↓	↓	↓	↓	↓	↓	↓	↓	↓	↓	↓		↓	↓	1 3	2 4	3 6	5 8	9 12	14 17	22 25	32 37	49 55	↑	↑
	Sixth	2	12	↓	↓	↓	↓	↓	↓	↓	↓	↓	↓	↓	↓		↓	↓	1 3	3 5	4 6	7 9	12 14	18 20	27 29	40 43	61 64	↑	↑
	Seventh	2	14	↓	↓	↓	↓	↓	↓	↓	↓	↓	↓	↓	↓		↓	↓	2 3	4 5	6 7	9 10	14 15	21 22	32 33	48 49	72 73	↑	↑
E	First	3	3	↓	↓	↓	↓	↓	↓	↓	↓	↓	↓	↓	•	↓	↓	# 2	# 2	# 3	# 4	0 4	0 6	1 8	3 10	6 15	↑	↑	↑
	Second	3	6	↓	↓	↓	↓	↓	↓	↓	↓	↓	↓	↓		↓	↓	# 2	0 3	0 3	1 5	2 7	3 9	6 12	10 17	16 25	↑	↑	↑
	Third	3	9	↓	↓	↓	↓	↓	↓	↓	↓	↓	↓	↓		↓	↓	0 2	0 3	1 4	2 6	4 9	7 12	11 17	17 24	26 36	↑	↑	↑
	Fourth	3	12	↓	↓	↓	↓	↓	↓	↓	↓	↓	↓	↓		↓	↓	0 3	1 4	2 5	3 7	6 11	10 15	16 22	24 31	37 46	↑	↑	↑
	Fifth	3	15	↓	↓	↓	↓	↓	↓	↓	↓	↓	↓	↓		↓	↓	1 3	2 4	3 6	5 8	9 12	14 17	22 25	32 37	49 55	↑	↑	↑
	Sixth	3	18	↓	↓	↓	↓	↓	↓	↓	↓	↓	↓	↓		↓	↓	1 3	3 5	4 6	7 9	12 14	18 20	27 29	40 43	61 64	↑	↑	↑
	Seventh	3	21	↓	↓	↓	↓	↓	↓	↓	↓	↓	↓	↓		↓	↓	2 3	4 5	6 7	9 10	14 15	21 22	32 33	48 49	72 73	↑	↑	↑
F	First	5	5	↓	↓	↓	↓	↓	↓	↓	↓	↓	↓	•	↓	↓	# 2	# 2	# 3	# 4	0 4	0 6	1 8	↑	↑	↑	↑	↑	↑
	Second	5	10	↓	↓	↓	↓	↓	↓	↓	↓	↓	↓		↓	↓	# 2	0 3	0 3	1 5	2 7	3 9	6 12	↑	↑	↑	↑	↑	↑
	Third	5	15	↓	↓	↓	↓	↓	↓	↓	↓	↓	↓		↓	↓	0 2	0 3	1 4	2 6	4 9	7 12	11 17	↑	↑	↑	↑	↑	↑
	Fourth	5	20	↓	↓	↓	↓	↓	↓	↓	↓	↓	↓		↓	↓	0 3	1 4	2 5	3 7	6 11	10 15	16 22	↑	↑	↑	↑	↑	↑
	Fifth	5	25	↓	↓	↓	↓	↓	↓	↓	↓	↓	↓		↓	↓	1 3	2 4	3 6	5 8	9 12	14 17	22 25	↑	↑	↑	↑	↑	↑
	Sixth	5	30	↓	↓	↓	↓	↓	↓	↓	↓	↓	↓		↓	↓	1 3	3 5	4 6	7 9	12 14	18 20	27 29	↑	↑	↑	↑	↑	↑
	Seventh	5	35	↓	↓	↓	↓	↓	↓	↓	↓	↓	↓		↓	↓	2 3	4 5	6 7	9 10	14 15	21 22	32 33	↑	↑	↑	↑	↑	↑
G	First	8	8	↓	↓	↓	↓	↓	↓	↓	↓	↓	•	↓	↓	# 2	# 2	# 3	# 4	0 4	0 6	1 8	↑	↑	↑	↑	↑	↑	↑
	Second	8	16	↓	↓	↓	↓	↓	↓	↓	↓	↓		↓	↓	# 2	0 3	0 3	1 5	2 7	3 9	6 12	↑	↑	↑	↑	↑	↑	↑
	Third	8	24	↓	↓	↓	↓	↓	↓	↓	↓	↓		↓	↓	0 2	0 3	1 4	2 6	4 9	7 12	11 17	↑	↑	↑	↑	↑	↑	↑
	Fourth	8	32	↓	↓	↓	↓	↓	↓	↓	↓	↓		↓	↓	0 3	1 4	2 5	3 7	6 11	10 15	16 22	↑	↑	↑	↑	↑	↑	↑
	Fifth	8	40	↓	↓	↓	↓	↓	↓	↓	↓	↓		↓	↓	1 3	2 4	3 6	5 8	9 12	14 17	22 25	↑	↑	↑	↑	↑	↑	↑
	Sixth	8	48	↓	↓	↓	↓	↓	↓	↓	↓	↓		↓	↓	1 3	3 5	4 6	7 9	12 14	18 20	27 29	↑	↑	↑	↑	↑	↑	↑
	Seventh	8	56	↓	↓	↓	↓	↓	↓	↓	↓	↓		↓	↓	2 3	4 5	6 7	9 10	14 15	21 22	32 33	↑	↑	↑	↑	↑	↑	↑
H	First	13	13	↓	↓	↓	↓	↓	↓	↓	↓	•	↓	↓	# 2	# 2	# 3	# 4	0 4	0 6	1 8	↑	↑	↑	↑	↑	↑	↑	↑
	Second	13	26	↓	↓	↓	↓	↓	↓	↓	↓		↓	↓	# 2	0 3	0 3	1 5	2 7	3 9	6 12	↑	↑	↑	↑	↑	↑	↑	↑
	Third	13	39	↓	↓	↓	↓	↓	↓	↓	↓		↓	↓	0 2	0 3	1 4	2 6	4 9	7 12	11 17	↑	↑	↑	↑	↑	↑	↑	↑
	Fourth	13	52	↓	↓	↓	↓	↓	↓	↓	↓		↓	↓	0 3	1 4	2 5	3 7	6 11	10 15	16 22	↑	↑	↑	↑	↑	↑	↑	↑
	Fifth	13	65	↓	↓	↓	↓	↓	↓	↓	↓		↓	↓	1 3	2 4	3 6	5 8	9 12	14 17	22 25	↑	↑	↑	↑	↑	↑	↑	↑
	Sixth	13	78	↓	↓	↓	↓	↓	↓	↓	↓		↓	↓	1 3	3 5	4 6	7 9	12 14	18 20	27 29	↑	↑	↑	↑	↑	↑	↑	↑
	Seventh	13	91	↓	↓	↓	↓	↓	↓	↓	↓		↓	↓	2 3	4 5	6 7	9 10	14 15	21 22	32 33	↑	↑	↑	↑	↑	↑	↑	↑
J	First	20	20	↓	↓	↓	↓	↓	↓	↓	•	↓	↓	# 2	# 2	# 3	# 4	0 4	0 6	1 8	↑	↑	↑	↑	↑	↑	↑	↑	↑
	Second	20	40	↓	↓	↓	↓	↓	↓	↓		↓	↓	# 2	0 3	0 3	1 5	2 7	3 9	6 12	↑	↑	↑	↑	↑	↑	↑	↑	↑
	Third	20	60	↓	↓	↓	↓	↓	↓	↓		↓	↓	0 2	0 3	1 4	2 6	4 9	7 12	11 17	↑	↑	↑	↑	↑	↑	↑	↑	↑
	Fourth	20	80	↓	↓	↓	↓	↓	↓	↓		↓	↓	0 3	1 4	2 5	3 7	6 11	10 15	16 22	↑	↑	↑	↑	↑	↑	↑	↑	↑
	Fifth	20	100	↓	↓	↓	↓	↓	↓	↓		↓	↓	1 3	2 4	3 6	5 8	9 12	14 17	22 25	↑	↑	↑	↑	↑	↑	↑	↑	↑
	Sixth	20	120	↓	↓	↓	↓	↓	↓	↓		↓	↓	1 3	3 5	4 6	7 9	12 14	18 20	27 29	↑	↑	↑	↑	↑	↑	↑	↑	↑
	Seventh	20	140	↓	↓	↓	↓	↓	↓	↓		↓	↓	2 3	4 5	6 7	9 10	14 15	21 22	32 33	↑	↑	↑	↑	↑	↑	↑	↑	↑

↓ = Use first sampling plan below arrow (refer to continuation of table on following page, when necessary). If sample size equals or exceeds lot or batch size, do 100 percent inspection.
↑ = Use first sampling plan above arrow.
Ac = Acceptance number
Re = Rejection number
• = Use corresponding single sampling plan (or alternatively, use multiple sampling plan below, where available).
++ = Use corresponding double sampling plan (or alternatively, use multiple sampling plan below, where available).
# = Acceptance not permitted at this sample size.

TABLE 10.9 *(concluded)*

Master Table for Tightened Inspection—Multiple Sampling (Mil. Std. 105D, Table IV–B, Continued)

Sample size code letter	Sample	Sample size	Cumulative sample size	Acceptable Quality Levels (tightened inspection) 0.010	0.015	0.025	0.040	0.065	0.10	0.15	0.25	0.40	0.65	1.0	1.5	2.5	4.0	6.5	10	15	25	40	65	100	150	250	400	650	1000
				Ac Re	Ac Re	Ac Re	Ac Re	Ac Re	Ac Re	Ac Re	Ac Re	Ac Re	Ac Re	Ac Re	Ac Re	Ac Re	Ac Re	Ac Re	Ac Re	Ac Re	Ac Re	Ac Re	Ac Re	Ac Re	Ac Re	Ac Re	Ac Re	Ac Re	Ac Re
K	First	32	32	↓	↓	↓	↓	↓	↓	•	↓	↓	# 2	# 2	# 3	# 4	0 4	0 6	1 8	↑	↑	↑	↑	↑	↑	↑	↑	↑	↑
	Second	32	64										# 2	0 3	0 3	1 5	2 7	3 9	6 12										
	Third	32	96										0 2	0 3	1 4	2 6	4 9	7 12	11 17										
	Fourth	32	128										0 3	1 4	2 5	3 7	6 11	10 15	16 22										
	Fifth	32	160										1 3	2 4	3 6	5 8	9 12	14 17	22 25										
	Sixth	32	192										1 3	3 5	4 6	7 9	12 14	18 20	27 29										
	Seventh	32	224										2 3	4 5	6 7	9 10	14 15	21 22	32 33										
L	First	50	50	↓	↓	↓	↓	↓	•	↓	↓	# 2	# 2	# 3	# 4	0 4	0 6	1 8	↑	↑	↑	↑	↑	↑	↑	↑	↑	↑	↑
	Second	50	100									# 2	0 3	0 3	1 5	2 7	3 9	6 12											
	Third	50	150									0 2	0 3	1 4	2 6	4 9	7 12	11 17											
	Fourth	50	200									0 3	1 4	2 5	3 7	6 11	10 15	16 22											
	Fifth	50	250									1 3	2 4	3 6	5 8	9 12	14 17	22 25											
	Sixth	50	300									1 3	3 5	4 6	7 9	12 14	18 20	27 29											
	Seventh	50	350									2 3	4 5	6 7	9 10	14 15	21 22	32 33											
M	First	80	80	↓	↓	↓	↓	•	↓	↓	# 2	# 2	# 3	# 4	0 4	0 6	1 8	↑	↑	↑	↑	↑	↑	↑	↑	↑	↑	↑	↑
	Second	80	160								# 2	0 3	0 3	1 5	2 7	3 9	6 12												
	Third	80	240								0 2	0 3	1 4	2 6	4 9	7 12	11 17												
	Fourth	80	320								0 3	1 4	2 5	3 7	6 11	10 15	16 22												
	Fifth	80	400								1 3	2 4	3 6	5 8	9 12	14 17	22 25												
	Sixth	80	480								1 3	3 5	4 6	7 9	12 14	18 20	27 29												
	Seventh	80	560								2 3	4 5	6 7	9 10	14 15	21 22	32 33												
N	First	125	125	↓	↓	↓	•	↓	↓	# 2	# 2	# 3	# 4	0 4	0 6	1 8	↑	↑	↑	↑	↑	↑	↑	↑	↑	↑	↑	↑	↑
	Second	125	250							# 2	0 3	0 3	1 5	2 7	3 9	6 12													
	Third	125	375							0 2	0 3	1 4	2 6	4 9	7 12	11 17													
	Fourth	125	500							0 3	1 4	2 5	3 7	6 11	10 15	16 22													
	Fifth	125	625							1 3	2 4	3 6	5 8	9 12	14 17	22 25													
	Sixth	125	750							1 3	3 5	4 6	7 9	12 14	18 20	27 29													
	Seventh	125	875							2 3	4 5	6 7	9 10	14 15	21 22	32 33													
P	First	200	200	↓	↓	•	↓	↓	# 2	# 2	# 3	# 4	0 4	0 6	1 8	↑	↑	↑	↑	↑	↑	↑	↑	↑	↑	↑	↑	↑	↑
	Second	200	400						# 2	0 3	0 3	1 5	2 7	3 9	6 12														
	Third	200	600						0 2	0 3	1 4	2 6	4 9	7 12	11 17														
	Fourth	200	800						0 3	1 4	2 5	3 7	6 11	10 15	16 22														
	Fifth	200	1000						1 3	2 4	3 6	5 8	9 12	14 17	22 25														
	Sixth	200	1200						1 3	3 5	4 6	7 9	12 14	18 20	27 29														
	Seventh	200	1400						2 3	4 5	6 7	9 10	14 15	21 22	32 33														
Q	First	315	315	↓	•	↓	↓	# 2	# 2	# 3	# 4	0 4	0 6	1 8	↑	↑	↑	↑	↑	↑	↑	↑	↑	↑	↑	↑	↑	↑	↑
	Second	315	630					# 2	0 3	0 3	1 5	2 7	3 9	6 12															
	Third	315	945					0 2	0 3	1 4	2 6	4 9	7 12	11 17															
	Fourth	315	1260					0 3	1 4	2 5	3 7	6 11	10 15	16 22															
	Fifth	315	1575					1 3	2 4	3 6	5 8	9 12	14 17	22 25															
	Sixth	315	1890					1 3	3 5	4 6	7 9	12 14	18 20	27 29															
	Seventh	315	2205					2 3	4 5	6 7	9 10	14 15	21 22	32 33															
R	First	500	500	•	↑	↓	# 2	# 2	# 3	# 4	0 4	0 6	1 8	↑	↑	↑	↑	↑	↑	↑	↑	↑	↑	↑	↑	↑	↑	↑	↑
	Second	500	1000				# 2	0 3	0 3	1 5	2 7	3 9	6 12																
	Third	500	1500				0 2	0 3	1 4	2 6	4 9	7 12	11 17																
	Fourth	500	2000				0 3	1 4	2 5	3 7	6 11	10 15	16 22																
	Fifth	500	2500				1 3	2 4	3 6	5 8	9 12	14 17	22 25																
	Sixth	500	3000				1 3	3 5	4 6	7 9	12 14	18 20	27 29																
	Seventh	500	3500				2 3	4 5	6 7	9 10	14 15	21 22	32 33																
S	First	800	800			# 2																							
	Second	800	1600			# 2																							
	Third	800	2400			0 2																							
	Fourth	800	3200			0 3																							
	Fifth	800	4000			1 3																							
	Sixth	800	4800			1 3																							
	Seventh	800	5600			2 3																							

↓ = Use first sampling plan below arrow. If sample size equals or exceeds lot or batch size, do 100 percent inspection.
↑ = Use first sampling plan above arrow (refer to preceding page, when necessary).
Ac = Acceptance number
Re = Rejection number
• = Use corresponding single sampling plan (or alternatively, use multiple sampling plan below, where available).
 # = Acceptance not permitted at this sample size.

TABLE 10.10
Limit Numbers for Reduced Inspection (Mil. Std. 105D, Table VIII)

Number of sample units from last 10 lots or batches	Acceptable Quality Level																									
	0.010	0.015	0.025	0.040	0.065	0.10	0.15	0.25	0.40	0.65	1.0	1.5	2.5	4.0	6.5	10	15	25	40	65	100	150	250	400	650	1000
20 - 29	*	*	*	*	*	*	*	*	*	*	*	*	*	*	*	0	0	2	4	8	14	22	40	68	115	181
30 - 49	*	*	*	*	*	*	*	*	*	*	*	*	*	*	0	0	1	3	7	13	22	36	63	105	178	277
50 - 79	*	*	*	*	*	*	*	*	*	*	*	*	*	0	0	2	3	7	14	25	40	63	110	181	301	
80 - 129	*	*	*	*	*	*	*	*	*	*	*	*	0	0	2	4	7	14	24	42	68	105	181	297		
130 - 199	*	*	*	*	*	*	*	*	*	*	*	0	0	2	4	7	13	25	42	72	115	177	301	490		
200 - 319	*	*	*	*	*	*	*	*	*	*	0	0	2	4	8	14	22	40	68	115	181	277	471			
320 - 499	*	*	*	*	*	*	*	*	*	0	0	1	4	8	14	24	39	68	113	189						
500 - 799	*	*	*	*	*	*	*	*	0	0	2	3	7	14	25	40	63	110	181							
800 - 1249	*	*	*	*	*	*	*	0	0	2	4	7	14	24	42	68	105	181								
1250 - 1999	*	*	*	*	*	*	0	0	2	4	7	13	24	40	69	110	169									
2000 - 3149	*	*	*	*	*	0	0	2	4	8	14	22	40	68	115	181										
3150 - 4999	*	*	*	*	0	0	1	4	8	14	24	38	67	111	186											
5000 - 7999	*	*	*	0	0	2	3	7	14	25	40	63	110	181												
8000 - 12499	*	*	0	0	2	4	7	14	24	42	68	105	181													
12500 - 19999	*	0	0	2	4	7	13	24	40	69	110	169														
20000 - 31499	0	0	2	4	8	14	22	40	68	115	181															
31500 - 49999	0	1	4	8	14	24	38	67	111	186																
50000 & Over	2	3	7	14	25	40	63	110	181	301																

* Denotes that the number of sample units from the last 10 lots or batches is not sufficient for reduced inspection for this AQL. In this instance more than 10 lots or batches may be used for the calculation, provided that the lots or batches used are the most recent ones in sequence, that they have all been on normal inspection, and that none has been rejected while on original inspection.

TABLE 10.11
Average Outgoing Quality Factors for Normal Inspection (Single Sampling) (Mil. Std. 105D, Table V–A)

Code Letter	Sample Size	Acceptable Quality Level																									
		0.010	0.015	0.025	0.040	0.065	0.10	0.15	0.25	0.40	0.65	1.0	1.5	2.5	4.0	6.5	10	15	25	40	65	100	150	250	400	650	1000
A	2															18			42	69	97	160	220	330	470	730	1100
B	3														12			28	46	65	110	150	220	310	490	720	1100
C	5													7.4			17	27	39	63	90	130	190	290	430	660	
D	8												4.6			11	17	24	40	56	82	120	180	270	410		
E	13											2.8			6.5	11	15	24	34	50	72	110	170	250			
F	20										1.8			4.2	6.9	9.7	16	22	33	47	73						
G	32									1.2			2.6	4.3	6.1	9.9	14	21	29	46							
H	50								0.74			1.7	2.7	3.9	6.3	9.0	13	19	29								
J	80							0.46			1.1	1.7	2.4	4.0	5.6	8.2	12	18									
K	125						0.29			0.67	1.1	1.6	2.5	3.6	5.2	7.5	12										
L	200					0.18			0.42	0.69	0.97	1.6	2.2	3.3	4.7	7.3											
M	315				0.12			0.27	0.44	0.62	1.00	1.4	2.1	3.0	4.7												
N	500			0.074			0.17	0.27	0.39	0.63	0.90	1.3	1.9	2.9													
P	800		0.046			0.11	0.17	0.24	0.40	0.56	0.82	1.2	1.8														
Q	1250	0.029			0.067	0.11	0.16	0.25	0.36	0.52	0.75	1.2															
R	2000			0.042	0.069	0.097	0.16	0.22	0.33	0.47	0.73																

Note: For the exact AOQL, the above values must be multiplied by $\left(1 - \frac{\text{Sample size}}{\text{Lot or batch size}}\right)$; see paragraph 11.4 of the standard.

TABLE 10.12
Average Outgoing Quality Limit Factors for Tightened Inspection (Single Sampling) (Mil. Std. 105D, Table V–B)

Code letter	Sample size	Acceptable Quality Level																									
		0.010	0.015	0.025	0.040	0.065	0.10	0.15	0.25	0.40	0.65	1.0	1.5	2.5	4.0	6.5	10	15	25	40	65	100	150	250	400	650	1000
A	2																			42	69	97	160	260	400	620	970
B	3															12			28	46	65	110	170	270	410	650	1100
C	5														7.4			17	27	39	63	100	160	250	390	610	
D	8													4.6			11	17	24	40	64	99	160	240	380		
E	13												2.8			6.5	11	15	24	40	61	95	150	240			
F	20											1.8			4.2	6.9	9.7	16	26	40	62						
G	32										1.2			2.6	4.3	6.1	9.9	16	25	39							
H	50									0.74			1.7	2.7	3.9	6.3	10	16	25								
J	80								0.46			1.1	1.7	2.4	4.0	6.4	9.9	16									
K	125							0.29			0.67	1.1	1.6	2.5	4.1	6.4	9.9										
L	200						0.18			0.42	0.69	0.97	1.6	2.6	4.0	6.2											
M	315					0.12			0.27	0.44	0.62	1.0	1.6	2.5	3.9												
N	500				0.074			0.17	0.27	0.39	0.63	1.0	1.6	2.5													
P	800			0.046			0.11	0.17	0.24	0.40	0.64	0.99	1.6														
Q	1250		0.029			0.067	0.11	0.16	0.25	0.41	0.64	0.99															
R	2000	0.018			0.042	0.069	0.097	0.16	0.26	0.40	0.62																
S	3150			0.027																							

Note: For the exact AOQL, the above values must be multiplied by $\left(1 - \frac{\text{Sample size}}{\text{Lot or batch size}}\right)$; see paragraph 11.4 of the standard.

TABLE 10.13

Limiting Quality (in Percent Defective) for which P_a = 10 Percent (for Normal Inspection, Single Sampling) (Mil. Std. 105D, Table VI–A)

Code letter	Sample size	Acceptable Quality Level															
		0.010	0.015	0.025	0.040	0.065	0.10	0.15	0.25	0.40	0.65	1.0	1.5	2.5	4.0	6.5	10
A	2															68	
B	3														54		
C	5													37			58
D	8												25			41	54
E	13											16			27	36	44
F	20										11			18	25	30	42
G	32									6.9			12	16	20	27	34
H	50								4.5			7.6	10	13	18	22	29
J	80							2.8			4.8	6.5	8.2	11	14	19	24
K	125						1.8			3.1	4.3	5.4	7.4	9.4	12	16	23
L	200					1.2			2.0	2.7	3.3	4.6	5.9	7.7	10	14	
M	315				0.73			1.2	1.7	2.1	2.9	3.7	4.9	6.4	9.0		
N	500			0.46			0.78	1.1	1.3	1.9	2.4	3.1	4.0	5.6			
P	800		0.29			0.49	0.67	0.84	1.2	1.5	1.9	2.5	3.5				
Q	1250	0.18			0.31	0.43	0.53	0.74	0.94	1.2	1.6	2.3					
R	2000			0.20	0.27	0.33	0.46	0.59	0.77	1.0	1.4						

With two exceptions, if a plan from the standard is used without provision for shifts to tightened inspection under the prescribed circumstances, it cannot be said that Mil. Std. 105D is being employed. The principal purpose of the standard is to put pressure on the producer to turn out material that is of quality at least as good as the designated AQL and this purpose is not attained in the absence of the provision for tightened inspection. The provision for reduced inspection is optional and is not essential for the operation of the standard. The two exceptions referred to above are the selection of a lot tolerance plan and a special AOQL plan. Thus it is possible (1) to extract a normal plan from the standard that will give designated protection against accepting an isolated bad lot, and (2) it is possible to select a tightened inspection plan for which the AOQL equals approximately the AQL. These, however, are special uses of the standard.[3]

The meaning of the AQL should also be clearly understood. It is essentially a brief way of characterizing the inspection procedure adopted. It is an aspect of the "judicial" process. It is not an aspect of specification or production. Product specifications and production procedures are not based directly on the AQL. For a producer who is faced with inspection of his product under Mil. Std. 105D, the AQL is simply a notice that if the product specifications and production procedures are such that the fraction defective of the process is no greater than the AQL, preferably less, a large percentage of his lots will be accepted.

Before discussing special features of the standard some of the technical aspects of its structure should be noted. Most of these stem from the symmetric character of Table 10.3 (Table II–A of the Standard). This table is the basic table; other tables such as Tables 10.4–10.9 are based on it. It is important then for understanding the standard to note the implications of the symmetry of this basic table. Let us review these implications.

The OC curves presented in the standard are all of Type B since they are based on the assumption that a sampling scheme will be applied to a stream of lots and that we are in essence sampling from a process. The binomial distribution is thus the correct distribution for constructing the OC curves. This means that with the small values of p' usually involved, the probabilities can be reasonably well approximated by the Poisson distribution. When the sampling scheme is applied to defects per hundred units, the Poisson distribution applies directly. With a Poisson distribution the probability of acceptance is determined by $u' = p'n$ and c, the acceptance number. Thus if p' and n are varied in inverse proportion, P_a will be unchanged. Hence if the AQLs in Table 10.3 vary in the same ratios as the n's, then with a constant acceptance number c down a diagonal of the table, the P_a at p' equal to the AQL will remain constant. Actually the AQLs and n's are rounded off to simple numbers so that exact ratios are not maintained

[3] Table 10.11 (Mil. Std. 105D, Table V–A) is primarily for informational purposes. It might preferably have been omitted to avoid confusion.

and an exact value of P_a is not adhered to. The P_a's remain roughly constant, however.

Since the Poisson distribution is used to calculate the various P_a's, approximately at least, Figure 4.9 can be used for this purpose. In Figure 4.9, however, the scale on which $p'n$ is measured is logarithmic. The distance (in logs) from 0.1 to 1.0 is thus the same as the distance from 1.0 to 10.0, and so on. Consequently, with a fixed n, the quantity AQL $\times$ n will vary by equal logarithmic units (i.e., by equal distances on the $p'n$ scale) if the AQL's progress by a constant proportion or in other words maintain a constant ratio to each other. The actual distance or ratio will, of course, depend on how many AQLs we choose for a given cycle. Mil. Std. 105D selects 5 AQLs per cycle so that the logarithmic distance that will separate them will be $1/5 = 0.20$. A table of common logarithms shows that this is the equivalent of a constant ratio equal to 1.585 $(=(10)^{1/5})$. Thus if we start with an AQL = 1 percent, this will yield other AQLs equal to 1.585 percent, 2.512 percent, 3.981 percent, 6.310 percent, 10.00 percent, 15.85 percent, 25.12 percent, and so forth in the upward direction and 0.6310 percent, 0.3981 percent, 0.2512 percent, 0.1585 percent, 0.1000 percent, 0.06310 percent, 0.03981 percent, and so on in the downward direction. Mil. Std. 105D adopts this system[4] of selecting AQLs but rounds the figures off to 1 percent, 1.5 percent, 2.5 percent, 4.0 percent, 6.5 percent, 10.0 percent, 15 percent, 25 percent, and so forth, and to 0.65 percent, 0.40 percent, 0.25 percent, 0.15 percent, 0.10 percent, 0.065 percent, 0.04 percent, and so on.

As noted above, the values of n, the sample size in a single-sampling plan, are kept roughly in the same proportion as the AQLs. Those selected for Mil. Std. 105D were 2, 3, 5, 8, 13, 20, 32, 50, 80, 125, 200, 315, 500, 800, 1,250 and 2,000. *Down any diagonal* of Table 10.3, therefore, the product AQL $\times$ n is approximately constant. Since c is constant on each diagonal, then the probability of acceptance (P_a) at the AQL, as approximated by the Poisson distribution, is also constant and the exact (binomial) probabilities are roughly so. The particular values of the P_a will depend, of course, on the values of c adopted. For values of c from 0 to 5, as adopted in Mil. Std. 105D, and AQLs from 0.10 to 1.00, we have the P_a's given in Table 10.14.

A second important feature of Table 10.3 is how the P_a's for a given AQL vary with n. Even if we exclude the $c = 0$ plans, which have very special characteristics and are generally viewed as "mavericks," it will be noted from Table 10.14 that the P_a at a given AQL is seen to vary with increasing sample size from about 0.91 to about 0.99. This feature of the

---

[4] If a finer graduation of AQLs was desired, say k per cycle, then the ratio of AQLs would be given by a logarithmic difference of $1/k$ or a ratio equal to $(10)^{1/k}$. For $k = 6$, for example, and the starting point of 1 percent, the rounded off AQLs would be 1, 1.5, 2, 3, 4.5, 7, 10, and so on. The material in this paragraph and the next is based on the Rutgers Technical Report No. 10 by Harold F. Dodge. Also see I. D. Hill's paper in the *Journal of Quality Technology,* April 1973.

TABLE 10.14
Selected Probabilities of Acceptance at the AQL

	AQL						
n	*0.10*	*0.15*	*0.25*	*0.40*	*0.65*	*1.00*	c
13						0.8781	0
20					0.8781		
32				0.8794			
50			0.8825			0.9098	1
80		0.8869			0.9037	0.9526	2
125	0.8825			0.9098	0.9508	0.9617	3
200			0.9098	0.9526	0.9569	0.9834	5
315		0.9180	0.9544	0.9608	0.9817		
500	0.9098	0.9595	0.9617	0.9834			
800	0.9526	0.9662	0.9834				
1,250	0.9617	0.9875					
2,000	0.9834						

SOURCE: From Minutes of ABC Working Group.

standard was, and still is, a matter of controversy. At the drafting of Mil. Std. 105D, Harold F. Dodge and the Standards Committee of the American Society for Quality Control of which Dodge was then chairman, and also certain European quality control experts, argued for keeping the probability of acceptance at the AQL constant, at least roughly so. In justification of this point of view it will be noted that a P_a approximately constant at the AQL effects an overall simplification. If the P_a at the AQL were known to be fixed, a quality manager would immediately know what to expect as soon as the AQL was designated and this would facilitate negotiation of the AQL between producer and consumer. With these fixed P_a's, a single set of calculations would be necessary to determine the effects of the tightened inspection rule if production were at the AQL level. Comparison between sampling plans would also be facilitated. With the P_a at the AQL set at about 0.95, a special additional feature is that if tightened inspection consists simply of reducing the acceptance number by 1, the AOQL[5] under the tightened plan would be about equal to or slightly less than the AQL. Thus, with nondestructive testing, if a producer institutes screening of rejected lots when he is on tightened inspection, he will know that the AOQL will be within the desired limits. This last is a point that is emphasized by Dodge and the ASQC.

The ABC Working Group that prepared Mil. Std. 105D rejected the above arguments for a fixed P_a at the AQL as being less compelling than the arguments for a P_a that increased with sample size. Rejection of a large lot has more serious consequences for the producer than rejection of a small lot and a P_a at the AQL that increases with sample size reduces the risk

[5] See Chapter 16, Section 1.

of rejection of a large lot since in 105D sample sizes vary with lot sizes. A large sample will also give a steeper and more discriminating OC curve than a smaller sample, which means that the protection of the consumer against an isolated bad lot will be increased. Hence the argument runs, these benefits to the consumer should be shared by the producer. What was probably more important in the eyes of the government agencies that were writing the standard, was that whereas the sample sizes increased less rapidly than the lot sizes for which the sample sizes were recommended, the cost of inspection per unit would be reduced with larger samples. To Dodge the relatively small variation in the P_a at the AQL, as sample size increases (particularly for acceptance numbers of 2 on up), was likely in practice to be of little influence in affecting the size of an inspection lot and the claimed advantages would thus not materialize. A modification of Mil. Std. 105D more recently proposed by H. F. Dodge and supported by the Standards Group of the American Society for Quality Control would be to use sample sizes one step lower for all plans for which 105D now recommends the use of the acceptance numbers $c = 1$ (cf. Table 10.14). This would raise the probability of acceptance of these plans at the AQL to 0.959.

Having selected a set of sample sizes, the next step is to relate them to a recommended set of lot sizes. This is done in Table 10.2 (Mil. Std. 105D, Table 1). If the midpoint of each lot range is computed, it will be noted that when the sample size is plotted against the midpoint of the lot range on double log paper, the points follow roughly a straight line up to $n =$ 80 and thereafter another straight line with a flatter slope. This relationship goes back to the earlier Army Ordnance Sampling tables and is reported by H. F. Dodge to have been based empirically on Bell Telephone System experience. The lot ranges selected in this manner yield a very rapidly decreasing ratio of sample size to lot size. Thus for lots ranging from 26 to 50 the associated sample size is 8, which is 21.3 percent of the midpoint of the lot range. For lots ranging from 281 to 500, the associated sample size is 50, which is 12.8 percent of the midpoint of the lot range and for lots ranging from 10,001 to 35,000, the associated sample size is 315 which is 1.4 percent of the midpoint of the lot range. This shows the great economy in inspection costs per unit attained by the use of large lots.

As has been noted above, the requirement of a shift to tightened inspection when there is evidence of deterioration in quality is an essential part of the standard. This requirement has two components: one is the set of rules for shifting from normal to tightened inspection and back again and the other the set of plans constituting tightened inspection. These will be discussed in reverse order.

In his report to the ABC Working Group, Harold Dodge had advocated that plans for tightened inspection be derived simply by subtracting one from the acceptance number for normal inspection, keeping the sample size constant, i.e., c_T should equal $c_N - 1$. This was to be followed up to $c = 15$. Dodge argued for this system of tightened inspection plans since it did not put an excessive burden of rejections on the producer. Also, if the producer undertook to screen rejected lots when under these tightened plans, his AOQL would be no more than the AQL and would thus give adequate consumer protection. The reason for the cutoff at $c = 15$, was that normal inspection had AOQLs for $c \geq 15$ that were already within the limit of the AQL. This all assumes, of course, nondestructive testing, but Dodge had expressed the view that it is probably best to design another set of plans than Mil. Std. 105D when destructive tests had to be employed.

Contrary to the point of view adopted by Dodge, Professor E. P. Rossow of Berlin-Charlottenburg Technical University, who was a prominent member of the sampling committee of the European Organization for Quality Control, took a position[6] in favor of increasing the sample size to get better discrimination when there was evidence of deterioration of quality, instead of reducing the acceptance number. He was particularly opposed to the latter procedure when the criterion for shifting to tightened inspection was the two out of five rule which had been proposed by Dodge. This will be discussed below.

Dr. Hamaker, chairman of the EOQC Sampling Committee, questioned the advisability of *requiring* shifts to tightened inspection in the standard. His point was that there may be other means of bringing pressure to bear on the producer. If there was more than one source of supply, for example, the threat of ceasing to buy further from a supplier whose product shows signs of deterioration might be enough to cause him to take corrective action.

The ABC Working Group rejected all of the above proposals. A shift to tightened inspection was made compulsory. The increase in sample size, in lieu of a reduction in the acceptance number, was considered to have ancillary disadvantages. The variations in amount of inspection would be annoying. Furthermore, if inspection was being done by the consumer, to increase his inspection costs because a producer had poor results did not make sense. In some cases, under the consumer's contract, the producer was paid for the testing and could increase his profits by testing larger samples.

The ABC Working Group went along with Dodge's proposal part way. Tightened inspection was accomplished by reducing the acceptance number and keeping the sample size fixed. For normal c's of 1, 2, and 3 the reduction was only 1, as Dodge had urged, but for normal c's of 5, 7, 10, 14 the reduction was 2, and for a normal c of 21, it was 3. The ABC Working Group believed that a reduction of 1 throughout did not put enough pressure on the producer. Dodge's argument that under his system the AOQL under tightened in inspection would be equal to or less than the AQL did not seem to carry heavy weight with the Working Group. Whether this was because they wanted the standard to be useful for destructive as well as nondestructive testing was not clear. As it turns out, however, Table 10.12 (Table V–B of 105D) reveals that except for $c = 0$ plans, the AOQLs under tightened inspection are very close to the AQLs.

---

[6] Minutes of ABC Working Group.

The rule for shifting from normal to tightened inspection and back again has been even more controversial than the nature of the tightened plans themselves. Previous versions of Mil. Std. 105 had provided for keeping records of the fraction defective and when estimates of the process average fraction defective ran above a prescribed limit, a shift to tightened inspection was required. Dodge suggested the two out of five rule to simplify matters since he believed that the cumbersome procedures were keeping tightened inspection from being used when it should have. Dodge indicated, however, that the two out of five rule seemed appropriate as a *general* rule only if the P_a at the AQL was reasonably constant under normal inspection, which as seen above was not true for the plans finally adopted for Mil. Std. 105D. Some members of the Sampling Committee of EOQC (Mr. Mendizabal from Spain and Professor Rossow from Germany) pointed out that if the process average was in the neighborhood of the AQL, the chance vagaries of sampling would in itself cause an annoying shifting back and forth between normal, tightened, and even reduced inspection. A simulation with glass beads that was run by Mendizabal showed, for example, that when p' equaled an AQL of 4 percent, there were three shifts to reduced inspection and one shift to tightened inspection. Seventy-four of the total of 200 lots studied were inspected on reduced inspection and 7 on tightened inspection.[7]

The two out of five rule for shifting from normal to tightened inspection and a requirement of a run of five acceptances to go back to normal was the rule finally adopted by the ABC Working Group. Subsequently, the rule came under further criticism from Japanese quality control experts. Like Mendizabal they argued that there is a considerable amount of misswitching from normal to tightened or normal to reduced inspection when the process is actually producing material of AQL quality continuously. They pointed out that there is even a sizable chance that production would be discontinued. Since Japanese industrialists view being on tightened inspection as an indication of inferior quality and a mark of dishonor, the two out of five switching rule was subjected to considerable criticism in Japan. Consequently, the Japanese Industrial Standard Z 9015, *Sampling Inspection Plans by Attributes with Severity Adjustment (Acceptance Sampling Where a Consumer Can Select Suppliers)* goes back to a procedure under which a switch to tightened inspection is based on sample estimates of the process average. This Japanese switching rule differs from previous Mil. Std. 105 procedures in that the requirement for estimating the process average applies only to the past five lots after a lot has been rejected and was not a continuing requirement as under older versions of Mil. Std. 105. The Japanese modifications have been opposed by the Standards Group of the Standards Committee of the American Society for Quality Control as a step backward to a type of procedure that had caused the misuse of Mil. Std. 105 in the United States. It was also pointed out that the AQL was a *maximum* acceptable percent defective and the

---

[7] *Journal of Quality Technology,* April 1971, p. 93.

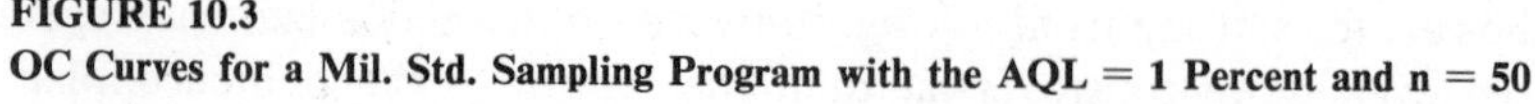

FIGURE 10.3
OC Curves for a Mil. Std. Sampling Program with the AQL = 1 Percent and n = 50

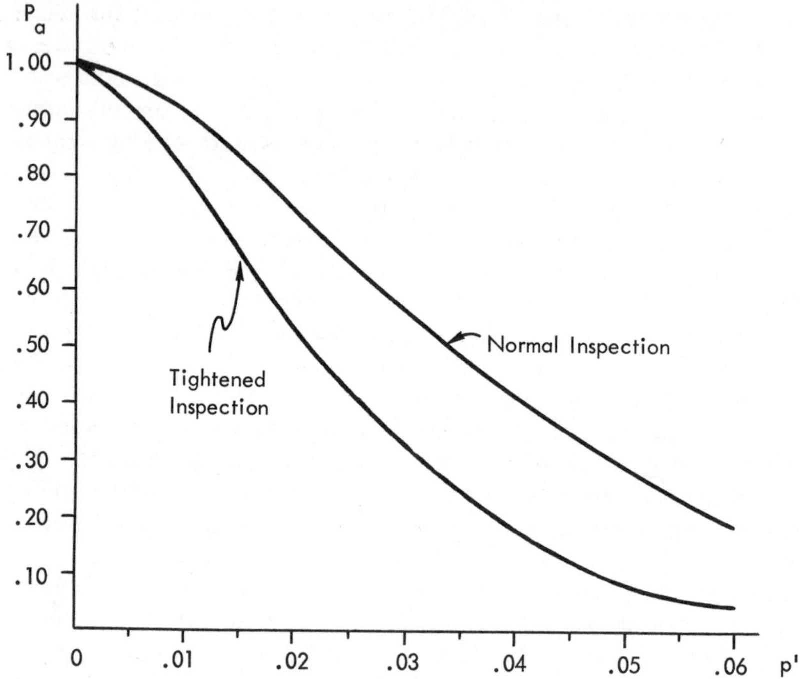

actual process average should preferably be less than the AQL. To encourage production *at* the AQL by *less* severe switching rules would reduce the protection to the consumer when the process average moved above the AQL.

This section may be fittingly concluded by reference to Figure 10.3 that shows the OC curve for normal inspection of a Mil. Std. 105D plan together with the OC curve for tightened inspection under the plan. The difference in the heights of the two curves at various levels of incoming quality reveals graphically the pressure that this particular plan puts on the producer to correct his process. Numerical values for the two curves will be found in Table X-H-1 of the standard.

4. ANSI/ASQC Z1.4–1981

ANSI/ASQC Z1.4–1981, *Sampling Procedures and Tables for Inspection by Attributes* is a revision of ANSI Z1.4–1971 which corresponded directly to Mil Std. 105D. The modifications made in Mil. Std. 105D are described in the Foreword* to Z1.4–1981 as follows:

---

* Copyright American Society for Quality Control, Inc. Reprinted by permission.

The present revision ANSI/ASQC Z1.4–1981 was undertaken to modernize terminology and to emphasize the system aspect of the procedure through incorporation of the operating characteristic curves and other measures computed for scheme performance reflecting the basic strategy including the switching rules.

All tables, table numbers, and procedures used in MIL-STD-105D were retained. The tables are unchanged to make the tabular content completely compatible with MIL-STD-105D. Modifications from the MIL-STD-105D format beyond editorial refinements include:

1. Substitution of the word "nonconformity" for "defect" throughout, in conformance with ANSI/ASQC A2–1978. Substitution of the word "nonacceptance" for "rejection" when it refers to a result of following the procedure. Forms of the word "reject" are retained when they refer to actions the customer may take. The term "rejection number" is retained when it refers to the nomenclature on Tables II, III, IV and X to be consistent with tables of the same numbers in MIL-STD-105D.
2. Presentation of the switching rules to put them in conformance with ANSI Z1.9–1980, the ANSI version of MIL-STD-414. [See Chapter 13 below.] This includes an option for reduced inspection without use of limit numbers (as in ANSI Z1.9–1980. Use without the limit numbers improves the performance of a scheme by accepting more lots at the AQL with no change in discrimination below the indifference quality level.
3. Introduction of the following tables:

   Table XI Average Outgoing Quality Limit Factors for ANSI Z1.4 Scheme Performance (Single Sampling)

   Table XII Limiting Quality for ANSI Z1.4 Scheme Performance for which $P_a = 10$ Percent (Single Sampling)

   Table XIII Limiting Quality for ANSI Z1.4 Scheme Performance for which $P_a = 5$ Percent (Single Sampling)

   Table XIV Average Sample Size Tables for ANSI Z1.4 Scheme Performance (Single Sampling)

   Table XV Scheme Performance with Switching Rules—for each Code Letter showing

   1. Operating Characteristic Curves for ANSI Z1.4 Scheme Performance
   2. Tabulated Values for Operating Characteristic Curves for ANSI Z1.4 Scheme Performance

An illustration of the scheme OC curves given in Z1.4–1981 is presented in Figure 10.4. It is so selected that a detailed comparison can be made with the OC curves for individual sampling plans presented in Figure 10.1,

FIGURE 10.4

Scheme Performance with Switching Rules. Chart XV-K Operating Characteristic Curves for ANSI Z1.4 Scheme Performance. Curves for double and multiple sampling are matched as closely as practicable.*

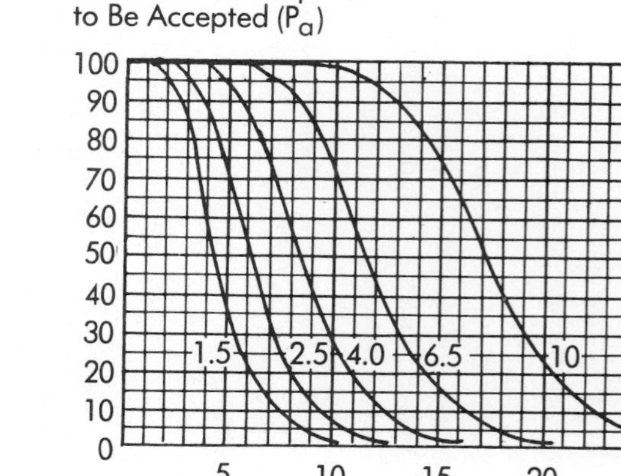

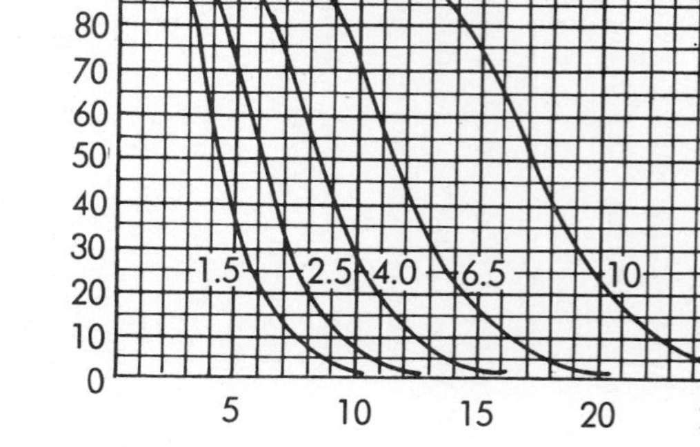

Quality of Submitted Lots (in percent defective for AQLs ⩽ 10; in defects per hundred units for AQLs > 10).

Note: Figures on curves are Acceptable Quality Levels (AQLs) for normal inspection.

TABLE 10.15
Tabulated Values for Operating Characteristic Curves for ANSI Z1.4 Scheme Performance. Table XV-K-1 of Z1.4.*

P_a	Acceptable Quality Levels (normal inspection)								
	0.10	0.40	0.65	1.0	1.5	2.5	4.0	6.5	10
	p (in percent nonconforming or nonconformities per hundred units)								
99.0	0.0167	0.153	0.455	0.738	1.49	2.43	4.01	6.34	10.3
95.0	0.0573	0.292	0.643	1.06	1.97	3.07	4.84	7.32	11.7
90.0	0.0916	0.392	0.771	1.28	2.21	3.43	5.33	7.96	12.5
75.0	0.178	0.586	1.02	1.65	2.59	3.96	6.08	8.96	13.7
50.0	0.359	0.873	1.42	2.21	3.11	4.70	7.08	10.3	15.2
25.0	0.694	1.35	2.16	3.14	4.10	5.94	8.65	12.2	17.4
10.0	1.15	1.94	3.11	4.26	5.34	7.42	10.4	14.2	19.8
5.0	1.50	2.37	3.79	5.04	6.20	8.41	11.5	15.6	21.4
1.0	2.31	3.32	5.31	6.73	8.04	10.5	13.9	18.3	24.5

p. 218. For example, for Code Letter K to which Figure 10.1 and Figure 10.4 both apply, it will be noted that for an AQL = 1 percent, the P_a = 0.99 point for the individual single sampling plan comes at $p' = 0.658$ percent, while the 0.99 point for the scheme OC curve comes at $p' = 0.738$ percent, showing that in this area the scheme OC curve is above the OC curve for the individual plan. This results from the scheme OC curve allowing for shifts to reduced inspection in this range of p' values. At the other end, the $P_a = 0.10$ point for the individual sampling plan falls at $p' = 5.35$ percent, while the $P_a = 0.10$ point for the scheme OC curve falls at $p' =$ 4.26 percent, which shows that the scheme OC curve lies below the OC curve for the individual sampling plan for the larger values of p'. This is because the scheme OC curve becomes heavily affected by switches to tightened inspection at these higher levels of p'. In fact, a comparison of Figures 10.1 and 10.4 shows that the 10 percent for the sampling scheme is the same as that listed in Table 10.1 for the 10 percent point of the tightened inspection plan, viz $p' = 4.26$ percent.

5. INTERNATIONAL ORGANIZATION FOR STANDARDIZATION (ISO) STANDARD 2859

As noted above, the ISO Std. 2859 that was initially issued in 1974 was essentially the same as Mil. Std. 105D. It is currently (1986) proposed that editorial revision of 2859 follow ANSI/ASQC Z1.4, in replacing the terms *defect* and *defective unit* by the terms *nonconformity* and *nonconforming unit.* The addition of three new parts is being considered. The old standard will be designated Part I. A proposed introduction to the standard will be designated Part 0. Part II, which has already been adopted, discusses sampling plans indexed by LQs and is essentially the same as the 1984 Supplement 1 to the British Standard 6001. (See Section 3 of Chapter XIV below.) A proposed Part III deals with skip-lot sampling and is based on the proposed ANSI-ASQC skip-lot standard described in the next section.

6. A PROPOSED ANSI/ASQC STANDARD FOR AN ATTRIBUTE SKIP-LOT SAMPLING PROGRAM

This chapter will be concluded with mention of a proposed ANSI/ASQC standard for a skip-lot sampling program that is designed for use with attribute lot-by-lot plans described in ANSI/ASQC Z1.4–1981. Because of the essential identity of the Z1.4 plans with those of Mil. Std. 105D and ISO Std. 2859, this proposed skip-lot standard is equally suitable for use with these other standards. In fact, as indicated above it may, possibly with some modification, become a part of the ISO standard. It will be noted that the proposed program is not the same as the skip-lot sampling plan (SkSP-1) proposed by H. F. Dodge[8] in 1955.

As described in the proposed standard its purpose is to provide a procedure for reducing the inspection effort on products submitted by those suppliers who have demonstrated their ability to control, in an effective manner, all facets of product quality and consistently produce superior quality material. The procedure is not to be applied to the inspection of product characteristics which involve the safety of personnel. It may be used in place of reduced inspection if it is more cost effective to do so.

In the proposed standard, "skip-lot sampling" takes the form of an acceptance sampling procedure in which some lots in a series are accepted without inspection (other than possible spot checks) when the sampling results from a stated number of immediately preceding lots meet specified criteria. When inspection is on a skip-lot basis, the lots to be inspected are chosen randomly in accordance with a stated frequency which is called the "skip-lot frequency." A skip-lot frequency of 1 lot in 2, for example, is explained to mean that the long-run average proportion of inspected lots is 50 percent. The procedure of this proposed standard is designed to protect against passing any significant quantity of poor quality product and was developed assuming that in order to qualify for skip-lot inspection, product quality is running at or better than one-half AQL. The statistical characteristics of the proposed standard

[8] See *Industrial Quality Control* 11, No. 5 (February 1955) pp. 3–5. A description of the Dodge SkSP-1 plan will also be found in Edward G. Schilling, *Acceptance Sampling in Quality Control,* pp. 443–45.

are described in the paper, "A Proposed Attribute Skip-Lot Sampling Program" by Burton S. Liebesman and Bernard Saperstein, published in the *Journal of Quality Technology* 15, No. 3, July 1983, pp. 130–40. Section 8 of the proposed standard gives a skip-lot bibliography.

7. PROBLEMS

10.1. *a.* Using Military Standard 105D, find a single-sample fraction-defective sampling plan for lots of about 15,000 and an AQL of 2.5 percent. use inspection level II.

b. Find the plans you would shift to under tightened inspection and under reduced inspection. Draw the OC curves for the three plans on the same chart.

10.2. *a.* A military agency wishes a sampling inspection plan for which the AQL is 1.5 percent. Lots run about 8,000 items. Find the plan it would adopt if it follows Military Standard 105D and uses single sampling. Use inspection level II.

b. Find the plans you would shift to under tightened inspection and under reduced inspection. Draw the OC curves for the three plans on the same chart.

10.3. Do Problem 10.1 for a product that is judged to require the use of—

a. Inspection level III.

b. Inspection level I.

10.4. Do Problem 10.2 for a product that is judged to require the use of—

a. Inspection level III.

b. Inspection level I.

10.5. *a.* Using Military Standard 105D, find a single-sample defects-per-unit sampling plan for lots of about 450 and an AQL of 4 defects per hundred units. Use inspection level II.

b. Find the plans you would shift to under tightened inspection and under reduced inspection. Draw the OC curves for the three plans on the same chart.

10.6. *a.* Using Military Standard 105D, find a single-sample defects-per-unit sampling plan for lots of about 1,000 and an AQL of 1.5 defects per hundred units. Use inspection level II.

b. Find the plans you would shift to under tightened inspection and under reduced inspection. Draw the OC curves for the three plans on the same chart.

10.7. *a.* Using Military Standard 105D, find a single-sample defects-per-unit sampling plan for lots of about 35 and an AQL of 15 defects-per-hundred units. Use special inspection level S–4.

b. Find the plans you would shift to under tightened inspection and under reduced inspection. Draw the OC curves for the three plans on the same chart.

10.8. Repeat Problem 10.1 for (1) a double-sampling plan and (2) a multiple-sampling plan, omitting the drawing of the OC curves.

10.9. Repeat Problem 10.2 for (1) a double-sampling plan and (2) a multiple-sampling plan, omitting the drawing of the OC curves.

10.10. Repeat Problem 10.3 for (1) a double-sampling plan and (2) a multiple-sampling plan, omitting the drawing of the OC curves.

10.11. Repeat Problem 10.4 for (1) a double-sampling plan and (2) a multiple-sampling plan, omitting the drawing of the OC curves.

8. SELECTED REFERENCES*

ABC Working Group (B '61), ASQC (B '81), British Standards Institution (B '84), Dodge (B '59), Grant and Leavenworth (B '80), Hald (B '81), Hald and Thyregod (P '65), Hill (P '73), ISO (B '74), Keefe (P '63), Koyama (P '79), Liebesman (P '79), Liebesman and Hawley (P '84), Liebesman and Saperstein (P '83), Pabst (P '63 and P '71), Schilling (P '82 and B '82), Statistical Research Group, Columbia University. (B '48), Stephens and Larson (P '67), U.S. Department of Defense (B: Mil. Std. 105D, '63; Supply and Logistics Handbook, '54; Quality Control and Reliability Handbook, '60).

---

* B and P refer to the Book and Periodical sections, respectively, of the Cumulative List of References in Appendix V.

11

Acceptable Sampling by Variables to Control the Fraction Nonconforming: Standard Deviation Known

When a quality characteristic is measurable on a continuous scale and is known to have a distribution of a specific type, say it is known to be normally distributed, it may be possible to use as a substitute for an attributes sampling plan a sampling plan based on sample measurements such as the mean of the sample or the mean and standard deviation of the sample. Such plans are called variables sampling plans.

1. THE ADVANTAGE OF VARIABLES SAMPLING PLANS

These variables sampling plans have the primary advantage that the same operating characteristic curve can be obtained with a smaller sample than is required by an attributes plan. The precise measurements required by a variables plan will probably cost more than the simple classification of items required by an attributes plan, but the reduction in sample size may more than offset this extra expense. For example, an attributes plan may require a sample of 50, but a comparable variables plan a sample of only 30. If the unit cost of measurement is less than 1.67 times that of simply classifying the items, the variables plan will in this instance effect a saving. Such saving may be especially marked if inspection is destructive and the item is expensive.

2. DISADVANTAGES OF VARIABLES SAMPLING PLANS

The principal disadvantage of variables sampling plans is that a separate plan must be employed for each quality characteristic that is being inspected. If an item is inspected for five quality characteristics, for example, it would be necessary to have five separate variables inspection plans, whereas acceptance or rejection of the lot as a whole could be based on a single attributes plan. It is also theoretically possible, although not very likely, that under a variables plan a lot will be rejected by the variables criteria, even though the actual sample contains no nonconforming items. A supplier unfamiliar with the nature of variables sampling would undoubtedly protest strongly against his product being rejected in this way. A third difficulty is that the distribution of the quality characteristic must be of the specified form.

The last difficulty mentioned should be explained in detail. In the plans usually applied, it is assumed in theory that sampling is from a normal universe. Thus with respect to Type B OC curves it is assumed that in the output of the process the measured characteristic is normally distributed. With respect to Type A OC curves it is assumed that the measured characteristic is normally distributed among the items of the lot. If the process or lot is nonnormal, the sampling distribution of means and standard deviations will not be the same as if the process or lot were normal, but the computations of probability will not be very seriously affected unless the departure from normality is very marked and the samples are small. All variables plans, however, assume some way of going from a mean and standard deviation to a fraction nonconforming. If the items are not normally distributed, estimates of the fraction nonconforming based on the mean and standard deviation will not be the same as if the items were normally distributed. Differences on this score may become especially serious when we are dealing with very small fractions nonconforming. For example, if the mean of a normal process or lot lies three standard deviations below a single upper specification limit, it will have no more than 0.00135 nonconforming. On the other hand, if in a nonnormal process or lot with considerable skewness and/or kurtosis say with $\gamma_1' = 1.00$, and γ_2' 1.5, the mean lies three standard deviations below

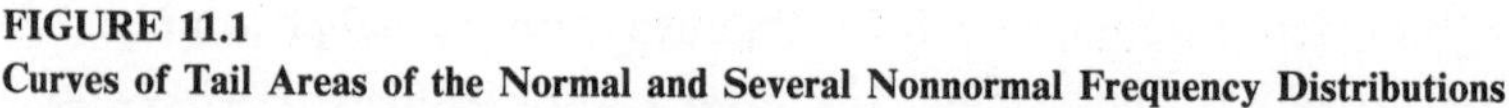
FIGURE 11.1
Curves of Tail Areas of the Normal and Several Nonnormal Frequency Distributions

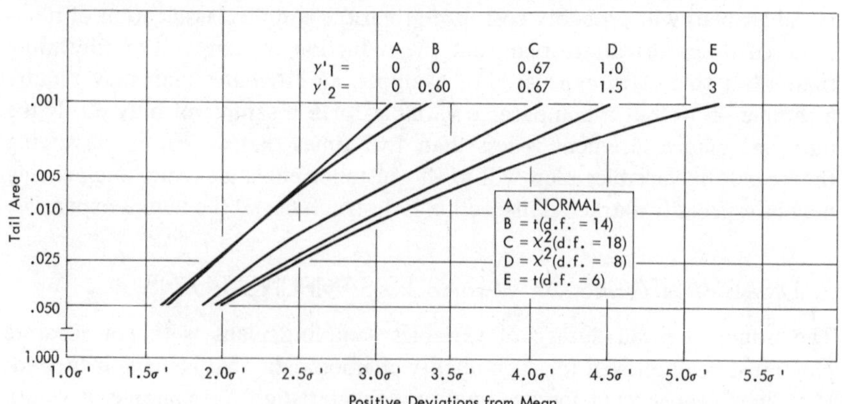

the specification limit, possibly 0.01000 of the items might be nonconforming or seven times that for a normal distribution. Compare Figure 11.1. A study of this figure suggests that when we are dealing with fractions nonconforming below 0.01, we must be exceptionally cautious on this score.

It is also to be noted that in strict theory the assumption of normality implies an infinite universe. For Type A OC curves this raises difficulties. In practice, if a lot is large, we can talk about it being normally distributed without serious difficulty. This may not be true for small lots. We shall assume here that if a small lot has a bell-shaped distribution with the same γ coefficients[1] as the normal distribution, viz, $\gamma_1' = 0$, $\gamma_2' = 0$, it will be satisfactory to consider it as normally distributed.

3. RELATIONSHIP BETWEEN THE MEAN AND STANDARD DEVIATION OF A NORMALLY DISTRIBUTED PROCESS OR LOT AND ITS FRACTION NONCONFORMING

If the items of a process or lot have a normal distribution, there exists an exact functional relationship between the fraction nonconforming and the mean and standard deviation. This relationship is given by the table of areas under the normal curve (Table A2 of Appendix II). For example, if the mean is $\bar{X}' = 100$ and the standard deviation is $\sigma' = 10$ and if there is a single lower limit $L = 82$, then the fraction nonconforming is found by computing z (i.e., the normal deviate) for the specification limit and finding the fraction of cases below that limit. Thus, for the given data,

$$z = \frac{L - \bar{X}'}{\sigma'} = \frac{82 - 100}{10} = -1.8$$

and Table A2 of Apendix II shows that the portion of the area below -1.8 is 0.0359. For a mean of 100 and a standard deviation of 10, the fraction nonconforming is thus 3.59 percent. If in another case the mean were 106 and the standard deviation were 12, we would have

$$z = \frac{82 - 106}{12} = -2.0$$

and Table A2 gives the fraction nonconforming as 0.0228. Hence, for any mean and any standard deviation and for a given specification limit, we can always find the process or lot fraction nonconforming, assuming that the items are normally distributed.

If the standard deviation is constant, the fraction nonconforming will vary directly with the mean. This is illustrated in Figure 11.2. It will be noticed that for $\bar{X}' > L$, the closer $\bar{X}'$ is to L, the larger the fraction nonconforming. There will in this case be a one-to-one correspondence between the mean and the fraction nonconforming.

---

[1] See Chapter 27, Section 3.

FIGURE 11.2
Illustrations of the Relationship between the Mean ($\bar{X}'$) of a Process or Lot and the Fraction Nonconforming (p'), with the Standard Deviation Assumed Constant

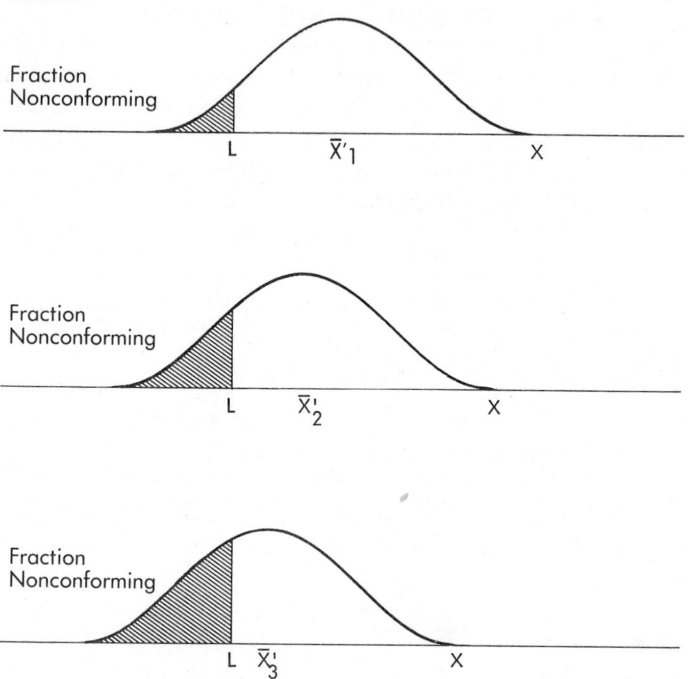

If σ' is not constant, the fraction nonconforming will be affected by both the mean and the standard deviation. If two processes or lots have different distributions, they may or may not have the same fraction nonconforming. If the standard deviation of A is the same as that of B, but the mean is closer to the lower specification limit L, the fraction nonconforming of A will be higher. If the standard deviation of A is the same as that of B but the mean is farther away from L, the fraction nonconforming of A will be lower. If the means of the two are the same but the standard deviation of A is larger than that of B, the fraction nonconforming of A will be higher. If the means are the same but the standard deviation of A is smaller than that of B, its fraction nonconforming will be lower (see Figure 11.3).

It will be noticed that the fraction nonconforming is found immediately from the normal table as soon as we know the z for the specification limit. Hence, instead of representing the quality of a process or lot by its fraction nonconforming, we can, if we wish, represent it by the value of z that gives that fraction nonconforming. (When we are given the fraction nonconforming first, we can find the corresponding value of z from the normal area table.) For processes or lots whose items are normally distributed, we have the results shown in Table 11.1. These are illustrated in Figure 11.4.

FIGURE 11.3
Illustration of How the Fraction Nonconforming Varies with the Mean and Standard Deviation of the Distribution of a Quality Characteristic

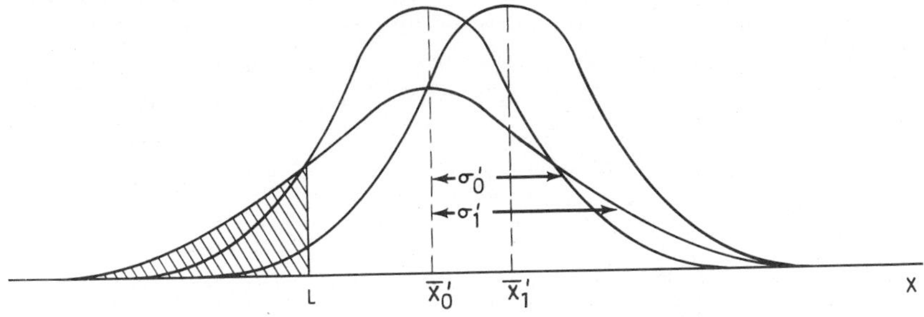

TABLE 11.1
Equivalent Values of z and the Fraction Nonconforming for a Normal Distribution*

Fraction Nonconforming	Value of $z_U = \frac{U - \bar{X}'}{\sigma'}$ or $z_L = \frac{\bar{X}' - L}{\sigma'}$
0.25	0.6745
0.20	0.8416
0.15	1.0364
0.10	1.2816
0.05	1.6449
0.02	2.0537
0.01	2.3263

* It will be noted that z_L is so defined that it has a positive sign for $p' < 0.50$.

FIGURE 11.4
Illustration of a Relationship between z_L and the Fraction Nonconforming

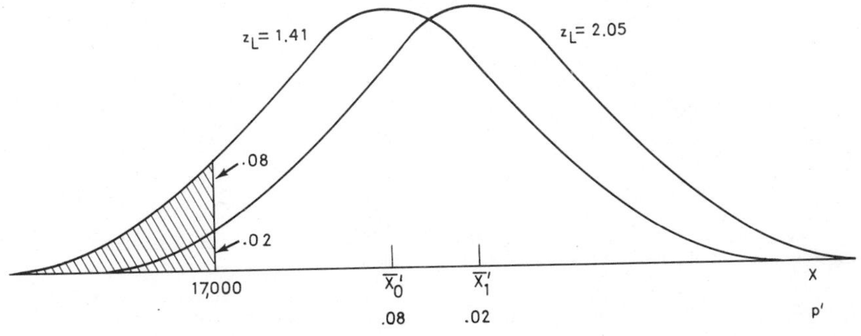

The reader will please note that the symbol z is used in Chapters 11 and 12 to mean simply "standard normal deviate." It is printed without a prime since no attempt is made to indicate whether the deviate is being computed from universe parameters or sample data, this being made clear from the context in each case. It will also be noted that "special" z's, such as z_L, z_U, z_1, z_2, and $z_{p'}$ are so defined that under normal circumstances they will be positive but in general z may be positive or negative. If the subscript of z is a fraction nonconforming, it pertains to the *tail* area or areas.

4. VARIABLES SAMPLING PLANS WHEN PROCESSES OR LOTS ARE NORMALLY DISTRIBUTED AND THE STANDARD DEVIATION IS KNOWN

Especially great economy is obtained by variables plans when a process or lot is normally distributed and its standard deviation is known. This situation is often reasonably well approximated in industry. Machine variability, for example, may vary only slightly from machine to machine and time to time,[2] but machine settings may differ considerably on occasion. Consequently, we may wish to sample machine output to see if the mean is such as to yield an acceptable fraction nonconforming, assuming that the standard deviation remains at its previously known value.

Under these conditions, we may proceed in several ways. If there is a single lower limit, we may use the sample mean to compute $z_L = \dfrac{\bar{X} - L}{\sigma'}$ and taking a lead from Section 3, estimate the fraction nonconforming of the process or lot as the area under the normal curve beyond z_L. If we wish a slightly better estimate we may treat the quantity $\dfrac{\bar{X} - L}{\sigma'}\sqrt{\dfrac{n}{n-1}}$ (hereafter designated by Q_L) as a normal deviate, since this will yield the minimum variance unbiased estimate[3] of the fraction nonconforming p'. However we estimate p', we may operate a sampling plan as follows. From a random sample of size n compute z_L or Q_L and estimate p'. If this estimate of p' exceeds a specified maximum value M, reject the lot; otherwise accept it.

We can, however, proceed in another way. We can translate the critical M into a critical value for z_L. Thus, if we represent this critical value for z_L by the symbol k, then k will be defined as that value of z_L such that the proportion of the cases in excess of this value of z_L, or if we prefer in excess of the normal variable

---

[2] Knowledge of the standard deviation might be obtained from a range or standard deviation control chart for the machine (see Chapter 21, Sections 2 and 3).

[3] See G. J. Lieberman and G. T. Resnikoff, "Sampling Plans for Inspection by Variables," *Journal of the American Statistical Association* 50 (1955), pp. 457–516.

$$Q_L = z_L \sqrt{\frac{n}{n-1}}$$

(in one direction) equals[4] M. Then we can operate a sampling plan as follows. From a random sample of size n, compute $z_L = \frac{\bar{X} - L}{\sigma'}$. If z_L is greater than or equal to k, accept; otherwise reject.

To keep our discussion in line with the Department of Defense standard for sampling inspection by variables,[5] we shall call the k-method Procedure 1 and the M-method Procedure 2, which is the reverse of the order in which they are discussed above. In Procedure 2 we shall in this chapter use the quantity Q_L to estimate p'.

4.1. Derivation of a Variables Plan with a Specified p_1', p_2', α, and β: Single Specification Limit

Let us assume (1) that the quality characteristic in question is measurable on a continuous scale and is normally distributed, (2) that the standard deviation is known, and (3) that there is but a single specification limit. Initially let the specification limit be a lower limit L.

The basic problem in designing a sampling plan is to find the sample size n and the acceptance criterion – the k or the M – that will yield the characteristics specified for the plan. Suppose, for example, that the lower specification limit on the tensile strength of a given product is 17,000 psi and it is desired to set up a sampling plan based on variables that, in terms of the fraction nonconforming, has a $p_1' = 0.01$, a $p_2' = 0.08$, $\alpha = 0.05$, and $\beta = 0.10$. Let the distribution of the tensile strength of the product be normal in form and let the standard deviation of its tensile strength be 800 psi. Such a distribution will have a fraction nonconforming of 0.01 if its mean $\bar{X}'$ is such that $\frac{\bar{X}' - 17{,}000}{800} = 2.3263$. For the probability that a normally distributed variable will fall short of its mean by $2.3263\sigma'$ is just 0.01. Likewise, the distribution will have a fraction nonconforming of 0.08 if its mean is such that $\frac{\bar{X}' - 17{,}000}{800} = 1.4053$. The first of these equations yields $\bar{X}' =$ 18,861 psi and the second $\bar{X}' = 18{,}124$ psi. Call these values of $\bar{X}'$ that will yield p_1' and p_2' quality the $\bar{X}_1'$ and $\bar{X}_2'$ of the plan and designate $\frac{\bar{X}_1' - L}{\sigma'}$

---

[4] In other words, using Q_L, k is defined by

$$\int_{k\sqrt{\frac{n}{n-1}}}^{\infty} \frac{e^{-y^2/2}}{\sqrt{2\pi}}\, dy = M$$

[5] See Chapter 13.

as z_1 and $\dfrac{\bar{X}'_2 - L}{\sigma'}$ as z_2. Hence, the desired plan will be obtained if its $z_1 = 2.3263$, its $z_2 = 1.4053$, $\alpha = 0.05$, and $\beta = 0.10$.

Under Procedure 1 we will accept a lot if

$$\frac{\bar{X} - L}{\sigma'} \geq k \tag{11.1}$$

After adding and subtracting $\dfrac{\bar{X}'}{\sigma'}$ this condition can also be written, accept if

$$\frac{\bar{X} - \bar{X}'}{\sigma'} + \frac{\bar{X}' - L}{\sigma'} \geq k$$

or if

$$\frac{\bar{X} - \bar{X}'}{\sigma'} \geq k - \frac{\bar{X}' - L}{\sigma'}$$

Multiplication of each side by $\sqrt{n}$ yields the condition that

$$\frac{\bar{X} - \bar{X}'}{\sigma'/\sqrt{n}} \geq \left(k - \frac{\bar{X}' - L}{\sigma'}\right)\sqrt{n}$$

Hence, if we note that $\dfrac{\bar{X}'_1 - L}{\sigma'} = z_1$ and $\dfrac{\bar{X}'_2 - L}{\sigma'} = z_2$, we see that the criteria for our plan will be met if n and k are such that

$$\text{Probability}\left(\frac{\bar{X} - \bar{X}'}{\sigma'/\sqrt{n}} \geq (k - z_1)\sqrt{n}\right) = 1 - \alpha$$

and

$$\text{Probability}\left(\frac{\bar{X} - \bar{X}'}{\sigma'/\sqrt{n}} \geq (k - z_2)\sqrt{n}\right) = \beta$$

Compare Figure 11.5. Since $\dfrac{\bar{X} - \bar{X}'}{\sigma'/\sqrt{n}}$ is normally distributed with zero mean and unit standard deviation,[6] it follows from the above that we must have

$$(k - z_1)\sqrt{n} = z_{1-\alpha}$$

and

$$(k - z_2)\sqrt{n} = z_\beta$$

where $z_{1-\alpha}$ and z_β are normal deviates the probability of exceeding which are $1 - \alpha$ and β, respectively. But if the probability of exceeding $z_{1-\alpha}$ is 1

[6] See Chapter 6, Section 1.2. This is exactly true if X is normally distributed. See Problem 6.30.

FIGURE 11.5
Illustrating the Relationship between the z's Involved in the Design of a Variables Sampling Plan (σ' Known)

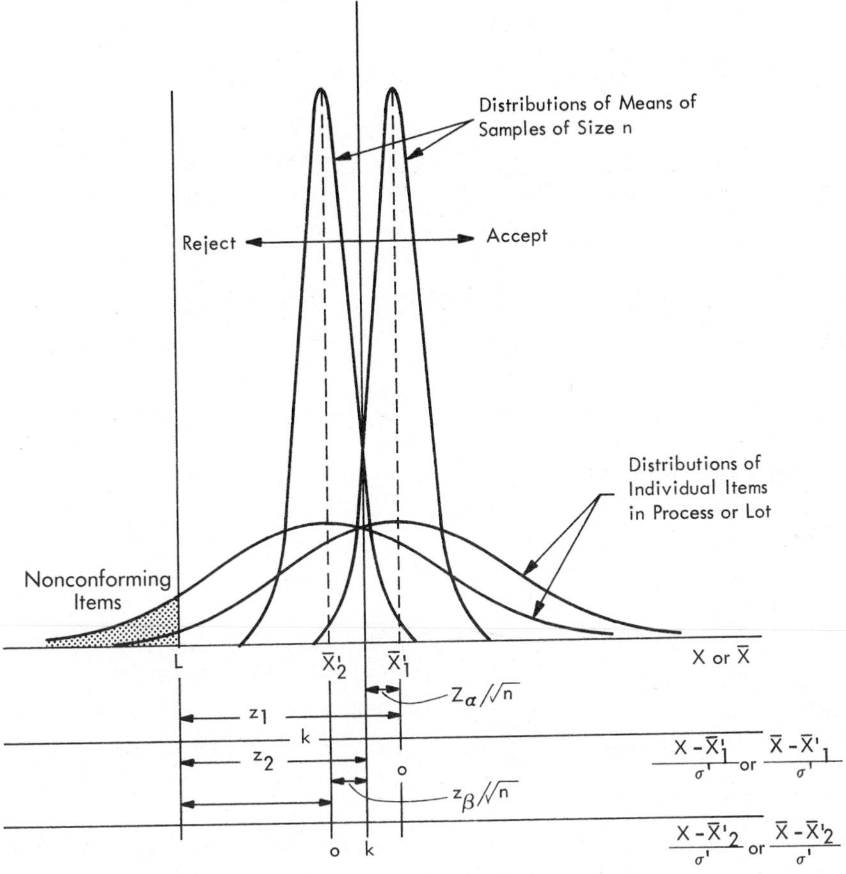

$-\alpha$, then $z_{1-\alpha}$ will equal $-z_\alpha$ where z_α is the normal deviate the probability of exceeding which is α. (Note here that for $\alpha < 0.50$, $z_{1-\alpha}$ will be negative and z_α positive.) Hence the criteria for our plan require that

$$(k - z_1)\sqrt{n} = -z_\alpha$$

$$(k - z_2)\sqrt{n} = z_\beta$$

These can be translated into

(11.2a)
$$k = z_1 - \frac{z_\alpha}{\sqrt{n}}$$

or

(11.2b)
$$k = z_2 + \frac{z_\beta}{\sqrt{n}}$$

and

(11.3)
$$n = \left(\frac{z_\alpha + z_\beta}{z_1 - z_2}\right)^2$$

For our example, we have[7] $z_\alpha = 1.6449$, $z_\beta = 1.2816$, $z_1 = z_{0.01} = 2.3263$, and $z_2 = z_{0.08} = 1.4053$. Equation (11.3) then yields

$$n = \left(\frac{1.6449 + 1.2816}{2.3263 - 1.4053}\right)^2$$
$$= 10.1 \doteq 10$$

Equation (11.2a) yields

$$k = 2.3263 - \frac{1.6449}{\sqrt{10}} = 1.806$$

which would be the value to use if we wish to have the α of our plan exactly as specified. Equation (11.2b) yields

$$k = 1.4053 + \frac{1.2816}{\sqrt{10}} = 1.811$$

which would be the value to use if we wish to have the β of our plan exactly as specified. For our example, however, we shall take k equal to the average of the two values derived above or

$$k = \frac{1.806 + 1.811}{2} = 1.809$$

Procedure 1 for our example will then run as follows. Take a random sample of 10 items from a lot and compute $z_L = \frac{\bar{X} - 17{,}000}{800}$. If $z_L \geq 1.809$, accept the lot; otherwise reject.

Under Procedure 2 we take the extra step of translating z values into estimates of fraction nonconforming. This will be done here by treating the quantity $Q_L = z_L\sqrt{\frac{n}{n-1}} = \frac{\bar{X} - L}{\sigma'}\sqrt{\frac{n}{n-1}}$ as a normal variable. The value of M will therefore be the proportion of the area under the normal curve beyond $k\sqrt{\frac{n}{n-1}}$. For our example, $k\sqrt{\frac{n}{n-1}} = 1.809\sqrt{\frac{10}{9}} = 1.907$ which yields $M = 0.0283$. Procedure 2 for our example will therefore work as follows:

[7] It seems desirable here to take z to at least four decimals. For an extended table of normal probabilities, see Truman L. Kelley, *The Kelley Statistical Tables.*

Take a random sample of 10 items, compute $Q_L = \dfrac{\bar{X} - 17{,}000}{800}\sqrt{\dfrac{10}{9}}$ and using this as a normal deviate, determine the area in excess of Q_L. Call this $\hat{p}_L$. If $\hat{p}_L \le 0.0283$, accept; otherwise reject.

There is nothing in either of the above procedures that requires the specification limit to be on the low side. If we have an upper specification limit, $U = 20{,}000$, say, then the $\bar{X}'$ corresponding to a fraction nonconforming of 0.01 would be given by $\dfrac{200{,}000 - \bar{X}'}{800} = 2.3263$ and the $\bar{X}'$ corresponding to a fraction nonconforming of 0.08 would be given by $\dfrac{20{,}000 - \bar{X}'}{800} = 1.4053$. The first yields $\bar{X}' = 18{,}139$ and the second $\bar{X}' = 18{,}876$. These would be the new $\bar{X}'_1$ and $\bar{X}'_2$, but the z_1 and z_2 would be the same as before. Hence the n and k will also be the same, viz $n = 10$, $k = 1.809$. (Note that k is here the critical value for z_U and is the same as the k for z_L.)

With the upper limit $U = 20{,}000$. Procedure 1 will therefore be: If $\dfrac{20{,}000 - \bar{X}}{\sigma'} \ge 1.809$, accept; otherwise reject. Procedure 2 will be this. Note that $k\sqrt{\dfrac{n}{n-1}} = 1.907$ as before, which again yields $M = 0.0283$. The procedure will be to accept a lot if the estimate of the fraction nonconforming $(\hat{p}_U)$ yielded by treating $Q_U = \dfrac{U - \bar{X}}{\sigma'}\sqrt{\dfrac{n}{n-1}}$ as a normal deviate is less than or equal to M. This is, accept if $\hat{p}_U \le 0.0283$; otherwise reject. It will be noted that except for substitution of U for L, Procedures 1 and 2 are exactly the same for an upper as for a lower limit.

4.2. OC Curves for Plans with a Single Specification Limit

The OC curve for the variables plan derived above for either a lower or upper limit is shown in Figure 11.6. *The OC curve is the same for the two Procedures 1 and 2.* It is more easily computed, however, with reference to Procedure 1.

If p' is the fraction nonconforming and $z_{p'}$ is the (positive) normal deviate corresponding to this value of p', the $\bar{X}'$ that will yield the fraction nonconforming p' will be

$$\bar{X}'_{p'} = L + z_{p'}(\sigma').$$

Then as noted in Section 4.1 above, the probability of acceptance will be

$$\text{Probability}\left(\frac{\bar{X} - \bar{X}'_{p'}}{\sigma'/\sqrt{n}} \ge (k - z_{p'})\sqrt{n}\right)$$

Since $\dfrac{\bar{X} - \bar{X}'_{p'}}{\sigma'/\sqrt{n}}$ will be normally distributed with zero mean and unit variance, this will be the same as the probability that a standard normal deviate z

FIGURE 11.6
Operating Characteristic Curve for the Variables Sampling Plan $n = 10$, $k = 1.809$ (standard deviation known)

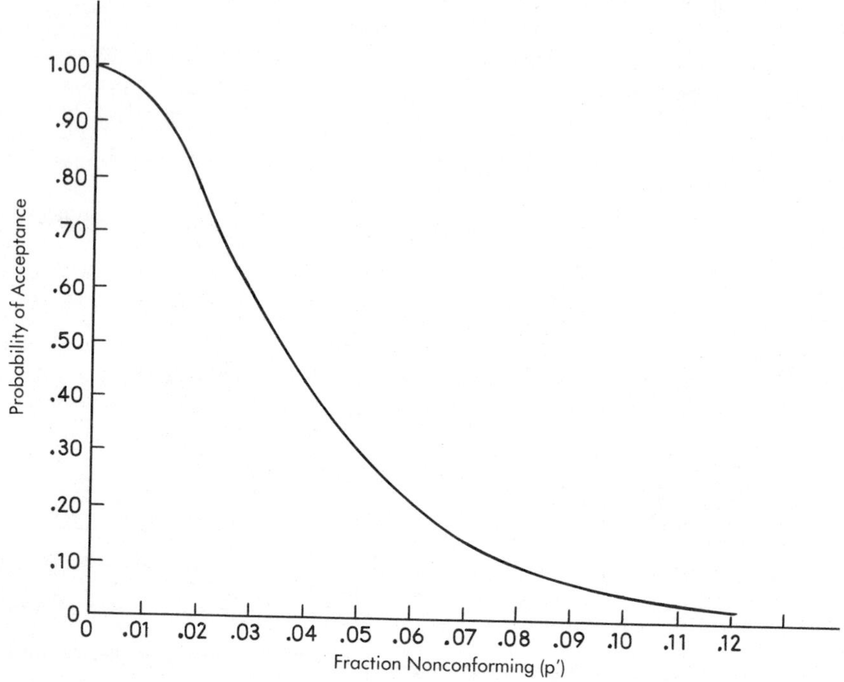

will exceed $(k - z_{p'})\sqrt{n}$. For example, if $p' = 0.03$, then $z_{0.03} = 1.881$, $(k - z_{p'})\sqrt{n} = (1.809 - 1.881)\sqrt{10} = -0.2277$, and the probability that a standard normal deviate should exceed -0.2277 is 0.59 (compare Figure 11.6). Other points on the OC curve can be computed in the same way.

4.3. Derivation of a Variables Plan with a Specified p_1', p_2', α, and β: Double Specification Limits

When there are *both* upper and lower specification limits, when the standard deviation is known, and when the process or lot is normally distributed, we can compute a lower limit for the fraction nonconforming without sampling a single item. For the minimum fraction nonconforming of a normally distributed process or lot with upper and lower specification limits is obtained when $\bar{X}'$ falls exactly between the specification limits. (See Figure 11.9). In this case the z deviate for each end will be $\dfrac{(U - \bar{X}')}{\sigma'} = \dfrac{-(L - \bar{X}')}{\sigma'} = \dfrac{(U - L)}{2\sigma'}$ and the proportion of normal area beyond $\pm z$ will be the fraction nonconforming.

When there are two specification limits, therefore, when the process or lot is normally distributed and when σ' is known, the first step is to note whether the area under a standard normal curve beyond $z = \pm\frac{U-L}{2\sigma'}$ is greater than what is acceptable. If it is, we may reject the lot without sampling. Consider an example. We are interested, let us suppose, in the width of a slot in a terminal block manufactured by a process the output of which is normally distributed. Specification limits on the slot are 0.8800 and 0.8780 inches and a process fraction nonconforming greater than 0.01 is deemed undesirable. The process standard deviation (σ') is known to be 0.0005. Then we have immediately $\frac{U-L}{2\sigma'} = \frac{0.8800 - 0.8780}{2(0.0005)} = 2$. This means that even if $\bar{X}'$ had its best possible value, namely, $\frac{U+L}{2} = \frac{0.8800 + 0.8780}{2} =$ 0.8790, the process fraction nonconforming would be no better than 2(0.0228) = 0.0456 (the area lying beyond the normal deviates ±2). The process is therefore obviously unacceptable and can be rejected without sampling.

If, however, the minimum possible fraction nonconforming is better than we seek, it may pay to sample, since $\bar{X}'$ may be either favorably or unfavorably located. Let us consider a case in which $(U - L)/2\sigma'$ is so large that if the process is centered, there will be practically no nonconforming material. Thus let us suppose that in the previous example specification limits are 0.8800 and 0.8750 inches. Then $\frac{U-L}{2\sigma'} = \frac{0.8800 - 0.8750}{2(0.0005)} = 5$ and we see that if $\bar{X}'$ were near the center of the specification range, the fraction nonconforming would be practically zero. The general relationship between the mean and fraction nonconforming in this case is as follows (see Figure 11.7):

If $\bar{X}$ Is at —	*Fraction Nonconforming Is —*
0.8750 or 0.8800	0.50
0.8754 or 0.8796	0.20
0.8756 or 0.8794	0.10
0.8758 or 0.8792	0.05
0.8762 or 0.8788	0.01

For widely spread specification limits such as just described, a variables sampling plan will consist of two single-limit plans, one for application at the lower specification limit, the other for application at the upper specification limit. For example, if we wished the overall plan to have a $p'_1 = 0.01$, a $p'_2 = 0.08$, $\alpha = 0.05$, and $\beta = 0.10$, the two single-limit plans would, in terms of Procedures 1 and 2, be identical with the upper and lower limit plans derived in Section 4.1. For it will be recalled that these plans were determined by the p'_1, p'_2, α, and β. Since the values prescribed for these quantities are the same here as in Section 4.1, the plans will also be the same. Thus, in this double-limit case, we will take $n = 10$, $k = 1.809$, and $M = 0.0283$,

FIGURE 11.7
Illustration of the Relationship between Variation in the Quality of a Product and the Fraction Nonconforming when There Is a Double Specification Limit

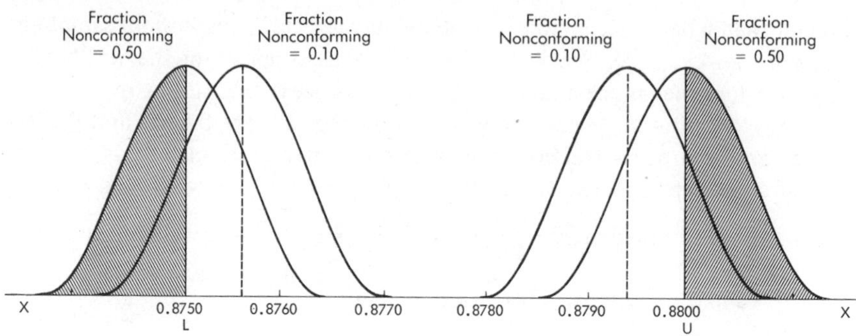

and Procedures 1 and 2 will be as follows. Procedure 1: Accept if $\frac{\bar{X}-L}{\sigma'} \geq 1.809$ and $\frac{U-\bar{X}}{\sigma'} \geq 1.809$ that is, if $\bar{X} \geq L + 1.809\sigma'$ and if $\bar{X} \leq U - 1.809\sigma'$; otherwise reject. (See Figure 11.8.) Procedure 2: Compute $Q_L = \frac{\bar{X}-L}{\sigma'}\sqrt{\frac{n}{n-1}}$, and find $\hat{p}_L$ as the normal area falling beyond Q_L. Likewise compute $Q_U = \frac{U-\bar{X}}{\sigma'}\sqrt{\frac{n}{n-1}}$ and find $\hat{p}_U$ as the normal area falling beyond Q_U. If either $\hat{p}_L$ or $\hat{p}_U$ exceeds M, reject; otherwise accept. It will be noted that in our example the p'_1 and p'_2 pertain to the process fraction nonconforming and that reference is throughout to a Type B OC curve.

It remains to discuss the case in which the two specification limits are not relatively far apart, say $\frac{U-L}{2} < 3\sigma'$, but are still not so close that the minimum fraction nonconforming to be obtained, if $\bar{X}'$ were centered between the two limits, is in excess of the acceptable level of 0.01. Suppose

FIGURE 11.8
Illustration of a Double-Limit Sampling Plan with Widely Spread Specification Limits. (Curves picture distributions of sample means.)

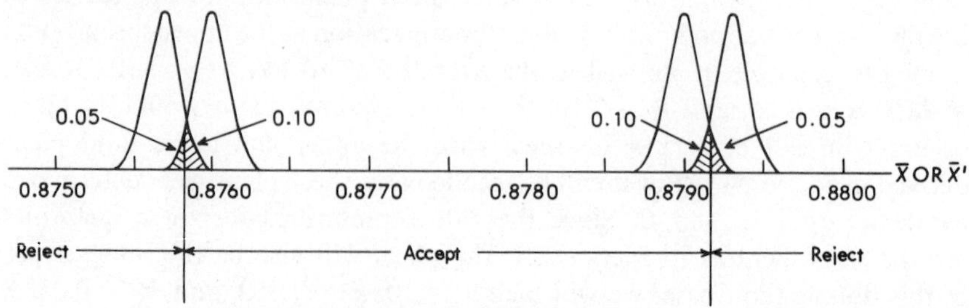

that in our example specification limits fall at 0.8800 and 0.8773. Then the minimum fraction nonconforming to be obtained (yielded by an $\bar{X}' = \frac{0.8800 + 0.8773}{2} = 0.87865$) is the proportion of cases beyond the normal deviates $\pm z = \pm\frac{0.8800 - 0.8773}{2(0.0005)} = \pm 2.70$. This is 0.0069 which is less than the acceptable level of 0.01 so that knowledge of σ' is not sufficient to reject without sampling.

The relationship between $\bar{X}'$ and p' will now be as follows (it will be remembered that $\sigma' = 0.0005$):

$\bar{X}'$	$\frac{L - \bar{X}'}{\sigma'}$	$\frac{U - \bar{X}'}{\sigma'}$	p'_L	p'_U	p'
0.87865	−2.70	2.70	0.00347	0.00347	0.0069
0.8787	−2.80	2.60	0.0026	0.0047	0.0073
0.8786	−2.60	2.80	0.0047	0.0026	0.0073
0.8788	−3.00	2.40	0.00135	0.0082	0.0096
0.8785	−2.40	3.00	0.0082	0.00135	0.0096
0.8789	−3.20	2.20	0.00069	0.0139	0.0146
0.8784	−2.20	3.20	0.0139	0.00069	0.0146

A graph is shown in Figure 11.9 on the next page.

From Figure 11.9 it may be estimated that a $p' = 0.01$ will be given by $\bar{X}' = 0.878485$ or $\bar{X}' = 0.878815$. But this will be distributed such that 0.0089 of the blocks will be nonconforming at one end and 0.0011 at the other. Hence, if the plan we want is one with a $p'_1 = 0.01$, $p'_2 = 0.08$, $\alpha = 0.05$, and $\beta = 0.10$, this can again be accomplished by setting up two single-limit plans such that if the fraction nonconforming is 0.0089 at one end, there will be a 0.95 chance of acceptance, whereas if the fraction nonconforming is 0.08 at that end, there will be only a 0.10 chance of acceptance. We shall thus have

$$n = \left(\frac{z_{0.05} + z_{0.10}}{z_{0.0089} - z_{0.0800}}\right)^2$$

$$= \left(\frac{1.6449 + 1.2816}{2.3698 - 1.4053}\right)^2 = 9.2 \doteq 9$$

and k will be given by the average of

$$\frac{1.2816}{\sqrt{9}} + 1.4053 = 1.8325$$

and

FIGURE 11.9
Relationship between the Mean and Fraction Nonconforming of a Normally Distributed Process or Lot: Close Specification Limits

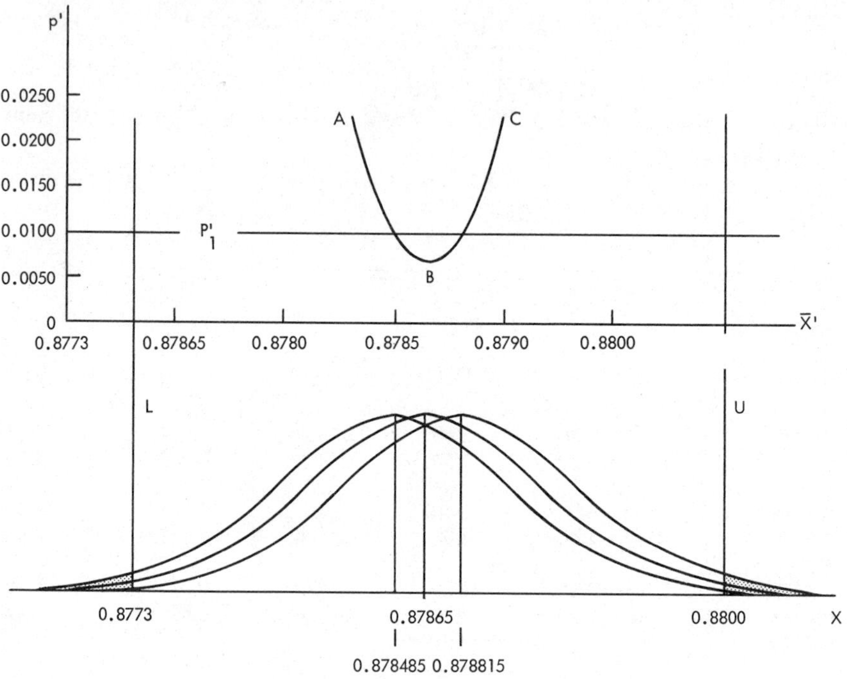

$$2.3698 - \frac{1.6449}{\sqrt{9}} = 1.8215$$

which yields

$$k = 1.827$$

We shall also have $k\sqrt{\frac{n}{n-1}} = 1.8727\sqrt{\frac{9}{8}} = 1.938$, which taken as a normal deviate yields $M_1 = 0.0263$, where M_1 is the acceptance limit for the fraction nonconforming at one end.

Procedure 1 will be to accept a lot if $\frac{\bar{X} - 0.8773}{\sigma'} \geq 1.827$ and if $\frac{0.8800 - \bar{X}}{\sigma'} \geq 1.827$; otherwise reject. Procedure 2 will be to accept if $Q_L = \frac{\bar{X} - 0.8773}{\sigma'}\sqrt{\frac{9}{8}}$ yields a $\hat{p}_L \leq 0.0263$ and if $Q_U = \frac{0.8800 - \bar{X}}{\sigma'}\sqrt{\frac{9}{8}}$ yields a $\hat{p}_U \leq 0.0263$; otherwise reject. The OC curves for either of the above procedures will be that for a single-limit plan and can thus be derived as described in Section 4.1 above.

It is to be noted that if the sample $\bar{X}$ is such that $\frac{\bar{X} - 0.8773}{0.0005} = 1.827$, it follows that $\bar{X} = 0.87821$. But then $\dfrac{U - \bar{X}}{\sigma'} = \dfrac{0.8800 - 0.87821}{0.0005} = \dfrac{0.00179}{0.0005}$ $= 3.58$ and $3.58\sqrt{\dfrac{9}{8}} = 3.80$. Taking this as a normal deviate, we have an estimate of the fraction nonconforming at the *upper* end equal to 0.00007. This suggests that we could set up an overall M equal to $0.0263 + 0.00007 = 0.02637$ and operate Procedure 2 as follows. Compute $Q_L = \dfrac{\bar{X} - 0.8773}{\sigma'}\sqrt{\dfrac{9}{8}}$ and determine $\hat{p}_L$. Also compute $\dfrac{0.8800 - \bar{X}}{\sigma'}\sqrt{\dfrac{9}{8}}$ and determine $\hat{p}_U$. If $\hat{p}_L + \hat{p}_U \leq 0.02637$, accept; otherwise reject. We thus have derived a unified double-limit plan. When we come to Chapter 12, where it will be assumed that σ' is unknown, this unified procedure will be the most satisfactory one.

In concluding this section, it should be noted that there is a continuum of sampling plans from the situation in which the spread between specifications is so wide that we need apply the p_1' and p_2' criteria only at one end, to the situation in which the specification limits are so close that the minimum fraction nonconforming yielded when $\bar{X}'$ is centered between L and U is high enough to lead to immediate rejection. In our example, it will be recalled that for specified p_1', p_2', α, and β, the first set of conditions called for a sample size of 10, the second for no sampling at all. It should therefore have been no surprise that an intermediate situation calling for the same p_1', p_2', α, and β yielded a sample size of nine. As the minimum point of the *ABC* curve in Figure 11.9 moves from 0 to above the p_1', the sample size of the plans needed to meet a constant set of values for the p_1', p_2', α, and β will continually decrease. There will likewise be a continual decrease in the M from that given by the single-limit plan down to the p_1'.[8]

5. A SPECIAL ATTRIBUTES PLAN AS AN ALTERNATIVE TO A VARIABLES PLAN WHEN σ' IS KNOWN

It is only fair to note at this point that, *under the conditions for which the variables plans of Sections 4.1 and 4.3 are applicable,* savings in inspection costs can also be attained by using an attributes plan with tighter specification limits than those prescribed by engineering design. This goes by the name of compressed limit gauging, or narrow limit gauging, or increased severity testing.

If we are willing to assume, for example, that items in a process or lot are normally distributed and that the standard deviation is known, then we can work with fictitious specification limits. With reference to the example of Section 4.1, we can increase the specification of minimum tensile strength to 17,500 psi. Under the assumed conditions, with $\sigma' = 800$, this will be

[8] A variation omitted from the above discussion is the case in which a different p_1' is specified for each of the specification limits. For plans providing for this, the reader is referred to Military Standard 414. (See Chapter 13.)

the equivalent of taking p'_1 as 0.0445 instead of 0.01 and p'_2 as 0.218 instead of 0.08. (For $\frac{17{,}500 - 18{,}861}{800} = -1.70$, and 4.45 percent of a normal distribution lies below $z = -1.70$. Likewise, $\frac{17{,}500 - 18{,}124}{800} = -0.78$, and 21.8 percent of a normal distribution lies below $z = -0.78$.) In other words, under the assumed conditions of normality and a standard deviation equal to 800, a process or lot that has 1 percent of its items below 17,000 is the equivalent of one that has 4.45 percent of its items below 17,500. Also a process or lot that has 8 percent of its items below 17,000 is the equivalent of one that has 21.8 percent of its items below 17,500. Now Table 7.1 indicates that an attributes single-sampling plan that will come close to having[9] a $p'_1 = 0.01$, a $p'_2 = 0.08$, $\alpha = 0.05$, and $\beta = 0.10$ is $n = 67$, $c = 2$. But as just noted, we can get a attributes single-sampling plan that is the equivalent of $n = 67$, $c = 2$, by finding a plan for which the $p'_1 = 0.0445$, $p'_2 = 0.218$, $\alpha = 0.05$, and $\beta = 0.10$. Such a plan is $n = 31$, $c = 3$ where now all items below 17,500 psi are taken as "nonconforming." The saving here is not so great as that obtained by the variables plan but it is still considerable.

Compressed limit gauging of this kind may be found useful when the original p'_1 and p'_2 are small and when variables sampling is costly or impracticable. It will be noted, however, that its usefulness, like that of the variables plans discussed in this chapter, depends on the distribution of items being normal, or at least of known form. For the same fictitious limits to be used for a series of lots, the process standard deviation and other distribution parameters must be constant if the OC curve is to be that desired.

6. VARIABLES SAMPLING FOR NONNORMAL LOTS OF KNOWN FORM

It is possible to employ variables plans based on $\bar{X}$ to nonnormal processes or lots, provided the form of the distribution is known and our knowledge of auxiliary characteristics is such that if we knew the mean we could immediately determine the fraction nonconforming. For the sampling distribution of means tends to the normal form rather rapidly with the sample size n and is approximately normal for most nonnormal universes met with in practice and most sample sizes employed in acceptance sampling.[10] All that is needed is sufficient knowledge of the process or lot distribution to go from the mean to a fraction nonconforming.

A procedure is given in Appendix I (27) for applying a variables sampling plan when the universe can be described by a Pearson Type III distribution, which in standardized form $\left(\text{i.e., the variable is taken as } \frac{X - \bar{X}'}{\sigma'}\right)$ depends only on the coefficient of skewness γ'_1. This is a moderately skewed distribu-

[9] The actual α will be 0.03 instead of $\alpha = 0.05$.

[10] See Chapter 6, Section 1.2.

tion. Tables of the Type III distribution prepared by L. R. Salvosa give, for specified values of the coefficient of skewness γ'_1, the fractions of product lying beyond certain listed distances from the mean expressed as multiples of the standard deviation. If the standard deviation and coefficient of skewness are known, the sample $\bar{X}$ can be used for accepting and rejecting lots in a manner similar to that applied to normally distributed processes or lots with known standard deviation. For details the reader is referred to Appendix I (27).

7. PROBLEMS

11.1 As a milk distributor you wish to control the butterfat content of the milk you purchase for distribution. Past research has indicated that under stable conditions the amount of butterfat per quart varies in accordance with a normal frequency distribution. The average butterfat per quart varies significantly, however, from time to time, but the standard deviation remains constant at about 0.15 ounces per quart. Cans with less than 1.000 ounce of butterfat per quart are considered by you to be below the standard of your trade (i.e., "nonconforming"). To control the quality of your purchases, you wish to find a variables sampling plan that will do the following: (1) accept 95 percent of the shipments when the cans are running only 1 out of 100 below standard; (2) accept only 10 percent of the shipments when the cans are running 4 out of 100 below standard. Find a variables sampling plan that will meet these criteria. Give two alternative procedures for carrying out the plan. Draw the OC curve for the plan. Find an attributes plan with the same p'_1, p'_2, α, and β as your variables plan and compare the sample sizes.

11.2 The illumination of a certain type of lamp varies from lamp to lamp in accordance with a normal frequency distribution. The standard deviation of the distribution is 168 end foot-candles. Variations in the manufacturing process usually affect the average illumination but not the standard deviation. Let the lower specification limit on the illumination of a lamp be 150 end foot-candles and all lamps giving less illumination than this be classed as nonconforming (cf. Harry G. Romig, *Allowable Average in Sampling Inspection,* p. 15).

Design a variables sampling plan that will have a $p'_1 = 0.005$, a $p'_2 = 0.05$, $\alpha = 0.05$, and $\beta = 0.10$. Give two alternative procedures for carrying out the plan. Draw the OC curve for the plan. Find a plan based on attributes with the same p'_1, p'_2, α, and β, and compare the sample size with that of your variables plan.

11.3 The current consumption of a certain lamp is normally distributed with a standard deviation of 1.75 milliampers. Variations in the manufacturing process usually affect the average current consumption but not the standard deviation. The upper specification limit on current consumption is 50 milliamperes, and the lower specification limit on current consumption is 35 milliamperes. Lamps the current consumption of which lies outside these limits are classed as nonconforming (cf. H. G. Roming, *Allowable Average in Sampling Inspection,* p. 15).

Design a variables sampling plan that will have a $p'_1 = 0.01$, a $p'_2 = 0.03$, $\alpha = 0.05$, and $\beta = 0.10$. Give two alternative procedures for carrying out the plan. Draw the OC curve for the plan. Find a plan based on attributes with the same p_1', p_2', α, and β, and compare the sample size with that of your variables plan.

11.4 Specifications on a piece part are 4.350 ± 0.005. The machine that manufactures these parts is known to yield approximately a normal variation in part dimensions with a standard deviation of 0.001. You are asked by the management to design a lot-by-lot variables sampling plan with a $p_1' = 0.02$, a $p_2' = 0.05$, $\alpha = 0.05$, $\beta = 0.10$. Give two alternative procedures for carrying out the plan and draw the OC curve. Find an attributes plan with the same p_1', p_2', α, and β and compare the sample size with that of your variables plan.

11.5 Suppose that in Problem 11.4, the standard deviation had been 0.003, and sketch the report you would make to the management in response to their request.

11.6 Specifications for a dimension are 10.50 ± 0.03. In the manufacturing process it has been found that variation in the dimension follows approximately a normal distribution with a constant standard deviation of 0.013. You are asked by your company to set up a lot-by-lot variables sampling plan with a $p_1' = 0.03$, a $p_2' = 0.07$, $\alpha = 0.05$, $\beta = 0.10$. Give the plan you would recommend, offering two alternative procedures for its execution. Draw the OC curve for the plan. Find an attributes plan with the same p_1', p_2', α, and β, and compare the sample size with that of your variables plan.

11.7 The resistance of a certain lamp is approximately distributed in accordance with a Pearson Type III distribution with a standard deviation of 3.000 ohms and $\gamma_1' = 0.5$. Variations in the manufacturing process usually affect the average resistance but not the other characteristics of the distribution. The upper tolerance limit is 90 ohms. Lamps whose resistance exceeds this limit are classed as nonconforming (cf. Harry G. Romig, *Allowable Average in Sampling Inspection,* p. 15).

Design a sampling plan based on $\bar{X}$ that will have a $p_1' = 0.008$, a $p_2' = 0.03$, $\alpha = 0.05$, and $\beta = 0.10$. Draw the OC curve for the plan. Find an attributes plan with the same p_1', p_2', α, and β, and compare the sample size with that of your variables plan. [See Appendix I (27).]

11.8 For solving Problem 11.4, set up attributes plans that employ compressed limit gauging. More specifically, given the fact of a normal distribution with $\sigma' = 0.001$, work out plans with "tightened" limits at ± 0.0048, ± 0.0045, and ± 0.0040 and note which gives the greatest saving with respect to the ordinary attributes plan having the same specified p_1', p_2', α, and β.

8. SELECTED REFERENCES*

Acceptance Sampling (B '50), Bowker and Goode (B '52), Burr (B '76), Clifford (P '47), Cowden (B '57), Dudding and Jennett (B '44), Enrick (B '71), Freeman (B '42), Grant and Leavenworth (B '80), Lieberman and Resnikoff (P '55), Ott and Mundel (P '54), Owen (P '66, P '67, and P '69), Peach (B '47), Romig (B '39), Schilling (B '82), Schwartz and Kaufman (P '51), Stevens (P '48), Tippett (B '50), and Zimmer and Burr (P '63).

---

* B and P refer to the Book and Periodical sections, respectively, of the Cumulative List of References in Appendix V.

12

Acceptance Sampling by Variables to Control the Fraction Nonconforming: Standard Deviation Unknown

When the process or lot standard deviation is unknown we proceed in much the same manner as we do when the standard deviation is known,[1] except that we substitute some sample estimate of σ' for σ' itself. This introduces greater uncertainty into the procedure, and to obtain an OC curve essentially the same as that obtained when the standard deviation is known, it is necessary to take a larger sample. This is the price we pay for not knowing σ'.

1. DERIVATION OF A VARIABLES PLAN WITH A SPECIFIED p'_1, p'_2, α, AND β

1.1. Normally Distributed Processes or Lots: Single Specification Limit

In Chapter 11 we saw that when the standard deviation is known, there are two procedures that can be followed in carrying out a variables sampling plan. These were called Procedures 1 and 2. When the standard deviation is unknown, we can continue to use these same two procedures, modified, however, to allow for our lack of knowledge of σ'.

1.1.1. Procedure 1. When σ' is unknown and there is a lower specification limit only, Procedure 1 runs as follows. Take a sample of n and compute $\bar{X} = \Sigma X/n$, $s = \sqrt{\dfrac{\Sigma(X-\bar{X})^2}{n-1}}$, and $z_L = \dfrac{\bar{X}-L}{s}$. If z_L is equal to or larger

[1] The reader will have a better understanding of Chapter 12 if he first studies Chapter 11. He should in particular note the warning given in Section 2 of Chapter 11 about the risks involved when processes and lots are not normally distributed.

than some predetermined limit k, accept a lot; if it is less than k, reject. Symbolically,

$$\text{if } \frac{\bar{X}-L}{s} \geq k, \text{ accept; if } \frac{\bar{X}-L}{s} < k, \text{ reject.}$$

When there is an upper specification limit only,

$$\text{if } \frac{U-\bar{X}}{s} \geq k, \text{ accept; if } \frac{U-\bar{X}}{s} < k, \text{ reject.}$$

To determine a variables sampling plan that follows Procedure 1 and has a specified p'_1, p'_2, α, and β, it is necessary as before to find an n and a k. For each n and k we select, we can derive an OC curve for the plan. The problem is to find an n and a k that will lead to an OC curve that satisfies the specified criteria.

Special formulas for finding n and k for given p'_1, p'_2, α, and β are offered by W. Allen Wallis. These are as follows.[2]

(12.1)

$$k = \frac{z_\alpha z_2 + z_\beta z_1}{z_\alpha + z_\beta}$$

$$n = \left(1 + \frac{k^2}{2}\right)\left(\frac{z_\alpha + z_\beta}{z_1 - z_2}\right)^2$$

where the z's are the normal deviates the probability of exceeding which are the p'_1, p'_2, α, and β. Resemblance to formula (11.3) will be readily noted. It is also readily seen that because we do not know the lot standard deviation, we must now take $\left(1 + \frac{k^2}{2}\right)$ times as many items in our sample to get the same OC curve. (The reader should note once again that z_L, z_U, z_1, z_2, and $z_{p'}$ are all so defined that under normal circumstances they will be positive, but that in general z may be positive or negative.)

Formulas (12.1) are based on the assumption that $\bar{X} \pm ks$ is approximately normally distributed with a mean[3] of $\bar{X}' \pm k\sigma'$ and a standard deviation equal[4] approximately to $\sigma'\sqrt{\frac{1}{n} + \frac{k^2}{2n}}$. For a single lower specification limit the probability of acceptance is the probability that $\frac{\bar{X}-L}{s} \geq k$ or $\bar{X} - ks \geq L$. If we subtract $\bar{X}' - k\sigma'$ from each side and divide by

[2] Statistical Research Group, Columbia University, *Techniques of Statistical Analysis* (New York: McGraw-Hill, 1947), p. 17. A more precise procedure is based on the use of the noncentral t distribution. See G. J. Resnikoff and G. L. Lieberman, *Tables of the Non-Central* t-*Distribution* (Stanford, Calif.: Stanford University Press, 1975), pp. 10–16.

[3] Cf. Appendix I (11). $E(\bar{X} + ks) = E(\bar{X}) + kE(s) \doteq \bar{X}' + k\sigma'$. See Table M of Appendix II.

[4] Cf. Appendix I (13). Var $(\bar{X} + ks)$ = Var $(\bar{X})$ + k^2 Var (s), because in samples from a normal universe, $\bar{X}$ and s are independent. Also see Chapter 6.

$\sigma'\sqrt{\dfrac{1}{n}+\dfrac{k^2}{2n}}$, this inequality-equality becomes

$$\frac{(\bar{X}-ks)-(\bar{X}-k\sigma')}{\sigma'\sqrt{\dfrac{1}{n}+\dfrac{k^2}{2n}}} \geq \frac{L-(\bar{X}'-k\sigma')}{\sigma'\sqrt{\dfrac{1}{n}+\dfrac{k^2}{2n}}} = \frac{\dfrac{L-\bar{X}'}{\sigma'}+k}{\sqrt{\dfrac{1}{n}+\dfrac{k^2}{2n}}}$$

Let $z_1 = \dfrac{\bar{X}'-L}{\sigma'}$ when $\bar{X}'$ and σ' are such as to yield a fraction nonconforming equal to p_1' and let $z_2 = \dfrac{\bar{X}'-L}{\sigma'}$ when $\bar{X}'$ and σ' are such as to yield a fraction nonconforming equal to p_2'. Then to meet the p_1', α criteria we must have

$$\frac{-z_1+k}{\sqrt{\dfrac{1}{n}+\dfrac{k^2}{2n}}} = -z_\alpha$$

and to meet the p_2', β criteria we must have

$$\frac{-z_2+k}{\sqrt{\dfrac{1}{n}+\dfrac{k^2}{2n}}} = z_\beta$$

These two equations yield formulas (12.1).

To illustrate the use of formulas (12.1), suppose the desired criteria are $p_1' = 0.01$, $p_2' = 0.08$, $\alpha = 0.05$, and $\beta = 0.10$. The formulas then yield

$$k = \frac{(1.6449)(1.4053)+(1.2816)(2.3263)}{(1.6449+1.2816)} = 1.809$$

$$n = \left(1+\frac{(1.809)^2}{2}\right)\left(\frac{1.6449+1.2816}{2.3263-1.4053}\right)^2$$

$$= 26.6 \doteq 27$$

Another method of deriving a single-limit variables sampling plan is provided by a nomograph constructed by L. J. Jacobson. This is shown in Figure 12.1 on the next page. The scale on the left of this nomograph is simply a graphical representation of the normal area table (Table A2, Appendix II). The left-hand markings are the tail areas or fractions nonconforming; the right-hand markings are values of z. The right-hand scale is also a graphical representation of the normal area table, but this shows the area from $-\infty$ to z. The left-hand markings give the area; the right-hand markings the z values. The use of the nomograph is illustrated at the bottom of the chart. Draw a line from p_2' on the left-hand scale to β on the right-hand scale, also a line from p_1' on the left-hand scale to $1-\alpha$ on the right-hand

FIGURE 12.1
Nomograph for Determining a Variables Sampling Plan Based on $\bar{X}$ with σ' Unknown—Given: p'_1, p'_2, α, and β; To Find: n and k

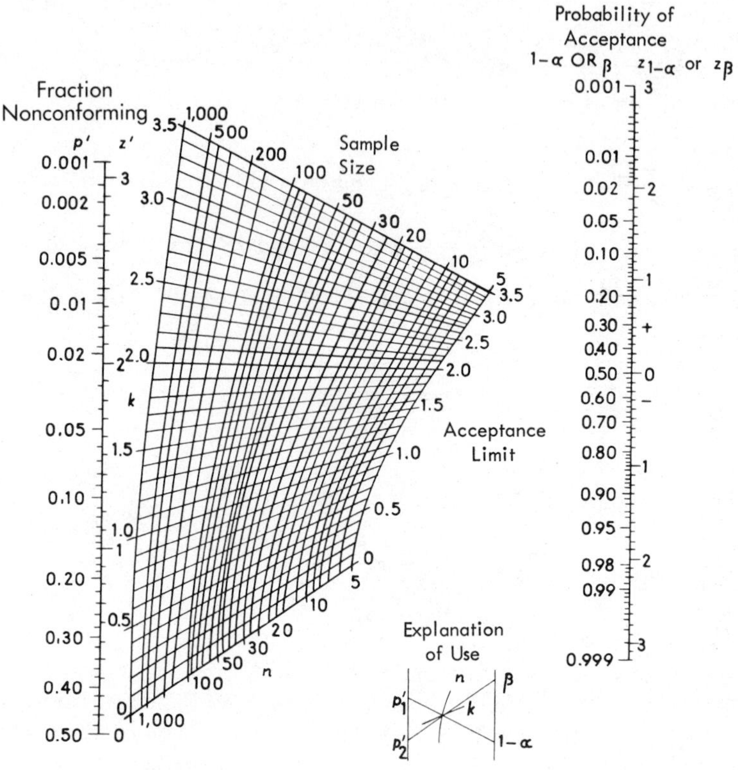

SOURCE: Leo J. Jacobson, "Nomograph for Determination of Variables Inspection Plan for Fraction Defective," *Industrial Quality Control,* November 1949, p. 24. Copyright American Society for Quality Control, Inc. Reprinted by permission.

scale. Find the intersection of these lines on the n, k grid. These are the values for the sampling plan desired. The nomograph gives only rough results and seems to have an upward bias in n and a downward bias in k. For example, for a $p'_1 = 0.01$, a $p'_2 = 0.08$, $\alpha = 0.05$, and $\beta = 0.10$, the nomograph yields $n = 30$, $k = 1.80$ compared with $n = 27$, $k = 1.809$ yielded by formula. The nomograph is very easy to use, however, and is of great help in exploratory work. If greater precision is wanted, the n and k of a plan should be derived from the formulas.

In addition to the Wallis formulas and the Jacobson nomograph, there is a table of values of k and n for various p'_1's from 0.001 to 0.05 and various p'_2's from 0.0015 to 0.40. This will be found in Statistical Research Group, Columbia University, *Techniques of Statistical Analysis,* pp. 22–25.

To find the OC curve for a one-sided variables plan with unknown σ', we can get approximate results as in formulas (12.1) by assuming that $\bar{X}$

$-\ ks$ has a normal distribution with a mean of $\bar{X}' - k\sigma'$ and a standard deviation equal to $\sigma' \sqrt{\frac{1}{n} + \frac{k^2}{2n}}$. Thus, if we have a single lower limit, the probability of acceptance, which is Prob. $\left(\left(\frac{\bar{X} - L}{s}\right) \geq k\right)$, is given by

$$\begin{aligned} &\text{Prob. } (\bar{X} - ks \geq L) \\ &= \text{Prob. } \frac{(\bar{X} - ks) - (\bar{X}' - k\sigma')}{\sigma' \sqrt{\frac{1}{n} + \frac{k^2}{2n}}} \geq \frac{L - (\bar{X}' - k\sigma')}{\sigma' \sqrt{\frac{1}{n} + \frac{k^2}{2n}}} \qquad \textbf{(12.2)} \\ &= \text{Prob. } (z \geq z_A) \end{aligned}$$

where

$$z_A = \frac{k - z_{p'}}{\sqrt{\frac{1}{n} + \frac{k^2}{2n}}} \qquad \textbf{(12.3)}$$

$$z_{p'} = \frac{\bar{X}' - L}{\sigma'}$$

and z represents a standardized normal variable. Since processes or lots are assumed to be normally distributed, we can go immediately[5] from p' to $z_{p'}$ and by (12.3) to z_A, and hence compute the probability of acceptance for product of p' quality.

Generally it is easier to compute the p' that will have a stated probability of acceptance. In this case the formula to use is the inverse of formula (12.3) or

$$z_{p'} = k - z_A \sqrt{\frac{1}{n} + \frac{k^2}{2n}} \qquad \textbf{(12.3a)}$$

To illustrate, for the plan $n = 27$, $k = 1.809$, the value of p' corresponding to a probability of acceptance of 0.10 (i.e., a $z_A = 1.282$) will be given by

$$z_{p'} = 1.809 - 1.282 \sqrt{\frac{1}{27} + \frac{(1.809)^2}{54}} = 1.408$$

which yields $p' = 0.08$. Thus we have verified one of the desired characteristics of the plan, namely that it should have a probability of acceptance of 0.10 for product of 0.08 quality.

For a rough sketch of the OC curve it is easier to use the Jacobson nomograph. Place a ruler on the chart so that it passes through the n, k point given by Figure 12.1 for the plan. Then, pivoting it around that point, read off various values of the probability of acceptance on the right-hand

[5] See Chapter 11, Section 3.

FIGURE 12.2
The Operating Characteristic Curve for the Variables Sampling Plan $n = 27, k = 1.809$ (standard deviation unknown)

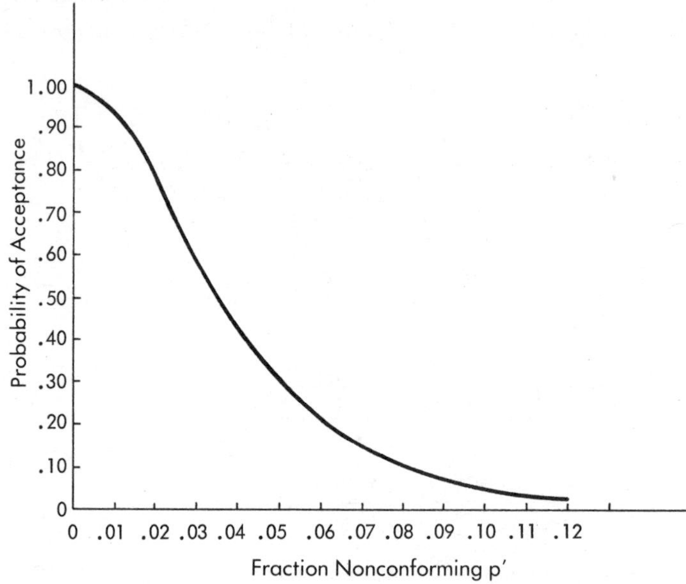

scale against various values of p' on the left-hand scale. The OC curve for the plan $n = 27$, $k = 1.809$ is shown in Figure 12.2.

1.1.2. Procedure 2. Procedure 2 which is the statistical equivalent of Procedure 1 when there is only a single specification limit runs as follows. From $z_L = \frac{\bar{X} - L}{s}$ or $z_U = \frac{U - \bar{X}}{s}$ make an estimate $(\hat{p})$ of the fraction nonconforming. If this exceeds a predetermined limit M, reject the lot; otherwise accept. At first glance this seems practically identical with Procedure 2 when σ' is known. There is this difference, however. When σ' is unknown the minimum-variance method of Lieberman and Resnikoff requires special tables and a special procedure for determining M, $\hat{p}_L$, and $\hat{p}_U$. Precise execution can only be attained with the use of such tables. It is possible, however, to attain resonably good results by using Figures 12.3 and 12.4. For specified values of n, Figure 12.3 gives estimates of $\hat{p}_L$ or $\hat{p}_U$ directly from $z_L = \frac{\bar{X} - L}{s}$ or $z_U = \frac{U - \bar{X}}{s}$. Figure 12.4 gives values of M as a function of k for specified values of n. In this text Figures 12.3 and 12.4 will be used, but if precise results are desired reference should be made to the Lieberman-Resnikoff tables.[6]

---

[6] See Cumulative List of References. These tables will also be found in Military Standard 414. See Chapter 13.

FIGURE 12.3
Chart for Determining $\hat{p}$ from z, Standard Deviation Plans

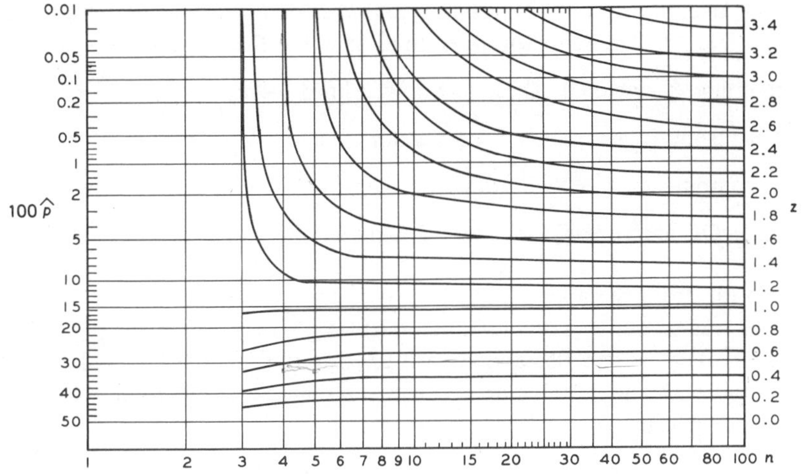

FIGURE 12.4
Chart for Determining the Maximum Allowable Fraction Nonconforming *M*

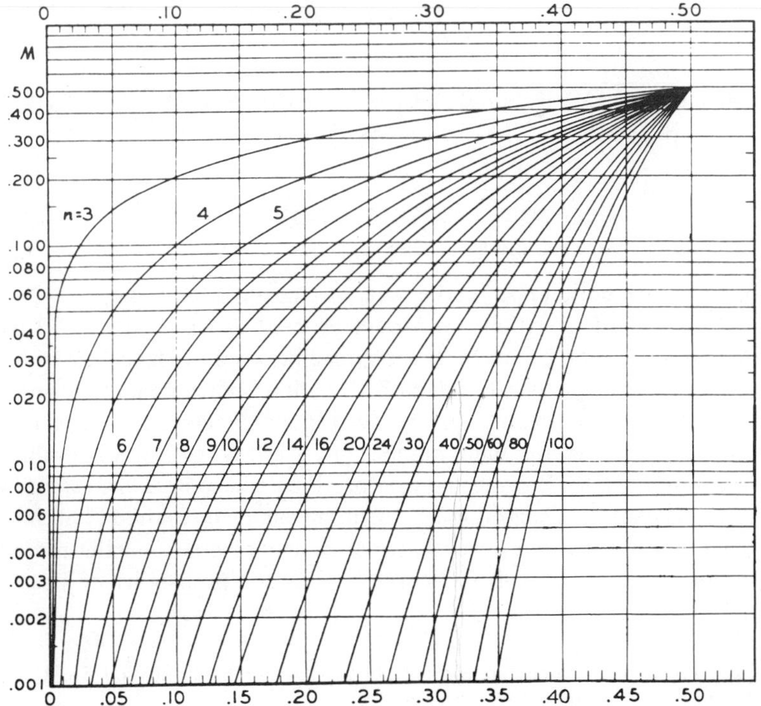

If there is a single specification limit, therefore, a variables sampling plan using Procedure 2 can be derived for a specified p'_1, p'_2, α, and β as follows. Use formulas (12.1) to find n and k as described above.[7] Use Figure 12.4 to find M. The plan will operate as follows. Take a random sample of n, compute $\bar{X}$ and s. If there is a lower limit, compute $z_L = \dfrac{\bar{X} - L}{s}$ and read $\hat{p}_L$ from Figure 12.3. Accept a lot if $\hat{p}_L \le M$. If there is an upper limit, compute $z_U = \dfrac{U - \bar{X}}{s}$ and read $\hat{p}_U$ from Figure 12.3. Accept if $\hat{p}_U \le M$. The OC curve for Procedure 2 will be the same as that for Procedure 1.

To illustrate the above let us return to the problem of page 277. There the aim was to derive a single-limit variables plan for which the $p'_1 = 0.01$, the $p'_2 = 0.08$, $\alpha = 0.05$, and $\beta = 0.10$. For this plan the n and k were found to be 27 and 1.809. For Procedure 2 the abscissa in Figure 12.4 will be taken equal to

$$\frac{1 - k\sqrt{n}/(n-1)}{2} = \frac{1 - \dfrac{1.809\sqrt{27}}{26}}{2} = 0.32$$

and we then read (for $n = 27$) $M = 0.033$. Suppose that $L = 17{,}000$ and a sample of 27 items yields $\bar{X} = 18{,}526$ and $s = 754$; we would then have

$$z_L = \frac{18{,}526 - 17{,}000}{754} = 2.02$$

and from Figure 12.3 we would read $\hat{p}_L = 0.019$. Since $0.019 < 0.033$, we would accept the lot.

1.2. Normally Distributed Processes and Lots: Double Specification Limits

When there are both upper and lower specification limits and we do not know the standard deviation, we cannot proceed as easily as we did when the standard deviation is known. If the spread between the specification limits is not a large multiple of the standard deviation, and how large a multiple it is we do not now know a priori, we can no longer find a one to one correspondence between a finite number of z's and a given fraction nonconforming. For when there can be nonconforming items at both ends simultaneously, any given total fraction nonconforming can be composed of an infinite combination of upper and lower fractions nonconforming. Furthermore, we can no longer derive a single OC curve relating the probability of lot accep-

[7] For a more precise procedure use tables of the noncentral t distribution. See Resnikoff-Lieberman reference in Appendix V (Book section).

tance under a specified variables plan to the fraction nonconforming, since the curve is partially dependent upon how the total fraction nonconforming is split between the specification limits. There is in this case a band of OC curves. Fortunately, however, investigation shows that this band of OC curves, which is widest in the neighborhood of p_1', is nevertheless not very wide at this point. A study by George L. Resnikoff[8] shows that the OC curve derived for a single-limit plan with specified p_1', p_2', α, and β is the lower limit of the band of OC curves for a two-sided specification plan with the same p_1', p_2', α, and β and for most practical cases can be taken as the OC curve for the two-sided plan.

1.2.1. Procedure 1. If we overlook this lack of precision in the OC curve, there is still the question of deciding upon the procedure to be used. Procedure 1 which uses $\frac{\bar{X}-L}{s}$ or $\frac{U-\bar{X}}{s}$ runs into difficulties when faced with two-sided specification limits. We can start with the application of a single-limit plan at each end. These single-limit plans would have the p_1', p_2', α, and β stipulated for the double-limit plan but would be applied to each end separately. When there are in fact two limits, however, we must watch for cases in which σ' is relatively large (suggested in practice by a relatively large s). For in this case, the lot might be accepted by application of the single-limit plan at each end, but, because there may be nonconforming items at *both* ends, the total may be greater than that desired. Because of this, Procedure 1 is usually modified when there are two-sided specifications by requiring that not only must the single-limit criteria be met but that also the sample standard deviation must not be too large. The criteria for acceptance under Procedure 1 are thus stated to be:

$$\frac{\bar{X}-L}{s} \geq k,$$

$$\frac{U-\bar{X}}{s} \geq k \text{ and} \tag{12.4}$$

$$s \leq \text{the } MSD,$$

where MSD stands for Maximum Standard Deviation. The MSD may be derived as follows.[9] From Figure 12.4 compute an M corresponding to the

[8] *A New Two-Sided Acceptance Region for Sampling by Variables,* Applied Mathematics and Statistics Laboratory, Stanford University (1952).

[9] An intuitive feeling for this procedure can be obtained from the following. As indicated in Chapter 11, if a process or lot is normally distributed, if its standard deviation σ' is known, and if there are both an upper and lower specification limit, then a minimum fraction nonconforming can immediately be computed. When the standard deviation is not known, but a p_1' is specified, a maximum value for σ' could be computed such that the minimum fraction nonconforming associated with that σ' would not exceed the p_1'. Under this procedure we would find the z associated with a fraction nonconforming equal to $p_1'/2$ and then find the maximum σ' by dividing $(U - L)/2$ by this z. The method in the text is similar to this except that it contains allowances for sampling fluctuations. Actually the MSD is the value of s at which

FIGURE 12.5
Illustration of Procedure 1 for Two-Sided Specifications

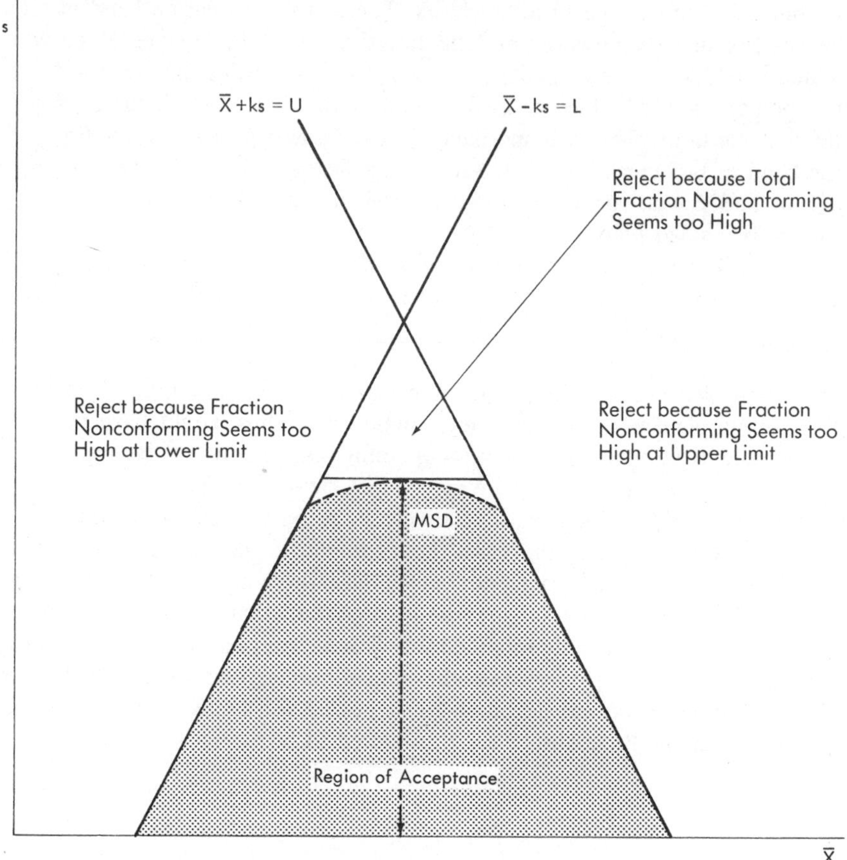

k and n of the single-limit plan. Divide this M by 2 and use Figure 12.4 in reverse to find a new k, which we will call k^*, corresponding to $M/2$. Take the MSD as equal to $(U - L)/2k^*$. For example, if $k = 1.700$ and $n = 20$, we would enter Figure 12.4 with an abscissa equal to

$$\frac{1 - \dfrac{(1.700)\sqrt{20}}{19}}{2} = 0.30$$

and M would be 0.04. Then $M/2$ would equal 0.02. For $n = 20$, the abscissa of Figure 12.4 corresponding to an ordinate of 0.02 would be 0.27 and k^*

---

the acceptance contour in the $\bar{X}$, s plane reaches its maximum when Procedure 1 is used. See United States Department of the Navy, Bureau of Ordnance, *Mil. Std.* 414 *Technical Memorandum* and Figure 12.5. Also see Figure 13.1 in Chapter 13.

would be $\frac{19[1-2(0.27)]}{\sqrt{20}} = 1.954$. The sample s would thus have to be such that $s \leq \frac{U-L}{3.908}$.

To illustrate the derivation of a double-limit variables sampling plan that uses Procedure 1 let us again design a plan for which the $p'_1 = 0.01$, $p'_2 = 0.08$, $\alpha = 0.05$, and $\beta = 0.10$. Suppose as in Chapter 11, that we are concerned with the width of a slot in a terminal block and let the lower specification limit on the slot be 0.8750 inches and the upper specification limit be 0.8800 inches as before.

We begin by designing a single-limit plan for these values. Application of formulas (12.1) yields, as previously, $k = 1.809$ and $n = 27$. The MSD is computed as follows. For $k = 1.809$ and $n = 27$, the quantity $\frac{1-\frac{k\sqrt{n}}{n-1}}{2} = 0.32$. For $n = 27$, the ordinate of Figure 12.4 at an abscissa equal to 0.32 is approximately 0.033. The abscissa for an ordinate of 0.033/2 = 0.0165 and an $n = 27$ is approximately 0.271. The equation

$$\frac{1-\frac{k^*\sqrt{27}}{26}}{2} = 0.271$$

yields $k^* = 2.292$. Hence

$$\text{MSD} = \frac{0.8800-0.8750}{2(2.292)} = 0.00109$$

Under Procedure 1, therefore, we would accept if in a sample of 27 blocks

$$\frac{\bar{X}-0.8750}{s} \geq 1.809$$

$$\frac{0.8800-\bar{X}}{s} \geq 1.809$$

and

$$s \leq 0.00109$$

Otherwise we would reject.

1.2.2. Procedure 2. Unlike Procedure 1, Procedure 2 can be applied to the double-limit case without modification. As in the single-limit case, we estimate $\hat{p}_L$ by entering Figure 12.3 with $z_L = \frac{\bar{X}-L}{s}$ and $\hat{p}_U$ by entering with $z_U = \frac{U-\bar{X}}{s}$. The two separate estimates are then summed. A lot is accepted if $\hat{p}_L + \hat{p}_U \leq M$, where M is the same M that would be derived for a single-limit plan. This comprehensiveness of Procedure 2, together with its optimum statistical efficiency, is the principal reason for its use.

With reference to the illustration of the previous section, we would first derive an n and a k for a single-limit plan with the same p'_1, p'_2, α, and β as the desired double-limit plan. By application of formulas (12.1) this was found to yield $n = 27$, $k = 1.809$. As before, the quantity $\dfrac{1 - \dfrac{k\sqrt{n}}{n-1}}{2}$ equals 0.32 and Figure 12.4 yields $M = 0.033$. To operate the plan we would compute

$$z_L = \frac{\bar{X} - 0.8750}{s} \text{ and } z_U = \frac{0.8800 - \bar{X}}{s}$$

and from Figure 12.3 find $\hat{p}_L$ and $\hat{p}_U$. We would accept if $\hat{p}_L + \hat{p}_U \leq 0.033$. Otherwise we would reject.

1.2.3. The OC Curve. The operating characteristic curve for the double-limit plan would, with the qualifications noted at the beginning of Section 1.2, be essentially the same as that for the single-limit plan with the same p'_1, p'_2, α, and β. For the illustration discussed above, the OC curve would be that shown in Figure 12.2.

1.3. Normally Distributed Processes and Lots: Use of the Range

In some cases, quality control engineers may prefer to work with the sample range *(R)* rather than with the sample standard deviation *(s)*. This may be done in variables sampling at the cost of reduced efficiency. In other words, a range plan can be designed that will have the same risks (same OC curve) as a standard deviation plan, but will require a somewhat larger sample size. This contravenes the fundamental purpose of variables sampling, but the diminished gain may be partly offset by greater ease of administration.

If the sample is broken down into subgroups of equal size[10] and the average range computed, the statistic $\dfrac{\bar{R}}{d_2^*}$ can be used in place of s in the variables plans.[11] To get approximately the same OC curve as obtained by using s, we look in Table D3, Appendix II, for a value of v that is close to $n - 1$. This will yield a new sample size $n_R = gm$. If there is a choice, we select the combination with highest v for equal gm. A study by Grubbs and Weaver[12] suggests that when gm is divisible by 8, $m = 8$ will be the

[10] Military Standard 414 requires that the subgroups be formed in the order in which the measurements are made. This would be well to adopt as standard procedure.

[11] Cf. Chapter 6, Section 6.

[12] F. E. Grubbs and C. L. Weaver, "The Best Unbiased Estimate of Population Standard Deviation Based on Group Ranges," *Journal of the American Statistical Association* 42 (1947), pp. 224–41.

optimum subgroup size. Also $m = 7$ or 9 will give good results. This can be verified by a study of Table D3.

If Procedure 1 is to be used, accept a lot if $\dfrac{\bar{X} - L}{\bar{R}/d_2^*} \geq k$ and there is only a lower specification limit or accept if $\dfrac{U - \bar{X}}{\bar{R}/d_2^*} \geq k$ and there is only an upper specification limit. Otherwise reject. If there are double specification limits, add the criterion that $\bar{R}/d_2^* \leq$ MSD, computed in the same manner as before. The procedure is exactly the same as when s is employed except that now $\dfrac{\bar{R}}{d_2^*}$ is used in its stead.

If Procedure 2 is to be employed, we proceed as when s is employed, except that (1) in using Figure 12.4 the abscissa is taken equal to $\dfrac{1 - k/\sqrt{v}}{2}$ and in the figure n is taken equal to $v + 1$, (2) the indexes used to derive $\hat{p}_L$ and/or $\hat{p}_U$ are $z_L = \dfrac{\bar{X} - L}{\bar{R}/d_2^*}$ and $z_U = \dfrac{U - \bar{X}}{\bar{R}/d_2^*}$, and (3) Figure 12.6 is used to derive $\hat{p}_L$ and $\hat{p}_U$ from z_L and z_U instead of Figure 12.3. As before, we accept if $\hat{p}_L \leq M$ or $\hat{p}_U \leq M$ in the case of single-limit plans, or if $\hat{p}_L + \hat{p}_U \leq M$ in the case of double-limit plans.

As an illustration let us consider the design of the double-limit plan discussed in the previous section. Here the desired plan called for a $p_1' = 0.01$, a $p_2' = 0.08$, $\alpha = 0.05$, and $\beta = 0.10$. To meet these criteria it was found that we must have $k = 1.809$ and $n = 27$. To derive a range plan with approximately the same OC curve we note that $n - 1 = v = 27 - 1 = 26$. In Table D3 the gm combination that yields a v closest to 26 is $g = 14$, $m = 3$. The combination $g = 5$, $m = 7$, however, requires less measure-

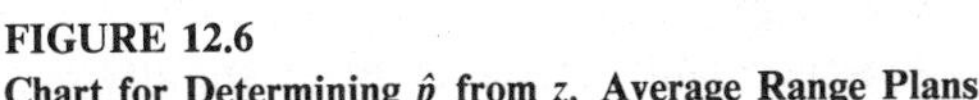

FIGURE 12.6
Chart for Determining $\hat{p}$ from z, Average Range Plans

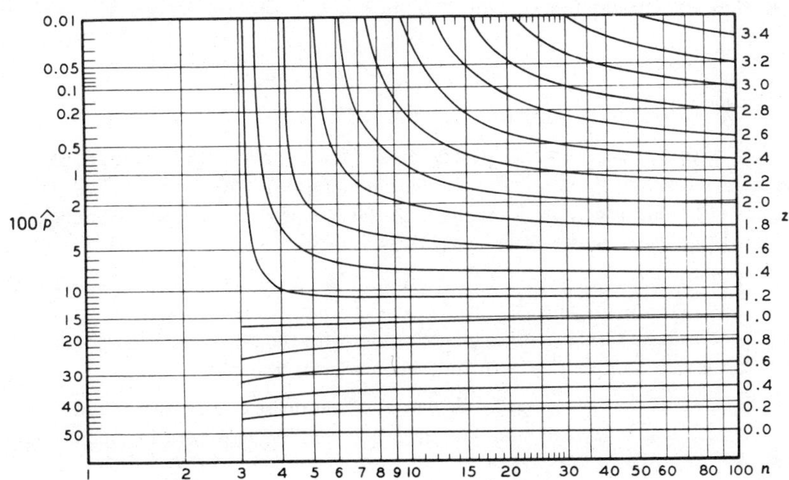

ments and yields a slightly higher ν, viz $\nu = 26.6$. The latter combination will therefore be adopted.

Under Procedure 1 the plan would operate as follows: Take a random sample of 35 items and subdivide it into 5 subgroups of 7. For each subgroup find the range R and compute the average range $\bar{R}$. Accept if

$$\frac{\bar{X} - L}{\bar{R}/d_2^*} \geq 1.809$$

$$\frac{U - \bar{X}}{\bar{R}/d_2^*} \geq 1.809 \text{ and}$$

$$\frac{\bar{R}}{d_2^*} \leq \text{MSD}$$

where MSD is the maximum standard deviation computed from Figure 12.4 as explained above except that now we take abscissa values $= \dfrac{1 - k/\sqrt{\nu}}{2}$. This last condition could also be written $\bar{R} \leq$ MAR, where MAR is the maximum average range and is defined as (MDS)d_2^*.

Under Procedure 2 we compute M. This can be done by finding for $n = \nu + 1$ the ordinate in Figure 12.4 corresponding to an abscissa equal to $\dfrac{1 - k/\sqrt{\nu}}{2}$, which for our problem has a value of $\dfrac{1 - 1.809/\sqrt{26.6}}{2} = 0.325$, yielding $M = 0.039$. Procedure 2 will then operate as follows. Take a random sample of 35 items and subdivide it into 5 subgroups of 7. For each subgroup find the range R and compute $\bar{R}$. Compute $z_L = \dfrac{\bar{X} - L}{\bar{R}/d_2^*} = \dfrac{\bar{X} - 0.8750}{\bar{R}/2.73}$ and $z_U = \dfrac{U - \bar{X}}{\bar{R}/d_2^*} = \dfrac{0.8800 - \bar{X}}{\bar{R}/2.73}$. From Figure 12.6 estimate $\hat{p}_L$ and $\hat{p}_U$. If $\hat{p}_L + \hat{p}_U \leq 0.039$ accept the lot; otherwise reject.

The OC curve for the range plan will, of course, be essentially the same as that for the corresponding s plan, since it was so designed.

2. PROBLEMS

12.1. Assume that the standard deviation of the process is unknown and do Problem 11.1 using *(a)* the standard deviation method *(b)* the range method. If formulas (12.1) are used, check results with the Jacobson nomograph. Compare the new sample size with that when σ' was known.

12.2. Assume that the standard deviation of the process is unknown and do Problem 11.2 using *(a)* the standard deviation method and *(b)* the range method. If formulas (12.1) are used, check results with the Jacobson nomograph. Compare the new sample size with that when σ' was known.

12.3. Assume that the standard deviation of the process is unknown and do Problem 11.3 using *(a)* the standard deviation method and *(b)* the range method. If formulas (12.1) are used, check results with the Jacobson nomograph. Compare the new sample size with that when σ' was known.

12.4. Assume that the standard deviation of the process is unknown and do Problem 11.4 using *(a)* the standard deviation method and *(b)* the range method. If formulas (12.1) are used, check results with the Jacobson nomograph. Compare the new sample size with that when σ' was known.

12.5. Assume that the standard deviation of the process is unknown and do Problem 11.6 using *(a)* the standard deviation method and *(b)* the range method. If formulas (12.1) are used, check results with the Jacobson nomograph. Compare the new sample size with that when σ' was known.

3. SELECTED REFERENCES*

Acceptance Sampling (B '50), Bowker and Goode (B '52), Bowker and Lieberman (B '72), Chernoff and Lieberman (P '57), Cowden (B '57), Duncan, A. J. (P '58), Fertig and Mann (P '77 and P '83), Hamaker (P '79), Jennett and Welch (P '39), Lieberman and Resnikoff (P '55), Nelson (P '81), Owen and Chou (P '83), Resnikoff and Lieberman (B '57), Schilling (B '82), Statistical Research Group, Columbia University (B '47), U.S. Dept. of the Navy, Mil. Std. 414, *Technical Memorandum,* and Zimmer and Burr (P '63).

---

* B and P refer to the Book and Periodical sections, respectively, of the Cumulative List of References in Appendix V

13

Acceptance Sampling by Variables to Control the Fraction Nonconforming: The U.S. Department of Defense Mil. Std. 414 and Related Variables Standards

The U.S. Department of Defense Mil. Std. 414, *Sampling Procedures and Tables for Inspection by Variables for Percent Defective* was issued in 1957 for use as an alternative to the attributes standard Mil. Std. 105A which had been issued in 1950. As noted in Chapter 10 a significant revision of the attributes standard was issued in 1963 as Mil. Std. 105D. Mil. Std. 414 has not been revised by the U.S. Department of Defense, however, to bring it in line with the revised attributes standard, but various revised forms of Mil. Std. 414 have been issued by other organizations.

The initiative for revision of Mil. Std. 414 was undertaken by the United Kingdom with the issuance in 1974 of the U.K. Standard 05–30/Issue 1, *Sampling Procedures and Charts for Inspection by Variables.* Subsequently the British were directed by the international group representing the armies of the United States, the United Kingdom, Canada, and Australia, known as the ABCA group, to develop recommendations for a quadripartite variables standard that would match Mil. Std. 105D better than Mil. Std. 414. This quadripartite variables standard was eventually issued as QSTAG 330 a recent version of which is dated 1985. The United Kingdom also became the Secretariat of a Working Group of Technical Committee 69 of the International Organization for Standardization (ISO) to undertake the writing of an ISO variables standard. ISO Std. 3951, *Sampling procedures and charts for inspection by variables for percent defective* was circulated in draft form among its member bodies in 1976 and was finally adopted[1] as in ISO standard in 1981. In the United States Mil. Std. 414 was adopted intact in 1972 as the American National Standard ANSI Z1.9. A revision of this by an ASQC working committee resulted in the ANSI/ASQC Standard Z1.9–1980.

It is to be noted in passing that Mil. Std. 414, like Mil. Std. 105D, uses the term *defect* in the broad general sense of a *nonconformity* to specifications, and the term *defective unit* is used simply to mean a *nonconforming unit.* On the other hand, ANSI/ASQC Z1.9–1980 has gone over to the use of the terms *nonconformity* and *nonconforming unit.* In our discussion here, we will use the terms interchangeably as seems appropriate.

Since Mil. Std. 414 is the basic standard, it will be discussed here at some length. At the end of the chapter the modifications of Mil. Std. 414 that have been made by the more recent variables standards will be illustrated by the changes in it made by ANSI/ASQC Std. Z1.9–1980 and ISO Std. 3951. How the newer variables standards will match Mil. Std. 105D is indicated in Table 13.10 below.

1. DESCRIPTION OF MILITARY STANDARD 414

Within the range 0.04 to 15 the variables standard is indexed for the same AQLs as the attributes standard,[2] but the lot size classes are different. There are five general levels of inspection and level IV is designated as "normal." Like the attributes standard, it uses sample size code letters, but the same letter does not mean the same sample size in both standards. It gives operating characteristic curves, but offers single-sampling plans only.

Military Standard 414 is divided into four sections. Section A is a general description of the sampling plans. It explains the various terms used in the Standard, defines various classes of defects, and indicates the method of sample selection. A special part of Section A provides for mixed attributes and variables sampling when lots have been previously screened to meet specifications and in other cases. Section B of the Standard gives variables plans based on the sample standard deviation for the case in which the process standard deviation is unknown. Section C gives variables plans based on the sample range, or on the average of ranges of sample subgroups of 5 items each, when the process standard deviation is unknown, and Section D gives variables plans based on the sample mean for the case when the process standard deviation is known. Operating characteristic curves are given in Section A only for plans of Section B, but the plans of Section C and D are so matched

---

[1] An interim revision was issued in 1985. This was primarily devoted, however, to the correction of errors and did not change the basic features of the initial documents. In our discussion the standard will be simply referred to as ISO Std. 3951.

[2] Higher AQL classes in Military Standard 105D are for defects per hundred units and not percent defective. The variables standard is not offered as a means of controlling defects per unit.

with those of Section B that their OC curves are approximately the same. No OC curves are given for mixed sampling plans.

Variables sampling plans are given for both single and double specification limits, and in the latter case the possibility of different AQLs at each limit is provided for. In the case of a single limit, the k and M methods (i.e., Procedures 1 and 2 of Chapters 11 and 12) are offered in each section as alternative approaches. (In the Standard these are referred to as Form 1 and Form 2 of the acceptability criterion.) In the case of double specification limits, the M (Form 2) method only is offered. For the operation of the M method the Lieberman-Resnikoff tables for estimating the process percent defective are reproduced in full. The Standard also contains tables of factors (F and f) that could be used in calculating the maximum standard deviation and maximum average range, but it does not make these quantities part of the acceptability criteria.

In all cases, provision is made for the computation of a process average and a shift to tightened or reduced inspection when warranted. The process average is taken in every instance as equal to the mean of the sample estimates of percent defective computed for lots submitted for *original* inspection. If the k or Form 1 method is employed, these percent defectives must be computed from the special tables provided (normal tables using Q_L or Q_U as a normal deviate when the process standard deviation is known or tables corresponding to Figures 12.3 and 12.6 when the standard deviation is unknown). In the M method they are computed in the application of the method itself. Usually the averages are figured for the preceding 10 lots.

Tightened inspection is to be instituted when the process average exceeds the AQL and at the same time more than a certain number T of the lots on which the process average is based have estimates of the percent defective that exceed the AQL.[3] Reduced inspection is substituted when (1) the 10 preceding lots have been under normal inspection and none have been rejected, (2) when the estimated percent defective for each of these preceding lots is less than a specified lower limit for which a special table is provided or, for certain plans, when the estimated percent defective is equal to zero for a specified number of consecutive lots,[4] and (3) production is at a steady rate.

---

[3] T is obtained as follows: Let p_0' be the AQL and let P_0 equal the probability that the estimate of p' based on $\bar{X}$ and s will exceed p_0'. Then, the probability that in c or more out of n lots from a process producing product of p_0' quality, we shall obtain estimates of p' in excess of p_0' is given by the binomial

$$\alpha = \sum_{x=c}^{x=n} \frac{n!}{x!(n-x)!} (P_0)^x (1-P_0)^{n-x}$$

T is set equal to $c - 1$ where c is so determined that α is close to 0.005.

The lower limits for reduced inspection are set such that if each lot of the group is from a process producing AQL quality, the probability of going to reduced inspection is approximately equal to 0.005. For full details see United States Department of the Navy, Bureau of Ordnance, *Mil. Std. 414 Technical Memorandum.*

[4] See footnote 3.

2. TABLES FROM THE STANDARD

The more important tables from Mil. Std. 414 are reproduced in Tables 13.1 to 13.9. Table 13.1 gives sample size code letters for various lot sizes and various levels of inspection. Tables 13.2 and 13.3 and Tables 13.6 and 13.7 cover the standard deviation and range methods when the standard deviation of the lot or process is unknown. Tables 13.4 and 13.8 give data for use in determining when tightened inspection is to be employed and Tables 13.5 and 13.9 give factors for computing the maximum allowable standard deviation *(s)* and the maximum allowable average range $(\bar{R})$. For tables relating to the case when the standard deviation of the process is known and for tables relating to going to reduced inspection, the reader is referred to the Standard itself.

The tables are used as follows. The producer and consumer agree upon an AQL. The lot size to be used and the inspection level adopted lead via Table 13.1 to a sample size code letter. Entry of this into the appropriate one of Tables 13.2, 13.3, 13.6, or 13.7 yields a specific sample size and an acceptance criterion.

For process variability unknown, and a method of analysis that uses the sample standard deviation and Procedure 1 (Form 1 of the Standard), Table 13.2 gives the acceptability criteria to be used with a single specification limit. A lot is accepted if, with an upper specification limit, the statistic *(U*

TABLE 13.1
Sample Size Code Letters* (Table A–2, Mil. Std. 414)

Lot Size	*Inspection Levels*				
	I	*II*	*II*	*IV*	*V*
3 to 8	B	B	B	B	C
9 to 15	B	B	B	B	D
16 to 25	B	B	B	C	E
26 to 40	B	B	B	D	F
41 to 65	B	B	C	E	G
66 to 110	B	B	D	F	H
111 to 180	B	C	E	G	I
181 to 300	B	D	F	H	J
301 to 500	C	E	G	I	K
501 to 800	D	F	H	J	L
801 to 1,300	E	G	I	K	L
1,301 to 3,200	F	H	J	L	M
3,201 to 8,000	G	I	L	M	N
8,001 to 22,000	H	J	M	N	O
22,001 to 110,000	I	K	N	O	P
110,001 to 550,000	I	K	O	P	Q
550,001 and over	I	K	P	Q	Q

* Sample size code letters given in subsequent tables are applicable when the indicated inspection levels are to be used.

TABLE 13.2

Master Table for Normal and Tightened Inspection for Plans Based on Variability Unknown (standard deviation method) (single specification limit—Form 1) (Table B–1, Mil. Std. 414)

Sample size code letter	Sample size	Acceptable Quality Levels (normal inspection)													
		.04	.065	.10	.15	.25	.40	.65	1.00	1.50	2.50	4.00	6.50	10.00	15.00
		k	k	k	k	k	k	k	k	k	k	k	k	k	k
B	3	↓	↓	↓	↓	↓	↓	↓	▼	▼	1.12	.958	.765	.566	.341
C	4	↓	↓	↓	↓	↓	↓	↓	1.45	1.34	1.17	1.01	.814	.617	.393
D	5	↓	↓	↓	↓	↓	↓	1.65	1.53	1.40	1.24	1.07	.874	.675	.455
E	7	↓	↓	↓	↓	2.00	1.88	1.75	1.62	1.50	1.33	1.15	.955	.755	.536
F	10	↓	↓	↓	2.24	2.11	1.98	1.84	1.72	1.58	1.41	1.23	1.03	.828	.611
G	15	2.64	2.53	2.42	2.32	2.20	2.06	1.91	1.79	1.65	1.47	1.30	1.09	.886	.664
H	20	2.69	2.58	2.47	2.36	2.24	2.11	1.96	1.82	1.69	1.51	1.33	1.12	.917	.695
I	25	2.72	2.61	2.50	2.40	2.26	2.14	1.98	1.85	1.72	1.53	1.35	1.14	.936	.712
J	30	2.73	2.61	2.51	2.41	2.28	2.15	2.00	1.86	1.73	1.55	1.36	1.15	.946	.723
K	35	2.77	2.65	2.54	2.45	2.31	2.18	2.03	1.89	1.76	1.57	1.39	1.18	.969	.745
L	40	2.77	2.66	2.55	2.44	2.31	2.18	2.03	1.89	1.76	1.58	1.39	1.18	.971	.746
M	50	2.83	2.71	2.60	2.50	2.35	2.22	2.08	1.93	1.80	1.61	1.42	1.21	1.00	.774
N	75	2.90	2.77	2.66	2.55	2.41	2.27	2.12	1.98	1.84	1.65	1.46	1.24	1.03	.804
O	100	2.92	2.80	2.69	2.58	2.43	2.29	2.14	2.00	1.86	1.67	1.48	1.26	1.05	.819
P	150	2.96	2.84	2.73	2.61	2.47	2.33	2.18	2.03	1.89	1.70	1.51	1.29	1.07	.841
Q	200	2.97	2.85	2.73	2.62	2.47	2.33	2.18	2.04	1.89	1.70	1.51	1.29	1.07	.845
		.065	.10	.15	.25	.40	.65	1.00	1.50	2.50	4.00	6.50	10.00	15.00	
		Acceptable Quality Levels (tightened inspection)													

All AQL values are in percent defective.

↓ Use first sampling plan below arrow, that is, both sample size as well as k value. When sample size equals or exceeds lot size, every item in the lot must be inspected.

TABLE 13.3

Master Table for Normal and Tightened Inspection for Plans Based on Variability Unknown (standard deviation method) (double specification limit and Form 2—single specification limit) (Table B–3, Mil. Std. 414)

Sample size code letter	Sample size	Acceptable Quality Levels (normal inspection)													
		.04	.065	.10	.15	.25	.40	.65	1.00	1.50	2.50	4.00	6.50	10.00	15.00
		M	M	M	M	M	M	M	M	M	M	M	M	M	M
B	3	↓	↓	↓	↓	↓	↓	↓	▼	▼	7.59	18.86	26.94	33.69	40.47
C	4								1.53	5.50	10.92	16.45	22.86	29.45	36.90
D	5							1.33	3.32	5.83	9.80	14.39	20.19	26.56	33.99
E	7					0.422	1.06	2.14	3.55	5.35	8.40	12.20	17.35	23.29	30.50
F	10				0.349	0.716	1.30	2.17	3.26	4.77	7.29	10.54	15.17	20.74	27.57
G	15	0.099	0.186	0.312	0.503	0.818	1.31	2.11	3.05	4.31	6.56	9.46	13.71	18.94	25.61
H	20	0.135	0.228	0.365	0.544	0.846	1.29	2.05	2.95	4.09	6.17	8.92	12.99	18.03	24.53
I	25	0.155	0.250	0.380	0.551	0.877	1.29	2.00	2.86	3.97	5.97	8.63	12.57	17.51	23.97
J	30	0.179	0.280	0.413	0.581	0.879	1.29	1.98	2.83	3.91	5.86	8.47	12.36	17.24	23.58
K	35	0.170	0.264	0.388	0.535	0.847	1.23	1.87	2.68	3.70	5.57	8.10	11.87	16.65	22.91
L	40	0.179	0.275	0.401	0.566	0.873	1.26	1.88	2.71	3.72	5.58	8.09	11.85	16.61	22.86
M	50	0.163	0.250	0.363	0.503	0.789	1.17	1.71	2.49	3.45	5.20	7.61	11.23	15.87	22.00
N	75	0.147	0.228	0.330	0.467	0.720	1.07	1.60	2.29	3.20	4.87	7.15	10.63	15.13	21.11
O	100	0.145	0.220	0.317	0.447	0.689	1.02	1.53	2.20	3.07	4.69	6.91	10.32	14.75	20.66
P	150	0.134	0.203	0.293	0.413	0.638	0.949	1.43	2.05	2.89	4.43	6.57	9.88	14.20	20.02
Q	200	0.135	0.204	0.294	0.414	0.637	0.945	1.42	2.04	2.87	4.40	6.53	9.81	14.12	19.92
		.065	.10	.15	.25	.40	.65	1.00	1.50	2.50	4.00	6.50	10.00	15.00	
		Acceptability Quality Levels (tightened inspection)													

All AQL and table values are in percent defective.

↓ Use first sampling plan below arrow, that is, both sample size as well as M value. When sample size equals or exceeds lot size, every item in the lot must be inspected.

TABLE 13.4

Values of T for Tightened Inspection (standard deviation method) (Table B–6, Mil. Std. 414)

Sample size code letter	*Acceptable Quality Levels (in percent defective)*														*Number of Lots*
	.04	*.065*	*.10*	*.15*	*.25*	*.40*	*.65*	*1.0*	*1.5*	*2.5*	*4.0*	*6.5*	*10.0*	*15.0*	
B	*	*	*	*	*	*	*	*	*	2 4 5	3 5 6	4 6 8	4 7 9	4 8 11	5 10 15
C	*	*	*	*	*	*	*	2 3 5	2 4 6	3 5 7	3 6 8	4 7 9	4 7 10	4 8 11	5 10 15
D	*	*	*	*	*	*	2 4 5	3 4 6	3 5 7	3 6 8	4 6 9	4 7 10	4 7 10	4 8 11	5 10 15
E	*	*	*	*	2 4 5	3 4 6	3 5 6	3 5 7	4 6 8	4 6 9	4 7 9	4 7 10	4 8 11	4 8 11	5 10 15
F	*	*	*	3 4 6	3 5 6	3 5 7	3 6 8	4 6 8	4 6 9	4 7 9	4 7 10	4 8 11	4 8 11	4 8 11	5 10 15
G	3 4 6	3 5 6	3 5 6	3 5 7	3 6 7	4 6 8	4 6 9	4 7 9	4 7 9	4 7 10	4 7 10	4 8 11	4 8 11	4 8 11	5 10 15
H	3 5 6	3 5 7	3 5 7	3 6 8	4 6 8	4 6 9	4 7 9	4 7 9	4 7 10	4 7 10	4 8 11	4 8 11	4 8 11	4 8 11	5 10 15
I	3 5 7	3 6 7	4 6 8	4 6 8	4 6 9	4 7 9	4 7 9	4 7 10	4 7 10	4 7 10	4 8 11	4 8 11	4 8 11	4 8 11	5 10 15
J	3 6 8	4 6 8	4 6 8	4 6 9	4 7 9	4 7 9	4 7 10	4 7 10	4 7 10	4 8 11	4 8 11	4 8 11	4 8 11	4 8 11	5 10 15
K	4 6 8	4 6 8	4 6 9	4 6 9	4 7 9	4 7 9	4 7 10	4 7 10	4 8 10	4 8 11	4 8 11	4 8 11	4 8 11	4 8 11	5 10 15
L	4 6 8	4 6 9	4 6 9	4 7 9	4 7 9	4 7 10	4 7 10	4 7 10	4 8 10	4 8 11	4 8 11	4 8 11	4 8 11	4 8 11	5 10 15
M	4 6 9	4 7 9	4 7 9	4 7 9	4 7 10	4 7 10	4 7 10	4 7 10	4 8 11	4 8 11	4 8 11	4 8 11	4 8 11	4 8 11	5 10 15
N	4 7 9	4 7 9	4 7 10	4 7 10	4 7 10	4 7 10	4 8 11	4 8 11	4 8 11	4 8 11	4 8 11	4 8 11	4 8 11	3 8 11	5 10 15
O	4 7 10	4 7 10	4 7 10	4 7 10	4 7 10	4 8 11	4 8 11	4 8 11	4 8 11	4 8 11	4 8 11	4 8 11	4 8 11	4 8 11	5 10 15

* There are no sampling plans provided in this Standard for these code letters and AQL values.

TABLE 13.4 *(concluded)*
Values of T for Tightened Inspection (standard deviation method) (Table B–6, Mil. Std. 414)

Sample size code letter	*Acceptable Quality Levels (in percent defective)*														*Number of Lots*
	.04	*.065*	*.10*	*.15*	*.25*	*.40*	*.65*	*1.0*	*1.5*	*2.5*	*4.0*	*6.5*	*10.0*	*15.0*	
	4	4	4	4	4	4	4	4	4	4	4	4	4	4	5
P	7	7	7	8	8	8	8	8	8	8	8	8	8	8	10
	10	10	10	10	11	11	11	11	11	11	11	11	11	12	15
	4	4	4	4	4	4	4	4	4	4	4	4	4	4	5
Q	7	8	8	8	8	8	8	8	8	8	8	8	8	8	10
	10	11	11	11	11	11	11	11	11	11	11	11	11	12	15

The top figure in each block refers to the preceding 5 lots, the middle figure to the preceding 10 lots and the bottom figure to the preceding 15 lots.

Tightened inspection is required when the number of lots with estimates of percent defective above the AQL from the preceding 5, 10, or 15 lots is greater than the given value of T in the table, and the process average from these lots exceeds the AQL.

$-\bar{X})/s$ equals or exceeds k or if, with a lower specification limit, the statistic $(\bar{X} - L)/s$ equals or exceeds k.

For process variability unknown and a method of analysis that uses the sample standard deviation and Procedure 2 (Form 2 of the Standard), Table 13.3 gives the acceptability criteria to be used with either a single or double specification limit. A lot is accepted if, with a single lower specification limit, $\hat{p}_L \leq M$ or if, with a single upper specification limit, $\hat{p}_U \leq M$ or if, with a double specification limit, $\hat{p}_L + \hat{p}_U \leq M$. The Standard contains tables for determining $\hat{p}_L$ and $\hat{p}_U$ from the sample values of $\bar{X}$ and s. Graphical estimates of $\hat{p}_L$ and $\hat{p}_U$ can be obtained from Figure 12.3 above by setting z equal to $(\bar{X} - L)/s$ and $(U - \bar{X})/s$, respectively. For double specification limits Table 13.5 gives factors for computing the maximum allowable sample standard deviation. If the sample standard deviation exceeds this maximum value, the lot is rejected.

Tables 13.6, 13.7, and 13.9 are similar tables to be used when the sample range is used in place of the sample standard deviation. For samples of size $n = 7$ or less, the statistics $(\bar{X} - L)/R$ and $(U - \bar{X})/R$ are used in place of the statistics $(\bar{X} - L)/s$ and $(U - \bar{X})/s$. For samples of size $n =$ 10 or more, the statistics $(\bar{X} - L)/\bar{R}$ and $(U - \bar{X})/\bar{R}$ are used in place of the statistics $(\bar{X} - L)/s$ and $(U - \bar{X})/s$, where $\bar{R}$ is the average of the ranges of subsets of five items each taken in the order of production. With Tables 13.6 and 13.7, Figure 12.6 can be used to estimate $\hat{p}_L$ and $\hat{p}_U$ with z equal to $\dfrac{(\bar{X} - L)}{\bar{R}/d_2^*}$ and $\dfrac{(U - \bar{X})}{\bar{R}/d_2^*}$, respectively. (See Section 1.3 of Chapter 12.) For double specification limits, Table 13.9 gives factors for computing the maximum allowable average range. If the sample $\bar{R}$ exceeds this maximum, the lot is rejected.

Tables 13.4 and 13.8 give the values of T to be employed in deciding

TABLE 13.5

Values of F for Maximum Standard Deviation (MSD) (Table B–8, Mil. Std. 414)

Sample Size Code Letter	Sample Size	*Acceptable Quality Levels (in Percent Defective)*													
		.04	*.065*	*.10*	*.15*	*.25*	*.40*	*.65*	*1.00*	*1.50*	*2.50*	*4.00*	*6.50*	*10.00*	*15.00*
B	3										.436	.453	.475	.502	.538
C	4								.339	.353	.374	.399	.432	.472	.528
D	5							.294	.308	.323	.346	.372	.408	.452	.511
E	7					.242	.253	.266	.280	.295	.318	.345	.381	.425	.485
F	10				.214	.224	.235	.248	.261	.276	.298	.324	.359	.403	.460
G	15	.182	.188	.195	.202	.211	.222	.235	.248	.262	.284	.309	.344	.386	.442
H	20	.177	.183	.190	.197	.206	.216	.229	.242	.255	.277	.302	.336	.377	.432
I	25	.174	.180	.187	.193	.203	.212	.225	.238	.251	.273	.297	.331	.372	.426
J	30	.173	.179	.185	.192	.201	.210	.223	.236	.249	.270	.295	.328	.369	.423
K	35	.170	.176	.183	.189	.198	.208	.220	.232	.245	.266	.291	.323	.364	.416
L	40	.169	.176	.182	.188	.198	.207	.219	.232	.245	.266	.290	.323	.363	.416
M	50	.166	.172	.178	.184	.194	.203	.214	.227	.241	.261	.284	.317	.356	.408
N	75	.162	.168	.174	.181	.189	.199	.211	.223	.235	.255	.279	.310	.348	.399
O	100	.160	.166	.172	.179	.187	.197	.208	.220	.233	.253	.276	.307	.345	.395
P	150	.158	.163	.170	.175	.185	.193	.206	.216	.230	.249	.271	.302	.341	.388
Q	200	.157	.163	.168	.175	.183	.193	.203	.215	.228	.248	.269	.302	.338	.386

The MSD may be obtained by multiplying the factor F by the difference between the upper specification limit U and lower specification limit L. The formula is MSD = F(U − L). The MSD serves as a guide for the magnitude of the estimate of lot standard deviation when using plans for the double specification limit case, based on the estimate of lot standard deviation of unknown variability. The estimate of lot standard deviation, if it is less than the MSD, helps to insure, but does not guarantee, lot acceptability.

Note: There is a corresponding acceptability constant in Table B–1 (Table 13.2 of this text) for each value of F.

TABLE 13.6

Master Table for Normal and Tightened Inspection for Plans Based on Variability Unknown (range method) (single specification limit—Form 1) (Table C–1, Mil. Std. 414)

Sample size code letter	Sample size	Acceptable Quality Levels (normal inspection)													
		.04	.065	.10	.15	.25	.40	.65	1.00	1.50	2.50	4.00	6.50	10.00	15.00
		k	k	k	k	k	k	k	k	k	k	k	k	k	k
B	3	↓	↓	↓	↓	↓	↓	↓	↓	↓	.587	.502	.401	.296	.178
C	4	↓	↓	↓	↓	↓	↓	↓	.651	.598	.525	.450	.364	.276	.176
D	5	↓	↓	↓	↓	↓	↓	.663	.614	.565	.498	.431	.352	.272	.184
E	7	↓	↓	↓	↓	.702	.659	.613	.569	.525	.465	.405	.336	.266	.189
F	10	↓	↓	↓	.916	.863	.811	.755	.703	.650	.579	.507	.424	.341	.252
G	15	1.09	1.04	.999	.958	.903	.850	.792	.738	.684	.610	.536	.452	.368	.276
H	25	1.14	1.10	1.05	1.01	.951	.896	.835	.779	.723	.647	.571	.484	.398	.305
I	30	1.15	1.10	1.06	1.02	.959	.904	.843	.787	.730	.654	.577	.490	.403	.310
J	35	1.16	1.11	1.07	1.02	.964	.908	.848	.791	.734	.658	.581	.494	.406	.313
K	40	1.18	1.13	1.08	1.04	.978	.921	.860	.803	.746	.668	.591	.503	.415	.321
L	50	1.19	1.14	1.09	1.05	.988	.931	.893	.812	.754	.676	.598	.510	.421	.327
M	60	1.21	1.16	1.11	1.06	1.00	.948	.885	.826	.768	.689	.610	.521	.432	.336
N	85	1.23	1.17	1.13	1.08	1.02	.962	.899	.839	.780	.701	.621	.530	.441	.345
O	115	1.24	1.19	1.14	1.09	1.03	.975	.911	.851	.791	.711	.631	.539	.449	.353
P	175	1.26	1.21	1.16	1.11	1.05	.994	.929	.868	.807	.726	.644	.552	.460	.363
Q	230	1.27	1.21	1.16	1.12	1.06	.996	.931	.870	.809	.728	.646	.553	.462	.364
		.065	.10	.15	.25	.40	.65	1.00	1.50	2.50	4.00	6.50	10.00	15.00	
		Acceptable Quality Levels (tightened inspection)													

All AQL values are in percent defective.

↓ Use first sampling plan below arrow, that is, both sample size as well as k value. **When sample size equals or exceeds lot size, every item in the lot must be inspected.**

TABLE 13.7

Master Table for Normal and Tightened Inspection for Plans Based on Variability Unknown† (range method) (double specification limit and for Form 2—single specification limit) (Table C–3, Mil. Std. 414)

Sample size code letter	Sample size	d_2^* factor	Acceptable Quality Levels (normal inspection)													
			.04	.065	.10	.15	.25	.40	.65	1.00	1.50	2.50	4.00	6.50	10.00	15.00
			M	M	M	M	M	M	M	M	M	M	M	M	M	M
B	3	1.910	↓	↓	↓	↓	↓	↓	↓	↓	↓	7.59	18.86	26.94	33.69	40.47
C	4	2.234	↓	↓	↓	↓	↓	↓	↓	1.53	5.50	10.92	16.45	22.86	29.45	36.90
D	5	2.474	↓	↓	↓	↓	↓	↓	1.42	3.44	5.93	9.90	14.47	20.27	26.59	33.95
E	7	2.830	↓	↓	↓	↓	.28	.89	1.99	3.46	5.32	8.47	12.35	17.54	23.50	30.66
F	10	2.405	↓	↓	↓	.23	.58	1.14	2.05	3.23	4.77	7.42	10.79	15.49	21.06	27.90
G	15	2.379	.061	.136	.253	.430	.786	1.30	2.10	3.11	4.44	6.76	9.76	14.09	19.30	25.92
H	25	2.358	.125	.214	.336	.506	.827	1.27	1.95	2.82	3.96	5.98	8.65	12.59	17.48	23.79
I	30	2.353	.147	.240	.366	.537	.856	1.29	1.96	2.81	3.92	5.88	8.50	12.36	17.19	23.42
J	35	2.349	.165	.261	.391	.564	.883	1.33	1.98	2.82	3.90	5.85	8.42	12.24	17.03	23.21
K	40	2.346	.160	.252	.375	.539	.842	1.25	1.88	2.69	3.73	5.61	8.11	11.84	16.55	22.38
L	50	2.342	.169	.261	.381	.542	.838	1.25	1.60	2.63	3.64	5.47	7.91	11.57	16.20	22.26
M	60	2.339	.158	.244	.356	.504	.781	1.16	1.74	2.47	3.44	5.17	7.54	11.10	15.64	21.63
N	85	2.335	.156	.242	.350	.493	.755	1.12	1.67	2.37	3.30	4.97	7.27	10.73	15.17	21.05
O	115	2.333	.153	.230	.333	.468	.718	1.06	1.58	2.25	3.14	4.76	6.99	10.37	14.74	20.57
P	175	2.331	.139	.210	.303	.427	.655	.972	1.46	2.08	2.93	4.47	6.60	9.89	14.15	19.88
Q	230	2.330	.142	.215	.308	.432	.661	.976	1.47	2.08	2.92	4.46	6.57	9.84	14.10	19.82
			.065	.10	.15	.25	.40	.65	1.00	1.50	2.50	4.00	6.50	10.00	15.00	
			Acceptable Quality Levels (tightened inspection)													

All AQL and table values are in percent defective.

↓ Use first sampling plan below arrow, that is, both sample size as well as M value. When sample size equals or exceeds lot size, every item in the lot must be inspected.

† *Military Standard 414* uses *c* to represent d_2^*.

on whether tightened inspection should be instituted. Normally, the process average is estimated for the last 10 lots submitted for original inspection. This will be the average of the $\hat{p}_L$ or the $\hat{p}_U$ if a single specification limit is involved, or the average of the sums $\hat{p}_L + \hat{p}_U$ if a double specification limit is involved. (The approximate values of the $\hat{p}_L$ and $\hat{p}_U$ may be derived from Figures 12.3 and 12.6, as indicated above.) If the average for the last 5, 10 or 15 lots exceeds the AQL and at the same time more than T of the individual lots have estimates of the percent defective that exceed the AQL, tightened inspection is to be instituted.

3. DISCUSSION OF MIL. STD. 414

For cases where it is applicable, Mil. Std. 414 was developed as a substitute for Mil. Std. 105D that would, through smaller sample sizes, significantly reduce the cost of inspection. It must be used with great caution, however. The operating characteristics given in Section A of Mil. Std. 414 assume a basic normal distribution and cannot be taken to be valid if the distribution of a process is not normal. The difficulty is not so much a matter of how the distribution of sample statistics is affected by nonnormality, as it is a matter of what fraction defective is yielded by a given process mean and standard deviation. It was seen in Section 2 of Chapter 11 that the effect of nonnormality on this fraction defective can be quite marked. It can be especially pronounced when the AQL is exceptionally small. Unfortunately, this is a situation that would call for large sample sizes when sampling by attributes and thus becomes the very occasion when the pressure to go to variables sampling is the greatest. This does not mean that Mil. Std. 414 cannot be applied to nonnormal processes. It simply means that the risks involved will be different than those shown by the OC curves presented in the Standard.

It has been suggested that a test for normality be made part of the standard, at least included in an appendix, but this again must be used with great caution. If the test rejects the normal hypothesis, this would, of course, be a warning against using the given OC curves to determine risks. It would not, however, reveal what the true risks are. But if the normal hypothesis passes the test, can we then proceed with confidence in the risks given by the OC curves of the Standard? Probably not. The test for normality would be likely to be applied to the sample data in hand and the sample sizes of Mil. Std. 414 are relatively small. With such small samples many nonnormal situations might go undetected. A more serious question is that we are not interested in the overall normality of the distribution of the process as much as we are in whether the normal assumption is valid for converting a mean and standard deviation into a fraction defective *on the tails of the distribution.* What is thus really needed is a test of normality on the tails. If the fractions defective involved are small, determination of whether the distribution is

TABLE 13.8
Values of T for Tightened Inspection (range method) (Table C–6, Mil. Std. 414)

Sample Size Code Letter	*Acceptable Quality Levels (in Percent Defective)*														Number of Lots
	.04	*.065*	*.10*	*.15*	*.25*	*.40*	*.65*	*1.0*	*1.5*	*2.5*	*4.0*	*6.5*	*10.0*	*15.0*	
										2	3	4	4	4	5
B	*	*	*	*	*	*	*	*	*	4	5	6	7	8	10
										5	6	8	9	11	15
								2	2	3	3	4	4	4	5
C	*	*	*	*	*	*	*	3	4	5	6	7	7	8	10
								5	6	7	8	9	10	11	15
							2	3	3	3	4	4	4	4	5
D	*	*	*	*	*	*	4	4	5	6	6	7	7	8	10
							5	6	7	8	9	10	10	11	15
					2	2	3	3	3	4	4	4	4	4	5
E	*	*	*	*	3	4	4	5	5	6	7	7	7	8	10
					4	5	6	6	7	8	9	10	10	11	15
				2	3	3	3	3	4	4	4	4	4	4	5
F	*	*	*	4	4	5	5	5	6	6	7	7	8	8	10
				5	5	6	7	7	8	9	9	10	11	11	15
	2	2	3	3	3	3	4	4	4	4	4	4	4	4	5
G	4	4	4	5	5	5	6	6	6	7	7	8	8	8	10
	5	5	6	6	7	7	8	8	9	9	10	11	11	11	15

H	3 5 6	3 5 7	3 5 7	3 6 7	4 6 8	4 6 8	4 7 9	4 7 9	4 7 10	4 7 10	4 7 11	4 8 11	4 8 11	4 8 11	5 10 15
I	3 5 7	3 5 7	3 6 7	4 6 8	4 6 8	4 6 9	4 7 9	4 7 9	4 7 10	4 7 10	4 8 11	4 8 11	4 8 11	4 8 11	5 10 15
J	3 5 7	3 6 7	4 6 8	4 6 8	4 6 9	4 7 9	4 7 9	4 7 10	4 7 10	4 7 10	4 8 11	4 8 11	4 8 11	4 8 11	5 10 15
K	3 6 7	4 6 8	4 6 8	4 6 8	4 7 9	4 7 9	4 7 10	4 7 10	4 7 10	4 8 11	4 8 11	4 8 11	4 8 11	4 8 11	5 10 15
L	4 6 8	4 6 8	4 6 9	4 6 9	4 7 9	4 7 9	4 7 10	4 7 10	4 7 10	4 8 11	4 8 11	4 8 11	4 8 11	4 8 11	5 10 15
M	4 6 8	4 6 9	4 7 9	4 7 9	4 7 9	4 7 10	4 7 10	4 7 10	4 8 11	4 8 11	4 8 11	4 8 11	4 8 11	4 8 11	5 10 15
N	4 7 9	4 7 9	4 7 9	4 7 10	4 7 10	4 7 10	4 7 10	4 8 11	4 8 11	4 8 11	4 8 11	4 8 11	4 8 11	4 8 11	5 10 15
O	4 7 9	4 7 10	4 7 10	4 7 10	4 7 10	4 7 10	4 8 11	4 8 11	4 8 11	4 8 11	4 8 11	4 8 11	4 8 11	4 8 11	5 10 15

* There are no sampling plans provided in this Standard for these code letters and AQL values.

TABLE 13.9
Values of f for Maximum Average Range (MAR) (Table C–8, Mil. Std. 414)

Sample Size Code Letter	*Sample Size*	*Acceptable Quality Levels (in Percent Defective)*													
		.04	*.065*	*.10*	*.15*	*.25*	*.40*	*.65*	*1.00*	*1.50*	*2.50*	*4.00*	*6.50*	*10.00*	*15.00*
B	3										.833	.865	.907	.958	1.028
C	4								.756	.788	.836	.891	.965	1.056	1.180
D	5							.730	.764	.801	.857	.923	1.011	1.118	1.263
E	7					.695	.727	.765	.804	.846	.910	.985	1.086	1.209	1.374
F	10				.529	.553	.579	.610	.642	.677	.730	.793	.876	.977	1.112
G	15	.444	.460	.477	.493	.517	.542	.572	.602	.637	.688	.748	.830	.928	1.058
H	25	.416	.432	.447	.463	.486	.509	.537	.567	.600	.649	.707	.785	.879	1.004
I	30	.411	.426	.442	.457	.480	.503	.531	.560	.593	.642	.699	.776	.870	.993
J	35	.408	.423	.438	.454	.476	.499	.527	.556	.588	.637	.694	.771	.864	.987
K	40	.402	.417	.432	.447	.469	.492	.519	.548	.580	.628	.684	.761	.852	.968
L	50	.396	.411	.426	.441	.463	.486	.503	.542	.573	.621	.676	.752	.843	.963
M	60	.390	.405	.419	.434	.455	.478	.505	.533	.564	.608	.666	.740	.830	.949
N	85	.382	.398	.412	.427	.448	.470	.497	.525	.555	.602	.656	.729	.818	.934
O	115	.378	.392	.406	.421	.442	.464	.490	.517	.548	.594	.648	.720	.808	.923
P	175	.371	.384	.399	.413	.434	.455	.481	.508	.538	.584	.637	.708	.794	.908
Q	230	.369	.384	.397	.412	.432	.454	.480	.507	.536	.582	.633	.706	.792	.906

The MAR may be obtained by multiplying the factor f by the difference between the upper specification limit U and lower specification limit L. The formula is MAR = f(U − L). The MAR serves as a guide for the magnitude of the average range of the sample when using plans for the double specification limit case, based on the average range of the sample of unknown variability. The average range of the sample, if it is less than the MAR, helps to insure, but does not guarantee, lot acceptability.

Note: There is a corresponding acceptability constant in Table C–1 (Table 13.6 of this text) for each value of f.

normal on the tails will take a relatively large set of sample data. For AQLs of 0.01, several hundred cases should be used; for AQLs of 0.001, several thousand. This probably means that in most real situations an extensive pilot study would be required before the assumption of normality could be judged valid for assessing the risks involved in using the Standard.

Although Mil. Std. 414 is intended to be an alternative to Mil. Std. 105 in appropriate circumstances, it is out of line with the attributes standard in several respects. This is partly due to Mil. Std. 105 being revised since Mil. Std. 414 was issued. One difference is the list of lot size classes. The lot size classes do not have the property that the ratio of sample size to lot size (n/N) always decreases as lot size increases. Mil. Std. 105D does have this property.

The rules for going to tightened inspection are very complicated in Mil. Std. 414. In Mil. Std. 105D, the rule for going to tightened inspection is simply that 2 out of 5 consecutive lots fail to pass on original inspection. This replaced a much more complicated procedure in earlier versions of Mil. Std. 105 that was similar to that required by Mil. Std. 414.

The probability of acceptance[5] at the AQL does not rise as rapidly with larger sample sizes in Mil. Std. 414 as does this probability in Mil. Std. 105D. It is argued by some that this has discouraged the use of Mil. Std. 414. [cf. Kao, 1971] On the other hand, those who argued against this feature of Mil. Std. 105D (see above Chapter 10) and in favor of having all plans such that the probability of acceptance at the AQL is approximately 0.95, would like to see this probability equal to approximately 0.95 for all Mil. Std. 414 plans. The issues were discussed above in Chapter 10, Section 3.

4. DEVIATIONS FROM MIL. STD. 414 IN THE NEWER VARIABLES STANDARDS: ANSI/ASQC Z1.9-1980 AND ISO STD. 3951[6]

While the newer variables standards ANSI/ASQC Z1.9–1980 and ISO Std. 3951 differ from Mil. Std. 414 in various respects, as described in this section, it is to be carefully noted that except for the deletion of the Mil. Std. 414 Code-Letter Groups J and L, the deletion of plans for the extreme AQLs of 0.04 percent, 0.065 percent, and 15 percent and, in the case of ISO Std. 3951, the deletion of plans based on the sample range, the individual sampling plans of the newer standards are the same as the corresponding individual sampling plans of Mil. Std. 414. The deviations from Mil. Std. 414 pertain basically to the structure and scheme procedures of the standard rather than to differences in the individual plans. These deviations are described in the sections that follow.

---

[5] With the assumption of a normal distribution, of course.

[6] A review of these two particular standards is sufficient to indicate the kinds of changes that have been made in general in newer variables standards.

4.1. Changes in Lot Size Ranges

Both ANSI/ASQC Z1.9–1980 and ISO Std. 3951 have adopted the lot size ranges of the attributes standard Mil. Std. 105D, which are the same as the lot-size ranges of ANSI/ASQC Z1.4 and ISO Std. 2859. A comparison of Table 10.2 and Table 13.1 will show the difference in lot size ranges.

4.2. Changes in Assignment of Code Letter Groups

To facilitate further the coordination of the variables and attributes standards, the newer variables standards modify the assignment of Code-Letter Groups. More specifically the plans in Mil. Std. 414 labelled J and L are not included in the newer standards and the plans retained are relettered alphabetically dropping the symbol O. Thus, for the unknown standard deviation method, normal sampling, we have the following:

Sample Size	*Mil. Std. 414 Code Letter*	*ANSI/ASQC Z1.9–1980 and ISO Std. 3951 Code Letter*†
3	B	B
4	C	C
5	D	D
7	E	E
10	F	F
15	G	G
20	H	H
25	I	I
30	J	*
35	K	J
40	L	*
50	M	K
75	N	L
100	O	M
150	P	N
200	Q	P

* No plan for this sample size.

† From Table B–1 of ANSI/ASQC Std. Z1.9–1980. Copyright American Society for Quality Control, Inc. Reprinted by permission.

The change in lot size ranges and the above rearrangement of Code Letter Groups means that General Inspection Levels for ANSI/ASQC Z1.9–1980 and ISO Std. 3951 differ from Mil. Std. 414 levels given in Table 13.1. Thus the inspection levels III, IV, and V in Table 13.1 are replaced in the newer standards by "Normal Inspection Levels" as follows:

	Normal Inspection Levels†		
Lot Size	*I*	*II*	*III*
3–8	B	B	C
9–15	B	B	D
16–25	B	C	E
26–50	C	D	F
51–90	D	E	G
91–150	E	F	H
151–280	F	G	I
281–500	G	H–I*	J
501–1200	H	J	K
1201–3200	I	K	L
3201–10,000	J	L	M
10,001–35,000	K	M	N
35,001–150,000	L	N	P
150,001–500,000	M	P	P
500,001 and over	N	P	P

* H for lot sizes 281–400 and I for lot sizes: 401–500.
† From Table A2 of ANSI/ASQC Std. Z1.9–1980. Copyright American Society for Quality Control, Inc. Reprinted by permission.

For Special Inspection Levels the reader is referred to the individual standards.

4.3. Changes in the Switching Rules

ANSI/ASQC Z1.9–1980 and ISO Std. 3951 have simpler switching rules. The Mil. Std. 105D rule for going from normal to tightened inspection when on original inspection two out of five or less consecutive lots have been rejected has been adopted by the newer variables standards. If five consecutive lots have been accepted under tightened inspection, return may be made to normal inspection. If tightened inspection has had to be maintained for 10 consecutive lots, sampling inspection is to be discontinued until the process has been improved.

ANSI/ASQC Z1.9–1980 and ISO Std. 3951 differ somewhat with respect to the provisions for going to reduced inspection. Neither has the set of limit numbers prescribed by Mil. Std. 414. ANSI/ASQC Z1.9–1980 permits reduced inspection to be instituted provided:

1. The last 10 lots have been on normal inspection and none has been rejected,
2. production is at a steady rate, and
3. reduced inspection is considered desirable by the responsible authority and is permitted by the contract or specification.

ISO Std. 3951 allows reduced inspection to be instituted after 10 successive lots have been accepted under normal inspection provided that:

1. they would have been acceptable if the AQL had been one step tighter,
2. the process is in statistical control, and
3. reduced inspection is considered desirable by the responsible authority.

4.4. Use of Acceptance Charts when There Are Double Specification Limits

Whereas ANSI/ASQC Z1.9–1980 follows the Form 2 procedures of Mil. Std. 414 in employing elaborate tables for determining the sample percent nonconforming from the quality indexes $(U - \bar{X})/s$ and $(\bar{X} - L)/s$ in the case of double specification limits and from this percentage determines lot acceptability, ISO 3951 contains charts that may be entered directly with the sample values of $\bar{X}$ and s (or $\bar{X}$ and σ' in the known σ' case) to determine lot acceptability. An acceptance chart of this kind is shown in Figure 13.1. In the chart, boundaries are given for various AQLs. If a sample point falls within or on the appropriate boundary, the lot is accepted.

4.5. Plans Based on the Sample Mean Range $\bar{R}$

ANSI/ASQC Z1.9–1980, like Mil. Std. 414, contains sampling plans based on the average sample range $\bar{R}$. ISO Std. 3951 does not.

4.6. Switch to the "Known" σ' Method

If, as inspection proceeds, a control chart for s or $\bar{R}$ indicates that the variability of the process is in control, ISO Std. 3951 allows the responsible authority to permit a switch to the "known" σ' method using for σ' the estimate given by the control chart. The continued maintenance of the control chart for s or $\bar{R}$ is required. Mil. Std. 414 and ANSI/ASQC Z1.9–1980 contain no explicit provision of this kind, but presumably such a switch could be made under either standard provided it was allowed by the responsible authority.

5. HOW WELL THE NEWER VARIABLES STANDARDS MATCH THE ATTRIBUTES STANDARDS

As noted above, a principal reason for the changes in Mil. Std. 414 that are effected in ANSI/ASQC Z1.9 and ISO Std. 3951 is to make for a better match between the variables standards and the attributes standards. How well this is accomplished is shown in Table 13.10, which is a reproduction of Table 4 of ANSI/ASQC Z1.9. This gives the difference in percentage points between the qualities of submitted product for the probabilities of

FIGURE 13.1
Example of an Acceptance Chart for a Variables Sampling Plan (for a plan using Mil. Std. 414 "s" method sample size 7)

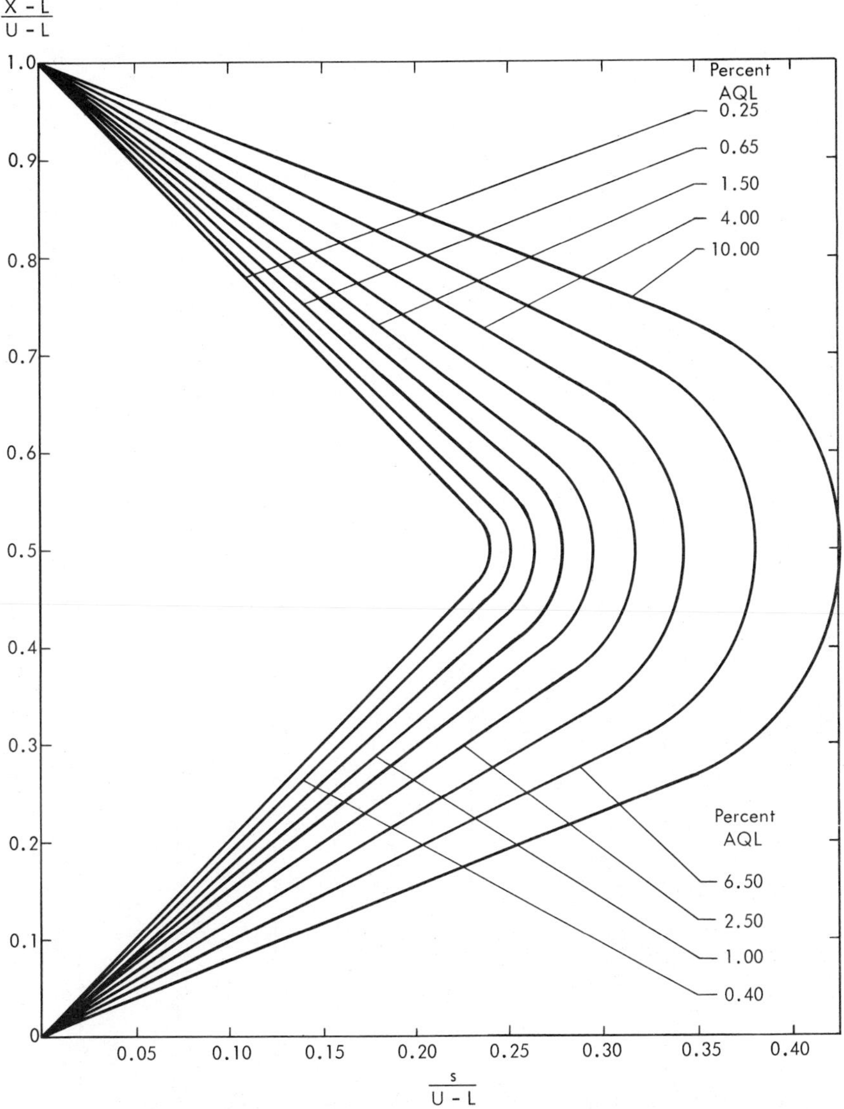

SOURCE: This chart is reproduced with permission from ISO Std. 3951, *Sampling procedures and charts for inspection by variables for percent defective,* p. 44, copyright 1981 by the International Organization for Standardization. Copies of the standard may be purchased in the U.S. from American National Standards Institute at 1430 Broadway, New York, N.Y. 10018.

TABLE 13.10
Difference in (Abscissa) Percentage Points—ANSI/ASQC Z1.9–1980 Minus Z1.4 (105D) (Table 4 of Z1.9†)

Probability of Acceptance (Percent)	*Z1.4 Code Letter*	*Z1.9–1980 Code Letter*	*Acceptable Quality Level*										
			.10	*.15*	*.25*	*.40*	*.65*	*1.00*	*1.50*	*2.50*	*4.00*	*6.50*	*10.00*
95.0	B	B									.19		
50.0											−.30		
10.0											−.77		
95.0	C	C								.30			−.78
50.0										1.54			−2.79
10.0										4.25			−2.85
95.0	D	D							.13			1.66	−3.99
50.0									1.41			.65	−5.70
10.0									6.24			3.95	−3.58
95.0	E	E						.136			−.16	−2.06	−3.84
50.0								.99			1.06	−1.89	−3.97
10.0								4.89			5.13	1.28	−1.15
95.0	F	F					.104			−.15	−1.39	−2.29	−6.19
50.0							.36			.37	−1.41	−2.19	−6.81
10.0							2.33			3.30	1.16	.59	−4.52
95.0	G	G				.089			−.04	−.68	−1.30	−3.20	−4.69
50.0						.06			.13	−.76	−1.25	−3.23	−4.45
10.0						.64			1.78	.97	.78	−1.34	−2.47

95.0					.067			.078	−.45	−.70	−1.95	−2.51	−4.02
50.0	H	H			.000			.33	−.50	−.44	−1.79	−1.71	−2.89
10.0					.23			1.67	.84	1.35	.14	.61	−.40
95.0								.138	−.38	−.54	−1.73	−2.22	−3.63
50.0	H	I						.06	−.84	−.76	−2.18	−2.20	−3.46
10.0								.41	−.57	−.09	−1.46	−1.16	−2.28
95.0				.066			.156	−.09	−.33	−94	−1.26	−1.70	−2.25
50.0	J	J		−.103			.12	−.28	−.52	−1.08	−1.14	−1.20	−1.48
10.0				−.55			.32	−.02	−.09	−.45	−.09	.11	.03
95.0			.059			.116	−.014	−.07	−.60	−.68	−.90	−.88	−1.90
50.0	K	K	−.364			.03	−.20	−.18	−.86	−.66	−.63	−.25	−1.30
10.0			−.48			−.04	−.23	−.02	−.70	−.19	.09	.62	−.52
95.0					.092	.021	.017	−.25	−.41	−.47	−.44	−.64	
50.0	L	L			−.019	−.15	−.10	−.41	−.50	−.31	−.04	.04	
10.0					−.21	−.29	−.10	−.36	−.31	.12	.60	.84	
95.0				.068	.031	.037	−.089	−.14	−.30	−.21	−.42		
50.0	M	M		−.022	−.078	−.05	−.16	−.12	−.21	.14	.12		
10.0				−.18	−.19	−.06	−.08	.07	.12	.72	.89		
95.0			.059	.026	.037	−.043	−.026	−.05	−.12	−.16			
50.0	N	N	−.016	−.075	−.044	−.13	−.05	.01	.03	.16			
10.0			−.158	−.21	−.13	−.18	−.01	.18	.31	.62			
95.0			.041	.039	.017	.036	.069	.09	.00				
50.0	P	P	−.013	−.014	−.026	.041	.15	.25	.25				
10.0			−.094	−.072	−.044	.097	.29	.50	.60				

acceptance of 0.95, 0.50, and 0.10 respectively under the attributes standard Z1.4 and the qualities of submitted product for which the probablistics of acceptance are also 0.95, 0.50, and 0.10 under the variables standard Z1.9. The same set of differences would, of course, also characterize a conparison of Mil. Std. 105D and ISO Std. 2859 with Z1.9 or ISO Std. 3951.

6. SPECIAL WARNING

It is well to conclude this chapter with repetition of the warning given in Chapter 11 on the use of variables sampling plans to give assurance with respect to very small fractions nonconforming. The OC curves worked out in Chapters 11 and 12 and all those presented in Military Standard 414, assume *random* sampling from a *normal* universe. The OC curves for both attributes and variables plans assume random sampling, but the assumption of a specific distributional form is a special feature of variables plans. Before a reader uses such plans it is strongly advised that he reread Section 2 of Chapter 11 and Section 3 of this chapter.

7. PROBLEMS

13.1. As an inspector for the U.S. Navy, describe a variables sampling plan you would use for an AQL of 1.5 percent and lots of 7,000 items. Give two alternative plans each with two alternative procedures, as follows:

a. Standard Deviation Method

Form 1

Form 2

b. Range Method

Form 1

Form 2

How do the sample sizes compare with those that would have been employed under Military Standard 105D?

13.2. As an inspector for the U.S. Army describe a variables sampling plan you would use for an AQL of 1 percent and lots of 15,000 items. Give two alternative plans each with two alternative procedures, as follows:

a. Standard Deviation Method

Form 1

Form 2

b. Range Method

Form 1

Form 2

13.3. A military agency decides to use a variables sampling plan for lots of 200 and for an AQL = 4.0. Give two alternative plans that it might use each with two alternative procedures as follows:

a. Standard Deviation Method

Form 1

Form 2

b. Range Method

Form 1

Form 2

13.4. Using the information given in Section 4.2 of this chapter, find the variables sampling plan that would be given by ANSI/ASQC Z1.9–1980 in answer to Problem 13.1*a* above.

13.5. Using the information given in Section 4.2 of this chapter, find the variables sampling plan that would be given by ANSI/ASQC Z1.9–1980 in answer to Problem 13.2*a* above.

8. SELECTED REFERENCES*

ANSI/ASQC Z1.9 (B '80), Bender (P '75), Elder and Muse (P '82), ISO 3951 (B '81), Kao (P '71), Schilling (B '82), United Kingdom (B '73), U.S. Dept. of Defense, *Mil.-Std.-414* and U.S. Dept. of the Navy, *Mil.-Std.-414 Technical Memorandum.*

* "B" and "P" refer to the "Book" and "Periodical" sections, respectively, of the Cumulative List of References in Appendix V.

14

Some Special Topics Pertaining to Lot Sampling

1. COMPARISON OF SAMPLING AND ADMINISTRATIVE EFFICIENCY OF VARIOUS TYPES OF LOT SAMPLING PLANS

With such a variety of sampling plans to choose from—single, double, multiple, item-by-item sequential, and variables—the quality control engineer may be puzzled as to which type to use in a given situation. Usually a final decision depends upon administrative considerations as well as sampling efficiency, and these will be noted in Section 1.3 below. For the moment let us consider some theoretical aspects of comparison of sampling plans.

1.1. Identity of OC Curves

Sampling plans of different types can generally be designed so that for practical purposes they have roughly the same OC curves. *The risk involved in sampling is thus not a point of difference in the comparison of various types of plans. Meaningful comparisons are only made between plans that have roughly the same OC curve.*

In actual comparisons, however, the practical question arises as to how we determine when the OC curves of two different plans are "roughly the same." Since the OC curve for a single-sampling plan, for example, is not the same mathematical function as the OC curve for a double-sampling plan, the two curves cannot be made to coincide at every point. In what sense than can we say that the two have roughly the same OC curve?

Several answers have been given to this question. The most common is to match curves approximately at the p'_1 and p'_2 points and then assume that other parts of the curves do not deviate greatly. A second method due to H. C. Hamaker is to match (1) the 50 percent (or indifference) points and (2) the relative slopes of the curves at this common point. A third method due to the Army Chemical Corps is to match the points of inflection and the slopes at these points. For single-sampling attributes plans, these last two methods yield identical results.

A detailed discussion of the advantages and disadvantages of various methods of matching OC curves will not be attempted here. Whenever it is necessary in this text to match OC curves, the method of matching p_1''s and p_2''s is followed. For certain theoretical purposes, however, e.g., the classification of plans of a certain type or the comparison of relative efficiencies, the indifference point method and inflection point method have special advantages. For matching the OC curves of all kinds of plans—single, double, multiple, sequential, and variables—the inflection point method has particularly good properties when especially close matching is desired in the neighborhood of p_1'. For further discussion of the matching of OC curves, the reader is referred to the references by Hamaker and the Army Chemical Corps at the end of this chapter.

1.2. Comparison of Efficiencies

If the OC curves of two acceptance sampling plans match reasonably well, their relative efficiency *with respect to the amount of sampling required* may be determined by comparing the average sample numbers (ASNs) at p_1'. It is assumed here that plans will be operating in most instances in the vicinity of p_1', but comparison may also be made at other points.

In general, in the neighborhood of p_1', the relative efficiency of various types of comparable acceptance-sampling plans runs, in inverse order, as follows:

1. Single sampling by attributes.
2. Double sampling by attributes.
3. Multiple sampling by attributes.
4. Item-by-item sequential sampling.
5. Single sampling by attributes using fictitious specification limits and assuming a known distributional form with known parameters other than $\bar{X}'$.
6. Variables sampling, depending on the relative cost of measurement.

As noted previously, Wald and Wolfowitz have shown by mathematical proof that an item-by-item SPR sampling plan is more or at least as efficient at the p_1' and p_2' points as any other attributes acceptance-sampling plan with these same two points on the OC curve. The gain in comparison with single sampling may run something like 50 percent. It is to be noted, however, that the relative efficiency of acceptance-sampling plans for product qualities intermediate to p_1' and p_2' may not be those listed above. If a plan is selected on the basis of its efficiency at the p_1' point, it is therefore important that most incoming product be of that quality.

To illustrate the relative sampling efficiency of various types of plans, let us review those derived in Chapters 7–12. These all had a $p'_1 = 0.01$ and a $p'_2 = 0.08$ with $\alpha = 0.05$ and $\beta = 0.10$. They ran as follows:

Type of Plan	*Sample Size or ASN without Curtailment* at* p'_1
1. Single sampling by attributes	$n = 67$ (this is slightly tighter than warranted because the $\alpha = 0.03$ instead of the prescribed 0.05)
2. Double sampling by attributes	ASN at $p'_1 \doteq 45$
3. Multiple sampling by attributes	ASN at $p'_1 \doteq 41$
4. Item-by-item SPR sampling by attributes	ASN at $p'_1 \doteq 38$
5. Single sampling by variables, unknown σ', average range method	$n = 35$
6. Single sampling by attributes using compressed limits, known σ'	$n = 31$
7. Single sampling by variables unknown σ', standard deviation method	$n = 27$
8. Single sampling by variables, known σ'	$n = 10$

* At p'_1 the difference between the ASN with and without curtailment is not very great. See, for example, Figure 8.2.

Several things may be noted from the above comparison. It is readily apparent that sequential sampling pays off well. In this case it reduces sampling costs by $\left(1 - \frac{38}{67}\right) = 43$ percent as compared with single sampling. The comparison is not strictly fair since the single-sampling plan is a little more discriminating, its α being 0.03 instead of 0.05, but the fact of considerable cost reduction is clearly evident. Double sampling also yields a marked reduction in sampling costs compared with single sampling (33 percent), and multiple sampling does almost as well as an item-by-item SPR plan.[1]

It is also clear from the above comparison that additional knowledge, such as knowledge of distributional form and standard deviation, yields large dividends. If we know that a process or lot is normally distributed, but do not know its standard deviation, we can, through sampling by variables, get a saving in this case of $\left(1 - \frac{27}{67}\right) = 60$ percent as compared with single sampling by attributes. In comparison with item-by-item SPR sampling the saving is $\left(1 - \frac{27}{38}\right) = 29$ percent. If we know a distribution is normal and also know its standard deviation, we can get a still greater saving with variables sampling. In comparison with single sampling by attributes the cost reduction is $\left(1 - \frac{10}{67}\right) = 85$ percent, and in comparison with item-by-item sequential sampling, the reduction is 74 percent. If variables measurements are considerably more costly than simple classification as nonconforming or conforming,

[1] The multiple plan was a little tighter than required. See Chapter 9, Section 4.

we can still capitalize on knowledge of distributional form and standard deviation by using an attributes plan with compressed limits. The saving in this case amounts to $\left(1-\frac{31}{67}\right)=54$ percent, as compared with single sampling and $\left(1-\frac{31}{38}\right)=18$ percent as compared with item-by-item SPR sampling.

Finally, the above comparison makes it clear that in terms of sampling efficiency it is expensive to use the range. Although the tendency may be to use the range if electronic computer facilities are not readily available for calculating the sample standard deviation, it should be recognized that use of the range costs money. It may be that its administrative facility more than offsets the larger amount of sampling required, but the quality control engineer should at least recognize what he is paying for its use. In our example, the range plan costs $\left(\frac{35}{27}-1\right)=29$ percent more than a comparable variables plan using the standard deviation. In other examples, the difference might not be as great, but it is likely to be considerable. Just how good or how bad range plans can be relative to standard deviation plans will be revealed by a study of Table D3 in Appendix II. The comparison is between the ν given in the table for given values of g and m and $gm - 1$.

1.3. Comparison of Administrative Features

The above comparisons of efficiency are based solely on the number of items to be tested and inspected to get a given degree of quality assurance. Other factors also enter into overall cost. It is to be noted, for example, that variables plans are likely to involve higher costs per unit because of the more precise measurements that are required. Other administrative features are summarized in Table 14.1. These indicate clearly that single sampling has a number of administrative advantages that in a given situation might override the sampling inefficiency noted in the previous section.

2. THE USE OF PRIOR INFORMATION ABOUT THE PROCESS

If a close relationship exists between producer and consumer (they may actually be two different departments of the same organization), prior information about the process producing the lots submitted for inspection may be useful in designing minimum cost inspection procedures.

2.1. Lot Break-Even Quality Levels

Before discussing how prior information about the process may be used, let us consider the lot break-even quality level which we will represent by $(X/N)_r$.

The concept of a lot break-even quality level plays a key role in the

TABLE 14.1
Administrative Factors Influencing Choice of Sampling Plan*

Factor	Type of Sampling Plan: Single	Double	Sequential and Multiple
Variability of inspection load	Constant	Variable	Variable
Sampling costs when all items can be taken as needed	Most expensive	Intermediate	Least expensive
Sampling costs when all items must be drawn at once	Least expensive	Most expensive	Intermediate
Accurate estimation of lot quality†	Best	Intermediate	Worst
Sampling costs when dependent on the number of samples drawn	Least expensive	Intermediate	Most expensive
Amount of record keeping	Least	Intermediate	Most expensive
Psychological: "give supplier more than one chance"	Worst	Intermediate	Best

* Cf. U.S. Navy, *Standard Sampling Inspection Procedures,* D4.03.02/14.
† If estimate is based on a large number of lots, differences from one type of sampling to another may not matter.

development of minimum cost sampling inspection plans in A. Hald's *Statistical Theory of Sampling Inspection by Attributes*[2] and in the discussion of optimum inspection procedures in W. Edwards Deming's *Out of the Crisis.*[3] Hald discusses the lot break-even quality level with reference to the linear cost model[4]

$$k_r = R_1 + R_2(X/N)$$
$$k_a = A_1 + A_2(X/N) \tag{14.1}$$

where k_r equals the cost per unit of rejecting and screening a lot and k_a equals the cost per unit of accepting a lot without inspection. The first parameter in each of these equations describes the cost due to rejecting or accepting an item without regard to quality, whereas the second gives the additional cost per nonconforming item. The lot break-even quality is given by equating k_r to k_a which yields

$$(X/N)_r = \frac{R_1 - A_1}{A_2 - R_2} \tag{14.2}$$

[2] A. Hald, *Statistical Theory of Sampling Inspection by Attributes* (New York: Academic Press, Inc., 1981).

[3] W. Edwards Deming, *Out of the Crisis* (Cambridge, Mass., Massachusetts Institute of Technology Center for Advanced Engineering Study, 1986).

[4] Hald, *Statistical Theory of Sampling Inspection by Attributes,* p. 12.

Deming considers the testing of incoming parts that are to go into an assembly.[5] He defines the lot break-even quality level as the fraction nonconforming that equals the ratio k_1/k_2 where k_1 is the cost to input one part and k_2 is the cost to test an assembly that fails because a nonconforming part was put into the production line.

To bring Deming's discussion of lot break-even quality into line with Hald's we shall set $A_1 = 0$ in equation (14.2), as Hald indicates is often done and we shall assume that the vendor replaces all nonconforming parts found by the purchaser's inspection by good parts at no cost to the purchaser.[6] This means that we can also take $R_2 = 0$. Under these conditions the Hald formulation of the lot break-even quality becomes $(X/N)_r = R_1/A_2$ and is identical with Deming's formulation with $k_1 = R_1$ and $k_2 = A_2$. We shall assume these simplifying conditions in the discussion that follows.

2.2. An Optimum Procedure when Lots Are from a Process in a State of Statistical Control at a Constant Level

If a control chart for fraction nonconforming (see Chapter 19 below) shows that a process has been in a state of statistical control at a constant fraction p nonconforming, if the lots to be inspected can be viewed as random samples from the process, and if a lot break-even quality level $(X/N)_r$ can be determined, then an optimum procedure would be to accept all lots without inspection if $p \leq (X/N)_r$ and to reject and screen all lots if $p > (X/N)_r$. For in the situation described, the lot fraction nonconforming would vary from lot to lot in accordance with the binomial distribution and the fraction nonconforming in a random sample from a given lot would be uncorrelated (from lot to lot) with the fraction nonconforming in the remainder of the lot.[7] Hence, sampling inspection would not provide useful additional information for making a decision to accept or reject a lot and would thus not be an effective procedure. As soon as there is evidence, however, that the process is no longer in control, a sampling inspection plan should be applied or lots should be inspected 100 percent. It should also be noted that if the process level p is such that 100 percent inspection is the optimum procedure, it would be desirable to check intermittently on how well the screening procedure is being carried out.

2.3. Optimum Procedures when a Process Is not in a State of Statistical Control at a Constant Level

The determination of an optimum procedure to follow when a study of the process indicates that it is not in a state of statistical control at a constant

[5] Deming, *Out of the Crisis,* Chapter 15. Sections 2.2, 2.3.1–2.3.3, 2.3.5 and 2.3.6 below. Follow closely Deming's discussion. Deming also provides examples with more complicated functions.

[6] Deming makes a somewhat different assumption than this, but the difference does not affect the argument developed here.

[7] A. M. Mood, "On the dependence of sampling inspection plans upon population distributions," *Annals of Mathematical Statistics* 14 (1943), pp. 415–25.

level depends on what the study reveals and what is known about the process in general.

2.3.1. Lot Fractions Nonconforming All to One Side of the Break-Even Quality Point. If in a study of past lots from a process all lots are found to have fractions nonconforming below the lot break-even quality level, the optimum procedure will be to accept current lots without inspection. If past lot fractions nonconforming all fall above the lot break-even quality level, the optimum procedure will be to inspect current lots 100 percent. If in the above cases the accept-without-inspection policy is adopted, the process should still be sampled periodically to see if a tendency has developed to produce some lots with fractions nonconforming above $(X/N)_r$.

2.3.2. Slight Straddle. If a study of past lots reveals that only a small percentage (say 1 in 20) had fractions nonconforming that exceeded $(X/N)_r$ it will still probably be near optimum to accept current lots without inspection. See p. 324 below for further discussion of a slight straddle.

2.3.3. Mixtures of Binomials. A review of incoming parts may reveal that they come from several different sources of supply. There may be several different vendors or within a plant several different combinations of operator and equipment may have produced the parts. If the quality of the output of each source is in statistical control at a constant level and if it is administratively feasible to adopt separate policies for each source of supply, we have simply a multiplication of the principle stated in Section 2.2 above. Accept without inspection all lots from source i if $p_i \leq (X/N)_r$ or inspect the lots from this source 100 percent if $p_i > (X/N)_r$.

If production of the various sources of supply are in a state of statistical control, but it is not administratively feasible to adopt different inspection policies for each source, we have a situation that is referred to as a "mixture of binomials." If the process levels $p_1, p_2, \ldots, p_k$ at which the various processes are in statistical control are all either above or below the lot break-even quality level $(X/N)_r$, we proceed as in the single-source case (Section 2.3.1). The problem arises when some of the p_i are above $(X/N)_r$ and some are below. If the different sources of supply are different vendors, a long-run solution might be to stop buying from those vendors with p_i's above $(X/N)_r$. If the different sources of supply are different operator-equipment combinations within a company's own plant, a solution may be retraining of the poorer operators or possibly the purchase of new equipment. If the situation cannot be corrected, in the short-run at least, the problem is what to do. If jumping from one source of supply to another is haphazard, the overall process might be judged to be in a "state of chaos," the treatment of which is discussed in Section 2.4 below.

2.3.4. A Predictable Distribution for Lot Quality. If the study of the quality of lots received in the past from a given source of supply does not indicate a state of statistical control at a constant level p, there may nevertheless be a suggestion that the distribution of lot quality follows a consistent pattern and a state of statistical control exists with respect to that pattern. Presumably the hypothesis of a state of statistical control at a constant level is rejected because a control chart shows numerous points exceeding the 3σ control limits based on the assumption of a constant p equal to the average lot quality for the period under study and no special assignable causes can be found to explain these outlying points. (See Chapter 19.) If the past output studied is divided into two periods, however, the shape of the distribution of lot quality may appear to be about the same in both periods with approximately the same standard deviation. If lots are of a constant size N, then the stability of the variation in lot quality could be determined as follows. Take the average lot quality for the first period ($\bar{p}_1$) and the standard deviation of lot quality in that period (s_1) as "standard values" for a control chart analysis of lot quality (X/N) in the second period. (See Chapter 19.) The central line on this chart would be $\bar{p}_1$ and 3-sigma limits would be set at $\bar{p}_1 \pm 3s_1$. If the fraction nonconforming (X/N) for each lot of the second period is plotted on this chart and if all points fall within the control limits and there are no long runs or other indication of nonrandom variation, it would seem reasonable to assume that the variation in lot quality follows a consistent pattern and that production is in control with respect to that pattern. The pattern would be estimated by the distribution of lot quality for the whole period of past production that was studied.

If the distribution of lot quality falls entirely below the lot break-even quality level, we would currently accept lots without inspection except for a small sample used to maintain the control chart. If the distribution of lot quality falls entirely above the lot break-even quality level, we would currently inspect all lots 100 percent. (See above p. 320.) If the distribution of lot quality straddles the lot break-even quality level by more than an insignificant amount[8] and if our general knowledge of the production process for the given source of supply supports the hypothesis that the variability in lot quality will continue to follow the pattern found from the study of past records, an economic procedure may be[9] to find a sampling plan that would minimize total costs with respect to that distribution. We will refer to such a plan as a Bayesian[10] minimum cost sampling plan since it is based on a "prior distribution" of lot quality. Theoretical discussion of this situation is best explained mathematically.

2.3.4.1. Mathematical Models for the Prior Distribution of Lot Quality. When a process is in a state of statistical control at constant level p, and lots may be viewed as random samples from the process, the distribution

---

[8] It is suggested that at least 5 percent of the distribution fall above the lot break-even quality level. Cf. discussion, p. 320.

[9] See, however, discussion on p. 328 below.

[10] The name comes from the work of Thomas Bayes. See Section 9 of Chapter 2 above. Bayes' original paper is listed in the Cumulative Bibliography in Appendix V. See Hald, *Statistical Theory,* for an extensive discussion of Bayesian sampling plans.

of the number of nonconforming units (X) in lots of size N will have the binomial distribution

(14.3) $$b(X,N,p) = \frac{N!}{X!(N-X)!} p^X (1-p)^{N-X}$$

If p is not in control at a constant level, however, but with the production of each lot varies at random in accordance with a designated distribution $f(p)$, the equation (14.3) becomes a conditional distribution dependent on the value of p and the distribution of X in lots of size N from the process in general will be the sum of the binomial probabilities (14.3) weighted by the probabilities of the various values of p. For continuous variation in p, the distribution of the number of nonconforming units X in lots of size N will be given by the integral

(14.4) $$\int_0^1 b(X,N,p)f(p)dp$$

Hald discusses the case in which the distribution of p takes the form of a beta distribution.[11] The density function for this is

(14.5) $$f(p,s,t) = p^{s-1}(1-p)^{t-1}/\beta(s,t)$$

where s and t are distribution parameters, s being referred to as the "shape parameter," and $\beta(s,t)$ is the beta function $\int_0^1 u^{s-1}(1-u)^{t-1}du$ which can be shown[12] to be equal to $\Gamma(s)\Gamma(t)/\Gamma(s+t)$, the gamma function $\Gamma(m)$ being defined by the integral

(14.6) $$\Gamma(m) = \int_0^\infty y^{m-1}e^{-y}dy$$

which equals $(m - 1)!$ if m is an integer. The beta distribution can be bell-shaped or J-shaped, and its mean and variance are[13]

(14.7) $$\bar{p} = s/(s+t)$$

and

(14.8) $$V\{p\} = \frac{\bar{p}(1-\bar{p})}{(s+t+1)} = \frac{\bar{p}^2(1-\bar{p})}{(s+\bar{p})}.$$

When $f(p)$ is a beta distribution, equation (14.4) above becomes

(14.9) $$\text{Prob. }(X,N,p,s) = \int_0^1 \frac{p^{s-1}(1-p)^{t-1}}{\beta(s,t)} \frac{N!}{X!(N-X)!} p^X(1-p)^{N-X}dp$$

---

[11] Hald, *Statistical Theory*, pp. 128 ff.

[12] Ibid.

[13] Ibid.

$$= \frac{N!}{X!(N-X)!} \int_0^1 \frac{p^{X+s-1}(1-p)^{N-X+t-1}}{\beta(s,t)} dp = \frac{N!}{X!(N-X)!} \frac{\beta(X+s, N-X+t)}{\beta(s,t)}$$

which becomes

(14.10)

$$\text{Prob. } (X,N,p,s) = \frac{\Gamma(N+1)}{\Gamma(X+1)\Gamma(N-X+1)} \frac{\Gamma(X+s)\Gamma(N-X+t)}{\Gamma(N+s+t)} \frac{\Gamma(s+t)}{\Gamma(s)\Gamma(t)}.$$

Equations (14.9) and (14.10) are formulas for the "beta-binomial" distribution and represent the distribution of lot quality (X/N) when the process average p is randomly distributed in accordance with the parameters of a beta distribution which are s and t. The mean[14] of the beta-binomial distribution is $E(X/N) = \bar{p}$, the variance is $\text{Var}(X/N) = \{\bar{p}\bar{q}/N\}\ (s + N\bar{p})/(s + \bar{p})$ where $\bar{q} = 1 - \bar{p}$ and s is the shape parameter.

Equation (14.10) can be evaluated by a computer program for the gamma function or by tables of the gamma function if N is small enough. Hald[15] states that in his own experience it has usually been possible to obtain a good fit to observed quality distributions by using a beta-binomial model.

If lot quality is measured by the number of nonconformities per unit area of opportunity and if a process is in statistical control at a constant average number λ of nonconformities per unit, the distribution of the number of nonconformities (C) in lots that may be viewed as random samples of N units from the process will be given by the Poisson distribution

(14.11)

$$g(C,N,\lambda) = \frac{(\lambda N)^C e^{-\lambda N}}{C!}$$

If the process average λ is not constant, however, but varies randomly from lot to lot in accordance with the distribution $h(\lambda)$, then the distribution of nonconformities C in lots of N units from the process in general will be given by the integral

(14.12)

$$\int_0^\infty g(C,N,\lambda)h(\lambda)d\lambda$$

Hald[16] discusses the case in which $h(\lambda)$ takes the form of a gamma distribution. Skipping over mathematical developments, we note[17] that taking $h(\lambda)$ in equation (14.12) as the gamma distribution yields

(14.13)

$$P(C,N\bar{\lambda},s) = \frac{\Gamma(C+s)}{C!\Gamma(s)} \cdot \frac{(N\bar{\lambda}/s)^C}{(1+N\bar{\lambda}/s)^{C+s}}$$

---

[14] Ibid., p. 129.

[15] Ibid., p. 133.

[16] Ibid., pp. 133 ff. The density function for the gamma distribution is $f(\lambda, \bar{\lambda}, s) = e^{v} v^{s-1} dv$ where $v = s\lambda/\bar{\lambda}$.

[17] Cf. Hald, *Statistical Theory,* p. 134.

as the distribution of the number of nonconformities C in lots of size N from the process. It is known as the "gamma-Poisson distribution." Its mean is $E\{C/N\} = \bar{\lambda}$, its variance is $\text{Var}\{C/N\} = (\bar{\lambda}/N)(s + N\bar{\lambda})/s$ and s is its shape parameter. Setting

$$\theta = \frac{s}{s + N\bar{\lambda}} = \frac{1}{1 + N\bar{\lambda}/s} \tag{14.14}$$

in equation (14.13) yields the negative binomial distribution

$$P(C, N\bar{\lambda}, s) = \frac{\Gamma(C + s)}{C!\Gamma(s)}\, \theta^s (1 - \theta)^C \tag{14.15}$$

as an alternate form for the gamma-Poisson distribution.

A useful characteristic of the gamma-Poisson (negative binomial model) is that by replacing $N\bar{\lambda}$ in equation (14.13) by $N\bar{p}$ and letting X represent the number of nonconforming units, we will get a reasonably good approximation to the beta-binomial distribution that would apply if lot quality were measured by the fraction nonconforming. The restraints are that $\bar{p}$ be less than 0.1 and $\bar{p}/s < 0.2$. In most practical cases we can thus work with the gamma-Poisson distribution whether lot quality is measured by the number of nonconforming units or number of nonconformities per unit.

If in a given situation we decide that a gamma-Poisson distribution is a suitable mathematical model to use, we can fit it to a set of observed data by equating the theoretical mean to the mean of the observed data and the theoretical variance to the variance of the observed data. If the relative frequencies yielded by the fitted mathematical model agree well with the relative frequencies of this histogram of past lots so that we can take the mathematical model as a smoothing formula for the observed data, it may be a practical way of representing long-run expectations.

A second use of the mathematical model would be to estimate the percent of the *process* prior distribution that exceeds the lot break-even quality level. Thus if we follow Hald in taking p_r and $\bar{\lambda}_r$ to represent lot break-even quality levels and set $\rho = \bar{p}/p_r$ or $\bar{\lambda}/\bar{\lambda}_r$ as the situation calls for, we can enter Hald's Table 38 with the value of ρ and the estimated value of s and find the estimated fraction (w_r) of the process prior distribution that exceeds the lot break-even quality level. If this estimated $w_r < 0.05$, Hald states that acceptance of all lots without inspection will be reasonably close to yielding minimum cost. Figure 14.1 of this text, which is derived from Hald's Table 38, shows the combinations of ρ and s for which $w_r < 0.05$.

If $w_r > 0.05$, a third use of the model would be to determine mathematically a Bayesian minimum cost sampling plan. The reader is referred to Hald for this procedure. A substitute procedure would be to find an approximate minimum cost plan from the Hald-Møller tables discussed in the next section using the values of ρ and s estimated from the observed data and referring to the Hald-Møller $w_r = 0.05$ or the $w_r = 0.10$ table as Hald's

Table 38 suggests is the more appropriate. (See the example at the end of the next section.)

2.3.4.2. The Hald-Møller Tables of Bayesian Minimum Cost Sampling Plans. If a gamma-Poisson distribution, either in its own right when quality is measured by the number of nonconformities per unit or as an approximation to the beta-binomial distribution when quality is measured by the fraction of nonconforming units, is judged to be an appropriate model for an observed distribution of lot quality, the Hald-Møller tables of sampling plans, presented in Tables 34 and 35 in Hald's book, may be found useful in deriving an approximate minimum cost plan. For a gamma-Poisson distribution of lot quality, the tables are a set of minimum cost plans indexed by nominal values of ρ and w_r (with associated values of s). In Hald's Table 34, w_r is taken equal to 0.05 and in his Table 35 it is taken equal to 0.10. The columns in these tables are headed by nominal values of $\rho = \bar{p}/p_r$ or $\bar{\lambda}/\bar{\lambda}_r$ that in his Table 34 run from $\rho = 0.60$ to 0.20 in steps of 0.10 and in his Table 35 run from $\rho = 0.70$ to 0.40 in steps of 0.10. With each listed value of ρ the value of the shape parameter s is given that, in combination with that value of ρ, yields the nominal value of $w_r = 0.05$ or $w_r = 0.10$ by which the table is headed. As an illustration, Table 14.2 of this text gives a section of the $w_r = 0.10$ table for which $\rho = 0.70$ and $s = 9.64$. If a given study of past data together with a stated lot break-even point yields an estimated value of ρ and s and from Hald's Table 38 a value of w_r is estimated from these values of ρ and s, furthermore if the gamma-Poisson model appears to be an appropriate model for the observed distribution of lot quality, then the Hald-Møller plan that has values of w_r, ρ, and s near those estimated from past data can be taken as a sampling plan that will probably yield expected costs not too far from minimum costs.

Let us consider the example given by Hald.[18] For a given set of observations of lot quality, he reports that the gamma-Poisson distribution fitted to the data (as an approximation to the beta-binomial distribution) has a mean of $\bar{p} = 0.0553$ and an $s = 7.70$. The break-even quality level, which he represents by p_r, is 0.0792. Thus $\rho = \bar{p}/p_r = 0.0553/0.0792 = 0.698 \doteq 0.70$. for $\rho = 0.70$ and $s = 7.7$, Hald's Table 38 indicates that w_r is about 12 percent. Lots consist of 30,000 items, so M_c which is defined as Np_r, equals 2,376. If we enter Hald's Table 35 (for which $w_r = .10$) with $M_c = 2{,}376$, we find that the associated acceptance number is $c = 35$. (See Table 14.2 on p. 326.) For $\rho = 0.70$, $s = 9.64$, which are reasonably close to the $\rho = 0.70$, $s = 7.7$ of the gamma-Poisson distribution fitted to the empirical data, the minimum cost plan for $c = 35$ has m, which is defined as equal to np_r, equal to 31.4 (the m_l and m_u are the same in this case, so no interpolation is necessary). Hence the optimum sample size for the tabulated plan will be $31.4/0.0792 = 396$. For the values of $\bar{p}$ and s obtained from the empirical data and the given lot size and break-even quality level, the nearest

[18] Hall, *Statistical Theory,* p. 150.

TABLE 14.2
An Excerpt from the Hald-Møller Tables of Bayesian Minimum Cost Sampling Plans ($w_r = 0.10$)

	$\rho = 0.70,\ s = 9.64$	
c	M_c	$m_l - m_u$
16	950	Accept
17	1000	13.4 − 13.4
18	1050	14.4 − 14.4
19	1110	15.4 − 15.4
20	1170	16.4 − 16.4
21	1230	17.4 − 17.4
22	1300	18.4 − 18.4
23	1360	19.4 − 19.4
24	1430	20.4 − 20.4
25	1500	21.4 − 21.4
26	1580	22.4 − 22.4
27	1650	23.4 − 23.4
28	1730	24.4 − 24.4
29	1810	25.4 − 25.4
30	1900	26.4 − 26.4
31	1980	27.4 − 27.4
32	2070	28.4 − 28.4
33	2160	29.4 − 29.4
34	2250	30.4 − 30.4
35	2340	31.4 − 31.4
36	2440	32.4 − 32.4
37	2540	33.4 − 33.4
38	2640	34.4 − 34.4
39	2740	35.4 − 35.4

SOURCE: Reproduced with permission from p. 487 of A. Hald, *Statistical Theory of Sampling Inspection by Attributes*, New York: Academic Press, 1981.

optimum tabulated plan is thus $n = 396$, $c = 35$. On the assumption that lot quality follows a gamma-Poisson distribution, the actual minimum cost Bayesian plan for the given N, ρ, and s (apparently derived by direct mathematical calculation) is given by Hald[19] to be $n = 457$, $c = 39$.

*2.3.4.3. **A Precautionary Note.*** Whan a Bayesian minimum cost sampling plan is used, a control chart should be maintained to check on the stability of the distribution of lot quality assumed in instituting the plan. This could be a continuation of the chart that was initially set up to decide whether the process was in control with respect to a given distribution of lot quality. (See p. 321 above.) If the currently maintained chart frequently

[19] Ibid.

goes out of control and frequent revisions of the plan become necessary, the use of a Bayesian plan may become administratively uneconomic. For further discussion of this see p. 328 below.

2.3.5. A Process in a "State of Chaos." A study of the quality of lots produced by a given process may reveal that lot quality wanders about in an unpredictable manner, is chaotic. In other words, there may be no evidence that the process is in a state of statistical control at a constant level nor evidence that it is in control with respect to a stable distribution of lot quality—conditions that have been discussed in Sections 2.2 and 2.3.4 above. Hopefully in the long run the process can be brought into some state of statistical control. The problem is what to do in the short run if the variation in lot quality straddles the lot break-even point. One solution is to apply 100 percent inspection. This may be more costly, however, than other procedures, two of which we will consider here.

2.3.5.1. Anscombe's Sequential Sampling Procedure. A procedure that may be applied in a short-run state of chaos and is near minimum cost is F. J. Anscombe's sequential sampling plan.[20] The initial sample taken under this plan would be $n = 0.375\sqrt{Nk_2/k_1}$ where N is the lot size and k_1 and k_2 are cost factors defined on p. 319 above, the ratio k_1/k_2 being the lot break-even quality level. Subsequent samples would be of size $n = k_2/k_1$. Sampling would continue, until either the total number of nonconforming units found is one less than the number of samples inspected, or the whole lot has been inspected. This plan has the disadvantage that it may be difficult to administer.

2.3.5.2. Joyce N. Orsini's Rules[21]. If lot quality is measured in terms of the fraction nonconforming, a simple procedure for dealing with a state of chaos that may be less costly than 100 percent inspection when lots are relatively large is to apply Joyce N. Orsini's rules. For the conditions indicated, they are:

$k_2 \geq 1000\, k_1$	Inspect the incoming lots 100%.
$1000\, k_1 > k_2 \geq 10\, k_1$	Test a random sample of 200 items from each lot. Accept the remainder of the lot if you find no nonconforming items in the sample. Screen the remainder if you encounter one or more nonconforming items in the sample.
$k_2 < 10\, k_1$	No inspection.

Joyce Orsini admits the possibility of a Bayesian solution but suggests that it may run into administrative difficulties that would produce a state

[20] Frank J. Anscombe, "Rectifying inspection of lots," *Journal of the American Statistical Association* 56 (1961), pp. 807–23. See Chapter 16, Section 2, for further discussion of Anscombe's sequential sampling procedure.

[21] See Joyce N. Orsini, "Simple Rule to Reduce the Total Cost of Inspection and Correction of Product in State of Chaos." Ph.D diss., New York University, 1982. Copyright J. Orsini.

of administrative chaos. In apparent reference to a Bayesian solution she writes:[22]

> For each example with known prior distribution, a minimum theoretical value for total cost can be calculated through arithmetic, use of calculus or by techniques of dynamic programming. . . .
>
> Unfortunately, each solution is unique to the particular problem, and a manufacturer would need to catalog sampling plans for each vendor for each part. Whenever a shipment of parts came in, the "correct" sampling plan would be instituted for that specific vendor and part. No two plans would be alike. Inspectors would change their procedure with each new lot of goods that came in.
>
> Costs of inspection and correction of product, or quality of incoming parts, may all change and necessitate calculation of a new sampling plan. Formulations of the new plan would require a specialist. Books of tables prepared in advance to cover changes would be voluminous and impractical. So, even if the prior distribution is known, the vigil would be constant.

Joyce Orsini offers her rules as an inspection procedure that might not yield minimum total cost but would be less costly than 100 percent inspection. The argument for her rules is based upon a study of their application to a spectrum of distributions of incoming lot quality that she considers to have a good chance of covering most of the worst states of chaos to be met in practice. She includes for comparison some intermediate distributions. The distributions studied are a decreasing right triangular distribution, an increasing right triangular distribution, a rectangular (uniform) distribution, a two-point distribution, three varieties of a V-shape distribution and three beta distributions. In her study of these various distributions, all of which have an average of 0.01, she found that except for extreme values of the lot break-even point her proposed rules resulted in substantial saving over 100 percent inspection. For a lot break-even quality level of 0.005 she found that in the examples studied the savings ranged from about 20 percent to 50 percent of the cost of 100 percent inspection. An investigation of results obtained with sample sizes of 100 and 300 instead of a sample size of 200, lead to the conclusion that over the range of distributions and values of the lot break-even quality level encountered in practice, it was difficult to improve on a sample of $n = 200$. Joyce Orsini warns that no claim is made that her suggested $n = 200$, $c = 0$ sampling plan "achieves minimum total cost, though in many cases, it comes close to minimum total cost."[23] It is offered as a simple, less costly substitute for 100 percent inspection when the process average varies relatively widely about the lot break-even quality level.

The sampling plan $n = 200$, $c = 0$ is presumably to be employed only while the process is in a "state of chaos" (in lieu of 100 percent inspection) and until the process can be brought into a state of statistical control at a

---

[22] Ibid., p. 9. Quoted with permission.

[23] Ibid., p. 62.

TABLE 14.3[24]
Type B Operating Characteristic Curve for the Sampling Plan $n = 200$, $c = 0$

Process Average Fraction Nonconforming	*Probability of Lot Acceptance*
0.000051	0.99
0.000256	0.95
0.000525	0.90
0.00144	0.75
0.00347	0.50
0.00693	0.25
0.0115	0.10
0.0150	0.05
0.0230	0.01

constant level, hopefully less than the lot break-even quality level $(X/N)_r$. If the steps taken to correct the process seem to be reaching this goal, it might be well to replace the $n = 200$, $c = 0$ plan by a sampling scheme that is less drastic with respect to rejections.

As indicated in Table 14.3 above, the sampling plan $n = 200$, $c = 0$ is a very tight plan. If the average fraction nonconforming in the "state of chaos" is 0.01, as in Joyce Orsini's examples, and if the lot break-even quality $(X/N)_r$ is 0.01, to bring the process into a state of statistical control at a level just less than 0.01 would be an acceptable solution that when established would allow acceptance without inspection. As this state is being approached, however, and tested for its stability, the continued application of the sampling plan $n = 200$, $c = 0$ would yield an excessively high number of rejections. It would seem that in this transitory period it might be more satisfactory to use the Mil. Std. 105D (ANSI Z1.4) sampling scheme listed for Code Letter L and an AQL = 0.65 percent, the normal sampling plan for which is $n = 200$, $c = 3$. This would require less 100 percent inspection for a situation for which acceptance-without-inspection is apparently about to be invoked.

In concluding this section the reader should be reminded that the Orsini plan $n = 200$, $c = 0$ is intended only for application to large lots. In her calculation of probabilities of acceptance she uses the binomial distribution which implies that she expects lots to be at least of size 2,000 (= 10 × 200). It may be noted in passing that Mil. Std. 105D lists the sampling plan $n = 200$, $c = 0$ for use with lot sizes 3,201–10,000. It could be that for smaller lots, say of size 800–1,000, the sampling plan $n = 80$, $c = 0$ would be a satisfactory substitute for 100 percent inspection. This is only a suggestion the truth of which would need to be demonstrated.[25]

---

[24] From Table X-L-1 of Mil. Std. 105D, column headed AQL = 0.065%.

[25] For further discussion along these lines see E. G. Schilling, "An Overview of Acceptance Control," *ASQC 37th Annual Quality Congress Transactions* (1983), p. 65.

2.3.6. Destructive Testing. With reference to the above discussion the question arises of what to do if testing is destructive. In this instance 100 percent inspection is impossible. In situations in which 100 percent inspection would be an optimum procedure if testing were nondestructive, the best answer is probably to devote special attention to keeping the process in a state of statistical control at a constant acceptable level.

3. SAMPLING PLANS INDEXED BY LIMITING QUALITY VALUES

3.1. Information on Limiting Quality Values Provided in Mil. Std. 105D and ANSI Std. Z1.4

Supplementary information on the process levels of percent nonconforming (LQLs) for which the probabilities of lot acceptance just equal 0.10 and 0.05 is given for the various plans of Mil. Std. 105D and ANSI Z1.4 in Tables VI–A and VI–B of these standards. Similar information for the process number of nonconformities per 100 units is given in Tables VII–A and VII–B. (See Table 10.7 of this text for Table VI–A of Mil. Std. 105D.) Like the main tables, however, these supplementary tables are indexed by AQLs and not LQLs. The consumer's risk of accepting undesirable lots can also be read from the OC curves given in Table X of these standards, but again the curves are tagged by AQLs and not LQLs. The supplementary information provided by Mil. Std. 105D and ANSI Z1.4 is thus not conveniently organized for those who seek to select a sampling plan with a specified LQL.

3.2 A Grouping of Mil. Std. 105D Plans by Nominal LQLs

The deficiency of Mil. Std. 105D and ANSI Z1.4 with respect to indexing by LQLs led to the issuance in 1984 of Supplement 1 to the British Standard 6001. In Tables 3 to 12 of this supplement, selected plans from the AQL indexed set that constitute the main body of the standard (plans with $c = 0$ are excluded) are grouped by nominal limiting quality levels of 0.5 percent, 0.8 percent, 1.25 percent, 2.00 percent, 3.15 percent, 5.0 percent, 8.0 percent, 12.5 percent, 20.0 percent, and 32.0 percent. OC curves are given for the plans of each group[26] and in Table 14 of the supplement the range of the process quality levels is indicated for the plans in each group for which the probabilities of lot acceptance are exactly 0.10 and 0.05. All of this pertains to the situation in which the sampling plan is presumably being applied to a continuing series of lots. For the consumer, however, who wishes to know the risks involved in applying a listed sampling plan to a single lot in isolation, Tables 3 to 12 give for each lot-size range the minimum and maximum hypergeometric probability of accepting a lot of nominal LQ quality. See,

[26] These OC curves give an insight as to how increasing the sample size while holding the LQL, P_a point constant affects the producer's risk. See Figure 14.1.

FIGURE 14.1
Operating Characteristic Curves for Sampling Plans of Table 14.5 (curves are identified by sample size code letter)*

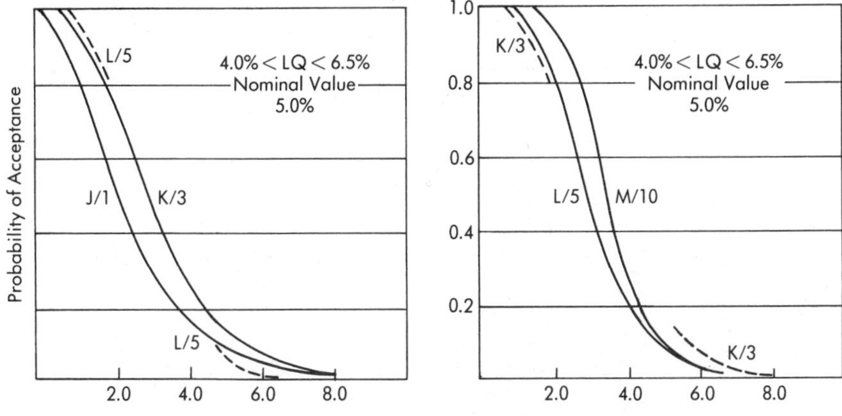

SOURCE: Figure 14.1 is from BS 6001: Supplement 1:1984 and is reproduced by permission of the British Standards Institution. Complete copies of the standard can be obtained from BSI at Linford Wood, Milton Keyes, MK14 6LE, England or national standards bodies.

for example, Table 14.4 of this text (next page). In 1985 a Part II to ISO Std. 2859 was adopted that contains the same information as this supplement 1 to BS 6001.

3.3 A Directly Derived Set of Plans Indexed by Nominal LQLs

A table of plans indexed by process limiting quality levels and designed for application to a series of lots was published by Acheson J. Duncan, August B. Mundel, A. Blanton Godfrey, and Valerie A. Partridge in the January 1980 issue of the *Journal of Quality Technology* and is reproduced here as Table 14.5. For the listed nominal limiting quality levels, this table gives plans for which the probability of acceptance is approximately 10 percent for lots from a process of average quality equal to the nominal LQL. The actual probabilities of lot acceptance at the nominal LQL are given in the table separately for each plan. The process level for which the probability of lot acceptance is 0.95 is also given for each plan so that the producer's risk is clearly indicated. A significant feature of the plans is that the lot-size ranges and sample sizes are compatible with those of Mil. Std. 105D, ANSI Z1.4, British Standard 6001, ISO Std. 2859 and other attributes standards that follow Mil. Std. 105D. They can thus be used in conjunction

TABLE 14.4
Single Sampling Plans for Nominal Limiting Quality 5.0 Percent (Table 8 of Supplement 1 to BS 6001)*

Lot Sizes for Given Inspection Levels					*BS 6001 Single Sampling Plan† (Normal Inspection)*			*Code Letter*	*Tabulated Values of Submitted Quality Accepted with Designated Probabilities† (Quality as Percent Nonconforming)*						*Acceptance Probabilities** for LQ (Based on Hypergeometric)*	
S-1 to S-3	*S-4*	*I*	*II*	*III*	*AQL*	*n*	*Ac*		*0.95*	*0.90*	*0.50*	*0.10*	*0.05*	*LQ*	*Max.*	*Min.*
> 80‡	81‡ to 500,000	81‡ to 10,000	81‡ to 1,200	81‡ to 500	0.65	80	1	J	0.444	0.666	2.09	4.78	5.80	5.0	0.086	0.000
	> 500,000	10,001 to 35,000	1,201 to 3,200	501 to 1,200	1.00	125	3	K	1.09	1.40	2.94	5.35	6.20	5.0	0.124	0.092
		35,001 to 150,000	3,201 to 10,000	1,201 to 3,200	1.00	200	5	L	1.31	1.58	2.84	4.64	5.26	5.0	0.062	0.048
		> 150,000	> 10,000	> 3,200	1.50	315	10	M	1.96	2.23	3.39	4.89	5.38	5.0	0.081	0.072

* Probability calculated by the Poisson approximation for Code letter K, L, and M. Binominal distribution for Code letter J.
** Maximum and minimum hypergeometric probabilities for permitted lot sizes for each plan.
† These are also Mil. Std. 105D plans.
‡ For fewer than 81 in the lot, 100 percent inspect the lot.

SOURCE: Table 14.4 is from BS 6001: Supplement 1: 1984 and is reproduced by permission of the British Standards Institution. Complete copies of the standard can be obtained from BSI at Linford Wood, Milton Keyes, MK 14 6LE, England or national standards bodies.

with plans from these standards for inspecting other characteristics for which the AQL is of primary interest.

3.4 Plans for Inspection of Isolated Lots

In addition to Tables 3 to 12 discussed in Section 3.2 above, Supplement 1 to BS 6001 also presents a table of lot LQ indexed plans that are recommended for use when both the supplier and consumer wish to regard a lot as isolated. This table is reproduced here as Table 14.6. It will be noted that it is based on random sampling from finite lots for both consumer and producer risks. It will also be noted that, like the tables discussed in Section 3.2.1 above, the lot-size ranges are the same as those of the AQL indexed plans of Mil. Std. 105D, ANSI Z1.4, and ISO Std. 2859. Many of the plans of Table 14.6 (pp. 336–37), are the same as the plans of Table 14.5 and it is interesting to note the differences between probabilities of acceptance based on the binominal or Poisson distributions calculated for lots from a process of nominal LQL quality and the probabilities of acceptance based on the hypergeometric distribution calculated for isolated lots of the same nominal limiting quality. When we make such a comparison we find, for example, that for the lot-size range 501–1,200 and sampling plan 80, 1 listed under a nominal limiting quality of 5 percent the binominal probability of acceptance for lots of that quality is 0.086 and the maximum hypergeometric probability over the lot-size range is 0.079, a difference of 0.007. For larger lot sizes the difference is somewhat less.

3.5. The Dodge-Romig LTPD Indexed Sampling Tables

If compatibility with the ranges of lot-size classes of Mil. Std. 105D, is not of special interest, it may be that the Dodge-Romig set of LTPD indexed plans will be found useful.[27] LTPD refers to the Lot Tolerance Percent Defective and is the same as what is designated here as the lot limiting quality (LQ). The nominal values of LTPD listed in the Dodge-Romig tables are: 0.5 percent, 1.0 percent, 2.0 percent, 3.0 percent, 4.0 percent, 5 percent, 7 percent, and 10 percent and all plans are designed to have a consumer's risk of 10 percent of accepting a given lot the quality of which equals the listed LTPD. Both single and double sampling plans are presented. The plans are indexed not only by the LTPD, but also by lot size ranges and ranges of the estimated process average. The last is an element in the selection of a plan, since the set of plans is so designed that they will minimize average total inspection for the designated process average.

---

[27] Harold F. Dodge and Harry G. Romig, *Sampling Inspection Tables* (2d ed.; New York: John Wiley & Sons, 1959).

TABLE 14.5
A Set of Directly Derived LQL Indexed Sampling Plans

Lot Size	Sample Size	*LQL (in percent)*													
		0.125	*0.2*	*0.315*	*0.5*	*0.8*	*1.25*	*2.0*	*3.15*	*5.0*	*8.0*	*12.5*	*20*	*31.5*	*50*
19–25	C 5														0 .03 .08 1.0 1.0
26–50	D 8													0 .05 .08 .64 .64	1 .09 .09 4.6 4.4
51–90	E 13												0 .06 .07 .39 .40	1 .05 .08 2.8 2.7	3 .05 .11 11.3 10.5
91–150	F 20											0 .07 .08 .26 .26	1 .07 .09 1.8 1.8	3 .08 .13 7.1 6.8	5 .02 .07 13.9 13.1
151–280	G 32										0 .07 .08 .16 .16	1 .08 .09 1.1 1.1	3 .09 .12 4.4 4.3	5 .04 .06 8.5 8.2	10 .03 .08 20.6 19.3
281–500	H 50									0 .08 .08 .10 .10	1 .08 .09 .72 .71	3 .11 .13 2.8 3.3	5 .05 .07 5.4 5.2	10 .05 .09 12.9 12.3	18 .03 .09 26.5 24.9
501–1,200	J 80								0 .08 .08 .06 .06	1 .09 .09 .45 .44	3 .11 .12 1.7 1.7	5 .06 .07 3.3 4.9	10 .06 .08 7.9 7.7	18 .05 .09 16.1 15.6	31 .03 .09 30.8 29.1
1,201–3,200	K 125							0 .08 .08 .04 .04	1 .09 .10 .29 .28	3 .12 .13 1.1 1.1	5 .06 .07 2.1 2.1	10 .08 .09 5.0 6.2	18 .07 .09 10.2 10.0	31 .06 .10 19.3 18.6	52 .04 .10 34.9 33.3

3,201–10,000	L 200						0 .08 .08 .03 .03	1 .09 .09 .18 .18	3 .12 .13 .69 .68	5 .06 .07 1.3 1.3	10 .07 .08 3.1 3.1	18 .08 .09 6.3	31 .06 .09 11.9 11.7	52 .05 .09 21.4 20.8	
10,001–35,000	M 315					0 .08 .08 .02 .02	1 .10 .10 .11 .11	3 .13 .13 .44 .43	5 .07 .07 .83 .83	10 .08 .09 2.0 2.0	18 .08 .09 4.0 4.0	31 .09 .10 7.5 7.4	52 .07 .09 13.4 13.2		
35,001–150,000	N 500				0 .08 .08 .01 .01	1 .09 .09 .07 .07	3 .13 .13 .27 .27	5 .07 .07 .52 .52	10 .08 .09 1.2 1.2	18 .09 .09 2.5 2.5	31 .08 .09 4.7 4.7	52 .09 .10 8.4 8.3			
150,001–500,000	P 800			0 .08 .08 .01 .01	1 .09 .09 .04 .04	3 .12 .12 .17 .17	5 .07 .07 .33 .33	10 .08 .08 .77 .77	18 .08 .09 1.6 1.6	31 .08 .09 2.9 2.9	52 .06 .07 5.2 5.2				
500,000	Q 1250		0 .08 .08 .004 .004	1 .10 .10 .03 .03	3 .13 .13 .11 .11	5 .07 .07 .21 .21	10 .09 .09 .49 .49	18 .09 .09 1.0 1.0	31 .10 .10 1.9 1.9	52 .09 .10 3.3 3.3					
	R 2000	0 .08 .08 .003 .003	1 .09 .09 .02 .02	3 .13 .13 .07 .07	5 .07 .07 .13 .13	10 .08 .08 .31 .31	18 .09 .09 .62 .62	31 .08 .09 1.2 1.2	52 .09 .09 2.1 2.1						

Key

AN.	
a	b
c	d

AN—acceptance number

a—binomial P_a for process average percent nonconforming = LQL

b—Poisson P_a for process average number of nonconforming per 100 units = LQL

c—Process average percent nonconforming for which binomial $P_a = 0.95$

d—Process average number of nonconformities per 100 units for which Poisson $P_a = 0.95$

SOURCE: Reprinted with permission from the *Journal of Quality Technology,* 12, no. 1 (January 1980), p. 43. Copyright American Society for Quality Control, Inc.

TABLE 14.6
Single Sampling Plans for Use when Both Consumer and Producer View a Lot as Isolated (Table 2 of Supplement 1 to BS 6001–1984)

Key:

Sample size/Acceptance no. (n/Ac)	Probability of acceptance at the nominal quality* (PLQ)
Percent defective, *(p)*	Probability of acceptance of p percent defective* (P_a)

Each cell in the table shows the producer's and consumer's risks P_a and PLQ, together with the acceptable quality *(p)* as indicated.

Lot Size	*Nominal Limiting Quality in Percent (LQ)*									
	0.5	*0.8*	*1.25*	*2.0*	*3.15*	*5.0*	*8.0*	*12.5*	*20.0*	*32.0*
16 to 25						Inspect every item	17/0 0.094 0 1.0	13/0 0.082 0 1.0	9/0 0.082 0 1.0	6/0 0.070 0 1.0
26 to 50				Inspect every item	Inspect every item	28/0 0.085 0 1.0	22/0 0.089 0 1.0	15/0 0.090 0 1.0	10/0 0.083 0 1.0	6/0 0.085 0 1.0
51 to 90			Inspect every item	50/0 † 0 1.0	44/0 0.094 0 1.0	34/0 0.103 0 1.0	24/0 0.098 0 1.0	16/0 0.094 0 1.0	10/0 0.094 0 1.0	8/0 0.040 0 1.0
91 to 150		Inspect every item	90/0 † 0 1.0	80/0 0.099 0 1.0	55/0 0.100 0 1.0	38/0 0.103 0 1.0	26/0 0.092 0 1.0	18/0 0.077 0 1.0	13/0 0.048 0 1.0	13/1 0.041 2.67 0.96
151 to 280	Inspect every item	170/0 0.102 0 1.0	130/0 0.095 0 1.0	95/0 0.089 0 1.0	65/0 0.090 0 1.0	42/0 0.097 0 1.0	28/0 0.086 0 1.0	20/0 0.062 0 1.0	20/1 0.062 1.79 0.96	13/1 0.044 2.86 0.95
281 to 500	280/0 0.089 0 1.0	220/0 0.097 0 1.0	155/0 0.095 0 1.0	105/0 0.092 0 1.0	80/0 0.61 0 1.0	50/0 0.067 0 1.0	32/0 0.068 0 1.0	32/1 0.071 1.0 0.97	20/1 0.065 1.80 0.95	20/3 0.072 7.20 0.95

501 to 1,200	380/0 0.101 0 1.0	255/0 0.098 0 1.0	170/0 0.100 0 1.0	125/0 0.069 0 1.0	125/1 0.081 0.250 0.97	80/1 0.079 0.417 0.96	50/1 0.078 0.667 0.96	32/1 0.075 1.08 0.95	32/3 0.090 4.42 0.95	32/5 0.029 8.50 0.95
1,201 to 3,200	430/0 0.099 0 1.0	280/0 0.095 0 1.0	200/0 0.074 0 1.0	200/1 0.083 0.188 0.95	125/1 0.088 0.281 0.95	125/3 0.119 1.13 0.95	80/3 0.106 1.75 0.95	50/3 0.112 2.78 0.95	50/5 0.047 5.38 0.95	50/10 0.042 12.9 0.95
3,201 to 10,000	450/0 0.099 0.010 0.96	315/0 0.076 0.010 0.97	315/1 0.091 0.110 0.96	200/1 0.087 0.180 0.95	200/3 0.120 0.690 0.95	200/5 0.061 1.32 0.95	125/5 0.058 2.12 0.95	80/5 0.055 3.32 0.95	80/10 0.056 7.92 0.95	80/18 0.041 16.1 0.95
10,001 to 35,000	500/0 0.080 0.009 0.96	500/1 0.089 0.071 0.95	315/1 0.094 0.111 0.95	315/3 0.123 0.437 0.95	315/5 0.066 0.834 0.95	315/10 0.080 1.97 0.95	200/10 0.069 3.99 0.95	125/10 0.077 5.01 0.95	125/18 0.069 10.2 0.95	80/18 0.041 16.1 0.95
35,001 to 150,000	800/1 0.090 0.044 0.95	500/1 0.090 0.071 0.95	500/3 0.128 0.273 0.95	500/5 0.065 0.524 0.95	500/10 0.083 1.23 0.95	500/18 0.086 2.50 0.95	315/18 0.077 3.99 0.95	200/18 0.078 6.31 0.95	125/18 0.069 10.2 0.95	80/18 0.041 16.1 0.95
150,001 to 500,000	800/1 0.091 0.044 0.95	800/3 0.118 0.170 0.95	800/5 0.060 0.328 0.95	800/10 0.075 0.771 0.95	800/18 0.082 1.56 0.95	500/18 0.086 2.50 0.95	315/18 0.077 3.99 0.95	200/18 0.078 6.31 0.95	125/18 0.069 10.2 0.95	80/18 0.041 16.1 0.95
$>$ 500,001	1,250/3 0.129 0.109 0.95	1,250/5 0.066 0.209 0.95	1,250/10 0.089 0.494 0.95	1,250/18 0.090 1.00 0.95	800/18 0.082 1.56 0.95	500/18 0.086 2.50 0.95	315/18 0.077 3.99 0.95	200/18 0.078 6.31 0.95	125/18 0.069 10.2 0.95	80/18 0.041 16.1 0.95

* The stated probability of acceptance is the maximum for lots of nominal limiting quality in the size range and the minimum for lots with percent defective *p*. These probabilities are based on the hypergeometric distribution. The producer's risk point is (p, P_a), the consumer's risk point is (LQ, PLQ).

† No lot in this size range can be of nominal quality since this implies a fractional number of defectives.

SOURCE: Table 14.6 is from BS 6001: Supplement 1:1984 and is reproduced by permission of the British Standards Institution. For directions for obtaining complete copies of the standard see footnote to Figure 14.1.

TABLE 14.7
Values of $D = Np_L$ Corresponding to f

f	.00	.01	.02	.03	.04	.05	.06	.07	.08	.09
0.9	1.0000	.9562	.9117	.8659	.8184	.7686	.7153	.6567	.5886	.5000
0.8	1.4307	1.3865	1.3428	1.2995	1.2565	1.2137	1.1711	1.1286	1.0860	1.0432
0.7	1.9125	1.8601	1.8088	1.7586	1.7093	1.6610	1.6135	1.5667	1.5207	1.4754
0.6	2.5129	2.4454	2.3797	2.3159	2.2538	2.1933	2.1344	2.0769	2.0208	1.9660
0.5	3.3219	3.2278	3.1372	3.0497	2.9652	2.8836	2.8047	2.7283	2.6543	2.5825
0.4	4.5076	4.3640	4.2270	4.0963	3.9712	3.8515	3.7368	3.6268	3.5212	3.4196
0.3	6.4557	6.2054	5.9705	5.7496	5.5415	5.3451	5.1594	4.9836	4.8168	4.6583
0.2	10.3189	9.7682	9.2674	8.8099	8.3902	8.0039	7.6471	7.3165	7.0093	6.7231
0.1	21.8543	19.7589	18.0124	16.5342	15.2668	14.1681	13.2064	12.3576	11.6028	10.9272
0.0	*	229.1053	113.9741	75.5957	56.4055	44.8906	37.2133	31.7289	27.6150	24.4149

* For values of $f < .01$ use $f = \frac{2.303}{D}$; for infinite lot size use sample size $n = \frac{2.303}{P_L}$.

SOURCE: From E. G. Schilling, "A Lot Sensitive Sampling Plan for Compliance Testing and Acceptance Inspection," *Journal of Quality Technology* 10 (April 1978), p. 48.

4. LOT SENSITIVE SAMPLING PLANS FOR HIGH QUALITY PRODUCT

E. G. Schilling's paper on "A Lot Sensitive Sampling Plan for Compliance Testing and Acceptance Inspection" in the April 1978, issue of the *Journal of Quality Technology* does not offer a set of plans indexed by LQs, but it does present a table (Table 14.7 of this text) for going from a selected lot LQ and a given lot size to a fraction of the lot that should be inspected to yield the specified LQ with a consumer's risk of 0.10. A requirement of the plans is that the acceptance number should be zero. These plans are intended for inspection of isolated lots and are recommended by Schilling for cases in which a high degree of safety is desired for the product being inspected. They are very tough on the producer, however, unless his product is near perfect. Thus, if a lot contains just one defective item, the probability of lot acceptance equals $1 - n/N$ which in most cases is relatively low. The 0.95 point on the OC curve comes at lot quality $= 0.0223\ p_L$, the 0.50 point comes at lot quality $= 0.301\ p_L$ and the 0.25 point at lot quality $= 0.602\ p_L$, where p_L is the selected limiting lot quality.

To illustrate the use of Table 14.7 suppose a limiting lot quality (p_L) of 0.01 is designated. For a lot size $N = 1{,}000$, Table 14.7 indicates that with $Np_L = 1{,}000\ (0.01) = 10$, 19 percent of the lot should be inspected and the sample size should thus be 190. In general enter the table at the nearest value of $D = Np_L$ and read the corresponding value of f as the sum of the associated row and column headings.

5. PROBLEMS

14.1. Select one of the multiple-sampling plans of Table 9.1 and find a single-sampling attributes plan, a double-sampling attributes plan, an item-by-item SPR sampling plan, and a variables plan that have "essentially the same" p_1', p_2', α, and β. As far as possible make all plans pass through the p_1', α point and compute the value of P_a for each plan at the values of p' for which the P_a of the multiple sampling plan is 0.50 and 0.10. Note differences. Determine the ASN for each of these plans at the p_1' and rank your plans as to "efficiency" at this point.

14.2. A process has been found to be in a state of statistical control at an average percent nonconforming of 2 percent.

a. If the lot break-even quality level for the given product equals 2.5 percent, what inspection procedure would you use? Justify your answer.

b. If the lot break-even quality level equals 1.5 percent and, the lot size is about 3,000 items, what inspection procedure would you use? Justify your answer.

14.3. A study of past production shows that the distribution of lot quality is reasonably well approximated by a gamma-Poisson distribution with average quality equal to 4.7 percent, and an $s = 9.1$. The break-even quality level is 6.8 percent. Referring to Table 14.6 find an approximately minimum cost Bayesian sampling plan for lots of size 20,000.

14.4. A lot of 2,000 items for which a high degree of safety is desired is to be sample-inspected. A limiting quality of 0.4 percent is deemed acceptable. What sample size would you use in inspecting the lot and what would your acceptance number be?

6. SELECTED REFERENCES*

British Standards Institution (B '84), Deming (B '86), Dodge and Romig (B '59), Duncan (P '80), Enell (P '84), Enters and Hamaker (B '51), Guild and Raka (P '80), Hamaker (P '50, P '53, and P '58), Hamaker and Van Strik (P '55), Hawkes (P '79), ISO (revision of B '74 expected in '86), Lauer (P '78), Martz (P '74), Mood (P '43), Orsini (B '82), Papadakis (P '85), Schilling (P '78 and P '83), Stephens and Dodge (P '76), U.S. Dept. of the Army, Chemical Corps Engineering Agency (B '53, ENASR No. PR-7, and '54, ENASR Nos. PR-12 and ES-1), and Wetherill and Chiu (P '74).

* "B" and "P" refer to the "Book" and "Periodical" sections, respectively, of the Cumulative List of References in Appendix V.

15

Acceptance Sampling by Variables to Give Assurance Regarding the Mean or Standard Deviation of a Process or Lot

The variables sampling plans discussed in this chapter are concerned with the average quality of a material or with the variability in the quality of the material and not with a fraction nonconforming. They are most likely to be used in the sampling of bulk material that comes in bags, boxes, drums, or the like.

As in the previous chapters, decisions to accept or reject will be based upon such sample statistics as the mean and standard deviation, but the OC curves will now give the probability of acceptance as a function of a process or lot mean or process or lot standard deviation. In the case of Type B OC curves it will be assumed as before that a process is operating in a random manner so that a series of lots can be viewed as random samples from the process. This will mean in turn that the samples taken from the lots under a given sampling plan can be viewed mathematically as if they came directly from the process. The probability of acceptance given by the Type B OC curve would then be the proportion of the randomly produced lots that would on the average be accepted by the sampling plan. Type A OC curves will, as previously, relate to a series of identical lots. Type A_L OC curves will relate to lots that are relatively large, say at least 10 times the sample. When lots are relatively small, they will be called Type A_S OC curves. Type A_L OC curves will be practically identical with Type B OC curves, except that in the case of the Type B OC curves the abscissa scale will refer to a process mean or process standard deviation while the Type A OC curve will refer to a lot mean or lot standard deviation. The abscissa scale will be marked $\bar{X}'$ or σ' in either case. When we refer to "the OC curve" of a sampling plan we shall mean the Type B OC curve or its practical mathematical equivalent, the Type A_L OC curve. Problems involving Type A_S OC curves will be considered only for the simplest models.

For strict application, the sampling plans discussed in this chapter all assume that the items of the process, or in the case of Type A_L OC curves the items in the lot, are normally distributed with respect to the quality characteristic in question. Minor departures from normality are not likely to affect Type B or Type A_L OC curves seriously if the sampling plan seeks to give assurance regarding a mean value. There must be greater caution if the sampling plan pertains to process or lot variability. There is also again the question of what is meant by a finite lot being normally distributed. In this chapter it will be sufficient to assume it has a bell-shaped distribution with the same γ coefficients as the normal distribution,[1] viz $\gamma_1' = 0$, $\gamma_2' = 0$. In general, if the process or lot distribution is distinctly nonnormal, it is recommended that professional advice be sought, especially before inaugurating a sampling plan to give assurance regarding variability.

1. LOT-BY-LOT SAMPLING PLANS TO GIVE ASSURANCE REGARDING THE MEAN QUALITY OF A MATERIAL: PLANS BASED ON THE SAMPLE MEAN WITH KNOWN STANDARD DEVIATION

1.1. Single-Sampling Plans

1.1.1. An Example. Consider the following example. A company is interested in the average bulk density of a chalk that it uses. To assure itself regarding the quality of the chalk, the company employs the following sampling plan. The chalk is shipped in bags. From each shipment a random sample of nine bags is taken and the bulk density of the contents of each bag is determined.[2] The mean of these nine measurements is computed, and if this falls below 0.1317 *gm/cc* the shipment is rejected. Otherwise it is accepted. The value 0.1317 is called the acceptance limit and is designated by the symbol $\bar{X}_a$.

The protection given by the above plan is revealed by its OC curve, which is shown in Figure 15.1. This gives the probability that in a random sample of nine bags from a lot that comes at random from a process, the mean density of which is $\bar{X}'$ *gm/cc,* the mean of the sample will be equal to or greater than 0.1317 *gm/cc.* It is based on the assumption that the standard deviation of bulk density runs about 0.006 *gm/cc,* regardless of the mean quality. Presumably the value of the standard deviation would be obtained from the company's past experience.

---

[1] See Chapter 27, Section 3.

[2] The "bulk density of a bag" will itself probably be a mean of several samples from the bag. This will not concern us here except to note that the method of determination will have a direct bearing on the standard deviation (σ') of such determinations.

FIGURE 15.1

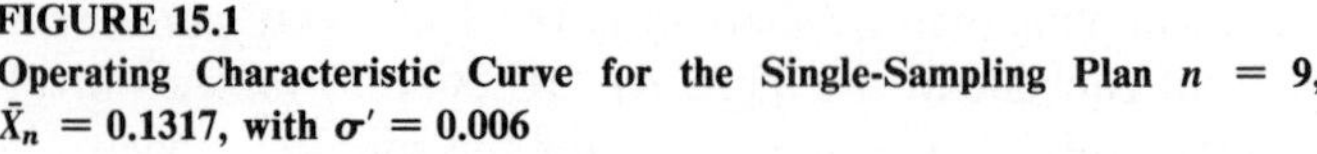

Operating Characteristic Curve for the Single-Sampling Plan $n = 9$, $\bar{X}_n = 0.1317$, with $\sigma' = 0.006$

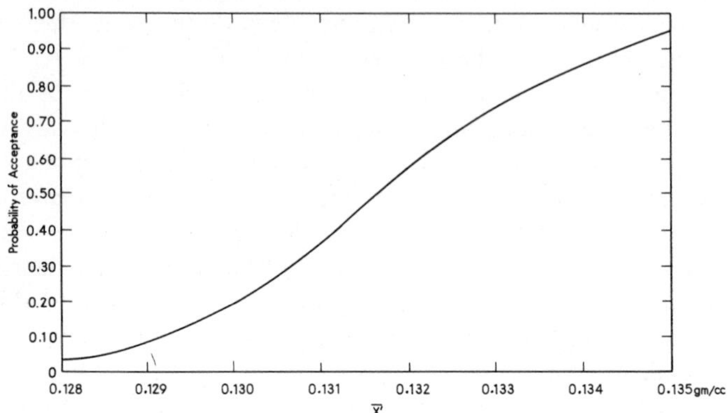

The computation of a point on the OC curve for the above plan can be illustrated by calculating the ordinate at $\bar{X}' = 0.1350$. The procedure consists of computing the quantity $\dfrac{\bar{X}_a - \bar{X}'}{\sigma'/\sqrt{n}}$ and looking in a table of the normal distribution to find the probability of a standardized normal deviate being greater than or equal to this value. In our example, $\bar{X}_a = 0.1317$, $\bar{X}' = 0.1350$, and we have $\dfrac{0.1317 - 0.1350}{0.006/\sqrt{9}} = -1.645$. From Table A2 of Appendix II we note that the probability of a standardized normal deviate being greater than or equal to -1.645 is 0.95. Hence the ordinate of the OC curve at $\bar{X}' = 0.1350$ is 0.95. In the same way it will be found that for $\bar{X}' = 0.1291$ the probability of acceptance is about 0.10. (See Figure 15.1.)

1.1.2. Derivation of a $\bar{X}$ Single-Sampling Plan with Specified $\bar{X}'_1$, $\bar{X}'_2$, α, and β. It is relatively simple to design a single-sampling plan based on $\bar{X}$ with a specified $\bar{X}'_1$, $\bar{X}'_2$, α, and β. The procedure will be explained with reference to the following example, which is similar to that of the previous section. As with every plan in Section 1, it will be assumed that σ' is known and constant.

Suppose that a good grade of a specified chalk averages 0.1350 *gm/cc* and a company wishes to have a plan that will accept lots of such chalk 95 times out of 100. That is, the company wants a plan with an $\bar{X}'_1 = 0.1350$ and $\alpha = 0.05$. Suppose, further, that, in contrast to the previous illustration, the company wants a plan that will accept, no more than 10 percent of the time, chalk that has a mean bulk density of 0.1300 *gm/cc.* In other words, it wants a plan with an $\bar{X}'_2 = 0.1300$ and $\beta = 0.10$. Suppose, finally, that the standard deviation of bulk density is known to be 0.006 *gm/cc,* as we assumed in the previous case.

To derive the desired plan, proceed as follows. Let n be the size of the

FIGURE 15.2
Operating Characteristic Curve for the Single-Sampling Plan $n = 12$, $\bar{X}_a = 0.1372$, with $\sigma' = 0.0063$

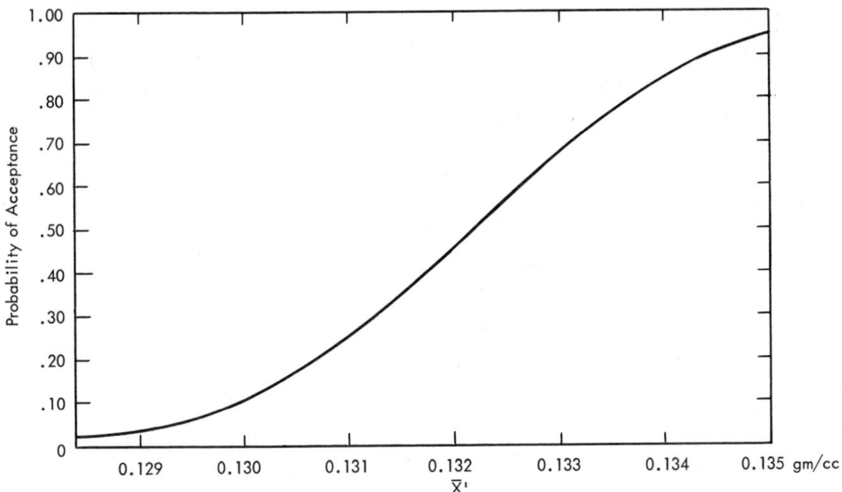

sample to be used and $\bar{X}_a$ the acceptance limit for the mean of the sample tests. If a process or lot actually has a mean grade of 0.1350 *gm/cc,* means of samples of n will have[3] a normal distribution with a mean of 0.1350 *gm/cc* and a standard deviation of $0.006/\sqrt{n}$. Hence, if lots of this kind of product are to have a 0.95 chance of being accepted, we must have

$$\frac{\bar{X}_a - 0.1350}{0.006/\sqrt{n}} = 1.645$$

Similarly, if the product actually has a mean grade of 0.1300 *gm/cc* and if lots of this kind are to have a 0.10 chance of being accepted, we must have

$$\frac{\bar{X}_a - 0.1300}{0.006/\sqrt{n}} = +1.282$$

These are two equations in $\bar{X}_a$ and n, the simultaneous solution of which yields the sampling plan $\bar{X}_a = 0.1372$ and $n = 12$. This is the desired plan.

The OC curve for the above plan is shown in Figure 15.2. It is derived in the manner described in Section 1.1.1.

1.1.3. Derivation of an $\bar{X}$ Single-Sampling Plan with Upper and Lower $\bar{X}'_{L2}$ and $\bar{X}'_{U2}$. The plan designed in Section 1.1.2. had only one $\bar{X}'_2$. There are cases, however, in which the quality characteristic can become unacceptable because it is either too high or too low. The problem will be considered here of how to design a plan with two $\bar{X}'_2$.

---

[3] Cf. above, Chapter 6, Section 1. Normality is exact only if the process or lot is normal.

Consider the following example: Suppose that the tensile strength of a certain steel varies with the heat treatment of this steel. The greater the heat, the greater the tensile strength. At higher temperatures, however, a certain brittleness develops that is undesirable. As a result, steel with high tensile strength is considered as unacceptable as steel with low tensile strength.

An airplane company wants a sampling plan that will reject steel with too high, as well as too low, tensile strength. Suppose that it considers 180,000 psi and 200,000 psi to be at the limits of tolerance and it wishes to accept steel with tensile strength outside these limits no more than 10 times out of 100. Let the standard deviation of tensile strength be 5,000 psi.

To derive a plan that will have the desired specifications we set $\bar{X}'_{L2} =$ 180,000 and $\bar{X}'_{U2} = 200{,}000$ and $\beta = 0.10$. We shall also set $\bar{X}'_1 = 190{,}000$ and as before take $\alpha = 0.05$. It will be noted that the procedure is valid only for plans in which $\bar{X}'_1$ lies halfway between $\bar{X}'_{L2}$ and $\bar{X}'_{U2}$. When this is not the case, the derivation of a plan with two $\bar{X}'_2$ is more difficult.

Let n be the number of tests to be made and let $\bar{X}_{Ua}$ and $\bar{X}_{La}$ be the upper and lower acceptance limits for the mean tensile strength of the n tests. We can assume that the sample means will be normally distributed. Hence the requirements for our plan will give us the following equations:

$$\frac{\bar{X}_{Ua} - 200{,}000}{5{,}000/\sqrt{n}} = -1.282$$

$$\frac{\bar{X}_{La} - 180{,}000}{5{,}000/\sqrt{n}} = 1.282$$

$$\frac{\bar{X}_{Ua} - 190{,}000}{5{,}000/\sqrt{n}} = 1.960$$

$$\frac{\bar{X}_{La} - 190{,}000}{5{,}000/\sqrt{n}} = -1.960$$

The last equation is redundant, since it can be derived from the other three.

The first and second of the above equations give

$$\bar{X}_{Ua} + \bar{X}_{La} = 380{,}000$$

The first and third give

$$\sqrt{n} = \frac{(1.282 + 1.960)(5{,}000)}{10{,}000} = 1.621$$

and $n = 2.6$, or approximately 3. The substitution of $n = 3$ in the first equation gives

$$\bar{X}_{Ua} - 200{,}000 = -1.282 \frac{(5{,}000)}{\sqrt{3}}$$

or

$$\bar{X}_{Ua} = 196{,}300$$

and this yields

$$\bar{X}_{La} = 380{,}000 - 196{,}300 = 183{,}700$$

Hence a plan that comes close to that desired is $n = 3$, $X_{Ua} = 196{,}300$, and $\bar{X}_{La} = 183{,}700$.

The plan just derived will have a $\beta = 0.10$ as desired, but the value of α will be 0.029 instead of 0.05. If we determine $\bar{X}_{Ua}$ from the third equation instead of the first, we will get $\bar{X}_{Ua} = 195{,}650$ and $\bar{X}_{La} = 184{,}350$, instead of 196,300 and 183,700. This second plan will yield $\alpha = 0.05$, but the β will be 0.0655, instead of 0.10. The second plan, $n = 3$, $\bar{X}_{Ua} = 195{,}650$, and $\bar{X}_{La} = 184{,}350$, errs on the conservative side, while the first plan, $n = 3$, $\bar{X}_{Ua} = 196{,}300$, and $\bar{X}_{La} = 183{,}700$, errs on the liberal side. The company can take its pick.

The OC curve for the more conservative of the two plans is shown in Figure 15.3. This is computed by finding the probability of a mean of a sample of 3 falling within the limits 195,650 and 184,350. Several points on the curve are easily obtained. When $\bar{X}' = 190{,}000$, the ordinate of the

FIGURE 15.3
Operating Characteristic Curve for the Double-Limit Single-Sampling Plan $n = 3$, $\bar{X}_{La} = 184{,}350$, and $\bar{X}_{Ua} = 195{,}650$, with $\sigma' = 5{,}000$

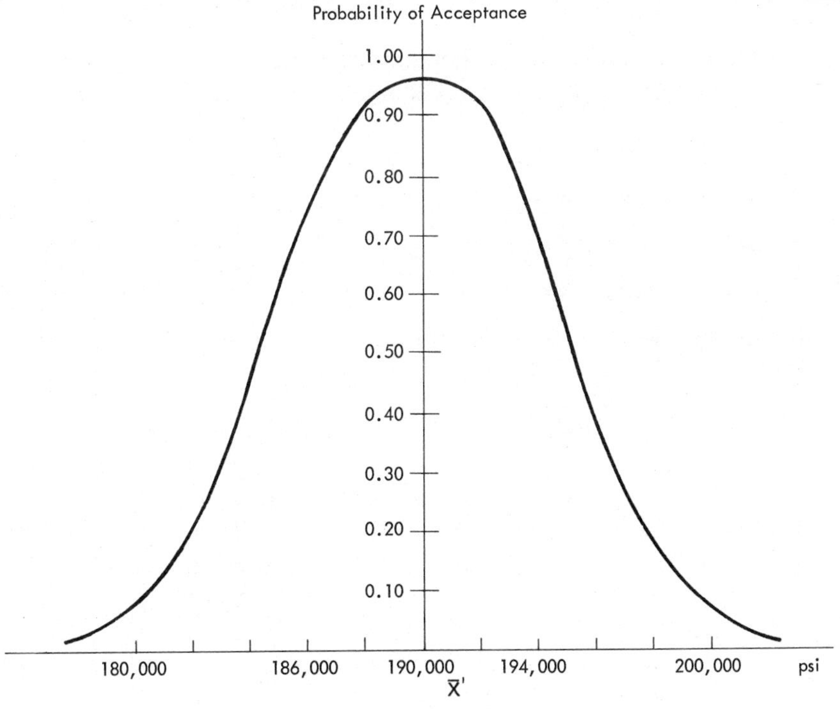

OC curve equals $1 - \alpha = 0.95$. When $\bar{X}' = 200{,}000$ or $180{,}000$, the ordinate equals 0.0655, as indicated above. When $\bar{X}' = 195{,}650$ or $184{,}350$, the ordinate equals 0.50. Other points are also symmetrical about 190,000. For example, if $\bar{X}' = 189{,}000$ we have, treating $\dfrac{\bar{X}_a - \bar{X}'}{\sigma'/\sqrt{n}}$ as a normally distributed variable,

$$\frac{184{,}350 - 189{,}000}{5{,}000/\sqrt{3}} = -1.61$$

and

$$\frac{195{,}650 - 189{,}000}{5{,}000/\sqrt{3}} = 2.30$$

Table A2 of Appendix II shows that the probability that a normally distributed variable (z) will lie between -1.61 and 2.30 is $0.9893 - 0.0537 = 0.9356$. The ordinate of the OC curve at 189,000 will thus be 0.9356. Because of the symmetry of the curve, it will also be the ordinate at 191,000.

1.1.4. Derivation of an $\bar{X}$ Single-Sampling Plan with Specified $\bar{X}'_1$, $\bar{X}'_2$, α, and β for Type A_s OC Curves. The derivation of an $\bar{X}$ plan for a Type A OC curve with small lots and known standard deviation is not a difficult matter. The procedure is exactly the same as that described in Section 1.1.2 except that the standard deviation of sample means is now taken as[4] $\sigma'\sqrt{(N-n)/n(N-1)}$ where N is the size of the lot and σ' is the lot standard deviation. The latter is assumed to be the same for all lots regardless of their mean values.

To illustrate let us use the same data as in Section 1.1.2, but assume that lots number 30 bags. The equations that will give us $\bar{X}_a$ and n are now

$$\frac{\bar{X}_a - 0.1350}{0.006\sqrt{(30-n)/n(29)}} = -1.645$$

$$\frac{\bar{X}_a - 0.1300}{0.006\sqrt{(30-n)/n(29)}} = +1.282$$

These yield an n of almost exactly 9 and an $\bar{X}_a = 0.1322$.

1.2. Sequential-Probability-Ratio Sampling Plans Based on ΣX

1.2.1. Description of an SPR Sampling Plan Based on ΣX. When sequential sampling is practicable in a plant, it may be profitable to use a sequential-probability-ratio plan based on the variable ΣX. For as noted in Chapter 8, in the neighborhood of the fitted points (here $\bar{X}'_1$ and $\bar{X}'_2$), sequential

---

[4] See J. Newman, "Contributions to the Theory of Small Samples Drawn from a Finite Population," *Biometrika,* 17 (1925), pp. 472–79.

sampling will lead to considerable saving in the amount of sampling inspection. Since $\bar{X} = \Sigma X/n$, a sampling plan based on ΣX is equivalent to a plan based on $\bar{X}$.

When a quality characteristic is normally distributed and when the standard deviation is known, a sequential-probability-ratio plan based on ΣX takes the following form. A unit is taken from a lot and a measurement X is made of the given quality characteristic. The result is plotted on a chart similar to Figure 15.4. If X falls on or above the line AB and high values of X are desirable, the lot is accepted immediately. If it falls on or below the line CD, the lot is rejected immediately. If it falls in between AB and CD, a second sample item is measured. The sum of the two measurements $(X_1 + X_2)$ is then plotted on the chart opposite $n = 2$ and the three decisions again considered. If the sum falls on or above AB, the lot is accepted; if it falls on or below CD, the lot is rejected; but if it falls between AB and CD, a third sample item is measured and the process repeated.

In a sampling plan of this kind, it is possible for the acceptance and rejection lines to be the opposite of those just described, i.e., the lower line may be the acceptance line and the upper line the rejection line. This would be true if low values of X are desirable. If both positive and negative deviations from the $\bar{X}'_1$ are undesirable, a slightly different type of plan is used. This is discussed below in Section 1.2.5.

It is to be noticed that the above plan can be used for fluids, blocks of material, and mixtures, as well as for lots of individual items. In the case of bulk material a unit would be simply a precisely defined portion of the total volume being inspected. With fluids, blocks of material, and mixtures, extra precautions might have to be taken to be assured that sampling is "at random," but otherwise the theory is valid.

1.2.2. Derivation of an SPR Sampling Plan for Specified Values of $\bar{X}'_1$, $\bar{X}'_2$, α, and β. For specified values of $\bar{X}'_1$, $\bar{X}'_2$, α, and β, and a given σ',

FIGURE 15.4
A Sequential-Probability-Ratio Sampling Chart for Σ(X − 0.1300): with a Single Lower $\bar{X}'_2$ and σ' Known

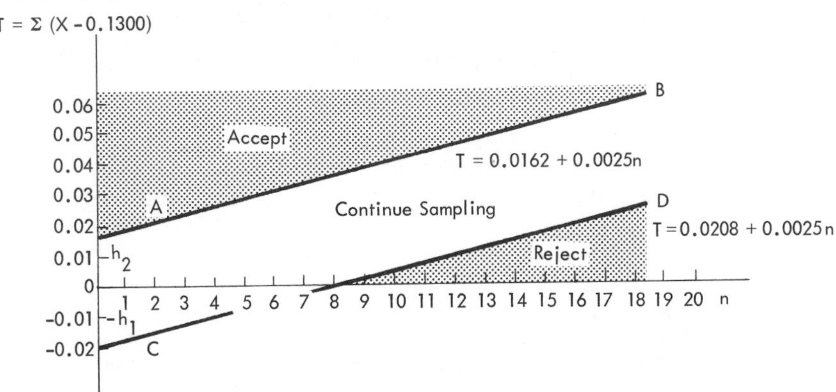

the limit lines for a sequential-probability-ratio plan based on ΣX with a single $\bar{X}'_2$ are[5]

(15.1)
$$T = h_2 + sn$$
$$T = -h_1 + sn$$

where for

A Lower $\bar{X}'_2$	An Upper $\bar{X}'_2$
$h_1 = \dfrac{b\sigma'^2}{\bar{X}'_1 - \bar{X}'_2}$	$h_1 = \dfrac{b\sigma'^2}{\bar{X}'_2 - \bar{X}'_1}$
$h_2 = \dfrac{a\sigma'^2}{\bar{X}'_1 - \bar{X}'_2}$	$h_2 = \dfrac{a\sigma'^2}{\bar{X}'_2 - \bar{X}'_1}$
(15.2) $s = \dfrac{\bar{X}'_1 + \bar{X}'_2}{2}$	$s = \dfrac{\bar{X}'_1 + \bar{X}'_2}{2}$
$a = 2.3026 \log_{10}\left(\dfrac{1-\alpha}{\beta}\right)$	$a = 2.3026 \log_{10}\left(\dfrac{1-\beta}{\alpha}\right)$
$b = 2.3026 \log_{10}\left(\dfrac{1-\beta}{\alpha}\right)$	$b = 2.3026 \log_{10}\left(\dfrac{1-\alpha}{\beta}\right)$

It will be noted that the spread between the lines, but not the slope, depends on the value of σ'.

To illustrate, let us consider again the chalk problem of Section 1.1.2. There we derived a single-sampling plan with $\bar{X}'_1 = 0.1350$ *gm/cc*, $\bar{X}'_2 = 0.1300$ *gm/cc*, $\alpha = 0.05$, $\beta = 0.10$, and it was given that $\sigma' = 0.006$ *gm/cc*. For the corresponding SPR sampling plan we would have

$$a = 2.3026 \log_{10}(0.95/0.10) = 2.2513$$

$$b = 2.3026 \log_{10}(0.90/0.05) = 2.8904$$

$$h_1 = \frac{2.8904(0.006)^2}{(0.1350 - 0.1300)} = 0.0208$$

$$h_2 = \frac{2.2513(0.006)^2}{(0.1350 - 0.1300)} = 0.0162$$

$$s = \frac{(0.1350 + 0.1300)}{2} = 0.1325$$

and the limit lines would be

---

[5] Statistical Research Group, Columbia University, *Sequential Analysis of Statistical Data: Applications* (New York: Columbia University Press, 1945), sec. 4. It will be noted that in Section 1.2, the symbol s is used for the slope of the acceptance lines and should in no way be taken to represent a standard deviation.

$$T = 0.0162 + 0.1325n$$
$$T = -0.0208 + 0.1325n$$

The above results are not graphically useful, however, since the h's are very small compared with s and the limit lines will rise steeply and be close together. It is best, therefore, to code the data. Let us substract 0.1300 from each observation. This will leave h_1 and h_2 unchanged, but will change s to s-coded = 0.0025. The SPR chart will then appear as in Figure 15.4. It is also to be noted that a tabular SPR form such as Figure 15.5 could be used instead of an SPR chart, the limits being calculated from the equations

(15.1a)
$$\Sigma(X - 0.1300) = 0.0162 + 0.0025n$$
$$\Sigma(X - 0.1300) = -0.0208 + 0.0025n$$

It is possible to code the data by subtracting s from each item. Since $s = \frac{\bar{X}'_1 + \bar{X}'_2}{2}$, this would make the coded s equal to zero and the limit lines would be horizontal, the upper line at level h_2 and the lower at level $-h_1$. In this case we would simply tabulate $(\Sigma X - s)$. If this exceeds h_2 we would accept in the case $\bar{X}'_2 < \bar{X}'_1$ (reject in the case $\bar{X}'_2 > \bar{X}'_1$); if it falls below $-h_1$, we would reject in the case $\bar{X}'_2 < \bar{X}'_1$ (accept in the case $\bar{X}'_2 > \bar{X}'_1$). In our example, we would accept if $\Sigma(X - 0.1325)$ was greater than or equal to 0.0162 and reject if it was algebraically less than or equal to -0.0208.

1.2.3. The OC Curve for an SPR Plan Based on ΣX: Single $\bar{X}'_2$. The operating characteristic curve for an SPR sampling plan based on ΣX with

FIGURE 15.5
Acceptance and Rejection Limits for the SPR Sampling Plan for $\Sigma(X - 0.1300)$ with $\bar{X}'_1 = 0.1350$, $\bar{X}'_2 = 0.1300$, $\alpha = 0.05$, $\beta = 0.10$, and σ' Given Equal to 0.006

N	*Rejection Limit*	*$T = \Sigma(X - 0.1300)$*	*Acceptance Limit*
1 . . .	−0.0188	__________	0.0187
2 . . .	−0.0158	__________	0.0212
3 . . .	−0.0133	__________	0.0237
4 . . .	−0.0108	__________	0.0262
5 . . .	−0.0083	__________	0.0287
6 . . .	−0.0058	__________	0.0312
7 . . .	−0.0033	__________	0.0337
8 . . .	−0.0008	__________	0.0362
9 . . .	0.0017	__________	0.0387
10 . . .	0.0042	__________	0.0412
11 . . .	0.0067	__________	0.0437
12 . . .	0.0092	__________	0.0462
13 . . .	0.0117	__________	0.0487
14 . . .	0.0142	__________	0.0512

a single $\bar{X}'_2$ is given by the following formulas.[6] Let $P_a(\bar{X}')$ be the probability of acceptance under the plan when the true mean of the lot is $\bar{X}'$. Also let

(15.3)
$$K = \frac{2(s - \bar{X}')}{\sigma'^2}$$

$$t_1 = h_1 K$$

$$t_2 = (h_1 + h_2)K$$

Then

(15.4)
$$P_a(\bar{X}') = \frac{e^{t_1} - 1}{e^{t_2} - 1}$$

Values of e^{t_1} and e^{t_2} can be easily obtained from tables of e^x. If these are not available, we can determine e^t from the equation

(15.5)
$$\log_{10} e^t = 0.4343t$$

A rough sketch of the OC Curve can be made from the fact that

$$P_a(\text{at } \bar{X}' = \bar{X}'_1) = 0.95$$

and

$$P_a(\text{at } \bar{X}' = \bar{X}'_2) = 0.10$$

according to the specifications of the plan and from the equation[7]

(15.6)
$$P_a(\text{at } \bar{X}' = s) = \frac{h_1}{h_1 + h_2}$$

To illustrate the computation of P_a for an intermediate point, consider the plan derived in the previous section. For $\bar{X}' = 0.1320$ we would have

$$K = \frac{2(0.1325 - 0.1320)}{(0.006)^2} = 27.8$$

$$t_1 = (0.0208)(27.8) = 0.5782$$

$$t_2 = (0.0162 + 0.0208)(27.8) = 1.029$$

$$\log_{10} e^{t_1} = 0.4343t_1 = 0.2511$$

$$\log_{10} e^{t_2} = 0.4343t_2 = 0.4469$$

$$e^{t_1} = 1.783 \qquad e^{t_2} = 2.798$$

$$P_a(\text{at } \bar{X}' = 0.1320) = \frac{1.783 - 1}{2.798 - 1} = \frac{0.783}{1.798} = 0.4355$$

---

[6] Ibid., p. 4.19.

[7] Ibid.

FIGURE 15.6
Operating Characteristic Curve for SPR Sampling Plan of Section 1.2.2 (σ' given equal to 0.006)

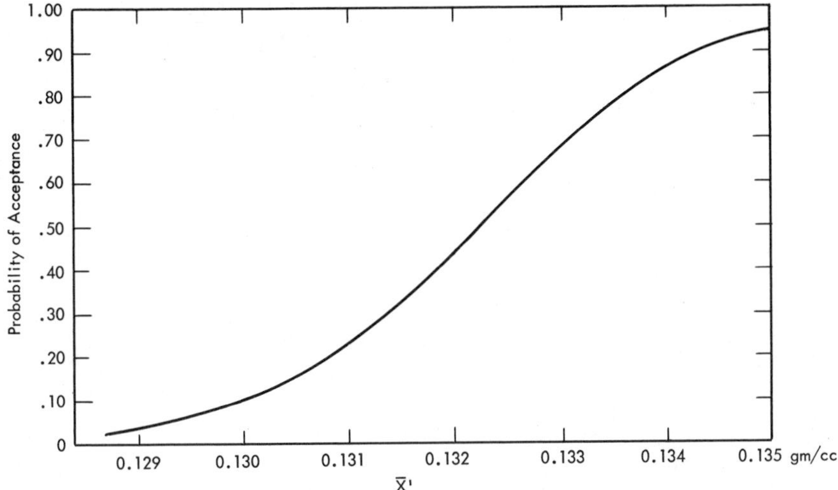

Similar computations yield Figure 15.6. This OC curve is almost the same as that of Figure 15.2, which is what we would have expected since they were both designed to have the same $\bar{X}'_1$, $\bar{X}'_2$, α, and β.

1.2.4. The ASN Curve: Single $\bar{X}'_2$. The average sample number curve for a sequential-probability-ratio sampling plan based on ΣX can be derived from the formula[8]

(15.7)
$$\text{ASN} = \frac{P_a(\text{at } \bar{X}')(h_1 + h_2) - h_1}{\bar{X}' - s}$$

Special values for the ASN, from which the curve can be sketched, are[9]

	With a Lower $\bar{X}'_2$		With an Upper $\bar{X}'_2$
$\text{ASN (at } \bar{X}'_2) =$	$\dfrac{(1-\beta)b - \beta a}{\dfrac{(\bar{X}_1 - \bar{X}'_2)^2}{2\sigma'^2}}$	$=$	$\dfrac{(1-\alpha)b - \alpha a}{\dfrac{(\bar{X}'_2 - \bar{X}'_1)^2}{2\sigma'^2}}$
(15.7a) $\text{ASN (at } s) =$	$\dfrac{h_1 h_2}{\sigma'^2}$	$=$	$\dfrac{h_1 h_2}{\sigma'^2}$
$\text{ASN (at } \bar{X}'_1) =$	$\dfrac{(1-\alpha)a - \alpha b}{\dfrac{(\bar{X}'_1 - \bar{X}'_2)^2}{2\sigma'^2}}$	$=$	$\dfrac{(1-\beta)a - \beta b}{\dfrac{(\bar{X}'_2 - \bar{X}'_1)^2}{2\sigma'^2}}$

[8] Ibid., p. 4.20.

[9] Ibid., pp. 4.10, 4.20.

For the plan derived in Section 1.2.2, we have

$$\text{ASN (at } \bar{X}' = 0.1300) = \frac{(0.90)(2.8904) - 0.10(2.2513)}{\dfrac{(0.1350 - 0.1300)^2}{2(0.006)^2}} = 6.8$$

$$\text{ASN (at } \bar{X}' = s = 0.1325) = \frac{(0.0208)(0.0162)}{(0.006)^2} = 9.4$$

$$\text{ASN (at } \bar{X}' = 0.1350) = \frac{(0.95)(2.2513) - 0.05(2.8904)}{\dfrac{(0.1350 - 0.1300)^2}{2(0.006)^2}} = 5.7$$

and for the point at which P_a was computed in the previous section,

$$\text{ASN (at } \bar{X}' = 0.1320) = \frac{(0.4355)(0.0162 + 0.0208) - (0.0208)}{(0.1320 - 0.1325)} = 9.4$$

Compare these results with Figure 15.7.

In sequential sampling by variables, there seems to be little advantage in using group sequential sampling or multiple sampling since, unlike attribute sequential sampling, a favorable or unfavorable decision is possible on each measurement.

1.2.5. A Sequential-Sampling Plan when There Are Both Upper and Lower $\bar{X}_2'$. When we have a situation such as described in Section 1.1.3 where the quality is considered less desirable if we move in either direction

FIGURE 15.7
Average Sample Number Curve for the SPR Sampling Plan of Section 1.2.2 (σ' given equal to 0.006)

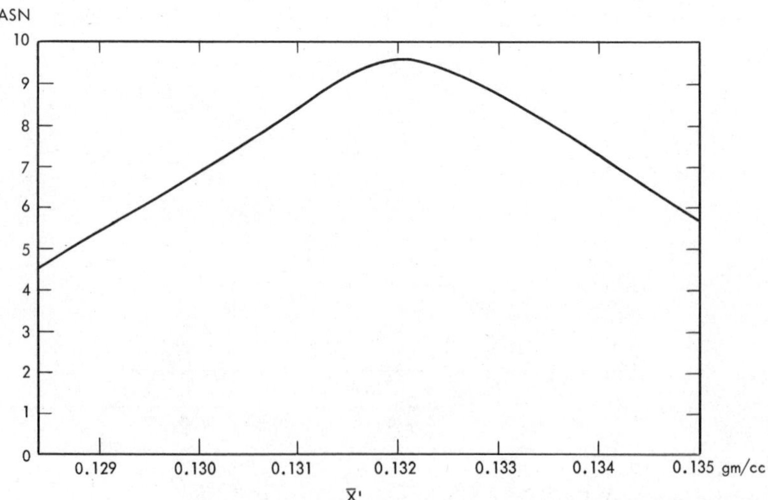

from the $\bar{X}'_1$, we proceed in a slightly different manner.[10] Instead of working now with ΣX, it is more convenient to use $\Sigma(X - \bar{X}'_1)$. Also we split the α so that the risk of falsely rejecting the lot is evenly divided between getting too high a sample result and getting too low a sample result. We then apply two separate single $\bar{X}'_2$ procedures. We set

(15.8)

$$a^* = 2.3026 \log_{10}[(1-\beta)/(\alpha/2)]$$

$$b^* = 2.3026 \log_{10}[(1-(\alpha/2))/\beta]$$

Then for each of the separate plans, we shall have

(15.9)

$$h_1 = \frac{b^* \sigma'^2}{(\bar{X}'_{U2} - \bar{X}'_{L2})/2}$$

$$h_2 = \frac{a^* \sigma'^2}{(\bar{X}'_{U2} - \bar{X}'_{L2})/2}$$

$$s = \frac{\bar{X}'_{U2} + \bar{X}'_{L2}}{4}$$

This assumes that $\bar{X}'_1$ falls precisely midway between the upper and lower $\bar{X}'_2$.

To illustrate, let us consider again the problem of Section 1.1.3. Here $\bar{X}'_1 = 190{,}000$ psi, $\bar{X}'_{L2} = 180{,}000$ psi, $\bar{X}'_{U2} = 200{,}000$ psi, $\sigma' = 5{,}000$ psi, $\alpha = 0.05$, and $\beta = 0.10$. For these data we have

$$a^* = 2.3026 \log_{10}(0.90/0.025) = 3.5835$$

$$b^* = 2.3026 \log_{10}(0.975/0.10) = 2.2773$$

$$h_1 = \frac{(2.2773)(5{,}000)^2}{(200{,}000 - 180{,}000)/2} = 5{,}693$$

$$h_2 = \frac{(3.5835)(5{,}000)^2}{(200{,}000 - 180{,}000)/2} = 8{,}959$$

$$s = \frac{200{,}000 - 180{,}000}{4} = 5{,}000$$

Although the two individual single-limit procedures will both have these same h's and s, the limit lines for each are plotted on the sequential-sampling chart inversely to each other. Call the individual procedures A and B. Then h_1 and h_2 for A are plotted normally with s positive. But h_1 and h_2 for B are plotted so that h_1 is marked off positively from zero and h_2 negatively and the limit lines are drawn with negative s. This is shown in Figure 15.8.

[10] This is a two-sided test proposed by Barnard and is slightly different from the Wald SPR test for two $\bar{X}'_2$. See references at end of this chapter. Also see Owen L. Davies (ed.), *The Design and Analysis of Industrial Experiments* (London: Oliver & Boyd, Ltd., 1954), p. 71. Irving W. Burr uses the Wald SPR procedure for approving a machine or process setting. See Cumulative List of References.

FIGURE 15.8
A Sequential-Sampling Chart with Both Upper and Lower $\bar{X}'_2$

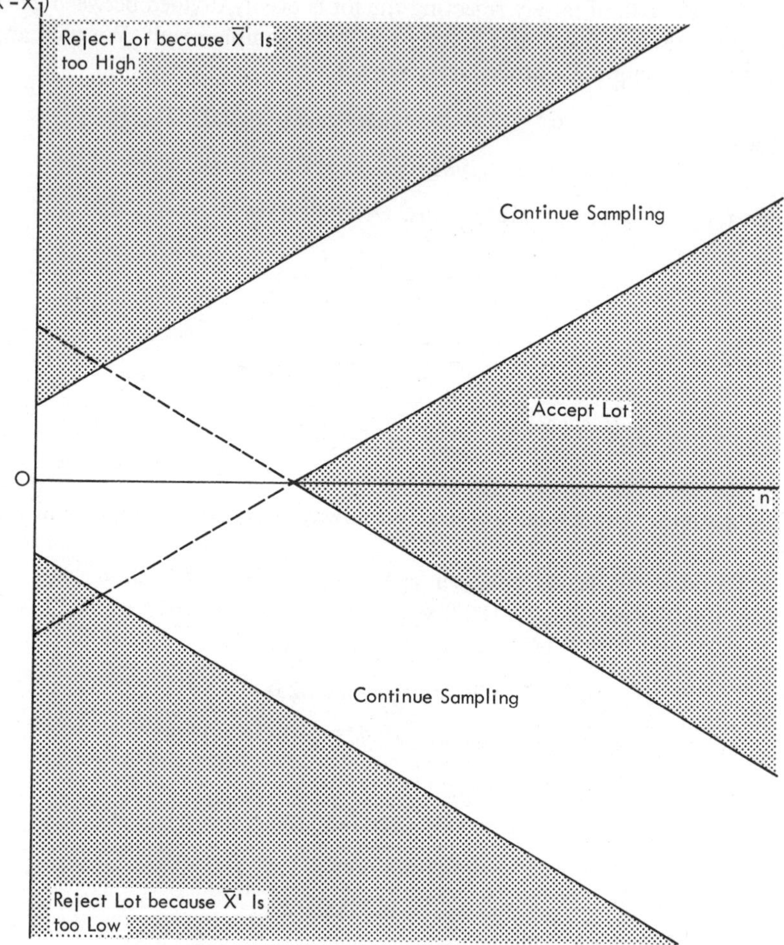

2. LOT-BY-LOT SAMPLING PLANS TO GIVE ASSURANCE REGARDING MEAN QUALITY: UNIVERSE STANDARD DEVIATION UNKNOWN

In the plans of Section 1 it was assumed that the standard deviation of the process or lot was known. Here this assumption will be removed and plans will be discussed that do not assume that the standard deviation is known. Because of lack of knowledge of the standard deviation, the plans of this section will require somewhat larger sample sizes to get OC curves equivalent to those of the plans based on a known standard deviation.

2.1. Single-Sampling Plans Based on the Statistic $\frac{\bar{X} - \bar{X}'}{s/\sqrt{n}}$

2.1.1. An Example. Consider again the chalk illustration of Section 1.1.1. The problem, it will be recalled, concerned a company that made a preliminary sample test of the chalk it was interested in buying. It was assumed in the earlier discussion that the company knew the standard deviation of bulk density and that this was constant. It will now be assumed that the company does not know the standard deviation of bulk density of the process or lot and that for Type A OC curves the standard deviation may vary from lot to lot.

Under the above conditions the company could use a sampling plan based on the statistic $t = \frac{\bar{X} - \bar{X}'_1}{s/\sqrt{n}}$ in which $s = \sqrt{\frac{\Sigma(X - \bar{X})^2}{n - 1}}$. For example, if it considers 0.1350 *gm/cc* as an acceptable quality, it could adopt the following plan. It could take a sample of 15 bags from a lot and compute both $\bar{X}$ and s. From these results it could compute the statistic $t = \frac{(\bar{X} - 0.1350)}{s/\sqrt{15}}$. If this had a positive value, or a negative value that was numerically equal to or less than 1.761, it would accept the lot. If it had a negative value that was numerically greater than 1.761, it would reject the lot.[11]

2.1.2. The OC Curve for a Plan Based on the Statistic $\frac{(\bar{X} - \bar{X}')}{s/\sqrt{n}}$. The OC curve for the above sampling plan is among those presented in Figure 15.9. This shows, for example, that if the actual average bulk density of a process or lot deviates negatively from 0.1350 *gm/cc* ($=\bar{X}'_1 = \bar{X}'_0$ of the chart) by as much as one σ', then with a sample of 15 the probability of accepting a lot is about 0.025.

The unfortunate part about the OC curve for a test using the t statistic is that the probability of acceptance is expressed as a function of standardized units and will not be of much practical use unless some estimate can be made of the process or lot standard deviation. Furthermore, according to our assumptions, with Type A OC curves the standard deviation may vary from lot to lot, so that the OC curve for the test may also vary from lot to lot. Probably the best we can do under these circumstances is to make a rough estimate as to what the average standard deviation might be for various kinds of lots. We can then use this in deriving an "average" OC curve for our tests. This curve will at least give a crude measure of the risks we run in our testing procedure and will provide a rough guide for the design of a sampling plan.

2.1.3. Derivation of a Sampling Plan Based on the Statistic $\frac{(\bar{X} - \bar{X}')}{s/\sqrt{n}}$ with a Given $\bar{X}'_1$, $\bar{X}'_2$, α, and β. To illustrate the derivation of a plan with

[11] The reader might do well at this point to review Section 5 of Chapter 6.

FIGURE 15.9

Operating Characteristics Curves for Single-Limit Sampling Plans Based on the Statistic $t = \dfrac{\bar{X} - \bar{X}'_0}{s/\sqrt{n}}$ with $\alpha = 0.05$*

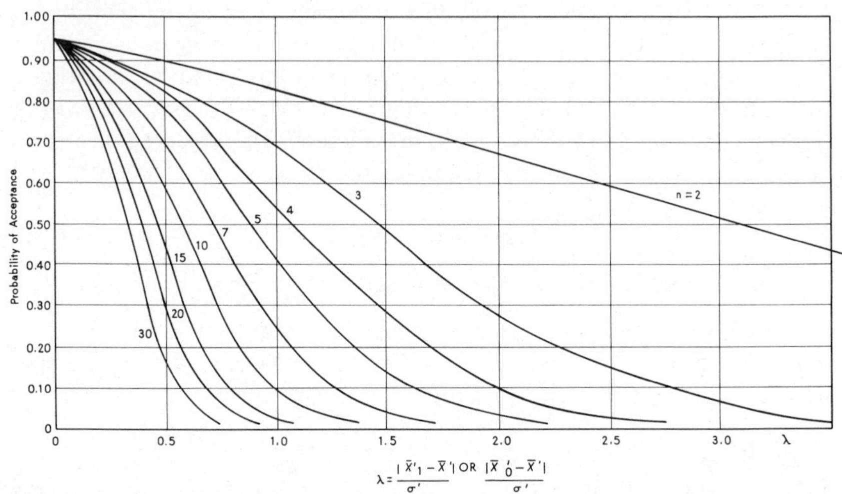

* In acceptance sampling, $\bar{X}'_0$ = the $\bar{X}'_1$ of the plan and $\bar{X}'$ = any other lot or process quality, n = size of sample. The lot or process is assumed to be normally distributed or approximately normally distributed. The sampling plan has only one acceptance limit. Source of original data: J. Neyman and B. Tobarska, "Errors of the Second Kind in Testing 'Student's' Hypothesis," *Journal of the American Statistical Association* 31, pp. 318–26.

a given $\bar{X}'_1$, $\bar{X}'_2$, α, and β, consider once more the chalk example. Suppose as previously that our company wants a sampling plan that will accept chalk of the 0.1350 *gm/cc* grade 95 times out of 100 (i.e., $\bar{X}'_1 = 0.1350$, $\alpha = 0.05$). It also wants to accept only 10 times out of 100 chalk that is of the grade 0.1300 *gm/cc* (i.e., $\bar{X}'_2 = 0.1300$, $\beta = 0.10$.) In this instance the company does not actually know the process or lot standard deviation (otherwise it would use the test of Section 1.1.2), but it makes a rough estimate that it will run about 0.005 *gm/cc* and decides to use this figure in designing its sampling plan. (It is not to be used, however, in the actual operation of the plan.)

From Figure 15.9 it is seen that the sampling plan that would meet these criteria is one for which n is about 10. The reasoning is as follows. The abscissa of Figure 15.9 at $\bar{X}' = \bar{X}'_2$ is $\lambda = \dfrac{\bar{X}'_1 - \bar{X}'_2}{\sigma'}$. For the given criteria and our estimate of σ' we have $\lambda = \dfrac{0.1350 - 0.1300}{0.005} = 1.00$, and at this value of λ we should have $P_a = 0.10$. Figure 15.9 shown that $n = 10$ comes close to satisfying these criteria. Hence the plan that would have the desired characteristic would be to take a sample of 10 and compute

$$t = \frac{(\bar{X} - 0.1350)}{s/\sqrt{10}}$$

If t has a positive value, or a negative value numerically equal to or less than 1.833, accept the lot. If it has a negative value numerically greater than 1.833, reject. (Note that 1.833 is the 0.05 point of a t distribution for $\nu = 9$.)

If the true standard deviation is greater than the assumed value of 0.005 *gm/cc*, the adopted plan will have a higher probability than 0.10 of accepting lots of grade 0.1300 *gm/cc*. For example, if the true standard deviation were 0.006 *gm/cc* (the value taken for σ' in Section 1) and not 0.005 as estimated here, then the probability of accepting a product of 0.1300 *gm/cc* quality with the sampling plan adopted here will be (as seen from Figure 15.9) approximately 0.15 instead of the desired 0.10 (note when $\sigma' = 0.006$, $\lambda = \dfrac{|0.1350 - 0.1300|}{0.006} = 0.83$). If the true standard deviation is less than the assumed value, the plan will have a lower probability than 0.10 of accepting product of grade 0.1300 *gm/cc*. In other words, if we err on the low side in our estimate of σ', our plan will be more liberal in accepting substandard product. If we err on the high side, our plan will be more conservative. It should be noted once again that the distribution of items in the process or lot is assumed to be normal, or approximately normal, in form.

2.2 Single-Sampling Plans Based on $\dfrac{\bar{X} - \bar{X}'}{\bar{R}}$ and $\dfrac{\bar{X} - \bar{X}'}{(\bar{R}/d_2^*)/\sqrt{n}}$

The plans of the previous section are based on a t statistic that uses $s = \sqrt{\dfrac{\Sigma(X - \bar{X})^2}{n - 1}}$ as a measure of variability. If the sample is small, say not greater than 12, the sample range can be used in lieu of s without much loss in sensitivity. If the sample is larger than 12, it can be divided into subsamples of equal size and the average subsample range can be used in lieu of s. The use of the range may simplify the administration of the sampling plan, but will require a larger sample to attain the same OC curve yielded by the t statistic of the previous sections. For discussion of these plans the reader is referred to the second edition of this text, pages 258–63.

2.3. Sequential-Sampling Plans

Sequential sampling with an unknown lot standard deviation follows much the same procedure as when the lot standard deviation is known. Given an $\bar{X}'_1$ and a single tolerance grade $\bar{X}'_2$, we can find a sequential plan that will have specified values for α and β. The statistic used in the plan is

(15.10)
$$U = \frac{\Sigma(X - \bar{X}'_1)}{\sqrt{\Sigma(X - \bar{X}'_1)^2}}$$

but unfortunately there is no simple formula for the limit lines. For special values, however, of $D = (\bar{X}'_1 - \bar{X}'_2)/\sigma'$ and α and β, tables have been

TABLE 15.1
Limiting Factors for a Sequential-Sampling Plan with Unknown Lot Standard Deviation, $\alpha = 0.05$, $\beta = 0.05$, and
$$D = \frac{\bar{X}'_1 - \bar{X}'_2}{\sigma'} = 1$$

n	*Lower Limit Factors*	*Upper Limit Factors*
2	(−2.14)	(2.13)
4	(−0.53)	(2.03)
6	0.03	2.04
8	0.37	2.09
10	0.63	2.16
15	1.11	2.34
20	1.47	2.52
25	1.76	2.70
30	2.02	2.88
35	2.24	3.05
40	2.45	3.21
45	2.64	3.36
50	2.82	3.50
60	3.16	3.77
70	3.45	4.03
80	3.73	4.27
90	3.99	4.49
100	4.24	4.70
150	5.27	5.65
200	6.15	6.48

SOURCE: Reproduced with permission from Owen L. Davies (ed.), *The Design and Analysis of Industrial Experiments* (London: Oliver and Boyd, 1954), p. 621.

prepared from which limits for U for each value of n can be worked out. A selection from these tables is reproduced in Table 15.1.

Let us apply a sequential test to the following set of individual measurements:

0.131	0.127	0.135	0.131	0.128	0.133	0.132
0.135	0.136	0.135	0.130	0.130	0.136	0.131
0.136	0.132	0.132	0.132	0.131	1.132	0.130
0.130	0.126	0.131	0.136	0.130	1.134	0.133
0.127	0.131	0.134	0.131	0.135	0.130	0.135

We shall assume that the order of making the measurements is from the top down and from left to right. As soon as a decision is reached, subsequent data in the set will be disregarded. With sequential sampling these surplus measurements would not have been made in the first place.

FIGURE 15.10

A Sequential-Sampling Chart for $U = \frac{\Sigma(X - \bar{X}'_1)}{\sqrt{\Sigma(X - \bar{X}'_1)^2}}$, **Single** $\bar{X}'_2$

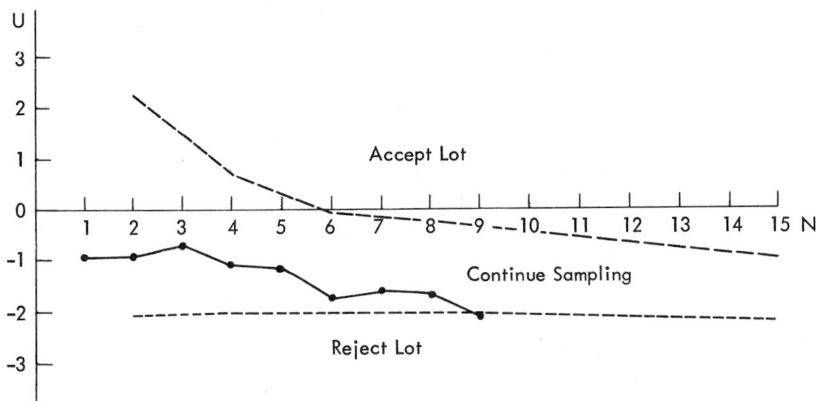

To design the test we need some knowledge of the value of σ'. Whether the value we take is close to or far away from the true value will affect the β risk and the ASN, but it will not affect the α, since the σ' itself will not be used in the test. Let us take $\sigma' = 0.005$, let $\bar{X}'_1 = 0.1350$, and let $\bar{X}'_2 = 0.1300$, which yields $D = \frac{0.1350 - 0.1300}{0.005} = 1$. Table 15.1 will then give us the limits for the sequential test, but the signs of the limits must be changed since $\bar{X}'_2$ is below $\bar{X}'_1$. The limits are shown graphically in Figure 15.10.

In tabular form the test proceeds as follows. For convenience the data are coded by multiplying by 1,000.

n	$(X - \bar{X}'_1)$	$\Sigma(X - \bar{X}'_1)$	$\Sigma(X - \bar{X}'_1)^2$	$\sqrt{\Sigma(X - \bar{X}'_1)^2}$	U
1	−4	−4	16	4	−1.00
2	0	−4	16	4	−1.00
3	+1	−3	17	4.1	−0.73
4	−5	−8	42	6.5	−1.23
5	−8	−16	106	10.3	−1.55
6	−8	−24	170	13.0	−1.85
7	+1	−23	171	13.1	−1.76
8	−3	−26	180	13.4	−1.94
9	−9	−35	261	16.2	−2.16

With this procedure the lot would be rejected after the 9th sample measurement.

If there are both and upper an lower $\bar{X}'_2$ equally spaced about the $\bar{X}'_1$, a satisfactory sequential test may be run by using two single $\bar{X}'_2$ tests and

FIGURE 15.11

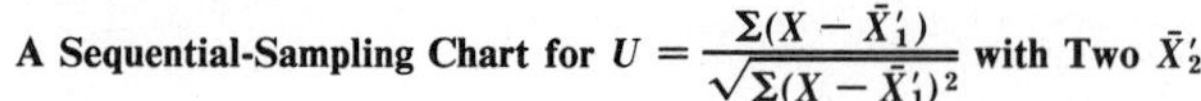

A Sequential-Sampling Chart for $U = \frac{\Sigma(X - \bar{X}'_1)}{\sqrt{\Sigma(X - \bar{X}'_1)^2}}$ **with Two** $\bar{X}'_2$

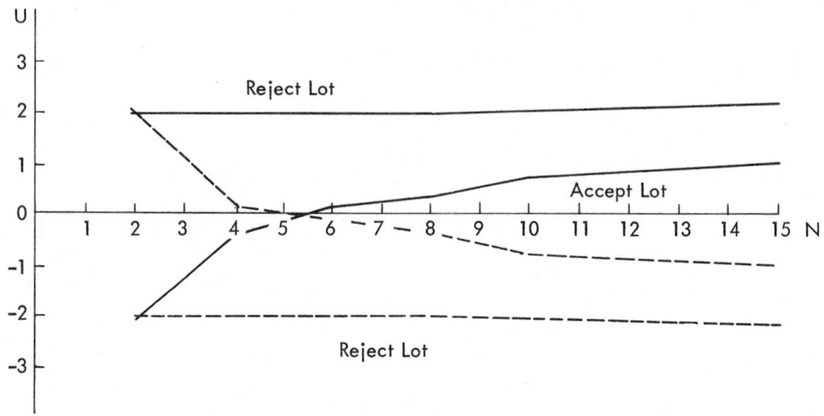

splitting the α. To be able to use Table 15.1 we would have to take $D = 1$ and an overall $\alpha = 0.10$. For the tensile strength problem of Section 1.1.3, the limit lines are shown in Figure 15.11 for $\alpha = 0.10$, $\beta = 0.05$, and $D = 1$. For other values of α, β, and D the reader is referred to the extensive table published in Owen L. Davies (ed.), *The Design and Analysis of Industrial Experiments,* 1954 ed. (pp. 617–24).

3. LOT-BY-LOT SAMPLING PLANS TO GIVE ASSURANCE REGARDING THE VARIABILITY OF A PROCESS OR LOT

3.1. Single-Sampling Plans Based on *s*

Occasionally the quality characteristic of a material that is of interest is not its average value but the variability around the average. In this case, a sampling inspection plan can be based on the statistic *s*.

Suppose, for example, that a manufacturer was willing to accept, a large percentage of the time, material the standard deviation of which was 25. In other words, 25 might be taken as the σ'_1 for his sampling plan. Suppose, further, that he did not wish to accept, more than 10 percent of the time, material whose standard deviation was 50 or more (i.e., 50 would be taken as σ'_2). Then he might adopt the following plan: He might take a sample of 10 from the lot. If the standard deviation of this sample is equal to or less than 34.3, he will accept the lot; if it exceeds 34.3, he will reject the lot. Let us consider how such a plan is derived.

To derive a sampling plan that is based on *s* and is required to have a specified σ'_1, σ'_2, α, and β, proceed as follows. Enter (or construct)[12] a

[12] See p. 361 below.

FIGURE 15.12
Operating Characteristic Curves for Single-Limit Sampling Plans Based on s, with $\alpha = 0.05$*

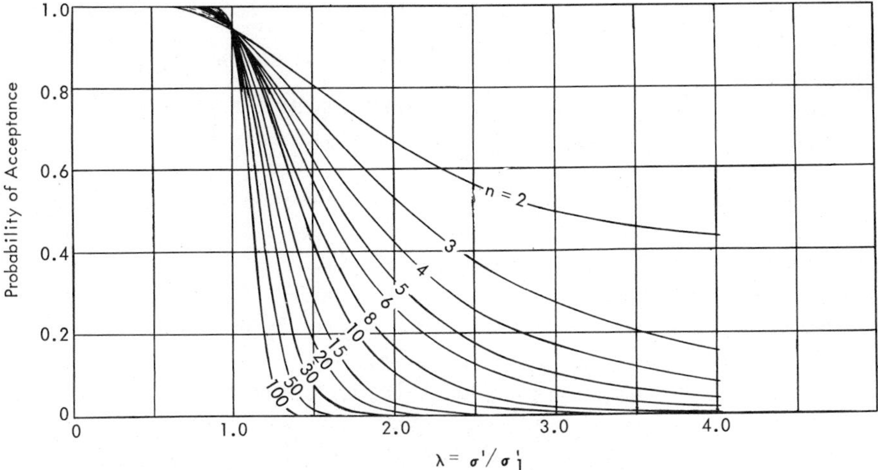

* n = size of sample. Based on the assumption that the process or lot or, in general, the "universe" is normal or approximately normal. Plan has a single acceptance limit.

SOURCE: Chart reproduced from D. C. Ferris, F. E. Grubbs, and C. L. Weaver, "Operating Characteristics of the Common Statistical Tests of Significance," *Annals of Mathematical Statistics,* 17 (1946), pp. 178–97.

chart like Figure 15.12 for the "α significance level." Since $\alpha = 0.05$ is the most commonly used "level of significance," Figure 15.12 is constructed on that basis. For a similar chart[13] based on a 0.01 level of significance, see Statistical Research Group, Columbia University, *Techniques of Statistical Analysis,* p. 275. To use Figure 15.12, proceed along a horizontal line at the β level on the P_a scale and find the intersection of this line with the vertical erected at the point $\lambda = \sigma_2'/\sigma_1'$. Find or interpolate for the n of the curve that comes the closest to this intersection point. This will be the size of the sample to be employed.[14] To find the acceptance value for the sample s enter a χ^2 table[15] with $\nu = n - 1$ and locate the $P = \alpha$ point. Call this χ^2_α. Then set $\dfrac{(n-1)s_a^2}{\sigma_1'^2} = \chi^2_\alpha$ and find s_a. This is the acceptance value for s.

To illustrate this procedure, let us show how the plan of our example was derived. Here the criteria called for $\sigma_1' = 25$, $\sigma_2' = 50$, $\alpha = 0.05$, and $\beta = 0.10$. Then $\lambda = \sigma_2'/\sigma_1' = 2$. To find the desired plan, we therefore enter Figure 15.12 at the level $P_a = 0.10$ and move to the right until we

---

[13] In this chart note that $\nu = n - 1$ when the chart is used to derive a sampling plan of the kind considered here.

[14] To be conservative, take the larger n of the two curves that bracket the intersection point.

[15] Such as Table C of Appendix II. In the discussion that follows a more extended table of χ^2 was used than Table C.

are immediately above $\lambda = 2$. We then note that the curve $n = 10$ is the one closest to our position. Hence we take 10 as the size of the sample.[16]

To find s_a we note from a χ^2 table that the $\chi^2_{0.05}$ point for $\nu = 10 - 1 = 9$ is 16.919. Hence the critical value for s is yielded by $\frac{9(s_a^2)}{(25)^2} = 16.919$ or $s_a = 34.3$. Therefore, our plan is to take a sample of 10 from a lot. If the sample s exceeds 34.3, reject the lot. If it equals or is less than 34.3, accept the lot.

Figure 15.12 presents "standardized" OC curves for tests of this kind. To find the OC curve for a particular test, we need only select the curve for the sample size used and convert the λ scale to a scale in terms of the lot standard deviation (σ'). For example, in our case we would take the curve $n = 10$ and change the λ scale to a σ' scale by multiplying λ by $\sigma'_1 = 25$.

If there is no curve in Figure 15.12 that matches the value of n of the plan or if $\alpha \neq 0.05$, so we cannot use Figure 15.12, we can always construct an OC curve as follows: To find the abscissa of the curve for which the ordinate is P_a enter a χ^2 table with the P of the table equal to $1 - P_a$ and find the χ^2 for $\nu = n - 1$. Also find from the χ^2 table the point for which $P = \alpha$ for $\nu = n - 1$. Then take λ equal to the square root of the ratio of the second χ^2 to the first. For example, let $n = 13$ and $\alpha = 0.05$. To find the abscissa point corresponding to $P_a = 0.10$, locate the column in the χ^2 table headed $P = 0.90$. Go down to the line $\nu = 13 - 1 = 12$ and note that $\chi^2_{0.90} = 6.304$. Also note from the table that $\chi^2_{0.05}$ for $\nu = 12$ equals 21.026. Then we have $\lambda^2 = 21.026/6.304$ and $\lambda = 1.83$. Again, if $n = 13$, but $\alpha = 0.01$, we find for $\nu = 12$, $\chi^2_{0.90} = 6.304$, and $\chi^2_{0.01} = 26.217$. Hence, in this case λ equals $\sqrt{26.217/6.304} = 2.04$.

3.2. Single-Sampling Plans Based on R

If we are willing to use small samples, we can adopt a sampling plan for testing variability that is similar to that described in Section 3.1 but which is based on the sample range instead of the sample standard deviation. Our plan in this case will consist of a sample size n and an acceptance limit for the sample R which we may designate as R_a.

Suppose that we wish to derive a sampling plan based on R that will have specified σ'_1, σ'_2, α, and β. To find such a plan, we use a chart like Figure 15.13 for the α level of significance. Figure 15.13 is based on $\alpha = 0.05$. To use Figure 15.13, we proceed along a horizontal line at the β level on the P_a scale and find the intersection of this line with the vertical erected at the point $\lambda = \sigma'_2/\sigma'_1$. Then we find or interpolate for the n curve that comes the closest to this intersection point. This will be the size of the sample

---

[16] $n = 11$ would have given us a more conservative plan, but our point came so close to the $n = 10$ curve that we decided to take that value.

FIGURE 15.13
Operating Characteristic Curves for Single-Limit Sampling Plans Based on R, with $\alpha = 0.05$

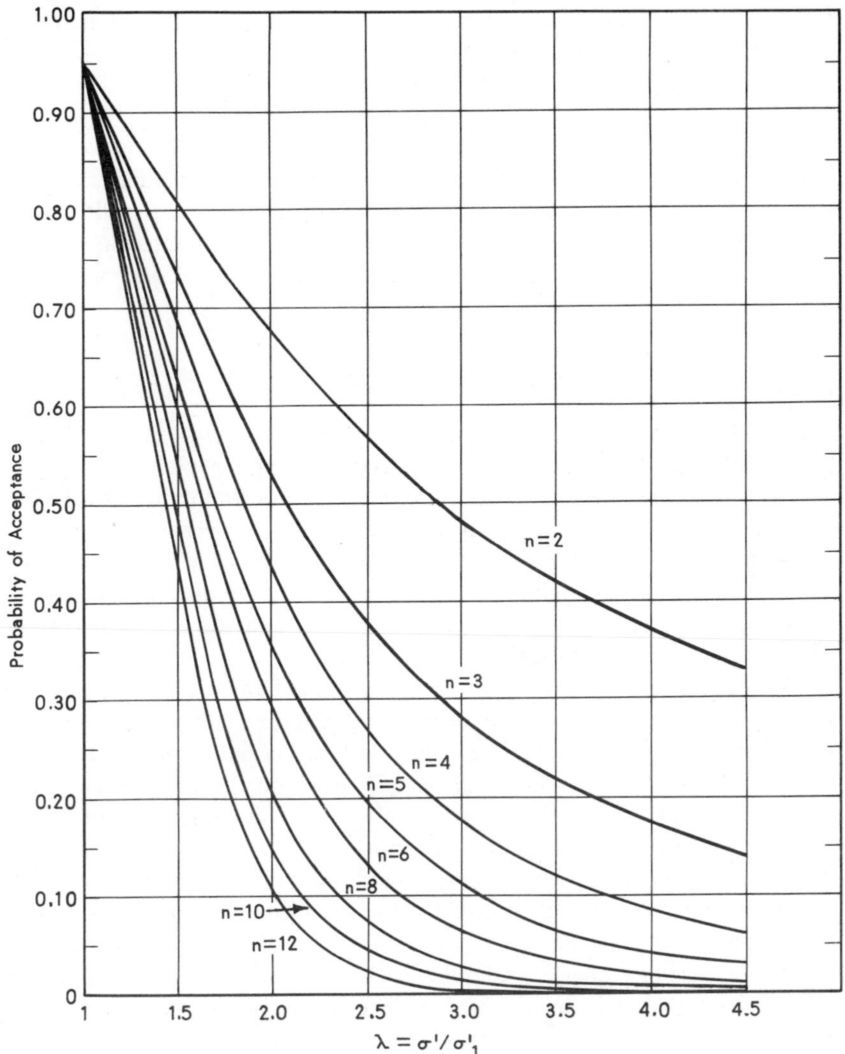

to be employed. To find the acceptance value for the sample R, enter Table D1 of Appendix II and locate the value of w for which, for the given value of n, the probability of an equal or greater value is just equal to α. When $\alpha = 0.05$, this is $w_{0.05}$. Then $R_a = (\sigma'_1)w_{0.05}$. (See Section 3 of Chapter 6).

If the sample is so large that the use of the range of the sample itself is inefficient (say $n > 12$), it is still possible to subdivide the sample into g subgroups of m each and compute an $\bar{R}/d_2^*$. The critical value for $\bar{R}/d_2^*$, call it $(\bar{R}/d_2^*)_a$, *will be found by setting* $\nu(\bar{R}/d_2^*)_a^2/\sigma'^2$ equal to the $\chi^2_{0.05}$ value for degrees of freedom equal to ν where ν is given by Table D3. In designing a sampling plan, the procedure would be to find an n from Figure 15.12. Then look for a combination[17] of g and m in Table D3 that will yield degrees of freedom close to $n - 1$.

3.3. Sequential Plans

It is also possible to derive sequential-probability-ratio sampling plans for testing variability. The plans are similar in many respects to the SPR plans described in Section 1.2 and will not be discussed here. The interested reader is referred[18] to Statistical Research Group, Columbia University, *Sequential Analysis of Statistical Data: Applications,* Section 6.

4. TIGHTENED AND REDUCED INSPECTION

It is unlikely that a program of rectifying inspection, such as described in Part 3 below, will become part of a variables sampling plan, the purpose of which is to give assurance regarding the mean or variability of a lot. In such plans, therefore, it is especially important to have provisions for tightened inspection that will put pressure on the supplier to provide higher grade material. In discussing tightened inspection, we should distinguish between situations in which there is but one $\bar{X}'_2$ (or σ'_2) and those in which there are both an upper and lower $\bar{X}'_2$ (or σ'_2).

When a single-sampling plan is employed and there is a single $\bar{X}'_2$ (or σ'_2), a provision for tightened inspection might run as follows. The program would be based on the sampling records. If the grand mean (or standard deviation) of 10 lots, say, falls significantly above (or below) the $\bar{X}'_1$ or σ'_1) and this is also true of the majority of the individual sample means (or sample s's), then the acceptance criterion ($\bar{X}_a$, t_a, or s_a) could be made more severe. If subsequently the grand mean (or standard deviation) becomes better than the acceptable quality level, return could be made to normal inspection.

In the case of sequential sampling with a single $\bar{X}'_2$, a special sequential chart could be used in which, beginning at a specified point, ΣX could be computed and plotted *over a series of lots.* Such a chart would have an α say of 0.01 or 0.001 instead of the 0.05 used for individual lots. If and when this cumulated ΣX reaches the zone of acceptance, a new ΣX would be started. If, however, it reached the region of rejection at any time, a shift would be immediately made to tightened inspection if in addition more

---

[17] Cf. Section 6 of Chapter 6.

[18] Also see Gayle W. McElrath and Jacob E. Bearman, "Sampling by Variables III: Protection on Variability," *National Convention Transactions, 1957,* American Society for Quality Control, pp. 447–62.

than half, say, of the preceding 10 lots had also been rejected. Tightened inspection would continue until the ΣX on this special chart again fell in the region of acceptance.

For two $\bar{X}'_2$ (or σ'_2), tightened inspection could be instituted with single sampling if the grand mean (or standard deviation) of say 10 lots differed significantly *in either direction* from the $\bar{X}'_1$ (or σ'_1) and more than half, say, of the last 10 lots had been rejected. Under sequential sampling, a two-sided sequential-sampling chart could be set up along the lines described in the previous paragraph for a one-sided chart, again taking the α considerably smaller than the standard 0.05 and cumulating ΣX from lot to lot.

When there is but a single $\bar{X}'_2$ (or σ'_2), it would also be possible to provide for a program of reduced inspection if the grand mean (or grand s) differed significantly from the $\bar{X}'_1$ (or σ'_1) in the direction of better quality. This would also be possible for two-sided plans in which the $\bar{X}'_2$ (or σ'_2) were relatively far apart, such that one-sided plans could be applied at either end. When the $\bar{X}'_2$ (or σ'_2) were relatively close, there would be little or no opportunity for resorting to reduced inspection.

5. PROBLEMS

15.1. The average breaking strength of cotton yarn is one measure of its quality. Let the standard deviation of the breaking strength of the particular grade of cotton yarn be 11 pounds and design a single-sampling plan based on sample averages that will yield a 0.95 chance of accepting yarn of 90-pound quality but a maximum chance of 0.10 of accepting yarn of 80-pound quality or less. Draw the OC curve for the plan. Assume a Type B curve.

15.2. Suppose that the shear strength of spot welds is found to be distributed with a standard deviation equal to 20 pounds regardless of the average shear strength. For a certain use, spot welding with an average shear strength of 375 pounds per weld is considered an acceptable quality level, while spot welding with an average shear strength of 365 pounds per weld is the lower tolerance limit. Design a single-sampling plan based on averages that will yield a 0.95 chance of accepting spot welding with an average shear strength of 375 pounds per weld but a maximum chance of 0.10 of accepting spot welding with an average shear strength of 365 pounds per weld or less. Draw the OC curve for this plan. Assume a Type B curve.

15.3. Weld shear strength is found to be correlated with weld diameter. Hence the former can be controlled by controlling the latter. Suppose that the distribution of weld diameters is found to have a constant standard deviation of 0.1 inches regardless of the average of the distribution. For a certain use, welds whose diameters average 0.6 inches are deemed to constitute an acceptable quality level. Welds whose diameters average only 0.4 inches are deemed to constitute a lower tolerance limit. Design a single-sampling plan based on averages that will yield a 0.95 chance of accepting welding the weld diameters of which average 0.6 inches but will yield a maximum chance of 0.10 of accepting welding the weld diameters of which average 0.4 inches or less. Draw the OC curve for your plan. Assume a Type B curve.

15.4. A quality characteristic of cotton yarn is the fineness of its fibers. Generally, fine fibers make for strong yarn. Very fine fibers, however, tend to reduce the rate of processing. Suppose that, for a given purpose, an average fineness of 4.9 micrograms per inch of fiber is an acceptable quality level but that an average of 3.7 or less (very fine fibers) or an average of 6.1 or more (coarse fibers) is deemed undesirable. The standard deviation of fineness for the given type of cotton is 0.7 micrograms. Design a double-limit, single-sampling plan based on sample averages that will yield a 0.95 chance of accepting cotton with average fiber fineness of 4.9 micrograms but a maximum chance of 0.10 of accepting cotton whose average fiber fineness is equal to or less than 3.7 micrograms or equal to or more than 6.1 micrograms. Draw the OC curve for this plan. Assume a Type B curve.

15.5. A whiskey distillery is required by law to have the contents of its bottles up to the quantity marked on the bottle. At the same time it is anxious not to put much more than the stated amount in each bottle. Suppose that, in a given case, bottles are marked as containing 32 fluid ounces. A statistical study shows that variations in the filling process plus variations in the dimensions of the bottles give rise to a standard deviation in contents of 0.05 ounces. The distillery sets 32.15 ounces as the lower tolerance limit for the average contents of its bottles. For its own protection, it sets 32.35 ounces as the upper tolerance limit. Bottles are tested in lots as they are filled.

Design a single-sampling plan that will have a 0.95 chance of passing as acceptable all lots of bottles whose average contents is 32.25 ounces but will have a maximum chance of 0.10 of passing as acceptable all lots whose average contents is 32.15 ounces or less or 32.35 ounces or more. Draw the OC curve for the plan. Assume a Type B curve.

15.6. Assume that in Problem 15.1 the standard deviation of the breaking strength of cotton yarn is unknown, and design a sampling plan based on $\bar{X}$ and s. Sketch the OC curve for the plan. Assume a Type B curve.

15.7. Assume that in Problem 15.2 the standard deviation of shear strength is unknown, and design a sampling plan based on $\bar{X}$ and s. Sketch the OC curve for the plan. Assume a Type B curve.

15.8. Assume that in Problem 15.3 the standard deviation of weld diameters is unknown, and design a sampling plan based on $\bar{X}$ and s. Sketch the OC curve for the plan. Assume a Type B curve.

15.9. Assume that in Problem 15.4 the standard deviation of fineness is unknown and design a sampling plan based on $\bar{X}$ and s. Sketch the OC curve for the plan. Assume a Type B curve.

15.10. Assume that in Problem 15.5 the standard deviation of contents is unknown, and design a sampling plan based on $\bar{X}$ and s. Sketch the OC curve for the plan. Assume a Type B curve.

15.11. In Problem 15.1 design a sequential-sampling plan instead of a single-sampling plan. Compute the ASN at $\bar{X}' = \bar{X}'_1$ and compare with the single-sampling plan.

15.12. In Problem 15.2 design a sequential-sampling plan instead of a single-sampling plan. Compute the ASN at $\bar{X}' = \bar{X}'_1$ and compare with the single-sampling plan.

15.13. In Problem 15.3 design a sequential-sampling plan instead of a single-sampling plan. Compute the ASN at $\bar{X}' = \bar{X}'_1$ and compare with the single-sampling plan.

15.14. In Problem 15.4 design a sequential-sampling plan instead of a single-sampling plan. Compute the ASN at $\bar{X}' = \bar{X}'_1$ and compare with the single-sampling plan.

15.15. In Problem 15.5 design a sequential-sampling plan instead of a single-sampling plan. Compute the ASN at $\bar{X}' = \bar{X}'_1$ and compare with the single-sampling plan.

15.16. In Problem 15.2 assume that the standard deviation of shear strength is unknown, but *guess* it to be equal to 10. Then design a sequential-sampling plan (σ' unknown) for the data of the problem. (Take $\beta = 0.05$.)

15.17. In Problem 15.3 assume that the standard deviation of weld diameters is unknown, but *guess* it to be equal to 0.2. Then design a sequential-sampling plan (σ' unknown) for the data of the problem. (Take $\beta = 0.05$.)

15.18. Derive a sampling plan based on the standard deviation that will have $\sigma'_1 = 0.015$, $\sigma'_2 = 0.045$, $\alpha = 0.05$, and $\beta = 0.10$. Draw the OC curve for the plan. Assume a Type B curve.

15.19. Derive a sampling plan based on the standard deviation that will have $\sigma'_1 = 1$, $\sigma'_2 = 1.50$, $\alpha = 0.05$, and $\beta = 0.10$. Draw the OC curve for the plan. Assume a Type B curve.

15.20. Derive a sampling plan based on the standard deviation that will have $\sigma'_1 = 2$, $\sigma'_2 = 2.50$, $\alpha = 0.05$, and $\beta = 0.10$. Draw the OC curve for the plan. Assume a Type B curve.

15.21. Derive a sampling plan based on the standard deviation that will have $\sigma'_1 = 1$, $\sigma'_2 = 1.50$, $\alpha = 0.01$, and $\beta = 0.10$. Draw the OC curve for the plan. Assume a Type B curve.

15.22. As a manufacturer of gauges, you wish your product to be such that the standard deviation of measurements made from it will be about 0.001. At the worst, you do not want the gauges to have a higher variability than 0.003. To check on your gauges, you select one at random from every 20 gauges and have a series of measurements made with it, the errors of which can be computed. In this sampling plan how many measurements would you make on each gauge and what would be the acceptance limit for the standard deviation of measurements?

15.23. Derive a sampling plan based on the range or average range that will have the same criteria as Problem 15.18. Sketch the OC curve for the plan. Assume a Type B curve.

15.24. Derive a sampling plan based on the range or average range that will have the same criteria as Problem 15.19. Sketch the OC curve for the plan. Assume a Type B curve.

6. SELECTED REFERENCES*

Acceptance Sampling (B '50), Armitage (B '61), Barnard (P '46), Barraclough and Page (P '59), Burr (B '76 and P '49), Curtiss (P '47), Davies (B '54), DeGroot and Nadler (P '58), Freeman (B '42), McElrath and Bearman (P '57), Mood and Graybill (B '63), Peach (B '47), and Schwartz and Kaufman (P '51).

---

* B and P refer to the Book and Periodical sections, respectively, of the Cumulative List of References in Appendix V.

Part 3

Rectifying Inspection

16

Rectifying Inspection for Lot-by-Lot Sampling

The acceptance sampling discussed in Part 2 was primarily concerned with (1) the protection that a sampling inspection program gives a consumer against accepting individual bad lots and (2) the incentive that may be given a supplier to produce acceptable quality through high rates of acceptance of good product and low rates of acceptance of poor product. No consideration was given to the *direct* effects on quality of the inspection process. In many cases, samples would be a small percentage of the lots inspected so that the direct improvements through removing nonconforming items from the sample would be negligible. In any event, the indirect effects on quality brought about by the pressure of lot rejections were deemed to be paramount.

Part 3 considers inspection programs in which inspection itself plays an important role in affecting the final quality of the outgoing product. In lot-by-lot sampling such schemes generally call for corrective inspection of rejected lots. In general such programs have the intention of correcting or eliminating through inspection, if necessary, a sufficient number of nonconforming items to attain a specified quality objective. These are called rectifying inspection programs.

Rectifying inspection programs were among the earliest of proposed sampling inspection schemes. The plans developed by Harold F. Dodge and Harry G. Romig at the Bell Telephone Laboratories before World War II[1] and subsequently published in their *Sampling Inspection Tables* were rectifying inspection plans. In more recent years Dodge's and Romig's ideas on rectifying inspection have been extended by G. A. Barnard and F. J. Anscombe.[2] Initially programs of rectifying inspection were associated with

[1] See Chapter 1.

[2] See references at end of this chapter.

lot-by-lot acceptance sampling, but in 1943 Dodge also proposed a rectifying scheme for continuous production. This chapter will be devoted to discussion of rectifying inspection for lot-by-lot sampling plans, and the next will discuss rectifying inspection for sampling from a continuous flow of production.

A rectifying inspection scheme will be of interest to a manufacturer that wishes to know about the average quality of product that is likely to result at a given stage of manufacture from the combination of production, sampling inspection, and rectification of rejected lots. It could be applied effectively to incoming inspection, in-process inspection, or final inspection to give assurance regarding the average quality of material used in the next stage of manufacture or regarding the average quality of the final product shipped to the consumer. Even if a large consumer only applies an acceptance sampling program, he may be interested in the average quality that is likely to result from the combination of his sampling inspection and the rectification procedures of his supplier. In other words, it may be of interest to view an overall inspection procedure as a single program even though various parts of the program are under separate managerial control.

1. RECTIFYING INSPECTION PLANS CALLING FOR 100 PERCENT INSPECTION OF REJECTED LOTS

Most rectifying inspection plans for lot-by-lot sampling call for 100 percent inspection of rejected lots. The criticism that was made in Chapter 7 of final 100 percent inspection as a general inspection program is not likely to apply here. In rectifying inspection, 100 percent inspection is restricted to rejected lots and this will in most cases be a small percentage of all product submitted for inspection. The volume of inspection, therefore, will likely be much less than that under final 100 percent inspection. Second, if the burden of inspecting rejected lots is placed on the "producer," whether this be the production department of the same company or an outside supplier, there will be a direct incentive to turn out a good product initially in order to hold to a minimum the amount of 100 percent inspection of the rejected product.

In the theoretical discussion that follows we shall assume for simplicity that the 100 percent inspection of rejected product is perfect inspection, just as in the theory of sampling inspection and in the discussion of probabilities and OC curves it is implied that the inspection of a sample is perfect inspection. We shall also assume that nonconforming items found during both sampling and 100 percent inspection are replaced by good ones. These assumptions will keep the analysis reasonably uncomplicated.

With 100 percent inspection of rejected lots, two features of rectifying inspection become of principal importance. One relates to the average quality of the material turned out by the combination of sampling and 100 percent inspection. The other relates to the average amount of inspection required by the program. Let us consider each of these in turn for various kinds of sampling plans.

1.1. Single-Sampling Plans for Inspection by Attributes

1.1.1. The Average Outgoing Quality Curve. The concept of an average quality that results from application of a scheme of rectifying inspection is that of an expected value—the value that would be attained in a long series of lots turned out by a process having some particular *process* average fraction nonconforming p'. In the discussion that follows it will be assumed that lot quality is binomially distributed[3] with a mean equal to this process average p'. The OC curves that will be used in the analysis will thus all be of Type B, and p' will stand for the average quality of the product turned out by the process and not for the quality of an individual lot. We shall now show how average outgoing quality is functionally related to the quality p' of incoming product.

As noted, the probability of lot acceptance for product quality equal to p' will be given by a Type B OC curve. Let this probability be called P_a. Then in lots of size N passed by the inspection plan, we will have

1. The n items taken in the sample
2. The $N - n$ items which are either
   a. Inspected 100 percent, or
   b. Accepted without further inspection

In category 1 all nonconforming items are replaced by good ones, so this category will contain no nonconforming items. The same is true of category 2*a*. In category 2*b* there will, on the average, be $p'(N - n)$ nonconforming items. Since category 2*b* will occur P_a proportion of the time, the sampling plan will, *on the average,* turn out lots that contain $P_a p'(N - n)$ nonconforming items.[4] Expressing this average outgoing quality (AOQ) as a fraction nonconforming, we have[5]

$$\text{AOQ} = \frac{P_a p'(N - n)}{N} \tag{16.1}$$

which for large lot sizes and relatively small sample sizes becomes approximately

$$\text{AOQ} = P_a p' \tag{16.1a}$$

To give a numerical illustration of the derivation of this formula, suppose that our sampling plan is $n = 100$, $c = 2$, and suppose that lots contain 10,000 items (i.e., $N = 10{,}000$). Let all rejected lots be inspected 100 percent

---

[3] See Chapter 7. It is, of course, possible to talk about an average outgoing quality with respect to a series of identical lots, all of quality p', but this is not as meaningful as assuming that p' is a process average. This second point of view is the one that will be taken in Section 2, since it simplifies the mathematics.

[4] Cf. Chapter 4, Section 4.1.

[5] If the nonconforming items are thrown out and not replaced, the formula becomes

$$\text{AOQ} = \frac{P_a p'(N - n)}{N - p'n - (1 - P_a)p'(N - n)}$$

and let all items that are found to be nonconforming be replaced by good ones. Then, if product quality is 0.05, the OC curve tells us that 12 percent of the lots will be accepted and 88 percent rejected.[6] If 1,000 lots are inspected, then, according to the theory of probability, we would expect the following: About 120 lots will be accepted and about 880 lots will be rejected. In the samples of 100 taken from each lot, all nonconforming items will be replaced by good ones. In the 9,900 items in each of the 880 lots rejected, all nonconforming items will be replaced by good ones. In the 9,900 items in each of the 120 lots accepted without further inspection, $0.05(9{,}900) = 495$ items will on the average be nonconforming. Hence, out of the total of 10,000,000 items turned out by the inspection department, about $495(120) = 59{,}400$ will be nonconforming. This is an average fraction nonconforming in outgoing lots of 0.00594.

Direct substitution in the formula yields the same result, viz

$$AOQ = (0.12)(0.05)\left(\frac{10{,}000 - 100)}{10{,}000}\right) = 0.00594$$

which is approximately equal to

$$AOQ \doteq (0.12)(0.05) = 0.0060$$

In many cases sufficiently good results can be obtained by using the approximate formula. This means that the AOQ curve of a sampling plan will be obtained approximately by multiplying the ordinate of the OC curve by the corresponding abscissa. In other words, if lots are relatively large, the ordinates of the AOQ curve are approximately the areas of the successive rectangles that can be inscribed under the OC curve.

As indicated by the above formula, the average outgoing quality will vary with the fraction nonconforming of the incoming material. If the latter is very small, most of the lots will be accepted, and the outgoing material will have a low fraction nonconforming because the incoming material had a low fraction nonconforming. On the other hand, if the incoming material has a high fraction nonconforming, most of the lots will be rejected and inspected 100 percent. Because of the latter, the outgoing quality will be good. Hence, for both low and high fractions nonconforming of incoming material, the quality of outgoing material will be good. In between, there will be a point at which the outgoing material will be at its worst. The AOQ curve pictures this relationship between incoming quality and outgoing quality. The AOQ curve for the plan $n = 100$, $c = 2$ is shown in Figure 16.1. It will be noticed that the curve shows the fraction nonconforming of the outgoing material and therefore fluctuates inversely with quality. The maximum ordinate of the curve measures the worst possible quality of material turned out by the plan and is known as the average outgoing quality limit

[6] See Chapter 7, Section 2.1.1.

FIGURE 16.1
Average Outgoing Quality Curve for the Sampling Inspection Plan $n = 100$, $c = 2$

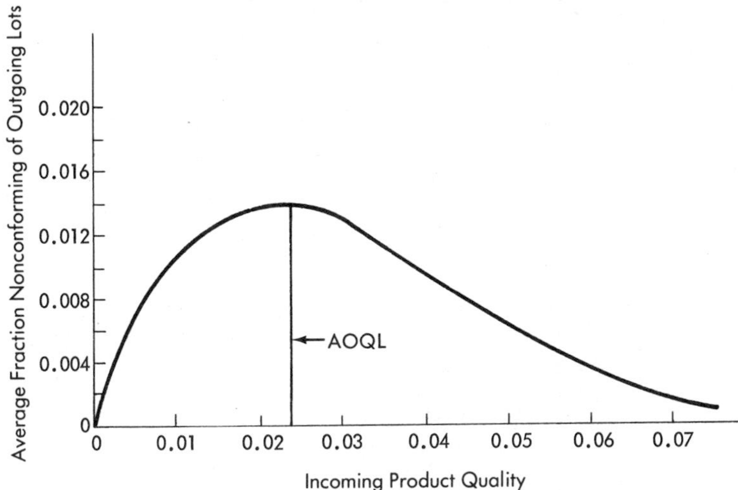

or briefly the AOQL. It is to be noted once again that this is an average value over many lots.

Table 16.1 gives[7] values of $y = \dfrac{n(\text{AOQL})}{1 - n/N}$ for various values of c. If we find the value of c for a plan and multiply the table value of y by $(1 - n/N)/n$, we get the value of the AOQL. Thus for the plan $n = 100$, $c = 2$, with $N = 10{,}000$, the AOQL is $1.372(0.99)/100 = 0.014$.

In concluding the discussion of average outgoing quality it will be noted that the AOQ and AOQ curve can be markedly affected by misclassification of items in inspection. See the 1975 paper by Case, Bennett, and Schmidt in the Cumulative List of References in Appendix D.

1.1.2. The Total Inspection Curve. In discussing the total amount of inspection called for on the average by a program of rectifying inspection, it will be assumed that rejected lots that have been screened are accepted for use or sale without further inspection. If a supplier has to do the screening of his rejected lots, it may be that the consumer will want to reinspect these screened lots before he finally accepts them. If this is done, it means the number of items sampled in this reinspection program must be added to the average amount of total inspection. This special situation is not considered in the analysis that follows, but modification can easily be made to take care of it.[8]

---

[7] See p. 376.

[8] If P_a is the probability of lot acceptance for product of original p' quality, then assuming the screening to be perfect, we need simply add to the average total inspection the quantity $n_r(1 - P_a)$ where n_r is the size of the sample used in the reinspection of resubmitted lots.

TABLE 16.1
Values of $y = P_a p'_M n$

c	$P_a p'_M n$	c	$P_a p'_M n$
0 . . .	0.3679	21 . . .	14.66
1 . . .	0.8400	22 . . .	15.43
2 . . .	1.371	23 . . .	16.20
3 . . .	1.942	24 . . .	16.98
4 . . .	2.544	25 . . .	17.76
5 . . .	3.168	26 . . .	18.54
6 . . .	3.812	27 . . .	19.33
7 . . .	4.472	28 . . .	20.12
8 . . .	5.146	29 . . .	20.91
9 . . .	5.831	30 . . .	21.70
10 . . .	6.528	31 . . .	22.50
11 . . .	7.233	32 . . .	23.30
12 . . .	7.948	33 . . .	24.10
13 . . .	8.670	34 . . .	24.90
14 . . .	9.398	35 . . .	25.71
15 . . .	10.13	36 . . .	26.52
16 . . .	10.88	37 . . .	27.33
17 . . .	11.62	38 . . .	28.14
18 . . .	12.37	39 . . .	28.96
19 . . .	13.13	40 . . .	29.77
20 . . .	13.89		

SOURCE: Reproduced with permission from Table 2–3 of the Dodge-Romig, *Sampling Inspection Tables—Single and Double Sampling* (2d ed; New York: John Wiley & Sons, 1959).

The average total amount of inspection called for under a program of 100 percent inspection of rejected lots will depend on the quality of the material submitted. If the material contains no nonconforming items, there will be no rejections, and the amount of inspection per lot will be n. If the items are all nonconforming, every lot will be submitted to 100 percent inspection, in which case the amount of inspection per lot will be N, the size of the lot. If material is between 0 nonconforming and 100 percent nonconforming, the average amount of inspection per lot will be between n and N. If the product is of quality p' and the probability of lot acceptance is P_a, then, *on the average,* the amount of inspection per lot will be

(16.2) $$\text{ATI} = n + (1 - P_a)(N - n)$$

where ATI stands for average total inspection. For example, if $N = 10{,}000$, $n = 100$, $c = 2$, and if the fraction nonconforming of material submitted for inspection (p') is $= 0.05$, then, from the OC curve, $P_a = 0.12$ and $1 - P_a = 0.88$. Hence,

$$\text{ATI} = 100 + (0.88)(10{,}000 - 100) = 8{,}812$$

If $p' = 0.01$, then $P_a = 0.92$, $1 - P_a = 0.08$, and

$$\text{ATI} = 100 + (0.08)(10{,}000 - 100) = 892$$

Of course, if a particular lot is accepted, only n items are inspected, or if a particular lot is rejected, N items are inspected. What the ATI represents is the average of n's and N's over many lots from a process of average p' quality. Curves of the average total inspection per lot are shown in Figure 16.2 for lots of size 1,000, 5,000, and 10,000. In each case $n = 100$ and $c = 2$.

FIGURE 16.2
Average Total Inspection Curves for the Sampling Plan $n = 100$, $c = 2$: Lots of 1,000, 5,000, and 10,000

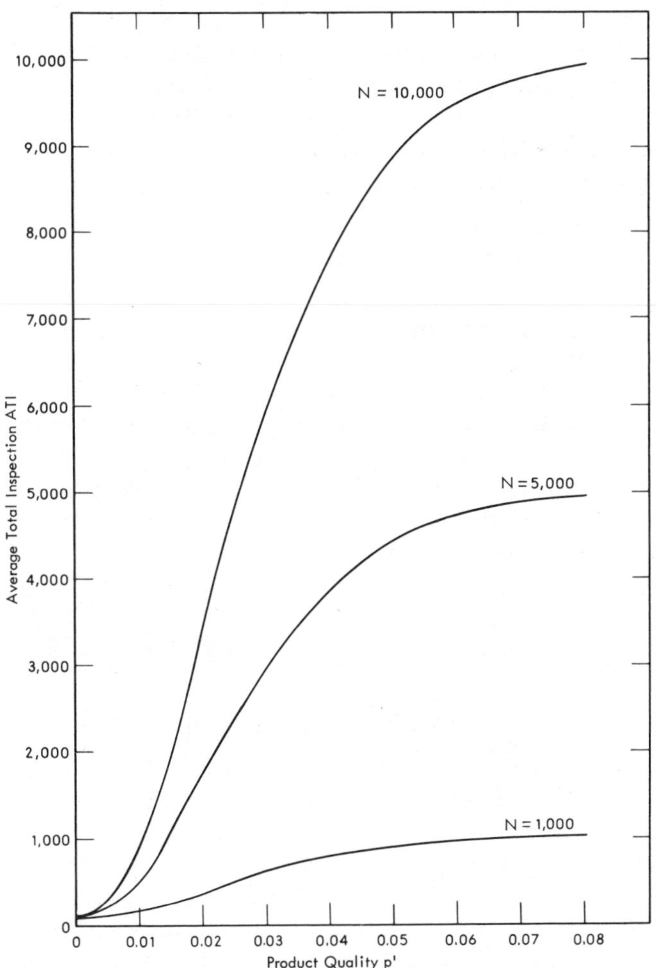

1.1.3. Design of a Rectifying Inspection Plan with a Specified AOQL. Since the AOQL of a rectifying inspection plan is likely to be its most important characteristic, it is often of interest to design a plan with a specified AOQL. Specification of this characteristic is not sufficient, however, to determine a unique plan. Frequently, therefore, plans that are designed to have a specified AOQL are in addition designed to yield minimum ATI at the most likely incoming quality level which we shall call the process average.

The theoretical procedure for finding a plan that has a specified AOQL and minimizes average total inspection for a given process average is straightforward. From equation (16.1) we have the relationship

(16.1)
$$\text{AOQ} = P_a p' \frac{(N-n)}{N}$$

Since the probability of acceptance P_a depends on the n and the c of a plan and on product quality p', once these are known, the AOQ can be found. The average outgoing quality limit (AOQL) is the maximum value we get for AOQ as p' varies. Let us designate by p'_M the value of p' which yields the AOQL. Hence we have

(16.1b)
$$\text{AOQL} = P_a p'_M \frac{(N-n)}{N}$$

where it is to be remembered that P_a is to be computed for $p' = p'_M$. If, now, the AOQL is specified, equation (16.1b) gives a relationship between n and c, since both P_a and p'_M are determined when n and c are known. Hence, to secure the solution of our problem, we can give c arbitrary values, find the corresponding values of n that satisfy equation (16.1b), and then determine the average amounts of total inspection yielded by these values of n and c. The combination which yields the smallest ATI for p' equal to the process average will be the plan that satisfies the specified criteria.

To carry out this procedure readily, we must have an easy way of solving equation (16.1) for n when we have given c an arbitrary value. This is provided by Table 16.1. For note that equation (16.1b) may be put in the form

(16.1c)
$$\text{AOQL} = \frac{y}{n}(1 - n/N)$$

where $y = P_a p'_M n$ and Table 16.1 gives values of y for various values of c. Thus, for any selected value of c, we can find y, and substitution of this in equation (16.1c) will yield the value of n.

Let us illustrate the procedure with reference to a concrete example. Suppose that we wish to find a single-sampling rectifying inspection plan that has an AOQL of 0.015 and minimizes average total inspection when the process average is 0.01. Suppose further that lots contain 5,000 items.

Let us start with $c = 2$. For $c = 2$, Table 16.1 gives $y = 1.371$. Hence,

$$\text{AOQL} = \frac{1.371}{n}(1 - n/N)$$

$$n = \frac{1.371N}{N(\text{AOQL}) + 1.371}$$

which for $N = 5{,}000$ and AOQL $= 0.015$ yields

$$n = \frac{1.371(5{,}000)}{5{,}000(0.015) + 1.371} \doteq 90$$

For $n = 90$, $c = 2$, and $p' = 0.01$, we have[9] $P_a = 0.94$. Hence

$$\text{ATI} = n + (1 - P_a)(N - n) = 90 + (0.06)(5{,}000 - 90) = 385$$

Next try $c = 4$. For $c = 4$, Table 16.1 gives $y = 2.544$. Hence

$$n = \frac{2.544(5{,}000)}{5{,}000(0.015) + 2.544} \doteq 164$$

For $n = 164$, $c = 4$, and $p' = 0.01$, we have $P_a = 0.973$. Hence

$$\text{ATI} = 164 + (0.027)(5{,}000 - 164) = 295$$

We have gained by increasing c and n, although the gain is not very great.

Now try $c = 5$. For $c = 5$, Table 16.1 gives $y = 3.168$. Hence

$$n = \frac{3.168(5{,}000)}{5{,}000(0.015) + 3.168} \doteq 203$$

For $n = 203$, $c = 5$, and $p' = 0.01$, we have $P_a = 0.983$. Hence

$$\text{ATI} = 203 + (0.017)(5{,}000 - 203) = 285$$

We continue to gain, but the amount of the gain is small, which indicates we are near the minimum point.

Try $c = 6$. For $c = 6$, Table 16.1 gives $y = 3.812$. Hence

$$n = \frac{3.812(5{,}000)}{5{,}000(0.015) + 3.812} \doteq 242$$

For $n = 242$, $c = 6$, and $p' = 0.01$, we have $P_a = 0.988$. Hence

$$\text{ATI} = 242 + (0.012)(5{,}000 - 242) = 299$$

This time we lost. The amount of inspection is higher with $n = 242$, $c = 6$, than with $n = 203$, $c = 5$. The minimum plan we are seeking is therefore $n = 203$, $c = 5$.

---

[9] This may be read from Figure 4.9 or, if greater precision is necessary, from Table E of Appendix II or from Molina's tables. It is assumed here that p' and n will be such that the Poisson distribution will give a reasonably good approximation to the binomial distribution.

1.1.4. The Dodge-Romig Sampling Inspection Tables. Fortunately, the Dodge-Romig *Sampling Inspection Tables* make it generally unnecessary to do all this work to find a plan that will minimize average total inspection for a given AOQL and a specified process average p'. A section of these tables is reproduced in Figure 16.3. In the majority of cases, we can find a good approximation to the plan we want without undertaking tedious calculations. For example, in our case the Dodge-Romig tables state that for an AOQL = 0.015 and a process average ranging from 0.0091 to 0.0120, the minimum inspection plan is $n = 205$, $c = 5$, which is practically the same as the plan that we laboriously derived above. The Dodge-Romig tables give minimum inspection plans for specified AOQL values of 0.0010, 0.0025, 0.0050, 0.0075, 0.0100, 0.0150, 0.0200, 0.0250, 0.0300, 0.0400, 0.0500, 0.0700, and 0.1000. In each case, six classes of values for the process average are specified. Figure 16.3 of this text shows part of the section of the Dodge-Romig tables for an AOQL = 0.0100.

FIGURE 16.3
Section of the Dodge-Romig Single-Sampling Lot Inspection Tables Based on Stated Values of the Average Outgoing Quality Limit.
Average Outgoing Quality Limit = 1.0 Percent

	Process Average 0.41 to 0.60%			*Process Average 0.61 to 0.80%*			*Process Average 0.81 to 1.00%*		
Lot Size	n	c	p_t%	n	c	p_t%	n	c	p_t%
1–25	All	0	. . . .	All	0	. . . .	All	0	. . . .
26–50	22	0	7.7	22	0	7.7	22	0	7.7
51–100	27	0	7.1	27	0	7.1	27	0	7.1
101–200	32	0	6.4	32	0	6.4	32	0	6.4
201–300	33	0	6.3	33	0	6.3	65	1	5.0
301–400	70	1	4.6	70	1	4.6	70	1	4.6
401–500	70	1	4.7	70	1	4.7	70	1	4.7
501–600	75	1	4.4	75	1	4.4	75	1	4.4
601–800	75	1	4.4	75	1	4.4	120	2	4.2
801–1,000	80	1	4.4	120	2	4.3	120	2	4.3
1,001–2,000	130	2	4.0	130	2	4.0	180	3	3.7
2,001–3,000	130	2	4.0	185	3	3.6	235	4	3.3
3,001–4,000	135	2	3.9	185	3	3.6	295	5	3.1
4,001–5,000	190	3	3.5	245	4	3.2	300	5	3.1
5,001–7,000	190	3	3.5	305	5	3.0	420	7	2.8
7,001–10,000	245	4	3.2	310	5	3.0	430	7	2.7
10,001–20,000	250	4	3.2	435	7	2.7	635	10	2.4
20,001–50,000	380	6	2.8	575	9	2.5	990	15	2.1
50,001–100,000	445	7	2.6	790	12	2.3	1,520	22	1.9

SOURCE: Reproduced with permission of Bell Telephone Laboratories, Inc., from H. F. Dodge and H. G. Romig, *Sampling Inspection Tables—Single and Double Sampling* (2d ed.; New York: John Wiley & Sons, 1959), p. 200.

Several things are to be noted about the Dodge-Romig tables. First, they apply strictly to programs that submit the rejected lots to 100 percent inspection. Second, for the most effective use of the tables, some knowledge must be had of the average fraction nonconforming of the incoming material. When new material is first being inspected, the quality of the incoming material will have to be determined from a preliminary sample or from data provided by the producer. What is actually needed is the average results of the manufacturing and inspection processes of the supplier. After the program has been in effect for a while, the inspection office may be able to estimate this process average reasonably well from the data it has collected. These subsequent calculations may lead to the adoption of a new plan.

Suppose, for example, that in our illustration the material supplied turns out, for the most part, to have a fraction nonconforming of 0.014 instead of 0.010, as we originally estimated it to have. The plan we adopted was $n = 205$, $c = 5$. (Our own calculations gave $n = 203$, $c = 5$, but the table values which we would normally have used are $n = 205$, $c = 5$.) For a process average of 0.014, $P_a = 0.927$ and

$$\text{ATI} = 205 + (0.073)(5{,}000 - 205) = 555$$

If we had designed our plan originally for $p' = 0.014$, we would have taken $n = 280$, $c = 7$. For a process average of 0.014, this would have yielded a $P_a = 0.954$, and an

$$\text{ATI} = 280 + (0.046)(5{,}000 - 280) = 497$$

Thus a mistake in assuming that the average fraction nonconforming of the material supplied us is 0.01 when it actually is 0.014 would mean that we would, on the average, inspect 555 items per lot instead of 497 items. The cost of our mistake is about 10 percent of total inspection costs. Of course, as soon as we become aware of our error in estimating the true product quality being supplied us, we can shift to the plan which will give us minimum cost.

Note may be made here of another point. If we are willing to allow a moderate increase in our AOQL, we can secure a considerable reduction in our inspection costs. For example, suppose that we are willing to allow an AOQL of 0.020 instead of requiring an AOQL of 0.015. Then our minimum inspection plan for a process average of 0.01 and lots of size 5,000 would be $n = 125$, $c = 4$. For $p' = 0.01$, this yields $P_a = 0.991$ and ATI $= 125 + (0.009)(5{,}000 - 125) = 169$. This is about a 40 percent reduction in inspection costs. Table 16.2 shows the relationship between the AOQL, p', and the ATI for lots of 5,000. It illustrates further the reduction in inspection costs to be obtained by slight changes in the AOQL.

1.1.5. Design of a Rectifying Inspection Plan with a Specified p_t'. Although a company may employ a sampling inspection plan that calls for 100 percent inspection of rejected lots, its interest in the AOQL may possibly be secondary to its interest in the "consumer's risk" or lot tolerance percent

TABLE 16.2
Variation in Inspection Costs with the AOQL (inspection lots of 5,000)

	Average Amount of Inspection per Lot				
AOQL	p′ = *0.0005*	p′ = *0.0020*	p′ = *0.005*	p′ = *0.01*	p′ = *0.02*
0.0010	934				
0.0025	485	855			
0.0050	257	358	823		
0.0075		202	435		
0.0100			274	676	
0.0150				287	
0.0200				169	530
0.0250					320
0.0300					198

SOURCE: Based on the Dodge-Romig tables.

nonconforming. In this case it may wish to design a rectifying plan that for lots of quality p'_t has a specified consumer's risk and at the same time minimizes the average total amount of inspection for the most likely incoming quality or process average.

It should be carefully noted at this point that the criterion relating to the lot tolerance fraction nonconforming is an individual lot criterion and the specified consumer's risk will run in terms of a Type A OC curve. On the other hand, the criterion relating to minimum inspection at a given process average is a process criterion and probabilities will run in terms of a Type B OC curve. Thus if the consumer's risk is P, the p'_t criterion requires the relationship

(16.3)
$$\sum_{X=0}^{c} C_{n-X}^{(1-p'_t)N} C_{X}^{p'_t N} \Big/ C_n^N = P$$

which makes use of the hypergeometric distribution. The minimum inspection criterion requires a minimum value for

(16.2a)
$$\text{ATI} = n + (N-n)\left[1 - \sum_{X=0}^{c} \frac{n!}{X!(n-1)!} p'^{X}(1-p')^{n-X}\right]$$

where p' is the process average and probabilities are given by the binomial distribution. The n and c that satisfy these two criteria give the desired sampling plan.

The criterion (16.3) is difficult to work with so we will use the binomial approximation to the hypergeometric given by equation (4.24). A practical substitute for (16.3) will thus be

(16.3a)
$$\sum_{X=0}^{c} \frac{(p'_t N)!}{X!(p'_t N - X)!}\left(\frac{n}{N}\right)^{X}\left(1 - \frac{n}{N}\right)^{p'_t N - X} = P$$

The values of p' are also likely to be less than 0.10 in most problems, and n is likely to be sufficiently large to use the Poisson as a reasonable approximation to the binomial distribution in (16.2a). Consequently this can be written

$$\textbf{(16.2b)} \qquad \text{ATI} = n + (N - n)\left[1 - \sum_{X=0}^{c} (p'n)^X e^{-p'n}/X!\right]$$

To illustrate the use of the above relationships to design a sampling plan let us consider another numerical example. Let the desired lot tolerance fraction nonconforming be $p_t' = 0.05$ with the consumer risk $P = 0.10$. Let $N = 500$ and let the expected value of incoming material (the "process average") equal 0.02. We shall again work by trial and error.

Let us start with a 10 percent sample or $n = 50$. With $n = 50$, $N = 500$, and $p_t'N = (0.05)(500) = 25$, the binomial probability tables suggest that c should equal 0. For if we use Table 2 of the National Bureau of Standards *Tables of the Binomial Probability Distribution,* we will find that for n of the NBS tables $= p_t'N = 25$ and p of the NBS tables $= n/N = 50/500 = 0.10$, the probability of r of the NBS tables $= c + 1$ or more is approximately 0.90 (in this case 0.928) only if $c = 0$. For $c = 0$ we have[10]

$$\text{ATI} = 50 + (450)(1 - 0.38) = 329$$

that is, for $n = 50$, $c = 0$, the average total inspection at a process average of 0.02 is 329.

Let us try next $c = 2$. The binomial tables then show that n/N should equal about 0.20 or $n = 100$. With $n = 100$, we have

$$\text{ATI} = 100 + (400)(1 - 0.67) = 232$$

which is a sizable gain.

Take next $c = 4$. For this the binomial tables show that n/N should equal about 0.30 or $n = 150$. With $n = 150$, we have

$$\text{ATI} = 150 + (350)(1 - 0.82) = 213$$

Since the gain is less, it looks as if we may be nearing the minimum.

Take $c = 5$. For this the binomial tables show that n/N should equal about 0.34 or $n = 170$. With $n = 170$, we have

$$\text{ATI} = 170 + (330)(1 - 0.87) = 213$$

These last answers agreed to the third figure so it is obvious that we are now at the minimum. A more precise calculation yields 214.6 for the former and 212.2 for the latter, so $n = 170$, $c = 5$ is apparently the minimum plan we seek. A check on $c = 6$ yields $n = 190$ and an ATI $= 218$, so n

[10] Probabilities involved in calculating the ATI are read from Figure 4.9 until we get near the minimum when the need for greater accuracy requires reference to tabular values.

= 170, c = 5 is the minimum plan. It will be noticed, however, that the plans n = 125, c = 3; n = 150, c = 4; and n = 190, c = 6 also yield ATIs that are very close to the minimum value.

The Dodge-Romig *Sampling Inspection Tables* give lot tolerance plans as well as AOQL plans. A sample table is given in Figure 16.4. This is for $p_t' = 0.05$. It will be noticed that for the data of the example given above, the Dodge-Romig tables give the minimum plan as n = 125, c = 3. As noted, the ATI for this is very close to that of the true minimum plan (215 versus 213). It will be noted that the Dodge-Romig tables for p_t' plans also give the AOQLs for these plans, just as their AOQL plans also give p_t' values, these being in this case simply the 10 percent points of the OC curves for the plans. The Dodge-Romig tables give exact minimum plans only for values of the process average that fall at the midpoints of the ranges specified and hence cannot be expected to give true minimum plans for other values

FIGURE 16.4
Section of the Dodge-Romig Single-Sampling Lot Inspection Tables Based on Stated Values of the Lot Tolerance Percent Nonconforming and Consumer's Risk = 0.10.
Lot Tolerance Percent Nonconforming = 5.0 Percent

	Process Average 0.51 to 1.00%			*Process Average 1.01 to 1.50%*			*Process Average 1.51 to 2.00%*		
Lot Size	n	c	*AOQL %*	n	c	*AOQL %*	n	c	*AOQL %*
1–30	All	0	0	All	0	0	All	0	0
31–50	30	0	0.49	30	0	0.49	30	0	0.49
51–100	37	0	0.63	37	0	0.63	37	0	0.63
101–200	40	0	0.74	40	0	0.74	40	0	0.74
201–300	70	1	0.92	70	1	0.92	95	2	0.99
301–400	70	1	0.99	100	2	1.0	120	3	1.1
401–500	100	2	1.1	100	2	1.1	125	3	1.2
501–600	100	2	1.1	125	3	1.2	150	4	1.3
601–800	100	2	1.2	130	3	1.2	175	5	1.4
801–1,000	105	2	1.2	155	4	1.4	180	5	1.4
1,001–2,000	130	3	1.4	180	5	1.6	230	7	1.7
2,001–3,000	135	3	1.4	210	6	1.7	280	9	1.9
3,001–4,000	160	4	1.5	210	6	1.7	305	10	2.0
4,001–5,000	160	4	1.5	235	7	1.8	330	11	2.0
5,001–7,000	185	5	1.7	260	8	1.9	350	12	2.2
7,001–10,000	185	5	1.7	260	8	1.9	380	13	2.2
10,001–20,000	210	6	1.8	285	9	2.0	425	15	2.3
20,001–50,000	235	7	1.9	305	10	2.1	470	17	2.4
50,001–100,000	235	7	1.9	355	12	2.2	515	19	2.5

SOURCE: Reproduced with permission of Bell Telephone Laboratories, Inc., from H. F. Dodge and H. G. Romig, *Sampling Inspection Tables—Single and Double Sampling* (2d ed.; New York: John Wiley & Sons, 1959), p. 184.

of the process average in the stated ranges. The differences, however, are not very great, as we have just seen.

1.2. Double-Sampling Plans for Inspection by Attributes

1.2.1. The AOQ Curve. If nonconforming items are replaced by good ones in both sampling and 100 percent inspection, the AOQ of a double-sampling plan for inspection by attributes that calls for 100 percent inspection of rejected lots will be given by

$$\text{AOQ} = \frac{[P_{a1}(N-n_1) + P_{a2}(N-n_1-n_2)]p'}{N} \tag{16.4}$$

For $P_{a1}(N - n_1)$ is the average or expected number of items that will be passed per lot after taking only a sample of n_1, and $P_{a2}(N - n_1 - n_2)$ will be the expected number of items passed per lot after taking a sample of $n_1 + n_2$. These items will over many lots have an average fraction nonconforming of p' so that the ratio of nonconforming to total items passed by the inspection plan has an overall average given by formula (16.4).

If N is large relative to n_1, formula (16.4) will reduce approximately to

$$\text{AOQ} = P_a p' \tag{16.4a}$$

When applied to the double-sampling plan $n_1 = 50$, $n_2 = 100$, $c_1 = 2$, $c_2 = 6$, the AOQ curve is that shown in Figure 16.5.

FIGURE 16.5
Average Outgoing Quality Curve for the Double-Sampling Plan $n_1 = 50$, $n_2 = 100$, $c_1 = 2$, $c_2 = 6$

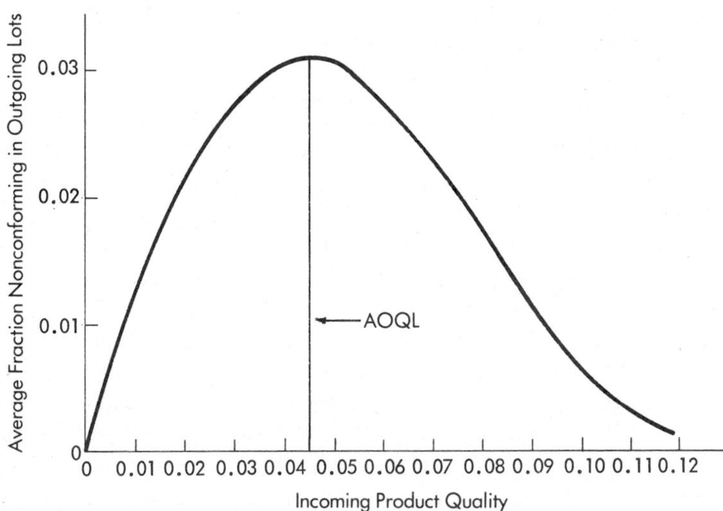

1.2.2. The Average Total Inspection Curve. Average total inspection under a double-sampling plan in which rejected lots are inspected 100 percent is given by the formula

$$\text{ATI} = n_1 P_{a1} + (n_1 + n_2) P_{a2} + N(1 - P_a) \tag{16.5}$$

in which P_{a1} is the probability of acceptance on the first sample, P_{a2} is the probability of acceptance on the second sample, P_a is the probability of final acceptance, and $1 - P_a$ is the probability of final rejection. Since $P_a = P_{a1} + P_{a2}$, the above can be written

$$\text{ATI} = n_1 + n_2(1 - P_{a1}) + (N - n_1 - n_2)(1 - P_a) \tag{16.6}$$

The reasoning underlying formula (16.5) is as follows:

1. n_1 items only will be inspected if the lot is accepted on the first sample, and the chance of this is P_{a1}.
2. $(n_1 + n_2)$ items will be inspected if the lot is accepted on the second sample, and the chance of this is P_{a2}.
3. N items will be inspected if the lot is rejected, and the chance of this is $1 - P_a$.

To illustrate the use of formula (16.6) for computing average total inspection, consider the computation of average total inspection for the plan $n_1 = 50$, $n_2 = 100$, $c_1 = 2$, $c_2 = 6$ and lot size $N = 1{,}600$. Take $p' = 0.06$. Then, from Molina's tables, we find that the probability of acceptance on the first sample (i.e., probability of 2 or less nonconforming items out of 50) is 0.423, and the probability of rejection on the first sample (i.e., probability of more than 6 nonconforming items out of 50) is 0.034. The probability of final rejection was calculated in Chapter 8 to be 0.532. Hence, for $p' = 0.06$, the average total inspection is

$$\begin{aligned}\text{ATI} &= 50 + 100(1 - 0.423) + (1{,}600 - 150)(0.532)\\ &= 50 + 58 + 772 = 880\end{aligned}$$

This gives one point on the ATI curve. The whole curve is shown in Figure 16.6. The figure also shows the ATI curve for a single-sampling plan that has approximately the same OC curve as the given double-sampling plan. It will be noticed that the double-sampling plan has less total inspection than the single-sampling plan at low and high values of the process average, while at intermediate values total inspection under single-sampling is slightly less than under double-sampling.

1.2.3. Design of Double-Sampling Rectifying Inspection Plans. The Dodge-Romig *Sampling Inspection Tables* described in Section 1.1.4 also give double-sampling plans that will have specified AOQLs and will minimize average total inspection for a given "process average." Likewise they give double-sampling plans for specified p_t''s that will again minimize average total inspection for a given process average. It will be noted that in deriving

FIGURE 16.6
Comparison of ATI Curves for Single- and Double-Sampling Plans that Have Approximately the Same OC Curves

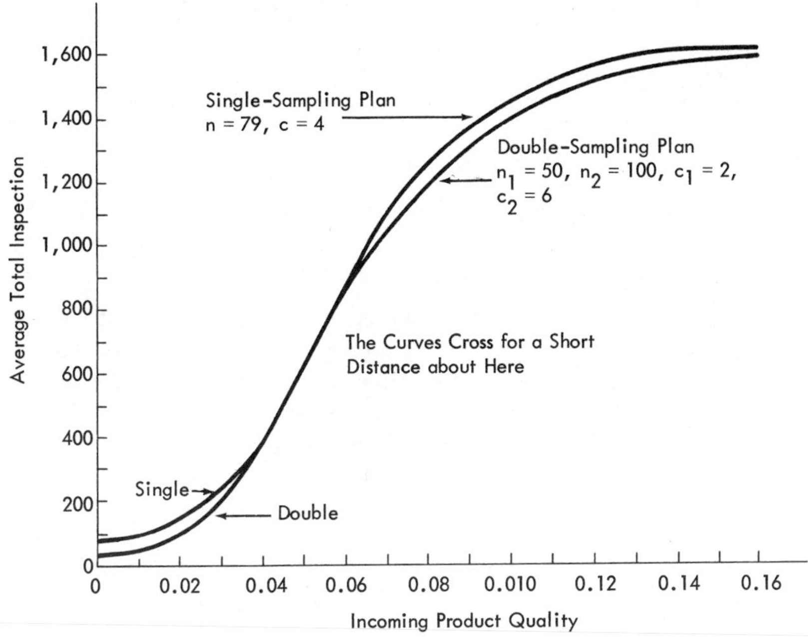

these double-sampling plans no attempt was made to keep n_1 and n_2 in any preassigned ratio. They all, however, take $c_2 = c_3$. (See Chapter 8, Section 1.)

1.3. Multiple-Sampling Plans for Inspection by Attributes

The formulas given in the previous section for double-sampling plans can be easily extended to multiple sampling. Thus the formula for the AOQ will be

(16.7) $$\text{AOQ} = \frac{[P_{a1}(N - n_1) + P_{a2}(N - n_1 - n_2) + \ldots + P_{ak}(N - n_1 - n_2 - \ldots - n_k)]p'}{N}$$

which if N is relatively large, will reduce again to

(16.7a) $$\text{AOQ} \doteq P_a p'$$

Similarly the formula for average total inspection will be

(16.8)
$$\text{ATI} = n_1 P_{a1} + (n_1 + n_2)P_{a2} + \ldots (n_1 + n_2 + \ldots n_k)P_{ak} + N(1 - P_a)$$

1.4. Item-by-Item SPR Sampling Plans

As in other cases the AOQ for an item-by-item SPR sampling plan where rejected lots are inspected 100 percent is given approximately by

(16.1a) $$\text{AOQ} \doteq P_a p'$$

The average total inspection curve will be given by the formula

(16.9) ATI (at p') =

$$\frac{P_a \log \dfrac{\beta}{1-\alpha}}{p' \log (p_2'/p_1') + (1-p') \log [(1-p_2')/(1-p_1')]} + (1-P_a)N$$

when N is the lot size, p_1' is the value of p' for which the probability of lot acceptance is $1 - \alpha$, p_2' is the value of p' for which the probability of lot acceptance is β, and P_a is the probability of lot acceptance for product of p' quality computed by formulas (8.5). The reasoning goes as follows. Formula (8.7) gives the average sample number as

$$E(n) = P_a \frac{A}{C} + (1 - P_a)\frac{B}{C}$$

where

$$A = \log \frac{\beta}{1-\alpha}$$

$$B = \log \frac{1-\beta}{\alpha}$$

$$C = p' \log \left(\frac{p_2'}{p_1'}\right) + (1-p') \log [(1-p_2')/(1-p_1')]$$

The amount of sampling n thus behaves like a random variable that takes on the value A/C when the lot is accepted and the value B/C when it is rejected. Under the rectifying inspection considered here, however, a rejected lot is inspected 100 percent. Hence, the average total inspection can be written

$$\text{ATI} = P_a \frac{A}{C} + (1 - P_a)N$$

which is formula (16.9).

1.5. Single-Sampling Plans for Inspection by Variables

The general principles developed in Section 1.1 may be applied to variables plans as well as to attributes plans. The formulas for the AOQ and ATI of single-sampling attributes plans will apply directly to the AOQ and ATI of variables plans used to control fraction nonconforming. The design of a variables plan to meet certain rectifying criteria, however, does require special discussion.

1.5.1. Design of a Variables Sampling Plan with a Specified AOQL and Minimum ATI for a Given p': Process σ' Known and Constant. Let it be assumed, as in sampling inspection by attributes, that all nonconforming items found in sampling or screening are replaced by good ones. Then, as previously, the average outgoing quality, i.e., the average fraction nonconforming of outgoing lots, will be given by

$$\text{AOQ} = P_a p' \frac{N-n}{N} \tag{16.1}$$

where p' is the average fraction nonconforming of incoming lots, N is the size of the lots, and n is the sample size.

The average outgoing quality limit is, by definition, the maximum value of AOQ. Thus, if p'_M is the value of p' at which the AOQ attains its maximum, then

$$\text{AOQL} = P_a p'_M \frac{(N-n)}{\text{N}} \tag{16.1b}$$

Let us suppose that Procedure 1 of Chapter 11 is used in carrying out the variables sampling plan. Then for a given AOQL, equation (16.1b) defines a relationship between n and k that must be satisfied by the plan, for both P_a and p'_M are functions of n and k. If an n is arbitrarily selected, the k can be found which satisfies this equation. From this n and k, a P_a can be found for the specified process average and the ATI can be found from the equation

$$\text{ATI} = n + (1 - P_a)(N - n) \tag{16.2}$$

These calculations can be repeated until the n and k are found that yield the minimum ATI.

The theory is thus straightforward enough, but the difficulty comes in finding a convenient method of deriving k from n by use of equation (16.1b). In general, this has to be done by trial and error, and the work is extensive. A special set of plans based on the work of H. G. Romig is presented in Table 16.3.

2. RECTIFYING INSPECTION PLANS WITH LESS THAN 100 PERCENT INSPECTION OF REJECTED LOTS

Sampling followed by 100 percent inspection of rejected lots is not the only possible type of rectifying inspection. Another scheme developed by F. J. Anscombe[11] offers overall AOQL protection and also lot tolerance protection

[11] F. J. Anscombe, "Tables of Sequential Inspection Schemes to Control Fraction Defective," *Journal of the Royal Statistical Society* Ser. A, 112 Part II (1949), pp. 180–206.

TABLE 16.3
Sampling Inspection Plans Based on $\bar{X}$ (with σ' Known) that Have an AOQL = 0.01 and Minimize ATI for Specified p' (plans assume the process is normally distributed)

	Process Average or p'									
	0.002		*0.004*		*0.006*		*0.008*		*0.010*	
Lot Size	n	k	n	k	n	k	n	k	n	k
100	6	2.044	8	2.024	10	2.011	12	2.004	13	2.001
500	10	2.046	15	2.050	21	2.059	29	2.069	35	2.073
1,000	11	2.052	18	2.062	28	2.081	41	2.095	50	2.103

SOURCE: Based on Table V of H. G. Romig, *Allowable Average in Sampling Inspection.*

for each lot passed without the necessity of going to 100 percent inspection. It runs as follows.

From a lot of N items, take at random an original sample equal to f_1N, where f_1 is a fraction less than 1. If there are no nonconforming items in this sample, accept the lot immediately. If there are 1 or more nonconforming items, take another random sample equal to f_2N, where f_2 is another fraction less than 1 and usually different from f_1. If there is only 1 nonconforming item in the combined samples, accept the lot; otherwise, take a third sample of f_2N. Continue on in this way taking samples of size f_2N. If at the completion of the rth sample, there are $r - 1$ or less nonconforming items in the cumulated samples, stop inspection and accept the lot; otherwise take an $(r + 1)$th sample. Continue on until either the partially rectified lot is accepted or until the whole lot has been inspected. All nonconforming items that are found are to be removed and replaced by conforming items.

Under such a plan it is possible to guarantee that the chance that the number of nonconforming items *(d)* remaining in a partially rectified lot will equal or exceed d_t will not be more than a specified β risk, and it is also possible to determine an average d and its upper limit or what amounts to an AOQL.[12] For $f_1 + f_2D < 1$, the average d is given by

$$\bar{d} = D\left[1 - \frac{f_1}{1 - f_2D} + \frac{f_1f_2}{(1 - f_2D)^3}\right] \tag{16.10}$$

where D is the number of nonconforming items in the lot, i.e., $D = p'N$, where N is the size of the lot. The maximum of $\bar{d}$ as D varies is

$$\text{Maximum average } d = \frac{(1 - \sqrt{f_1})^2}{f_2} + \frac{1}{\sqrt{f_1}} - 1 \tag{16.11}$$

[12] This AOQL is with reference to a stream of lots, all of which are of identical quality, p'. It thus stands in contrast to the AOQLs of Section 1 which were with reference to a stream of lots that are binomially distributed with mean quality p'. The assumption that D varies binomially from lot to lot would lead to a somewhat different $\bar{d}$ for the whole set of lots than that given by formula (16.10).

and the average amount of inspection is

$$\text{(16.12)} \qquad \text{ATI} = N\left[\frac{f_1}{(1-f_2D)} - \frac{f_1 f_2^2 D}{(1-f_2D)^3}\right]$$

For this scheme Anscombe gives tables of f_1 and f_2 that for specified values of $d_t = 5, 10, \ldots, 100$ and $\beta = 0.10$ and 0.01, will yield specified AOQLs. The tables also list the average amount of inspection as a function of lot quality. A selection from one of his tables is shown in Figure 16.7.

It will be seen from Figure 16.7, for example, that if we take an initial sample of 524 items from a lot of 1,000 and then succeeding samples of 46 items each, following the rectifying rules laid down above, we can guarantee with a 0.90 chance of being right that the lot contains less than 5 nonconforming items, i.e., the partially rectified lot contains less than 0.5 percent nonconforming items. Under this scheme the maximum average number of nonconforming items passed per lot will be 2. Furthermore, for lots equal to the AOQL in quality (i.e., for lots that contain only 2 nonconforming items) the average amount of inspection per lot will be 57 percent. For lots with 4 nonconforming items per lot we shall have to inspect 64 percent of each lot on the average, and for lots with 8 or more nonconforming items per lot we shall have to inspect 80 percent of each lot on the average.

The saving that may be attained by sequential rectifying inspection will be illustrated by an example. Let lots contain 1,000 items each and let the process average equal 0.008. For an AOQL of 0.01, the single-sampling plan recommended by the Dodge-Romig tables is $n = 120$, $c = 2$. With 100 percent inspection of rejected lots, this will yield a minimum average total inspection of 184 items per lot and a p_t' of 0.043. The corresponding Dodge-Romig double-sampling plan is $n_1 = 70$, $n_2 = 150$, $c_1 = 0$, $c_2 = 4$. For this the ATI, minimized at 0.008, is 158 and the p_t' is 0.038. On the other hand, Anscombe's tables show[13] that sequential rectifying inspection will guarantee an AOQL of 0.0094 with a p_t' of 0.040 and average total inspection

FIGURE 16.7
A Selection from F. J. Anscombe's Tables of Sequential Rectifying Inspection Plans

Scheme		*Average Percentage of Lot Inspected, for* D *Equal to*									*Max. Av.* d =
f_1	f_2	D = 0	*1*	2	3	4	5	6	8	*10*	*AOQL*
0.3690	0.1900	*0.37*	*0.44*	0.54	0.67						1.4
0.4238	0.0982	0.42	0.47	*0.52*	*0.58*	0.65	0.74				1.8
0.4773	0.0639	0.48	0.51	0.54	0.58	*0.63*	0.68	0.74	0.89		1.9
0.5241	0.0459	0.52	0.55	0.57	0.60	0.64	0.67	0.71	0.80	0.92	2.0

Guarantee: Chance is at most 0.10 that 5 or more nonconforming items remain in any lot after inspection.

SOURCE: F. J. Anscombe, "Tables of Sequential Inspection Schemes to Control Fraction Defective," *Journal of the Royal Statistical Society* Ser. A, 112 Part II (1949), p. 202. Italicized figures are the minima for the various lot qualities.

[13] Anscombe, "Tables of Sequential Inspection Schemes," p. 203.

will be only 133 items per lot. Thus the sequential rectifying inspection gives a better guarantee at a saving of 28 percent in inspection costs compared with Dodge-Romig single sampling and almost as good a guarantee at a saving of 16 percent compared with Dodge-Romig double sampling.

3. PROBLEMS

16.1. Find a single-sample fraction-nonconforming sampling inspection plan that has an AOQL = 0.0075 and for lots of 7,000 minimizes total inspection when the process average is 0.005. What quality product has a 0.95 chance of lot acceptance? A 0.10 chance?

16.2. Find a single-sample fraction-nonconforming sampling inspection plan that has an AOQL = 0.02 and for lots of 15,000 minimizes total inspection when the process average is 0.01. What quality product has a 0.95 chance of lot acceptance? A 0.10 chance?

16.3. Find a single-sample fraction-nonconforming sampling inspection plan that has an AOQL = 0.02 and for lots of 1,500 minimizes total inspection when the process average is 0.01. What quality product has a 0.95 chance of lot acceptance? A 0.10 chance? How does the "sampling ratio" (ratio of sample size to lot size) for this plan compare with that found in Problem 16.2?

16.4. Find a single-sample fraction-nonconforming sampling inspection plan that has an AOQL = 0.035 and for lots of 15,000 minimizes total inspection when the process average is 0.01. What quality product has a 0.95 chance of lot acceptance? A 0.10 chance?

16.5. Find a single-sample fraction-nonconforming sampling inspection plan that has an AOQL = 0.06 and for lots of 1,500 minimizes total inspection when the process average is 0.01. What quality product has a 0.95 chance of lot acceptance? A 0.10 chance?

16.6. *a.* Suppose that the true process average is 0.02. How much inspection would you have to do under the plan derived in Problem 16.2? How much does this differ from the amount of inspection you would have obtained if you had designed your plan so as to minimize total inspection at $p' = 0.02$ instead of 0.01?

b. How much would you save in total inspection if in Problem 16.2 you were willing to allow an AOQL of 0.025 instead of 0.02?

16.7. *a.* Suppose that the true process average is 0.005. How much inspection would you have to do under the plan derived in Problem 16.3? How much does this differ from the amount of inspection you would have obtained if you had designed your plan so as to minimize total inspection at $p' = 0.005$, instead of 0.01?

b. How much would you save in total inspection if in Problem 16.3 you were willing to allow an AOQL of 0.03 instead of 0.02?

16.8. Given $p_t' = 0.07$, find a single-sample fraction-nonconforming sampling inspection plan that will have $P_a = 0.10$ when $p' = 0.07$ and for lots of 8,000 minimizes total inspection when the process average is at 0.02. What is the AOQL of your plan? What quality lots have a 0.95 chance of being accepted?

16.9. Given $p_t' = 0.10$, find a single-sample fraction-nonconforming sampling inspection plan that will have $P_a = 0.10$ when $p' = 0.10$ and for lots of 5,000 minimizes total inspection when the process average is 0.005. What is the AOQL of your plan? What quality lots have a 0.95 chance of being accepted?

16.10. Given $p_t' = 0.05$, find a single-sample fraction-nonconforming sampling inspection plan that will have $P_a = 0.10$ when $p' = 0.05$ and for lots of 7,000 minimizes total inspection when the process average is 0.01. What is the AOQL of your plan? What quality lots have a 0.95 chance of being accepted?

16.11. Given $p_t' = 0.06$, find a single-sample fraction-nonconforming sampling inspection plan that will have $P_a = 0.10$ when $p' = 0.06$ and for lots of 10,000 minimizes total inspection when the process average is 0.01. What is the AOQL of your plan? What quality lots have a 0.95 chance of being accepted?

16.12. Given $p_t' = 0.025$, find a single-sample fraction-nonconforming sampling inspection plan that will have $P_a = 0.10$ when $p' = 0.025$ and for lots of 10,000 minimizes total inspection when the process average is 0.01. What is the AOQL of your plan? What quality lots have a 0.95 chance of being accepted?

16.13. Suppose in Problem 16.8 that the true process average is 0.01 instead of the expected 0.02. How much will your total inspection actually be? How much will this differ from the total inspection you would have obtained if you had originally designed your plan to give minimum inspection for $p' = 0.01$?

16.14. Suppose in Problem 16.9 that the true process average is 0.01 instead of 0.005 as expected. How much will your total inspection actually be? How much will this differ form the total inspection you would have obtained if you had originally designed your plan to give minimum inspection for $p' = 0.01$?

16.15. Given a lot tolerance number of nonconformities per 100 units = 7. Find a single-sample nonconformities-per-unit sampling inspection plan that has $P_a = 0.10$ for $u' = 7$ and for lots of 8,000 minimizes total inspection when the "process average" is 2 nonconformities per 100 units. (*Hint:* This is worked in exactly the same way as Problem 16.8 and has the same answer except that c represents the maximum total number of allowable nonconformities. Use Dodge-Romig tables with $p_t' = 0.07$ and $p' = 0.02$.) Assume that rejected lots are inspected 100 percent and that all nonconformities are repaired or removed or nonconforming units replaced by effective units and compute the AOQL for your plan in terms of average number of defects per 100 untis. What quality product has a 0.95 chance of lot acceptance?

16.16. Given a lot tolerance nonconformities per 100 units = 10, find a single-sample nonconformities-per-unit sampling inspection plan that has $P_a = 0.10$ for $u' = 10$ and for lots of 5,000 minimizes total inspection when the process average is 0.5 nonconformities per 100 units. Assume that rejected lots are inspected 100 percent and all nonconforming units are made conforming, and compute the AOQL for your plan. What quality product has a 0.95 chance of lot acceptance? (Compare Problem 16.9.)

16.17. Find a single-sample nonconformities-per-unit sampling inspection plan that has an AOQL of 2 nonconformities per 100 units and for lots of 15,000 minimizes total inspection when the process average is 1 nonconformity per 100 units. (*Hint:* Use the Dodge-Romig tables with AOQL = 0.02 and $p' = 0.01$.) What quality product has a 0.95 chance of lot acceptance? A 0.10 chance?

16.18. Find a single-sample nonconformities-per-unit sampling inspection plan that has an AOQL of 4 nonconformities per 100 units and for lots of 10,000 minimizes total inspection when the process average is 2 nonconformities per 100 units. What quality product has a 0.95 chance of lot acceptance? A 0.10 chance?

16.19. For lots of 5,000 make graphs of the average total inspection curves for the double-sampling plans of *(a)* Problem 8.1; *(b)* Problem 8.2.

16.20. Given $p_t' = 0.07$, find a double-sampling plan that has a $P_a = 0.10$ for $p' = 0.07$ and minimizes total inspection for lots of size 11,000 and a process average of 0.02. What is the AOQL of this plan? Find a single-sampling plan that has the same $p_{0.10}'$ and also has minimum total inspection for lots of size 11,000 and a process average of 0.02.

16.21. Given $p_t' = 0.05$, find a double-sampling plan that has $P_a = 0.10$ for $p' = 0.05$ and minimizes total inspection for lots of size 5,000 and a process average equal to 0.01. What is the AOQL of this plan? Find a single-sampling plan that has the same $p_{0.10}'$ and also minimizes total inspection for lots of size 5,000 and a process average of 0.01.

16.22. Find a double-sampling plan that has an AOQL = 0.01 and minimizes total inspection for lots of size 11,000 and a process average of 0.005. What is the $p_{0.10}'$ of the plan? Find the single-sampling plan that has the same AOQL and minimizes total inspection for lots of size 11,000 and a process average of 0.005.

16.23. Find a double-sampling plan that has an AOQL = 0.02 and minimizes total inspection for lots of size 11,000 and a process average of 0.01. What is the $p_{0.10}'$ of the plan? Find the single-sampling plan that has the same AOQL and minimizes total inspection for lots of size 11,000 and a process average of 0.01.

16.24. Compute the AOQ for the multiple-sampling plan of Problem 9.1 at the point $p' = 0.10$. Assume lots of size 8,000.

16.25. Compute the AOQ for the multiple-sampling plan of Problem 9.2 at the point $p' = 0.005$. Assume lots of size 5,000.

16.26. Sketch the AOQ curve for the sequential-sampling plan of *(a)* Problem 8.11 and *(b)* Problem 8.12. Assume lots of size 10,000.

16.27. Compute the ordinate of the average total inspection curve of the multiple-sampling plan of Problem 9.1 at the point $p' = 0.10$. Take $N = 8{,}000$.

16.28. Compute the ordinate of the average total inspection curve of the multiple-sampling plan of Problem 9.2 at the point $p' = 0.005$. Take $N = 5{,}000$.

16.29. For lots of 10,000 make a graph of the average total inspection curve for the sequential-sampling plan of *(a)* Problem 8.11 and *(b)* Problem 8.12.

16.30. A given quality characteristic has a lower specification limit of 900. The lot size is 500, and the items in the lot are normally distributed. The standard deviation is about the same from lot to lot and is estimated at 100. The lot fraction nonconforming usually runs about 0.004. Find a sampling plan based on $\bar{X}$ that will have an AOQL of 0.01 and will minimize total inspection when the process fraction defective is 0.004.

16.31. A given quality characteristic has an upper specification limit of 200. The lot size is 1,000, and the items in the lots are normally distributed. The standard deviation is about the same from lot to lot and is estimated at 20. The lot fraction nonconforming usually runs about 0.008. Find a sampling plan based on $\bar{X}$ that will have an AOQL of 0.01 and will minimize total inspection when the process fraction nonconforming is 0.008.

16.32. You wish to use "sequential" rectifying inspection. You want to be able to guarantee with a 0.90 chance of being right that less than 5 nonconforming items remain in the lot when screening is stopped and the remainder of the lot is accepted. Determine the initial fraction of the lot (f_1) that will be inspected and the subsequent fractions (f_2) such that for a process average of 2 nonconforming items per lot, the average amount of inspection will be a minimum. State what this minimum average inspection will be if the process average of 2 per lot is realized. What will the average amount of inspection be if the process average should actually be 4 nonconforming items per lot?

4. SELECTED REFERENCES*

Anscombe (P '49 and P '61), Bowker and Lieberman (B '55), Burr (B '76), Case, Bennett and Schmidt (P '75), Cowden (B '57), Dodge and Romig (B '59), Grant and Leavenworth (B '80), Hall (P '79), Sackrowitz (P '75), Steck and Owen (P '59), and Wortham and Mogy (P '70).

---

* B and P refer to the Book and Periodical sections, respectively, of the Cumulative List of References in Appendix V.

17

Sampling Plans for Continuous Production

1. DIFFICULTIES WITH LOT-BY-LOT PLANS WHEN PRODUCTION IS CONTINUOUS

The sampling plans discussed in previous chapters were generally lot-by-lot plans. In Part 2 they were systematic arrangements for making decisions regarding the acceptance or rejection of lots. In Chapter 16 of Part 3 the additional rectifying procedures discussed were also based on lots. In all these plans lotting was an essential ingredient. Many assembly and other manufacturing operations, however, may be conveyorized or performed on a continuous moving line. When applied to continuous production of this type in which lots are not normally created as a natural result of the production process, lotting creates special problems.[1]

Under conditions of continuous production, lotting requires either that items be allowed to accumulate at given points of production or that a "lot" be artifically marked off as a given segment of the items in the production line or of the items about to be produced. The first procedure, which creates banks of items at various production points, requires extra space, increases inventories, and in the case of explosive material produces additional safety hazards. With the second procedure various special difficulties may arise. If a continuous flow of product is arbitrarily marked off into "lots," it is possible in a lot-by-lot plan to reject material not yet produced. If a lot is ultimately rejected and a 100 percent inspection of the lot is called for, it may be necessary to recall product from subsequent operations. Disassembly,

[1] See, for example, United States Department of the Navy, Bureau of Ordnance, NAVORD OSTD 81, *Sampling Procedures and Tables for Inspection by Attributes on a Moving Line* (1952), p. 1; and L. Storer, "Continuous Sampling Inspection," *1956 Middle Atlantic Conference Transactions,* American Society for Quality Control, p. 16.

sometimes partially destructive, may be necessary. These and other difficulties with lotting may interfere with efficient production operations.

2. CONTINUOUS-SAMPLING PLANS FOR INSPECTION BY ATTRIBUTES

For the above reasons, special sampling-inspection plans have been developed for application to continuous production. These are commonly called continuous-sampling plans in contrast to "lot-by-lot" plans. They are all "rectifying" in character in that through partial screening the quality of the product is improved.

2.1. The Dodge Continuous-Sampling Plans

Continuous-sampling plans were first proposed by Harold F. Dodge of the Bell Telephone Laboratories. Writing in the *Annals of Mathematical Statistics* in 1943,[2] he described the following scheme for application to nondestructive inspection on a go-no-go basis of a continuous flow of individual units of product offered to the inspector in order of production.

2.1.1. CSP-1. Dodge's initial plan has come to be called CSP-1. At the start of this plan all product is inspected 100 percent.[3] As soon as i consecutive units of product are found to be free of nonconformities, 100 percent inspection is discontinued and only a fraction (f) of the units are inspected. These individual sample units are to be selected one at a time at random from the flow of product. If a sample unit is found nonconforming, reversion is immediately made to 100 percent inspection and the cycle is completed. All nonconforming units found are corrected or replaced with good ones.

There may be associated with each CSP-1 plan an overall AOQL, the specific value, of course, depending on the values of i and f. The same AOQL can be attained by different i and f combinations. To find the i and f combination that will yield a stipulated AOQL, Dodge prepared a special chart that is reproduced in Figure 17.1. This charts shows, for example, that an AOQL of 1 percent will be yielded by a plan for which $f = 0.05$ and $i = 150$. The same AOQL will be yielded by a plan for which $f = 0.25$ and $i = 60$.

In 1961 John S. White published a chart for CSP-1 plans from which values of i and f can more easily be read for a specified AOQL than they can be read from the Dodge chart. This is reproduced here as Figure 17.2. In the case of both the Dodge and White charts it is assumed that in computing the AOQL, nonconforming items discovered in inspection are replaced by conforming items. A third chart, very similar to White's, that does not allow for replacement of nonconforming items in the computation of the AOQL

---

[2] Vol. 14 (1943), pp. 264–79. He also applied it to continuous flows of "sub-lots."

[3] The plan could begin with sampling inspection.

FIGURE 17.1
Curves for Determining Values of f and i of a Continuous-Sampling (*CSP*-1) Plan for a Given AOQL

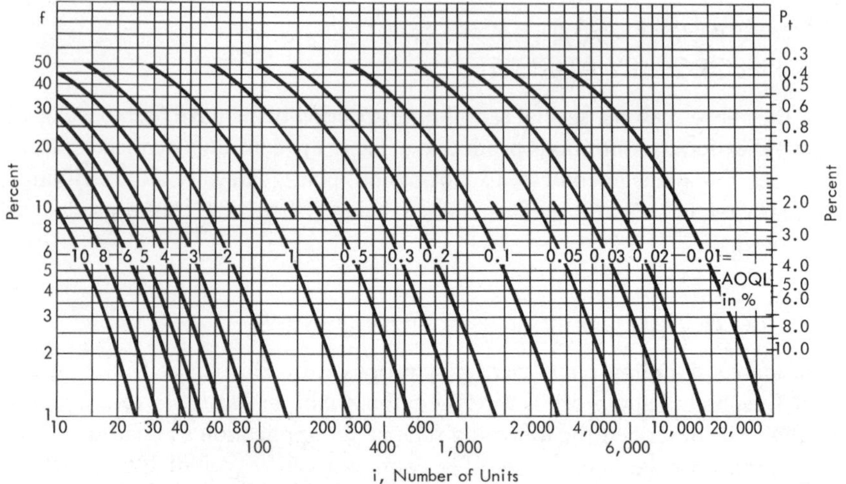

p_t in percent = the value of percent nonconforming, in a consecutive run of N = 1,000 product units for which the probability of acceptance, P_a, is 0.10 for a sample size of f percent.

SOURCE: Reproduced from the *Annuals of Mathematical Statistics,* 14 (1943), p. 274.

was presented in 1970 by F. L. Abraham in a Quality Evaluation Memorandum, "A New Graphical Method of Parameter Selection for CSP-1, CSP-2 and CPS-R," issued by the Concepts Branch, U.S. Ammunition Procurement and Supply Agency, Joliet, Illinois. Table 17.3 below, which is derived from the White chart, gives numerical values of i and f for selected nominal AOQLs.

The actual selection of f and i is usually based on practical considerations. For example, if the inspector is at the end of the production line, it may be best to have i no greater than some small multiple of the number of units on the production line at any one time. Also f may be influenced by the ordinary work loads of the inspector and operators on the line.[4] Dodge points out, however, that the protection against spotty quality in a continuous run of product becomes quite poor if f is less than 2 percent.[5] In the operation of the plan, Dodge suggests that the best results may be obtained by putting the burden of 100 percent inspection on the production department, while leaving the sampling inspection to the ordinary inspectors.[6] This gives a constant load for the inspection department.

---

[4] See H. F. Dodge, "A Sampling Inspection Plan for Continuous Production," *Annals of Mathematical Statistics* 14 (1943), p. 273.

[5] Ibid., p. 274. The p_t scale in Figure 17.1 will give some guide to the protection afforded by the plan against "spotty" production. The p_t is percent nonconforming in a run of 1,000 consecutive product units, for which the probability of acceptance by sample is 0.10 for a percentage sample equal to the corresponding f value shown on the chart.

[6] Ibid., pp. 274–75. The sampling inspectors, however, should carry on a verification inspection of the screening operations.

FIGURE 17.2
John S. White's Chart of AOQL Values for CSP-1

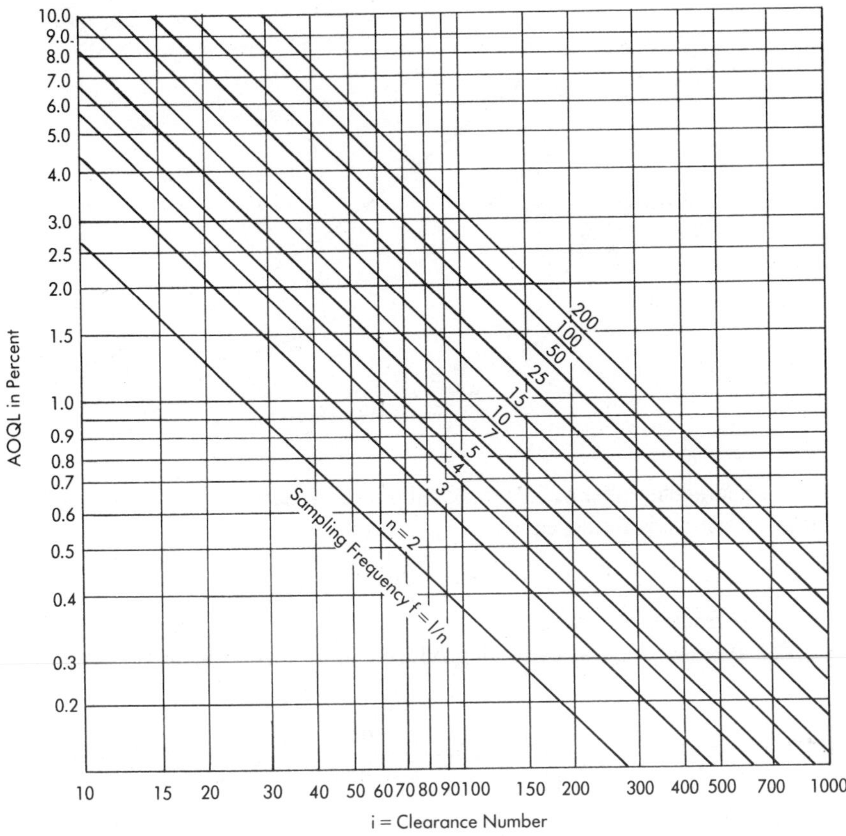

SOURCE: Reproduced with permission from *Industrial Quality Control* 17, May 1961, p. 79. Copyright American Society for Quality Control, Inc.

Some of the formulas pertaining to this type of sampling plan are as follows.[7]

The average number of pieces inspected in a 100 percent screening sequence following the finding of a nonconformity equals

(17.1)
$$u = \frac{1 - q'^{i}}{p'q'^{i}}$$

where $q' = 1 - p'$ and p' is the fraction nonconforming when the process is operating in control at level p'.[8]

---

[7] Dodge, "A Sampling Inspection Plan for Continuous Production," *Annals of Mathematical Statistics* 14 (1943), pp. 264–79.

[8] See Chapters 18 and 19 below.

FIGURE 17.3
Operating Characteristic Curves for Continuous-Sampling Plans, CSP-1

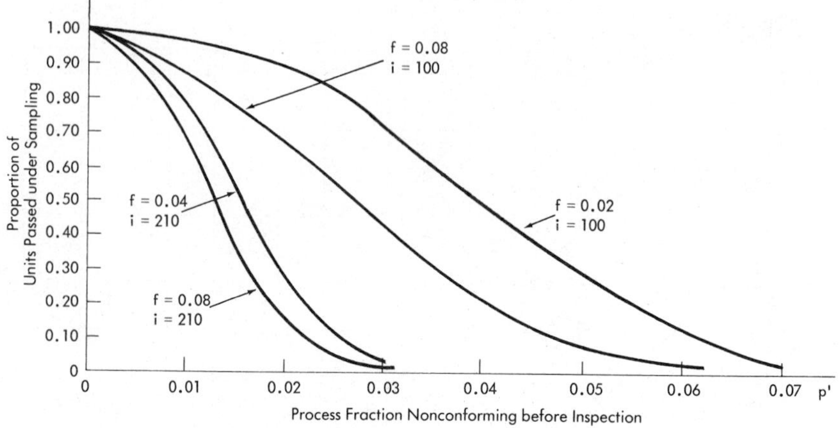

The average number of pieces passed under the sampling procedure before a nonconformity is found equals

(17.2)
$$v = 1/fp'$$

where f = fraction inspected.

The average fraction of total produced units inspected in the long run equals

(17.3)
$$\text{AFI} = \frac{u + fv}{u + v}$$

The average fraction of produced units passed under the sampling procedure is

(17.4)
$$P_a = \frac{v}{u + v}$$

When P_a is plotted as a function of p', we get what may be called an operating characteristic curve. Whereas an OC curve for a lot-by-lot acceptance-sampling plan gives the percentage of lots that would be passed under sampling inspection, so the P_a for a continuous-sampling plan gives the percentage of items passed under sampling inspection. Graphs of operating characteristic curves for various values of f and i are shown in Figure 17.3. It will be noted that for moderate values of f, i has much more effect on the curve than f does.

To illustrate the use of these formulas, suppose[9] that $p' = 0.015$, $f = 0.10$, and $i = 50$. Then the average number of pieces inspected in a 100

[9] In a letter to the author Dodge suggests that the process average p' should be about two thirds of the AOQL for a continuous-sampling plan to be economical.

percent screening sequence following the finding of a nonconforming item is

$$u = \frac{1-(0.985)^{50}}{(0.015)(0.985)^{50}} = \frac{1-0.470}{(0.015)(0.470)} = 75.2$$

The average number of pieces inspected in a period of sampling inspection is

$$v = \frac{1}{(0.10)(0.015)} = \frac{1}{0.0015} = 667$$

The average fraction of total product inspected in the long run is

$$\text{AFI} = \frac{75.2 + 66.7}{742} = 0.19$$

The average fraction of units passed under the sampling procedure is

$$P_a = \frac{667}{742} = 0.90$$

This would be the height of the OC curve at the point $p' = 0.015$.

If there is more than one class of nonconformities, say major and minor nonconformities, CSP-1 may be applied to each class separately. More specifically, there will be an i major and i minor, but it will be convenient to use the same f. Under this joint procedure, it will be possible for production to be under 100 percent inspection for major nonconformities but under sampling inspection for minor nonconformities, or vice versa. When they are both under sampling inspection, however, the same units of product will be sampled for inspection for major nonconformities as for minor nonconformities. Special note should be made of the following. If production is under 100 percent inspection for major nonconformities, for example, but under sampling inspection for minor nonconformities, discovery of a unit of product with a minor nonconformity *among the nonsampled units being inspected 100 percent for major nonconformities* should not be cause for shifting from sampling to 100 percent inspection of minor nonconformities, and similarly when minor nonconformities are under 100 percent inspection and major under sampling inspection.

In conclusion of the discussion of the Dodge plan, two comments are in order. The first concerns the selection of the sample items. Originally Dodge simply specified that the sample should be drawn in such a manner as to assure an unbiased sample. In subsequent conversations with the author, he expressed the view, shared strongly by the author, that, if possible, random numbers should be used. For each group of $1/f$ items, for example, one number should be selected from a group of random numbers running between 0 and $\frac{1}{f} - 1$. Thus, if $f = 0.10$, we could, for each group of 10 items, select a digit (a 0, 1, 2, . . . , or 9) from a table of random numbers, and this would indicate the item to be selected. Another method would be to decide by random numbers whether *each item* should be inspected or passed. Since, on a continuous production line, items occur in a definite order, it would seem that this would be the easiest of all places to use random numbers.

A final word should be said about the AOQL of a Dodge plan. If a process is in control at a level p', or more specifically, if the probability of occurrence of a nonconforming item has a constant value p', the application of CSP-1 will guarantee an average outgoing quality (AOQ) which will be less than p'. For small values of p', the AOQ will be small because p' is small, and for relatively large values of p', the AOQ will also be small because of the frequent application of 100 percent inspection under the plan. For intermediate values of p', the AOQ will be a maximum for some value of p'. This will be the AOQL of the process. If a process is not in control but can be looked upon as consisting of a series of production segments, for each of which the probability of occurrence of a nonconforming item is a constant p', it is very likely that the AOQL computed for a controlled process will still serve as an upper limit on the average outgoing quality. For, if the AOQL for a controlled process is the AOQ reached at $p' = p'_M$, then a series of production segments operating at levels, some of which are different from p'_M, will under the sampling plan generally have an AOQ that is less than the AOQL for a controlled process.[10]

It is theoretically possible for rare exceptions to occur. If the process is artificially manipulated, for example, such that p' shifts in precise sympathy with the phases of the sampling plan, e.g., $p' = 1$ when the plan calls for sampling at the rate f but becomes zero as soon as the plan calls for 100 percent inspection, then an AOQ would result that would be higher than the AOQL for a controlled process. If $i = 150$, $f = 0.05$, for example, the AOQL for a controlled process would be 1 percent. If sampling is systematic, however, i.e., every 20th item is selected under the sampling regimen, then under the type of shifting in p' just specified, the AOQ would be 19/170 which equals 11 out of 100 instead of 1 out of 100.

It is extremely unlikely that such perfectly coordinated shifts in p' would occur in practice. Even though the production operators know when inspection is on sampling and on 100 percent inspection, they are not likely to let down in their efforts to maintain quality if they are required to perform the 100 percent inspection when a nonconformity is found. It would seem justifiable, therefore, to speak generally of the AOQL computed for a CSP-1 plan on the assumption the process is in control as being the AOQL of the plan, whether or not the process is actually in control.[11] The same point

---

[10] Cf., however, C. Derman, M. V. Johns, Jr., and G. J. Lieberman, "Continuous Sampling Procedures without Control," *Annals of Mathematical Statistics* 30 (1959), pp. 1175–91.

[11] In a letter to the author, Dodge notes that the AOQ's actually experienced usually are considerably less than the AOQLs of the plans used.

applies to the variations of the basic Dodge plan that are discussed in the next section.

2.1.2. Variations in the Dodge Plan. Variations in the original Dodge plan have followed two principal lines. One is to meet the feeling among plant inspectors that the occurrence of a single isolated nonconforming unit does not generally warrant return to 100 percent inspection, especially when dealing with minor nonconformities. To meet this objection, Dodge and Torrey proposed[12] CSP-2 and CSP-3. Under CSP-2, 100 percent inspection will not be resorted to when production is under sampling inspection until two nonconforming sample units have been found within a space of k sample units of each other. It is likely that k will be taken equal to i. CSP-3 follows CSP-2 but gives additional protection against spotty production. It requires that after a nonconforming unit has been found in sampling inspection, the *immediately following* four units should be inspected. If any of these four are nonconforming, return is made immediately to 100 percent inspection. If none are nonconforming, the plan proceeds as under CSP-2.

A second variation is aimed at avoiding the abrupt changes in inspection inherent in the previously described plans. This is accomplished by allowing for several "levels" of sampling inspection. Multilevel sampling plans, as these are called, were originally proposed by G. J. Lieberman and H. Solomon.[13] They begin with 100 percent inspection as does CSP-1, and go to a fraction f as soon as a run of i conforming units has been found. When, however, under sampling inspection at rate f a run of i consecutive sample units has been found free of nonconformities, then resort is made to sampling at the rate f^2. Again, if a run of i consecutive units is found free of nonconformities, resort is made to sampling at the rate f^3, and so forth as far as the designer of the plan wishes to go. If at any time when under sampling inspection a nonconforming unit is found, return is immediately made to the next lower level of sampling. The effect of this multilevel plan is obviously to lighten the inspection load when the process is turning out a relatively low percentage of nonconforming units and to increase it when the process is not doing as well. The purpose is to do this without too abrupt changes in the inspection load. For constant p', Lieberman and Solomon have derived AOQL contours, similar to Figure 17.1, for two-level plans and have provided approximate formulas for plans with three or more levels. J. A. Greenwood[14] has derived multilevel plans similar to the Lieberman-Solomon plans, except that they call for 100 percent inspection of all product units represented by any sample unit that proves to be nonconforming.

---

[12] "Additional Continuous Sampling Inspection Plans," *Industrial Quality Control,* November 1951, pp. 5–9.

[13] "Multi-Level Continuous Sampling Plans," *Annals of Mathematical Statistics* 26 (1955), pp. 686–704.

[14] Unpublished memorandum, Bureau of Aeronautics, Navy Department.

2.2. Other Continuous-Sampling Plans

A form of continuous sampling somewhat different from the Dodge procedure was suggested by A. Wald and J. Wolfowitz[15] in 1945. Their SPC plan proceeds as follows:

1. Demark segments of the process numbering N_0 items each.
2. Begin by inspecting one item drawn at random from each successive group of $1/f$ items, where f is the fraction sampled.
3. If at any time prior to the inspection of fN_0 sample units the number of nonconforming sample items equals or exceeds $M^* = \dfrac{N_0 L}{\left(\dfrac{1}{f} - 1\right)}$, where L is the specified AOQL, stop sampling and, beginning with the first unit that follows the group in which the last of the M^* nonconforming units was found, inspect 100 percent of the remaining units in N_0.
4. Start inspecting each segment of N_0 units afresh.

For example, if we take $N_0 = 1{,}000$, $f = 0.10$, and desire[16] an AOQL of 0.018, then $M^* = \dfrac{1{,}000(0.018)}{\left(\dfrac{1}{0.10} - 1\right)} = \dfrac{18}{9} = 2.$

This SPC plan will guarantee the AOQL, whether the process is in control or not. Wald and Wolfowitz argue that it is likely to be less expensive than Dodge's CSP-1 when N_0 is large (say 1,000 or more when the AOQL = 0.045) and more expensive when N_0 is small (say 400 when the AOQL = 0.045), but in the first case it does not give as good protection against spotty material as in the second case.

M. A. Girschick has proposed a continuous-sampling plan along the Wald-Wolfowitz line that has good statistical properties. (See references at end of this chapter.) The plan runs as follows:

1. Begin with sampling at the rate of one item selected at random from each consecutive group of m items. The number of items found to be nonconforming are recorded and cumulated. Inspection terminates when a total of d nonconforming items has been found.

2. If at the time the dth nonconforming item has been found, the number of groups examined, designated by g, is greater than or equal to some predetermined number G, then the material passed is deemed satisfactory and a new cycle of sampling inspection is begun.

---

[15] "Sampling Inspection Plans for Continuous Production which Insure a Prescribed Limit on the Outgoing Quality," *Annals of Mathematical Statistics* 14 (1945), pp. 30–49.

[16] The AOQL desired might be 0.02, but 0.018 is selected since it makes M^* equal to an integer.

3. If at the time the dth nonconforming item has been found, $g < G$, then beginning with the next group of m items a total of $G - g$ groups [$= m(G - g)$ items] is inspected 100 percent. Upon conclusions of this 100 percent inspection, a new cycle of sampling inspection is begun. A diagram of the procedure follows:

> *Beginning Stage.* Select 1 item at random from each consecutive group of m items. Cumulate number of nonconforming items found (x) and number of groups sample inspected (g). Stop when number of nonconforming items (x) equals d.

If $g < G$	*If* $g \geq G$
Inspect the next $G - g$ groups 100%. Then start cycle over again.	Start cycle over again.

If in the above procedure all nonconforming items are replaced by good ones, the plan will guarantee an AOQL equal to

(17.5)
$$\text{AOQL} = \frac{m-1}{m}\frac{d}{G}$$

If nonconforming items are not replaced, then

(17.6)
$$\text{AOQL} = \frac{d}{G}$$

These AOQLs are valid regardless of the pattern of nonconforming items in the process. This plan has been used successfully in at least one industrial company.[17]

Girschick gives a formula for an OC curve of his plan that shows how the probability that a given inspection will terminate without resort to 100 percent inspection varies with the process average when the process is in a state of statistical control. Under this same condition he also gives formulas for the AOQ, the average fraction inspected, and the variance in the AOQ for a segment of a specified size. A special modification of the plan allows for two levels of sampling inspection which will make for a more effective inspection of a spotty process.

2.3. Comments on the Various Plans

If we brush aside the details of the various plans, it would appear that the Dodge-type plans, on the one hand, and the Wald-Wolfowitz-Girschick-type, on the other, differ in two fundamental respects. The Dodge plans come closer to an inspector's natural instincts. Thus under these plans resort is made to tightened inspection with the appearance of one or two closely spaced

[17] See Albrecht, Gulde, H., MacLean, A., and Thompson, P., "Continuous Sampling at Minneapolis-Honeywell," *Industrial Quality Control,* September 1955, pp. 4–9.

nonconforming items; whereas under the Wald-Wolfowitz-Girschick plans nonconformities are allowed to accumulate before resort is made to tightened inspection. Dodge plans also usually begin with 100 percent inspection,[18] while the Wald-Wolfowitz-Girschick plans start with sampling inspection, something that many inspectors would not be inclined to do.

The Dodge type of plan, such as a CSP or multilevel plan, does not offer a watertight guarantee for all possible conditions, but, as noted above, the AOQL for the state of control is very likely to be the AOQL for conditions met in practice. The Wald-Wolfowitz-Girschick plans assure an AOQL for any kind of a process. The principal unknown feature in the use of both the Dodge and Wald-Wolfowitz type of plan is how closely the AOQ for an infinite process is approached when the plan is applied to runs of production that are not very long. Girschick, as noted above, took the interesting step of computing for his plan the variance in the AOQ for segments of specified finite size from a controlled process.

2.4. Military Standard 1235B

The U.S. Department of Defense has issued a standard set of continuous sampling plans entitled *Single- and Multi-Level Continuous Sampling Procedures and Tables for Inspection by Attributes.* The current (1985) version bears the designation Mil. Std. 1235B and is dated December 10, 1981. The manual *Functional Curves of the Continuous Sampling Plans* that accompanies Mil. Std. 1235B has the designation Mil. Std. 1235A–1 (MU), which it had when it served as a supplement to the *A* version of Mil. Std. 1235. The date of issuance of the manual is June 28, 1974.

Mil. Std. 1235B provides five different types of continuous sampling plans for attributes,[19] namely,

a. CSP-1, a single-level continuous sampling procedure which provides for alternating between sequences of 100 percent inspection and sampling inspection.

b. CSP-F, a variation of the CSP-1 plans in that CSP-F plans are applied to a relatively short run of product, thereby permitting smaller clearance numbers to be used.

c. CSP-2, a modification of CSP-1 in that 100 percent inspection resumes only if the number of conforming sampled units that separate any two nonconforming sampled units is less than a prescribed number.

d. CSP-T, a multi-level continuous sampling procedure which provides for reducing the sampling frequency upon demonstration of superior product quality.

e. CSP-V, a single-level continuous sampling procedure which is an alter-

---

[18] As noted previously, however, Dodge plans can start with sampling, and in many cases are so applied.

[19] From the Foreword to the standard.

native to CSP-T in that these plans provide for reducing the clearance number in good quality situations where reduction of sampling frequency has no economic merit.

Section 7 of the Standard contains the definitions of terms of particular importance to the proper use of the Standard's provisions.

Table 17.1 below is a reproduction of Table 2-A of the standard. It gives the clearance number i and sampling frequency f for CSP-1 plans indexed in such a way that a CSP-1 plan can readily be selected as an alternative for a Mil. Std. 105D lot-by-lot attributes sampling scheme with a designated AQL. A Department of Defense contract ordinarily cites a specification and this will list the sampling standards that may be used. Thus a specification could list[20] Mil. Std. 1235 and Mil. Std. 105 as permissible sampling standards and specify an AQL (or AQLs if more than a single class of nonconformities is to be considered). If the production process and inspection requirements are such that continuous sampling may be effectively carried out, the contractor may currently (1985) use Mil. Std. 1235B. If he decides to apply a CSP-1 plan, the clearance number i for his plan will be taken from the column in Table 17.1 headed by the specified AQL. The actual i that will be used will be determined by the code letter or associated sampling frequency that is adopted. Table 17.1 provides permissible code letters based on the number of units in the production interval. The standard states[21] that a

> code letter and its associated sampling frequency should be selected after considering such influencing factors as inspection time per unit of product, production rate, and proximity to other inspection stations. When idle inspector time is a significant consideration, a plan with a higher sampling frequency and lower clearance number is usually preferred.

For example, suppose a contract lists Mil. Std. 105 and Mil. Std. 1235 as permissible sampling standards and prescribes an AQL of 1 percent. Then, if Mil. Std. 105D is used in the inspection of a continuing series of lots and a lot of 2,000 items is to be inspected under inspection level II, Code Letter K would be called for and for normal single sampling this would require the use of the sampling plan $n = 125$, $c = 3$. If Mil. Std. 1235B is used in the inspection of a continuing flow of product and sample frequency code letter F is judged to be appropriate (this need not be the same as the sample size code letter called for under the alternate 105D plan), then the CSP-1 plan to be used would have a clearance number $i = 89$ and a sampling frequency $f = 1/10$.

Users of Table 2–A of the standard should carefully note the warning given at the bottom of the table which states that the AQLs listed in the table are provided as indexes to simplify use of the table, but have no other

[20] Ordinarily the letters after the standard number are omitted, the ones in effect at the time of the bid invitation being considered authorized.

[21] Section 1.8.2.

TABLE 17.1
Values of *i* for CSP-1 Plans (Table 2–A of Mil. Std. 1235B)

Samp Freq Code Ltr	*f*	*AQL* in percent*															
		.010	*.015*	*.025*	*.040*	*.065*	*0.10*	*0.15*	*0.25*	*0.40*	*0.65*	*1.0*	*1.5*	*2.5*	*4.0*	*6.5*	*10.0*
A	1/2	1,540	840	600	375	245	194	140	84	53	36	23	15	10	6	5	3
B	1/3	2,550	1,390	1,000	620	405	321	232	140	87	59	38	25	16	10	7	5
C	1/4	3,340	1,820	1,310	810	530	420	303	182	113	76	49	32	21	13	9	6
D	1/5	3,960	2,160	1,550	965	630	498	360	217	135	91	58	38	25	15	11	7
E	1/7	4,950	2,700	1,940	1,205	790	623	450	270	168	113	73	47	31	18	13	8
F	1/10	6,050	3,300	2,370	1,470	965	762	550	335	207	138	89	57	38	22	16	10
G	1/15	7,390	4,030	2,890	1,800	1,180	930	672	410	255	170	108	70	46	27	19	12
H	1/25	9,110	4,970	3,570	2,215	1,450	1,147	828	500	315	210	134	86	57	33	23	14
I	1/50	11,730	6,400	4,590	2,855	1,870	1,477	1,067	640	400	270	175	110	72	42	29	18
J	1/100	14,320	7,810	5,600	3,485	2,305	1,820	1,302	790	500	330	215	135	89	52	36	22
K	1/200	17,420	9,500	6,810	4,235	2,760	2,178	1,583	950	590	400	255	165	106	62	43	26
		.018	*.033*	*.046*	*.074*	*.113*	*.143*	*.198*	*0.33*	*0.53*	*0.79*	*1.22*	*1.90*	*2.90*	*4.94*	*7.12*	*11.46*
		AOQL in percent															

* AQLs are provided as indexes to simplify use of this table, but have no other meaning relative to the plans.

meaning relative to the plans. The point cannot be emphasized too strongly. Continuous sampling plans are not AQL plans. The use of AQLs in the standard is simply to provide a means, as described above, of finding a CSP considered an acceptable alternate to a lot-by-lot attributes plan for which an AQL has been designated.

An important question is how a continuous sampling plan is judged to be an acceptable alternative for a lot-by-lot attributes sampling plan with a designated AQL. In Table 2–A of the standard all the CSP-1 plans listed under a given AQL index are designed to have an AOQL equal to that given at the bottom of the specified AQL column. This AOQL is the same as the maximum *scheme* AOQL of the Mil. Std. 105D plans with the designated AQL that in the case of single sampling do not have zero acceptance numbers. If all plans in Mil. Std. 105D with a designated AQL have zero acceptance numbers for single sampling, the AOQL of the CSP-1 plans with that AQL index is computed to be equal to the maximum AOQL for these Mil. Std. 105D plans with zero acceptance numbers.[22]

To illustrate the above procedure we note that for an AQL = 4 percent Table XI of ANSI-Z1.4 (the civilian version of Mil. Std. 105D) gives AOQL factors for *scheme* performance as follows:

Lot Size	*Sample Size for Normal Single Sampling*	*Sample Size Code Letter*	*AOQL Factor*
51–90	13	E	4.5%
91–150	20	F	4.9
151–280	32	G	4.9
281–500	50	H	5.1
501–1,200	80	J	5.0

A footnote indicates that a better approximation to the AOQL is obtained by multiplying the above factors by (1 − Normal Plan Sample Size/Lot Size). When this is done for the factors 5.1 percent and 5.0 percent above, using the maximum of the lot-size range, we get AOQLs = 0.051 (1 − 50/500) = 4.59 percent and 0.050 (1 − 80/1,200) = 4.67 percent. The figure used in Table 2–A of Mil. Std. 1235B is 4.94 percent. The deviation of the Mil. Std. 1235B figure is apparently due to a slightly different procedure for computing the AOQL of a mil. Std. 105D scheme.[23] In Mil. Std. 1235B,

[22] American-British-Canadian-Australian Armies Standardization Program, Quadripartite Advisory Publication 16, *The Mathematical Background of QSTAG 340,* pp. VI–2 and VI–3.

[23] In deriving Mil. Std. 1235B, the AOQLs of the Mil. Std. 105D schemes were calculated by allowing only for shifts between normal and tightened inspection. This had almost no material effect. What is more significant is that in the computation of the scheme AOQLs reported in ANSI Z1.4, it is assumed that after discontinuance of inspection under clause 8.4, sampling inspection is resumed using a *tightened inspection* plan while the computations on which Mil. Std. 1235B are based assume sampling inspection under a Mil. Std. 105D scheme is resumed using a normal inspection plan. See ABCA Quadripartite Advisory Publication 16, p. VI–2.

therefore, CSP-1 plans that have an AOQL of 4.94 percent are taken as acceptable alternatives for Mil. Std. 105D schemes with AQLs of 4 percent.

The philosophy of matching continuous sampling plans to lot sampling plans by equating AOQLs had been discussed by R. L. Storer in his paper[24] in *Industrial Quality Control* in 1956 and this procedure had been followed in Mil. Std. 1235 (Ord)—1962. Storer argued that in, "most instances, the conventional inspection system actually does have AOQL connotations, for the customary action taken by the contractor upon the return of a rejected lot is to screen the lot under the supervision of an Ordnance inspector, removing all defects of the kind causing rejection." Subsequently when revision of Mil. Std. 1235 (Ord) was being discussed, the Standards Group of the American Society for Quality Control, at the suggestion of Harold Dodge who was then a member, raised objections to the matching of AOQLs and advocated instead that the AOQLs of the continuous sampling plans be matched[25] to the AQLs of Mil. Std. 105D. It will be noted in Table 17.1 above that the AOQLs of the CSP-1 plans selected as acceptable alternative procedures are *larger* than the AQLs of the Mil. Std. 105D plans they are intended to match. It will be recalled that Mil. Std. 105D, through application of tightened inspection when quality deteriorates, seeks to bring pressure on a supplier to attain a process average *at least as good* as the AQL. If a CSP plan is adopted that has an AOQL equal to the AQL of a Mil. Std. 105D plan, the ASQC Standards Committee argued that the objectives of the two procedures would be brought into accord. Table 17.2 below gives CSP-1 plans that for a specified sampling frequency have AOQLs equal to the nominal AQLs used in Mil. Std. 105D.

In Mil. Std. 1235 (Ord)—1962 the AOQLs of the CSP were tied in with the AOQLs of the *individual* plans of Mil. Std. 105D using Code Letter M, normal single sampling for the matching. As described above, the AOQLs of Mil. Std. 1235B are tied in with the AOQLs of the Mil. Std. 105D sampling *schemes,* which run somewhat less than the AOQLs of *individual* normal Mil. Std. 105D plans (say, for example, 5 percent instead of 6 percent) and this has reduced somewhat the gap between CSP AOQLs and the AQLs of the related Mil. Std. 105D procedures. Army statisticians, however, chose not to go the whole way and equate the AOQLs of the CSP to Mil. Std. 105D AQLs. A comparison of Table 2–A of Mil. Std. 1235B (See Table 17.1 of this text) with Table 17.2 shows that the clearance values i run larger in Table 17.2 which means that the inspection cost involved in the use of these plans will run higher. For example, as noted above, under an AQL heading of 1 percent and a sampling frequency of 1/10, Table 2–A of Mil. Std. 1235B calls for an i value of 89, whereas the corresponding

---

[24] R. L. Storer, "The Use of Continuous Sampling in Ammunition Procurement," *Industrial Quality Control* 12, No. 11 (May 1956), pp. 48–53.

[25] *Quadripartite Advisory Publication* 16, pp. VI–3 and VI–4.

TABLE 17.2

CSP-1 Plans the AOQLs of which Equal the AQLs of Mil. Std. 105D* (clearance interval *i*)

f	*AOQL in percent*															
	.010	*.015*	*.025*	*.04*	*.065*	*.10*	*.15*	*.25*	*.40*	*.65*	*1.0*	*1.5*	*2.5*	*4.0*	*6.5*	*10.0*
1/2	2,800	1,850	1,150	700	430	280	187	115	70	42	27	18	11	—	—	—
1/3	4,600	3,100	1,850	1,150	710	463	310	185	115	70	46	30	18	11	—	—
1/4	6,000	4,000	2,400	1,500	920	600	410	235	150	93	60	40	23	14	—	—
1/5	7,100	4,700	2,850	1,800	1,100	710	470	290	180	110	71	47	29	17	10	—
1/7	8,800	5,800	3,500	2,200	1,400	890	580	355	225	138	89	59	35	21	13	—
1/10	10,800	7,200	4,300	2,700	1,700	1,100	720	430	270	170	110	72	43	26	16	10
1/15	13,300	8,800	5,300	3,300	2,050	1,330	880	535	333	208	134	89	53	33	20	12
1/25	16,700	11,000	6,700	4,200	2,550	1,650	1,100	665	415	255	165	110	66	40	24	15
1/50	21,000	14,000	8,400	5,300	3,250	2,100	1,400	840	530	330	210	140	84	52	31	20
1/100	26,500	17,700	10,500	6,600	4,100	2,600	1,700	1,050	660	405	265	175	104	65	39	24

* These plans were derived from Figure 1 of John S. White, "A New Graph for Determining CSP-1 Sampling Plans," *Industrial Quality Control* 17 (May 1961), pp. 18–19 (see Figure 17.2 of this chapter) and an extrapolation thereof. They were checked by reference to Figure 17.1 of this chapter and by the approximate inverse relationship that exists between the AOQL and i when f is constant. Numerical computations on an electronic computer indicate that with the exception of plans for 10.0 percent AOQLs and one 6.5 percent AOQL plan, the exact AOQLs for the plans of Table 17.2 are within ±5 in the next decimal place of the AOQL listed. The exact AOQL for $f = 1/15$, AOQL = 6.5 percent is about 6.44 percent and the exact AOQLs for those listed as 10.0 percent are within ±0.2 percent. Since in the calculation of the AOQLs of Table 17.1 nonconforming units were not replaced, whereas Dodge and White assume replacement, a fairer comparison between Table 17.1 and Table 17.2 is given by adding 1 to the i values of Table 17.2. Cf. Fred L. Abraham, *Journal of Quality Technology* 3 (Nov. 1, 1971), p. 4.

TABLE 17.3
Values of *S* for CSP-1 Plans (Table 2–B of Mil. Std. 1235B)

Samp Freq Code Ltr	*f*	*AQL* in percent*															
		.010	*.015*	*.025*	*.040*	*.065*	*0.10*	*0.15*	*0.25*	*0.40*	*0.65*	*1.0*	*1.5*	*2.5*	*4.0*	*6.5*	*10.0*
A	1/2	1,850	925	721	451	295	273	197	119	75	55	36	22	17	11	10	6
B	1/3	4,080	1,950	1,600	993	649	579	442	268	166	120	78	52	36	24	19	16
C	1/4	6,010	2,915	2,360	1,460	1,010	926	699	421	262	177	115	79	57	36	28	20
D	1/5	8,320	3,890	3,100	1,930	1,390	1,150	975	589	367	258	165	109	76	45	40	27
E	1/7	11,400	5,670	4,660	2,895	1,980	1,750	1,355	813	507	376	244	154	109	63	54	34
F	1/10	16,900	7,590	6,640	4,120	2,800	2,595	1,985	1,245	624	543	352	221	164	90	82	51
G	1/15	24,400	11,300	9,250	5,760	4,020	3,820	2,960	1,810	922	856	524	327	241	141	138	75
H	1/25	35,500	16,900	13,900	8,640	5,950	5,740	4,560	2,760	1,390	1,350	839	524	390	212	189	105
I	1/50	59,800	26,900	23,000	14,300	10,300	10,100	8,440	5,070	3,170	2,445	1,590	913	733	368	334	212
J	1/100	96,000	39,800	36,400	23,300	16,900	16,500	14,300	8,710	6,020	3,980	2,600	1,640	1,360	642	601	352
K	1/200	148,100	63,700	58,000	36,000	29,000	28,500	25,400	15,200	9,470	8,030	4,365	2,835	2,150	1,080	1,025	636
		.018	*.033*	*.046*	*.074*	*.113*	*.143*	*.198*	*0.33*	*0.53*	*0.79*	*1.22*	*1.90*	*2.90*	*4.94*	*7.12*	*11.46*
		AOQL in percent															

* AQLs are provided as indices to simplify use of this table, but have no other meaning relative to the plans.

plan in Table 17.2 has an i value[26] of 110. What may be needed is a study of several examples of the comparative *overall* costs of the two procedures.

In Mil. Std. 1235B the application of CSP-1 plans is accompanied by the requirement that inspection be discontinued if a run of 100 percent inspection exceeds a specified quantity S. Values of S for the CSP-1 plans of Table 2–A are given here in Table 17.3 (Table 2–B of the standard). This provision has the same intent as the provision for discontinuance of inspection in Mil. Std. 105D if sampling remains on tightened inspection for more than 10 consecutive lots (Section 8.4 of the standard).

For tables of other types of continuous sampling plans contained in Mil. Std. 1235B and discussion thereof, the reader is referred to the standard itself. A relatively extensive discussion of Mil. Std. 1235A, which is the same[27] as Mil. Std. 1235B, will be found in Edward G. Schilling, *Acceptance Sampling in Quality Control,* pp. 426 ff.

3. PROBLEMS

17.1. Find a Dodge continuous-sampling plan (a CSP-1) that has an AOQL of 2 percent. If the process fraction nonconforming is 0.01, what is the average number of items that will have to be inspected following the finding of a nonconforming item and what is the average number of items passed under the sampling procedure before a nonconforming item is found? Sketch the OC curve for your plan.

17.2. Find a Dodge continuous-sampling plan (a CSP-1) that has an AOQL of 3 percent. If the process fraction nonconforming is 0.02, what is the average number of items that will have to be inspected following the finding of a nonconforming item, and what is the average number of items passed under the sampling procedure before a nonconforming item is found? Sketch the operating characteristic curve for your plan.

17.3. An AQL of 0.25 percent is specified in a military contract and Mil. Std. 1235B is listed as a permissible sampling standard to use. State the CSP-1 plan the contractor would use if he judged Sample Size Code Letter H to be appropriate.

17.4. An AQL of 2.5 percent is specified in a military contract and Mil. Std. 1235B is listed as a permissible sampling standard to use. State the CSP-1 plan the contractor would use if he judged Sample Size Code Letter E to be appropriate.

4. SELECTED REFERENCES*

Albrecht, Gulde, MacLean, and Thompson (P '55), Anscombe (P '58), Beattie (P '62), Blackwell (P '77), Bowker (P '56), Derman, Johns, and Lieberman (P '59),

---

[26] Since Table 17.1 does not assume replacement of nonconforming units, a fairer comparison is given by adding 1 to the i values of Table 17.2. See the footnote to that table.

[27] The B reference was assigned when the U.S. Navy also adopted the standard and it became a Department of Defense standard. The contents were not changed.

* B and P refer to the Book and Periodical sections, respectively, of the Cumulative List of References in Appendix V.

Derman, Littauer, and Solomon (P '57), Dodge (P '43, P '55, and P '69–'70), Dodge and Torrey (P '51), Girschick (B '54), Hasssn (B '65), Hillier (P '64), Lieberman and Solomon (P '55), Murphy (P '59), Read and Beatie (P '61), Romig (P '53), Savage, I. R. (P '59), Schilling, E. G. (B '82), Shahani (P '79), Stephens (P '81), U.S. Department of Defense (B '74 and B '81), Wald and Wolfowitz (P '45) and Yang (P '83).

Part 4

Control Charts

18

The General Theory of Control Charts

A control chart is a statistical device principally used for the study and control of repetitive processes. Dr. Walter A. Shewhart, its originator,[1] suggests that the control chart may serve, first, to define the goal or standard for a process that the management might strive to attain; second, it may be used as an instrument for attaining that goal; and, third, it may serve as a means of judging whether the goal has been reached. It is thus an instrument to be used in specification, production, and inspection and, when so used, brings these three phases of industry into an interdependent whole.[2]

1. CAUSES OF VARIATION IN QUALITY

At the basis of the theory of control charts is a differentiation of the causes of variation in quality. With the adoption of the statistical point of view in industry, it has come to be recognized that certain variations in the quality of product belong to the category of chance variations about which little can be done other than to revise the process. This chance variation is the sum of the effects of the whole complex of chance causes. In this complex of causes the effect of each cause is slight, and no major part of the total variation can be traced to a single cause. The set of chance causes that produces variation in the quality of a manufactured product is thus like the set of forces that causes a penny to turn up heads or tails when it is tossed in a random manner.

Besides chance variations in quality, there are variations produced by "assignable causes." These are relatively large variations that are attributable to special causes. For the most part, assignable causes consist of—

[1] See Chapter 1 for a brief review of the history of statistical quality control.

[2] W. A. Shewhart, *Statistical Method from the Viewpoint of Quality Control* (Washington, D.C.: Graduate School, Department of Agriculture, 1939), chap. i.

1. Differences among machines.
2. Differences among workers.
3. Differences among materials.
4. Differences in each of these factors over time.
5. Differences in their relationships to one another.

2. THE THEORY OF CONTROL CHARTS

If chance variations are ordered in time or possibly on some other basis, they will behave in a random manner. They will show no cycles or runs or any other defined pattern. No specific variation to come can be predicted from knowledge of past variations.

On the other hand, variation produced by chance causes follows statistical laws. For example, if 10 pennies are tossed in a random manner, the relative frequencies with which 0, 1, 2, . . . , 10 heads occur will tend, as the tossing is continued, to approach the frequencies of a binomial distribution. Likewise, in random samples of n units each from a process that is affected only by chance causes, the probabilities of getting 0, 1, 2, . . . , n nonconforming units will also be given by the binomial distribution. The variation produced by a system of chance causes can thus be predicted for mass phenomena, although the effect of any particular cause cannot be determined, nor can the variation of specific individuals be predicted.

Knowledge of the behavior of chance variations is the foundation on which control chart analysis rests. If a group of data is studied and it is found that their variation conforms to a statistical pattern that might reasonably be produced by chance causes, then it is assumed that no special assignable causes are present. The conditions which produced this variation are, accordingly, said to be under control. They are under control in the sense that, if chance causes are alone at work, then the amount and character of the variation may be predicted for large numbers, and it is not possible to trace the variation of a specific instance to a particular cause. On the other hand, if the variations in the data do not conform to a pattern that might reasonably be produced by chance causes, then it is concluded that one or more assignable causes are at work. In this case the conditions producing the variation are said to be out of control.

As noted above, a control chart is a device for describing in concrete terms what a state of statistical control is; second, a device for attaining control; and, third, a device for judging whether control has been attained. The nature of a control chart may be explained as follows.

Suppose that samples of a given size are taken from a process at more or less regular intervals and suppose that for each sample some statistic X is computed. This might be the sample fraction nonconforming, the sample mean, or the sample range. Being a sample result, X will be subject to sampling fluctuations. If no assignable causes are present, these sampling fluctuations in X will be distributed in a definite statistical pattern such as that pictured

FIGURE 18.1
Distribution of Chance Variations in a Sample Measure of Quality

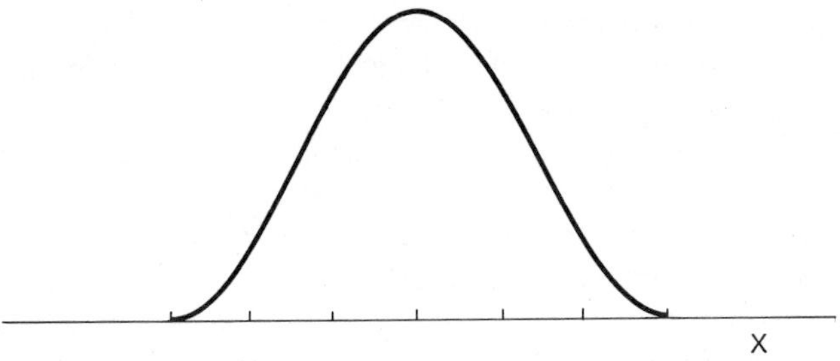

in Figure 18.1. If enough samples are taken, it is possible to estimate the mean and certain extreme points of this distribution. Suppose, for example, theory suggests that the sampling distribution of X is normal in form. Then, from the mean of the samples it would be possible to estimate the mean of the distribution of X and from the within-sample variation the standard deviation of X and, from these, to determine 0.001 probability points. If the vertical scale of a chart is calibrated in units of X and the horizontal scale marked with respect to time or some other basis for ordering X and if horizontal lines are drawn through the estimated mean of X and through an extreme value on the upper and lower tail of the distribution of X (see Figure 18.2), the result is a control chart for X.

FIGURE 18.2
Illustration of the Theoretical Basis for a Control Chart

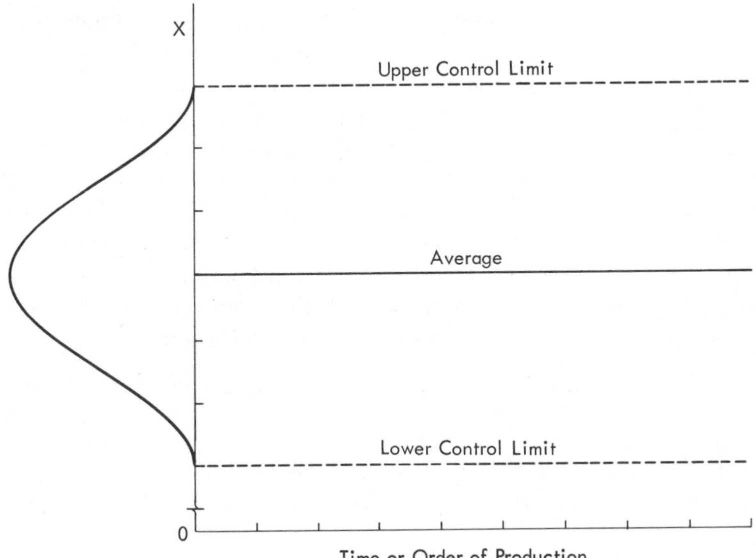

If sample values of X are plotted for a significant range of output and time and if these values all fall within the control limits and show no cycles or runs above or below average or runs up or down, then it can be said that the process is in a state of statistical control at the level designated with respect to the given measure of quality. In the use of control charts it is the goal of management to reduce fluctuations in a process until they are in a state of statistical control at the level desired.

A control chart may thus be used to specify the goal of management. It is also an instrument for attaining that goal. To see whether a process is in control, past data pertaining to the process are plotted on a control chart. If the data conform to a pattern of random variation within the control limits, the process will be judged as being in control at a level equal to the mean line on the chart. If the data do not conform to this pattern, as is almost always the case in the beginning, then departures from the pattern are investigated and assignable causes tracked down. If an exceptional cause of variation is on the unfavorable side, the effort is made to eliminate the special cause of this variation. If the exceptional variation is on the favorable side, an effort may be made to extend and perpetuate the cause producing it. In this way the process may eventually be brought close to a state of satistical control at a desirable level. After a condition of control has been satisfactorily approximated, departure from the condition may be quickly detected by maintaining a control chart on current output.

But, as noted above, a control chart is not only a device for specifying and attaining a state of statistical control. It is also a device for judging whether the state of control has been attained; for a control chart is constructed in conformity with statistical theory and can consequently be used to test the hypothesis of control. Thus, if sample values of X, when ordered in time or on some other basis, all fall within the control limits without varying in a nonrandom manner within the limits, then the process may be judged to be within control at the level indicated by the chart. Likewise, if a process has been judged to be in control and new sample results from the same process are plotted on the old control chart, then, if these new points continue to fall within the limits on the chart and do not lead to nonrandom fluctuations in the group of points as a whole, the process may be judged to be continuing in a state of statistical control at the given level.

Dr. Walter A. Shewhart emphasizes strongly that a process should not be judged to be in control unless the pattern of random variation has persisted for some time and for a sizable volume of output. He writes:

> This potential state of economic control can be approached only as a statistical limit even after the assignable causes of variability have been detected and removed. *Control of this kind can not be reached in a day. It can not be reached in the production of a product in which only a few pieces are manufactured. It can, however, be approached scientifically in a continuing mass production.*[3]

[3] Ibid., p. 46.

More specifically, he states:

> It has also been observed that a person would seldom if ever be justified in concluding that a state of statistical control of a given repetitive operation or production process has been reached until he had obtained, under presumably the same essential conditions, a sequence of *not less than twenty-five samples of four that satisfied Criterion I.*[4]

The reader should also note the following: If no points fall outside the control limits and if there is no evidence of nonrandom variation within the limits, it does not mean that assignable causes are not present. It simply means that the hypothesis that chance causes are alone at work is a tenable hypothesis and that it is likely to be unprofitable to look for special assignable causes. This is merely using the scientific principle of keeping our explanation of phenomena as simple as possible.[5] If chance can reasonably explain our results, we look no further. Thus, to repeat once again, if a control chart shows a process is "in control," it means that the hypothesis of random variation is a reasonable one to adopt for managerial purposes. When the chart fails to show control, then other action is reasonable.

Figures 18.3 and 18.4, 19.1, and 21.1 and 21.2 illustrate various kinds of control charts. Figures 18.3 and 19.1 show the fractions nonconforming of output taken in order of production. On each chart there is a central line and certain limits called control limits. In Figure 18.3 all points but one lie within the limits. If it were not for that point, it might be said that the process is "under control" at the level indicated by the central line on the chart. Figure 19.1 shows a number of points outside the control limits and illustrates a process that is obviously not in control. In Figure 19.4 this same process appears to have been brought under control at a considerably lower level. Figures 21.1 and 21.2 show the averages and ranges of samples of output taken in order of production and plotted on charts with central lines and control limits.[6] The points all fall nicely within the limits, so that the charts suggest that the process is in control at the levels indicated, with respect both to its average and to its general variability. Figure 18.4 illustrates a control chart on which the ordering of the samples is other than on a time basis.

3. CONTROL LIMITS

The limits on the control chart pictured in Figure 18.2 are 0.001 probability limits. They were determined so that, if chance causes alone were at work,

---

[4] Ibid., p. 37. Dr. Shewhart's Criterion I is that the points lie within the control limits (see his *Economic Control of Quality of Manufactured Product* [New York: D. Van Nostrand & Co., Inc., 1931], p. 304). His example of 25 samples of 4 each refers to an $\bar{X}$-chart. See Chapter 21. Also see Proschan and Savage (P '60). This suggests that less than 25 samples might do.

[5] In general, a single explanation of phenomena is deemed preferable to a set of special explanations: *Entia non sunt multiplicandum practer necessitatem.*

[6] The range chart of Figure 21.2 has only an upper limit.

FIGURE 18.3
An Example of a Fraction-Nonconforming Control Chart (a p-Chart)*

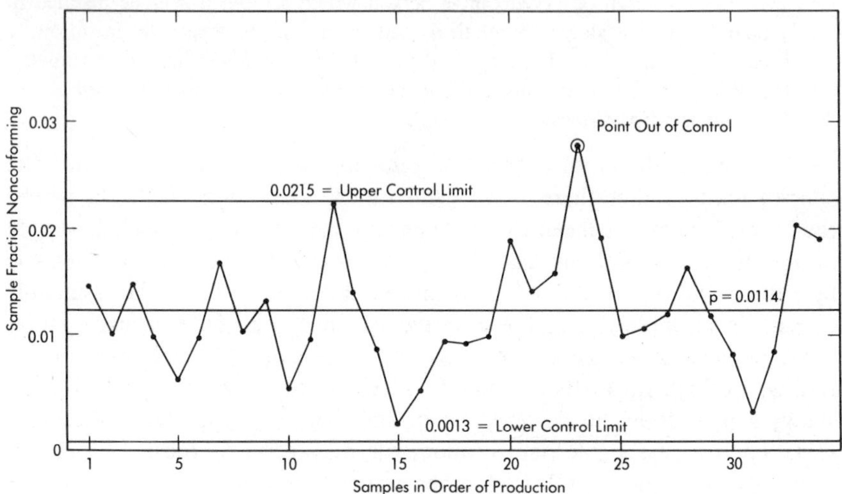

* See Figure 18.8. In constructing a control chart for the data of Figure 18.8, it was assumed that the sample size was 1,000. See James R. Crawford and Preston C. Hammer, "Statistical Quality Control at Lockheed," Quality Control Reports, No. 9, p. 7.

the probability of a point's falling above the upper limit would be one out of a thousand and the probability of a point's falling below the lower limit would be one out of a thousand. Since a hunt will be made for an assignable cause if a point falls outside these limits, they determine the risk of making such a hunt when actually there is no assignable cause of variation. Since two out of a thousand is a very small risk, the 0.001 limits may be said to

FIGURE 18.4
Fractions Nonconforming of Samples of 200 from 20 Different Machines Plotted on a Control Chart

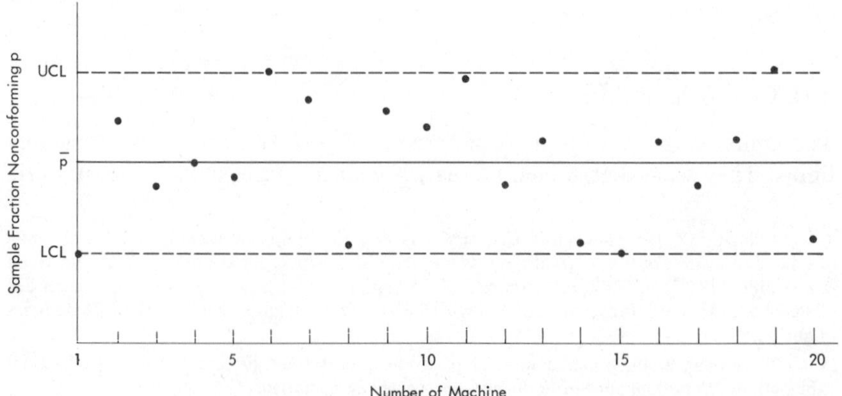

give practical assurance that, if a point falls outside, the variation was produced by an assignable cause. Two out of a thousand is a purely arbitrary figure; there is no reason why the probability of exceeding the limits by chance could not be set at one out of a hundred or higher. The decision would in any case depend on the risks the management of the quality control program wishes to run. It is customary to use limits that approximate the 0.002 standard, but this is not necessary.

If the system of chance causes produces a variation in X that follows the normal curve, the 0.001 probability limits are practically equivalent to $3\sigma'$ limits; for under a normal curve the probability that a deviation from the mean will exceed $3\sigma'$ in one direction is 0.00135 or in both directions is 0.0027. For normal variation, therefore, $3\sigma'$ limits are the practical equivalent of 0.001 probability limits.

In the United States, whether X is normally distributed or not, it is customary to base the control limits upon a multiple of the standard deviation. Usually this multiple is 3 and the limits are called 3-sigma limits. This term is used whether the standard deviation actually involved is the universe standard deviation, some estimate thereof, or simply a "standard value" for control chart purposes. Usually it is readily inferred from the context precisely what standard deviation is involved. We shall follow this practice except where it might be confusing.

If the distribution is very skewed in a positive direction, the 3-sigma limit will fall short of the upper 0.001 limit, while the lower 3-sigma limit will be below the lower 0.001 limit. In such instances the use of 3-sigma limits means that the risk of looking for assignable causes of positive variation, when none exists, will be greater than one out of a thousand. The risk of looking for an assignable cause of negative variation, when none exists, will be reduced. The net result, however, will be an increase in the risk of a chance variation beyond the control limits. How much this risk will be increased will depend on the degree of skewness. It may not increase the risk very much. If variations in quality should follow a Poisson distribution,[7] for example, for which $p'n = 0.8$, the risk of exceeding the upper limit by chance would be raised by the use of 3-sigma limits[8] from 0.001 to 0.009. On the other hand, the risk of exceeding the lower limit is reduced from 0.001 to 0. The net difference is 0.008.[9] The difference between 3-sigma limits and 0.001 probability limits for an R-chart is shown in Figure 18.5.

The use of 3-sigma limits is justified on the grounds that in the case of variables (as contrasted with attributes) the sampling distribution of X is

---

[7] See Chapter 4, Section 5.2.

[8] For a Poisson distribution the mean and σ'^2 both equal $p'n$. Hence the upper 3-sigma limit would come at $0.8 + 3\sqrt{0.8} = 3.48$. For $p'n = 0.8$, the probability of getting 3 or less is 0.991.

[9] Also see *Control Chart Method of Controlling Quality During Production* (ANSI Std. Z1.3) (New York: American National Standards Institute, 1985).

FIGURE 18.5
Illustration of an *R*-Chart Showing Both 3-Sigma and 0.001 Probability Limits*

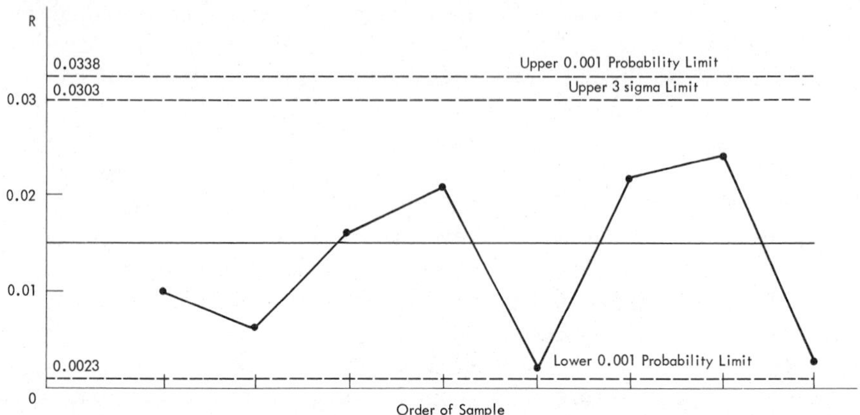

* There is no lower 3-sigma limit, since $\bar{R} - 3\sigma'_{Rn}$ is less than zero in this case.

frequently not known well enough to compute probability limits and that 3-sigma limits have been found to give good practical results.[10] It will be pointed out in Chapter 21, however, that in certain circumstances 2-sigma and even 1.5-sigma limits may be more economical than 3-sigma limits.

It is to be noted that the limits on a control chart need not be constant. In fact, if the size of the sample inspected varies from one time to another and if the vertical scale on the chart measures absolute variations in quality, such as fraction nonconforming, mean density, or the like, the control limits will necessarily be different for each occasion. For in such cases both probability limits and 3-sigma limits will vary with the size of the sample. To avoid confusion on the part of the operating staff, constant limits are sometimes drawn on the basis of an average sample size, and the exact limit for any case is figured only when the sample result falls close to this average limit. Another procedure is to change the vertical scale so that it measures deviations from average quality in terms of standard deviation units and thus becomes independent of the size of the sample. For examples of control charts with samples of varying size see Figures 19.5, 197.7, and 19.9.

British writers suggest the use of inner or *warning* limits in addition to the ordinary *action* limits. These are essentially 2-sigma limits. (The British prefer probability limits to sigma limits and actually use 0.001 and 0.025

[10] If 3-sigma limits are used, the Camp-Meidel variation of Tchebychev's inequality (see Chapter 4, above) will give a maximum figure for the probability of exceeding a control limit, whatever the statistical law of variation in quality. For the Camp-Meidel formula says that under fairly general conditions the probability that a variable will deviate from its mean by more than $k\sigma'$ is less than $1/(2.25k^2)$. Hence the probability of exceeding a 3-sigma control limit is less than 1/20.25 or less than 0.050. This is somewhat greater than 0.001 but is still not very large.

limits instead of 3-sigma and 2-sigma limits.) If a point falls outside or near the inner limit, they advise additional data be immediately taken and the combined results compared to limits calculated for the enlarged sample size.[11]

4. THE OPERATING CHARACTERISTIC CURVE FOR A CONTROL CHART

When a control chart is used to pass judgment on whether a process was or is currently in control at a given level, the chart has associated with it an operating characteristic curve. This curve shows how the control chart operates under varying process conditions. An OC curve for a fraction-non-conforming chart used to control current ouput is shown in Figure 18.6.

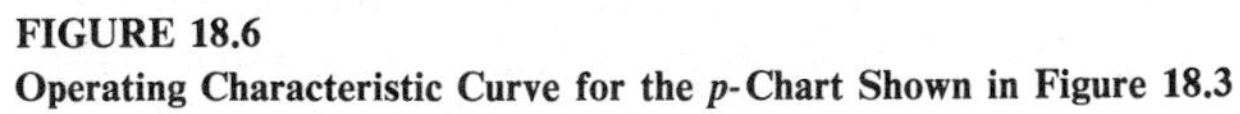

FIGURE 18.6
Operating Characteristic Curve for the *p*-Chart Shown in Figure 18.3

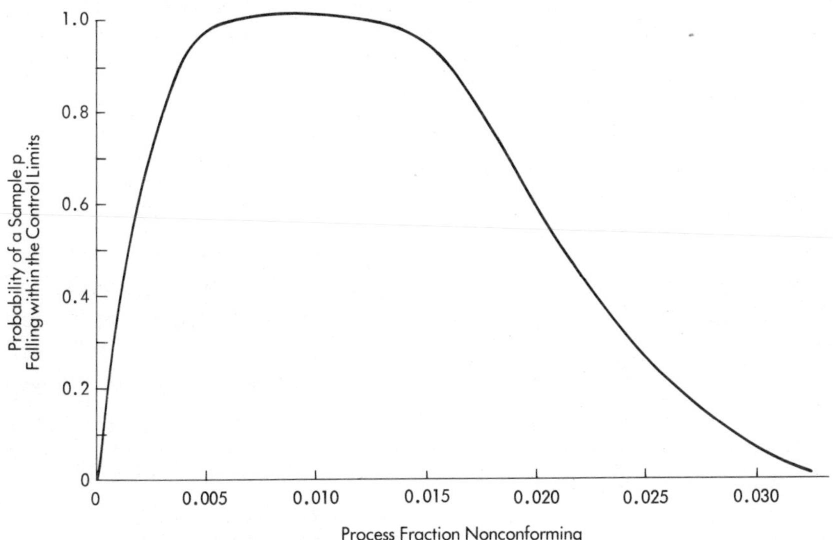

This gives the probability of a single-sample result's falling within the control limits when the process fraction nonconforming is actually above or below the central line shown in the chart. It shows the risk of saying that the process is in control at the designated level because a point falls within the control limits, when actually the process is operating at a different level. The OC curve depicted in Figure 18.6 shows that in the use of the control chart of Figure 18.3 there is a risk of 0.30 that, when a sample point is

[11] See, for example, B. P. Dudding and W. J. Jennett, *Quality Control Charts* (British Standard 600 R: 1942) (London: British Standards Institution, 1942).

plotted, it will fall within the control limits, even though the actual process average is at 0.024 and not 0.0114. In other words, the OC curve for a control chart used to control current output shows the chance of *not* catching a shift in the process average on the first sample taken after the shift has occurred. Since successive samples are presumably independent of one another if the process is operating in a random manner, the probability computed from the OC curve can be used to determine the chance of not catching a shift of a specified amount within 2, 3, or h samples taken after the shift has occurred. For if P_a is the probability of not catching a shift on the first sample, P_a^h is the probability of not catching it within h successive samples.

The OC curve for a control chart used to study past output would show the probability of all n points falling inside the limits for the given set of sample data studied, expressed as a function of the actual process characteristics. This is usually a complicated function, and great care must be taken in specifying the conditions to which it applies. The important point for the reader to note here is that the OC curve for a chart used to control current output is quite a different thing from an OC function that pertains to a test of the hypothesis of control applied to past data.

5. THE SIZE OF SAMPLE AND FREQUENCY OF SAMPLING FOR CURRENT CONTROL

If the output of a process is inspected 100 percent, then the sample of output that occurs in an hour or a day or a week may be the sample taken for control chart purposes. In this case, the sample size is determined by the rate of production and will vary from period to period. If a control chart is based on a sample especially selected for control purposes, both the size of the sample and the frequency of sampling will have to be determined by the control authorities.

Before considering the practical solutions of this problem, let us consider how the size of the sample affects a control chart and its OC curve. A larger sample will cause the limits of a control chart to be closer to the control line on the chart. This is because the standard deviations of such statistics as p, $\bar{X}$, and R, vary inversely with $\sqrt{n}$. Hence the larger the n, the smaller the standard deviation, and the closer 3-sigma or 0.001 probability limits will be to the central line on the chart. The effect of these tighter control limits is to pull in the OC curve and thus reduce the risk of not catching a shift in the process (see Figure 18.7).

However, we must also consider the frequency of sampling. Large samples taken at short intervals would, of course, give the best protection against shifts in the process, but this will be expensive. The practical problem, then, is whether to take larger samples at less frequent intervals or smaller samples at more frequent intervals. A complete solution of this problem would require knowledge of the costs involved in not catching shifts when they occur,

FIGURE 18.7
Operating Characteristic Curves for Two *p*-Charts with Samples of Different Size
(both charts have the same central line)

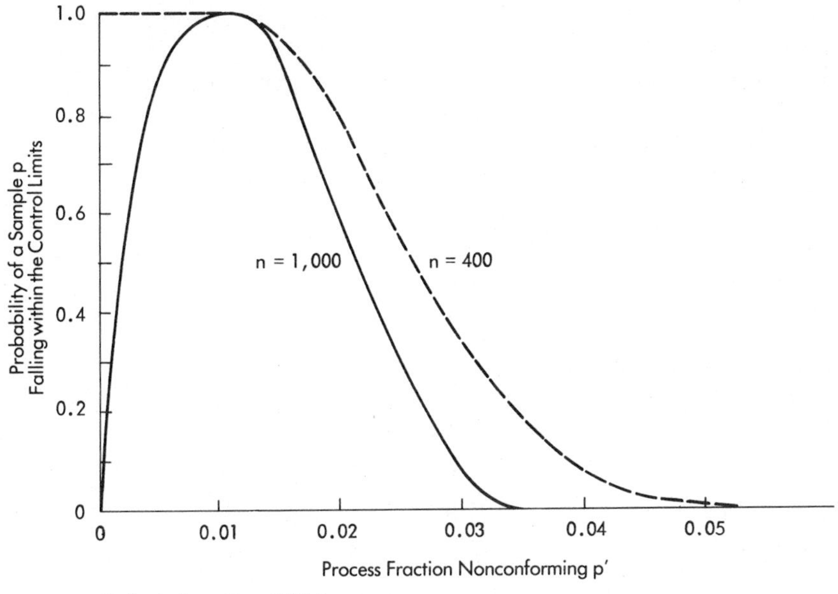

costs connected with the amount and frequency of the inspection process, and the a priori probabilities of certain shifts occurring. Special theoretical solutions to this problem have been offered and will be discussed in subsequent chapters. Also practical procedures that have been suggested will be discussed as the various kinds of control charts are taken up.

6. SUBGROUPING

It is important to note that the samples on a control chart should represent subgroups of output that are as homogeneous as possible. In other words, the subgroups should be such that if assignable causes are present, they will show up in differences between the subgroups rather than in differences between the members of a subgroup. A natural subgroup, for example, would be the output of a given shift. It would not make sense to take as a subgroup the ouput for an arbitrarily selected period of time, especially if this overlapped two or more shifts; for a difference in shifts might be an assignable cause that in the second case might not be detected by the variation from sample to sample. Similarly, if a process used four machines, say, it would be better to take a separate sample from the output of each machine than to have each sample made up of items from all four machines. For, again, differences between machines may be an assignable cause of variation that it is the object of the control chart analysis to detect.

Subgrouping also ties in closely with determination of control limits. In the case of control charts for means and ranges, for example, the control limits for variation *between sample means* are based directly on the average range of variation *within samples,* i.e., limits for variation between subgroups are based on variation within subgroups. For it is when the variation between subgroups exceeds that which may be predicted on the basis of the variation within subgroups that the process is declared to be out of control.

Nothing is more important in the setting up of a control chart than the careful determination of subgroups.

7. THE THEORY OF RUNS IN RANDOM DATA

In analyzing a control chart, the theory of runs will be found very useful. Although the discussion that is introduced here is oriented toward control charts, it can be applied to any series of data.

7.1. Sampling Distribution of Runs

A run is a succession of items of the same class. When the output of a process, for example, is classified as nonconforming and conforming, a succession of nonconforming units would be a "run" of this class of units. As another example, consider an ordered series plotted about its average, as shown in Figure 18.8. A point above the average may be considered as belong-

FIGURE 18.8
A Fraction-Nonconforming Chart for the Punch Press Section of an Aircraft Factory, 1944–45

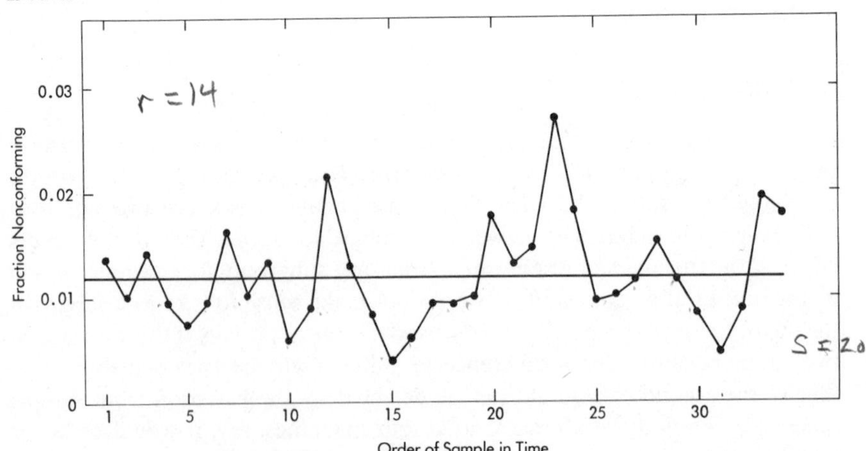

SOURCE: Adapted with slight modifications from Figure I on page 7 of OPRD Quality Control Reports, No. 9, *Statistical Quality Control at Lockheed,* by J. H. Crawford and P. C. Hammer.

ing to one class, and a point below the average (ignoring points exactly equal to the average) may be considered as belonging to another class. In such cases we speak of a run above average or a run below average. There are also "runs up" and "runs down." A succession of increases in value is a run up and a succession of decreases is a run down. The run, it will be noted, refers to the increases or decreases, not to the numbers which produce the increases or decreases. Thus, in the series 5, 4, 6, 8, 10, 12, 11, there is a run up of 4, since there are four increases in a row. Likewise, 7, 10, 8, 6, 5, 4, 3, 2, 4, illustrates a run down of 6.

The runs appearing in a series of data may be studied in various ways. Two ways that have been fruitful are (1) to count the total number of runs of any given class and (2) to note the length of the longest run of a given type. For *random* series, an investigation has been made of both the distribution of the total number of runs and the distribution of the lengths of runs of various kinds. Tables N1 to Q of Appendix II give various probability points of several of these distributions. They may be used to test the hypothesis of randomness in an ordered series of data. A discussion of these tables and their use follows. The discussion will be restricted to cases in which the data are divided into only two classes; e.g., nonconforming and conforming, above average and below average, increases and decreases.

7.2. Distribution of the Total Number of Runs

Tables N1 and N2 of Appendix II give the 0.005 and 0.05 points of the distribution of the total number of runs of two classes of elements in a purely random series. To employ these tables, we must first note the total number of items in each class of elements. Call the smaller number of items r and the larger number of items s. Then look under the proper r and s entries and find the limiting value for the total number of runs. If the given number of runs is less than that shown in Table N1, the probability is less than 0.005 that it could have been produced by a random process. If the given number of runs is less than that shown in Table N2, the probability is less than 0.05 that it could have been produced by a random process.

These tables are taken from S. Swed and C. Eisenhart, "Tables for Testing Randomness of Grouping in a Sequence of Alternatives," *Annals of Mathematical Statistics,* 14 (1943), pp. 66–82. The basic theory will be found in W. L. Stevens, "Distribution of Groups in a Sequence of Alternatives," *Annals of Eugenics,* 9, Part I (1939), pp. 10–17, and A. Wald and J. Wolfowitz, "On a Test Whether Two Samples Are from the Same Population," *Annals of Mathematical Statistics,* 11 (1940), pp. 147–62.

To illustrate the use of Tables N1 and N2, consider Figure 18.8. This shows the fraction nonconforming in a sample of the weekly output of a punch press section of an aircraft factory. The two classes of elements are the fractions nonconforming above the central line on the chart and the fractions nonconforming below the central line. The number of points above the line is 14; the number below the line is 20. Hence $r = 14$ and $s = 20$. Above the line the number of runs is

Runs of 1	5
Runs of 2	2
Runs of 5	1
Total	8

Below the line the number of runs is

Runs of 1	2
Runs of 2	1
Runs of 3	2
Runs of 4	1
Runs of 6	1
Total	7

The total number of runs of both kinds is $8 + 7 = 15$. In Tables N1 and N2 we find that for $r = 14$ and $s = 20$, the 0.005 limiting value is 9 and the 0.05 limiting value is 12. Hence we cannot conclude from this analysis that the number of runs is less than would be expected on the assumption of randomness.

We can apply the same analysis to the number of runs up and down. The number of changes in the whole series consists of 15 decreases, 17 increases, and 1 no change. We ignore the latter and take $r = 15$ and $s = 17$.

The runs down are

Runs down of 1	5
Runs down of 2	2
Runs down of 3	2
Total	9

The runs up are

Runs up of 1	2
Runs up of 2	6
Runs up of 3	1
Total	9

The total number of runs up and down is 18. For $r = 15$ and $s = 17$, Table N1 gives the 0.005 limiting value as 9 and Table N2 gives the 0.05 limiting value as 11. Since 18 is greater than either of these values, we conclude that, on the basis of the runs up and down, there is no reason to reject the hypothesis of randomness.

Table N3 of Appendix II is a continuation of Tables N1 and N2 for values of $r = s$ up to 100. This is useful in analyzing a long series of data when the dividing line between one class and another is taken as the median. For if there is an odd number of cases, there will be as many cases above the middle value (the median) as below it. If there is an even number of cases, the median will fall between the two middle values, and again there will be as many cases above the median as below it. The exceptions to the above occur when there is more than one case with the median value. In this instance, some of the cases having the median value must be assigned to the side with the smaller number of cases, so as to make an equal number above and below the median. In practice, it is probably best to assign all except possibly one of the median values to one side or the other of the line. However the assignment is made, we should follow the conservative rule of trying to increase and not decrease the number of runs.

For example, Figure 18.9 presents measurements of the inner diameters of 78 landing gear trunions. For the sake of discussion, it will be assumed that the trunions are presented on the chart in the order of their production, so that Figure 18.9 may be viewed as a graph of how the process operated over time. Of these 78 measurements, there are 11 which have the median value of 1.7495 inches. There are 36 above the median value and 31 below. In order to use the total runs above and below the median as a statistic for testing randomness, we shall need to add 8 of the median values to the lower side and 3 to the upper side. To be conservative, we have done this so as to increase the number of runs. The points we have elected to put above the median are marked with an *a;* those that we have decided to put below the median are marked with a *b.* The final result is to put 39 points above the median and 39 points below it. If there had been an odd

FIGURE 18.9
Variations in the Inner Diameter of Landing Gear Trunions

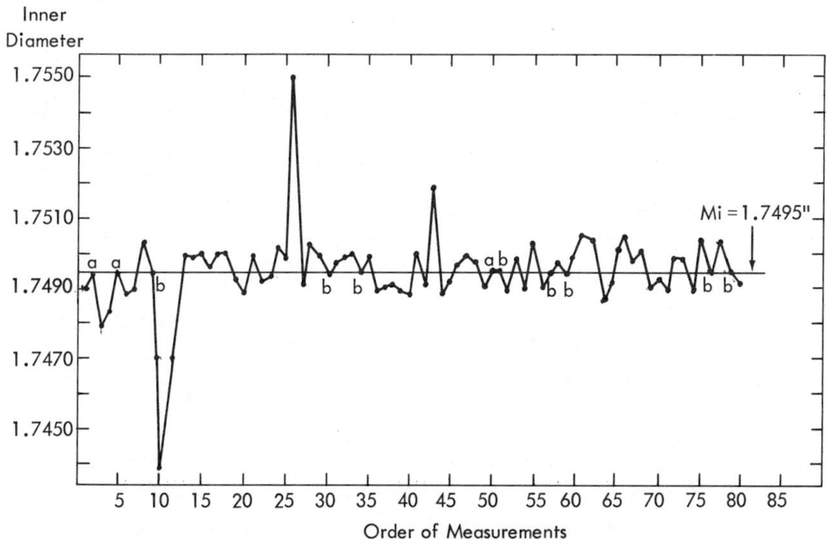

SOURCE: Taken from Quality Control Reports, No. 9, *Statistical Quality Control at Lockheed,* p. 10.

number of cases, we would have disregarded one of the median values. A good one for this purpose would have been the third from the beginning (the first marked b), since neglecting it would not have decreased the number of runs.

When we count the number of runs in Figure 18.9, we find there are 21 runs above the median and 21 below the median, or a total of 42 runs. Table N3 shows that for $r = s = 39$, the 0.005 point is 28 and the 0.05 point is 32. Hence, at either level, there is no indication that nonrandom influences are at work to cause runs in the series.

7.3. Distribution of the Lengths of Runs of Any One Class

Table P of Appendix II gives the 0.05, 0.01, and 0.001 points of the cumulative distribution of lengths of runs on either side of the median of n cases. This is useful in testing randomness by considering the lengths of runs. The procedure may be carried out as follows.

Divide the total set of data into two equal sections, those above the median and those below it. If this cannot be done because there are several cases exactly equal to the median, then one or more of the latter can be assigned to the side that has the smaller number of cases, so that there will be equal numbers on either side. Again, the conservative rule is to make the assignment as far as possible so that the short runs are increased in size rather than the long runs. If there is an odd number of cases, one of the median values must be ignored.

After having divided the data into two groups of $n/2$ each, we determine the longest run on either side of the median and note whether this is greater than one of the limiting values listed in Table P. If our sample contains a run greater than a selected one of these limits for the given value of n, we conclude that some nonrandom influences have been at work.

For example, consider the 49 points plotted in Figure 18.10. The median of these points is 0.10, and there are two values equal to the median. Since there are 24 points above the line and 23 points below it, one of the median values must be added to the latter group. The second median value is the best to pick in this case, since it merely makes a run of 1 into a run of 2. When this adjustment is completed, we notice that the longest run above or below the median line is 5. According to Table P, the limiting value for a set of 50 points is a run of 10 for a 0.05 level of significance and runs of 11 and 12 for the 0.01 and 0.001 levels. The lengths of the runs in this case give no evidence of nonrandom influences.

It might be noted at this point that there is a rule of thumb often followed by quality control engineers that a run of 7 or more indicates nonrandom influences. Table P indicates that for 20 items the 0.05 limiting value is a run of 7. For more items or for lower significance levels (e.g., 0.01 or 0.001) the limiting values are higher than 7. In some cases, therefore, the rule of 7 may lead to the search for nonrandom influences when none exists more often than one is usually willing to take such chances in quality control.

FIGURE 18.10
Variations in the Fraction of Bellows Rejected at Final Inspection

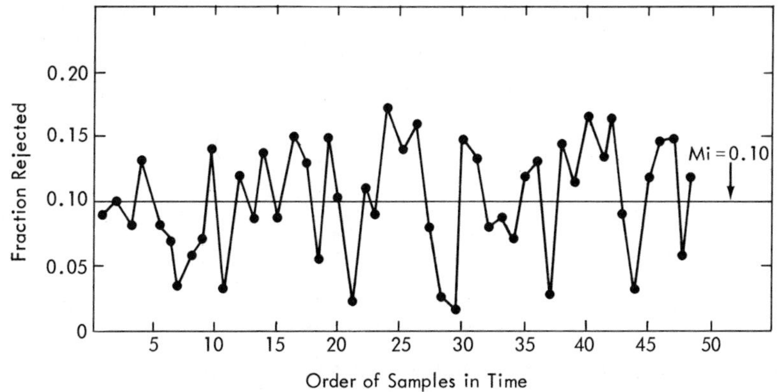

SOURCE: Adapted with permission from Wilbur L. Burns, "Quality Control Proves Itself in Assembly," *Industrial Quality Control,* January 1948, pp. 12–17.

Table Q gives the limiting values of runs up and down in a series of *n* numbers. More precisely, the table lists the shortest run for which the probability of an equal or greater run does not exceed 0.0032 in one case and 0.0567 in the other case. These may be considered limiting values for randomness of a series. For example, consider again the 49 points plotted in Figure 18.10. The longest run up or down in the series is a run of 3. The 0.05 and 0.003 limiting values are 6 and 7. There is, therefore, no evidence of nonrandom influences.

7.4. The Average Run Length (ARL) Curve for a Control Chart Used to Control Current Output

The discussion of the distribution of the lengths of runs in the previous section was related to criteria for determining whether a process is in control or out of control. The theory of run lengths also provides an important measure of the effectiveness of a chart that is being used to control the quality of current output. This measure is the average length of a run of in-control points that follows immediately after there has been a specified change in the process. A plot of the average run length (ARL) against the changed state of the process to which it is related is referred to as the ARL curve of the chart for the given type of change.

An ARL curve for a chart used to control the quality of current output can be derived if we can calculate (1) the probability that a given change in the process, say an increase in the process average, will be detected on the first sample after the change has occurred, (2) the probability it will not be detected on the first sample, but will be detected on the second sample, (3) the probability it will not be detected on the first or second sample, but will be detected on the third sample, and so on. For example, in the case of the p-chart shown in Figure 19.4 of the next chapter, the individual samples are assumed to be independent of each other. The chart has a single (upper) control limit (UCL) and for a specified increase in the process fraction nonconforming the probabilities of various run lengths will be:

Run Length (r)	*Probability*
1	$(1-P_a)$
2	$P_a(1-P_a)$
3	$P_a^2(1-P_a)$
⋮	⋮
k	$P_a^{k-1}(1-P_a)$
⋮	⋮

where P_a is the probability that a single sample fraction nonconforming will be less than or equal to the UCL. The average run length (ARL) will thus be given by the sum of the infinite series

$$\textbf{(18.1)}\quad (1)(1-P_a)+(2)P_a(1-P_a)+(3)P_a^2(1-P_a)+\ldots + (k)P_a^{k-1}(1+P_a)+\ldots=(1-P_a)\sum_{r=1}^{\infty}(r)P_a^{r-1}$$

Now it is well known that $\sum_{r=0}^{\infty} q^r = 1/(1-q) = 1/p$ where $p = (1-q)$.

If we differentiate both sides with respect to q, we have $\sum_{r=1}^{\infty} rq^{r-1} = 1/(1-q)^2 = 1/p^2$, and $p\sum_{r=1}^{\infty} rq^{-1} = 1/p$. The ARL given by the sum of the infinite series (18.1) will thus equal $1/(1-P_a)$. A plot of the ARL given by the sum of (18.1) against the increased process average to which it pertains will yield the ARL curve for the p-chart of Figure 19.4. This is shown in Figure 19.15 of Chapter 19.

8. SUMMARY OF OUT-OF-CONTROL CRITERIA

It might be helpful to the reader to summarize at this point the various criteria that have been mentioned above as suggesting that a process is out of control. These are as follows:

1. One or more points outside the limits on a control chart.
2. One or more points in the vicinity of a warning limit. This suggests the need for immediately taking more data to check on the possibility of the process being out of control.
3. A run of 7 or more points. This might be a run up or run down or simply a run above or below the central line on the control chart.
4. Cycles or other nonrandom patterns in the data. Such patterns may be of great help to the experienced operator. Other criteria that are sometimes used are the following:

5. A run of 2 or 3 points outside of 2-sigma limits.
6. A run of 4 or 5 outside of 1-sigma limits.

It is to be noted that the use of a multiplicity of criteria will increase the risk of looking for trouble when none exists. If there are b independent tests in the "battery of tests," and if α_i is the probability that the ith test will falsely indicate an out-of-control situation, then the overall risk of getting a false indication of lack of control is

$$\alpha_{\text{overall}} = 1 - \prod_{i=1}^{b} (1 - \alpha_i) \tag{18.2}$$

There is some question as to whether all of the above criteria are independent of each other so formula (18.2) must be used with care. Hilliard and Lasater[12] show in a simulation study that when a 3-standard-deviation-limits test, a 0.05 length of runs test, and a 0.05 number-of-runs test are used in combination on either the $\bar{X}$-chart or the R-chart or both, with 25 samples of 4 items each, no standard given, the probability of one or more false indications of nonrandomness is about 0.25. The authors note that this risk can be reduced by cutting down the risks on individual tests (allowing a longer run above the central line, for example), but they also suggest that maybe a high rate of false alarms is not bad, since it is likely to be more important to detect real out-of-control situations. Also they point out that the combination of tests has worked well in practice.

9. PROBLEMS

18.1. Analyze the first 100 items of Table 3.2 (Figure 4.16) for evidence of runs.

18.2. Analyze the data of Problem 3.3*a*(i) for runs, going down each column.

18.3. Analyze the data of Problem 3.3*a*(ii) for runs, going down each column.

18.4. Analyze the data of Problem 3.3*a*(iv) for runs, going down each column.

18.5. Analyze the data of Problem 3.3*a*(v) for runs, going down each column.

18.6. Analyze the data of Figure 19.2 for runs. (See text, page 440.)

18.7. Analyze the data of Figure 19.4 for runs. (See text, page 442.)

10. SELECTED REFERENCES*

American Society for Quality Control (B: ANSI/ASQC Std. Z1.1–1985), Barnard (P '59), Barton (P '67), Box and Jenkins (P '62), Dudding and Jennett (B '42), Grubbs (P '83), Hilliard and Lasater (P '66), Juran (B '63), Montgomery (P '80), Mosteller (P '41), Mullet (P '77), Ott (B '75), Shewhart (B '31, B '39 and P '39), Swed and Eisenhart (P '43), Tippett (B '50), Vance (P '83), Vance and McDonald (P '79), Wallis and Moore (B '41), and Wilks (B '48 and B '62).

---

[12] See J. E. Hilliard and H. Alan Lasater, "Type One Risks when Several Tests Are Used Together on Control Charts for Means and Ranges," *Industrial Quality Control,* August 1966, pp. 56–61.

* B and P refer to the Book and Periodical sections, respectively, of the Cumulative List of References in Appendix V.

19

Control Charts for Fraction Nonconforming (p-Charts)

A p-chart shows variations in the fraction nonconforming of output. The central line on the chart is an estimate of the fraction nonconforming of the process, and the upper and lower control limits are usually 3-sigma limits. If p' is the process fraction nonconforming on the assumption that the process is under control, then σ'_p is $\sqrt{\frac{p'(1-p')}{n}}$. The control limits will, therefore, be constant only when n is constant. A picture of a p-chart for constant n is shown in Figure 18.3.[1] In this chapter we shall first show how a p-chart can be set up and then discuss its use in attaining control and in detecting subsequent departures from control.

1. SETTING UP A p-CHART — n CONSTANT

Consider the setting up of a p-chart with reference to the production of railway car side frames. Suppose that these frames are in continuous production in a foundry and that a random sample of 50 frames is taken from each day's output and inspected. Let us say that the results of 28 days of past operations are as follows:[2]

---

[1] See above, p. 422.

[2] These and subsequent data have been adapted from the chart on p. 15 of the paper by William M. Armstrong, "Foundry Applications of Quality Control," *Industrial Quality Control,* May 1946, pp. 12–16.

Date	Number Rejected	Date	Number Rejected
April 27	4	May 11	4
28	9	12	3
29	10	14	11
30	11	15	8
May 1	13	16	14
2	30	17	21
3	26	18	25
4	13	19	18
5	8	21	10
6	23	22	8
7	34	23	18
8	25	24	19
9	18	25	4
10	12	26	8

Our job is to set up a control chart for controlling the fraction nonconforming of railway car side frames.

We must begin by seeing whether the process is already in control. To do this we set up a preliminary chart based on the given data. Our first task is to estimate the average fraction nonconforming of the existing process estimate. This is done by finding the total number of nonconforming items for the period for which the preliminary analysis is to be undertaken and dividing by the total number inspected. For the above data, we have $\bar{p} = 407/1{,}400 = 0.291$. If the process is under control, this 0.291 becomes an estimate of the process fraction nonconforming p' and is taken as the central line on our control chart. We shall follow the standard procedure and set up 3-sigma limits. Using $\bar{p} = 0.291$ as an estimate of p', we can take as an estimate of σ'_p the quantity[3]

(19.1) $$\hat{\sigma}_p = \sqrt{\frac{\bar{p}(1-\bar{p})}{n}} = \sqrt{\frac{(0.291)(0.709)}{50}} = 0.064$$

Then $3\hat{\sigma}_p = 0.192$. This added to and subtracted from 0.291 gives 0.483 as the upper limit for our chart and 0.099 as the lower limit. The control chart with its central line and upper and lower control limits is shown in Figure 19.1, together with the past results plotted as fractions nonconforming. It will be noted that five points are outside the limits on the high side and four on the low side. The process is obviously not in control.

What is now the next step? Our ultimate objective is to get the process under control. To do this, we must, as the work proceeds, investigate points which are out of control, so that we can remove or guard against any special causes of poor quality and possibly perpetuate special causes of a genuinely high quality. In this case, let us suppose that investigation reveals the follow-

[3] The placing of a ^ over σ indicates that it is an estimated value.

FIGURE 19.1

A p-Chart for Analyzing Past Data

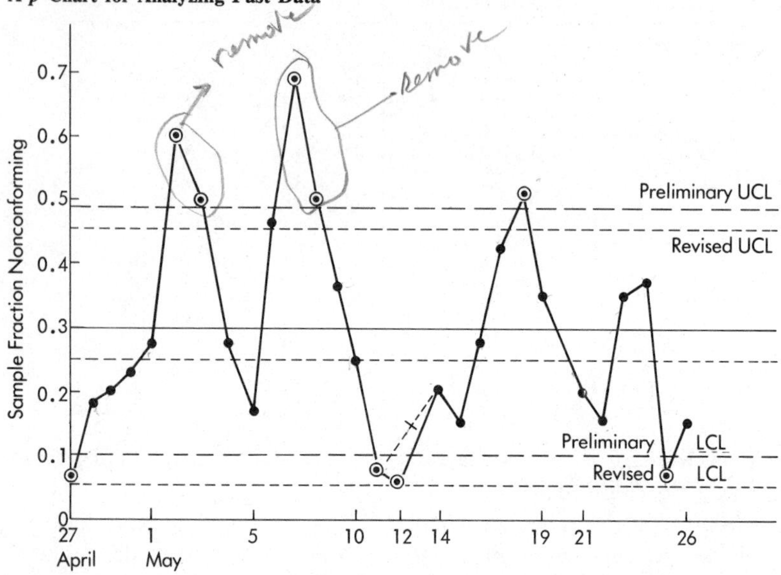

ing. On May 2 and 7, several new workers have been added to the force, and the foreman attributes the high fractions nonconforming on May 2 and 3 and May 7 and 8 to the breaking in of these men. Also, a checkback reveals that the correct figure for May 12 was 7, not 3, the latter being a mistake in copying the original records. No assignable causes were found for other points outside the limits. As a result of this investigation, it is decided to have all new men given a special training course before being placed on the job. It is also decided that the whole force might profit from extra training, and such a program is set up on May 28.

In the meantime the data are analyzed further to derive a basis for setting up a control chart for controlling future production. The first step is to throw out all data for which assignable causes have been found and to correct the figure for May 12. Then $\bar{p}$, UCL, and LCL are recomputed for these revised data. The results are as follows:

$$\bar{p} = \frac{296}{1200} = 0.247$$

$$\text{UCL} = 0.247 + 3\sqrt{\frac{(0.247)(0.753)}{50}} = 0.430$$

$$\text{LCL} = 0.247 - 3\sqrt{\frac{(0.247)(0.753)}{50}} = 0.064$$

When we draw in the new control limits, we note that all the points are within the limits except those for May 6 and May 18. This indicates that there is still some lack of control, but, before we make any further decisions, let us examine the new chart for runs.

There are 24 points altogether; 14 of them are below the central line and 10 above it. There are 5 runs below the line and 4 above it, or a total of 9 runs above or below the central line. For $r = 10$ and $s = 14$, the 0.05 limit on the number of runs is 8 and the 0.005 limits is 6 (see Tables N1 and N2 in Appendix II). Hence the total number of runs above and below average is clearly within the limits of chance. Again, there are 11 runs up and down. For $r = 9$ and $s = 13$, this is even farther within the limits expected by chance (see Tables N1 and N2).

The median of the 24 cases is 0.22, and there are two cases having this value. Assign the first of these to the group of cases above the median, and the second to the group below the median. Examination then shows that there is no long run above the median or below the median. Once more there is no indication of nonrandom forces (see Table P in Appendix II).

Finally, we note that the longest run up or down in the whole series of 24 points is a run up of 4. For $n = 24$, Table Q, Appendix II, indicates that a run up of 4 is not outside the chance limits. (For $n = 20$ the probability of a run up or run down of 4 or more is greater than 0.0567.)

In whatever manner we analyze the 24 points for runs we find no evidence of nonrandom influences.

With only two points beyond the new control limits and with no evidence of nonrandom influences within the limits, further refinement seems unnecessary. Let us therefore take the central line and upper and lower control limits of the revised chart as standards for controlling production in the next month. For the p-chart to control current production, we thus have the following standard values:[4]

$$p'' = 0.247$$

$$\text{UCL} = 0.430$$

$$\text{LCL} = 0.064$$

The above is a solution of a particular problem in which the assignable causes found for the points outside the control limits affected these points only. It is possible, however, that the assignable cause unearthed by investigation might reasonably have affected other points that were themselves not outside the control limits. For example, a substitute workman might have been employed for a whole week. During this week only one day's output was outside the control limits, say; but, since this workman was employed on the other days as well, the process fraction nonconforming on these days was probably higher than it should have been Consequently, in recomputing $\bar{p}$, it would be best to discard the data for the whole week.

---

[4] A symbol with a double prime refers to a standard value. In contrast, p' is the actual process value and p a sample value.

It is also possible that an assignable cause may affect the whole last part of the data. For example, suppose that some new raw material had been used in the last two weeks and had caused one or more of the points during that period to be below the lower control limit. If this same raw material was going to be used during the next month, then $\bar{p}$ and UCL and LCL for the control chart for the next month should be based on the last two weeks' operations rather than on the operations of the previous weeks. The object in setting up the next month's chart is to determine limits within which the results may be expected to fall if the process is operating in a random manner. This means that the central line should be set as closely as possible to the level at which the process will actually operate if it does operate in a random manner.[5]

In concluding this section, special note may be made of Figure 19.3, a graphical device for computing 3-sigma limits given p', p'', or $\bar{p}$. With the help of this chart we need no longer compute 3-sigma by formula. We need only enter the abscissa scale of Figure 19.3 with the given value of n, proceed vertically until we hit the line for which p' equals its given value or the value estimated from $\bar{p}$, and then read off 3-sigma on the right-hand scale.

2. OPERATION AND REVISION OF THE p-CHART

For the rest of May and the whole of June the operations of the foundry are as follows:

Date	*Nonconforming Frames in Sample of 50*	*Date*	*Nonconforming Frames in Sample of 50*
May 28	14	June 14	4
30	5	15	5
31	8	16	2
June 1	8	18	3
3	9	19	5
4	4	20	5
5	8	21	4
6	4	22	5
7	10	23	4
8	6	25	6
9	5	26	6
10	9	27	3
11	7	28	6
12	4	29	7
13	2	30	5

[5] If the management is not satisfied with the level attained by the process, it is probably best to revise the whole process rather than artificially to set a lower standard and then seek to force the process down to that standard by investigating each point above the artificially set upper limit. According to theory, such a procedure is not likely to be very successful.

FIGURE 19.2
A *p*-Chart in Operation, Showing Original and Revised Limits

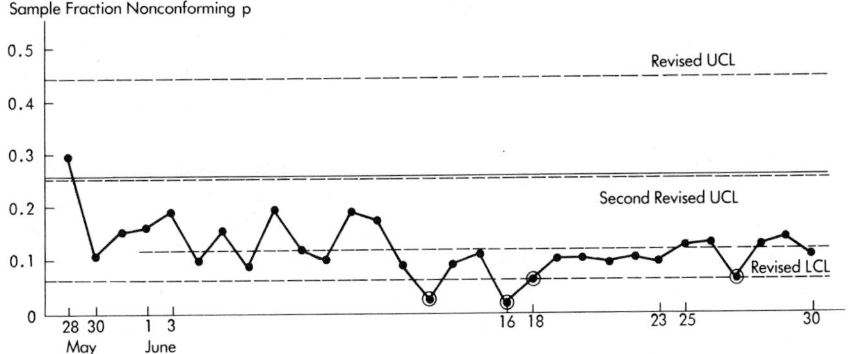

In Figure 19.2 these results are shown plotted on the control chart set up at the beginning of the new period.[6] The striking thing is the new level around which the fraction defective is now obviously fluctuating. That something happened to change the process became definitely clear on June 13, when one of the points fell below the lower control limit. That the average fraction defective had shifted to a lower level was suggested prior to that by the long run of points below the central line on the control chart.

Let us suppose that investigation on June 13 following the occurrence of a point outside the lower control limit indicated that the only apparent change in the process was in the attitude and interest of the operators. For several days following May 28, all the operators had been given extra training, the new quality control system had been explained, and charts were posted in the foundry shop on which the fraction nonconforming was plotted from day to day. The increased interest of the men in the quality of output plus their improved knowledge of the process seemed to be the only assignable cause of the favorable shift in the process fraction nonconforming.

Toward the end of June it is decided to revise the control chart. June 1 seems to be the logical point at which to start, since by that date the new program of education and orientation might be thought to have had its full effect. From June 1 to June 30, the average fraction nonconforming was $146/1{,}350 = 0.108$. This is taken tentatively as the new p''. From Figure 19.3 $3\hat{\sigma}_p$ is computed, and an upper control limit is determined from the formula

$$\text{UCL} = \bar{p} + 3\hat{\sigma}_p = 0.108 + 0.132 = 0.240$$

There is now no LCL, since application of the $3\hat{\sigma}_p$ formula yields a negative result.

---

[6] See immediately below.

The new lines are plotted on the chart to see how well they explain the results for the month of June (see Figure 19.2). It is noted that no points fall outside the new limits. Altogether, 27 samples were taken in the month of June; 11 of these are seen to have a fraction nonconforming greater than the new p'' and 16 a fraction nonconforming less than the new p''. The total number of runs above and below average are 12, which is above the 0.05 chance limit for $r = 11$ and $s = 16$ (see Table N1, Appendix II). The median of the 27 points is 0.10, and there are 6 points with this value. If these 6 points are conservatively assigned to the classes of points above and below the median, there are no long runs above or below the median. There are also no long runs up or down. On all scores, therefore, the process appears to be under control at the new level. The values $p'' = 0.108$ and UCL = 0.240 are therefore adopted as standard values for control purposes for July.

The results for July are as follows:

Date	*Number of Nonconforming Items in Sample of 50*	*Date*	*Number of Nonconforming Items in Sample of 50*
July 1	7	July 18	1
2	8	19	3
3	7	20	4
4	8	21	7
5	3	23	2
6	3	24	7
7	6	25	4
10	4	26	4
11	4	27	2
12	3	28	7
13	4	29	0
14	2	30	5
17	2	31	6

Figure 19.4 shows how these results look on the control chart set up at the beginning of the month.

It will be noted that none of the points in Figure 19.4 fall above the upper control limit. Hence it may be concluded that nothing happened during the period to increase the fraction nonconforming. The chart has no lower limit, so we have to turn to other means of determining whether assignable causes have at any time caused exceptionally low fractions nonconforming. Incidentally, this difficulty might have been avoided at the beginning of the month by taking a sample large enough to give a lower limit on the chart.

According to the rule[7] of 7, the long run below the central line beginning on July 10 suggests that possibly something happened about that time to

[7] See Chapter 18, Section 7.3, for a discussion of this rule.

FIGURE 19.3
Nomograph for Computing Sigma or 3-Sigma, Given p', p'', or $\bar{p}$

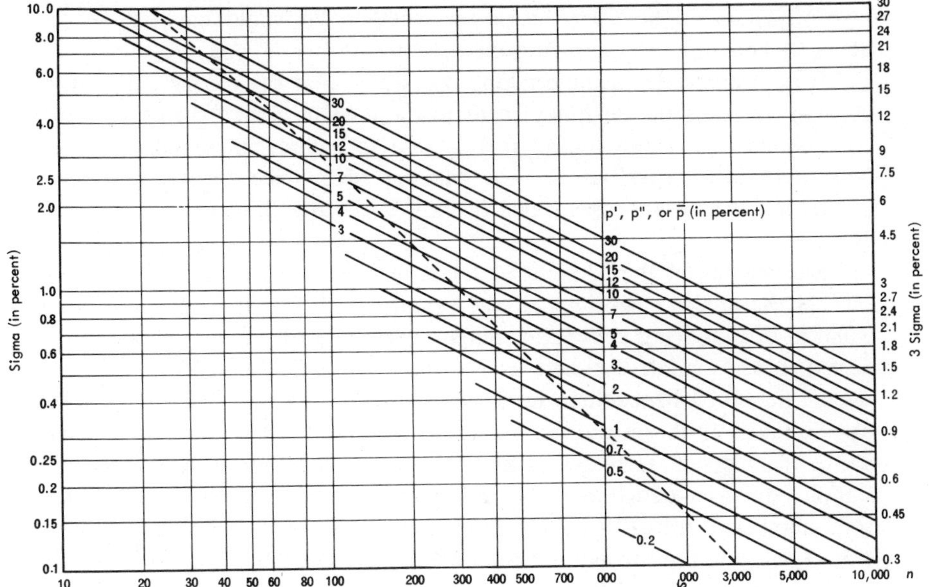

Note: Dotted line indicates sample size for which lower limit will be zero.

SOURCE: Reproduced with permission from John M. Howell. "Aids for Computing Limits for p-Charts." *Industrial Quality Control,* July 1949, pp. 20–23. Howell's original chart is larger and has more guidelines than Figure 19.3. The latter permits only a rough interpolation between the scale markings. The original should be referred to whenever greater accuracy is desired.

FIGURE 19.4
A p-Chart in Operation: A Continuation of Figure 19.2

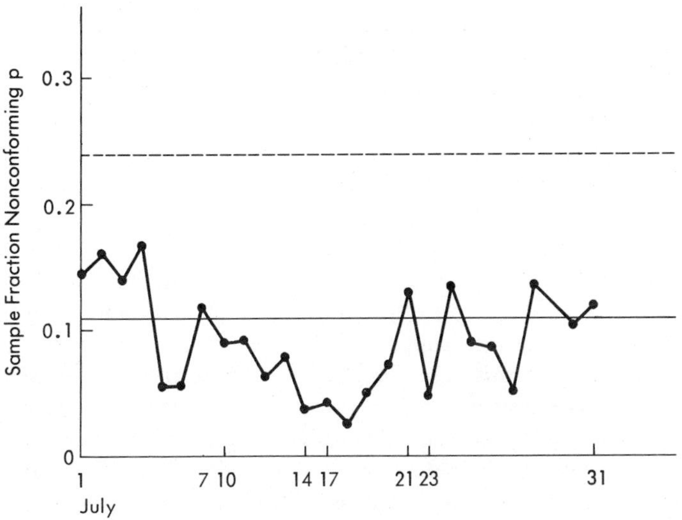

lower the fraction nonconforming. If we followed this rule of thumb, we would have looked for an assignable cause beginning on July 19. For the sake of argument, however, let us see whether we would have arrived at the same conclusion by applying the accepted theory of runs.

It is to be noted, in undertaking our study, that the theory outlined in Chapter 18 refers the lengths of runs to the median, not to the central line on the chart. It is also logical in this case to include in our analysis not only the data for July but also the data for June, since the hypothesis we are testing is that the operations in July are really the same as those in June.

The median of the 53 samples taken in June and July is 0.10. For the sake of the analysis consider all the samples with fraction nonconforming equal to 0.10 as lying above the median, except that for June 9, which may be temporarily ignored. (There are 26 cases below 0.10 and 27 equal to or greater than 0.10; hence there are 53 cases altogether. One of the 0.10 samples can be taken as the median sample, and the rest must be put in the class above the median. Here we picked the sample for June 9 as the median sample, since its omission would not produce any long run or affect the total number of runs.) With this adjustment, it is noted that there is a long run of 9 below the median, beginning July 10. The 0.05 limit for $n = 50$, however, is a run of 10, so that even this long run below the median is not sufficient to indicate an assignable cause. With respect to the total number of runs above and below the median, we have the following: Altogether there are 27 runs above and below the median. For $r = s = 26$, the 0.005 and 0.05 limiting values are 17 and 20 (see Table N3, Appendix II), both of which are much less than the number obtained. In general, therefore, the theory of runs does not suggest the existence of assignable causes producing nonrandom variation. It does not in this instance verify the conclusion derived from applying the rule of 7.

The reader may argue that, although the theory of runs gives us no reason to suspect assignable causes, it still is true that the level of points in July is less than the level in June. The real question, of course, is whether it is significantly lower, that is, whether the difference in levels in the two months might be such as could reasonably occur as a result of chance. So far we are not equipped to answer this question. It is shown below in Chapter 26, however, that there is a reasonable probability that the observed difference could have occurred by chance.

Although the rule of 7 and possibly the general impression of the reader suggest that something is causing a lower fraction nonconforming, standard statistical tests do not lend support to this conclusion. If we follow the guidance of the latter in this case, we shall not revise the control chart at the end of July on the basis of the July figures only. If we wish, we can combine the June and July figures and determine a value of $\bar{p}$ based on the two months' operations. Whether we do this or not, it would be advisable to recommend, for the next month at least, that the size of the sample be increased from 50 to 100, so that the p-chart will have a lower limit (cf. Figure 19.2).

3. p-CHARTS WITH VARIABLE n

The analysis of Sections 1 and 2 was greatly simplified because we took a constant sample of 50 frames. In many cases a fraction-nonconforming control chart is based on 100 percent inspection of output, and the latter varies from day to day. In this case we can proceed in one of three ways.

The straightforward procedure is to compute separate control limits for each sample result as illustrated in Figure 19.6. For if we use 3-sigma limits, our σ'_p varies inversely with $\sqrt{n}$. Hence with a different n we shall have

FIGURE 19.5
A p-Chart with Samples of Varying Size*

* These data pertain to the fraction nonconforming of battery adapters used in Handie-Talkies. See F. A. Palumbo and Edward S. Strugala, "Fraction Defective of Battery Adapter Used in Handie-Talkie." *Industrial Quality Control,* November 1945, pp. 6–8.

FIGURE 19.6
Illustration of a Work Sheet for a p-Chart when n Varies* ($p'' = 0.042$ and $3\sqrt{p''(1-p'')} = 0.6015$; for remainder of data see Figure 19.5)

Date July	*Sample Size*	*Number of Rejects*	*Sample Fraction Nonconforming*	$\frac{3\sqrt{p''(1-p'')}}{\sqrt{n}}$	*UCL*†
20	90	0	0	0.063	0.105
21	105	0	0	0.059	0.101
23	105	4	0.038	0.059	0.101
24	155	8	0.052	0.048	0.090
25	155	2	0.013	0.048	0.090
26	155	0	0	0.048	0.090
27	210	4	0.020	0.042	0.084
28	155	7	0.045	0.048	0.090
30	155	5	0.032	0.048	0.090

* These data pertain to the fraction nonconforming of battery adapters. See F. A. Palumbo and Edward S. Strugala "Fraction Defective of Battery Adapter Used in Handie-Talkie," *Industrial Quality Control,* November 1945, p. 8.

† $UCL = p'' + 3\sqrt{p''(1-p'')/n} = 0.042 + 0.6015/\sqrt{n}$.

different limits. Since σ'_p varies inversely with $\sqrt{n}$, the larger n is, the smaller σ'_p is, and the narrower the limits. On the other hand, the smaller n is, the larger σ'_p is, and the wider the limits. Figure 19.5 shows a p-chart with varying limits and Figure 19.6 shows a section of the worksheet from which Figure 19.5 was derived. In this case there is no lower limit, so only the upper limit is shown.

A second procedure is to express our variable in standard deviation units. In other words, instead of plotting the sample fraction nonconforming in each case, we plot $\frac{p - p''}{\sigma''_p}$ or $\frac{p - p''}{\sqrt{\frac{p''(1 - p'')}{n}}}$. This "stabilizes" our variable, and the resulting chart is called a stabilized p-chart. On this chart the limits are, of course, constant and occur at −3 and +3. Figure 19.7 shows Figure 19.5 in stabilized form and Figure 19.8 shows a section of the worksheet that was used to derive Figure 19.7.

The third procedure is to base our limits on an average output. This will give us constant limits, which in most cases will be fairly close to the exact limits for each sample. If the variation in sample size is especially great in a given case or if a sample point falls close to the approximate limits, so that we need to know the precise limits to determine whether the point is in or out of control, we can take the trouble to compute exact limits for that particular case. Figure 19.9 illustrates this third procedure. Since the last method involves the least work, it is the one most commonly used.

When the sample size varies, we must be especially careful about the interpretation of runs up and down on a p-chart; for, if the sample size varies, a given variation in the sample fraction nonconforming does not have the same significance in each case. For example, if $p'' = 0.10$, a fraction nonconforming of 0.12 represents a sample result 2 point above average, and a sample fraction nonconforming of 0.14 represents a sample result 4 points above average. But, if the first sample consists of 400 items and the second of 25, the standard deviation of the second will be 4 times as large

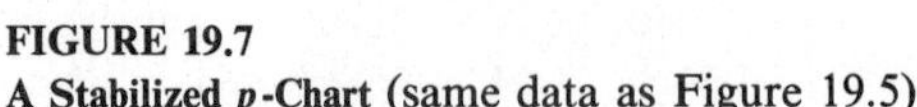
FIGURE 19.7
A Stabilized *p*-Chart (same data as Figure 19.5)

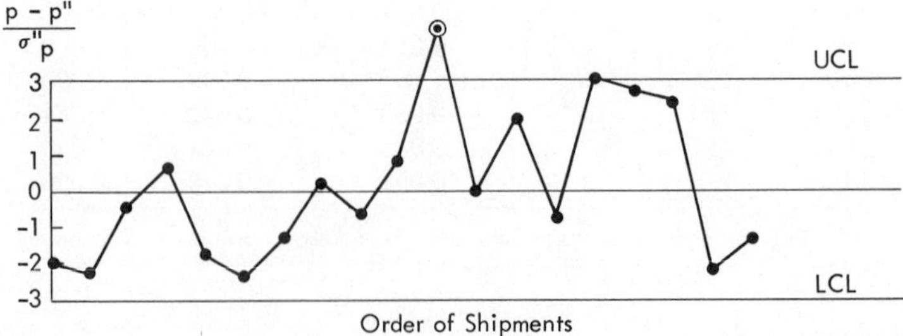

FIGURE 19.8
Illustration of a Work Sheet for a Stabilized *p*-Chart* ($p'' = 0.042$; $\sqrt{p''(1-p'')} = 0.2005$; UCL = 3.0; LCL = −3.0.)

Date July	*Sample Size* n	*Number of Rejects* c	*Sample Fraction Nonconforming* $p = c/n$	$\sqrt{\frac{p''(1-p'')}{n}} = \sigma''_p$	$p - p''$	$\frac{p-p''}{\sigma''_p}$
20	90	0	0	0.021	−0.042	−2.0
21	105	0	0	0.020	−0.042	−2.1
23	105	4	0.038	0.020	−0.004	−0.2
24	155	8	0.052	0.016	+0.010	+0.6
25	155	2	0.013	0.016	−0.029	−1.8
26	155	0	0	0.016	−0.042	−2.6
27	210	4	0.020	0.014	−0.022	−1.6
28	155	7	0.045	0.016	+0.003	+0.2
30	155	5	0.032	0.016	−0.010	−0.6

* Based on Figure 19.6. Also see Chapter 23, Section 4, for a discussion of how to use binomial probability paper in constructing a stabilized *p*-chart.

FIGURE 19.9
A *p*-Chart with a UCL Based on an Average Sample Size (same data as Figure 19.5)

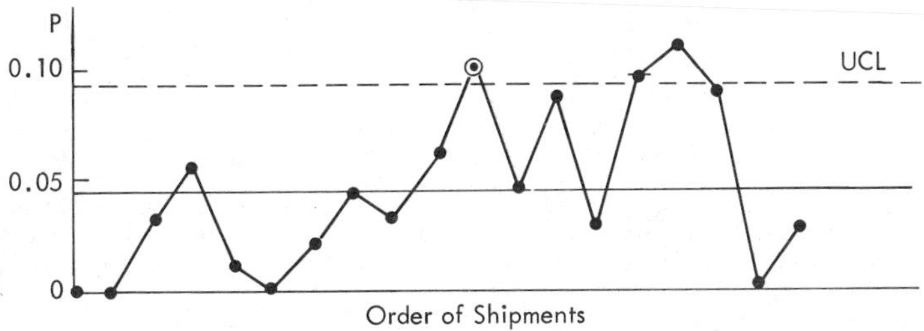

as the standard deviation of the first. Hence, in standard deviation units, the first will represent a deviation twice as large as the second. Whereas on an ordinary *p*-chart there would be a rise from the first to the second sample, on a stabilized *p*-chart there would be a fall. Points 15 and 16 in Figures 19.5 and 19.7 illustrate this phenomenon. Since, on the stabilized *p*-chart, deviations of equal size have the same probability significance, the stabilized chart would appear to be the proper type of chart on which to study runs up and down when the sample size varies.[8]

[8] Cf. Acheson J. Duncan, "Detection of Non-random Variation when Size of Sample Varies," *Industrial Quality Control,* January 1948, pp. 9–12.

4. THE OC CURVE FOR A p-CHART USED TO CONTROL CURRENT OUTPUT

The OC curve for a p-chart used to control current output is found by computing the probability of a point's falling within the limits when the actual fraction nonconforming is some given value p'.

The procedure is shown in Figure 19.10, in which the OC curve for the control chart of Figure 19.4 is worked out. This is an OC curve for a single-limit control chart. The procedure for computing an OC curve for a double-limit p-chart is illustrated in Figure 19.11, in which the OC curve is derived for the original control chart of Figure 19.2. The two OC curves are shown in Figures 19.12 and 19.13.

FIGURE 19.10

Illustration of a Work Sheet for Deriving the OC Curve for a Single-Limit p-Chart (UCL $= 0.240$; $n = 50$. For $p' \leq 0.10$, use Poisson distribution; for $p' > 0.10$, use normal distribution)

p'	$p'n$	$\sigma'_p = \sqrt{\frac{p'(1-p')}{n}}$	$0.240 - p'$	$\frac{0.240 - p'}{\sigma'_p}$	*Probability of Sample Fraction Nonconforming* ≤ 0.240
0 . . .	0	. . . .	. . . .	. . . .	1.000
0.10 . . .	5	. . . .	. . . .	. . . .	0.99856
0.15 . . .	. . . . . .	0.0505	0.09	1.78	0.9625
0.20 . . .	. . . . . .	0.0566	0.04	0.71	0.7612
0.24 . . .	. . . . . .	. . . .	0	0	0.5000
0.28 . . .	. . . . . .	0.0635	−0.04	−0.63	0.2643
0.30 . . .	. . . . . .	0.0648	−0.06	−0.93	0.1762
0.35 . . .	. . . . . .	0.0674	−0.11	−1.63	0.0515
0.40 . . .	. . . . . .	0.0693	−0.16	−2.31	0.0104

5. THE ARL CURVE FOR A SINGLE-LIMIT p-CHART USED TO CONTROL CURRENT OUTPUT

In Section 7.4 of Chapter 18 it was shown that if a p-chart has only an upper control limit (UCL) and if sample results plotted on the chart are independent of each other, the number of samples that will have to be taken on the average before the chart will detect an increased process average is given by $1/(1 - P_a)$, where P_a is the probability of a single sample result being less than or equal to the UCL. It was noted that a plot of these average run lengths (ARLs) against the increased process averages to which they pertained would yield an ARL curve for the chart.

For the p-chart of Figure 19.4, the central line of which falls at 0.108 and the upper control limit at 0.240, selected points of the ARL curve are

FIGURE 19.11
Illustration of a Work Sheet for Deriving the OC Curve for a Double-Limit *p*-Chart (UCL = 0.430; LCL = 0.064; n = 50)

For $p' \leq 0.10$ use the Poisson distribution.

p′	p′n	*Prob.* (c < 0.43n) − *Prob.* (c ≤ 0.064n)
0	0	0
0.01	0.5	0.002
0.02	1.0	0.02
0.03	1.5	0.07
0.04	2.0	0.14
0.05	2.5	0.24
0.06	3.0	0.35
0.08	4.0	0.57
0.10	5.0	0.73

For $p' > 0.10$ use the normal distribution.

p′	$\sqrt{p'(1-p')}$	$\sigma'_p = \sqrt{\frac{p'(1-p')}{n}}$	$\frac{0.430 - p'}{\sigma'_p}$	$\frac{0.064 - p'}{\sigma'_p}$	*Prob.* $\left[\frac{0.064 - p'}{\sigma'_p} \leq z \leq \frac{0.430 - p'}{\sigma'_p}\right]$
0.15	0.357	0.0505	5.54	−1.70	0.95
0.25	0.433	0.0612	2.94	−3.04	0.997
0.30	0.458	0.0648	2.01	−3.64	0.98
0.35	0.477	0.0675	1.19	....	0.88
0.40	0.490	0.0693	0.43	....	0.67
0.43	....	....	0	....	0.50
0.50	0.500	0.0707	−0.99	....	0.16
0.55	0.498	0.0704	−1.70	....	0.04
0.60	0.490	0.0693	−2.45	....	0.01

calculated in Figure 19.14. The curve itself is shown in Figure 19.15. This indicates that if the process average persists at the level of the central line of the chart, there will be long runs, say of 600 to 700 samples on the average, without any false alarms. In the case of a relatively small upward shift in the process average, say to 0.15, the average run length will be about 27 samples. If the process average fraction nonconforming increases to 0.24 or more, the chart will pick up the change on the average on the first or second sample taken after the shift has occurred. If management believes it is better to detect the smaller shifts more rapidly, the sample size n should be increased, which, of course, will lower the UCL. The general question of the economic design of a p-chart used to control current output is discussed in the next section.

FIGURE 19.12
Operating Characteristic Curve for the Single-Limit *p*-Chart; *UCL* = 0.240, *n* = 50 (see Figure 19.4)

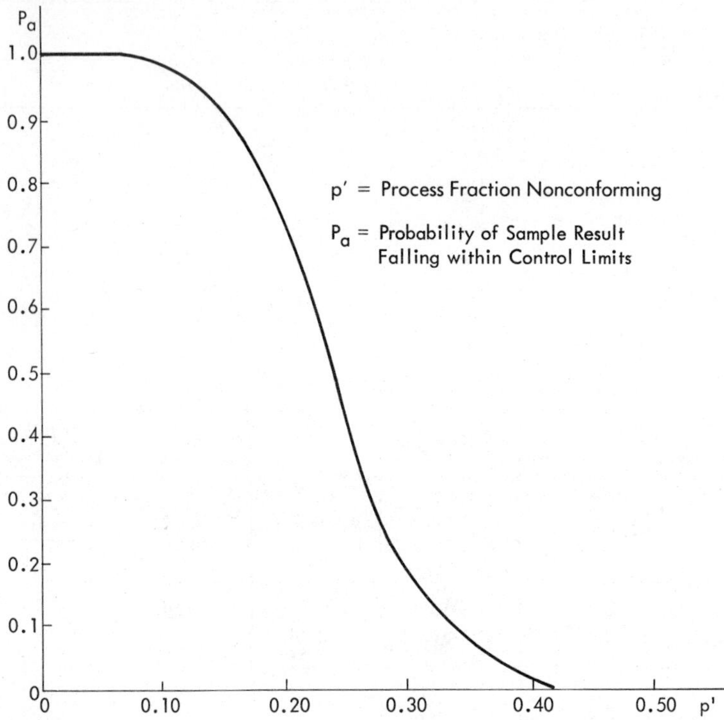

FIGURE 19.13
Operating Characteristic Curve for the Double-Limit *p*-Chart: *LCL* = 0.064, *UCL* = 0.430, *n* = 50 (see Figure 19.2)

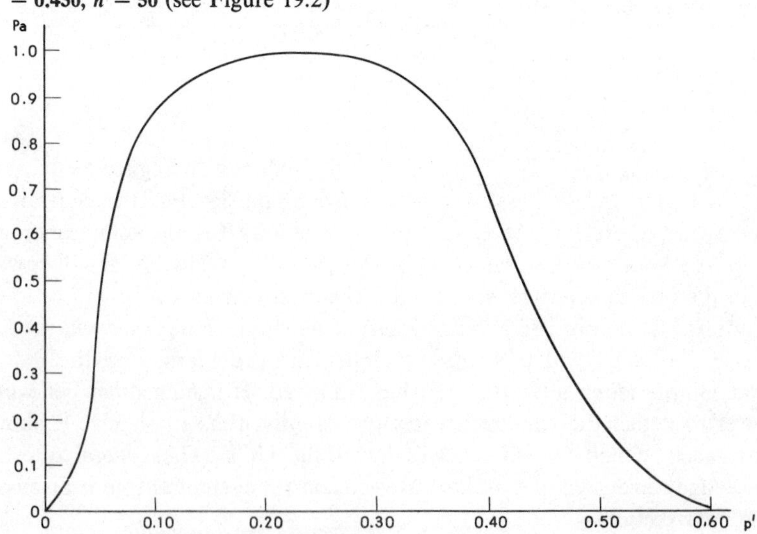

FIGURE 19.14
Illustration of a Work Sheet for Deriving the ARL Curve for a Single-Limit p-Chart (data pertain to Figure 19.4 for which $\bar{p} = 0.108$, UCL = 0.240, and $n = 50$.)

p′	*Probability* (P_a) *of a Sample Fraction Nonconforming* ≤ 0.240	$1 - P_a$	$ARL = 1/(1 - P_a)$
0	1.000	0	—
0.10	0.99856	0.00144	694.44
0.15	0.9625	0.0375	26.67
0.20	0.7612	0.2388	4.19
0.24	0.5000	0.5000	2.00
0.28	0.2643	0.7357	1.36
0.30	0.1762	0.8238	1.21
0.35	0.0515	0.9485	1.05
0.40	0.0104	0.9896	1.01

* From Figure 19.10.

6. SIZE OF SAMPLE AND FREQUENCY OF SAMPLING FOR CURRENT CONTROL

As noted in the previous chapter, many fraction-nonconforming control charts are based on 100 percent inspection of output, and the fraction nonconforming for the output of each hour, day, or week is plotted on the chart. In this case no separate decision as to the frequency of the sample is necessary, since this is most likely to be based on administrative convenience and the general rate of production.

When the p-chart is based on a sample of the total output for any given period, the quality control engineer will have to decide on both the size and the frequency of the sample and also the best limits to use. Theoretical studies that suggest some good answers are reviewed below. First, however, let us consider certain practical considerations. If p' is small, n should be large enough so that we have a good chance of getting some nonconforming items in our sample. Otherwise we might have a situation in which the occurrence of just one nonconforming item might indicate lack of control. One writer[9] has suggested that n be large enough to get one or more nonconforming items per sample at least 90 out of 100 times. For example, if $p' = 0.01$, then, if the probability of $c \geq 1$ is to equal 0.90 (i.e., $P[c < 1] = 0.10$), we would have to have $p'n = 2.3$, which means that n should be 230.

Another way of determining the sample size is to reason as follows: Suppose we decide, for example, that the control chart should be such that it would have a chance of 0.50 or more of catching on a single sample an

---

[9] William B. Rice, *Control Charts in Factory Management* (New York: John Wiley & Sons, 1947), p. 82.

FIGURE 19.15
The Average Run Length Curve for the Single-Limit *p*-Chart Shown in Figure 19.4 (data from Figure 19.14)

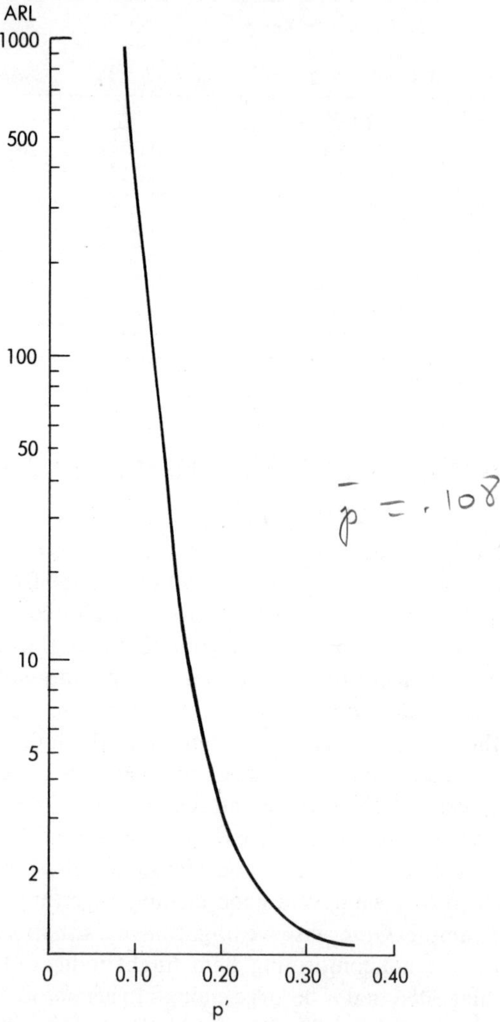

increase in the process fraction nonconforming of as much as, say 0.05. For instance, if $p'' = 0.07$, we may wish to have a chance of 0.50 or more of catching on a single sample a shift to a process fraction nonconforming of $p' = 0.12$ or more. If $p'n \geq 5$, as it frequently will be, we can use the normal curve to compute probabilities. This means that the n should be such that the upper control limit equals p'' plus the increase in the process average we wish a 0.50 chance of detecting on a single sample (for, when

$p' =$ UCL there will be a 0.50 chance of a point's being above the UCL if the distribution of p is normal). Hence in this case we can derive n from the relationship $3\sqrt{\frac{p''(1-p'')}{n}} = d$, where d is the increase in p' that we wish a 0.50 chance of catching on a single sample. Thus, for the above example, we would have $3\sqrt{\frac{(0.07)(0.93)}{n}} = 0.05$, which yields $n = 234$. If $p'n$ should turn out to be smaller than 5, we could make some adjustment with the help of the Poisson distribution or the binomial distribution itself (see Chapter 21, Section 8). If we wish the chart to have a 0.90 chance of catching on a single sample an increase of d in p', then n would be given by the relationship $(1.282 + 3)\sqrt{\frac{p''(1-p'')}{n}} = d$ since the probability is just 0.90 of a standardized normal variable falling above -1.282. For our example we would have $(1.282 + 3)\sqrt{\frac{(0.07)(0.93)}{n}} = 0.05$, which yields $n = 478$.

The theoretical studies mentioned above are based on designing a chart so that it will minimize the costs and losses involved in operating the chart. If pertinent costs can be estimated, a relatively simple procedure can probably be worked out for designing a specific chart that will be approximately optimum.[10] Studies by the author suggest some general conclusions. Unless the cost of tracking down false alarms is very high relative to the unit cost of inspection, say 500 times or more, the control limits should be such that two or more nonconforming items in a sample would indicate lack of control. Even when this cost is high, acceptance numbers generally run low, say 3–5. Sample sizes are mostly determined by the degree of shift in the process that is expected. The author found that for a chart designed to pick up a shift in the process average from 0.01 to 0.06 sample sizes of about 30 to 70 were optimum, the larger sizes occurring when the acceptance number was 2 or 3 as against 1. To pick up a shift from 0.01 to 0.02, sample sizes, with an acceptance number of 1, ran about 80 to 120. For an acceptance number of 5, sample sizes were in the neighborhood of 300. With a higher initial level of 0.05, and a shift to 0.07, optimum sample sizes ran about 30 to 50 with an acceptance number of 1. With n being largely determined by the size of the shift to be detected, adjustment to variation in the loss rate and inspection costs was largely effected by variation in the frequency of sampling. With a high rate of loss from operating in an out-of-control state, the interval between samples would be small; with high inspection costs, the interval would be large.

---

[10] Compare the method developed by W. K. Chiu and G. B. Wetherill for $\bar{X}$-charts in their paper "A Simplified Scheme for the Economic Design of $\bar{X}$-Charts," *Journal of Quality Technology,* Vol. 6, April 1974, pp. 63–69.

FIGURE 19.16

Operating Characteristic Curves for p-Charts for Past Data (chart gives probability of failing to detect a single *slippage* from a state of control at level p'_1 to level p'_2: $g = 20$, $n = 50$)

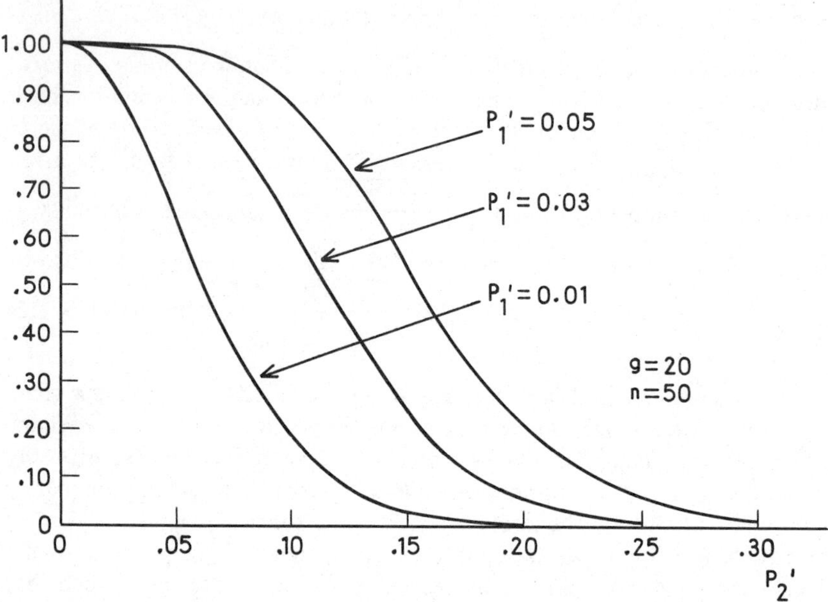

SOURCE: See Olds (B '56) in Cumulative List of References.

7. THE OC FUNCTION FOR A p-CHART USED TO ANALYZE PAST DATA

Professor Edwin G. Olds[11] has made an extensive study of the power of a p-chart used to study past data. In particular he has concerned himself with the following situation. He supposes that for $g - 1$ occasions on which a process has been sampled it has been in control at level p'_1, but on one occasion it was operating at a higher level p'_2.

For the situation just discussed, Olds has determined the probability that the sample taken on the occasion when $p' = p'_2$ will yield a fraction nonconforming p falling above the upper 3-sigma control limit computed from all g subgroups. For 20 subgroups of 50 items each, i.e., for $g = 20$, and $n = 50$, the probability of failing to detect such a single slippage (which is equal to one minus the probability of detecting it) is shown in Figure 19.16 for values of $p'_1 = 0.01$, 0.03, and 0.05. The chart itself expresses the probability of acceptance as a function of p'_2. It will be noted that the curves

[11] See Olds (B '56) in Cumulative List of References.

FIGURE 19.17
Upper and Lower Bounds for Risk of a Single Point Falling beyond Upper 3-Sigma Control Limit on a p-Chart when Process Was Actually in Control at Level p' at Each Sampling: $g = 20$, $n = 50$

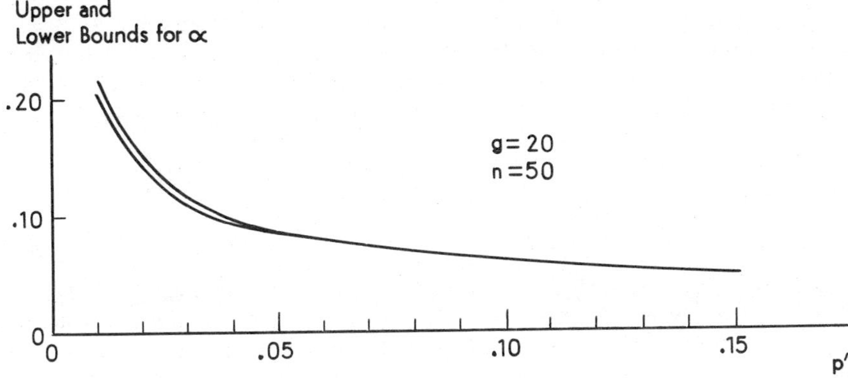

SOURCE: See Olds (B '56) in Cumulative List of References.

are dependent upon specified values of p'_1 and p'_2 and not simply on their difference. Olds' study shows that for a given number of subgroups g, the probability of failing to detect the slippage decreases directly with increases in the sample size n. It also shows that for a given n, it decreases with increases in g.

The risk of finding a point above the upper 3-sigma control limit and falsely concluding there is lack of control when in fact the process is in control at level $p' = p'_1$ for all g samples may be called the α risk associated with a particular p-chart and specified level p'_1. The upper and lower bounds for this risk when $g = 20$, $n = 50$ are shown in Figure 19.17 as a function of p'. For a given g, this risk decreases as n increases, but for a given n it increases as g increases. For a total of 200 sample items, Olds points out that a g of 10 and an n of 20 is better than a g of 20 and an n of 10 because in the latter instance the α risk is lower while the power of the chart to detect differences is greater. For the comparison to be valid, however, the production spans would have to be equal for the two charts.

8. p-CHARTS AND χ^2 TESTS

When analyzing a set of past samples for lack of control, it is possible, as an alternative to a p-chart, to set up the data in the form of a $2 \times n$ contingency table and run a χ^2 test of homogeneity. This alternative procedure is discussed in Chapter 28.

The difference between a p-chart analysis of past data and a χ^2 test lies mainly in the kind of deviation that will lead to rejection of the hypothesis of control or homogeneity. In the former, a single marked deviation will lead to rejection, whereas in the latter, rejection is based primarily upon average departures from hypothetical expectations. The OC curves for the two analyses bring out this difference. The Olds analysis in Section 7 expressed the probability of acceptance as a function of the difference between a general control level p'_1 and a single slippage to p'_2. The OC curve for a χ^2 test runs in terms of squared differences between a common hypothetical p' and alternative p'_j, which may be different for each sample. In industry we will often be more interested in a single marked slippage than in an overall average difference.

9. PROBLEMS

19.1. Using the following data, construct a p-chart for controlling production in the next month. Assume that assignable causes are found for all points outside the limits on the control chart. Draw the OC curve for this p-chart. Derive the ARL curve that will indicate the effectiveness of the chart to be used next month in picking up an increase in the process average.

Day of Month	*Number Inspected*	*Number Nonconforming*
1	200	6
2	200	6
4	200	6
5	200	5
6	200	0
7	200	0
8	200	6
10	200	14
11	200	4
12	200	0
13	200	1
14	200	8
15	200	2
17	200	4
18	200	7
19	200	1
20	200	3
21	200	1
22	200	4
24	200	0
25	200	4
26	200	15
27	200	4
28	200	1

19.2. Assume that the fractions nonconforming of Problem 3.3a(iii) are all based upon a sample of 180. Consider that the first 24 of these figures represent the output for the past month, and proceed to set up control limits for controlling the output of the next month. (Assume that assignable causes are found for every point outside the limits on the initial chart.) Consider the remaining data as the output for the next month and plot them on the chart set up for that month. Is the second month's output in control at the level set at the beginning of the month? Would you recommend any change in the p-chart to be used for the third month?

19.3. Set up a control chart for next month's operations based upon the following results for the past month. (Assume assignable causes for every point out of control.)

Day of Month	*Number Inspected*	*Number Nonconforming*
1	196	39
2	210	25
4	210	28
5	210	43
6	210	15
7	174	6
8	180	14
10	196	3
11	181	10
12	184	25
13	130	15
14	190	6
15	180	22
17	180	8
18	196	8
19	208	9
20	204	5
21	210	5
22	180	7
24	180	15
25	190	19
26	180	15
27	185	2
28	190	8

19.4. The following are the fractions nonconforming of shaft and washer assemblies during the month of April in samples of 1,500 each.[12]

[12] Read from Figure IV of George V. Herrold, "The Introduction of Quality Control at Colonial Radio Corporation," *Industrial Quality Control,* July 1944, p. 8.

Date April	Fraction Nonconforming	Date April	Fraction Nonconforming
1	0.11	17	0.04
3	0.06	18	0.07
4	0.10	19	0.04
5	0.11	20	0.04
6	0.14	21	0.04
7	0.11	22	0.03
10	0.14	24	0.06
11	0.03	25	0.06
12	0.02	26	0.04
13	0.03	27	0.03
14	0.03	28	0.04
15	0.03		

From these data construct a p-chart for controlling quality during the next month.

19.5. Below are data on the production of a piece of electrical equipment. Is the manufacturing process under control with respect to its fraction nonconforming?

Date Produced May	Number of Items in Lot	Number Nonconforming
1	1,361	161
2	1,278	118
3	1,328	220
4	710	67
5	735	85
6	726	82
8	803	78
9	850	74
10	700	60
11	670	45
12	680	54
13	801	64
15	717	53
16	722	52
17	691	51
18	756	53
19	652	46
20	701	62
22	741	66
23	754	46
24	703	38
25	682	47
26	719	44
27	713	57

19.6. In samples of 90 bellows each, the following are the fractions nonconforming of output from April 14 to June 30[13] (cf. Figure 6.13):

0.09	0.06	0.13
0.10	0.15	0.14
0.08	0.10	0.03
0.13	0.02	0.15
0.08	0.12	0.12
0.07	0.09	0.17
0.03	0.17	0.14
0.06	0.14	0.17
0.07	0.16	0.09
0.14	0.08	0.03
0.03	0.03	0.12
0.12	0.02	0.15
0.09	0.15	0.15
0.14	0.13	0.06
0.09	0.08	0.12
0.15	0.09	
0.13	0.07	

Is the process represented by these data under control with respect to its fraction nonconforming?

19.7. The process average has been shown to be 0.03. Your control chart calls for taking daily samples of 800 items. What is the chance that if the process average should suddenly shift to 0.06, you would catch the shift on the first sample taken after the shift? On the first or second sample after the shift? Derive the ARL curve indicating the effectiveness of the chart in picking up an increase in the process average.

19.8. The process average has been shown to be 0.03. Your control chart calls for taking a daily sample of 400 items. What is the chance that, if the process average should suddenly shift to 0.06, you would catch the shift on the first sample taken after the shift? On the first or second sample after the shft? Compare your answer with that of Problem 19.7. Derive the ARL curve indicating the effectiveness of the chart in picking up an increase in the process average.

19.9. The process average has been shown to be 0.07. Your control chart calls for taking daily samples of 400 items. What is the chance that, if the process average should suddently shift to 0.10, you would catch the shift on the first sample taken after the shift? On the first or second sample after the shift? Compare your answer with that of Problem 19.8. Derive the ARL curve indicating the effectiveness of the chart in picking up an increase in the process average.

19.10. The process average has been shown to be 0.12. You wish a shift to a process average of 0.15 to have approximately a 0.50 chance of being caught on the

[13] Adapted from William L. Burns, "Quality Control Proves Itself in Assembly," *Industrial Quality Control*, January 1948, p. 14.

first sample after the shift. How large should your daily sample be for control chart purposes?

19.11. The process average has been shown to be 0.07. You wish a shift to a process average of 0.10 to have approximately a 0.50 chance of being caught on the first sample after the shift. How large should your daily sample be for control chart purposes? Suppose you wanted a 0.90 chance of detecting the shift on the first sample, what would your answer be?

10. SELECTED REFERENCES*

American Society for Quality Control (B: ANSI/ASQC Stds. Z1.1, Z1.2 and Z1.3 1985 and P '46–), American Society for Testing Materials (B '76), Armstrong (P '46), Burr (B '76), Cowden (B '57), Duncan (P '78), Gibra (P '78 and P '81), Grant and Leavenworth (B '80), Kennedy (B '48), Knowler (P '46), Nelson (P '83), Olds (B '56), Palumbo and Strugala (P '45), Peach (B '47), Reynolds (P '71), Rice (B '47), Schrock (B '50), Simon (B '41), Smith, E. S. (B '47), U.S. War Production Board, Office of Production Research and Development (OPRD), *Quality Control Reports,* 1945–, and Williams, Looney and Peters (P '85).

---

* B and P refer to the Book and Periodical sections, respectively, of the Cumulative List of References in Appendix V.

20

Control Charts for Number of Nonconformities per Unit (c-Charts and u-Charts)

1. EXAMPLES OF USE

In many cases it is necessary, or at least more convenient, to work with the number of nonconformities per unit than with the fraction nonconforming. If we were producing oilcloth that was inspected for blemishes, the natural thing would be to take some unit of area, say, 100 square yards, and count the number of blemishes per unit. The variable in this case would be the number of blemishes per 100 square yards. We could, if we wished, look upon the 100 square yards of oilcloth as a sample of 900 square feet, and we could count the fraction of square feet that are blemished. If a blemish does not cover more than 1 square foot and there is only one blemish per square foot, then the number of blemishes per 100 square yards and the fraction of square feet blemished would be equivalent variables. The first, however, would be the more natural one.

As another example, consider cases in which the product is a large unit, say, a radio, where there may be nonconformities at innumerable points, although any one unit rarely has more than a few nonconformities. In these cases it is again natural to count the number of nonconformities per unit. If it is possible to determine all the points at which the unit may be nonconforming, we could, if we wished, determine the fraction of nonconforming points in each unit. But again this seems a more arbitrary way of proceeding; and it may be impossible, or at least very impracticable, to determine all the points at which a radio may be nonconforming.

2. THE SAMPLING DISTRIBUTION OF THE NUMBER OF NONCONFORMITIES PER UNIT

Experience with variation in the number of nonconformities per unit, e.g., number of blemishes per 100 square yards or number of nonconformities per radio, indicates that the distribution of the variable follows very closely the form of the Poisson distribution. This fact, which may be established empirically, is also supported by certain theoretical considerations.

A theoretical argument for using the Poisson distribution to measure sampling fluctuations in the number of nonconformities per unit can always be made whenever the number of nonconformities per unit can be translated into a fraction nonconforming. Examples of how this might be done have already been given. As noted above, if a blemish is limited to 1 square foot and if only one blemish occurs per square foot, the number of blemishes per 100 square yards can be expressed as a fraction nonconforming in a total of 900 square feet. Again, if we can count the total number of points at which a radio may be nonconforming, then the number of nonconformities per radio can also be expressed as the fraction of nonconforming points. In situations such as this, we have already seen that when the fraction nonconforming for the whole process is very small, say, less than 0.10, and the sample size (the 900 square feet or number of possible nonconforming points in a radio) is large, then sampling fluctuations in the sample fraction nonconforming is approximated closely by the Poisson distribution. Usually the condition of a small p' and a large n are satisfied when we have to do with the number of nonconformities per unit. The only theoretical difficulty is whether we can always translate the number of nonconformities per unit into an equivalent fraction nonconforming. When this can be done, we are on sound theoretical grounds in using the Poisson distribution. When it cannot be done, we can still justify the use of the Poisson distribution on the empirical basis that it seems to work well.

The formula for the Poisson distribution may be written $P(c) = \dfrac{c'^{c} e^{-c'}}{c!}$, *where* c is the number of nonconformities in a sample consisting of a single unit and c' the average number of nonconformities per unit for the universe. The standard deviation of c is $\sqrt{c'}$. It will be recalled that Figure 4.9 gives the sum of the ordinates of the Poisson distribution from 0 to c, i.e., the probability of c nonconformities or less, and Molina's tables[1] give the ordinates of the distribution as well as the sum of the ordinates from c to infinity, i.e., the probability of c or more. When the number of nonconformities can be translated into an equivalent fraction nonconforming, we have[2] $c' = p'n$, where p' is the universe fraction nonconforming, and $\sigma'_c = \sqrt{p'n}$.

---

[1] E. C. Molina, *Poisson's Exponential Binomial Limit* (New York: D. Van Nostrand Co., Inc., 1949). See also Table E of Appendix II.

[2] See Chapter 4, Sections 5.2 and 7.2.

In applying the Poisson formula, it is to be noted that an inspection unit may be a unit of any kind. It may be a single item like a radio, or possibly a set of radios, or it may be a unit of length, area, volume, weight, or even time. However defined, it is important that the inspection unit be kept constant throughout the analysis. It must always represent the same "area of opportunity" for the occurrence of nonconformities. This does not mean that we cannot inspect more than one unit at a time. As a matter of fact, in some cases we may determine the average number of nonconformities per unit for a sample of several units, and the number of units may vary from sample to sample. This special case will be taken up subsequently. All that need be noted here is that an inspection unit, however defined, retains that definition throughout the analysis.

3. CONSTRUCTION AND OPERATION OF A c-CHART WITH A CONSTANT-SIZE SAMPLE

The following data may be used to illustrate the construction of a c-chart. They show the number of nonconformities per group in 25 successive groups of 5 radio sets each.[3]

Group Number	*Number of Nonconformities per Group*	*Group Number*	*Number of Nonconformities per Group*
1	77	14	87
2	64	15	40
3	75	16	22
4	93	17	92
5	45	18	89
6	61	19	55
7	49	20	25
8	65	21	54
9	45	22	22
10	77	23	49
11	59	24	33
12	54	25	20
13	41		

The total number of nonconformities for the 25 groups is 1,393, and the average number per unit (i.e., per group of 5 radios) is $\bar{c} = 1{,}393/25 = 55.7$. As a preliminary estimate, we take the process average number of nonconformities per unit (c') as equal to $\bar{c}$. We therefore take as an estimate of σ'_c the quantity $\hat{\sigma}_c = \sqrt{\bar{c}} = 7.5$. Hence $3\hat{\sigma}_c = 3\sqrt{\bar{c}} = 22.5$, and the

[3] Mason E. Wescott, "Attribute Charts in Quality Control," *Conference Papers, First Annual Convention, American Society for Quality Control and Second Mid-west Quality Control Conference, June 5 and 6, 1947* (Chicago: John S. Swift Co., Inc., 1947), p. 67.

FIGURE 20.1
A c-Chart for Analyzing Past Data: Variation in Number of Nonconformities per Group of Five Radios

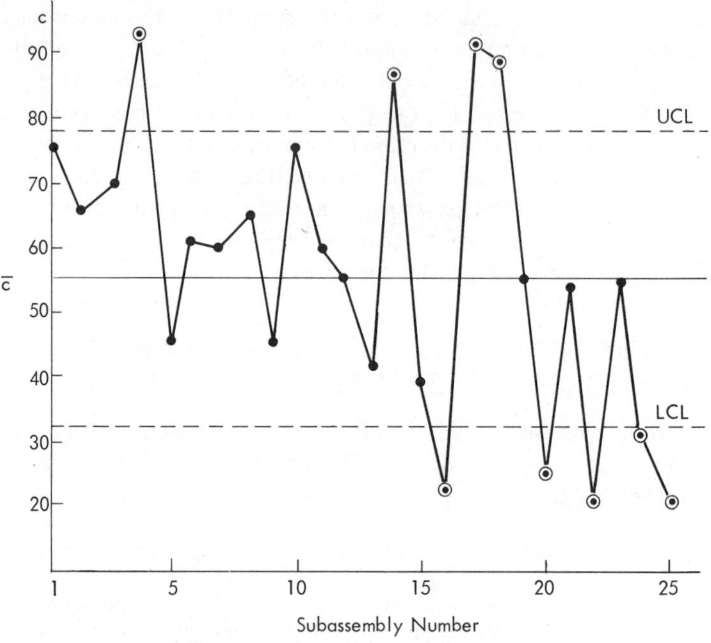

upper control limit becomes $\bar{c} + 3\sqrt{\bar{c}} = 55.7 + 22.5 + 78.2$ and the lower control limit becomes $\bar{c} - 3\sqrt{\bar{c}} = 55.7 - 22.5 = 33.2$. These give the central line and upper and lower limits for our preliminary control chart.[4] This chart with the individual points from which it was derived is shown in Figure 20.1.

It is obvious from Figure 20.1 that the manufacturing and assembly processes are not under control. Four points are above the upper limit and five points are below the lower limit. Let us suppose that assignable causes are found for all but one of these points. Say that high points 17 and 18 are found to be due to certain faulty material used in the production of these sets; high point 4 was traced to a particularly poor workman; and high point 14 to an old machine that did not function properly. Let all these special causes of poor quality be eliminated. The low points 16, 20, 22, and 25, we shall say, are traced to one inspector who was just learning the job. Assume that he missed many of the actual nonconformities in the radios he inspected, and exclude these low points from the rest. No assignable cause was found for point 24, so that this will be the only one of the points outside the control limits to be kept.

---

[4] If we wished, we could use probability limits instead of 3-sigma limits. For a comparison, see *Control Chart Method of Controlling Quality During Production* (ANSI Std. Z1.3) (New York: American National Standards Institute, 1985).

After omitting points 4, 14, 16, 17, 18, 20, 22, and 25, $\bar{c}$ for the remaining cases is found to be 943/17 = 55.5. This is so close to the old value that we can take the old central line and the old control limits as standards for the operation of the control chart in the next period. Let the number of nonconformities in each of the next 25 groups be as follows.

Group Number	*Number of Nonconformities per Group*	*Group Number*	*Number of Nonconformities per Group*
26	26	39	46
27	23	40	32
28	9	41	46
29	15	42	49
30	63	43	31
31	39	44	36
32	58	45	41
33	61	46	49
34	59	47	39
35	51	48	49
36	33	49	43
37	40	50	43
38	40		

Figure 20.2 shows how these results look when plotted. Let us assume that the first four points represent radios inspected by the new inspector and that they should therefore be disregarded. Suppose, for example, that the foreman decided that the new man would probably never be a first-class inspector and consequently removed him from the job.

On the 36th point and again on the 40th and 43d points the process is seen to have gone below the lower control limit. Suppose that an investigation was made immediately after point 36 occurred and it was found that some new material had begun to be used with point 35. The succession of points following this, all of which are below the old average, show the beneficial effects of using this material. It was consequently ordered that the new material be employed in all radio sets from that time on.

Suppose that at the end of the 50th group it is decided to revise the limits, basing the new average on points 35–50. The result is $\bar{c} = 668/17 = 41.7$. This gives:

$$\hat{\sigma}_c = \sqrt{\bar{c}} = \sqrt{41.7} = 6.46$$

$$UCL = \bar{c} + 3\sqrt{\bar{c}} = 41.7 + 19.4 = 61.1$$

$$LCL = \bar{c} - 3\sqrt{\bar{c}} = 41.7 - 19.4 = 22.3$$

Let 41.7 be the central line and 61 and 22 be the upper and lower control limits for the chart to be used in the next period. Suppose that the standard they represent is considered satisfactory by the manager, so that no major overhauling of the process is deemed necessary.

FIGURE 20.2
A *c*-Chart in Operation: Variation in Number of Nonconformities per Group of Five Radios: A Continuation of Figure 20.1

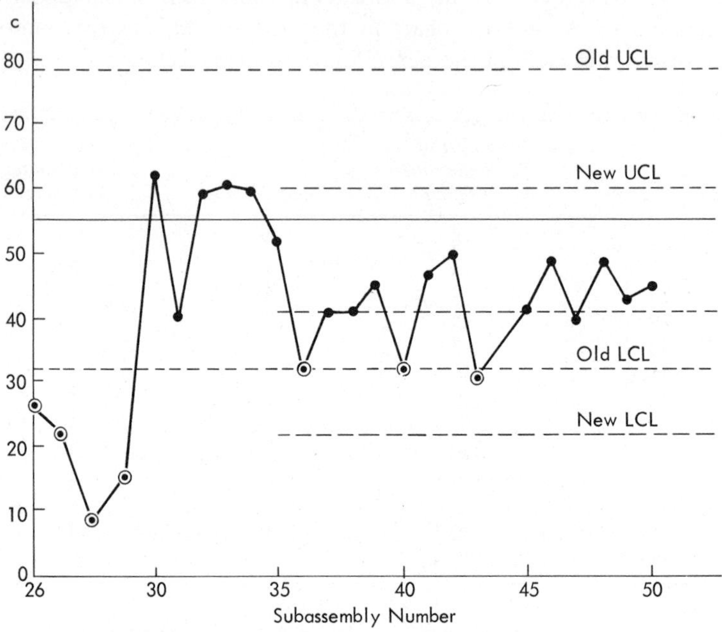

4. THE OC CURVE FOR A *c*-CHART WITH A CONSTANT-SIZE SAMPLE

The OC curve for a *c*-chart gives the probability of a point's falling on or within the limits as a function of the process average number of nonconformities per unit. It is easily derived from tables or charts of the Poisson distribution. For the final *c*-chart adopted above, the OC curve is derived as follows:[5]

u'	*Prob.* of $c \leq 61$	−	*Prob.* of $c \leq 22$	=	*Prob.* $(22 < c \leq 61)$
15 . . .	1.0000	−	0.9673	=	0.0327
18 . . .	1.0000	−	0.8551	=	0.1449
20 . . .	1.0000	−	0.7206	=	0.2794
22 . . .	1.0000	−	0.5564	=	0.4436
27 . . .	1.0000	−	0.1952	=	0.8048
30 . . .	1.0000	−	0.0806	=	0.9194
42 . . .	0.9977	−	0.0005	=	0.9972
50 . . .	0.9443	−	0.0000	=	0.9443
55 . . .	0.8111	−	0.0000	=	0.8111
61 . . .	0.5340	−	0.0000	=	0.5340
70 . . .	0.1545	−	0.0000	=	0.1545
75 . . .	0.0560	−	0.0000	=	0.0560

[5] Computed from Molina's *Poisson's Exponential Binomial Limit,* Table II.

FIGURE 20.3
Operating Characteristic Curves for c-Charts with Samples of Different Size

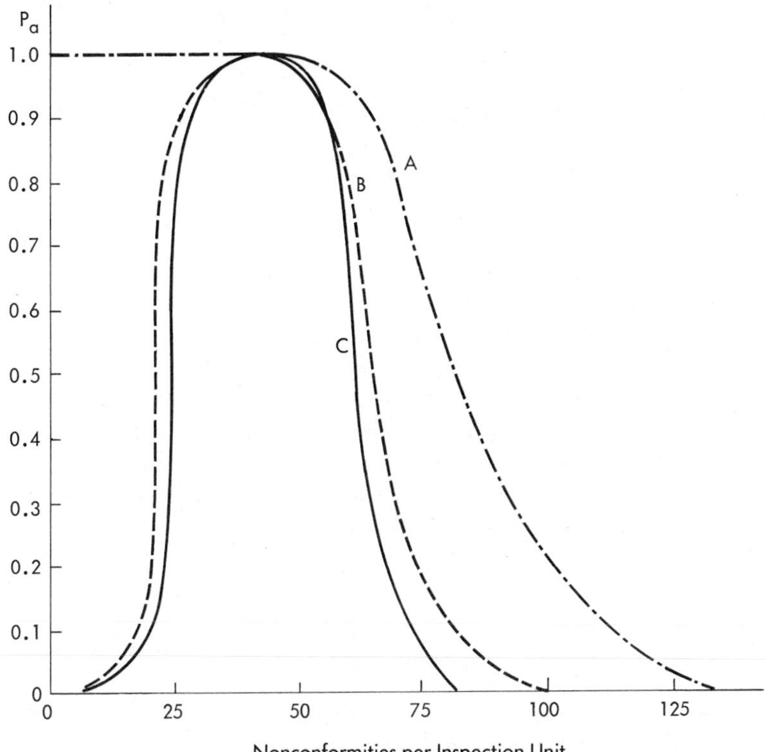

A = OC curve for a sample consisting of 0.2 of an inspection unit; B = OC curve for a sample consisting of 0.6 of an inspection unit; C = OC curve for a sample consisting of 1 inspection unit. Take c' = nonconformities per inspection unit times number of inspection units.

This shows that the true average number of nonconformities might go as high as 70 or as low as 18 and we would still have more than a 0.10 chance of not catching the shift on a single unit inspected (also see Figure 20.3).

5. THE ARL CURVE FOR A SINGLE-LIMIT c-CHART WITH A CONSTANT-SIZE SAMPLE USED TO CONTROL CURRENT OUTPUT

If a c-chart with a constant-size sample has only an upper control limit and is used to detect upward shifts in the process average of current production, we can, as in the case of p-charts, easily derive an ARL curve for the chart. If only the upper limit of the c-chart adopted in Section 3 above is retained and if successive samples are independent of each other, we can easily derive the ARL curve for the chart from the data pertaining to the OC curve presented in the previous section. We will thus have the following:

c'	*Prob.** (P_a) *of* $c \le 61$	$1 - P_a$	$ARL = 1/(1 - P_a)$
42	0.9977	0.0023	434.8
50	0.9443	0.0557	17.9
55	0.8111	0.1889	5.3
61	0.5340	0.4660	2.1
70	0.1545	0.8455	1.2
75	0.0560	0.9440	1.1

* From Section 4.

This shows that if the process average number of nonconformities per unit remains in the neighborhood of the central line on the chart (=41.7), a run of about 435 samples will be inspected on the average without the occurrence of a false alarm. If the process number of nonconformities per unit increases to 55, the chart will on the average detect the change in 5 to 6 samples. If the process average increases to 61 nonconformities per unit (the UCL) or to a greater number, the chart will on the average detect the change on the first or second sample after the shift has occurred (see Figure 24.4, p. 470).

6. DETERMINATION OF THE SIZE OF THE SAMPLE FOR PURPOSES OF CURRENT CONTROL

The sample used in connection with the c-chart of the previous section consisted of a single "inspection unit" of 5 radios. It might equally well have consisted of 3 or 1 or any number of radios.

If the "inspection unit" is defined as a set of 5 radios, then a sample of 3 radios is a sample amounting in size to three-fifths of an inspection unit, and a sample of 1 radio a sample amounting to one-fifth of an inspection unit. If the inspection unit is defined as a single radio, then a sample of 5 radios is a sample of 5 inspection units.

In general, the inspection unit is the "accounting" or statistical unit with respect to which the records are kept. The number of nonconformities plotted on the c-chart are with respect to this unit, and the central line and limits are also with respect to it. As noted above, when once the "inspection unit" has been defined, it is essential to keep this definition throughout the analysis. Since the inspection unit is the "accounting" unit, it is well to measure the size of the sample in terms of the number of inspection units it contains.

The inspection unit is determined by accounting convenience. The size of the sample, however, should be determined with reference to statistical principles.

One criterion that has been proposed is that the sample should be at least large enough so that there is a lower limit above zero,[6] for one purpose of a control chart is to catch any lowering in inspectors' efficiency.

Another approach to this problem is via the OC curve. Figure 20.3 shows, for each of three different cases, how the probability of not catching a shift of a specified size varies with the degree of the shift. Curve *A* is the case in which we take one radio as our sample, curve *B* the case in which we take three radios, and curve *C* the case in which we take five radios. In case *A* there is no lower limit, so there is little probability of catching a shift in the negative direction. The probability of catching an upward shift on a single sample is also much less in case *A* than in case *B* or case *C*. Case *B*, in which the sample consists of three radios, gives some protection against lowering of inspection standards, which would mean "favorable" shifts in the number of nonconformities, but it does not give as much protection as case *C*. It also does more poorly than case *C* on the upward side.

Assume that we have adopted an inspection unit of five radios. Then, in choosing the size of sample to employ, we could select one that would yield a probability of 0.50 or more of catching, on a single inspection, a shift of as much as 25 nonconformities per inspection unit (i.e., five nonconformities per radio) in either direction. Figure 20.4 shows that a sample consisting of three radios will give us approximately this amount of protection when the central line on the control chart is at 40 nonconformities per inspection unit (i.e., 8 nonconformities per radio).

A quick way of solving this problem in general is to note that the 0.50 point on the OC curve comes close to the upper and lower control limits. When a sample consists of k inspection units (k being less than, equal to, or greater than 1), these limits will be

$$\begin{aligned} UCL &= \bar{c} + 3\sqrt{\bar{c}/k} \\ LCL &= \bar{c} - 3\sqrt{\bar{c}/k} \end{aligned} \tag{20.1}$$

If the shift against which we want 0.50 protection is one of d nonconformities per inspection unit, then we can choose k so that $3\sqrt{\bar{c}/k}$ is equal approximately to d. For example, suppose that $\bar{c} = 40$ and $d = 25$. Then we can find k from the equation $d = 3\sqrt{\bar{c}/k}$. For the given data this yields $25 = 3\sqrt{40/k}$ or $k = \frac{9(40)}{625} = \frac{360}{625} = 0.58$, or approximately 0.6 of an inspection unit (i.e., three radios), which is what we saw directly from our charts. If d is set at 10, then we would have $10 = 3\sqrt{40/k}$ or $k = \frac{9(40)}{100} = \frac{360}{100} = 3.6$. In other words, we would use a sample consisting of 3.6 inspection units or 18 radios.

[6] Westcott, "Attribute Charts in Quality Control," p. 68. (See Footnote 3 above.)

FIGURE 20.4
The Average Run Length Curve for the Revised *c*-Chart of Figure 20.2 (data on p. 468)

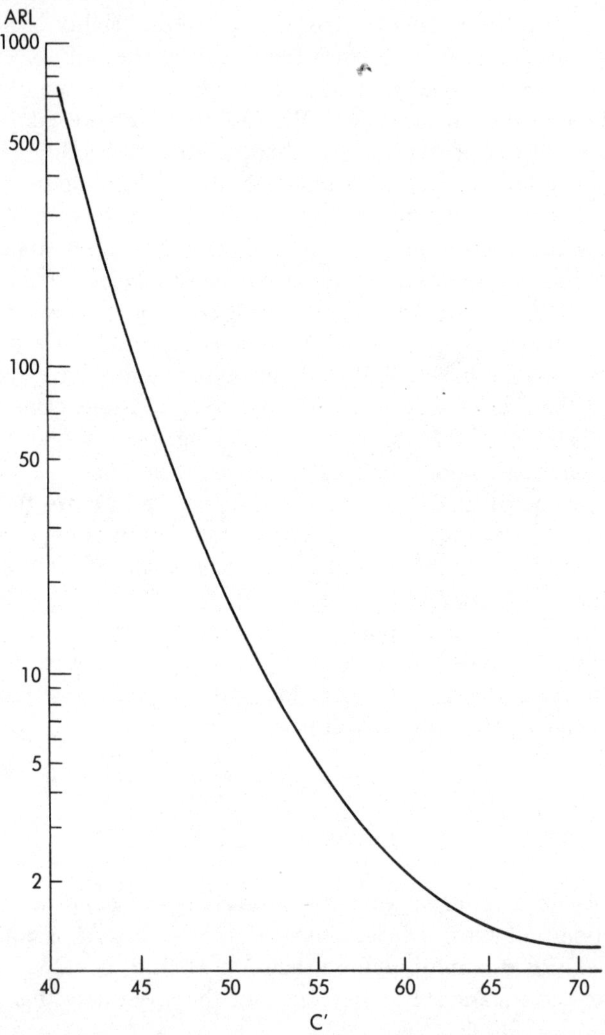

7. *u*-CHARTS

Having found the desired sample size, we should, if we wish to use a *c*-chart, redefine our inspection unit so that the sample size and inspection unit become identical. Thus, in terms of the examples just discussed, we should change our inspection unit from five radios to 3 radios or to 18 radios. In this case, we would on each occasion plot the total sample number of nonconformities directly on our *c*-chart. If we choose a new inspection unit of 3 radios, the central line in the c-chart will become $40(0.6) = 24$; if we choose a new inspection unit of 18 radios, the central line will become $40(3.6) = 144$.

If after determining the sample size we do not redefine the inspection unit, it becomes necessary to set up a control chart in terms of the *average* sample number of nonconformities per inspection unit. In other words, if c is the total number of nonconformities found in any sample and k is the number of inspection units in a sample, we set up a control chart on which we would plot the quantity $u = c/k$. Such charts are called u-charts.

When the sample size varies from sample to sample, it is necessary to use a u-chart in order to have a constant central line. The control limits, however, will vary. This is explained in the next section.

8. A u-CHART WITH A VARYING-SIZE SAMPLE

Sometimes u-charts, like p-charts, are based on 100 percent inspection of production. In these instances the number of units composing a sample will undoubtedly vary from sample to sample. Suppose, for example, that we had the following results:

Inspection Lot Number	*Square Yards Oilcloth Inspected*	*Number of Nonconformities in Total Lot*	*Number of Nonconformities per 100 Square Yards*
1..........	200	5	2.5
2..........	250	7	2.8
3..........	100	3	3.0
4..........	90	2	2.2
5..........	120	4	3.3
6..........	80	1	1.3

Here each inspection lot contains a different number of square yards of oilcloth. If we take 100 square yards as the unit that we shall use on our u-chart, then in each ease, to get comparability, we must convert the number of nonconformities per inspection lot to the number of nonconformities per unit of 100 square yards. However, when these results are plotted on the control chart, we must take into consideration that each figure is based on a different number of units. The control limits will thus vary from sample to sample. The larger the number of units in a sample, the narrower the limits.

If each sample consists of a single unit, the limits for the u-chart are given by $\text{UCL} = \bar{u} + 3\sqrt{\bar{u}}$, $\text{LCL} = \bar{u} - 3\sqrt{\bar{u}}$. But if the number of units in a sample is k, then the standard deviation of the average number of nonconformities per unit is $\sqrt{\bar{u}}/\sqrt{k} = \sqrt{\bar{u}/k}$. Hence, for a sample of k units, the limits are given by

(20.1a)

$$\text{UCL} = \bar{u} + 3\sqrt{\bar{u}/k}$$
$$\text{LCL} = \bar{u} - 3\sqrt{\bar{u}/k}$$

To illustrate, suppose that we wish to construct a u-chart for controlling the production of oilcloth and decide to base it upon the above data (actually, we would use a much larger volume of production than this).

The first step is to compute the average number of nonconformities per 100 square yards (our unit) for the six inspection lots. This is easily done by adding up the total number of nonconformities in each lot and dividing by the total number of 100-square-yard units inspected. This gives $\bar{u} = 22/8.40 = 2.62$. Note that we do not take a simple average of the number of nonconformities per 100 square yards in each lot. If we wish to work with these figures, we must take a weighted average. That is, we would have to weight 2.5 by 2.00, 2.8 by 2.50, 3.0 by 1.00, and so on. The result would then be $\bar{u} = 2.62$ as above.

The central line on our chart will be $\bar{u}$. The limits for each sample will be given by (20.1a) above which yields the table on top of p. 473 (also see Figure 20.5):

FIGURE 20.5
A u-Chart with Samples of Varying Size (upper control limit only)

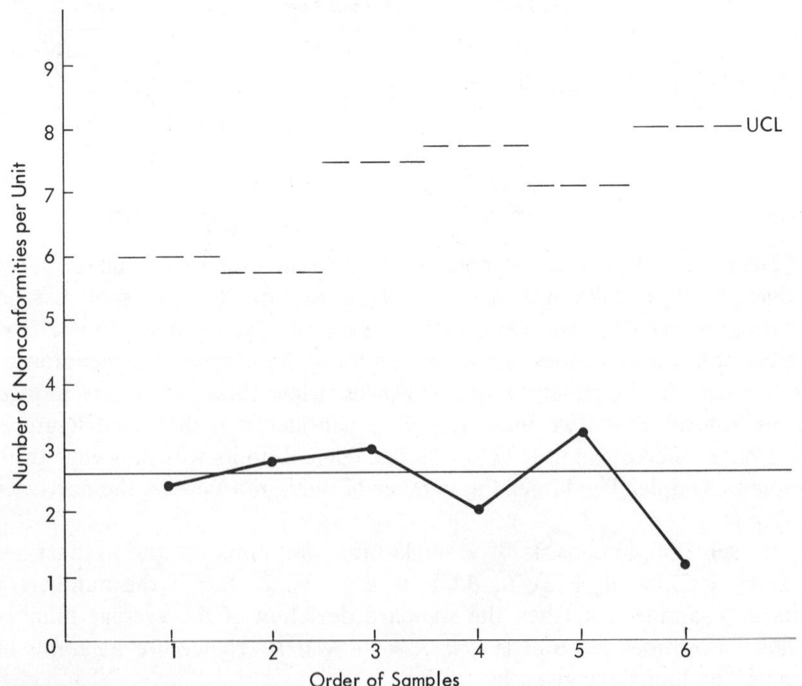

Inspection Lot Number	k	$3\sqrt{\bar{u}/k}$	$UCL = \bar{u} + 3\sqrt{\bar{u}/k}$	$LCL = \bar{u} - 3\sqrt{\bar{u}/k}$
1	2.0	3.43	6.05	—
2	2.5	3.07	5.69	—
3	1.0	4.86	7.48	—
4	0.9	5.12	7.74	—
5	1.2	4.43	7.05	—
6	0.8	5.43	8.05	—

An OC curve can be constructed only for a u-chart with a constant size sample. When the number of units varies from sample to sample, we can, if we wish, construct an OC curve for the average number of units. This would be the OC curve for the control chart with average limits.

9. c-CHARTS, u-CHARTS, AND χ^2 TESTS

As with p-charts, χ^2 tests offer an alternative approach to the problems attacked by c-charts and u-charts. This is discussed in Chapter 28.

10. PROBLEMS

20.1 The following are the number of nonconformities found at subassembly inspection of radio sets during a given period of operation (6 sets = 1 inspection unit):

Subassembly Number	*Number of Nonconformities*	*Subassembly Number*	*Number of Nonconformities*
1	70	14	40
2	64	15	21
3	81	16	56
4	105	17	91
5	40	18	70
6	62	19	65
7	53	20	50
8	48	21	28
9	82	22	24
10	90	23	60
11	110	24	75
12	54	25	25
13	88		

a. From these data set up a c-chart for use in the next period. Assume that assignable causes are found for all points outisde the preliminary control limits. Derive the ARL curve that will indicate the effectiveness of the chart to be used next period in picking up an increase in the process average.

b. The number of nonconformities per inspection unit during the next period is as follows:

Subassembly Number	*Number of Nonconformities*	*Subassembly Number*	*Number of Nonconformities*
26	35	39	49
27	14	40	37
28	21	41	51
29	33	42	54
30	40	43	45
31	63	44	33
32	62	45	41
33	55	46	57
34	65	47	50
35	70	48	63
36	45	49	48
37	38	50	49
38	38		

Plot these data on the control chart derived in *a*. Indicate which points are outside the control limits. Would you recommend revision of your control chart?

20.2. The following are the number of breakdowns in successive lengths of 10,000 feet each of rubber-covered wire.[7]

Length Number	*Number of Breakdowns*	*Length Number*	*Number of Breakdowns*
1	1	16	20
2	1	17	1
3	3	18	6
4	7	19	12
5	8	20	4
6	1	21	5
7	2	22	1
8	6	23	8
9	1	24	7
10	1	25	9
11	10	26	2
12	5	27	3
13	0	28	14
14	19	29	6
15	16	30	8

Do these data come from a controlled process?

20.3. A large corporation has the following accident record for the years 1930–31:

[7] *ASTM Manual on Presentation of Data* (Philadelphia: American Society for Testing and Materials, 1940), p. 66. For further analysis of these data, see the later ASTM publication, *Manual on Presentation of Data and Control Chart Analysis.* STP 15D (Philadelphia: American Society for Testing and Materials, 1976), Table XIII.

Date	Number of Disabling Injuries	Million Man-Hours of Exposure
1930 January	11	0.175
February	4	0.178
March	5	0.175
April	8	0.180
May	4	0.183
June	4	0.198
July	9	0.210
August	12	0.212
September	2	0.210
October	6	0.211
November	6	0.195
December	7	0.200

Date	Number of Disabling Injuries	Million Man-Hours of Exposure
1931 January	7	0.201
February	3	0.202
March	10	0.203
April	8	0.215
May	9	0.220
June	11	0.227
July	17	0.237
August	9	0.233
September	16	0.236
October	7	0.232
November	2	0.216
December	7	0.202

Would you conclude from the above data that the variation from month to month in the number of disabling injuries per million manhours of exposure was due entirely to chance fluctuations, or would you conclude that "assignable causes" were at work that from time to time caused fundamental changes in accident-producing conditions?

11. SELECTED REFERENCES*

American Society for Quality Control (B: ANSI/ASQC Stds. Z1.1, Z1.2 and Z1.3–1985 and P '46–), American Society for Testing and Materials (B '76), Burr (B '76), Cowden (B '57), Dodge (P '28), Grant and Leavenworth (B '80), Howell and Johnson (P '47), Reynolds (P '71), and Simon (B '41).

* B and P refer to the Book and Periodical sections, respectively, of the Cumulative List of References in Appendix V.

21

Variables Control Charts

Control charts based upon measurements of quality characteristics are often found to be a more economical means of controlling quality than control charts based on attributes. Occasionally, variables charts are the only kind that can be used. The variables control charts that are most commonly used are average, or $\bar{X}$-charts, range, or R-charts, and standard deviation, or charts. Other variables charts, such as control charts for individuals, are also sometimes employed.

If the output of a process forms a normal frequency distribution, this distribution will be completely described when its mean and standard deviation are known. Even when the distribution of output is not normal, the mean and standard deviation are important measures of the distribution. Significant changes in either the mean or the standard deviation are an indication of significant changes in the process, and if specification limits are close to the existing mean, these changes may cause significant changes in the fraction nonconforming.[1] When control is undertaken by using variables instead of attributes, it usually takes the form of employing an $\bar{X}$-chart to control the average of the process and an R-chart or s-chart to control the general variability of the process. The combination of $\bar{X}$-chart and R-chart or an $\bar{X}$-chart and s-chart will give reasonably good control of the whole process.

The sample range is simpler than the sample standard deviation and can be easily employed whether computer facilities are readily available or not. In small samples of 10 or less from a normal distribution it is also almost as efficient as the sample standard deviation. For these reasons the discussion that follows will be primarily devoted to the use of $\bar{X}$-charts and R-charts; s-charts are discussed briefly in Section 10.2 below.

[1] Compare Chapter 11.

1. $\bar{X}$-CHARTS

An $\bar{X}$-chart shows variations in the averages of samples. On it are a central line and upper and lower control limits. If the chart is being used to study past data, the central line is set at $\bar{\bar{X}}$, the average of all the data. The UCL and LCL usually are $3\sigma'_{\bar{X}}$ limits or rather $3\hat{\sigma}_{\bar{X}}$ limits, where $\hat{\sigma}_{\bar{X}}$ is an estimate of $\sigma'_{\bar{X}}$ derived from the data. If the chart is being used to control current output, the central line may be a standard $\bar{X}''$ derived from past data or selected by the management to attain certain objectives. The limits may also be based on a standard value for the process standard deviation (σ'') that has been derived from past data or selected by the management. The formulas for the control limits are

(21.1)
$$\left.\begin{aligned} \text{UCL} &= \bar{\bar{X}} + \frac{3\hat{\sigma}}{\sqrt{n}} \\ \text{LCL} &= \bar{\bar{X}} - \frac{3\hat{\sigma}}{\sqrt{n}} \end{aligned}\right\} \textit{For charts for analyzing past data}$$

(21.2)
$$\left.\begin{aligned} \text{UCL} &= \bar{X}'' + \frac{3\sigma''}{\sqrt{n}} \\ \text{LCL} &= \bar{X}'' - \frac{3\sigma''}{\sqrt{n}} \end{aligned}\right\} \textit{For charts for attaining current control}$$

Here $\hat{\sigma}$ is an estimate of σ', which we shall see shortly is usually taken as equal to $\bar{R}/d_2$. This is explained below.

If the $\bar{X}$ chart is being used to control current production, a sample of n items is taken from the process every so often and a quality measurement made of each item. The average of these measurements is then computed and plotted on the chart. So long as the sample averages do not fall outside the control limits or do not show any nonrandom variation within the limits, the process is deemed to be in control with respect to its central tendency. If a point falls outside the control limits or nonrandom variation occurs within the limits, the process is deemed to be out of control with respect to its central tendency, and an investigation is undertaken to find the assignable cause of this extreme variation. Figure 21.1 is an example of an $\bar{X}$-chart.

2. R-CHARTS

An R-chart shows variations in the ranges of samples. On it is a central line and an upper and lower control limit. If the chart is being used to analyze past data, the central line will be the average of the sample ranges *($\bar{R}$)*. If it is being used to control current output, the central line may be some standard derived from past data or adopted by the management to attain certain objectives. The control limits are usually $3\hat{\sigma}_R$ limits, where $\hat{\sigma}_R$ is estimated from the data or $3\sigma''_R$ limits where σ''_R is computed from

FIGURE 21.1
An $\bar{X}$-Chart for Analyzing Past Data (see Table 21.1 and Section 3)

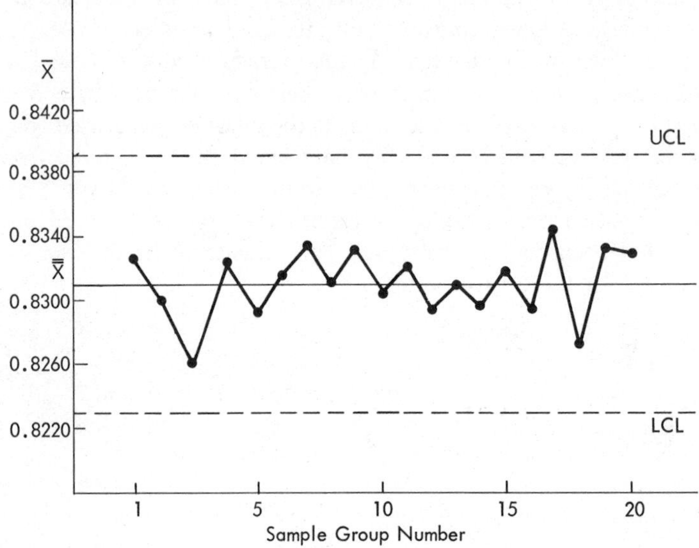

a standard value for the process standard deviation (σ''). Since the distribution of sample ranges is rather skewed, there is a tendency in some quarters to use probability limits instead of $3\hat{\sigma}_R$ or $3\sigma_R''$ limits. These are also based on $\bar{R}$ or σ'', as will be explained below. All formulas for R-charts are based on the assumption that the universe of individual values is normally dis-

FIGURE 21.2
An R-Chart for Analyzing Past Data* (see Table 21.1 and Section 3)

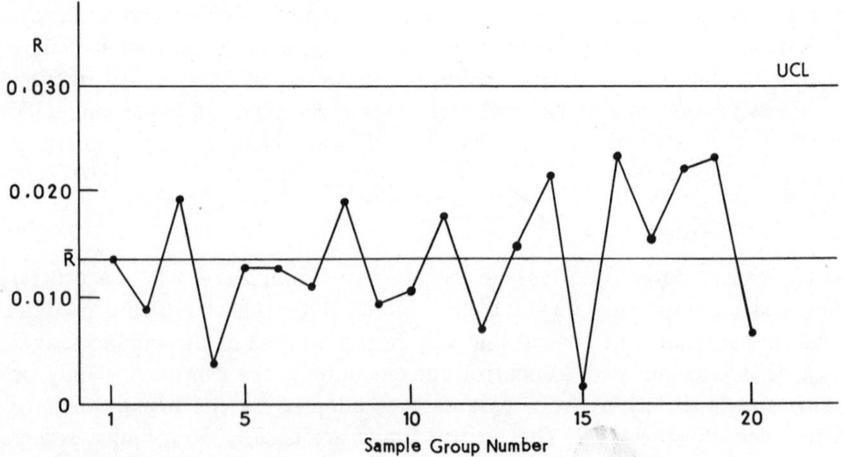

* This chart has only an upper control limit; other R-charts may have both upper and lower control limits.

tributed. Good results have been obtained, however, by the use of the normal procedure when the universe is not normal.

When an R-chart is being used to control current output, the range of a sample of n items is computed and plotted on the R-chart. This accompanies the plotting of the mean of the sample on the $\bar{X}$-chart. Usually the R-chart is shown directly below the $\bar{X}$-chart. If the sample range does not fall outside the control limits and if there is no evidence of nonrandom variation within the limits, the process is deemed to be under control with respect to its variability. If a point falls outside the control limits or if there is evidence of nonrandom variation within the limits, the process is deemed to be out of control with respect to its variability. In this case, an investigation is undertaken to locate assignable causes. Figure 21.2 is an example of an R-chart.

3. CONSTRUCTION AND OPERATION OF $\bar{X}$- AND R-CHARTS

3.1. Charts Based on Observed Data

The general procedure for setting up $\bar{X}$- and R-charts from observed data may be illustrated with reference to the heights of fragmentation bomb bases listed[2] in Table 3.2. These heights are arranged in 29 groups of 5 each and are, in fact, 29 daily samples taken from the production process. The means and ranges of these 29 samples are listed in Table 21.1. For the purposes of our illustration, let us suppose that we have the data for the first 20 days only and that we have the problem of setting up control charts.

It is best to begin with the construction of an R-chart because the limits of the $\bar{X}$-chart will depend upon the estimate of the process standard deviation and this should properly be made from the R-chart. *It is not correct to estimate the process standard deviation from all the data* $\left(\text{e.g., } s = \sqrt{\frac{\Sigma(X - \bar{X})^2}{n - 1}}\right)$ *and use this in setting up limits for the* $\bar{X}$*-chart. For the estimate so computed will be affected by the variation between the sample means and is therefore not an independent measuring rod for detecting extreme variations in these means.* The estimate of the process standard deviation to be used in setting up limits for the $\bar{X}$-chart must be computed from the within-sample variation to the exclusion of the between-sample variation. If the R-chart shows that the variability of the process is in control, such an estimate may be derived from the average of the sample ranges. If the R-chart shows that the variability of the process is generally out of control, it is better not to set up an $\bar{X}$-chart until control of process variability has been attained. If only a point or two are outside control limits, however, and if assignable causes can be found for these points, we might venture to set up limits for the $\bar{X}$-chart on the basis of the remaining points.

---

[2] See p. 479.

TABLE 21.1
Overall Heights of Fragmentation Bomb Bases ($\bar{X}$ and R for samples of five measurements)*

Group Number	$\bar{X}$	R
1	0.8324	0.014
2	0.8306	0.008
3	0.8262	0.020
4	0.8326	0.004
5	0.8290	0.013
6	0.8316	0.013
7	0.8336	0.012
8	0.8310	0.020
9	0.8336	0.010
10	0.8306	0.011
11	0.8332	0.018
12	0.8288	0.006
13	0.8310	0.016
14	0.8294	0.023
15	0.8322	0.003
16	0.8288	0.025
17	0.8344	0.016
18	0.8270	0.023
19	0.8338	0.025
20	0.8332	0.007
	$\bar{\bar{X}}$ 0.8312	$\bar{R}$ 0.01435
21	0.8310	0.012
22	0.8304	0.009
23	0.8328	0.005
24	0.8332	0.016
25	0.8282	0.022
26	0.8306	0.004
27	0.8354	0.023
28	0.8330	0.024
29	0.8312	0.008

* See Table 3.2 for original data.

With respect to the first 20 samples of Table 21.1, we may proceed to set up an R-chart as follows. First, we compute the average sample range $\bar{R}$. For the given data $\bar{R} = 0.01435$. This becomes the central line on the R-chart. Let us follow the standard practice and compute $3\hat{\sigma}_R$ limits where $\hat{\sigma}_R$ is an estimate of the universe standard deviation of sample ranges ($\hat{\sigma}'_R$) for the given sample size. To do this we note from Table D1 of Appendix II that $\sigma'_w = 0.864$ for samples of 5, w being the ratio R/σ'. Hence,[3] σ'_R

[3] Cf. Problem 3.20.

$= 0.864\sigma'$. But σ' can itself be estimated[4] as equal to $\bar{R}/d_2$. Hence we can estimate σ'_R by using $\bar{R}/d_2$ for σ'. This gives

(21.3) $$\hat{\sigma}_R = \sigma'_w \frac{\bar{R}}{d_2}$$

or to use the quality control notation,

(21.3a) $$\hat{\sigma}_R = d_3 \frac{\bar{R}}{d_2}$$

where d_3 is defined as equal to σ'_w. Thus for samples of 5 we have $\hat{\sigma}_R = \frac{0.864}{2.326}\bar{R} = 0.3715\bar{R}$, and, for the given value of $\bar{R}$, $\hat{\sigma}_R = 0.3715(0.01435) = 0.00533$. The upper control limit for our R-chart is thus

$$\text{UCL} = \bar{R} + 3\hat{\sigma}_R = 0.01435 + 3(0.00533) = 0.03034$$

The lower control limit would be computed from the formula

$$\text{LCL} = \bar{R} - 3\hat{\sigma}_R$$

but its use in this case leads to a negative answer. We therefore say there is no lower limit.

As indicated above, $3\hat{\sigma}_R$ is taken equal to $\frac{3d_3\bar{R}}{d_2}$, so that the upper control limit can be written

$$\text{UCL} = \left(1 + \frac{3d_3}{d_2}\right)\bar{R}$$

Thus, if we set $D_4 = \left(1 + \frac{3d_3}{d_2}\right)$, we have

(21.4) $$\text{UCL} = D_4\bar{R}$$

Likewise, the lower control limit can be written

$$\text{LCL} = \left(1 - \frac{3d_3}{d_2}\right)\bar{R}$$

and if we set $D_3 = \left(1 - \frac{3d_3}{d_2}\right)$, we have

(21.5) $$\text{LCL} = D_3\bar{R}$$

Values of D_3 and D_4 for various values of n will be found in Table M of Appendix II.

As an alternative to 3-sigma limits, we could have computed probability limits for the R-chart. To do this, we make use of Table D1 of Appendix

---

[4] See Chapter 6, Section 3.

II. This gives the 0.001 and 0.999 points of the distribution of the relative range $w = R/\sigma'$. To get the 0.001 and 0.999 points for R, we need only multiply the 0.001 and 0.999 points for w by σ'. If σ' is estimated as equal to $\bar{R}/d_2$, the 0.001 and 0.999 points for R are estimated as equal to $\frac{w_{0.001}}{d_2}\bar{R}$ and $\frac{w_{0.999}}{d_2}\bar{R}$. For our present problem we have $\frac{w_{0.001}}{d_2}\bar{R}$ $= \frac{0.37}{2.326}(0.01435) = 0.00228$, and $\frac{w_{0.999}}{d_2}\bar{R} = \frac{5.48}{2.326}(0.014350 = 0.0338)$. The probability limits are thus higher than the corresponding $3\hat{\sigma}_R$ limits.

It is to be noted that the quantities $w_{0.001}/d_2$ and $w_{0.999}/d_2$ are also referred to as $D_{0.001}$ and $D_{0.999}$, so that, when probability limits are used, we have

(21.6) and

$$\text{UCL} = D_{0.999}\bar{R}$$
$$\text{LCL} = D_{0.001}\bar{R}$$

A table of $D_{0.001}$ and $D_{0.999}$ may be found in Grant and Leavenworth, *Statistical Quality Control,* 5th ed. (New York: McGraw-Hill, 1980), page 292, for values of n from 2 to 10.

The preliminary values for the central line and the $3\hat{\sigma}_R$ limits for our R-chart are shown in Figure 21.2, together with the sample R's for the first 20 days. None of the points falls outside the control limit or shows other indications of assignable causes. Hence the variability of the process is deemed to be under control in this preliminary period. Let the central line and control limit of the preliminary R-chart be taken as the standard for use in the next period.

Having found the process in control with respect to variability, we can now safely proceed to set up the $\bar{X}$-chart. Our estimate of σ' is $\hat{\sigma} = \bar{R}/d_2$ $= \frac{0.01435}{2.326} = 0.006169$. The preliminary upper and lower limits on our $\bar{X}$-chart will therefore be

$$\text{UCL} = \bar{\bar{X}} + \frac{3\hat{\sigma}}{\sqrt{n}} = 0.8312 + \frac{3(0.006169)}{\sqrt{5}} = 0.8395$$

and

$$\text{LCL} = \bar{\bar{X}} - \frac{3\hat{\sigma}}{\sqrt{n}} = 0.8312 - \frac{3(0.006169)}{\sqrt{5}} = 0.8229$$

These $\bar{X}$ control limits may also be derived from the equations:

(21.7) $$\text{UCL} = \bar{\bar{X}} + A_2\bar{R} \qquad \text{LCL} = \bar{\bar{X}} - A_2\bar{R}$$

where $A_2 = \frac{3}{d_2\sqrt{n}}$. Thus, in the present instance, we would have

$$\text{UCL} = 0.8312 + 0.577(0.01435) = 0.8395$$
$$\text{LCL} = 0.8312 - 0.577(0.01435) = 0.8229$$

Values of A_2 have been tabulated for various values of n and are presented in Table M of Appendix II.

The preliminary values for the central line and upper and lower control limits of our $\bar{X}$-chart are set up in Figure 21.1 and the data for the first 20 days are plotted on the chart. It will be noted that once again all the points fall within the limits and there is no evidence of nonrandom variation within the limits. The process appears to be well under control with respect to its average.

In Figures 21.3 and 21.4 the central values and control limits of Figures 21.1 and 21.2 are taken as standard values for the following month's operations. When the results for the next nine days are plotted on the $\bar{X}$- and R-charts, both charts indicate that the process remains under control at the average level and with the variability specified.

It is to be noted that when a point is out of control on a $\bar{X}$-chart and is subsequently discarded on the finding of an assignable cause for the occasion, it may be desirable also to discard the corresponding point on the R-chart before calculating central lines and limits for the next period. Whether it is advisable to do this or not will depend on the nature of the assignable cause. If the assignable cause discovered subsequent to the investigation of a point out of control on an X-chart is of such a character that it suggests the likelihood of a different variability for the occasion, then the corresponding point should also be discarded in determining the R-chart to be used in controlling quality in the next period. Or, if the assignable cause discovered subsequent to the investigation of a point out of control on an R-chart is

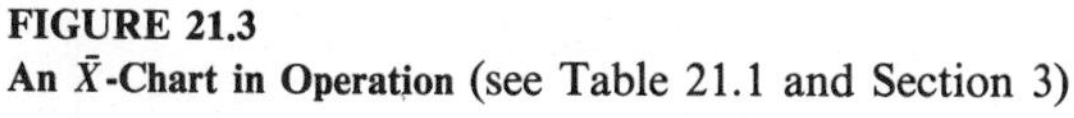

FIGURE 21.3
An $\bar{X}$-Chart in Operation (see Table 21.1 and Section 3)

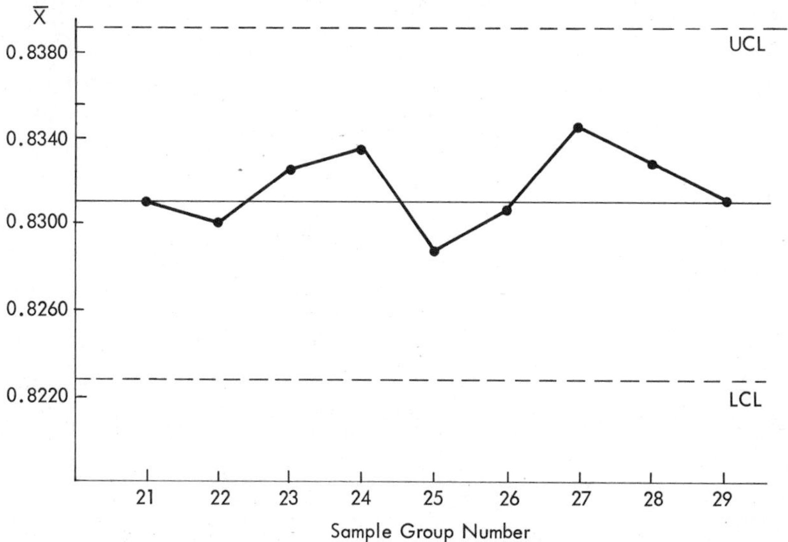

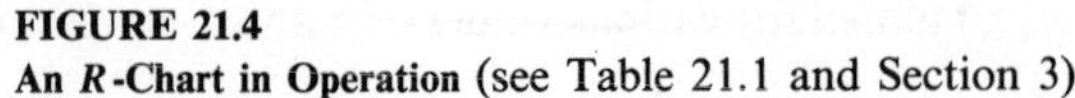

FIGURE 21.4
An *R*-Chart in Operation (see Table 21.1 and Section 3)

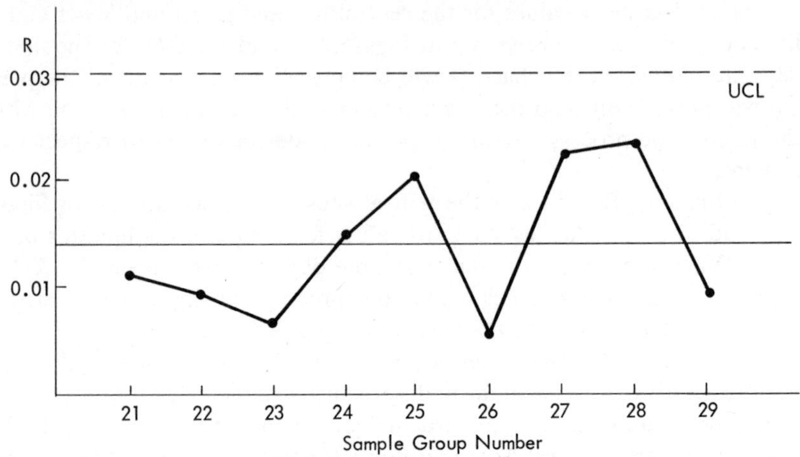

of such a character that it suggests the likelihood of a different mean for the occasion, then the corresponding point should also be discarded in determining the $\bar{X}$-chart to be used in controlling quality in the next period. In fact, the points (if any) that are to be used in computing central lines and control limits for the two charts for the next period will depend largely on the types of assignable causes discovered in the investigation of out-of-control points. This is a matter of engineering knowledge and intuition, not a matter of statistical theory.

3.2 Charts Based on Standard Values

If the management from the start provides standard values at which it is expected the process can be controlled, the construction of $\bar{X}$- and R-charts is simplied. If the standard values are represented by $\bar{X}''$ and σ'', the central line on the $\bar{X}$-chart is $\bar{X}''$, and the control limits are given by the equations

(21.8)

$$\text{UCL} = \bar{X}'' + \frac{3\sigma''}{\sqrt{n}}$$

$$\text{LCL} = \bar{X}'' - \frac{3\sigma''}{\sqrt{n}}$$

The central line on the R-chart is obtained by multiplying σ'' by d_2 for the sample size employed,[5] and upper and lower control limits are given by

[5] See Chapter 6, Section 3.

(21.9)
$$\text{UCL} = d_2\sigma'' + 3d_3\sigma''$$
$$\text{LCL} = d_2\sigma'' - 3d_3\sigma''$$

where d_2 and d_3 are given by Table D1 of Appendix II. To get further simplification, it is customary in quality control to set $A = 3/\sqrt{n}$, $D_1 = (d_2 - 3d_3)$, and $D_2 = (d_2 + 3d_3)$. The control limits for $\bar{X}$ then become

(21.2a)
$$\text{UCL} = \bar{X}'' + A\sigma''$$
$$\text{LCL} = \bar{X}'' - A\sigma''$$

and the control limits for R become

(21.10)
$$\text{UCL} = D_2\sigma''$$
$$\text{LCL} = D_1\sigma''$$

Values of A, D_1, and D_2 for various values of n are listed in Table M of Appendix II.

The standard values provided by the management may be values that experience with similar processes in the past suggests can be attained. Also, if points fall outside the control limits on the $\bar{X}$-chart simply because the machines involved are not set at the right levels and if adjustment of the settings will bring the points within the control limits, then the management's selection of a standard value for the central line on the $\bar{X}$-chart (i.e., the value of $\bar{X}''$) becomes essentially the management's directive for setting the machines. It is possible that process variability may likewise be primarily a matter of machine adjustment, in which case the management's standard for the process standard deviation (i.e., the value of σ'') may again be essentially its directive for machine settings.

The quality control engineer, however, must be on his guard against arbitrarily selected values of $\bar{X}''$ and σ''. This is especially true of arbitrary values of σ''. If an R-chart shows control when the limits are based upon an estimate of σ' derived from actual experience with the process, then, if the management arbitrarily sets a value for σ'' that is significantly less than the actual standard deviation of the process, increased expense will be incurred in trying to find assignable causes, when only chance causes may be at work. This does not mean that more assignable causes may not be uncovered; it simply means that the risk of fruitless searches is increased. Possibly a less expensive course would be to use the estimate of σ' based upon the actual data but to increase the size of the sample taken for control chart purposes. This would enhance the chance of finding assignable causes without increasing the risk of fruitless searches.[6] A third course would be to revise the whole process, say, buy new machines. This is likely to be initially the most expensive, but it might in the long run be less expensive than trying to reduce the variability of the process by arbitrarily setting a low value for σ'' and, as

[6] See Section 5 below on operating characteristics of $\bar{X}$- and R-charts.

a result, continually looking for "assignable causes" that cannot be found.

In concluding the discussion of $\bar{X}$- and R-charts, *it is to be especially noted that specification limits should never be shown on an $\bar{X}$-chart since it pertains to means not individual items.*

4. SUPPLEMENTARY CHARTS OF INDIVIDUAL ITEMS

Often $\bar{X}$- and R-charts can be more readily interpreted if they are accompanied by a tier chart of individual items. Such a chart is shown in Figure 21.5. There the individual data of Table 3.2 (from which Table 21.1 was derived) are plotted by samples, the items for each sample being plotted vertically in order of size. If a point falls outside the limits on the $\bar{X}$-chart, the chart for individual items will show, for example, whether this was caused by a single item or whether the whole group had moved. Similarly, if a point falls above the upper limit on the R-chart, the chart for individual items may indicate which items are especially responsible for the increased variation and thus aid in discovering the assignable cause. It is also possible to draw specification limits on a chart for individuals, whereas as noted in the previous section these should not be shown on a control chart for means.

If standard values $\bar{X}''$ and σ'' are available and *if the process is exactly normally distributed,* then it is possible to place $3\sigma''$ limits on the chart for individual items and in this way make it into a control chart for items. For in this case one or more items outside the control limits will suggest departure from the previous state of control. Such a chart by itself, however,

FIGURE 21.5
Chart of Individual Heights of Bomb Bases by Sample Groups (data from Table 3.2)

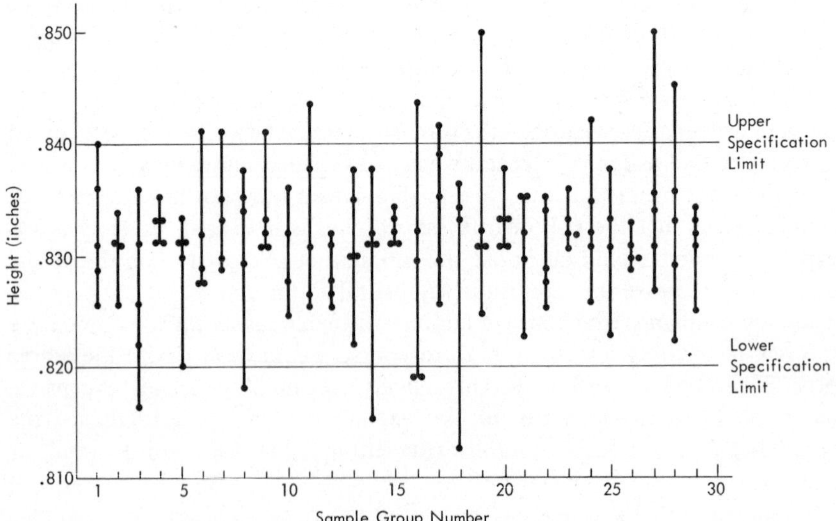

must be used with great caution. For points outside the limits of the items chart may indicate that the assumption of normality is wrong rather than that the process means or process standard deviation has changed. If a chart for individual items is constructed for past data and the standard deviation that is used to set the limits is computed from the data as a whole $\left(\text{i.e., } s = \sqrt{\frac{\Sigma(X - \bar{X})^2}{n-1}}\right)$ rather than from an R-chart based on proper subgrouping, little information is provided as to whether or not the process is in a state of control. Also, unless the data are very numerous, little information is given as to normality.

5. OPERATING CHARACTERISTICS OF $\bar{X}$- AND R-CHARTS

5.1. Charts Used to Control Current Output

The derivation and interpretation of the operating characteristic functions of $\bar{X}$- and R-charts that are used to control current output are essentially the same as the derivation and interpretation of the OC functions for p-charts and c-charts, although the procedure is somewhat more involved. What is required is simply a function that gives the probability of the points' falling within the control limits when the process has actually changed as to either its central tendency or its variability or both. As before, the OC function gives the probability of not catching on a single sample a specified change in the process.

In deriving the OC functions, we shall have to make one basic assumption, viz. that the distribution of the process is normal. This is necessary because it is the one case adequately covered by statistical theory. It is not likely to be a serious limitation, however, since measurable quality characteristics are often normally distributed or at least approximately so.

When the process is normally distributed, we can derive an OC curve for an R-chart that is independent of the means of the process. The OC function for an $\bar{X}$-chart is unfortunately dependent upon both the mean and the standard deviation of the process. This makes the OC curve for an R-chart the simpler to derive, and we shall therefore consider it first.

5.1.1. The OC Curve for an R-Chart Having a $3\hat{\sigma}_R$ Upper Limit. As an example, consider the R-chart with $3\hat{\sigma}_R$ limits derived in Section 3. For this, $n = 5$, $\bar{R} = 0.01435$, $UCL = 0.03034$, and there is no lower limit. Let the standard value taken for sigma be $\sigma'' = \bar{R}/d_2 = 0.01435/2.326 = 0.006169$, and suppose that the process standard deviation was originally equal to this σ'' but subsequently increases by 20 percent, i.e., σ' goes from 0.006169 to 0.00740. What is the chance that such a shift would not be caught by our R-chart on a single sample? To answer this we have simply to compute the probability that a sample of five items will have a range less than or equal to $UCL = 0.03034$. This is readily done with the help of Table D1 of Appendix II, which gives the sampling distribution of $w =$

R/σ'. For the probability that R will be less than or equal to UCL is the same as the probability that w will be less than or equal to UCL/σ', which in the given case equals $\frac{0.03034}{0.00740} = 4.10$. Table D1 shows that this is between 0.950 and 0.975. (Straight-line interpolation gives as a rough answer 0.968.) This yields one point on our OC curve. Suppose that σ' is now 50 percent above the original value of 0.006169, i.e., $\sigma' = 0.00925$. Then the probability

FIGURE 21.6
Operating Characteristic Curves for R-Charts Using $3\sigma''_R$ Limits

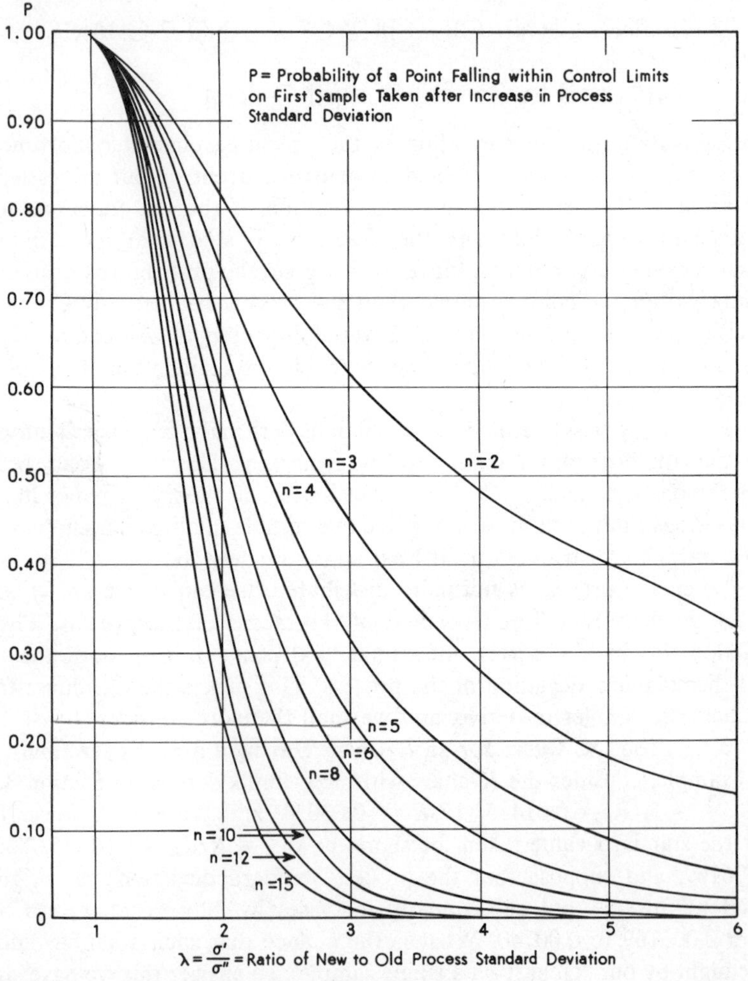

P_a = Probability of a point falling within control limits on first sample taken after increase in process standard deviation.

SOURCE: Reproduced from Acheson J. Duncan, "Operating Characteristics of R Charts," *Industrial Quality Control,* March 1951, pp. 40–41. For a more general chart, see Henry Scheffe, "Operating Characteristics of Average and Range Charts," *Industrial Quality Control,* May 1949, pp. 13–18.

that R will be less than or equal to 0.03034 is equal to the probability that w is equal to or less than $\frac{0.03034}{0.00925}$ or 3.28. From an extension of Table D1 this is computed to be approximately 0.86. This gives us another point on the OC curve.

Calculations of the kind just described yield the results presented in Figure 21.6. This gives the OC curves for R-charts for n varying from 2 to 15. The abscissa of the chart is the ratio of the new standard deviation (σ') to the old (taken as equal to the standard for the chart, viz, σ''), and the ordinate is the probability that the shift in σ' will not be caught by the R-chart on a single sample. The chart does not show the OC curves to the left of $\sigma'/\sigma'' = 1$, since in nearly every case we are concerned about increases in σ' rather than decreases. Figure 21.7 shows the full OC curves for $n = 5$ and $n = 10$. The OC curves for $n \leq 6$ do not drop to the left but continue getting closer to $P = 1$, since in these cases there is no lower limit. For $n > 6$, an R-chart making use of $3\hat{\sigma}_R$ limits does have a lower limit, and the OC curves for such charts are similar in shape to that for $n = 10$.

FIGURE 21.7
The Full Operating Characteristic Curves for Two *R*-Charts

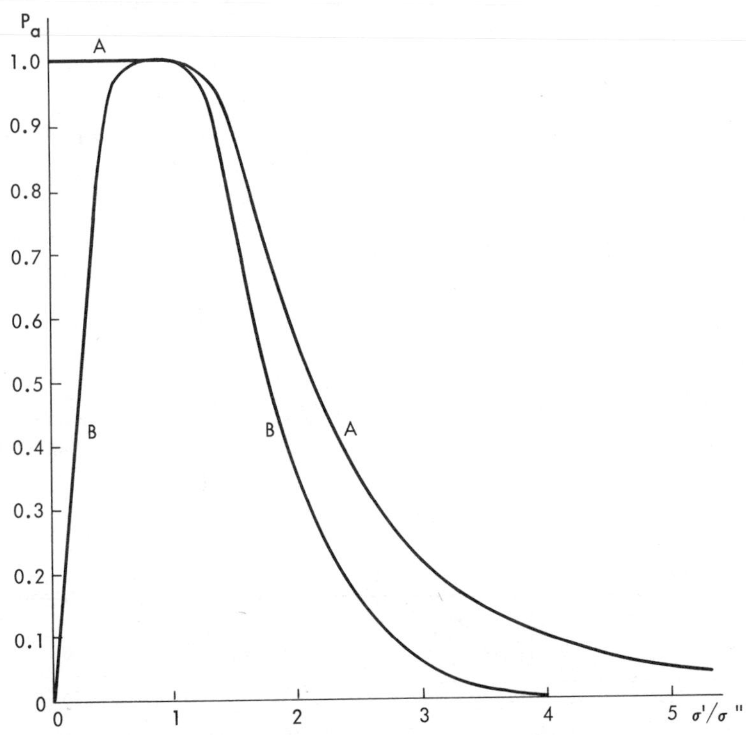

P = probability of not catching designated shift in σ' on a single sample.

In concluding this section, it is of interest to compare the OC curves for samples of various sizes. It will be noted, for example, that, with a doubling of the process standard deviation, the probability of not catching the shift on a single sample is as high as 0.60 for a sample of 5 but is about 60 percent of this for a sample of 10. For an increase in σ' of 50 percent, the two probabilities are 0.85 and 0.75, respectively.

5.1.2. The OC Curve for an $\bar{X}$-Chart Using 3-Sigma Control Limits when σ' Is Constant. If σ' remains constant, the OC curve for an $\bar{X}$-chart is easy to construct. We simply compute the probability that a sample $\bar{X}$ will fall within the control limits as a function of the process average $\bar{X}'$. Let us suppose that σ' equals σ'', the standard value used in constructing the chart. Then, when $\bar{X}'$ equals the standard for the chart, i.e., $\bar{X}' = \bar{X}''$,

FIGURE 21.8
Operating Characteristic Curve for the $\bar{X}$-Chart of Figure 21.3 when σ' Is Constant at 0.006169

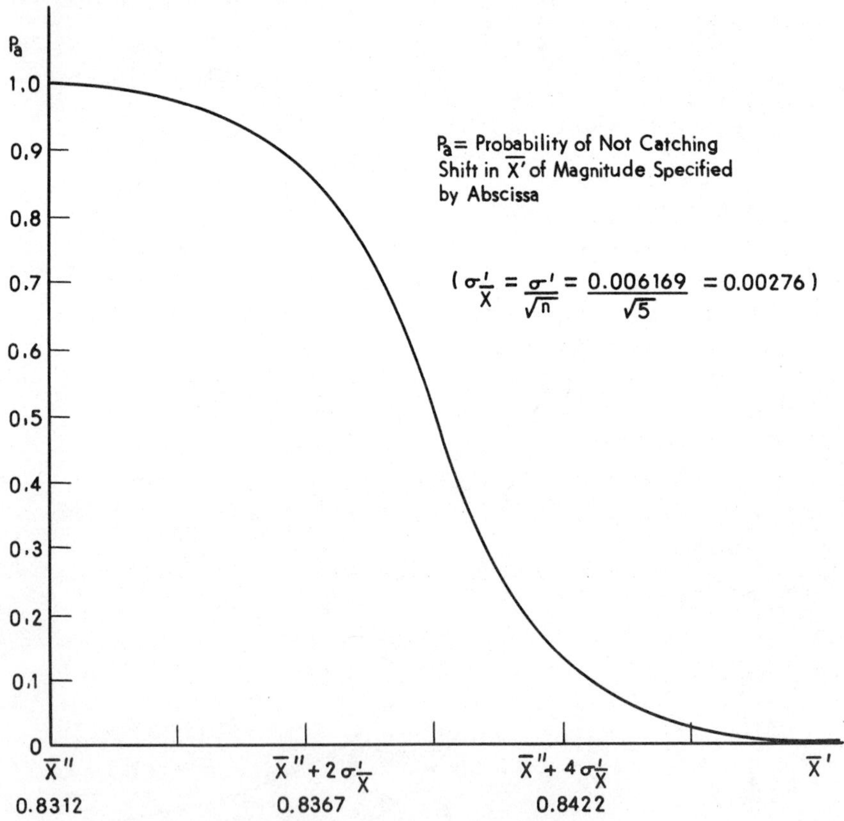

P_a = probability of not catching shift in $\bar{X}'$ of magnitude specified by abscissa.

$$\left(\sigma'_{\bar{X}} = \frac{\sigma'}{\sqrt{n}} = \frac{0.006169}{\sqrt{5}} = 0.002761\right)$$

the probability of a point's falling within the control limits is 0.9973. If $\bar{X}'$ shifts to $\bar{X}'' + 3\sigma''_{\bar{X}}$ or $\bar{X}'' - 3\sigma''_{\bar{X}}$, the probability of a point's falling within the control limits will be[7] 0.50. In general, if $\bar{X}'$ shifts to $\bar{X}'' + k\sigma''_{\bar{X}}$, the probability of a point's falling within the control limits will be the probability that a variable normally distributed with a standard deviation equal to $\sigma''_{\bar{X}}(= \sigma''/\sqrt{n})$ will fall above $\bar{X}'' + k\sigma''_{\bar{X}} - \bar{X}'' - 3\sigma''_{\bar{X}} = (k - 3)\sigma''_{\bar{X}}$. Thus, if $k = 2$, we have $P =$ Prob. $\bar{X} > -\sigma''_{\bar{X}} = 0.84$; if $k = 4$, we have $P =$ Prob. $\bar{X} > \sigma''_{\bar{X}} = 0.16$, and so on. A graph of the OC curve for the $\bar{X}$-chart of Figure 21.3 with $\sigma'' = 0.006169$ is presented in Figure 21.8.

5.1.3. The OC Function for an $\bar{X}$ Chart when σ' Is Variable. If σ' is variable, the probability of a point's falling within the control limits on an $\bar{X}$-chart will depend on both the process mean and the process standard deviation. Hence the OC function for the $\bar{X}$-chart must be represented by a surface, not a curve.

Details of the construction of the probability contours of such a surface are described in the revised (2d) edition of this text.[8] It will suffice here to note that if there is a large change in the process standard deviation, there is some chance it may throw a point outside the limits on the $\bar{X}$-chart. The increase in σ' would have to be four- to fivefold, however, if without any change in $\bar{X}$, there is to be a 0.5 chance of picking up the change in σ' on the $\bar{X}$ chart. This is the reason why we must have an R-chart to accompany our $\bar{X}$-chart. On a R-chart with $n = 5$, σ' would have to increase only 2.2 times to have a 0.5 chance of catching it.[9] This is not especially good, but it is much better than the $\bar{X}$-chart will do.

5.2. Charts Used to Study Past Data

The operating characteristics of $\bar{X}$-charts of past data have been studied by E. P. King. The analysis is beyond the scope of this book. Suffice it to report that when the variation in the process means is of the order of one standard deviation, an $\bar{X}$-chart with 25 samples of 5 each has a very high probability of discovering such variation.[10]

6. AVERAGE RUN LENGTH (ARL) CURVES FOR $\bar{X}$- AND R-CHARTS USED TO CONTROL CURRENT OUTPUT

Our discussion of ARL curves for $\bar{X}$- and R-charts will pertain to charts based on standard values. As explained in Section 3.2 above,[11] the central

---

[7] The probability of a point's falling beyond the opposite limit is practically zero.

[8] Acheson J. Duncan, *Quality Control and Industrial Statistics* (2d ed.: Homewood, Ill.: Richard D. Irwin, 1959), pp. 396–402.

[9] See Figure 21.6 for $n = 5$.

[10] Edgar P. King, "The Operating Characteristic of the Control Chart for Sample Means," *Annals of Mathematical Statistics,* 23 (1925), pp. 384–95.

[11] Also see the discussion on pp. 433–34.

line on the $\bar{X}$-chart will be set at the standard value ($\bar{X}''$) selected for the process average and the standard value (σ'') selected for the process standard deviation will be used in setting the control limits. For an R-chart the standard value σ'' will be used in setting both the central line and the control limits. It will be assumed that 3-sigma limits are used on both charts. The ARL curve for an $\bar{X}$-chart will pertain to its effectiveness in detecting an upward shift in the process average under conditions in which the process standard deviation remains fixed at σ''. The ARL curve for an R-chart will pertain to its effectiveness in detecting an upward shift in the process standard deviation. In both cases the increases in the process parameters will be taken to be of such a magnitude that the probability of a sample result falling below the lower control limit will be sufficiently small to be disregarded in the analysis. If in practice a sample result does fall below the lower control limit when there has been an increase in the process mean or process standard deviation, it will be assumed that the wrong causes will be investigated and the conclusion of the research will be that the sample in question is one of those that sometimes fall below the LCL simply as a result of chance causes. For sample sizes of 6 or less an R-chart will not have a lower 3-sigma control limit.

6.1. The ARL Curve for an $\bar{X}$-Chart Using 3-Sigma Control Limits and a Constant Standard Value σ'' for the Process Standard Deviation

To illustrate the derivation of an ARL curve for an $\bar{X}$-chart that uses 3-sigma control limits and a constant standard value σ'' for the process standard deviation, we will derive the ARL curve for the $\bar{X}$-chart of Figure 21.3 with respect to its effectiveness in detecting an increase in the process average. Figure 21.3 is a chart that has a central line set at a standard value ($\bar{X}''$) equal to 0.8312 and, in setting 3-sigma control limits, uses a constant standard value of $\sigma'' = 0.006169$ for the process standard deviation. The sample size (n) is 5, which yields an upper control limit $\bar{X}'' + 3\sigma''/\sqrt{n}$ equal to $0.8312 + 3(0.006169)/\sqrt{5} = 0.8395$ and a lower control limit $\bar{X}'' - 3\sigma''/\sqrt{n}$ equal to $0.8312 - 3(0.006169)/\sqrt{5} = 0.8229$. We will assume that sample $\bar{X}$'s are normally distributed, which is likely to be at least approximately true in most cases.

If in fact the process average $\bar{X}'$ equals $\bar{X}''$ and the process standard deviation σ' equals σ'' so that we have $\sigma'_{\bar{X}} = \sigma''/\sqrt{n}$, the probability that a sample $\bar{X}$ will fall within the control limits will equal the probability that a standard normal variable z will lie between -3 and $+3$ which equals 0.9973. In other words, the probability of a false alarm in either direction will equal $1 - 0.9973 = 0.0027$, which is the equivalent of an average run length of $1/(0.0027) = 370.4$ without a false alarm.

If in the use of the $\bar{X}$-chart of Figure 21.3 the process standard deviation σ' remains equal to σ'' but the process average $\bar{X}'$ increases from $\bar{X}'' =$

FIGURE 21.9
A Work Sheet for Deriving the ARL Curve for an $\bar{X}$-Chart Using 3-Sigma Control Limits and a Constant Standard Value σ'' for the Process Standard Deviation (given $\bar{X}'' = 0.8312$, $\sigma'' = 0.006169$, $n = 5$)

k	*Process Average* $=\bar{X}'' + k\sigma''/\sqrt{n}$ $= 0.8312 + 0.002759$ k	k − 3	*Prob** (P_a) $\bar{X} \le UCL = 0.8395$	*ARL* $= 1/(1 - P_a)$
0	0.8312	−3	0.99865	740.7†
0.2	0.8318	−2.8	0.99744	390.6
0.4	0.8323	−2.6	0.99534	214.6
0.6	0.8329	−2.4	0.99180	122.0
0.8	0.8334	−2.2	0.98610	71.9
1.0	0.8340	−2.0	0.97725	44.0
1.2	0.8345	−1.8	0.96407	27.8
1.4	0.8351	−1.6	0.94520	18.2
1.6	0.8356	−1.4	0.91924	12.4
2.0	0.8367	−1.0	0.84134	6.3
3.0	0.8395	0	0.50000	2.0

* P_a is the probability of a standard normal variable $\ge k - 3$. Taken from *Biometrica Tables for Statisticians,* Vol. I.

† This is the average run length if there is only an UCL. With both an UCL and a LCL, the average run length without a false alarm is 370.4. See text.

0.8312 to $\bar{X}'' + k\sigma''/\sqrt{n} = 0.8312 + k(0.006169)/\sqrt{5} = 0.8312 + 0.002759k$, then as noted above, the probability that a single sample mean will not exceed the UCL of 0.8395 will, under the normality assumption, equal the probability that a normal variable with standard deviation $\sigma''/\sqrt{n}$ will be greater than or equal to $\bar{X}'' + k\sigma''/\sqrt{n} - (\bar{X}'' + 3\sigma''/\sqrt{n}) = (k - 3)\sigma''/\sqrt{n}$. These probabilities are given in Figure 21.9 for specified values of k. If the probability for a given k is P_a, then on the assumption that sample $\bar{X}$'s are independent of each other, the average run length associated with the specified k will equal[12] $1/(1 - P_a)$. The ARLs for the various k's are given in Figure 21.9 and are shown graphically in Figure 21.10.

If management is dissatisfied with the computed ARL curve, a new chart could be adopted using a larger size sample or less than 3-sigma control limits or both. It is to be noted, however, that a tightening of the chart will increase the probability of picking up smaller, possibly unimportant shifts in the process average and may lead to a constant overadjustment and under-adjustment of the process that the management will consider undesirable.[13] The standard value σ'' should not be changed unless there are good reasons for believing it is not equal[14] to the actual process standard deviation.

---

[12] See pp. 434 above. Also note the discussion at the beginning of this section about a sample result falling below the LCL.

[13] A smaller σ' might be obtained, of course, by improving the process.

[14] See comments of W. Edwards Deming in *Quality, Productivity and Competitive Position,* p. 121.

FIGURE 21.10
The Average Run Length Curve for the $\bar{X}$-Chart Shown in Figure 21.3 Using Only the Upper Control Limit (data from Figure 21.9)

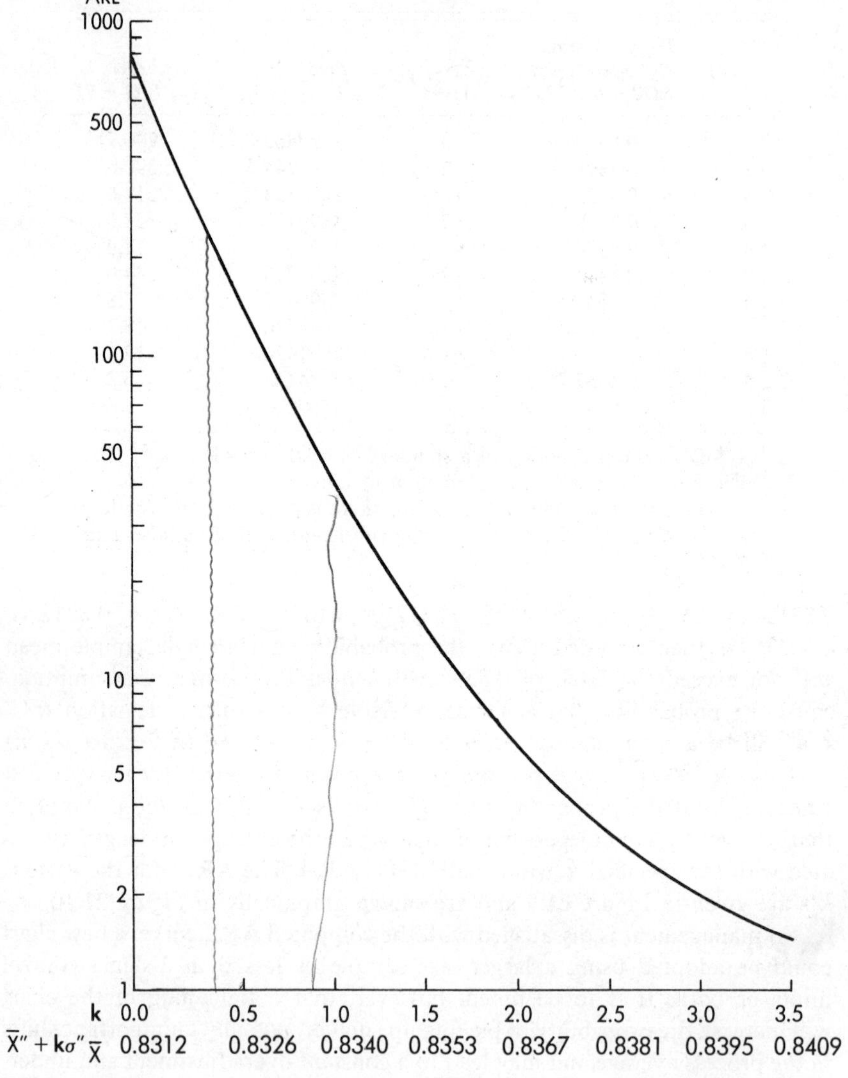

6.2. The ARL Curve for an R-Chart Having a $3\sigma''_R$ Upper Control Limit

For an R-chart with a $3\sigma''_R (= 3d_3\sigma'')$ upper control limit that is used to control current output, a curve of average run lengths for specified increases in the process standard deviation can be derived approximately from the

OC curves given in Figure 21.6. For as in the case of p-charts, c-charts, and $\bar{X}$-charts, if sample results are independent of each other, the ARL with which an R-chart using a $3\sigma''_R$ upper control limit will detect a specified increase in the process standard deviation will equal $1/(1 - P_a)$, the P_a being the probability of not detecting the specified increase in σ' given by Figure 21.6. Thus, if $n = 5$, Figure 21.6 shows that there will be a probability of about 0.60 that an R-chart with a $3\sigma''_R$ upper control limit will not detect on a single sample a doubling of the process standard deviation. Hence the ARL will in that case equal $1/(1 - 0.60) = 1/0.40 = 2.5$. For $n = 5$, other points on the ARL curve may be estimated by noting the abscissa values of the points of intersection of the OC curve for $n = 5$ with the horizontal lines through the P_a values marked on the vertical scale of Figure 21.6. A work sheet for the procedure is given in Figure 21.11.

FIGURE 21.11
Illustration of a Work Sheet for Deriving from Figure 21.6 the ARL Curve for an R-Chart Using a $3\sigma''_R$ Upper Control Limit and $n = 5$

Ratio of the New Process Standard Deviation (σ') to the Old (σ''), Read from Figure 21.6	*Prob. (P_a) of Not Detecting the Increase*	*ARL = $1/(1 - P_a)$*
1.30	0.90	10.0
1.60	0.80	5.0
1.80	0.70	3.3
2.00	0.60	2.5
2.23	0.50	2.0
2.48	0.40	1.7
3.15	0.20	1.25
4.80	0.10	1.11

A smoothing of the ARLs computed in Figure 21.11 will yield an approximation to the ARL curve for an R-chart with a $3\sigma''_R$ upper control limit and $n = 5$. This is presented in Figure 21.12. It shows that if the new process standard deviation is approximately 2.23 times as large as the old process standard deviation or larger, the increase in the variability of the process will be detected on the first or second sample taken after the increase has occurred. If management does not find this satisfactory, an R-chart could be adopted that uses a larger size sample or less than 3-sigma control limits or both. As noted above, for $n \geq 7$, an R-chart with 3-sigma control limits will have a lower as well as an upper control limit, but for important increases in the process standard deviation the probaiblity of a point falling below the LCL is taken to be negligible. For a chart that uses less than 3-sigma control limits, this may not be true.

FIGURE 21.12
The Average Run Length Curve for the *R*-Chart Shown in Figure 21.4 (data from Figure 21.11)

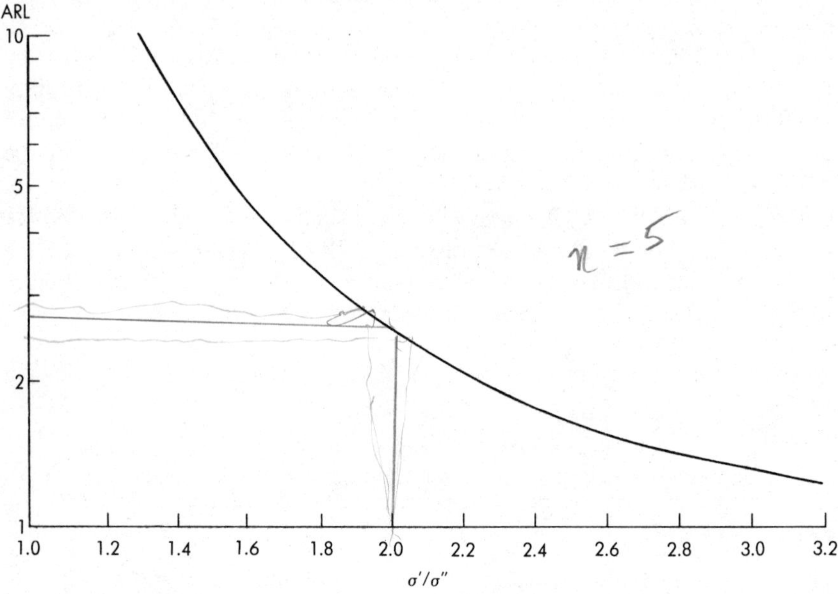

7. THE DESIGN OF $\bar{X}$- AND R-CHARTS FOR CURRENT CONTROL

The determination of sample size and frequency of sampling for $\bar{X}$ and R-charts used to maintain current control is a difficult problem of statistical theory. Its complete solution depends not only on the various risks inherent in the sampling process but on the cost of inspection and of scrap and research. In finding the final solution, quality control engineers need the help of both the accountant and the mathematical statistician.

7.1. Arguments for Samples of 4 or 5

Although the problem of the proper sample size and frequency of sampling has not been completely solved, solutions of one sort or another have been worked out in practice. Most often, samples consist of 4 or 5 items taken fairly frequently. The arguments given for samples of this size are as follows. Each is briefly discussed, and in some cases the general applicability of the argument is questioned.

1. For the same frequency of sampling, samples of 4 or 5 cost less than larger samples. Both the amount of inspection and the amount of computation are less. If samples are larger than 10, the R-chart does not work so well

and it would probably be desirable to divide the sample into subsamples of 4 or 5 and to set up an $\bar{R}$-chart. Such a chart is discussed below in Section 10.1. The use of an $\bar{R}$-chart, however, would increase the amount of work and introduce a "strange" chart.

2. With samples of 4 or 5 there is less chance of a change occurring during the taking of the sample than with larger samples. As W. A. Shewhart puts it, "If the cause system is changing, the sample size should be as small as possible so that the averages of samples do not mask the change."[15] How important this factor is depends on the type of product being produced. If a machine stamps out a ring every 5 seconds, the time involved in collecting a sample of 10 or 20 will be only 1 or 2 minutes.

3. Samples of 4 make it easy to compute the square root of *n;* samples of 5 permit the computing of an average by simply doubling the total and moving the decimal point (?). Samples of 10 or 16, however, would be equally easy in this respect.

4. Samples of 4 or 5 taken every 15 minutes, say, will be better than samples of 16 or 20 taken every hour, because they will catch more quality changes that occur during the hour (?). This argument needs to be carefully considered. Suppose that production is at a continuous rate with a mean of $\bar{X}'$ and a standard deviation of σ'. Let our control charts be based on samples of 4 taken every 15 minutes on the quarter hour. The upper limit of the $\bar{X}$-chart will thus be at $\bar{X}'' + \frac{3\sigma''}{\sqrt{4}}$. Suppose, now, that at 10 minutes after the hour the mean of the process $\bar{X}'$ increases from $\bar{X}' = \bar{X}''$ (the standard) to $\bar{X}' = \bar{X}'' + 1.5\sigma''$ without any change in the process standard deviation. With the $\bar{X}$-chart as set up, there will be a 0.5 chance of not catching this shift on any one sample. The chance of not catching the shift on any of the next three samples will therefore be $(0.5)^3$ or 0.125. On the other hand, if our control charts are based on samples of 16 every hour on the hour, the upper limit on our $\bar{X}$-chart will be at $\bar{X}'' + \frac{3}{\sqrt{16}}\sigma''$. With this chart, a shift of $1.5\sigma''$ in the process average at 10 minutes after the hour will have a chance of only 0.00135 of not catching the shift on the next sample. In fact, in this particular example, we would have to take 10 successive samples of 4 before the probability of catching the shift is more than the probability of catching it with a single sample of 16.

If the principal loss, however, that arises from not catching the shift consists of the volume of nonconforming product that is allowed to pass, then if the shift occurs at 10 minutes after the hour, the loss will, *on the average,* be less with samples of 4 every 15 minutes than with samples of 16 every hour. Suppose *a* is the average number of nonconforming units

---

[15] W. A. Shewhart, *Economic Control of Quality of Manufactured Product* (New York: D. Van Nostrand Co., Inc., 1931), p. 314.

produced per quarter hour after the shift has occurred. Then if the shift occurs 10 minutes after the hour, the average (expected) losses for the two types of control charts will be as follows:

Chart using samples of 4 on the quarter hour—

$$(0.5)(0.33a)+(0.5)^2(1.33a)+(0.5)^3(2.33a)$$
$$+(0.5)^4(3.33a)+(0.5)^5(4.33a)+(0.5)^6(5.33a)$$
$$+(0.5)^7(6.33a)+\ldots=1.33a$$

Chart using samples of 16 on the hour—

$$(0.99865)(3.33a)+(0.99865)(0.00135)(7.33a)$$
$$+(0.99865)(0.00135)^2(11.33a)+\ldots\doteq 3.34a$$

Thus, with respect to the average number of nonconforming items produced before the shift is caught, samples of 4 on the quarter hour do better in this particular case than samples of 16 on the hour.

If the shift in the process takes place 10 minutes before, rather than after, the hour, samples of 16 on the hour will obviously catch the shift with less average loss. If it may be presumed that process changes occur at random, then an average time for the occurrence of a shift is just before or just after the half hour. If they occur just after the half hour, the average loss incurred with samples of 4 on the quarter hour will be $4a$, while that incurred with samples of 16 on the hour will be approximately $2a$. If the shift occurs just before the quarter hour, the average loss incurred with samples of 4 on the quarter hour will be $2a$, while that incurred with samples of 16 on the hour will be slightly more than $2a$. In general, the average over all possible times of occurrence yields a smaller loss for samples of 16 on the hour than for samples of 4 on the quarter hour.

7.2. Theoretical Studies

If pertinent cost data are available, W. K. Chiu and G. B. Wetherill offer[16] a relatively simple procedure for designing specific $\bar{X}$-charts. A general study of the design of $\bar{X}$-charts has been undertaken by the author from which some interesting conclusions may be drawn.[17] The study was a theoretical one based on minimizing overall cost in respect to a specific model. The model was reasonably comprehensive, however, and it is believed that the conclusions are likely to be generally valid. At least they are suggestive of the proper direction in which to move if changes in design are contemplated.

The principal conclusions are as follows:

1. The customary sample sizes of 4 or 5 are close to optimum if the shifts to be detected are relatively large, e.g., if the assignable cause produces

[16] W. K. Chiu and G. B. Wetherill, "A Simplified Scheme for the Economic Design of $\bar{X}$-Charts," *Journal of Quality Technology,* 6, April 1974, pp. 63–69.

[17] Acheson J. Duncan, "The Economic Design of $\bar{X}$ Charts Used to Maintain Current Control of a Process," *Journal of the American Statistical Association,* 51 (1956), pp. 228–42.

a shift of $2\sigma'$ or more in the process average. If it is the aim of the chart to detect shifts in the process average as small as one σ', sample sizes of 15 to 20 are more economical than sample sizes of 4 or 5.

2. If a shift in the process average causes a high rate of loss, i.e., high relative to the cost of inspection, it is better to take small samples quite frequently than large samples less frequently. For example, when the rate of loss is high, samples of 4 or 5 taken every half hour are better than samples of 8 or 10 taken every hour.

3. Under certain circumstances charts using 2-sigma or even 1.5-sigma limits are more economical than charts using the conventional 3-sigma limits. This is true if it is possible to decide very quickly and inexpensively that nothing is wrong with the process when a point (just by chance) happens to fall outside the control limits, i.e., when the cost of looking for trouble when none exists is low. Contrariwise, it will be more economical to use charts with 3.5-sigma to 4-sigma limits if the cost of looking for trouble is very high.

4. If the unit cost of inspection is relatively high, the most economical design is one that takes small samples (say samples of 2) at relatively long intervals (say every 4 to 8 hours) with narrow control limits, say 1.5-sigma.

8. $\bar{X}$- AND R-CHARTS VERSUS p-CHARTS

It is to be noted than an attribute control chart can always be substituted for a set of variables charts if it is deemed desirable. If process variability exceeds the range between the specification limits, the process inevitably produces some nonconforming units. If process variability is less than the range between the specification limits, the process may not produce any nonconforming product. In the latter case, however, it is possible to set up fictitious specification limits which are tighter than the true limits and thus create a fictitious percent nonconforming.[18] In every case, therefore, it is possible to employ a p-chart in lieu of an $\bar{X}$-chart and R-chart. This being true, it is desirable to consider the advantages and disadvantages of the $\bar{X}$-chart–R-chart combination as compared with a p-chart.

A p-chart has the advantage that it merely calls for determining whether or not the product is "nonconforming." An $\bar{X}$-chart–R-chart combination can be used only for a single quality characteristic, but a p-chart may be based on a number of quality characteristics. In the latter instance, it may serve to summarize the general quality of the output, and in all cases it gives management a record of quality history.[19] On the other hand, a p-chart will usually require a much larger sample than an equally satisfactory

[18] This is sometimes called compressed limit gauging or 'narrow limit gauging.' Cf. N. L. Enrick, *Quality Control* (2d ed.; New York: Industrial Press, 1954), pp. 66 ff.; and Ellis R. Ott and A. B. Mundel, "Narrow Limit Gauging," 10, *Industrial Quality Control,* March 1954, pp. 21–28. Also see Chapter 11, Section 5.

[19] Cf. Eugene L. Grant, *Statistical Quality Control* (3d ed.; New York: McGraw-Hill, 1964), p. 234.

combination of $\bar{X}$- and R-charts,[20] say, a sample of 50 as against a sample of 10. Unless the cost of determining effectiveness under an attribute plan is much less than the cost of numerical measurement under a variables plan, the advantage of economy is likely to be on the side of the variables plan. Variables control also has the advantage that it enables the engineer to locate the cause of the trouble much more readily. The use of a chart for controlling central tendency and another for controlling variability in itself gives him an insight into the basic operations of the process. Furthermore, the variables chart helps him to estimate the extent of the departure from previous norms. All this facilitates the tracing of assignable causes of variation.

The advantage of variables charts with respect to size of sample may be illustrated by an example. Suppose that a process has a standard deviation of 10 and that the specification limits are 340 and 380. Then, if the quality characteristic in question is normally distributed and if the standard deviation remains constant, the minimum fraction nonconforming will be obtained when the process average is 360. This minimum fraction nonconforming will be 0.0454 (the probability of exceeding $2\sigma'$ in either direction). If the process average shifts to 350 (i.e., by $1\sigma'$), the fraction nonconforming rises to 0.1601 (the probability of exceeding $1\sigma'$ in one direction and $3\sigma'$ in the other direction). To have a 0.50 chance of catching such a shift on the first sample occurring after the shift, the size of the sample used on the variables charts should be such that the lower $3\sigma'$ limit of the $\bar{X}$-chart falls at 350. For the data in question this requirement is that $360 - \frac{3(10)}{\sqrt{n}} = 350$. Hence, to get the desired result, n should equal $\left[\frac{3(10)}{360 - 350}\right]^2 = 9$.

A p-chart that would give the same degree of protection would have to be based on a much larger sample than 9. This may be determined as follows:

If the process average is 0.0454, the upper control limit would be given by $0.0454 + 3\sqrt{\frac{(0.0454)(0.9546)}{n}}$. If a shift to a process average of 0.1601 is to have a chance of 0.50 of being caught on the first sample after the shift, then, on the assumption that the normal distribution gives a rough approximation to the probabilities involved, we must have

$$0.0454 + 3\sqrt{\frac{(0.0454)(0.9546)}{n}} = 0.1601$$

This yields $n = 30$. Tables of the binomial distribution show that a sample of about 50 will give results closer to those desired.[21] If the chance of catching

---

[20] Another disadvantage of a p-chart is that as the quality gets better, larger sample sizes are needed.

[21] If $n = 47$ and the upper control limit on the original p-chart is taken as 7, the probability of 7 or less is approximately 0.9986 when $p = 0.0454$ and approximately 0.5151 when $p = 0.1601$. (These results are based on straight-line interpolation, which involves some error.)

the shift to 350 on the first sample following the shift had been set at 0.10 instead of 0.50, then a sample of 3 on the $\bar{X}$-chart would have been sufficient. Approximately equivalent protection would then be given by a p-chart employing a sample of 12. It is obvious that if the cost of variables inspection does not greatly exceed the cost of attribute inspection, the variables plan will be definitely cheaper.

9. CONTROL CHART FOR INDIVIDUALS

If a process turns out a series of individual items that do not fall naturally into subgroups or the process is too slow to form special clusters over time, we can still use *a control chart for individuals.* This is accomplished by first constructing a chart for the moving range of 2 successive items. From the average moving range of 2, we can estimate σ' as equal to $\bar{R}/d_2$ for an n of 2, for such an estimate is likely to be only slightly affected by variations in individuals due to assignable causes. We can thus take the control limits equal to

$$\bar{X} \pm \frac{3\bar{R}}{d_2} \tag{21.11}$$

For the standard values $\bar{X}''$ and σ'', the control chart for individuals will have the limits

$$\bar{X} \pm 3\sigma'' \tag{21.12}$$

It should be noted that control charts for individuals must be very carefully interpreted if the process shows evidence of marked departure from normality. In such cases the multiples of sigma used to set control limits might better be derived from a Pearson Type III distribution or some other distribution for which the percentage points have been computed. See, for example, Table 42, of the *Biometrika Tables for Statisticians,* 2d ed. Another possibility is to chart some transformation of the data instead of the original data.[22]

Table 21.2 gives data on the strength of an organic chemical used in the manufacture of certain dyestuffs. These may be used to illustrate a control chart for individuals and a chart for the moving range. For these data $\bar{X} = 47.45$ and $\bar{R} = 4.175$. For the R-chart we have the UCL $= D_4\bar{R} = 3.267(4.175) = 13.64$, and for the X-chart we have the UCL $= \bar{X} + 3\frac{\bar{R}}{d_2} = 47.45 + \frac{3(4.175)}{1.128} = 57.55$, and the LCL $= 47.45 - s\frac{3(4.175)}{1.128} = 37.35$. (See Figure 21.13.)

[22] Cf. Dudley J. Cowden, *Statistical Methods in Quality Control* (Englewood Cliffs, N.J.: Prentice-Hall, 1957), chap. xxi on "Control Charts for Nonnormal Distributions."

TABLE 21.2
Percent Strengths of Samples of an Organic Paste

Batch Number	*Mean Strength of Three Samples*	*Moving Range of Two*
1	50.0	...
2	43.9	6.1
3	41.2	2.7
4	49.9	8.7
5	48.6	1.3
6	46.5	2.1
7	43.8	2.7
8	55.7	11.9
9	48.7	7.0
10	50.3	1.6
11	46.5	3.8
12	44.3	2.2

FIGURE 21.13
Control Charts for Individuals and for the Moving Range of Two (data from Table 21.2)

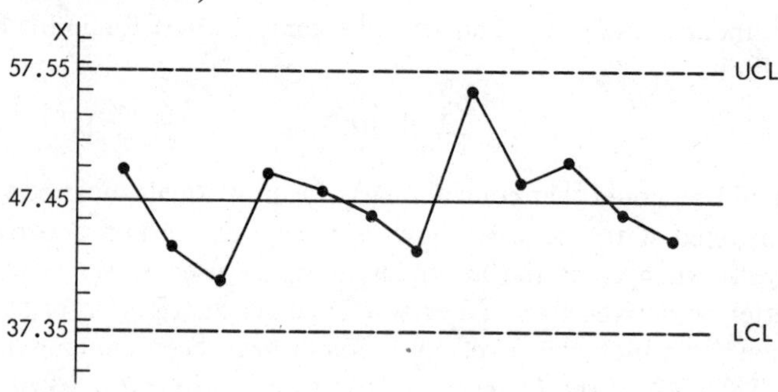

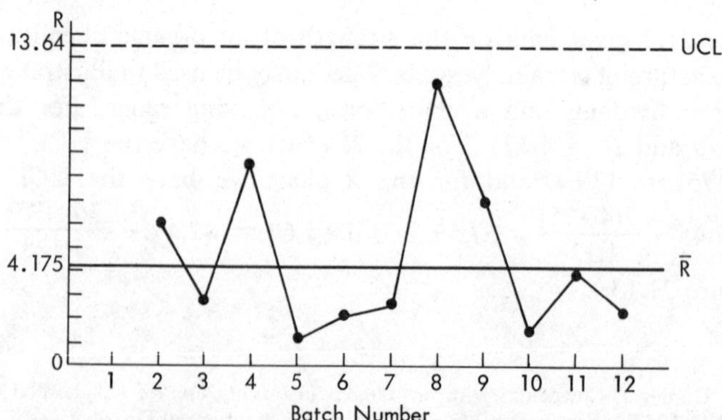

10. OTHER VARIABLES CONTROL CHARTS

10.1. The $\bar{R}$-Chart

When samples are large, say $n > 12$, the range loses rapidly in efficiency and it is best not to use an R-chart to control process variability. As noted above, however, it is possible to split a large sample into subsamples each of size 4 or 5 and run an $\bar{R}$-chart. For g subsamples of size m, yielding a total sample of gm items, the central line on an $\bar{R}$-chart for past data would be at $\bar{\bar{R}}$ and 3-sigma limits would fall at $\left(1 \pm \frac{3d_3}{d_2\sqrt{g}}\right)\bar{\bar{R}}$ where d_2 and d_3 are factors given in Table D1 and Table M, Appendix II, for $n = m$. For a standard value σ'', the central line would be $d_2\sigma''$ and the control limits $\left(d_2 \pm \frac{3d_3}{\sqrt{g}}\right)\sigma''$. Because of the division of the large sample into small subsamples and the loss in degrees of freedom due to ignoring the variation between subsamples within each sample, the $\bar{R}$ chart is not so efficient as other charts that make use of the whole of the sample information.

10.2. The s-Chart

When a standard value is not given for σ', and all we have is an ordered set of s's, we construct an s-chart as follows:

Compute the mean s and take the central line on the control chart as equal to $\bar{s}$.

As shown in footnote 19 of Chapter 6, if X is normally distributed, the standard deviation of s is $\sigma'\sqrt{1-c_4^2}$. If X is normally distributed, we also have $E(s) = c_4\sigma'$, so that $\bar{s}/c_4$ will give an unbiased estimate of σ'. Hence for a normal universe, the standard deviation of s would be estimated as equal to $\frac{\bar{s}}{c_4}\sqrt{1-c_4^2}$ and the control limits for an s-chart would be estimated to be

(21.13)

$$\text{LCL} = \bar{s} - 3\frac{\bar{s}}{c_4}\sqrt{1-c_4^2} = \left(1 - \frac{3}{c_4}\sqrt{1-c_4^2}\right)\bar{s} = B_3\bar{s}$$

$$\text{UCL} = \bar{s} + 3\frac{\bar{s}}{c_4}\sqrt{1-c_4^2} = \left(1 + \frac{3}{c_4}\sqrt{1-c_4^2}\right)\bar{s} = B_4\bar{s}$$

where

(21.14) $$B_3 = \left(1 - \frac{3}{c_4}\sqrt{1-c_4^2}\right) \text{ and } B_4 = \left(1 + \frac{3}{c_4}\sqrt{1-c_4^2}\right)$$

Values of B_3 and B_4 are given in Table M for values of n from 2 to 25.

While formulas (21.13) for control limits for an s-chart were derived from the assumption that X is normally distributed, they can be taken as control limits for s when X is not normally distributed. In such a situation they become simply arbitrary limits for engineering decisions and great care must be exercised in assigning probabilities to their use.

When a standard value σ'' is available, the central line on the s-chart becomes $c_4\sigma''$ and control limits for s are given by

(21.15)
$$\begin{aligned} \text{LCL} &= c_4\sigma'' - 3\sigma''\sqrt{1-c_4^2} = (c_4 - 3\sqrt{1-c_4^2})\,\sigma'' = B_5\sigma'' \\ \text{UCL} &= c_4\sigma'' + 3\sigma''\sqrt{1-c_4^2} = (c_4 + 3\sqrt{1-c_4^2})\,\sigma'' = B_6\sigma'' \end{aligned}$$

where

(21.16) $$B_5 = c_4 - 3\sqrt{1-c_4^2} \text{ and } B_6 = c_4 + 3\sqrt{1-c_4^2}$$

Values of B_5 and B_6 are also tabulated in Table M. Formulas (21.15) and (21.16) are again based on the assumption of a normal universe, but they can be used for nonnormal universes if great care is exercised in assessing the probabilities associated with the limits they yield.

10.3. The s^2-Chart

In contrast to engineering practice, mathematical statisticians advocate the use of an s^2-chart to control process variability when samples are large. The quantity s^2 is an unbiased statistic, and for a normal universe probability limits for it can be readily derived from the χ^2-distribution. Thus, for

$$s_j^2 = \frac{\sum_i (X_{ij} - \bar{X}_j)^2}{n-1},$$

the central line on the control chart would be $\bar{s}_j^2$, and 0.001 probability limits would fall at $\frac{\bar{s}_j^2}{(n-1)}\chi^2_{0.999}$ and $\frac{\bar{s}_j^2}{(n-1)}\chi^2_{0.001}$ for $n - 1$ degrees of freedom. Table C of Appendix II gives values of $\chi^2_{0.001}$, but not $\chi^2_{0.999}$. The latter can be interpolated from values given in Table 7 of *Biometrika Tables for Statisticians,* Vol. 1, 2d ed., pp. 122 ff. Table C could be used for 0.005 limits, however.

10.4. Other Charts

Other variables control charts have been offered as variations in the standard charts or as devices to handle special situations. Midrange and median charts are simple substitutes for $\bar{X}$-charts. Charts have also been developed for largest and smallest values. Details on the nature and construction of these special charts will be found in the revised (2d) edition of this text[23] and in the references quoted at the end of this chapter.

---

[23] Duncan, *Quality Control and Industrial Statistics,* pp. 378–81.

11. ALTERNATIVES TO VARIABLES CONTROL CHART ANALYSIS

In concluding the discussion of variables control charts it should be noted that there are alternative statistical procedures for finding answers to questions that have in this chapter been sought by control charts. This is particularly true when a chart is used to answer questions about past data. Such procedures are analysis of variance, Lambda tests, studentized-range tests, and so forth. These procedures are discussed in Chapters 29 to 31. To many statisticians these more sophisticated methods of analysis are preferable to the use of control charts. For further discussion of this the reader is referred to Chapters 29 to 31.

For personnel with little training in statistical theory, it may be noted that Ellis R. Ott's[24] "Analysis of Means" may be an alternate procedure that is more meaningful than analysis of variance since it deals directly with means rather than variances. To illustrate Ott's procedure suppose we have standard values $\bar{X}''$ and σ'' for the process mean and process standard deviation. Then a factor Z_α can be computed for which the probability is $1 - \alpha$ that a set of k independently and normally distributed means of samples of size n will all fall within the limits $\bar{X}'' \pm Z_\alpha \sigma''/\sqrt{n}$. Thus for $k = 3$, Z_α will equal[25] 2.39 and if, for example, $\bar{X}'' = 0.8321$, $\sigma'' = 0.006169$, and $n = 5$, the limiting values for three subsequently obtained sample means will be $0.8321 \pm 2.39(0.006169)/\sqrt{5}$ or 0.8231 and 0.8393. Hence, if the sample means obtained should equal 0.8310, 0.8354, and 0.8328, we would conclude that the process continues in control at the given standard values. For a comprehensive discussion of the analysis of means in general, the reader is referred to Ott's paper noted above, to his book *Process Quality Control, Trouble Shooting and Interpretation of Data* and to the January 1983 issue of the *Journal of Quality Technology,* which contains articles on the analysis of means by other authors. The reader is also referred to the papers by Edward G. Schilling and Lloyd S. Nelson listed in Appendix V of this text.

12. PROCESS CAPABILITY STUDIES

After $\bar{X}$- and R-charts show that a process is in control, a study of process capabilities is often undertaken. This is to find out whether the process can meet specifications and, if not, to estimate the fraction nonconforming. If the control charts are based on considerable data, $\bar{X}'$ and σ' are estimated from the charts. Thus we have

---

[24] "Analysis of Means—A Graphical Procedure," *Industrial Quality Control,* 24 (1967), pp. 101–09, reproduced in *Journal of Quality Technology,* 15 (1983), pp. 10–18.

[25] If p is the probability that a standard normal deviate will lie between $\pm Z_\alpha$, the probability that three such deviates will independently lie within the limits is p^3. If p^3 is to equal $1 - 0.05 = 0.95$, then p must equal 0.98304 and the normal tables will give $Z_\alpha = 2.39$.

$$\text{Estimate of } \bar{X}' = \bar{\bar{X}}$$

$$\text{Estimate of } \sigma' = \frac{\bar{R}}{d_2}$$

If we assume the process is normally distributed, we can immediately determine the percent nonconforming from the given specifications.

Let us examine the data of Section 3.1. Here we had $\bar{\bar{X}} = 0.8312$, $\hat{\sigma} = \bar{R}/d_2 = 0.01435/2.326 = 0.006169$, and Table 3.2 gives the specifications as 0.840 and 0.820. Thus the spread between the specifications is less than 6 times our estimate of σ', so if the process is normal, we shall almost certainly have some nonconforming material. Thus, from our sample data we have

$$z_U = \frac{0.840 - 0.8312}{0.006169} = 1.43$$

$$z_L = \frac{0.820 - 0.8312}{0.006169} = -1.82$$

The first suggests that the upper percent nonconforming is 7.63 percent and the second that the lower percent nonconforming is 3.44 percent, a total of 11.07 percent nonconforming. (See Figure 21.14.) Even if the process were centered in the middle of the specification range, 10.52 percent nonconforming would be produced (for then $z = \frac{0.0100}{0.006169} = 1.63$).

This method of estimating the percent nonconforming is very crude and should only be used when the basic data are numerous. If the process is not in control for $\bar{X}$, the estimate for the percent nonconforming is purely

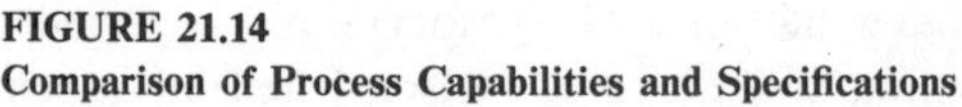
FIGURE 21.14
Comparison of Process Capabilities and Specifications

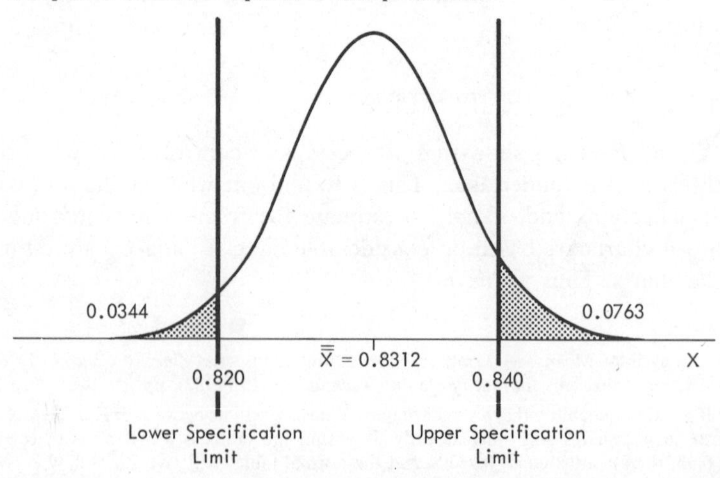

hypothetical and pertains solely to what the percentage would be if the process were brought into control for $\bar{X}$. For the analysis to have any meaning, the R-chart must be in control. It is also important to remember the assumption of a normal distribution of product. When the estimate of the percent nonconforming is very small, say less than 0.01, this assumption must be accepted with great caution. In general, when we are concerned with very small fractions nonconforming, it is safer to use methods that are free from assumptions as to distributional form. Wilks' statistical tolerances, for example, are very appropriate for this, although it will be quickly recognized that large samples, say of the order of 1,500 items, will be necessary to have a high degree of confidence in the conclusions.[26] The reader is referred to Section 4 of Chapter 6.

13. PROBLEMS

21.1. The following are means and ranges of 20 samples of 5 each. The data pertain to the overall lengths of a fragmentation bomb base manufactured during the war by the American Stove Company.[27] The measurements are in inches.

Group Number	$\bar{X}$	*R*	*Group Number*	$\bar{X}$	*R*
1	0.8372	0.010	11	0.8380	0.006
2	0.8324	0.009	12	0.8322	0.002
3	0.8318	0.008	13	0.8356	0.013
4	0.8344	0.004	14	0.8322	0.005
5	0.8346	0.005	15	0.8304	0.008
6	0.8332	0.011	16	0.8372	0.011
7	0.8340	0.009	17	0.8282	0.006
8	0.8344	0.003	18	0.8346	0.005
9	0.8308	0.002	19	0.8360	0.004
10	0.8350	0.006	20	0.8374	0.006

a. From these data, set up an $\bar{X}$-chart and an R-chart to control the lengths of bomb bases produced in the future. Derive ARL curves for these charts showing their effectiveness in picking up an increase in the process average and the process standard deviation respectively.

b. The above samples were taken every 15–20 minutes in order of production after changing fixtures. The production rate was 350–400 per hour, and the specification limits were 0.820 and 0.840 inches.

On the assumption that lengths of bomb bases are normally distributed, what percent of the bomb bases would you estimate to have lengths outside the specification limits when the process is under control at the levels indi-

[26] See Acheson J. Duncan, "Process Capability Studies," *Proceedings, Middle Atlantic Conference,* American Society for Quality Control, February 1958.

[27] See Lester A. Kauffman, *Statistical Quality Control at the St. Louis Division of American Stove Company* (OPRD, Quality Control Reports, No. 3, August 1945), p. 12.

cated by the above data? In other words, what is the percent nonconforming of the above process operating at the levels indicated? Could the percent nonconforming be reduced to zero by changing the process average?

What would happen to the percent nonconforming if the process average should shift to 0.8370? What is the probability that you would catch such a shift on your control chart on the first sample following the shift? How many samples would you have to take to have a chance of approximately 0.95 of catching the shift on at least one of these samples? If this shift is not caught before the eighth sample, approximately how many nonconforming bomb bases will have been produced in the interval?

c. If the lengths of the bomb bases had been simply checked with a "Go Not-Go" gauge and the control of production exercised by a p-chart, approximately what size sample would have given about the same protection against shifts in the process average as that given by your $\bar{X}$- and R-chart?

21.2. The following are means and ranges of samples of 5. The data pertain to the depth to shoulder of fragmentation bomb heads manufactured during the war by the American Stove Company.[28] The measurements are in inches.

Group Number	$\bar{X}$	R	*Group Number*	$\bar{X}$	R
1	0.4402	0.015	16	0.4362	0.015
2	0.4390	0.018	17	0.4380	0.019
3	0.4448	0.018	18	0.4350	0.008
4	0.4432	0.006	19	0.4378	0.011
5	0.4428	0.008	20	0.4384	0.009
6	0.4382	0.010	21	0.4392	0.006
7	0.4358	0.011	22	0.4378	0.008
8	0.4440	0.019	23	0.4362	0.016
9	0.4366	0.010	24	0.4348	0.009
10	0.4368	0.011	25	0.4338	0.005
11	0.4360	0.011	26	0.4366	0.014
12	0.4402	0.007	27	0.4346	0.009
13	0.4332	0.008	28	0.4374	0.015
14	0.4356	0.017	29	0.4339	0.024
15	0.4314	0.010	30	0.4368	0.014

a. From the first 20 samples, set up an $\bar{X}$-chart and an R-chart. Plot the next 10 samples on these charts to see if the process continues "under control" both as to average and range.

b. The above samples were taken every 15–20 minutes in order of production. The production rate was 350–400 per hour, and the specification limits were 0.430 and 0.460 inches.

On the assumption that depth-to-shoulder measurements of fragmentation bomb heads are normally distributed, what percent of the bomb heads would you estimate to have depth-to-shoulder measurements outside the specification limits when the process is under control at the levels indicated by the above data? In other words, what is the percent nonconforming of

[28] Ibid., pp. 14–15.

the above process operating at the levels indicated? Could the percent nonconforming be reduced to zero by changing the process average?

What would happen if the process average should shift to 0.4315? What is the probability that you would catch such a shift on your control chart on the first sample following the shift? How many samples would you have to take to have a chance of approximately 0.95 of catching the shift on at least one of these samples? If this shift is not caught before the eighth sample, approximately how many nonconforming bomb heads could have been produced in the meantime?

c. If the depth-to-shoulder dimensions had been controlled on an attribute basis by means of a p-chart, what size sample would have given about the same protection against shifts in the process average as that given by your $\bar{X}$- and R-charts?

21.3. The following data pertaining to incandescent lamps show the average life and the range for 32 samples of 5 lamps each. The data are in hours.[29]

Sample Number	$\bar{X}$	R	*Sample Number*	$\bar{X}$	R
1	1,080	420	17	1,270	420
2	1,390	670	18	1,580	470
3	1,460	180	19	1,560	650
4	1,380	320	20	750	580
5	1,090	70	21	1,200	590
6	1,230	690	22	1,080	360
7	1,370	950	23	1,730	190
8	1,310	380	24	1,170	310
9	1,630	1,080	25	1,260	760
10	2,120	350	26	1,420	340
11	1,230	580	27	1,290	160
12	1,600	680	28	1,500	360
13	2,290	470	29	1,210	940
14	2,050	270	30	1,940	540
15	1,580	170	31	760	670
16	1,510	670	32	1,150	480

Are these data from a controlled process? Base your answer upon $\bar{X}$- and R-charts.

21.4. The following are measurements of inside diameters.[30] The data represent the number of 0.0001 inches above 0.7500 inches. The measurements are taken in sample groups of 5 each.

---

[29] Cf. Warren R. Purcell, "Saving Time in Testing Life," *Industrial Quality Control,* March 1947, p. 16.

[30] Cf. Irving W. Burr, *Some Experiments Illustrating Principles of Quality Control* (OPRD, Quality Control Reports, No. 12), p. 4.

Sample	X_1	X_2	X_3	X_4	X_5
1	15	11	8	15	6
2	14	16	11	14	7
3	13	6	9	5	10
4	15	15	9	15	7
5	9	12	9	8	8
6	11	14	11	12	5
7	13	12	9	6	10
8	10	15	12	4	6
9	8	12	14	9	10
10	10	10	9	14	14
11	13	16	12	15	18
12	7	10	9	11	16
13	11	7	16	10	14
14	11	7	10	10	7
15	13	9	12	13	17
16	17	10	11	9	8
17	4	14	5	11	11
18	8	9	6	13	9
19	9	10	7	10	13
20	15	10	12	12	16

a. Set up $\bar{X}$- and R-charts for controlling this dimension in the future.

b. Set up an s-chart to be used in lieu of the R-chart.

c. Set up an s^2-chart to be used in lieu of the R-chart.

21.5. Specifications on a dial standoff were 0.3200 and 0.3220 inches. Samples of 5 were taken every 45 minutes with the following results[31] (measured as deviations from 0.3210 in 0.0001 inches).

	Item Number				
Sample Number	*1*	*2*	*3*	*4*	*5*
1	1	9	6	9	9
2	9	4	3	0	3
3	0	9	0	3	2
4	1	1	0	2	1
5	−3	0	−1	0	−4
6	−7	2	0	0	2
7	−3	−1	−1	0	−2
8	0	−2	−3	−3	−2
9	2	0	−1	−3	−1
10	0	−2	−1	−2	−2
11	−3	−2	−1	−1	2
12	−16	2	0	−4	−1
13	−6	−3	0	0	−8
14	−3	−5	5	0	5
15	−1	−1	−1	−2	−1

[31] H. R. Harrison, "Statistical Quality Control Will Work on Short-Run Jobs," *Industrial Quality Control,* September 1956, pp. 8–11.

a. Using $\bar{X}$- and R-charts indicate whether the above data show the process to be in control? If so, project the limits into the next period. If not, throw out points outside control limits and set up tentative limits for the next period. Do the following data fall within the extended limits?

	Item Number				
Sample Number	*1*	*2*	*3*	*4*	*5*
16	−1	−2	−2	0	−4
17	−2	2	−1	−1	0
18	0	4	0	0	0
19	1	2	1	1	−3
20	0	−3	3	3	−1
21	1	2	1	2	1
22	−1	0	2	−1	2
23	0	−1	0	0	0
24	1	0	−1	1	0
25	2	2	2	1	1
26	−3	2	0	1	−1

b. Use an s-chart in lieu of the R-chart.

c. Use an s^2-chart in lieu of the R-chart.

21.6. A particular department of a paper manufacturing company is expected to produce a standard number of container units per machine-hour of operation. The variations that occurred from this standard over a 30-day period were as follows:

Day	*Variation from Standard*
1	−12
2	26
3	−36
4	−30
5	34
6	8
7	34
8	−24
9	40
10	−26
11	69
12	−38
13	0
14	84
15	32
16	7
17	24
18	5
19	−3
20	−44

Day	Variation from Standard
21	−45
22	1
23	−12
24	18
25	−27
26	−48
27	−57
28	23
29	98
30	−32

Taking $\bar{X}'$ as zero and using a moving range of 2, set up a control chart for the daily variations from standard.

14. SELECTED REFERENCES*

Abraham and Box (P '79), American Society for Quality Control (B '85 and P '46–), American Society for Testing and Materials (B '76), Bauer and Hackl (P '78 and P '80), Chiu and Cheung (P '77), Chiu and Wetherell (P '74), Bowker and Lieberman (B '72), Burr (B '76), Cowden (B '57), Dudding and Jennett (B '42), Duncan, A. J. (P '56, P '58, and P '71), Ferrell (P '53), Folks, Pierce and Stewart (P '65), Grant and Leavenworth (B '80), Hilliard and Lasater (P '66), Howell (P '49), Johnson and Counts (P '79), Keen and Page with appendix by Hartley (P '53), Kennedy (B '48), Knowler (P '46), Kurtz, Link, Tukey, and Wallace (P '66), Mitten and Sanoh (P '61), Moore (P '58), Nelson (P '74 and P '82), Olds (P '56), OPRD Reports (B #3), Page (P '55), Peach (B '47), Proschan and Savage (P '60), Reynolds (P '71), Rice (B '47), Roberts (P '58), Robinson and Ho (P '78), Rutherford (B '48), Schilling (Ps '73), Schrock (B '50), Shewart (B '31), Simon (B '41), Smith, E. S. (B '47), and Wheeler (P '83).

* B and P refer to the Book and Periodical sections, respectively, of the Cumulative List of References in Appendix V.

22

Cumulative Sum Control Charts

The control charts discussed in previous chapters can all be classed as Shewhart charts, after the name of the man who originated this type of chart. The basic rule in using such charts is to take action when a point falls outside the "control limits," usually 3-sigma limits. Other rules have been developed from time to time, however. As noted above,[1] inner limits at $\pm$ 2-sigma have been used as warning limits. The rule also has been suggested[2] of taking *two* points in succession outside the "inner limits" as a signal for action. The use of a run of 7 above the central line on the chart as an action signal has been noted.[3] From these developments in the use of the Shewhart chart, it was a natural step to adopt a rule for action that was based on *all* the data and not only the last few samples. This was done by the use of a cumulative sum chart. Such charts were proposed in 1954 by a British statistician, E. S. Page,[4] and developed by him and other British statisticians. (See references at end of this chapter.)

1. THE ADVANTAGE OF PLOTTING CUMULATIVE SUMS

The cumulative sum control chart, or "cusum chart" as it has come to be called, is used primarily to maintain current control of a process. Its advantage over the ordinary Shewhart chart is that it may be equally effective at less expense. This stems from the possibility of the cusum chart picking up a sudden and persistent change in the process average more rapidly than a comparable Shewhart chart, especially if the change is not large. It may also possibly locate the time of change more sharply.

---

[1] See Chapter 18, Section 3.

[2] Cf. E. S. Page, "Control Charts with Warning Lines," *Biometrika,* 42 (1955), p. 243.

[3] See Chapter 18, Section 7.3.

[4] See E. S. Page, "Continuous Inspection Schemes," *Biometrika,* 41 (1954), pp. 100–115.

To illustrate the above, let us look at Figure 22.1. The top shows a plot of the number of red beads obtained in successive drawings of 100 beads at random from a box containing 5 percent red beads. These data are plotted on an ordinary Shewhart p-chart with 3-sigma control limits. After 20 samples of beads were drawn, a slight increase was made in the number of red beads in the box. Despite this change in percentage of red beads in the box, not a single sample point subsequently fell outside the control limits on the Shewhart chart nor was there any run above the mean line to indicate the change. It is difficult, indeed, to see any difference in the Shewhart chart before and after the change in the percentage of red beads in the box.

In the lower part of Figure 22.1 the same data are plotted as a cumulative sum. Actually a plot is made of the algebraic sum of the number of red beads in each sample minus 5, the expected number per sample. Since the original proportion of red beads in the box was 5 percent, the cumulative sum would initially be expected to hover around zero. That it actually has a negative drift is simply evidence of a bias in the method of sampling. The beads were stirred by hand each time and a sample drawn from the box by a paddle. Previous experiments with sampling from the box suggested that something in the character of the red beads caused them to drift to

FIGURE 22.1
Variation in Number of Red Beads in a Sample of 100

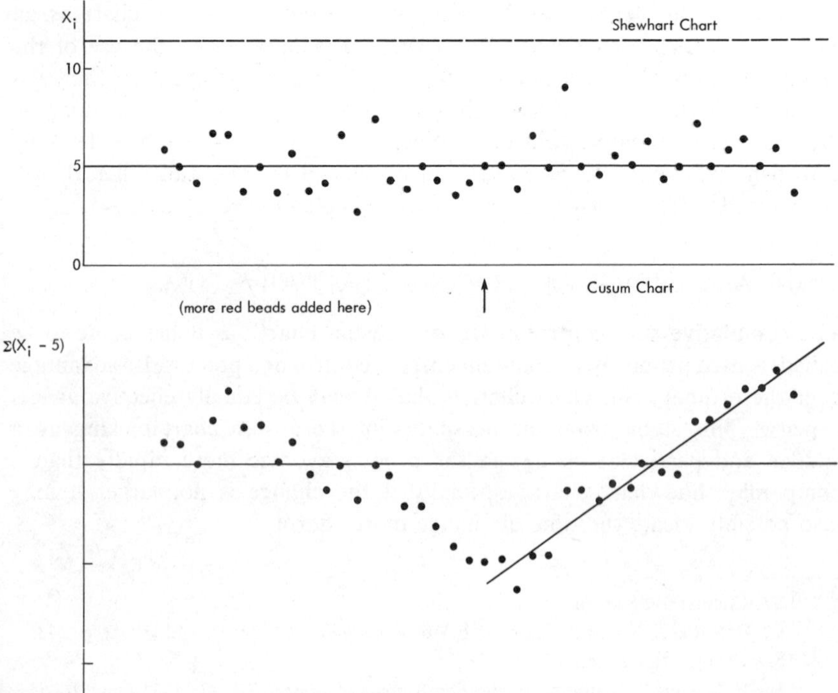

the top of the box, and when samples were taken by sliding the paddle in along the bottom, they showed a downward bias in the number of red beads obtained. This sampling bias is indicated in Figure 22.1 by the initial downward trend of the cumulative sum. After the number of red beads in the box was increased, however, the cusum sum changed its downward trend and started upward. It continued to move in this direction until the end of the sampling. The time of the change in the number of red beads in the box is indicated clearly by the reversal in trend of the cumulative sum. Figure 22.1 thus illustrates in a general way the possible advantage of a cusum chart. A more careful comparison of Shewhart charts and cusum charts is undertaken in Section 5. Prior to this it is necessary to discuss the operation and design of cusum charts.

2. A ONE-SIDED DECISION PROCEDURE

2.1. A Simple Cusum Chart for Protection against a Shift in the Process Average in a Single Direction

Cumulative sum charts were first devised for one-sided schemes. The original procedure was to plot the cumulative sum of "scores" taken to represent the quality measurements. When the chart was to be used to detect an upward shift in the process average, the rule for deciding when such a change had occurred was to compare the last point plotted with the lowest point previously plotted. If the difference exceeded a specified quantity h, the conclusion was that a change had occurred. The system of scoring quality was so chosen that the mean sample path on the chart was downwards when the quality was satisfactory and upwards when unsatisfactory. If the primary purpose of the cusum chart is to detect an upward shift in the process average, however, and there is no interest in using it for record keeping, this early cusum procedure for one-sided schemes has certain practical difficulties. If quality remains good, the trend of the cusum chart is negative and frequent revision of the chart downward is necessary. A further difficulty is that the chart may pick up small shifts in the process average that may not be of any importance.

For the above reasons, the original cusum procedure for one-sided schemes on the upper side has been modified by the adoption of a reference value k. Under this modified procedure the quantities plotted on the chart are the cumulative sums of $\bar{X} - k$ where k is a value greater than the target for $\bar{X}$. Only positive values and the first negative value of such sums are used, however. As soon as the cumulative sum falls below zero, a new series is begun. When the cumulative sum of $\bar{X} - k$ exceeds the decision interval h, it is decided that the process average has shifted above k. (See Figure 22.2). This modified procedure eliminates much unnecessary paper work. It also calls attention only to those upward shifts in the process average that (through the selection of k) are considered to be of importance.

FIGURE 22.2
Plot of the Cusum ($\bar{X} - k$) (plot starts afresh as soon as cusum falls below zero)

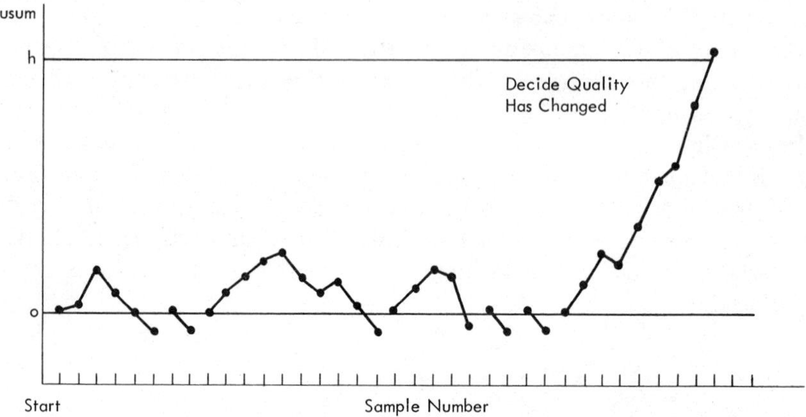

A one-sided scheme can, of course, be used to detect shifts in the negative instead of the positive direction. In this case k will be less than the target value and the test will be whether $\Sigma(\bar{X} - k)$ has a negative value that falls below $-h$.

2.2. The Average Run Length (ARL) Curve for a One-Sided Cusum Procedure

In the case of Shewhart p-charts, c-charts, $\bar{X}$-charts, and R-charts, we found that the average run length (ARL) curve was a meaningful way of describing their effectiveness. For a cusum sum chart the ARL curve is the primary device for presenting its effectiveness.

To determine the ARL curve for a cusum scheme is not a simple matter. For continuous variables, direct derivation will usually require the use of an electronic computer, either to solve the integral equations that arise in the analytical formulation of the problem or through simulation. For discrete variables like difficulties arise. Fortunately a nomogram has been constructed by K. W. Kemp for finding approximate values of the ARLs of some one-sided cusum schemes when the variable plotted is normally distributed. This is reproduced in Figure 22.3.

Let us note how this nomograph is used. Suppose that there is a quality level m_a that we consider to be an acceptable quality level and another quality level m_r greater than m_a that we consider to be barely tolerable, any worse quality being rejectable. Let the reference value k be between m_a and m_r and let σ' be known. Then for a one-sided cusum scheme with a specified reference value k, decision interval h and sample size n, we can find the ARLs at m_a and m_r as follows. In Figure 22.3 note the point on the $h\sqrt{n}/\sigma'$ line yielded by the given values for h and n. Call this

FIGURE 22.3
Nomogram from which ARL Values Can Be Determined when X Is Normally Distributed

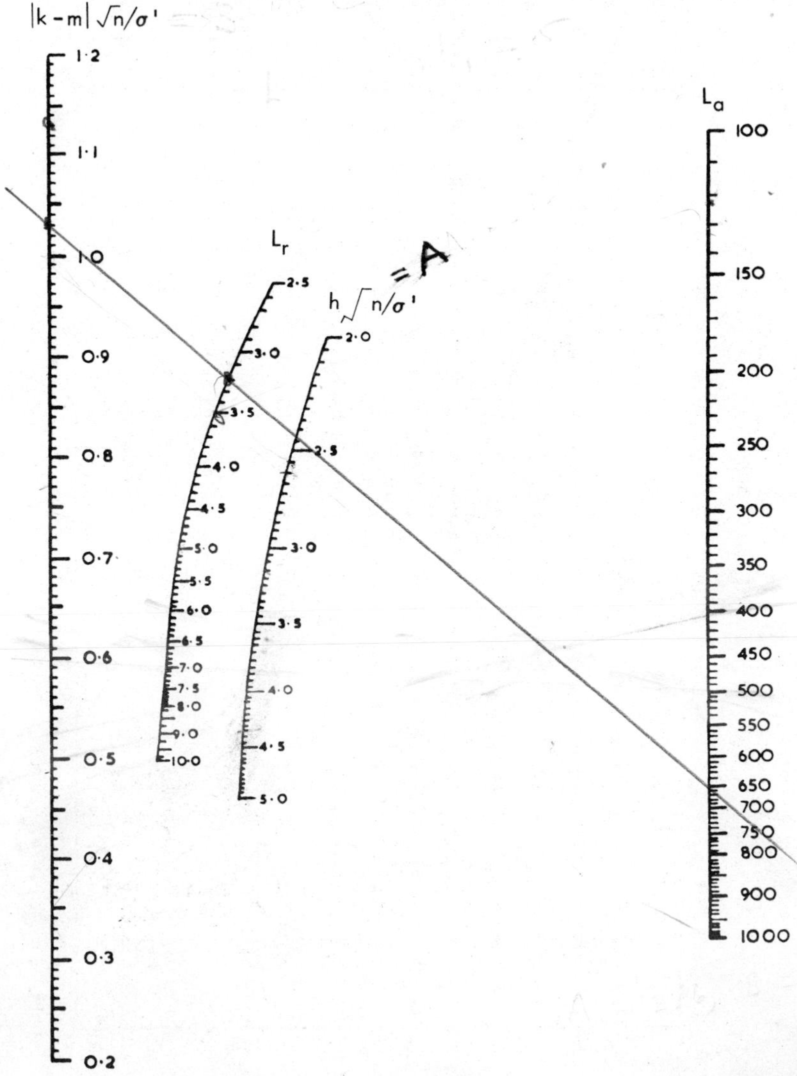

SOURCE: Reproduced with permission from *Applied Statistics,* 11 (1962), p. 23. For another nomogram see Amrit L. Goel and S. M. Wu. "Determination of A.R.L. and a Contour Nomogram for Cusum Charts to Control Normal Mean," *Technometrics,* 13 (1971), pp. 221–30.

point A. To find the ARL at m_a find the point on the $|k - m|\sqrt{n}/\sigma'$ line given by setting $m = m_a$. Call this point B. Note where a line through the points A and B cuts the line L_a. This point on L_a is the ARL at m_a. To find the ARL at m_r, find the point on the $|k - m|\sqrt{n}/\sigma'$ line given by setting $m = m_r$. Call this point C. Note where a line through the points

FIGURE 22.4
Sketch of the ARL Curve for the One-Sided Cusum Scheme Given by $k = 105$, $h = 13$, and $n = 4$, with σ' Known to Be 10 and the ARL Curve for an $\bar{X}$-Chart with the Same n and σ'

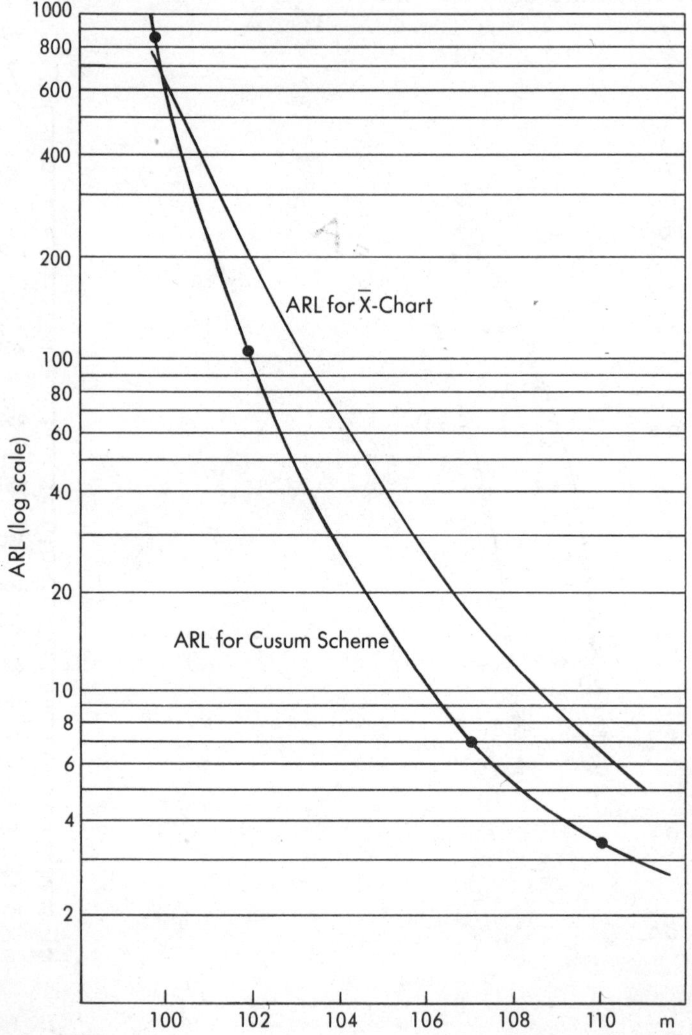

A and C cuts the line L_r. This point on L_r is the ARL at m_r. The same procedure can be used to find the ARLs for other values of m, using the L_a line when m is less than k and the L_r line when m is greater than k. A similar procedure can be used to sketch the ARL curve for a negative one-sided cusum chart when the barely tolerable (or just rejectable) quality level m_r is below m_a and the decision interval is $-h$.

To illustrate the sketching of the ARL curve for a positive one-sided cusum scheme, let $m_a = 100$, $m_r = 110$, and $\sigma' = 10$. Suppose that the cusum scheme has $k = 105$, $h = 13$, and $n = 4$. Then $h\sqrt{n}/\sigma' = 13(2)/10 = 2.6$, $|k - m_a|\sqrt{n}/\sigma' = (5)(2)/10 = 1$ and the ARL at $m = 100$ is about 830. Further with $|k - m_r|\sqrt{n}/\sigma' = (5)(2)/10 = 1$, the ARL at $m = 110$ is about 3.5. To find the ARL at $m = 102$ we compute $|105 - 102|\sqrt{4}/10 = (3)(2)/10 = 0.6$ and note that a line connecting the point 0.6 on the $|k - m|\sqrt{n}/\sigma'$ line with the point 2.6 on the $h\sqrt{n}/\sigma'$ line cuts the line L_a at about 104. This is approximately the ARL at $m = 102$. Again to find the ARL at $m = 107$, we compute $|105 - 107|\sqrt{4}/10 = 0.4$ and note that the line connecting the point 0.4 on the $|k - m|\sqrt{n}/\sigma'$ line with the point 2.6 on the $h\sqrt{n}/\sigma'$ line cuts the L_r line at about 7. This is approximately the ARL at $m = 107$. From these results we can sketch the ARL curve for the cusum scheme $k = 105$, $h = 13$, and $n = 4$ with σ' given equal to 10. The results are graphed in Figure 22.4. It will be noted that the ARL is plotted on semi-log paper in order to make a presentable chart. The ARL curve for the $\bar{X}$-chart also shown in Figure 22.4 is discussed below in Section 5.

3. TWO-SIDED CUMULATIVE SUM SCHEMES AND V-MASKS

3.1. Use of a Pair of One-Sided Cusum Charts

To attain two-sided protection with cusum procedures it is possible to run concurrently two one-sided schemes of the kind discussed above—one with an upper reference value k_1, the other with a lower reference value k_2. The two procedures can be concisely combined, however, by the use of a **V**-shaped mask. Such a mask is shown in Figure 22.5.

3.2. V-Masks

A **V**-mask is used as follows. Let a sample of n items be taken periodically from a process and let the mean $\bar{X}$ be computed. Let $\bar{X}''$ be the target for the process and let the data plotted on the cumulative sum chart be the cumulative sums of $\bar{X} - \bar{X}''$. To determine whether the process average has shifted from or is different from $\bar{X}''$, each point on the cusum chart is tested as it is plotted. This is done by placing the point P of the mask immediately over the last cusum point plotted on the chart with the mask leveled so that the line PO is horizontal. If any of the previously plotted points is covered by the mask, it is an indication that the process had changed. So long as the cumulative sums stay within the angle of the V of the chart, the process is viewed as being in control. The distance from the vertex of the V, i.e., from the point O, to the point P is called the *lead distance* of the mask and is usually represented by the symbol d. Half the angle of the V is generally represented by the symbol θ. A given mask is thus determined by its d and θ.

FIGURE 22.5
A V-Shaped Mask for Use on a Cusum Chart

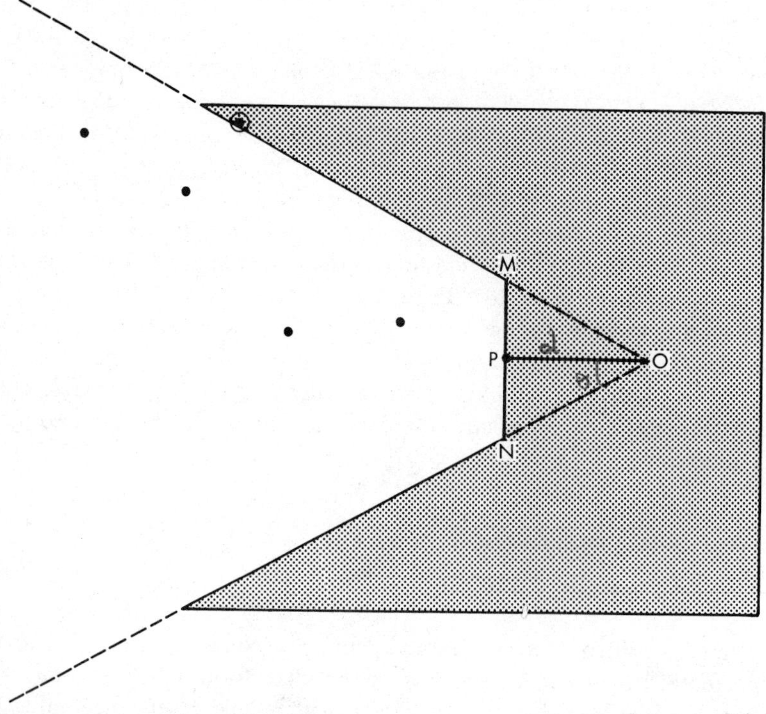

Note: Covered point indicates a downward shift.

It is interesting to note that the mask can also be used to indicate the amount of the change in the process average. This is done by pivoting the mask around the point *P* placed on the last plotted cumulative sum. If a position can be found such that the last *t* cusum points fall within the *V* of the mask when these *t* points at least include the points covered when the mask was horizontal, then the direction of the line *OP* in the new position can be used to measure the increase in the process average. See Figure 22.6.

3.3. The Equivalence of the Two Procedures

If w is the horizontal distance between successive points on a cusum chart measured in terms of unit distance on the vertical scale, then two one-sided schemes for which $(k_1 + k_2)/2 = \bar{X}''$ and the **V**-mask scheme will be equivalent if

(22.1)

$$k_1 - \bar{X}'' = \bar{X}'' - k_2 = w \tan \theta, \quad \text{and}$$

$$h = d \tan \theta = \frac{d(k_1 - \bar{X}'')}{w}$$

FIGURE 22.6
Use of V-Mask to Estimate Change in Process Average

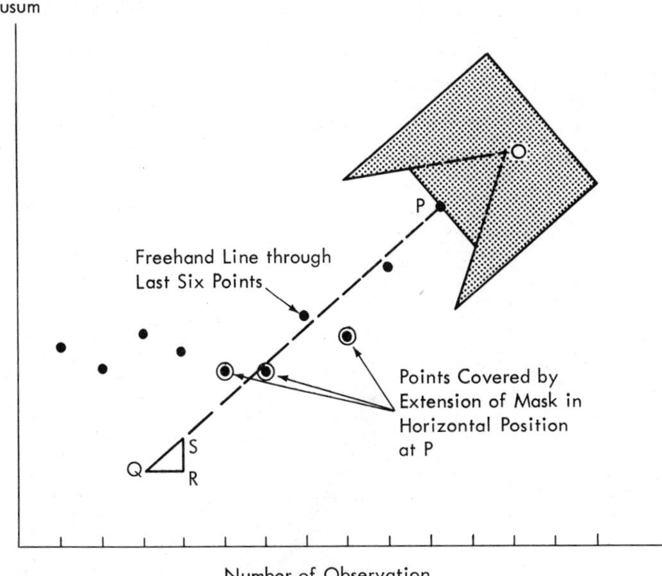

Kenneth W. Kemp demonstrates this equivalence as follows:[5]

[Figure 22.7] shows a mask with limbs inclined at an angle θ to the horizontal. The cumulative sum at A, which is the last plotted point, is S_n and at C is S_{n-r}. O is the vertex of the **V**-mask.

Extend OA and let the perpendicular from C meet it at B. Then

$$CB = S_n - S_{n-r}$$

$$BA = rw$$

The path of the plotted points will cross the lower limb of the ***V*** when

$$BC \geq BO \tan \theta$$

that is when

$$S_n - S_{n-r} \geq (rw + d) \tan \theta$$

or

$$\sum_{i=n-r+1}^{i=n} (\bar{X} - \bar{X}'' - w \tan \theta) \geq d \tan \theta$$

---

[5] Kenneth W. Kemp, "The Use of Cumulative Sums for Sampling Inspection Schemes," *Applied Statistics,* 11 (1962), p. 20. In the quotation $\bar{X}$ has been substituted for x and $\bar{X}''$ for m_a.

FIGURE 22.7
Cumulative Sum of ($\bar{X} - \bar{X}''$) Plotted against Number of Samples

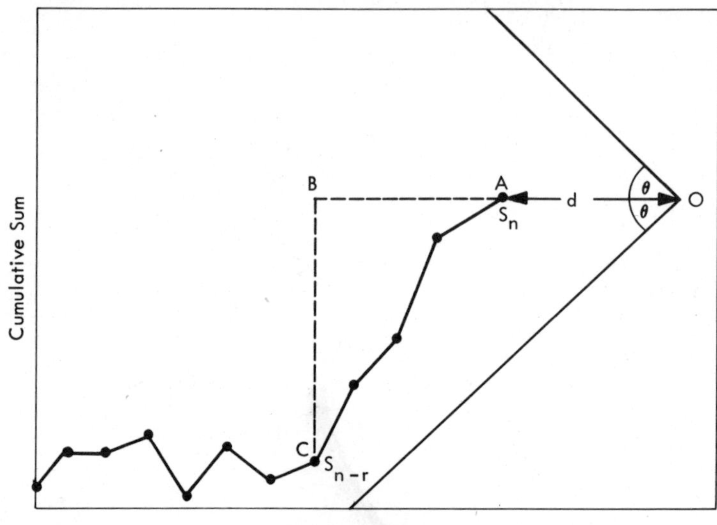

SOURCE: Reproduced with permission from Kenneth W. Kemp, "The Use of Cumulative Sums for Sampling Inspection Schemes," *Applied Statistics,* 11 (1962), p. 20.

This is equivalent to cumulating the deviations of $\bar{X}_i$ from a reference value $k_1 + w \tan \theta$ and using a decision interval $h = d \tan \theta$.

Similar use of the upper limb of the mask[6] would be equivalent to cumulating the deviations of $\bar{X}_i$ from a reference value $k_2 = \bar{X}'' - w \tan \theta$ and taking $h = -d \tan \theta$.

3.4. The ARL for a Two-Sided Scheme

The ARL for a two-sided scheme can readily be derived from the ARLs for the two one-sided schemes of which it is composed. K. W. Kemp has shown[7] that if $L_1(m)$ is the ARL at m of a one-sided scheme with reference value $k_1(> m)$ and decision interval h, and if $L_2(m)$ is the ARL at m of a one-sided scheme with reference value $k_2(< m)$ and decision interval $-h$, then if $L_d(m)$ is the ARL at m of the two-sided scheme obtained by using both of these simultaneously, we have

$$\frac{1}{L_d(m)} = \frac{1}{L_1(m)} + \frac{1}{L_2(m)} \tag{22.2}$$

[6] A **V**-mask can thus be used for a one-sided scheme if it is so desired as well as for a two-sided scheme.

[7] See Kenneth W. Kemp, "The Average Run Length of the Cumulative Sum Chart When a **V**-Mask Is Used," *Journal of the Royal Statistical Society,* Ser. B, 23 (1961), pp. 149–53.

Equation (22.2) immediately gives us a way of sketching the ARL for a **V**-mask with specified d and θ. For from equations (22.1) we have $k_1 = \bar{X}'' + w \tan \theta$, $k_2 = \bar{X}'' - w \tan \theta$, and $h = d \tan \theta$. It will be noted that at the target value m_a the ARL for a symmetrical two-sided scheme (or equivalent **V**-mask) will be just half that of either of the one-sided schemes of which it is composed. At the barely tolerable (or just rejectable) quality level m_{r1}, however, the ARL for the two-sided scheme will be practically equal to $L_1(m_{r1})$ since $1/L_2(m_{r1})$ will be negligible. Similarly $L_d(m_{r2})$ will equal $L_2(m_{r2})$ since $1/L_1(m_{r2})$ will be negligible.

4. DESIGN OF ORDINARY CUSUM CHARTS AND V-MASKS

4.1. A Normally Distributed Variable

Just as an acceptance-sampling plan can be designed so that its OC curve will pass approximately through two given points (the p_1', α and p_2', β points) so a cusum scheme can be designed such that its ARL curve will pass approximately through two designated points. If m_a again represents acceptable quality and m_r represents barely tolerable (or just rejectable) quality, a cusum scheme can be found that will have specified values for the ARL at m_a and m_r. As we shall see, this can be done (within limits) by use of Figure 22.3. It will be assumed that the σ' of the process is known and constant and that $\bar{X}$ is normally distributed.

With σ' given, there are three parameters at the disposal of the designer. These are the reference value k, the decision interval h, and the sample size n (or for a **V**-mask the lead distance d, the angle θ, and the sample size n). Now it turns out that three parameters are a greater number than is needed to meet the specified conditions, with the result that more than one cusum scheme can be derived such that its ARL curve will pass through the designated points. Further study reveals,[8] however, that for a given h, the k value that leads to minimum sample size n is a value half way between m_a and m_r. It has therefore been suggested that k be taken as approximately equal to $(m_a + m_r)/2$. The h and n can then be varied to secure the desired ARL at m_a and m_r.

To illustrate let us seek a one-sided cusum scheme for which the ARL will be about 500 when the process average is in the neighborhood of 100 (the acceptable quality level) and about 5 when the process average is in the neighborhood of 110 (the tolerable or just rejectable quality level). Let the standard deviation of the process be 10. The suggestion given above for the selection of k yields $k = \dfrac{110 + 100}{2} = 105$. We now seek values of h

[8] Kemp, in "The Use of Cumulative Sums for Sampling Inspection Schemes," *Applied Statistics,* 11 (1962), p. 24, notes that the curve for n is very flat near its minimum point and that any convenient value of k in the vicinity of $(m_a + m_r)/2$ will be satisfactory.

and n that will give us the desired ARLs. This is very easily done [since $(k - m_a) = (m_r - k)$] by placing a ruler on Figure 22.3 such that it connects the point 5 on the L_r line with the point 500 on the L_a line and noting where the ruler intersects the $|k - m|\sqrt{n}/\sigma'$ scale and the $h\sqrt{n}/\sigma'$ scale. The $|k - m|\sqrt{n}/\sigma'$ scale yields

$$|k - m|\sqrt{n}/\sigma' = 0.742$$

or

$$\sqrt{n} = \frac{0.742\sigma'}{|k - m|} = \frac{0.742(10)}{5} = 1.484$$

or

$$n = 2.35$$

If we are conservative, we shall round n upwards to 3. This gives

$$|k - m|\sqrt{n}/\sigma' = \frac{|105 - 100|\sqrt{3}}{10} = 0.866$$

and a ruler laid on this point and the point 500 on the L_a scale will intersect the $h\sqrt{n}/\sigma'$ line at the point

$$h\sqrt{n}/\sigma' = 2.72$$

or

$$h = \frac{2.72\sigma'}{\sqrt{n}} + \frac{2.72(10)}{1.732} = 15.7$$

Our scheme will thus be to cumulate $\bar{X} - 105$, where $\bar{X}$ is the mean of a sample of 3. If this cusum becomes negative, we start afresh. If it exceeds 15.7, we conclude that the process is out of control.

The ruler that connects 500 on the L_a line and the point 2.72 on the $h\sqrt{n}/\sigma'$ line will also simultaneously intersect the L_r line at 3.9. The scheme $k = 105$, $h = 15.7$, and $n = 3$ thus yields the desired ARL at $m = 100$ and a somewhat better ARL at $m = 110$ than that originally called for. By laying the ruler on the point 0.866 on the $|k - m|\sqrt{n}/\sigma'$ line and the point 5 on the L_r line, we could have found a scheme ($k = 105$, $h = 21.1$, and $n = 3$) that would have kept the L_r at the designated value of 5 but yielded an L_a far in excess of that specified. (The actual L_a is off the chart.)

If we take $n = 2$ instead of 3, we have

$$|k - m|\sqrt{n}/\sigma' = \frac{|105 - 100|\sqrt{2}}{10} = 0.707$$

and a ruler connecting this point with the point 500 on the L_a line yields

$$h\sqrt{n}/\sigma' = 3.35$$

or

$$h = \frac{3.35\sigma'}{\sqrt{n}} = \frac{3.35(10)}{\sqrt{2}} = 23.7$$

The ruler also intersects the L_r line at the point 5.5. This is reasonably close to the desired ARL of 5 at $m = 110$. Thus the plan $k = 105$, $h = 23.7$, and $n = 2$ apparently comes closer to the desired goals than does the scheme in which $n = 3$.

If our object is to design a **V**-mask such that the ARL at $m = 100$ is 250 and the ARL at $m = 110$ is 5, we would take $d = h/\tan\theta$ and $\theta = \tan^{-1}(|k - m|/w) = \tan^{-1}(5/w)$, where w is the ratio of a unit distance on the horizontal scale to the unit distance on the vertical scale. Thus, if $w = 2$, we would have $\tan\theta = 2.5$, $\theta = 68.2°$, and $d = 9.48$.

TABLE 22.1
Values of $|m_a - m_r|\sqrt{n}/\sigma'$ for Cumulative Sum Schemes with a Central Reference Value for a Variety of L_a and L_r

	L_a		
L_r	*250*	*500*	*1,000*
2.50 . . .	2.10	2.34	2.54
5.00 . . .	1.36	1.48	1.60
7.50 . . .	1.03	1.14	1.25
10.00 . . .	0.86	0.96	1.04

Table 22.1 is a reproduction of a table by K. W. Kemp that, for a central reference value k and given σ', will immediately give the sample n required for a one-sided cusum scheme with specified L_a and L_r. The value of h can then be obtained from Figure 22.3. To illustrate let us take $L_a = 500$ and $L_r = 5$, as above, then Table 22.1 gives

$$|m_a - m_r|\sqrt{n}/\sigma' = 1.48$$

which for $m_a = 100$, $m_r = 110$, and $\sigma' = 10$ yields

$$n = \left(\frac{1.48\sigma'}{m_a - m_r}\right)^2 = \left(\frac{1.48(10)}{10}\right)^2 \doteq 2.2$$

which is essentially the same as the value 2.35 we obtained by direct use of Figure 22.3. This will lead to adoption of $n = 2$, $k = 105$, and $h = 23.7$ as before. Table 22.1 may be found very helpful for a quick determination of the sample size required by a specified cusum scheme.

4.2. A Poisson Variable

When we seek a cusum chart to control a proportion, say a proportion nonconforming, and when this proportion is sufficiently small to use the Poisson distribution to approximate the binomial distribution which theoretically applies, we may be able to find the chart we seek from another table prepared by K. W. Kemp. This is reproduced in Table 22.2. Let p'_a and p'_r represent the acceptable and barely tolerable (or just rejectable) qualities, respectively, and let $R = p'_r/p'_a$. Then, selecting the L_a and L_r in the table that come the closest to those we desire, we can find the n, the k, and the h for the cusum chart. The last two are given directly by the table, and n is found from the relationship $np'_a = m_a$ or $n = m_a/p'_a$, where the m_a is provided by the table.

To illustrate suppose that we have $p'_a = 0.01$ and $p'_r = 0.04$ and let us be satisfied with an $L_a = 500$ and a $L_r = 7.50$. Then for $R = \dfrac{0.04}{0.01} = 4$, we have $n(0.01) = 0.24$ or $n = 24$. The table also gives $k = 0.60$ and $h = 2.75$. The control procedure will thus be to plot $\Sigma(X - 0.60)$ and to conclude the process is out of control whenever this exceeds 2.75. Here X is the number of nonconforming items in the sample of 24. Whenever the sum becomes negative we start the summation anew.

Control of an increase in p' could likewise be exercised by using the lower limb of a **V**-mask with $\theta = \tan^{-1}(k - m_a)/w = \tan^{-1}(0.60 - 0.24)/w$ and $d = w(2.75)/(0.60 - 0.24)$. If we take w equal to 2, this yields tan

TABLE 22.2
Values of m_a, R, h, and k for Fraction-Nonconforming Sampling Schemes

R	L_a	L_r = 5.00			L_r = 7.50			L_r = 10.00		
		m_a	k	h	m_a	k	h	m_a	k	h
2.50	500	1.18	2.00	5.00	0.64	1.20	3.75	0.50	0.90	3.75
	250	0.93	1.50	4.50	0.52	0.90	3.50	0.42	0.80	3.00
	125	0.71	1.20	3.75	0.47	0.70	3.25	0.32	0.60	2.25
3.00	500	0.66	1.20	4.00	0.46	0.90	3.50	0.32	0.70	3.00
	250	0.56	0.90	3.00	0.40	0.80	3.00	0.27	0.60	2.50
	125	0.48	0.80	3.00	0.31	0.60	3.00	0.15	0.30	2.00
3.50	500	0.54	1.20	3.00	0.35	0.80	3.00	0.24	0.60	2.75
	250	0.41	0.90	2.50	0.27	0.60	2.50	0.18	0.40	2.50
	125	0.34	0.70	2.25	0.18	0.40	2.00	0.13	0.30	1.75
4.00	500	0.38	0.90	2.75	0.24	0.60	2.75	0.16	0.40	2.50
	250	0.32	0.80	2.25	0.21	0.60	2.00	0.12	0.30	2.00
	125	0.28	0.70	1.75	0.16	0.40	1.75	0.07	0.20	1.50

$\theta = 0.18$, $\theta = 10.2°$, and $d = 16.4$. The upper limb of the mask could not be used to control a decrease in p', since the Poisson distribution will be skewed and the table was prepared for an upper one-sided cusum chart.

5. A COMPARISON OF CUSUM CHARTS AND SHEWHART CHARTS

It will be meaningful at this point to compare cusum charts with Shewhart charts.

One way of comparing the two control procedures is to compare the ARL curve for a Shewhart $\bar{X}$-chart that has only an UCL with the ARL curve for a one-sided cusum scheme that has the same target value $\bar{X}''$, uses the same standard value σ'' for the process standard deviation and employs the same sample size n. Such a comparison is presented in Figure 22.4 above. It will be noted that the cusum scheme has a longer run length than the Shewhart $\bar{X}$-chart and thus fewer false alarms when the process is on target and has shorter run lengths when there have been sizable increases in the process average.

Another way of comparing the two procedures is to match the ARL curve for a cusum scheme to the ARL curve for a given $\bar{X}$-chart and note the reduction that can be attained in the sample size. To illustrate this type of comparison we shall assume that the change in the process average that is to be picked up by the charts is a sudden jump in the process average that persists until it is detected. One measure of effectiveness will thus be the quickness or average run length with which the change will be detected. With this measure an appropriate procedure would be to match the two types of charts so that they have approximately the same ARLs at given m_a and m_r and then to compare the sample sizes. We shall assume that $\bar{X}$ is normally distributed with known σ' and shall restrict ourselves to one-sided situations.

To illustrate such a comparison, let us pick the $\bar{X}$-chart of Figure 21.3 as the Shewhart chart to which we shall match a cusum chart. Figure 21.8 indicates that for a shift in the process mean from 0.8312 to 0.8381 the probability of this chart detecting the change on a single sample ($= 1 - P_a$) is about 0.25. Hence at this value of m, the ARL for the Shewhart chart is $1/0.25 = 4$. When the process is in control, i.e., the process mean $= 0.8312$, the probability of a point going outside the upper control limit is 0.00135 (since the chart uses 3-sigma limits) which means an average run length of about 740.

To set up a "comparable" cusum chart, we seek one that will have about the same ARL at $\bar{X}' = 0.8312$ as the Shewhart chart and also about the same ARL at $\bar{X}' = 0.8381$. Figure 22.3 shows that with an $L_a = 740$ and an $L_r = 4$, $|k - m_a|\sqrt{n}/\sigma' = 0.895$ and $h\sqrt{n}/\sigma' = 2.85$. We shall take k as centrally located between 0.8312 and 0.8381 or as having the

it is a one-sided procedure = 1/.00135 because

value (0.8312 + 0.8381)/2 = 0.83465. Then n for the cusum chart will be given by

$$|0.83465 - 0.8313|\sqrt{n}/0.006169 = 0.895$$

or

$$\sqrt{n} = (0.006169)(0.895)/0.00345) = 1.60$$

and

$$n = 2.56$$

To be conservative we shall use $n = 3$ for the cusum chart. With $n = 3$ we will have

$$|k - m_a|\sqrt{n}/\sigma' = \frac{(0.00345)(1.732)}{0.006169} = 0.969$$

If we match the two curves exactly at the L_r, this will yield (from Figure 22.3)

$$h\sqrt{n}/\sigma' = \frac{h(1.732)}{0.006169} = 3.00$$

or

$$h = 0.006169(3.00)/1.732 = 0.01069$$

Our "matching" cusum procedure will thus be to plot $\Sigma(\bar{X} - 0.83465)$ where $\bar{X}$ is the mean of a sample of 3 from the process. Whenever this sum becomes negative, we start a new sum. Whenever it exceeds 0.01069, we conclude that the process has gone out of control. This procedure will pick up shifts to 0.8381 just as quickly as will the Shewhart chart and will give "false alarms," when the process is really in control, much less frequently than the one in 740 samples for the Shewhart chart. (The ARL at $\bar{X}' = 0.8312$ for the cusum scheme is in excess of 1,000. See Figure 22.3.) We thus get a saving of 40 percent in our sampling costs together with less risk of looking

TABLE 22.3
Ratio of the Sample Size Required for a Shewhart Scheme to that Required for a Cumulative Sum Scheme with Equivalent Run Lengths at AQL and RQL

	L_a		
L_r	*250*	*500*	*1,000*
2.50 . . .	1.3	1.3	1.3
5.00 . . .	1.8	1.9	2.0
7.50 . . .	2.2	2.4	2.5
10.00 . . .	2.5	2.8	3.0

for trouble when it does not exist. Table 22.3 gives the results of similar comparisons between the standard Shewhart chart and matching cusum schemes. In defense of the Shewhart chart it is only fair to note that the above comparisons are between cusum charts and *standard* 3-sigma Shewhart charts.

An extensive comparison of Shewhart control charts and cumulative sum control charts was carried out by A. L. Goel in his 1968 Ph.D. thesis at the University of Wisconsin on "A Comparative and Economic Investigation of $\bar{X}$ and Cumulative Sum Control Charts" (available in microfilm). Besides comparisons of sample sizes for matched ARLs at m_a and m_r, similar to that above, he studied the ARL when the two charts were given the same ARL at m_a and the charts individually designed to have minimum ARL at m_r. He found the minimum ARL at m_r for the cusum chart was much less than that for the $\bar{X}$-chart. With an expected shift in the process mean of $2\sigma'$, for example, the ARL at this new process level was 62 percent smaller for the cusum chart. On the other hand, if the actual shift in the process mean turns out to be much larger or much smaller than that for which the charts were designed, then Goel notes that the $\bar{X}$-chart would have been a better choice. This is because for much larger shifts the $\bar{X}$-chart has smaller ARLs. For much smaller shifts the cusum chart has considerably smaller ARLs, but this is a disadvantage since it means that there will be more searches for relatively unimportant assignable causes than is presumably desirable.

The most meaningful of Goel's comparisons is between the two charts when each have the minimum cost design for the same cost data and technical factors. He found in this case that there was very little difference. One example yielded a loss-cost value of \$401.38 for the $\bar{X}$-chart and \$400.93 for the cusum chart. He notes, however, that the loss-cost difference in favor of the cusum chart becomes relatively large when a smaller than optimum sample size is used. When the shifts in the process mean are actually smaller or larger than that for which the charts are designed, there is also a minor economic advantage in favor of the cusum chart, say about 5 percent as a maximum. Goel concludes that an optimally designed cusum chart is just slightly better than an optimum $\bar{X}$-chart, but that if a nonoptimized $\bar{X}$-chart is used, the difference could be much larger in favor of the cusum chart.

The above comparisons are all based on operating efficiency and cost. They do not include other considerations such as the more meaningful graphic interpretations offered by the cusum chart in the event of a sudden shift in the process mean. (See Section 1 above.) It should also be noted that all comparisons relate to sudden shifts and none to a slowly changing process.

6. OTHER CUSUM CHARTS

Most of the above discussion of cusum charts has pertained to charts for variables, such as $\bar{X}$, that are normally distributed or to charts for Poisson variables. Charts have also been worked out, however, to control other types of variables such as a standard deviation for a normal distribution and the mean and standard deviation combined. For descriptions of these special applications of cusum procedure, the reader is referred to the references cited, in particular to the 1960 *Biometrika* paper by Ewan and Kemp and the book by Johnson and Leone.

7. SPECIAL FORMS OF CUSUM MASKS AND MORE INVOLVED CUSUM SCHEMES

In conclusion we will consider special forms of **V**-masks and more involved cusum schemes that have been developed.

7.1. Truncated V-Masks

In the use of the **V**-mask shown in Figure 22.5, the point O at the vertex of the mask does not enter into the discussion as to whether the process is in control with respect to the given target. In practice, therefore, much of the mask to the right of the line MPN can be cut off and the mask will take less space. A segment of the line PO should be retained to aid in the use of the mask to estimate a change in the process average as illustrated in Figure 22.6. For the truncated **V**-mask the design elements will be the decision interval $h = MP = NP$ and the tangent f of the angle θ that the limit lines make with the horizontal.

7.2. A "Standardized" V-Mask

If cusum charts are being run concurrently on several variables, it may be useful to measure variations in standard deviation units and to use a single "standardized" form of a **V**-mask for all variables. Consider cusum charts for sample means and let the variable to be cumulated be the deviation $(\bar{X}_i - \bar{X}'')$ of a sample mean from the target value $\bar{X}''$. Let a unit distance on the vertical scale equal the standard deviation of the sample means $(\sigma'/\sqrt{n})$ where σ' is the assumed "known" standard deviation of the process. In practice σ' can be taken as the standard deviation of the process given by an R-chart or s-chart that in a preliminary analysis has been brought into a state of statistical control. Let the unit distance or plotting interval on the horizontal scale equal two of the vertical scale units, i.e., let[9] $w = 2$. Assume that an acceptable "standardized" **V**-mask will be one that will yield an average run length of 370 when the process is in control at the target value (which is the ARL for a zero displacement of the process average for a standardized 3-sigma Shewhart chart[10]) and assume further, somewhat

[9] See p. 520 above.

[10] For a normally distributed variable the probability of exceeding 3-sigma limits in either direction is 0.0027 and the ARL = 1/.0027 = 370.

arbitrarily, that the standardized **V**-mask will yield a small average run length of, say, 5 to 6, when the process average has shifted by as much as 1.5 vertical scale units, i.e., by $1.5\sigma'/\sqrt{n}$.

A study of Figure 22.3 reveals what the parameters for the standardized **V**-mask should be to approximate these specified design criteria. Note first that we are cumulating means of deviations from the target value $\bar{X}''$ so that m_a will be zero. Note further that a unit distance on the vertical scale of the cusum chart will equal $\sigma'/\sqrt{n}$ so that in Figure 22.3 $\sigma'/\sqrt{n}$ will equal one. The left-hand scale of Figure 22.3 will thus be simply $|k - m|$ and the barely tolerable value m_r will be simply 1.5. We will then find that if we set the parameter $k = 0.5$, measured in $\sigma'/\sqrt{n}$ units, we will have $|k - m_a| = |0.5 - 0| = 0.5$ and $|k - m_r| = |0.5 - 1.5| = 1$. If $\sigma'/\sqrt{n} = 1$, the h-parameter scale directly measures h in $\sigma'/\sqrt{n}$ units. It is then seen from Figure 22.3 that a "standardized" **V**-mask with parameters $k = 0.5$ and $h = 4.2$, both measured in $\sigma'/\sqrt{n}$ units, will yield approximately the desired ARLs. For a line through the points $|k - m_a| = 0.5$ and $h = 4.2$ in Figure 22.3, standardized by taking $\sigma'/\sqrt{n} = 1$, will cut the L_a scale at approximately an ARL = 370 and a line through $|k - m_r| = |0.5 - 1.5| = 1$ and $h = 4.2$ will cut the L_r scale at an ARL = 5.1. The standardized values of $k = 0.5$ and $h = 4.2$ will thus yield the desired criteria for the "standardized" **V**-mask.[11] Since k/w is the tangent of the angle θ that a guide line makes with the horizontal, we will have $\tan \theta = 0.5/2 = 0.25$ for the "standardized" **V**-mask and the decision interval will be $4.2\ \sigma'/\sqrt{n}$ units. A picture of a truncated "standardized" **V**-mask is shown in Figure 22.8. Not only can this "standardized" **V**-mask be used for simultaneous control of several cusum charts for sample means, but it can also be used as a reference for comparison with other control procedures. Since Figure 22.3 was used in deriving the "standardized" **V**-mask, the one requirement is that the sample means being cumulated must be normally distributed, a requirement that is usually met in practice, at least approximately.

7.3. Snub-Nosed Cusum Masks

Although an ordinary cusum chart will detect a small shift in the process mean more rapidly than a comparable Shewhart chart, the latter will detect a sudden sharp change more rapidly. This disadvantage of an ordinary cusum procedure can be remedied, however, by the use of a combination of two parallel cusum schemes. One scheme could be that provided by the "standardized" **V**-mask discussed in the previous section and a supplementary scheme could be one effected by a **V**-mask with a smaller decision interval h, and a larger reference value k. The former will do well in picking up small shifts in the process mean and the latter will do better at picking up sudden

---

[11] Cf. James M. Lucas, "Cumulative Sum (CUSUM) Control Schemes," American Society for Quality Control, *39th Annual Quality Congress Transactions*, p. 369.

FIGURE 22.8
A Standardized V-Mask

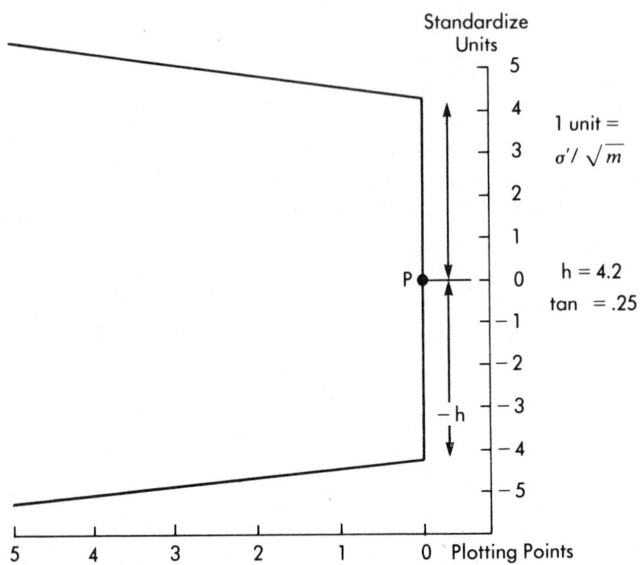

large shifts. A graph of such combined cusum schemes is shown in Figure 22.9, in which the supplementary **V**-mask has an $h = 1.6$ and a $k = 2.0$ standardized units. The decision line that is effective at any point is the inner line and the mask for the combined scheme is called a snub-nosed cusum mask.[12] A parabolic cusum mask that is a sort of smoothed snub-nosed mask was proposed by James M. Lucas[13] in 1973.

7.4. Two More-Involved Cusum Schemes

Besides noting special forms of **V**-masks, mention should be made of two, somewhat involved cusum schemes that have been found useful. One is a cusum scheme for a fast initial response (a FIR scheme) which is recommended for use at start up or following a readjustment of a process after an out-of-control signal.[14] The other is a cusum scheme that puts Shewhart control limits on an ordinary cusum chart.[15] The cusum feature of this combination will quickly detect small shifts from the target value while the addition of Shewhart limits increases the speed of detecting large shifts. It is thus

[12] ISO Draft Proposal 7871, "Introduction to Cumulative Sum Charts."

[13] "A Modified **V**-Mask Control Scheme," *Technometrics,* 15 (1973) pp. 833–47.

[14] See James M. Lucas and Ronald B. Crozier, "Fast Initial Response for CUSUM Quality Control Schemes: Give your CUSUM a Head Start," *Journal of Quality Technology,* 24 (1982), pp. 199–205.

[15] James M. Lucas, "Combined Shewhart-CUSUM Quality Control Schemes," *Journal of Quality Technology,* 14 (1982), pp. 51–59.

FIGURE 22.9
A Snub-Nosed Cusum Mask

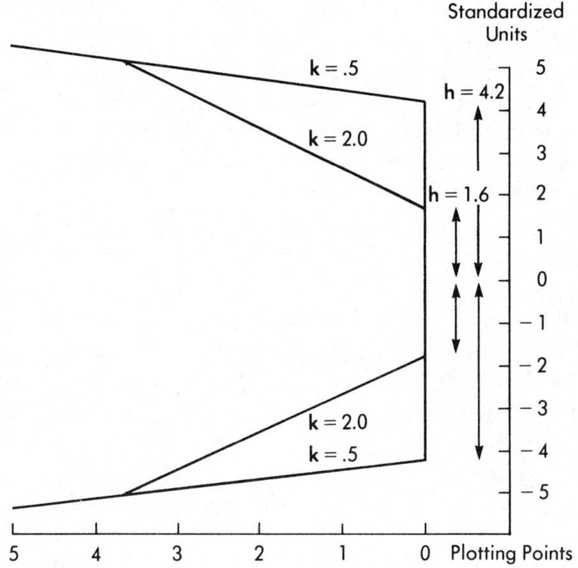

similar in performance to the snub-nosed **V**-mask and parabolic cusum mask described in the previous section.

8. PROBLEMS

22.1. Subtract 0.4380 from each of the $\bar{X}$ values of Problem 21.2. Cumulate the results by group number and show them on a chart. What inference can you make?

22.2. Subtract 21.5 from each of the values in the last column of Table 26.1 headed "average." Cumulate the results over time and show them on a chart. What inference can you make?

22.3. *a.* A one-sided cusum chart has a k of 110, an h of 9.5. The sample size is 9, and the standard deviation of individual values is given as 10. The chart was originally designed for an acceptable quality level *(m_a)* of 107 and a barely tolerable (rejectable) quality level *(m_r)* of 113. Sketch the ARL curve for the chart. Use semilog paper.

b. Sketch the ARL curve for a *pair* of one-sided schemes, one upper and one lower, such as described in *(a)* with $k_1 = 110$, $k_2 = 104$, and $\pm h = \pm 9.5$. Assume that $m_a = 107$, $m_{r1} - 113$, and $m_{r2} = 101$. Take the sample size to be 9 and the standard deviation of individual values to be 10 as in *(a)*.

c. Find the values of d and θ for a **V**-mask that would be the equivalent of the two one-sided schemes given in *(b)*. Sketch its ARL curve.

22.4. *a.* A one-sided cusum scheme has a k of 0.8330 and an h of 0.0120. The sample size is 4, and the standard deviation of individual values is given as 0.0060. The chart was originally designed for an acceptable quality level *(m_a)* of 0.8310 and a barely tolerable (rejectable) quality level *(m_r)* of 0.8350. Sketch the ARL curve for the chart. Use semilog paper.

b. Sketch the ARL curve for a *pair* of one-sided schemes, one upper and one lower, such as described in *(a)* with $k_1 = 0.8330$, $k_2 = 0.8290$, and $\pm h = \pm 0.0120$. Assume that $m_a = 0.8310$, $m_{r1} = 0.8350$, and $m_{r2} = 0.8270$. Take the sample size to be 4 and the standard deviation of individual values to be 0.0060 as in *(a)*.

c. Find the values of d and θ for a **V**-mask that would be the equivalent of the two one-sided schemes given in *(b)*. Sketch its ARL curve.

22.5. An acceptable quality level *(m_a)* is set at 6.50 and a barely tolerable (rejectable) quality level *(m_r)* at 2.00. The standard deviation of individual units is given as 5. What sample size should be used for a one-sided cusum scheme if L_a is to be about 1,000 and L_r about 2.5? (*Hint:* Use Table 22.1.) What will be the values of k and h to be used in operating the scheme? (*Hint:* To find h use Figure 22.3.)

22.6. An acceptable quality level *(m_a)* is set at 0.8310 and a barely tolerable (rejectable) quality level *(m_r)* at 0.8350. The standard deviation of individual values is given as 0.0050. What sample size should be used for a one-sided cusum scheme if L_a is to be about 500 and L_r about 7.5? (*Hint:* Use Table 22.1.) What will be the values of k and h to be used in operating the scheme? (*Hint:* To find h use Figure 22.3.)

22.7. An acceptable quality level *(m_a)* is set at 25 and a barely tolerable (rejectable) quality level *(m_r)* at 10. The standard deviation of individual values is given as 15. What sample size should be used for a one-sided cusum scheme if L_a is to be about 800 and L_r about 3? (*Hint:* Adopt a central reference value and use Figure 22.3) What will be the k and h for the cusum scheme? (*Hint:* Use Figure 22.3 to find h.)

22.8. An acceptable quality level *(m_a)* is set at 1,000 and a barely tolerable (rejectable) quality level *(m_r)* at 800. The standard deviation of individual values is given as 200. What sample size should be used for a one-sided cusum scheme if L_a is to be about 400 and L_r about 6? (*Hint:* Adopt a central reference value and use Figure 22.3.) What will be the k and h for the cusum scheme? (*Hint:* Use Figure 22.3.) What will be the k and h for the cusum scheme? (*Hint:* Use Figure 22.3 to find h.)

22.9. Suppose that we set up another cusum chart that is exactly like that derived in Problem 22.5 except the k is symmetrically placed on the other side of m_a and we use h instead of $-h$ as our decision interval. Suppose that these two charts are operated simultaneously to provide a two-sided cusum scheme. What then will be the ARL at an $m_a = 6.50$? at an $m_r = 2.00$? at an $m_r = 11.00$? Derive the d and θ for a **V**-mask that will be the equivalent to this pair of one-sided charts?

22.10. Suppose that we set up another cusum chart that is exactly like that derived in Problem 22.6 except that k is symmetrically placed on the other side of m_a and we use $-h$ instead of h as our decision interval. Suppose that these two charts are operated simultaneously to provide a two-sided cusum scheme. What then will be the ARL at an $m_a = 0.8310$? at an $m_r = 0.8350$? at an

$m_r = 0.8270$? Derive the d and θ for a **V**-mask that will be the equivalent of this pair of one-sided charts.

22.11. An acceptable quality level is set at 50 and barely tolerable (rejectable) quality levels at 45 and 55. The standard deviation of individual units is given as 5. What sample size should be used for a cusum scheme using a **V**-mask if L_a is to be about 250 and L_r about 7.5? (*Hint:* Take L_a for a "component" one-sided scheme to be double 250 and use Table 22.1.) Find the d and θ for the mask. (*Hint:* Find first the k and h as in Problems 22.5 and 22.6.)

22.12. An acceptable quality level is set at 30 and barely tolerable (rejectable) quality levels at 20 and 40. The standard deviation of individual values is given as 20. What sample size should be used for a cusum scheme using a **V**-mask if L_a is to be about 500 and L_r about 10? (*Hint:* Take L_a for a "component" one-sided scheme to be double 500 and use Table 22.1.) Find the d and θ for the mask. (*Hint:* Find first the k and h as in Problems 22.5 and 22.6.)

22.13. For a given process the acceptable quality level for the proportion nonconforming (p'_a) is taken as 0.02 and the barely tolerable (rejectable) quality level (p'_r) at 0.05. Management is satisfied to have an L_a of 250 and an L_r of 7.50. Find the sample size (n) and the k and h to be used in operating a cusum chart to control the process. (*Hint:* Use Table 22.2.) Indicate precisely what will be plotted on your cusum chart.

22.14. For a given process the acceptable quality level for the proportion nonconforming (p'_a) is taken as 0.02 and the barely tolerable (rejectable) quality level (p'_r) at 0.07. Management is satisfied to have an L_a of 500 and an L_r of 5.00. Find the sample size (n) and the k and h to be used in operating a cusum chart to control the process. (*Hint:* Use Table 22.2.) Indicate precisely what will be plotted on the cusum chart.

22.15. It is noted in the text that if a truncated "standardized" **V**-mask is designed to have an $h = 4.2$ and a $k = 0.5$, both measured in $\sigma'/\sqrt{n}$ units, then this "standardized" **V**-mask will have an ARL of 5.1 when the process average has shifted by $1.5\sigma'/\sqrt{n}$ units. For a shift of the same magnitude what will be the ARL of a 3-sigma Shewhart $\bar{X}$-chart the vertical scale of which is also measured in $\sigma'/\sqrt{n}$ units? What is the answer for a 2-sigma Shewhart $\bar{X}$-chart?

22.16. Draw a snub-nosed **V**-mask one component of which is the "standardized" **V**-mask described in the text and the other is a truncated **V**-mask with $k = 1.8$ and $h = 2$.

9. SELECTED REFERENCES*

Bagshaw and Johnson (P '75), Bissell (P '84), Elder, Provost and Ecker (P '81), Ewan (P '63), Ewan and Kemp (P '60), Freund (P '62), Goel (B '68), Goldsmith and Whitfield (P '61), Johnson (P '61), Johnson and Leone (P '62, and B '64), Kemp (P '58, P '61 and P '62), Khan (P '79 and P '81), Lucas (P '76, P '82 and P '85), Lucas and Crosier (Ps '82), Marquardt (P '84), Mumford (P '80), Neube and Woodall (P '84), Roberts (P '59), and Truax (P '61).

---

* B and P refer to the Book and Periodical sections, respectively, of the Cumulative List of References in Appendix V.

23

Special Devices and Procedures

1. ACCEPTANCE CONTROL CHARTS

1.1. Nature of Acceptance Control Charts

In the preceding chapters we discussed the use of a control chart as a device for determining whether a process has been or is in control at some specified or computed level. Specification limits were considered only with reference to the fraction nonconforming of the process. In discussing $\bar{X}$-charts it was tacitly assumed that specification limits were close to the standard level at which the process was being controlled and that a shift in the process average from this standard level would increase the fraction nonconforming; hence the need for control.

The natural dispersion of a process, however, is sometimes much less than the dispersion permitted by the specifications. Specification limits on a given dimension, for example, might be $L = 0.1220$ and $U = 0.1250$ inches, but the tool forming this dimension may, over a short interval, produce parts the distribution of which will have a standard deviation of only 0.0002 inches.[1] The spread between specification limits would, in this case, be $15\sigma'$. Under such conditions there is considerable room for the process average to shift within the specification limits without producing nonconforming items.

When the engineering tolerance on a dimension greatly exceeds the natural dispersion of the manufacturing process, it may be desirable to use another kind of $\bar{X}$-chart which is aimed directly at control of the fraction nonconforming of the process. For such a "control chart" the limits will be based on the specification limits for the product rather than on some standard level for the process average. So long as the sample means fall within these limits,

[1] Cf. J. Manuele, "Control Chart for Determining Tool Wear," *Industrial Quality Control,* May 1945, pp. 7–10.

the process will be deemed to be acceptable. There will no longer be any presumption, however, that the process average remains stable. It can vary considerably within the specification limits and so be "out of control" in the sense of varying in a nonrandom manner, but there will be no interest in picking up such variations until they cause the process to produce an undesirable fraction of nonconforming items. $\bar{X}$-charts of this kind are called acceptance control charts.[2]

1.2. Design of an Acceptance Control Chart

1.2.1. Designs Based on a Rejectable Process Level and β Risk. In our initial discussion let us assume that the sample size n is given and we simply have the problem of determining how close the control limits should come to the specification limits L and U. We shall throughout Section 1.2 assume that the process standard deviation (σ') is known.

We shall begin by selecting a Rejectable Process Level (RPL) which will be a process fraction nonconforming that we can barely tolerate. We shall also select a probability (β) with which we are willing to accept a process with a fraction nonconforming equal to the RPL. The limits for the control chart will then be set to yield this value of β at the RPL.

To illustrate the procedure let the RPL $= 0.025$ and $\beta = 0.10$. Let σ' be the known standard deviation of the process and assume that the product is normally distributed. Then the limits for the control chart will fall at $U - 1.960\sigma' - 1.282\sigma'/\sqrt{n}$ and $L + 1.960\sigma' + 1.282\sigma'/\sqrt{n}$. For if the process average is at $U - 1.960\sigma'$ or $L + 1.960\sigma'$ and if the process is normal, the fraction nonconforming will be 0.025. Furthermore, if the process average is at $U - 1.960\sigma'$, the probability of a sample mean falling below the upper control limit $U - 1.960\sigma' - 1.282\sigma'/\sqrt{n}$ is 0.10; similarly for the lower control limit. (See Figure 23.1.)

1.2.2 Designs Based on an APL and α Risk. We can also design an acceptance control chart with reference to an acceptable process level (APL) and a probability (α) of *rejecting* the process when the average is in fact equal to the APL. If we take n as given, the procedure is almost exactly the same as discussed in the preceding section. The difference is that the control limits are now on the "outer side" of the $\bar{X}$ values that yield a fraction nonconforming equal to the APL whereas they were on the "inner side" of the $\bar{X}$ values that yielded a fraction nonconforming equal to the RPL. Thus, if the process is normal, if the APL $= 0.01$ and $\alpha = 0.001$, and if σ' is known, then the control limits will fall at $U - 2.326\sigma' +$

[2] See Richard A. Freund, "Acceptance Control Charts," *Industrial Quality Control,* October 1957, pp. 13–23, for a fuller discussion of the use of control charts for acceptance puposes. Also see David Hill, "Modified Control Limits," *Applied Statistics,* Vol. V (1956), pp. 12–19. The "modified control charts" discussed in the earlier editions of this text are the same as the acceptance control charts described in Section 1.2.2.

FIGURE 23.1
Limits on an Acceptance Control Chart, RPL = 0.025 and β = 0.10

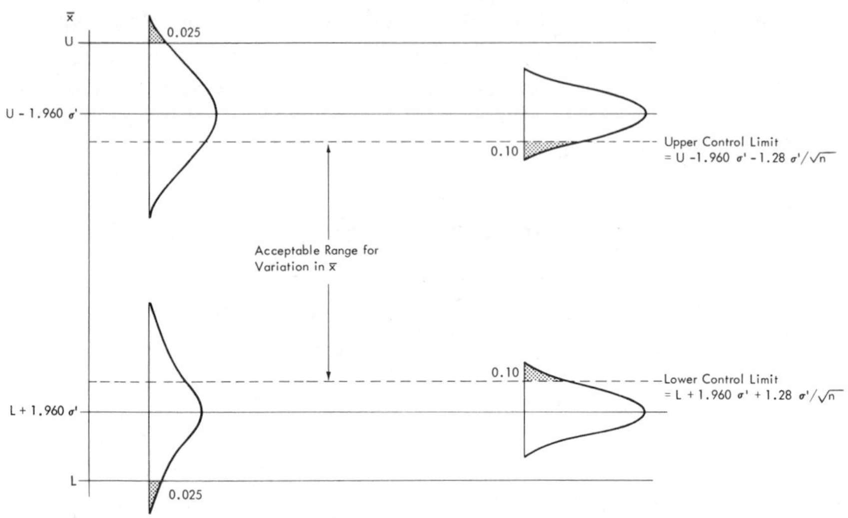

$3.090\sigma'/\sqrt{n}$ and $L + 2.326\sigma' - 3.090\sigma'/\sqrt{n}$. For this case it will be left to the reader to draw the figure that corresponds to Figure 23.1.

1.2.3. Designs Based on Both an RPL and β and an APL and α. The design of Section 1.2.1 ignores the α risk of rejecting a process of APL quality, although it can be computed, however, as soon as n is known. Likewise the design of Section 1.2.2 ignores the β risk of accepting a process of RPL quality, although it again can be figured as soon as n is known. It is obvious that a third possibility is to adjust n so that both sets of criteria can be met simultaneously. This assumes that n is not seriously restricted by cost[3] and other considerations.

To illustrate, suppose as before that the process is normal and the standard deviation σ' is known and let the RPL = 0.025 and β = 0.10 and the APL = 0.01 and α = 0.001. Then the n that will attain these goals will be given by the equation

$$U - 1.960\sigma' - 1.282\sigma'/\sqrt{n} = U - 2.326\sigma' + 3.090\sigma'/\sqrt{n}$$

Thus

$$\sqrt{n} = \frac{3.090 + 1.282}{2.326 - 1.960} = 11.9$$

and n lies between 141 and 142. Either could be used to approximate the desired goals. The control limits will fall at $U - 1.960\sigma' - 1.282\sigma'/\sqrt{n}$ and $L + 1.960\sigma' + 1.282\sigma'/\sqrt{n}$.

[3] The best design would in fact be based on minimizing overall costs, but this is beyond what we wish to take up here.

1.3. Estimation of σ'

To operate an acceptance control chart we must have a value of σ'. This is probably best determined from an R-chart. The procedure is to take a number of samples from the process, all of a given size; $n = 8$ is considered to be a good figure.[4] Plot the ranges of these samples on an R-chart and throw out any that are outside the control limits for which assignable causes can be found. If more than one or two samples are outside the limits, it is probably better to try first to bring the process under control with respect to its general variability before proceeding further with setting up an $\bar{X}$-chart. If the variability appears to be under control, then we can compute $\bar{R}$ and estimate σ' from the formula $\sigma' = \bar{R}/d_2$. It is well to keep this R-chart running currently with the acceptance control chart.

It will be noted that a basic assumption of the acceptance control chart is that the variance around the process average is stable although the process average itself may fluctuate.

1.4. Tool Wear

An acceptance control chart may be particularly useful when a process is subject to tool wear.[5] Under these conditions the natural spread of the process at any one time will be much less than the spread over the whole life of the tool, and special control limits will be applicable. Figure 23.2 illustrates an acceptance control chart that is used to control a process subject to tool wear.

It will be noticed that if no nonconforming items are tolerated, then the lower specification limit plus $3\sigma'$, i.e., $L + 3\sigma'$ or $L + 3\bar{R}/d_2$, is the proper initial setting for the tool, since that is the lowest possible position that will produce practically no nonconforming items if the process is under control. If $100p'$ percent nonconforming items is tolerated, then the proper initial setting is $L + k\sigma'$ or $L + k\bar{R}/d_2$, where k is the value of z in the normal table such that it is exceeded in a positive direction by $100p'$ percent of the cases. The upper control limit could be set symmetrically with respect to the upper specification limit, although an RPL approach at this end might be more appropriate.

It will also be noticed that the life of the tool will be determined not only by the normal rate of tool wear but also by two other factors. The first is the decision as to the percentage of nonconforming items that will be tolerated for any short interval of production, and the second is the natural dispersion of the process for short intervals (i.e., σ'). If the tolerance percent

---

[4] See F. E. Grubbs and C. L. Weaver, "The Best Unbiased Estimate of the Population Standard Deviation Based on Group Ranges," *Journal of the American Statistical Association*, 42 (1947), pp. 224–41.

[5] See Manuele, "Control Chart for Determining Tool Wear."

FIGURE 23.2
Illustration of an Acceptance Control Chart for Controlling Tool Wear*

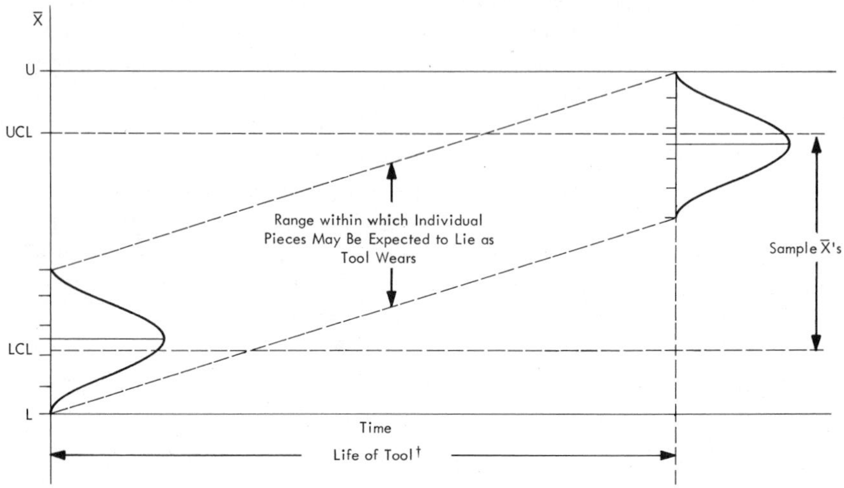

* When sample $\bar{X}$ goes above UCL, the life of the tool is ended. If tools are uniform enough, a diagram such as the above may be established for the given class of tools. The average life then becomes the standard. Tools are replaced at the end of that time. Under these conditions, UCL and LCL merely give protection against unusual events during the life of the tool. See J. Manuele, "Control Chart for Determining Tool Wear." *Industrial Quality Control,* May 1945, p. 9.

† If no nonconforming items are tolerated.

nonconforming is increased or σ' reduced, the life of the tool will be longer, and vice versa.

1.5. Acceptance Control Charts and Acceptance Sampling Plans[6]

Statistically an acceptance control chart and a double-limit variables sampling plan are practically identical. There is a definite administrative difference however. An acceptance sampling plan is applied to lots, and the action taken is to accept or reject a lot. An acceptance control chart is applied to a process, and the action taken is to do something about the process. Furthermore, under an acceptance sampling plan the items should be sampled at random from the whole lot. With an acceptance control chart it is likely to be desirable to take the sample from the process all at once. If it is spread out over a segment of the process, a change that has taken place in the process during the production of that segment may be covered up by the averaging of the sample results.

[6] Alfred J. Winterhalter, "Development of Reject Limits for Measurements," *Industrial Quality Control,* January 1945, pp. 12–15, and March 1945, pp. 12–13; and Norbert L. Enrick, "Operating Characteristics of Reject Limits for Measurements," *Industrial Quality Control,* September 1945, pp. 9–10.

2. DIFFERENCE CONTROL CHARTS[7]

It sometimes happens that variability in test results in significantly affected by testing conditions which cannot easily be controlled. Under such circumstances an $\bar{X}$-chart may show lack of control, which may not, in reality, be lack of control of quality but only lack of control of testing techniques. Suppose, for example, that a sample is inspected only once a day and that weather conditions significantly affect the inspection tests. Under these conditions, the effects of variations in weather become confused with any real variations in quality, and the control chart does not give a true indication of control or lack of control in the production process. To overcome this difficulty, the difference control chart has been devised.

To use a difference control chart, it is necessary to have a standard lot which is known to have an output controlled at the desired level. This is called the reference lot. Such a lot could be taken as a part of the output that had been produced under controlled conditions, or it might be made up through 100 percent inspection and artificial selection of the items.

To operate a difference control chart, we proceed as follows: We take a sample from each day's output and also a sample from the reference lot. In each case we calculate the mean of the sample. Then we take the difference between the average of the standard sample and the average of the current sample and plot the difference on an $\bar{X}$-difference chart. The central line on the chart will be 0. The control limits on the $\bar{X}$-difference chart will fall at $\pm \frac{3}{\sqrt{n}} \sqrt{\hat{\sigma}_r^2 + \hat{\sigma}_c^2}$, where $\hat{\sigma}_r$ is an estimate of the standard deviation of the reference lot and σ_c is the estimated standard deviation of the current output. These both might be estimated, for example, by computing $\bar{R}$ from m samples of n each—say, 20 samples of 5 each. Then we could take $\hat{\sigma}_r = \frac{\bar{R}_r}{d_2\sqrt{n}}$ and $\hat{\sigma}_c = \frac{\bar{R}_c}{d_2\sqrt{n}}$, and the control limits on the $\bar{X}$-difference chart would be

$$\text{UCL} = +\frac{3}{d_2\sqrt{n}} \sqrt{\bar{R}_r^2 + \bar{R}_c^2} = +A_2\sqrt{\bar{R}_r^2 + \bar{R}_c^2} \tag{23.1}$$

$$\text{LCL} = -\frac{3}{d_2\sqrt{n}} \sqrt{\bar{R}_r^2 + \bar{R}_c^2} = -A_2\sqrt{\bar{R}_r^2 + \bar{R}_c^2}$$

If points fall outside the above limits and assignable causes are found, we say the process is out of control; if no points fall outside the limits and there is no evidence of nonrandom variation within the limits, we say the process is under control with respect to its average. We can do this now, since variability in test results due to variations in testing conditions from

[7] See F. E. Grubbs, "The Difference Control Chart with an Example of Its Use," *Industrial Quality Control,* July 1946, pp. 22–25.

day to day have been eliminated by taking differences. The ordinary R-chart is kept on the current process as usual, since it is assumed that variations in testing conditions from day to day do not affect test variations. If this were not true, we would also have to have an R-difference chart, constructed along lines similar to the $\bar{X}$-difference chart. It is also well to keep an R-chart on sampling from the reference lot, to be sure that the sampling is at random.

3. THE LOT PLOT METHOD

3.1. Description of Method

A special type of variables sampling plan developed by Dorian Shainin is known as the lot plot method.[8] This runs as follows:

1. A random sample of 50 items[9] is taken from the lot, and a frequency distribution is made of these items in the manner indicated below.
2. First, a subsample of 5 is taken from the original sample of 50, and the $\bar{X}$ and R are computed for this subsample.
3. The mean of the subsample of 5 is taken as a convenient arbitrary origin for constructing the frequency distribution, and the size of the class interval is determined so that twice the value of R computed in (2) includes from 7 to 16 intervals. The intervals are numbered 0, 1, 2, −1, −2, and so on, taking the 0 interval as that containing the arbitrary origin (compare Chapter 3, Section 4.1.1.2).
4. The initial subsample of 5 cases is distributed among the various intervals determined in (3), each item being indicated on the frequency chart by a "1."
5. The remaining 45 cases are divided into groups of 5 items each, and the items in each group are entered on the chart by recording the number of their group (see Figure 23.4).
6. The sum of the class interval deviations and the range in class interval units is entered for each group in a supplementary table. From this table the mean of the whole 50 items ($\bar{\bar{X}}$) and the average range for the 10 groups ($\bar{R}$) are computed, both in class interval units. Then σ' (in class interval units) is estimated by dividing $\bar{R}$ by $d_2 = 2.326$.
7. Upper and lower "control limts" are laid off by adding and subtracting $3\bar{R}/d_2$ (in class interval units) from $\bar{\bar{X}}$ (in class interval units).
8. If the distribution of the 50 items approximates the normal form and if the "control limits" fall within the specification limits, the lot is accepted.

---

[8] Dorian Shainin, "The Hamilton Standard Lot Plot Method of Acceptance Sampling by Variables," *Industrial Quality Control,* July 1950, pp. 15–34.

[9] Dorian Shainin found 50 to be a good number. The plan can be used for larger size samples, however, if desired.

9. If the distribution of the 50 sample items is significantly skewed or shows significant kurtosis and if the "control limits" are near the specification limits, the data are studied further. The steps taken are discussed below.

10. If the distribution of the 50 items approximates the normal form but the "control limits" exceed the specification limits in either or both directions, indicating that a fraction of the lot will not conform to specifications, the data are passed on to a salvage board for review and further action.

11. In some cases one or more points will fall beyond the "control limits." In such cases an attribute plan is employed. If no points lie beyond specification limits out of the total of 50, the lot is accepted. If the number exceeds 3, the lot is submitted to 100 percent inspection.

3.2. The OC Curve for the Lot Plot Method

If we assume (1) that the lot is large and is normally distributed, (2) that nonnormal looking samples are treated like normal looking samples instead of by the special procedures called for by the plan, and (3) that the spread between specifications is considerably greater than six times the lot standard deviation, then a type A_L OC curve can be derived as follows:

Under the above conditions an OC curve will give the probability of a "control limit" not exceeding a specification limit for designated values of the lot fraction nonconforming. By the lot plot method the upper "control limit" is given by $\bar{\bar{X}} + 3\bar{R}/d_2$, or for a sample of 50 subdivided into 10 groups of 5 each, by $\bar{\bar{X}} + 3\bar{R}/2.326$. Since both $\bar{\bar{X}}$ and $\bar{R}$ are sample values, the quantity $\bar{\bar{X}} + 3\bar{R}/2.326$ will not give the actual $3\sigma'$ limit for the lot but will be subject to sampling errors. To construct an OC curve for given conditions we have to find the probability that $\bar{\bar{X}} + 3\bar{R}/2.326$ will not exceed the upper specification limit when the actual mean and standard deviation of the lot (assumed to be normally distributed) is such as to produce a fraction of nonconforming units beyond the upper specification limit equal to p'. In computing this probability, we shall assume that $\bar{\bar{X}} + 3\bar{R}/2.326$ is normally or approximately normally distributed, that its average value in many samples is $\bar{X}' + 3\sigma'$, and its variance in many samples is[10]

$$\textbf{(23.2)} \qquad \sigma'^2_{\bar{\bar{X}}} + \left(\frac{3}{2.326}\,\sigma'_{\bar{R}}\right)^2$$

With $n = 50$, we have $\sigma'^2_{\bar{\bar{X}}} = \sigma'^2/50 = 0.02\sigma'^2$ and $\sigma'^2_{\bar{R}} = \sigma'^2_R/10 = \sigma'^2 d_3^2/10$, since[11] $d_3 = \sigma'_R/\sigma'$. But Table D1 of Appendix II shows that $d_3 = 0.864$ for samples of 5. Hence, in this particular case, we have

---

[10] It will be recalled that $\sigma'^2_{x+y} = \sigma'^2_x + \sigma'^2_y$, if x and y are independent (see Chapter 4, Section 10.1, Theorem 4.2). If the lot is normally distributed, $\bar{\bar{X}}$ and $\bar{R}$ are independent, since $\bar{X}$ and R for each group of 5 are independent.

[11] See Chapter 6, Section 3, and Chapter 21, Section 3.1.

$$\frac{3}{2.326}\sigma'_{\bar{R}} = \frac{3}{2.326}\frac{0.864}{\sqrt{10}}\sigma' = 0.352\sigma'$$

and

$$\sqrt{\sigma'^2_{\bar{\bar{X}}} + \left(\frac{3}{2.326}\sigma'_{\bar{R}}\right)^2} = \sqrt{0.02\sigma'^2 + 0.124\sigma'^2} = 0.38\sigma'$$

Let the lot fraction nonconforming on the upper side be 0.004. This means that, if the lot is normally distributed, the specification limit comes at $\bar{X}' + 2.65\sigma'$. Hence the probability of accepting the lot on the upper side is the probability that $\bar{\bar{X}} + 3\bar{R}/2.326$ will be less than $\bar{X}' + 2.65\sigma'$. Since $\bar{\bar{X}} + 3\bar{R}/2.326$ is assumed to be normally distributed about a mean equal to $\bar{X}' + 3\sigma'$ with a standard deviation equal to $0.38\sigma'$, the aforesaid probability is that of a normal variable z exceeding

$$\frac{(\bar{X}' + 2.65\sigma') - (\bar{X}' + 3\sigma')}{0.38\sigma'} = -\frac{0.35}{0.38} = -0.92$$

in the negative direction. The normal tables give this as 0.18. Similar calculations give the following results. These results are plotted in Figure 23.3.

(1) p'	(2) $z_{p'} = \dfrac{\text{Spec. Limit} - \bar{X}'}{\sigma'}$	(3) $\dfrac{(\bar{X}' + z_{p'}\sigma') - (\bar{X}' + 3\sigma')}{0.38\sigma'}$	$P[z < (3)]$
0.0001	3.61	0.61/0.38 = 1.60	0.945
0.0005	3.27	0.27/0.38 = 0.71	0.761
0.0010	3.08	0.08/0.38 = 0.21	0.583
0.00135	3.00	0	0.500
0.002	2.87	−0.13/0.38 = −0.34	0.367
0.004	2.65	−0.35/0.38 = −0.92	0.179
0.008	2.41	−0.59/0.38 = −1.55	0.061
0.012	2.257	−0.743/0.38 = −1.95	0.026
0.020	2.054	−0.946/0.38 = −2.49	0.006

C. C. Craig and L. E. Moses have worked out OC curves for the "normal analysis" when the spread between specifications is about six times the lot standard deviation or less.[12] Under these conditions, when the lot mean falls midway between the specification limits, the OC curve is several percentage points higher than that shown in Figure 23.3.

[12] See Cecil C. Craig, "Some Remarks Concerning the Lot Plot Plan, *Industrial Quality Control,* September 1953, pp. 41–48; and Lincoln E. Moses, "Some Theoretical Aspects of the Lot Plot Sampling Inspection Plan," *Journal of the American Statistical Association,* 51 (1956), pp. 84–107.

FIGURE 23.3
Operating Characteristic Curve for the Lot Plot Method when Lot Distribution Is Normal and $n = 50$

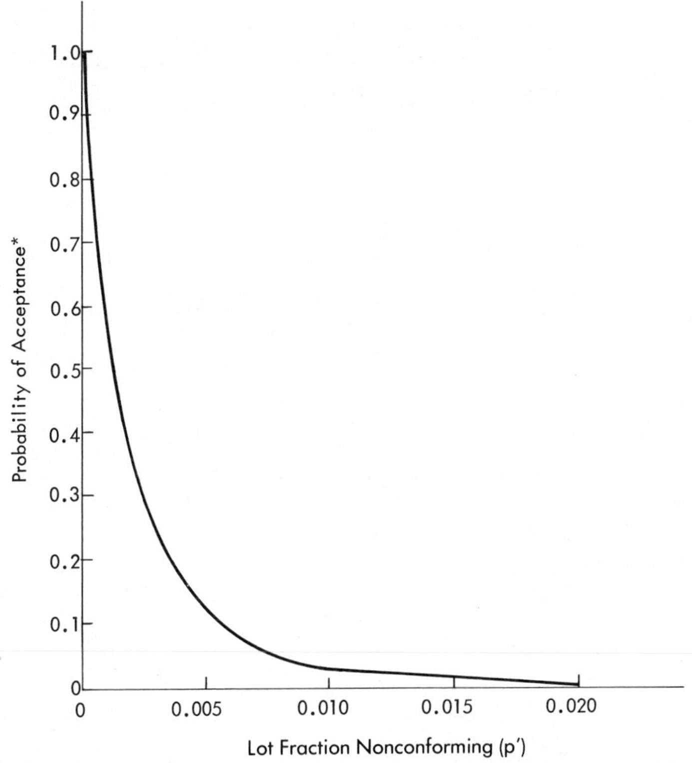

* Refers to probability that the upper (lower) "control limit" for a sample will fall below (above) the upper (lower) specification limit when, in fact, $100p$ percent of the lot lies above (below) the upper (lower) specification limit.

3.3. Action when Sample Distribution Is Nonnormal

When the distribution of the 50 cases is significantly skewed or multimodal or truncated or shows features markedly different from a normal distribution, the lot plot plan calls for special procedures of analysis. In the case of a highly skewed distribution standard deviations are computed separately for the cases above and below the mode. For a multimodal distribution, the control limits are computed from the smallest and largest modes, respectively. Further details as to these special procedures will be found in the references at the end of this chapter.

3.4. An Example

Let us suppose that the first 50 items of Table 3.2 are a random sample from a lot of 500 items and let us apply the lot plot method. The steps are as follows:

1. Since the data are presumably drawn at random, the first 5 items are a convenient subgroup with which to begin the analysis. This group has a mean of 0.8324 and a range of 0.014. We, therefore, take a class interval of 0.003 [2 × 0.014 divided by 0.003 yields 9 to 10 intervals which lie between 7 and 16, as directed] and take 0.8320 as the arbitrary origin, since it is the nearest "whole unit" to 0.8324. Class limits are now located at 0.8305, 0.8275, and so on, and 0.8335, 0.8365, 0.8395 and so forth, which

FIGURE 23.4
Lot Plot of Overall Height of Fragmentation Bomb Bases

Class Limits	*Interval Deviations**	*Distribution of Cases*	
0.8455—	5		——UCL
0.8425—	4		
0.8395—	3	1 6 7 9	——Spec. Limit
0.8365—	2	8	
0.8335—	1	1 2 3 4 7 8 8 10	
0.8305—	0	1 2 2 2 3 4 4 4 4 5 5 5 6 7 9 9 9 9 10 10	
0.8275—	−1	1 5 6 6 6 7 7 8 10	
0.8245—	−2	1 2 3 10	
0.8215—	−3	3	
0.8185—	−4	5	——Spec. Limit
0.8155—	−5	3 8	----LCL

* Origin = 0.8320.

Supplementary Table

Group Number	*Sum*	*Average*	*Range*
1	1	0.2	5
2	−1	−0.2	3
3	−9	−1.8	6
4	1	0.2	1
5	−5	−1.0	4
6	0	0	4
7	2	0.4	4
8	−2	−0.4	7
9	3	0.6	3
10	−2	−0.4	3
Total		−2.4	40
Average		−0.24	4

will make the arbitrary origin the midpoint of an interval and cause the class limits to fall between measured values.[13] Thus, with this scaling, each item can be clearly assigned to one, and only one, interval. Finally, the various class intervals are numbered up and down from the class interval containing the arbitrary origin, which itself is numbered 0.

2. We now enter the various items on the lot plot diagram by writing the number of the subgroup to which each belongs to the right of the interval in which it falls (see Figure 23.4).

3. From this we construct the supplementary table in which we enter the sum and average of the class interval deviations and the range of these deviations for each group. At the bottom of this table we compute the grand average and the average range (both in class interval units).

4. Then, as the last step, we compute the "control limits" in terms of class interval units. Thus

$$\bar{\bar{X}} + \frac{3\bar{R}}{d_2} = -0.24 + \frac{3}{2.326}(4) = 4.92$$

$$\bar{\bar{X}} - \frac{3\bar{R}}{d_2} = -0.24 - \frac{3}{2.326}(4) = -5.40$$

The lot plot diagram (Figure 23.4) shows that for our data the "control limits" are beyond the specification limits. The lot, therefore, would not be accepted but would be sent to a salvage board for review.[14]

3.5 Appraisal of the Lot Plot Plan

The lot plot plan may be appraised briefly as follows:

1. The procedure requires relatively simple computations and is reasonably well understood by plant personnel. The latter find the "picture" of the lot given by the histogram helpful in interpreting lot quality and the provision for always taking the same sample size is administratively convenient.

2. Its use in many plants has been accompanied by improved quality and lower inspection costs. It is intended as a substitute for 100 percent inspection and in the normal case, at least, yields a very tight OC curve. (See Figure 23.3.)

3. When lots are highly skewed or leptokurtic, however, operating characteristics of the plan may not be as tight as for normal lots. The study of

---

[13] If special gauges are used, they can be "zeroed" so that the markings will come at the midpoints of the intervals.

[14] If we assume the distribution is normal with a mean of approximately 0.8313 [i.e., 8.8320 − 0.24(0.003)] and a standard deviation of 0.0052 [i.e., (4/2.326)(0.003)], it can be estimated that about 6.2 percent of the bomb bases lie beyond the specification limits. This is the fact on which the salvage board would have to base its decision.

L. E. Moses[15] indicates that a highly leptokurtic lot that is 2.5 percent nonconforming may have as high as a 0.10 chance of acceptance. A similar conclusion is drawn for highly skewed lots where the fraction nonconforming occurs all at one end.

4. On the other hand, the lot plot plan may under some circumstances be excessively tight. For example, Moses found that certain kinds of nonnormal lots containing no nonconformities at all have a high chance of being rejected.

5. In general, the decision of the inspector to use "normal analysis" or special procedures depends upon the inspector's personal inclinations, and this subjective element makes it impossible to work out theoretical OC curves for the lot plot plan in all its phases.

6. "The provision," Moses writes,[16] "that though an inspector may *accept* the lot it requires the Salvage Review Board to dispose of one in any other way should have important administrative advantages in many settings; economic factors will tend to be systematically weighed at the time of disposition of the lot; if the Salvage Review Board contains representation from the departments which (in addition to inspection) are concerned, then, fewer dispositions of lots should result in difficulties such as production flow problems."

4. BINOMIAL PROBABILITY PAPER (BIPP)

A device that has many uses in statistical analysis and quality control is binomial probability paper, invented by F. Mosteller and J. W. Tukey.[17] A picture of this paper is shown in Figure 23.5. It will be noted that the scales on each axis are square root scales.

On this paper a universe probability is plotted as a straight line through the origin, the angle being determined as a $100(1 - p)$, $100p$ split. For example, a 10 percent line could be plotted as a 90, 10 split, i.e., we move to the right 90 units and up 10 and then connect that point with the origin. We could, if we wished, also plot a 10 percent line as a 10, 90 split.

A sample result is plotted as a triangle. Thus, if we have n_1 of one category and n_2 of another, we plot the three points (n_1, n_2), $(n_1 + 1, n_2)$ and $(n_1, n_2 + 1)$. If n_1 and n_2 are small, these form a sizable triangle and are plotted as such. For example, see point 3, 1 in Figure 23.5. If n_1 is small and n_2 large, the triangle reduces to a horizontal line. If n_1 is large and n_2 small, it plots as a vertical line. If both are large, the triangle plots as a point. If either n_1 or n_2 is very large, it may be necessary to take the scales as $\frac{1}{10}$ of the regular scales. That is, read n_1 as $n_1/10$ or n_2 as $n_2/10$ or both.

---

[15] Lincoln E. Moses, "Some Theoretical Aspects of the Lot Plot Sampling Inspection Plan," *Journal of the American Statistical Association,* 51 (1956), pp. 84–107.

[16] Ibid., p. 106.

[17] See "The Uses and Usefulness of Binomial Probability Paper," *Journal of American Statistical Association,* 44 (1949), pp. 174–212.

FIGURE 23.5
Binomial Probability Paper (BIPP)*

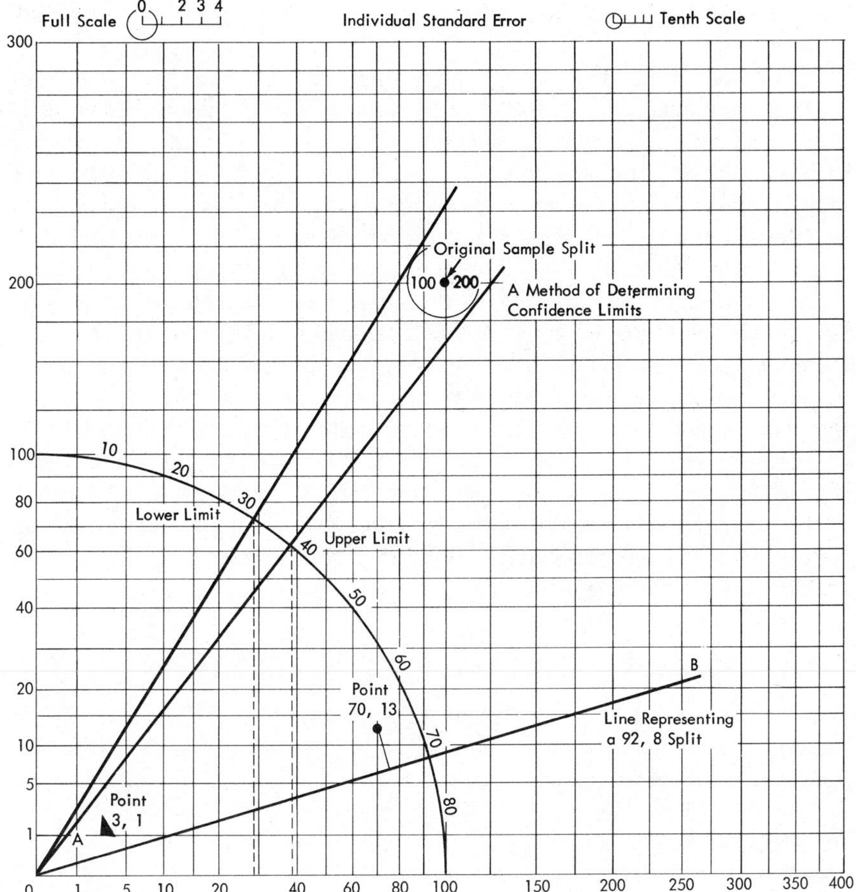

* Binomial probability paper was designed by F. Mosteller and J. W. Tukey and is sold by the Codex Book Co., Inc., Norwood, Massachusetts.

One of the uses of binomial probability paper is to determine the probability of a given sample result. To do this, the universe probability line is first drawn on the chart. For example, if $p' = 0.08$, the universe line could be plotted as line AB on the chart. The next step is to plot the sample point. If this were $n_1 = 70$, $n_2 = 13$, for example, it would be plotted as the triangle with corners (70, 13), (71, 13), and (70, 14) (see Figure 23.5). To find the probability of such a sample or one that deviates further from the universe line, we take the nearest corner of the sample triangle, or take the sample "point" itself, if the triangle is practically a point, and measure its distance from the universe line. Next we lay off this distance on the standard deviation scale in the upper part of the chart and determine the number of standard deviation units that the sample is away from the universe value.

Finally, we enter Table A2 with this result and read off the desired probability. For the example given, the distance of the sample "point" from the universe line is about 12 mm. or $2.4\sigma'$, which indicates a probability of being exceeded of about 0.01.

Binomial probability paper is very useful in constructing a stabilized p-chart. The standard value p'' or the average value $\bar{p}$ may be drawn as a "universe line" on the chart, and the sample points may be plotted as triangles or line segments about this line. The distances of the samples from the line may be measured and converted into standard deviation units. These standardized deviations may then be plotted on the regular stabilized p-chart. Binomial paper thus avoids the calculations of Figure 19.8 of Chapter 19. In using the binomial paper in this way, if the number of nonconforming items is very small compared with the number of conforming items, we can plot the total number as if it were the number of conforming items. It also might be more convenient in this case to take the number of conforming items as the abscissa and the number of nonconforming items as the ordinate. In this case the triangles become vertical lines. Mosteller and Tukey plot these vertical lines rather than their midpoints on the stabilized p-chart, each located on the chart at its vertical distance from the average line on the binomial paper.[18] For the many other uses of BIPP the reader is referred to the original article by Mosteller and Tukey.

5. MOVING AVERAGE AND MOVING RANGE CHARTS

There are occasions when moving averages and moving ranges are more in accord with industrial practice than are ordinary averages and ranges. To note what is meant by a "moving average," consider the 10 cases 5, 4, 6, 3, 2, 9, 5, 4, 3, 4. We can, if we wish, break up these 10 cases into two subgroups of 5 each and calculate the average of the first group and the average of the second group. These would be ordinary averages. If we applied moving averages, however, we would average the first 5; then drop the first figure, add the sixth, and average this new set of 5; next drop the second figure, add the seventh, and average this third set of 5; and so on. Numerically we would have

$$\bar{X}_1 = \frac{5+4+6+3+2}{5} = 4.0$$

$$\bar{X}_2 = \frac{4+6+3+2+9}{5} = 4.8$$

$$\bar{X}_3 = \frac{6+3+2+9+5}{5} = 5.0$$

etc.

[18] Ibid., p. 200.

The above are moving averages of 5 items. Moving ranges of 5 items would be the ranges of these same groups. Thus,

R_1 would be the range of 5, 4, 6, 3, 2, which is 4

R_2 would be the range of 4, 6, 3, 2, 9, which is 7

R_3 would be the range of 6, 3, 2, 9, 5, which is 7

etc.

Moving average and range charts are used to control current processes. They are set up from past data, however, in the same way as ordinary charts are. Thus we would first compute the grand average of the past data and the average range of individual group ranges, as was done in Chapter 21. Then we would derive the central line and control limits in the ordinary way. When we came, however, to plot new data, for purposes of current control, we would plot moving averages and ranges.

The method of moving averages and ranges is particularly suitable for lines of production in which it takes some time to produce a single item. By adding to our "sample" each new item as it is produced and dropping the nth item previous, our moving average and moving range will give not only the usual control but also an up-to-date picture of the process.

A point out of control on a moving average or moving range chart has the same significance as a point out of control on an ordinary average or range chart. Since we are dealing with "moving averages" and "moving ranges," however, the points on the chart are not independent.[19] Hence several points in a row outside or near the control limits do not have the significance that they would on an ordinary chart on which the points are independent.

6. PROBLEMS

23.1. The results from the use of a cutting tool were as follows. All dimensions are the number of inches above 0.250 inches. Specification limits are 0.255 and 0.265 inches.

Time					
9 A.M.	0.005	0.006	0.006	0.007	0.008
10 A.M.	0.005	0.006	0.007	0.008	0.008
11 A.M.	0.007	0.007	0.008	0.008	0.009
12 M.	0.008	0.008	0.009	0.010	0.011
1 P.M.	0.009	0.010	0.010	0.010	0.012
2 P.M.	0.010	0.012	0.012	0.013	0.014
3 P.M.	0.013	0.013	0.014	0.014	0.015

[19] When a moving average of 5 items is used, points must be 5 intervals apart before they become independent.

Machine stopped and cutting edge ground

4 P.M.	0.005	0.005	0.006	0.007	0.008
5 P.M.	0.006	0.006	0.007	0.007	0.008
6 P.M.	0.006	0.007	0.008	0.009	0.010
7 P.M.	0.007	0.007	0.008	0.009	0.011
8 P.M.	0.008	0.009	0.010	0.010	0.012
9 P.M.	0.010	0.010	0.011	0.012	0.013
10 P.M.	0.012	0.012	0.013	0.014	0.015

Machine stopped and cutting edge ground

11 P.M.	0.005	0.006	0.007	0.008	0.009
12 P.M.	0.006	0.006	0.007	0.007	0.009
1 A.M.	0.006	0.007	0.008	0.009	0.009
2 A.M.	0.007	0.008	0.008	0.008	0.010
3 A.M.	0.009	0.009	0.009	0.010	0.011
4 A.M.	0.010	0.010	0.011	0.011	0.012
5 A.M.	0.011	0.011	0.012	0.014	0.015

From these results set up an R-chart and an $\bar{X}$ acceptance control chart. Assume that no nonconforming product will be tolerated.

23.2. A sample of 5 items is taken from a process every hour with the following results:

Sample	$\bar{X}$	R
1	0.7540	0.0011
2	0.7542	0.0014
3	0.7542	0.0009
4	0.7546	0.0010
5	0.7550	0.0008
Machine readjusted		
6	0.7539	0.0009
7	0.7541	0.0012
8	0.7543	0.0011
9	0.7547	0.0007
10	0.7549	0.0015
Machine readjusted		
11	0.7541	0.0017
12	0.7542	0.0010
13	0.7545	0.0011
14	0.7548	0.0009
15	0.7551	0.0012

Specifications are 0.7538 and 0.7553. Set up an R-chart. Take the RPL = 0.05 and $\beta = 0.10$ and derive control limits for an $\bar{X}$ acceptance control chart. Assume X is normally distributed.

23.3. Consider the last 50 cases of Table 3.2 as a random sample from a large lot and apply the lot plot method to a study of the data. If the specification limits are 0.820 and 0.840 inches, what would be your recommendation regarding the disposition of the lot?

23.4. Consider the first 50 cases of Problem 3.3*a* (i) as a random sample from a large lot and apply the lot plot method to a study of the data. If the lower specification on tensile strength is 80 pounds, what would be your recommendation regarding the disposition of the lot?

23.5. Consider the first 50 cases of Problem 21.4 as a random sample from a large lot and apply the lot plot method to a study of the data. If the specification limits on the dimension are 0.7500 and 0.7520 inches, what would be your recommendation regarding the disposition of the lot?

23.6. A sample of 15 is taken from a large lot, the fraction nonconforming of which is 0.28. Use BIPP to determine the probability of getting 3 or less nonconforming items in the sample. Check your answer by numerical calculations or the use of binomial tables.

23.7. A sample of 100 is taken from a large lot, the fraction nonconforming of which is 0.08. Use BIPP to determine the probability of getting 5 or less nonconforming items in the sample. Check your results with Figure 4.9 or Molina's tables.

23.8. Check the results of Figure 19.8 by using BIPP.

23.9. Use BIPP to work Problem 19.3.

23.10. Use BIPP to work Problem 19.5.

7. SELECTED REFERENCES*

Burr (B '76), Cowden (B '57), Craig (P '53), Dudding and Jennett (B '42), Enrick (P '45), Freund (P '57), Grant and Leavenworth (B '80), Grubbs (P '46), Grubbs and Weaver (P '47), Hicks (P '56), Hill (P '56), Indianapolis Section, ASQC (P '52), Manuele (P '45), Moses (P '56), Mosteller and Tukey (P '49), Satterthwaite (P '56 I.Q.C.), Shainin (P '50), Winterhalter (P '45), and Woods (P '76).

---

* B and P refer to the Book and Periodical sections, respectively, of the Cumulative List of References in Appendix V.

Part 5

Some Statistics Useful in Industrial Research

24

Estimation of Lot and Process Characteristics

1. POINT ESTIMATION

In previous discussions we have talked about estimation of a lot or process average or a lot or process standard deviation without much theorizing as to how this should be done. We shall undertake here a brief discussion of statistical theories of estimation.

1.1. Posterior Probability

One of the oldest methods of estimation is to use Bayes' theorem to compute posterior probabilities and to adopt the estimate that has the maximum posterior probability. Let us explain the procedure with reference to a simple industrial example.

Let us suppose that five-tenths of the time a manufacturer's process turns out 7 percent nonconforming units; two tenths of the time it turns out 6 percent nonconforming units; and three tenths of the time it turns out 5 percent nonconforming units. You receive a shipment from the manufacturer and inspect a sample of 100 items. Six are found nonconforming. What is the grade of the material which the manufacturer sent you? Is it the 7 percent grade, the 6 percent grade, or the 5 percent grade? To answer this, you note that, if it is the 7 percent grade, the probability of getting 6 nonconforming units out of 100 is[1] 0.1490; if it is the 6 percent grade, the probability

[1] Computed from E. C. Molina, *Poisson's Exponential Binomial Limit.*

of getting 6 nonconforming units is 0.1606; and if it is the 5 percent grade, the probability of getting 6 nonconforming units is 0.1462. Hence the posterior probabilities of the various grades are as follows:[2]

For the 7 percent grade, (0.5)(0.1490)/0.15048 = 0.495

For the 6 percent grade, (0.2)(0.1606)/0.15048 = 0.213

For the 5 percent grade, (0.3)(0.1462)/0.15048 = 0.291

The 7 percent grade has the highest posterior probability so this will be the estimate of lot quality.

Let us consider this same problem in a more general framework. If a lot has a fraction nonconforming p, then the probability that a sample of n items will contain exactly X nonconforming items is for relatively large lots, approximately

(24.1)
$$P(X|p) = \frac{n!}{X!(n-X)!} p^X (1-p)^{n-X}$$

Suppose now that p is itself a random variable that has a beta distribution, in other words, its density function is[3]

(24.2)
$$f(p) = \frac{(\alpha + \beta + 1)!}{\alpha!\beta!} p^{\alpha}(1-p)^{\beta}$$

Then the joint probability (1) that for a given lot p falls in the interval p to $p + dp$ and (2) that a sample of n taken from the lot contains X nonconforming items is

$$P(X, p) = P(X|p)f(p)\, dp$$

According to Bayes' Theorem, the posterior probability that the p that produced the given X lies in the interval p to $p + dp$ is

(24.3)
$$P(p|X) = \frac{P(X|p)f(p)\, dp}{\int_0^1 P(X|p)f(p)\, dp}$$

Inserting (24.1) and (24.2) in (24.3) and adjusting the integral gives

---

[2] The total probability of 6 nonconforming units is (0.5)(0.1490) + (0.2)(0.1606) + (0.3)(0.1462) = 0.15048. For discussion of posterior probability, see Chapter 2, Section 9.

[3] See A. M. Mood and F. A. Graybill, *Introduction to the Theory of Statistics* (2d ed.; New York: McGraw-Hill, 1963), p. 129. Since the quality is assumed to vary from time to time, the prime notation is not used here. It will be noted that if α is set equal to $s - 1$ and β to $t - 1$, we will obtain equation 14.5 of Chapter 14 above. Different notations are used in the two cases to facilitate the use of the reference given in each instance.

$$P(p|X)=\frac{\dfrac{n!}{X!(n-X)!}p^X(1-p)^{n-X}\dfrac{(\alpha+\beta+1)!}{\alpha!\beta!}p^\alpha(1-p)^\beta dp}{\dfrac{n!}{X!(n-X)!}\dfrac{(\alpha+\beta+1)!(\alpha+X)!(\beta+n-X)!}{\alpha!\beta!\qquad(\alpha+\beta+n+1)!}}$$

(24.4)

$$\int_0^1\frac{(\alpha+\beta+n+1)!p^{\alpha+X}(1-p)^{\beta+n-X}}{(\alpha+X)!(\beta+n-X)!}dp$$

$$=\frac{(\alpha+\beta+n+1)!}{(\alpha+X)!(\beta+n-X)!}p^{\alpha+X}(1-p)^{\beta+n-X}dp,\ \text{since the integral}=1$$

This is recognized as another beta distribution. The estimate of lot quality p that maximizes this posterior probability is the same as the value of p that maximizes the log of its density function. Thus, setting the derivative of the log equal to zero yields[4]

$$\frac{d}{dp}[(\alpha+X)\text{ lot }\hat{p}+(\beta+n-X)\log(1-\hat{p})]$$

$$=(\alpha+X)/\hat{p}-(\beta+n-X)/(1-\hat{p})=0$$

or

$$(\alpha+X)(1-\hat{p})=(\beta+n-X)\hat{p}$$

or

(24.5) $$\hat{p}=\frac{\alpha+X}{\alpha+\beta+n}$$

This result shows clearly how the "prior distribution" affects the results. If α and β are both small relative to X and n, $\hat{p}$ does not differ much from X/n. If $\alpha=\beta=0$, which are the parameters for a uniform prior distribution, $\hat{p}=X/n$. A uniform prior distribution means that all values of p between 0 and 1 are equally likely. This is the prior distribution we

[4] Although this book has not undertaken extensive discussions of general cost minimizing procedures, it should be pointed out here that modern Bayesian estimates are not the same as these maximum posterior probability estimates. In current "Bayesian" thinking reference is made to a loss function $l(\hat{\theta},\theta)$, which indicates how much is lost if we choose $\hat{\theta}$ as an estimate when in fact the true parameter value is θ. If $P(\theta_i|E)$ are the posterior probabilities, the Bayesian procedure is to chose $\hat{\theta}$ such that the "risk"

$$\sum_i l(\hat{\theta},\theta_i)P(\theta_i|E)$$

is a minimum. Cf. A. M. Mood and F. A. Graybill, *Introduction to the Theory of Statistics,* p. 187. Cf. also discussion of minimum cost Bayesian sampling plans in Section 2 of Chapter 14 above.

are likely to assume when we are completely ignorant of how p varies since it "distributes our ignorance equally."

In some instances prior distributions that are used to compute posterior probabilities are distributions of actual occurrence. More often, however, they represent our judgment about a situation. In these cases the posterior probabilities are no longer purely objective, but become partly subjective in character.

1.2. The Method of Maximum Likelihood

Because we do not often have positive knowledge regarding the distribution of prior probabilities, other methods of estimation have been developed that do not depend on this. One of the most popular of these is the method of maximum likelihood. This method says that we should take as our estimate of a universe parameter the value that maximizes the probability of producing the sample result. For example, a lot the fraction nonconforming of which is 0.15 has a much smaller chance of producing a sample with a fraction nonconforming of 0.086 than has a lot the fraction nonconforming of which is 0.10. The lot that would have the greatest chance of producing a sample with a fraction nonconforming of 0.086 is a lot the fraction nonconforming of which is[5] also 0.086. That the sample statistic is the maximum likelihood estimate of the corresponding universe parameter is frequently but not always the case. For example, the mean and the root-mean-square-deviation of a sample from a normal universe are joint maximum likelihood estimates of the mean and standard deviation of the universe but the sample mean and sample standard deviation are not joint maximum likelihood estimates. (See Problem 24.29.)

The method of maximum likelihood is favored by many statisticians because the estimates it yields have good properties. Thus,

1. Maximum likelihood estimates are *consistent*. More precisely, the larger the sample, the less the probability that the estimate deviates from the universe value by any predetermined amount.

---

[5] For relatively large lots, the probability of getting X nonconforming items out of n is approximately

$$P\left(\frac{X}{n}\right) = \frac{n!}{X!(n-X)!} p'^{X}(1-p')^{n-X}$$

(see Chapter 5, Section 1.1.1). This will be a maximum when $\log P\left(\frac{X}{n}\right)$ is a maximum. Hence for a given n and X, the value of p' that will maximize $P\left(\frac{X}{n}\right)$ is the value that maximizes $\log P\left(\frac{X}{n}\right)$. But $\log P\left(\frac{X}{n}\right)$ will be a maximum for p' if $\frac{d}{dp'}\log P\left(\frac{X}{n}\right) = 0$ or if $\frac{X}{p'} - \frac{n-X}{1-p'} = 0$, or if $(1-p')X - (n-X)p' = 0$ or if $p' = \frac{X}{n}$. Hence, $\frac{X}{n}$ is the maximum likelihood estimate of p'.

2. Maximum likelihood estimates are approximately *normally distributed* when the samples are large.

3. Maximum likelihood estimates have *maximum efficiency* in large samples. More precisely, no other estimate which is approximately normally distributed in large samples has a sampling variance which is less than that of the maximum likelihood estimate. For example, the mean of a sample from a normal universe is the maximum likelihood estimate of the mean of the universe. The median of the sample could also be used, and it is approximately normally distributed in large samples. The sampling variance of the median, however, is about 1.57 times that of the mean. The median of a sample from a normal universe is thus said to have an efficiency in large samples of about 64 percent. The maximum likelihood estimate (in this case, the mean) has the smallest possible sampling error in large samples and always has 100 percent efficiency in large samples.

4. Maximum likelihood estimates are "*sufficient.*" This means that no other estimate that is based on the same data and is independent of the maximum likelihood estimate is able to yield any further information about the universe parameter to be estimated. In more technical language, if $\hat{\theta}$ is used to estimate the universe parameter θ, and θ^* is any other statistic not a function of $\hat{\theta}$, then if for each of the statistics θ^*, the conditional distribution of θ^* given $\hat{\theta}$ does not contain θ, the statistic $\hat{\theta}$ is said to be a sufficient statistic. Maximum likelihood estimates are sufficient statistics or are at least functions of the minimal sufficient statistic. This property is valid for any sample size.

5. Finally, if t is a maximum likelihood estimate of the universe parameter T, it is generally true that t^2 is a maximum likelihood estimate of T^2, $\sqrt{t}$ is a maximum likelihood estimate of $\sqrt{T}$, and, in general, $F(t)$ is a maximum likelihood estimate of $F(T)$. Because of these good properties, maximum likelihood estimates are often called optimum estimates or best estimates.

Maximum likelihood estimates have the disadvantage that they are not always easy to find mathematically. Their derivation also depends on knowledge of the mathematical form of the distribution of the universe, e.g., whether it is normal or not. Maximum likelihood estimates are also not always unbiased, although multiplication by a simple constant may often make them unbiased.

1.3. Unbiased Estimates

The property of an estimate being unbiased is viewed as so important that it is often made a basis for estimation. An unbiased estimate of a universe parameter is one whose mean value, in an infinitely large set of samples, is the universe parameter.[6] For example, it was seen in Chapter 6 that the

[6] That is, the estimate is "mean-unbiased." G. W. Brown, "On the Small Sample Estimation," *Annals of Mathematical Statistics,* (1947), pp. 582–85, suggests that a "median-unbiased" estimate might be better than the usual "mean-unbiased" estimate.

mean of sample means is the mean of the universe, and that the mean of sample variances is the variance of the universe.[7] For that reason, the sample mean is said to be an unbiased estimate of the universe mean, and the sample variance an unbiased estimate of the universe variance. On the other hand, it was seen that the mean value of sample standard deviations is not the standard deviation of the universe.[8] The standard deviation of a sample is therefore a "biased estimate" of the universe standard deviation. If s is divided by the c_4 factor, however, we get an unbiased estimate of σ'. It is thus seen that if t^2 is an unbiased estimate of T^2, t is not necessarily an unbiased estimate of T. This is one of the main differences between maximum likelihood estimates and unbiased estimates. It should also be noticed that the property of being unbiased usually is independent of the sample size n, unless it is stated to be otherwise.

In statistical quality control, unbiased estimates are more frequently used than are maximum likelihood estimates. Unbiased estimates of important process values are shown in the accompanying tabulation:

Process Value		*Unbiased Estimate*
p'		p
$\bar{X}'$		$\bar{X}$
σ'	Single sample	s/c_4
	Group of k samples all of same size	$\frac{1}{c_4}\frac{s_1+s_2+\cdots+s_k}{k}$
		$\bar{R}/d_2$
σ'^2		s^2

We sometimes speak of best unbiased estimates. We usually mean by this that they have minimum sampling variance for any size sample and are thus better than other unbiased estimates. For example,

$$\frac{1}{c_4}\frac{s_1+s_2+\cdots+s_k}{k}$$

is a better estimate of σ' than $\bar{R}/d_2$. The reason why the latter is used so often is that it is easier to work with and the difference in efficiency is not great if the sample is small, say 10 or less.

2. CONFIDENCE INTERVALS[9]

Although we often want a single "best" estimate of some lot or process value, it is also frequently important to determine an interval that will have a high probability of including the universe value. For example, we may

[7] See Chapter 6, Sections 1.2 and 2.3.

[8] See Chapter 6, Section 2.5.

[9] Based largely on S. S. Wilks, *Elementary Statistical Analysis* (Princeton, N.J.: Princeton University Press, 1943), chap. X.

find that a random sample of 100 items from a given lot contains 15 nonconforming items. From this result, what can we say about the fraction nonconforming in the lot? Would the interval from 14 to 16 percent have a high probability of including the lot value? Or should we make it 13 to 17 percent or 10 to 20 percent? Again a sample of 20 rods from a given manufacturing process has an average length of 4.504 inches and a standard deviation of 0.0083 inches. What can we say in general about the average length of rods turned out by the process, assuming that the process is operating in a random manner?[10] Is there a high probability that the interval 4.500 to 4.508 includes the process value? These questions are answered by the theory of confidence intervals.

2.1. Confidence Intervals for the Universe Mean when σ' Is Known

2.1.1. Determination of 0.95 Confidence Intervals. Let us consider first a problem concerned with a mean value. Let a random sample of n values have a mean of $\bar{X}$ and assume that the standard deviation of the universe from which the sample was taken is known to have the value σ'. What interval can we set up that will have a 0.95 chance of including the unknown universe mean?

To solve this problem, consider Figure 24.1. The ordinate of this figure is the value of the universe mean ($\bar{X}'$). The abscissa is the value of the sample mean ($\bar{X}$). A point plotted on this chart represents a set of $\bar{X}',\bar{X}$ values. The line bisecting the angle between the axes is the line on which $\bar{X}'$ and $\bar{X}$ have equal values.

Now suppose that the universe mean is $\bar{X}'_1$. Then the means of all possible samples of n from the universe will be distributed around $\bar{X}'_1$ as an average, with a standard deviation equal to $\sigma'/\sqrt{n}$, where σ' is the standard deviation of the universe.[11] The distribution will be normal or approximately normal in form.[12] Hence, if we lay off $1.96\sigma'/\sqrt{n}$ to the left and right of $\bar{X} = \bar{X}_1$, we will get the points a_1 and b_1. It can thus be said that, if $\bar{X}' = \bar{X}'_1$, the probability that a sample mean $\bar{X}$ will lie between a_1 and b_1 is[13] 0.95. Similarly, if the universe mean were $\bar{X}'_2$, we can say that the chance is 0.95 that a sample mean of n cases will lie between a_2 and b_2, where $a_2 = \bar{X}_2 - 1.96\sigma'/\sqrt{n}$ and $b_2 = \bar{X}_2 + 1.96\sigma'/\sqrt{n}$. It will be noticed that the loci of all such points as a_1 and a_2 will be a straight line parallel to the bisecting line OD and at a distance $1.96\sigma'/\sqrt{n}$ to the left of this line. Similarly, the loci of all such points as b_1 and b_2 will be a straight line

---

[10] That is, the process is under control; see Chapter 2, Section 5.

[11] See Chapter 6, Section 1.

[12] Ibid.

[13] See Table A2 of Appendix II.

FIGURE 24.1
Illustration of How the Distribution of Sample Means Varies with the Universe Mean

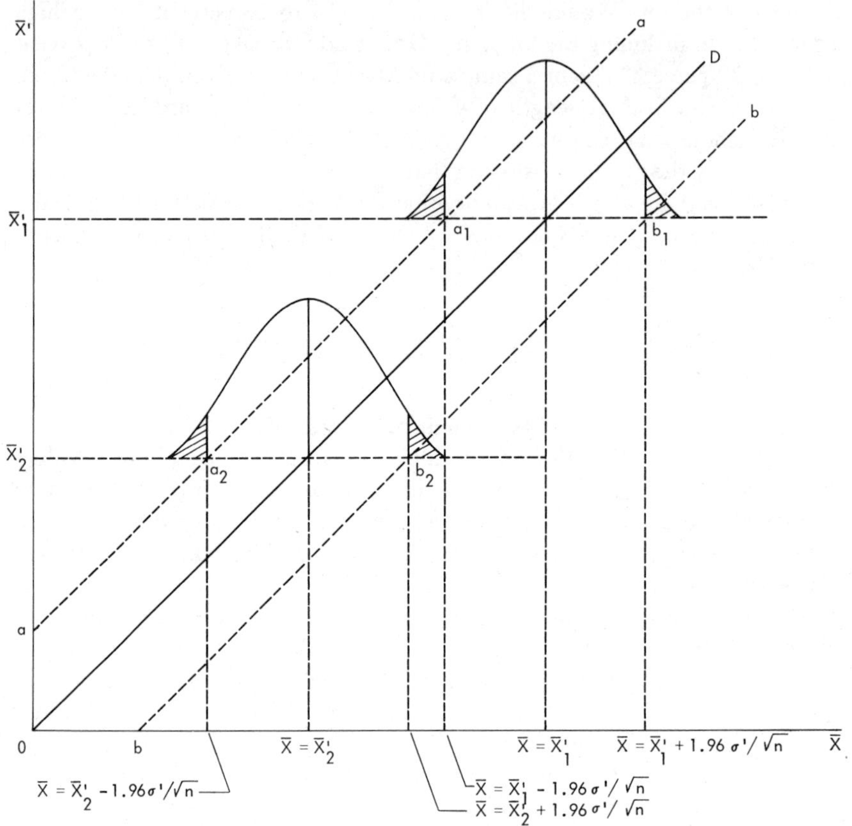

parallel to OD and at a distance of $1.96\sigma'/\sqrt{n}$ to the right of OD. Figure 24.1 shows these two loci.

It will be argued now that, if we know the mean of a sample $(\bar{X})$ and know the universe standard deviation (σ'), we can find an interval that will have a 0.95 chance of including the universe mean $\bar{X}'$. This will be called a 0.95 confidence interval for $\bar{X}'$. Such an interval can be found by adding and subtracting from $\bar{X}$ the quantity $1.96\sigma'/\sqrt{n}$. Suppose, for example, that we have a sample mean equal to $\bar{X}_3$ and that the mean of the universe from which this came is $\bar{X}'_3$, i.e., we have the point P_3 in the Figure 24.2. The procedure just described will be the equivalent of moving up and down from P_3 to the points a_3 and b_3 on the a and b loci and then finding the ordinates of these a and b points. The ordinate for a_3 will give the upper confidence limit for $\bar{X}'_3$, and the ordinate for b_3 will give the lower confidence limit for $\bar{X}'_3$.

In the case $\bar{X} = \bar{X}_3$, it will be seen that the confidence interval includes

FIGURE 24.2
Illustration of How Confidence Limits for $\bar{X}'$ Are Determined

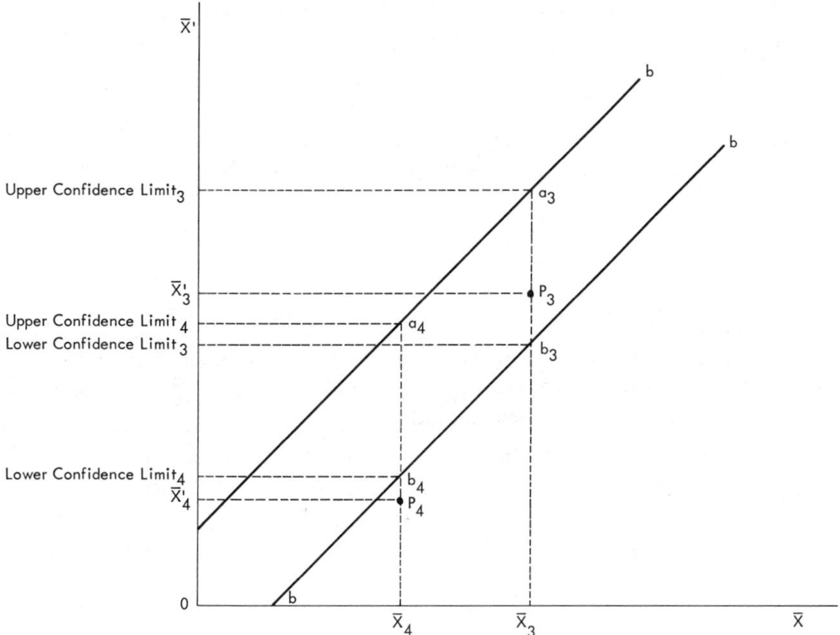

the universe mean $\bar{X}'_3$. Consider, now, the case in which the sample mean has the value $\bar{X}_4$ and the mean of the universe from which this sample came is $\bar{X}'_4$, i.e., we have the point P_4. If we move vertically from P_4, we will hit the a and b loci at a_4 and b_4, and when we find the ordinates of these a and b points we get (upper confidence limit)$_4$ and (lower confidence limit)$_4$. These are the confidence limits for $\bar{X}'_4$ based on the sample mean $\bar{X}_4$. It will be noticed that this second confidence interval does *not* include the universe mean $\bar{X}'_4$.

From the foregoing illustrations it should be clear that our confidence intervals will include the universe value whenever we get points P (i.e., combinations of $\bar{X}'$ and $\bar{X}$ values) that lie on or between the a and b loci, and they will not include the universe value whenever the P points lie outside these loci. But, from the construction of the diagram, for any given value of the universe mean $\bar{X}'$, we shall secure points outside a and b loci only 5 times out of 100. Hence, 95 times out of 100, samples drawn from this universe will lead to confidence intervals that will include the universe value. Since this applies to any given universe mean, it applies to the procedure in general. If this procedure is followed, 95 percent of the time our confidence intervals will include the universe value.

The foregoing argument was geometrical in character. An algebraic formulation is equally simple and has the advantage that it can easily be applied to many other types of problems. This algebraic formulation is as follows:

If the universe mean is $\bar{X}'$, the probability that the sample mean will lie between $\bar{X}' \pm 1.96\sigma'/\sqrt{n}$ is equal to 0.95. This may be written:

$$\text{(I)} \qquad \text{Prob. } (\bar{X}' - 1.96\sigma'/\sqrt{n} \leq \bar{X} \leq \bar{X}' + 1.96\sigma'/\sqrt{n}) = 0.95$$

But this is the equivalent of saying that the probability that $\bar{X} + 1.96\sigma'/\sqrt{n}$ lies above $\bar{X}'$ and $\bar{X} - 1.96\sigma'/\sqrt{n}$ lies below $\bar{X}'$ equals 0.95, or, symbolically,

$$\text{(II) Prob. } (\bar{X}' \leq \bar{X} + 1.96\sigma'/\sqrt{n} \quad \text{and} \quad \bar{X} - 1.96\sigma'/\sqrt{n} \leq \bar{X}') = 0.95$$

The last inequalities can be combined to read as follows:

$$\text{(III)} \qquad \text{Prob. } (\bar{X} - 1.96\sigma'/\sqrt{n} \leq \bar{X}' \leq \bar{X} + 1.96\sigma'/\sqrt{n}) = 0.95$$

The argument is that statement III is true if statement I is. Statement III, however, is the confidence interval statement. For if $\bar{X} - 1.96\sigma'/\sqrt{n}$ is taken as the lower limit of the confidence interval and $\bar{X} + 1.96\sigma'/\sqrt{n}$ is taken as the upper limit, then statement III merely says that the interval so determined has a 0.95 chance of including the universe value.

To illustrate the calculation of a confidence interval for $\bar{X}'$: Let $n = 16$, $\bar{X} = 0.8320$, and $\sigma' = 0.0030$. Then a 0.95 confidence interval for $\bar{X}'$ is

$$0.8320 \pm 1.96\frac{(0.0030)}{\sqrt{16}} = \begin{cases} 0.83053 \\ 0.83347 \end{cases}$$

2.1.2. Intervals with Other Confidence Coefficients. The confidence coefficient can be any value. It need not be 0.95, but can be 0.90 or 0.99 or 0.999, just as we please. If this is changed, however, then the multiple of $\sigma'/\sqrt{n}$ that is added and subtracted from $\bar{X}$ must also be changed. For 0.90 confidence intervals, the multiple is 1.645; for 0.99 confidence intervals, it is 2.576; and for 0.999 confidence intervals, it is 3.291. These all follow from the fact that the distribution of sample means is normal or approximately normal.[14] They do not apply to statistics which do not have a normal distribution.

It will be noted that the higher the confidence coefficient, the wider the confidence interval. In other words, for a given size sample, greater confidence in our estimate can be bought only at the expense of less preciseness.

2.1.3. Single-Limit Confidence Intervals. The confidence intervals we have described have all been two-limit intervals, with an equal probability of making mistakes in either direction. Thus the 0.95 confidence interval we derived has a 0.025 chance of not including the universe value because the interval fell below it and a 0.025 chance of not including the universe

[14] If it is only approximately normal, the confidence coefficients yielded by the given multiples will not be exactly 0.95, 0.99, and so on.

value because the interval fell above it. Thus, in a probability sense, our confidence interval was a "balanced" interval.

It is perfectly possible to derive confidence intervals for which the probability of not covering the universe value in one direction is not the same as the probability of not covering the universe value in the other direction. An important class of this type of interval is a single limit interval. The limit may be either an upper or a lower limit. A 0.95 upper limit interval would be obtained by adding $1.645\sigma'/\sqrt{n}$ to the sample mean $\bar{X}$. Thus it can be said that the interval $-\infty$ to $\bar{X} + 1.645\sigma'/\sqrt{n}$ has a 0.95 chance of including the universe value. Likewise, it can be said that the interval $\bar{X} - 1.645\sigma'/\sqrt{n}$ to $+\infty$ has a 0.95 chance of including the universe mean $\bar{X}'$. This would be a 0.95 lower limit interval. When the investigator is interested in variation in one direction only, these single limit intervals are the most appropriate ones to use.

2.1.4. Confidence Intervals and the Size of the Sample. It is easily seen from the formulas for confidence intervals that the larger the sample size, the narrower the interval. In other words, the larger the sample, the better our estimate of the universe value. In more exact language it may be said that the preciseness of our estimate increases with $\sqrt{n}$.

2.2. Confidence Intervals for the Population Mean when σ' Is Unknown

The foregoing argument has been based on the assumption that we know the standard deviation of the universe. If we do not know the standard deviation of the universe, we can use the standard deviation of the sample if the sample is large—say, greater than 30. If the sample is less than 30, however, and σ' is unknown, we should use the t distribution in place of the normal distribution.[15]

For a 0.95 two-limit equal-probability confidence interval, the argument runs as follows: From Section 5 of Chapter 6 we know that the statistic $\sqrt{n}(\bar{X} - \bar{X}')/s$ is distributed in the form of a t distribution with $\nu = n - 1$. Hence,

$$\text{Prob.}\left(-t_{0.025} \leq \frac{\sqrt{n}(\bar{X} - \bar{X}')}{s} \leq t_{0.025}\right) = 0.95,$$

where $t_{0.025}$ is the 0.025 point of the t distribution for $\nu = n - 1$. But this may be written:

$$\text{Prob.}\ (\bar{X}' \leq \bar{X} + t_{0.025}s/\sqrt{n} \text{ and } \bar{X} - t_{0.025}s/\sqrt{n} \leq \bar{X}') = 0.95,$$

or

$$\text{Prob.}\ (\bar{X} - t_{0.025}s/\sqrt{n} \leq \bar{X}' \leq \bar{X} + t_{0.025}s/\sqrt{n}) = 0.95.$$

---

[15] See Chapter 6, Section 5.

Hence our confidence limits become $\bar{X} - t_{0.025}s/\sqrt{n}$ and $\bar{X} + t_{0.025}s/\sqrt{n}$, where $t_{0.025}$ is the 0.025 point of the t distribution for $\nu = n - 1$. It will be noted that $t_{0.025}$ is the value of t listed in Table B of Appendix II under the heading $P = 0.05$, since that table gives the probability of exceeding $\pm t$. These last confidence intervals are the ones to be used if n is small and the standard deviation of the universe is unknown. They are based on the assumption that the universe is normal but may be used with reasonable accuracy when the universe is moderately nonnormal.

As an example, suppose in Section 2.1.1. that σ' is unknown and all the information we have is that the s of the sample $= 0.0031$. Then a 0.95 confidence interval is given by

$$0.8320 \pm 2.131 \frac{(0.0031)}{\sqrt{16}} = \begin{cases} 0.83035 \\ 0.83365 \end{cases}$$

Note that absence of knowledge of σ' causes the interval to be slightly wider than when we knew its value. Compare Section 2.1.1.

2.3. Confidence Intervals for a Universe Variance or Standard Deviation

2.3.1. Small Samples from a Normal Universe. Confidence limits for a universe variance or standard deviation are most easily obtained when the universe is normal. For, in this case, we saw in Chapter 6, Section 2.3, that sample values of $\frac{(n-1)s^2}{\sigma'^2}$ form a χ^2 distribution with $\nu = n - 1$. Hence we have the following:

$$\text{Prob.}\left(\chi^2_{0.975} \le \frac{(n-1)s^2}{\sigma'^2} \le \chi^2_{0.025}\right) = 0.95$$

But this can be written:

$$\text{Prob.}\left(\sigma'^2 \le \frac{(n-1)s^2}{\chi^2_{0.975}} \quad \text{and} \quad \frac{(n-1)s^2}{\chi^2_{0.025}} \le \sigma'^2\right) = 0.95$$

or

$$\text{Prob.}\left(\frac{(n-1)s^2}{\chi^2_{0.025}} \le \sigma'^2 \le \frac{(n-1)s^2}{\chi^2_{0.975}}\right) = 0.95$$

Hence, $\frac{(n-1)s^2}{\chi^2_{0.025}}$ and $\frac{(n-1)s^2}{\chi^2_{0.975}}$ can be taken as the lower and upper 0.95 confidence limits for σ'^2, where $\chi^2_{0.025}$ and $\chi^2_{0.975}$ are the 0.025 and 0.975 points of the χ^2 distribution for $\nu = n - 1$.

To illustrate the procedure, suppose that the standard deviation of 10 tests of the tensile strength of a given grade of cotton yarn is 11 pounds. Then 0.95 confidence limits for the universe variance are as follows:

$$\text{Lower limit} = \frac{9(11)^2}{\chi^2_{0.025}(\nu=9)} = \frac{9(121)}{19.0} = 57.3 \text{ and}$$

$$\text{Upper limit} = \frac{9(11)^2}{\chi^2_{0.975}(\nu=9)} = \frac{9(121)}{2.70} = 403.3$$

and the limits for the universe standard deviation are $\sqrt{57.3}$ and $\sqrt{403.3}$, or 7.57 and 20.1. It will be noticed that these limits are not symmetrical around the sample value of 11, although the confidence limits are balanced in a probability sense.

2.3.2. Large Samples from a Normal Universe. When the universe is normal and n is large, say, >30, we can assume[16] that sample standard deviations are approximately normally distributed about the universe standard deviation as a mean, with a standard deviation (i.e., σ'_σ) equal approximately to $\sigma'/\sqrt{2n}$. Hence we have

$$\text{Prob. } (\sigma' - 1.96\sigma'/\sqrt{2n} \leq s \leq \sigma' + 1.96\sigma'/\sqrt{2n}) = 0.95$$

which may be written

$$\text{Prob.}\left(\sigma' \leq \frac{s}{(1-1.96/\sqrt{2n})} \quad \text{and} \quad \frac{s}{(1+1.96/\sqrt{2n})} \leq \sigma'\right) = 0.95$$

or

$$\text{Prob.}\left(\frac{s}{(1+1.96/\sqrt{2n})} \leq \sigma' \leq \frac{s}{(1-1.96/\sqrt{2n})}\right) = 0.95$$

Hence, confidence limits for σ' are[17]

$$\frac{s}{(1+1.96/\sqrt{2n})} \quad \text{and} \quad \frac{s}{(1-1.96/\sqrt{2n})}$$

For example, if in our previous illustration n had equaled 50 instead of 10, we would have had the following:

$$\text{Lower confidence limit} = \frac{11}{(1+1.96/\sqrt{100})} = 9.2$$

$$\text{Upper confidence limit} = \frac{11}{(1-1.96/\sqrt{100})} = 13.7$$

---

[16] See Chapter 6, Sections 2.4 and 2.5.

[17] R. A. Fisher suggests that we take $\sqrt{2\chi^2} - \sqrt{2\nu - 1}$ as being normally distributed with 0 mean and unit variance (cf. Chapter 6, Section 2.3). This yields the confidence limits

$$\frac{s}{\sqrt{1-\frac{3}{(2n-2)}} + 1.96/\sqrt{2n-2}} \quad \text{and} \quad \frac{s}{\sqrt{1-\frac{3}{(2n-2)}} - 1.96/\sqrt{2n-2}}$$

2.4. Confidence Intervals for a Universe Proportion or Fraction

2.4.1. A Diagram of Confidence Belts. Because of the discrete nature of the data, the loci demarking the various confidence belts for a universe proportion or fraction are not smooth lines, such as the loci for a universe mean shown in Figure 24.2, but are a set of unconnected points. Suppose, for example, that we have taken a sample of 10 from a large lot and have determined the fraction of nonconforming items in the sample. We wish to determine 0.95 confidence limits for the lot fraction nonconforming. Since only 11 different sample results are possible, only 11 different confidence intervals can be obtained. Figure 24.3 shows that, if we get 0 nonconforming items, then the confidence interval for p' is 0 to 0.31; if we get 1 nonconforming item, the confidence interval is 0 to 0.45; if we get 2 nonconforming items, the confidence interval is 0.03 to 0.56; and so on. In the diagram the various possible confidence intervals are shown by heavy dots. The general trend

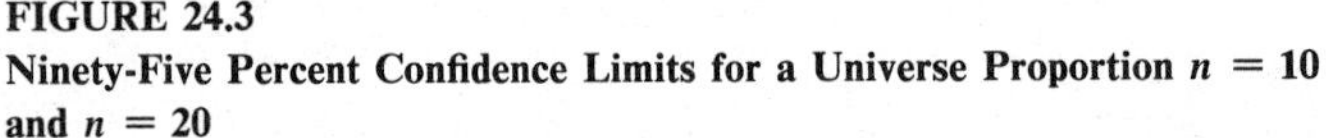

FIGURE 24.3
Ninety-Five Percent Confidence Limits for a Universe Proportion $n = 10$ and $n = 20$

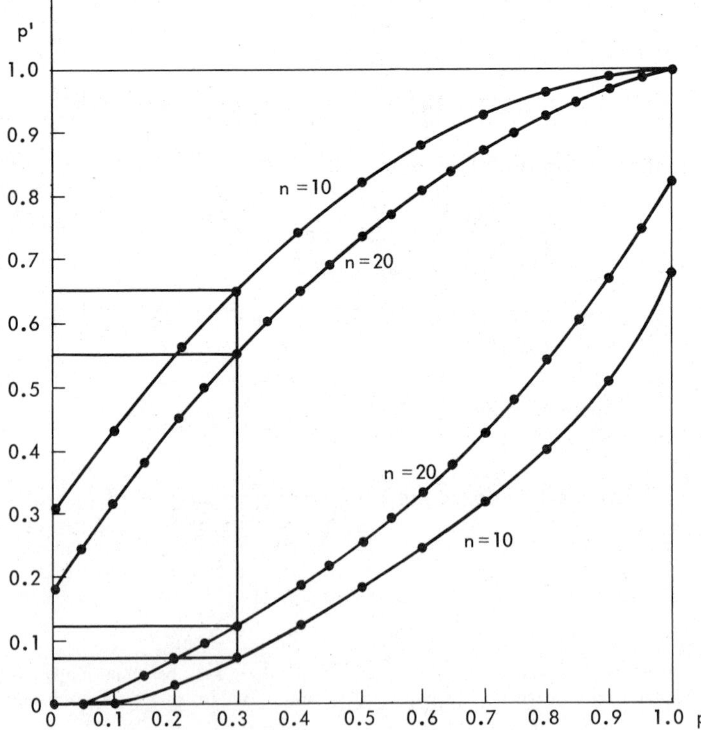

SOURCE: Based on Table 1.1 of George W. Snedecor, *Statistical Methods* (Ames, Iowa: Iowa State College Press, 1946). Also compare C. J. Clopper and E. S. Pearson, "The Use of Confidence or Fiducial Limits," *Biometrika;* 26 (1934), pp. 404–13.

of these dots is shown by a smooth curve drawn through the dots, but actually the dots themselves are the only points of any significance. If we had taken a sample of 20 instead of 10, there would have been 21 possible sample results and 21 possible confidence intervals. The dots in Figure 24.3 for $n = 20$ are therefore not so far apart as those for $n = 10$, but the dots themselves are still the only points of significance.

Figure 24.4 gives the smooth curves that trace out the confidence limits for p' for selected values of n from 10 to 1,000. The confidence coefficient associated with this diagram is 0.95. To use the diagram, note the position of the sample fraction on the abscissa scale and read off on the ordinate scale the values of the two limiting lines for the given value of n. For example, if X/n equals 0.3 and $n = 100$, then the confidence limits are 0.22 and 0.40. Figure 24.4 may be used whenever a rough determination of a confidence interval is all that is desired. When more accurate methods are judged desirable, one of the methods described in Sections 2.4.2 to 2.4.4 may be employed.

FIGURE 24.4
Chart for Determining 0.95 Confidence Limits for a Universe Proportion for Samples of Size 10 to 1,000

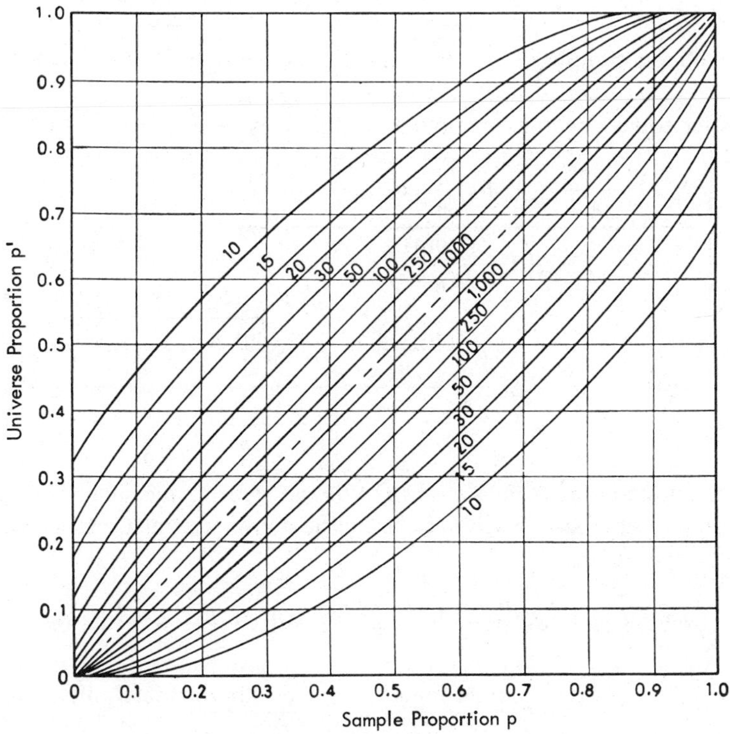

SOURCE: Reproduced with permission from Figure 4 of C. J. Clopper and E. S. Pearson, "The Use of Confidence or Fiducial Limits," *Biometrika*, 26 (1934), p. 410. More precise results can be obtained from the chart (Table 41) in Pearson and Hartley, *Biometrika Tables for Statisticians*, Vol. I, 2d ed.

2.4.2 Confidence Intervals Derived from Tables of the Binomial Distribution. If n is small and p is between 0.01 and 0.50, the National Bureau of Standards tables can be used to establish confidence limits for p'. Thus, if $n = 25$ and the sample $p = 0.08$ (i.e., 2 out of 25), then we can find the upper 0.95 confidence limit for p' by selecting the column $n = 25$, $X = 3$, of Table 2 of the binomial tables (actually the table column reads $n = 25$, $r = 3$), and moving down to the point at which the probability of 3 or more is close to 0.9750. The result is p' equals approximately 0.26. The lower 0.95 confidence limit is given by using the column $n = 25$, $X = 2$, and moving down to the point at which the probability of 2 or more is about 0.025. The result is approximately 0.01. Hence the confidence interval for p' is approximately 0.01 to 0.26.

2.4.3. Confidence Intervals Based on the Normal Approximation to the Binomial Distribution. When n is large and p' is not less than 10 percent, we have seen that the binomial distribution can be approximated reasonably well by a normal distribution with a mean equal to p' and a standard deviation equal to $\sqrt{\frac{p'(1-p')}{n}}$. When this is true, we can consider the sample p as being practically a continuous variable and can find confidence limits by the algebraic method employed in previous sections. Thus we may write

$$\text{Prob.}\left(p' - 1.96\sqrt{\frac{p'(1-p')}{n}} \leq p \leq p' + 1.96\sqrt{\frac{p'(1-p')}{n}}\right) \doteq 0.95$$

After some algebraic manipulation, this may be rewritten

$$\text{Prob.}\left\{\frac{2p + \frac{(1.96)^2}{n} - 1.96\sqrt{\frac{4p}{n}(1-p) - \frac{(1.96)^2}{n^2}}}{2\left(1 + \frac{(1.96)^2}{n}\right)} \leq p' \right.$$

$$\left. \leq \frac{2p - \frac{(1.96)^2}{n} + 1.96\sqrt{\frac{4p}{n}(1-p) - \frac{(1.96)^2}{n^2}}}{2\left(1 + \frac{(1.96)^2}{n}\right)}\right\} \doteq 0.95$$

If we drop the terms $(1.96)^2/n$ and $(1.96)^2/n^2$ as being too small to be given practical consideration when n is large, the above statement can be written in the simple form

$$\text{Prob.}\left(p - 1.96\sqrt{\frac{p(1-p)}{n}} \leq p' \leq p + 1.96\sqrt{\frac{p(1-p)}{n}}\right) \doteq 0.95$$

Hence, 0.95 confidence limits for p' are given approximately by adding to and subtracting from p the quantity $1.96\sqrt{\frac{p(1-p)}{n}}$.

For example, suppose that we have a sample of 75 and suppose that the number of nonconforming items in the sample is 15, i.e., $p = \frac{15}{75} = 0.20$. Then, 0.95 confidence limits for p' are

$$p' \doteq 0.20 \pm 1.96\sqrt{\frac{(0.2)(0.8)}{75}} \doteq \begin{cases} 0.29 \\ 0.11 \end{cases}$$

If BIPP is available,[18] confidence limits for a universe proportion can be readily obtained by plotting n_1 and n_2 as a triangle or point on the paper and then constructing two universe lines (two "splits") whose short distances from the triangle or point correspond to the two-sided level of confidence desired. Thus, if 0.95 confidence limits are required, short distances of 10 mm. should be used. The coordinates of the intersection of the two splits with the quarter-circle on the paper give the confidence limits desired.[19]

To illustrate, let us suppose that a sample of 300 contains 100 nonconforming items. To find a 0.95 confidence interval, we plot the triangle (100, 200), (101, 200), (100, 201), which can in this case be approximated by a point. (It could equally well have been plotted as 200, 100; but 100, 200 was used to prevent confusion with other lines in Figure 23.5.) We then draw arcs of circles of radius 10 mm. about the extreme vertices of the triangle or, in this case, simply draw a circle of radius 10 mm. about the sample "point." Finally, on each side, we draw tangents to these arcs through the origin. The intersections of the two tangents with the quarter-circle give on the abscissa scale (on the ordinate scale if the sample point had been plotted as 200, 100) the 0.95 confidence limits as 0.28 and 0.39 (see Figure 23.5).

2.4.4. Confidence Intervals Based on the Poisson Approximation to the Binomial Distribution. If p is small but n is large, the Poisson distribution may be used to set up confidence limits for p'. This can be done with the help of Figure 4.9, or from Molina's table of cumulated probabilities. From Figure 4.9 the upper 0.95 confidence limit may be found by entering the chart at the 0.025 ordinate and proceeding horizontally to the right until the curve X = sample number of nonconforming or other specified type of units is reached. The abscissa value of this point gives $p'n$, and $p'n$ divided by n gives the upper limit for p'. To get the lower 0.95 confidence limit, enter the chart at the 0.975 ordinate; proceed horizontally to X = sample number of nonconforming units minus 1, and read off $p'n$ on the abscissa scale. The $p'n$ divided by n will give the lower confidence limit for p'.

For example, suppose that $p = 0.006$ and $n = 2{,}700$, i.e., there were 16 nonconforming units out of 2,700. To get the upper 0.95 confidence limit for p', enter Figure 4.9 at the vertical ordinate 0.025, proceed to the curve $X = 16$, and read off on the abscissa scale $p'n = 26$. Then the upper limit for p' would be 26/2,700 or 0.0096. To get the lower 0.95 confidence limit

---

[18] See Chapter 23, Section 4.

[19] See F. Mosteller and J. W. Tukey, "The Uses and Usefulness of Binomial Probability Paper," *Journal of the American Statistical Association,* 44 (1949), p. 186.

for p', enter Figure 4.9 at the vertical ordinate 0.975, proceed to the curve $X = 15$, and read off on the abscissa scale $p'n = 9$. Then the lower limit would be 9/2,700 or 0.0033. The interval 0.0033 to 0.0096 would thus form a 0.95 confidence interval for the value of p', i.e., this interval may be said to have a probability of 0.95 of covering the universe value.

If the sample number of nonconforming units (X) is greater than 19, Figure 4.9 cannot be used for determining 0.95 confidence limits; or, if X is greater than 14, the figure cannot be used to determine 0.998 confidence limits or those with higher coefficients. In these cases, resort must be had to Molina's Table II.[20]

The use of Molina's tables for determining confidence limits can be best explained by means of an illustration. Suppose that a sample of 2,700 items from a large lot contains 22 nonconforming units. This would mean a sample fraction nonconforming of 0.0082. To establish an upper 0.95 confidence limit for the lot fraction nonconforming, enter Molina's Table II at the line $X = 23$ and go across the table until the probability of X or more nonconforming units is equal to 0.975. Actually, it is found that for $p'n$ (Molina's a) = 33, the probability of 23 or more is 0.9719, and for $p'n = 34$, it is 0.9809. Hence, by linear interpolation, the $p'n$ for a probability of 23 or more = 0.975 is equal to $33 + \dfrac{0.0031}{0.0090} = 33.3$. This $p'n$ divided by n, i.e., 33.3/2,700 = 0.0123, is the upper 0.95 confidence limit. To establish a lower 0.95 confidence limit for the true fraction nonconforming, enter Molina's Table II at the line $X = 22$ and go across the table until the probability of X or more nonconforming units is equal to 0.025. Actually, it is found that for $p'n = 13.7$ the probability of 22 or more is 0.0252. Linear interpolation yields $p'n = 13.79$. This $p'n$ divided by n, i.e., 13.79/2,700 = 0.0051, is the lower 0.95 confidence limit. Hence the whole confidence interval runs from 0.0051 to 0.0123, and this interval has a chance of 0.95 of including the universe value.

2.4.5. Confidence Intervals when the Universe Is Relatively Small. If the universe is relatively small, so we have to use the hypergeometric distribution to measure probabilities,[21] but if both the sample and universe are large enough to use the normal distribution as an approximation to the hypergeometric distribution, then we can determine confidence limits for p' in much the same way as with the normal approximation to the binomial distribution. The upper and lower 0.95 confidence limits will be given approximately by

$$p'_U = p + 1.96\sqrt{\frac{p(1-p)}{n}\left(1-\frac{n}{N}\right)}$$

and

[20] Or Table E of Appendix II. If $X > 30$, the normal distribution may be the easiest to use.

[21] See Chapter 4, Section 5.3.

$$p'_L = p - 1.96\sqrt{\frac{p(1-p)}{n}\left(1-\frac{n}{N}\right)}$$

when N is the size of the universe.

To illustrate, suppose that a sample of 1,000 is taken from a lot of 3,000 parts and suppose 12 percent of the sample, i.e., 120 parts, is found to be nonconforming. Then the lower 0.95 confidence limit for the lot fraction nonconforming is approximately

$$\begin{aligned} p'_L &= 0.120 - 1.96\sqrt{\frac{0.12(0.88)}{1{,}000}\left(1-\frac{1{,}000}{3{,}000}\right)} \\ &= 0.120 - 1.96(0.0084) \\ &= 0.104 \end{aligned}$$

and the upper 0.95 confidence limit is approximately

$$\begin{aligned} p'_U &= 0.120 + 1.96\sqrt{\frac{0.12(0.88)}{1{,}000}\left(1-\frac{1{,}000}{3{,}000}\right)} \\ &= 0.120 + 1.96(0.0084) \\ &= 0.136 \end{aligned}$$

The interval 0.104 to 0.136 may thus be said to have a 0.95 chance of covering the lot fraction nonconforming.[22]

If the sample is a large proportion of the universe, but both the sample and universe are small, say 100 or less, we can use the Lieberman-Owen *Tables of the Hypergeometric Distribution* to find a confidence interval for p'. Suppose, for example, that we had a sample of 40 from a lot of 100 and found 4 nonconforming items in the sample. Search in the Lieberman-Owen tables under a sample size of 40 from a universe of 100 reveals that a universe with 4 nonconforming items will yield a probability of 0.023 of getting exactly 4 in the sample. (Usually we would look for 4 or more, but in this case there are only 4 in the universe.) Search also shows that if the universe contains 21 nonconforming items, the probability of getting 4 or less in the sample is 0.023. Hence 0.954 confidence limits for the universe fraction nonconforming would be 0.04 to 0.21.

2.4.6 The Stated Confidence Coefficient as a Lower Limit. Because of the discrete nature of the binomial and hypergeometric distributions, the confidence coefficient of 0.95 used in the preceding paragraphs is a lower limit for the actual confidence coefficient involved. If we refer back to Figure 24.3, it will be noted that, when $p' = 0.30$ and $n = 20$, confidence intervals derived from this diagram will include the universe value if $2 \le X \le 10$, and the probability of this is 0.975. Again, if $p' = 0.50$ and $n = 20$, confidence

[22] Since both the sample and the universe are fairly large in this case, the limits differ little from those yielded by the binomial distribution.

intervals derived from the diagram will include the universe value if $6 \leq X \leq 14$, and the probability of this is 0.959.

2.4.7. The Quality Measurement Plan (QMP). A method for estimating the quality of current output that is of special interest is the Quality Measurement Plan (QMP) developed by Bruce Hoadley of the Bell Telephone Laboratories. A comprehensive discussion of the plan is beyond the scope of this book. For full details the interested reader is referred to Hoadley's paper on the subject in the *Bell System Technical Journal* 60, No. 2, February

FIGURE 24.5
A QMP Location Summary

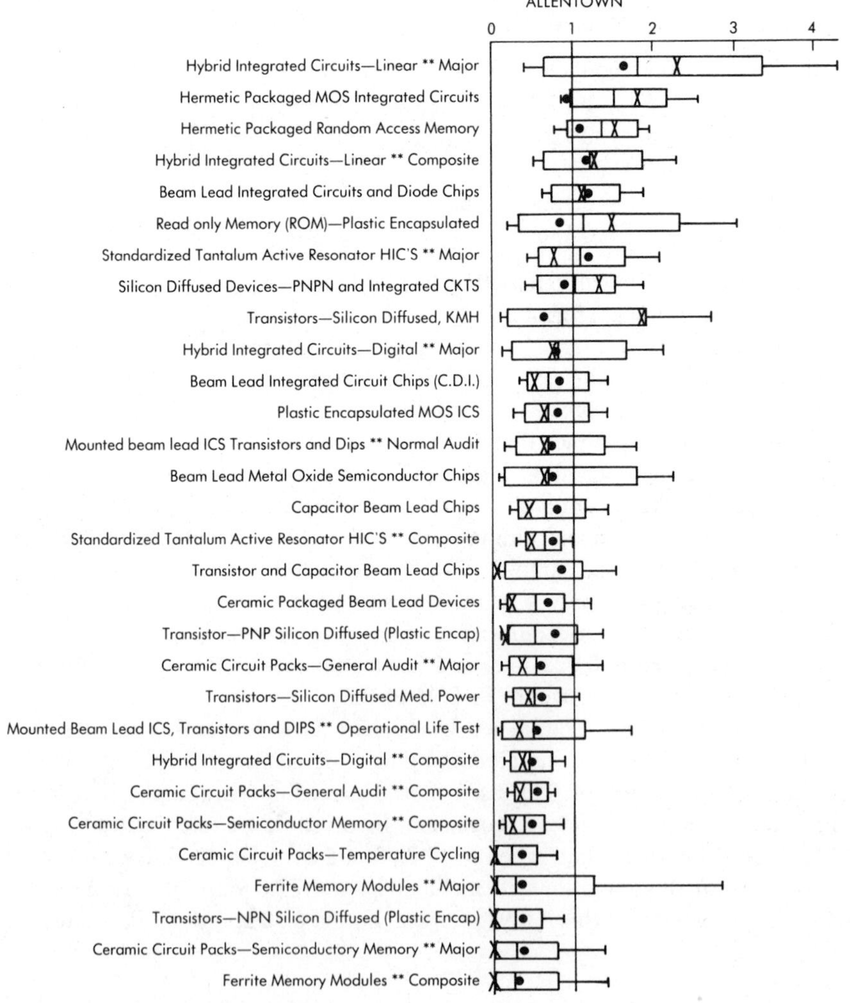

SOURCE: Reproduced with permission from Bruce Hoadley, "The Quality Measurement Plan (QMP)," *The Bell System Technical Journal* 60 (1981), Fig. 6, p. 237.

1981, pp. 215–73. We will simply note here that a QMP presents confidence intervals for the quality of current items of production that are, "computed from current and past data and are derived from a new Bayesian approach to the empirical Bayes problem for Poisson observations"[23] —a "Bayes-empirical Bayes procedure," as it is sometimes called. The confidence intervals are presented as a set of box-and-whisker plots[24] as illustrated in Figure 24.5, which is taken from Hoadley's paper, p. 237. The key for a QMP box-and-whisker plot for an individual item is shown in Figure 24.6, which

FIGURE 24.6
Key to a QMP Box Plot

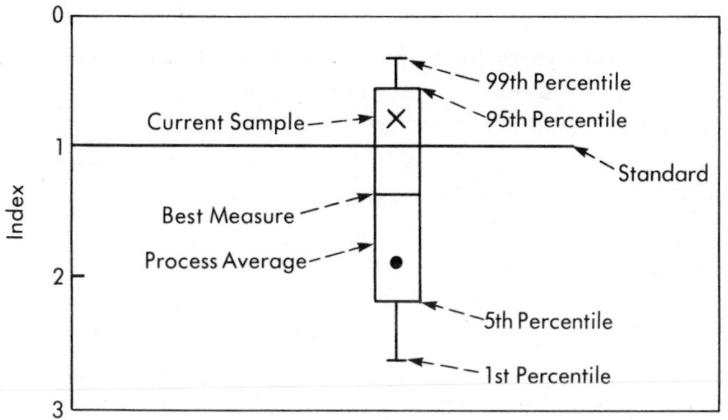

SOURCE: Adapted with permission from Bruce Hoadley, "The Quality Measurement Plan (QMP)," *The Bell System Technical Journal* 60 (1981), Fig. 4, p. 235.

is based on Hoadley's Figure 4. With respect to this key, Hoadley writes that a QMP box-and-whisker plot is

> a graphical representation of the posterior distribution for current production given the six most recent periods of audit data. The whiskers display the 99th and 1st percentiles and the box displays the 95th and 5th percentiles. The Best Measure is the posterior mean or Bayes estimate. It is a weighted average of the process average ("dot") and the current sample ("X"). The weight is the ratio of sampling variance to total variance. If all the variance is due to sampling, then the production is stable and the process average is the Best Measure of current quality. If the sampling variance is zero, then the current sample is the Best Measure.[25]

---

[23] Bruce Hoadley, "The Quality Measurement Plan (QMP)," *Bell System Technical Journal,* 60, No. 2, February 1981, p. 215.

[24] See Section 5.3 of Chapter 3.

[25] Quoted with permission from Bruce Hoadley, "The Quality Measurement Plan (QMP), p. 237.

3. PROBLEMS

24.1. Your past records suggest that the quality of lots received from a given supplier has varied randomly somewhat as follows:

Percent Nonconforming	*Proportion of Lots*
0.01	0.03
0.02	0.13
0.03	0.26
0.04	0.28
0.05	0.19
0.06	0.08
0.07	0.02
0.08	0.01

In a current sample of 100 items from a lot of 5,000 items you find 8 nonconforming items. Estimate the percent nonconforming in the lot by the method of maximum posterior probability.

24.2. Your past records suggest that the quality of lots received from a given supplier has varied randomly somewhat as follows:

Percent Nonconforming	*Proportion of Lots*
0.01	0.24
0.02	0.18
0.03	0.15
0.04	0.12
0.05	0.09
0.06	0.07
0.07	0.05
0.08	0.04
0.09	0.03
0.10	0.02
0.11	0.01

In a current sample of 100 items from a lot of 5,000 items, you find 8 nonconforming items. Estimate the percent nonconforming in the lot by the method of maximum posterior probability. Compare your answer with that to Problem 24.1.

24.3. In Chapter 3 the mean of the heights of 145 bomb bases was found to be 0.8314 inches, and their standard deviation was 0.00593 inches. Make a single estimate of the mean and standard deviation of the process, assuming that it is operating in a random manner. Determine 0.95 confidence limits for the mean and standard deviation of the process.

24.4. In Problem 3.7 you were asked to compute the mean of the data of Problem 3.3*a* (i) and in Problem 3.22 you were asked to compute the standard deviation. Estimate the mean and standard deviation of the process, assuming it is operating in a random manner. Determine 0.95 confidence limits for the mean and standard deviation of the process.

24.5. In Problem 3.7 you were asked to compute the mean of the data of Problem

3.3*a*(iii) and in Problem 3.22 you were asked to compute the standard deviation. Estimate the mean and standard deviation of the process, assuming that it is operating in a random manner. Determine 0.95 confidence limits for the mean and standard deviation of the process.

24.6. Work Problem 24.3, taking $n = 15$.

24.7. Work Problem 24.4, taking $n = 10$.

24.8. Work Problem 24.5, taking $n = 20$.

24.9. Work Problem 24.3, using a confidence coefficient of 0.90. A confidence coefficient of 0.99.

24.10. Work Problem 24.4, using a confidence coefficient of 0.90. A confidence coefficient of 0.99.

24.11. Work Problem 24.5, using a confidence coefficient of 0.90. A confidence coefficient of 0.99.

24.12. In Chapter 19, of 1,300 railway car frames, 113 were found to be nonconforming. Assume the manufacturing process to be operating in a random manner and estimate its fraction nonconforming. Determine 0.95 confidence limits (cf. Section 2.4 above).

24.13. After working Problem 19.5, determine 0.95 confidence limits for the process fraction nonconforming.

24.14. After working Problem 19.6, determine 0.95 confidence limits for the process fraction nonconforming.

24.15. The inspection unit for inspecting a given type of cloth is 100 square yards. The total number of nonconformities found after inspecting 10,000 square yards is 216. Determine 0.95 confidence limits for the average number of nonconformities per 100 square yards for the process, assuming that it is operating in a random manner.

24.16. The total number of nonconformities found in the inspection of 200 radios is 140. Determine 0.95 confidence limits for the average number of nonconformities per radio for the process, assuming that it is operating in a random manner.

24.17. Two hundred rounds of ammunition from a large lot yielded an average muzzle velocity of 1,750 feet per second. The standard deviation in muzzle velocity was 15 feet per second. Determine 0.95 confidence limits for the average muzzle velocity of the lot.

24.18. You wish to estimate the average outside diameter of a bushing produced by a given manufacturing process. You take a random sample of 20 bushings and find that the average outside diameter of the 20 bushings is 1.509 inches and their standard deviation is 0.008 inches. Determine 0.95 confidence limits for the true average of the process.

24.19. In Table 3.2 the standard deviation of the first 10 tests is 0.0043. Establish 0.95 confidence limits for the universe standard deviation. What is an unbiased estimate of the universe standard deviation?

24.20. In Table 3.2 the standard deviation of the first 75 tests is 0.0053. Establish confidence limits for the true standard deviation, using a 0.95 confidence coefficient. What is an unbiased estimate of the universe variance?

24.21. Ten samples of 12 each have the ranges 41, 35, 43, 41, 29, 36, 31, 39, 27, 34. Estimate the standard deviation of the universe from which these samples were drawn.

24.22. Examination of 500 items reveals that 214 are nonconforming. Assume that these are a random sample from a given manufacturing process that appears to be well under control and estimate the process fraction nonconforming. Also determine 0.95 confidence limits and 0.90 confidence limits for the process fraction nonconforming. Would the same limits be closer or farther apart if a sample of 1,000 items had been taken and 428 had been found nonconforming?

24.23. Examination of 4,000 items reveals that 49 are nonconforming. Estimate the process fraction nonconforming and determine 0.95 confidence limits for the process fraction nonconforming.

24.24. A sample of 30 from a lot of 60 items contains 3 nonconforming items. Derive confidence limits for the lot fraction nonconforming. Make your confidence coefficient as close to 0.95 as possible.

24.25. A sample of 25 from a lot of 80 contains 2 nonconforming items. Derive confidence limits for the lot fraction nonconforming. Make your confidence coefficient as close to 0.95 as possible.

24.26. A preliminary estimate shows that your process average is "about" 7 percent nonconforming when the process is under control. How many cases should you have in total to estimate the process fraction nonconforming within 1 percent either way? 0.2 percent either way? Assume that the confidence coefficient associated with the prescribed interval is 0.998 in the first case and 0.95 in the second case.

24.27. A preliminary estimate shows that your process fraction nonconforming is "about" 0.13 when the process is under control. How many cases should you have to estimate the process fraction nonconforming within 0.5 percent either way? Assume a 0.95 confidence coefficient.

24.28.* Six samples from Poisson distribution, $P(X) = \frac{u'^X e^{-u'}}{X!}$, have the results X_1, X_2, X_3, X_4, X_5, and X_6, respectively. Show that the maximum likelihood estimate of u' is $\Sigma X_i/6$. (*Hint:* Likelihood equals $L = \frac{u'^{\Sigma X_i} e^{-6u'}}{\Pi X_i!}$ and the u' that maximizes log L also maximizes L.)

24.29.* Five samples from a normal universe with density function

$$f(X) = \frac{e^{-(X-a)^2/2b}}{\sqrt{2\pi b}}$$

have the results X_1, X_2, X_3, X_4, and X_5. Show that the joint maximum likelihood estimates of a and b are

$$\hat{a} = \sum X_i/5$$

$$\hat{b} = \sum (X_i - \hat{a})^2/5$$

---

* Students with training in the calculus will find starred problems will give them knowledge of statistical theory that will be very useful in reading the text.

Note that a and b are the mean and variance of the universe and $\hat{a}$ and $\hat{b}$ are the sample mean and sample mean square deviation.

4. SELECTED REFERENCES†

Bayes (P 1763), Bowker and Lieberman (B '55), Bowley (B '26), Chandra and Hahn (P '81), Cheng and Iles (P '83), Chou and Owen (P '84), Clunies-Ross (P '58), Cramer (B '46), Crow and Gardner (P '59), Deely and Lindley, Deming (B '50), Fisher (B '56), Ghosh (P '79), Hall (P '82), Hoadley (P '81), Jennison and Turnbull (P '83), Kendall (B '48), Mainland, Herrara, and Sutcliffe (B '56. Contains a very extensive table of confidence intervals for p'.), Memon (P '63), Mood and Graybill (B '63), Morris (P '83), Munford (P '80), Nelson (P '78), Pachores (P '60), Satterthwaite (P '57), Simon (B '41. Appendix contains discussion of posterior probability applied to estimation of a fraction nonconforming.), Smith, J. G. and Duncan (B '45), Statistical Research Group, Columbia University (B '47. Chapter X contains confidence interval charts for p' with confidence coefficients of 0.99, 0.95, 0.90, and 0.80), Tomshy and Iwashi (P '79), and Wilks (B '48 and B '62).

† B and P refer to the Book and Periodical sections, respectively, of the Cumulative List of References in Appendix V.

25

Tests of Hypotheses Pertaining to Proportions, Means, and Variances

An important concept in industrial research is that of testing hypotheses. We wish to know, for example, whether a new method A is better than some standard method S; whether product A is better than product B; whether the outputs of machines I, II, and III form a homogeneous mass of product; whether a quality characteristic of a product is independent of a given condition of production; and so forth. To answer such questions we employ the statistical theory of testing hypotheses.[1]

This chapter will discuss tests of hypotheses pertaining to a universe proportion, tests of hypotheses pertaining to a universe mean, and tests of hypotheses pertaining to a universe variance. Subsequent chapters will be concerned with tests of differences between two proportions and two means, tests of the ratio of two variances, and more involved tests. In most instances, the design of such tests to meet specified criteria will also be discussed.

1. THE THEORY OF TESTING HYPOTHESES

To illustrate the testing of hypotheses, consider the following game. Let a die be shaken in a box and rolled on the table. If an even number turns up, Player A pays Player B \$1.00. If an odd number turns up, Player B pays Player A \$1.00. Umpire C does the shaking and rolling of the die. If

[1] As the reader proceeds in this chapter, he will find that the elementary theory of testing hypotheses is the equivalent of the theory of acceptance sampling developed in Part 2 above. He will discover, for example, that the hypothetical value to be tested is the equivalent of the p'_1 of a sampling plan, the risk of rejecting the hypothesis when it is true is the equivalent of the α of the sampling plan, and the OC curve for the test is the same as the OC curve for the sampling plan. Accordingly, this chapter is to some extent a repetition of what has already been discussed in Part 2.

the die is symmetrical and rolled in a random manner, the theory of probability tells us that, in the long run, we can expect an equal number of even and odd numbers. Under these conditions, the game will be a fair one.

Before the game begins, it is very desirable to test the fundamental hypothesis of symmetry and randomness. To do this A and B agree to have C roll the die, say, 100 times. If the results agree with those expected, they will play the game as agreed upon; if not, they will make the proper modifications. The basic question, of course, is how close an agreement with expectations is necessary before the hypothesis of symmetry and randomness can be accepted. Consideration of this question leads us to the theory of testing hypotheses.

In testing a hypothesis we may make two kinds of errors. The first kind of error is to reject a hypothesis when it is actually true. With reference to our illustration, we may conclude after C rolls the die 100 times, that it is unsymmetrical or is thrown in a biased manner, when actually this is not true. The second kind of error is to accept a hypothesis when it is not true. Thus we may say that the die is symmetrical and thrown in a random manner when actually it is not. The risk of the first kind of error can always be determined by the statistician when he designs his statistical test. He can arrange, for example, that the chance of his saying the die is unsymmetrical or is thrown in a biased manner, when actually it is not, will equal 5 out of 100. The risk of an error of the first kind is usually represented by α. If the coefficient of risk is α, the test is said to be conducted at the α significance level. In many tests α is set at 0.05.

The risk of an error of the second kind varies with the actual conditions that exist and can be determined by the statistician only as a function of those conditions. Thus the chance of saying that the die is symmetrical and thrown in an unbiased manner when actually it is either unsymmetrical or thrown in a biased manner will vary with the degree of asymmetry or bias. If the asymmetry or bias is such as to make the true probability of an even number equal to 0.55, for example, the risk of concluding that this probability is 0.50 is greater than if the true probability of an even number were, say, 0.75. The relationship between the chance of accepting a hypothesis and the actual conditions that exist is given by the operating characteristic or OC function for the test. Fortunately, statisticians have worked out formulas or derived tables or found approximations to the OC functions of many statistical tests, although these are not always in the form most desired.

2. TESTS OF HYPOTHESES PERTAINING TO A FRACTION NONCONFORMING

Testing whether a die, when shaken in a box and rolled on a table, will tend on the average to yield an equal number of even and odd numbers is similar to testing whether an industrual process is operating in a random manner to produce a specified fraction of nonconforming units. Assume that an even number is a conforming item, an odd number a nonconforming item; make the specified fraction nonconforming 0.5, assume that the operation of the process is similar to the rolling of the die, and we have an exact parallel.

2.1. A Two-Tail Test

2.1.1. Description of Test. Let us consider the die problem first. The hypothesis is that it is symmetrical and thrown in a random manner or, what amounts to the same thing, that the probability of an even number is 0.5. Set the risk of an error of the first kind at 0.05, and let us agree that we shall test the die by rolling it 100 times. Take as the statistic for testing the hypothesis the relative frequency of even numbers in the 100 rolls of the die. If this falls short of the lower acceptance limit p_{La} or exceeds the upper acceptance limit p_{Ua}, we shall reject the hypothesis; if it equals or falls between these limiting values, we shall accept the hypothesis. The quantities p_{La} and p_{Ua} have to be determined so that the risk of an error of the first kind (α) equals 0.05.

2.1.2. Determination of Acceptance Limits. To determine p_{La} and p_{Ua}, note that, if our hypothesis is true, the chances of getting various relative frequencies in our sample will be given approximately by the normal distribution. More precisely, the chance that the relative frequency will deviate from 0.5 by as much as $k\sigma_p'$ is given approximately by the area under the normal curve above $z = k$ and below $z = -k$ (see Figure 25.1). The standard deviation (σ_p') is given by $\sigma_p' = \sqrt{\frac{(0.5)(0.5)}{100}} = 0.05$. Hence the risk of making an error of the first kind will just equal 0.05, if we accept the hypothesis

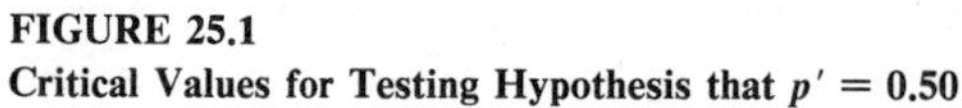

FIGURE 25.1
Critical Values for Testing Hypothesis that $p' = 0.50$

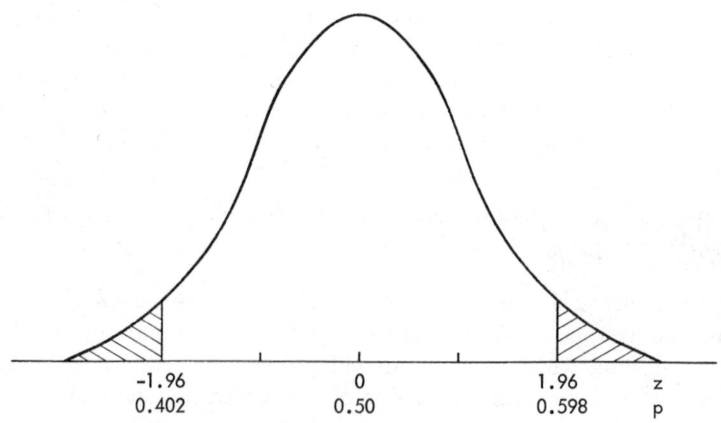

when the relative frequency of the sample equals or lies between 0.5 ± 1.96 (0.05) and reject it when it lies outside these limits.[2] Hence for our problem,

$$p_{La} = 0.500 - 1.96(0.05) = 0.402$$

$$p_{Ua} = 0.500 + 1.96(0.05) = 0.598$$

In other words, if we get 40 or less or 60 or more even numbers from 100 rolls of the die, we shall reject the hypothesis of symmetry and unbiased rolling of the die. If we get 41 to 59 even numbers, we shall accept the hypothesis.

The risk of rejecting our hypothesis when it is true is 0.05. This was determined by the design of the test. We could have made it 0.10 by taking

$$p_{La} = 0.500 - 1.645(\sigma'_p)$$

$$p_{Ua} = 0.500 + 1.645(\sigma'_p)$$

or we could have made it 0.01 by taking

$$p_{La} = 0.500 - 2.576(\sigma'_p)$$

$$p_{Ua} = 0.500 + 2.576(\sigma'_p)$$

2.1.3. The OC Curve for the Test. The risk of making an error of the second kind can be determined only as a function of the actual conditions. Let us suppose, for a moment, that the die is asymmetrical or is thrown with a bias. This is the same as assuming that the true probability of an even number is not 0.5 but some other fraction, p'. The risk of an error of the second kind is, under these conditions, the chance of getting 41 to 59 even numbers in 100 rolls when the true probability of an even number is p'. This risk is shown in Figure 25.2 as a function of p'. It is the OC curve for the test under the assumed conditions.

To derive this OC curve, we merely compute the probability of getting a z lying between

$$\frac{0.402 - p'}{\sqrt{\frac{p'(1-p')}{100}}} \text{ and } \frac{0.598 - p'}{\sqrt{\frac{p'(1-p')}{100}}}$$

For various values of p' we have the results given on the next page. These are the values plotted in Figure 25.2. For small p' we can use Figure 4.9 to compute the OC curve. This is illustrated in Section 2.2 below.

The curve just derived gives us the risks we run in accepting the hypothesis when the true probability is other than 0.5. Such a condition, it was assumed, could arise if the die was asymmetrical. It could also arise if the die were thrown in a random, but biased, manner. No simple curve can be derived for describing the risks involved in accepting the hypothesis when the rolling

[2] See Table A2 of Appendix II.

p'	$\sigma'_p = \sqrt{\frac{p'(1-p')}{n}}$	$p_{La} - p'$	$p_{Ua} - p'$	$\frac{p_{La} - p}{\sigma'_p}$	$\frac{p_{Ua} - p'}{\sigma'_p}$	$P_a = Prob. \left[\frac{p_{La} - p'}{\sigma_p} \leq z \leq \frac{p_{Ua} - p'}{\sigma'_p}\right]$
0.5	0.0500	−0.098	0.098	−1.96	1.96	0.9750 − 0.0250 = 0.9500
0.55	0.0497	−0.148	0.048	−2.98	0.97	0.8340 − 0.0014 = 0.8326
0.45	0.0497	−0.048	0.148	−0.97	2.98	0.9986 − 0.1660 = 0.8326
0.6	0.0490	−0.198	−0.002	−4.04	−0.04	0.4840 − 0.0000 = 0.4840
0.4	0.0490	0.002	0.198	0.04	4.04	1.0000 − 0.5160 = 0.4840
0.7	0.0458	−0.298	−0.102	−6.51	−2.23	0.0129 − 0.0000 = 0.0129
0.3	0.0458	0.102	0.298	2.23	6.50	1.0000 − 0.9871 = 0.0129

FIGURE 25.2
Operating Characteristic Curve for a Two-Tail Test of the Hypothesis that $p' = 0.50$ with α Taken as 0.05

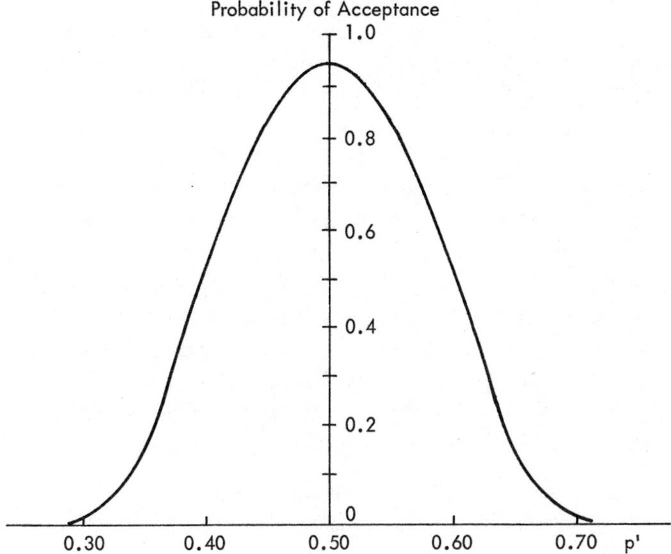

is done in a nonrandom manner. In this case the results are not chance results but are controlled by the thrower. Hence what we get will depend on how they are controlled. If the thrower rolls the die so as to alternate with even and odd numbers, we will be sure to accept the hypothesis (falsely, of course). If he throws the die so as to produce an even number every third throw, then we shall get 33 even numbers out of 100, and we shall be sure to reject the hypothesis. The point here is that our original test is not sufficient to protect us against accepting our hypothesis when random conditions do not exist, but leaves us at the mercy of the thrower. Every test of this kind should therefore be supplemented by a test for randomness. This could be done by studying runs in the data.[3]

2.2. A One-Tail Test

2.2.1. Description of Test. Let us consider a one-tail test applied to an industrial process. Suppose that a machine has been turning out parts that have, in past operations, run 4 percent nonconforming. An adjustment is made in the machine, and you are interested in seeing whether this has lowered the fraction nonconforming. To find the answer, you undertake to test the hypothesis that there has been no change. This will be done by taking a sample from the current process.

[3] See Chapter 18, Section 7.

The test you set up is as follows:

1. Hypothesis: The current process is operating in a random manner with a fraction nonconforming of 0.04.

2. Size of Sample: Arbitrarily take $n = 500$. A rational basis for selecting n is discussed below in Section 5.

3. Statistic: The fraction nonconforming in the sample is to be the statistic used in making the test.

4. Risk of an Error of the First Kind: You decide to run a risk of 0.05 of rejecting the hypothesis when it is true.

5. Acceptance Limit: In the present instance you decide to have only one acceptance limit p_a, which is less than 0.04. If the sample fraction nonconforming is less than p_a, you will reject the hypothesis and conclude that the adjustment has lowered the process fraction nonconforming. If the sample fraction nonconforming is equal to or greater than p_a you will accept the hypothesis. The quantity p_a is to be determined so that the probability that the sample fraction nonconforming will be less than p_a is just 0.05 when the process fraction nonconforming is 0.04.

2.2.2. Determination of Acceptance Limit $\mathbf{p_a}$. To determine the acceptance limit p_a, note that the distribution of sample fractions nonconforming will, because p' is small, be given approximately by the Poisson distribution. With $n = 500$ and $p' = 0.04$, Figure 4.9 shows us that the chance of $500p_a$ nonconforming units or less is about[4] 0.05 when $500p_a = 13$. Hence we take $p_a = 13/500 = 0.026$. If our sample fraction nonconforming is less than 0.026, we reject the hypothesis that the process fraction nonconforming is 0.04.

2.2.3. The OC Curve for the Test. An OC curve can be constructed to show the probability of accepting the given hypothesis when the process is a random one with a fraction nonconforming other than 0.04, say, p'. This is done readily with the help of Figure 4.9. To get P_a for a given p', we compute $p'n$ and find the probability of np_a or more nonconforming items in our sample. The results are plotted in Figure 25.3. It will be noticed that in this case we have an **S**-shaped OC curve. The reason for this is that we have taken only one rejection value. In the previous example, we got a bell-shaped curve because we took rejection values above and below the value specified by the hypothesis.

It is to be noted that we used Figure 4.9 (i.e., the Poisson approximation) to calculate probabilities in this problem because we were dealing primarily with small values of p', not because we were running a one-tail test. The normal approximation or the Poisson approximation can both be used in either a one-tail or a two-tail test.

---

[4] Because of the discrete nature of the Poisson distribution, we cannot in this case get an α exactly equal to 0.05.

FIGURE 25.3
Operating Characteristic Curve for a One-Tail Test of the Hypothesis that $p' = 0.04$, with α Taken as 0.05

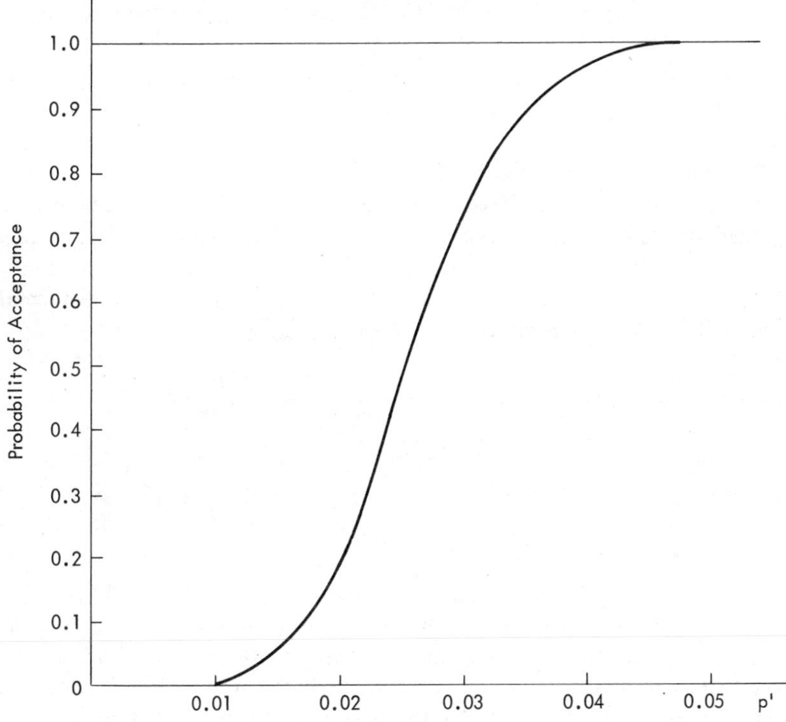

3. TESTS OF HYPOTHESES PERTAINING TO A UNIVERSE MEAN

3.1. When σ' Is Known

In testing hypotheses pertaining to a universe mean the procedure is simplest when the standard deviation of the universe is known. For in this case we can[5] treat the sample mean $\bar{X}$ as having a normal distribution with a mean equal to the universe mean $\bar{X}'$ and a standard deviation (or standard error as it is frequently called) equal to $\sigma'/\sqrt{n}$. This will be exactly true if the universe is normal and approximately true if the universe is moderately non-normal.

Let us consider an example. Suppose a manufacturer of synthetic rubber claims that the average hardness of his rubber is 64.3° Shore. As a consumer of rubber, we wish to test his claim. This can be done as follows:

1. Hypothesis: Let us suppose that a control chart on the range indicates that process variability is in control with a standard deviation equal to 2°

---

[5] See above, Chapter 6, Section 1.

Shore. Our hypothesis is that the sample we take for test purposes is a random sample from a normal or approximately normal universe the mean of which is 64.3° Shore.

2. Size of Sample: Take n equal to 25. This we shall suppose is determined by one of the methods described in Section 5.

3. The statistic to be used for the test will be the mean of the sample. We could use the median, but this is not so "efficient" a statistic as the mean.[6]

4. Risk of an Error of the First Kind: Let this equal 0.05.

5. Acceptance Limit: Derive a single acceptance limit $\bar{X}_a$ below 64.3°. This will be determined so that the probability of getting a sample mean less than $\bar{X}_a$ is just 0.05 if the hypothesis is true. Since under our assumption of a normal or approximately normal universe sample means will be normally distributed, $\bar{X}_a$ will be determined from the equation

$$\bar{X}_a = 64.3 - 1.645\sigma'_{\bar{X}} = 64.3 - 1.645\frac{(2)}{\sqrt{25}} = 63.64$$

Hence, our test will reject the hypothesis if the sample mean is less than 63.64° Shore and will accept it otherwise. As in the previous example, we use a single acceptance limit because we are not displeased if the hardness of the manufacturer's rubber is, in general, greater than claimed.

6. The OC Curve: The OC curve for the above test is obtained by computing the probability of getting a mean of a random sample equal to or greater than 63.64° when the mean of the universe is other than 64.3°, say, $\bar{X}'$. The universe standard deviation is assumed throughout to be 2° Shore. The ordinates of the OC curve for the above test are obtained as follows:

True Mean $\bar{X}'$	$\bar{X}_a - \bar{X}'$	$\frac{(\bar{X}_a - \bar{X}')}{\sigma'_{\bar{X}}}$	*Prob.* $\left(z \geq \frac{(\bar{X}_a - \bar{X}')}{\sigma'_{\bar{X}}}\right)$
62.5	63.64 − 62.5 = 1.14	1.14/0.4 = 2.85	0.0022
63.0	63.64 − 63.0 = 0.64	0.64/0.4 = 1.60	0.0548
63.5	63.64 − 63.5 = 0.14	0.14/0.4 = 0.35	0.3632
64.3	63.64 − 64.3 = −0.66	−0.66/0.4 = −1.645	0.9500
65.0	63.64 − 65.0 = −1.36	−1.36/0.4 = −3.4	0.9997

A graph of the OC curve is shown in Figure 25.4.

It is to be noted that the normal test described above is the correct procedure whenever the standard deviation is known, regardless of whether we have a large or small sample.

[6] See Chapter 24, Section 1.2.

FIGURE 25.4
Operating Characteristic Curve for a One-Tail Test of the Hypothesis that $\bar{X}'$ = 64.3, Given $\sigma' = 2$ and α Taken as 0.05

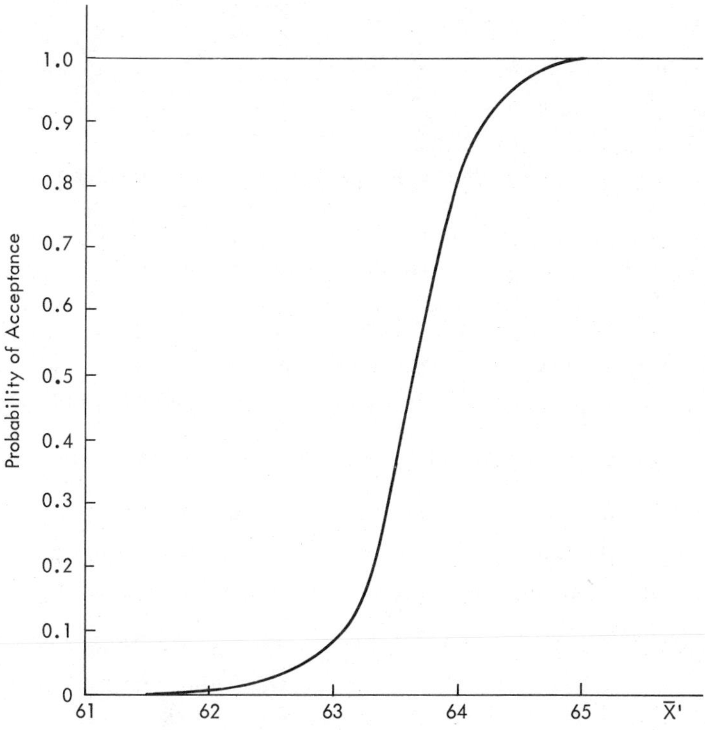

3.2. When σ' Is Unknown

When the standard deviation of the universe is unknown, the correct procedure is to run a t test. This means that we use the statistic $t = \dfrac{\bar{X} - \bar{X}'}{s/\sqrt{n}}$ instead of the normal statistic $z = \dfrac{\bar{X} - \bar{X}'}{\sigma'/\sqrt{n}}$. The statistic $t = \dfrac{\bar{X} - \bar{X}'}{s/\sqrt{n}}$, it will be recalled,[7] follows the t distribution, the percentage points of which are given in Table B of Appendix II. It will be noted that there is a different t distribution for each value of the degrees of freedom ν, which in the testing of a hypothesis pertaining to a single mean equals $n - 1$.

Consider again an example. If in the problem of the previous section we did not know the universe standard deviation, the procedure would run as follows:

[7] See Chapter 6, Section 5.

1. Hypothesis: The new hypothesis is that our sample comes from a normal universe the mean of which is 64.3° and the standard deviation of which is unknown.

2. Size of Sample: Let $n = 25$ as before.

3. Statistic: Now use the statistic $t = \frac{\bar{X} - \bar{X}'}{s/\sqrt{n}}$. This means that we shall have to compute both the mean and the standard deviation of our sample.

4. Risk of an Error of the First Kind: Let $\alpha = 0.05$.

5. Acceptance Limit: Since the statistic being used to make the test is different from that used in the case of known σ' and since this will be distributed in accordance with the t distribution and not the normal distribution, we shall need a new acceptance criterion. If we are interested only in a one-tail test, the critical value will be $-t_{0.05}$ for $\nu = n - 1$ (t for $P = 0.10$ in Table B of Appendix II, since the table gives both tail areas). If we are interested in a two-tail test, the critical values will be $\pm t_{0.025}$ for $\nu = n - 1$. For the present problem, a one-tail test seems the more suitable. For $\nu = 25 - 1 = 24$, the t table shows that $-t_{0.05}$ is -1.711. Hence our test will be to take a sample of 25 pieces of rubber, compute the mean and standard deviation of the sample, derive the value of $\frac{(\bar{X} - \bar{X}')}{s/\sqrt{n}}$, and note how it stands with reference to -1.711. If it is algebraically less than -1.711 (i.e., numerically greater in a negative direction), we shall reject the hypothesis. If it is equal to, or algebraically greater than, -1.711, we shall accept the hypothesis.

6. The OC Curve: The OC curve for the test will be found or interpolated among the OC curves presented in Figure 15.9. To use this figure we take $\bar{X}'_0$ as the hypothetical value being tested and $\bar{X}'$ as the actual value of the universe mean. The use of Figure 15.9 requires making some guess as to the value of σ'. If we guess a value higher than the true value, the OC curve we draw will be higher than it should be, which means we shall overestimate the risk of errors of the second kind. If we guess a value for σ' that is lower than the true value, the OC curve will be lower than it should be, which means that we shall underestimate the risk of errors of the second kind. Let us guess $\sigma' = 2°$. Then the OC curve for the test outlined in steps (1) to (5) will be that shown in Figure 25.5. This has been interpolated from Figure 15.9 and may be in slight error, but it is probably good enough for our purposes.

Because $n = 25$, the OC curve for the case in which σ' is unknown but is guessed at 2 differs but little from the OC curve for the case in which σ' is known to be 2. It will be noticed, however, that the curve for the case σ' unknown is slightly higher throughout most of the range than is the curve for the case σ' known. Consequently, for the same size sample we run greater risks of errors of the second kind. For example, if the true

FIGURE 25.5
Operating Characteristic Curve for a One-Tail Test of the Hypothesis that $\bar{X}' = 64.3°$, σ' Guessed at 2° and α Taken as 0.05

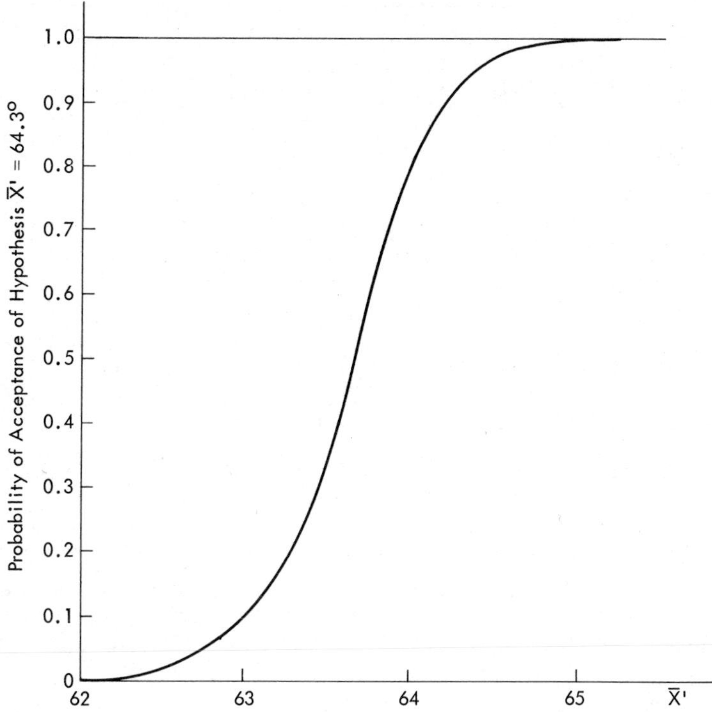

mean deviates from the hypothetical mean by as much as $0.4\sigma'$, the probability of accepting the hypothetical value is about 0.36 for the case in which σ' is known and about 0.40 for the case in which σ' is unknown, but guessed correctly.

Figure 25.6 gives OC curves for two-tail t tests with an α of 0.05. It should be used in lieu of Figure 15.9 when two-tail and not one-tail tests are employed.

4. TESTS OF HYPOTHESES PERTAINING TO A VARIANCE OR STANDARD DEVIATION

The last of the elementary tests of hypotheses that will be considered in this chapter is the test of a hypothesis pertaining to a universe variance. Since the variance is the square of the standard deviation, the test of the hypothesis $\sigma'^2 = \sigma_0'^2$ is the same as the test of the hypothesis $\sigma' = \sigma_0'$.

FIGURE 25.6
Operating Characteristic Curves* for Two-Tail t Tests, α Taken as 0.05 (for one-tail t tests, see Figure 15.9)

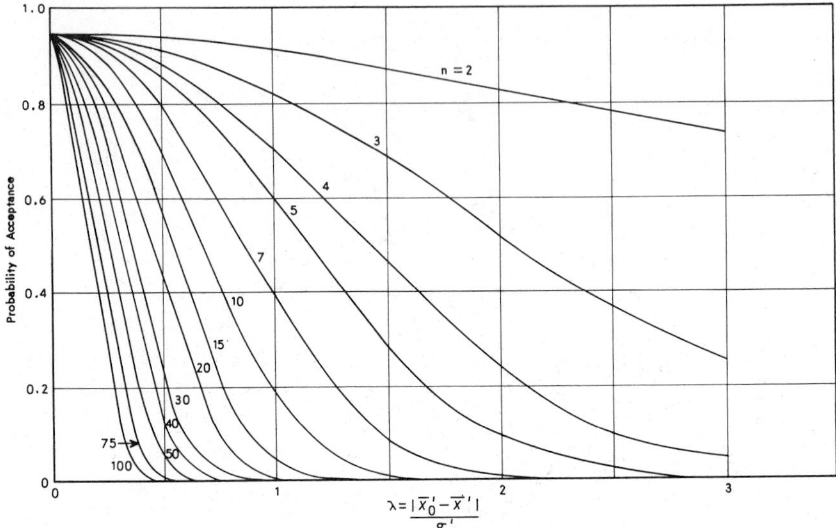

* In Figure 25.6 the OC curves for the larger values of n (say 20 to 100) become less and less reliable the higher we go on the chart. For probabilities greater than 0.5 they should be used for only very rough computations.

SOURCE: Reproduced from Figure 7 of C. D. Ferris, F. E. Grubbs, and C. L. Weaver, "Operating Characteristics for the Common Statistical Tests of Significance," *Annals of Mathematical Statistics,* 17 (1946), p. 195.

4.1. When $n > 30$

When $n > 30$, we can assume that s is normally distributed about σ' as a mean, with a standard deviation equal to[8] $\sigma'/\sqrt{2n}$. Thus, to test the hypothesis that $\sigma' = \sigma'_0$, we take as critical values $\sigma'_0 \pm 1.96\,\sigma'_0/\sqrt{2n}$ if we are interested in a two-tail test with $\alpha = 0.05$, or the value $\sigma'_0 + 1.645\sigma'_0/\sqrt{2n}$ (or possibly $\sigma'_0 - 1.645\sigma'_0/\sqrt{2n}$) if we are interested in a one-tail test with $\alpha = 0.05$.

Suppose, for example, that a process has been running for some time with a variability of $\sigma' = 10$. Some change is made, and we wish to determine whether the variability has been affected. To test the null hypothesis that $\sigma' = 10$, we take a random sample of 100 items from the modified process. If we take $\alpha = 0.05$ and use a two-tail test, the critical values for rejecting this hypothesis will be $s_{La} = 10 - 1.96\dfrac{(10)}{\sqrt{2(100)}} = 8.615$ and $s_{Ua} = 10 + 1.96\dfrac{(10)}{\sqrt{2(100)}} = 11.385$. If our sample s falls outside these limits, we shall

[8] Cf. Chapter 6, Section 2.5. R. A. Fisher suggests that a slightly more refined test will be given by taking the mean at $\sqrt{(2n-3)/2n}\ \sigma'$ (*Statistical Methods for Research Workers,* Table III). For $n = 30$ this equals $0.975\sigma'$ and gets closer to σ' as n is increased.

reject the null hypothesis; otherwise we shall accept it. The OC curve for this test will be obtained as follows:

(1) Universe Standard Deviation σ'	(2) $s_{La} - \sigma' =$ $8.615 - \sigma'$	(3) $s_{Ua} - \sigma' =$ $11.385 - \sigma'$	(4) $\frac{s_{La} - \sigma'}{\sigma'/\sqrt{2n}} = \frac{8.615 - \sigma'}{\sigma'/\sqrt{200}}$	(5) $\frac{s_{Ua} - \sigma'}{\sigma'/\sqrt{2n}} = \frac{11.385 - \sigma'}{\sigma'/\sqrt{200}}$	(6) Prob. (4) $\leq z \leq$ (5)
7	1.615	4.385	3.26	8.85	0.0006
8	0.615	3.385	1.09	6.00	0.14
8.615					0.50
9	−0.385	2.385	−0.61	3.76	0.73
10	−1.385	1.385	−1.96	1.96	0.95
11	−2.385	0.385	−3.06	0.49	0.69
11.385					0.50
12	−3.385	−0.615	−3.98	−0.72	0.24
13	−4.385	−1.615	−4.76	−1.76	0.04

A graph of these figures is shown in Figure 25.7.

FIGURE 25.7
Operating Characteristic Curve for a Two-Tail Test of the Hypothesis $\sigma' = 10$, α Taken as 0.05 and $n = 100$

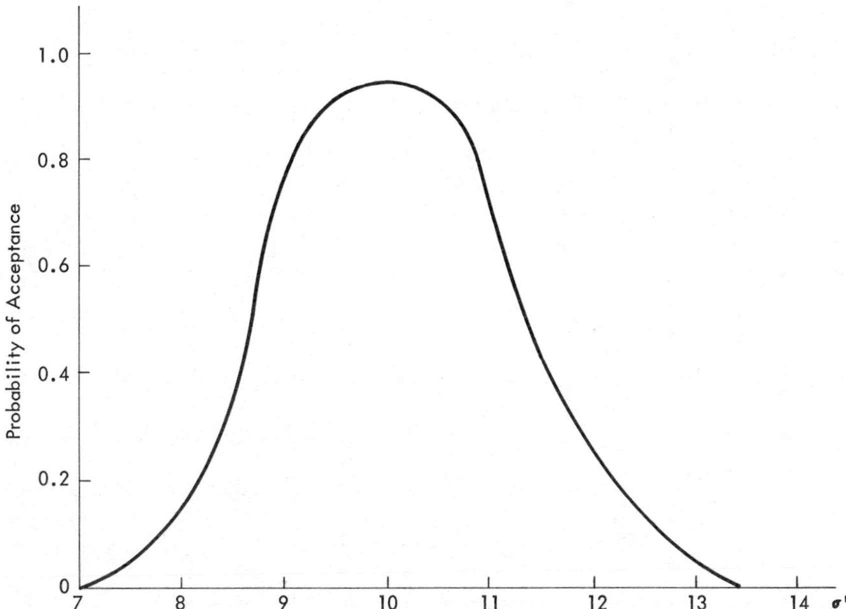

4.2 When $n \leq 30$

When $n \leq 30$, we must use more exact procedures. For n of any size we know[9] that, when $\sigma'^2 = \sigma_0'^2$, the statistic $(n - 1)s^2/\sigma_0'^2$ is distributed in the form of the χ^2 distribution with $\nu = n - 1$. For a one-tail test with $\alpha = 0.05$, a critical value will be given by either $\chi^2_{0.05}$ or $\chi^2_{0.95}$, depending on which tail we use. For a two-tail test with $\alpha = 0.05$, critical values will be given by $\chi^2_{0.025}$ and $\chi^2_{0.975}$. The test will consist of computing s^2 for the sample and finding the value of $(n - 1)s^2/\sigma_0'^2$. If this exceeds the critical value (or values) of χ^2 for $\nu = n - 1$, we reject the hypothesis $\sigma'^2 = \sigma_0'^2$; otherwise we accept it.

For example, suppose that we wish to test the hypothesis that the standard deviation of a process is 10 units (i.e., $\sigma'^2 = 100$). If we take a sample of 15 and decide on a one-tail test with $\alpha = 0.05$, the critical value for $(n - 1)s^2/\sigma_0'^2$ will be $\chi^2_{0.05}$ for $\nu = 15 - 1 = 14$. Table C of Appendix II shows this to be 23.7. Suppose, now, that we take 15 items at random from the process and find the standard deviation of this sample to be 16. Then we would have $(n - 1)s^2/\sigma_0'^2 = (14)(16)^2/(10)^2 = 35.8$, and we would reject the hypothesis $\sigma' = 10$.

OC curves for the test just described are given for various values of n in Figure 15.12, Here λ should be taken as σ'/σ_0'. Consider, for example, the test discussed above, in which $n = 15$ and the critical value for $(n - 1)s^2/\sigma_0'^2$ is 23.7. Figure 15.12 shows that, if the σ' is actually 1.5 times σ_0', i.e., $\sigma' = 15$ and not 10, the probability of accepting the hypothesis $\sigma' = 10$ is approximately 0.30.

5. LIST OF STEPS IN TESTING HYPOTHESES

We offer here a checklist of the steps to be taken in testing hypotheses. The reader may find it useful in doing the problems.

1. State hypothesis (preferably in equation form).
2. State whether test is to be one-tail (which tail) or two-tail.
3. State assumptions.
4. State statistic to be used; compute its value and state degrees of freedom.
5. State value of α to be used and why it was selected to be high or low.
6. State limits. These must be consistent with 2.
7. State results and/or decision or action.
8. Find OC curve for test.

In the expository material presented earlier in this chapter α was generally taken to be 0.05, but this was an arbitrary decision. Tables and charts of

[9] See Chapter 6, Section 2.3. The universe, of course, must be normal.

OC curves are generally worked out for $\alpha = 0.05$ and/or 0.01, and this may lead us to adopt one of these values for our test. If a test is designed with reference to the costs involved in making wrong decisions, α will become a variable that can be selected so as to minimize cost. This broader point of view will not be taken in this text, however. In the next section we shall seek the sample size that will give us stated risks of errors of the first and second kind, but these risks will be selected more or less arbitrarily.

6. DETERMINATION OF THE SIZE OF THE SAMPLE

In the discussion so far we have purposely avoided the question, How large a sample shall be taken to carry out a test? In all practical problems, however, this must be answered.

There are various ways of selecting the size of the sample. If a fixed sum of money has been allotted for a test, then we might take n as large as we can for the money allotted.

A method that is often used in other problems is to specify the size of the standard error that is desired, and then calculate n from this. For example, if we are testing a hypothesis pertaining to a proportion p', the standard error of a sample proportion is $\sigma'_p = \sqrt{\dfrac{p'(1-p')}{n}}$ when the universe proportion is p'. If the hypothetical value for p' is 0.5 and if we wish σ'_p to be no more than 0.02 when $p' = 0.5$, then we find n by solving the equation

$$0.02 = \sqrt{\frac{(0.5)(0.5)}{n}}$$

which yields $n = 625$.

The above method can also be used to determine sample size in tests of hypotheses pertaining to means and variances. Thus the standard error of a sample mean is $\sigma'_{\bar{X}} = \sigma'/\sqrt{n}$. Hence, if σ' is known or can be estimated and if it is decided that $\sigma'_{\bar{X}}$ should be a, say, then we have $n = \sigma'^2/a^2$. Likewise, the standard error of a sample standard deviation from a normal universe is approximately $\sigma'_s \doteq \sigma'/\sqrt{2n}$. Hence, if σ' is known or can be roughly estimated and if it is decided that σ'_s, should be b, say, then we have $n = \sigma'^2/2b^2$.

A third method of determining sample size that may in many cases be the equivalent of the above is to derive n from decisions as to risks of errors of the first and second kind, i.e., the risk of rejecting the given hypothesis when it is true and the risk of accepting it when it is not true. Suppose, for example, that we wish to test the hypothesis that $p' = 0.04$. We are willing to run a risk of 0.05 of rejecting this hypothesis when it is true (i.e., α is set equal to 0.05), but we do not wish to run a risk greater than 0.10 of accepting the hypothesis $p' = 0.04$ if p' is actually equal to or less than 0.02. From these criteria we can derive a sample size n and an acceptance limit p_a as follows:

The conditions are that n and p_a must be such that, when $p' = 0.02$, the probability of getting a sample proportion equal to or greater than p_a must be 0.10; and, when $p' = 0.04$, the probability of getting a sample proportion equal to or greater than p_a must be 0.95. To find an n and p_a that will meet these two conditions simultaneously, we can use the nomograph of Figure 7.5. Since this gives the probability of c or less, we draw a line through 0.02 on the left-hand scale and 0.90 on the right-hand scale. A second line is drawn through 0.04 on the left and 0.05 on the right. Their intersection yields $n = 600$ and $c = 16$. Since we are working with probabilities of rejection in this case, we take $p_a = (c + 1)/n = 17/600 = 0.028$. If the procedure carries us off the chart, we can use a trial-and-error method with Molina's tables, or we can use the normal distribution as described below.

If p_a had been on the upper side of the hypothetical value being tested, i.e., if we were using a one-tail test on the upper side, then the procedure for finding n and p_a would be exactly the same as that for finding a single-sampling inspection plan. For example, if we wish to find a plan for testing the hypothesis that $p' = 0.04$ with $\alpha = 0.05$, but do not wish a risk of more than 0.10 of accepting this hypothesis when $p' = 0.08$, we proceed exactly as we would in designing a sampling inspection plan for which the p_1' is to equal 0.04, the $p_2' = 0.08$, $\alpha = 0.05$, and $\beta = 0.10$. The n for our test will be the n for the sampling inspection plan and the p_a for our test will be c/n. Any of the methods described in Section 2.3 of Chapter 7 can be used.

If the hypothetical value of p' that is being tested (designate it by p_0') is not small—say it is greater than 10 percent—we may have to use the normal distribution for designing the test. Suppose, for example, that $p_0' = 0.20$ and we wish to accept the given hypothesis not more than 10 times out of 100 if the true p' is equal to or greater than 0.30. Let $\alpha = 0.05$ as before. In this case we would have to find n and p_a as follows: If α is to be 0.05, we must have

$$\frac{p_a - p_0'}{\sqrt{\dfrac{p_0'(1-p_0')}{n}}} = \frac{p_a - 0.2}{\sqrt{\dfrac{0.2(1-0.2)}{n}}} = 1.645$$

Likewise, if the probability of acceptance is to be 0.10 or less if $p' \geq 0.30$, we must have

$$\frac{p_a - p'}{\sqrt{\dfrac{p'(1-p')}{n}}} = \frac{p_a - 0.3}{\sqrt{\dfrac{0.3(1-0.3)}{n}}} = -1.282$$

These two equations become

$$\frac{p_a - 0.2}{1.645\sqrt{0.2(1-0.2)}} = \frac{1}{\sqrt{n}}$$

and

$$\frac{p_a - 0.3}{-1.282\sqrt{0.3(1-0.3)}} = \frac{1}{\sqrt{n}}$$

which, when solved simultaneously, yield $p_a = 0.252$ and $n = 161$. The test then is to take a sample of 161 and reject the hypothesis $p' = 0.20$ if the sample proportion is greater than 0.252.

If we wish to employ a two-tail test with symmetrical acceptance limits, we can come close to getting what we want by deriving n and p_a for a one-tail test with α equal to half that for the two-tail test. Let the lower limit obtained in this way be p_{La}. Then p_{Ua} will equal $p_0' + (p_0' - p_{La})$. If $p_0'n$ is reasonably large—say, 5 or more—this method will give results reasonably close to what we wish.

Suppose, for example, we wish to set up a two-tail test to test the hypothesis $p' = 0.15$ and suppose that we do not wish to accept this hypothesis more than 10 times out of 100 if the universe p' is either as low as 0.10 or as high as 0.20. Let $\alpha = 0.05$. Then we can find p_{La} and p_{Ua} and n for this test by finding p_a and n for a one-tail test that yields a probability of acceptance of 0.10 or less when $p' \geq 0.20$ and $\alpha = 0.025$. Thus, using the normal distribution method as described above, we have

$$\frac{p_{Ua} - 0.15}{\sqrt{\frac{(0.15)(1-0.15)}{n}}} = 1.96$$

$$\frac{p_{Ua} - 0.20}{\sqrt{\frac{(0.20)(1-0.20)}{n}}} = -1.282$$

which yields $p_{Ua} = 0.179$ and $n = 581$. Hence p_{La} would equal $0.15 - (0.179 - 0.15) = 0.121$. If we had worked from the other end, we would have had the equations

$$\frac{p_{La} - 0.15}{\sqrt{\frac{(0.15)(1-0.15)}{n}}} = -1.96$$

$$\frac{p_{La} - 0.10}{\sqrt{\frac{(0.10)(1-0.10)}{n}}} = 1.282$$

which yield $p_{La} = 0.118$, $p_{Ua} = p_0' - (p_{La} - p_0') = 0.15 - (0.118 - 0.15) = 0.182$, and $n = 478$. An intermediate plan that would be a good compromise would be to take $n = \frac{478 + 581}{2} = 530$, $p_{La} = 0.12$, and $p_{Ua} = 0.18$.

Tests pertaining to means and variances can also be designed with reference to risks of errors of the first and second kind.[10] Consider, first, the case in which σ' is known and let us design a test pertaining to a mean. Suppose that σ' is known to be 2 and we wish to test the hypothesis[11] that $\bar{X}' = 64.3$. We set $\alpha = 0.05$, which means that we are willing to run a risk of 0.05 of rejecting the hypothesis when it is true. On the other hand, we decide that we do not want a probability of more than 0.10 of accepting the hypothesis when the universe $\bar{X}'$ is 63 or less. Our test, we shall assume, is to be a one-tail test. Then the n and acceptance limit $\bar{X}_a$ which will yield the results desired are given by the equations

$$\frac{\bar{X}_a - \bar{X}'_0}{\sigma'/\sqrt{n}} = \frac{\bar{X}_a - 64.3}{2/\sqrt{n}} = -1.645$$

and

$$\frac{\bar{X}_a - \bar{X}'}{\sigma'/\sqrt{n}} = \frac{\bar{X}_a - 63.0}{2/\sqrt{n}} = 1.282$$

which yield $n = 21$, $\bar{X}_a = 63.7$.

If σ' is not known and we are to use a one-tail t-test, we use Figure 15.9 to design the test.[12] In this case, we calculate $\lambda = \dfrac{|\bar{X}'_0 - \bar{X}'|}{\sigma'}$, where $\bar{X}'_0$ is the hypothetical value being tested, $\bar{X}'$ is given, say, the tolerance value that we are willing to accept no more than 10 times out of 100, and σ' is guessed as best we can. Then we find the curve which comes the closest to intersecting the 0.10 line at this value of λ. The n for this curve is the sample size to be used. For example, suppose in the previous problem we did not know that $\sigma' = 2$ but were willing to make a rough guess that it was approximately 2. Then we would have

$$\lambda = \frac{|64.3 - 63.0|}{2} = \frac{1.3}{2} = 0.65$$

and from Figure 15.9 we may estimate n at 23. We derive the critical t by finding the $t_{0.05}$ point in a t table for $\nu = n - 1$. Thus for $n = 23$, we have $t = -1.717$. Our test will thus be to take a sample of 23 and compute[13] $\dfrac{\bar{X} - 64.3}{s/\sqrt{23}}$. If this has a negative value numerically greater than -1.717, we reject the hypothesis.

---

[10] The procedures are identical with those of Chapter 15 for designing variables sampling plans.

[11] Cf Section 3.1 above.

[12] This again is identical with the derivation of a variables sampling plan when σ' is unknown (see Chapter 15).

[13] It will be recalled that $s = \sqrt{\dfrac{\Sigma(X - \bar{X})^2}{n - 1}}$

If we wish to use a two-tail t test, we proceed in a similar way but use Figure 25.6 instead of Figure 15.9.

We can design a test pertaining to a variance or standard deviation in a like manner. Suppose, for example, that we wish to test the hypothesis that $\sigma' = 10$ and prefer a one-tail test. We are willing to run a risk of 0.05 of rejecting this hypothesis when it is true, but do not wish to run a risk of more than 0.10 of accepting it if the universe σ' is 15 or more. To find n and the critical limit s_a that will give us these results, we use the equations

$$\frac{s_a - \sigma'_0}{\sigma'_0/\sqrt{2n}} = \frac{s_a - 10}{10/\sqrt{2n}} = 1.645$$

$$\frac{s - \sigma'}{\sigma'/\sqrt{2n}} = \frac{s - 15}{15/\sqrt{2n}} = -1.282$$

These yield the results $n = 26$ and $s_a = 12.3$.

The above analysis yields an $n < 30$. Consequently, it may be well to check the results by using the χ^2 distribution. If we select the statistic $(n - 1)s^2/\sigma_0'^2$ to carry out a one-tail test, the critical value would be $\chi^2_{0.05}$ for $v = n - 1$. To find the n which will give us the desired results, we use Figure 15.12. That is, we compute $\lambda = \sigma'/\sigma'_0$, where σ'_0 is the hypothetical value being tested and σ' is given the value, say, that we are willing to accept not more than 10 times out of 100. Then we find the n of the curve that intersects the 0.10 level closest to this value of λ. Thus, for the given plan, we shall have $\lambda = 15/10 = 1.50$. It appears from Figure 15.12 that the curve $n = 27$ comes the closest to intersecting the 0.10 line at $\lambda = 1.5$. This is close to the value of n found by the large-sample method and may be taken as the n desired. Our test will thus be to take a sample of 27 and compute.

$$\chi^2 = \frac{(n - 1)s^2}{\sigma_0'^2} = \frac{26s^2}{(10)^2}$$

If this exceeds $\chi^2_{0.05}$ for $v = n - 1 = 27 - 1 = 26$, i.e., if it exceeds 38.9, we reject the hypothesis that $\sigma' = 10$.

7. A POINT OF LOGIC

It is to be noted that in Part 2 on "Lot Acceptance Sampling Plans" we were concerned with making decisions, and "to accept a lot" after it had been submitted to sampling inspection meant exactly what the words implied. The lot was accepted.

In the theory of testing hypotheses, however, our acceptance and rejection of hypotheses may in many instances be merely an intellectual exercise associated with no course of action. In such a situation "to accept a hypothesis" may simply mean that it is not rejected. It may mean that the hypothesis is tentatively accepted or simply that we reserve judgment. Such tentative acceptance is most likely to occur when the original test has not been carefully planned and it is believed that the risks of an error of the second kind are much higher than desirable. In such an instance, the hypothesis is tentatively accepted subject to possible later revision when more data are available. In the future we shall talk about accepting a hypothesis without always pointing out that acceptance may mean only tentative acceptance.

8. PROBLEMS

25.1. The fraction nonconforming of a certain controlled process is 0.17. A supplier of a new basic material claims that the use of his material will reduce the fraction nonconforming. You make a trial run with this new material and, out of an output of 400, you find 56 nonconforming units. Would you accept the claim that the fraction nonconforming of the process will be reduced by the new material? What is the hypothesis you test? Draw the OC curve for your test.

25.2. Manufacturer X claims that his product is only 2 percent nonconforming. Being a purchaser of this type of product, you direct your research department to test the manufacturer's claim. The department checks a sample of 200 items bought in the market and finds 7 nonconforming units. Would you or would you not feel warranted in concluding that the manufacturer's claim was false? Draw the OC curve for your test.

25.3. A large company wants to check the attitude of its employees on the adoption of a contributory health insurance plan. It takes a random sample of 200 persons from the total working force of 4,000 and finds that 110 of these persons favor this measure. The company does not wish to propose the plan unless the sample poll gives a "high degree of assurance" that at least 50 percent of the working force is in favor of it. Would you conclude that the above results give such assurance? If your answer is in the negative, how many out of the 200 persons sampled would have to favor the proposal before you would consider the company to be sufficiently assured to take action? Justify your answer.

25.4. A shoe manufacturer claims that he can supply, at the same price now being paid, shoes that will give longer service than the shoes now being used for garrisoned troops of the United States Army. Army records show that, for such troops, the average life of shoes is 10 months, with a standard deviation of two months.

To test the manufacturer's claim, the Army issues 100 pairs of the new shoes at random among garrisoned troops. On the average, these 100 pairs wear out in 10½ months. Would you or would you not accept the manufacturer's claim? Draw the OC curve for your test.

25.5. A supplier of cotton yarn claims that his product has an average breaking strength of 90 pounds. The company for which you work is interested in buying this yarn, but decides first to test his claim. If the test leads to acceptance of the claim, it will buy the yarn.

a. The management tells you to take a sample of 16 pieces of yarn, but leaves the rest of the test to your discretion. The inspection department informs you that, from their past experience with this kind of cotton yarn, the standard deviation of breaking strength may be taken as 12 pounds. On the basis of the management's directive and the inspection department's estimate of the standard deviation, draw up the details of the test as you would apply it. Draw the OC curve for the test.

b. What is the risk you run under your test of advising the company to buy the yarn when its average breaking strength is actually only 82 pounds. If you do not like this result, what recommendation would you make to the management regarding the test?

c. If the yarn has actually a breaking strength of 90 pounds, what is the risk under your test that you might reject the claim?

d. If you cannot get any prior information regarding the standard deviation of the breaking strength of this kind of cotton yarn, what recommendation would you make to the management regarding the test?

e. You carry out the test as designed in *a* and find that the average breaking strength for the sample of yarn tested is 86 pounds. Would you or would you not advise the management to buy the yarn?

25.6. A milk distributor claims that the milk sold by his company contains, on the average, 0.0110 pounds of butterfat per quart. A consumers' research organization makes a check on his claim by taking a random sample of 16 quart bottles of the milk. The amount of butterfat in each bottle is measured, and the arithmetic mean of the amount of butterfat in a quart of milk is found to be 0.0105 pounds; the standard deviation of the measurements of quantity of butterfat in the sample of 16 bottles was found to be 0.0012 pounds.

a. Could the consumers' research organization justifiably conclude that the milk distributor's claim was false? Sketch the OC curve for the test.

b. Suppose, instead of taking a sample of 16 quart bottles, that the consumers' research organization had taken 100 quart bottles and found that the arithmetic mean was 0.0105 pounds of butterfat and the standard deviation 0.0012. Under these conditions could the consumers' research organization justifiably conclude that the milk distributor's claim was false? Draw the OC curve for the test.

25.7. A producer claims that the lengths of the parts he manufactures has a standard deviation of 0.04 inches. You measure a sample of 10 parts and find that their standard deviation is 0.07 inches. Could you reasonably reject the producer's claim? Draw the OC curve for your test.

25.8. A manufacturer of gauges claims that the standard deviation in the use of his gauge is 0.0003. Unknown to an analyst who is using this gauge, you have him measure the same item 8 times during the course of his regular testing duties and find that the standard deviation of the 8 measurements is 0.0006. In view of these results would you be justified in rejecting the manufacturer's claim? Sketch the OC curve for your test.

25.9. The standard value for the central line on a *p*-chart is $p'' = 0.075$. The chart requires the plotting of the fraction nonconforming in samples of 100. For 5 samples plotted during a week's operations the average fraction nonconforming is 0.060. Would you conclude from this that the process average is significantly less than the standard of 0.075? Draw the OC curve for your test.

25.10. The standard value for the central line on an $\bar{X}$-chart is $\bar{X}'' = 0.8314$ inches, and the chart requires the plotting of means of samples of 5. The standard deviation of the process is estimated from past experience to be 0.0060 inches. If the average of 20 successive samples of five each is 0.8330 inches, would you conclude that the process average is significantly greater than the standard? Draw the OC curve for your test.

25.11. The standard value for the central line on an R-chart is 0.0140 inches, and the chart calls for the plotting of ranges of samples of 5. If 25 successive samples have an average range $(\bar{R})$ of 0.0163 inches, would you infer that the variability of the process was significantly different from the standard? (*Hint:* Treat $\bar{R}$ as a normally distributed variable with $\sigma'_{\bar{R}} = \sigma'_R/\sqrt{n}$.) Draw the OC curve for your test.

25.12. A manufacturer claims that his product is 1 percent nonconforming. You wish to test this claim. You do not want to run a risk of more than 0.05 of rejecting his claim if it is true. On the other hand, you do not wish to run a risk of more than 0.10 of accepting his claim if his product is actually as much as 3 percent nonconforming. How large a sample of his product would you test? What would be your acceptance limit?

25.13. The standard value for the central line on a p-chart is 0.055. The chart requires the plotting of the fractions nonconforming of samples of 200. At the end of m samples you want to revise the standard if the cumulated sample results indicate that the process average is significantly different from the old standard. You do not wish to run a risk of more than 0.05 of changing the standard if the process average is not actually different from the standard. On the other hand, you do not want to run a risk of more than 0.10 of not changing the standard, if the process average differs from the standard by as much as 0.01. How many samples should you cumulate before making the test?

25.14. *a.* Design a test for determining whether the hypothesis $\bar{X}' = 500$ is valid, given that $\sigma' = 50$ and that the "tolerance grade" for the material is 520.

b. Design the above test if σ' is not known but is guessed to be approximately 50.

25.15. The standard value for an $\bar{X}$-chart is 0.8314, and the chart requires the plotting of means of samples of 5. The standard deviation of the process is estimated from past data to be 0.0060. At the end of m samples you want to revise the central line of the $\bar{X}$-chart if the cumulated sample results indicate that the process average is significantly different from the old standard. You do not wish to run a risk of more than 0.05 of changing the standard when the process average is actually not different from the old standard. On the other hand, you do not want to run a risk of more than 0.10 of not changing the standard when the process average has shifted by as much as 0.001. How many samples would you cumulate before making the test?

25.16. You wish to test whether the standard deviation of measurements made by a certain thermometer is significantly greater than 0.010°, given a "tolerance σ'" of 0.040°. How many measurements would you make and what would be your acceptance limit?

9. SELECTED REFERENCES*

Bowker and Lieberman (B '55 and B '72), Brownlee (B '60), Davies (B '57), Dixon and Massey (B '69), Fisher (B '56), Freund (B '67), Johnson (B '40), Mood and Graybill (B '63), Neyman (B '50), Resnikoff and Lieberman (B '57), Smith, J. G. and Duncan (B '45), Srivastava (P '58), Tippet (B '50), and Tukey (P '60).

---

* B and P refer to the Book and Periodical sections, respectively, of the Cumulative List of References in Appendix V.

26

Tests Pertaining to the Difference between Two Proportions, Means, or Variances

1. TESTS PERTAINING TO THE DIFFERENCE BETWEEN TWO PROPORTIONS

A common problem is to determine whether two percentages or proportions are significantly different. For example, experiments may be run on two processes and the results may be compared to see if they differ with respect to their fractions nonconforming. The sample results will undoubtedly differ, but the primary question is: Are the two processes different? In other words, how much must two sample proportions differ before we say the universes are different?

1.1. Test for Large Samples

1.1.1. Theory. The theory underlying tests of differences between two proportions may be outlined as follows.

Suppose that we have many independent pairs of samples, the first sample in each pair consisting of n_1 cases from universe I and the second sample in each pair consisting of n_2 cases from universe II. Let p_1 be the proportion with a certain attribute in the first sample of a pair, say, p_1 is the fraction nonconforming, and let p_2 be the proportion in the second sample. Let p'_1 and p'_2 be the two universe proportions. Then if $p'_1 n_1$ and $p'_1 n_2$ are sufficiently large,[1] the differences $p_1 - p_2$ among the many pairs of samples will tend to have a sampling distribution that is normal in form, with a mean equal

[1] See Chapter 4, Section 7.1.

to the difference between the universe proportions $p'_1 - p'_2$ and a standard deviation equal to

$$\text{(26.1)} \qquad \sigma'_{p_1 - p_2} = \sqrt{\sigma'^2_{p_1} + \sigma'^2_{p_2}} = \sqrt{\frac{p'_1(1-p'_1)}{n_1} + \frac{p'_2(1-p'_2)}{n_2}}$$

It should be emphasized that the samples must be independent of each other.

The hypothesis that we are interested in testing is that the universe proportions are the same, i.e., that $p'_1 = p'_2$. If we reject this, we accept the alternative hypothesis.

To test the hypothesis that the universe proportions are the same, we estimate the supposed common proportion by taking a weighted average of the two sample proportions. Thus we estimate the common proportion as equal to

$$\text{(26.2)} \qquad \hat{p} = \frac{n_1 p_1 + n_2 p_2}{n_1 + n_2}$$

and use this in the standard error formula (26.1). In other words, we take

$$\text{(26.3)} \qquad \hat{\sigma}_{p_1 - p_2} = \sqrt{\hat{p}(1-\hat{p})\left(\frac{1}{n_1} + \frac{1}{n_2}\right)}$$

as an estimate of $\sigma'_{p_1 - p_2}$. If we want a two-tail test and are willing to run a risk of 0.05 of rejecting the hypothesis $p'_1 = p'_2$ when it is true, then we will reject or accept our hypothesis, depending on whether or not $\dfrac{p_1 - p_2}{\hat{\sigma}_{p_1 - p_2}}$ falls outside of the limits ± 1.96. If we wish to test for a difference in one direction only—say, that the alternative we are interested in is that $p'_1 > p'_2$—then we take $+1.645$ as the critical value. If $\dfrac{p_1 - p_2}{\hat{\sigma}_{p_1 - p_2}}$ exceeds this, we reject the hypothesis $p'_1 = p'_2$. Otherwise we accept it.

1.1.2. An Example. To illustrate the above test, consider the following data.[2] In June of a given year a foundry turned out 1,350 car side frames, of which 146 were found to be nonconforming. In the following July it turned out 1,300 car side frames, of which 113 were found to be nonconforming. If the production process was operating in a random manner in both months, can it be concluded that the level at which it was operating in July was significantly different from that in June, or can the apparent difference be reasonably attributed to chance? Let us suppose that tests of this kind are made once a month and the management is interested in a shift in either direction. A two-tail test is therefore appropriate. Let us agree to run a risk of 0.05 of rejecting the hypothesis $p'_1 = p'_2$ when it is true, so that our critical values will be -1.96 and $+1.96$.

[2] Cf. Chapter 19, Section 2.

Since the hypothesis to be tested is that the process is the same in both months, we pool our two samples and estimate the assumed common value of p' as equal to

$$\hat{p} = \frac{146 + 113}{1{,}350 + 1{,}300} = \frac{259}{2{,}650} = 0.098$$

Hence

$$\hat{\sigma}_{p_1 - p_2} = \sqrt{(0.098)(0.902)\left(\frac{1}{1{,}350} + \frac{1}{1{,}300}\right)} = 0.0116$$

and

$$\frac{p_1 - p_2}{\hat{\sigma}_{p_1 - p_2}} = \frac{146/1{,}350 - 113/1{,}300}{0.0116} = \frac{0.108 - 0.087}{0.0116} = 1.81$$

Since +1.81 lies between the limits ±1.96, we accept the hypothesis $p_1' = p_2'$ and conclude that at the 0.05 level there is not a significant difference between the two fractions nonconforming.

1.1.3. A Chi-Square Test of a Difference between Two Proportions. In concluding this section, it should be noted that the chi-square test described in Chapter 28 offers an alternative procedure for testing the difference between two proportions when n_1 and n_2 are large. It is the statistical equivalent of the procedure developed above in Section 1.1.1, and has the advantage that tables are available for selected points on the OC curve and a formula can be derived for determination of sample sizes.

1.2. Test for Small Samples

When $p_1' n_1$ or $p_2' n_2$ is small, an exact test should be applied. This is described in Appendix I (31).

1.3. Confidence Limits for a Difference between Two Proportions

If two proportions are found to be significantly different, it might be of interest to determine confidence limits for the difference. If the samples are large, this can be done in a rough way by laying off around the sample difference some multiple of the estimated standard error of the difference. Thus, 0.95 confidence limits for $p_1' - p_2'$ will be given approximately by

$$(p_1 - p_2) \pm 1.96\sqrt{\frac{p_1(1 - p_1)}{n_1} + \frac{p_2(1 - p_2)}{n_2}}$$

Thus, if $p_1 = 0.11$, $p_2 = 0.07$, $n_1 = 1{,}000$, and $n_2 = 2{,}000$, 0.95 confidence limits for $p_1' - p_2'$ are given as approximately equal to

$$(0.11 - 0.07) \pm 1.96\sqrt{\frac{0.11(1-0.11)}{1{,}000} + \frac{0.07(1-0.07)}{2{,}000}}$$

$$= 0.04 \pm 0.022 = \begin{cases} 0.018 \\ 0.062 \end{cases}$$

2. TESTS PERTAINING TO THE DIFFERENCE BETWEEN TWO MEANS

Testing the difference between two sample means is an even more common problem than testing the difference between two proportions. As in the case of testing a hypothesis pertaining to a single mean, there are two cases to be considered: one in which the standard deviations of the two universes are known and the other when they are unknown. In the latter case, there are two subcases: one in which the standard deviations are known to be equal but the common value is unknown, and the other when we have no knowledge of equality.

2.1. Testing the Difference between Two Means when σ_1' and σ_2' Are Known

2.1.1. Theory. When the standard deviations of the two universes are known, the test of a difference between two means is a normal test, regardless of the sample size.[3] For if the universes are normal, means of samples therefrom are normally distributed and likewise differences between them. Even if the universes are moderately nonnormal, means and differences of means will still tend to be normally distributed. From Theorem 4.1 of Chapter 4, the mean of the difference between the sample means is the difference between the universe means, and if the samples are independent, Theorem 4.2 indicates that the variance of the difference is the sum of the variances of the individual sample means. Thus we have $\bar{X}_1 - \bar{X}_2$ normally or approximately normally distributed with

(26.4) $$E(\bar{X}_1 - \bar{X}_2) = E(\bar{X}_1) - E(\bar{X}_2) = \bar{X}_1' - \bar{X}_2'$$

$$\sigma'^2_{\bar{X}_1 - \bar{X}_2} = \sigma'^2_{\bar{X}_1} + \sigma'^2_{\bar{X}_2} = \frac{\sigma_1'^2}{n_1} + \frac{\sigma_2'^2}{n_2} \text{ if } \bar{X}_1 \text{ and } \bar{X}_2 \text{ are independent.}$$

It thus follows that the statistic

$$z = \frac{(\bar{X}_1 - \bar{X}_2) - (\bar{X}_1' - \bar{X}_2')}{\sqrt{\dfrac{\sigma_1'^2}{n_1} + \dfrac{\sigma_2'^2}{n_2}}}$$

will have a normal distribution with a zero mean and a unit standard deviation.

[3] Cf. Section 3.1 of Chapter 25.

It is to be emphasized that the two samples being compared must be independent. We must be particularly careful in this regard about paired data. For example, if we are comparing two devices for testing tensile strength, we might have the following:

Piece of Wire	*Device A*	*Device B*
No. 1	90	92
No. 2	80	80
No. 3	74	73

In this case the A and B data vary together from piece to piece and are thus not independent. Consequently, the methods outlined in this section should not be applied. The correct procedure is to compute individual differences (e.g., 90 − 92, 80 − 80, 74 − 73, and so on) and then test whether the mean difference is significantly different from zero. This reduces the problem of paired data to a single sample problem of the type discussed in the previous chapter.

2.1.2. An Example. Suppose that we wish to determine whether machine I and machine II have the same setting. It is known from past experience that the standard deviation of machine I is 0.053 inches and that of machine II is 0.047 inches. Let six measurements be made from the output of machine I and five from machine II, as follows:

Machine I	*Machine II*
0.37	0.33
0.32	0.31
0.41	0.38
0.35	0.26
0.38	0.39
0.31	

The hypothesis to be tested is: $\bar{X}'_1 = \bar{X}'_2$. It is known that $\sigma'_1 = 0.053$ and $\sigma'_2 = 0.047$. Let us take $\alpha = 0.05$ and use a two-tail test. The critical ratios will therefore be ± 1.960. We have $n_1 = 6$, $n_2 = 5$. Compute

$$\bar{X}_1 = 0.357,$$

$$\bar{X}_2 = 0.334,$$

$$\sigma'_{\bar{X}_1 - \bar{X}_2} = \sqrt{\frac{(0.053)^2}{6} + \frac{(0.047)^2}{5}} = 0.030$$

and

$$z = \frac{\bar{X}_1 - \bar{X}_2}{\sigma'_{\bar{X}_1 - \bar{X}_2}} = \frac{0.357 - 0.334}{0.030} = 0.77$$

This is less than 1.96, so we conclude that the settings are not different.

2.1.3. The Operating Characteristics of the Test. Figure 26.1 gives an idea as to how effective a test such as described in the previous section will be in detecting differences between two universe means. It answers the question as to what the probability of accepting the hypothesis $\bar{X}'_1 = \bar{X}'_2$ would be if the two samples came from universes the standard deviations of which are both equal to an average standard deviation σ^*, this being defined as the square root of the weighted average of the two originally specified variances. In other words, Figure 26.1 assumes a common standard deviation σ^* given by

$$\sigma^* = \sqrt{\frac{n_2\sigma_1'^2 + n_1\sigma_2'^2}{n_1 + n_2}}$$

This is a logical choice for a common standard deviation since we will have

$$\sigma^* \sqrt{\frac{1}{n_1} + \frac{1}{n_2}} = \sqrt{\frac{\sigma_1'^2}{n_1} + \frac{\sigma_2'^2}{n_2}}$$

and the standard deviation of the difference between the mean of a sample of n_1 and the mean of a sample of n_2 will be the same for the case of equal universe variances as it is for the case of unequal variances discussed in the previous section.

FIGURE 26.1
Operating Characteristic Curve for a Two-Tail Test of the Difference between Two Means, α Taken as 0.05

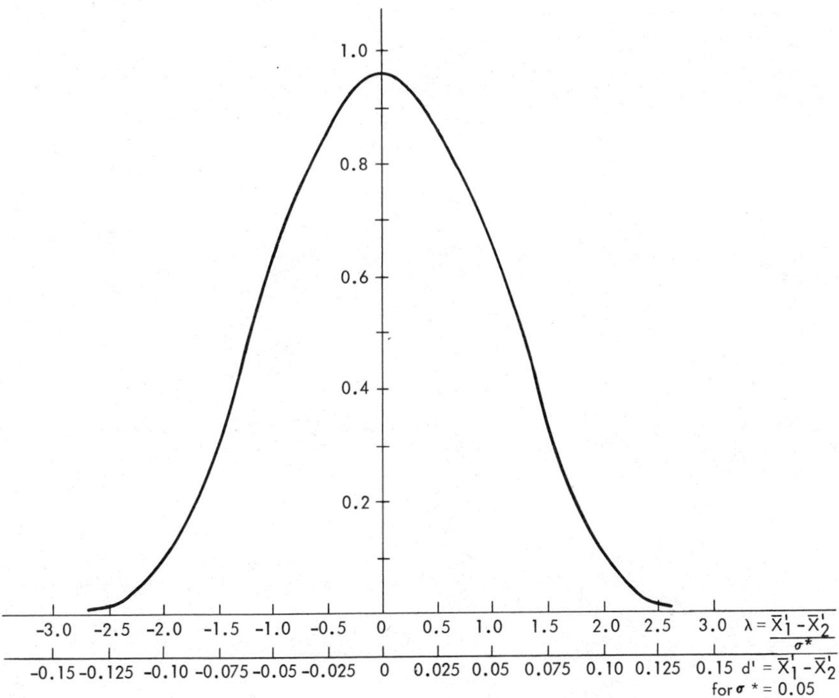

The upper half of the OC curve of Figure 26.1 is derived as follows:

(1) $\lambda = d'/\sigma^*$	(2) $d' = \bar{X}'_1 - \bar{X}'_2$	(3) $d'/\sigma'_{\bar{X}_1-\bar{X}_2}$	(4) 1.96 − (3)	(5) −1.96 − (3)	(6) Prob. [(5) ≤ z ≤ (4)]
0	0	0	1.96	−1.96	0.95
0.30	0.015	0.50	1.46	−2.46	0.92
0.60	0.030	1.00	0.96	−2.96	0.83
0.90	0.045	1.50	0.46	†	0.68
1.20	0.060	2.00	−0.04	†	0.48
1.50	0.075	2.50	−0.54	†	0.30
1.80	0.090	3.00	−1.04	†	0.15
2.10	0.105	3.50	−1.54	†	0.06
2.40	0.120	4.00	−2.04	†	0.02
2.70	0.135	4.50	−2.54	†	0.01

† Taken as $-\infty$ in computation (6).

FIGURE 26.2
Operating Characteristic Curve for a One-Tail Test of the Difference between Two Means, α Taken as 0.05

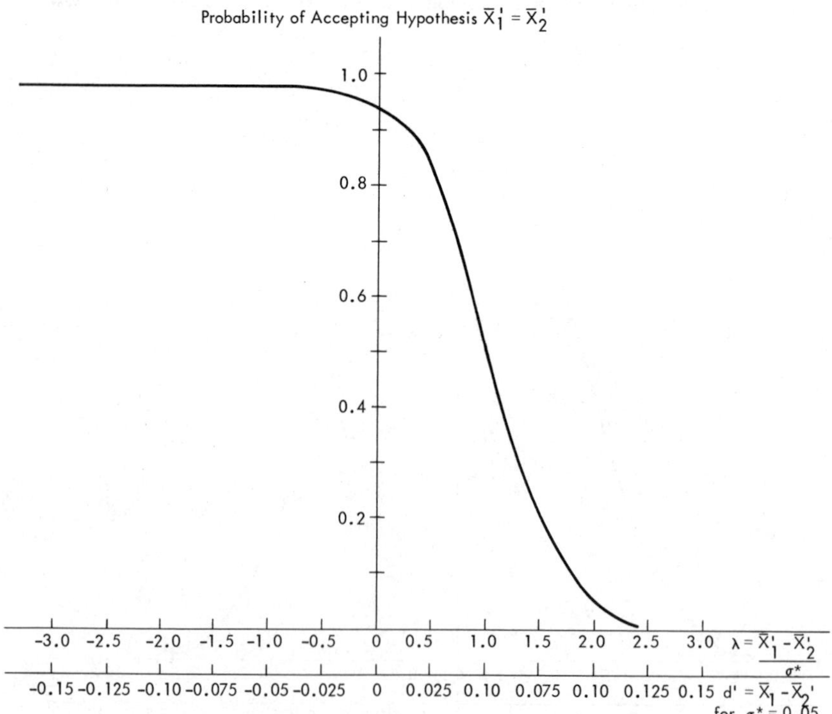

The lower half is derived in a similar manner, the curve being symmetrical around zero.

It will be noted that the OC curve is most easily derived by arbitrarily selecting values in column (3) and computing the values in columns (2) and (1) from those in (3) by multiplying by $\sigma'^2_{\bar{X}_1-\bar{X}_2}$ and $\sqrt{\frac{1}{n_1}+\frac{1}{n_2}}$, respectively. It will also be noted that since the OC curve is symmetrical in this case, the above values apply to the left-hand tail as well.

If our test had used only one critical value, say, 1.645, then the OC curve would have been derived as follows:

(1) $\lambda = d'/\sigma^*$	*(2)* $d' = \bar{X}'_1 - \bar{X}'_2$	*(3)* $d'/\sigma'_{\bar{X}_1-\bar{X}_2}$	*(4)* *1.645 − (3)*	*(5)* *Prob.* [$z \leq$ *(4)*]
−0.30	−0.015	−0.50	2.145	0.98
0	0	0	1.645	0.95
0.30	0.015	0.50	1.145	0.87
0.60	0.030	1.00	0.645	0.74
0.90	0.045	1.50	0.145	0.56
1.20	0.060	2.00	−0.355	0.36
1.50	0.075	2.50	−0.855	0.20
1.80	0.090	3.00	−1.355	0.09
2.10	0.105	3.50	−1.855	0.03
2.40	0.120	4.00	−2.355	0.01

A picture of this one-tail OC curve is given in Figure 26.2.

2.2. Testing the Difference between Two Means when σ'_1 and σ'_2 Are Unknown but Equal

2.2.1. Theory. When σ'_1 and σ'_2 are unknown but *are known to be equal,* we use a t statistic for testing the difference between two means. The t test is an exact test if the universes are normal and have a common variance. If they are moderately nonnormal, but have common variances, the t test will give fairly good results.

If the universes are normal with a common variance, then the statistic

(26.5)
$$t = \frac{d - d'_0}{s\sqrt{\frac{1}{n_1}+\frac{1}{n_2}}}$$

has a t distribution with $\nu = n_1 + n_2 - 2$, where $d = \bar{X}_1 - \bar{X}_2$, d'_0 equals the hypothetical value for $\bar{X}'_1 - \bar{X}'_2$ that is being tested,

(26.6)
$$s = \sqrt{\frac{(n_1 - 1)s_1^2 + (n_2 - 1)s_1^2}{n_1 + n_2 - 2}} = \sqrt{\frac{\Sigma x_1^2 + \Sigma x_2^2}{n_1 + n_2 - 2}}$$

and $x_1 = X_1 - \bar{X}_1$ and $x_2 = X_2 - \bar{X}_2$. This means that in employing this test, if we take $\alpha = 0.05$, the critical values will be $\pm t_{0.025}$ for $\nu = n_1 + n_2 - 2$, if we use a two-tail test, or $t_{0.05}$ for $\nu = n_1 + n_2 - 2$, if we use a one-tail test. (These will be $t_{0.05}$ and $t_{0.10}$ of Table B of Appendix II, because of the table's doubling of the tail area.)

2.2.2. An Example. The following are data on the drained weight after filling of the contents of size No. 2½ cans of Standard Grade Tomatoes in Puree:[4]

Sample Taken in the Morning	*Sample Taken in the Afternoon*
22.0 oz.	22.5 oz.
22.5	19.5
22.5	22.5
24.0	22.0
23.5	21.0

Between the morning and afternoon production runs, the filling machine had been reset and there was some question as to whether the averages were the same for both periods. There is no reason, however, to believe that the variation in weight was different for the morning and afternoon runs.

Statistically we wish to test the hypothesis that $\bar{X}'_1 = \bar{X}'_2$ with the assumption that $\sigma'_1 = \sigma'_2$. The work is as follows:
First compute

$$\bar{X}_1 = \frac{22.0 + 22.5 + 22.5 + 24.0 + 23.5}{5} = 22.9$$

$$\bar{X}_2 = \frac{22.5 + 19.5 + 22.5 + 22.0 + 21.0}{5} = 21.5$$

For convenience pick 20 as an arbitrary origin and set $\zeta = X - 20$. Then compute

$$\Sigma x_1^2 = \Sigma \zeta_1^2 - (\Sigma \zeta_1)^2/n_1 = 2^2 + 2.5^2 + 2.5^2 + 4^2 + 3.5^2 - \frac{(14.5)^2}{5}$$

$$= 2.70$$

$$\Sigma x_2^2 = \Sigma \zeta_2^2 - (\Sigma \zeta_2)^2/n_2 = 2.5^2 + 0.5^2 + 2.5^2 + 2^2 + 1^2 - \frac{(8.5)^2}{5}$$

$$= 3.30$$

Finally compute

[4] A larger set of data on weights of cans of tomatoes are given in Table 26.1. Data are from E. L. Grant and R. S. Leavenworth, *Statistical Quality Control* (4th ed.; New York: McGraw-Hill, 1972), p. 41.

$$s = \sqrt{\frac{\Sigma x_1^2 + \Sigma x_2^2}{n_1 + n_2 - 2}} = \sqrt{\frac{2.70 + 3.30}{5 + 5 - 2}} = 0.866$$

If we take $\alpha = 0.05$ and use a two-tail test, we have the critical values ± 2.306. Computing t for our sample, we get

$$t = \frac{22.9 - 21.5}{0.866\sqrt{\frac{1}{5} + \frac{1}{5}}} = \frac{1.4}{0.547} = 2.56$$

Since this falls above the upper critical value, we reject the hypothesis $\bar{X}'_1 = \bar{X}'_2$.

2.2.3. The OC Curve for the Test. OC curves for two-tail t tests are given in Figure 25.6 and OC curves for one-tail t tests are given in Figure 15.9. In both figures $\alpha = 0.05$. When these are used with respect to a test of the difference between two means, we must take as the value of λ the quantity

$$\lambda = \frac{|d_0 - d'|}{\sigma'} \sqrt{\frac{n_1 n_2/(n_1 + n_2)}{n}} \tag{26.7}$$

where d_0 is the hypothetical difference between means that is being tested and d' is the actual difference, σ' is the common standard deviation of the two universes, and $n = n_1 + n_2 - 1$. Since in both cases we must guess σ', the accuracy of our OC curves with respect to the assumed differences $d' = \bar{X}'_1 - \bar{X}'_2$ will depend on the accuracy of our estimate of σ'. If σ' is overestimated, we will be on the conservative side, in that the OC curve based on it will yield a result higher than the true probability of not detecting a given difference.

To illustrate the use of Figure 25.6, suppose that the common universe standard deviation is guessed to be 1 ounce, and let $n_1 = n_2 = 5$. Then the probability of accepting the hypothesis $\bar{X}'_1 = \bar{X}'_2$, when actually $\bar{X}'_1 - \bar{X}'_2$ is as great as 1 ounce and a two-tail test is used, is found by computing

$$\lambda = \frac{|\bar{X}'_1 - \bar{X}'_2|}{\sigma'} \sqrt{\frac{n_1 n_2/(n_1 + n_2)}{n_1 + n_2 - 1}} = \frac{1}{1}\sqrt{\frac{25/10}{9}} = 0.527$$

and noting that the height of the curve $n = 9$ at $\lambda = 0.527$ is approximately 0.7. Thus the chance of detecting a difference of 1 ounce with samples of 5 and a common universe standard deviation of 1 is about $1 - 0.7 = 0.3$.

2.2.4. Use of the Range in Testing the Difference between Two Means. If samples are small and equal in size and if it is known that the universe standard deviations are equal, a quick test of a difference between two means can be effected by using the average range. Such a test will not be as efficient as one using the standard deviation, but the speed and ease of calculation may more than offset the loss.

Consider again the example of Section 2.2.2. Here the range of the morning sample of cans was 2 ounces and that of the afternoon sample was 3 ounces. Following the Patnaik procedure described in Section 6 of Chapter 6, we can under the hypothesis that $\bar{X}'_1 = \bar{X}'_2$ treat $\dfrac{\bar{X}_1 - \bar{X}_2}{\left(\dfrac{\bar{R}}{d_2^*}\right)\sqrt{\dfrac{2}{n}}}$ as if it had a t distribution with degrees of freedom given by Table D3. For the given example, we have

$$t = \frac{22.9 - 21.5}{\left(\dfrac{2.5}{2.5}\right)\sqrt{\dfrac{2}{5}}} = \frac{1.4}{0.66} = 2.12$$

The critical t will be that for 7.5 degrees of freedom since this is the value of ν given by Table D3 for $g = 2$ and $m = 5$. For $\alpha = 0.05$ and a two-tail test, this will be a critical t of about 2.33. Since the sample t is less than this critical t, we will accept the hypothesis and conclude that the process is not actually running lighter in the afternoon that in the morning. The same conclusion was reached in Section 2.2.2, but the two tests do not have to agree.

2.3. Testing the Difference between Two Sample Means when σ'_1 and σ'_2 Are Unknown and May Not Be Equal

2.3.1. Theory. In many problems we do not know σ'_1 and σ'_2 nor do we know that they are equal. In this instance an excellent way of proceding is to apply a special test devised by B. L. Welch. This test is based on a more exact procedure developed by Alice Aspin and is consequently known as the Aspin-Welch test.[5]

The Aspin-Welch test consists in treating

$$t = \frac{d - d'_0}{\sqrt{\dfrac{s_1^2}{n_1} + \dfrac{s_2^2}{n_2}}}$$

as if it had a t distribution with degrees of freedom given by

$$\nu = \frac{1}{\dfrac{c^2}{n_1 - 1} + \dfrac{(1-c)^2}{n_2 - 1}} \tag{26.8}$$

where

[5] See particularly *Biometrika* 36 (1949), pp. 290–96.

(26.9)

$$c = \frac{s_1^2/n_1}{s_1^2/n_1 + s_2^2/n_2},$$

$$s_1^2 = \frac{\Sigma(X_1 - \bar{X}_1)^2}{n_1 - 1} \quad \text{and} \quad s_2^2 = \frac{\Sigma(X_2 - \bar{X}_2)^2}{n_2 - 1}$$

Mrs. Aspin gives tables for a more exact test of the above statistic. These show that when $c = 0.5$, which if $n_1 = n_2$ suggests that $\sigma'_1 = \sigma'_2$, the procedure of Mrs. Aspin yields a t test with degrees of freedom at least as large as those obtained when we *know* that $\sigma'_1 = \sigma'_2$ and run the ordinary t test described in Section 2.2. On the other hand, when c is small or large, say, less than 0.2 or greater than 0.8, the degrees of freedom approach those yielded by estimating σ'_1 and σ'_2 separately, viz. $n_1 - 1$ and $n_2 - 1$. The general conclusion is that if we know that $\sigma'_1 = \sigma'_2$, it is better to use the test of Section 2.2. If we do not have this knowledge, it is better to use the Aspin-Welch test.

Some statisticians have based the use of the t test of Section 2.2 on the preliminary test[6] and acceptance of the hypothesis that $\sigma'_1 = \sigma'_2$. With the Aspin-Welch test this is not necessary since the sample s_1^2 and s_2^2 enter into and determine the degrees of freedom of the test.

2.3.2. An Example. Suppose that in our tomato problem, we had the following results:

Morning sample: 22.0, 22.5, 22.5, 24.0, and 23.5 ounces
Afternoon sample: 24.5, 19.5, 25.5, 20.0, 18.0, 21.5, and 21.5 ounces

and suppose we are unwilling to assume that the process has the same variability in the afternoon as it has in the morning. Then we can test the hypothesis that $\bar{X}'_1 = \bar{X}'_2$ as follows:

We have

$$\bar{X}_1 = \frac{22.0 + 22.5 + 22.5 + 24.0 + 23.5}{5} = 22.9$$

$$\bar{X}_2 = \frac{24.5 + 19.5 + 25.5 + 20.0 + 18.0 + 21.5 + 21.5}{7} = 21.5$$

Also, if we pick 20 as an arbitrary origin, we have[7] for $\zeta = X - 20$,

$$s_1^2 = \frac{\Sigma(X_1 - \bar{X}_1)^2}{n_1 - 1} = \frac{\Sigma(\zeta_1 - \bar{\zeta}_1)^2}{n_1 - 1} = \frac{\Sigma\zeta_1^2 - (\Sigma\zeta_1)^2/n_1}{n_1 - 1}$$

$$= \frac{2^2 + 2.5^2 + 2.5^2 + 4^2 + 3.5^2 - (14.5)^2/5}{4}$$

$$= 0.675$$

[6] See Section 3.1 below.

[7] See Section 2.2.2 above.

$$s_2^2 = \frac{\Sigma(X_2 - \bar{X}_2)^2}{n_2 - 1} = \frac{\Sigma(\zeta_2 - \bar{\zeta}_2)^2}{n_2 - 1} = \frac{\Sigma\zeta_2^2 - (\Sigma\zeta_2)^2/n_2}{n_1 - 1}$$

$$= \frac{4.5^2 + (-0.5)^2 + 5.5^2 + 0^2 + (-2)^2 + 1.5^2 + 1.5^2 - (10.5)^2/7}{6}$$

$$= 7.25$$

$$\frac{s_1^2}{n_1} = \frac{0.675}{5} = 0.135 \quad \text{and} \quad \frac{s_2^2}{n_2} = \frac{7.25}{7} = 1.036$$

and

$$c = \frac{0.135}{0.135 + 1.036} = 0.115$$

Hence, the degrees of freedom are given by

$$\frac{1}{\nu} = \left[\frac{(0.15)^2}{4} + \frac{(0.885)^2}{6}\right] = [0.0035 + 0.123] = 0.127$$

or

$$\nu = 7.9 \doteq 8$$

Hence the test is

$$t = \frac{\bar{X}_1 - \bar{X}_2}{\sqrt{\dfrac{s_1^2}{n_1} + \dfrac{s_2^2}{n_2}}} = \frac{22.9 - 21.5}{\sqrt{0.135 + 1.037}} = 1.3$$

which is less than the $t_{0.025}$ point 2.306. We, therefore, accept the hypothesis $\bar{X}'_1 = \bar{X}'_2$. Here we lost 2 degrees of freedom as compared with the ordinary t test.

2.3.3. The OC Curve. As far as the writer knows, the OC curve for an Aspin-Welch test has not been worked out. It can probably be roughly approximated, however, by one of the curves in Figure 15.9 or Figure 25.6, if σ' is taken equal to the average of the guesses as to σ'_1 and σ'_2 and formula (26.7) is used with n equal to one more than the degrees of freedom ν.

2.4. The Design of a Test of a Difference between Two Means

A test of a difference between two means may be designed to meet various criteria. Let us consider the following.

Suppose that in comparing two means we want a $1 - \beta$ chance of detecting a difference of δ between $\bar{X}'_1$ and $\bar{X}'_2$ and wish to run a two-tail test at the α level of significance. Then for a known σ'_1 and σ'_2 the test procedure

FIGURE 26.3
A Geometrical Representation of a Two-Tail Test of a Difference between Two Means

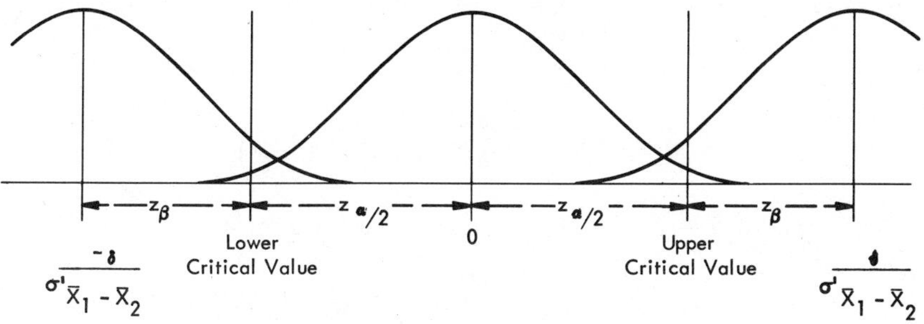

would be such that the geometrical relationship shown in Figure 26.3 would hold true. This indicates we must have

$$\frac{\delta}{\sigma'_{\bar{X}_1-\bar{X}_2}}=z_{\alpha/2}+z_\beta$$

Here $z_{\alpha/2}$ and z_β are the standard normal deviates the probability of exceeding which are $\alpha/2$ and β, respectively. The above yields

(26.10) $$\sigma'_{\bar{X}_1-\bar{X}_2}=\frac{\delta}{z_{\alpha/2}+z_\beta}$$

Now if measurement of X_1 costs $\$c_1$ per unit and measurement of X_2 $\$c_2$ per unit, then for a given precision, $\sigma'_{\bar{X}_1-\bar{X}_2}$, the total cost ($=c_1n_1+c_2n_2$) will be a minimum[8] if n_1 is chosen equal to $n_2\sqrt{\frac{c_2\sigma_1'^2}{c_1\sigma_2'^2}}$. To meet the desired risk criteria with minimum cost, we should, therefore, have

---

[8] Mathematically we minimize $c_1n_1+c_2n_2$ subject to the condition $\frac{\sigma_1'^2}{n_1}+\frac{\sigma_2'^2}{n_2}=K$. This yields $c_1\frac{\partial n_1}{\partial n_2}+c_2=0$, where $\frac{\partial n_1}{\partial n_2}$ is given by

$$-\frac{\sigma_1'^2}{n_1^2}\frac{\partial n_1}{\partial n_2}-\frac{\sigma_2'^2}{n_2^2}=0$$

or

$$\frac{\partial n_1}{\partial n_2}=-\frac{\sigma_2'^2n_1^2}{n_2^2\sigma_1'^2}$$

Hence for minimum cost we must have

$$-c_1\frac{\sigma_2'^2n_1^2}{\sigma_1'^2n_2^2}+c_2=0$$

or

$$n_1=n_2\sqrt{\frac{c_2\sigma_1'^2}{c_1\sigma_2'^2}}$$

$$\frac{\sigma_1'^2}{n_2\sqrt{\frac{c_2\sigma_1'^2}{c_1\sigma_2'^2}}}+\frac{\sigma_2'^2}{n_2}=\left(\frac{\delta}{z_{\alpha/2}+z_\beta}\right)^2$$

or

$$n_2=\frac{(z_{\alpha/2}+z_\beta)^2}{\delta^2}\left(\sigma_1'\sigma_2'\sqrt{\frac{c_1}{c_2}}+\sigma_2'^2\right) \tag{26.11}$$

For example, if $\delta=2$, $\alpha=0.05$, and $\beta=0.10$, we must have

$$n_2=\frac{(1.960+1.282)^2}{4}\left(\sigma_1'\sigma_2'\sqrt{\frac{c_1}{c_2}}+\sigma_2'^2\right)$$

and

$$n_1=n_2\sqrt{\frac{c_2\sigma_1'^2}{c_1\sigma_2'^2}}$$

From here on the decision about n_1 and n_2 will depend on our knowledge of σ_1', σ_1', c_1, and c_2. If we know these, n_1 and n_2 are immediately determined. If we guess them, n_1 and n_2 may be larger or smaller than we need to attain the desired criteria. If we have no knowledge at all about σ_1', σ_2', c_1, and c_2, it might be necessary to run a small "pilot study" before we plan the complete test.

If we have reason to believe that $\sigma_1'=\sigma_2'$ and if our guess as to this common σ' yields an n_1 and an n_2 such that $n_1+n_2-2<30$, then it is wise to check the results with the OC curves for t tests. For example, if $c_1 = c_2$ and $\sigma_1'=\sigma_2'$ ($=\sigma'$ say), and if we take $\alpha=0.05$, $\beta=0.10$, and $\delta = 1$ then n_1 will equal n_2 ($=n$ say), and if we guess $\sigma'=0.5$, then n will be 5. To see what the probability is of detecting a difference of 1 when $\sigma'^2 = 0.25$ and $n_1=n_2=5$, we compute

$$\lambda=\frac{\bar{X}_1'-\bar{X}_2'}{\sigma'}\sqrt{\frac{n_1n_2/(n_1+n_2)}{n_1+n_2-1}}=\frac{1}{0.5}\sqrt{\frac{25/10}{9}}=1.054$$

and note in Figure 25.6 that the height of the curve $n=9$ at $\lambda=1.054$ is approximately 0.25 instead of the 0.10 that we aimed for.

It is desirable, therefore, to redesign a test whenever the n turns out to be small and we have to use the t test. This redesigning will have to be done by trial and error. For example, in our present instance $n_1=n_2=5$ gave too high a result. Let us try $n_1=n_2=10$. Then for $\sigma'^2=0.25$ and $\bar{X}_1'-\bar{X}_2'=1$, we get $\lambda=\frac{1}{0.5}\sqrt{\frac{100/20}{19}}=1.026$ and find from Figure 25.6 that for $n=19$ the probability of acceptance is about 0.02. We have gone too far. It looks like $n=7$ will be right. Let us see. Using $n_1=n_2 = 7$, we have $\lambda=\frac{1}{0.5}\sqrt{\frac{49/14}{13}}=1.038$, which yields $P_\alpha=0.10$, or the characteristic we want.

If we guess σ'_1, σ'_2 and these are not equal, and if for estimated c_1 and c_2, formula (26.11) yields small n_1 and n_2, a safe way of proceeding would seem to be as follows. Average the guessed σ'_1 and σ'_2 and temporarily taking $n_1 = n_2 = n$ find the n for a t test with a common σ' equal to this average σ'. Double this n and compute n_1 and n_2 from the equations $n_2 + n_2\sqrt{\frac{c_2\sigma_1'^2}{c_1\sigma_2'^2}} = 2n$ and $n_1 = n_2\sqrt{\frac{c_2\sigma_1'^2}{c_1\sigma_2'^2}}$. The sample sizes derived in this way will yield an Aspin-Welch test with an OC curve that will probably be steeper than originally desired, but departure will at least be on the conservative side.

2.5. Confidence Limits for $\bar{X}'_1 - \bar{X}'_2$

Formulas for confidence limits for $\bar{X}'_1 - \bar{X}'_2$ will, as in the test procedure, depend upon our knowledge of σ'_2 and σ'_2.

If we know σ'_1 and σ'_2, 0.95 confidence limits for $\bar{X}'_1 - \bar{X}'_2$ are given by

(26.12)
$$\bar{X}_1 - \bar{X}_2 \pm 1.960\sqrt{\frac{\sigma_1'^2}{n_1} + \frac{\sigma_1'^2}{n_2}}$$

If we do not know σ'_1 and σ'_2 but know that they are equal, 0.95 confidence limits for $\bar{X}'_1 - \bar{X}'_2$ are given by

(26.13)
$$\bar{X}_1 - \bar{X}_2 \pm t_{0.025} s\sqrt{\frac{1}{n_1} + \frac{1}{n_2}}$$

where $t_{0.025}$ is the 0.025 point of a t distribution with degrees of freedom equal to $n_1 + n_2 - 2$ (the 0.05 point in Table B) and

$$s = \sqrt{\frac{\Sigma(X_1 - \bar{X}_1)^2 + \Sigma(X_2 - \bar{X}_2)^2}{n_1 + n_2 - 2}}$$

If we do not know σ'_1 and σ'_2 and cannot assume them to be equal, 0.95 confidence limits for $\bar{X}'_1 - \bar{X}'_2$ may be obtained from the Welch formula[9]

(26.14)
$$\bar{X}_1 - \bar{X}_2 + 1.960\sqrt{\frac{s_1^2}{n_1} + \frac{s_2^2}{n_2}}\left[1 + \frac{1 + (1.960)^2}{4}\frac{\left(\frac{s_1^4}{n_1^2(n_1 - 1)} + \frac{s_2^4}{n_2^2(n_2 - 1)}\right)}{\left(\frac{s_1^2}{n_1} + \frac{s_2^2}{n_2}\right)^2}\right]$$

where $s_1^2 = \dfrac{\Sigma(X_1 - \bar{X}_1)^2}{n_1 - 1}$ and $s_2^2 = \dfrac{\Sigma(X_2 - \bar{X}_2)^2}{n_2 - 1}$.

[9] B. L. Welch, *Biometrika* 34 (1947), p. 35.

3. TESTS PERTAINING TO THE RATIO OF TWO VARIANCES

A problem that often arises is to determine whether the variability of one set of data is significantly different from the variability of another set. It may be important to know, for example, whether process A yields a more uniform product than process B. Or, again, a process may be modified, and we might like to know whether the change has reduced or increased its variability.

3.1. Theory

Whereas in comparing two proportions or two means, we took arithmetic differences, in comparing two variances, it is easier mathematically to take the ratio of the two variances. The mathematical statisticians have found that when both universes are normal and the samples are independent, sample values of the ratio $s_1^2\sigma_2'^2/s_2^2\sigma_1'^2$ are distributed among many samples in the form of a distribution called the F distribution. In other words, the F distribution is the sampling distribution of the statistic $s_1^2\sigma_2'^2/s_2^2\sigma_1'^2$. Here s_1^2 =

FIGURE 26.4
Graph Illustrating the *F* Distribution

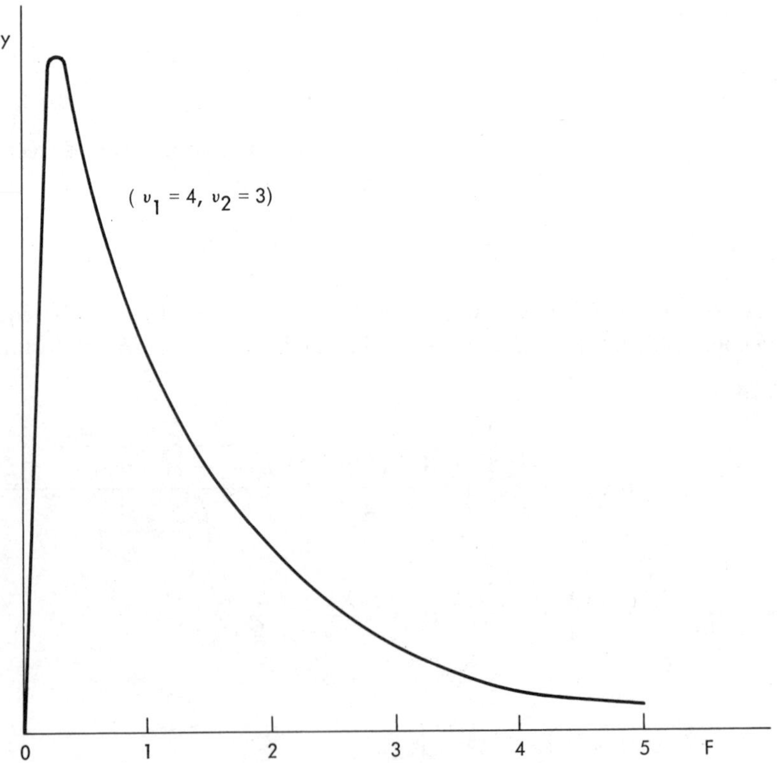

$\frac{\Sigma(X_1 - \bar{X}_1)^2}{n_1 - 1}$ and is an unbiased estimate of $\sigma_1'^2$, while $s_2^2 = \frac{\Sigma(X_2 - \bar{X}_2)^2}{n_2 - 1}$ and is an unbiased estimate of $\sigma_2'^2$.

A picture of the F distribution is shown in Figure 26.4, and its various percentage points are given in Table J of Appendix II. The distribution depends on two parameters ν_1 and ν_2, and in the present problem we have $\nu_1 = n_1 - 1$ and $\nu_2 = n_2 - 1$. For example, if we have many samples of 10 items each from universe I and also many samples of 5 items each from universe II, and if we pair these samples at random, computing the ratio $s_1^2\sigma_2'^2/s_2^2\sigma_1'^2$ for each pair, these sample ratios will form a distribution that will have the shape of the F distribution with $\nu_1 = 9$ and $\nu_2 = 4$. Consequently, it can be estimated that if the universes are normal, 95 percent of the ratios will be less than 6.00 and 99 percent less than 14.7 (see Table J).

3.2. An Example

Consider the data of Table 26.1 dealing again with weights of cans of tomatoes. For these data, $n_1 = n_2 = 40$, $\bar{X}_1 = 21.78$, $\bar{X}_2 = 20.71$, $s_1 = 1.76$, and $s_2 = 1.55$. It might be of interest to determine whether the morning variability

TABLE 26.1
Drained Weight after Filling of Contents of Size No. 2½ Cans of Standard Grade Tomatoes in Puree (weight in ounces)

Day	*Hour*	*Sample*					*Average*
	First Morning Tests						
Sept. 21	9:30	22.0	22.5	22.5	24.0	23.5	22.9
22	10:00	23.0	23.5	21.0	22.0	20.0	21.9
23	9:30	21.0	22.5	20.0	22.0	22.0	21.5
25	8:00	19.5	20.5	21.0	20.5	21.0	20.5
27	7:30	21.5	25.0	21.0	19.0	21.0	21.5
28	7:15	22.0	17.5	21.0	22.0	23.5	21.2
29	9:00	22.5	21.5	21.0	21.5	23.5	22.0
30	9:30	22.5	24.5	25.5	20.0	21.0	22.7
	Last Afternoon Tests						
21	5:25	22.5	19.5	22.5	22.0	21.0	21.5
22	5:00	21.5	20.5	19.0	19.5	19.5	20.0
23	3:30	19.0	21.0	21.0	21.0	20.5	20.5
25	5:30	21.0	20.5	19.5	22.0	21.0	20.8
27	4:40	19.5	22.5	15.5	20.0	22.5	20.0
28	3:30	21.0	19.5	22.0	20.0	20.0	20.5
29	5:20	21.5	22.0	21.5	20.5	22.5	21.6
30	3:30	22.5	19.5	21.5	20.5	20.0	20.8

SOURCE: Data reproduced with permission from E. L. Grant and R. S. Leavenworth, *Statistical Quality Control* (4th ed.; New York: McGraw-Hill, 1972), p. 41.

is in fact greater than the afternoon variability as suggested by the sample standard deviations.

As in previous problems, our hypothesis will be that there is no difference in universe variances, or, in other words, that $\sigma_1'^2 = \sigma_2'^2$. Hence, according to the previous section, s_1^2/s_2^2 will be distributed in the form of an F distribution with $\nu_1 = n_1 - 1$ and $\nu_2 = n_2 - 1$. If we take $\alpha = 0.05$, we can test the hypothesis $\sigma_1'^2 = \sigma_2'^2$ by computing s_1^2/s_2^2 for our sample and noting its relationship to the value of F in Table J for $\nu_1 = 39$ and $\nu_2 = 39$. Thus, we have $s_1^2 = 3.10$, $s_2^2 = 2.40$, and $s_1^2/s_2^2 = 3.10/2.40 = 1.29$.

Let us suppose that there may be some reason for believing that if there is any difference, the larger variance will occur in the morning. It might be argued, for example, that it takes a little time for the men and machines to "settle down" and that by afternoon the variability in results is significantly less than in the morning. Under these conditions the test should be a one-tail test, and, with $\alpha = 0.05$, the critical value for our ratio is the $F_{0.05}$ point for $\nu_1 = 39$ and $\nu_2 = 39$. Table J does not give this value of F, but it does give $F_{0.05}$ for $\nu_1 = 30$, $\nu_2 = 30$, as equal to 1.84; $F_{0.05}$ for $\nu_1 = 30$, $\nu_2 = 60$, as equal to 1.65; $F_{0.05}$ for $\nu_1 = 60$, $\nu_2 = 30$, as equal to 1.74; and $F_{0.05}$ for $\nu_1 = 60$, $\nu_2 = 60$, as equal to 1.53. $F_{0.05}$ for $\nu_1 = 39$, $\nu_2 = 39$, lies between these limits. Hence, the sample value of s_1^2/s_2^2, viz. 1.29, is not above the $F_{0.05}$ for $\nu_1 = 39$, $\nu_2 = 39$; and we therefore accept the hypothesis that $\sigma_1' = \sigma_2'$. In other words, we conclude that the variances are the same in the morning and afternoon.

If we prefer a two-tail test, we can find a lower critical value from Table J by taking $F_{1-\alpha/2}$ for $\nu_1 = n_1 - 1$ and $\nu_2 = n_2 - 1$ as equal to $1/F_{\alpha/2}$ for $\nu_1 = n_2 - 1$ and $\nu_2 = n_1 - 1$. In other words, to get the lower percentage point $F_{1-\alpha/2}$, we take the reciprocal of $F_{\alpha/2}$ for the degrees of freedom in reverse. For example, $F_{0.99}$ for $\nu_1 = 6$, $\nu_2 = 10$, equals the reciprocal of $F_{0.01}$ for $\nu_1 = 10$, $\nu_2 = 6$. Thus, $F_{0.99}$ for $\nu_1 = 6$, $\nu_2 = 10$, is equal to 1/7.87, or 0.127. (See Table J and note that F is cumulated from the upper end.)

To illustrate a two-tail test, suppose in the above instance that we wanted to test merely whether the variances of the weights of the cans of tomatoes in the morning and afternoon were different. To do this, let $\alpha = 0.05$. Our critical values will then be $F_{0.975}$ and $F_{0.025}$. Since $\nu_1 = \nu_2 = 39$ in this case, we would take $F_{0.975}$ as equal to $1/F_{0.025}$. Linear interpolation on the reciprocals[10] of ν_1 and ν_2 yields $F_{0.025}$ as equal approximately to 1.89. Hence, $F_{0.975}$ will equal approximately $1/1.89 = 0.529$. The two critical values are thus approximately 1.89 and 0.529. If s_1^2/s_2^2 exceeds 1.89 or falls short of 0.529, we shall reject the hypothesis that $\sigma_1' = \sigma_2'$.

It is to be noted that in running a two-tail test an equivalent procedure to that described above is to follow the rule of not deciding which sample variance to designate as s_1^2 and which s_2^2 until we have seen the data. Then

[10] See notes on Table J in Appendix II.

we call the *larger* variance s_1^2 and ν_1 its degrees of freedom. That is, we form the sample F-ratio by putting the larger s^2 on top. The ratio is then referred to $F_{\alpha/2}$.

3.3. The OC Curve for a One-Tail Test

The OC curve for a comparison of two variances gives the probability of accepting the hypothesis $\sigma_1' = \sigma_2'$ as a function of the actual ratio between the two universe variances. The OC curve, of course, will depend on the value of α and will be different for a one-tail test than for a two-tail test.

The OC curve for a one-tail test can be derived as follows. As pointed out above the ratio $s_1^2\sigma_1'^2/s_2^2\sigma_1'^2$ is distributed in the form of an F distribution. For a one-tail test with $\alpha = 0.05$, the critical value of F is taken as the 0.05 point in the F table for $\nu_1 = n_1 - 1$, and $\nu_2 = n_2 - 1$. Call this $F_{0.05}$. Thus, when $\sigma_1'^2 = \sigma_2'^2$ or $\sigma_1'^2/\sigma_2'^2 = 1$, we have Prob. $(s_1^2/s_2^2 \leq F_{0.05})$ $= 0.95$. This gives us one point on the OC curve. To locate other points we find the Prob. $(s_1^2/s_2^2 \leq F_{0.05})$ when[11] the ratio $\lambda = \sigma_1'/\sigma_2' > 1$; in other words, we find[12] the Prob. $(s_1^2/s_2^2 \leq F_{0.05} | \sigma_1'/\sigma_2' = \lambda)$, where $\lambda > 1$. If, however, we multiply both sides of the inequality by $\sigma_2'^2/\sigma_1'^2$, this probability is equal to Prob. $(s_1^2\sigma_2'^2/s_2^2\sigma_1'^2 \leq F_{0.05}\sigma_2'^2/\sigma_1'^2 | \sigma_1'/\sigma_2' = \lambda)$ or to Prob. $(s_1^2\sigma_2'^2/s_2^2\sigma_1'^2 \leq F_{0.05}/\lambda^2 | \sigma_1'/\sigma_2' = \lambda)$. Since $s_1^2\sigma_2'^2/s_2^2\sigma_1'^2$ is distributed in the form of an F distribution, this last probability can theoretically be found from an F table. We simply compute $F_{0.05}/\lambda^2$ for given values of λ and then from an F table find the probability of F's exceeding $F_{0.05}/\lambda^2$ for the given values of $\nu_1 = n_1 - 1$ and $\nu_2 = n_2 - 1$.

Interpolation in the F tables is not easy, however; so in practice it is simpler to work in the reverse direction. Set the probability of acceptance, i.e., Prob. $(s_1^2\sigma_2'^2/s_2^2\sigma_1'^2 \leq F_{0.05}/\lambda^2 | \sigma_1'/\sigma_2' = \lambda)$ equal to β and note that $F_{0.05}/\lambda^2$ is the equivalent of $F_{1-\beta}$. Thus we have $F_{0.05}/F_{1-\beta} = \lambda^2$. Hence, if we start with β, we can find $F_{1-\beta}$ in the F table (i.e., the value of F which has the probability of $1 - \beta$ of being exceeded) and then compute λ. In other words, we find the abscissa of our OC curve for a selected ordinate instead of finding the ordinate for a selected abscissa.

To illustrate, let us find the OC curve for the test described in the previous section. There $\nu_1 = \nu_2 = 39$. From Table J linear interpolation on the reciprocals[13] of ν_1 and ν_2 yields $F_{0.05}$ for $\nu_1 = 39$, $\nu_2 = 39$, as equal to 1.70.

We know that when $\lambda = 1$, $\beta = 0.95$. This gives us one point on the OC curve. Let us now find λ for $\beta = 0.05$. If $\beta = 0.05$, $1 - \beta = 0.95$,

---

[11] Since we are dealing with a one-tail test, we are concerned only with values of $\sigma_1'/\sigma_2' > 1$.

[12] Note that Prob. $(A|B)$ means probability of A given B. Hence, Prob. $(s_1^2/s_2^2 \leq F_{0.05} | \sigma_1'/\sigma_2' = \lambda)$ means the probability that s_1^2/s_2^2 should be less than or equal to $F_{0.05}$, given $\sigma_1'/\sigma_2' = \lambda$.

[13] See notes on Table J in Appendix II.

FIGURE 26.5
Operating Characteristic Curves for One-Tail Tests Comparing Two Variances, α Taken as 0.05 and $n_1 = n_2$*

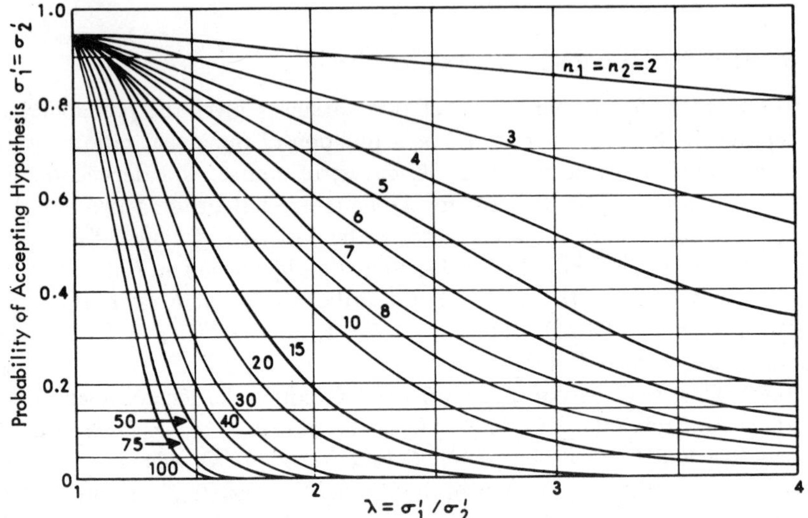

* Probabilities computed from Figure 26.5 may be off by as much as 0.03.
SOURCE: Reproduced from Figure 3 of C. D. Ferris, F. E. Grubbs, and C. L. Weaver, "Operating Characteristics for the Common Statistical Tests of Significance," *Annals of Mathematical Statistics* 17 (1946), p. 185.

and, as pointed out above, the 0.95 point of the F distribution is given by $1/F_{0.05}$ with the degrees of freedom reversed. Since $F_{0.05}$ for $\nu_1 = \nu_2 = 39$ is 1.70, then $F_{0.95}$ for $\nu_1 = \nu_2 = 39$ equals $1/1.70 = 0.588$. Hence, $\lambda^2 = \frac{F_{0.05}}{F_{0.95}} = \frac{1.70}{0.588} = 2.89$ and $\lambda = 1.70$. In other words, at $\lambda = 1.70$ the ordinate of the OC curve is 0.05. In like manner, we can find the abscissa for other values of β.

Fortunately, the work of drawing OC curves involving the ratio of two variances has already been done for us.[14] A chart that can be used when $n_1 = n_2$ is given in Figure 26.5. For cases in which $n_1 \neq n_2$, the reader is referred to the references given in footnote 14.

Thus, instead of computing the 0.05 point on the OC curve for the problem above, we could have read it directly from Figure 26.5. For $\lambda = 1.70$, for example, Figure 26.5 shows that the ordinate of the OC curve for $n_1 = n_2 = 40$ is approximately 0.05.

---

[14] See C. D. Ferris, F. E. Grubbs, and C. L. Weaver, "Operating Characteristics for the Common Statistical Tests of Significance," *Annals of Mathematical Statistics* 17 (1946), pp. 178–97; and Statistical Research Group, Columbia University, *Techniques of Statistical Analysis* (New York: McGraw-Hill, 1947), chap. VIII.

3.4. The OC Curve for a Two-Tail Test

Since two-tail tests are used less frequently in comparing variances than one-tail tests, the references of footnote 14 do not give OC curves for these tests. They can be derived in the same way as OC curves for one-tail tests, however. Thus, with $\alpha = 0.05$, we have, for $\sigma_1' = \sigma_2'$, Prob. $(F_{0.975} \leq s_1^2/s_2^2 \leq F_{0.025}) = \text{Prob.}\left(\frac{1}{F_{0.025}(\nu_2, \nu_1)} \leq \frac{s_1^2}{s_2^2} \leq F_{0.025}(\nu_1, \nu_2)\right) = 0.95$, which will give us one point on the OC curve.[15] To find other points, it is necessary to find

$$\text{Prob.}\left(\frac{1}{F_{0.025}(\nu_2, \nu_1)} \leq \frac{s_1^2}{s_2^2} \leq F_{0.025}(\nu_1, \nu_2) | \sigma_1'^2/\sigma_2'^2 = \lambda^2\right)$$

But this equals

$$\text{Prob.}\left(\frac{\sigma_2'^2}{\sigma_1'^2 F_{0.025}(\nu_2, \nu_1)} \leq \frac{s_1^2\sigma_2'^2}{s_2^2\sigma_1'^2} \leq \frac{\sigma_2'^2 F_{0.025}(\nu_1, \nu_2)}{\sigma_1'^2} | \sigma_1'^2/\sigma_2'^2 = \lambda^2\right)$$

Since $s_1^2\sigma_2'^2/s_2^2\sigma_1'^2$ is distributed in the form of an F distribution, this last probability can be derived from an F table. Thus, if $\sigma_1'^2/\sigma_2'^2 = \lambda^2$, we can compute $F_{0.025}(\nu_1, \nu_2)/\lambda^2$ and $1/F_{0.025}(\nu_2, \nu_1)\lambda^2$ and find from an F table the probability of

$$\frac{1}{F_{0.025}(\nu_2, \nu_1)\lambda^2} \leq F \leq \frac{F_{0.025}(\nu_1, \nu_2)}{\lambda^2}$$

For $n_1 = n_2$, charts of OC curves for two-tail F-tests will be found in the Bowker-Lieberman handbook.

3.5 The Design of a Test

For samples of equal size, Figure 26.5 will help us design a one-tail test with $\alpha = 0.05$. Thus, if we wish a test that will not fail to detect a λ ratio of 2.0 more than 10 times out of 100, we will find immediately from looking at Figure 26.5 that n_1 and n_2 should each be 20.

If sample tests do not cost the same and therefore make it inadvisable to have $n_1 = n_2$, we can use Table L of Appendix II to find a combination of n_1 and n_2 that for a given β and λ will yield minimum cost. For example, Table L shows[16] that for $\beta = 0.10$ and for λ^2 in the neighborhood of 4, samples should be such that $n_1 - 1$ and $n_2 - 1$ are (60, 12) or (24, 15) or (15, 24) or (12, 40) or (10, 120). If sample I tests cost double sample II tests, i.e., if $c_1 = 2c_2$, then the costs for these various tests would be $135c_2$,

---

[15] Note that $F_\alpha(\nu_1, \nu_2)$ means the F_α point for ν_1 and ν_2.

[16] For use of Table L in this connection take $\phi^2 = \lambda^2$.

$66c_2$, $57c_2$, $67c_2$, and $143c_2$, respectively. Of the group, the third would be the cheapest. Hence, to minimize our cost for the desired protection, we would take $n_1 = 16$, $n_2 = 25$.

3.6. Confidence Limits for the Ratio of Two Variances

Ninety-five percent confidence limits for the ratio of the two universe variances can be derived by computing $s_1^2F_{0.025}(\nu_2, \nu_1)/s_2^2$ and $s_1^2/F_{0.025}(\nu_1, \nu_2)s_2^2$. The first gives an upper limit, the second a lower limit for $\sigma_1'^2/\sigma_2'^2$.

As an example, consider the problem of Section 3.2 and suppose that $s_1^2 = 7.20$ instead of 3.10, as was actually the case. Then, 0.95 confidence limits for $\sigma_1'^2/\sigma_2'^2$ would be

$$\frac{7.20(1.70)}{2.40} = 5.10$$

$$\frac{7.20}{2.40(1.70)} = 1.76$$

It will be noticed that the F factor is the same (1.70) in both cases because $n_1 = n_2$. If $n_1 \neq n_2$, the F factors would be different.

4. PROBLEMS

26.1. You wish to determine whether lot B has a larger fraction nonconforming than lot A. A random sample of 250 items from lot A is found to contain 10 nonconforming items. Another random sample of 300 items from lot B is found to contain 18 nonconforming items. What conclusion do you draw from these results?

26.2. During July the fraction nonconforming in an output of 1,125 items is 0.025. During August, after some changes had been introduced, the fraction nonconforming in an output of 1,250 was 0.018. Could we reasonably infer from these data that the process had been improved by the changes that were made? Justify your answer.

26.3. You wish to determine whether the night shift differs from the day shift with respect to the fraction nonconforming of output. After a week's operations you find the night shift turned out 542 items, of which 41 were nonconforming. During the same time the day shift turned out 632 items, of which 45 were nonconforming. What are your conclusions?

26.4. *a.* From past experience it is known that machine A has a standard deviation of 0.003 and machine B a standard deviation of 0.004. The settings of the two machines have recently been changed, but both were presumably set alike. A sample of 10 items from machine A yields a mean of 2.534, whereas a sample of 15 items from machine B yields a mean of 2.542. Could you reasonably infer from these data that the machines were not set alike, i.e., that the process means were different? Assume that the setting of a machine has no effect on the standard deviation.

b. Sketch the OC curve for the test.

26.5. *a.* The standard deviation of tests for determining percentage of nitrogen in mixed fertilizer is known to be 0.06 percent. In a certain experiment, the sample of fertilizer taken from a bag is ground to a powder and put into two different bottles. One bottle is retained by the fertilizer company; the other is sent to a state laboratory for test. At each place three determinations are made of the percentage of nitrogen. The results are as follows:

Company Laboratory	*State Laboratory*
4.42%	4.39%
4.43	4.48
4.58	4.31

Could you reasonably conclude from these results that the method of determining percentage of nitrogen used by the state laboratory has a downward bias relative to that used by the company?

b. Sketch the OC curve for the test.

26.6. *a.* A sample of 500 rounds of ammunition supplied by manufacturer A yielded a mean muzzle velocity of 1,752 feet per second, with a standard deviation of 102 feet per second. A sample of 400 rounds of ammunition supplied by manufacturer B yielded a mean muzzle velocity of 1,713 feet per second, with a standard deviation of 90 feet per second. Would you conclude from these results that the two lots of ammunition differed with respect to their mean velocities? Justify your answer.

b. Sketch the OC curve for your test.

c. Determine 0.95 confidence limits for the difference between the universe means.

26.7. *a.* During a given week a process turned out 220 items of product, the mean weight of which was 2.46 pounds, with a standard deviation of 0.57 pounds. During the next week a different raw material was used, and the mean weight of the 205 items turned out was 2.55 pounds, with a standard deviation of 0.48 pounds. Would you conclude from these results that the use of the new raw material had increased the mean weight of the product? Justify your answer.

b. Sketch the OC curve for the test.

c. Determine 0.95 confidence limits for the difference in mean weight.

26.8. *a.* The following are 16 independent determinations of the melting point of hydroquinone, 8 made by analyst I and 8 made by analyst II. Data are in degrees centigrade.[17]

[17] See Grant Wernimont, "Quality Control in the Chemical Industry, II. Statistical Quality Control in the Chemical Laboratory," *Industrial Quality Control,* May 1947, p. 8.

Analyst I	*Analyst II*
174.0	173.0
173.5	173.0
173.0	172.0
173.5	173.0
171.5	171.0
172.5	172.0
173.5	171.0
173.5	172.0

Would you conclude from these data that there was a tendency for one analyst to get higher results than the other? Justify your answer.

b. Sketch the OC curve for your test.

c. Use another test to answer *a.* Do your results agree?

d. Determine 0.95 confidence limits for the difference.

26.9. *a.* It is argued that the resistance of wire A is greater than the resistance of wire B. You make tests on each wire with the following results:

Wire A	*Wire B*
0.140 ohms	0.135 ohms
0.138	0.140
0.143	0.142
0.142	0.136
0.144	0.138
0.137	0.140

What conclusions do you draw? Justify your answer.

b. Sketch the OC curve for your test.

c. Use another test to answer *a.* Do your results agree?

d. Determine 0.95 confidence limits for the difference.

26.10. *a.* It is argued that the skein strength of yarn A is greater than that of yarn B. Tests yield the following:

Yarn A	*Yarn B*
99 pounds	93 pounds
93	94
99	75
97	84
90	91
96	
93	
88	
89	

What conclusions would you draw? Justify your answer.

b. Sketch the OC curve for your test.

c. Determine 0.95 confidence limits for the difference.

26.11. *a.* Two operators are making fragmentation bomb bases. The means of samples of 5 each from the output of the two operators are reported as follows:[18]

Operator A	*Operator B*
0.8372 inches	0.8336 inches
0.8324	0.8356
0.8318	0.8338
0.8344	0.8346
0.8346	0.8334
0.8332	0.8348
0.8340	0.8342
0.8344	0.8320
0.8308	0.8326
0.8350	0.8326
0.8380	0.8332
0.8322	0.8342
0.8356	0.8358
0.8322	
0.8304	
0.8372	
0.8282	
0.8346	
0.8360	
0.8374	

Do the outputs of the two operators differ with respect to their variability? Justify your answer.

b. Sketch the OC curve for your test.

c. Determine 0.95 confidence limits for the ratio of the two variances.

26.12. *a.* With reference to Problem 26.10 it is argued that the variability of yarn A is less than the variability of yarn B. Do the data of that problem substantiate this contention? Justify your answer.

b. Sketch the OC curve for your test.

c. Determine 0.95 confidence limits for the ratio of the two variances.

26.13. *a.* You wish to determine whether the skein strength of yarn A is different from that of yarn B. The standard deviation in skein strength is known to be somewhere near 10 pounds for both types of yarn. You are willing to run a maximum risk of 0.05 of saying that the yarns are different when they are actually the same. On the other hand, you do not wish to run a risk of more than 0.10 of saying that they are the same, when actually the average skein strengths differ by as much as 5 pounds. The cost of determining the strength of both yarns is the same. Determine the sample sizes and acceptance limits you would use in your test.

b. Suppose that tests of A cost twice as much as tests of B; what then would your answer be?

---

[18] Lester A. Kauffman, *Statistical Quality Control at the St. Louis Division of the American Stove Company* (OPRD Quality Control Reports, No. 3, August 1945), pp. 12–13.

26.14. You wish to determine whether operator B is turning out a product that is more variable than the product of operator A. You are willing to run a maximum risk of 0.05 of saying that B's product is more variable than A's when actually it is not. On the other hand, you do not wish to run a risk of more than 0.10 of saying that the variability of B's product is equal to or less than that of A's when it is actually one and a half times greater than A's. Determine the sample sizes and acceptance limit you would use in your test.

26.15. You wish to determine whether product A is more variable in quality than product B. You are willing to run a maximum risk of 0.05 of concluding that product A is more variable than product B when actually it is not. On the other hand, you do not wish to run a risk of more than 0.10 of concluding that product A is equally variable or less variable than product B when actually it is 1.75 times more variable. Tests of quality of product A cost three times as much as tests of quality of product B. Determine the sample sizes and the acceptance limit you would use in your test.

26.16. Suppose that the data of Problem 26.8 represent measurements on eight different thermometers, each analyst making one measurement on each thermometer. Rework the problem in the light of this additional information. Is your conclusion any different?

26.17.**a.* Find the joint density function for two variables X and Y that are independently distributed as χ^2 with ν_1 and ν_2 degrees of freedom, respectively.

b. Set up the transformation

$$F = \frac{X/\nu_1}{Y/\nu_2}, \qquad G = Y$$

Note that the Jacobian is $\nu_1 G/\nu_2$ and write the joint density function for F and G.

c. Set $U = \left(\frac{\nu_1 F}{\nu_2} + 1\right) G$, multiply and divide by $\left(\frac{\nu_1 + \nu_2 - 2}{2}\right)!$, and integrate U from 0 to ∞. (This will turn out to be the integral of a χ^2 density function over the whole range and will be 1.) The remainder is the density function for F. Show that this equals

$$f(F) = \frac{\left(\frac{\nu_1 + \nu_2 - 2}{2}\right)!}{\left(\frac{\nu_1 - 2}{2}\right)!\left(\frac{\nu_2 - 2}{2}\right)!}\left(\frac{\nu_1}{\nu_2}\right)^{\nu_1/2} \frac{F^{(\nu_1-2)/2}}{\left(1 + \frac{\nu_1 F}{\nu_2}\right)^{(\nu_1+\nu_2)/2}}$$

You have shown that the ratio of a χ^2 variable divided by its degrees of freedom to another independently distributed χ^2 variable divided by its degrees of freedom has an F distribution.

d. How does Section 3.1 of this chapter follow from what has just been proved? (*Hint:* See Section 2.3 of Chapter 6.)

---

* Students with training in the calculus will find starred problems will give them knowledge of statistical theory that will be very useful in reading the text.

5. SELECTED REFERENCES*

Aspin (P '49), Benjamini (P '83), Bowker and Lieberman (B '55. This contains OC curves for many tests of differences between means and ratios of variances.), Brownlee (B '60), Ferris, Grubbs, and Weaver (P '46), Fisher (B '58), Freeman (B '42), Goldman (P '63), Gurland and McCollough (P '62), Johnson (B '49), Lee and Garland (P '75), Mood and Graybill (B '63), Pearson and Hartley (B '58), Smith, J. G. and Duncan (B '45), and Statistical Research Group, Columbia University (B '47).

---

* B and P refer to the Book and Periodical sections, respectively, of the Cumulative List of References in Appendix V.

27

Tests of Normality

Many statistical tests are based on the assumption that the universe from which a sample is drawn is normally distributed. It is therefore wise, if possible, to test this assumption of normality. Various procedures are available for making such a test.

1. GRAPHIC METHODS

If the sample is large enough to construct a histogram, a normal curve with the same mean and same standard deviation as the given data can be plotted with the histogram to see how well the normal curve fits. This is a nonnumerical comparison and is probably not much good as a test of normality. In extreme cases a decision can just as easily be made without plotting the curve, whereas in borderline cases a visual comparison of this kind provides no criteria on which to base a decision.

Another graphic method is to plot the cumulated distribution on normal probability paper and see how well it is fitted by a straight line. Figure 27.1 shows the bomb-base data of Table 3.6 plotted on normal probability paper. The data show some departure from a straight line, but whether this is significant or not cannot be told from the graph.

2. THE χ^2 TEST OF GOODNESS OF FIT

A more commonly used method of testing normality is to employ a χ^2 test of goodness of fit. A χ^2 test is a test that compares a set of sample frequencies with a set of frequencies that would be expected on the basis of some hypothesis. If the two sets compare well, the hypothesis is accepted; if they compare badly, the hypothesis is rejected. Since the distribution on which the decision

FIGURE 27.1
Bomb-Base Data of Table 3.6 Plotted on Normal Probability Paper

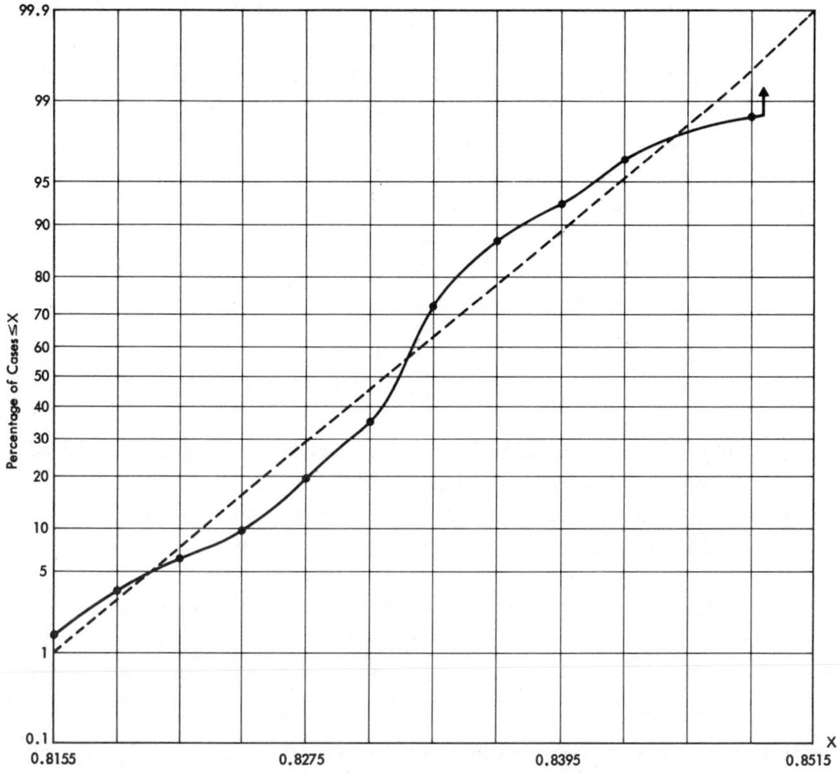

to accept or reject is based is the χ^2 distribution,[1] the test is called a χ^2 test. It can be used to test the fit of any frequency curve.

The precise formulation of a χ^2 test is as follows: Let $F_1, F_2, \ldots,$ F_k be sample frequencies of k classes, and let $f_1, f_2, \ldots, f_k$ be the frequencies that would be expected on the basis of hypothesis H_0. Then, if H_0 is true, sample values of the quantity

$$\sum_{i=1}^{k} \frac{(F_i - f_i)^2}{f_i}$$

will tend[2] to form a χ^2 distribution with ν equal to k minus the number of relations between the F_i and the f_i that are used to determine the f_i.

In applying a χ^2 test to determine the normality of a distribution, the actual frequencies of a sample histogram are compared with the theoretical frequencies obtained by assuming that the universe is normally distributed.

[1] See Chapter 6, Section 2.3.

[2] The use of the χ^2 distribution is only an approximation.

The normal curve is "fitted" by giving it the same mean and same standard deviation as the sample histogram. The theoretical and actual frequencies are compared as described by the formula above, and the χ^2 table is entered with $\nu = k - 3$.

The value of ν is taken as $k - 3$ because of the method of fitting. For, in addition to taking the total number of cases the same for both theoretical and actual frequencies, the theoretical mean is taken the same as the actual mean and the theoretical variance the same as the actual variance. The fitting process thus imposes the three conditions:

$$\sum f_i = \sum F_i$$

(27.1)
$$\bar{X}' = \sum F_i X_i / n$$

$$\sigma'^2 = \sum F_i x_i^2 / n$$

In the language of mathematical statistics these three conditions are said to cause a loss of three "degrees of freedom."

To illustrate a χ^2 test of normality, consider Table 3.4 of Chapter 3, which gives the frequency distribution of the overall heights of 145 fragmentation bomb bases. The mean of this sample distribution is 0.8314 inches, and its standard deviation is 0.00593 inches.[3] In Table 27.1 there is computed the number of cases that would be expected to fall in each interval on the assumption that the bomb bases are, in fact, normally distributed with a mean equal to 0.8314 inches and a standard deviation equal to 0.0059 inches.[4] These are compared with the actual frequencies, and the quantity $\sum \frac{(F_i - f_i)^2}{f_i}$ is computed as described above. The result is 35.54. The $\chi^2_{0.05}$ value for $\nu = 9 - 3 = 6$ is 12.6. Since the sample result is greater than this, we reject the hypothesis of normality. We must conclude that the distribution of heights of bomb bases is not normal.

It is to be noted that in carrying out this test it is the conservative rule to have the theoretical frequencies in any row of column (5) equal to 5 or more. If this is not the case, two or more groups should be combined. Thus, in our example, it might have been better to have combined the group 0.8395 to 0.8425 with the group 0.8425 to ∞. This was not done because the combination of intervals reduces the "degrees of freedom" (ν) and hence the sensitivity of the test, and in any case 4.3 is close to 5 and also close to the actual number for that interval.[5]

---

[3] See Chapter 3, Section 4.1.1.2 and Section 4.2.4.4.

[4] Note that in Table 27.1 the standard deviation has been rounded off to two significant figures.

[5] For a further discussion of this point, see Chap. 28, Sec. 2.

TABLE 27.1
A Table Illustrating the χ^2 Test of Normality

(1)		(2)	(3)	(4)	(5)	(6)	(7)	(8)	(9)
Upper Limit of Class Interval X	$x = X - \bar{X}$	*Limits in Standardized Units* $z = (X - \bar{X})/\sigma'$	*Relative Frequency from* $-\infty$ *to* z	*Relative Frequency of Cell*	*Absolute Theoretical Frequency* (f)	*Actual Frequency* (F)	F − f	$(F - f)^2$	$\frac{(F-f)^2}{f}$
0.8215	−0.0099	−1.68	0.0465	0.0465	6.8	9	2.2	4.84	0.71
0.8245	−0.0069	−1.17	0.1210	0.0745	10.8	5	−5.8	33.64	3.11
0.8275	−0.0039	−0.66	0.2546	0.1336	19.4	14	−5.4	29.16	1.50
0.8305	−0.0009	−0.15	0.4404	0.1858	26.9	21	−5.9	34.81	1.29
0.8335	0.0021	0.36	0.6406	0.2002	29.0	55	26.0	676.00	23.31
0.8365	0.0051	0.86	0.8051	0.1645	23.9	23	−0.9	0.81	0.03
0.8395	0.0081	1.37	0.9147	0.1096	15.9	7	−8.9	79.21	4.98
0.8425	0.0111	1.88	0.9700	0.0553	8.0	6	−2.0	4.00	0.50
∞	∞	∞	1.0000	0.0300	4.3	5	0.7	0.49	0.11
					145.0	145			$35.54 = \sum \frac{(F_i - f_i)^2}{f_i}$

3. USE OF MEASURES OF SKEWNESS AND KURTOSIS

A third way of determining nonnormality is to compute certain measures of nonnormal characteristics such as skewness and kurtosis.

3.1. Moments

The use of moments to measure skewness and kurtosis was noted in Chapters 3 and 4. For grouped data the kth moment about the mean is defined as follows:

For Sample Data: *For a Universe:*

(27.2a)
$$\mu_k = \frac{\sum_j F_j(X_j - \bar{X})^k}{n} \qquad \mu'_k = \sum_j P_j(X_j - \bar{X}')^k$$

where X_j is the midpoint of the jth interval and all cases in an interval are assumed to be concentrated at its midpoint. For continuous data the kth moment is defined by

(27.2b)
$$\mu'_k = \int_{-\infty}^{\infty} (X - \bar{X}')^k f(X)dX$$

In general,[6] $\mu_1 = 0$ and $\mu_2 = \text{RMSD}^2$.

The kth moment around any arbitrary origin A is defined by the formulas:

For Sample Data: *For a Universe:*

(27.3)
$$\mu_{ak} = \frac{\sum_j F_j(X_j - A)^k}{n} \qquad \mu'_{ak} = \sum_j P_j(X_j - A)^k$$

or

$$\mu'_{ak} = \int_{-\infty}^{\infty} (X - A)^k f(X)dX$$

3.2. The Third Moment as a Measure of Skewness

If the distribution of a universe is symmetrical, its third moment about the mean will be zero. If the distribution is skewed to the right, the third moment about the mean will have a positive value, because the large size of the X_j on the long tail will more than offset the greater number of smaller X_j on the shorter tail of the distribution. Hence, for a positively skewed distribution, μ'_3 will be positive. Likewise, for a negatively skewed distribution (i.e., one with the long tail to the left), μ'_3 will be negative. For these reasons the

---

[6] See Chapter 3, footnote 7 and Section 4.2.3.

third moment around the mean is taken as a measure of the absolute skewness of a distribution. Similarly, the ratio

$$\gamma_1' = \frac{\mu_3'}{\sigma'^3} \tag{27.4}$$

is taken as a measure of relative skewness.[7]

If a distribution is symmetrical, γ_1' will be zero; but it should be noted that, if $\gamma_1' = 0$, it does not necessarily follow that a distribution is symmetrical.[8]

The sample measure γ_1 will be taken as an estimate of γ_1' for the universe.

3.3. Measures of Kurtosis

A measure of the kurtosis of a universe is given by

$$\gamma_2' = \frac{\mu_4'}{\mu_2'^2} - 3 \tag{27.5}$$

For, if μ_4' is large relative to μ_2', it indicates relatively large tails. Hence $\mu_4'/\mu_2'^2$ is a measure of kurtosis.

For the normal distribution,[9] $\mu_4'/\mu_2'^2$ has the value 3. Since the normal distribution arises very frequently and is often used as a basis of reference for distributions that are not normal, the quantity γ_2' is defined so that it will be 0 when a distribution has the normal degree of kurtosis. Thus $\gamma_2' > 0$ means that a distribution has a sharper peak, thinner shoulders, and fatter tails than the normal distribution;[10] $\gamma_2' < 0$ means that a distribution has a flatter peak, fatter shoulders, and thinner tails than the normal distribution.

A second method of measuring kurtosis is afforded by the ratio a'. For a discrete variable,

$$a' = \sum_j P_j |X_j - \bar{X}'|/\sigma' \tag{27.6}$$

for a continuous variable,

$$a' = \int_{-\infty}^{\infty} |X - \bar{X}| f(X) dX/\sigma' \tag{27.6a}$$

For a normal distribution this should be 0.7979. For distributions with less than normal kurtosis, $a' > 0.7979$. For distributions with greater than normal kurtosis, $a' < 0.7979$. The sample measures γ_2 and $a = \dfrac{\text{AD}}{\text{RMSD}}$ will be taken as estimates of γ_2' and a' for the universe.

---

[7] In some texts $\beta_1' = \mu_3'^2/\mu_2'^3$ is taken as a measure of relative skewness. Thus $\gamma_1' = \sqrt{\beta_1'}$.

[8] W. A. Shewhart, *Economic Control of Quality of Manufactured Product* (New York: D. Van Nostrand Co., Inc., 1931), p. 75.

[9] See Chapter 4, Section 6.1.

[10] Cf. Figure 3.8, Chapter 3.

3.4. A Short Method of Computing γ_1 and γ_2

The short methods for computing a sample mean and a sample standard deviation[11] may be extended to the computation of the sample γ coefficients. In fact, if γ_1 and γ_2 are to be computed as part of a general frequency distribution analysis, it is probably best to compute $\bar{X}$, s, γ_1, and γ_2 all from a single work sheet, such as shown in Table 27.2. The directions are briefly as follows:

Fill in columns (1)–(3) of the work sheet as indicated. Select an arbitrary origin at the midpoint of one of the intervals near the center of the distribution.[12] In column (4) write the deviations from the arbitrary origin measured in class interval units (ξ). In column (5) write the product of the frequencies times the deviations from the arbitrary origin. Obtain columns (6), (7), and (8) by multiplication of immediately preceding columns by ξ. Use columns (9) and (10) as check columns in the manner indicated at the bottom of the table. Compute the following moments around the arbitrary origin measured in class interval units:

(27.3a)

$$\mu_{a\,1} = \frac{\sum_j F_j\,(\xi_j)}{n} \qquad \mu_{a\,2} = \frac{\sum_j F_j\,(\xi_j)^2}{n}$$

$$\mu_{a\,3} = \frac{\sum_j F_j\,(\xi_j)^3}{n} \qquad \mu_{a\,4} = \frac{\sum_j F_j\,(\xi_j)^4}{n}$$

From these derive the moments around the mean measured in class interval units as follows:[13]

(27.7)

$$\mu_2 = \mu_{a\,2} - (\mu_{a\,1})^2$$

$$\mu_3 = \mu_{a\,3} - 3\mu_{a\,2}\mu_{a\,1} + 2(\mu_{a\,1})^3$$

$$\mu_4 = \mu_{a\,4} - 4\mu_{a\,3}\mu_{a\,1} + 6\mu_{a\,2}(\mu_{a\,1})^2 - 3(\mu_{a\,1})^4$$

It will be noted that measuring the moments in class interval units simplifies the computations, and the moments around the mean in terms of original units can be obtained easily by multiplication by the class interval raised to the kth power.

We now have all we need for computing the final measures. Thus

$$s^2 = \frac{n}{n-1}\mu_2$$

$\gamma_1 = \mu_3/\mu_2^{3/2}$ (valid whether μ_3 and μ_2 are measured in original or class interval units)

---

[11] See Chapter 3, Sections 4.1.1.2 and 4.2.4.4.

[12] See Chapter 3, Section 4.1.1.2, for a discussion of how to pick A.

[13] See Appendix I (3), (4), and (5).

TABLE 27.2
Computation of Moments for Distribution of Table 3.4

(1) X Lower Class Limits	(2) Midpoint	(3) F	(4) ξ	(5) $F(\xi)$	(6) $F(\xi)^2$	(7) $F(\xi)^3$	(8) $F(\xi)^4$	(9) $(1+\xi)^4$	(10) $F(1+\xi)^4$
0.8125 –		2	−6	−12	72	−432	2,592	625	1,250
0.8155 –		3	−5	−15	75	−375	1,875	256	768
0.8185 –		4	−4	−16	64	−256	1,024	81	324
0.8215 –		5	−3	−15	45	−135	405	16	80
0.8245 –		14	−2	−28	56	−112	224	1	14
0.8275 –		21	−1	−21	21	−21	21	0	0
0.8305 –	0.8320	55	0	0	0	0	0	1	55
0.8335 –		23	1	23	23	23	23	16	368
0.8365 –		7	2	14	28	56	112	81	567
0.8395 –		6	3	18	54	162	486	256	1,536
0.8425 –		2	4	8	32	128	512	625	1,250
0.8455 –		1	5	5	25	125	625	1,296	1,296
0.8485 –		2	6	12	72	432	2,592	2,401	4,802
		145	. . .	−27	567	−405	10,491	5,655	12,310

Check $\Sigma_{10} = \Sigma_3 + 4\Sigma_5 + 6\Sigma_6 + 4\Sigma_7 + \Sigma_8$

$12,310 = 145 - 108 + 3,402 - 1,620 + 10,491$

$$\gamma_2 = (\mu_4/\mu_2^2) - 3 \quad \text{(valid whether } \mu\text{'s are measured in original or class interval units)}$$

$$\mathbf{a} = \frac{\text{AD}}{\text{RMSD}} = \frac{\text{AD}}{\sqrt{\mu_2}}$$

3.5. Computations Illustrated

These calculations may be illustrated with reference to the data of Table 3.4 of Chapter 3, the major computations for which are summarized in Table 27.2. Class interval units are employed until close to the end of the computations. First,

$$\mu_{a\,1} = \frac{\sum_j F_j\,(\xi_j)}{n} = \frac{-27}{145} = -0.1862 \text{ (class interval units)}$$

Hence,

$$\bar{X} = A + \mu_{a\,1} \text{ (in class interval units) } (i) = 0.8320 + \left(\frac{-27}{145}\right)(0.003) = 0.8314$$

inches.

Second, continuing to use class interval units, we have

$$\mu_{a\,2} = \frac{\Sigma F(\xi)^2}{n} = \frac{567}{145} = 3.9103$$

and

$$\mu_2 = \mu_{a\,2} - (\mu_{a\,1})^2 = 3.9103 - (-0.1862)^2 = 3.8756$$

Hence RMSD $= \sqrt{\mu_2} = \sqrt{3.8756} = 1.9686$ class interval units or 1.9686 (0.003) $= 0.0059$ inches.

Third, we also have

$$\mu_{a\,3} = \frac{\sum_j F_j\,(\xi_j)^3}{n} = \frac{-405}{145} = -2.7931$$

and

$$\begin{aligned} \mu_3 &= \mu_{a\,3} - 3\mu_{a\,2}\mu_{a\,1} + 2(\mu_{a\,1})^3 \\ &= (-2.7931) - 3(3.9103)(-0.1862) + 2(-0.1862)^3 \\ &= -0.60 \text{ in class interval units cubed.} \end{aligned}$$

Hence

$$\gamma_1^2 = \frac{(\mu_3)^2}{(\mu_2)^3} = \frac{(-0.60)^2}{(3.8756)^3} = 0.0062$$

and

$$\gamma_1 = -0.079$$

Finally,

$$\mu_{a\,4} = \frac{\sum_j F_j\,(\xi_j)^4}{n} = \frac{10{,}491}{145} = 72.3517$$

and

$$\mu_4 = \mu_{a\,4} - 4\mu_{a\,3}\mu_{a\,1} + 6\mu_{a\,2}(\mu_{a\,1})^2 - 3(\mu_{a\,1})^4$$
$$= 72.3517 - 4(-2.7931)(-0.1862) + 6(3.9103)(-0.1862)^2$$
$$-3(-0.1862)^4 = 71.08 \text{ in class interval units to the fourth power}$$

Hence

$$\gamma_2 = \frac{\mu_4}{\mu_2^2} - 3 = \frac{71.08}{(3.8756)^2} - 3 = 1.73$$

3.6. The Significance of Measures of Skewness and Kurtosis

Tables F to H of Appendix II give the 0.05 and 0.01 points of the sampling distributions of γ_1, γ_2, and a. If a given sample has a value of γ_1 or γ_2 or a beyond the 0.05 point for that statistic, the population may be deemed to be nonnormal. A stricter test would use the 0.01 points. For example, the data of Table 27.2 pertaining to the heights of bomb bases has[14] $\gamma_1 = 0.08$, $\gamma_2 = 1.73$, and $a = \dfrac{0.0038}{0.0059} = 0.644$. The size of the sample is 145. For samples of 200, Table G shows that the upper 0.05 point for γ_2 is 0.57 and the upper 0.01 point is 0.98. The corresponding points for $n = 145$ will be somewhat higher than this, but they will obviously be much less than the sample value of 1.73. The sample value of γ_2 thus indicates departure from normality. The sample value of a is 0.644, which is considerably below the lower 0.05 and 0.01 points for a given in Table H for $n = 100$ and $n = 200$, also indicating departure from normality. Both γ_2 and a measure nonnormality on the basis of the kurtosis of the distribution.[15] The

---

[14] The average deviation was computed from the mean, and it was assumed that all cases in any interval had the value of the midpoint of the interval. The actual method of computation employed the following formula,

$$\text{AD} = \frac{\Sigma F|\xi| + (n_1 - n_2)\mu_{a\,1}}{n}\,(i)$$

in which n_1 is the number of cases below the mean, n_2 is the number above the mean, and $\mu_{a\,1}$ is measured in class interval units. Thus

$$\text{AD} = \frac{187 + (107 - 80)(-0.1862)}{145}(0.003) = 0.0038$$

To use this formula the arbitrary origin (A) must be in the same interval as the mean.

[15] See above, Section 3.3.

γ_1, which equals -0.08, indicates no departure from normality on account of skewness. It will be noted that a universe can have $\gamma_1 = 0$ or $\gamma_2 = 0$ without their being normal. Sample values close to these normal values do not therefore prove normality; they merely do not lead to the rejection of the normal hypothesis.

4. PROBLEMS

27.1. Determine whether the data of Problem 3.3*a* (i) could reasonably have come from a normal universe by—

a. Plotting a cumulative distribution of the data on normal probability paper.

b. Running a χ^2 test of goodness of fit.

c. Computing γ_1, γ_2, and $\boldsymbol{a}$.

27.2. Determine whether the data of Problem 3.3*a* (ii) could reasonably have come from a normal universe by—

a. Plotting a cumulative distribution of the data on normal probability paper.

b. Running a χ^2 test of goodness of fit.

c. Computing γ_1, γ_2, and $\boldsymbol{a}$.

27.3. Determine whether the data of Problem 3.3*a* (v) (without the extreme value 106) could reasonably have come from a normal universe by—

a. Plotting a cumulative distribution of the data on normal probability paper.

b. Running a χ^2 test of goodness of fit.

c. Computing γ_1, γ_2, and $\boldsymbol{a}$.

27.4. A sample of 200 items has $\gamma_1 = 0.28$, $\gamma_2 = 0.6$, and an $\boldsymbol{a} = 0.82$. Would you infer from these results that the data are normally distributed?

27.5. To test the distribution of brittle spots in nylon bars, a set of 270 bars are submitted to bending tests. Each bar was bent in five places, and the number of places in which each bar broke was recorded. These results were compared with what would have been expected if the brittleness of the bars was evenly distributed throughout the bar (i.e., the chance of a break was the same for every bend). The results were as follows:

Breaks per Bar	*Actual Number Bars*	*Number Expected on Hypothesis of Uniform Distribution of Brittleness**
0	157	130
1	69	108
2	35	36
3 or more	19	6

* Calculated from the binomial distribution. Cf. W. L. Gore, "Quality Control in the Chemical Industry. IV. Statistical Methods in Plastics Research and Development," *Industrial Quality Control,* September 1947, p. 6.

Would you infer from these data that the actual number of bars and theoretical number of bars match sufficiently well to justify the hypothesis of a uniform distribution of brittleness? Justify your answer.

[Note that this problem is similar to the problem of testing normality. Here we are testing whether the *binomial distribution* is a good fit to the data. Since the total number of cases is the same for theoretical and actual frequencies and since the chance of a break is presumably estimated from the data, we lose *two* degrees of freedom (compare Section 2, above). Otherwise the procedure is the same as that described for χ^2 tests of normality.]

27.6. It is suggested in Problem 4.19 that among the employees of a certain public utility the distribution of the number of absences per year because of disability follows a Poisson distribution. Fit a Poisson distribution to the data of this problem by estimating the universe mean from the sample and determine whether the Poisson hypothesis could be accepted. (How many degrees of freedom are lost in this case?)

5. SELECTED REFERENCES*

Anscombe (P '63, a Bayesian approach), Anscombe and Glynn (P '83), Chernoff and Lieberman[16] (P '54), Cowden (B '57), D'Agostino (P '70), D'Agostino and Rosman (P '74), Fisher (B '58), Folks and Blankenship (P '67), Foster (P '62), Gastwirth and Owens (P '77), Geary and Pearson (B '38), Johnson (B '49), Kendall (B '48), Kimball (P '60), Massey[17] (P '51),Mulholland (P '77), Nelson (P '81), Pearson, D'Agostino and Bowman (P '77), Pettitt and Stephens (P '77), Royston (P '82), Shapiro and Wilk[18] (P '65), Shapiro, Wilk and Chen (P '68), Smith, J. G. and Duncan (B '44 and B '45), Spiegelhalter (P '77), Watson (P '57, P '58, and P '59).

---

* B and P refer to the Book and Periodical sections, respectively, of the Cumulative List of References in Appendix V.

[16] This article shows that when n is small, it is good to plot individual points on normal probability paper at $\left(X_1, \frac{1}{2n}\right), \left(X_2, \frac{3}{2n}\right), \ldots, \left(X_n, \frac{2n-1}{2n}\right)$ where $X_1 \leq X_2 \leq \ldots \leq X_n$. This procedure will make the abscissa of the 0.50 point a good estimate of $\bar{X}'$ and the distance between the abscissa of the 0.8413 point and the abscissa of the 0.50 point a good estimate of σ'.

[17] The Kolmogorov-Smirnov test discussed by Massey has certain advantages as a "goodness of fit test." For sample distributions of more than 35 cases, the 0.05 critical value of the maximum absolute difference between the sample and universe cumulative distributions is $1.36/\sqrt{n}$.

[18] The Shapiro-Wilk test for normality, which is based on a linear function of order statistics, is comparatively quite sensitive to a wide range of nonnormality, even with samples as small as $n = 20$. Their paper contains an interesting summary of how various tests for normality compare.

28

Contingency Tables and Chi Square Tests

Statisticians have developed three powerful techniques for the study of relationships between various factors. These are contingency tables, the analysis of variance, and regression analysis. Contingency tables reveal associations between classifications. The analysis of variance reveals, and to a certain extent measures, the relationship between a given (preferably a normally distributed) variable and one or more classifications. It also analyzes the total variance of the variable into its component parts. Regression analysis reveals and measures functional relationships between two or more variables. Each of these techniques will be explained and illustrated in turn—contingency tables in this chapter, analysis of variance and regression analysis in the following chapters.

1. 2 × 2 TABLES

Table 28.1 illustrates the simplest form of a contingency table. It is a two-way classification table with respect to the toughness of nylon bars. The horizontal classification divides the bars into two classes according to their brittleness. The vertical classification divides the bars into two classes according to the length of the heat treatment. The contingency table itself cross-classifies the bars with respect to these two bases of classification.

The problem that a two-way contingency table seeks to solve is whether one classification is independent of the other. For example, Table 28.1 aims to give an answer to the question: Does the length of the heating cycle have any effect on the toughness of nylon bars? The test associated with a two-way contingency table is thus a test of independence.[1]

It will be recalled[2] that when an item is cross-classified as in Table 28.1,

[1] Cf. Chapter 2, Section 8.2.

[2] See Chapter 2, Section 8.2.

TABLE 28.1
Effect of Length of Heating Cycle on Toughness of Nylon Bars

Length of Heating Cycle (Seconds)	*Number Classed as Brittle*	*Number Classed as Tough*	*Total*
30	77 (127)	323 (273)	400
90	177 (127)	223 (273)	400
Total	254	546	800

SOURCE: W. L. Gore, "Quality Control in the Chemical Industry. IV. Statistical Methods in Plastics Research and Development," *Industrial Quality Control*, September 1947, p. 7.

if the frequencies of each row are proportional to those of all other rows or, what amounts to the same thing, if the frequencies of each column are proportional to those of all other columns, then the two classifications are independent of each other. If a universe of items, however, shows the proportionality of frequencies characteristic of independence, a random sample from that universe will very likely not have the same properties, because of the uncertainties of sampling. A cross-classification of sample data that does not have exactly proportional frequencies is therefore not a proof that the universe frequencies are not proportional and the classification therefore dependent. The question that has to be answered by the statistician is: By how much must a contingency table of sample data depart from the proportional pattern before it is reasonable to conclude that in the universe as a whole the classifications are dependent?

The statistical device that is used to test the hypothesis of independence is the χ^2 test, the same test that was used to test for normality.[3] Frequencies are obtained for each cell of a contingency table that would be expected for that cell if the hypothesis of independence were true and if the universe proportion for each marginal classification were the same as the sample proportion. These theoretical or expected frequencies are then compared with the actual sample frequencies by means of the χ^2 test.

The process may be illustrated with reference to Table 28.1. This shows the cross-classification of 800 nylon bars with respect to their brittleness. The frequencies are not in proportion; but the question is: Do they deviate sufficiently from proportional frequencies to justify the conclusion that the variation in heat treatment affects the brittleness of nylon bars, or can the deviation from proportionality be reasonably attributed to sampling? To compute the frequency expected for each cell on the hypothesis of independence, we take the total for each column and redistribute the total among the various

[3] See Chapter 27, Section 2.

rows in the same proportions as the grand total is distributed in the row total column. In other words, we get a set of cell frequencies that are proportional in the same way that the row or column totals are proportional. In Table 28.1, for example, the row totals are in a 1 to 1 ratio (400 to 400). Hence the cell frequencies in the first row that we would expect on the assumption of independence should bear a 1 to 1 ratio with the cell frequencies of the second row, column by column. Thus in Table 28.1 we have

$$\frac{127}{127} = \frac{273}{273} = \frac{400}{400}$$

If we wish, we can work the other way. The ratio of the column totals is $\frac{254}{546} = 0.465$. Hence, the cell frequencies in the first column that we would expect on the assumption of independence should be 0.465 times the cell frequencies in the second column, row by row. Thus, in Table 28.1, we have

$$\frac{127}{273} = \frac{127}{273} = \frac{254}{546}$$

It is to be emphasized that the theoretical frequencies obtained in this way are those that would be expected on the assumption of independence, together with the assumption that the ratios of the row and column totals in the universe are the same as those in the given sample. It also may be noted that the practical way to get the theoretical frequency for any given cell is to compute the ratio of the total for the row (column) in which the cell occurs to the grand total and then multiply this ratio by the total of the column (row) in which the cell occurs. Thus, we have

$$\frac{400}{800} \times 254 = 127$$

or

$$\frac{254}{800} \times 400 = 127$$

To test the hypothesis of independence for Table 28.1, we compute

$$\Sigma \frac{(F_i - f_i)^2}{f_i} = \frac{(77 - 127)^2}{127} + \frac{(177 - 127)^2}{127}$$

$$+ \frac{(323 - 273)^2}{273} + \frac{(223 - 273)^2}{273} = 57.6$$

and compare this result with the $\chi^2_{0.05}$ point for the proper value of v, assuming that we have set the risk of rejecting the hypothesis when it is true at 0.05.

As noted in Section 2 of Chapter 27, the v of a χ^2 test is associated, as the mathematical statisticians say, with the "degrees of freedom" involved in the testing procedure.

In a test of independence the number of degrees of freedom are determined by counting the number of independent restrictions placed upon the theoretical frequencies when they are being determined and subtracting this number from the number of comparisons made between theoretical and actual frequencies. In our case we required that the expected cell frequencies add by rows to 400 and 400, respectively, and by columns to 254 and 546, respectively. That is, we imposed the following restrictions:

$$f_{11}+f_{12}=\text{Total for row 1}$$

$$f_{21}+f_{22}=\text{Total for row 2}$$

$$f_{11}+f_{21}=\text{Total for column 1}$$

$$f_{12}+f_{22}=\text{Total for column 2}$$

But one of these can be obtained from the other three, so only three of the restrictions are independent. Altogether, theoretical frequencies were determined for 4 cells. Since they were subjected to 3 independent restrictions, the degrees of freedom (ν) are $4 - 3 = 1$.

It therefore follows that for a 2×2 contingency table, the critical value of χ^2 (assuming $\alpha = 0.05$) is the $\chi^2_{0.05}$ point for one degree of freedom (i.e., for $\nu = 1$), viz. 3.84. In our example, this is much less than the sample result (57.6). We therefore reject the hypothesis of independence and conclude that variations in heat treatment do have some effect on the toughness of nylon bars.

It is to be noted that a χ^2 test is an approximate one. For a 2×2 contingency table, an improvement in the approximation yielded by the χ^2 distribution is obtained by subtracting[4] 0.5 from the absolute value of the difference $F_i - f_i$ for each cell before squaring it. In other words, somewhat better results are obtained if we use the statistic

$$\sum \frac{(|F_i - f_i| - 0.5)^2}{f_i}$$

When the frequencies are large, this correction makes little difference. (In our example, the result is affected only in the third place.) When the frequencies run smaller, however, it improves the approximation.

It is to be noted that the χ^2 test as applied to a 2×2 contingency table is the equivalent of the test described in Section 1.1 of Chapter 26 for testing the difference between two proportions. With reference to Table 28.1, for example, it is seen that the fraction of brittle nylon bars produced with a 30-second heating cycle is $77/400 = 0.1925$, whereas the fraction of

[4] This is the so-called correction for continuity. It is the same type of adjustment described above for the approximation of a set of normal ordinates by an area under the normal curve (see Chapter 4, Section 7.1).

brittle nylon bars produced with a 90-second heating cycle is 177/400 = 0.4425. To test the hypothesis that the variations in heat treatment have no effect on the brittleness of nylon bars, we can test the hypothesis that in the universe the fraction of brittle nylon bars under the 30-second heating cycle was the same as the fraction of brittle nylon bars under the 90-second heating cycle. The two tests are mathematically equivalent and subject to the same limitations.

One of the limitations in using the normal distribution to test the difference between two proportions is that p' should not be too small. The same applies to the χ^2 test. If one of the marginal row or column proportions of a 2×2 contingency table is small—say, low enough to make the expected frequency for any cell less than 5 (cf. Section 7.1 of Chapter 4)—it is the conservative rule to resort to some other method. A direct procedure of this kind is described in Appendix I (31).

2. $r \times c$ TABLES

Tests of independence can be carried out for two-way contingency tables that have r groups in one classification and c groups in the other. The only difference is that the computations are longer.

Consider Table 28.2. This classifies rejects of metal castings by causes and weeks of production. The question is: Does the distribution of rejects by causes vary significantly from week to week? To obtain the theoretical frequencies (based on the assumption of independence), we apply the marginal percentages for the whole three-week period to the totals for each week. Thus, the expected frequency of sand rejects for the first week is $\frac{299}{573}(180) = 93.9$; the expected frequency of misrun rejects in the first week is $\frac{27}{573}(180) = 8.5$; the expected frequency of corebreak rejects in the second week is $\frac{82}{573}(213) = 30.5$; and so on.

TABLE 28.2
Causes of Rejection of Metal Castings

	1st Week	*2d Week*	*3d Week*	*Total*
Sand	97	120	82	299
Misrun	8	15	4	27
Shift	18	12	0	30
Drop	8	13	12	33
Corebreak	23	21	38	82
Broken	21	17	25	63
Other	5	15	19	39
Total	180	213	180	573

SOURCE: George A. Hunt, "A Training Program Becomes a 'Clinic,'" *Industrial Quality Control*, January 1948, p. 26.

The various frequencies expected on the assumption of independence are listed in Table 28.3. From this table we have

$$\sum \frac{(F_i - f_i)^2}{f_i} = \frac{(97 - 93.9)^2}{93.9} + \frac{(8 - 8.5)^2}{8.5} + \frac{(18 - 9.4)^2}{9.4}$$

$$+ \frac{(8 - 10.4)^2}{10.4} + \frac{(23 - 25.7)^2}{25.7} + \frac{(21 - 19.8)^2}{19.8} + \frac{(5 - 12.3)^2}{12.3}$$

$$+ \frac{(120 - 111.1)^2}{111.1} + \frac{(15 - 10.0)^2}{10.0} + \frac{(12 - 11.2)^2}{11.2} + \frac{(13 - 12.3)^2}{12.3}$$

$$+ \frac{(21 - 30.5)^2}{30.5} + \frac{(17 - 23.4)^2}{23.4} + \frac{(15 - 14.5)^2}{14.5} + \frac{(82 - 93.9)^2}{93.9}$$

$$+ \frac{(4 - 8.5)^2}{8.5} + \frac{(0 - 9.4)^2}{9.4} + \frac{(12 - 10.4)^2}{10.4}$$

$$+ \frac{(38 - 25.7)^2}{25.7} + \frac{(25 - 19.8)^2}{19.8} + \frac{(19 - 12.3)^2}{12.3} = 45.06$$

It is proved in statistical theory that the quantity $\sum \frac{(F_i - f_i)^2}{f_i}$ is distributed approximately in accordance with the χ^2 distribution with $\nu = (r - 1)(c - 1)$, i.e., number of rows minus 1 times the number of columns minus 1. Hence, for Table 28.2, $\nu = (r - 1)(c - 1) = (7 - 1)(3 - 1) = 12$ and the $\chi^2_{0.05}$ point for $\nu = 12$ is 21.0. Since 45.53 is greater than this, we conclude that there is a difference in the distribution of rejects from week to week. Indeed, the difference is so great that we would have rejected the hypothesis of independence at the 0.001 level of significance, since the 0.001 point of the χ^2 distribution for $\nu = 12$ is 32.9.

In $r \times c$ contingency tables ($r > 2$ or $c > 2$ or both), we do not use the particular correction for continuity described in the previous section.

TABLE 28.3
Expected Frequencies for Table 28.2 on the Assumption of Independence

	1st Week	*2d Week*	*3d Week*	*Total**
Sand	93.9	111.1	93.9	299.0
Misrun	8.5	10.0	8.5	27.0
Shift	9.4	11.2	9.4	30.0
Drop	10.4	12.3	10.4	33.0
Corebreak	25.7	30.5	25.7	82.0
Broken	19.8	23.4	19.8	63.0
Other	12.3	14.5	12.3	39.0
Total	180.0	213.0	180.0	573.0

* Because of the rounding-off of the expected frequencies to make the columns add to the proper totals, the actual sum across a row may differ slightly from the total brought down from Table 28.2.

Usually it is unnecessary to make such a correction. If in special cases a refinement of this kind is deemed desirable, the rule to follow is this: "Calculate χ^2 by the usual formula. Find the next lowest possible value of χ^2 to the one to be tested, and use the tabular probability for a value of χ^2 midway between the two."[5]

In $r \times c$ contingency tables, there is more likely to be need of guarding against a low expected frequency for a cell. The conservative rule is to have at least a theoretical frequency of 5 for any one cell. If this is not obtained with the table as originally set up, then it is the practice to combine several cells until a theoretical frequency of 5 is obtained for the combined cells. This procedure cuts down the sensitivity of the χ^2 test, however. W. G. Cochran points out that the rule of 5 is quite conservative and suggests that, if we are willing to run an error of 20 percent in probabilities read from the χ^2 table (an error up to 1 percent at the 5 percent level and up to 0.2 percent at the 1 percent level), we can let our smallest expected frequency run as low as 2 if there are 6 degrees of freedom; as low as 1 if there are 10 degrees of freedom; and as low as 0.5 if there are 25 degrees of freedom—all provided there is only a single small frequency in the table.[6]

3. TESTING FOR HOMOGENEITY OF A SET OF PERCENTAGES

The χ^2 tests discussed in the previous section are a means of testing for homogeneity of a set of percentages. Suppose, for example, that we have r samples for each of which the fraction nonconforming has been determined, and suppose that we wish to decide whether they could reasonably have come from a common universe. This is essentially the same problem for which, in Chapter 19, the p-chart for past data was offered as a solution. Here we shall show how this problem can be solved by a χ^2 test.

Let us consider first the case in which we assume by hypothesis that the samples come from a common universe but the p' (e.g., the fraction nonconforming of the universe) is not specified. In this case, we have essentially an $r \times c$ contingency table in which the r categories of one classification are the r samples being compared and the c categories of the other are the two classes, nonconforming and conforming, into which the items of each sample are classified. For example, the data, from which the p-chart of Figure 19.1 was constructed, can be arranged as indicated in Table 28.4. (Disregard the last column for the moment.)

For these data the common percent nonconforming may be estimated at $407/1{,}400 = 29.07$ percent. When the number of nonconforming items expected in each cell on the basis of this estimate of p' (viz. 29.07 percent

[5] Quoted from W. G. Cochran, "The χ^2-Correction for Continuity," *Iowa State College Journal of Science* 16 (1942), p. 423. Cochran adds that if the possible χ^2's are closely spaced together, the probabilities given by the corrected and uncorrected χ^2's may differ only by a negligible amount, in which case the correction can be ignored.

[6] Ibid., p. 433.

TABLE 28.4
Comparison of the Nonconforming Quality of a Series of Lots

Sample for:	*Number of Items Rejected as Nonconforming*	*Number of Items Accepted*	*Total*	*Contribution to* $\sum \frac{(F-f)^2}{f}$
April 27	4	46	50	10.71
28	9	41	50	2.94
29	10	40	50	1.97
30	11	39	50	1.19
May 1	13	37	50	0.22
2	30	20	50	23.34
3	26	24	50	12.85
4	13	37	50	0.22
5	8	42	50	4.10
6	23	27	50	7.02
7	34	16	50	36.93
8	25	25	50	10.71
9	18	32	50	1.19
10	12	38	50	0.61
11	4	46	50	10.71
12	3	47	50	12.85
14	11	39	50	1.19
15	8	42	50	2.94
16	14	36	50	0.03
17	21	29	50	4.10
18	25	25	50	10.71
19	18	32	50	1.19
21	10	40	50	1.97
22	8	42	50	4.10
23	18	32	50	1.19
24	19	31	50	1.97
25	4	46	50	10.71
26	8	42	50	4.10
	407	993	1,400	$181.76 = \sum \frac{(F-f)^2}{f}$

of 50, or 14.5) is compared with the actual number nonconforming, *and* the number of conforming items expected in each cell (viz. 35.5) is compared with the actual number,[7] the quantity $\sum \frac{(F-f)^2}{f}$ is found to be beyond $\chi^2_{0.001}$ for 27 degrees of freedom (i.e., 181.76 is beyond 55.5). It is to be concluded, therefore, that the samples do not come from a common universe or, in other words, that the process is not in control.

[7] Beginners sometimes make the mistake of comparing only the actual and expected number of *nonconforming units* and overlook the comparison of actual and expected number of conforming units.

Since the samples are ordered in time, the hypothesis of control could not have been accepted even if $\sum \frac{(F-f)^2}{f}$ had been less than $\chi^2_{0.001}$ until the χ^2 test had been supplemented by a run test. Tests of randomness over time are needed here just as they were when the data were analyzed by a p-chart. The p-chart, however, had the advantage that the data were already set up in a form to which a run test could easily be applied.

With a p-chart it was also readily clear which points were out of control. Here the "out-of-control" points can be noted by reviewing the row-by-row contributions to $\sum \frac{(F-f)^2}{f}$. It will be noted that the heavy contributions are samples for April 27 and May 2, 3, 7, 8, 11, 12, 18, and 25, which are also the same points that were outside the control limits in Figure 19.1. Theoretically, for the contingency table and p-chart procedures to be equivalent, we should use a critical χ^2 for the contingency table the probability of exceeding which is 0.0027 instead of 0.001. Table C of Appendix II, however, does not provide us[8] with such values of χ^2.

If we wished to compare the given sample percentages with some standard value of p', we would compute the frequencies expected in each class on the basis of this standard p' and find $\sum \frac{(F-f)^2}{f}$ as before. In this case, however, we would use a critical χ^2 for 28 degrees of freedom instead of 27, since the sum of the actual and expected frequencies need no longer agree by columns but only by rows. This contingency analysis would compare with a p-chart for which a standard value p'' was given.

A special case of comparing a set of percentages is the comparison of *two* percentages. In this instance we are back to a 2×2 contingency table, which means that we should use

$$\sum \frac{(|F-f|-0.5)^2}{f}$$

to give the proper correction for continuity. This χ^2 method of testing for the difference between two percentages (without the correction for continuity) is the arithmetic equivalent of the method described in Section 1.1 of Chapter 26. Whether we use one or the other is a matter of indifference. The χ^2 procedure leads more readily, however, into the OC curve for the test. (See Section 5 below.)

4. COMPARISON OF NONCONFORMITIES PER UNIT AND OTHER POISSON VARIABLES

Sometimes data come to us in the form of number of occurrences per unit. Such, for example, would be number of nonconformities per radio, number

[8] See, however, A. J. Duncan, "A Chi-Square Chart for Controlling a Set of Percentages," *Industrial Quality Control,* November 1950, p. 12, where values of $\chi^2_{0.0027}$ are given.

of blemishes per table top, and number of accidents per 100,000 manhours. In these instances we do not have a cross-classification of frequencies that might be set up in a contingency table. The frequencies involved, however, will in random samples follow a Poisson distribution, and because of this, comparison can be made by a χ^2 test. Thus, if the number of occurrences per unit ran $c_1, c_2, \ldots, c_k$ for k sample units, we could test the homogeneity of the units (i.e., we could test the hypothesis that the k units all came from populations with the same mean number of occurrences per unit) by computing the statistic $\sum_i \frac{(c_i - \bar{c})^2}{\bar{c}}$ and noting whether this exceeded a critical χ^2 for $k - 1$ degrees of freedom. If the hypothetical value of the common mean number of occurrences is specified as equal to u' (and not estimated from the data), we would compute the statistic $\sum \frac{(c_i - u')^2}{u'}$ and compare this with the critical χ^2 for k degrees of freedom.

It will be noted that the χ^2 test just described offers an alternative procedure to the construction of a c-chart. To illustrate this let us apply the procedure to the data of Section 3 of Chapter 30. For these data $\bar{c} = 55.7$. We therefore have

$$
\begin{aligned}
\sum_i \frac{(c_i - \bar{c})^2}{\bar{c}} = \frac{1}{55.7}\,[&(77 - 55.7)^2 + (64 - 55.7)^2 \\
&+ (75 - 55.7)^2 + (93 - 55.7)^2 + (45 - 55.7)^2 \\
&+ (61 - 55.7)^2 + (49 - 55.7)^2 + (65 - 55.7)^2 \\
&+ (45 - 55.7)^2 + (77 - 55.7)^2 + (59 - 55.7)^2 \\
&+ (54 - 55.7)^2 + (41 - 55.7)^2 + (87 - 55.7)^2 \\
&+ (40 - 55.7)^2 + (22 - 55.7)^2 + (92 - 55.7)^2 \\
&+ (89 - 55.7)^2 + (55 - 55.7)^2 + (25 - 55.7)^2 \\
&+ (54 - 55.7)^2 + (22 - 55.7)^2 + (49 - 55.7)^2 \\
&+ (33 - 55.7)^2 + (20 - 55.7)^2] = 216.6
\end{aligned}
$$

Since this is much greater than $\chi^2_{0.001}$ for 24 degrees of freedom, we must conclude that the data are not homogeneous, or in other words, the process is not in control.[9] Again those terms with large contributions to $\sum \frac{(c - \bar{c})^2}{\bar{c}}$ will indicate the principal sources of nonhomogeneity or lack of control.

The above procedure can be used to compare two as well as several units. In this case the statistic to be used would be

---

[9] As in the previous section, the hypothesis of control cannot be accepted until the χ^2 test has been supplemented by a run test.

$$\frac{(c_1-\bar{c})^2+(c_2-\bar{c})^2}{\bar{c}}$$

which would be compared with a critical χ^2 for one degree of freedom.

The procedure can also be employed to sets of sample units in which the number of units per sample differ from sample to sample. In this case, we would need to compute the weighted average number of occurrences per unit $(\bar{\bar{c}})$, and the statistic to be employed would be

$$\sum_i \frac{(X_i - k_i\bar{\bar{c}})^2}{k_i\bar{\bar{c}}}$$

where X_i is the total number of nonconformities occurring in the ith sample, k_i is the number of units in the ith sample, and $\bar{\bar{c}} = \sum_i X_i \Big/ \sum_i k_i$.

It is to be noted once again that the use of χ^2 in these tests of independence and homogeneity is an approximate procedure that may not work well if numerous expected values $(k_i\bar{\bar{c}})$ are very small. We must be especially on our guard against this when we are comparing only two samples. For an exact procedure for comparing two Poisson variables the reader is referred to the *Biometrika Tables for Statisticians,* Vol. I, 2d ed., p. 72.

5. THE OPERATING CHARACTERISTICS OF χ^2 TESTS

The ability of a test to detect differences is given by the OC curve for the test. For χ^2 tests of independence and homogeneity, departures from the condition of independence or homogeneity are measured by

(28.1)
$$\lambda = \sum \left[\frac{(m_i' - m_i)^2}{m_i}\right]$$

where m_i is the frequency of the ith cell that is expected on the assumption of independence or homogeneity and m_i' is the frequency expected on some arbitrary alternative basis. Both the m_i and m_i' must, of course, sum to the marginal totals given in the problem. For this measure of departure from hypothesis, Evelyn Fix has tabulated abscissa and ordinates for OC curves for degrees of freedom from 1 to 100. A selected set of her results are presented in Table 28.5.

Let us determine now the 0.10 point of the OC curve for the test of toughness of nylon bars discussed in Section 1. In any 2 × 2 table such as Table 28.1, designated differences between m_i' and m_i for the given cells must be the same for all cells, since as noted above m_i' and m_i must sum to the same row and column totals (i.e., the $m_i' - m_i$ must sum to zero by rows and columns). Call this common difference d. Then for a 2 × 2 table we have

(28.1a)
$$\lambda = d^2 \sum \left(\frac{1}{m_i}\right)$$

TABLE 28.5
Values of λ for which the Probability of Falsely Accepting the Hypothesis of Independence Is 0.10* (probability of falsely rejecting hypothesis of independence = α)

Degrees of Freedom	λ	
	α = 0.05	*α = 0.01*
1	10.509	14.879
2	12.655	17.427
3	14.172	19.248
4	15.405	20.737
5	16.470	22.033
6	17.419	23.187
7	18.284	24.238
8	19.083	25.211
9	19.829	26.122
10	20.532	26.981
20	26.132	33.852
30	30.379	39.074
40	33.941	43.461
50	37.069	47.312
70	42.48	53.99
100	49.29	62.39

* Smaller λ's yield higher probabilities of accepting the hypothesis of independence.

SOURCE: Abridged from Evelyn Fix, "Tables of Non-Central χ^2," *University of California Publications in Statistics,* Vol. I No. 2 (1949), pp. 15–19.

To find the d (for the given m_i) that has a 0.90 chance of being detected, we set $d^2\,\Sigma(1/m_i)$ equal to the $\lambda_{0.10}$ for one degree of freedom. If we take a significance level of 0.05, this gives us

$$10.509 = d^2 \sum \left(\frac{1}{m_i}\right)$$

For the data of Table 28.1, which gives the m_i for the nylon bar problem, we have

$$10.509 = d^2 \left[\frac{1}{127} + \frac{1}{273} + \frac{1}{127} + \frac{1}{273}\right]$$

from which

$$d^2 = \frac{10.509}{0.023074} = 455.45$$

and $d = 21.3$, or approximately 21. Hence, if the true expectations (the m_i') were as shown in Table 28.6 (note that $m_i' = m_i \pm d$), there would still be a chance of 0.10 of not detecting such a difference.

TABLE 28.6
Alternative Expected Frequencies of Nylon Bars (see Table 28.1)

Length of Heating Cycle	*Number of Bars Classed as Brittle*	*Number of Bars Classed as Tough*	*Total*
30 seconds	106	294	400
90 seconds	148	252	400
Total	254	546	800

If we viewed the above problem as a comparison of two percentages, Table 28.6 would mean that in reality the percentages could be 26.5 percent for bars given the 30-second heating and classed as brittle and 37.0 percent for bars given the 90-second heating and classed as brittle, a difference of 10.5 percentage points, and we would still have a chance of 0.10 of saying that there was no difference. If we should feel that a difference of 5 percentage points, for example, was important enough to detect, then the above test is definitely not adequate since the risk of failing to detect a difference of 5 percentage points is greater than 0.10.

Let us consider now the question of what total sample size would be adequate to detect a difference in percentages equal to a specified value. Suppose that we wish a 0.90 chance of detecting a difference of δ percentage points (expressed as a fraction, say, 0.10), and suppose that we estimate the average percentage of the class in question to be p_0. If we let ρ and $(1 - \rho)$ be the estimated proportional split in the other classification, then the formula for determining n is

$$\textbf{(28.2)} \qquad n = \frac{10.509\left(\dfrac{1}{\rho} + \dfrac{1}{1-\rho}\right)^2}{\delta^2\left[\dfrac{1}{\rho p_0} + \dfrac{1}{\rho(1-p_0)} + \dfrac{1}{(1-\rho)p_0} + \dfrac{1}{(1-\rho)(1-p_0)}\right]}$$

It is to be noted that this formula is not valid if it should yield a very small n. This, however, is rarely the case for the values of δ that we are usually interested in.

To illustrate suppose we estimate that $\rho = 0.5$ and that $p_0 = 31.75$ percent. Then to have a 0.90 chance of detecting a difference of 5 percentage points, we should take

$$n = \frac{10.509\left(\dfrac{1}{0.5} + \dfrac{1}{0.5}\right)^2}{(0.05)^2\left[\dfrac{1}{0.5(0.3175)} + \dfrac{1}{0.5(0.6825)} + \dfrac{1}{0.5(0.3175)} + \dfrac{1}{0.5(0.6825)}\right]}$$

$$\doteq 3{,}600$$

The reader should not confuse d and δ. They are related as follows. Let the cell frequencies expected on the basis of independence or no difference in percentages be

m_1	m_2	ρn
m_3	m_4	$(1-\rho)n$
		n

Then for independence or no difference we must have

$$\frac{m_1}{\rho n} = \frac{m_3}{(1-\rho)n} \text{ and } \frac{m_2}{\rho n} = \frac{m_4}{(1-\rho)n}$$

Let the alternative to independence be given by the frequencies $m_i \pm d$, i.e., by

m_1+d	m_2-d	ρn
m_3-d	m_4+d	$(1-\rho)n$
		n

The difference in percentages for the row classification would be

$$\frac{m_1+d}{\rho n} - \frac{m_3-d}{(1-\rho)n} = \frac{d}{\rho n} + \frac{d}{(1-\rho)n} = \frac{d}{\rho(1-\rho)n}$$

It is this difference that is specified to be δ. We thus have

$$d = \rho(1-\rho)n\,\delta$$

When we are concerned with an $r \times c$ contingency table for which r or c or both are greater than two, formula (28.1a) does not apply and we must go back to formula (28.1). In this instance simplification in terms of some meaningful quantity may not be possible and the best we can probably do is to try out various values of m_i' for given m_i and note how close the λ's come to $\lambda_{0.10}$.

6. PROBLEMS

28.1. In April 1944, a new assembly crew B turned out 4,554 tanks, of which 352 showed front discharge silver solder leaks. An old crew A turned out in the same period 4,109 tanks, of which 217 showed front discharge silver solder leaks (cf. OPRD Quality Control Reports, No. 3, p. 19). Is it reasonable to conclude from these data that crew B is operating in a way that causes a larger fraction of nonconforming tanks than crew A? Work the problem in two ways: (1) as a 2 × 2 contingency table and (2) as a difference between two proportions.

28.2. Two different types of coning guides are used in producing a given yarn item. Production from each type of guide was sampled, and the items were classified as follows:[10]

Guides	*Nonconforming*	*Conforming*	*Total*
A	20	180	200
B	12	88	100
Total	32	268	300

a. Could you infer from these data whether there was an "interaction" between the number of nonconforming items produced and the type of guide? Work this problem in two ways.

b. Suppose that the frequencies in *a* were all multipled by 10, i.e., they were 200, 120, 1,800, and 880. How would your answer be affected?

28.3. The following is the classification of 95 "reject bobbins" according to the type of nonconformity and "buffing cycle" in which it occurred:[11]

Buffing Cycle	*Niched Top Flange*	*Niched Bottom Flange*	*Total*
No. 1	32	12	44
No. 2	14	22	36
No. 3	6	9	15
Total	52	43	95

Can you infer from these data whether type of nonconformity is independent of the cycle in which it occurred? Justify your answer by statistical analysis.

28.4. A study was made of tire wear. After a given mileage test, the tires scrapped were distributed as follows:[12]

	Front	*Rear*
Right	115	65
Left	125	95

Would you conclude from these data: *(i)* That front tires get more wear than rear tires? *(ii)* That left tires get more wear than right tires? *(iii)* That

[10] See Harry E. Robbins, "Some Applications of Standard Chi Square Significance Tests to Plant Problems," American Society for Quality Control, *1956 Middle Atlantic Conference Transactions,* pp. 43–51.

[11] Ibid.

[12] Cf. A. W. Swan, "The χ^2 Significance Test—Expected vs. Observed Results," *The Engineer,* December 31, 1948, pp. 679.

there is any "interaction" between the various positions? (Work this part of the problem in two ways.)

28.5. A study was made of the effect of the time of work on the quality of work. It was the practice in a plant for a crew to change shifts once a month. A study of three months' operations by one crew that was intact for the period showed the following:

Shift	*Nonconforming*	*Conforming*
8:00–4:00	52	921
4:00–12:00	61	902
12:00–8:00	73	851

Would you conclude from these data that the time of work affected the quality of the results? Justify your answer.

28.6. Use a χ^2 test to answer Problem 2.5, Chapter 2.

28.7. In the development of the VT fuse during World War II, a study was made of the causes of failure of vacuum tubes. The following data illustrate the procedure.[13] The data are the number of tubes that failed for the reasons indicated.

Position in Shell	*Type of Failure*		
	A	*B*	*C*
Top block	75	10	15
Bottom block . . .	40	30	10

Would you conclude from these data that the distribution of failures by type was the same for both shell positions? Justify your answer.

28.8. Use a χ^2 test of homogeneity to test whether the data of Problem 19.5 came from a controlled process.

28.9. Use a χ^2 test to determine whether the data of Problem 20.3 came from a controlled process.

28.10. Two years ago the number of accidents in your plant was 153. Last year you introduced some safety devices and the number of accidents dropped to 145. Could you reasonably assume that the drop was due to the new safety measures? Assume that other conditions in the plant, particularly the number of employees, remained unchanged.

28.11. You wish to determine whether the percent nonconforming of lot A is significantly different from that of lot B. From past experience you judge that the fraction nonconforming will run about 10 percent and you wish to have a 0.90 chance of detecting a difference between the lots of 2 percentage points. Assume an α risk of 0.05 and determine the number of items you will inspect

[13] Cf. Besse B. Day, "Application of Statistical Methods to Research and Development in Engineering," *Review of the International Statistical Institute,* 1949, Nos. 3, 4.

from each lot. Assume that the cost of inspection is the same for both A and B items and that both lots are large, say in excess of 500,000.

28.12. You wish to determine whether machine X can turn out a better quality product than machine Y. You judge that the general level of the fraction nonconforming for either machine will be in the neighborhood of 6 percent and you wish to have a 0.90 chance of detecting a difference between the fractions nonconforming of the two machines of 1 percentage point. Assume an α risk of 0.05 and determine the number of items you will inspect from each machine. Assume that the cost of inspection is the same for the output of each machine.

7. SELECTED REFERENCES*

Bennett and Hsu (P '60), Broffitt and Randles (P '77), Brown (P '74), Brownlee (B '48 and B '60), Cochran (P '42, P '52, and P '55. The '52 paper should be read by all those who actually employ χ^2-tests. This also has a comprehensive list of references.), Craddock and Flood (P '70), Day (P '49), Fisher (B '58), Fix (P '49), Haber (P '80), Hald (P '52), Hommell (P '78), Johnson, N. S. (P '75), Kendall (B '48), Lawal (P '80), Mainland, Herrera, and Sutcliffe (B '56), Mathai and Katiyar (P '79), contains tables of 1% and 5% points, Metha and Patel (P '83), Pagano and Halvorsen (P '81), Smith, J. G. and Duncan (B '45), Statistical Research Group, Columbia University (B '47), Swan (P '48), and Yule and Kendall (B '49. A classic in the field of contingency tables).

---

* B and P refer to the Book and Periodical sections, respectively, of the Cumulative List of References in Appendix V.

29

Analysis of Variance

1. ANALYSIS OF VARIANCE FOR A ONE-WAY CLASSIFICATION

One of the most powerful tools of statistical analysis is what is known as analysis of variance. Basically it consists of classifying and cross-classifying statistical results and testing whether the means of a specified classification differ significantly. In this way it is determined whether the given classification is important in affecting the results. For example, the output of a given process might be cross-classified by machines and operators (each operator having worked on each machine). From this cross-classification it could be determined whether the mean qualities of the outputs of the various operators differed significantly. Also it could *independently* be determined whether the mean qualities of the outputs of the various machines differed significantly. Such a study would determine, for example, whether uniformity in quality of output could be increased by standardizing the procedures of the operators (say through special training) and similarly whether it could be increased by standardizing the machines (say through resetting).

The statistical procedure for testing differences between several means will be discussed in this chapter and the next. This will be done for various degrees and kinds of classification and cross-classification. Subsequent chapters will be concerned with such topics as general tests of homogeneity, multiple comparisons of means, estimation of the components of variance and analysis of covariance.

1.1. A Problem

The analysis of variance for a one-way classification may be illustrated with reference to the data of Table 29.1. This gives three determinations of the melting point of hydroquinone on each of four thermometers. The question

TABLE 29.1
Determinations of the Melting Point of Hydroquinone in Degrees Centrigrade

Thermometers			
A	*B*	*C*	*D*
174.0	173.0	171.5	173.5
173.0	172.0	171.0	171.0
173.5	173.0	173.0	172.5

SOURCE: See Grant Wernimont, "Quality Control in the Chemical Industry. II. Statistical Quality Control in the Chemical Laboratory," *Industrial Quality Control,* May 1947, p. 8.

to be put to these data is: Do the thermometers read differently, or can the variations between thermometers be attributed primarily to chance errors?

In carrying out the analysis, we shall suppose that other controllable factors that might affect the measurement of temperature are held constant so that the variations in readings on the same thermometer are purely "experimental errors." This may give a constant bias to all readings, but it will not affect the analysis of variance. How an experiment is designed depends in part on the object of the experiment, and questions pertaining to this are discussed in Chapter 35.[1] For the moment we shall be concerned only with the technique of analysis of variance.

The computations of analysis of variance can become rather complicated, so we attempt to simplify them as much as possible. A useful preliminary step is to "code" the data.[2] In our present problem we make our work easier by subtracting 172 from each figure in Table 29.1. The results are given in Table 29.1A. In cases where there are decimals, the computations may be further facilitated by multiplying by some multiple of 10. Neither subtraction of a constant nor multiplication by a constant will affect the ratio of variances that will subsequently be used in the analysis. If it is desired to estimate the mean readings of each thermometer, however, or to determine the components of variance,[3] some decoding may be necessary.

1.2. Theory

In a one-way analysis of variance the hypothesis to be tested is the "null hypothesis" that the items in the various classes come from universes the means of which are equal. More precisely, the mathematical model for the analysis is

$$X_{ij} = \mu + \theta_j + \epsilon_{ij} \tag{29.1}$$

[1] See especially Section 1.

[2] Cf. Section 4.1.1.2 of Chapter 3.

[3] See Chapter 31.

TABLE 29.1A
Data of Table 29.1 Coded to an Arbitrary Origin of 172

	Thermometers				
	A	*B*	*C*	*D*	
	2.0	1.0	−0.5	1.5	
	1.0	0.0	−1.0	−1.0	
	1.5	1.0	1.0	0.5	
Σ	4.5	2.0	−0.5	1.0	Grand total = 7.0
$\bar{X}_j$	1.5	0.67	−0.17	0.33	$\bar{X} = 0.58$

where μ is a constant, the θ_j are class differentials, and ϵ_{ij} is a random, normal deviate with mean zero and variance σ'^2, these being the same for all classes. The hypothesis to be tested is that the θ_j are zero for all j.

With reference to the thermometer example, the null hypothesis is that all the thermometers read alike. This means that in an infinity of melting-point determinations with each thermometer the mean would be the same for all thermometers. It is assumed that the variances of the experimental errors are identical for all thermometers.

The procedure for testing the null hypothesis in a one-way analysis of variance is based on some simple theoretical notions. These may be outlined as follows:

Let us consider Table 29.1A. Here we have three determinations with each of four thermometers. If these thermometers can all be used without bias [i.e., in equation (29.1) the θ_j are all 0], then we really have four random samples of determinations from universes that may be taken to be identical. Let us deviate from the trend of the argument for the moment and suppose that under the assumption of unbiased thermometers we wish to estimate the common universe variance. To do this we can proceed in two ways. We can recall[4] that means of samples of size r from a single universe tend to have a variance that equals the variance of the universe divided by r. Hence, if we wish to estimate the common variance of the universes of melting-point determinations, we can compute the variance of the four thermometer means around their grand mean and *multiply* by the sample size r. For if in an infinity of samples from identical universes

$$\sigma'^2_{\bar{X}} = \sigma'^2/r,$$

then from four samples we can estimate σ'^2 as equal to r times the variance of the four sample means.[5] Thus

[4] Cf. above, Chapter 6, Section 1.2.

[5] See Chapter 6, Section 2.2 and Appendix I (22) and (33). If the number of cases in each column are unequal, the numerator of $s^2_{\bar{C}}$ would be $\sum_j n_j(\bar{X}_j - \bar{X})^2$.

$$\text{Estimate of } \sigma'^2 = \frac{r \sum_j (\bar{X}_j - \bar{X})^2}{c - 1}$$

[A bar implies a mean over all suppressed subscripts; e.g., here $\bar{X}_j$ is a mean over the ith subscript and $\bar{X}$ the grand mean over both i and j.]

Call this estimate s_C^2 since it is based on the column means of Table 29.1.A.

It is also possible to estimate σ'^2 by computing separate sample variances for each of the four thermometers and then averaging these sample results. To estimate σ'^2 we take[6]

Estimate of σ'^2

$$= \frac{s_1^2 + s_2^2 + \cdots + s_c^2}{c} = \frac{1}{c}\left[\frac{\Sigma(X_1 - \bar{X}_1)^2}{(r-1)} + \frac{\Sigma(X_2 - \bar{X}_2)^2}{(r-1)} \cdots + \frac{\Sigma(X_c - \bar{X}_c)^2}{(r-1)}\right] = \frac{\sum_i \sum_j (X_{ij} - \bar{X}_j)^2}{c(r-1)}$$

Call this estimate s_e^2 since it is an estimate of the error variance.

Now under the hypothesis that the three determinations on each of the four thermometers are really random samples from identical normal universes, it can be shown that the estimates s_C^2 and s_e^2 are independent of each other.[7] This means that in many such samples with the four thermometers, the values of s_C^2 and s_e^2 computed from each sample will show no association.

This brings us back to the basic argument. We have been talking about making independent estimates of the common universe variance, but what we want is a procedure for testing whether the means of the various classes (thermometers) differ significantly. The theory we have just developed, however, about the estimates s_C^2 and s_e^2 gives us the tool we need for testing for differences between the thermometer means. For if the null hypothesis is true, i.e., if the four universes of determinations are normal universes with identical means, then on the assumption that the universe standard deviations are all equal, it follows that ratios such as s_C^2/s_e^2 computed from an infinity of samples (each consisting of r readings from the same c thermometers) will tend to form an F distribution[8] with degrees of freedom given by $\nu_1 = c - 1$ and $\nu_2 = c(r - 1)$. Consequently, if in the case of a single sample we wish to test the hypothesis that

$$\bar{X}'_1 = \bar{X}'_2 = \cdots = \bar{X}'_c,$$

or in terms of model (29.1) that $\theta_j = 0$ for all j, we can compute the sample statistic s_C^2/s_e^2 and compare it with F_α for the selected value of α.

---

[6] See Appendix I (33). When the number of cases in each column are unequal, the denominator of s_e^2 is $\sum_j (n_j - 1)$.

[7] See H. B. Mann, *Analysis and Design of Experiments* (New York: Dover Publications, 1949), chap. ii and iii.

[8] See Chapter 26, Section 3.1. Also see Mann, *Analysis and Design of Experiments.*

It will be noted that the analysis of variance test is always a "one-sided" test, since a low sample value of s_C^2/s_e^2 would mean that the "fit" of the sample means to the hypothesis $\bar{X}'_1 = \bar{X}'_2 = \ldots = \bar{X}'_c$ is exceptionally good. An extremely low sample s_C^2/s_e^2 might cast doubt on the randomness of the results but not on the equality of the means.

For the data of Table 29.1A, the two estimates s_C^2 and s_e^2 may be derived as follows:

$$s_C^2 = 3\left[\frac{(1.5-0.58)^2+(0.67-0.58)^2+(-0.17-0.58)^2+(0.33-0.58)^2}{4-1}\right]$$

$$= \frac{3(1.47)}{3} = 1.47$$

$$s_e^2 = \frac{1}{4(3-1)}[(2.0-1.5)^2+(1.0-1.5)^2+(1.5-1.5)^2$$
$$+(1.0-0.67)^2+(0.0-0.67)^2+(1.0-0.67)^2$$
$$+(-0.5-(-0.17))^2+(-1.0-(-0.17))^2+(1.0-(-0.17))^2$$
$$+(1.5-0.33)^2+(-1.0-0.33)^2+(0.5-0.33)^2]$$

$$= \frac{6.50}{8} = 0.81$$

The value of the sample s_C^2/s_e^2 is thus

$$s_C^2/s_e^2 = 1.47/0.81 = 1.82$$

For $\nu_1 = 3$ and $\nu_2 = 8$, the critical $F_{0.05}$ point is 4.07. Since the sample s_C^2/s_e^2 is less than this critical value, we do not reject the null hypothesis and tentatively conclude that the thermometers really read alike.

1.3. Short-Cut Computations

The computations of the previous section are longer than are necessary. When we have an equal number of cases in each class, it has become customary to use a short-cut procedure.

The short-cut procedure distinguishes a "column sum of squares," a "total sum of squares," and an "error sum of squares." The column sum of squares is merely the number of items in each class *(r)* times the sum of the squares of the deviations of the class means from the grand mean. It is symbolized by

$$\text{Column SS} = r\sum_j (\bar{X}_j - \bar{X})^2$$

The total sum of squares is the sum of the squares of the deviations of the individual items from the grand mean and is symbolized by

$$\text{Total SS} = \sum_i \sum_j (X_{ij} - \bar{X})^2$$

The error sum of squares is the sum of the squares of the deviations of the individual items in each class from the mean of that class summed for all classes. It is symbolized by

$$\text{Error SS} = \sum_i \sum_j (X_{ij} - \bar{X}_j)^2$$

The short-cut method relies on the mathematical identity that

$$\text{Total SS} = \text{Column SS} + \text{Error SS}$$

The mathematical derivation of the short-cut procedure will be found in Appendix I, Sections (32) and (34).

For the data of Table 29.1A the computations of the short-cut procedure run as follows:

1. Sum each column, square these column sums, sum and divide by the number of items in the columns. Thus,

$$\frac{4.5^2 + 2^2 + (-0.5)^2 + 1^2}{3} = 8.50$$

2. Square each individual item and sum. Thus,

$$2^2 + 1^2 + 1.5^2 + 1^2 + 0^2 + 1^2 + (-0.5)^2 + (-1)^2 + 1^2 + 1.5^2 + (-1)^2 + 0.5^2 = 15.00$$

3. Sum all the cases, square this sum, and divide by the number of cases. This is usually called the correction factor. Thus,

$$\frac{(2 + 1 + 1.5 + 1 + 0 + 1 - 0.5 - 1 + 1 + 1.5 - 1 + 0.5)^2}{12} = 4.08$$

4. The column sum of squares equals

$$(1) - (3) = 8.50 - 4.08 = 4.42$$

5. The total sum of squares equals

$$(2) - (3) = 15.00 - 4.08 = 10.92$$

6. The error sum of squares equals

$$(5) - (4) = 10.92 - 4.42 = 6.50$$

These results check with those obtained in the preceding section.

The various sums of squares are now brought together in an analysis of variance table such as Table 29.2.

The letters "df" in Table 29.2 stand for "degrees of freedom."[9] The degrees of freedom for the column sum of squares are the number of column totals that may be assigned arbitrarily without affecting the grand total of

[9] Compare Chapter 27, Section 2 and Chapter 28, Section 1.

TABLE 29.2
Analysis of Variance Table for a One-Way Classification

Source of Variation	*Sum of Squares*	*df*	*Mean Square*	*Variance Ratio*
Thermometers (columns)	4.42	3	$s_C^2 = 1.47$	1.82
Experimental error (residuals)	6.50	8	$s_e^2 = 0.81$	
Total	10.92	11	....	....

the data. In general, it equals the number of columns minus 1. In Table 29.1A it equals $4 - 1 = 3$. The number of degrees of freedom for residuals is the number of items for each column that might be filled in arbitrarily without affecting the total for the column, multiplied by the number of columns.[10] In Table 29.1A it equals $4(3 - 1) = 8$. Division of the sums of squares by the corresponding degrees of freedom yields a mean square for columns and an error mean square. The mean square for columns is s_C^2, and the error mean square is s_e^2. For Table 29.2, $s_C^2/s_e^2 = 1.47/0.81 = 1.82$, which is less than $F_{0.05}$ ($\nu_1 = 3$, $\nu_2 = 8$) $= 4.07$. As noted above, we do not reject the null hypothesis.

1.4. OC Curves for Analysis of Variance with a One-Way Classification

Like the test of any hypothesis, the test of the null hypothesis in analysis of variance will involve both errors of the first and second kind. The risk of the latter will as usual be given by the operating characteristic curve for the test.

The OC curve for a one-way analysis of variance will depend upon the hypotheses alternative to the null hypothesis. With reference to the mathematical model (29.1), the null hypothesis may be stated in the form $X_{ij} = \mu + \epsilon_{ij}$, where ϵ_{ij} is a normally distributed random variable with zero mean and standard deviation σ' and μ is a constant. This merely says that the variable X_{ij} is normally distributed around a constant μ. Let $X_{ij} = \mu + \theta_j + \epsilon_{ij}$ be an alternative hypothesis in which the μ and ϵ_{ij} are defined as in the null hypothesis and (1) the θ_j have specified values other than 0 and are restricted to these given values, or (2) the θ_j are a random sample from a universe of θ's, the variance of which is $\sigma_\theta'^2$. With reference to our thermometer problem, these alternatives may be explained as follows. If the four thermometers used in making the measurements are the only four in our laboratory, we would be interested in the first alternative hypothesis. Here the θ_j's would be the fixed relative biases of each thermometer. On the other

[10] If the number of items in a column varies from column to column, $\nu_2 = n - c$, where n equals the total number of cases.

hand, if these four thermometers are a random sample of thermometers used and to be used in the laboratory, we would be interested in the second alternative hypothesis. In this case the relative biases of our four thermometers (θ_j) would be viewed as a random sample of thermometer biases in general.

If we make the assumption that the θ_j's are fixed values, the OC curve for the test of the null hypothesis is expressed as a function of the quantity $\lambda = \sqrt{\Sigma \theta_j^2 / c\sigma'^2}$. If we consider θ as a random variable, then the OC curve for the test of the null hypothesis is expressed as a function of $\lambda = \sigma'_\theta / \sigma'$, in which σ'_θ is the standard deviation of the θ universe.

The OC curves for fixed θ_j's have been worked out by P. C. Tang,[11] and tables have been reproduced in H. B. Mann, *Analysis and Design of Experiments.*[12] The OC curves for the fixed θ_j's can also be easily derived from E. S. Pearson and H. O. Hartley's charts of the power function for analysis of variance F tests.[13] Charts I and II in Appendix II, give the 0.10 and 0.50 points of the Pearson and Hartley curves for $\alpha = 0.05$ and 0.01. In using these charts and Mann's tables it is to be noted that

$$\phi^2 = \frac{r \sum_j \theta_j^2}{c\sigma'^2} = r\lambda^2 \tag{29.2}$$

Figure 29.1 gives the OC curves for fixed θ_j's for selected values of r and c. For example, if the means of the universes of determinations to be made with the four thermometers vary among themselves such that the average of the squared deviations from their own average is 25 percent of the experimental error variance (σ'^2), the value of λ will be $\sqrt{0.25} = 0.5$. Figure 29.1 shows that if three measurements are made with each thermometer, the probability of *not* detecting such differences will be approximately 0.80. If, with three measurements for each of four thermometers, the variation in the universe means is as much as 100 percent of the experimental variance (i.e., $\lambda = 1$), then the probability of not detecting such a variation is about 0.36. It will be instructive for the reader to check Figure 29.1 with Chart I of Appendix II.

When θ is viewed as a random variable, the OC curves for the test can be derived from the OC curves for the ratio of two ordinary variances.[14] Tables of the OC function for an ordinary variance ratio are given in Statistical Research Group, Columbia University, *Techniques of Statistical Analysis.*[15] Table L[16] of Appendix II, which is taken from this source, gives the 10

---

[11] *Statistical Research Memoirs,* Vol. II (1938).

[12] New York: Dover Publications, 1949.

[13] *Biometrika* 38 (1951), pp. 112–30.

[14] Compare Chapter 26, Section 3.3. The ratio is that of two variables each of which has independently a χ^2 distribution. When the θ_j's are fixed, the numerator of the variance ratio is distributed as noncentral χ^2.

[15] Tables 8.3 and 8.4.

[16] See p. 1024.

FIGURE 29.1
Operating Characteristic Curves for Single-Factor Analysis of Variance Tests: Alternative Hypothesis Having the Form $X_{ij} = \mu + \theta_j + \epsilon_{ij}$, Where the θ_j Have Fixed Values (c columns, r in a column, and α taken as 0.05)

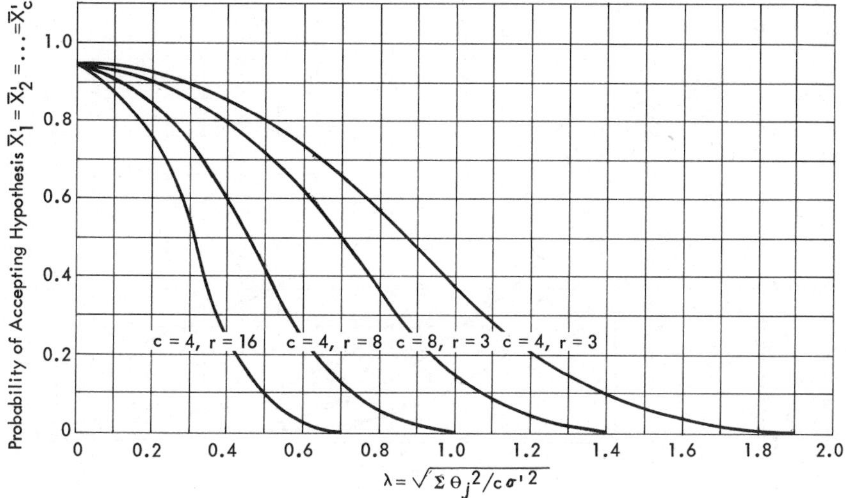

percent points of these OC curves for $\alpha = 0.05$. In using Table L and other tables from the *Techniques of Statistical Analysis,* it is to be noted that the ϕ^2 in Table L and the ϕ in the *Techniques* tables[17] must be taken equal to $(\sigma'^2 + r\sigma_\theta'^2)/\sigma'^2 = 1 + r\lambda^2$, where $\lambda^2 = \sigma_\theta'^2/\sigma'^2$. For example, with $c = 4$, $r = 3$, we have $\nu_1 = c - 1 = 3$ and $\nu_2 = c(r - 1) = 8$. From Table L we then have ϕ^2 for $\nu_1 = 3$, $\nu_2 = 8$, equal to 21.35. Therefore $1 + r\lambda^2 = 1 + 3\lambda^2 = 21.35$, or $\lambda^2 = 20.35/3 = 6.78$, or $\lambda = 2.6$. Hence, the chance is just 0.10 of not detecting a difference in thermometers when σ_θ' for the thermometers is 2.6 times the experimental standard deviation σ'. Particular OC curves based on the *Techniques* tables are shown in Figure 29.2. These curves are for the same values of r, c, and α as shown in Figure 29.1 and thus permit a comparison between the OC curves for fixed θ_j's and those for random θ.

It will first be noted that the OC curves for fixed variation are generally lower than those for random variations. For example, suppose that $c = 4$, $r = 3$. If $\lambda = 2.0$, a specified fixed difference is almost certain to be detected but a random variation corresponding to $\lambda = 2$ has still a chance of 0.20 of not being detected. A similar conclusion follows from a comparison of the other curves. In general, fixed variations have a higher chance of being detected than equal random variations. This agrees with our intuition; for in the second case we have two reasons for failing to detect the variation.

[17] Note carefully that ϕ^2 In Table L equals ϕ in the *Techniques* tables.

FIGURE 29.2
Operating Characteristic Curves for Single-Factor Analysis of Variance Tests: Alternative Hypothesis Having the Form $X_{ij} = \mu + \theta_j + \epsilon_{ij}$, Where θ_j Is a Random Variable with Variance σ'^2_θ (*c* columns, *r* in a column, and α taken as 0.05)

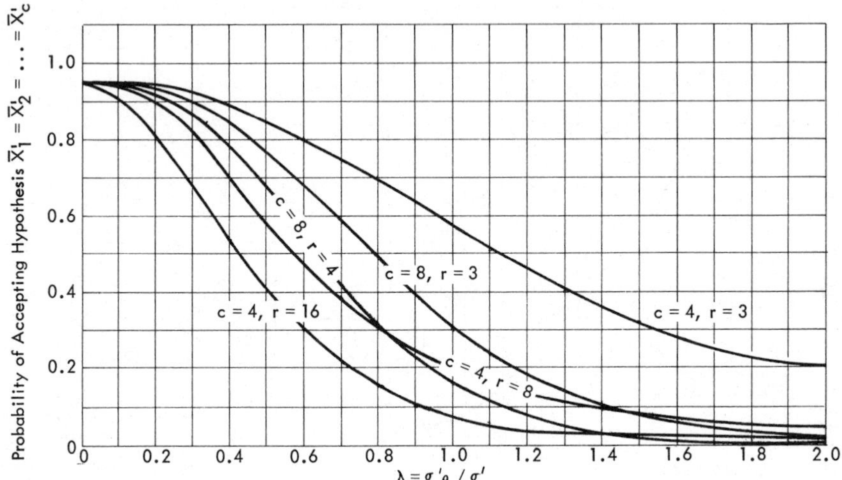

With reference to our problem, if the thermometers to which the data pertain are a random sample of thermometers in general, then it is possible that the particular four thermometers selected might just happen to show little variation. But even if we happen to get four thermometers which vary considerably, we might still get sample measurements from each thermometer, the averages of which vary little from thermometer to thermometer.

1.5. The Design of Analysis of Variance for a One-Way Classification

When dealing with fixed variation (e.g., a specified set of thermometers), the only element of design that is free for the statistician to determine is the number of cases to be taken for each class of the factor (i.e., the number of measurements per themometer). In other words, c is fixed, but r may be chosen by the statistican. Figure 29.1 shows, for example, that for $c = 4$, considerable improvement in sensitivity is gained by increasing r from 3 to 8 and again from 8 to 16. Thus if $\lambda = 1$, the $c = 4$, $r = 3$ test has a chance of about 0.36 of not detecting the difference, whereas the $c = 4$, $r = 8$ test is almost sure to detect it. Again, a difference corresponding to a $\lambda = 0.7$ will almost certainly be detected by the $c = 4$, $r = 16$ test, but with the $c = 4$, $r = 3$ test there is a chance of 0.66 of not detecting it.

A similar improvement in the sensitivity of the test with increases in r may be noted in the case of random variation (e.g., the case in which the thermometers are a random sample of thermometers). In this instance, however, we have an additional element of design in that both c and r can be determined by the statistician. In the case of random variation a test with a larger number of factor classes (e.g., a larger number of thermometers) and smaller number of cases per class (small number of measurements per thermometer) may in certain instances do better in detecting differences than a test with a smaller number of factor classes (thermometers) and a larger number of cases per class (measurements per thermometer). Thus Figure 29.2 shows that the test $c = 4$, $r = 8$ does better than the test $c = 8$, $r = 4$ for values of λ smaller than 0.83, but after that point the second test does better.

A rough rule often followed by those who are not interested in precise computations of risks is that in an F test it is well to have the error mean square based on at least 6 degrees of freedom. For a brief survey of the F table will show that critical values of F diminish rather rapidly up to about $\nu_2 = 6$ but less rapidly after that.

1.6. The Studentized Range Test

It may be noted in passing that instead of running an ordinary one-way analysis of variance it is possible to test the null hypothesis of no difference between $\bar{X}'_1, \bar{X}'_2, \ldots, \bar{X}'_c$ by running a "studentized range test." This was mentioned in Section 6 of Chapter 6.

In the comparison of a set of means the studentized range test consists simply in taking the ratio of the difference between the highest and lowest sample means to the estimated standard deviation of a sample mean. Thus we compute

$$q = \frac{\bar{X}_H - \bar{X}_L}{s/\sqrt{r}}$$

where s is the square root of the error mean square given in an analysis of variance table and r is the number of cases upon which each sample mean is based, assumed to be the same for each mean. The quantity q has a studentized range distribution with the number of groups g equal to the number of means compared ($=c$) and the degrees of freedom for s equal to $c(r - 1)$. (See Table D2 of Appendix II.)

Applying the above to the data of Table 29.1A and using the s given by Table 29.2, we have

$$q = \frac{1.5 - (-0.17)}{\sqrt{0.81}/\sqrt{3}}$$

Table D2 indicates that the 0.05 point for $q(g, \nu) = q(4, 8)$ is 4.53 and since this is greater than 3.21, the studentized range test does not lead to a rejection of the null hypothesis. This is the same conclusion as was reached by ordinary analysis of variance.[18]

It is to be noted that the studentized range test and the ordinary analysis of variance test need not agree. Ranges are seriously affected by individual variations, while variances water down these extremes by averaging them with other less extreme variations. It is thus possible for the range test to show significance and the variance-ratio method not to show significance, and vice versa. It is likely that in many industrial analyses, the range method would be preferred to the variance-ratio method just because it is more sensitive to a single extreme variation.

The studentized range test is introduced here not simply as an alternative procedure that might be more applicable in certain instances, but primarily because it has become the basis for certain procedures of grouping means after they have been shown to be significantly different. This is discussed in Sections 1.2 and 1.3 of Chapter 31.

1.7. Analysis of Individual Degrees of Freedom

A very important approach to analysis of variance is effected by the analysis of so-called individual degrees of freedom. This approach gives a good insight into the analysis of the data and lays a foundation for much of the theory of design of experiments. Although our discussion of the theory underlying this approach is a little lengthy, the reader is urged to take the time to master it.

1.7.1. The Argument in Brief. In the thermometer problem of previous sections, 3 degrees of freedom were assigned to the variation between the four thermometers. This is in fact the number of independent comparisons that can be made between four things. For example, suppose that thermometers A and B are of American make and thermometers C and D are of German make. Then we could select our comparisons as follows. We could compare the two American thermometers with the two German thermometers; we could compare one American thermometer with the other; and we could compare one German thermometer with the other. If there were a single observation (X) on each thermometer, the first comparison would be, say, $X_{A1} + X_{A2} - X_{G1} - X_{G2}$; the second would be $X_{A1} - X_{A2}$; and the third, $X_{G1} - X_{G2}$. Each such comparison could be said to account for one

[18] A quicker but less sensitive procedure might be to use $\bar{R}/d_2^*$ as a substitute for s in computing q. (See Section 6 of Chapter 6.) The $\bar{R}$ would be simply the average of the within group ranges. The quantity d_2^* and the degrees of freedom for $\bar{R}/d_2^*$ to be used in evaluating q would be given by Table D3. In the given problem, for example, $g = c = 4$, $m = r = 3$, $d_2^* = 1.75$, and the degrees of freedom for $\bar{R}/d_2^*$ would be 7.5. For the data $\bar{R} = 1.625$ and

$$q = \frac{1.67}{\left(\frac{1.625}{1.75}\right) \Big/ \sqrt{3}} = 3.12.$$

The critical $q_{0.05} = 4.61$ and again we would not reject the null hypothesis.

of the three degrees of freedom allowed us in the comparison of four things. Now if we set

$$W_1 = \frac{1}{2}[X_{A1} + X_{A2} + X_{G1} + X_{G2}]$$

$$W_2 = \frac{1}{2}[X_{A1} + X_{A2} - X_{G1} - X_{G2}]$$

$$W_3 = \frac{1}{\sqrt{2}}[X_{A1} - X_{A2}]$$

$$W_4 = \frac{1}{\sqrt{2}}[X_{G1} - X_{G2}]$$

(the coefficients being such as to make the transformation orthogonal), it is easy to show that

$$\sum_{1=1}^{4} W_i^2 = \sum_{1}^{4} X_i^2$$

or that

$$W_2^2 + W_3^2 + W_4^2 = \sum_{i=1}^{4} X_i^2 - W_1^2 = \sum_{i=1}^{4} (X_i - \bar{X})^2$$

Thus there is a relationship between the sum of the squares of the individual W_i which are related to the individual comparisons, and the sum of the squares of the deviations of the original data from their mean. In an analysis of variance we could test the sum of squares on the right (with 3 degrees of freedom) or we could test the individual W_i^2 on the left (each with a single degree of freedom).

1.7.2. The Fuller Argument. Let us now look at the argument in more detail. Let $X_1 \ldots, X_n$ be n independently derived quantities and let

$$\begin{aligned} W_1 &= a_{11}X_1 + a_{12}X_2 + \ldots + a_{1n}X_n \\ W_2 &= a_{21}X_1 + a_{22}X_2 + \ldots + a_{2n}X_n \\ &\vdots \\ W_n &= a_{n1}X_1 + a_{n2}X_2 + \ldots + a_{nn}X_n \end{aligned} \tag{29.3}$$

where

$$\begin{aligned} &\sum_j a_{ij}^2 = 1 \text{ for all } i \text{ and} \\ &\sum_j a_{ij}a_{kj} = 0 \text{ for all } i \text{ and } k \text{ where } i \neq k \end{aligned} \tag{29.4}$$

This means that W is an orthogonal transformation of X. Such a transformation merely revolves the axes, but leaves the angles between the axes and scales along the axes unchanged. Since the origin is not changed by the transformation and the scales remain the same, it follows immediately that the distance between any point and the origin will be the same in either coordinate system. Hence we will have

(29.5) $$\sum_1^n W_i^2 = \sum_1^n X_i^2$$

Consider now another set of variables Z_i. Let

(29.6)
$$\begin{aligned}
Z_1 &= b_{11}X_1 + b_{12}X_2 + \ldots + b_{1n}X_n \\
Z_2 &= b_{21}X_2 + b_{22}X_2 + \ldots + b_{2n}X_n \\
&\;\;\vdots \\
Z_n &= b_{n1}X_1 + b_{n2}X_2 + \ldots + b_{nn}X_n
\end{aligned}$$

where the b's are integers such that

(29.7)
$$\begin{aligned}
&b_{11} = b_{12} = \ldots = b_{1n} = 1, \text{ and} \\
&\sum b_{ij}b_{kj} = 0 \text{ for all } i \text{ and } k \text{ where } i \neq k
\end{aligned}$$

The conditions (29.7) also require $\sum_j b_{ij} = 0$ for $i = 2, \ldots, n$.

If we set

(29.8) $$D_i = \sum_j b_{ij}^2$$

it is easy to see that

(29.9) $$W_i = Z_i/\sqrt{D_i}$$

In setting up the transformation it is easier to work with the Z_i, but the analysis generally runs in terms of the W_i.

In individual degrees of freedom analysis a transformation such as (29.6) is usually set up in tabular form. An example of such a transformation applied to the thermometer data of Table 29.1A is shown in Table 29.3. This table says that

$$\begin{aligned}
Z_1 &= X_1 + X_2 + X_3 + X_4 + X_5 + X_6 + X_7 + X_8 + X_9 + X_{10} \\
&\quad + X_{11} + X_{12} \\
Z_2 &= X_1 + X_2 + X_3 + X_4 + X_5 + X_6 - X_7 - X_8 - X_9 - X_{10} \\
&\quad - X_{11} - X_{12} \\
Z_3 &= X_1 + X_2 + X_3 - X_4 - X_5 - X_6
\end{aligned}$$

$Z_4 = \qquad X_7 + X_8 + X_9 - X_{10}$
$\quad - X_{11} - X_{12}$

$Z_5 = X_1 - X_2$

$Z_6 = X_1 + X_2 - 2X_3$

Etc.

The last column of this table is simply a list of the D_i.

There is no rule for deriving the coefficients for a table like Table 29.3. Mathematically the only requirement is that they satisfy the restrictions (29.7). In practice the coefficients are so selected that the individual rows will represent meaningful comparisons among the data. Furthermore, as we shall see, the selection may be made in different ways, dependent on what comparisons we are interested in. The comparisons that make up a set cannot be selected arbitrarily, however, for we must not forget the restrictions (29.7). After one has had some practice, it becomes an easy matter to select coefficients.

Let us look at the comparisons of Table 29.3 from the point of view of the thermometer problem. Z_1 is the total of all the determinations. Z_2 is the difference between the total for the American thermometers and the total for the German thermometers, and affords a comparison between these two groups. (It will be noted that readings with American thermometers all have the same coefficient +1 and readings with German thermometers all have coefficients of −1.) Z_3 is a comparison between thermometer A and thermometer B, and Z_4 is a comparison between thermometer C and thermometer D. Z_5 is a comparison between the first determination with thermometer A and the second determination, and Z_6 is a comparison between the sum of determinations 1 and 2 with thermometer A and double determination 3 with thermometer A, and so forth for the remaining Z_i's.

Consider now the analysis of variance model

(29.10) $$X_{ij} = \mu + \theta_j + \epsilon_{ij} \quad \begin{cases} i = 1, 2, \ldots, r \\ j = 1, 2, \ldots, c \end{cases}$$

in which the ϵ_{ij}'s are normally and independently distributed with zero mean and variance σ'^2, this being the same for all ϵ_{ij}. The quantity μ is the grand mean and the θ_j are differentials for the existence of which we are testing. For the sake of simplicity we shall assume that $\mu = 0$. This can be done without affecting the argument since the scaling of a variable is arbitrary. Also for the moment let us assume that the hypothesis we are testing is true and that the θ_j are all individually 0. The model then becomes simply $X_{ij} = \epsilon_{ij}$.

In the rest of the argument we shall assume that we are talking about the 12 X_i listed in Table 29.3. The argument, however, is in essence quite general. We shall make the following statements, some of which will be given without proof.

TABLE 29.3
A Transformation of Melting Points

	Thermometer A			*Thermometer B*			*Thermometer C*			*Thermometer D*			
Z_i	X_1 (2.0)	X_2 (1.0)	X_3 (1.5)	X_4 (1.0)	X_5 (0.0)	X_6 (1.0)	X_7 (−0.5)	X_8 (−1.0)	X_9 (1.0)	X_{10} (1.5)	X_{11} (−1.0)	X_{12} (0.5)	*Divisor* D_i
Z_1	1	1	1	1	1	1	1	1	1	1	1	1	12
Z_2	1	1	1	1	1	1	−1	−1	−1	−1	−1	−1	12
Z_3	1	1	1	−1	−1	−1	0	0	0	0	0	0	6
Z_4	0	0	0	0	0	0	1	1	1	−1	−1	−1	6
Z_5	1	−1	0	0	0	0	0	0	0	0	0	0	2
Z_6	1	1	−2	0	0	0	0	0	0	0	0	0	6
Z_7	0	0	0	1	−1	0	0	0	0	0	0	0	2
Z_8	0	0	0	1	1	−2	0	0	0	0	0	0	6
Z_9	0	0	0	0	0	0	1	−1	0	0	0	0	2
Z_{10}	0	0	0	0	0	0	1	1	−2	0	0	0	6
Z_{11}	0	0	0	0	0	0	0	0	0	1	−1	0	2
Z_{12}	0	0	0	0	0	0	0	0	0	1	1	−2	6

1. If the X's are submitted to transformation (29.3), the coefficients being those of Table 29.3 divided by their respective $\sqrt{D_i}$, then each of the W_i's is a linear function of the ϵ_{ij} and hence is normally distributed.[19] (Cf. Problem 4.48.)

2. The mean of each W_i will be zero. The first is zero because μ is zero. All the rest will be zero because of the orthogonal nature of the transformation, whether μ is zero or not.

3. By the corollary to Theorem 4.2 the variance of each W_i will be σ'^2, for the orthogonality of the transformation yields $\sum_j a_{ij}^2 = 1$ for every i and the W_i are independent. (See Statement 5 below.)

4. The above means that each of the W_i/σ' will have a standard normal distribution, i.e., the mean will be zero and variance will be one.

5. Since the ϵ_{ij} are independent, the W_i's are also independent of each other. The argument here is a simple one. Since the ϵ's are all independent their joint probability is obtained by multiplying the individual probabilities, which means that the density function for the joint distribution is of the form $Ce^{-\Sigma\epsilon^2/2\sigma'^2}$. But under the given assumptions $\Sigma\epsilon^2 = \Sigma X^2 = \Sigma W^2$, so that the joint density function of the W_i's is[20] $Ce^{-\Sigma W^2/2\sigma'^2}$. This will factor into parts each of which is dependent only upon a single W_i. This means that the W_i are independent, for their marginal distributions will be the same as their conditional distributions. (See Chapter 2, Section 8.2, and Chapter 4, Section 4.4.)

6. Since each W_i/σ' has a normal distribution with a zero mean and unit variance, then each W_i^2/σ'^2 will have a χ^2 distribution with one degree of freedom. (Cf. Problem 6.34.)

7. Since the W_i are all independent, the sum of any of the W_i^2/σ'^2 will have a χ^2 distribution with degrees of freedom equal to the number of the W_i^2/σ'^2 summed.[21] (Cf. Problem 6.35.)

8. Form mutually exclusive subsets of the W_i^2/σ'^2 and let one subset contain ν_1 such quantities and let another subset contain ν_2. If S_1 is the sum of the W_i^2/σ'^2 for the first subset and S_2 is the sum of the W_i^2/σ'^2 for the second subset, then the ratio $\frac{S_1}{\nu_1} \Big/ \frac{S_2}{\nu_2}$ has an F distribution with ν_1 and ν_2 degrees of freedom.[22] (Cf. Problem 26.17.)

All the above was on the basis of the assumption that the hypothesis to be tested was true, viz. $\theta_j = 0$ for all j. It will still hold for W_5 to

[19] Cf. A. M. Mood and F. A. Graybill, *Introduction to the Theory of Statistics* (2d ed.; New York: McGraw-Hill, 1963), p. 211.

[20] The orthogonality of the transformation yields a "Jacobian" of 1, so the density function goes over without modification. (Cf. Problem 6.36.)

[21] See D. A. S. Fraser, *Statistics: An Introduction,* (New York: John Wiley & Sons, 1958), p. 193.

[22] See Mood and Graybill, *Introduction to the Theory of Statistics,* p. 231.

W_{12}, however, even if the hypothesis is not true, since these will be the same functions of the ϵ's whether the hypothesis is true or not. For when the model (29.10) is substituted in these W_i's, the μ and θ_j's all cancel out. It will not hold for W_2 or W_3 or W_4 if the hypothesis is not true. This immediately indicates a way for testing each of the comparisons W_2 to W_4. If the hypothesis is true, $\dfrac{W_2^2}{\sum_5^{12} W_i^2/8}$ will have an F distribution with 1 and 8 degrees of freedom. If a sample value of this statistic yields a sufficiently large F, we reject the hypothesis and claim that comparison W_2 shows a significant difference. We can proceed in the same way with W_3 and W_4. If we are interested simply in an overall test of the θ_j differentials, we can use the statistic $\dfrac{\sum_2^4 W_i^2/3}{\sum_5^{12} W_i^2/8}$, which if the hypothesis is true, will have an F distribution with 3 and 8 degrees of freedom.

In the thermometer problem we have

$$\sum W_i^2 = Q_1 + Q_2 + Q_3 + Q_4 + Q_5$$

where

$$Q_1 = W_1^2 = Z_1^2/D_1 = \text{Correction for sample mean}$$

$$Q_2 = W_2^2 = Z_2^2/D_2 = \text{German-American comparison}$$

$$Q_3 = W_3^2 = Z_3^2/D_3 = \text{Within-American comparison}$$

$$Q_4 = W_4^2 = Z_4^2/D_4 = \text{Within-German comparison}$$

$$Q_5 = \sum_{i=5}^{12} W_i^2 = \sum_{i=5}^{12} Z_i^2/D_i = \text{Error SS}$$

It follows that if we wish to test whether American thermometers have a significant bias relative to German thermometers, we can compute $\dfrac{Q_2/1}{Q_5/8}$ and compare it with $F_{0.05}$ for $\nu_1 = 1$, $\nu_2 = 8$. Again to test whether American thermometer A has a bias relative to American thermometer B, we can compute $\dfrac{Q_3/1}{Q_5/8}$ and compare it with $F_{0.05}$ for $\nu_1 = 1$, $\nu_2 = 8$. Similarly a test of a difference between thermometers C and D is given by $\dfrac{Q_4/1}{Q_5/8}$. To test whether the thermometers as a group differ among themselves (the test originally carried out in Section 1.2), we can compute $\dfrac{(Q_2 + Q_3 + Q_4)/3}{Q_5/8}$.

1.7.3. Computations. Computations for individual degrees of freedom analysis can be carried out in direct conformance with theory. If we use the data of Table 29.1A we have

$$X_1 = 2.0 \quad X_4 = 1.0 \quad X_7 = -0.5 \quad X_{10} = 1.5$$
$$X_2 = 1.0 \quad X_5 = 0.0 \quad X_8 = -1.0 \quad X_{11} = -1.0$$
$$X_3 = 1.5 \quad X_6 = 1.0 \quad X_9 = 1.0 \quad X_{12} = 0.5$$

It follows from Table 29.3 then that

$$W_1^2 = \frac{Z_1^2}{D_1} = \frac{(2 + 1 + 1.5 + 1 + 0 + 1 - 0.5 - 1 + 1 + 1.5 - 1 + 0.5)^2}{12}$$
$$= \frac{(7)^2}{12} = 4.08$$

$$W_2^2 = \frac{Z_2^2}{D_2} = \frac{(2 + 1 + 1.5 + 1 + 0 + 1 + 0.5 + 1 - 1 - 1.5 + 1 - 0.5)^2}{12}$$
$$= \frac{(6)^2}{12} = 3$$

$$W_3^2 = \frac{Z_3^2}{D_3} = \frac{(2 + 1 + 1.5 - 1 - 0 - 1)^2}{6} = \frac{(2.5)^2}{6} = 1.04$$

$$W_4^2 = \frac{Z_4^2}{D_4} = \frac{(-0.5 - 1 + 1 - 1.5 + 1 - 0.5)^2}{6} = \frac{(-1.5)^2}{6} = 0.38$$

$$W_5^2 = \frac{Z_5^2}{D_5} = \frac{(2 - 1)^2}{2} = \frac{(1)^2}{2} = 0.5$$

$$W_6^2 = \frac{Z_6^2}{D_6} = \frac{(2 + 1 - 2(1.5))^2}{6} = 0$$

$$W_7^2 = \frac{Z_7^2}{D_7} = \frac{(1 - 0)^2}{2} = \frac{(1)^2}{2} = 0.5$$

$$W_8^2 = \frac{Z_8^2}{D_8} = \frac{(1 + 0 - 2(1))^2}{6} = \frac{(-1)^2}{6} = 0.167$$

$$W_9^2 = \frac{Z_9^2}{D_9} = \frac{(-0.5 + 1)^2}{2} = \frac{(0.5)^2}{2} = 0.125$$

$$W_{10}^2 = \frac{Z_{10}^2}{D_{10}} = \frac{(-0.5 - 1 - 2(1))^2}{6} = \frac{(-3.5)^2}{6} = 2.042$$

$$W_{11}^2 = \frac{Z_{11}^2}{D_{11}} = \frac{(1.5 + 1)^2}{2} = \frac{(2.5)^2}{2} = 3.125$$

$$W_{12}^2 = \frac{Z_{12}^2}{D_{12}} = \frac{(1.5 - 1 - 2(0.5))^2}{6} = \frac{(-0.5)^2}{6} = 0.042$$

Note that the above agrees exactly with the results of Section 1.3. We have

$$W_1^2 = \text{Correction Factor} = 4.08,$$

$$W_2^2 + W_3^2 + W_4^2 = \text{Column SS} = 3 + 1.04 + 0.38 = 4.42, \text{ and}$$

$$\sum_{i=5}^{12} W_i^2 = \text{Error SS} = 0.5 + 0 + 0.5 + 0.167 + 0.125 + 2.042 + 3.125$$

$$+ 0.042 = 6.50$$

As before, let $Q_2 = W_2^2$, $Q_3 = W_3^2$, $Q_4 = W_4^2$, and $Q_5 = \sum_{i=5}^{12} W_i^2$. Then to test American versus German thermometers, we have

$$\frac{Q_2/1}{Q_5/8} = \frac{3/1}{6.5/8} = \frac{3}{0.81} = 3.70$$

But $F_{0.05}$ $(\nu_1 = 1, \nu_2 = 8) = 5.32$, so the difference is not significant. Likewise to compare the two American thermometers we have

$$\frac{Q_3/1}{Q_5/8} = \frac{1.04/1}{6.5/8} = 1.28$$

and to compare the two German thermometers we have

$$\frac{Q_4/1}{Q_5/8} = \frac{0.38/1}{6.5/8} = 0.47$$

Neither of these variance ratios is significant. Finally we have

$$\frac{(Q_2 + Q_3 + Q_4)/3}{Q_5/8} = \frac{4.42/3}{6.5/8} = \frac{1.47}{0.81} = 1.81$$

and $F_{0.05}$ $(\nu_1 = 3, \nu_2 = 8) = 4.07$. The variation between thermometers is therefore not significant. This is the same test run in Section 1.2.

It will be noted that the computation of individual components for the error term is not necessary. We can compute W_1^2, W_2^2, W_3^2, and W_4^2 and then subtract the total of these components from ΣX_i^2.

The above numerical analysis is summarized in Table 29.4.

1.7.4. Special Analysis for Equally Spaced Quantitative Factors. If the previous sections had not been concerned with a qualitative factor such as thermometers but with an *equally spaced quantitative factor,* the analysis might have taken a special slant. Suppose, for example, that A, B, C, and D no longer stand for different thermometers but represent different levels of impurity. Specifically, let the melting points in column A of Table 29.1A represent those obtained (after coding, of course) when the percentage of impurities is 0.1, let those in column B be the results obtained when the percentage of impurities is 0.2, those in column C when it is 0.3, and those in column D when it is 0.4. Note once again that the impurity factor is quantitative and *equally spaced.*

In this kind of problem, we may not only be interested in whether the

TABLE 29.4
Analysis of Variance Table Showing Individual Degrees of Freedom for Comparison of Column Means

Source of Variation	Sum of Squares	df	Mean Square	Variance Ratio
Thermometers:				
A + B versus C + D	3.00	1	3.00	3.70
A versus B	1.04	1	1.04	1.28
C versus D	0.38	1	0.38	0.47
Total Thermometers	4.42	3	1.47	1.81
Error (residual)	6.50	8	0.81	
Total Deviations from Sample Mean	10.92	11		
Sample mean (correction factor)	4.08	1		
ΣX_i^2	15.00	12		

TABLE 29.5
Another Transformation of Melting Points

	0.1% Impurity			0.2% Impurity			0.3% Impurity			0.4% Impurity			
Z_i	X_1 (2.0)	X_2 (1.0)	X_3 (1.5)	X_4 (1.0)	X_5 (0.0)	X_6 (1.0)	X_7 (−0.5)	X_8 (−1.0)	X_9 (1.0)	X_{10} (1.5)	X_{11} (−1.0)	X_{12} (0.5)	*Divisor* D_i
Z_1	1	1	1	1	1	1	1	1	1	1	1	1	12
Z_2	−3	−3	−3	−1	−1	−1	1	1	1	3	3	3	60
Z_3	1	1	1	−1	−1	−1	−1	−1	−1	1	1	1	12
Z_4	−1	−1	−1	3	3	3	−3	−3	−3	1	1	1	60
Z_5	1	−1	0	0	0	0	0	0	0	0	0	0	2
Z_6	1	1	−2	0	0	0	0	0	0	0	0	0	6
Z_7	0	0	0	1	−1	0	0	0	0	0	0	0	2
Z_8	0	0	0	1	1	−2	0	0	0	0	0	0	6
Z_9	0	0	0	0	0	0	1	−1	0	0	0	0	2
Z_{10}	0	0	0	0	0	0	1	1	−2	0	0	0	6
Z_{11}	0	0	0	0	0	0	0	0	0	1	−1	0	2
Z_{12}	0	0	0	0	0	0	0	0	0	1	1	−2	6

percentage of impurity affects the melting point but also whether there is any linear, quadratic, or higher order trend in these effects. Fortunately, in our individual degrees of freedom analysis we can associate a separate degree of freedom with each of the trend components. Thus, if we replace Table 29.3 with Table 29.5, which is the same as Table 29.3 except for Z_2, Z_3, and Z_4, the new Z_2, Z_3, and Z_4 will represent the linear, quadratic, and cubic components of the variation in column means. For the mathematics underlying this, the reader is referred to J. G. Smith and A. J. Duncan, *Elementary Statistics and Applications,* chap. xxii.

That Z_2, Z_3, and Z_4, measure the linear, quadratic, and cubic effects, respectively, may be seen intuitively by noting the following. Z_2 measures the difference in melting points between levels of impurity that are equally spaced around the mean level of impurity. It is the effect you would get if you drew a straight line between an average of determinations at a low level of impurity and an average of determinations at a high level. Z_3 measures the quadratic effect or elementary curvature of the relationship between impurity and melting point by taking the difference between the sum of determinations at high and low levels of impurity and determinations at middle levels. Z_4 measures the cubic effect by comparing the determinations at low-intermediate and high levels of impurity, on the one hand, with high intermediate and low levels of impurity, on the other hand. No effects higher than the cubic effect can be tested with only four levels of impurity. Again it will be noted that in practice it is not necessary to compute any more of the Z components than those desired.

The Z coefficients that give the linear, quadratic, cubic, and higher order effects vary with the number of levels of the quantitative factor. Table 29.6 presents the linear, quadratic, and cubic coefficients for levels of 3, 4, 5, and 6. These came from Fisher and Yates, *Statistical Tables,* Table XXIII, which contains orthogonal coefficients as high as the fifth degree up to 75 levels.

The general idea underlying these orthogonal coefficients may be explained as follows. If we have a third-degree equation in x, say

$$y = a + bx + cx^2 + dx^3,$$

we can transform this into a linear equation in p_1, p_2, and p_3 of the form

$$y = A + Bp_1 + Cp_2 + Dp_3 \tag{29.11}$$

where p_1 is a linear function of x, p_2 is a quadratic function of x, and p_3 is a cubic function of x, and where $\Sigma p_i p_j = 0$ for all i, j. If this is done, it turns out that when the p-form of the equation is "fitted" to a set of equally spaced data by the method of least squares (see Chapters 32 and 33), then estimates of B, C, and D, the linear, quadratic, and cubic coefficients, are given by

TABLE 29.6
Orthogonal Coefficients for Determining Linear, Quadratic, and Cubic Components for Levels of the Quantitative Factor from 3 to 6

3 Levels		*4 Levels*			*5 Levels*			*6 Levels*		
L	*Q*	*L*	*Q*	*C*	*L*	*Q*	*C*	*L*	*Q*	*C*
								−5	+5	−5
					−2	+2	−1			
		−3	+1	−1				−3	−1	+7
−1	+1				−1	−1	+2			
		−1	−1	+3				−1	−4	+4
0	−2				0	−2	0			
		+1	−1	−3				+1	−4	−4
+1	+1				+1	−1	−2			
		+3	+1	+1				+3	−1	−7
					+2	+2	+1			
								+5	+5	+5

SOURCE: Reproduced with permission from Table XXIII of R. A. Fisher and F. Yates, *Statistical Tables for Biological, Agricultural and Medical Research* (London: Oliver & Boyd, Ltd., 1953).

(29.12)

$$B = \sum p_1 y \Big/ \sum p_1^2$$
$$C = \sum p_2 y \Big/ \sum p_2^2$$
$$D = \sum p_3 y \Big/ \sum p_3^2$$

The coefficients in Table 29.6 are the p_1, p_2, etc. In terms of Table 29.5, for example, this means that $B = Z_2/D_2$, $C = Z_3/D_3$, and $D = Z_4/D_4$.

Applying this special analysis to the numerical data, we obtain

$$\text{New } W_2^2 = \text{New } Z_2^2/D_2 = \frac{[-3(2) - 3(1) - 3(1.5) - 1 - 0 - 1 - 0.5 - 1 + 1 + 3(1.5) + 3(-1) + 3(0.5)]^2}{60} = \frac{169}{60} = 2.82$$

$$\text{New } W_3^2 = \text{New } Z_3^2/D_3 = \frac{[2 + 1 + 1.5 - 1 - 0 - 1 + 0.5 + 1 - 1 + 1.5 - 1 + 0.5]^2}{12} = \frac{16}{12} = 1.33$$

$$\text{New } W_4^2 = \text{New } Z_4^2/D_4 = \frac{[-2 - 1 - 1.5 + 3(1) + 3(0) + 3(1) - 3(-0.5) - 3(-1) - 3(1) + 1.5 - 1 + 0.5]^2}{60} = \frac{16}{60} = 0.27$$

TABLE 29.7
Analysis of Variance Table Showing Linear, Quadratic, and Cubic Components of Column Means

Source of Variation	*Sums of Squares*	*df*	*Mean Square*	*Variance Ratio*
Percentage impurities:				
Linear component	2.82	1	2.82	3.48
Quadratic component	1.33	1	1.33	1.64
Cubic component	0.27	1	0.27	0.33
Error (residual)	6.50	8	0.81	
Total	10.92	11	...	...

Again these results are in agreement with those previously obtained since the Column Sum of Squares equals 2.82 + 1.33 + 0.27 = 4.42. The data are tabulated in Table 29.7.

None of the trend components is significant, for

$$\frac{\text{New } W_2^2/1}{Q_5/8} = \frac{2.82/1}{6.50/8} = 3.48$$

$$\frac{\text{New } W_3^2/1}{Q_5/8} = \frac{1.33/1}{6.50/8} = 1.64$$

$$\frac{\text{New } W_4^2/1}{Q_5/8} = \frac{0.27/1}{6.50/8} = 0.33$$

all of which are less than $F_{0.05}$ $(\nu_1 = 1, \nu_2 = 8) = 5.32$.

2. ANALYSIS OF VARIANCE FOR A TWO-WAY CLASSIFICATION WITH A SINGLE CASE IN EACH CLASS

2.1. The New Problem

In Section 1 we were concerned with several measurements of the melting point of hydroquinone made with each of four thermometers. In taking these measurements it was assumed that all other controllable factors were held constant and that variations in determinations with the same thermometer were the result of "experimental errors."

Let us now change the conditions of the problem and assume that each of the three determinations of the melting point of hydroquinone using a given thermometer were made by different analysts. This two-way classification of the results is indicated in Table 29.8, and the coded data are given in Table 29.8A.

Two questions can now be put to the data: (1) Is there any analyst bias in determining the melting point of hydroquinone? (2) Do all the thermometers read alike?

TABLE 29.8
Determinations of the Melting Point of Hydroquinone in Degrees Centigrade: Further Classified

	Thermometer			
Analyst	*A*	*B*	*C*	*D*
I	174.0	173.0	171.5	173.5
II	173.0	172.0	171.0	171.0
III	173.5	173.0	173.0	172.5

SOURCE: Grant Wernimont, "Quality Control in the Chemical Industry. II. Statistical Quality Control in the Chemical Laboratory," *Industrial Quality Control,* May 1947, p. 8.

TABLE 29.8A
Data of Table 29.8 Coded to an Arbitrary Origin of 172

	Thermometer					
Analyst	*A*	*B*	*C*	*D*	Σ	$\bar{X}$
I	2.0	1.0	−0.5	1.5	4.0	1.00
II	1.0	0.0	−1.0	−1.0	−1.0	−0.25
III	1.5	1.0	1.0	0.5	4.0	1.00
Σ	4.5	2.0	−0.5	1.0	7.0	0.58
$\bar{X}$	1.50	0.67	−0.17	0.33	0.58	....

2.2. Theory

The basic model for a two-way analysis of variance with a single case in each class is that the observed value (melting point) X_{ij} in each cell is the algebraic *sum* of—

1. An overall mean (true melting point) μ;
2. A row effect (analyst bias) τ_i $(i = 1, 2, \ldots r)$;
3. A column effect (thermometer bias) θ_j $(j = 1, 2, \ldots c)$; and
4. A random residual ϵ_{ij}, which is normally distributed with zero mean and standard deviation σ';

with possibly either the τ_i's or the θ_j's or both being all zero. If by hypothesis the τ_i's are all zero, then, as in the case of a one-way classification, it is possible to make independent estimates of σ'^2 from (1) the variation between row means (between analyst means) and (2) the residual variance derived as explained below, and then to run a variance-ratio F test, this being possible whatever the values of θ_j. Likewise, if by hypothesis the θ_j's are all zero (thermometers all read alike), it is possible to make independent estimates of σ'^2 from (1) the variations between column means (between thermometer means) and (2) the residual variance, and then run a variance-ratio F test, this again being possible whatever the values of the τ_i's.

The estimate of σ'^2 based on row means is

$$s_R^2 = \frac{c \sum_i (\bar{X}_i - \bar{X})^2}{r - 1} \tag{29.13}$$

and the estimate of σ'^2 based on column means is

$$s_C^2 = \frac{r \sum_j (\bar{X}_j - \bar{X})^2}{c - 1} \tag{29.14}$$

The residual variance is derived from the sum of the squares of the deviations of the actual values X_{ij} from the "average" value for each cell given by $\bar{X} + (\bar{X}_i - \bar{X}) + (\bar{X}_j - \bar{X}) = \bar{X}_i + \bar{X}_j - \bar{X}$, and the estimate of σ'^2 based on the residual variance is

$$s_{\text{res.}}^2 = \frac{\sum_i \sum_j (X_{ij} - \bar{X}_i - \bar{X}_j - \bar{X})^2}{(r-1)(c-1)} \tag{29.15}$$

To test the hypothesis that $\tau_1 = \tau_2 = . . . = \tau_r = 0$, we compute $s_R^2/s_{\text{res.}}^2$ and compare it with F_α for $\nu_1 = r - 1$, $\nu_2 = (r - 1)(c - 1)$. To test the hypothesis that $\theta_1 = \theta_2 = . . . \theta_c = 0$, we compute $s_C^2/s_{\text{res.}}^2$ and compare it with F_α for $\nu_1 = c - 1$ and $\nu_2 = (r - 1)(c - 1)$. It will be noted that in this model we speak of a residual rather than an error variance because the residual may contain random interactions as well as experimental errors. See the next chapter.

In carrying out the mathematical analysis it is assumed that if the τ_r's and θ_j's are fixed factors, then $\Sigma\tau_i = 0$ and $\Sigma\theta_j = 0$; if the τ_i's and θ_j's are random selections from a universe of τ's and a universe of θ's, respectively, it is assumed that $\bar{\tau}' = 0$ and $\bar{\theta}' = 0$. This means that we can never estimate the absolute values of τ or θ, but merely their differential values. For example, any bias common to all four thermometers could not be detected by any analysis of variance, and any estimates of thermometer biases would be simply "differential" biases. If a common bias existed, it would be statistically confounded with μ.

The foregoing algebraic foundation may be made clearer by numerical illustration with the data of Table 29.8A. Thus

$$s_R^2 = \frac{4[(1.00 - 0.58)^2 + (-0.25 - 0.58)^2 + (1.00 - 0.58)^2]}{3 - 1} = 2.09$$

$$s_C^2 = \frac{3[(1.5 - 0.58)^2 + (0.67 - 0.58)^2 + (-0.17 - 0.58)^2 + (0.33 - 0.58)^2]}{4 - 1}$$

$$= 1.47$$

$$
\begin{aligned}
s^2_{\text{res.}} &= \frac{1}{(4-1)(3-1)}[(2-1.5-1.00+0.58)^2 \\
&+(1-1.5+0.25+0.58)^2+(1.5-1.5-1+0.58)^2 \\
&+(1-0.67-1+0.58)^2+(0-0.67+0.25+0.58)^2 \\
&+(1-0.67-1+0.58)^2+(-0.5+0.17-1+0.58)^2 \\
&+(-1+0.17+0.25+0.58)^2+(1+0.17-1+0.58)^2 \\
&+(1.5-0.33-1+0.58)^2+(-1-0.33+0.25+0.58)^2 \\
&+(0.5-0.33-1+0.58)^2] \\
&=0.39
\end{aligned}
$$

To test the hypothesis that the analysts have no biases in determining the melting point of hydroquinone, we compute 2.09/0.39 = 5.36 and note that this exceeds $F_{0.05}$ ($\nu_1 = 2$, $\nu_2 = 6$) = 5.14. We therefore reject the null hypothesis in this case and conclude that analyst biases do exist.

To test the hypothesis that the thermometers all read alike, using $\alpha = 0.05$, we compute 1.47/0.39 = 3.77 and note that this is less than $F_{0.05}$ ($\nu_1 = 3$, $\nu_2 = 6$) = 4.76. Again the null hypothesis pertaining to thermometers is not rejected.

Although this last result agrees with that of Section 2, it is possible that it might have differed, even though the data are the same. The reason is that in Section 1 the variation between analysts was included in the residual error, whereas here the residual error is free of analyst bias. This separation of analyst variation reduces the degrees of freedom for the residual variance and hence increases the critical F, but it might nevertheless so reduce the residual error as to make the thermometer variation significant.

2.3. Short-Cut Computations

As in the case of a one-way classification, it is possible to shorten greatly the computations by using special methods. The steps in the short-cut computations for a two-way classification are as follows:

1. Sum each row, square these row sums, add, and divide by the number in each row. Thus

$$\frac{4^2+(-1.0)^2+4^2}{4}=8.25$$

2. Sum each column, square these column sums, add, and divide by the number in each column. Thus

$$\frac{4.5^2+2.0^2+(-0.5)^2+(1.0)^2}{3}=8.50$$

3. Square each individual item and sum. Thus

$$2^2 + 1^2 + 1.5^2 + 1^2 + 0^2 + 1^2 + (-0.5)^2 + (-1)^2 + 1^2 + 1.5^2 + (-1)^2 + (0.5)^2 = 15.00$$

4. Sum all the cases, square this sum, and divide by the number of cases. Thus

$$\frac{(2 + 1 + 1.5 + 1 + 0 + 1 - 0.5 - 1 + 1 + 1.5 - 1 + 0.5)^2}{12} = 4.08$$

This is the "correction factor."

5. The row sum of squares equals

$$(1) - (4) = 8.25 - 4.08 = 4.17$$

6. The column sum of squares equals

$$(2) - (4) = 8.50 - 4.08 = 4.42$$

7. The total sum of squares equals

$$(3) - (4) = 15.00 - 4.08 = 10.92$$

8. The residual sum of squares equals

$$(7) - (5) - (6) = 10.92 - 4.17 - 4.42 = 2.33$$

The above analysis is set up in tabular form in Table 29.9. The degrees of freedom are now:

degrees of freedom for analysts equals number of analysts minus 1, i.e., $(r - 1)$;

degrees of freedom for thermometers equals number of thermometers minus 1, i.e., $(c - 1)$;

degrees of freedom for residual sum of squares equals $(r - 1)(c - 1)$.

If each sum of squares is divided by its degrees of freedom, we get the mean squares listed in the table. Variance ratios that are significant at the 0.05 level are starred.

TABLE 29.9
Analysis of Variance Table for a Two-Way Classification with a Single Case in Each Class

Source of Variation	*Sum of Squares*	*df*	*Mean Square*	*Variance Ratio*
Analysts (row)	4.17	2	$s_R^2 = 2.09$	5.36*
Thermometers (columns)	4.42	3	$s_C^2 = 1.47$	3.77
Residual	2.33	6	$s_{res.}^2 = 0.39$	
Total	10.92	...	....	....

2.4. Pooling

A two-way analysis of variance raises for the first time the question of pooling. In the thermometer-analyst problem we have just seen that the thermometer variation is not significant. If we conclude then that this variation is due purely to chance, we may be led to pool the thermometer sum of squares with the original residual sum of squares to get a new residual sum of squares based on $6 + 3 = 9$ degrees of freedom. This will give us a new residual mean square of $6.75/9 = 0.75$. We can now test the analyst variation against the new residual mean square, having 9 degrees of freedom for error instead of 6. The critical $F_{0.05}$ will thus become 4.26 instead of 5.14, and we have a more precise test.

The danger with pooling the thermometer variation with the original residual sum of squares is that the "thermometer variation" may not be purely chance variation but may contain some real thermometer variation that is too small to detect by our statistical test. Pooling the sums of squares in this case may increase the residual mean square and make for a less sensitive test. Actually in Table 29.9 the new variance ratio would be $2.09/0.75 = 2.8$, and we would now conclude that the analyst variation was not significant.

The problem of pooling[23] can be met in various ways. We can adopt the policy of never pooling, in which case we may in some cases fail to get as precise a test as would be possible. (The OC curve will be poorer.) We can always pool, in which case the risks of falsely rejecting or incorrectly accepting the null hypothesis will have to be modified since the basis for our tests is much more complicated than if we never pool. Finally, we can pool sometimes and not others.

If the degrees of freedom for the original residual mean square are less than 6, we have much to gain by pooling. On the other hand, if the variance ratio for the sum of square that is to be pooled is high, although not high enough to be significant, the danger of incorporating nonchance variability in our residual term may be fairly high. The author is inclined to favor the following rule. If the degrees of freedom for the original residual mean square are 6 or more, do not pool. If the degrees of freedom for the original residual mean square are less than 6, pool if the variance ratio for the sum of squares to be pooled is not greater than 2. This policy will have slightly different risks than the 100 percent nopooling rule but seems a reasonably practical procedure.

2.5. OC Curves for a Two-Way Analysis of Variance

For fixed factors, operating characteristic curves for two-way analyses of variance are derived in exactly the same way as those for one-way analyses

[23] See F. Mosteller, "On Pooling Data," *Journal of the American Statistical Association*, 43 (1948), pp. 231–42.

of variance with an equal number in each class. This is true for both row and column factors.

If we set

(29.16) $$\lambda^2 = \frac{\Sigma\theta_j^2}{c\sigma'^2}$$

then in Mann's tables or in Charts I and II of Appendix II,

(29.17) $$\phi = \lambda\sqrt{r}$$

or if we set

(29.18) $$\lambda^2 = \frac{\Sigma\tau_i^2}{r\sigma'^2}$$

then

(29.19) $$\phi = \lambda\sqrt{c}$$

It must be remembered, however, that the degrees of freedom associated with the residual variance will now be $(r - 1)(c - 1)$ and not $c(r - 1)$ or $r(c - 1)$.

For random factors, the tables of the OC curves for F tests given in the Statistical Research Group's *Techniques of Statistical Analysis* can be used as before. Table L of Appendix II gives the 0.10 points of these curves. It is to be noted that if we set

(29.20) $$\lambda^2 = \frac{\sigma_\theta'^2}{\sigma'^2}$$

then

(29.21) ϕ^2 in Table L and ϕ in S.R.G.'s *Techniques of Statistical Analysis*

$$= \frac{(\sigma'^2 + r\sigma_\theta'^2)}{\sigma'^2} = 1 + r\lambda^2$$

or if we set

(29.22) $$\lambda^2 = \frac{\sigma_\tau'^2}{\sigma'^2}$$

(29.23) ϕ^2 in Table L and ϕ in S.R.G.'s *Techniques of Statistical Analysis*

$$= \frac{(\sigma'^2 + c\sigma_\tau'^2)}{\sigma'^2} = 1 + c\lambda^2$$

Again it must be remembered that the degrees of freedom associated with the residual mean square are now $(r - 1)\ (c - 1)$ not $c(r - 1)$ or $r(c - 1)$.

If one factor is fixed and the other random, we can use fixed factor procedures for getting the OC curve for the fixed factor test and random factor procedures for getting the OC curve for the random factor test.

Some examples will illustrate the above. Suppose that in the thermometer-analyst problem we consider both factors fixed. Let the actual analyst variance be 2 times the true residual variance, i.e., let $\lambda^2 = \left(\frac{\Sigma\tau_i^2}{r}\right)/\sigma'^2 = 2$ and $\lambda = 1.41$. We shall then have $\phi = 1.41\sqrt{4} = 2.82$. From Chart I of Appendix II, it is seen that with $\nu_1 = 2$, $\nu_2 = 6$ the chance of not detecting a variation of this magnitude is less than 0.10. (For $\nu_1 = 2$, $\nu_2 = 6$ the ϕ for 0.10 is about 2.7, or slightly less than the given ϕ of 2.82.)

If we consider the thermometers as random, then if the variance of thermometer biases in the universe of thermometers is three times that of the residual variance, i.e., $\lambda^2 = \sigma_\theta'^2/\sigma'^2 = 3$, then $\phi^2 = 1 + 3\lambda^2 = 1 + 9 = 10$. From Table L we note that for $\nu_1 = 3$ and $\nu_2 = 6$, the probability of not detecting such a thermometer variance is greater than 0.10 (for 25.14 given in Table L is greater than 10). Table 8.3 in the *Techniques of Statistical Analysis* gives the probability at about 0.30.

2.6. Individual Degrees of Freedom

The analysis of individual degrees of freedom is easily extended to a two-way classification with a single case in each class. In Table 29.3, for example, two degrees of freedom may be taken from the error sum of squares and used to represent comparisons among analysts. One set of individual comparisons is shown in Table 29.10. The Z_1 to Z_4 are the same as in Table 29.3,

TABLE 29.10
A Transformation of Melting Points for a Two-Way Classification

	Thermometer A			*Thermometer B*			*Thermometer C*			*Thermometer D*			
	Analyst			*Analyst*			*Analyst*			*Analyst*			
	I	*II*	*III*	*I*	*II*	*III*	*I*	*II*	*III*	*I*	*II*	*III*	*Divisor*
Z_i	X_1 (2.0)	X_2 (1.0)	X_3 (1.5)	X_4 (1.0)	X_5 (0.0)	X_6 (1.0)	X_7 (−0.5)	X_8 (−1.0)	X_9 (1.0)	X_{10} (1.5)	X_{11} (−1.0)	X_{12} (0.5)	D_i
Z_1	1	1	1	1	1	1	1	1	1	1	1	1	12
Z_2	1	1	1	1	1	1	−1	−1	−1	−1	−1	−1	12
Z_3	1	1	1	−1	−1	−1	0	0	0	0	0	0	6
Z_4	0	0	0	0	0	0	1	1	1	−1	−1	−1	6
Z_5	1	0	−1	1	0	−1	1	0	−1	1	0	−1	8
Z_6	1	−2	1	1	−2	1	1	−2	1	1	−2	1	24
Z_7	1	0	−1	1	0	−1	−1	0	1	−1	0	1	8
Z_8	1	−2	1	1	−2	1	−1	2	−1	−1	2	−1	24
Z_9	1	0	−1	−1	0	1	0	0	0	0	0	0	4
Z_{10}	1	−2	1	−1	2	−1	0	0	0	0	0	0	12
Z_{11}	0	0	0	0	0	0	1	0	−1	−1	0	1	4
Z_{12}	0	0	0	0	0	0	1	−2	1	−1	2	−1	12

TABLE 29.11
Analysis of Variance Table Showing Individual Degrees of Freedom for Comparison of Row and Column Means

Source of Variation	*Sum of Squares*	*df*	*Mean Square*	*Variance Ratio*
Thermometers:				
A + B versus C + D	3.00	1	3.00	7.69*
A versus B	1.04	1	1.04	2.67
C versus D	0.38	1	0.38	0.97
Total Thermometers	4.42	3	1.47	3.77
Analysts:				
I versus III	0.00	1	0	0
I + III versus II	4.17	1	4.17	10.7*
Total Analysts	4.17	2	2.09	5.36*
Residual	2.33	6	0.39	
Total Deviations from Sample Mean	10.92	11		
Sample mean (correction factor)	4.08	1		
ΣX_1^2	15.00	12		

but Z_5 and Z_6 are now comparisons among analysts. Z_7 to Z_{12} remain residuals, although they take a different form than in Table 29.3. It is to be remembered that the transformation must meet requirements (29.7) throughout.

The computations of Z_i^2/D_i will be the same as before for Z_1 to Z_4 and need not be repeated here.[24] For Z_5 and Z_6 we have

$$\frac{Z_5^2}{D_5}=\frac{[2-1.5+1-1+(-0.5)-1+1.5-0.5]^2}{8}=\frac{0}{8}=0$$

$$\frac{Z_6^2}{D_6}=\frac{[2-2+1.5+1-2(0)+1+(-0.5)-2(-1)+1+1.5-2(-1)+0.5]^2}{24}$$

$$=\frac{100}{24}=4.17$$

The residual sum of squares can be computed directly from Table 29.10 in the same way or it can be obtained by subtracting $\sum_{i=1}^{6} Z_i^2/D_i$ from ΣX_i^2. The results are given in Table 29.11. Again variance ratios that are significant at this 0.05 level are starred.

It will be noted that the significance of the variation among analysts is due entirely to the deviation of analyst II from analysts I and III. This is

[24] See Section 1.7.3 above.

obvious in the given example since the mean determination for analyst I is the same as that for analyst III. It will also be noted that the comparison A + B verses C + D turns out to be a significant individual comparison.

If either the row or column factor is quantitative and equally spaced, we can analyze it for significant trend components, i.e., for linear components, quadratic components, and so on. This can be done independently for the row and column factors in precisely the same manner as described in Section 1.7.4.

In a 3-row, 4-column analysis, for example, we can identify Z_2, Z_3, and Z_4 with trend components for the column factor and we can *independently* identify Z_5 and Z_6 with trend components for the row factor. Also one factor can be analyzed for trend components without a similar analysis being applied to the other. In fact, one factor could be qualitative. It must be remembered, however, that as long as we recognize both factors, the residual variation will be based on Z_7 to Z_{12} and will have only 6, not 8, degrees of freedom for this variation.

A word of caution is due at this point. If the comparisons in individual degrees of freedom analysis have been selected prior to the running of the experiment, i.e., they have been built into the experiment, then each can be tested against the error term at the 0.05 level as long as we are willing to falsely accept as significant 5 comparisons out of 100. From this point of view the comparisons are looked upon as independent subexperiments and our error rate is on a *per comparison* basis.

What we must be on our guard against is a "fishing expedition" in which we run every and all possible comparisons at the 0.05 level. The danger is that we shall conclude that the overall experiment has shown some significant result if one or more of these comparisons turns out significant, and fail to realize that we shall be drawing false conclusions in more than 5 *out of 100 experiments.*

3. PROBLEMS

29.1. Five examiners are checked on the time they take to use a certain gauge. The record of gauging times in seconds is as follows:[25]

Examiners				
1	*2*	*3*	*4*	*5*
13	14	14	15	13
14	14	15	14	16
12	13	13	14	14
14	14	15	14	14

[25] Cf. B. A. Griffith, A. E. R. Westman, and B. H. Lloyd, "Analysis of Variance," *Industrial Quality Control,* May 1948, p. 19.

a. Use an analysis of variance to determine whether the examiners differ significantly in their gauging speed.

b. Use a studentized range test to determine whether the examiners differ significantly in their gauging speed.

c. Estimate the error variance in speed of gauging, assuming that it is the same for every examiner.

d. Assuming that the examiners are a random sample of examiners in general, sketch the OC curve for your test.

e. Assuming that these five examiners are the only five used in the plant, sketch the OC curve for your test.

f. Discuss the design of such a test as this, if you were free to pick the number of examiners and number of measurements.

g. Suppose that the last reading of examiner 2 and the last two readings of examiner 4 became lost, so that in effect only the first three readings of examiner 2 and the first two readings of examiner 4 were available for analysis. Carry out the analysis of variance for this case of unequal column numbers. (See footnotes 5 and 6 of this chapter.)

29.2. The tensile strength of a certain rubber vulcanizate shows the following variation with the accelerator used.[26] Data are in pounds per square inch.

Accelerators

A	*B*	*C*
3,900	4,300	3,700
4,100	4,200	3,900
4,000	4,300	3,600

a. Use an analysis of variance to determine whether the accelerators have any effect on the tensile strength of the rubber vulcanizate.

b. Use a studentized range test to determine whether the accelerators have any effect on the tensile strength of the rubber vulcanizate.

c. Estimate the random variance in the vulcanizate, assuming it is the same for each accelerator.

d. Assuming that these accelerators are the only ones used, sketch the OC curve for the test.

e. How would your OC curve be affected if you had had 6 cases for each accelerator?

29.3. Four thermocouples were placed vertically one above the other at equal distances so that variation in temperature from layer to layer of a process could be studied.

---

[26] Cf. Hugh M. Smallwood, "Design of Experiments in Industrial Research," *Industrial and Engineering Chemistry—Analytical Chemistry* 19 (1947), p. 952.

At random intervals of time six readings were made with each thermocouple with the following results.[27] (Temperatures are all measured in degrees Fahrenheit.)

Determination Number	Thermocouple Number			
	1	*2*	*3*	*4*
1	440	510	525	530
2	550	535	530	550
3	480	470	530	580
4	520	500	550	555
5	495	505	550	515
6	500	530	495	530

a. Use an analysis of variance to determine whether there is a significant variation from layer to layer (i.e., from thermocouple to thermocouple). Do both as a one-way and two-way problem.

b. Use a studentized range test to determine whether there is a significant variation from thermocouple to thermocouple.

c. Is there a significant linear trend in layer temperature? A significant quadratic variation?

29.4. *a.* In Table 3.2 are listed sample measurements of the heights of fragmentation bomb bases. Each sample contains 5 cases. Take the first 20 of these samples and determine by an analysis of variance whether the means vary significantly from sample to sample. Check the conclusions you draw with the control chart analysis presented in Chapter 21 above.

b. Determine whether the means in *a* are significantly different by using a studentized range test.

c. Is there a significant linear trend in the means of the samples? A significant quadratic variation?

d. Estimate the variance of the individual items turned out by the process.

e. Sketch the OC curve for your test in *a.*

f. Would the OC curve have been much poorer if you had taken 10 samples instead of 20?

29.5. The tensile strength of a rubber vulcanizate shows the following variations with different conditions of production.[28] Data are in pounds per square inch.

---

[27] Adapted from data presented by Wyatt H. Lewis in *Industrial Quality Control,* May 1953, p. 81.

[28] Cf. Smallwood, "Design of Experiments in Industrial Research."

Cure at 140° (Minutes)	Accelerators A	B	C
40	3,900	4,300	3,700
60	4,100	4,200	3,900
80	4,000	4,300	3,600

a. Use an analysis of variance to determine whether the accelerators have any effect on the tensile strength.

b. Does the type of cure have any effect on the tensile strength?

c. Estimate the residual variance.

d. Compare your results with those obtained in Problem 29.2.

29.6. A textile company was interested in studying variations in dynamic absorption tests of its "Water Repellant Cotton Oxford."[29] They were interested especially in a laundering effect and a testing effect. Four laboratories (A, B, C, and D) were prevailed to join in the experiment. A strip of cloth 24 yards in length was considered to be homogeneous with respect to dynamic absorption. A 24-yard strip was thus cut into four pieces of 6 yards each, and one was assigned at random to each of the four laboratories for laundering. Then after laundering, each of the 6-yard strips of cloth were cut into four 1½-yard strips and one of these strips was assigned at random to each of the four laboratories for determination of dynamic absorption. The results were as follows. All figures are in coded form and represent percentage gain by weight minus 42 percent.

Launderings	Dynamic Absorption Tests A	B	C	D
I	8.13	7.76	14.75	8.11
II	2.27	8.74	6.51	2.90
III	2.44	3.39	5.11	2.16
IV	1.82	3.26	5.39	2.52

What conclusions would you draw from an analysis of variance of the results of this experiment?

29.7. The following represents the yield of a certain chemical under varying conditions.[30] The results are in pounds.

[29] Norman R. Garner, "Studies in Textile Testing," *Industrial Quality Control,* May 1956, pp. 44–46.

[30] Smallwood, "Design of Experiments in Industrial Research."

Temperature	*Concentration of Inert Solvent* 40%	50%	60%
50°	45.1	45.7	44.9
60°	44.8	45.8	44.7

a. Does the concentration of inert solvent have a significant effect on the yield?

b. Is the linear component of the concentration significant? the quadratic component?

c. Does the temperature have a significant effect on the yield?

d. Estimate the experimental variance.

29.8. A 24-hour cumulative sample of city gas is collected from a main and tested for calorific value each day from Monday to Saturday, inclusive, for nine weeks. The coded results of each test are as follows:[31]

Day	*Week Number* 1	2	3	4	5	6	7	8	9
Mon.	5	1	−4	5	−13	−8	−2	−4	−10
Tues.	3	6	−10	−2	−7	−2	−4	2	2
Wed.	8	4	−14	−3	3	0	5	−11	−12
Thurs. . . .	8	10	−5	−1	4	−2	4	1	−12
Fri.	4	−1	7	−5	5	−3	−7	−3	−6
Sat.	3	−9	3	−8	−6	0	−3	8	−1

a. Is there a significant variation from week to week?

b. Is there a significant linear component in the week-to-week variation? A significant quadratic component? A significant cubic component? (See Fisher and Yates, *Statistical Tables,* for orthogonal polynomial coefficients.)

c. Is there a significant variation from day to day?

d. Is there a significant linear component in the day-to-day variation? A significant quadratic component? A significant cubic component?

e. Estimate the residual variance.

f. Assuming that the 9 weeks are a random sample of weeks in general, sketch the OC curve for your test in *a.*

[31] Cf. Griffith, Westman, and Lloyd, "Analysis of Variance," p. 20.

4. SELECTED REFERENCES*

(For other references on "Analysis of Variance" see Chapter 30.)

DeLury (P '54), Davies (B '57), Fisher (B '58. Biological illustrations. Modern developments in analysis of variance are mainly due to Fisher.), Fisher and Yates (B '53), Freeman (B '42), Gabriel (P '63), Grandage (P '58), Hicks (P '56), Mood and Graybill (B '63), Pearson, E. S. and Hartley (B '58), Schwarz, G. (P '60), Smith J. G. and Duncan (B '44. Chapter xxii has a simple mathematical account of orthogonal polynomials.), Snedecor and Cochran (B '67. Contains an extensive, nonmathematical discussion of analysis of variance. Illustrations mostly from agricultural research.), Srivastava (P '59), Tang (P '38), Tippett (B '50), Tukey (P '51), and Wishart and Metakides (P '53).

---

* B and P refer to the Book and Periodical sections, respectively, of the Cumulative List of References in Appendix V.

30

Analysis of Variance (continued)

1. TWO-WAY CLASSIFICATION, WITH MORE THAN ONE BUT AN EQUAL NUMBER OF CASES IN EACH CLASS

1.1. An Extension of the Thermometer-Analyst Problem

In Section 2 of the previous chapter we were concerned with a two-way classification problem in which there was but a *single* case in each class. In the example selected for illustrative purposes, three analysts each made one determination of the melting point of hydroquinone with each of four different thermometers. There was but a single result for each combination of analyst and thermometer.

Let us now extend the analyst-thermometer illustration to the case in which each analyst makes two determinations of the melting point of hydroquinone with each of the four thermometers. The data with which we shall work will be those given in raw form in Table 30.1 and in coded form in Table 30.1A. It will be the latter that we shall use in our actual computations.

1.2. Theory

The theoretical model underlying the analysis assumes that each determination of the melting point of hydroquinone X is the algebraic *sum* of

1. An overall mean ("true" melting point) μ;
2. A row effect (analyst bias) τ_i, $(i = 1, 2, \ldots, r)$;
3. A column effect (thermometer bias) θ_j, $(j = 1, 2, \ldots, c)$;
4. An "interaction" effect (analyst-thermometer joint bias) ψ_{ij}; and
5. A random residual (experimental error) ϵ_{ijk}, which is normally distributed with zero mean and standard deviation σ', $(k = 1, 2, \ldots, g)$;

TABLE 30.1
Determinations of the Melting Point of Hydroquinone in Degree Centigrade: Additional Data

	Thermometer			
Analyst	*A*	*B*	*C*	*D*
I	174.0	173.0	171.5	173.5
	173.5	173.5	172.5	173.5
II	173.0	172.0	171.0	171.0
	173.0	173.0	172.0	172.0
III	173.5	173.0	173.0	172.5
	173.0	173.5	173.0	173.0

SOURCE: See Grant Wernimont, "Quality Control in the Chemical Industry. II. Statistical Quality Control in the Chemical Laboratory," *Industrial Quality Control,* May 1947, p. 8.

with possibly any of the effects (2), (3), and (4) all being zero. The number of cases in each cell is g. In equation form, we have

(30.1) $$X_{ijk} = \mu + \tau_i + \theta_j + \psi_{ij} + \epsilon_{ijk}$$

In analyzing this extended model we need to distinguish carefully whether the factors are both fixed, or both random, or one fixed and the other random. The three cases may be called the fixed-effects model, the random-effects model, and the mixed-effects model. Let us consider first the fixed-effects model.

1.2.1. The Fixed-Effects Model. If both column and row factors are "fixed," we make the added assumption that whenever the τ_i, θ_j, or ψ_{ij} exist, then $\sum_i \tau_i = 0$, $\sum_j \theta_j = 0$, $\sum_i \psi_{ij} = 0$ for all j, and $\sum_j \psi_{ij} = 0$ for all i. This means that none of these effects can be measured absolutely but only as differential deviations (the θ_j as deviations from μ, the τ_i as deviations from μ and the ψ_{ij} as deviations from $\mu + \theta_j + \tau_i$).

If we set up the "null" hypothesis that the τ_i's are each individually zero, it is possible to make independent estimates of the experimental error variance σ'^2 from

1. The variation among row (analyst) means, and
2. The residual variance based on the sum of the squares of the deviations of the cell elements from their individual cell means,

and then, whatever the values of θ_j or ψ_{ij}, to test the null hypothesis by running a variance-ratio F test.

If we set up the "null" hypothesis that the θ_j's are each individually zero, it is possible to make independent estimates of the experimental or error variance σ'^2 from

TABLE 30.1A
Data of Table 30.1 Coded to an Arbitrary Origin of 172

	Thermometer									
Analyst	*A*		*B*		*C*		*D*		Σ	$\bar{X}$
I	2.0		1.0		−0.5		1.5			
	1.5	*3.5*	1.5	*2.5*	0.5	*0.0*	1.5	*3.0*	*9.0*	*1.12*
II	1.0		0.0		−1.0		−1.0			
	1.0	*2.0*	1.0	*1.0*	0.0	*−1.0*	0.0	*−1.0*	*1.0*	*0.12*
III	1.5		1.0		1.0		0.5			
	1.0	*2.5*	1.5	*2.5*	1.0	*2.0*	1.0	*1.5*	*8.5*	*1.06*
Σ		*8.0*		*6.0*		*1.0*		*3.5*	*18.5*	*0.77*
$\bar{X}$		*1.33*		*1.00*		*0.17*		*0.58*	*0.77*	

1. The variation among column (thermometer) means, and
2. The residual variance based on the sum of the squares of the deviations of the cell elements from their individual cell means,

and then, whatever the values of τ_i or ψ_{ij}, to test the null hypothesis by running a variance-ratio F test.

Finally, if we set up the null hypothesis that the ψ_{ij}'s are individually zero, it is possible to make independent estimates of the experimental variance σ'^2 from

1. The sum of the squares of the deviations of the cell means from the "average" value given by $\bar{X}_j + \bar{X}_i - \bar{X}$, and
2. The residual variance as defined above,

and then, whatever the values of θ_j or τ_i, to test the null hypothesis by running a variance-ratio F test.

Algebraically the various estimates of σ'^2 are defined as follows:

(1) Estimate of σ'^2 based on row means, $s_R^2 = \dfrac{cg \sum_i (\bar{X}_i - \bar{X})^2}{r-1}$

(30.2) (2) Estimate of σ'^2 based on column means, $s_C^2 = \dfrac{rg \sum_j (\bar{X}_j - \bar{X})^2}{c-1}$

(3) Estimate of σ'^2 based on cell means, $s_{RC}^2 =$

$$\frac{g \sum_i \sum_j (\bar{X}_{ij} - \bar{X}_j - \bar{X}_i + \bar{X})^2}{(r-1)(c-1)}$$

and

(4) Estimate of σ'^2 based on residual variance, $s_e^2 =$

$$\frac{\sum_i \sum_j \sum_k (X_{ijk} - \bar{X}_{ij})^2}{rc(g-1)}$$

To test the hypothesis that $\tau_1 = \tau_2 = \cdots = \tau_r = 0$, we compute s_R^2/s_e^2 and compare it with F_α for $\nu_1 = r - 1$, $\nu_2 = rc(g - 1)$. To test the hypothesis that $\theta_1 = \theta_2 = \cdots = \theta_c = 0$, we compute s_C^2/s_e^2 and compare it with F_α for $\nu_1 = c - 1$, $\nu_2 = rc(g - 1)$. To test the hypothesis that $\psi_{ij} = 0$ for all ij, we compute s_{RC}^2/s_e^2 and compare it with F_α for $\nu_1 = (r - 1)(c - 1)$, $\nu_2 = rc(g - 1)$.

Except for a factor, g, s_R^2, s_C^2, and s_{RC}^2 are exactly what we would get if we reduced Table 30.1A to a two-way classification with a single case in each class by computing averages for each cell and analyzing it by the methods of Section 2.2 of Chapter 29. They are therefore not new concepts, although the reader at the moment may have difficulty in identifying the "interaction mean square" (s_{RC}^2) with the "residual mean square" $(s_{\text{res.}}^2)$ of the previous chapter. But more of this later. Numerically s_R^2, s_C^2, and s_{RC}^2 are g times the s_R^2, s_C^2, and $s_{\text{res.}}^2$ of Chapter 29 applied to the cell averages of Table 30.1A.

The new residual mean square s_e^2 may be illustrated numerically with respect to Table 30.1A as follows:

$$
\begin{aligned}
s_e^2 = {} & \frac{1}{(3)(4)(2-1)}[(2-1.75)^2+(1.5-1.75)^2+(1-1.25)^2+(1.5-1.25)^2 \\
& +((-0.5)-0)^2+(0.5-0)^2+(1.5-1.5)^2+(1.5-1.5)^2+(1-1)^2 \\
& +(1-1)^2+(0-0.5)^2+(1-0.5)^2+(-1-(-0.5))^2+(0-(-0.5))^2 \\
& +(-1-(-0.5))^2+(0-(-0.5))^2+(1.5-1.25)^2+(1-1.25)^2 \\
& +(1-1.25)^2+(1.5-1.25)^2+(1-1)^2+(1-1)^2+(0.5-0.75)^2 \\
& +(1-0.75)^2] \\
= {} & \frac{2.625}{12} = 0.22
\end{aligned}
$$

Other numerical calculations are carried out in Section 1.3.

In concluding this section, a few words should be said in explanation of the "interaction" elements ψ_{ij}. If there are no main effects, i.e, $\tau_i = 0$ for all i and $\theta_j = 0$ for all j and no interactions, i.e., $\psi_{ij} = 0$ for all ij, then the determination of the melting point of hydroquinone (X_{ijk}) will fluctuate around a constant level μ (see Figure 30.1a). If there are analyst biases but no thermometer biases or interactions, i.e., not all the τ_i equal zero

FIGURE 30.1
Average Levels for a Two-Way Analysis

a.

No Main Effects
No Interaction

X_{ijk}

μ

A B C D
Thermometers

No Main Effects
No Interaction

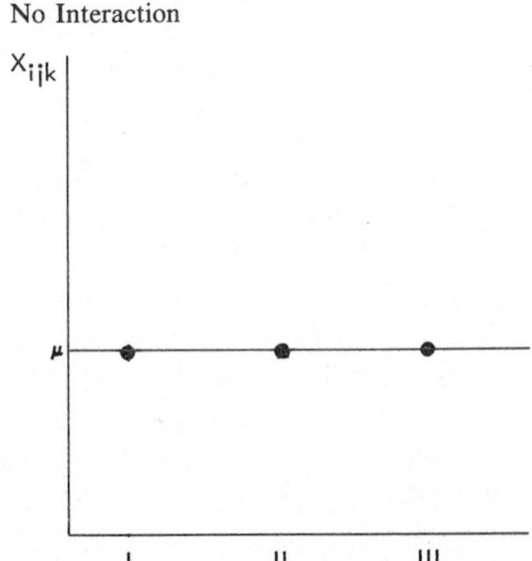

FIGURE 30.1 *(concluded)*

b.

Analyst Effects Only

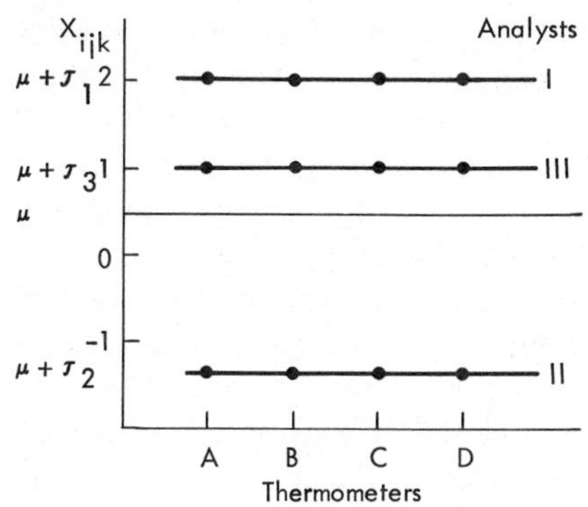

Thermometer Effects Only

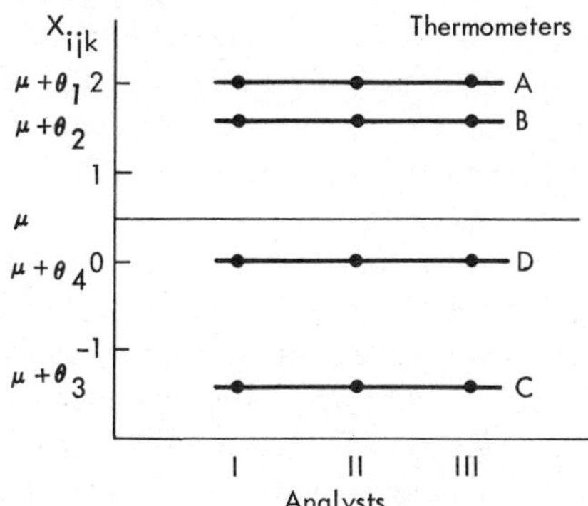

c.

Analyst and Thermometer Effects Only

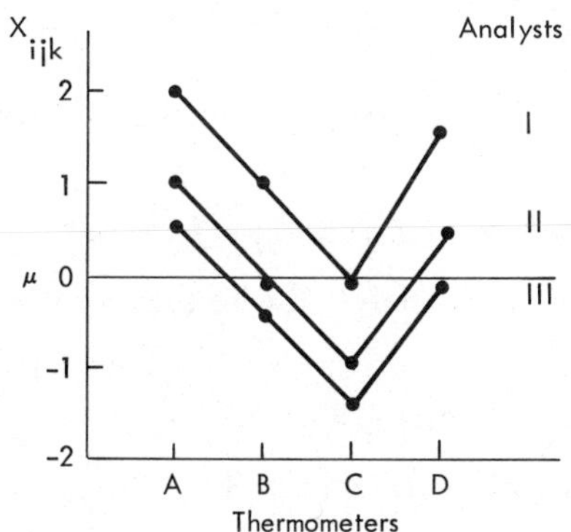

Analyst and Thermometer Effects Only

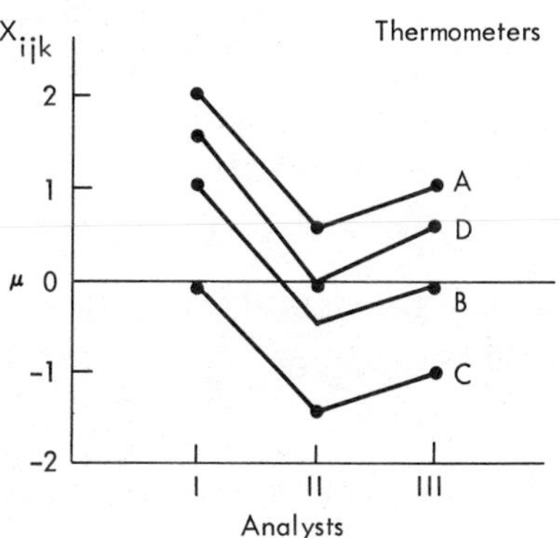

d.

Main Effects and Interaction

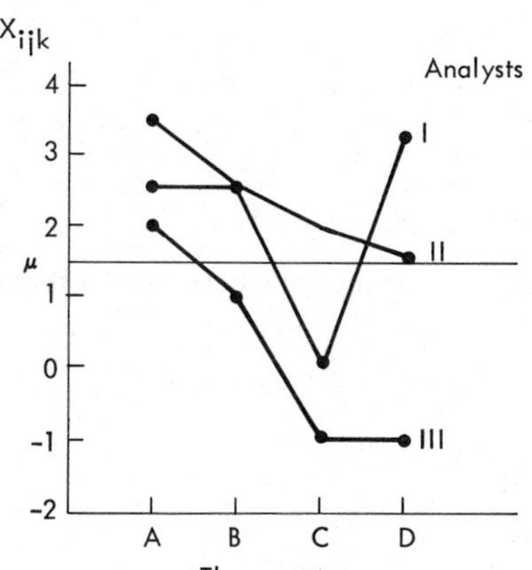

Main Effects and Interaction

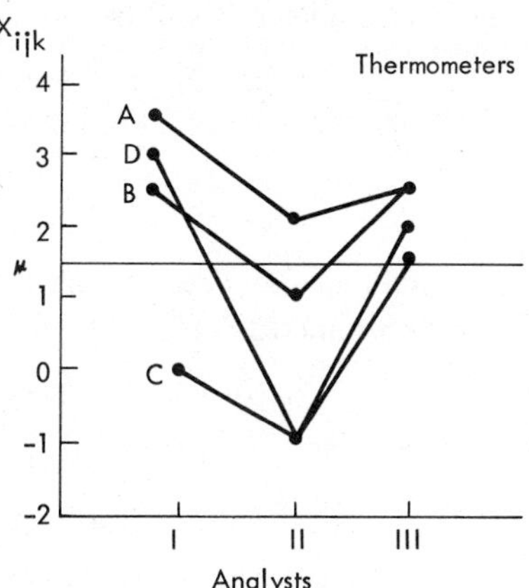

but all the θ_j and ψ_{ij} equal zero, then the X_{ijk} will fluctuate around different levels for the different analysts (see Figure 30.1b). If there are both analyst and thermometer biases but no interactions, i.e., not all the τ_i nor all the θ_j are zero but all the ψ_{ij} are zero, the X_{ij} will fluctuate around different levels for different combinations of analysts and thermometers, but these levels will vary from thermometer to thermometer in precisely the same way (i.e., the arithmetic differences will be the same) for every analyst, and the levels will vary from analyst to analyst in precisely the same way for every thermometer. In Figure 30.1c, for example, the lines of comparison will be parallel. If there are main effects and interactions, i.e., not all τ_i, nor all θ_j, nor all ψ_{ij} are zero, then the differences between thermometer levels will vary from analyst to analyst and the differences between analyst levels will vary from thermometer to thermometer (see Figure 30.1d).

Algebraically the argument runs as follows: If X_{111} and X_{112} are the determinations by analyst I with thermometer A and $\bar{X}_{11.}$ is the mean of these, then

$$\begin{aligned}\bar{X}_{11.} &= \frac{1}{2}[\mu + \tau_1 + \theta_1 + \psi_{11} + \epsilon_{111} + \mu + \tau_1 + \theta_1 + \psi_{11} + \epsilon_{112}] \\ &= \mu + \tau_1 + \theta_1 + \psi_{11} + \bar{\epsilon}_{11.}\end{aligned}$$

Likewise if $\bar{X}_{12.}$ is the mean of the two determinations by analyst I with thermometer B, we have

$$\bar{X}_{12.} = \mu + \tau_1 + \theta_2 + \psi_{12} + \bar{\epsilon}_{12.}$$

Similarly, the mean of the two determinations by analyst II with thermometer A will be

$$\bar{X}_{21.} = \mu + \tau_2 + \theta_1 + \psi_{21} + \bar{\epsilon}_{21.}$$

and the mean of the two determinations with thermometer B will be

$$\bar{X}_{22.} = \mu + \tau_2 + \theta_2 + \psi_{22} + \bar{\epsilon}_{22.}$$

The difference between analyst I's mean determinations with thermometer A and thermometer B will be

$$\begin{aligned}\bar{X}_{11.} - \bar{X}_{12.} &= \mu + \tau_1 + \theta_1 + \psi_{11} + \bar{\epsilon}_{11.} \\ &\quad - \mu - \tau_1 - \theta_2 - \psi_{12} - \bar{\epsilon}_{12.} \\ &= \theta_1 - \theta_2 + \psi_{11} - \psi_{12} + \bar{\epsilon}_{11.} - \bar{\epsilon}_{12.}\end{aligned}$$

and the difference between analyst II's mean determinations with thermometer A and thermometer B will be

$$\begin{aligned}\bar{X}_{21.} - \bar{X}_{22.} &= \mu + \tau_2 + \theta_1 + \psi_{21} + \bar{\epsilon}_{21.} \\ &\quad - \mu - \tau_2 - \theta_2 - \psi_{22} - \bar{\epsilon}_{22.} \\ &= \theta_1 - \theta_2 + \psi_{21} - \psi_{22} + \bar{\epsilon}_{21.} - \bar{\epsilon}_{22.}\end{aligned}$$

The difference between these differences is

$$\begin{aligned}(\bar{X}_{11.} - \bar{X}_{12.}) - (\bar{X}_{21.} - \bar{X}_{22.}) &= \theta_1 - \theta_2 + \psi_{11} - \psi_{12} + \bar{\epsilon}_{11.} - \bar{\epsilon}_{12.} \\ &\quad - \theta_1 + \theta_2 - \psi_{21} + \psi_{22} - \bar{\epsilon}_{21.} + \bar{\epsilon}_{22.} \\ &= (\psi_{11} - \psi_{12} - \psi_{21} + \psi_{22}) \\ &\quad + (\bar{\epsilon}_{11.} - \bar{\epsilon}_{12.} - \bar{\epsilon}_{21.} + \bar{\epsilon}_{22.})\end{aligned}$$

and is seen to depend only on the interaction elements (ψ_{ij}) and an error term. If the ψ_{ij} were all zero, this difference between differences would be zero except for an error term.

The usual procedure is to test the interaction first. If this is found to be significant, we know immediately that the factors under study affect the results and, in many cases, there may be no interest in proceeding further to test the main effects. In the present problem, for example, if we should find a significant interaction, we might turn immediately to a study of the individual thermometers and individual analysts (cf. Figure 30.1d). We could then assign to each analyst that thermometer which when used by him showed the smallest deviation from the grand average $\bar{X}$. Thus, in the present case, we would assign thermometer A to analyst II, thermometer B to analyst I, and thermometer D to analyst III. Under these conditions we would not be interested in the averages of the various analysts with different thermometers nor in the averages of the various thermometers over the three analysts. There may be cases, however, in which we are interested in the main effects even though there is a significant interaction.

1.2.2. The Random-Effects Model. If both column and row factors are viewed as being "random" factors (i.e., the analysts are viewed as a random sample from an infinite universe of analysts, and thermometers are viewed as a random sample from an infinite universe of thermometers), then we assume that in the universe of τ_i's, $\bar{\tau}'_i = 0$, in the universe of θ_j's, $\bar{\theta}'_j = 0$, and in the joint (τ_i, θ_j) universe the mean interaction element $\bar{\psi}'_{ij} = 0$, with variances σ'^2_τ, σ'^2_θ, and σ'^2_ψ, respectively. This means that in the universe of τ's, θ's, and ψ's, we are again dealing with differentials, not absolute values.

If we set up the null hypothesis that the τ_i are all zero (i.e., $\sigma'^2_\tau = 0$), in other words, that none of the analysts in the universe of analysts has a mean bias, then it turns out that s^2_R [see equations (30.2)] is[1] an unbiased estimate of $g\sigma'^2_\psi + \sigma'^2$, whereas under the same hypothesis s^2_{RC} is also an unbiased estimate of $g\sigma'^2_\psi + \sigma'^2$. Hence to test this null hypothesis we compute s^2_R/s^2_{RC} and compare[2] it with F_α for $\nu_1 = r - 1$ and $\nu_2 = (r - 1)(c - 1)$. Note that we compare the row mean square with the *interaction* mean square, not the error mean square.

---

[1] See Table 31.3 below.

[2] Here and in subsequent tests we assume that all the random variables are normally distributed.

If we set up the null hypothesis that the θ_j are all zero in the universe (i.e., $\sigma'^2_\theta = 0$), in other words, that none of the thermometers in the universe of thermometers has a mean bias, then it turns out that s^2_C is an unbiased estimate of $g\sigma'^2_\psi + \sigma'^2$, whereas under the same hypothesis s^2_{RC} is an unbiased estimate of $g\sigma'^2_\psi + \sigma'^2$. Hence to test the given hypothesis, we compute s^2_C/s^2_{RC} and compare it with F_α for $\nu_1 = c - 1$ and $\nu_2 = (r - 1)(c - 1)$. It will be noted again that we compare the column mean square with the *interaction* mean square not the error mean square.

Finally, if we set up the hypothesis that all the ψ_{ij} equal zero in the universe of ψ_{ij} (i.e., $\sigma'^2_\psi = 0$), in other words, that there is no interaction between analysts and thermometers in general, then it turns out that s^2_{RC} is an unbiased estimate of σ'^2, whereas the error mean square s^2_e is an independent unbiased estimate of σ'^2. Hence to test this null hypothesis we compute s^2_{RC}/s^2_e and compare it with F_α for $\nu_1 = (r - 1)(c - 1)$ and $\nu_2 = rc(g - 1)$. Note that here the interaction mean square is compared with the *error* mean square.

Although the tests are carried out differently in this case, the arithmetical calculation of the sums of squares is the same as in the fixed case and will be discussed in Section 1.3 below.

In concluding this section it should be noted that in the random model we usually go from the basic significance tests to a study of the components of variance. It might be, for example, that in the course of a year many different analysts make determinations of the melting point of hydroquinone and many different thermometers are used. Interest would then center on how much of the overall variation in the determinations is due to analyst differences in general and how much to thermometer differences in general. If either or both proved relatively large, then an attempt might be made through a training program or through a carefully written set of instructions to reduce the analyst variation, or again pressure might be put on the suppliers of thermometers to produce a more uniform quality. The procedure for analyzing components of variance will be discussed in Chapter 31.

1.2.3. The Mixed-Effects Model. Suppose that in our thermometer-analyst problem, we view the analysts as being fixed (i.e., we are interested in the three given analysts only) but the thermometers are viewed as being a random sample from an "infinite" universe of thermometers. We then have a mixed model.

In our thinking about this mixed model we must be very careful about our assumptions since there have been many points of view as to how the model should be interpreted. We shall proceed as follows:[3]

Let $m(i, t)$ represent the "true" determination by analyst i with thermometer t, "true" in the sense that it is free from experimental error. Let

[3] This section is based largely on Henry Scheffé's article "Alternative Models for the Analysis of Variance," *Annals of Mathematical Statistics* 27 (1956), pp. 251–71. The symbol t is used here instead of j since reference is to a thermometer in the *universe* of thermometers instead of one in the *sample* of thermometers actually used in the experiment.

$$\mu = m(.\ ,\ .)$$

where dots represent means, i.e., μ is the grand mean of true determinations for the three analysts over the universe of thermometers. Let

$$\tau_i = \mu(i,\ .) - m(.\ ,\ .)$$

i.e., τ_i is the difference between the mean of the ith analyst's determinations with the universe of thermometers and the grand mean, and let

$$\theta_t = m(.\ ,\ t) - m(.\ ,\ .)$$

i.e., θ_t equals the difference between the mean of the three analysts determinations with thermometer t and the grand mean. Finally let

$$\begin{aligned}\psi_{it} &= m(i, t) - [m(i\ ,\ .) - m(.\ ,\ .) + m(.\ ,\ t) - m(.\ ,\ .) + m(.\ ,\ .)] \\ &= m(i, t) - m(i\ ,\ .) - m(.\ ,\ t) + m(.\ ,\ .)\end{aligned}$$

i.e., ψ_{it} is the difference between the "true" determination of the ith analyst on thermometer t and the value that would be expected from addition of the thermometer differential $\theta_t = m(.\ ,\ t) - m(.\ ,\ .)$, the analyst bias $\tau_i = m(i\ ,\ .) - m(.\ ,\ .)$ and the grand mean $m(.\ ,\ .)$.

These definitions give the model equation

$$m(i, t) = \mu + \tau_i + \theta_t + \psi_{it}$$

If we add to this a random experimental error ϵ that is normally distributed with zero mean and standard deviation σ', we have the final equation:

$$X_{itk} = \mu + \tau_i + \theta_t + \psi_{it} + \epsilon_{itk}$$

It is easily seen from the above that $\sum_i \tau_i = 0$, mean of θ_t in the universe of θ_t's is zero, $\sum_i \psi_{it} = 0$ for all[4] t, and mean of ψ_{it} equals zero for all i.

With this mixed model, it turns out that if the θ_t are all zero, i.e., $\sigma'^2_\theta = 0$, in the universe of θ_t, in other words, none of the thermometers are biased, then s_C^2 is an unbiased estimate of σ'^2 and the hypothesis $\theta_t = 0$ for all t can be tested by computing s_C^2/s_e^2 and comparing the result with F_α for $\nu_1 = c - 1$ and $\nu_2 = rc(g - 1)$.

Similarly, if the ψ_{it} are all zero, then s_{RC}^2 is an unbiased estimate of σ'^2 and the hypothesis $\psi_{it} = 0$ for all i and t can be tested by computing s_{RC}^2/s_e^2 and comparing it with F_α for $\nu_1 = (r - 1)(c - 1)$, $\nu_2 = rc(g - 1)$. Note that in both these cases, comparison is with the *error* mean square s_e^2.

On the other hand, if the $\tau_i = 0$ for all i, in other words, none of the analysts have any bias, then s_R^2 is an unbiased estimate[5] of the same quantity

[4] $\sum_i m(i, t) - \sum_i m(i\ ,\ .) - \sum_i m(.\ ,\ t) + \sum_i m(.\ ,\ .) = rm(.\ ,\ t) - rm(.\ ,\ .) - rm(.\ ,\ t) + rm(.\ ,\ .) = 0$. ($r$ is the number of analysts.)

[5] See Section 3.3 of Chapter 31.

as s_{RC}^2. For this reason, the hypothesis $\tau_i = 0$ for all i is often tested by computing s_R^2/s_{RC}^2 and comparing the result with F_α for $\nu_1 = r - 1$, $\nu_2 = (r-1)(c-1)$, i.e., the s_R^2 is compared with the *interaction* mean square. It has been shown[6] by H. Scheffé, however, that if the random elements ψ_{it} are correlated, the F distribution does not strictly hold and the exact procedure is to use a Hotelling T^2 test. The latter is explained in Appendix II (35). It will be found that the Hotelling T^2 test involves considerable work, however. An approximate procedure using an F test with an adjustment in the degrees of freedom has been worked out by J. P. Imhoff.[7] He suggests we use the regular F statistic but that the degrees of freedom be taken as h and $h(J-1)$ where J is the number of "levels" of the random factor ($= c$ for our example) and

(30.3) $$h = (I-1)[1 + (I-2)(1-\theta)^2]^{-1}$$

where I is number of "levels" of the fixed factor ($= r$ for our example) and

(30.4) $$\theta = \text{the smaller of} \begin{Bmatrix} 1 \\ s_e^2/s_{RC}^2 \end{Bmatrix}$$

Imhoff gives a table of the 0.05 and 0.01 points for $F_{h,mh}$, where $m = J - 1 = c - 1$ in our example. Imhoff's table is reproduced in this text as Table K, Appendix II. It will be noted that when $I = 2$, the procedure reduces to the ordinary F test.

1.3. Computations

To compute the various sums of squares needed in a two-way analysis of variance with more than one but an equal number in each class, we employ short-cut procedures as before. For the data of Table 30.1A, the computations run as follows:

1.3.1. Sums of Squares and Mean Squares. To compute the various sums of squares:

1. Sum the items in each row (i.e., for each analyst), square these sums, and add. Divide the total by the number of items in a row. Thus,

$$\frac{9^2 + 1^2 + 8.5^2}{8} = 19.281$$

2. Sum the items for each column (i.e., for each thermometer), square these sums, and add. Divide the total by the number of items in a column. Thus,

---

[6] Scheffé, "Alternative Models for the Analysis of Variance." Also see reference of footnote 7. No difficulty arises if in the variance-covariance matrix of the $m(i, t)$ the covariances are all equal.

[7] "Testing the Hypothesis of No Fixed Main-Effects in Scheffe's Mixed Model," *Annals of Mathematical Statistics* 33 (1962), pp. 1085–94.

$$\frac{8^2 + 6^2 + 1^2 + 3.5^2}{6} = 18.875$$

3. Sum the items for each cell (i.e., for each analyst-thermometer combination), square these sums, and add. Divide the total by the number of items in a cell. Thus,

$$\frac{3.5^2 + 2^2 + 2.5^2 + 2.5^2 + 1^2 + 2.5^2 + 0^2 + (-1)^2 + 2^2 + 3^2 + (-1)^2 + 1.5^2}{2} = 26.625$$

4. Square each individual item and add. Thus,

$$2^2 + 1.5^2 + 1^2 + 1^2 + 1.5^2 + 1^2 + 1^2 + 1.5^2 + 0^2 + 1^2 + 1^2 + 1.5^2 + (-0.5)^2 + 0.5^2 + (-1)^2 + 0^2 + 1^2 + 1^2 + 1.5^2 + 1.5^2 + (-1)^2 + 0^2 + 0.5^2 + 1^2 = 29.250$$

5. Sum all the data, square this sum, and divide by the total number of cases. Thus,

$$\frac{(18.5)^2}{24} = 14.260$$

This is the correction factor.

6. The row sum of squares equals

$$(1) - (5) = 19.281 - 14.260 = 5.021$$

7. The column sum of squares equals

$$(2) - (5) = 18.875 - 14.260 = 4.615$$

8. The interaction sum of squares equals

$$(3) - (5) - (6) - (7) = 26.625 - 14.260 - 5.021 - 4.615 = 2.729$$

9. The total sum of squares equals

$$(4) - (5) = 29.250 - 14.260 = 14.990$$

10. The residual or error sum of squares equals

$$(9) - (3) + (5) = 14.990 - 26.625 + 14.260 = 2.625$$

The above computations are summarized in Table 30.2. (In general, r is the number of rows, c the number of columns, and g the number of items in each cell.)

The degrees of freedom for analysts and thermometers are the same as in Chapter 29. The interaction sum of squares has $(r - 1)(c - 1)$ degrees of freedom, which for our data = (3)(2) = 6. Although the interaction sum of squares is made up 12 terms based on the sums of each cell, this is subject to 6 independent restrictions, in that the totals of the cell sums for each row and column must equal the row and column totals. Hence, the total

TABLE 30.2
Analysis of Variance Table for a Two-Way Classification with More Than One but an Equal Number of Cases in Each Class

Source	Sum of Squares	df	Mean Square	Variance Ratio		
				Fixed	Random	Mixed
Analysts (rows)	5.021	2	$s_R^2 = 2.51$	11.41**	5.58*	5.58(?)
Thermometers (columns) . . .	4.615	3	$s_C^2 = 1.54$	7.00**	3.42	7.00**
Analyst-thermometer interaction (cells)	2.729	6	$s_{RC}^2 = 0.45$	2.04	2.04	2.04
Experimental error (residual)	2.625	12	$s_e^2 = 0.22$			
Total	14.990					

(?) See Section 1.2.3 above.
* Significant at 0.05 level.
** Significant at 0.01 level.

degrees of freedom for the interaction sum of squares is 12 − 6 = 6. The degrees of freedom for the error sum of squares equals 12 since there are 12 restrictions on the 24 components in that each set of items for any one cell must have the sum for that cell. As previously noted, each sum of squares divided by its degrees of freedom gives the corresponding mean square.

1.3.2. Tests for Fixed-Effects Model. If all the factors are fixed, the variance-ratio tests run as follows. For all tests α is taken as 0.05.

Hypothesis: Analysts have no biases.

$$\frac{s_R^2}{s_e^2} = \frac{2.51}{0.22} = 11.41$$

$F_{0.05}(\nu_1 = 2, \nu_2 = 12) = 3.89$
Hypothesis rejected.

Hypothesis: Thermometers all read alike.

$$\frac{s_C^2}{s_e^2} = \frac{1.54}{0.22} = 7.00$$

$F_{0.05}(\nu_1 = 3, \nu_2 = 12) = 3.49$
Hypothesis rejected.

Hypothesis: There is no interaction between analysts and thermometers.

$$\frac{s_{RC}^2}{s_e^2} = \frac{0.45}{0.22} = 2.04$$

$F_{0.05}(\nu_1 = 6, \nu_2 = 12) = 3.00$
Hypothesis not rejected.

1.3.3. Tests for Random-Effects Model. If both factors are random, the variance-ratio tests run as follows. For all tests α is taken as 0.05.

Hypothesis: Analysts in universe have no mean biases ($\sigma_\tau'^2 = 0$).

$$\frac{s_R^2}{s_{RC}^2} = \frac{2.51}{0.45} = 5.58$$

$F_{0.05}(\nu_1 = 2, \nu_2 = 6) = 5.14$
Hypothesis rejected.

Hypothesis: Thermometers in the universe have no mean biases ($\sigma_\theta'^2 = 0$).

$$\frac{s_C^2}{s_{RC}^2} = \frac{1.54}{0.45} = 3.42$$

$F_{0.05}(\nu_1 = 3, \nu_2 = 6) = 4.76$
Hypothesis not rejected.

Hypothesis: There is no interaction between analysts and thermometers in general ($\sigma_\psi'^2 = 0$).

$$\frac{s_{RC}^2}{s_e^2} = \frac{0.45}{0.22} = 2.04$$

$F_{0.05}(\nu_1 = 6, \nu_2 = 12) = 3.00$
Hypothesis not rejected.

1.3.4. Tests for Mixed-Effects Model As noted above, if the analysts are viewed as being fixed and the thermometers are random, the main or average random effects and the interaction effects are tested by variance-ratio F tests. These run as follows. For all tests α is taken as 0.05.

Hypothesis: Thermometers in universe have no biases ($\sigma_\theta'^2 = 0$).

$$\frac{s_C^2}{s_e^2} = \frac{1.54}{0.22} = 7.00$$

$F_{0.05}(\nu_1 = 3, \nu_2 = 12) = 3.49$
Hypothesis rejected.

Hypothesis: There is no interaction between analysts and thermometers in general ($\sigma_\psi'^2 = 0$).

$$\frac{s_{RC}^2}{s_e^2} = \frac{0.45}{0.22} = 2.04$$

$F_{0.05}(\nu_1 = 6, \nu_2 = 12) = 3.00$
Hypothesis not rejected.

To test the main effects of analysts (the fixed factor), we use special procedures. The Hotelling T^2 test for these data is worked out in Appendix I (35) and leads to acceptance of the null hypothesis. Using the approximate procedure developed by Imhoff, we compute $s_R^2/s_{RC}^2 = 2.51/0.45 = 5.58$.

Then to get the degrees of freedom for running the F test, we use the Imhoff formula (30.3) to get

$$h = (3-1)[1 + (3-2)(1 - 0.22/0.45)^2]^{-1} = 1.59$$

For this value of h, Table K, Appendix II, indicates that the critical $F_{0.05}$ value is about 6.2. Since this is larger than the sample variance ratio, we again accept the null hypothesis which agrees with the exact Hotelling T^2 test. It will be noted that if we had used the ordinary procedure and compared 5.58 with the $F_{0.05}$ for 2 and 6 degrees of freedom (= 5.14), we would have rejected the null hypothesis.

1.4. Operating Characteristic Curves

For a *fixed-effects* model the operating characteristic curves for a two-way analysis of variance with r rows, c columns, and g elements in each cell can be derived in a manner similar to the OC curves for a one-way classification. Mann's table and Pearson and Hartley's power curves can be used as before. See Charts I and II in Appendix II.

(30.5) If λ^2 is taken equal to $\dfrac{\Sigma\tau_i^2}{r\sigma'^2}$, then $\phi = \lambda\sqrt{cg}$

(30.6) If λ^2 is taken equal to $\dfrac{\Sigma\theta_j^2}{c\sigma'^2}$, then $\phi = \lambda\sqrt{rg}$

(30.7) If λ^2 is taken equal to $\dfrac{\Sigma\psi_{ij}^2}{rc\,\sigma'^2}$, then $\phi = \lambda\sqrt{\dfrac{rcg}{(r-1)(c-1)+1}}$

It should be noted that the degrees of freedom associated with the error term is now $rc(g-1)$.

For a *random-effects* model, the OC curves can again be derived from Tables 8.3 and 8.4 of the Statistical Research Group's *Techniques of Statistical Analysis.* Table L of Appendix II gives the 0.10 points. In the test for interaction, if λ^2 is taken equal to $\sigma_\psi'^2/\sigma'^2$, then

(30.8) ϕ^2 in Table L and the *Techniques* $\phi = \dfrac{\sigma'^2 + g\sigma_\psi'^2}{\sigma'^2} = 1 + g\lambda^2$

The degrees of freedom for the denominator of F will be $rc(g-1)$. In the test for a row effect, if λ^2 is taken equal to $\dfrac{\sigma_\tau'^2}{\sigma'^2 + g\sigma_\psi'^2}$, then

(30.9) ϕ^2 in Table L and the *Techniques* $\phi =$

$$\frac{\sigma'^2 + g\sigma_\psi'^2 + cg\,\sigma_\tau'^2}{\sigma'^2 + g\sigma_\psi'^2} = 1 + cg\lambda^2$$

In the test for a column effect, if λ^2 is taken equal to $\dfrac{\sigma_\theta'^2}{\sigma'^2 + g\sigma_\psi'^2}$, then

(30.10) ϕ^2 in Table L and the *Techniques* $\phi=$

$$\frac{\sigma'^2+g\sigma_\psi'^2+rg\sigma_\theta'^2}{\sigma'^2+g\sigma_\psi'^2}=1+rg\lambda^2$$

For the *mixed-effects* model, the OC curves for random main effects and interactions are the same as for the random case. For the fixed factor, however, the situation becomes very complicated. Whether we use the exact Hotelling T^2 test or Imhoff's approximation, the abscissa for the OC curve must mathematically be expressed not only in terms of the actual τ_i's but also in terms of the variances and covariances of the ψ_{ij}. All that can be done practically is to derive the OC curve for the ordinary F test with $(r-1)$ and $(r-1)(c-1)$ degrees of freedom (this corresponds with equal covariances for the ψ_{ij}) and note that the OC curve for Imhoff's approximate F test is not likely to be any better than this and may be considerably (say 20–25 percent) worse. For small values of $(c-1)$, say 4 or less, the OC curve for the Hotelling T^2 test can be very much worse than that for the ordinary F test and worse even than the Imhoff approximation. The reader is referred to Imhoff's paper in the *Annals of Mathematical Statistics* 33 (1962), p. 1091.

1.5. Individual Degrees of Freedom

The principles underlying the analysis of individual degrees of freedom in a two-way analysis of variance with more than one but an equal number of items in each class are the same as those of simpler models. In all models, the basic requirement is that the transformation of the observed values X_{ijk} into the individual comparisons $(Z_i/\sqrt{D_i})$ be orthogonal.

Table 30.3 presents a transformation of the analyst-thermometer data. The Z_1 to Z_{12} are exactly the same as in Table 29.10 except that we now have a total of 24 items instead of 12. As always, Z_1^2/D_1 is the "correction factor." The Z_2 to Z_{12} are the same comparisons as in Table 29.10. In Table 29.10, however, Z_7 to Z_{12} were taken to represent residual variations which, in the model then appropriate, could be a combination of random interaction components and experimental error. In Table 30.3, Z_7 to Z_{12} represent random *or* fixed interaction variations plus error and we now have Z_{13} to Z_{24} which represent pure error variations. The quantity $\sum_{i=13}^{24} Z_i^2/D_i$ is the error sum of squares.

That Z_7 to Z_{12} are interaction comparisons is readily seen from a study of Table 30.3, Z_7, for example, compares the difference between analyst I and analyst III totaled over thermometers A and B with this same difference totaled over thermometers C and D. Z_8 compares the difference between analyst I and analyst III, on the one hand, and analyst II, on the other hand, totaled over thermometers A and B with this same difference totaled over thermometers C and D; and similarly for Z_9 to Z_{12}. It will be a good

TABLE 30.3

A Transformation of Melting Points for a Two-Way Classification with More Than One Case in a Class

Z_i	*Thermometer A*						*Thermometer B*						*Thermometer C*						*Thermometer D*						
	Analysts						*Analysts*						*Analysts*						*Analysts*						
	I		*II*		*III*		*I*		*II*		*III*		*I*		*II*		*III*		*I*		*II*		*III*		
	X_{111} (2.0)	X_{112} (1.5)	X_{121} (1.0)	X_{122} (1.0)	X_{131} (1.5)	X_{132} (1.0)	X_{211} (1.0)	X_{212} (1.5)	X_{221} (0.0)	X_{222} (1.0)	X_{231} (1.0)	X_{232} (1.5)	X_{311} (−0.5)	X_{312} (0.5)	X_{321} (−1.0)	X_{322} (0.0)	X_{331} (1.0)	X_{332} (1.0)	X_{411} (1.5)	X_{412} (1.5)	X_{420} (−1.0)	X_{422} (1.0)	X_{431} (0.5)	X_{432} (1.0)	D_i
Z_1	1	1	1	1	1	1	1	1	1	1	1	1	1	1	1	1	1	1	1	1	1	1	1	1	24
Z_2	1	1	1	1	1	1	1	1	1	1	1	1	−1	−1	−1	−1	−1	−1	−1	−1	−1	−1	−1	−1	24
Z_3	1	1	1	1	1	1	−1	−1	−1	−1	−1	−1	0	0	0	0	0	0	0	0	0	0	0	0	12
Z_4	0	0	0	0	0	0	0	0	0	0	0	0	1	1	1	1	1	1	−1	−1	−1	−1	−1	−1	12
Z_5	1	1	0	0	−1	−1	1	1	0	0	−1	−1	1	1	0	0	−1	−1	1	1	0	0	−1	−1	16
Z_6	1	1	−2	−2	1	1	1	1	−2	−2	1	1	1	1	−2	−2	1	1	1	1	−2	−2	1	1	48
Z_7	1	1	0	0	−1	−1	1	1	0	0	−1	−1	−1	−1	0	0	1	1	−1	−1	0	0	1	−1	16
Z_8	1	1	−2	−2	1	1	1	1	−2	−2	1	1	−1	−1	2	2	−1	−1	−1	−1	2	2	−1	−1	48
Z_9	1	1	0	0	−1	−1	−1	−1	0	0	1	1	0	0	0	0	0	0	0	0	0	0	0	0	8
Z_{10}	1	1	−2	−2	1	1	−1	−1	2	2	−1	−1	0	0	0	0	0	0	0	0	0	0	0	0	24
Z_{11}	0	0	0	0	0	0	0	0	0	0	0	0	1	1	0	0	−1	−1	−1	−1	0	0	1	1	8
Z_{12}	0	0	0	0	0	0	0	0	0	0	0	0	1	1	−2	−2	1	1	−1	−1	2	2	−1	−1	24
Z_{13}	1	−1	0	0	0	0	0	0	0	0	0	0	0	0	0	0	0	0	0	0	0	0	0	0	2
Z_{14}	0	0	1	−1	0	0	0	0	0	0	0	0	0	0	0	0	0	0	0	0	0	0	0	0	2
Z_{15}	0	0	0	0	1	−1	0	0	0	0	0	0	0	0	0	0	0	0	0	0	0	0	0	0	2
Z_{16}	0	0	0	0	0	0	1	−1	0	0	0	0	0	0	0	0	0	0	0	0	0	0	0	0	2
Z_{17}	0	0	0	0	0	0	0	0	1	−1	0	0	0	0	0	0	0	0	0	0	0	0	0	0	2
Z_{18}	0	0	0	0	0	0	0	0	0	0	1	−1	0	0	0	0	0	0	0	0	0	0	0	0	2
Z_{19}	0	0	0	0	0	0	0	0	0	0	0	0	1	−1	0	0	0	0	0	0	0	0	0	0	2
Z_{20}	0	0	0	0	0	0	0	0	0	0	0	0	0	0	1	−1	0	0	0	0	0	0	0	0	2
Z_{21}	0	0	0	0	0	0	0	0	0	0	0	0	0	0	0	0	1	−1	0	0	0	0	0	0	2
Z_{22}	0	0	0	0	0	0	0	0	0	0	0	0	0	0	0	0	0	0	1	−1	0	0	0	0	2
Z_{23}	0	0	0	0	0	0	0	0	0	0	0	0	0	0	0	0	0	0	0	0	1	−1	0	0	2
Z_{24}	0	0	0	0	0	0	0	0	0	0	0	0	0	0	0	0	0	0	0	0	0	0	1	−1	2

exercise for the reader to see for himself that these Z's are all differences between differences between column or row means and hence are "interactions."

Fortunately we can easily derive the Z coefficients that make up the interaction comparisons by line-by-line multiplication of the Z coefficients that constitute the main effects (comparisons between means). Thus, in Table 30.3 we have *coefficient-wise*

(30.11)

$$\begin{aligned} Z_7 &= Z_2 Z_5 \\ Z_8 &= Z_2 Z_6 \\ Z_9 &= Z_3 Z_5 \\ Z_{10} &= Z_3 Z_6 \\ Z_{11} &= Z_4 Z_5 \\ Z_{12} &= Z_4 Z_6 \end{aligned}$$

Numerical computations are carried out in exactly the same way as in previous analyses of individual degrees of freedom. It will be a good exercise for the reader to satisfy himself that the thermometer (column) sum of squares equals $\frac{Z_2^2}{D_2} + \frac{Z_3^2}{D_3} + \frac{Z_4^2}{D_4}$, the analyst (row) sum of squares equals $\frac{Z_5^2}{D_5} + \frac{Z_6^2}{D_6}$, the interaction sum of squares equals $\sum_{i=7}^{12} Z_i^2/D_i$, and finally the error sum of squares equals $\sum_{i=13}^{24} Z_i^2/D_i$.

When both row and column factors are quantitative and equally spaced, the analysis of individual degrees of freedom enables us to divide the overall interaction into individual interactions between the various trend components. Suppose in Table 30.3, for example, that we were not dealing with analysts and thermometers but with equally spaced quantitative factors. The columns might represent determinations with 0.1 percent, 0.2 percent, 0.3 percent, and 0.4 percent impurities of type A, and the rows might represent 0.2 percent, 0.3 percent, and 0.4 percent impurities of type B. Then we could set up new Z factors as indicated in Table 30.4 and compute the Z_i^2/D_i in which we are interested. For Table 30.4, the various numerical results are summarized in Table 30.5. Here we have seen fit to isolate the linear (Z_2), quadratic (Z_3), and cubic (Z_4) components of A and the linear (Z_5) and quadratic (Z_6) components of B. We have also noted the interaction (Z_7) between the linear component of A with the linear component of B and the interactions (Z_8 and Z_{10}) between the linear and quadratic components of A and B. The rest of the Z's (i.e., Z_9, Z_{11}, and Z_{12}) we have grouped as a remainder term, since it is thought (in this case) that these higher order interactions do not exist in reality and are a part of the error term. It will be seen from Table 30.4 that Z_7 is the interaction between linear components in the sense that it is the difference between the linear component of A when B is at level 1 and the linear component of A when B is at level 3, i.e., it

TABLE 30.4
Another Transformation of Melting Points for a Two-Way Classification with More Than One Case in a Class

Z_i	A_1						A_2						A_3						A_4						D_i
	B_1		B_2		B_3		B_1		B_2		B_3		B_1		B_2		B_3		B_1		B_2		B_3		
	(2.0)	(1.5)	(1.0)	(1.0)	(1.5)	(1.0)	(1.0)	(1.5)	(0.0)	(1.0)	(1.0)	(1.5)	(−0.5)	(0.5)	(−1.0)	(0.0)	(1.0)	(1.0)	(1.5)	(1.5)	(−1.0)	(0.0)	(0.5)	(1.0)	
Z_1	1	1	1	1	1	1	1	1	1	1	1	1	1	1	1	1	1	1	1	1	1	1	1	1	24
Z_2	−3	−3	−3	−3	−3	−3	−1	−1	−1	−1	−1	−1	1	1	1	1	1	1	3	3	3	3	3	3	120
Z_3	1	1	1	1	1	1	−1	−1	−1	−1	−1	−1	−1	−1	−1	−1	−1	−1	1	1	1	1	1	1	24
Z_4	−1	−1	−1	−1	−1	−1	3	3	3	3	3	3	−3	−3	−3	−3	−3	−3	1	1	1	1	1	1	120
Z_5	−1	−1	0	0	1	1	−1	−1	0	0	1	1	−1	−1	0	0	1	1	−1	−1	0	0	1	1	16
Z_6	1	1	−2	−2	1	1	1	1	−2	−2	1	1	1	1	−2	−2	1	1	1	1	−2	−2	1	1	48
Z_7	3	3	0	0	−3	−3	1	1	0	0	−1	−1	−1	−1	0	0	1	1	−3	−3	0	0	3	3	80
Z_8	−1	−1	0	0	1	1	1	1	0	0	−1	−1	1	1	0	0	−1	−1	−1	−1	0	0	1	1	16
Z_9	1	1	0	0	−1	−1	−3	−3	0	0	3	3	3	3	0	0	−3	−3	−1	−1	0	0	1	1	80
Z_{10}	−3	−3	6	6	−3	−3	−1	−1	2	2	−1	−1	1	1	−2	−2	1	1	3	3	−6	−6	3	3	240
Z_{11}	1	1	−2	−2	1	1	−1	−1	2	2	−1	−1	−1	−1	2	2	−1	−1	1	1	−2	−2	1	1	48
Z_{12}	−1	−1	2	2	−1	−1	3	3	−6	−6	3	3	−3	−3	6	6	−3	−3	1	1	−2	−2	1	1	240
Z_{13} to Z_{24} same as in Table 30.3																									

TABLE 30.5
An Analysis of Variance Table for Two Quantitative Factors

Source of Variation	*Sum of Squares*	*df*	*Mean Square*	*Variance Ratio†*
Main effects:				
Linear A	2.852	1	2.852	13.33**
Quadratic A	0.844	1	0.844	3.95
Cubic A	0.919	1	0.919	4.30
Total A	4.615	3	1.538	7.19**
Linear B	0.016	1	0.016	0.75
Quadratic B	5.005	1	5.005	23.40**
Total B	5.021	2	2.511	11.74**
Interactions:				
Linear A × Linear B	0.003	1	0.003	0.001
Quadratic A × Linear B ...	1.266	1	1.266	5.92*
Quadratic B × Linear A ...	0.876	1	0.876	4.10
Remainder	0.584	3	0.195	
Total Interaction	2.729	6	0.45	2.33
Experimental error	2.625	12	0.22	

* Significant at the 0.05 level.
** Significant at the 0.01 level.
† Based on a pooling of the interaction remainder and the experimental error, since the remainder term is presumably an error term and is not significantly different from the experimental error.

is the linear trend over B in the linear component of A. It will also be seen to be the linear trend over A in the linear component of B. Likewise, it is seen that Z_8 is the interaction between quadratic A and linear B in the sense that it is the difference between the quadratic component of A when B is at level 1 and the quadratic component of A when B is at level 3, i.e., it is the linear trend over B in the quadratic component of A. Again Z_{10} is the interaction between quadratic B and linear A in the sense that it is the linear trend over A in the quadratic component of B.

Table 30.5 shows a highly significant linear effect of A and a highly significant quadratic effect of B, and these are sufficient in both cases to account for significant main effects. A significant interaction was also found in that the quadratic component of A was found to vary linearly with B. In running these tests, the experimental error sum of squares was pooled with the remainder interaction sum of squares since the latter was believed a priori to represent error and the difference between the two mean squares was not significant. This gave us 15 degrees of freedom instead of 12, and a critical $F_{0.05} = 4.54$ instead of 4.75 for most of the tests.

2. ANALYSIS OF VARIANCE FOR A THREE-WAY CLASSIFICATION, A SINGLE CASE IN EACH CLASS

2.1. Further Extension of the Problem

Consider, now, an extension of the analysis to a three-way classification with a single case in each class. Let us suppose that in Tables 30.1 and 30.1A the first measurement by each analyst was made on June 1 and the second on June 2. In other words, we have two replications of the whole experiment. The data then become classified in three ways: by thermometer, analyst, and day. The questions of interest are: Do the analysts have any biases in reading the thermometers? Are any of the thermometers biased? Does the day have any effect on the measurement of the melting point of hydroquinone? Are there interactions between the various factors?

2.2. Theory

When we have three or more classifications it is well to use a mathematical notation that is easier to handle than the Greek letters that have been employed so far. Following a device of H. Scheffé, we shall use the letter "a" to stand for a component effect. To this letter we shall attach *superscripts* indicating the particular classification or classifications to which reference is made and *subscripts* indicating which class or classes are involved. If the effect is viewed as fixed, we shall use an ordinary a, if it is viewed as being random, we shall use an italic *a*. Interactions between fixed factors will be fixed and interactions between random factors will be random. An interaction between a fixed factor and a random factor will be called a mixed factor and represented by an *m* instead of an a.

2.2.1. The Fixed-Effects Model. If all factors are fixed, the basic equation for a three-way analysis of variance with a single case in each class is

(30.12) $$X_{ijk} = \mu + \mathrm{a}_i^A + \mathrm{a}_j^B + \mathrm{a}_k^C + \mathrm{a}_{ij}^{AB} + \mathrm{a}_{ik}^{AC} + \mathrm{a}_{jk}^{BC} + \epsilon_{ijk}$$

where

μ = overall average ("true" melting point)
a_i^A = a row effect (analyst bias), $i = 1, 2, \ldots, r$
a_j^B = a column effect (thermometer bias), $j = 1, 2, \ldots, c$
a_k^C = a group effect (day effect), $k = 1, 2, \ldots, g$
a_{ij}^{AB} = a row-column (analyst-thermometer) interaction
a_{ik}^{AC} = a row-group (analyst-day) interaction
a_{jk}^{BC} = a column-group (thermometer-day) interaction
ϵ_{ijk} = a random residual variation that is normally distributed with zero mean and variance σ'^2 for all classes.

It will be assumed that

(30.16)
$$\sum_i \mathrm{a}_i^A = \sum_j \mathrm{a}_j^B = \sum_k \mathrm{a}_k^C = 0$$
$$\sum_i \mathrm{a}_{ij}^{AB} = \sum_j \mathrm{a}_{ij}^{AB} = \sum_i \sum_j \mathrm{a}_{ij}^{AB} = 0$$
$$\sum_i \mathrm{a}_{ik}^{AC} = \sum_k \mathrm{a}_{ik}^{AC} = \sum_i \sum_k \mathrm{a}_{ik}^{AC} = 0$$
$$\sum_j \mathrm{a}_{jk}^{BC} = \sum_k \mathrm{a}_{jk}^{BC} = \sum_j \sum_k \mathrm{a}_{jk}^{BC} = 0$$

which means that only differential effects can be measured.

Let us define:

$$s_R^2 = \frac{cg \sum_i (\bar{X}_i - \bar{X})^2}{r - 1}$$

$$s_C^2 = \frac{rg \sum_j (\bar{X}_j - \bar{X})^2}{c - 1}$$

$$s_G^2 = \frac{rc \sum_k (\bar{X}_k - \bar{X})^2}{g - 1}$$

(30.14)
$$s_{RC}^2 = \frac{g \sum_i \sum_j (\bar{X}_{ij} - \bar{X}_i - \bar{X}_j + \bar{X})^2}{(r - 1)(c - 1)}$$

$$s_{RG}^2 = \frac{c \sum_i \sum_k (\bar{X}_{ik} - \bar{X}_i - \bar{X}_k + \bar{X})^2}{(r - 1)(g - 1)}$$

$$s_{CG}^2 = \frac{r \sum_j \sum_k (\bar{X}_{jk} - \bar{X}_j - \bar{X}_k + \bar{X})^2}{(c - 1)(g - 1)}$$

$$s_{\text{res.}}^2 = \frac{\sum_i \sum_j \sum_k (X_{ijk} - \bar{X}_{ij} - \bar{X}_{ik} - \bar{X}_{jk} + \bar{X}_i + \bar{X}_j + \bar{X}_k - \bar{X})^2}{(r - 1)(c - 1)(g - 1)}$$

Then in line with previous analysis, we can test the hypothesis $a_i^A = 0$ for all i by computing $s_R^2/s_{\text{res.}}^2$ and comparing it with F_α for $\nu_1 = r - 1$, $\nu_2 = (r - 1)(c - 1)(g - 1)$. Similarly, we can test the hypothesis $a_j^B = 0$ for all j and the hypothesis $a_k^C = 0$ for all k by computing $s_C^2/s_{\text{res.}}^2$ and $s_G^2/s_{\text{res.}}^2$, respectively. To test for the various interactions we compute $s_{RC}^2/s_{\text{res.}}^2$, $s_{RG}^2/s_{\text{res.}}^2$, and $s_{CG}^2/s_{\text{res.}}^2$.

2.2.2. The Random-Effects Model. If all factors are random, the basic mathematical model is

(30.15)
$$X_{ijk} = \mu + a_i^A + a_j^B + a_k^C + a_{ij}^{AB} + a_{ik}^{AC} + a_{jk}^{BC} + \epsilon_{ijk}$$

where in the various universes

(30.16)
$$\begin{aligned}
&\text{Mean of } a_i^A = 0 \text{ and variance of } a_i^A = \sigma_A'^2\\
&\text{Mean of } a_j^B = 0 \text{ and variance of } a_j^B = \sigma_B'^2\\
&\text{Mean of } a_k^C = 0 \text{ and variance of } a_k^C = \sigma_C'^2\\
&\text{Mean of } a_{ij}^{AB} = 0 \text{ and variance of } a_{ij}^{AB} = \sigma_{AB}'^2\\
&\text{Mean of } a_{ik}^{AC} = 0 \text{ and variance of } a_{ij}^{AC} = \sigma_{AC}'^2\\
&\text{Mean of } a_{jk}^{BC} = 0 \text{ and variance of } a_{jk}^{BC} = \sigma_{BC}'^2\\
&\text{Mean of } \epsilon_{ijk} = 0 \text{ and variance of } \epsilon_{ijk} = \sigma'^2
\end{aligned}$$

TABLE 30.6
Expected Values of Mean Squares in a Three-Way Analysis of Variance, Random Effects

Source of Variation	*Mean Square*	*An Unbiased Estimate of—*
Row (A)	s_R^2	$\sigma'^2 + g\sigma'^2_{AB} + c\sigma'^2_{AC} + cg\sigma'^2_{AC}$
Column (B)	s_C^2	$\sigma'^2 + g\sigma'^2_{AB} + r\sigma'^2_{BC} + rg\sigma'^2_{B}$
Group (C)	s_G^2	$\sigma'^2 + c\sigma'^2_{AC} + r\sigma'^2_{BC} + rc\sigma'^2_{C}$
Row × Column (AB)	s_{RC}^2	$\sigma'^2 + g\sigma'^2_{AB}$
Row × Group (AC)	s_{RG}^2	$\sigma'^2 + c\sigma'^2_{AC}$
Column × Group (BC)	s_{CG}^2	$\sigma'^2 + r\sigma'^2_{BC}$
Residual	$s_{\text{res.}}^2$	σ'^2

and the random variables are normally distributed and independent of each other. In this case the mean squares of formulas (30.14) are unbiased estimates of certain linear functions of the universe variances.[8] These are listed in Table 30.6.

In running the various F tests we first test the interactions. We compute $s_{RC}^2/s_{\text{res.}}^2$ and compare it with F_α for $\nu_1 = (r - 1)(c - 1)$ and $\nu_2 = (r - 1)(c - 1)(g - 1)$. We compute $s_{RG}^2/s_{\text{res.}}^2$ and compare it with F_α for $\nu_1 = (r - 1)(g - 1)$ and $\nu_2 = (r - 1)(c - 1)(g - 1)$, and we compute $s_{CG}^2/s_{\text{res.}}^2$ and compare it with F_α for $\nu_1 = (c - 1)(g - 1)$ and $\nu_2 = (r - 1)(c - 1)(g - 1)$.

If the interactions are all significant, we cannot proceed with the ordinary F tests of the main effects by computing such quantities as s_R^2/s_{RC}^2 or s_C^2/s_{CG}^2, since significance may be due to an "outside" interaction term rather than the variance of the main effect for which we are testing. One way to solve this difficulty is to take ratios of linear functions of the mean squares. For example, to test the hypothesis $\sigma_A'^2 = 0$, we compute $(s_R^2 + s_{\text{res.}}^2)/(s_{RC}^2 + s_{RG}^2)$ and compare it with F_α for the degrees of freedom derived as follows:

In general let L be a linear function of several independent estimates of variance, say,

$$L = k_a s_a^2 + k_b s_b^2 + k_c s_c^2 \tag{30.17}$$

Then Satterthwaite[9] has shown that $\dfrac{\nu L}{\text{Exp. Value of } \Sigma k_i s_i^2}$ will follow approximately a χ^2 distribution of which the degrees of freedom can be estimated from[10]

[8] For a further discussion of components of variance see Chapter 31, Section 3.

[9] F. E. Satterthwaite, "Synthesis of Variance," *Psychometrika* 6 (1941), pp. 309–16; and "An Approximate Distribution of Estimates of Variance Components," *Biometrics Bulletin* 2 (1946), pp. 110–14. Also see W. G. Cochran, "Testing a Linear Relationship among Variances," *Biometrics* 7 (1951), pp. 17–32.

[10] The original formula is in terms of the expected values of s^2, but since we do not know these we use the sample values.

(30.18)
$$\nu = \frac{L^2}{\dfrac{(k_a s_{\bar{a}}^2)^2}{\nu_a} + \dfrac{(k_b s_{\bar{b}}^2)^2}{\nu_b} + \dfrac{(k_c s_{\bar{c}}^2)^2}{\nu_c}}$$

where ν_a, ν_b, and ν_c are the degrees of freedom associated with $s_{\bar{a}}^2$, $s_{\bar{b}}^2$, and $s_{\bar{c}}^2$, respectively. In using this it is well to have the k's positive. According to this rule, the degrees of freedom for the statistic $F = (s_R^2 + s_{\text{res.}}^2)/(s_{RC}^2 + s_{RG}^2)$ will be

$$\nu_1 = \frac{(s_R^2 + s_{\text{res.}}^2)^2}{\dfrac{(s_R^2)^2}{(r-1)} + \dfrac{(s_{\text{res.}}^2)^2}{(r-1)(c-1)(g-1)}}$$

and

$$\nu_2 = \frac{(s_{RC}^2 + s_{RG}^2)^2}{\dfrac{(s_{RC}^2)^2}{(r-1)(c-1)} + \dfrac{(s_{RG}^2)^2}{(r-1)(g-1)}}$$

2.2.3. The Mixed-Effects Model. If some factors are fixed and others random, we have a mixed model. For example, suppose that A is fixed and B and C are random. In our illustration, analysts might be fixed and thermometers and days might be random. The model would then be

(30.19) $$X_{ijk} = \mu + a_i^A + a_j^B + a_k^C + m_{ij}^{AB} + m_{ik}^{AC} + a_{jk}^{BC} + \epsilon_{ijk}$$

where

(30.20) $$\sum_i a_i^A = 0; \ \sum_i m_{ij}^{AB} = 0; \ \sum_i m_{ik}^{AC} = 0$$

and in the various universes

(30.21)

Mean of $a_j^B = 0$ and variance of $a_j^B = \sigma_B'^2$

Mean of $a_k^C = 0$ and variance of $a_k^C = \sigma_C'^2$

Mean over B of $m_{ij}^{AB} = 0$ and variance over B of $m_{ij}^{AB} = \sigma_{AB}'^2$ *for each*[11] i

Mean over C of $m_{ik}^{AC} = 0$ and variance over C of $m_{ik}^{AC} = \sigma_{AC}'^2$ *for each*[11] i

Mean of $a_{jk}^{BC} = 0$ and variance of $a_{jk}^{BC} = \sigma_{BC}'^2$

Mean of $\epsilon_{ijk} = 0$ and variance of $\epsilon_{ijk} = \sigma'^2$

It is also assumed that the random components are all normally distributed and independent of each other within the restrictions laid down in (30.20).

---

[11] We exclude the possibility that $\sigma_{AB}'^2$ and $\sigma_{AC}'^2$ are different for the different levels of the fixed factor A.

TABLE 30.7
Expected Values of Mean Squares in a Three-Way Analysis of Variance, Mixed Effects
(A fixed, B and C random)

Source	*Mean Square*	*An Unbiased Estimate of—*
Row (A)	s_R^2	$\sigma'^2 + \frac{gr}{r-1}\sigma'^2_{AB} + \frac{cr}{r-1}\sigma'^2_{AC} + \frac{cgr}{r-1}\sigma'^2_A$
Column (B)	s_C^2	$\sigma'^2 + r\sigma'^2_{BC} + rg\sigma'^2_B$
Group (C)	s_G^2	$\sigma'^2 + r\sigma'^2_{BC} + rc\sigma'^2_C$
Column × row (AB)	s_{RC}^2	$\sigma'^2 + \frac{gr}{r-1}\sigma'^2_{AB}$
Row × group (AC)	s_{RG}^2	$\sigma'^2 + \frac{cr}{r-1}\sigma'^2_{AC}$
Column × group (BC)	s_{CG}^2	$\sigma'^2 + r\sigma'^2_{BC}$
Residual	$s_{\text{res.}}^2$	σ'^2

Under these conditions the mean squares of formulas (30.14) are unbiased estimates of the linear functions of variances listed in Table 30.7. (It will be noted that $\sigma_A'^2$ is defined as $\sum_i (a_i^A)^2/r$.)

The AB interaction can be tested by computing $s_{RC}^2/s_{\text{res.}}^2$ and comparing it with F_α for $\nu_1 = (r-1)(c-1)$ and $\nu_2 = (r-1)(c-1)(g-1)$. Likewise, the AC interaction can be tested by computing $s_{RG}^2/s_{\text{res.}}^2$ and comparing it with F_α for $\nu_1 = (r-1)(g-1)$ and $\nu_2 = (r-1)(c-1)(g-1)$ and the BC interaction can be tested by computing the ratio $s_{CG}^2/s_{\text{res.}}^2$ and comparing it with F_α for $\nu_1 = (c-1)(g-1)$ and $\nu_2 = (r-1)(c-1)(g-1)$.

The main or average effects of B and C can be tested by computing the ratios s_C^2/s_{CG}^2 and s_G^2/s_{CG}^2 and comparing them with F_α for $\nu_1 = c-1$, $\nu_2 = (c-1)(g-1)$ and $\nu_1 = g-1$, $\nu_2 = (c-1)(g-1)$, respectively. To test the A main effects we could compute $\frac{s_R^2 + s_{\text{res.}}^2}{s_{RC}^2 + s_{RG}^2}$, with degrees of freedom following Satterthwaite's rule given in the previous section. As discussed in Section 1.2.3, however, this last test would not be strictly valid and we would have to very cautious in drawing conclusions in such cases. The Imhoff procedure cannot be applied directly to this three-factor case. Precise results would require the use of a Hotelling T^2 test.

2.3. Computations

In carrying out the computations for a three-way analysis of variance we need three separate two-way tables. For the analyst-thermometer-day problem, the three tables required are the summary parts of Tables 30.1A, 30.8,

TABLE 30.8
Determinations of the Melting Point of Hydroquinone Classified by Days and Analysts

Days	Analysts I	II	III	Σ	$\bar{X}$
1	2.0	1.0	1.5		
	1.0	0.0	1.0		
	−0.5	−1.0	1.0		
	1.5	−1.0	0.5		
	4.0	−1.0	4.0	7.0	0.58
2	1.5	1.0	1.0		
	1.5	1.0	1.5		
	0.5	0.0	1.0		
	1.5	0.0	1.0		
	5.0	2.0	4.5	11.5	0.96
Σ	9.0	1.0	8.5	18.5	0.77
$\bar{X}$	1.12	0.12	1.06	0.77	. . .

TABLE 30.9
Determinations of the Melting Point of Hydroquinone Classified by Days and Thermometers

Days	Thermometers A	B	C	D	Σ	$\bar{X}$
1	2.0	1.0	−0.5	1.5		
	1.0	0.0	−1.0	−1.0		
	1.5	1.0	1.0	0.5		
	4.5	2.0	−0.5	1.0	7.00	0.58
2	1.5	1.5	0.5	1.5		
	1.0	1.0	0.0	0.0		
	1.0	1.5	1.0	1.0		
	3.5	4.0	1.5	2.5	11.5	0.96
Σ	8.0	6.0	1.0	3.5	18.5	0.77
$\bar{X}$	1.33	1.00	0.17	0.58	0.77	. . . .

and 30.9. A two-way table with the total of each cell is all that is needed. For the given data the computations run as follows:

1. In Table 30.1A sum the items for each analyst, square, and sum by rows. Divide by the number in a row. Thus,

$$\frac{9^2 + 1^2 + 8.5^2}{8} = 19.281$$

2. In Table 30.1A sum the items for each thermometer, square, and sum by columns. Divide by the number in a column. Thus,

$$\frac{8^2 + 6^2 + 1^2 + 3.5^2}{6} = 18.875$$

3. In Table 30.8 sum the items for days (rows of Table 30.8), square, and sum by days. Divide by the number of cases for a day. Thus,

$$\frac{7^2 + 11.5^2}{12} = 15.104$$

4. In Table 30.1A sum the items for each cell (each analyst-thermometer combination), square, and add for all cells. Divide by the number in a cell. Thus

$$\frac{3.5^2 + 2^2 + 2.5^2 + 2.5^2 + 1^2 + 2.5^2 + 0^2 + (-1)^2 + 2^2 + 3^2 + (-1)^2 + 1.5^2}{2} = 26.625$$

5. In Table 30.8 sum the items for each cell (each analyst-day combination), square, and add for all cells. Divide by the number in a cell. Thus

$$\frac{4^2 + (-1)^2 + 4^2 + 5^2 + 2^2 + 4.5^2}{4} = 20.563$$

6. In Table 30.9 sum the items for each cell (each thermometer-day combination), square, and add for all cells. Divide by the number in a cell. Thus

$$\frac{4.5^2 + 2^2 + (-0.5)^2 + 1^2 + 3.5^2 + 4^2 + 1.5^2 + 2.5^2}{3} = 20.750$$

7. Square each individual item and sum for all items. Thus

$$\begin{aligned}&2^2 + 1.5^2 + 1^2 + 1^2 + 1.5^2 + 1^2 + 1^2 + 1.5^2 + 0^2 + 1^2 + 1^2\\&+ 1.5^2 + (-0.5)^2 + 0.5^2 + (-1)^2 + 0^2 + 1^2 + 1^2 + 1.5^2 + 1.5^2\\&+(-1)^2 + 0^2 + 0.5^2 + 1^2 = 29.250\end{aligned}$$

8. Sum all the data, square this total, and divide by the total number of cases. Thus

$$\frac{(18.5)^2}{24} = 14.260$$

9. The row (analyst) sum of squares (reference is to Table 30.1A) equals

$$(1) - (8) = 19.281 - 14.260 = 5.021$$

10. The column (thermometer) sum of squares equals

$$(2) - (8) = 18.875 - 14.260 = 4.615$$

11. The group (day) sum of squares (reference is to Table 30.8) equals

$$(3) - (8) = 15.104 - 14.260 = 0.844$$

12. The row-column (analyst-thermometer) interaction sum of squares equals

$$(4) - (8) - (9) - (10) = 26.625 - 14.260 - 5.021 - 4.615 = 2.729$$

13. The row-group (analyst-day) interaction sum of squares equals

$$(5) - (8) - (9) - (11) = 20.563 - 14.260 - 5.021 - 0.844 = 0.438$$

14. The column-group (thermometer-day) interaction sum of squares equals

$$(6) - (8) - (10) - (11) = 20.750 - 14.260 - 4.615 - 0.844 = 1.031$$

15. The residual sum of squares equals

$$\begin{aligned}&(7) - (8) - (9) - (10) - (11) - (12) - (13) - (14)\\ &= 29.250 - 14.260 - 5.021 - 4.615 - 0.844 - 2.729 - 0.438 - 1.031\\ &= 0.312\end{aligned}$$

The above results are summarized in Table 30.10.

2.3.1. Fixed-Effects Tests. For the model in which analysts, thermometers, and days are all viewed as fixed, we test for the significance of each effect as follows. Note that comparison is always with the residual mean square. H_0 is the null hypothesis of no effect.

Effect	*Variance Ratio*	ν_1	ν_2	$F_{0.05}$	*Conclusion*
Analysts	$s_R^2/s_{\text{res.}}^2 = 2.52/.05 = 50.4$**	2	6	5.14	Reject H_0
Thermometers	$s_C^2/s_{\text{res.}}^2 = 1.54/.05 = 30.8$**	3	6	4.76	Reject H_0
Days	$s_G^2/s_{\text{res.}}^2 = 0.84/.05 = 16.8$**	1	6	5.99	Reject H_0
An. × Therm.	$s_{RC}^2/s_{\text{res.}}^2 = 0.45/.05 = 9.0$**	6	6	4.28	Reject H_0
An. × Days	$s_{RG}^2/s_{\text{res.}}^2 = 0.22/.05 = 4.4$	2	6	5.14	Do not reject H_0
Therm. × Days	$s_{CG}^2/s_{\text{res.}}^2 = 0.34/.05 = 6.8$*	3	6	4.76	Reject H_0

It would seem that all effects are significant except the analyst-day interaction.

2.3.2. Random-Effects Tests. When all the effects are random, the various tests may not be so simple. We begin by comparing the interaction mean squares with the residual mean square. This we have just seen yields significant interactions for analysts and thermometers, and thermometers and days, but not analysts and days. Since the analyst-day interaction is not significant,

Table 30.10
Analysis of Variance Table for Three Factors, One Case for Each Factor Combination

Source	Sum of Squares	df	Mean Square
Analysts	5.021	2	$s_R^2 = 2.52$
Thermometers . . .	4.615	3	$s_C^2 = 1.54$
Days	0.844	1	$s_G^2 = 0.84$
AT	2.729	6	$s_{RC}^2 = 0.45$
AD	0.438	2	$s_{RG}^2 = 0.22$
TD	1.031	3	$s_{CG}^2 = 0.34$
Residual	0.312	6	$s_{\text{res.}}^2 = 0.05$
Total	14.990	23	

we can test for analyst main effects by comparing its mean square with the mean square for the thermometer-analyst interaction. Thus

$$s_R^2/s_{RC}^2 = \frac{2.52}{0.45} = 5.6$$

For $\nu_1 = 2$, $\nu_2 = 6$, $F_{0.05} = 5.14$, so the analyst main effect is significant. Likewise, we can test for a day main effect by comparing the day mean square with the thermometer-day interaction.

Thus

$$s_G^2/s_{CG}^2 = \frac{0.84}{0.34} = 2.47$$

For $\nu_1 = 1$, $\nu_2 = 3$, $F_{0.05} = 10.1$, so the day main effect is not significant. Apparently days in general have some effect on individual thermometers (since AT is significant), but there is no average effect attributable to days.

To test for thermometer main effects we must proceed in a roundabout manner. According to Section 2.2, we can treat

$$\frac{s_C^2 + s_{\text{res.}}^2}{s_{RC}^2 + s_{CG}^2} = \frac{1.54 + 0.05}{0.45 + 0.34} = \frac{1.59}{0.79} = 2.0$$

as having approximately an F distribution with

$$\nu_1 = \frac{(1.59)^2}{\dfrac{(1.54)^2}{3} + \dfrac{(0.05)^2}{6}} \doteq 3$$

and

$$\nu_2 = \frac{(0.79)^2}{\dfrac{(0.45)^2}{6} + \dfrac{(0.34)^2}{3}} = \frac{0.62}{0.034 + 0.038} \doteq 8.6$$

This yields a critical $F_{0.05}$ in excess of 3.86. Since $2 < 3.86$, we do not reject the null hypothesis. Although thermometers interact with days and analysts, there is no indication here that there are average differences among thermometers in general.

2.3.3. Mixed-Effects Tests. Following the argument in Section 2.2.3, we can validly run the following F tests.

Hypothesis: Thermometers and days do not interact.

$$\frac{s_{CG}^2}{s_{\text{res.}}^2} = \frac{0.34}{0.05} = 6.8 \qquad F_{0.05}(\nu_1 = 3, \nu_2 = 6) = 4.76$$

Reject hypothesis. Thermometers and days do interact.

Hypothesis: Thermometers have no average biases.

$$\frac{s_C^2}{s_{CG}^2} = \frac{1.54}{0.34} = 4.54 \qquad F_{0.05}(\nu_1 = 3, \nu_2 = 3) = 9.28$$

Hypothesis not rejected.

Hypothesis: Days have no average biases.

$$\frac{s_G^2}{s_{CG}^2} = \frac{0.84}{0.34} = 2.46 \qquad F_{0.05}(\nu_1 = 1, \nu_2 = 3) = 10.1$$

Hypothesis not rejected.

Hypothesis: Analysts and thermometers do not interact.

$$\frac{s_{RC}^2}{s_{\text{res.}}^2} = \frac{0.45}{0.05} = 9 \qquad F_{0.05}(\nu_1 = 6, \nu_2 = 6) = 4.28$$

Hypothesis rejected. Analysts and thermometers would appear to interact.

Hypothesis: Analysts and days do not interact.

$$\frac{s_{RG}^2}{s_{\text{res.}}^2} = \frac{0.22}{0.05} = 4.4 \qquad F_{0.05}(\nu_1 = 2, \nu_2 = 6) = 5.14$$

Hypothesis not rejected.

The following F test could also be run but it will not be strictly valid and conclusions must be accepted with caution. See Section 1.2.3.

Hypothesis: There are no average analyst biases.

$$\frac{s_R^2 + s_{\text{res.}}^2}{s_{RC}^2 + s_{RG}^2} = \frac{2.52 + 0.05}{0.45 + 0.22} = \frac{2.57}{0.67} = 3.84$$

$$\nu_1 = \frac{(2.57)^2}{\dfrac{(2.52)^2}{1} + \dfrac{(0.05)^2}{6}} \doteq 1$$

$$\nu_2 = \frac{(0.67)^2}{\frac{(0.45)^2}{6} + \frac{(0.22)^2}{2}} \doteq 8$$

The critical F ratio is thus $F_{0.05} = 5.32$
Hypothesis not rejected.

2.4. Operating Characteristic Curves

For fixed-effects or random-effects tests, the operating characteristic curves for a three-way classification with a single case in each class are derived in the same way as in a two-way classification with more than one, but an equal number in each class. For tests of main effects and interactions, the remarks of Section 1.4 will apply in detail, except that the degrees of freedom for the residual variation will be $(r - 1)(c - 1)(g - 1)$ instead of $rc(g - 1)$.

2.5. Analysis of Individual Degrees of Freedom

The analysis of individual degrees of freedom for the three-way classification will proceed in exactly the same way as for a two-way classification. Table 30.3 can be extended to apply to the data of Tables 30.1A, 30.8, and 30.9, with only minor changes. All that is needed is to introduce degrees of freedom for the day effects and the day interactions. These will be the new Z_{12} to Z_{18} and will run as indicated in Table 30.11. The remaining six degrees of freedom will be residual terms and may be computed by getting the products ("coefficientwise") of $Z_2 \times Z_5 \times Z_{13}$, $Z_3 \times Z_5 \times Z_{13}$, $Z_4 \times Z_5 \times Z_{13}$, $Z_2 \times Z_6 \times Z_{13}$, $Z_3 \times Z_6 \times Z_{13}$, and $Z_4 \times Z_6 \times Z_{13}$. If we had more than one item in a cell, these would be three-factor interaction terms. With only one item in a cell, however, they represent simply a residual variation that may or may not contain a random three-factor interaction as well as experimental error.

3. NESTED CLASSIFICATIONS

So far we have considered only analysis of variance of cross-classifications. Each class of each category was combined with each class of all other categories. There are situations, however, in which the classes of a given category are "nested" within the classes of another category in such a way that they occur in only one class of the latter. For example, if we run interlaboratory tests using three, *but not the same,* technicians in each laboratory, then the technicians are nested within the laboratories. A given technician works in only one laboratory. Again if we draw 10 bales of wool at random from a shipment and take 3 cores of wool from each bale, then the cores are "nested"

TABLE 30.11
Individual Degrees of Freedom for the Day Effects and Day Interactions of Tables 30.8 and 30.9

	X_{111}	X_{112}	X_{121}	X_{122}	X_{131}	X_{132}	X_{211}	X_{212}	X_{221}	X_{222}	X_{231}	X_{232}	X_{311}	X_{312}	X_{321}	X_{322}	X_{331}	X_{332}	X_{411}	X_{412}	X_{421}	X_{422}	X_{431}	X_{432}	Di
Z_{13}	1	−1	1	−1	1	−1	1	−1	1	−1	1	−1	1	−1	1	−1	1	−1	1	−1	1	−1	1	−1	24
Z_{14}	1	−1	1	−1	1	−1	1	−1	1	−1	1	−1	−1	1	−1	1	−1	1	−1	1	−1	1	−1	1	24
Z_{15}	1	−1	1	−1	1	−1	−1	1	−1	1	−1	1	0	0	0	0	0	0	0	0	0	0	0	0	12
Z_{16}	0	0	0	0	0	0	0	0	0	0	0	0	1	−1	1	−1	1	−1	−1	1	−1	1	−1	1	12
Z_{17}	1	−1	0	0	−1	1	1	−1	0	0	−1	1	1	−1	0	0	−1	1	1	−1	0	0	−1	1	16
Z_{18}	1	−1	−2	2	1	−1	1	−1	−2	2	1	−1	1	−1	−2	2	1	−1	1	−1	−2	2	1	−1	48

TABLE 30.12
Data on Set Viscosity of Rubber Tread Stocks

	Mix I		*Mix II*	
Tread Formula	*Sample 1*	*Sample 2*	*Sample 3*	*Sample 4*
Orthex	798	752	786	822
Tonox-orthex	775	796	764	750
Steam processed	734	707	734	726
Reclaim	732	719	707	678
ONV	838	813	753	808
40 black-orthex	802	774	904	923

SOURCE: Adapted from D. S. Villars, "Replication Degeneracy," American Society for Quality Control, *Quality Control Convention Papers,* 1954, pp. 135–46.

within the bales. Nested classifications are also called *hierarchal* classifications.

Pure nested classifications are easy to analyze. Suppose that in the laboratory example each technician runs "duplicate" tests. Then assuming c laboratories, r technicians in each laboratory, and g duplicate tests by each technician. we would have[12] a mean square for laboratories

$$\text{(30.22)} \qquad s^2_{\text{Lab}} = \frac{rg \sum_j (\bar{X}_j - \bar{X})^2}{c-1} \quad (\text{In general} = s^2_{\text{main category}}),$$

a mean square for technicians

$$\text{(30.23)} \qquad s^2_{\text{Tech}} = \frac{g \sum_j \sum_j (\bar{X}_{ij} - \bar{X}_j)^2}{c(r-1)} \quad (\text{In general} = s^2_{\text{1st subcategory}}),$$

and an error mean square

$$\text{(30.24)} \qquad s^2_e = \frac{\sum_j \sum_j \sum_k (X_{ijk} - \bar{X}_{ij})^2}{cr(g-1)} \quad (\text{In general} = s^2_{\text{2d subcategory}})$$

To test for laboratory differences we would compute $s^2_{\text{Lab}}/s^2_{\text{Tech}}$ and compare it with F_α for $\nu_1 = c - 1$ and $\nu_2 = c(r - 1)$ degrees of freedom. To test for technician differences we would compute s^2_{Tech}/s^2_e and compare it with F_α for $\nu_1 = c(r - 1)$, $\nu_2 = cr(g - 1)$.

Computations run along the same line as in cross-classifications. Consider the data of Table 30.12. These represent results of a rubber tread experiment. Two mixes were made for each of six different tread formulas. Two samples were taken from the slab cured from each mix and duplicate tests run. The samples were thus nested in mixes and the mixes within tread formulas.

[12] For simplicity it is assumed here that sampling errors in selecting the material to be analyzed are included in the general error term.

TABLE 30.12A
Data on Set Viscosity of Rubber Tread Stocks Coded to 700

	Mix I			*Mix II*			*Tread Formula*
Tread Formula	S_1	S_2	*Total*	S_3	S_4	*Total*	*Totals*
Orthex	98	52	150	86	122	208	358
Tonox-orthex	75	96	171	64	50	114	285
Steam processed . . .	34	7	41	34	26	60	101
Reclaim	32	19	51	7	−22	−15	36
ONV	138	113	251	53	108	161	412
40 black-orthex	102	74	176	204	223	427	603
						Grand Total	1,795

We compute the tread formula sum of squares by squaring each tread formula total, summing and dividing by the number of items for each tread formula, and then subtracting the correction factor. Thus, for the coded data of Table 30.12A, we have, the tread formula sum of squares equals:

$$\frac{(358)^2+(285)^2+(101)^2+(36)^2+(412)^2+(603)^2}{4}-\frac{(1{,}795)^2}{24}$$

$$=188{,}560-134{,}251=54{,}309$$

Then, for *each* tread formula, square the mix totals, add, divide by the number in each mix, and subtract the "correction factor" for that tread formula (= the total for the tread formula squared and divided by the number of measurements for that tread formula). Sum these results for all tread formulas to get the within-tread-formula sum of squares. For Table 30.12A this yields

$$\frac{(150)^2+(208)^2}{2}-\frac{(358)^2}{4}$$
$$+\frac{(171)^2+(114)^2}{2}-\frac{(285)^2}{4}$$
$$+\frac{(41)^2+(60)^2}{2}-\frac{(101)^2}{4}$$
$$+\frac{(51)^2+(-15)^2}{2}-\frac{(36)^2}{4}$$
$$+\frac{(251)^2+(161)^2}{2}-\frac{(412)^2}{4}$$
$$+\frac{(176)^2+(427)^2}{2}-\frac{(603)^2}{4}$$
$$=20{,}609$$

Finally we compute the total sum of squares by squaring each individual item, summing and subtracting the correction factor. This is

$$(98^2 + 75^2 + 34^2 + 32^2 + 138^2 + 102^2 + 52^2 + 96^2 + 7^2 + 19^2 + 113^2$$
$$74^2 + 86^2 + 64^2 + 34^2 + 7^2 + 53^2 + 204^2 + 122^2 + 50^2 + 26^2$$
$$+ (-22)^2 + 108^2 + 223^2) - 134{,}251 = 80{,}191.$$

The "between-samples within-mixes" sum of squares is computed as a residual. The results are summarized in Table 30.13.

We test the tread-formula variation by computing

$$s_T^2/s_M^2 = \frac{10{,}862}{3{,}435} = 3.2$$

Since $F_{0.05}$ for $\nu_1 = 5$, $\nu_2 = 6$ equals 4.39, variation between tread formulas is not significant. We test the variation between mixes within tread formulas by computing

$$s_M^2/s_e^2 = \frac{3{,}435}{439} = 7.8$$

Since $F_{0.05}$ for $\nu_1 = 6$ and $\nu_2 = 12$ is 3.0, variation between mixes within tread formulas is significant.

In numerous cases nested classifications occur with cross-classifications. For these more complicated models the reader is referred to H. Leon Harter and Mary D. Lum, *Partially Hierarchal Models in the Analysis of Variance* (Wright Air Development Center Technical Report 55–33).

Generally a nested factor is a random factor and in the study of nested models interest is likely to center upon the various components of variance. The analysis of components of variance is discussed in Section 3 of Chapter 31.

TABLE 30.13
An Analysis of Variance Table for a Pure Nested Classification

Source of Variation	*Sum of Squares*	*df*	*Mean Square*	*Variance Ratio*
Between tread formulas	54,309	5	$s_T^2 = 10{,}862$	3.2
Between mixes within tread formulas	20,609	6	$s_M^2 = 3{,}435$	7.8**
Between samples within mixes	5,273	12	$s_e^2 = 439$	
Total	80,191			

** Significant at the 0.01 level.

4. ASSUMPTIONS UNDERLYING ANALYSIS OF VARIANCE[13]

Before closing our discussion of analysis of variance, it is well to note what happens when the assumptions underlying the analysis of variance are not satisfied. The basic assumptions are as follows:

1. The various effects are additive,
2. The "experimental" (residual) errors must be independent of the main effects and interactions and of each other,
3. The "experimental" errors must have a common variance, and
4. The "experimental" errors must be randomly and normally distributed.

We shall consider these in reverse order.

From various studies that have been made, it has been found that variance-ratio F tests are not much affected by a moderate lack of normality and are therefore "robust tests." Usually the direction of error is mainly in announcing too many significant results. The OC curves are also moderately raised, and the test is less efficient in detecting differences.

Heterogeneity of error variance may cause trouble. The error or residual sum of squares is essentially derived from a pooling of the residual sum of squares for various subgroups. This is most easily seen in a one-way classification, where the error sum of squares is obtained by pooling the "within" sums of squares for the various groups. If in some groups experimental errors run less than in other groups, this pooling might cover significant differences that exist between the means or show significant differences between the means where they did not exist. If our groups bear the letters A, B, C, D, and E and the true model is one for which error variances for A, B, and C run about half of those for groups D and E, then it is possible that an ordinary analysis of variance that pools the group sums of squares might fail to detect differences in the means of A, B, and C that might be discovered if an F test were applied to them alone with the error variance being obtained solely from these groups. On the other hand, the pooling might so reduce the overall error sum of squares that an F test would show significance because the mean of E differed from the others. If D and E were compared by themselves, however, using the error sum of squares obtained from these two groups alone, the mean of the E group might not be found to be different from the mean of the D group. In such clear-cut cases as this, the difficulty caused by nonhomogeneity of error variances can be met easily by subdividing the problem into separate ones. In other cases, the difficulties created by heterogeneity of variances can be met by weighting the observations in proportion to their estimated individual variances. If the data follow the Poisson

[13] This section is based largely on W. G. Cochran's paper, "Some Consequences when the Assumptions for the Analysis of Variance Are Not Satisfied," *Biometrics* 3 No. 1 (1947), pp. 22–38.

distribution for which the variance equals the mean, a transformation may solve the difficulty.

Positive correlation among the residual errors may lead to underestimation of the true error variance, and negative correlation may lead to overestimation. This situation can be met by special adjustments.[14]

If the effects are nonadditive, say they are multiplicative instead of additive, that is A is 20 percent more than B, but they are treated as additive, the true nonadditivity effects (= interaction effects) become merged with the error term and increase the estimate of error variance which makes for a less precise test. The loss on this score is not likely to be great, however, unless the true error variance is relatively small.

5. ANALYSIS OF VARIANCE VERSUS $\bar{X}$- and R-CHARTS

It may be recognized by the reader that the analysis of variance technique described in this chapter and the last will serve as a substitute for an $\bar{X}$-chart for determining whether a set of past data came from a "controlled process."[15] In fact, many statisticians tend to prefer the analysis of variance for this problem.

The pros and cons of the argument *analysis of variance vs. control charts,* have been excellently summarized by C. C. Craig as follows:

> It is nevertheless true that the two procedures are not exactly equivalent. The analysis of variance is a summary method based on the overall variation among sample means while the control chart for averages is based on the behavior of individual $\bar{X}$'s. It can happen that no one $\bar{X}$ will fall outside the limit lines and the analysis of variance will disclose that too many of them are near the limit lines. On the other hand a single point on the $\bar{X}$ chart might be well outside the control lines and yet because the remaining points are relatively near the center line the total sum of squares among averages will not be large enough for the analysis of variance to indicate lack of control. But in general, the two methods are nearly enough equivalent that both will disclose any clear case of lack of control among averages and when one says control is good the other will too.
>
> Enough has been said already to make it evident that for process control by variables, as in the example considered, the use of control charts for R's or σ's and $\bar{X}$'s does possess some advantages over the use of analysis of variance. In particular, we have noted that the R- or σ-chart[16] provides a means of judging whether or not the within sample variation is controlled so that one has justification for proceeding to estimate σ' from the data. But in using the analysis of variance, one in effect assumes that this control exists and by this method this

---

[14] Ibid., pp. 32–35.

[15] See Chapters 18 and 21.

[16] By definition Craig's σ equals the RMSD. His σ-chart is the equivalent of the s-chart discussed in Chapter 21, Section 9.2.

assumption is not checked.[17] This is important but in addition we must also point out other merits of the control chart system.

First as compared to the analysis of variance, it is both more readily understood and used. Second, and this is very important, the control chart is a visual device. Even people who have little grasp of the underlying principles can read in most cases the story told by $\bar{X}$- and R-charts. Third, the control chart is better adapted for maintaining a running check on the performance of a process which is, of course, almost essential to effective process control. Fourth, because we have limit lines drawn on the $\bar{X}$-chart, one identifies at once the out of control $\bar{X}$'s. After an analysis of variance reveals too much fluctuation among sample means, one still has to examine further the individual $\bar{X}$'s to try to determine at which point trouble made itself evident. Finally any experienced user of control charts reads from them a great deal more than that points are within limit lines or not. He gets information worth money to his shop that would be wholly or partly concealed in the purely tabular presentation of the results of an analysis of variance.

It may begin to appear that there is not much left that can be put on the credit side for the analysis of variance. But the latter has its points, too. A paramount feature is the fact that the analysis of variance is designed to get the maximum of information from a limited amount of data. If observations are costly, difficult, or time-consuming, then it becomes a prime consideration to use the most efficient statistical tools. There are two chief aspects of the analysis of variance which make it superior in such cases. Granting, as in many situations it can be, that the within sample variation is controlled, then the method of dividing the total sum of squares of deviations about the mean within samples by the number of degrees of freedom gives a better; that is, on the average closer, estimate of σ'^2, than the one obtained from $\bar{R}$ or $\bar{\sigma}$. This is less important in process control because it is usually easy to get plenty of observations. But of much more importance is the fact that our example has employed only the very simplest form of the analysis of variance. This was the proper form for data of that kind. We wished only to compare the variability due to a time effect with inherent process variability.

But suppose we were concerned with a dimension on machined parts which are coming off of each of the four spindles of an automatic screw machine. If it were suspected that there were spindle differences then control chart procedure is to keep a chart on each spindle. But if it could be assumed that each spindle has approximately the same uniformity, then a single analysis of variance can test for assignable causes due only to: (1) differences among the centers to which each spindle is working, and (2) to variation among samples taken at different times. The samples would consist of one or perhaps two pieces taken at the same time from each spindle and the method enables one to eliminate the time factor while studying the spindle effect and also to eliminate the spindle factor while looking for effects due to differences in time. The analysis of variance in this case would be a three-component breakdown of the total variation instead of a two-component one. The agricultural experimentalist, to illustrate a more complicated use of the same general method, gets yields for different plant varieties

[17] [*Author's Comment:*] Bartlett's or a similar test can and should be used to test the homogeneity of the variances. See Chapter 31, Section 2.

under different fertilizer treatments grown in different kinds of soil. If the variability of yields for any constant set of these three factors can be assumed very nearly the same, then he can study the effect of any one of these factors or of any two of them in combination for constant values of the others without having to perform the almost impossible number of separate experiments required if each one or two factors were held experimentally constant.

It may be gathered that the analysis can have important industrial application, especially in laboratory work. But for the situations in process control for which $\bar{X}$ and R or σ charts have been advocated by Shewhart and other competent authorities, these charts as tools for the job more than hold their own in comparison with the analysis of variance.[18]

6. PROBLEMS

30.1. The tensile strength of a rubber vulcanizate varies as follows. All data are in pounds per square inch.

Cure at 14° C (Minutes)	Accelerator		
	A	*B*	*C*
40	3,900	4,300	3,700
	3,600	3,700	4,100
60	4,100	4,200	3,900
	3,500	3,900	4,000
80	4,000	4,300	3,600
	3,800	3,600	3,800

a. Is there a significant interaction between accelerator and type of curing?

b. Is the "main" or "average" effect of the accelerators significant?

c. Is the "main" or "average" effect of the curing significant?

d. Estimate the experimental error variance.

e. Sketch the OC curve for (1) the test used in *a* and (2) the test used in *c.*

f. Compare your results with those of Problem 29.5.

30.2. The flexural strength of sheet castings of a polymer shows the following variations.[19] Data are in pounds per square inch.

---

[18] Quoted from Cecil C. Craig, "Control Charts versus the Analysis of Variance in Process Control by Variables," *Industrial Quality Control,* January 1947, pp. 14–16.

[19] Cf. W. L. Gore, "Statistical Techniques for Research and Development," *Conference Papers, First Annual Convention, American Society for Quality Control and Second Midwest Quality Control Conference, June 5 and 6, 1947* (Chicago: John L. Swift Co., Inc., 1947), p. 150.

Temperature of Bath	Time in Polymerization Bath with 1 Percent Catalyst	
	20 Minutes	*60 Minutes*
100°	9,500	11,500
	10,650	11,650
	9,700	11,250
	9,950	11,250
	10,100	11,900
120°	11,300	10,900
	11,750	11,500
	11,600	11,850
	11,650	11,700
	11,700	11,650

a. Is there a significant interaction between time of polymerization and temperature?

b. Is the "main" or "average" effect of the time of polymerization significant?

c. Is the "main" or "average" effect of the temperature significant?

d. Estimate the experimental error variance.

e. Sketch the OC curve for (1) the test used in *a* and (2) the test used in *b.*

30.3. The flexural strength of sheet castings of a polymer shows the following variations.[20] Data are in pounds per square inch.

a. Is there a significant interaction between time of polymerization and temperature?

Temperature of Bath	Time in Polymerization Bath with 2 Percent Catalyst	
	20 Minutes	*60 Minutes*
100°	11,800	11,900
	11,750	11,850
	11,800	11,850
	11,950	12,000
	11,900	12,100
120°	10,550	9,900
	11,000	10,150
	11,100	9,400
	11,350	9,800
	11,200	9,900

b. Is the "main" or "average" effect of the time of polymerization significant?

c. Is the "main" or "average" effect of the temperature significant?

[20] Ibid.

d. Estimate the experimental error variance.

e. Sketch the OC curve for (1) the test used in *a* and (2) the test used in *b.*

30.4. A study[21] was undertaken of the factors affecting variation in the lengths of steel bars. The lengths of four bars from each of four screw machines and two heat treatments were measured with the following results. (Data pertain to measurements less 4.380 times 1,000.)

Machine	Heat Treatment W	Heat Treatment L
A	6	4
	9	6
	1	0
	3	1
B	7	6
	9	5
	5	3
	5	4
C	1	−1
	2	0
	0	0
	4	1
D	6	4
	6	5
	7	5
	3	4

a. Is there a significant interaction between the four given machines and heat treatments?

b. Does heat treatment have a significant "main" or "average" effect over the four given machines?

c. Do the four given machines have a significant "main" or "average" effect?

d. Estimate the experimental error variance.

e. Sketch the OC curve (1) for the test used in *a* and (2) the test used in *c.*

f. View the four machines as a random sample of screw machines and determine (1) whether the heat treatments W and L have a "main" or "average" effect over all machines; (2) whether screw machines in general affect the length of steel bars; and (3) whether in general there is any interaction between screw machines and heat treatments. In which respects if any do your tests differ from those employed in *a, b,* and *c?*

g. Sketch the OC curve for the test of the machine effects in *f.*

30.5. Experimental work on rocket nozzles was carried out on a special testing facility which simulated the temperature environment the nozzle would actually

---

[21] Cf. William D. Baten, "An Analysis of Variance Applied to Screw Machines," *Industrial Quality Control,* April 1956, p. 8.

encounter in flight. Tests were conducted on specimens two inches in diameter. Percentage weight loss of the test specimen was selected as the yardstick of performance, a lower value indicating an improvement. The effects of four resin additives were investigated, the resin with no additive being retained as a "control." Each material was tested twice in succession on each of three different days, the order of testing the different materials being randomized over each day. The results were as follows:[22]

Type of Resin	*Day I*		*Day II*		*Day III*	
With additive A	Test 1	5.1	Test 11	7.4	Test 21	8.6
	Test 2	4.9	Test 12	7.9	Test 22	8.1
With additive B	Test 3	8.0	Test 13	10.2	Test 23	8.0
	Test 4	8.1	Test 14	10.5	Test 24	7.7
With additive C	Test 5	6.4	Test 15	5.5	Test 25	4.0
	Test 6	6.4	Test 16	5.1	Test 26	4.6
With additive D	Test 7	5.7	Test 17	5.5	Test 27	4.0
	Test 8	6.2	Test 18	4.7	Test 28	3.9
With no additive	Test 9	5.5	Test 19	5.8	Test 29	5.7
	Test 10	5.7	Test 20	5.7	Test 30	6.0

a. Is there a significant interaction between type of resin and days of experimentation?

b. Is there a significant difference between the various types of resin?

c. Is there a significant difference between days?

d. Indicate clearly what assumptions you are making about varibility in your answers to questions *a, b,* and *c.*

e. Estimate the test variance.

30.6. The following are yields of a certain chemical under varying conditions.[23] The results are in pounds.

	Concentration of an Inert Solvent		
Temperature	*40%*	*50%*	*60%*
Catalyst A:			
50°	45.1	45.7	44.9
60°	44.8	45.8	44.7
Catalyst B:			
50°	33.0	45.7	53.8
60°	32.6	45.5	54.2

[22] Adapted from S. Roy Wood, "Analysis of a Rocket Nozzle Ablation Experiment," American Society for Quality Control, *Annual Convention Transactions,* 1963, pp. 329–32. The data for even number tests are fictitious.

[23] Cf. Hugh M. Smallwood, "Design of Experiments in Industrial Research," *Industrial and Engineering Chemistry—Analytical Chemistry* 19 (1947), p. 952.

a. Is there any interaction between concentration of solvent and catalyst?

b. Is there any interaction between concentration of solvent and temperature?

c. Is there any interaction between temperature and catalyst?

d. Does the concentration of solvent have any "main" or "average" effect on the yield?

e. Is the linear component of concentration significant? The quadratic component?

f. Does the catalyst have any "main" or "average" effect on the yield?

g. Does the temperature have any "main" or "average" effect on the yield?

h. Estimate the variance of the residual variation.

i. Compare your results with those of Problem 29.7.

j. Sketch the OC curve for (1) the test used in *a* and (2) the test used in *d*.

30.7 *a.* Combine the data of Problems 30.2 and 30.3 and determine whether the catalyst has a significant "main" or "average" effect.

b. Is there a significant interaction between catalyst and temperature?

c. Is there a significant interaction between catalyst and time in polymerization bath?

d. Is there a significant three-factor interaction?

e. Estimate the variance of the residual variation over the whole set of data.

30.8. A study was made of the variations involved in sampling and testing bags of fertilizer. The manufacturing company was interested in comparing variations between mixes, between bags within mixes, between samples from the same bag, and between chemical determinations for the same sample. The results of the experiment were as follows. (Measurements pertain to the nitrogen content of the fertilizer and are in percentages.)

Mix A			
Bag 1		*Bag 2*	
Sample 1	*Sample 2*	*Sample 3*	*Sample 4*
$D_1 = 7.26$ $D_2 = 7.28$	$D_3 = 7.29$ $D_4 = 7.31$	$D_5 = 7.30$ $D_6 = 7.29$	$D_7 = 7.32$ $D_8 = 7.30$

Mix B			
Bag 3		*Bag 4*	
Sample 5	*Sample 6*	*Sample 7*	*Sample 8*
$D_9 = 7.38$ $D_{10} = 7.35$	$D_{11} = 7.40$ $D_{12} = 7.37$	$D_{13} = 7.34$ $D_{14} = 7.34$	$D_{15} = 7.31$ $D_{16} = 7.34$

a. Is there a significant variation between samples from the same bag?

b. Is there a significant variation between bags from the same mix?

c. Is there a significant variation between mixes?

d. Estimate the variance of chemical determinations of percent nitrogen in the given kind of fertilizer.

30.9. A given industrial product[24] comes in lots containing many boxes, each with a large number of items. A sample of 2 boxes is taken from a lot and a sample of 3 items from each box. Two tests are made on each item selected. For two lots we have the following results:

Lot I					
Box A			*Box B*		
Item	*Test 1*	*Test 2*	*Item*	*Test 1*	*Test 2*
#1	15.5	15.0	#4	12.8	13.3
#2	17.0	17.0	#5	13.0	12.9
#3	13.0	12.9	#6	16.0	16.1

Lot II					
Box C			*Box D*		
Item	*Test 1*	*Test 2*	*Item*	*Test 1*	*Test 2*
#7	12.5	12.7	#10	10.4	10.2
#8	10.5	10.8	#11	10.6	10.2
#9	11.5	11.3	#12	12.0	11.9

a. Is there a significant difference between items from the same box?

b. Is there a significant difference between boxes from the same lot?

c. Is there a significant difference between lots?

d. Estimate the test variance.

e. Indicate clearly what assumptions you are making about variability in your answers to questions *a* to *d.*

30.10. In a textile mill, data were collected on a quality characteristic known as the "yarn number." This is the number of 840-yard lengths of yarn per pound weight. For three weeks four specimens were taken from each of four different machines. The test results on the 48 specimens are given on the following page.

[24] Adapted from Gerald J. Desmond, "Sampling Frozen Beef Using a Variable Sampling Plan for Percent Fat," American Society for Quality Control, *Annual Convention Transactions,* 1960, pp. 273–76. Data for Test 2 are fictitious.

Yarn Numbers for 48 Specimens of Yarn.*

WEEK 1

Machine 43		*Machine 27*		*Machine 5*		*Machine 16*	
Specimen	*Test Result*	*Specimen*	*Test Result*	*Specimen*	*Test Result*	*Specimen*	*Test Result*
#1	55.5	#5	51.8	#9	56.0	#13	52.0
#2	50.2	#6	54.2	#10	57.1	#14	56.1
#3	54.8	#7	50.1	#11	55.2	#15	54.1
#4	52.1	#8	54.1	#12	51.1	#16	53.0

WEEK 2

Machine 8		*Machine 39*		*Machine 7*		*Machine 12*	
Specimen	*Test Result*	*Specimen*	*Test Result*	*Specimen*	*Test Result*	*Specimen*	*Test Result*
#17	50.3	#21	59.9	#25	55.5	#29	58.8
#18	58.8	#22	58.2	#26	50.2	#30	56.1
#19	54.0	#23	56.3	#27	51.6	#31	59.2
#20	53.8	#24	57.0	#28	54.7	#32	57.1

WEEK 3

Machine 21		*Machine 29*		*Machine 41*		*Machine 32*	
Specimen	*Test Result*	*Specimen*	*Test Result*	*Specimen*	*Test Result*	*Specimen*	*Test Result*
#33	50.4	#37	55.4	#41	58.8	#45	58.6
#34	52.1	#38	56.3	#42	57.3	#46	59.2
#35	51.6	#39	56.1	#43	56.8	#47	55.7
#36	53.5	#40	58.1	#44	53.9	#48	57.1

* Taken from Norbert L. Enrick, "Variation Flow Analysis for Process Improvement," American Society for Quality Control, *Annual Convention Transactions,* 1961, pp. 113–25.

a. Is there a significant variation between machines within weeks?
b. Is there a significant variation from week to week?
c. Estimate the variance between specimens from the same machine.
d. Indicate clearly what assumptions you are making about variability in your answers to questions *a* to *c*.

7. SELECTED REFERENCES*

(For other references on "Analysis of Variance," see Chapter 29.)

Bennett and Franklin (B '54. This discusses the case of random factors from small finite universes.), Bishop and Dudewicz (P '78 and P '81, the latter contains the mathematics for the former), Brown (P '75), Brownlee (B '48 and B '60), Cochran (P '47 and P '51. P '47 should be read by all who actually apply analysis of variance.), Cochran and Cox (B '57), Eisenhart (P '47. A review of the assumptions underlying analysis of variance and analysis of components of variance.), Gaylor and Hopper (P '69), Goulden (B '39), Harter and Lum (B '55. An excellent compendium of analysis of variance models. May be obtained from Office of Technical Services, U.S. Department of Commerce, Washington, D.C.), Hotelling (P '31, the original article on the Hotelling T^2 test.), Hsu (P '38), Hudson and Krutchkoff (P '68), Huntsberger (P '55), Johnson (B '49. Illustrations in the field of education.), Kempthorne (B '52), Kendall (B '48), Leone and Nelson (P '66), Leone, Nelson, Johnson, and Eisenstat (P '68), Mann (B '49. A comprehensive mathematical discussion of analysis of variance), Patnaik (P '49), Paull (P '50), Satterthwaite (P '41 and P '46), Scheffé (P '56 and B '59), Snee (P '83), Wilks (B '62), and Yates (P '34).

---

* B and P refer to the Book and Periodical sections, respectively, of the Cumulative List of References in Appendix V.

31

Analyses Supplementary to and Associated with Analysis of Variance

1. ADDITIONAL ANALYSES PERTAINING TO MEANS

1.1. Individual Allowances for Fixed Effects

1.1.1. Confidence Intervals for Individual Means. In evaluating the means of several independent samples, we often want to determine confidence intervals for one or more of the separate means. This is frequently undertaken supplementary to analysis of variance. It applies to fixed effects; for random effects we seek components of variance (Section 3).

In computing a confidence interval for a single mean in a set of several means, we usually estimate the residual variance from the whole set of data rather than from the individual items from which the mean in question is computed. This assumes, of course, that the hypothesis of homogeneity of variance is accepted. For the latter the reader is referred to Section 2 below.

1.1.1.1. One Way Classification. The 0.95 confidence limits for the mean of a given class in a one-way classification with c classes and r individuals per class are given by[1]

(31.1)
$$\bar{X}_j \pm t_{0.025}\,[\text{for } \nu = c(r-1)]s/\sqrt{r}$$

where s^2 is the error variance defined by the formula $s^2 = \dfrac{\sum\limits_i \sum\limits_j (X_{ij} - \bar{X}_j)^2}{c(r-1)}$.

For example, with reference to the thermometer problem of Chapter 29 and in particular with reference to Table 29.1, 0.95 confidence limits for the mean of thermometer A are

[1] It will be recalled that $\sigma_{\bar{X}}'^2 = \sigma_X'^2/n$. See Chapter 6, Section 1, Theorem 6.2.

$$173.5 \pm 2.306\sqrt{\frac{0.81}{3}} = \begin{cases} 174.7 \\ 172.3 \end{cases}$$

It will be recognized that in terms of the model equation, $X_{ij} = \mu + \theta_j + \epsilon_{ijk}$, we are getting confidence limits for $\mu + \theta_j$.

1.1.1.2. Two-Way Classification, One Case in Each Class For a two-way classification with r rows and c columns, and one item in each class, 0.95 confidence limits for the mean of a row are given by

(31.2) $$\bar{X}_i \pm t_{0.025}\,[\text{for } \nu = (r-1)(c-1)]s/\sqrt{c}$$

where s^2 is the residual variance and is given by the formula $s^2 = \sum_i \sum_j (X_{ij} - \bar{X}_i - \bar{X}_j + \bar{X})^2/(r-1)(c-1)$. The same formula will apply to the mean of a column $(\bar{X}_j)$, except that $\sqrt{r}$ will replace $\sqrt{c}$. In the first case we are getting confidence limits for $\mu + \tau_i$, and in the second case for $\mu + \theta_j$. For example, for the data of Table 29.8, 0.95 confidence limits for the mean of the readings of analyst I are

$$173.0 \pm 2.447\sqrt{\frac{0.39}{4}} = \begin{cases} 173.76 \\ 172.24 \end{cases}$$

(Note: df of t always equals those of the s used to set limits.)

1.1.1.3. Two Way Classification, More Than One Case in Each Class. For a two-way classification with r rows, c columns, and g elements in a class, 0.95 confidence limits for the mean of a row are given by

(31.3) $$\bar{X}_i \pm t_{0.025}\,[\text{for } \nu = rc(g-1)]s/\sqrt{cg}$$

where s^2 is the error variance and is given by the formula $s^2 = \dfrac{\sum_i \sum_j \sum_k (X_{ijk} - \bar{X}_{ij})^2}{rc(g-1)}$. The same formula will apply to the mean of a column $(\bar{X}_j)$ with $\sqrt{rg}$ replacing $\sqrt{cg}$ and to the mean of a class $(\bar{X}_{ij})$ with $\sqrt{g}$ replacing $\sqrt{cg}$. In the first case we are getting confidence limits for $\mu + \tau_i$, in the second case for $\mu + \theta_j$, and in the third case for $\mu + \theta_j + \tau_i + \psi_{ij}$. For example, for Table 30.1, 0.95 confidence limits for the mean of the readings of analyst I made with thermometer B are

$$173.25 \pm 2.179\sqrt{\frac{0.22}{2}} = \begin{cases} 173.97 \\ 172.53 \end{cases}$$

(Note: Denominator of radical equals number of items on which mean is based.)

1.1.1.4. Three-Way Classification. For a three-way classification with r rows, c columns, and g groups and a single case in each three-way class, 0.95 confidence intervals may be determined as follows. For the different notation see Chapter 30, Section 2.2.

For the mean of a row, i.e., for $\mu + a_i^A$, 0.95 confidence limits are given by

(31.4a) $\bar{X}_i \pm t_{0.025}$ [for $\nu = (r-1)(c-1)(g-1)$]$s/\sqrt{cg}$

For the mean of a column, i.e., for $\mu + a_j^B$, 0.95 confidence limits are given by

(31.4b) $\bar{X}_j \pm t_{0.025}$ [for $\nu = (r-1)(c-1)(g-1)$]$s/\sqrt{rg}$

For the mean of a group, i.e, for $\mu + a_k^C$, 0.95 confidence limits are given by

(31.4c) $\bar{X}_k \pm t_{0.025}$ [for $\nu = (r-1)(c-1)(g-1)$]$s/\sqrt{rc}$

For the mean of a two-way class, e.g., for $\mu + a_i^A + a_j^B + a_{ij}^{AB}$, 0.95 confidence limits are given by

(31.5a) $\bar{X}_{ij} \pm t_{0.025}$ [for $\nu = (r-1)(c-1)(g-1)$]$s/\sqrt{g}$

For $\mu + a_j^A + a_k^C + a_{ik}^{AC}$, 0.95 confidence limits are given by

(31.5b) $\bar{X}_{ik} \pm t_{0.025}$ [for $\nu = (r-1)(c-1)(g-1)$]$s/\sqrt{c}$

For $\mu + a_j^B + a_k^C + a_{jk}^{BC}$, 0.95 confidence limits are given by

(31.5c) $\bar{X}_{jk} \pm t_{0.025}$ [for $\nu = (r-1)(c-1)(g-1)$]$s/\sqrt{r}$

For the mean of any three-way class, i.e., for $\mu + a_i^A + a_j^B + a_k^C + a_{ij}^{AB} + a_{ik}^{AC} + a_{jk}^{BC}$, 0.95 confidence limits are given by

(31.6) $\bar{X}_{ij} + \bar{X}_{ik} + \bar{X}_{jk} - \bar{X}_i - \bar{X}_j - \bar{X}_k + \bar{X} \pm t_{0.025}$
[for $\nu = (r-1)(c-1)(g-1)$]h

where $h = [(rc + rg + cg - r - c - g + 1)/rcg]^{1/2}s$

In equations (31.4a) to (31.6), s^2 is the residual variance given by

$$s^2 = \frac{\sum_i \sum_j \sum_k (X_{ijk} - \bar{X}_{ij} - \bar{X}_{ik} - \bar{X}_{jk} + \bar{X}_i + \bar{X}_j + \bar{X}_k - \bar{X})^2}{(r-1)(c-1)(g-1)}$$

To illustrate, we shall compute 0.95 confidence limits for the mean determination of the melting point of hydroquinone by analyst I using thermometer B. (See Table 30.1 and Table 30.10.) These are

$$173.25 \pm 2.477\sqrt{0.05/2} = \begin{cases} 173.64 \\ 172.86 \end{cases}$$

1.1.1.5. Pure Nested Classification. With pure nest classifications each classification is nested in that next higher up. Generally the nested classifications are those for random variables. In such cases, the mean square (i.e., the s^2) for a given class is an unbiased estimate of the universe variance for elements of that class. For example, consider the model

(31.7) $$X_{ijk} = \mu + \alpha_i + \gamma_{ij} + \delta_{ijk} + \epsilon_{ijkl} \begin{cases} i = 1, 2 \ldots n_\alpha \\ j = 1, 2 \ldots n_\gamma \\ k = 1, 2 \ldots n_\delta \\ l = 1, 2 \ldots n_\epsilon \end{cases}$$

where γ is nested in α, δ in γ, and ϵ in δ, and γ, δ, and ϵ are all independently distributed random variables with zero means and variances $\sigma_{\gamma}'^2$, $\sigma_{\delta}'^2$, and $\sigma_{\epsilon}'^2$, respectively. If we define[2]

(31.8)

$$s_{\gamma}^2 = \frac{n_{\delta}n_{\epsilon}\sum_i\sum_j(\bar{X}_{ij..}-\bar{X}_{i...})^2}{n_{\alpha}(n_{\gamma}-1)}$$

$$s_{\delta}^2 = \frac{n_{\epsilon}\sum_i\sum_j\sum_k(\bar{X}_{ijk.}-\bar{X}_{ij..})^2}{n_{\alpha}n_{\gamma}(n_{\delta}-1)}$$

$$s_{\epsilon}^2 = \frac{\sum_i\sum_k\sum_k\sum_l(X_{ijkl}-\bar{X}_{ijk.})^2}{n_{\alpha}n_{\gamma}n_{\delta}(n_{\epsilon}-1)}$$

where the dots in the subscripts denote the elements over which means are taken, then

(31.9)

$s_{\lambda}^2/n_{\gamma}n_{\delta}n_{\epsilon}$ is an unbiased estimate of the variance of $\bar{X}_{i}\ldots$

$s_{\delta}^2/n_{\delta}n_{\epsilon}$ is an unbiased estimate of the variance of $\bar{X}_{ij}\ldots$

$s_{\epsilon}^2/n_{\epsilon}$ is an unbiased estimate of the variance of $\bar{X}_{ijk}\ .$

A proof of the second statement of (31.9) is as follows. Consider the variance of $\bar{X}_{ij...}$ From the model we have[3]

$$\bar{X}_{ij..} = \frac{n_{\delta}n_{\epsilon}\mu + n_{\delta}n_{\epsilon}\alpha_i + n_{\delta}n_{\epsilon}\gamma_{ij} + n_{\epsilon}\sum_k\delta_{ijk} + \sum_k\sum_l\epsilon_{ijkl}}{n_{\delta}n_{\epsilon}}$$

and since the variables are independent, we have[4]

$$\sigma_{\bar{X}ij..}'^2 = \frac{n_{\epsilon}^2n_{\delta}\sigma_{\delta}'^2 + n_{\epsilon}n_{\delta}\sigma_{\epsilon}'^2}{n_{\delta}^2n_{\epsilon}^2} = \frac{\sigma_{\delta}'^2}{n_{\delta}} + \frac{\sigma_{\epsilon}'^2}{n_{\delta}n_{\epsilon}}$$

But[5]

$$E(s_{\delta}^2) = \sigma_{\epsilon}'^2 + n_{\epsilon}\sigma_{\delta}'^2$$

and

$$E(s_{\epsilon}^2) = \sigma_{\epsilon}'^2$$

so that an unbiased estimate of $\sigma_{\delta}'^2$ is given by $\dfrac{s_{\delta}^2 - s_{\epsilon}^2}{n_{\epsilon}}$. Hence we have an unbiased estimate of $\sigma_{\bar{X}ij..}'^2$ given by $\dfrac{s_{\epsilon}^2}{n_{\delta}n_{\epsilon}} + \dfrac{s_{\delta}^2 - s_{\epsilon}^2}{n_{\delta}n_{\epsilon}} = \dfrac{s_{\delta}^2}{n_{\delta}n_{\epsilon}}$. The other statements of (31.9) may be proved in the same way.

---

[2] Compare Chapter 30, Section 3.

[3] $\bar{X}_{ij..}$ is the mean, it will be recalled, of $n_{\delta}n_{\epsilon}$ cases all of which have the same ij subscripts. Only the δ and ϵ components vary.

[4] See Chapter 4, Section 10.1, Corollary of Theorem 4.2.

[5] See Appendix I (33). Also see Section 3 below.

From the above it may be concluded that 0.95 confidence limits for the various means will be given as follows:

For $\bar{X}'_{i...}$, confidence limits are given by

$$\bar{X}_{i...} \pm t_{0.025} \,[\text{for } \nu = n_\alpha(n_\gamma - 1)] \frac{s_\gamma}{\sqrt{n_\gamma n_\delta n_\epsilon}} \tag{31.10a}$$

For $\bar{X}'_{ij..}$, 0.95 confidence limits are given by

$$\bar{X}_{ij..} \pm t_{0.025} \,[\text{for } \nu = n_\alpha n_\gamma(n_\delta - 1)] \frac{s_\delta}{\sqrt{n_\delta n_\epsilon}} \tag{31.10b}$$

For $\bar{X}'_{ijk.}$, 0.95 confidence limits are given by

$$\bar{X}_{ijk.} \pm t_{0.025} \,[\text{for } \nu = n_\alpha n_\gamma n_\delta(n_\epsilon - 1)] \frac{s_\epsilon}{\sqrt{n_\epsilon}} \tag{31.10c}$$

1.1.2. Confidence Intervals for "Differentials." ***1.1.2.1. One-Way Classification.*** Sometimes we may be interested in estimating the "differential" or "bias" of a given class. Thus, in a one-way classification, we may be interested in estimating θ_j instead of $\mu + \theta_j$. This may be done by computing the difference between the mean of that class and the grand mean, i.e., differential or bias equals $\bar{X}_j - \bar{X}$. For example, with reference to the data of Table 29.1A, the "bias" of thermometer A may be estimated to be 1.50° − 0.58° = 0.92° and the biases of the other thermometers at 0.67° − 0.58° = 0.09°, −0.17° − 0.58° = −0.85°, and 0.33° − 0.58° = −0.25°, respectively. These are estimates of the universe "differentials" θ_1, θ_2, θ_3, and θ_4.

For a one-way classification 0.95 confidence limits for the universe differentials or biases (θ_j) are given by[6]

$$\bar{X}_j - \bar{X} \pm t_{0.025} \,[\text{for } \nu = c(r-1)] s \sqrt{\frac{c-1}{rc}} \tag{31.11}$$

where $s^2 = \sum_i \sum_j (X_{ij} - \bar{X}_j)^2/c(r-1)$. Thus we have 0.95 confidence limits for the bias of thermometer A equal to $0.92 \pm 2.306(0.90)\sqrt{3/12}$ $= \begin{cases} 1.94. \\ -0.12. \end{cases}$

1.1.2.2. Two-Way Classification, a Single Case in Each Class. For this model, row differentials are given by $\bar{X}_i - \bar{X}$ and column differentials by $\bar{X}_j - \bar{X}$. In each case, 0.95 confidence limits are as follows:

For τ_i, 0.95 confidence limits are given by

---

[6] Note that $\bar{X}_1 - \bar{X} = \dfrac{(c-1)\bar{X}_1 - \bar{X}_2 - \ldots - \bar{X}_c}{c}$ and by the corollary of Theorem 4.2 (See Chapter 4, Section 10.1.), $\sigma'^2_{\bar{X}_1 - \bar{X}} = \dfrac{[(c-1)^2 + (c-1)]}{c^2} \sigma'^2_{\bar{X}j} = \dfrac{c-1}{c} \sigma'^2_{\bar{X}j} = \dfrac{c-1}{rc} \sigma'^2$.

(31.12a) $$\bar{X}_i - \bar{X} \pm t_{0.025}\,[\text{for } \nu = (r-1)(c-1)]s\sqrt{\frac{r-1}{rc}}$$

For θ_j, 0.95 confidence limits are given by

(31.12b) $$\bar{X}_j - \bar{X} \pm t_{0.025}\,[\text{for } \nu = (r-1)(c-1)]s\sqrt{\frac{c-1}{rc}}$$

In each case, s^2 is the residual variance and is given by

$$s^2 = \sum_i \sum_j (X_{ij} - \bar{X}_i - \bar{X}_j + \bar{X})^2/(r-1)(c-1)$$

1.1.2.3. Two-Way Classification, More Than One Case in a Class. Here, row differentials are given by $\bar{X}_i - \bar{X}$, column differentials by $\bar{X}_j - \bar{X}$, and interaction differentials by $\bar{X}_{ij} - \bar{X}_j - \bar{X}_i + \bar{X}$. For this case, 0.95 confidence limits are as follows:

For τ_i, 0.95 confidence limits are given by

(31.13a) $$\bar{X}_i - \bar{X} \pm t_{0.025}\,[\text{for } \nu = rc(g-1)]s\sqrt{\frac{r-1}{rcg}}$$

For θ_j, 0.95 confidence limits are given by

(31.13b) $$\bar{X}_j - \bar{X} \pm t_{0.025}\,[\text{for } \nu = rc(g-1)]s\sqrt{\frac{c-1}{rcg}}$$

For ψ_{jj}, 0.95 confidence limits are given by

(31.13c) $$\bar{X}_{ij} - \bar{X}_i - \bar{X}_j + \bar{X} \pm t_{0.025}\,[\text{for } \nu = rc(g-1)]s\sqrt{\frac{(r-1)(c-1)}{rcg}}$$

In each case s^2 is the error variance and is given by

$$s^2 = \sum_i \sum_j \sum_k (X_{ijk} - \bar{X}_{ij})^2/rc(g-1).$$

1.1.3. Error Rates. It should be noted that in setting up 0.95 confidence intervals for individual means or biases, as described above, the error rate is 0.05 *per determination.* In other words, in 100 such determinations, whether related to the same experiment or not, we shall on the average make 5 false statements about the universe parameters. The error rate is on a determination basis, not an experiment basis.

It should also be noted that these confidence limits can be set up without preliminary F tests and are valid whether the F tests show significance or not.

1.2. Allowances for Differences between Pairs of Means

In problems involving classification and cross-classification of data we may want to establish limits for differences between various pairs of means. In doing this we must be very careful in our statement of the risks involved in regard to the conclusions we draw. We have no right, for example, to pick the largest sample mean and the smallest sample mean of a set and lay off the usual 0.95 confidence limits for the difference between the means of these classes *expecting the risk of error to be 0.05 for comparisons of this special type,* i.e., selection of the highest and lowest sample means. For the risk of an incorrect statement will obviously be much higher than 0.05 since the particular comparison on each occasion is selected after seeing the data. If the two classes to be compared, however, were selected prior to seeing the data, then the standard procedure would give the correct risks. Or if we ran all comparisons, including the two extreme pairs, the risk would be 0.05 of a wrong statement on a per comparison basis. In other words, the use of standard procedures for setting up 0.95 confidence limits for the difference between two means selected a priori or between all possible pairs of means will lead to 5 misstatements *per hundred such comparisons.*

It is also to be noted that these confidence limits can be properly established subject to the stated risks regardless of whether an F test indicates a significant difference or not. Conversely, if 0.95 confidence limits for some comparisons do not include zero (indicating significant differences for those pairs), it does not follow that an overall F test will or should show a significant result at the 0.05 level. For a 0.05 F test is run on the basis of drawing wrong conclusions *in 5 out of 100 experiments.*

If we wish a procedure for establishing confidence limits for differences between means that will lead to misstatements in only 5 out of 100 experiments, we can resort to the studentized range, q. If s, the estimate of the standard deviation to be used in setting up confidence limits, is based on ν degrees of freedom and is independent of the class means to be compared, and if these class means are each computed from m cases and selected from a group of g means, then the 0.05 allowance for any comparison on an *experiment* basis will be $\pm q_{0.05}(g, \nu)s/\sqrt{m}$. That is we can say the limits

$$\bar{X}_1 - \bar{X}_2 \pm q_{0.05}(g, \nu)s/\sqrt{m} \tag{31.14}$$

will include the true universe difference and be wrong in no more than 5 *experiments* in 100.

Proof of the foregoing is as follows. Set up the deviations of the sample means $(\bar{X})$ from the universe means $(\bar{X}')$. Number in order of size; thus,

$$(\bar{X}_i - \bar{X}_i')_L \quad . \; . \; . \quad (\bar{X}_j - \bar{X}_j') \quad . \; . \; . \quad (\bar{X}_k - \bar{X}_k')_H$$

The quantity $\dfrac{(\bar{X}_k - \bar{X}_k')_H - (\bar{X}_i - \bar{X}_i')_L}{s/\sqrt{m}}$ will have a studentized range distribution, so that the probability is 0.95 that

$$(\bar{X}_k - \bar{X}_k')_H - (\bar{X}_i - \bar{X}_i')_L \leq q_{0.05}\frac{s}{\sqrt{m}}$$

The difference between any two other deviations will be less than the difference between the extremes. Thus,

$$\text{Probability } [|(\bar{X}_1 - \bar{X}'_1) - (\bar{X}_2 - \bar{X}'_2)| \leq q_{0.05}s/\sqrt{m}]$$

is equal to or greater than 0.95. In other words, the probability is at least 0.95 that

$$-q_{0.05}\frac{s}{\sqrt{m}} \leq (\bar{X}_1 - \bar{X}'_1) - (\bar{X}_2 - \bar{X}'_2) \leq +q_{0.05}\frac{s}{\sqrt{m}}$$

or that

$$(\bar{X}_1 - \bar{X}_2) - q_{0.05}\frac{s}{\sqrt{m}} \leq (\bar{X}'_1 - \bar{X}'_2) \leq (\bar{X}_1 - \bar{X}_2) + q_{0.05}\frac{s}{\sqrt{m}}$$

which is the confidence interval statement to be derived.

We can illustrate the above by reference to the data of Tables 30.1 and 30.1A. These yield

Mean for thermometer A = 173.33

Mean for thermometer B = 173.00

Mean for thermometer C = 172.17

Mean for thermometer D = 172.58

and $s = \sqrt{0.22} = 0.47$ based on 12 degrees of freedom.

The 0.95 confidence limits for all possible comparisons on *a per comparison basis* are based on the t distribution as follows:

$A - B$: $$173.33 - 173.00 \pm 2.179(0.47)\sqrt{\frac{1}{6} + \frac{1}{6}}$$

$$= 0.33 \pm 0.59 = \begin{cases} 0.92 \\ -0.26 \end{cases}$$ [Note that $t_{0.025}(12) = 2.179$.]

$A - C$: $$173.33 - 172.17 \pm 2.179(0.47)\sqrt{\frac{1}{6} + \frac{1}{6}}$$

$$= 1.16 \pm 0.59 = \begin{cases} 1.75 \\ 0.57 \end{cases}$$

$A - D$: $$173.33 - 172.58 \pm 2.179(0.47)\sqrt{\frac{1}{6} + \frac{1}{6}}$$

$$= 0.75 \pm 0.59 = \begin{cases} 1.34 \\ 0.16 \end{cases}$$

$B - C$: $$173 - 172.17 \pm 2.179(0.47)\sqrt{\frac{1}{6} + \frac{1}{6}}$$

$$= 0.83 \pm 0.59 = \begin{cases} 1.42 \\ 0.24 \end{cases}$$

$$B - D: \quad 173.00 - 172.58 \pm 2.179(0.47)\sqrt{\frac{1}{6}+\frac{1}{6}}$$

$$= 0.42 \pm 0.59 = \begin{cases} 1.01 \\ -0.17 \end{cases}$$

$$C - D: \quad 172.17 - 172.58 \pm 2.179(0.47)\sqrt{\frac{1}{6}+\frac{1}{6}}$$

$$= -0.41 \pm 0.59 = \begin{cases} 0.18 \\ -1.00 \end{cases}$$

The allowance in each case is $\pm 2.179(0.47)\sqrt{\frac{1}{6}+\frac{1}{6}} = \pm 0.59$.

On an *experiment basis,* the 0.95 allowance is

$$\pm q_{0.05}(4, 12)s/\sqrt{6} = \pm(4.20)(0.47)/\sqrt{6} = \pm 0.81$$

and the 0.95 confidence limits are

$$A - B: \quad 0.33 \pm 0.81 = \begin{cases} 1.14 \\ -0.48 \end{cases}$$

$$A - C: \quad 1.16 \pm 0.81 = \begin{cases} 1.97 \\ 0.35 \end{cases}$$

$$A - D: \quad 0.75 \pm 0.81 = \begin{cases} 1.56 \\ -0.06 \end{cases}$$

$$B - C: \quad 0.83 \pm 0.81 = \begin{cases} 1.64 \\ 0.02 \end{cases}$$

$$B - D: \quad 0.42 \pm 0.81 = \begin{cases} 1.23 \\ -0.39 \end{cases}$$

$$C - D: \quad 0.41 \pm 0.81 = \begin{cases} 1.22 \\ -0.40 \end{cases}$$

It will be noted that some of the confidence intervals based on the studentized-range allowance do not include zero, which is in line with the previous 0.05 F test that indicated the thermometer means of Table 30.1 were significantly different.

Although references are made to agreement or nonagreement with F tests, it is to be emphasized that the point of view in this section is not that of testing hypotheses but simply that of reporting results with allowances for experimental error. If errors are to be allowed at a 5 percent rate on a *per comparison* basis, then we use the "least significant difference"[7]

---

[7] Cf. John W. Tukey, "The Problem of Multiple Comparisons," unpublished dittoed notes, Princeton University, 1953. Also see H. Leon Harter, "Error Rates and Sample Sizes for Range Tests in Multiple Comparisons," *Biometrics* 13 (1957), pp. 511–36.

$t_{0.025}\, s\sqrt{2/m}$. If errors are to be allowed only in 5 percent of the *experiments,* then we use the "wholly significant difference"[8] $q_{0.05}\, s/\sqrt{m}$.

1.3. Subgrouping and Ranking the Means of a Set

If a group of g means are shown to be significant by an F test or studentized-range test, it may be of interest to analyze the group further. If we divide the g means into g subgroups containing $g - 1$ means each, we may wish to determine whether the means of each subgroup differ significantly. To do this we can apply further F or range tests to each subgroup. If a subgroup shows no significant difference, the analysis of that group stops. If a significant difference is found, then such subgroups can be further subdivided. The general rule will be that the means of no subgroup will be declared significantly different unless the means of all larger groups containing this particular subgroup are significantly different. Ultimately the analysis will yield a partial ranking of the means of the set.

It will be noted that the number of different subgroups of size $g - 1$, $g - 2$, and so on, that can be made from g means rises rapidly as g increases. Thus some orderly procedure is necessary. The usual approach is to arrange the g means in order of size and then, working first from one end and then the other, to consider subgroups of $g - 1$ and 1, $g - 2$ and 2, and so forth. If at any time a test of a subgroup indicates no significant difference, then the analysis stops for that subgroup.

As an example, let us analyze the following set of 6 means:

$$A = 502 \qquad D = 498$$
$$B = 528 \qquad E = 600$$
$$C = 564 \qquad F = 470$$

These are taken[9] from a 5×6 one-way analysis of variance for which the error mean square, based on 24 degrees of freedom is equal to 2,451. The studentized-range coefficient for $g = 6$, $\nu = 24$ is 4.37 (see Table D2, Appendix II), and the range factor for testing the above set of means is $4.37\sqrt{\dfrac{2{,}451}{5}} = 96.7$. The range of the 6 means in question is $600 - 470 = 130$ which is greater than 96.7, so we conclude that the 6 means differ significantly.

Arranged in order of size we have

F	D	A	B	C	E
470	498	502	528	564	600

---

[8] Tukey, "The Problem of Multiple Comparisons."

[9] See David B. Duncan, *Multiple Range and Multiple F Tests,* Department of Statistics and Statistical Laboratory, Virginia Polytechnic Institute, Technical Report No. 6 (1953), p. 3.

From the previous analysis we conclude that F is less than E. Next look at the subgroup F, D, A, B, and C. For the moment let us be ultraconservative regarding the Type I error (with resultant loss in Type II error) and let us use the studentized-range allowance for *the whole set* of g means to separate significantly different means in subgroups containing *less than* g means. We shall call this the Tukey method.[10] Applying this procedure, we note that the range of the 5 means F, D, A, B, and C is 94 and that this is less than the allowance 96.7. Hence we do not conclude that the means of this subgroup differ significantly. Analysis in this direction stops. Start now at the other end. Here we have D, A, B, C, and E with a range of 102. We therefore conclude that the means of this group are significantly different and further analysis at this end is warranted. Consider next the 4 mean group A, B, C, and E. The range of these means is 98 and again this exceeds 96.7, so we conclude that the means of this group also differ significantly. A further subdivision is in order. The next subgroup of means is B, C, and E which has a range of 72; but 72 is less than 96.7 so we conclude that these means do not differ significantly.

We can summarize the results in the following diagram:

$F \qquad D \quad A \qquad B \qquad C \qquad E$

(underline from F to C; underline from B to E)

Thus we conclude that F, D, and A are less than E, but that B and C may be the same as E. We also conclude that C, B, A, and D may be the same as F.

From the way in which the previous analysis was run we can be assured that we shall not falsely conclude that differences exist in more than 5 *experiments* out of 100. Some statisticians argue, however, that the procedure is more conservative than necessary with respect to the Type I error and that because of this, it has a relatively high Type II error. In other words, the conservatism in making false statements about differences that do not exist is bought at the expense of failing to make true statements about differences that do exist. Some take the position that in comparing means within any subgroup, it is sufficient to use the α-level range factor *appropriate to a group of that size.* We shall call this the Newman-Keuls method,[11] after the men especially responsible for it.

Consider again the given example. When we come to compare subsets of 5 means let us base the allowance factor upon the 0.05 studentized-range coefficient for 5 means and not 6, and when we compare subsets of 4 means let us use the 0.05 studentized-range coefficient for subsets of 4, and so on. This means that we will compare the range of F, D, A, B, and C (i.e., 94)

---

[10] Cf. Tukey's "The Problem of Multiple Comparisons." Also see Harter, "Error Rates and Sample Sizes for Range Tests in Multiple Comparisons."

[11] See David B. Duncan, "Multiple Range and Multiple F Tests," *Biometrics* 11 (1955), pp. 1–42.

with $q_{0.05}$ (for $g = 5$, $\nu = 24$) $\sqrt{\frac{s^2}{m}}$ or $4.17 \sqrt{\frac{2{,}451}{5}} = 92.3$ and in this case will conclude that F is less than C. Continuing further, the 0.05 studentized-range coefficient for $g = 4$, $\nu = 24$ is 3.90 and the factor for testing for a difference between 4 means is $3.90 \sqrt{\frac{2{,}451}{5}} = 86.4$. The actual range of the means F, D, A, and B is 58, so the analysis from the right will stop here. From the other end we will have $D < E$ (since $102 > 92.3$) and $A < E$ (since $98 > 86.4$), but B is not less than E (since $72 < q_{0.05}$ ($g = 3$, $\nu = 24$) $\frac{s}{\sqrt{m}} = 78.5$). Thus using this less conservative procedure we will conclude that

$$F < E$$

$$D < E$$

$$A < E$$

$$F < C$$

In making these conclusions, however, we will run a greater risk (than 5 experiments per 100) in concluding that differences exist when they do not.

David B. Duncan argues[12] that a still better procedure is to make an additional adjustment in the level of α in accordance with the number of means in the set being analyzed. The rule he offers runs as follows. For sets of p means let the proportion of sets in which wrong statements will be tolerated be put at $1 - (1 - \alpha[2])^{p-1}$, where $\alpha[2]$ is the α to be used for comparing sets of 2 means in the absence of other means, say, $\alpha[2] = 0.05$. For comparable $\alpha[2]$'s, this method will yield a higher overall risk than the Tukey or Newman-Keuls method of concluding, on an experiment basis, that differences exist when they do not. It has a greater probability, however, of finding differences that do exist, and Duncan argues that it balances nicely the Type I and Type II risks. Duncan offers special tables for the use of his method with $\alpha[2]$ equal to 0.01 and 0.05.

The most liberal method, on an experiment basis, is to use $t_{0.025} \sqrt{\frac{2s^2}{m}}$ as an allowance for separating means. All means that differ by this amount will be considered to be significantly different, and this can also be used to separate the means into "nonsignificantly different" groups. This is called the least-significant-difference method. It is not as acceptable as the other methods, since for relatively large groups of means, the error rate on an experiment basis is judged by some statisticians to be too high.[13]

Another method that might be mentioned is the "modified Tukey

[12] Ibid.

[13] Ibid.

method." This bases the allowance factor on an average of the studentized-range factor for the whole set and that for the number of means currently analyzed.

All the above methods have been reviewed by H. Leon Harter in *Biometrics* 13 (1957), pp. 511–36. Harter gives tables that permit comparisons between the Type I error rates of the tests and also tables showing the operating characteristics or Type II errors of the tests.

1.4. Outliers

Occasionally in a set of individual values or a set of means there is one value that appears to differ considerably from the others. The tendency among those not trained in statistics is to discard these exceptional values as not belonging to the set. The statistician prefers to follow a policy in such cases for which he can calculate the probability of wrong decisions. Of course, if a value is *known* to have some characteristic that differentiates it sharply from the rest other than being simply an extreme value of the set, then this value can be discarded on the grounds that its inclusion will destroy the homogeneity of the set. Such decisions must be based on sound knowledge and must be exercised with great caution. If there is any doubt, it is best to follow some statistical rule.

If the extreme value is a mean, one way of testing whether it belongs to the group is to run a subgrouping analysis, such as described in the previous section, carrying the test down to a comparison between the extreme mean and the mean adjacent to it. If we are interested solely in a single extreme value, be it a mean or an individual value, use can be made of the "extreme studentized deviate."[14] Thus, if an independent estimate of σ' is available, say an estimate based on within-group variation in the case of a mean outlier or previous data *(not included in the set under study)* in the case of an individual item, then the extreme value may be tested by computing

(31.15a) $$t_e = \frac{(\bar{X}_e - \bar{X})}{s_{\bar{X}}} \text{ for an extreme mean } \bar{X}_e$$

or

(31.15b) $$t_e = \frac{(X_e - \bar{X})}{s_X} \text{ for an extreme individual } X_e$$

If t_e exceeds the critical value for the specified α given in Table U in Appendix II, it may be concluded that the extreme value does not belong.

---

[14] See K. R. Nair, *Biometrika* 35 (1948), p. 118; and 39 (1952), p. 189. Also see H. A. David, "Revised Upper Percentage Points of the Extreme Studentized Deviate from the Sample Mean," *Biometrika* 43 (1956), p. 85. For a readily available discussion, see E. S. Pearson and H. O. Hartley (eds.), *Biometrika Tables for Statisticians* (2d ed.; London: Cambridge University Press, 1958), Vol. I, p. 50.

This extreme studentized-deviate test can be employed to detect an outlier mean whether or not an analysis of variance F test indicates significance.

Other tests of an outlier that make use only of the data in hand have been worked out by W. J. Dixon.[15] These make use of the statistics

$$r_{10} = \frac{X_2 - X_1}{X_n - X_1} \quad \text{or} \quad \frac{X_n - X_{n-1}}{X_n - X_1}$$

$$r_{11} = \frac{X_2 - X_1}{X_{n-1} - X_1} \quad \text{or} \quad \frac{X_n - X_{n-1}}{X_n - X_2}$$

$$r_{21} = \frac{X_3 - X_1}{X_{n-1} - X_1} \quad \text{or} \quad \frac{X_n - X_{n-2}}{X_n - X_2}$$

$$r_{22} = \frac{X_3 - X_1}{X_{n-2} - X_1} \quad \text{or} \quad \frac{X_n - X_{n-2}}{X_n - X_3}$$

where the X_i represent either individual values or means arranged in order of size from X_1 to X_n. Dixon's table of critical values for r_{10}, r_{11}, r_{21}, and r_{22} is reproduced as Table W of Appendix II. When this method is used to test an extreme mean, the samples from which the means are computed should all have the same size. It should also be noted that the distribution of $\bar{X}$ or X is assumed to be normal.

Let us apply this rule to the 6 means recorded in Section 1.3 above. Here $\bar{X}_1 = 470$ and $\bar{X}_n = 600$. To test whether $E(= \bar{X}_6)$ belongs with the others of the group we compute

$$r_{10} = \frac{|\bar{X}_6 - \bar{X}_5|}{|\bar{X}_6 - \bar{X}_1|} = \frac{|600 - 564|}{|600 - 470|} = \frac{36}{130} = 0.28$$

Since this is less than the critical value (0.56) for $\alpha = 0.05$ (see Table W), we conclude that E should not be judged to be different from the others. Note that this conclusion is based solely upon the 6 means and does not make use of an outside measure of error as did the analysis of Section 1.3.

1.5. Selecting the Maximum or Minimum Group

Often when we compare the means of several different groups, we are interested primarily in picking the group that has the largest or smallest (universe) mean. Fortunately the test procedure for this is very simple. If we have g sample means, we select that group as optimum that has the optimum sample mean.[16]

---

[15] W. J. Dixon, "Processing Data for Outliers," *Biometrics* 9 (1953), p. 74.

[16] D. B. Duncan points out that the α of this test is 100 percent (*Multiple Range and Multiple F Tests,* p. 36) when $\bar{X}'_1 = \bar{X}'_2, \ldots = \bar{X}'_g$, but the author does not feel this is a serious criticism since rarely are means exactly equal in the universes. Duncan considers the Bechhofer test a special case of the ranking analysis of Section 1.3.

The contribution that statisticians have made to this problem is to help us decide what is the proper sample size for the test. Robert E. Bechhofer has prepared tables from which the sample size can be computed that will give a specified assurance that the truly optimum group will be selected if it differs by as much as a preassigned amount from the next to the optimum group. A few select values from his table are reproduced in Table Y of Appendix II of this text.

Suppose, for example, that with a 0.90 assurance of being right, we wish to determine the largest of 5 universe means when the difference between the largest and the next to the largest universe mean is one half of the universe standard deviation (i.e., $\lambda = 0.5$). It is assumed that this standard deviation is the same for all universes. (For instance, variation may be due to experimental errors.) To find the solution we look in Table Y opposite $g = 5$, $t = 1$, and note that for $P_\alpha = 0.90$, $\lambda\sqrt{n} = 2.5997$. Hence $\sqrt{n} = 2.5997/0.5 = 5.1984$ and n equals approximately 25. In other words, we take a sample of 25 from each group.

It will be noted that the difference that we wish to detect must be expressed as a multiple of the (common) known universe standard deviation, but Bechhofer, Dunnett, and Sobel[17] have worked out tables for a two-sample procedure to be used when the universe variance is not known.

Bechhofer has also given tabular values for sample size when it is desired to rank correctly the t largest means. The ranking procedure is to select as the highest t ranking universe groups the samples with the t highest means. To have an assurance of 0.90, 0.95, or 0.99 of being right, we enter Table Y with the given values of g and t and read off the value of $\lambda\sqrt{n}$. Here λ is the ratio of the smallest difference between the means of the t truly highest ranking groups to the common universe standard deviation. To find the desired n we divide by the assumed λ. For example, if $\lambda = 1.0$, $g = 12$, and $t = 3$, and we wish an assurance of 0.99 of being right, we have $\lambda\sqrt{n} = 4.7039$, $\sqrt{n} = 4.7039/1.0 = 4.7039$, and $n \doteq 22$. In other words, we should take 22 measurements in each of the 12 groups if we wish to select the truly 3 highest groups when the difference between any two of the universe means of these groups is not less than one standard deviation.

2. TESTS OF HOMOGENEITY OF VARIANCES

Occasionally we wish to determine whether a set of variances could all have come from the same universe. For example, if we have a set of samples taken from a process at different intervals, we might like to know whether the variability of the process has remained constant during the interval. Or we may have several lots and may be interested in knowing whether the

[17] R. E. Bechhofer, C. W. Dunnett, and M. Sobel, "A Two-Sample Multiple Decision Procedure for Ranking Means of Normal Populations with a Common Unknown Variance," *Biometrika* 41 (1954), pp. 170–76.

variances of the items within the lots differ significantly from lot to lot. Or, again, we may be carrying out an analysis of variance with several cases for each class and may wish to test the assumption that the variances for every class are the same—one of the basic assumptions in analysis of variance.

One way of determining whether a set of g variances is homogeneous is to set up an s^2-chart or possibly an R-chart. (See Chapter 21.) Another method, usually preferred to s^2 or R-charts by statisticians, is to use the quantity

$$M = 2.3026\,[\nu \log (\Sigma \nu_i s_i^2/\nu) - \Sigma \nu_i \log s_i^2] \tag{31.16}$$

where the s_i^2 are the independent estimates of variance to be compared, based respectively on ν_i degrees of freedom, and where $\nu = \Sigma \nu_i$. M is proportional to the log of the ratio of the arithmetic mean to the geometric mean of the g sample estimates of variance and is approximately distributed as χ^2 with $g - 1$ degrees of freedom. M. S. Bartlett has shown that if C is defined by the equation

$$C = 1 + \frac{1}{3(g-1)}\left\{\sum_i \frac{1}{\nu_i} - \frac{1}{\nu}\right\}, \tag{31.17}$$

then the distribution of M/C is somewhat more closely approximated by a χ^2 distribution with $g - 1$ degrees of freedom than that of M alone. M/C, however, is not always adequate if some ν_i are 1, 2, or 3. More precise tests can be made with the help of a table of maximum and minimum approximations to the 0.05 and 0.01 percentage points of M published as Table 32 of *Biometrika Tables for Statisticians,* Vol. I.[18]

In comparing a set of variance estimates each based on the same number of degrees of freedom, a quick and easy test is to compute the ratio $\frac{\text{Max. } s_i^2}{\text{Min. } s_i^2}$. Limiting points for the distribution of this ratio are given in Table T of Appendix II.

If we have a series of small samples, an even quicker procedure is to take the ratio $\frac{\text{Max. } (R/d_2^*)^2}{\text{Min. } (R/d_2^*)^2}$ and use Table T with the degrees of freedom given in Table D3 for m equal to the size of the samples and $g = 1$.

It must be remembered that the above tests are all based on the assumption that the random variation within each of the g groups follows the normal law. If this is not true, a significant value of M or the ratio $\frac{\text{Max. } s_i^2}{\text{Min. } s_i^2}$ may indicate departure from normality rather than heterogeneity of variance. "Tests of this kind are, indeed, more sensitive to departure from normality than the ordinary tests of the analysis of variance."[19]

---

[18] Also see Carl A. Bennett and Norman L. Franklin, *Statistical Analysis in Chemistry and the Chemical Industry* (New York: John Wiley & Sons, 1954), p. 198.

[19] *Biometrika Tables for Statisticians,* p. 57.

To illustrate Bartlett's test, consider the eight samples of five each that constitute the first half of Table 26.1. These, it will be recalled, are samples of cans of tomatoes turned out in the morning. The s_i^2 for the weights of the cans in the eight samples of five are as follows:[20]

Sample Number	s_i^2	*Sample Number*	s_i^2
1	0.675	5	4.750
2	2.050	6	5.075
3	1.000	7	1.000
4	0.375	8	5.325

The question of interest is this: Is the variability of the process constant, or do some of the samples show significantly greater variability than the others? To answer this, let us set up the null hypothesis that the variability of the process is constant and apply Bartlett's test.

We first derive $\sum_i \nu_i \log s_i^2 =$

$$
\begin{aligned}
4\log(0.675) &= 4(9.8293-10)\\
4\log(2.050) &= 4(0.3118)\\
4\log(1.000) &= 4(0)\\
4\log(0.375) &= 4(9.5740-10)\\
4\log(4.750) &= 4(0.6767)\\
4\log(5.075) &= 4(0.7055)\\
4\log(1.000) &= 4(0)\\
4\log(5.325) &= \underline{4(0.7263)}\\
&\ 4(21.8236-20)\\
&= 7.2944
\end{aligned}
$$

Then we compute $\nu \log(\Sigma \nu_i s_i^2/\nu) = 32 \log(\Sigma s_i^2/8) = 32(0.4033) = 12.9056$

$$M = 2.3026(12.9056 - 7.2944) = 12.92$$

$$C = 1 + \frac{1}{21}\left(\frac{8}{4} - \frac{1}{32}\right) = 1.094$$

$$\frac{M}{C} = \frac{12.92}{1.094} = 11.81$$

[20] $s^2 = \Sigma(X - \bar{X})^2/(\nu - 1)$. For further information on computation of s^2, see Chapter 3, Section 4.2.4.2.

The $\chi^2_{0.05}$ value for $df = 8 - 1 = 7$ is 14.1. Since our M/C is less than this, the null hypothesis that the variances all come from the same universe is not rejected. We conclude that the variability of the process is constant.

The ratio Max. s_i^2/Min. s_i^2 is very easy to apply. For our data, this ratio is $\frac{5.325}{0.375} = 14.2$. Reference to Table T shows that the 0.05 point for this statistic for $g = 8$, $\nu_i = 4$ is 37.5, which again does not lead to rejection of the null hypothesis.

Using the range method, we have

Sample Number	*Range*	$(R/d_2^*)^2$	*Sample Number*	*Range*	$(R/d_2^*)^2$
1	2.0	0.650	5	6.0	5.852
2	3.5	1.992	6	6.0	5.852
3	2.5	1.016	7	2.5	1.016
4	1.5	0.366	8	4.5	3.294

These yield

$$\frac{\text{Max. } (R/d_2^*)^2}{\text{Min. } (R/d_2^*)^2} = \frac{5.852}{0.366} = 15.99$$

which is less than 0.05 point in Table T for $\nu_i = 3.8$. Once more we do not reject. In practice we would only have to compute $(R/d_2^*)^2$ for the maximum and minimum range.

3. ANALYSIS OF COMPONENTS OF VARIANCE

Often we are not so much interested in determining whether certain factors affect the results as we are in measuring how much they affect it. For example, we may have good reasons a priori for believing that the effects of the factors are significant, and we may wish simply to estimate the proportion each contributes to the overall variance. Such an analysis is called components of variance analysis.

3.1. Fixed-Effects Models

3.1.1. One-Way Classification. Let us begin by discussing a few basic principles. Usually components of variance analysis is applied to purely random models or mixed models, but let us look first at a fixed-effects model. For a one-way classification the equation for such a model is

(31.18)
$$X_{ij} = \mu + \theta_j + \epsilon_{ij}$$

where ϵ_{ij} is a normally distributed random variable with constant variance σ'^2, but θ_j is a fixed factor that takes on all of the given values θ_1, θ_2, . . . θ_c and μ, of course, is a constant. Since the θ_j are only "differentials,"

FIGURE 31.1
Illustration of Components of Variance in a One-Way Classification: Fixed Effects

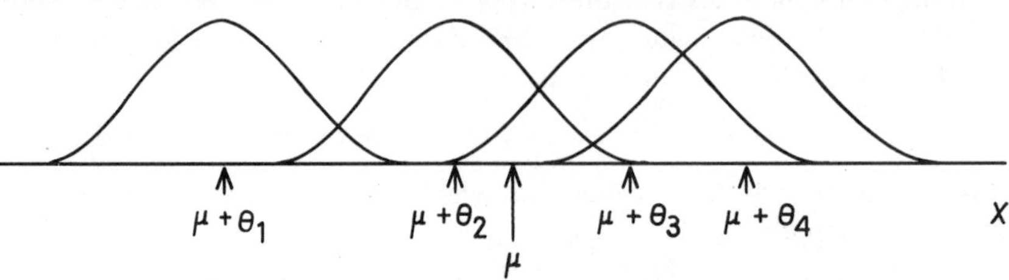

we have $\Sigma\theta_j = 0$; any excess that might exist is by definition put into μ. For $c = 4$, the variation can be represented by a diagram such as Figure 31.1. For example, the θ_j may be deviations in the settings of four machines from their grand mean (μ), and the variation of the product of each machine around its setting may be given by σ'^2.

Now in terms of the above example, if n items are produced such that $n/4$ are turned out by each machine, the variance of the overall product $\frac{\sum_i \sum_j (X_{ij} - \mu)^2}{n}$ will be given by[21] the sum of the between-machine variance, $\sigma_\theta'^2 = \frac{\sum_j (\mu + \theta_j - \mu)^2}{4} = \frac{\Sigma\theta_j^2}{4}$ plus the pooled within-machine variance, $\frac{\sum_i \sum_j (X_{ij} - \mu - \theta_j)^2}{n}$. If n is large enough, the latter will be close to σ'^2.

In brief, then, we have for large n

$$\sigma'^2_{\text{product}} \doteq \sigma_\theta'^2 + \sigma'^2 \tag{31.19}$$

where $\sigma_\theta'^2 = \frac{\Sigma\theta_j^2}{c}$. The proportion of total variation due to the variation in

[21] Proof: The total sum of squares can be written

$$\begin{aligned}\sum_i \sum_j (X_{ij} - \mu)^2 &= \sum_i \sum_j [(X_{ij} - \mu - \theta_j) + (\mu + \theta_j - \mu)]^2 \\ &= \sum_i \sum_j (X_{ij} - \mu - \theta_j)^2 + \frac{n}{4} \sum_j (\mu + \theta_j - \mu)^2 \\ &\quad + 2 \sum_i \sum_j (X_{ij} - \mu - \theta_j)(\mu + \theta_j - \mu) \\ &= \sum_i \sum_j (X_{ij} - \mu - \theta_j)^2 + \frac{n}{4} \sum_j \theta_j^2\end{aligned}$$

since $\sum_i (X_{ij} - \mu - \theta_j) = 0$ for all j. The statement follows when the total sum of squares is divided by n.

TABLE 31.1
Mean Squares as Unbiased Estimates of Linear Functions of Variance Components: Fixed Effects, One-Way Classification

Model	*Mean Square**	*An Unbiased Estimate of:†*
Model A: $X_{ij} = \mu + \theta_j + \epsilon_{ij}$; $i = 1,2, \cdots r$; $j = 1,2, \cdots c$; μ, θ—fixed; ϵ—random. Definition: $\sigma'^2_\theta = \frac{\Sigma \theta_j^2}{c}$	$s^2_{\text{bet. classes}} = \frac{r \sum_j (\bar{X}_j - \bar{X})^2}{c-1}$	$\sigma'^2 + \frac{rc}{c-1}\sigma'^2_\theta$
	$s^2_e = \frac{\sum_i \sum_j (X_{ij} - \bar{X}_j)^2}{c(r-1)}$	σ'^2

* For short-cut methods of computing mean squares, see Chapter 29, Section 1.3.
† See Appendix I(33). Note that $\Sigma\theta_j = 0$. Many statisticians prefer to define σ'^2_θ as equal to $\Sigma\theta_j^2/(c-1)$. When this is done, the expected value of $s^2_{\text{bet. classes}}$ becomes $\sigma'^2 + r\sigma'^2_\theta$.

machine settings is thus $\sigma'^2_\theta/\sigma'^2_{\text{prod.}}$, and the proportion due to within machine variability is $\sigma'^2/\sigma'^2_{\text{prod.}}$. The first is a measure of the reduction in product variability that might be attained if the four machines were set exactly alike so that the θ_j were all 0.

In a one-way classification such as the above, we may estimate the various components of variance by taking, say, r items at random from each of c classes. The relationships between the mean squares and components of variance will then be those given in Table 31.1. These yield the following unbiased estimates[22] of σ'^2 and σ'^2_θ:

(31.20)
$$\hat{\sigma}^2 = s^2_e$$
$$\hat{\sigma}^2_\theta = \frac{(s^2_{\text{bet. classes}} - s^2_e)}{rc/(c-1)}$$

Consider, for example, the problem of Section 1, Chapter 29, in which biases of thermometers are studied. Table 29.2 gives the thermometer mean square (s^2_C) equal to 1.47 and the error mean square (s^2_e) equal to 0.81. The error variance (σ'^2) is thus immediately estimated at $\hat{\sigma}^2 = 0.81$, and the thermometer variance (σ'^2_θ) is estimated as equal to $\hat{\sigma}^2_\theta = \frac{(1.47 - 0.81)}{12/3} = 0.165$. If each of the three thermometers is used equally often, the variance of an individual measurement may then be taken as equal to $\hat{\sigma}^2_\theta + \hat{\sigma}'^2 = 0.165 + 0.81 = 0.975$, and it may be estimated that thermometer variance

[22] A hat on a sample symbol means it is an estimate of the corresponding universe parameter.

accounts for $\frac{0.165}{0.975} = 17$ percent of the total measurement variance. In other words, we could reduce measurement error by only 17 percent if we were able to make the three thermometers read exactly alike. Since in fact, the thermometer variance was not significantly different from zero in this case, we could reasonably take the position that there is no thermometer variance anyway.

3.1.2. Two-Way Classification. The same principles may be applied to a two-way classification. Suppose that we have c machines and r operators and these are combined so that in the course of operations an equal amount of time is assigned to each machine-operator combination. Then, if the machine effects (θ_j), the operator effects (τ_i) and the interaction elements (ψ_{ij}) combine additively, i.e., if the model is

$$X_{ijk} = \mu + \tau_i + \theta_j + \psi_{ij} + \epsilon_{ijk}, \tag{31.21}$$

it is easy to show[23] that variance of product

$$\sigma'^2_{\text{product}} = \sigma'^2_{\tau} + \sigma'^2_{\theta} + \sigma'^2_{\psi} + \sigma'^2 \tag{31.22}$$

If now we take equal samples of size g from the output of each factor combination, then the various components of variance can be estimated from the relationships given in Table 31.2. Thus unbiased estimates of σ'^2, σ'^2_{ψ}, σ'^2_{θ} and σ'^2_{τ} will be given by

$$\begin{aligned} \hat{\sigma}^2 &= s^2_{\text{res.}} \\ \hat{\sigma}^2_{\psi} &= \frac{(c-1)(r-1)}{gcr}(s^2_{RC} - s^2_{\text{res.}}) \\ \hat{\sigma}^2_{\theta} &= \frac{c-1}{rgc}(s^2_{C} - s^2_{\text{res.}}) \\ \hat{\sigma}^2_{\tau} &= \frac{r-1}{cgr}(s^2_{R} - s^2_{\text{res.}}) \end{aligned} \tag{31.23}$$

From these results the proportion of the total variance attributable to each factor can be measured, on the assumption, of course, that every combination of machine and operator is used equally over time. On the assumption that machine and operator differences could be eliminated or reduced by resetting of the machines or remedial training of operators, the above analysis would indicate how much could be gained by such action.

Consider the data of Table 30.2 of Chapter 30. Here we have

Mean square for analysts $s(^2_R) = 2.51$

Mean square for thermometers $(s^2_C) = 1.54$

[23] The reasoning is the same as in a one-way classification.

$$\text{Interaction mean square } (s^2_{RC}) = 0.45$$

$$\text{Residual mean square } (s^2_{\text{res.}}) = 0.22$$

The experimental error variance (σ'^2) may thus be estimated at

$$\hat{\sigma}^2 = 0.22;$$

the interaction variance (σ'^2_ψ) may be estimated at

$$\hat{\sigma}^2_\psi = \frac{(3)(2)}{24}(0.45 - 0.22) = 0.058$$

the thermometer variance (σ'^2_θ) may be estimated at

$$\hat{\sigma}^2_\theta = \frac{(4-1)}{24}(1.54 - 0.22) = 0.165$$

TABLE 31.2
Mean Squares as Unbiased Estimates of Linear Functions of Variance Components: Fixed Effects, Two-Way Classification

Model	*Mean Square**	*An Unbiased Estimate of—†*
Model B: $X_{ijk} = \mu + \tau_i + \theta_j + \psi_{ij} + \epsilon_{ijk}$	$s^2_R = \dfrac{cg\,\Sigma(\bar{X}_i - \bar{X})^2}{r-1}$	$\sigma'^2 + \dfrac{cgr}{r-1}\sigma'^2_\tau$
$i = 1, 2, \ldots, r$; $j = 1, 2, \ldots, c$	$s^2_C = \dfrac{rg\,\Sigma(\bar{X}_j - \bar{X})^2}{c-1}$	$\sigma'^2 + \dfrac{rgc}{c-1}\sigma'^2_\theta$
$k = 1, 2, \ldots, g$; μ, τ, θ, ψ—fixed; ϵ—random	$s^2_{RC} = \dfrac{g\,\Sigma\Sigma(\bar{X}_{ij} - \bar{X}_i - \bar{X}_j - \bar{X})^2}{(r-1)(c-1)}$	$\sigma'^2 + \dfrac{gcr}{(c-1)(r-1)}\sigma'^2_\psi$
Definitions: $\sigma'^2_\tau = \dfrac{\Sigma\tau_i^2}{r}$; $\sigma'^2_\theta = \dfrac{\Sigma\theta_j^2}{c}$; $\sigma'^2_\psi = \dfrac{\Sigma\Sigma\psi_{ij}^2}{rc}$	$s^2_{\text{res.}} = \dfrac{\Sigma\Sigma\Sigma(X_{ijk} - \bar{X})^2}{rc(g-1)}$	σ'^2

* For short methods of computation, see Chapter 30, Section 1.3.

† For the nature of the theoretical analysis, see Appendix I (33). The general procedure is to substitute the model in the formula for each mean square, make note of such relationships as $\sum_j \theta_j = 0$ and $\sum_i \tau_i = 0$, and then apply the mean-value operator E. As noted in Table 31.1, many statisticians prefer to define σ'^2_τ, σ'^2_θ, and σ'^2_ψ in such a way as to simplify the variance estimates. This simplification can be accomplished by defining σ'^2_τ as equal to $\Sigma\tau_i^2/(r-1)$, $\sigma'^2_\theta = \Sigma\theta_j^2/(c-1)$, and $\sigma'^2_\psi = \dfrac{\Sigma\Sigma\psi_{ij}^2}{(r-1)(c-1)}$. This practice was not followed here since it was believed to be more confusing than that adopted. See Section 3.4, however.

and the analyst variance ($\sigma_\tau'^2$) may be estimated at

$$\hat{\sigma}_\tau^2 = \frac{(3-1)}{24}(2.51 - 0.22) = 0.191$$

If in the course of measurement each thermometer was used equally often by each analyst, the measurement variance may be estimated at[24]

$$\hat{\sigma}_\tau^2 + \hat{\sigma}_\theta^2 + \hat{\sigma}_\psi^2 + \hat{\sigma}^2 = 0.191 + 0.165 + 0.058 + 0.22 = 0.634$$

In this instance much could be gained by training the three analysts to use thermometers alike and by making the four thermometers to read alike. Since the interaction variance is not significantly different from zero and in any case is small, little could be gained by taking the trouble of assigning particular thermometers to particular analysts.

3.2. Random-Effects Models (Infinite Universes)

When all variable effects are random, the components of variance are not variances of a fixed set of effects but of the universe of effects from which the given set being analyzed are a random sample. In this case the various mean squares in an analysis of variance are unbiased estimates of certain linear functions of these universe variance components. The functions are given in Table 31.3 for four different models. The universes are assumed in all cases to be infinite.

Unbiased estimates of the individual variance components can be obtained as follows:

One-Way Classification (Model C): Unbiased estimates of σ'^2 and $\sigma_\theta'^2$ are given by

(31.24)

$$\hat{\sigma}^2 = s_e^2$$

$$\hat{\sigma}_\theta^2 = \frac{s_{\text{bet. classes}}^2 - s_e^2}{r}$$

Two-Way Classification (Model D): Unbiased estimates of σ'^2, $\sigma_\theta'^2$, and $\sigma_\tau'^2$ are given by

(31.25)

$$\hat{\sigma}^2 = s_{\text{res.}}^2$$

$$\hat{\sigma}_\theta^2 = \frac{s_C^2 - s_{\text{res.}}^2}{r}$$

$$\hat{\sigma}_\tau^2 = \frac{s_R^2 - s_{\text{res.}}^2}{c}$$

---

[24] Since the interaction variance is not significantly different from zero, it could be omitted from the estimate if so desired.

TABLE 31.3

Expected Values for Various Mean Squares in Four Analysis of Variance Models: All Variable Effects Random, Universes Infinite

Model	*Mean Square**	*An Unbiased Estimate of—* †
One-Way Classification		
Model C:		
$X_{ij} = \mu + \theta_j + \epsilon_{ij}$	$s^2_{\text{bet. classes}} = \dfrac{r\Sigma(\bar{X}_j - \bar{X})^2}{c-1}$	$\sigma'^2 + r\sigma'^2_\theta$
$i = 1, 2, \ldots, r$; $j = 1, 2, \ldots, c$	$s^2_e = \dfrac{\Sigma\Sigma(X_{ij} - \bar{X}_j)^2}{c(r-1)}$	σ'^2
μ—fixed; θ, ϵ—random		
Two-Way Classification		
Model D:		
$X_{ij} - \mu + \tau_i + \theta_j + \epsilon_{ij}$	$s^2_R = \dfrac{c\Sigma(\bar{X}_i - \bar{X})^2}{r-1}$	$\sigma'^2 + c\sigma'^2_\tau$
$i = 1, 2, \ldots, r$; $j = 1, 2, \ldots, c$	$s^2_C = \dfrac{r\Sigma(\bar{X}_j - \bar{X})^2}{c-1}$	$\sigma'^2 + r\sigma'^2_\theta$
μ—fixed; τ, θ, ϵ—random	$s^2_{\text{res.}} = \dfrac{\Sigma\Sigma(X_{ij} - \bar{X}_i - \bar{X}_j + \bar{X})^2}{(r-1)(c-1)}$	σ'^2
Model E:		
$X_{ijk} = \mu + \tau_i + \theta_j + \psi_{ij} + \epsilon_{ijk}$	$s^2_R = \dfrac{cg\Sigma(\bar{X}_i - \bar{X})^2}{r-1}$	$\sigma'^2 + g\sigma'^2_\psi + cg\sigma'^2_\tau$
$i = 1, 2, \ldots, r$	$s^2_C = \dfrac{rg\Sigma(\bar{X}_j - \bar{X})^2}{c-1}$	$\sigma'^2 + g\sigma'^2_\psi + rg\sigma'^2_\theta$
$j = 1, 2, \ldots, c$; $k = 1, 2, \ldots, g$; μ—fixed	$s^2_{RC} = \dfrac{g\Sigma\Sigma(\bar{X}_{ij} - \bar{X}_i - \bar{X}_j + \bar{X})^2}{(r-1)(c-1)}$	$\sigma'^2 + g\sigma'^2_\psi$
τ, θ, ψ, ϵ —random	$s^2_e = \dfrac{\Sigma\Sigma\Sigma(X_{ijk} - \bar{X}_{ij})}{rc(g-1)}$	σ'^2
Three-Way Classification		
Model F:		
$X_{ijk} = \mu + a^A_i + a^B_j + a^C_k + a^{AB}_{ij} + a^{AC}_{ik} + a^{BC}_{jk} + \epsilon_{ijk}$	$s^2_R = \dfrac{cg\Sigma(\bar{X}_i - \bar{X})^2}{r-1}$	$\sigma'^2 + c\sigma'^2_{AC} + g\sigma'^2_{AB} + cg\sigma'^2_A$
$i = 1, 2, \ldots, r$	$s^2_C = \dfrac{rg\Sigma(\bar{X}_j - \bar{X})^2}{c-1}$	$\sigma'^2 + r\sigma'^2_{BC} + g\sigma'^2_{AB} + rg\sigma'^2_B$
$j = 1, 2, \ldots, c$; $k = 1, 2, \ldots, g$	$s^2_G = \dfrac{rc\Sigma(\bar{X}_k - \bar{X})^2}{g-1}$	$\sigma'^2 + r\sigma'^2_{BC} + c\sigma'^2_{AC} + rc\sigma'^2_C$
μ—fixed	$s^2_{RC} = \dfrac{g\Sigma\Sigma(\bar{X}_{ij} - \bar{X}_i - \bar{X}_j + \bar{X})^2}{(r-1)(c-1)}$	$\sigma'^2 + g\sigma'^2_{AB}$
A, B, C, AB, AC, BC, ϵ	$s^2_{RG} = \dfrac{c\Sigma\Sigma(\bar{X}_{ik} - \bar{X}_i - \bar{X}_k + \bar{X})^2}{(r-1)(g-1)}$	$\sigma'^2 + c\sigma'^2_{AC}$
—random	$s^2_{CG} = \dfrac{r\Sigma\Sigma(\bar{X}_{jk} - \bar{X}_j - \bar{X}_k + \bar{X})^2}{(c-1)(g-1)}$	$\sigma'^2 + r\sigma'^2_{BC}$
	$s^2_{\text{res.}} = \dfrac{\Sigma\Sigma\Sigma(X_{ijk} - \bar{X}_{ij} - \bar{X}_{ik} - \bar{X}_{jk} + \bar{X}_i + \bar{X}_j + \bar{X}_k - \bar{X})^2}{(c-1)(r-1)(g-1)}$	σ'^2

* For short-cut methods of computation, see Chapter 29, Sections 1.3 and 2.3, and Chapter 30, Sections 1.3 and 2.3.

† For the nature of the theoretical analysis, see Appendix I (33). The procedure is to substitute the model in each mean square and apply the mean-value operator E. Also see footnote † to Table 31.2.

Two-Way Classification (Model E): Unbiased estimates of σ'^2, σ'^2_ψ, σ'^2_θ, and σ'^2_τ are given by

$$\hat{\sigma}^2 = s_e^2$$

$$\hat{\sigma}^2_\psi = \frac{s_{RC}^2 - s_e^2}{g}$$

(31.26)

$$\hat{\sigma}^2_\psi = \frac{s_{RC}^2 - s_e^2}{g}$$

$$\hat{\sigma}^2_\theta = \frac{s_C^2 - s_{RC}^2}{rg}$$

$$\hat{\sigma}^2_\tau = \frac{s_R^2 - s_{RC}^2}{cg}$$

Three-Way Classification (Model F): Unbiased estimates of σ'^2, σ'^2_{BC}, σ'^2_{AC}, σ'^2_{AB}, σ'^2_C, σ'^2_B, and σ'^2_A are given by:

$$\hat{\sigma}^2 = s^2_{\text{res.}}$$

$$\hat{\sigma}^2_{BC} = \frac{s_{CG}^2 - s^2_{\text{res.}}}{r}$$

$$\hat{\sigma}^2_{AC} = \frac{s_{RG}^2 - s^2_{\text{res.}}}{c}$$

(31.27)

$$\hat{\sigma}^2_{AB} = \frac{s_{RC}^2 - s^2_{\text{res.}}}{g}$$

$$\hat{\sigma}^2_C = \frac{s_G^2 + s^2_{\text{res.}} - s_{RG}^2 - s_{CG}^2}{rc}$$

$$\hat{\sigma}^2_B = \frac{s_C^2 + s^2_{\text{res.}} - s_{RC}^2 - s_{CG}^2}{rg}$$

$$\hat{\sigma}^2_A = \frac{s_R^2 + s^2_{\text{res.}} - s_{RC}^2 - s_{RG}^2}{cg}$$

The above estimates could be combined in certain circumstances to estimate the percentage of the overall variability due to each of the component factors. Thus let X_i represent the individual units of product turned out in a given interval of time k by a random selection of the production factors from the respective factor universe, then the total variance of the individual items X_i, over a large number of intervals, k, will be given approximately by

(31.28) $$\hat{\sigma}_X'^2 = \hat{\sigma}_A'^2 + \hat{\sigma}_B'^2 + \hat{\sigma}_C'^2 + \hat{\sigma}_{AB}'^2 + \hat{\sigma}_{AC}'^2 + \hat{\sigma}_{BC}'^2 + \hat{\sigma}'^2$$

and the proportion of this total variance contributed by each factor can be computed by dividing by $\sigma_X'^2$.

For a numerical illustration, let us return once again to the thermometer-analyst problem. Consider the data of Table 30.10. Here we have a three-factor model with

Analyst mean square = 2.52

Thermometer mean square = 1.54

Day mean square = 0.84

Analyst-thermometer interaction mean square = 0.45

Analyst-day interaction mean square = 0.22

Thermometer-day interaction mean square = 0.34

Residual mean square = 0.05

The experimental variance (σ'^2) can thus be estimated at

$$\hat{\sigma}^2 = 0.05$$

the *TD* interaction variance ($\sigma_{TD}'^2$) may be estimated at

$$\hat{\sigma}_{TD}^2 \frac{0.34 - 0.05}{3} = 0.0967$$

The *AD* interaction variance ($\sigma_{AD}'^2$) may be estimated at

$$\hat{\sigma}_{AD}^2 = \frac{0.22 - 0.05}{4} = 0.0425$$

the *AT* interaction variance ($\sigma_{AT}'^2$) may be estimated at

$$\hat{\sigma}_{AT}^2 = \frac{0.45 - 0.05}{2} = 0.200$$

the day variance ($\sigma_D'^2$) may be estimated at

$$\hat{\sigma}_D^2 = \frac{0.84 + 0.05 - 0.22 - 0.34}{(3)(4)} = 0.0275$$

the thermometer variance ($\sigma_T'^2$) may be estimated at

$$\hat{\sigma}_T^2 = \frac{1.54 + 0.05 - 0.45 - 0.34}{(3)(2)} = 0.1333$$

and the analyst variance ($\sigma_A'^2$) may be estimated at

$$\hat{\sigma}_A^2 = \frac{2.52 + 0.05 - 0.45 - 0.22}{(4)(2)} = 0.2375$$

If we assume that a measurement is made by selecting an analyst at random from a large number of such analysts (an unlikely assumption), a thermometer at random from a large number of thermometers, and that day-to-day effects occur at random over time, then the variance of measurements may be estimated at

$$\hat{\sigma}_X^2 = 0.2375 + 0.1333 + 0.0275 + 0.2000 + 0.0425 + 0.0967 + 0.0500 = 0.7875$$

It may be concluded that considerably greater precision of measurement could be attained by training analysts to work alike and by more careful zeroing of thermometers.

3.3. Mixed Models

The analysis of components of variance for mixed models is a composite of the analyses for fixed and random models. Table 31.4 presents the expected values of the various mean squares for several mixed models. Unbiased estimates of the individual components can be obtained as follows:

Model G: Unbiased estimates of σ'^2, $\sigma_\theta'^2$, and $\sigma_\tau'^2$ and are given by

$$\begin{aligned} \hat{\sigma}^2 &= s_{\text{res.}}^2 \\ \hat{\sigma}_\theta^2 &= \frac{s_C^2 - s_{\text{res.}}^2}{r} \qquad (31.29) \\ \hat{\sigma}_\tau^2 &= \frac{r-1}{cr}(s_R^2 - s_{\text{res.}}^2) \end{aligned}$$

Model H: Unbiased estimates of σ'^2, $\sigma_\psi'^2$, $\sigma_\theta'^2$, and $\sigma_\tau'^2$ are given by

$$\begin{aligned} \hat{\sigma}^2 &= s_e^2 \\ \hat{\sigma}_\psi^2 &= \frac{r-1}{rg}(s_{RC}^2 - s_e^2) \\ \hat{\sigma}_\theta^2 &= \frac{s_C^2 - s_e^2}{rg} \qquad (31.30) \\ \hat{\sigma}_\tau^2 &= \frac{r-1}{cgr}(s_R^2 - s_{RC}^2) \end{aligned}$$

Model J: Unbiased estimates of σ'^2, $\sigma_{BC}'^2$, $\sigma_{AC}'^2$, $\sigma_{AB}'^2$, $\sigma_C'^2$, $\sigma_B'^2$, and $\sigma_A'^2$ are given by

$$\begin{aligned} \hat{\sigma}^2 &= s_{\text{res.}}^2 \\ \hat{\sigma}_{BC}^2 &= \frac{g-1}{rg}(s_{CG}^2 - s_{\text{res.}}^2) \end{aligned}$$

$$\hat{\sigma}^2_{AC} = \frac{(r-1)(g-1)}{crg}(s^2_{RG} - s^2_{\text{res.}})$$

(31.31)
$$\hat{\sigma}^2_{AB} = \frac{r-1}{gr}(s^2_{RC} - s^2_{\text{res.}})$$

$$\hat{\sigma}^2_{C} = \frac{g-1}{rcg}(s^2_{G} - s^2_{CG})$$

$$\hat{\sigma}^2_{B} = \frac{(s^2_{C} - s^2_{\text{res.}})}{rg}$$

$$\hat{\sigma}^2_{A} = \frac{r-1}{cgr}(s^2_{R} - s^2_{RC})$$

It will be noted that estimates (31.29) to (31.31) will be unbiased despite the failure of certain variance ratios to follow strictly the F distribution.[25]

To illustrate the above, suppose that in the thermometer problem we assume that both analysts and days are fixed factors, but thermometers are random. Then referring to the data of Table 30.10, which is relisted in Section 3.2 above, we would estimate the various components of variance as follows:

The residual variance σ'^2 can be estimated at

$$\hat{\sigma}^2 = 0.05$$

the thermometer-day interaction variance (σ'^2_{TD}) may be estimated at

$$\hat{\sigma}^2_{TD} = \frac{1}{(3)(2)}(0.34 - 0.05) = 0.048$$

the analyst-day interaction variance (σ'^2_{AD}) may be estimated at

$$\hat{\sigma}^2_{AD} = \frac{(2)(1)}{(4)(3)(2)}(0.22 - 0.05) = 0.014$$

the analyst-thermometer interaction variance (σ'^2_{AT}) may be estimated at

$$\hat{\sigma}^2_{AT} = \frac{2}{(2)(3)}(0.45 - 0.05) = 0.133$$

the day variance (σ'^2_{D}) may be estimated at

$$\hat{\sigma}^2_{D} = \frac{1}{(3)(4)(2)}(0.84 - 0.34) = 0.021$$

the thermometer variance (σ'^2_{T}) may be estimated at

$$\hat{\sigma}^2_{T} = \frac{1}{(3)(2)}(1.54 - 0.05) = 0.248$$

[25] See Chapter 30, Section 1.2.3.

TABLE 31.4
Expected Values of Mean Squares in Some Analysis of Variance Models: Mixed Effects, Random Factors from Infinite Universes

Model	*Mean Square**	*Unbiased Estimate of—†*
Two-Way Classification		
Model G:		
$X_{ij} = \mu + \tau_i + \theta_j + \epsilon_{ij}$	$s_R^2 = \dfrac{c\,\Sigma(\bar{X}_i - \bar{X})^2}{r-1}$	$\sigma'^2 + \dfrac{cr}{r-1}\sigma_\tau'^2$
μ, τ—fixed		
θ, ϵ—random	$s_C^2 = \dfrac{r\,\Sigma(\bar{X}_j - \bar{X})^2}{c-1}$	$\sigma'^2 + r\sigma_\theta'^2$
$i = 1, 2, \ldots, r$		
$j = 1, 2, \ldots, c$	$s_{\text{res.}}^2 = \dfrac{\Sigma\Sigma(X_{ij} - \bar{X}_i - \bar{X}_j + \bar{X})^2}{(r-1)(c-1)}$	σ'^2
Special Definition:		
$\sigma_\tau'^2 = \dfrac{\Sigma\tau_i^2}{r}$		
Model H:		
$X_{ijk} = \mu + \tau_i + \theta_j + \psi_{ij} + \epsilon_{ijk}$	$s_R^2 = \dfrac{cg\,\Sigma(\bar{X}_i - \bar{X})^2}{r-1}$	$\sigma'^2 + \dfrac{gr}{r-1}\sigma_\psi'^2 + \dfrac{cgr}{r-1}\sigma_\tau'^2$
μ, τ—fixed		
θ, ϵ—random		
ψ—mixed	$s_C^2 = \dfrac{rg\,\Sigma(\bar{X}_j - \bar{X})^2}{c-1}$	$\sigma'^2 + rg\sigma_\theta'^2$
$i = 1, 2, \ldots, r$		
$j = 1, 2, \ldots, c$		
$k = 1, 2, \ldots, g$	$s_{RC}^2 = \dfrac{g\,\Sigma\Sigma(\bar{X}_{ij} - \bar{X}_i - \bar{X}_j + \bar{X})^2}{(r-1)(c-1)}$	$\sigma'^2 + \dfrac{gr}{r-1}\sigma_\psi'^2$
Special Definitions:		
$\sigma_\tau'^2 = \dfrac{\Sigma\tau_i^2}{r}$	$s_e^2 = \dfrac{\Sigma\Sigma\Sigma(\bar{X}_{ijk} - \bar{X}_{ij})^2}{rc(g-1)}$	σ'^2
$\sigma_\psi'^2$ = variance over θ of ψ_{ij} (Assumed same for all i)		

Model	Mean square*	Expected value†
Three-Way Classification		
Model J:		
$X_{ijk} = \mu + a_i^A + a_j^B + a_k^C + m_{ij}^{AB} + a_{ik}^{AC} + m_{jk}^{BC} + \epsilon_{ijk}$	$s_R^2 = \frac{cg\,\Sigma(\bar{X}_i - \bar{X})^2}{r-1}$	$\sigma'^2 + \frac{gr}{r-1}\sigma'^2_{AB} + \frac{cgr}{r-1}\sigma'^2_A$
μ, A, C, AC—fixed B, ϵ—random AB, BC—mixed	$s_C^2 = \frac{rg\,\Sigma(\bar{X}_j - \bar{X})^2}{c-1}$	$\sigma'^2 + rg\,\sigma'^2_B$
$i = 1, 2, \ldots, r$ $j = 1, 2, \ldots, c$ $k = 1, 2, \ldots, g$	$s_G^2 = \frac{rc\,\Sigma(\bar{X}_k - \bar{X})^2}{g-1}$	$\sigma'^2 + \frac{rg}{g-1}\sigma'^2_{BC} + \frac{rcg}{g-1}\sigma'^2_C$
Special Definitions: $\sigma'^2_A = \frac{\Sigma(a_i^A)^2}{r}$	$s_{RC}^2 = \frac{g\,\Sigma\Sigma(\bar{X}_{ij} - \bar{X}_i - \bar{X}_j + \bar{X})^2}{(r-1)(c-1)}$	$\sigma'^2 + \frac{gr}{r-1}\sigma'^2_{AB}$
$\sigma'^2_C = \frac{\Sigma(a_k^C)^2}{g}$	$s_{RG}^2 = \frac{c\,\Sigma\Sigma(\bar{X}_{ik} - \bar{X}_i - \bar{X}_k + \bar{X})^2}{(r-1)(g-1)}$	$\sigma'^2 + \frac{crg}{(r-1)(g-1)}\sigma'^2_{AC}$
$\sigma'^2_{AC} = \frac{\Sigma\Sigma(a_{ik}^{AC})^2}{rg}$	$s_{CG}^2 = \frac{r\,\Sigma\Sigma(\bar{X}_{jk} - \bar{X}_j - \bar{X}_k + \bar{X})^2}{(c-1)(g-1)}$	$\sigma'^2 + \frac{rg}{g-1}\sigma'^2_{BC}$
σ'^2_{AB} = Variance over B of m_{ij}^{AB} (Assumed same for all i)	$s_{\text{res.}}^2 = \frac{\Sigma\Sigma\Sigma(X_{ijk} - \bar{X}_{ij} - \bar{X}_{ik} - \bar{X}_{jk} + \bar{X}_i + \bar{X}_j + \bar{X}_k - \bar{X})^2}{(c-1)(r-1)(g-1)}$	σ'^2
σ'^2_{BC} = Variance over B *of* m_{jk}^{BC} (Assumed same for all k)		

* For short-cut methods of computation, see Chapter 29, Section 2.3 and Chapter 30, Sections 1.3 and 2.3.

† For the nature of the theoretical reasoning, see Appendix 1 (33). The general procedure is to substitute the model in each mean square, make use of the fact that fixed and mixed elements sum to zero over the levels of the fixed factor, and apply the mean-value operator *E*. Also see footnote † of Table 31.2.

and the analyst variance ($\sigma_A'^2$) may be estimated at

$$\hat{\sigma}_A^2 = \frac{2}{(4)(2)(3)}(2.52 - 0.45) = 0.173$$

3.4. Finite Universe Considerations

In a fixed-factor model, the "levels" of each factor make up a complete factor universe. In the random models so far considered, variable effects were viewed as random samples from *infinite* universes. The mixed models were a combination of the fixed and (infinite) random models.

Suppose now that all factors are random variables from *finite* universes. Then, if the variances of these universes are defined in special ways, the sample mean squares become unbiased estimates of functions of the variances in a manner illustrated in Table 31.5.

At first glance the introduction of finite universes may seem a further complication of an already rather complicated analysis. Actually, however, a study of the results for finite universes leads to simplification and unification of our thinking. For if a universe becomes infinite, then a finite universe factor such as $\left(1 - \frac{r}{r'}\right)$ approaches unity. Thus if in Table 31.5 all the *A*, *B*, and *C* universes become infinite, Model K becomes Model F of Table 31.3. If, on the other hand, all "samples" becomes complete universes, then such factors as $\left(1 - \frac{r}{r'}\right)$ become zero. Thus, if $r = r'$, $c = c'$, and $g = g'$, and if the special definitions of Table 31.5 are adopted, then Model K reduces to a straight fixed-factor model, similar to Model B of Table 31.2. (Note that $\sigma_A'^2$, $\sigma_{AB}'^2$, etc., are defined differently in Table 31.5 than in Table 31.2.) Finally, if some "samples" comprise the whole universe and the factors become fixed factors, while other samples are from infinite universes, then Model K becomes a mixed model. If $r = r'$ and $g = g'$, for example, and c' approaches infinity, then Model K becomes Model J of Table 31.4. (Note again the differences in definitions.)

The finite-universe model is thus the master model. The method of forming the coefficients of this model is suggested by Table 31.5. For further details the reader is referred to the book by Bennett and Franklin referenced in the footnote to the table.

3.5. Nested Classifications

For pure nested classifications, the various components of variance are readily found by subtracting one mean square from that immediately above it and dividing by the appropriate constant. The expected values of the various mean squares are shown in Table 31.6 for a relatively simple nested model.

TABLE 31.5
Expected Values of Mean Squares in a Finite Universe Model: All Variable Factors Random

Model	*Mean Square*	*Unbiased Estimate of—**
Model K:	$s_R^2 = \frac{cg\Sigma(\bar{X}_i - \bar{X})^2}{r-1}$	$\sigma'^2 + g\left(1 - \frac{c}{c'}\right)\sigma'^2_{AB} + c\left(1 - \frac{g}{g'}\right)\sigma'^2_{AC} + cg\sigma'^2_A$
$X_{ijk} = \mu + a_i^A + a_j^B + a_k^C + a_{ij}^{AB} + a_{ik}^{AC} + a_{jk}^{BC} + \epsilon_{ijk}$	$s_C^2 = \frac{rg\Sigma(\bar{X}_j - \bar{X})^2}{c-1}$	$\sigma'^2 + g\left(1 - \frac{r}{r'}\right)\sigma'^2_{AB} + r\left(1 - \frac{g}{g'}\right)\sigma'^2_{BC} + rg\sigma'^2_B$
$i = 1, 2, \ldots, r$ $j = 1, 2, \ldots, c$ $k = 1, 2, \ldots, g$	$s_G^2 = \frac{rc\Sigma(\bar{X}_k - \bar{X})^2}{g-1}$	$\sigma'^2 + c\left(1 - \frac{r}{r'}\right)\sigma'^2_{AC} + rc\left(1 - \frac{c}{c'}\right)\sigma'^2_{BC} + rc\sigma'^2_C$
Universes of size r', c', and g' Special Definitions:	$s_{RC}^2 = \frac{g\Sigma\Sigma(\bar{X}_{ij} - \bar{X}_i - \bar{X}_j + \bar{X})^2}{(r-1)(c-1)}$	$\sigma'^2 + g\sigma'^2_{AB}$
$\sigma'^2_A = \frac{\sum_{i=1}^{r'} (a_i^A)^2}{r'-1}$	$s_{RG}^2 = \frac{c\Sigma\Sigma(\bar{X}_{ik} - \bar{X}_i - \bar{X}_k + \bar{X})^2}{(r-1)(g-1)}$	$\sigma'^2 + c\sigma'^2_{AC}$
$\sigma'^2_B = \frac{\sum_{j=1}^{c'} (a_j^B)^2}{c'-1}$	$s_{CG}^2 = \frac{r\Sigma\Sigma(\bar{X}_{jk} - \bar{X}_j - \bar{X}_k + \bar{X})^2}{(c-1)(g-1)}$	$\sigma'^2 + r\sigma'^2_{BC}$
$\sigma'^2_C = \frac{\sum_{k=1}^{g'} (a_k^C)^2}{g'-1}$	$s_{\text{res.}}^2 = \frac{\Sigma\Sigma\Sigma(\bar{X}_{ijk} - \bar{X}_{ij} - \bar{X}_{ik} - \bar{X}_{jk} + \bar{X}_i + \bar{X}_j + \bar{X}_k - \bar{X})^2}{(c-1)(r-1)(g-1)}$	σ'^2
$\sigma'^2_{AB} = \frac{\sum_{i=1}^{r'}\sum_{j=1}^{c'} (a_{ij}^{AB})^2}{(r'-1)(c'-1)}$		
$\sigma'^2_{AC} = \frac{\sum_{i=1}^{r'}\sum_{k=1}^{g'} (a_{ik}^{AC})^2}{(r'-1)(g'-1)}$		
$\sigma'^2_{BC} = \frac{\sum_{j=1}^{c'}\sum_{k=1}^{g'} (a_{jk}^{BC})^2}{(c'-1)(g'-1)}$		

* See Carl A. Bennett and Norman L. Franklin, *Statistical Analysis in Chemistry and the Chemical Industry* (New York: John Wiley & Sons, 1954), p. 394. The derivation will be found in this reference, pp. 474–77. For three-factor interactions we have the product of two finite universe correction factors, e.g., $k\left(1 - \frac{c}{c'}\right)\left(1 - \frac{r}{r'}\right)\sigma'^2_{ABC}$ would be the three-factor interaction term in the expression for the *A* mean square.

TABLE 31.6
Expected Values of Mean Squares in a Pure Nested Model: All Effects Random except μ

*Model**	*Mean Square*	*Unbiased Estimate of—*
$X_{ijk} = \mu + \tau_i + \theta_{j(i)} + \epsilon_{ijk}$ $i = 1, 2, \ldots, r$	$s_R^2 = \dfrac{cg\Sigma(\bar{X}_i - \bar{X})^2}{r-1}$	$\sigma'^2 + g\sigma_\theta'^2 + cg\sigma_\tau'^2$
$j = 1, 2, \ldots, c$ $k = 1, 2, \ldots, g$	$s_{C(R)}^2 = \dfrac{g\Sigma\Sigma(\bar{X}_{ij} - \bar{X}_i)^2}{r(c-1)}$	$\sigma'^2 + g\sigma_\theta'^2$
	$s_e^2 = \dfrac{\Sigma\Sigma\Sigma(X_{ijk} - \bar{X}_{ij})^2}{rc(g-1)}$	σ'^2

* $\theta_{j(i)}$ means that the j classification is nested in the i classification.

For this model, we would have the following unbiased estimates of σ'^2, $\sigma_\theta'^2$, and $\sigma_\tau'^2$:

(31.32)
$$\hat{\sigma}^2 = s_e^2$$
$$\hat{\sigma}_\theta^2 = \frac{s_{C(R)}^2 - s_e^2}{g}$$
$$\hat{\sigma}_\tau^2 = \frac{s_R^2 - s_{C(R)}^2}{cg}$$

The analysis of other nested models follows a similar pattern. For a comprehensive catalogue of nested and partially nested models, the reader is referred to H. Leon Harter and Mary D. Lum, *Partially Hierarchal Models in the Analysis of Variance.* (See references in Appendix V.)

3.6 Confidence Limits for Components of Variance: Random Effects, Infinite Universes[26]

The estimates of the components of variance given in preceding sections are, of course, all subject to sampling errors, and unfortunately these errors are likely to be large since the degrees of freedom on which these estimates are based are usually very small. It is also unfortunate that simple exact methods are not available for determining confidence limits for the universe values. Standard error formulas have been available for some time, but the distributions of the statistics themselves are not known precisely, even in the case of normal variation of the basic measurements. The following is a method that is offered by Davies *et alia*[27] and is adopted here because (1)

[26] Confidence limits for components of variance for fixed and mixed factors are beyond the scope of this text.

[27] Owen L. Davies (ed.), *Statistical Methods in Research and Production with Special Reference to the Chemical Industry* (3d ed.; London: Oliver & Boyd, Ltd., 1957), p. 103.

it is relatively simple and (2) it makes a clear-cut distinction as to just how far the method is exact and where it becomes inexact.

The essence of the Davies method is to set up confidence limits for ratios of variances and linear functions of variances. This can be done with known risks in the same manner as in Section 3.6 of Chapter 26 confidence limits were established for the ratio of two universe variances. When the ratio of two linear functions of variances contains other variances than those in which we are immediately interested, the Davies method is to substitute their sample estimates. At this point the confidence limits for the ratio of the two variances in question become inexact, but it is believed that the results obtained are nevertheless reasonably good.

We shall explain the procedure with reference to the random-effects model listed in Table 31.3 as Model F. Consider the simplest case first, viz, that of confidence limits for the residual variance σ'^2. Since $\frac{(c-1)(r-1)(g-1)s^2_{\text{res.}}}{\sigma'^2}$ has a χ^2 distribution with $(c-1)(r-1)(g-1)$ degrees of freedom, 0.95 confidence limits for σ'^2 will be given by

(31.33)

$$\frac{(c-1)(r-1)(g-1)s^2_{\text{res.}}}{\chi^2_{0.025}[\nu=(c-1)(r-1)(g-1)]} \le \sigma'^2 \le \frac{(c-1)(r-1)(g-1)s^2_{\text{res.}}}{\chi^2_{0.975}[\nu=(c-1)(r-1)(g-1)]}$$

Compare Section 2.3.1 of Chapter 24.

Consider next σ'^2_{AB}. Since the mean value of s^2_{RC} is $\sigma'^2 + g\sigma'^2_{AB}$ and the mean value of $s^2_{\text{res.}}$ is σ'^2 (see Table 31.3), the ratio $[s^2_{RC}/(\sigma'^2 + g\sigma'^2_{AB})]/[s^2_{\text{res.}}/\sigma'^2]$ has an F distribution with $\nu_1 = (r-1)(c-1)$ and $\nu_2 = (c-1)(r-1)(g-1)$. Hence 0.95 confidence limits for the ratio $(\sigma'^2 + g\sigma'^2_{AB})/\sigma'^2$ will be[28]

$$\frac{s^2_{RC}}{F_{0.025}(\nu_1, \nu_2)s^2_{\text{res.}}} \le \frac{\sigma'^2 + g\sigma'^2_{AB}}{\sigma'^2} \le \frac{s^2_{RC}F_{0.025}(\nu_2, \nu_1)}{s^2_{\text{res.}}}$$

and 0.95 confidence limits for σ'^2_{AB}/σ'^2 will be given by

(31.34)
$$\frac{s^2_{RC}}{gF_{0.025}(\nu_1, \nu_2)s^2_{\text{res.}}} - \frac{1}{g} \le \frac{\sigma'^2_{AB}}{\sigma'^2} \le \frac{s^2_{RC}F_{0.025}(\nu_2, \nu_1)}{gs^2_{\text{res.}}} - \frac{1}{g}$$

---

[28] See Section 3.6 of Chapter 26. The underlying reasoning runs as follows: If $(s_1^2/\sigma_1'^2)/(s_2^2/\sigma_2'^2)$ has an F distribution, then

$$\text{Prob.}\left(F_{0.975} \le \frac{s_1^2/\sigma_1'^2}{s_2^2/\sigma_2'^2} \le F_{0.025}\right) = 0.95$$

But this is the same as saying that

$$\text{Prob.}\left(\frac{s_1^2}{F_{0.025}(\nu_1, \nu_2)s_2^2} \le \frac{\sigma_1^2}{\sigma_2'^2} \le \frac{s_1^2F_{0.025}(\nu_2, \nu_1)}{s_2^2}\right) = 0.95$$

which is the confidence interval statement. Note that it is a property of the F distribution that

$$F_{0.975}(\nu_1, \nu_2) = 1/F_{0.025}(\nu_2, \nu_1)$$

Given normal universes and random sampling, the above formulas are precise. If we are willing to accept rough approximations, we can take a step further and estimate 0.95 confidence limits for σ'^2_{AB} by multiplying the confidence limits for σ'^2_{AB}/σ'^2 by the unbiased estimate of σ'^2, viz. $s^2_{\text{res.}}$ Thus approximate 0.95 confidence limits for σ'^2_{AB} will be given by

$$\text{(31.35)}\qquad \frac{s^2_{RC} - s^2_{\text{res.}}F_{0.025}(\nu_1, \nu_2)}{gF_{0.025}(\nu_1, \nu_2)} \leq \sigma'^2_{AB} \leq \frac{s^2_{RC}F_{0.025}(\nu_2, \nu_1) - s^2_{\text{res.}}}{g}$$

In a similar way 0.95 confidence limits can be derived for the ratios σ'^2_{AC}/σ'^2 and σ'^2_{BC}/σ'^2 and approximate limits for σ'^2_{AC} and σ'^2_{BC}.

When we come to σ'^2_A, σ'^2_B, and σ'^2_C, we find that the precise analysis can be applied only to a ratio of two linear functions of variance components. Thus, following the preceding method of attack we find that 0.95 confidence limits for the ratio $(\sigma'^2 + c\sigma'^2_{AC} + g\sigma'^2_{AB} + cg\sigma'^2_A)/(\sigma'^2 + c\sigma'^2_{AC})$ will be

$$\frac{s^2_R}{F_{0.025}(\nu_1, \nu_2)s^2_{RG}} \leq \frac{\sigma'^2 + c\sigma'^2_{AC} + g\sigma'^2_{AB} + cg\sigma'^2_A}{\sigma'^2 + c\sigma'^2_{AC}} \leq \frac{s^2_R F_{0.025}(\nu_2, \nu_1)}{s^2_{RG}}$$

where $\nu_1 = r - 1$ and $\nu_2 = (r - 1)(g - 1)$. This can be put in the form

$$\text{(31.36)}\qquad \frac{s^2_R}{gF_{0.025}(\nu_1, \nu_2)s^2_{RG}} - \frac{1}{g} \leq \frac{\sigma'^2_{AB} + c\sigma'^2_A}{\sigma'^2 + c\sigma'^2_{AC}} \leq \frac{s^2_R F_{0.025}(\nu_2, \nu_1)}{gs^2_{RG}} - \frac{1}{g}$$

If now we use $s^2_{\text{res.}}$ as an unbiased estimate of σ'^2, $\dfrac{s^2_{RG} - s^2_{\text{res.}}}{c}$ as an unbiased estimate of σ'^2_{AC} and $\dfrac{s^2_{RC} - s^2_{\text{res.}}}{g}$ as an unbiased estimate of σ'^2_{AB}, then we can make rough estimates of the 0.95 confidence limits for σ'^2_A.

Let us illustrate the above by reference to the analysis of Section 3.2. This it will be recalled was based on the data of Table 30.10. Here an unbiased estimate of σ'^2 was given by $s^2_{\text{res.}}$, which equaled $\dfrac{0.312}{6} = 0.05$. Ninety-five percent confidence limits for σ'^2 will therefore be immediately given by

$$\frac{0.312}{14.4} \leq \sigma'^2 \leq \frac{0.312}{1.24}$$

or

$$0.022 \leq \sigma'^2 \leq 0.252$$

In Section 3.2 an unbiased estimate of σ'^2_{AT} was taken as 0.200. From (31.35) we have 0.95 confidence limits for σ'^2_{AT}/σ'^2 given by[29]

$$\frac{0.45}{2(5.82)(0.05)} - \frac{1}{2} \leq \frac{\sigma'^2_{AT}}{\sigma'^2} \leq \frac{(0.45)(5.82)}{2(0.05)} - \frac{1}{2}$$

---

[29] Note that since $\nu_1 = \nu_2$, $F(\nu_1, \nu_2) = F(\nu_2, \nu_1)$.

or

$$0.273 \le \frac{\sigma'^2_{AT}}{\sigma'^2} \le 25.690$$

Since σ'^2 is estimated at 0.05, rough 0.95 limits for σ'^2_{AT} will be

$$(0.273)(0.05) \le \sigma'^2_{AT} \le (25.690)(0.05)$$

or

$$0.014 \le \sigma'^2_{AT} \le 1.284$$

Again in Section 3.2 an unbiased estimate of σ'^2_{AD} was taken as 0.0425. From (31.35) we have 0.95 confidence limits for σ'^2_{AD}/σ'^2 given by

$$\frac{(0.22)}{2(7.26)(0.05)} - \frac{1}{2} \le \frac{\sigma'^2_{AD}}{\sigma'^2} \le \frac{(0.22)(39.3)}{2(0.05)} - \frac{1}{2}$$

or

$$0 \le \frac{\sigma'^2_{AD}}{\sigma'^2} \le 85.960$$

Rough 0.95 confidence limits for σ'^2_{AD} will be given by

$$0 \le \sigma'^2_{AD} \le (85.960)(0.05)$$

or

$$0 \le \sigma'^2_{AD} \le 4.298$$

Finally we can attack the problem of confidence limits for σ'^2_T, for example, as follows. Ninety-five percent confidence limits for

$$(\sigma'^2_{TD} + 2\sigma'^2_T)/(\sigma'^2 + 2\sigma'^2_{AT})$$

will be

$$\frac{1.54}{3(6.60)(0.45)} - \frac{1}{3} \le \frac{\sigma'^2_{TD} + 2\sigma'^2_T}{\sigma'^2 + 2\sigma'^2_{AT}} \le \frac{1.54(14.7)}{3(0.45)} - \frac{1}{3}$$

or

$$0 \le \frac{\sigma'^2_{TD} + 2\sigma'^2_T}{\sigma'^2 + 2\sigma'^2_{AT}} \le 16.436$$

From this, by using the estimates of σ'^2, σ'^2_{AT}, and σ'^2_{TD} derived in Section 3.2 we may take rough limits for σ^2_T as equal to

$$0 \le \sigma'^2_T \le \frac{(16.436)[0.05 + 2(0.20)] - 0.0967}{2}$$

or

$$0 \le \sigma'^2_T \le 3.65$$

3.7 Negative Values and Subsequent Estimation

In estimating components of variance and determining lower confidence limits by the method described above, the arithmetical computations sometimes result in negative values (as in Section 3.6). The usual practice in this case is to take the estimate or lower confidence limit as zero (as was done in Section 3.6).

When a component of variance is estimated at zero, the question then arises as to how we proceed in making estimates further up the table. For example, if in Table 31.6, $s_C^2 < s_{\text{res.}}^2$, we shall get a negative value for $s_C^2 - s_{\text{res.}}^2$ and we shall estimate $\sigma_\theta'^2$ as equal to zero. Then we may ask: How do we estimate $\sigma_\tau'^2$? Mathematically, the simplest procedure is to continue on as if we had secured a positive value for $\sigma_\theta'^2$ and estimate $\sigma_\tau'^2$ by subtracting s_C^2 from s_R^2 and dividing by cg. We could, however, argue that since $\sigma_\theta'^2$ is estimated at 0, both s_C^2 and $s_{\text{res.}}^2$ are estimates of σ'^2 and that $s_{\text{res.}}^2$ will be a better estimate than s_C^2 since it has a larger number of degrees of freedom. We could then estimate $\sigma_\tau'^2$ by subtracting $s_{\text{res.}}^2$ from s_R^2 and dividing by cg. This will give us a smaller estimate than the former one. A third procedure would be to take a weighted average of s_C^2 and $s_{\text{res.}}^2$ $\left(\text{equal to } \dfrac{r(c-1)s_C^2 + rc(g-1)s_{\text{res.}}^2}{r(c-1)+rc(g-1)}\right)$, subtract this from s_R^2, and divide by cg. This will yield an estimate intermediate to the other two.

The statistician has his choice of these three methods. The first we know is unbiased and is therefore to be recommended if the degrees of freedom for s_C^2 are reasonably large. The other methods will give biased results, but these may not be serious and we may prefer to employ them since they will have a greater number of degrees of freedom. W. A. Thompson, Jr., in the *Annals of Mathematical Statistics* 33 (1960), pp. 273ff., suggests optimum estimates that call for pooling mean squares whenever a difference is negative. This pooled estimate is used in subsequent calculations and is also used as an estimate of the common variance. When calculations yield negative estimates of variance, they are given the value zero.

4. ESTIMATION OF PRECISION OF MEASURING INSTRUMENTS AND PRODUCT VARIABILITY

An important problem closely allied to the analysis of components of variance is the separation of instrumental errors from product variation. If we are measuring a group of items with a single instrument, there is no way we can determine how much the variation in observed measurements is due to actual variation in the items and how much to errors of measurement. In such a case the variance of the observations is made up of the sum of inseparable components—one, the variance in the items measured; the other, the

TABLE 31.7
Burning Times of Thirty Fuses

Fuse Burning Times (Seconds)			
Observer I	*Observer II*	*Observer III*	*Mean Times (Seconds)*
10.10	10.07	10.07	10.080
9.98	9.90	9.90	9.927
9.89	9.85	9.86	9.867
9.79	9.71	9.70	9.733
9.67	9.65	9.65	9.657
9.89	9.83	9.83	9.850
9.82	9.75	9.79	9.787
9.59	9.56	9.59	9.580
9.76	9.68	9.72	9.720
9.93	9.89	9.92	9.913
9.62	9.61	9.64	9.623
10.24	10.23	10.24	10.237
9.84	9.83	9.86	9.843
9.62	9.58	9.63	9.610
9.60	9.60	9.65	9.617
9.74	9.73	9.74	9.737
10.32	10.32	10.34	10.327
9.86	9.86	9.86	9.860
10.01	lost	10.03	10.020
9.65	9.64	9.65	9.647
9.50	9.49	9.50	9.497
9.56	9.56	9.55	9.557
9.54	9.53	9.54	9.537
9.89	9.89	9.88	9.887
9.53	9.52	9.51	9.520
9.52	9.52	9.53	9.523
9.44	9.43	9.45	9.440
9.67	9.67	9.67	9.670
9.77	9.76	9.78	9.770
9.86	9.84	9.86	9.853

SOURCE: Leslie E. Simon, "On the Relation of Instrumentation to Quality Control," *Instruments* 19 (1946), pp. 654–56.

variance in the errors of measurement.[30] If we employ two or more instruments, however, it is possible to separate out these two components of variance.

The following is a method for separating product variation from instrument variation that has been suggested by Frank E. Grubbs.[31] It will be explained with reference to the data of Table 31.7. This contains fuse burning

---

[30] Cf. Frank E. Grubbs, "On Estimating Precision of Measuring Instruments and Product Variability," *Journal of the American Statistical Association* 43 (1948), p. 246.

[31] Ibid., pp. 243–64.

TABLE 31.8
Differences between Observers' Reports of Fuse Burning Times

I − II	*II − III*	*I − III*	*I − II*	*II − III*	*I − III*
+0.03	0	+0.03	+0.01	−0.01	0
+0.08	0	+0.08	0	−0.02	−0.02
+0.04	−0.01	+0.03	0	0	0
+0.08	+0.01	+0.09	—	—	−0.02
+0.02	0	+0.02	+0.01	−0.01	0
+0.06	0	+0.06	+0.01	−0.01	0
+0.07	−0.04	+0.03	0	+0.01	+0.01
+0.03	−0.03	0	+0.01	−0.01	0
+0.08	−0.04	+0.04	0	+0.01	+0.01
+0.04	−0.03	+0.01	+0.01	+0.01	+0.02
+0.01	−0.03	−0.02	0	−0.01	−0.01
+0.01	−0.01	0	+0.01	−0.02	−0.01
+0.01	−0.03	−0.02	0	0	0
+0.04	−0.05	+0.01	+0.01	−0.02	−0.01
0	−0.05	−0.05	+0.02	−0.02	0

SOURCE: Leslie E. Simon, "On the Relation of Instrumentation to Quality Control," *Instruments* 19 (1946), pp. 654–56.

times for 30 fuses as recorded by three different observers—I, II, and III. In the last column is the mean time for all three observers.

The first step in the analysis is to compute the differences in the observations between observers. This is done in Table 31.8, which records the differences between I's and II's observations, between I's and III's observations, and between II's and III's observations. Taking the difference between two observations on the same fuse eliminates the true burning time and gives simply the differences in the erros of observation. Thus, if X represents the true burning time of a fuse and e_1, e_2, and e_3 represent the errors of observers I, II, and III, respectively, then the difference between I's observation and II's observation is equal to $(X + e_1) - (X + e_2) = e_1 - e_2$; the difference between I's and III's observations is equal to $(X + e_1) - (X + e_3) = e_1 - e_3$; and the difference between II's and III's observations is equal to $(X + e_2) - (X + e_3) = e_2 - e_3$. The data of Table 31.8, therefore, contain only the errors of observation and do not contain any of the variation in the burning time of the fuses.

The second step is to compute unbiased estimates of variances for each of the three sets of differences in Table 31.8. If we set $d_{12} = e_1 - e_2$, $d_{13} = e_1 - e_3$, and $d_{23} = e_2 - e_3$, then an unbiased estimate of $\sigma'^2_{e_1-e_2}$ is given by

$$s^2_{e_1-e_2} = \frac{\Sigma d^2_{12} - (\Sigma d_{12})^2/n}{n-1} \tag{31.37}$$

where $n = 30$, the number of fuses. An unbiased estimate of $\sigma'^2_{e_1-e_3}$ is given by

(31.37a)
$$s^2_{e_1-e_3} = \frac{\Sigma d^2_{13} - (\Sigma d_{13})^2/n}{n-1}$$

and an unbiased estimate of $\sigma'^2_{e_2-e_3}$ is given by

(31.37b)
$$s^2_{e_2-e_3} = \frac{\Sigma d^2_{23} - (\Sigma d_{23})^2/n}{n-1}$$

For the data of Table 31.8 the results are

$$s^2_{e_1-e_2} = 0.0007030 \text{ sec.}^2$$
$$s^2_{e_1-e_3} = 0.0008878 \text{ sec.}^2$$
$$s^2_{e_2-e_3} = 0.0003108 \text{ sec.}^2$$

The third step is to estimate the standard deviations of the individual observer's errors. This can be done by the use of certain formulas that Grubbs has derived. The formulas are as follows:

Estimate of $\sigma'^2_{e_1}$ equals

(31.38)
$$\tfrac{1}{2}(s^2_{e_1-e_2} + s^2_{e_1-e_3} - s^2_{e_2-e_3})$$

Estimate of $\sigma'^2_{e_2}$ equals

(31.38a)
$$\tfrac{1}{2}(s^2_{e_1-e_3} + s^2_{e_1-e_2} - s^2_{e_1-e_3})$$

Estimate of $\sigma'^2_{e_3}$ equals

(31.38b)
$$\tfrac{1}{2}(s^2_{e_1-e_3} + s^2_{e_2-e_3} - s^2_{e_1-e_2})$$

For the given data we have

Estimate of $\sigma'_{e_1} = 0.0253$ sec.

Estimate of $\sigma'_{e_2} = 0.0079$ sec.

Estimate of $\sigma'_{e} = 0.0157$ sec.

The final step is to use another of Grubbs's formulas to estimate the variance in the burning time of the fuses. This is:

Estimate of σ'^2_X equals

(31.39)
$$s^2_{\frac{(X+e_1)+(X+e_2)+(X+e_3)}{3}} - \frac{1}{18}(s^2_{e_1-e_2} + s^2_{e_1-e_3} + s^2_{e_2-e_3})$$

The first term is simply the variance of the mean of the three observations, which can be computed from the last column of Table 31.7. Thus, if the mean of the three observations for the ith fuse is equal to $\bar{X}_i$, we have

(31.40)
$$s^2_{\frac{(X+e_1)+(X_2+e_2)+(X_3+e_3)}{3}} = \frac{\Sigma \bar{X}_i^2 - (\Sigma \bar{X}_i)^2/n}{n-1}$$

The other terms of the formula were computed in step 2.

Applying these formulas to the data, we get the estimates

$$\hat{\sigma}_X^2 = 0.046098 - \frac{1}{18}(0.0007030 + 0.0008878 + 0.0003108) = 0.04599$$

and

$$\hat{\sigma}_X = 0.2145$$

The standard errors for the above estimates are

$$\text{Standard error of } \hat{\sigma}^2_{e_1} = \sqrt{\frac{2}{n-1}\sigma'^4_{e_1} + \frac{1}{n-1}(\sigma'^2_{e_1}\sigma'^2_{e_2} + \sigma'^2_{e_1}\sigma'^2_{e_3} + \sigma'^2_{e_2}\sigma'^2_{e_3})}$$

$$\text{Standard error of } \hat{\sigma}^2_{e_2} = \sqrt{\frac{2}{n-1}\sigma'^4_{e_2} + \frac{1}{n-1}(\sigma'^2_{e_1}\sigma'^2_{e_2} + \sigma'^2_{e_1}\sigma'^2_{e_3} + \sigma'^2_{e_2}\sigma'^2_{e_3})}$$

(31.41)

$$\text{Standard error of } \hat{\sigma}^2_{e_3} = \sqrt{\frac{2}{n-1}\sigma'^4_{e_3} + \frac{1}{n-1}(\sigma'^2_{e_1}\sigma'^2_{e_2} + \sigma'^2_{e_1}\sigma'^2_{e_3} + \sigma'^2_{e_2}\sigma'^2_{e_3})}$$

and

$$\text{Standard error of } \hat{\sigma}^2_X = \sqrt{\frac{2}{n-1}\sigma'^4_X + \frac{1}{n-1}\left\{\frac{4}{9}(\sigma'^2_X\sigma'^2_{e_1} + \sigma'^2_X\sigma'^2_{e_2} + \sigma'^2_X\sigma'^2_{e_3}) + \frac{1}{9}(\sigma'^2_{e_1}\sigma'^2_{e_2} + \sigma'^2_{e_1}\sigma'^2_{e_3} + \sigma'^2_{e_2}\sigma'^2_{e_3})\right\}}$$

In practice we rarely know $\sigma'^2_{e_1}$, $\sigma'^2_{e_2}$, $\sigma'^2_{e_3}$, and σ'^2_X and hence have to use the estimates of these values derived from the sample.

For discussion of the procedures to be followed when there are n instruments or observers, and for other matters relating to Grubbs's method, the reader is referred to Grubbs's original article in the *Journal of the American Statistical Association* 43 (1948), pp. 243–64.

5. TESTS OF GENERAL HOMOGENEITY

5.1. Lambda Tests

As we have seen, an F test can be used to test the hypothesis that the means of several normal universes are equal, *given that their standard deviations are the same.* Likewise, Bartlett's test or the ratio Max. s_i^2/Min. s_i^2 can be used to test whether the variances of several normal universes are the same, regardless of whether the means are equal or not.

There may be cases, however, in which we simply want to know whether two normal universes are different and do not care whether it is due to a difference in means or a difference in variances. This would be a test of general homogeneity. Such a test is offered by Lambda tests. These tests permit a mathematical statistical decision pertaining to simultaneous variations in subgroup means and variances. If we have a set of g samples all of size m, we can test the joint hypothesis that $\bar{X}'_1 = \bar{X}'_2 = \ldots = \bar{X}'_g$ and $\sigma'_1 = \sigma'_2 = \ldots = \sigma'_g$ by computing the statistic

$$\Lambda_0 = \frac{\sigma_0^2}{[(\sigma_1^2)(\sigma_2^2) \ldots (\sigma_g^2)]^{1/g}} \tag{31.42}$$

where σ_0^2 is the second moment of the whole set of data given by

$$\sigma_0^2 = \frac{\sum_i \sum_j (X_{ij} - \bar{X})^2}{gm} = \frac{\sum_i \sum_j X_{ij}^2 - \left(\sum_i \sum_j X_{ij}\right)^2 \Big/ gm}{gm} \tag{31.43}$$

and σ_j^2 is the second moment of the jth sample, given by

$$\sigma_j^2 = \frac{\sum_i (X_{ij} - \bar{X}_j)^2}{m} = \frac{\sum_i X_{ij}^2 - \left(\sum_i X_{ij}\right)^2 \Big/ m}{m} \tag{31.44}$$

An alternative form is

$$\Lambda_0 = \frac{\sigma_0^2}{\frac{m-1}{m}[(s_1^2)(s_2^2) \ldots (s_g^2)]^{1/g}} \tag{31.45}$$

where

$$s_j^2 = \frac{\sum_i (X_{ij} - \bar{X}_j)^2}{m-1} \tag{31.46}$$

In (31.42) the denominator of Λ_0 will be recognized as the geometric mean of the g sample second moments.

Tables of $L_o = 1/\Lambda_0$ were originally computed by Mahalanobis and published in *Sankhyā: The Indian Statistical Journal* 1 (1963), p. 121. Tables of Λ_0 were computed by Dudley J. Cowden and published in his book on *Statistical Methods in Quality Control,* p. 687. Table S in Appendix II is a reproduction of Cowden's table. This gives the 0.05 points of the distribution of Λ_0 for various values of g and m and may be used to test at the 0.05 level the joint hypothesis $\bar{X}'_1 = \bar{X}'_2 = \ldots = \bar{X}'_g$ and $\sigma'_1 = \sigma'_2 = \ldots = \sigma'_g$.

As an example, consider again the eight samples of five each that constitute the first half of Table 26.1. The s^2 for each of these samples of cans of tomatoes was computed in Section 2 above, and σ_0^2 may be computed to be 3.028.

These values give us

$$\Lambda_0 = \frac{3.028}{\frac{4}{5}[(0.675)(2.050)(1.000)(0.375)(4.750)(5.075)(1.000)(5.325)]^{1/8}}$$

Taking logs, we have

$$\log \Lambda_0 = \log 3.028 - \log (4/5) - 1/8 \log [(0.675)(2.050)(1.000)(0.375)(4.750)(5.075)(1.000)(5.325)] = 0.4812 - (9.9031 - 10) - 1/8(1.8236) = 0.3501$$

and $\Lambda_0 = 2.24$. For $\alpha = 0.05$, $g = 5$, $m = 5$, Table S gives the critical Λ_0 as 2.22, and for $\alpha = 0.05$, $g = 10$, $m = 5$, it gives the critical Λ_0 as 2.06. The critical Λ_0 for $\alpha = 0.05$, $g = 8$, $m = 5$ lies between these tabular values. Hence the sample Λ_0 lies beyond the critical value, and we reject the hypothesis that universe means and variances are equal. In terms of the problem under discussion, it means that if we accept the assumption of a normal distribution of product, then we must conclude that there were changes in either the mean or the variance of the process over the period in question. Since a previous test indicated that the sample variances were not significantly different, it is to be inferred that the difference is due to significant variations in the mean.

In general, it is possible to use the Λ_0 test as a mathematical procedure for determining whether a normally distributed process shows evidence of lack of control.[32] It has the following disadvantages, however. It is a test based solely on the amount of variation and unlike a control chart does not give a picture of the pattern of variation. If a Λ_0 test shows lack of control, further analysis is necessary to find out whether the cause is a significant variation in the mean or variance or both. Finally, when there is no great assurance that the process is normally distributed, a significant Λ_0 may actually reveal a departure from normality rather than a significant variation in the mean or variance. An incidental disadvantage for use in process control is that tables do not give the 0.001 point which is nearer the value of α usually used in control analysis.

5.2. Multi-Vari Charts

A graphical method of studying joint homogeneity of means and variances is the multi-vari chart developed by Leonard Seder. This is a quick and penetrating procedure for analyzing variation when there is more than one case in a class. It is graphical, hence generally easy to follow. It is quick and simple, since tests based on it use the range.

The procedure may be illustrated with respect to the data of Table 31.9. Here are counts of a fastidious organism classified by day, technician, A.M. and P.M., and subsample. The multi-vari procedure is to plot the range of the subsample results as a line on a chart of actual values. For example, with reference to Figure 31.2, the results obtained by technician 1 on the morning of the third day, viz 35, 47, 73, are plotted at the time indicated,

[32] Dudley J. Cowden, *Statistical Methods in Quality Control* (Englewood Cliffs, N.J.: Prentice-Hall, 1957), chap. xi.

TABLE 31.9
Some Experimental Results on a Fastidious Organism (plate counts of organisms)

Sub-Sample	*Day 1*		*Day 2*		*Day 3*	
	Tech. 1	*Tech. 2*	*Tech. 1*	*Tech. 2*	*Tech. 1*	*Tech. 2*
A.M.						
A1	45,48,50	55,43,C	43, 36, 22	174, 90,134	35, 47, 73	209,190,188
A2	47,45,C*	50,44,52	104, 94,116	143,147,142	124,133,103	186,104,236
A3	48,51,41	65,56,C	118,120,108	144,130,132	100, 75,C	250,268,121
A4	37,40,48	47,55,58	83,104,C	110,119,137	86,111, 97	230,216,190
A5	46,43,39	54,60,46	120,101,106	C, C, C	159,172,139	233,228,C
A6	53,40,C	60,49,55	117,122, 97	120,138,142	156,122,125	206,204,200
P.M.						
P1	65,53,50	59,51,C	124,112, 99	117, C,C	124,156,C	36, 85, 82
P2	40,52,44	52,44,C	86,130,102	131,120,C	131,127, 74	180,172, 91
P3	50,64,55	59,44,C	97,122,C	122,113,126	166,145,C	200,134,148
P4	47,54,C	52,46,50	137,110,C	106, 99,C	82,106,105	174,205,186
P5	53,39,40	50,39,48	130,109,109	130, 97,108	136,118,182	144,121,136
P6	52,52,C	41,48,50	120,108,107	113,126,109	104,137,117	82, 95,101

* C = Contaminated plate.

as a line running from 35 to 73. Other subsamples are plotted in the same way, the dotted line distinguishing technician 1 from technician 2, and the lines being spaced horizontally to distinguish the day and time of day. Only subsamples of three are shown; those with missing data being omitted from the chart.

Figure 31.2 shows a very striking result. Without any further analysis it is immediately seen that the day-to-day counts vary markedly. It is also obvious that the within-sample variation also varies markedly from day to day. Finally, at certain times (say the morning of the third day), the results of the two technicians differ strikingly. All this is seen directly from the chart itself.

If numerical tests are desired, we can apply an F test based on ranges or a studentized-range test. Thus with respect to Figure 31.2 we may ask whether the within-sample variation differs from day 2 to day 3. To find the answer we average the sample ranges (not distinguishing technicians) on the two days and following Section 6 of Chapter 6, take $[(\bar{R}/d_2^*)^2$ for Day 3$]/[(\bar{R}/d_2^*)^2$ for Day 2$]$ as having an F distribution. Thus

$$\frac{(\bar{R}/d_2^*)^2 \text{ for Day 3}}{(\bar{R}/d_2^*)^2 \text{ for Day 2}} = \left(\frac{48.1}{1.70}\right)^2 \Big/ \left(\frac{23.8}{1.71}\right)^2 = 4.1$$

From an extension of Table D3 by constant differences, it may be determined that for $g = 20$, $m = 3$, the degrees of freedom are 36.6 and for $g = 17$, $m = 3$, the degrees of freedom are 31.1. The $F_{0.05}$ for $\nu_1 = 37$ and

FIGURE 31.2
A Multi-Vari Chart of Ranges of Three Plate Counts (see Table 31.9)

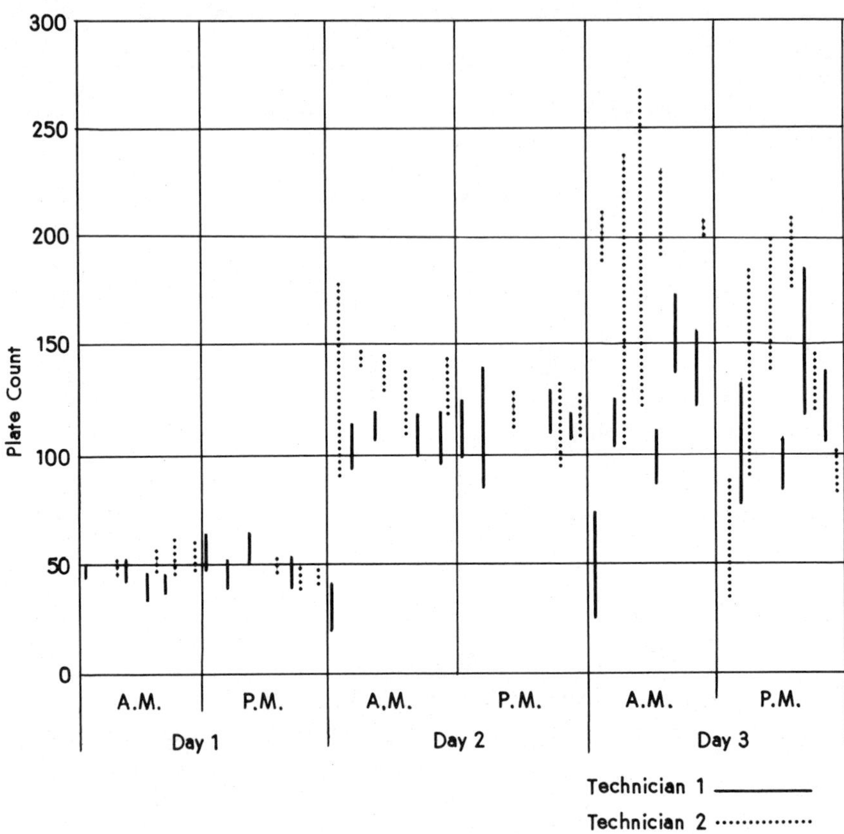

$\nu_2 = 31$ is in the neighborhood of 1.80. Hence the within-sample variations do differ as between the two days, 4.1 being greater than 1.80.

Another question that might be asked is: "Do the sample *levels* of technician 2 on the third day vary significantly?" From Table 31.9 it can be computed that the largest sample mean of technician 2 on the third day is 213 and the smallest is 67.7. The range of these means is $213 - 67.7 = 145.3$. In view of the within-sample variation, is this spread of the means significant? We answer this by a studentized-range test. For technicians 1 and 2 the average of the sample ranges on the third day is 48.1, and for $g = 20$, $m = 3$, an extension of Table D3 yields $d_2^* = 1.70$. Hence we have

$$q = \frac{145.3\sqrt{3}}{(48.1/1.70)} = \frac{251.7}{28.3} = 8.9$$

The $q_{0.05}$ point for $g = 11$, $\nu = 36$ (see Table D2) is approximately 4.86. The levels of technician 2 do differ significantly on the third day.

6. PROBLEMS

31.1. *a.* In Problem 29.2 determine 0.95 confidence limits for the tensile strength of the rubber vulcanizate produced with accelerator A.

b. In Problem 29.5 determine 0.95 confidence limits for the tensile strength of the rubber vulcanizate produced with accelerator A at 40 minutes.

c. In Problem 29.7 determine 0.95 confidence limits for the yield at temperature 50° and 50 percent concentration of inert solvent.

d. In Problem 30.1 determine 0.95 confidence limits for the rubber vulcanizate produced with accelerator B and the 60-minute cure.

e. In Problem 30.2 determine 0.95 confidence limits for the flexural strength of a casting produced at a temperature of 120° and 20 minutes in the polymerization bath.

f. In Problem 30.3 determine 0.95 confidence limits for the flexural strength of a casting produced at a temperature of 120° and 20 minutes in the polymerization bath. Compare your answer with that in *e.*

g. In Problem 30.4 determine 0.95 confidence limits for the lengths of steel bars produced by heat treatment W and machine C.

h. In Problem 30.6 determine 0.95 confidence limits for the yield with catalyst B at 60° and 60 percent inert solvent.

31.2. *a.* In Problem 29.2 determine 0.95 confidence limits for the "differential" effect of accelerator A.

b. In Problem 29.5 determine 0.95 confidence limits for the "differential" effect of (1) accelerator A and (2) the 40-minute cure.

c. In Problem 29.7 determine 0.95 confidence limits for the "differential" effect of (1) the 50° temperature and (2) the 50 percent concentration of inert solvent.

d. In Problem 30.1 determine 0.95 confidence limits for the "differential" effect of (1) accelerator B, (2) the 60-minute cure, and (3) the interaction between them.

e. In Problem 30.2 determine 0.95 confidence limits for the "differential" effect of (1) the 120° temperature, (2) the 20-minute bath, and (3) the interaction between them.

f. In Problem 30.3 determine 0.95 confidence limits for the "differential" effect of (1) the 120° temperature, (2) the 20-minute bath, and (3) the interaction between them.

g. In Problem 30.4 determine 0.95 confidence limits for the "differential" effect of (1) the heat treatment W, (2) the machine C, and (3) the interaction between them.

h. In Problem 30.6 determine 0.95 confidence limits for the "differential" effect of (1) catalyst B, (2) 60° temperature, (3) 60 percent inert solvent, (4) interaction between this catalyst and this temperature, (5) the interaction between this catalyst and this concentration of solvent, and (6) the interaction between this temperature and this concentration of solvent.

i. In Problem 30.7 determine 0.95 confidence limits for the interaction differential with Temp. = 120°, Time = 20 min., and Cat. = 1 percent.

31.3. *a.* In Problem 29.1 examiners 1 and 5 had been trained by different methods to use the given gauge. There are therefore a priori reasons for believing they might have different mean times. Do the data verify this expectation? State clearly the error rate that is associated with your test. Give the assumptions on which your analysis is based.

b. In Problem 29.1 assume no a priori beliefs and set up allowances for differences between pairs of means where the error rate will be (1) 5 wrong conclusions per 100 comparisons and (2) wrong conclusions in 5 percent of the experiments. Give the assumptions on which your analysis rests.

31.4. *a.* In Problem 29.3 there are a priori reasons for believing that temperature given by thermocouple 1 will be less than the temperature given by thermocouple 4. Do the data verify this expectation? State clearly the error rate that is associated with your test. Give the assumptions on which your analysis rests.

b. In Problem 29.3 assume no a priori beliefs and set up allowances for differences between pairs of means where the error rate will be (1) 5 wrong conclusions per 100 comparisons and (2) wrong conclusions in 5 percent of the experiments. Give the assumptions on which your analysis rests.

31.5. *a.* In Problem 29.5 there were a priori reasons for believing that on the average accelerator B would yield vulcanizate of tensile strength different from that which would be yielded by accelerator C. Do the data of the experiment verify this expectation? State clearly the error rate that is associated with your test. Give the assumptions on which your analysis rests.

b. In Problem 29.5 assume no a priori beliefs and set up allowances for differences between pairs of mean accelerator yields where the error rate will be (1) 5 wrong conclusions per 100 comparisons and (2) wrong conclusions in 5 percent of the experiments. Give the assumptions on which your analysis rests.

c. Repeat *(b)* for differences between mean yields for curing times.

31.6. Repeat Problem 31.5 for the data of Problem 30.1.

31.7. *a.* In Problem 30.4 there are a priori reasons for believing that machine B would lead to a mean length different from that which would be yielded by machine C. Do the data of the experiment verify this expectation? State clearly the error rate that is associated with your test. Give the assumptions on which your analysis rests.

b. In Problem 30.4 assume no a priori beliefs and set up allowances for differences between pairs of machine means where the error rate will be (1) 5 wrong conclusions per 100 comparisons and (2) wrong conclusions in 5 percent of the experiments. Give the assumptions on which your analysis rests.

c. Repeat *b* for pairs of machine means for heat treatment W only.

31.8. Separate the thermocouples of Problem 29.3 into subgroups of "like" mean values indicating which means belong to which subgroup. State clearly the error rates that are involved in your subgrouping and give the assumptions on which your analysis rests.

31.9. Separate the days of the week of Problem 29.8 into subgroups of "like" mean values indicating which means belong to which subgroups. State clearly the error rates that are involved in your subgrouping and give the assumptions on which your analysis rests.

31.10. Separate the machines of Problem 30.4 into subgroups of "like" mean values indicating which means belong to which subgroups. State clearly the error rates that are involved in your subgrouping and give the assumptions on which your analysis rests.

31.11. In Problem 29.3 determine whether thermocouple 4 can be classed as an outlier.

31.12. In Problem 29.6 determine whether laundering I can be classed as an outlier.

31.13. In Problem 30.4 determine whether machine C can be classed as an outlier.

31.14. *a.* Suppose that with a 0.95 assurance of being right, we wish to determine the largest of 7 universe means when the difference between the largest and next to the largest universe mean is equal to the known universe standard deviation. (Assume all universes have the same standard deviation.) Determine the number of observations to be made on each universe.

b. Repeat the problem for a 0.99 assurance.

31.15. *a.* Suppose that with a 0.95 assurance of being right, we wish to determine the largest of 4 universe means when the difference between the largest and next to the largest universe mean is equal to one fourth of the known universe standard deviation. (Assume all universes have the same standard deviation.) Determine the number of observations to be made on each universe.

b. Repeat the problem for a 0.99 assurance.

31.16. *a.* Suppose that with a 0.95 assurance of being right, we wish to determine and rank correctly the two largest of 7 universe means when both the difference between the largest and next largest and the difference between the second and third largest universe means are at least equal to half the known (common) standard deviation of the universes. Determine the number of observations to be made on each universe.

b. Repeat the problem for a 0.90 assurance.

31.17. In Problem 29.1 determine whether the error variance is the same for all examiners. Do this by two methods.

31.18. In Problem 29.2 determine whether the random variance in vulcanizate is the same for all accelerators. Do this by two methods.

31.19. In Problem 30.2 determine whether the within-class variation is the same for all classes. Do this by two methods.

31.20. In Problem 30.4 determine whether the within-class variation is the same for all classes. Do this by two methods.

31.21. *a.* In Problem 29.1 estimate the variance of the experimental error, assuming it is the same for all examiners, and determine 0.95 confidence limits for this variance.

b. Assume that the examiners are a random sample from an infinite universe of examiners and determine 0.95 confidence limits for the variance of examiner biases.

31.22. *a.* In Problem 29.3 assume that the readings on each thermocouple are independent of those made on other thermocouples and estimate the random variance in temperature. Assume that the universe variances are the same at each layer (thermocouple). Determine 0.95 confidence limits for this common variance.

b. In Problem 29.3 assume that temperatures on each of the six occasions are read "simultaneously" on all thermocouples and estimate the residual variance. Determine 0.95 confidence limits for this variance.

31.23. In Problem 29.6 assume that the four laboratories are a random sample of laboratories in general and the launderings a random sample of launderings. Estimate residual variance, laundering variance and laboratory variance, and determine 0.95 confidence limits for each.

31.24. In Problem 30.8 assume all variations are random and estimate the various components of variance. Determine approximate 0.95 confidence limits for each.

31.25. On Problem 30.9 assume all variations are random and estimate the various components of variance. Determine approximate 0.95 confidence limits for each.

31.26. In Problem 30.10 assume all variations are random and estimate the various components of variance. Determine approximate 0.95 confidence limits for each.

31.27. Study Problem 29.1 (1) by making a multi-vari chart and (2) by a Lambda test. What conclusions do you draw in each case?

31.28. Study the first ten groups of Table 3.2 (1) by making a multi-vari chart and (2) by a Lambda test. What conclusions do you draw in each case?

31.29. Study the data of Problem 30.1 by making a multi-vari chart. Compare your conclusions with those obtained by analysis of variance.

31.30. Study the data of Problem 30.2 by making a multi-vari chart. Compare your conclusions with those obtained by analysis of variance.

31.31. Study the data of Problem 30.4 by making a multi-vari chart. Compare your conclusions with those obtained by analysis of variance.

7. SELECTED REFERENCES*

Anderson and Bancroft (B '52), Andrews and Pregibdon (P '78)†, Anscombe (P '60)†, ASTM (B '76), Bainbridge (P '65), Bancroft and Han (P '83), Barnett (P '78)†, Bartlett (P '37)†, Basu (P '80), Bechhofer (P '54), Bechhofer, Dunnett, and Sobel (P '54), Beckman and Cook (P '83)†, Bennett and Franklin (B '54),

* B and P refer to the Book and Periodical sections, respectively, of the Cumulative List of References in Appendix V.

† References that specifically discuss "outliers."

Bliss, Cochran, and Tukey (P '56)†, Brown (P '75)†, Brownlee (B '48 and '60), Burdick and Sulken (P '78), Chao and Glaser (P '78), Cochran and Cox (B '57), Collett and Lewis (P '76)†, Conover, Johnson and Johnson (P '81), Cornfield (P '54), Cowden (B '57), David (P '56), David and Paulson (P '65)†, Davies (B '57), Dixon (P '53)†, Duncan, A. J. (P '55), Duncan, D. B. (P '55 and P '57), Eisenhart (P '47), Epstein (P '60), Feder (P '74), Federer (B '55), Gentleman and Wilk (P '75 and P '75)†, Glasser (P '76), Goldsmith and Gaylor (P '70), Grubbs (P '48 and P '69), Grubbs and Beck (P '72)†, Harter (P '57), Harter and Lum (B '55), Healy (P '61), Hicks (P '56), John (P '78)†, Keuls (P '52), Kitagawa (P '79)†, Krushall (P '60)†, Lam and Chiu (P '76), Ling (P '78), Mahalanobis (P '33), Newman (P '39), Ostle (B '54), Pearson and Hartley (B '58), Proschan (P '53)†, Rao, Kaplan and Cochran (P '81), Russell and Bradley (P '58), Samiuddin (P '76), Samiuddin, Hanif, and Asad (P '78), Samiuddin and Atiqullah (P '76), Scheffé (P '53 and B '59), Seder (P '50), Seder and Cowan (B '56), Shukla (P '73 and P '82), Simon (P '46), *Technometrics,* entire May 1965 issue devoted to multiple comparisons, Thompson (P '62 and P '63), Tietjen and Moore (P '72)†, Tukey (B '53), U.S. Dept. of Health and Human Services (B '83)†, and Wetherill and Afosu (P '74).

32

Regression and Correlation: Two Variables

An analysis of variance is independent of whether the factors involved are quantitative or qualitative in character. If all factors are quantitative, however, and if we are more interested in their general functional relationship than in their relationship at especially assigned levels, then a "regression" analysis becomes more appropriate than an analysis of variance.

The word "regression" is a technical term that has become part of the language of statistics primarily because of the work of Sir Francis Galton. In the 1880s Galton laid the foundations of modern correlation techniques in a study of the relationship between the heights of children and the heights of their parents. He plotted the average heights of the children against the average heights of their parents and found that, on the average, the offspring of tall parents were not so tall as their parents, while the offspring of short parents were not so short as their parents. He therefore concluded that human height tends to "regress" back to type. Since the time of Galton, a line or curve that shows how the mean[1] of the values of one variable associated with a given value of another variable changes with the value of the other variable has been called a line or curve of regression. In statistics, therefore, "regression" means simply average relationship. When we speak of the regression of X_1 on the X_2, we mean the relationship between (1) the average of the values of X_1 for a given value of X_2 and (2) the value of X_2. Likewise, the regression of X_2 on X_1 means the relationship between (1) the average of the values of X_2 for a given value of X_1 and (2) the value of X_1.

In industry we are interested in relationships between variables for several reasons. First, knowledge of relationships is useful in controlling quality of output. If we know which factors are the important ones in producing variabil-

[1] Galton used the median; see his *Natural Inheritance* (London: Macmillan & Co., 1889).

ity in the quality of output and if we learn to what extent variation in a factor causes variation in quality of output, then we may, by controlling the variation in the factor, control the variation in quality of output. Second, relationships between variables are useful in testing. Many tests of quality are destructive. If we can find indirect nondestructive tests that are highly correlated with direct destructive tests, then we can test for quality without destroying our product. For example, we can test for tensile strength by using a hardness test because the latter is highly correlated with the former. Third, a knowledge of relationships is useful in industrial research as it is in all research.

In most physical sciences, relationships are commonly determined by controlled experiments. In the social sciences and in certain physical sciences, like astronomy, controlled experiments may be impossible, or at least very difficult. Relationships must in such cases be discovered by analyzing the data as they come. The tool that was devised to accomplish this is the modern regression or correlation analysis. Often laboratory conditions cannot be set up that will exactly reproduce conditions in the plant. Consequently, the industrial engineer is frequently in the position of the social scientist and astronomer, in that he must take the data as he finds them. Regression analysis is thus a very useful tool of industrial research.

1. BIVARIATE FREQUENCY DISTRIBUTIONS

As noted above regression analysis was originally developed in the study of the joint distribution of two variables. Our discussion will thus begin with *bivariate* distributions.

1.1. A Bivariate Chart

In the study of a bivariate frequency distribution it is often helpful to plot the points first on a bivariate chart. Such a chart is shown in Figure 32.1. This shows in crude graphic form the relationship between the skein strength of a given type of cotton yarn and the length of the fibers that make up the yarn. Each point in the figure represents a pair of measurements for a particular piece of this yarn, the ordinate of the point (X_1) being its skein strength measured in pounds and the abscissa being its fiber length (X_2) measured in inches. This is the type of chart most of us would make, whether we knew any statistical theory or not.

1.2. The Bivariate Frequency Distribution

If X_1 and X_2 are continuous variables, we can divide both the X_1 scale and the X_2 scale into class intervals, and, by drawing horizontal and vertical lines through the class limits, we can subdivide the area of our bivariate chart into cells. Then our bivariate chart can be converted into a bivariate

FIGURE 32.1
Bivariate Chart Showing the Joint Distribution of Skein Strength and Fiber Length of 22s Cotton Yarn, United States Crop, 1944 (see Table 32.3)

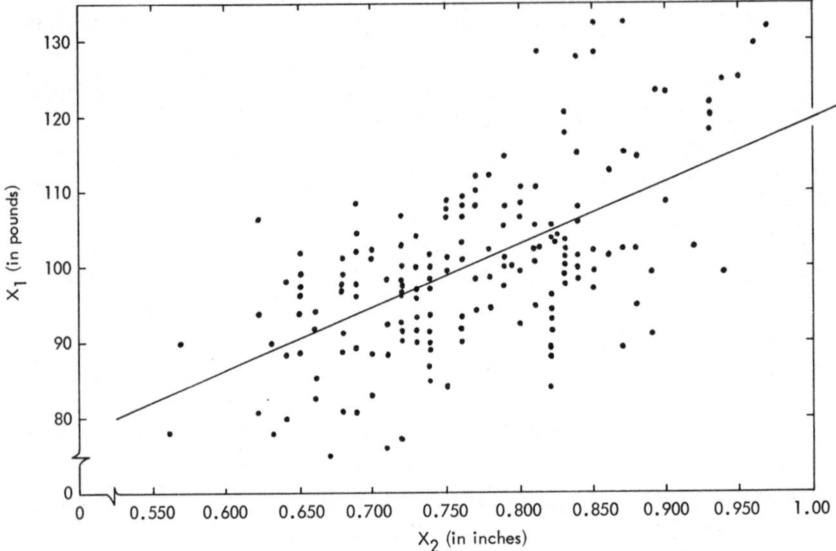

frequency distribution by recording the number of points that fall in each cell. When this is done for the data of Figure 32.1, we obtain Figure 32.2. This figure is the bivariate frequency distribution of 183 pieces of cotton yarn with respect to their skein strength and fiber length. It shows the frequency with which skein strength lies between certain specified limits at the same time that fiber length lies between certain other specified limits.

1.3. The Bivariate Histogram

If we represent a bivariate distribution by a three-dimensional model in which the frequencies of each cell are measured by the height of a rectangular parallelepiped erected on the cell, then we would have a bivariate histogram.[2] Such a model could easily be constructed by piling blocks on each cell, the number of blocks being the same as the number of cases in the cell. The height of a pile would thus become a measure of the frequency of the cell on which it was resting.

1.4. The Bivariate Frequency Surface

For a single continuous variable we moved from the histogram to the frequency curve.[3] For two continuous variables we can move from the bivariate

---

[2] See Chapter 3, Section 2.2.

[3] See Chapter 4, Section 3.2.

FIGURE 32.2
Bivariate Distribution of Skein Strength and Fiber Length of 22s Cotton Yarn, United States Crop, 1944

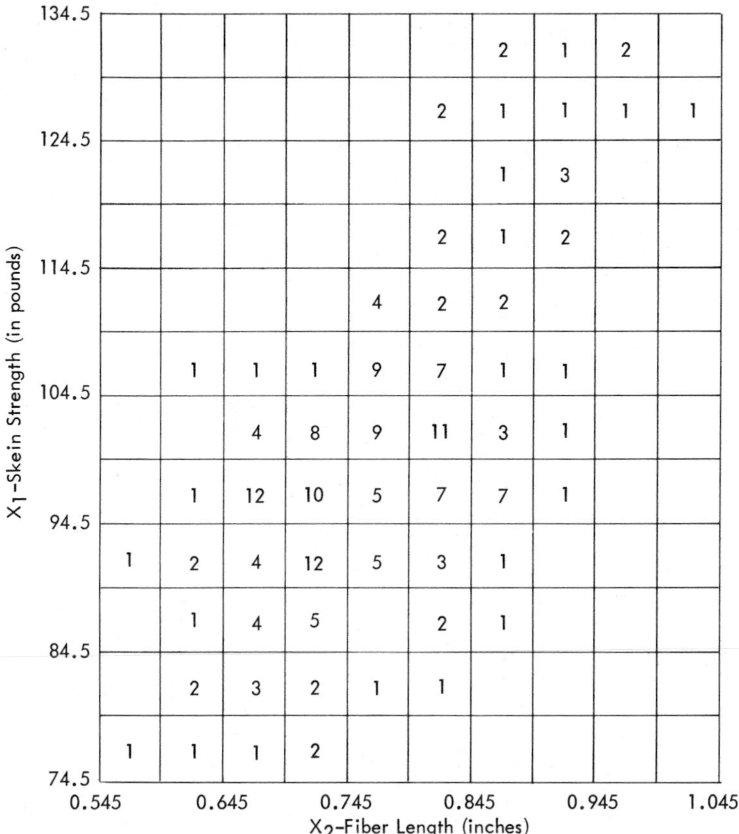

histogram to the bivariate frequency surface. If we express the joint frequencies in relative terms, i.e., we take F/n instead of n; if we use the volume of a pile of blocks to measure relative frequency; and, finally, if we increase the number of cases indefinitely and, at the same time, reduce the size of both the X_1 and X_2 class intervals (i.e., the size of our cells), then the tops of our piles of blocks will approach a limit that can be represented by a surface over the X_1, X_2 plane. This would be the bivariate frequency surface that would show the distribution of the universe of the given data. Just as the area under a frequency curve over a given range measures the relative frequency or probability of a case in that range, so the volume under a frequency surface over a given area of the X_1, X_2 plane measures the relative frequency or probability of a case in that area. In the future, when we refer to a universe of X_1 and X_2 values, X_1 and X_2 being continuous, it will be implied that the distribution of this universe may be represented by a frequency

FIGURE 32.3
Picture of a Normal Bivariate Frequency Surface with Correlation between the Variables

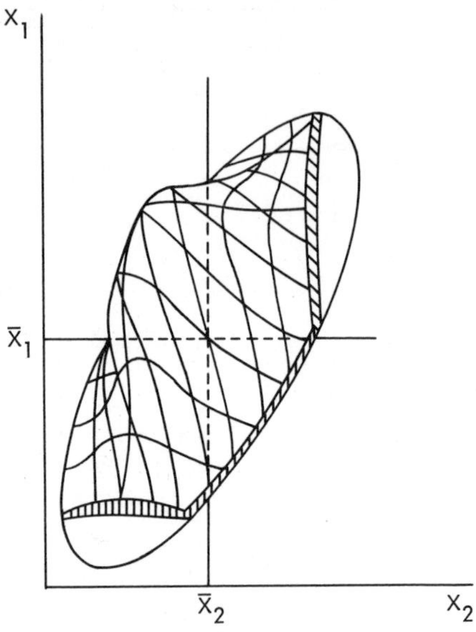

surface or "density function" of the kind just described. Figure 32.3 illustrates a *normal* bivariate density function.

2. IMPORTANT CHARACTERISTICS OF BIVARIATE FREQUENCY DISTRIBUTIONS AND THEIR MEASUREMENTS

2.1. Distributions of X_1 for Given Values of X_2

Even casual examination of a bivariate frequency distribution shows that for any given value of X_2, say, we do not have a single X_1 value but a distribution of X_1 values. Thus, for the 183 measurements studied in Figure 32.2, it will be noticed that, when fiber length lies between 0.695 and 0.745 inches, skein strength is distributed over 7 cells, ranging from a skein strength of 74.5 pounds to a skein strength of 109.5 pounds. This tendency for skein strength to be distributed over a range of values for any given value of fiber length not only is characteristic of a sample of data from the universe of all possible skein strength-fiber length measurements but also would be characteristic of the universe itself.

But why, the reader may ask, do we not get a single value of X_1 for any given value of X_2? Why do we have a distribution of X_1 values for each value of X_2? What physical phenomena produce such results?

The principal answer to these questions is that the dependent variable

X_1 is almost always affected by other variables than X_2. Refer for example to Figure 32.1. Here the points do not all fall on a line or curve primarily because fiber length is only one of many factors that determine skein strength. In the next chapter we shall consider two other factors, viz fiber tensile strength and fiber fineness, and shall see that all the factors that affect skein strength have not even then been exhausted. It is the variation of these other factors from case to case that causes a distribution of skein strength for a given fiber length. If the factors other than fiber length were few in number, we could use multiple regression analysis to measure the effect of each factor and thus reduce the final variation to errors of measurement. Other factors, however, are probably very numerous, with only a few of any great importance. Consequently, after we have taken account of the few important factors, we shall have left a residual variation that must be accepted and treated in much the same way as errors of measurement.

2.2. Variation in the Mean Value of X_1 with X_2

Although there is not a single value of X_1 for a given value of X_2, but a distribution of X_1 values, nevertheless in Figures 32.1 and 32.2 it will be seen that there is a "tendency" for X_1 values to be higher when X_2 is higher and lower when X_2 is lower. Thus, if we plot the means of X_1 for each X_2 interval (see Figure 32.4), we note that, with the exception of the last interval, in which there is only one case, the mean value of X_1 increases steadily with X_2. In the Universe of X_1, X_2 values it is likely that the locus of mean values of X_1 would trace out a smooth curve or, as seems probable in this case, a straight line, at least for the central values of X_2. As noted above,[4] it is this locus of mean values[5] that is called the regression of X_1 on X_2. If it is a straight line, the regression is said to be "linear." Since the regression of X_1 on X_2 gives the mean value of X_1 for a given value of X_2, it will help in estimating X_1 when we know X_2.

It is to be noted that in a bivariate distribution there is also a regression of X_2 on X_1 as well as a regression of X_1 in X_2. This would be the locus of mean values of X_2 for given values of X_1—a locus that is not the same as the locus of the mean values of X_1 for given values of X_2 (see Figure 32.4). When we are interested in using a regression to estimate one variable from another—tensile strength from hardness, say—the problem in hand will probably tell us which regression we want. The reader will note, however, that the regression of X_1 on X_2 should not be used to estimate the mean value of X_2 for a given value of X_1, nor should the regression of X_2 on X_1 be used to estimate the mean value of X_1 for a given value of X_2 (see

[4] See beginning of chapter.

[5] Other measures of central tendency could be used. As noted above, Sir Francis Galton used the median. The mean, however, is most commonly employed. If the distribution of X_1 values for a given X_2 is symmetrical, it makes no difference, of course, whether the mean or the median is used.

FIGURE 32.4
Chart Showing Changes in the Mean Value of X_1 with X_2 and Changes in the Mean Value of X_2 with X_1 for the Bivariate Distribution of Figure 32.2

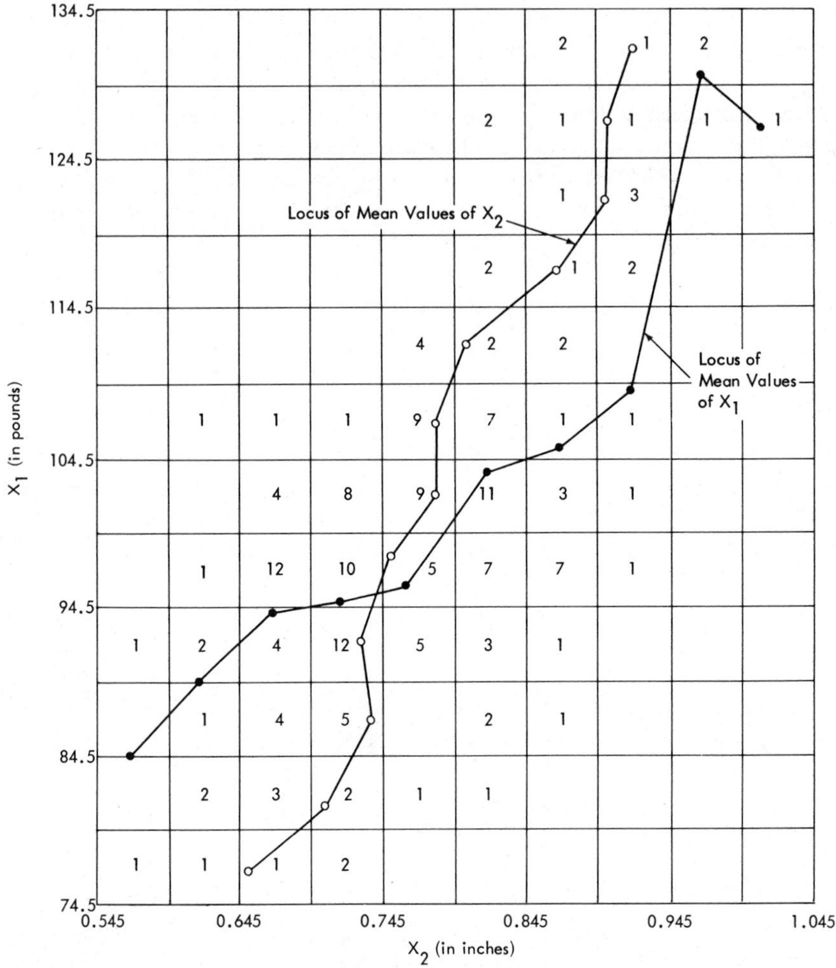

Figure 32.4). It is only when the distribution adheres very closely to a line or curve that the two regressions are close together.

A line of regression will be said to be "positive" or "negative" depending on whether it slopes upward or downward. If a regression is curvilinear, it may not be possible to speak of the regression as a whole being positive or negative.

2.3. The Standard Error of Estimate of a Line or Curve of Regression

If a line or curve of regression is used to estimate X_1 from X_2, it is highly desirable that we have some measure of the reliability of that estimate. Such a measure of reliability will be given by the standard deviation of the distribution of X_1 values for a given value of X_2. The regression value, it will be recalled, is the mean of X_1 values associated with a given X_2; what we want now is the standard deviation of these X_1 values. Then we can lay off ± 2 or ± 3 times this standard deviation from the regression value (the mean) and, in the case of most distributions met with in practice, obtain a range of values that will have a reasonably high probability of including the actual value of X_1. For example, if we are referring to a universe of X_1, X_2 values and if the distribution of X_1 values for the given X_2 value is normal in form, then twice the standard deviation laid off plus and minus from the regression value will yield a range that will include approximately 95 percent of the values of X_1. If a universe of X_1, X_2 values has the property that the standard deviation of X_1 values for a given X_2 is the same for all X_2 values, the bivariate distribution is said to be homoscedastic; if not, it is said to be heteroscedastic. Figure 32.5 illustrates a homoscedastic distribution in which the regression of X_1 on X_2 is linear.

The standard deviation of X_1 for a given X_2 is commonly called the standard error of estimate, since it measures the error involved in using the regression value to estimate X_1. The symbol that will be used in this text to represent the standard error of estimate for estimating X_1 from X_2 when we are referring to a universe of X_1, X_2 values will be $\sigma'_{1.2}$. Since the subsequent discussion will be concerned only with homoscedastic distributions, we shall not need to distinguish different values of $\sigma'_{1.2}$ for different values of X_2.

FIGURE 32.5
Illustration of a Linear Regression with Uniform Scatter (homoscedastic distribution)

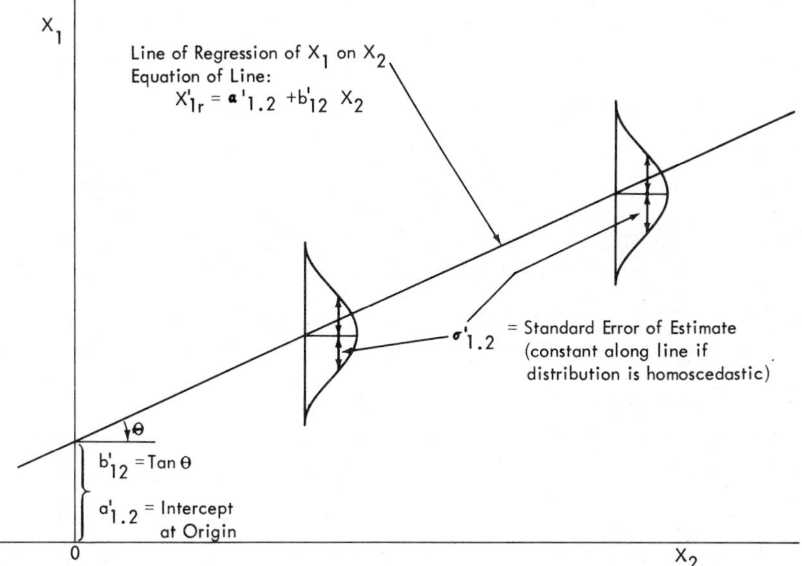

It is to be carefully noted that a standard error of estimate is not concerned with sampling fluctuations. It is very easy to become confused about this, since the term "standard error" is usually used when referring to a standard deviation of a sampling distribution.[6] When speaking of a standard error of estimate, therefore, it is advisable always to use the full expression. It is also to be noted that the standard error of estimate for the regression of X_1 on X_2 is not the same, except in special circumstances, as the standard error of estimate for the regression of X_2 on X_1. Thus, in general, $\sigma'_{1.2} \neq \sigma'_{2.1}$.

3. ESTIMATION OF A UNIVERSE REGRESSION AND STANDARD ERROR OF ESTIMATE FROM SAMPLE DATA

In practice, we usually have only a sample of X_1, X_2 values. Figure 32.1 shows the distribution of 183 pairs of skein strength and fiber length measurements; although it is a large sample, it is still only a sample. The universe would be the set of all possible measurements. Frequently, samples are as small as 10 or 20 cases. When we have only a sample of values and wish to estimate the regression of X_1 on X_2, say, and the standard error of estimate ($\sigma'_{1.2}$), the question arises as to the method of estimation.

3.1. Fitting a Line or Curve Freehand

In estimating a universe line or curve of regression from a set of sample data, one method is to plot the points on a scatter diagram such as Figure 32.1 and draw the curve or line in "freehand." The standard error of estimate can then be estimated by measuring graphically the deviations of the sample points from the fitted line and computing their standard deviation. The deviations should be measured vertically or horizontally, depending on whether the line is to be used to estimate X_1 from X_2 or X_2 from X_1.

This freehand procedure may not give bad results, especially if the points are numerous and cluster closely around the line or curve. Its principal disadvantage is that it is too personal. It is unlikely that two people would get exactly the same result. It also does not allow for formal statistical procedures, such as the determination of sampling errors, tests of hypotheses, and so forth. Nevertheless, in many cases it may be a satisfactory solution.

3.2. Estimating a Universe Line or Curve of Regression by the Method of Least Squares

3.2.1. Description of Method and Its Advantages. A more sophisticated approach to the estimation of a universe line or curve of regression is to use the method of least squares. This fits a line or curve to a set of sample

[6] See beginning of Chapter 6.

points such that the sum of the squares of the deviations of the sample points from the fitted line or curve is a minimum. Suppose, for example, that a universe line of regression is represented by the equation

$$X'_{1r} = a'_{1.2} + b'_{12}X_2$$

and let an estimate of this universe line be

$$X_{1r} = a_{1.2} + b_{12}X_2$$

where X_{1r}, $a_{1.2}$, and b_{12} are estimates of X'_{1r}, $a'_{1.2}$, and b'_{12}. Then, according to the method of least squares, $a_{1.2}$ and b_{12} are determined from the condition that

$$\Sigma(X_1 - X_{1r})^2 = \Sigma(X_1 - a_{1.2} - b_{12}X_2)^2$$

over all the sample points be a minimum.

The reader will note that the various quantities are now denoted by combinations of letters and subscripts. Thus $a'_{1.2}$ is the constant term of the universe line of regression that is used to estimate X_1 from X_2. Likewise, b'_{12} is the coefficient of X_2 in the universe line of regression that is used to estimate X_1 from X_2. The reason for the decimal point after the 1 in the subscript of $a_{1.2}$ will become clearer when we take up regressions involving several variables. This is done in the next chapter.[7] In the case of two variables the decimal point is not necessary. It is introduced here to facilitate later the generalization to more than two variables.

The method of least squares is mathematical and impersonal. If the variations around the universe regression are random, the method of least squares permits the computation of sampling errors and hence the determination of the reliability of estimates of X_1 made from the fitted line. Furthermore, if the distribution of points around the universe regression is not only random but normal in form, then the least-squares method gives the maximum likelihood estimate of the universe regression. For these reasons, lines and curves of regression are commonly estimated from sample data by the method of least squares.

The remainder of this chapter will be devoted to an exposition of the use of the method of least squares to estimate a linear regression and statistics associated therewith. Problems associated with curvilinear regression will be discussed in Section 3 of the next chapter.

3.2.2. The Least-Squares Equations for Estimating a Linear Regression of a Homoscedastic Distribution. If we assume that the universe regression is linear, say that

$$X'_{1r} = a'_{1.2} + b'_{12}X_2$$

and if we also assume that the residuals $u'_{1.2} = X_1 - X'_{1r}$ are randomly distributed with a zero mean and a constant standard deviation $\sigma'_{1.2}$, then,

[7] See especially Chapter 33, Sections 1.2.1 and 2.2.1.

to estimate $a'_{1.2}$ and b'_{12} from a set of sample data by the method of least squares, we use the equations[8]

$$\text{(32.1)}\quad \begin{aligned} \Sigma(X_1 - a_{1.2} - b_{12}X_2) &= \Sigma X_1 - na_{1.2} - b_{12}\Sigma X_2 = 0 \\ \Sigma(X_1 - a_{1.2} - b_{12}X_2)X_2 &= \Sigma X_1X_2 - a_{1.2}\Sigma X_2 - b_{12}\Sigma X_2^2 = 0 \end{aligned}$$

in which ΣX_1, ΣX_2, ΣX_1X_2, and ΣX_2^2 are the sums of the variables for the sample data. When these sums are computed and substituted in equations (21.1), we have two equations to be solved simultaneously for the two unknowns $a_{1.2}$ and b_{12}. The results so obtained will be the "least-squares estimates" of the universe values $a'_{1.2}$ and b'_{12} when the origin is the original origin for X_1 and X_2.

If we set up the transformation $x_1 = X_1 - \bar{X}_1$ and $x_2 = X_2 - \bar{X}_2$, the equation for the universe line of regression can be written

$$x'_{1r} = A' + b'_{12}x_2, \text{ where } A' = a'_{1.2} - \bar{X}_1 + \bar{X}_2 b'_{12}$$

If this is fitted by least squares, we will get for estimates of A' and b'_{12}

$$\text{(32.2)}\quad A = 0,\ b_{12} = \Sigma x_1x_2/\Sigma x_2^2$$

since $\Sigma x_1 = 0$ and $\Sigma x_2 = 0$. These estimates yield an estimate for $a'_{1.2}$ of

$$\text{(32.3)}\quad a_{1.2} = \bar{X}_1 - b_{12}\bar{X}_2$$

which could have been obtained directly[9] from the first of equations (32.1).

Since the universe distribution is assumed to be homoscedastic, the universe standard error of estimate will be the same at all points along the line of regression. In this case, an unbiased estimate of the square of the universe standard error of estimate may be obtained from the sample by computing $\Sigma u_{1.2}^2/(n - 2)$, where $u_{1.2} = X_1 - X_{1r} = X_1 - a_{1.2} - b_{12}X_2$ and n is the number of sample points. Hence, if we set

$$\text{(32.4)}\quad s_{1.2}^2 = \frac{\Sigma u_{1.2}^2}{n - 2}$$

then $s_{1.2}^2$ is an unbiased estimate of $\sigma'^2_{1.2}$.

Fortunately, if the line is fitted by least squares, we can calculate $\Sigma u_{1.2}^2$ from a relatively simple formula. This is[10]

$$\text{(32.5)}\quad \Sigma u_{1.2}^2 = \Sigma X_1^2 - a_{1.2}\Sigma X_1 - b_{12}\Sigma X_1X_2$$

When deviations are taken from the means of X_1 and X_2, we have the alternative formula

$$\text{(32.6)}\quad \Sigma u_{1.2}^2 = \Sigma x_1^2 - b_{12}\Sigma x_1x_2$$

---

[8] See Appendix I (36). It will be noted that the method gives a unique solution, i.e., there is only one pair of values for $a_{1.2}$ and b_{12} that satisfies the equations.

[9] Divide the first of equations (32.1) by n and note that $\bar{X} = \Sigma X/n$.

[10] See Appendix I (37).

It is to be noted that the above equations give an estimate of the universe line of regression of X_1 on X_2. The same method can be used to estimate the universe line of regression of X_2 on X_1. In this case we would minimize

$$\Sigma(X_2 - X_{2r})^2 = \Sigma(X_2 - a_{2.1} - b_{21}X_1)^2$$

With reference to a diagram such a Figure 32.1, it is often said that the line of regression of X_1 on X_2 is estimated by minimizing the sum of the squares of the "vertical" deviations and that the line of regression of X_2 on X_1 is estimated by minimizing the sum of the squares of the "horizontal" deviations.

4. AN EXAMPLE FOR UNGROUPED DATA

It is well at this point to work out an example. Table 32.1 gives the shear strength and weld diameter of 10 different welds, and Figure 32.6 suggests that there is some relationship between the two. We shall assume that this relationship is linear and fit a line of regression to the 10 points. We shall also assume that the universe distribution is homoscedastic.

The first step in the analysis is to compute the various sums of squares and cross-products that will be used in subsequent calculations. This is done in Table 32.2. It will be noted that the table contains three auxiliary columns for checking the computations. If a calculator is available, the various sums of squares and cross-products could be cumulated in the calculator and the table itself dispensed with. The checks, however, should always be carried out.

After calculating the various sums of squares and cross-products, we continue the computations as follows:

$$\bar{X}_1 = \frac{9{,}750}{10} = 975.0$$

$$\bar{X}_2 = \frac{2{,}239}{10} = 223.9$$

$$\sum x_1x_2 = \sum X_1X_2 - n\bar{X}_1\bar{X}_2 = 2{,}219{,}370 - 10(975.0)(223.9) = 36{,}345$$

TABLE 32.1
Shear Strength and Weld Diameter of Ten Spot Welds of 0.064″–0.064″ Steel

Shear Strength (X_1) *(Pounds)*	*Weld Diameter* (X_2) *(0.001 Inches)*	*Shear Strength* (X_1) *(Pounds)*	*Weld Diameter* (X_2) *(0.001 Inches)*
680	190	1,025	215
800	200	1,100	230
780	209	1,030	250
885	215	1,175	265
975	215	1,300	250

FIGURE 32.6
A Bivariate Chart of 10 Random Samples of Shear Strength (X_1) and Weld Diameter (X_2)

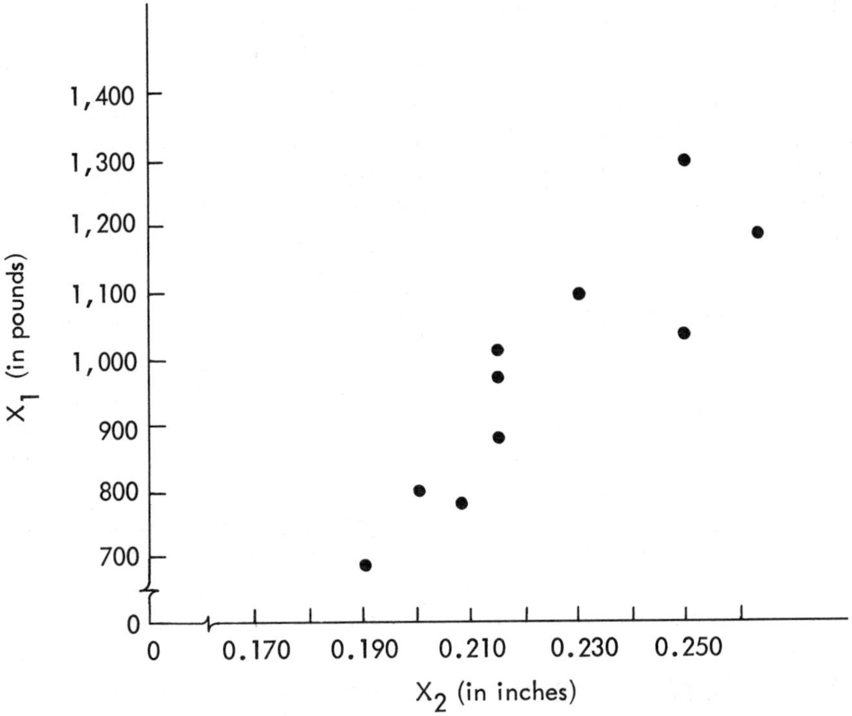

TABLE 32.2
Computation of Products and Sums for a Two-Variable Regression Analysis*

(1) X_1	(2) X_2	(3) X_1X_2	(4) X_1^2	(5) X_2^2	(6) (X_1+X_2)	(7) $(X_1+X_2)X_1$	(8) $(X_1+X_2)X_2$
680	190	129,200	462,400	36,100	870	591,600	165,300
800	200	160,000	640,000	40,000	1,000	800,000	200,000
780	209	163,020	608,400	43,681	989	771,420	206,701
885	215	190,275	783,225	46,225	1,100	973,500	236,500
975	215	209,625	950,625	46,225	1,190	1,160,250	255,850
1,025	215	220,375	1,050,625	46,225	1,240	1,271,000	266,600
1,100	230	253,000	1,210,000	52,900	1,330	1,463,000	305,900
1,030	250	257,500	1,060,900	62,500	1,280	1,318,400	320,000
1,175	265	311,375	1,380,625	70,225	1,440	1,692,000	381,600
1,300	250	325,000	1,690,000	62,500	1,550	2,015,000	387,500
9,750	2,239	2,219,370	9,836,800	506,581	11,989	12,056,170	2,725,951

Checks: 9,750 + 2,239 = 11,898 [ΣCol(1) + ΣCol(2) = ΣCol(6)]
2,219,370 + 9,836,800 = 12,056,170 [ΣCol(3) + ΣCol(4) = ΣCol(7)]
2,219,370 + 506,581 = 2,725,951 [ΣCol(3) + ΣCol(5) = ΣCol(8)]

* In some cases, it may facilitate the calculations if the original data are "coded" by referring them to an arbitrary origin and also possibly multiplying them by a constant factor to remove decimals. The values of Σx_1x_2 and b_{12} are not affected by the change of origin, but they are by multiplication by a constant. Final results will, in the latter case, have to be readjusted.

$$\sum x_2^2 = \sum X_2^2 - n\bar{X}_2^2 = 506{,}581 - 10(223.9)^2 = 5{,}269$$

$$b_{12} = \frac{\Sigma x_1 x_2}{\Sigma x_2^2} = \frac{36{,}345}{5{,}269} = 6.90$$

$$a_{1.2} = \bar{X}_1 - b_{12}\bar{X}_2 = 975 - 6.90(223.9) = -569.9$$

Hence, the estimated regression is

$$X_{1r} = -569.9 + 6.90X_2, \qquad \text{with origin at } X_1 = 0,\ X_2 = 0.$$

We also have

$$\sum x_1^2 = \sum X_1^2 - n\bar{X}_1^2 = 9{,}836{,}800 - 10(975)^2 = 330{,}550$$

and

$$\sum u_{1.2}^2 = \sum x_1^2 - b_{12} \sum x_1 x_2 = 330{,}550 - 6.90(36{,}345) = 79{,}770$$

$$s_{1.2}^2 = \frac{\Sigma u_{1.2}^2}{n-2} = \frac{79{,}770}{8} = 9{,}971.25$$

and

$$s_{1.2} = 99.9$$

5. AN EXAMPLE FOR GROUPED DATA

If the number of cases is large and an electronic computer is not available, the computations outlined in Table 32.2 will involve considerable work. To save labor, it is better in these instances to group the data and construct a bivariate frequency distribution for the sample data.

In Figure 32.7 the 183 measurements of skein strength and fiber length listed in Table 32.3 are cross-classified.[11] It will be noted that measurements of skein strength are now grouped into class intervals of 5 pounds and measurements of fiber length into class intervals of 0.05 inches. These class intervals are determined as described in Chapter 3, Section 2. The principal figure in each cell of Figure 32.7 is the frequency with which the cases fall in that cell, and these cell frequencies are totaled by rows and columns.

Computations are carried out with respect to two arbitrary origins, $A_1 = 102$ for X_1 and $A_2 = 0.770$ for X_2. The units are class interval (ξ) units. Successive multiplication of the frequency for each row or column by ξ gives $F\xi$ and $F\xi^2$ for each row or column, and the sum for all rows or columns gives the total for the data. From these sums we compute as shown in (32.7) on p. 813.

[11] Also see Figures 32.1 and 32.2.

FIGURE 32.7
Illustration of a Work Sheet for Computing the Regression of Skein Strength (X_1) on Fiber Length (X_2) of Cotton Yarn

X_1 \ X_2	0.545 –	0.595 –	0.645 –	0.695 –	0.745 –	0.795 –	0.845 –	0.895 –	0.945 –	0.995 –	Σ	ξ_1	$F\xi_1$	$F\xi_1^2$	$F\xi_1\xi_2$
129.5 –							2 $^{24}_{12}$	1 $^{18}_{18}$	2 $^{48}_{24}$		5	+6	+30	180	+90
124.5 –						2 $^{10}_{5}$	1 $^{10}_{10}$	1 $^{15}_{15}$	1 $^{20}_{20}$	1 $^{25}_{25}$	6	+5	+30	150	+80
119.5 –							1 $^{8}_{8}$	3 $^{36}_{12}$			4	+4	+16	64	+44
114.5 –						2 $^{6}_{3}$	1 $^{6}_{6}$	2 $^{18}_{9}$			5	+3	+15	45	+30
109.5 –					4 $^{0}_{0}$	2 $^{4}_{2}$	2 $^{8}_{4}$				8	+2	+16	32	+12
104.5 –		1 $^{-3}_{-3}$	1 $^{-2}_{-2}$	1 $^{-1}_{-1}$	9 $^{0}_{0}$	7 $^{7}_{1}$	1 $^{2}_{2}$	1 $^{3}_{3}$			21	+1	+21	21	+6
99.5 –			4 $^{0}_{0}$	8 $^{0}_{0}$	9 $^{0}_{0}$	11 $^{0}_{0}$	3 $^{0}_{0}$	1 $^{0}_{0}$			36	0	0	0	0
94.5 –		1 $^{3}_{3}$	12 $^{24}_{2}$	10 $^{10}_{1}$	5 $^{0}_{0}$	7 $^{-7}_{-1}$	7 $^{-14}_{-2}$	1 $^{-3}_{-3}$			43	−1	−43	43	+13
89.5 –	1 $^{8}_{8}$	2 $^{12}_{6}$	4 $^{16}_{4}$	12 $^{24}_{2}$	5 $^{0}_{0}$	3 $^{-6}_{-2}$	1 $^{-4}_{-4}$				28	−2	−56	112	+50
84.5 –		1 $^{9}_{9}$	4 $^{24}_{6}$	5 $^{15}_{3}$		2 $^{-6}_{-3}$	1 $^{-6}_{-6}$				13	−3	−39	117	+36
79.5 –		2 $^{24}_{12}$	3 $^{24}_{8}$	2 $^{8}_{4}$	1 $^{0}_{0}$	1 $^{-4}_{-4}$					9	−4	−36	144	+52
74.5 –	1 $^{20}_{20}$	1 $^{15}_{15}$	1 $^{10}_{10}$	2 $^{10}_{5}$							5	−5	−25	125	+55
Σ	2	8	29	40	33	37	20	10	3	1	183		−71	1033	+468
ξ_2	−4	−3	−2	−1	0	+1	+2	+3	+4	+5					
$F\xi_2$	−8	−24	−58	−40	0	37	40	30	12	5	−6				
$F\xi_2^2$	32	72	116	40	0	37	80	90	48	25	540				
$F\xi_1\xi_2$	+28	+60	+96	+66	0	+4	+34	+87	+68	+25	+468				

$$\bar{X}_1 = A_1 + \frac{\Sigma F \xi_1}{n}(i_1)$$

$$\bar{X}_2 = A_2 + \frac{\Sigma F \xi_2}{n}(i_2)$$

(32.7)
$$\sum Fx_1^2 = \left[\sum F\xi_1^2 - \frac{(\Sigma F\xi_1)^2}{n}\right] i_1^2$$

$$\sum Fx_2^2 = \left[\sum F\xi_2^2 - \frac{(\Sigma F\xi_2)^2}{n}\right] i_2^2$$

TABLE 32.3
Results of Fiber and Spinning Tests for Some Varieties of Upland Cotton Grown in the United States, Crop of 1944

X_1*	X_2†	X_1	X_2	X_1	X_2	X_1	X_2	X_1	X_2
99	0.85	91	0.72	97	0.83	78	0.56	134	0.90
93	0.82	102	0.74	106	0.88	90	0.57	132	0.87
99	0.75	98	0.78	102	0.81	91	0.74	103	0.83
97	0.74	108	0.76	99	0.80	109	0.76		
90	0.76	100	0.79	76	0.71	97	0.72		
96	0.74	99	0.75	77	0.72	132	0.85		
93	0.73	102	0.79	85	0.74	106	0.75		
130	0.96	101	0.84	89	0.69	110	0.81		
118	0.93	100	0.72	115	0.87	108	0.77		
88	0.70	100	0.73	108	0.84	107	0.79		
89	0.82	97	0.73	105	0.82	112	0.77		
93	0.80	108	0.90	106	0.84	127	0.84		
94	0.77	110	0.80	104	0.82	105	0.81		
75	0.67	101	0.75	89	0.87	81	0.68		
84	0.82	101	0.81	102	0.65	90	0.74		
91	0.76	114	0.79	99	0.65	87	0.74		
100	0.74	103	0.92	91	0.68	110	0.77		
98	0.71	92	0.82	97	0.68	102	0.69		
101	0.70	104	0.82	81	0.62	107	0.75		
80	0.64	99	0.88	88	0.64	93	0.76		
96	0.72	102	0.78	93	0.66	78	0.63		
102	0.70	102	0.79	98	0.64	83	0.70		
106	0.72	97	0.85	88	0.65	97	0.74		
89	0.74	98	0.84	96	0.69	81	0.69		
91	0.72	101	0.79	96	0.65	114	0.88		
90	0.72	96	0.88	97	0.65	125	0.95		
88	0.71	98	0.83	91	0.66	103	0.76		
83	0.70	103	0.82	90	0.63	100	0.79		
92	0.72	121	0.93	108	0.69	97	0.79		
91	0.73	95	0.81	99	0.68	97	0.82		
90	0.73	106	0.76	101	0.68	102	0.83		
92	0.71	105	0.79	104	0.69	115	0.92		

TABLE 32.3 (*Concluded*)

X_1*	X_2†	X_1	X_2	X_1	X_2	X_1	X_2	X_1	X_2
90	0.72	101	0.76	97	0.68	132	0.97		
108	0.75	88	0.82	97	0.69	125	0.94		
96	0.73	123	0.89	89	0.68	99	0.94		
115	0.84	97	0.68	102	0.85	128	0.85		
106	0.84	97	0.65	126	1.01	123	0.90		
101	0.86	99	0.68	104	0.73	102	0.81		
100	0.83	85	0.66	120	0.93	93	0.74		
99	0.85	83	0.66	113	0.86	98	0.72		
102	0.87	93	0.65	98	0.77	112	0.78		
96	0.87	99	0.68	94	0.78	106	0.80		
96	0.87	97	0.73	128	0.81	108	0.80		
91	0.89	106	0.62	102	0.72	84	0.75		
95	0.82	94	0.62	117	0.83	100	0.84		

* X_1 = skein strength of carded 22s cotton yarn, measured in pounds. Yarn strength is perhaps the most important single index of spinning quality. Good yarn strength not only increases the range of usefulness of a given cotton but it indicates good spinning and weaving performance. From the standpoint of strength, 22s yarn may be classified as follows: Excellent, above 110; Very Good, 101–10; Average, 93–100; Fair, 85–92; Poor, below 85.

† X_2 = Fiber length. Data here reported were determined by the fibrograph instrument, which is a photoelectric device for determining certain length values with respect to the fibers in a sample of cotton. The "upper half mean length" of a sample of cotton, as determined by this instrument, provides a measure of the average length of all fibers longer than the mean length, expressed in terms of decimal fractions of an inch. The "upper half mean length" is fairly closely related to the classer's designation of staple length. It may vary from that value, however, because of fiber characteristics other than length which may be taken into account by the classer. Mean length, which is the measure here chosen, is the average length of all fibers in the sample, excluding those shorter than ¼ inch.

For further details see source of data.

SOURCE: Data from U.S. Department of Agriculture, War Food Administration, Office of Marketing Services, *Results of Fiber and Spinning Tests for Some Varieties of Upland Cotton Grown in the United States, Crop of 1944* (Processed April 1945).

To calculate b_{12}, we need $\Sigma\Sigma F\xi_1\xi_2$. This is derived as follows. For each cell compute the product $\xi_1\xi_2$ for that cell. Enter this in the lower right-hand corner of the cell. Multiply the $\xi_1\xi_2$ product for the cell by the cell frequency and enter the results in the upper right-hand corner of the cell. Sum the cell values of $F\xi_1\xi_2$ by rows and columns, and then sum the row and column totals to get a grand total for the whole set of data. The answer obtained by summing the rows should be the same as the answer obtained by summing the columns. From the total $\Sigma\Sigma F\xi_1\xi_2$ we compute

$$\text{(32.8)} \qquad b_{12} = \frac{\Sigma\Sigma Fx_1x_2}{\Sigma Fx_2^2} = \frac{[\Sigma\Sigma F\xi_1\xi_2 - (\Sigma F\xi_1)(\Sigma F\xi_2)/n]i_1i_2}{\Sigma Fx_2^2}$$

and from this we have, as before,

$$\text{(32.9)} \qquad a_{1.2} = \bar{X}_1 - b_{12}\bar{X}_2$$

For the data grouped as in Figure 32.7 on p. 812, we have

$$\bar{X}_1 = 102 + \frac{(-71)}{(183)}(5) = 100.06$$

$$\bar{X}_2 = 0.770 + \frac{(-6)}{(183)}(0.05) = 0.7684$$

$$\sum Fx_1^2 = \left[1{,}033 - \frac{(-71)^2}{183}\right](5)^2 = 25{,}136.25$$

$$\sum Fx_2^2 = \left[540 - \frac{(-6)^2}{183}\right](0.05)^2 = 1.348$$

$$b_{12} = \frac{\Sigma\Sigma Fx_1x_2}{\Sigma Fx_2^2} = \frac{\left[468 - \frac{(-71)(-6)}{183}\right]}{1.348}(5)(0.05) = \frac{116.42}{1.348} = 86.36$$

$$a_{1.2} = 100.06 - 86.36(0.7684) = 33.70$$

$$s_{1.2}^2 = \frac{\Sigma v_{1.2}^2}{n-2} = \frac{25{,}136.25 - 86.36(116.42)}{181} = \frac{15{,}082.22}{181} = 83.33$$

$$s_{1.2} = \sqrt{83.33} = 9.1$$

The regression equation is thus $X_{1r} = 33.7 + 86.4X_2$, with origin at $X_1 = 0$, $X_2 = 0$.

6. USE OF THE LINE OF REGRESSION AND STANDARD ERROR OF ESTIMATE IN MAKING ESTIMATES

When we have several hundred cases and $\sigma'_{1.2}$ is relatively not very large, it is probably not necessary to consider sampling errors. Suppose this were true[12] of the example of Section 5. The equation

$$X_{1r} = 33.7 + 86.4X_2$$

might then be taken as a reasonably close approximation to the universe regression and $s_{1.2}$ as a reasonably good estimate of $\sigma'_{1.2}$. Hence, we could use them to make estimates of skein strength from knowledge of fiber length.

For example, if the fiber length were known to be 0.800 inches, the skein strength might be estimated at

$$X_{1r} = 33.7 + 86.4(0.800) = 102.8 \text{ pounds}$$

If we assumed that the skein strengths of individual pieces of yarn were normally distributed around X_{1r} as a mean, with a standard deviation of 9.1 pounds, then we might conclude that the chances were 0.95 that, when the fiber length was 0.800 inches, the skein strength of a piece of yarn would be between 102.8 ± 1.96(9.1) or between 85.0 and 120.6 pounds.

---

[12] Actually, the sampling variance of b_{12} in Section 5 is approximately 83.33/(1.348) = 61.80 and its standard error is $\sqrt{61.80} = 7.86$ which is about 9%. (See Section 7.)

7. SAMPLING ERRORS WHEN THE ORIGIN IS AT $X_2 = \bar{X}_2$, $X_1 = 0$

When the number of cases is small or $\sigma'_{1.2}$ is not relatively small, we need to proceed more carefully than we did in Section 6. The first step is to consider the samplig errors in the estimates of the constant term and the regression coefficient b'_{12}.

7.1. Sampling Distributions of $\bar{X}_1$ and b_{12}

If the origin is shifted to $X_2 = \bar{X}_2$ and $X_1 = 0$, the least-squares estimate of the constant term in the new universe regression equation is $\bar{X}_1$. (Cf. Section 3.2.2 above.) Statistical theory then provides us with information about the sampling error of $\bar{X}_1$ and b_{12} under the conditions (1) that the deviations from the universe line of regression are independent and normally distributed; (2) that the line is estimated by the method of least squares; and (3) that the sampling to which the theory applies is carried out in such a way that the sampling variation is variation in the deviations only and not in the regression values.[13] Under these conditions, it may be shown[14] that the average of many sample values of $\bar{X}_1$ will be the universe value $\bar{X}'_1$ and that the average of many sample values of b_{12} will be the universe value b'_{12}. It may also be shown that the distribution of sample values of $\bar{X}_1$ and b_{12} will be normal or approximately normal, and unbiased estimates of the squares of their standard deviations (standard errors) are $s^2_{\bar{X}_1} = \dfrac{s^2_{1.2}}{n}$ and $s^2_{b_{12}} = \dfrac{s^2_{1.2}}{s^2_2(n-1)}$. Furthermore, it may be shown that sample values of $\bar{X}_1$ and b_{12} are not correlated.

7.2. Confidence Limits for $\bar{X}'_1$ and b'_{12}

It follows from the above that, when the origin is at $X_2 = \bar{X}_2$, $X_1 = 0$, 0.95 confidence limits for $\bar{X}'_1$ are given by

(32.10)
$$\bar{X}_1 \pm t_{0.025}\frac{s_{1.2}}{\sqrt{n}}$$

and 0.95 confidence limits for b'_{12} are given by

(32.11)
$$b_{12} \pm t_{0.025}\frac{s_{1.2}}{s_2\sqrt{n-1}}$$

---

[13] This means that, in repeated experiments, we would always give X_2 the same set of values.

[14] For example, see M. G. Kendall, *The Advanced Theory of Statistics* (2d ed.; London: Charles Griffin & Co., Ltd., 1948), Vol. II, chap. xxii.

where $t_{0.025}$ is the 0.025 point of the t distribution for $\nu = n - 2$. If n is large, this is approximately equal to the normal 0.025 point, viz. 1.96. For example, for the welding data of Section 4, 0.95 confidence limits for $\bar{X}'_1$ are[15] $975.0 \pm 2.306 \dfrac{99.9}{\sqrt{10}} = \begin{cases} 902 \\ 1048 \end{cases}$ and 0.95 confidence limits for b'_{12} are 6.90 $\pm\, 2.306 \dfrac{99.9}{24.19\sqrt{9}} = \begin{cases} 10.1 \\ 3.7 \end{cases}$.

7.3. Testing Hypotheses Pertaining to $\bar{X}'_1$ and b'_{12}

If we wish to test a hypothesis pertaining to b'_{12}—say, we wish to test the hypothesis that $b'_{12} = B_{12}$—then we compute $\dfrac{b_{12} - B_{12}}{s_{b_{12}}}$ and look up the result in a t table for $\nu = n - 2$. If this exceeds $t_{\alpha/2}$ or t_α, depending on whether we use a two-tail or a one-tail test, we reject the hypothesis. OC curves for $\alpha = 0.05$ are given by Figure 15.9 for one-tail tests and Figure 25.6 for two-tail tests, except now we take

$$\lambda = \frac{|b'_{12} - B_{12}|}{\sigma'_{1.2}} \frac{\sqrt{\Sigma x_2^2}}{\sqrt{n-1}}$$

and read the n of the chart as equal to $n - 1$ of the sample. Of special interest is the hypothesis $b'_{12} = 0$. If this hypothesis is rejected, we say that b_{12} is significant at the α level.

We can likewise test the hypothesis that $\bar{X}'_1 = M_1$ by computing $\dfrac{\bar{X}'_1 - M_1}{s_{\bar{X}_1}}$ and looking up the result in a t table for $\nu = n - 2$. If this exceeds $t_{\alpha/2}$ or t_α, whichever is appropriate, we reject the hypothesis. OC curves for $\alpha = 0.05$ are given in Figure 15.9 for one-tail tests and Figure 25.6 for two-tail tests, except that we take $\lambda = \dfrac{|\bar{X}'_1 - M_1|}{\sigma'_{1.2}} \dfrac{\sqrt{n}}{\sqrt{n-1}}$ and read the n of the chart as equal to $n - 1$ of the sample. If we reject the hypothesis $\bar{X}'_1 = 0$, we say that $\bar{X}_1$ is significant at the α level. It must always be remembered that this particular analysis applies only when the origin is at $X_1 = \bar{X}_2$, $X_1 = 0$.

For the welding data of Section 4 we have

$$\frac{\bar{X}_1 - 0}{s_{\bar{X}_1}} = \frac{975.0 - 0}{99.9/\sqrt{10}} = \frac{975.0}{31.6} = 30$$

which greatly exceeds the $t_{0.025}$ point for $\nu = n - 2 = 10 - 2 = 8$ (viz 2.306). Hence $\bar{X}_1$ is clearly significant. We also have

[15] See Section 4 above.

$$\frac{b_{12}-0}{s_{b_{12}}}=\frac{6.90-0}{99.9/\sqrt{5{,}269}}=\frac{6.90}{1.38}=5$$

which is again greater than 2.306. Hence b_{12} is significant.

Although b_{12} turned out to be significant for the welding data, there was a sizable chance that, with as small a sample as 10, we might not secure significance. Consider the OC curve for a one-tail test of the hypothesis $b'_{12} = 0$. Let $\Sigma x_2^2 = 5{,}269$ and $\sigma'_{1.2} = 99.9$, as in the sample, then, if b'_{12} is actually as large as 5, we have

$$\lambda=\frac{|5-0|}{99.9}\frac{\sqrt{5{,}269}}{\sqrt{9}}=1.21$$

and Figure 15.9 shows that at this value of λ the ordinate of the curve $n = 9$ is about 0.08. Thus, with a sample of 10, there is still a chance of 8 out of 100 of saying that $b'_{12} = 0$ when actually it equals 5.

7.4. Confidence Limits for Regression Values

As noted above when the origin is at $\bar{X}_2$, 0, the least-squares estimate of the universe line of regression is $X_{1r} = \bar{X}_1 + b_{12}x_2$. Since $\bar{X}_1$ and b_{12} are both subject to sampling errors, the quantity X_{1r} will also be subject to sampling errors. From Theorem 4.1 of Chapter 4 we have the statement that the mean value of X_{1r} for a given value of x_2 will be $\bar{X}'_1 + b'_{12}x_2$ (the universe regression value), and from the corollary of Theorem 4.2 we have the sampling variance of X_{1r} for a given x_2 equal to the sampling variance of $\bar{X}_1$ plus the sampling variance of b_{12} multiplied by x_2^2, since $\bar{X}_1$ and b_{12} are independent. An unbiased estimate of this is $s_{\bar{X}_1}^2 + s_{b_{12}}^2x_2^2$. Furthermore, sample values of X_{1r} for a given value of x_2 will be approximately normally distributed. Hence, 0.95 confidence limits for X_{1r} are given by

(32.12) $$X'_{1r}=\bar{X}_1+b_{12}x_2\pm t_{0.025}\sqrt{s_{\bar{X}_1}^2+s_{b_{12}}^2x_2^2}$$

where $t_{0.025}$ equals the 0.025 point of the t distribution for $\nu = n - 2$. It will be noticed that these limits are a quadratic function of x_2 and thus form hyperbolic loci around the sample line of regression. These loci are 0.95 confidence limits for the values of X_{1r}, but they are not strictly confidence limits for the universe line as a whole (see S. N. Roy in periodical references in Appendix V).

The limiting loci for the welding problem of Section 4 are shown in Figure 32.8. The equations for these loci are

$$X_{1r}=975.0+6.90x_2\pm 2.306\sqrt{\frac{9{,}971}{10}+\frac{9{,}971}{5{,}269}x_2^2}$$

$$=975.0+6.90x_2\pm(2.306)(99.9)\sqrt{0.1+0.00019x_2^2}$$

FIGURE 32.8
Limit Values for Shear Strength (X_1) for Various Values of Weld Diameter (X_2)

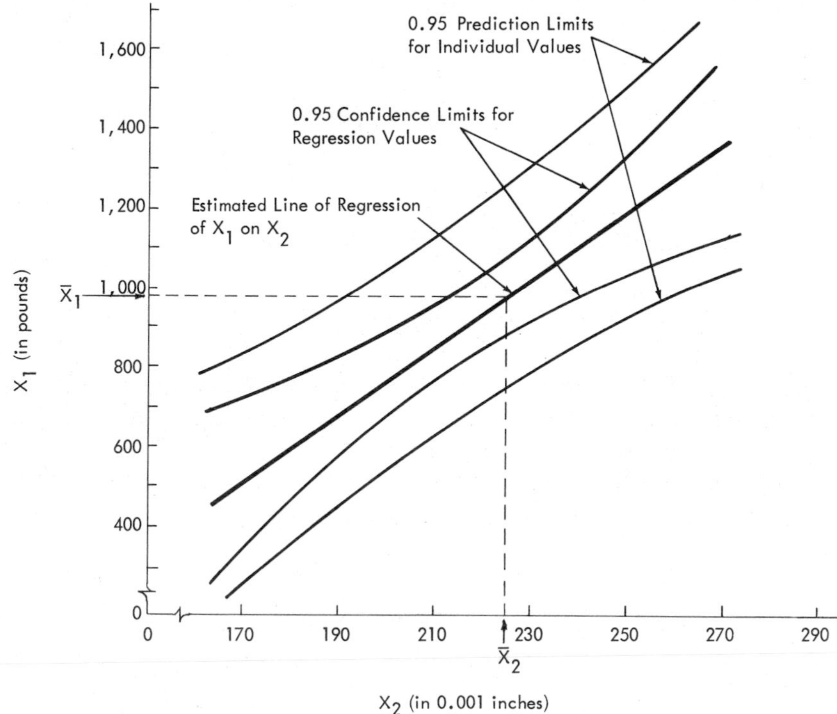

7.5. Limits for Individual Values

The above equations give the range of sampling error in the line of regression. If we use the sample line of regression to estimate a particular value of X_1, we must add to the error in the sample line of regression some measure of the possible deviation of the individual value from the regression value. On the assumption that the sampling error in X_{1r} and the deviation of the individual value from the regression value are independent of each other, the sampling variance of an individual estimate is $\sigma'^2_{\bar{X}_1} + \sigma'^2_{b_{12}} x_2^2 + \sigma'^2_{1.2}$, an unbiased estimate of which is $s^2_{\bar{X}_1} + s^2_{b_{12}} x_2^2 + s^2_{1.2}$. The limits

$$X_1 = \bar{X}_1 + b_{12} x_2 \pm t_{0.025} \sqrt{s^2_{\bar{X}_1} + s^2_{b_{12}} x_2^2 + s^2_{1.2}} \tag{32.13}$$

may consequently be viewed as "prediction limits"[16] for individual values. We saw above that, for a given x_2, the chances are 0.95 that $X_1 + b_{12} x_2 \pm t_{0.025} \sqrt{s^2_{\bar{X}_1} + s^2_{b_{12}} x_2^2}$ will include the universe regression value for that x_2. It can now be said that, for a given x_2, the chances are 0.95 that

[16] Compare Alexander M. Mood and Franklin A. Graybill, *Introduction to the Theory of Statistics* (2d ed.; New York: McGraw-Hill, 1963), p. 336.

$$\bar{X}_1 + b_{12}x_2 \pm t_{0.025}\sqrt{s^2_{\bar{X}_1} + s^2_{b_{12}}x_2^2 + s^2_{1.2}}$$

will include the actual X_1 that occurs on a single trial.

Figure 32.8 shows the prediction loci for individual values as well as the confidence loci for regression values. The prediction loci are given by

$$X_1 = 975 + 6.90x_2 \pm (2.306)(99.9)\sqrt{0.1 + 0.00019x_2^2 + 1}$$

Thus, if we find that a particular weld diameter is 250 thousandths of an inch (i.e., $x_2 = 250 - 223.9 = 26.1$), then we can say the chances are 0.95 that the limits

$$975 + (6.90)(26.1) \pm (2.306)(99.9)\sqrt{0.1 + 0.00019(26.1)^2 + 1} = \begin{cases} 1{,}410.3 \\ 899.8 \end{cases}$$

will include the shear strength of that particular weld. If we had not allowed for sampling errors in $\bar{X}_1$ and b_{12} and had assumed that the sample regression were the true regression, we would have said that, when $X_2 = 0.250$, the chances are 0.95 that an individual value of X_1 will lie between the limits

$$975 + 6.90(26.1) \pm (1.96)(99.9) = \begin{cases} 1{,}350.8 \\ 959.2 \end{cases}$$

This indicates the degree of error involved when the sample is small and allowance is not made for sampling errors in the determination of the line of regression.

It will be noted that the prediction limits for X_1 get wider as X_2 deviates from its mean, both positively and negatively. This means that predictions of the dependent variable are subject to the least error when the independent variable is near its mean, and are subject to the greatest error when the independent variable is distant from its mean. K. A. Brownlee recommends that we refuse to use a sample regression equation to make estimates of X_1 if the values of X_2 at which the estimates are made are beyond the range of X_2 values from which the regression equation was originally derived.[17]

8. THE COEFFICIENT OF CORRELATION

8.1. Definition and Meaning of the Coefficient of Correlation

With every linear regression there is associated a coefficient of correlation which measures the degree of association between the two variables. With reference to the line of regression of X_1 on X_2, the universe coefficient of correlation may be defined by the formula

[17] K. A. Brownlee, *Industrial Experimentation* (2d Amer. ed.; Brooklyn, N.Y.: Chemical Publishing Co., Inc., 1948), p. 71.

(32.14) $$r'_{12} = \pm\sqrt{1 - \frac{\sigma'^2_{1.2}}{\sigma'^2_1}}$$

where $\sigma'_{1.2}$ is the standard deviation of the deviations from the universe line of regression (i.e., the universe standard error of estimate). When the line of regression has a positive slope (i.e., X_1 and X_2 tend to increase and decrease together), r'_{12} is given a positive sign. When the line of regression has a negative slope (i.e., X_1 and X_2 tend to move in opposite directions), r'_{12} is given a negative sign. When $\sigma'_{1.2} = 0$ (i.e., when all points lie on the universe line of regression and there is no scatter), r'_{12} is $\pm$ 1. This means that there is perfect (either positive or negative) linear correlation between the two variables. In other words, the line of regression is a perfect fit. When $\sigma'_{1.2} = \sigma'_1$, i.e., the scatter about the line of regression is the same as the scatter about the line $X_1 = \bar{X}'_1$, then $r'_{12} = 0$, and there is no linear correlation between the two variables. In fact, in this case the horizontal line through $\bar{X}'_1$ and the line of regression are one and the same line. For, when there is no linear correlation between X_1 and X_2, the mean value of X_1 does not vary with X_2. Hence the regression value is always equal to $\bar{X}'_1$, whatever the value of X_2. It will be noted that r'^2_{12} cannot exceed 1, so that the limits for r'_{12} are $\pm$ 1.

For a bivariate *normal* distribution[18] r'_{12} is directly related to the slope (b'_{12}) of the universe line of regression by the equation $b'_{12} = \frac{\sigma'_1}{\sigma'_2} r'_{12}$. Thus when the slope (b'_{12}) is 0, r'_{12} also is 0. When $r'_{12} = \pm$ 1, $b'_{12} = \pm\frac{\sigma'_1}{\sigma'_2}$. If the variables are measured in standard deviation units, so that the standard deviations of the standardized units are both 1, then r'_{12} (for standardized variables) $= b'_{12}$ (for standardized variables). Thus, when standardized units are used, the correlation coefficient *is* the slope of the line of regression. This implies that, when the variables are standardized, the slope of the line of regression cannot exceed 1 in absolute value.

It is to be noted that the square of the coefficient of correlation may be viewed as measuring the relative amount of variation in the dependent variable that is "explained" by the independent variable. For it will be noted that $\frac{\sigma'^2_{1.2}}{\sigma'^2_1}$ is the fraction of the total variance of X_1 that is "unaccounted for" by the regression. Hence, r'^2_{12}, which equals $1 - \frac{\sigma'^2_{1.2}}{\sigma'^2_1}$, is the fraction of the total variance that is "accounted for" by the regression.

It will also be noted that r'_{12} is an index of how much we can improve our estimate of X_1 by employing knowledge of X_2. The standard error of estimate $\sigma'_{1.2}$ is a measure of the absolute error made in using X'_{1r} as an estimate of X_1. But σ'_1 can be looked upon as the "standard error of estimate"

[18] See Appendix I (38).

when we do not make use of the correlation of X_1 with X_2, and $\sigma'_{1.2}$ is the standard error of estimate when we make use of this correlation. From the definition of r'_{12} we have

$$\frac{\sigma'_{1.2}}{\sigma'_1} = \sqrt{1 - r'^2_{12}} \tag{32.15}$$

Hence it follows that r'_{12} is directly related to the reduction in the standard error of estimate through using the correlation of X_1 with X_2 to estimate X_1. Table 32.4 shows the improvement (i.e., $1 - \frac{\sigma'_{12}}{\sigma'_1}$) to be gained by using a line of regression and the accompanying standard error of estimate to measure variations in X_1 as compared with using $\bar{X}'_1$ and σ'_1.

8.2. Estimation of the Coefficient of Correlation from Sample Data

When we have sample data only, which is usually the case, we may estimate r'_{12} in various ways. For data from a bivariate *normal* universe, we may jointly estimate $\bar{X}'_1$, $\bar{X}'_2$, σ'_1, σ'_2, and r'_{12} in a way that maximizes the likelihood of getting the sample result.[19] These joint maximum likelihood estimates are

$$\begin{aligned} \bar{X}_1 &= \Sigma X_1/n \\ \hat{\sigma}_1 &= \text{RMSD}_1 = \sqrt{\frac{\Sigma(X_1 - \bar{X}_1)^2}{n}} \\ \bar{X}_2 &= \Sigma X_2/n \\ \hat{\sigma}_2 &= \text{RMSD}_2 = \sqrt{\frac{\Sigma(X_2 - \bar{X}_2)^2}{n}} \\ r_{12} &= \frac{\Sigma(X_1 - \bar{X}_1)(X_2 - \bar{X}_2)}{n\hat{\sigma}_1\hat{\sigma}_2} \end{aligned} \tag{32.16}$$

Because of this result, r_{12} is known as the product-moment coefficient of correlation. It can be shown that the above formula for r_{12} yields

$$\begin{aligned} r^2_{12} &= 1 - \frac{\Sigma u^2_{1.2}}{\Sigma(X_1 - \bar{X}_1)^2} = 1 - \frac{\Sigma u^2_{1.2}/n}{\Sigma(X_1 - \bar{X}_1)^2/n} \\ &= 1 - \frac{\hat{\sigma}^2_{1.2}}{\hat{\sigma}^2_1} \end{aligned} \tag{32.17}$$

which has the same form as the formula for r'^2_{12}. (Cf. (32.14) above.)

---

[19] See Section 1.2 of Chapter 24.

TABLE 32.4
Relationship between r'_{12} and Reduction in Standard Error of Estimate

Correlation Coefficient r'_{12}	*Percent Improvement in Using Line of Regression to Estimate* X_1 $\left(1-\frac{\sigma'_{1.2}}{\sigma'_1}\right)$	*Correlation Coefficient* r'_{12}	*Percent Improvement in Using Line of Regression to Estimate* X_1 $\left(1-\frac{\sigma'_{1.2}}{\sigma'_1}\right)$
0.60	20	0.90	56
0.65	24	0.92	61
0.70	29	0.94	66
0.75	34	0.96	72
0.80	40	0.98	81
0.85	47	1.00	100

SOURCE: Peter P. DiPaola, "Use of Correlation in Quality Control," *Industrial Quality Control,* July 1945, p. 14; see also Ralph J. Watkins, "A Criticism of 'Economics of Business Fluctuation in the United States, 1919–1925,' " *American Economic Review* 18 (1928), pp. 489–97.

Sometimes formula (32.17) is adjusted by replacing $\hat{\sigma}^2_{1.2}$ by $s^2_{1.2}$ and $\hat{\sigma}^2_1$ by s^2_1. We then have

(32.18) $$\tilde{r}^2_{12} = 1 - \frac{s^2_{1.2}}{s^2_1}$$

the "adjustment" causing $\tilde{r}_{12}$ to be less than r_{12}.

An *unbiased* estimate[20] of r'_{12} is given approximately by

(32.19) $$G(r_{12}) = r_{12}\left(1 - \frac{1 - r^2_{12}}{2(n-4)}\right)$$

Again the adjustment yields an estimate that is less than r_{12}.

If we apply the above formulas to the data of Section 4 on shear strength of weld diameters, we have the following. The joint maximum likelihood estimate of r'_{12} is:

$$r_{12} = \frac{36{,}345}{\sqrt{(330{,}550)(5{,}269)}} = 0.871$$

or by (32.17)

$$r_{12} = \sqrt{1 - \frac{79{,}770}{330{,}550}} = 0.871$$

The unbiased estimate of r'_{12} is

$$G(r_{12}) = 0.871\left(1 - \frac{1-(0.871)^2}{2(10-4)}\right) = 0.854$$

---

[20] See Olkin and Pratt, "Unbiased Estimation of Certain Correlation Coefficients," *Annals of Mathematical Statistics* 29 (1958), p. 203. Note that the Olkin and Pratt n is one less than the sample size.

If we use formula (32.18), which adjusts the sums of squares for degrees of freedom, we get

$$\tilde{r}_{12} = \sqrt{1 - \frac{79{,}770/8}{330{,}550/9}} = \sqrt{1 - 0.2715} = 0.854$$

which rounded off to the third decimal is in this case the same as $G(r_{12})$. Using the unbiased estimate $G(r_{12})$, we would conclude from the coefficient of correlation that about 73 percent ($= [G(r_{12})]^2$) of the variance in shear strength is accounted for by weld diameter.

For the cotton yarn data of Section 5 we have

$$r_{12} = \frac{116.42}{\sqrt{(1.348)(25{,}136.25)}} = 0.6323$$

$$G(r_{12}) = 0.6323\left(1 - \frac{1 - (.6323)^2}{2(183 - 4)}\right) = 0.6312$$

We can in this case estimate that approximately 36 percent of the variance in skein strength is accounted for by fiber length.

8.3. Testing Hypotheses Regarding the Value of r'_{12}

In a bivariate normal universe $b'_{12} = r'_{12}\frac{\sigma'_1}{\sigma'_2}$ and we can test the hypothesis that $r'_{12} = 0$ by testing the equivalent hypothesis that $b'_{12} = 0$. The latter was discussed in Section 7.3 above. Other hypotheses regarding r'_{12} can be tested by treating[21] $\tanh^{-1} r_{12}$ as if it were normally distributed with a mean of $\tanh^{-1} r'_{12}$ and a standard deviation of $\frac{1}{\sqrt{n-3}}$.

In conclusion it is to be noted that $r'^2_{12} = r'^2_{21}$ and all that has been said regarding r'_{12} and the line of regression of X_1 on X_2 applies *mutatis mutandis* to r'_{21} ($=r'_{12}$) and the line of regression of X_2 on X_1. It is also to be noted that r'_{12} measures linear and not curvilinear correlation.[22] It applies to a line of regression and not a curve of regression. It is true that, if X_1 and X_2 are completely independent, then r'_{12} will be zero; but, if $r'_{12} = 0$, it is still possible that X_1 and X_2 may be related in some way or other.

---

[21] See, for example, R. A. Fisher, *Statistical Methods for Research Workers* (13th rev. ed.; London: Oliver & Boyd, Ltd., 1958), Section 35, and J. G. Smith and A. J. Duncan, *Sampling Statistics and Applications* (New York: McGraw-Hill, 1945), pp. 300–301.

[22] If $\sigma'^2_{1.2}$ is the variance of the deviations from some curve of regression, then $I'^2_{12} = 1 - \frac{\sigma'^2_{1.2}}{\sigma'^2_1}$ is called the index of correlation pertinent to the given curvilinear regression. This is the same formula as (32.14), but I'_{12} is now used in place of r'_{12} to warn the user that $\sigma'_{1.2}$ is in this case measured from a nonlinear regression.

9. PROBLEMS

32.1. The following are data on the tensile strength and fineness of cotton fibers.[23] Plot these data on a bivariate chart such as Figure 32.1. Draw a line in "freehand" that seems to represent the line of regression of fiber tensile strength on fiber fineness. (X_3 = fiber tensile strength; X_4 = fiber fineness.)

X_3	X_4	X_3	X_4	X_3	X_4	X_3	X_4
76	4.4	78	4.9	78	3.8	78	4.1
78	4.2	77	4.1	78	4.6	82	3.8
73	4.2	76	4.4	78	4.2	81	4.5
72	4.4	81	3.8	85	3.8	80	5.6
73	4.3	76	5.1	82	4.2	82	4.1
69	4.6	78	4.8	80	4.2	90	4.2
69	4.6	82	4.9	81	3.7	82	4.8
80	3.6	85	5.2	80	4.0	82	4.4
78	3.6	83	4.3	78	4.9	81	5.1
73	3.7	77	4.5	85	3.6	93	5.0
71	4.6	80	4.2	81	3.4		
72	4.5	80	4.6	83	4.0		
76	4.2	77	5.0	86	4.6		
76	5.0	82	4.7	82	4.3		
70	4.8	80	4.3	76	4.3		
76	4.1	82	4.4	78	4.2		
78	3.1	93	3.8	78	4.2		
80	2.9	82	4.0	78	4.2		
83	3.9	90	4.8	73	4.1		
79	3.8	84	4.6	77	4.1		
83	4.0	90	3.8	74	4.2		
76	3.2	79	4.2	79	4.3		
78	3.4	79	4.0	79	4.2		
71	4.4	80	4.1	80	4.4		
74	5.0	83	4.4	69	5.1		
72	3.7	81	4.1	73	5.4		
69	4.2	80	3.9	72	5.0		
77	4.9	80	4.0	78	5.2		
78	3.9	83	4.2	83	4.1		
71	4.4	80	4.1	79	4.3		

32.2. The following are measurements of a chemical precipitate (X_1) and a reagent (X_2) used in obtaining this precipitate.

[23] Cf. U.S. Department of Agriculture, War Food Administration, Office of Marketing Services, "Results of Fiber and Spinning Tests for Some Varieties of Upland Cotton Grown in the United States, Crop of 1944" (Processed, April 1945).

X_1	X_2	X_1	X_2
9.4	8.2	12.2	7.5
6.5	5.8	13.3	7.9
7.3	6.4	14.2	8.6
7.9	5.9	13.8	9.0
9.0	6.5	11.3	8.1
9.3	7.1	8.1	5.7
10.6	7.8	6.1	5.3
11.4	8.1		

a. Plot these data on a bivariate chart such as Figure 32.1. Draw in a line freehand that seems to represent the line of regression of X_1 on X_2.

b. Estimate the line of regression of X_1 on X_2 by the method of least squares. Plot on your chart and compare with the line you drew in freehand.

c. If $X_2 = 6.5$, what would be your estimate of the mean value of X_1?

d. Estimate the standard error of estimate for the universe.

e. On the assumption that the values obtained in *(c)* and *(d)* are the true universe values and that deviations from the universe line of regression are normally distributed, determine 0.95 prediction limits for values of X_1 when $X_2 = 6.5$.

f. Derive limiting loci for the universe regression values.

g. On the assumption that deviations from the universe line of regression are normally distributed, determine 0.95 prediction limits for X_1, given $X_2 = 6.5$. Compare with your answer to *(e)*.

h. Estimate the universe coefficient of correlation. What percent of the variation in X_1 may be attributed to X_2?

i. Test the hypothesis that the increase in X_1 is, on the average, double the increase in X_2 (i.e., that $b'_{12} = 2$).

j. Derive the OC curve for the test employed in *(i)*.

k. Determine 0.95 confidence limits for the regression coefficient b'_{12}.

32.3. Two crews, A and B, are making fuel tanks. The percentage of tanks with leaks for each crew are as follows for a given month.[24]

Day of Month	*Crew A* (X_1)	*Crew B* (X_2)
1	57.6	75.5
2	55.7	61.4
4	56.4	46.8
5	58.7	52.8
6	50.5	59.5
7	50.5	45.9

[24] Cf. Lester A. Kauffman, *Statistical Quality Control at the St. Louis Division of American Stove Company* (OPRD Quality Control Reports, No. 3, August 1945), p. 17.

Day of Month	*Crew A* (X_1)	*Crew B* (X_2)
8	44.0	40.0
10	43.3	45.9
11	40.8	59.6
12	32.6	55.9
13	59.3	54.6
14	49.4	67.3
15	45.5	60.5
17	50.0	53.3
18	35.5	46.4
19	53.8	49.5
20	62.7	72.0
21	73.0	82.3
22	72.4	72.7
25	80.5	86.8
26	70.1	73.8
27	80.0	55.1
28	57.8	53.6

a. Estimate the universe coefficient of correlation for these data.

b. Determine whether this coefficient is significantly different from zero. Would you conclude that there was some common factor causing the percentage of tanks with leaks to vary similarly for the two crews?

32.4. In the manufacture of airplane fuselage frame sections, thousands of rivets are used to join aluminum sheets and frames. A study of the number of oversize rivet holes (X_1) and the number of minor repairs on a unit (X_2) reveals the following:[25]

X_1	X_2	X_1	X_2
45	22	38	17
52	26	56	36
49	21	65	29
60	28	43	17
67	33	41	25
61	32	61	27
70	33	72	40
54	25	34	14
52	34	69	41
67	35	60	30
52	30	64	36
70	37	70	31
59	33	47	13
40	13	53	20

[25] Cf. Peter P. DiPaola, "Use of Correlation in Quality Control," *Industrial Quality Control,* July 1945, p. 11.

X_1	X_2	X_1	X_2
32	12	50	25
41	18	62	31
48	18	44	19
49	24	63	25
55	23	64	33
63	36	42	16

a. Plot these data on a bivariate chart such as Figure 32.1. Draw a line freehand that seems to represent the line of regression of X_1 on X_2.

b. Estimate the line of regression of X_1 on X_2 by the method of least squares. Plot on your chart and compare with the line you drew in freehand.

c. The above study was undertaken to help estimate the average number of oversize rivet holes from the number of minor repairs on a unit. Suppose that a unit showed 30 minor repairs, what would you take as the expected number of oversize rivet holes?

d. Estimate the standard error of estimate for the universe.

e. Take the line of regression derived in *b* as a reasonably good estimate of the universe line of regression and the standard error of estimate derived in *d* as a reasonably good estimate of the universe standard error of estimate. Draw up a chart showing the line of regression and lines at vertical distances $\pm 3s_{1.2}$ from the regression line, where $s_{1.2}$ is the estimate of the standard error of estimate derived in *d*. These two lines may be viewed as control limits for X_1 when we know X_2. Consider values of X_1 only, and set up limits equal to $\bar{X}_1 \pm 3s_1/\sqrt{n}$. Note the difference between the regression line control chart and the ordinary control chart. Note several cases in which points would be out of control on the regression line control chart and not on the ordinary chart (cf. Peter P. DiPaola, "Use of Correlation in Quality Control," *Industrial Quality Control,* July 1945, pp. 10–14).

f. Assume that the deviations from the universe line of regression are normally distributed and, allowing for sampling errors in your estimate of the universe line of regression, derive 0.95 prediction limits for individual values of X_1. Plot these limits on a chart and compare with limits given by simply adding and subtracting $1.96s_{1.2}$ from the sample line of regression. Does a sample of 40 in this case seem sufficient to disregard sampling errors in the regression parameters? What factor other than sample size is important in determining these sampling errors?

32.5. *a.* Group the data of Problem 32.1 and estimate the universe line of regression of X_3 on X_4 by the method of least squares. Draw this line on the chart constructed in Problem 32.1. How does it compare with the freehand line you drew?

b. If $X_4 = 4.0$ micrograms, at what would you estimate the average tensile strength?

c. Estimate the universe standard error of estimate.

d. Taking the values derived in *(b)* and *(c)* as reasonably good estimates of the universe values (remember $n = 100$) and assuming that the deviations from the universe line of regression are normally distributed, estimate the probability that X_3 will exceed 85,000 pounds per square inch when X_4 equals 4.0 micrograms.

e. Test the hypothesis that $b'_{34} = -0.50$.

f. Derive the OC curve for the test employed in *(e)*.

g. Determine 0.95 confidence limits for the regression coefficient b'_{34}.

10. SELECTED REFERENCES*

Acton (B '59), American Society for Testing and Materials (B '62), Berkson (P '69), Bowden (P '68), Box (P '66), Box and Watson (P '62), Brown (P '75), Brownlee (B '60), Davies (B '57), Draper and Smith (B '67), Fisher (B '58), Halperin (P '61), Hahn (P '77), Harrison and McCabe (P '79), Hocking (P '83), Mallios (P '69), Mandel (P '84), Olkin and Pratt (P '58), Roy (P '54), Roy and Bose (P '53), Wilks (B '48), Williams (B '59 and P '69), and Youden (P '59).

---

* B and P refer to the Book and Periodical sections, respectively, of the Cumulative List of References in Appendix V.

33

Regression and Correlation: Three and More Variables

A dependent variable can in most cases be estimated with greater precision if we base the estimate upon several independent variables than if we base it on only one. We shall see, for example, that the skein strength of cotton yarn can be predicted with greater precision from knowledge of fiber length, fiber fineness, and fiber tensile strength than from fiber length alone. For, unless the new independent variables are highly correlated with the old, their addition to the regression equation helps account for some previously unaccounted-for variations in the dependent variable and thus reduces the standard error of estimate.

To understand the nature of a regression equation involving two or more independent variables, we must first consider multivariate frequency distributions. The simplest of these is a trivariate distribution.

1. TRIVARIATE REGRESSIONS

1.1. Trivariate Frequency Distributions and Some of Their Characteristics

When more than two variables are involved, we cannot easily represent their joint distribution by a graph. When there are only three variables, however, we can draw pictures that will show some of the characteristics of the trivariate distribution, and our imagination can do the rest. In the discussion that follows we shall guide the imagination by analogy with the two-variable case.

1.1.1. Trivariate Distributions. For given values of three variables—X_1, X_2, and X_3—we can "plot" a point in three-dimensional space just as for given values of two variables, X_1 and X_2, we can plot a point on a plane (cf. Figure 32.1). Numerous sets of such values will yield a collection of points scattered throughout space just as the points of Figure 32.1 are scattered throughout a plane. If we mark off class intervals on each of the three axes X_1, X_2, and X_3 and pass planes through the class limits of these intervals, these planes will intersect in space to form three-dimensional cells. If we record the number of points falling in each cell, we will have a "trivariate frequency distribution."

1.1.2. Trivariate Density Functions. We can, if we wish, express the frequency of each cell in relative form, i.e., we can compute F/n for each cell. This relative frequency becomes a measure of the density of the points in the cell. When we do not have a very large number of points, the variations in density from cell to cell will be abrupt, just as are the jumps from one rectangle to neighboring rectangles of a single variable histogram. If, however, we increase the number of cases indefinitely and continuously reduce the size of the class intervals (i.e., take smaller and smaller cells), the gradations of density from one cell to the next will become less and less abrupt, and we shall approach as a limit a continuously varying "density function." With three variables, we cannot draw a graph of this trivariate density function as we could of univariate and bivariate density functions, but we can, by analogy, imagine the existence of a higher dimensional surface that describes the distribution of the universe of points in three-dimensional space.

1.1.3. Trivariate Regressions. If we examine a trivariate distribution, we shall find that, for given values of X_2 and X_3, there will be a distribution of X_1 values. If there is an association, however, between X_1, on the one hand, and X_2 and X_3, on the other hand, we shall probably find that the mean value of X_1 for given values of X_2 and X_3 will tend to vary with X_2 and X_3. The plane or curved surface that shows how the mean value of X_1 varies with X_2 and X_3 is called the multiple regression of X_1 on X_2 and X_3. Figure 33.1 is a picture of a plane of regression of X_1 on X_2 and X_3 that slopes downward in both directions. For that particular case the mean value of X_1 for given values of X_2 and X_3 falls as X_2 increases and also as X_3 increases. In other cases the mean value of X_1 for given values of X_2 and X_3 might increase as X_2 increases and also as X_3 increases. In still other cases the mean value of X_1 for given values of X_2 and X_3 might increase as X_2 increases and decrease as X_3 increases, or it might decrease with X_2 and increase with X_3. In the case of a curved regression the surface of mean values of X_1 for given values of X_2 and X_3 might bend in an infinite variety of ways as X_2 and X_3 change.

For a trivariate distribution there are three regressions. There is the regression of X_1 on X_2 and X_3 that we have just discussed. There is also the regression of X_2 on X_1 and X_3 and the regression of X_3 on X_1 and X_2. If we wish to estimate X_i from X_j and X_k, then we use the regression of X_i on X_j and X_k.

1.1.4. The Standard Errors of Estimate. For any given point on the regression of X_i on X_j and X_k there will be a standard deviation of X_i

FIGURE 33.1
Illustration of a Plane of Regression of X_1 on X_2 and X_3

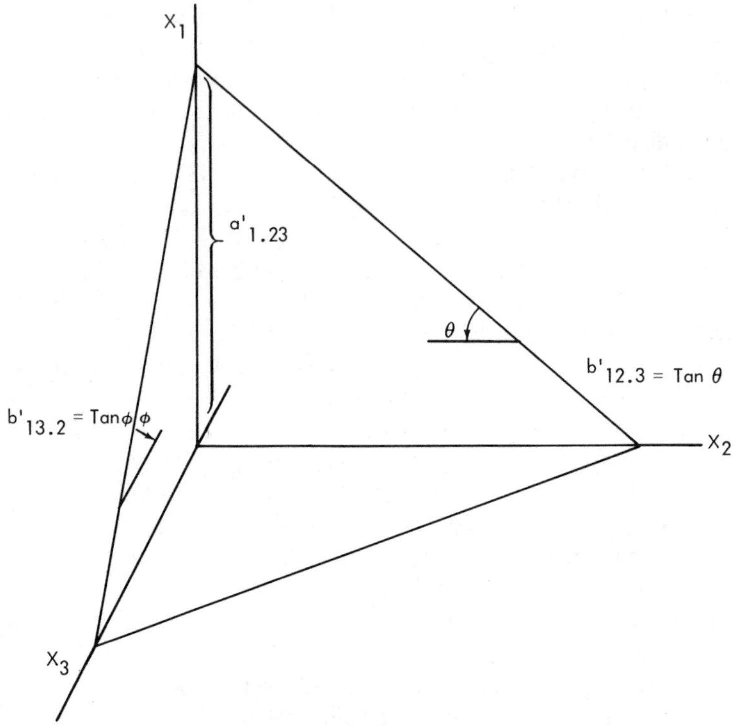

Equation of Plane: $x'_{1r} = a'_{1.23} + b'_{12.3}\, x_2 + b'_{13.2}\, x_3$.

values for the given X_j and X_k values. This will be the standard error of estimate of X_i for the given values of X_j and X_k. In the distributions we shall discuss, we shall assume that this standard error of estimate is the same for every point (i.e., the variation is homoscedastic) and shall therefore speak simply of the standard error of estimate for the given regression. For a universe distribution we shall represent the standard error of estimate for the regression of X_i on X_j and X_k by $\sigma'_{i.jk}$.

1.2. Estimate of Trivariate Linear Regression

As in the two-variable case, the practical problem in most cases is how to estimate the regression of an unknown universe from a sample set of data from that universe. As in the previous case, the method that is commonly used is the method of least squares. In this section we shall restrict ourselves to the estimation of a linear regression around which the variation is homoscedastic.

1.2.1. Estimation of the Plane of Regression of X_1 on X_2 and X_3 and

Its Standard Error of Estimate. Let the universe plane of regression of X_1 on X_2 and X_3 be represented by the equation

(33.1) $X'_{1r} = a'_{1.23} + b'_{12.3}X_2 + b'_{13.2}X_3$ origin at $X_1 = X_2 = X_3 = 0$

In this equation the subscripts indicate what the parameters are. Thus $a'_{1.23}$ is the constant term in the regression equation in which X_1 is the dependent variable and X_2 and X_3 are the independent variables. Also, $b'_{12.3}$ is the coefficient of X_2 in the equation in which X_1 is the dependent variable and X_3 is the other independent variable, and $b'_{13.2}$ is the coefficient of X_3 in the regression equation in which X_1 is the dependent variable and X_2 is the other independent variable. The placing of the dots should be obvious. In the case of $a'_{1.23}$ we are setting X_1 over against X_2 and X_3, whereas in the case of $b'_{12.3}$, say, we are linking together X_1 and X_2 with X_3 representing the extra variable.

Let the equation that is derived as an estimate of the universe equation be

(33.2) $X_{1r} = a_{1.23} + b_{12.3}X_2 + b_{13.2}X_3$ origin at $X_1 = X_2 = X_3 = 0$

If this is derived by the method of least squares,[1] we must have

$$\sum(X_1 - X_{1r})^2 \text{ a minimum}$$

or

$$\sum(X_1 - a_{1.23} - b_{12.3}X_2 - b_{13.2}X_3)^2 \text{ a minimum}$$

The necessary conditions for this are

$$\begin{aligned} &\sum(X_1 - a_{1.23} - b_{12.3}X_2 - b_{13.2}X_3) = 0 \\ \textbf{(33.3)} \qquad &\sum(X_1 - a_{1.23} - b_{12.3}X_2 - b_{13.2}X_3)X_2 = 0 \\ &\sum(X_1 - a_{1.23} - b_{12.3}X_2 - b_{13.2}X_3)X_3 = 0 \end{aligned}$$

which, when multiplied out, become

$$\begin{aligned} &\sum X_1 - na_{1.23} - b_{12.3}\sum X_2 - b_{13.2}\sum X_3 = 0 \\ \textbf{(33.4)} \qquad &\sum X_1X_2 - a_{1.23}\sum X_2 - b_{12.3}\sum X_2^2 - b_{13.2}\sum X_2X_3 = 0 \\ &\sum X_1X_3 - a_{1.23}\sum X_3 - b_{12.3}\sum X_2X_3 - b_{13.2}\sum X_3^2 = 0 \end{aligned}$$

These are the least-squares or "normal" equations for estimating the universe regression. The quantities ΣX_1, ΣX_2, ΣX_1X_2, and so on, can be computed from the given data, so that, after we have inserted their values, we have three simultaneous equations which can be solved for values of $a_{1.23}$, $b_{12.3}$, and $b_{13.2}$.

---

[1] See Appendix I (39). Also see Appendix I (37) for the method by which equation (33.7) is derived.

As in the two-variable case, it is usually easier in practice to use a slightly different set of equations from equations (33.4). These are equations in which the variables are referred to their mean values. For, in this case, the new constant term becomes zero, and the three least-squares equations reduce to a subset of two equations, viz.

(33.5)
$$b_{12.3}\sum x_2^2 + b_{13.2}\sum x_2 x_3 = \sum x_1 x_2$$
$$b_{12.3}\sum x_2 x_3 + b_{13.2}\sum x_3^2 = \sum x_1 x_3$$

To find $b_{12.3}$ and $b_{13.2}$, we compute Σx_2^2, Σx_3^2, $\Sigma x_2 x_3$, $\Sigma x_1 x_2$, and $\Sigma x_1 x_3$ from the data, insert their values in equations (33.5), and solve for the unknown parameters. Then $a_{1.23}$ is computed from the equation

(33.6)
$$a_{1.23} = \bar{X}_1 - b_{12.3}\bar{X}_2 - b_{13.2}\bar{X}_3$$

The actual computations are illustrated in the next section.

The variance around the universe plane of regression (33.1) is estimated from the equation

(33.7)
$$\sum u_{1.23}^2 = \sum X_1^2 - a_{1.23}\sum X_1 - b_{12.3}\sum X_1 X_2 - b_{13.2}\sum X_1 X_3$$

If deviations are taken from mean values, this reduces to

(33.8)
$$\sum u_{1.23}^2 = \sum x_1^2 - b_{12.3}\sum x_1 x_2 - b_{13.2}\sum x_1 x_3$$

which may in some problems be easier to evaluate than equation (33.7).

From $\Sigma u_{1.23}^2$ we can compute

(33.9)
$$s_{1.23}^2 = \frac{\Sigma u_{1.23}^2}{n - 3}$$

which is an unbiased estimate of the variance around the universe plane of regression. It will be noted that we divide by $n - 3$ now instead of by $n - 2$, since there are three parameters to be estimated instead of two.

1.2.2. Estimation of Other Planes of Regression and Their Standard Errors. Equations (33.5) to (33.9) give estimates of the plane of regression of X_1 on X_2 and X_3 and its associated standard error of estimate. Similar sets of equations will give the other two planes of regression and their standard errors of estimate. To get the plane of regression of X_2 on X_1 and X_3 and its standard error of estimate, interchange subscripts 1 and 2 in equations (33.5) to (33.9); and, to get the plane of regression of X_3 on X_1 and X_2 and its standard error of estimate, interchange subscripts 1 and 3 in these equations. In most cases we wish only one of these regressions, the choice being made obvious by the problem in hand. The variable to be estimated by the equation, it will be recalled, is always the dependent variable. Serious errors may result, for example, if we use the regression equation of X_1 and X_2 and X_3 to estimate X_2 or X_3.

1.3 An Example

1.3.1. The Problem. In Table 33.1 are listed 20 sets of measurements of four qualities of cotton yarn: the skein strength, fiber length, fiber tensile strength, and fiber fineness. We shall undertake to determine the regression

TABLE 33.1
Quality Characteristics of 20 Pieces of 22s Cotton Yarn

Piece Number	*Skein Strength (X_1) (Pounds)*	*Fiber Length (X_2) (0.01 Inch)*	*Fiber Tensile Strength (X_3) (1,000 Pounds per Square Inch)*	*Fiber Fineness (X_4) (0.1 Micrograms per Inch of Fiber)*
1	99	85	76	44
2	93	82	78	42
3	99	75	73	42
4	97	74	72	44
5	90	76	73	43
6	96	74	69	46
7	93	73	69	46
8	130	96	80	36
9	118	93	78	36
10	88	70	73	37
11	89	82	71	46
12	93	80	72	45
13	94	77	76	42
14	75	67	76	50
15	84	82	70	48
16	91	76	76	41
17	100	74	78	31
18	98	71	80	29
19	101	70	83	39
20	80	64	79	38

Variables X_1 and X_2 are described in the footnote to Table 32.3. Variables X_3 and X_4 may be described as follows:

X_3—fiber tensile strength, measured in 1,000 pounds per square inch, converted from Pressley index according to a formula. For most purposes fiber strength is a very important factor in determining yarn strength. Cottons with good fiber strength give less trouble in spinning than do weak-fibered cottons. The values reported under this caption may be classified as follows:

Excellent	90 or more
Very good	83–90
Average	78–82
Fair	72–77
Weak	Below 72

X_4—fiber fineness (weight per inch of fiber), in micrograms, expressed in terms of weight per unit of fiber length. The larger the figure reported, the coarser the fibers, and, conversely, the lower the figure, the finer the fiber. As a general rule, long cottons are fine-fibered and short cottons coarse-fibered. In general, fineness contributes to yarn strength. Very fine fibers, however, tend to make it more nappy and to reduce the rate of processing, so that the desirability of fiber fineness depends upon the specific end product or use. The following adjective ratings may be applied for purposes of comparison:

Fine	Below 4.0
Average	4.0–4.9
Slightly coarse	5.0–5.9
Coarse	6.0 and above

For further details see source of data.

SOURCE: United States Department of Agriculture, War Food Administration, Office of Marketing Services, *Results of Fiber and Spinning Tests for Some Varieties of Upland Cotton Grown in the United States, Crop of 1944* (processed April, 1945).

of skein strength (X_1) on fiber length (X_2) and fiber tensile strength (X_3). For the moment we shall ignore fiber fineness (X_4). The regression of skein strength on fiber length was studied in Section 5 of Chapter 32. The present analysis is an expansion of that study, although it covers a much smaller set of data. The principal objective of our analysis will be to derive an estimating equation for skein strength, so that from knowledge of fiber length and fiber tensile strength we can estimate skein strength. A supplementary objective will be to compare the relative importance of the two factors in determining skein strength.

1.3.2. Calculations. Table 33.2 shows the calculations required for computing Σx_2^2, Σx_3^2, and so on. The sums of the first nine columns yield quantities to be employed in subsequent computations. The last four columns help in checking the table. The checks are

$$\begin{aligned} &\sum \text{Col. (1)} + \sum \text{Col. (2)} + \sum \text{Col. (3)} = \sum \text{Col. 10} \\ &\sum \text{Col. (4)} + \sum \text{Col. (5)} + \sum \text{Col. (7)} = \sum \text{Col. 11} \\ &\sum \text{Col. (4)} + \sum \text{Col. (6)} + \sum \text{Col. (8)} = \sum \text{Col. 12} \\ &\sum \text{Col. (5)} + \sum \text{Col. (6)} + \sum \text{Col. (9)} = \sum \text{Col. 13} \end{aligned} \tag{33.10}$$

These provide enough overlapping so that, if any one of columns (4) to (9) is in error, this column will be immediately picked out. For example, if the second and fourth equations do not check but the third does, then it must be column (5), the only common term, that is in error. If a calculator is available, the computations of Table 33.2 can be cumulated on the calculator and the table itself dispensed with. The checks should always be carried out, however.

From Table 33.2 we have the following:

$$\bar{X}_1 = \frac{\Sigma X_1}{n} = \frac{1{,}908}{20} = 95.40$$

$$\bar{X}_2 = \frac{\Sigma X_2}{n} = \frac{1{,}541}{20} = 77.05$$

$$\bar{X}_3 = \frac{\Sigma X_3}{n} = \frac{1{,}502}{20} = 75.10$$

$$\sum x_1^2 = \sum X_1^2 - n\bar{X}_1^2 = 184{,}766 - 20(95.40)^2 = 2{,}742.80$$

$$\sum x_2^2 = \sum X_2^2 - n\bar{X}_2^2 = 119{,}951 - 20(77.05)^2 = 1{,}216.95$$

$$\sum x_3^2 = \sum X_3^2 - n\bar{X}_3^2 = 113{,}104 - 20(75.10)^2 = 303.80$$

$$\sum x_1x_2 = \sum X_1X_2 - n\bar{X}_1\bar{X}_2 = 148{,}344 - 20(95.40)(77.05) = 1{,}332.60$$

$$\sum x_1x_3 = \sum X_1X_3 - n\bar{X}_1\bar{X}_3 = 143{,}626 - 20(95.40)(75.10) = 335.20$$

$$\sum x_2x_3 = \sum X_2X_3 - n\bar{X}_2\bar{X}_3 = 115{,}754 - 20(77.05)(75.10) = 24.90$$

TABLE 33.2
Computation of Products and Sums for a Three-Variable Regression Analysis

(1) X_1	(2) X_2	(3) X_3	(4) X_1X_2	(5) X_1X_3	(6) X_2X_3	(7) X_1^2	(8) X_2^2	(9) X_3^2	(10) $(X_1+X_2+X_3)$	(11) $(X_1+X_2+X_3)X_1$	(12) $(X_1+X_2+X_3)X_2$	(13) $(X_1+X_2+X_3)X_2$
99	85	76	8,415	7,524	6,460	9,801	7,225	5,776	250	25,740	22,100	19,760
93	82	78	7,626	7,254	6,396	8,649	6,724	6,084	253	23,529	20,746	19,734
99	75	73	7,425	7,227	5,475	9,801	5,625	5,329	247	24,453	18,525	18,031
97	74	72	7,178	6,984	5,328	9,409	5,476	5,184	243	23,571	17,982	17,496
90	76	73	6,840	6,570	5,548	8,100	5,776	5,329	239	21,510	18,164	17,447
96	74	69	7,104	6,624	5,106	9,216	5,476	4,761	239	22,944	17,686	16,491
93	73	69	6,789	6,417	5,037	8,649	5,329	4,761	235	21,855	17,155	16,215
130	96	80	12,480	10,400	7,680	16,900	9,216	6,400	306	39,780	29,376	24,480
118	93	78	10,974	9,204	7,254	13,924	8,649	6,084	289	34,102	26,877	22,542
88	70	73	6,160	6,424	5,110	7,744	4,900	5,329	231	20,328	16,170	16,863
89	82	71	7,298	6,319	5,822	7,921	6,724	5,041	242	21,538	19,844	17,182
93	80	72	7,440	6,696	5,760	8,649	6,400	5,184	245	22,785	19,600	17,640
94	77	76	7,238	7,144	5,852	8,836	5,929	5,776	247	23,218	19,019	18,772
75	67	76	5,025	5,700	5,092	5,625	4,489	5,776	218	16,350	14,606	16,568
84	82	70	6,888	5,880	5,740	7,056	6,724	4,900	236	19,824	19,352	16,520
91	76	76	6,916	6,916	5,776	8,281	5,776	5,776	243	22,113	18,468	18,468
100	74	78	7,400	7,800	5,772	10,000	5,476	6,084	252	25,200	18,648	19,656
98	71	80	6,958	7,840	5,680	9,604	5,041	6,400	249	24,402	17,679	19,920
101	70	83	7,070	8,383	5,810	10,210	4,900	6,889	254	25,654	17,780	21,082
80	64	79	5,120	6,320	5,056	6,400	4,096	6,241	223	17,840	14,272	17,617
Σ 1,908	1,541	1,502	148,344	143,626	115,754	184,766	119,951	113,104	4,951	476,736	384,049	372,484

The least-squares equations are

$$1{,}216.95b_{12.3} + 24.90b_{13.2} = 1{,}332.60$$

$$24.90b_{12.3} + 303.80b_{13.2} = 335.20$$

These can be solved by substitution, elimination, or determinants, depending on the training of the investigator. For further discussion of the solution of least-squares equations, see Section 1.6 below. The solution itself is as follows:

$$b_{12.3} = 1.074$$

$$b_{13.2} = 1.015$$

This gives

$$\begin{aligned} a_{1.23} &= \bar{X}_1 - b_{12.3}\bar{X}_2 - b_{13.2}\bar{X}_3 \\ &= 95.40 - (1.074)(77.05) - (1.015)(75.10) \\ &= -63.58. \end{aligned}$$

Hence our estimate of the universe regression of X_1 on X_2 and X_3 is

(33.11) $$X_{1r} = -63.58 + 1.074X_2 + 1.015X_3$$

in which X_1 is measured in pounds, X_2 in hundredths of an inch, and X_3 in 1,000 pounds per square inch.

The standard error of estimate ($s_{1.23}$) is given by

$$\begin{aligned} s_{1.23}^2 = \frac{\Sigma u_{1.23}^2}{n-3} &= \frac{\Sigma x_1^2 - b_{12.3}\Sigma x_1x_2 - b_{13.2}\Sigma x_1x_3}{n-3} \\ &= \frac{2{,}742.80 - (1.074)(1{,}332.60) - 1.015(335.20)}{20-3} \\ &= 57.14 \end{aligned}$$

and

$$s_{1.23} = \sqrt{57.14} = 7.57$$

1.3.3. Use of Regression Equation in Making Estimates: No Allowance for Sampling Errors. Let us ignore sampling errors and take equation (33.11) as a good estimate of the universe regression. Then, if we assume that individual values of X_1 are normally distributed around X_{1r} as a mean, with a standard deviation equal to $s_{1.23}$, we can estimate skein strength from fiber length and fiber tensil strength and indicate the reliability of that estimate. For example, if our fibers average 0.800 inches in length and have an average tensile strength of 75,000 pounds per square inch, yarn made from such fibers can be estimated to have an average skein strength of

$$-63.58 + 1.074(80) + 1.015(75) = 98.46 \text{ pounds}$$

Furthermore, it may be estimated that the skein strength of individual pieces of yarn will, 95 percent of the time, lie between 98.46 ± 1.96(7.57) pounds, i.e., between 83.62 and 113.30 pounds.

1.3.4. Comparison with the Bivariate Regressions. It is interesting to compare these results with what would have been obtained if we had used only two variables, say, X_1 and X_2 or X_1 and X_3. Let us compare, first, the various standard errors of estimate. If we had correlated X_1 and X_2 alone, we would have obtained

$$s_{1.2}^2 = \frac{\Sigma x_1^2 - b_{12}\Sigma x_1 x_2}{n-2} = \frac{2{,}742.80 - (1.095)(1{,}332.60)}{18}$$

$$= \frac{2{,}742.80 - 1{,}459.20}{18} = 71.31$$

or

$$s_{1.2} = 8.45$$

Likewise, if we had correlated X_1 with X_3 alone, we would have obtained

$$s_{1.3}^2 = \frac{\Sigma x_1^2 - b_{13}\Sigma x_1 x_3}{n-2} = \frac{2{,}742{,}80 - (1.103)(335.20)}{18}$$

$$= \frac{2{,}742.80 - 369.73}{18} = 131.84$$

or

$$s_{1.3} = 11.48$$

Thus our estimate of the standard error of estimate for the multiple regression is less than our estimate of the standard error of estimate for either of the two simple regressions.

This relationship between the various estimates of the standard errors of estimate is a general one. If X_3 is correlated with X_1 and is not perfectly correlated with X_2, then the addition of X_3 to the regression equation will tend to reduce the standard error of estimate. In general, a multiple regression equation usually yields more precise estimates than a simple regression equation. In the above example, $s_{1.2} = 8.45$ was reduced to $s_{1.23} = 7.57$ and $s_{1.3} = 11.48$ to $s_{1.23} = 7.57$. In the first case there was a 10 percent increase in precision and in the second case a 34 percent increase.

The only exception to this general rule is the infrequent case in which the addition of a variable does not reduce the residual sum of squares (Σu^2) sufficiently to offset the loss of a degree of freedom (e.g., reduction of $n - 2$ to $n - 3$). It is to be noted, however, that this phenomenon is a feature of the estimation process. In the universe itself the standard error of estimate of a multiple regression is usually less than, or, at the most, equal to, the standard errors of estimate of regressions of lower order. In the universe, we can usually make more precise estimates—and always at least as precise estimates—from a multiple regression equation as we can from regressions of lower order.

Let us compare next the various regressions. If we had calculated the two-variable regressions, X_1 on X_2 and X_1 on X_3, we would have obtained

$$b_{12} = \frac{\Sigma x_1 x_2}{\Sigma x_2^2} = \frac{1{,}332.60}{1{,}216.95} = 1.095$$

and

$$b_{13} = \frac{\Sigma x_1 x_3}{\Sigma x_3^2} = \frac{335.20}{303.80} = 1.103$$

We also would have had

$$a_{1.2} = \bar{X}_1 - b_{12}\bar{X}_2 = 95.40 - 1.095(77.05) = 11.03$$

and

$$a_{1.3} = \bar{X}_1 - b_{13}\bar{X}_3 = 95.40 - 1.103(75.10) = 12.56$$

The two simple regression equations would have been

$$X_{1r} = a_{1.2} + b_{12}X_2 = 11.03 + 1.095X_2$$

with $s_{1.2} = 8.45$ and

$$X_{1r} = a_{1.3} + b_{13}X_3 = 12.56 + 1.013X_3$$

with $s_{1.3} = 11.48$.

Suppose that X_2 is known to be 0.800 and we use the regression of X_1 on X_2 to estimate X_1. We would get $X_{1r} = 11.03 + 1.095(80) = 98.63$ and, ignoring the sampling errors in estimates of $a_{1.2}$ and b_{12}, the 0.95 range of error for an individual value would be put at $98.63 \pm 1.96(8.45) = 82.07$ to 115.19. Now, if X_3 is known to be 75,000 psi and we use the regression equation of X_1 on X_3 to estimate X_1, we would get $X_{1r} = 12.56 + 1.103(75) = 95.29$, and, ignoring sampling errors in $a_{1.3}$ and b_{13}, the 0.95 range of error for an individual value would be put at $95.29 \pm 1.96(11.48) = 72.79$ to 117.79.

When the multiple regression equation was used, it will be recalled, we had

$$\begin{aligned} X_{1r} &= a_{1.23} + b_{12.3}X_2 + b_{13.2}X_3 \\ &= -63.58 + 1.074(80) + 1.015(75) = 98.46 \end{aligned}$$

and, ignoring sampling errors in $a_{1.23}$, $b_{12.3}$, and $b_{13.2}$, the 0.95 range of error for an individual value was 83.64 to 113.28. All three equations give us about the same estimate of X_1, but the estimate based on the multiple regression has the smaller range of error. The gain over the regression of X_1 on X_3 is considerable in this case. The gain over the regression of X_1

on X_2 is not very great; but some gain there is. This is the gain to be attributed to including both X_2 and X_3 in the regression equation.

Some comment should be made at this point on the relationship between the regression coefficient in a simple regression equation and the corresponding coefficient in the multiple regression equation. The coefficient $b'_{12.3}$ represents the effect on X_1 of a unit increase in X_2 *when* X_3 *is held constant.* It thus represents the *net* effect of X_2 on X_1. Likewise, $b'_{13.2}$ represents the effect on X_1 of a unit increase in X_3 *when* X_2 *is held constant.* It thus represents the *net* effect of X_3 on X_1. The coefficient b'_{12} represents the effect on X_1 of a unit increase of X_2 when X_3 is allowed to vary without restriction. It thus represents the total effect of X_2 on X_1, both through any direct link between the two factors and through any indirect links via other variables. Likewise, b'_{13} represents the *total* effect on X_1 of a unit increase in X_3, through both direct and indirect connections. With reference to the estimates derived, for example, it happens, probably because of the positive correlation between X_2 and X_3 (note that $\Sigma x_2 x_3$ is positive), that the net effects are slightly less than the total effects. No general rule can be laid down for this, however. The formulas connecting the least-squares estimates of the net and total regression coefficients are

(33.12)

$$b_{12.3} = \frac{b_{12} - \dfrac{\hat{\sigma}_1}{\hat{\sigma}_2} r_{13} r_{23}}{1 - r_{23}^2} \quad \text{where } r_{ij} = \frac{\Sigma x_i x_j}{n \hat{\sigma}_i \hat{\sigma}_j}$$

$$b_{13.2} = \frac{b_{13} - \dfrac{\hat{\sigma}_1}{\hat{\sigma}_3} r_{12} r_{23}}{1 - r_{23}^2}$$

where $\hat{\sigma}_i$ and r_{ij} are given by equation (32.16). Hence, $b_{12.3}$ may be greater or less than b_{12}, and $b_{13.2}$ may be greater or less than b_{13}, depending on the values of $\hat{\sigma}_1$, $\hat{\sigma}_2$, $\hat{\sigma}_3$, r_{12}, r_{13}, and r_{23}. If, however, $r_{23} = 0$, then $b_{12.3} = b_{12}$ and $b_{13.2} = b_{13}$.

1.4. The Multiple Correlation Coefficient for a Trivariate Regression

A multiple correlation coefficient associated with a given plane of regression measures the degree of association between the dependent variable and the function of the independent variables represented by the regression equation. For example, $R'_{1.23}$ is the correlation between X_1, on the one hand, and the linear expression $a'_{1.23} + b'_{12.3}X_2 + b'_{13.2}X_3$, on the other hand. If $\sigma'^2_{1.23}$ is the variance of the deviations from the universe plane of regression, then, by definition,

(33.13) $$R'^2_{1.23} = 1 - \sigma'^2_{1.23}/\sigma'^2_1$$

Since $\sigma'^2_{1.23}/\sigma'^2_1$ is the percentage of the variance of X_1 that is unexplained by the regression equation, $R'^2_{1.23}$ measures the percentage of "explained" variance. If X_2 and X_3 are uncorrelated,[2] $R'^2_{1.23} = r'^2_{12} + r'^2_{13}$. This says that, if X_2 and X_3 are uncorrelated, the percentage of variance of X_1 explained by X_2 and X_3 jointly is the sum of the percentages explained by each separately. This is not true if X_2 and X_3 are correlated.

The multiple correlation coefficient, like the simple correlation coefficient, is also directly related to the improvement in the precision of estimation to be obtained by using the multiple regression equation to estimate X_1. Thus equation (33.13) can be rewritten in the form

$$\text{(33.14)} \qquad \frac{\sigma'_{1.23}}{\sigma'_1} = \sqrt{1 - R'^2_{1.23}}$$

Hence there is a direct functional relationship between the ratio of $\sigma'_{1.23}/\sigma'_1$ and the value of $R'_{1.23}$. If we substitute $R'_{1.23}$ for r'_{12} in Table 32.4, we will obtain for selected values of $R'_{1.23}$ the percentage of improvement in precision[3] $(1 - \sigma'_{1.23}/\sigma'_1)$ to be gained through using the multiple regression equation to estimate X_1 instead of using simply $\bar{X}_1$.

It will be noted that a multiple correlation coefficient always pertains to a particular regression equation. Thus the multiple correlation coefficient $R'_{1.23}$ pertains to the plane of regression of X_1 on X_2 and X_3; the multiple correlation coefficient $R'_{2.13}$ pertains to the plane of regression of X_2 on X_1 and X_3; and the multiple correlation coefficient $R'_{3.12}$ pertains to the plane of regression of X_3 on X_1 and X_2. In the two-variable case, we had $r'_{12} = r'_{21}$. In the three-variable case, the multiple correlation coefficients are not necessarily equal. Thus, in general $R'_{1.23} \neq R'_{2.13} \neq R'_{3.12}$.

It is also to be noted that a multiple correlation coefficient, as defined by equation (33.13), measures linear correlation. Moreover, $R'_{i.jk}$ may be zero for a plane of regression, although there may be some nonlinear association between the variables.[4]

A universe multiple correlation coefficient can be estimated from the sample data just as was a simple correlation coefficient. If we take the sample multiple correlation as being defined by

$$\text{(33.15)} \qquad R_{i.jk} = \sqrt{1 - \frac{\Sigma u^2_{i.jk}}{\Sigma x^2_i}}$$

then an unbiased estimate of $R'^2_{i.jk}$ is given approximately by

---

[2] See Appendix I (41).

[3] Cf., Chapter 32, Section 8.1.

[4] If $\sigma'_{1.23}$ is the standard deviation of deviations from a curved plane or surface of regression, then, we can define an index of correlation by the formula

$$I'^2_{1.23} = \sqrt{1 - \sigma'^2_{1.23}/\sigma'^2_1}$$

This is the same formula as (33.13) but $I'_{1.23}$ is now used in place of $R'_{1.23}$ to warn the user that $\sigma'_{1.23}$ is in this case measured from a nonlinear regression.

(33.16)
$$I(R^2_{i.jk}) \doteq R^2_{i.jk} - \frac{2(1 - R^2_{i.jk})^2}{n - 1}$$

Another estimate frequently used is the "adjusted" sample multiple correlation coefficient,

(33.17)
$$R_{i.jk} = \sqrt{1 - \frac{\Sigma u^2_{i.jk}/(n - 3)}{\Sigma x^2_i/(n - 1)}}$$

For the cotton yarn data analyzed in the previous section we have

$$R_{1.23} = \sqrt{1 - \frac{971.36}{2742.8}} = 0.8037$$

and

$$I(R^2_{1.23}) = 0.6459 - \frac{2(1 - 0.6459)^2}{19} = 0.6327$$

which yields an estimate of $R'_{1.23} = 0.7954$. These results indicate that about 63 percent of the variation in skein strength is accounted for by variation in fiber length and fiber tensile strength. This stands in contrast to the 36 percent that was accounted for by fiber length alone. (See Section 8 of Chapter 32.) It also follows (see Table 32.4) that use of the plane of regression of skein strength on fiber length and fiber tensile strength to estimate skein strength improves the precision of the estimate by approximately 40 percent.

1.5 Sampling Errors of Trivariate Regression Coefficients

1.5.1. Standard Errors, Origin at $\bar{X}_2$, $\bar{X}_3$, 0. As in the two-variable case, when (1) deviations from the universe plane of regression are independent and normally distributed, (2) the sample plane is fitted by the method of least squares, (3) the independent variables have the same values in every sample, and (4) the independent variables are measured from their respective means, sample values of the parameters in a multiple regression equation are normally distributed about the universe parameters as means, and unbiased estimates of their sampling variances are as follows:

$$s^2_{\bar{X}_1} = \frac{s^2_{1.23}}{n} \text{ [note that with origin at } X_2 = \bar{X}_2,\ X_3 = \bar{X}_3,$$
$$X_1 = 0, \text{ the constant term equals } \bar{X}_1]$$

(33.18)
$$s^2_{b\,12.3} = \frac{s^2_{1.23}}{(n - 1)s^2_2(1 - r^2_{23})} = \frac{s^2_{1.23}}{\Sigma x^2_2 - (\Sigma x_2 x_3)^2/\Sigma x^2_3}$$

$$s^2_{b\,13.2} = \frac{s^2_{1.23}}{(n - 1)s^2_3(1 - r^2_{23})} = \frac{s^2_{1.23}}{\Sigma x^2_3 - (\Sigma x_2 x_3)^2/\Sigma x^2_2}$$

Having once computed $s_{1.23}$, s_2, s_3, and r_{23}, we can thus immediately estimate the standard errors of $\bar{X}_1$, $b_{12.3}$, and $b_{13.2}$. Testing hypotheses pertaining to these parameters and the establishment of confidence limits for the universe values proceed exactly as outlined in the two-variable case. The only difference is that we now enter the t table with $\nu = n - 3$ instead of $n - 2$.

1.5.2 Testing a Hypothesis Pertaining to a Multiple Regression Coefficient. To illustrate, let us test whether the regression coefficient of X_3 is significant (i.e., whether $b'_{13.2} = 0$). We have

$$s_{b_{13.2}} = \frac{s_{1.23}}{\sqrt{\Sigma x_3^2 - (\Sigma x_2 x_3)^2/\Sigma s_2^2}} = \frac{7.57}{\sqrt{303.80 - \dfrac{(24.90)^2}{1{,}216.95}}} = 0.434$$

Hence

$$\frac{b_{13.2} - 0}{s_{b_{13.2}}} = \frac{1.015}{0.434} = 2.34$$

The value of $t_{0.025}$ for $\nu = 20 - 3 = 17$ is 2.110. Hence, if $\alpha = 0.05$ and we use a two-tail test, the hypothesis that $b'_{13.2} = 0$ is rejected. In other words, we conclude that $b_{13.2}$ is significant.

1.5.3. The OC Curve for a Test. If we wish to derive the OC curve for a test such as that of Section 1.5.2, we can again use Figure 15.9 or Figure 25.6 with some adjustment. For a test of a hypothesis pertaining to a three-variable regression coefficient $b'_{ij.k}$, we set

$$\lambda = \frac{|B_{ij.k} - b'_{ij.k}|}{\sigma'_{i.jk}} \frac{\sqrt{(n-1)s_j^2(1 - r_{jk}^2)}}{\sqrt{n-2}}$$

$$= \frac{|B_{ij.k} - b'_{ij.k}|}{\sigma'_{i.jk}} \frac{\sqrt{\Sigma x_j^2 - (\Sigma x_j x_k)^2/\Sigma x_k^2}}{\sqrt{n-2}}$$

in which $B_{ij.k}$ is the hypothetical value of $b'_{ij.k}$ being tested, $b'_{ij.k}$ is the actual value of the universe regression coefficient, and $\sigma'_{i.jk}$ is the universe "standard error of estimate." In using Figure 15.9 or 25.6 with this value of λ, we read n on the chart as equal to $n - 2$ for the sample. Likewise, in testing a hypothesis pertaining to $\bar{X}'_1$ we set

$$\lambda = \frac{|\bar{X}'_1 - M_1|}{\sigma'_{1.23}} \frac{\sqrt{n}}{\sqrt{n-2}}$$

and again read the n of the chart as equal to $n - 2$ for the sample. The quantity M_1 is the hypothetical value being tested.

Let us, for example, derive a point on the OC curve for the test carried out above. In particular, we shall find the probability that we would have accepted the null hypothesis $b'_{13.2} = 0$, when actually $b'_{13.2} = 1.00$. We do not know $\sigma'_{1.23}$, but let us in this case take it equal to the sample estimate

TABLE 33.3
Analysis of Variance of Regression of X_1 on X_2 and X_3 (cotton yarn data of Table 33.1)

Source of Variation	*Sum of Squares*	*df*	*Mean Square*
Regression of X_1 on X_2	1,459.20	1	1,459.20
Addition of X_3	312.24	1	312.24
Joint regression X_1 on X_2 and X_3 ...	1,771.44	2	885.72
Remainder	971.36	17	57.14
Total	2,742.80	19	....

$s_{1.23} = 7.57$. If we had wished to derive the OC curve prior to the test, we would have had to make at least a rough guess as to the value of $\sigma'_{1.23}$. Using $\sigma'_{1.23} = 7.57$, we have

$$\lambda = \frac{|0 - 1.00|}{7.57} \frac{\sqrt{1{,}216.95 - \dfrac{(24.90)^2}{303.80}}}{\sqrt{20-2}} = 1.087$$

We then look for the curve in Figure 25.6 bearing the number 20 − 2 = 18 and find that the curve $n = 18$ has a height between 0.01 and 0.02 at $\lambda = 1.087$. Thus the probability of accepting the null hypothesis $b'_{13.2} = 0$ when actually $b'_{13.2} = 1.00$ (and $\sigma'_{1.23} = 7.57$) is between 1 and 2 chances out of 100. If $\sigma'_{1.23}$ is actually greater than 7.57, the true probability will be greater than 0.01 to 0.02. If $\sigma'_{1.23}$ is actually less than 7.57, the true probability will be less than 0.01 to 0.02. Anyway, unless our estimate of $\sigma'_{1.23}$ is far off, we can feel reasonably confident that our test would have declared $b_{13.2}$ significant when the actual $b'_{13.2}$ was as large as 1.00.

1.5.4. Analysis of Variance of the Regression Components. In testing the hypothesis $b'_{13.2} = 0$, it is possible to use an analysis of variance. For the data of Section 1.3 the total sum of squares for X_1 is $\Sigma x_1^2 = 2{,}742{,}80$. The sum of squares attributable to the regression of X_1 on X_2 and X_3 is equal to[5] $R^2_{1.23}\Sigma x_1^2 = \Sigma x_1^2 - \Sigma u^2_{1.23} = 2{,}742.80 - 971.36 = 1{,}771.44$. The sum of squares attributable to the regression of X_1 on X_2 is equal to[6] $r^2_{12}\Sigma x_1^2 = \Sigma x_1^2 - \Sigma u^2_{1.2} = 2{,}742.80 - 1{,}283.60 = 1{,}459.20$. Hence the sum of squares attributable to the addition of the term $b'_{13.2}X_3$ to the regression equation is the sum of squares attributable to the joint regression of X_1 on X_2 and X_3 minus the sum of squares attributable to the regression of X_1 on X_2, or $1{,}771.44 - 1{,}459.20 = 312.24$. We can then set up the analysis of variance table presented in Table 33.3. To test whether the X_3 term makes a significant

[5] See formula (33.15), $\Sigma u^2_{1.23}$ is the sum of squares attributable to the residual variation. Hence $\Sigma X_1^2 - \Sigma u^2_{1.23}$ is the sum of squares attributable to the regression.

[6] See equation (32.17).

addition to the regression, we take the ratio of its mean square to the remainder mean square and note whether the result exceeds $F_{0.05}$ (using $\alpha = 0.05$) for $\nu_1 = 1$, $\nu_2 = n - 3$. For the above data, this yields $\frac{312.24}{57.14} = 5.46$. But $F_{0.05}$ for $\nu_1 = 1$, $\nu_2 = 17$, is 4.45. Hence the X_3 term is a significant addition to the regression equation. This is another way of determining whether the net effect of X_3 on X_1 is significant.

An analysis of variance can also be used to test the significance of the multiple correlation coefficient (i.e., whether $R'_{1.23} = 0$). This is the same as testing in the above example whether the joint regression of X_1 on X_2 and X_3 is significant. To do this, we divide the mean square for the joint regression by the remainder mean square. The result is then compared with the $F_{0.05}$ point for $\nu_1 = 2$, $\nu_2 = n - 3$. Thus, in our example, we have $\frac{885.72}{57.14} = 15.5$. Since $F_{0.05}$ for $\nu_1 = 2$, $\nu_2 = 17$, is 3.59, the multiple correlation is clearly significant.

1.5.5 Confidence Limits for the Regression Coefficients. To determine 0.95 confidence limits for a universe regression coefficient, we simply add to and subtract from the sample value $t_{0.025}$ (for $\nu = n - 3$) times the estimated standard error of the coefficient. For example, for the cotton yarn data we have

$$b_{13.2} = 1.015$$

and

$$s_{b\,13.2} = 0.434$$

Hence 0.95 confidence limits for $b'_{13.2}$ are given by

$$b'_{13.2} = 1.015 \pm 2.111(0.434) = \begin{cases} 1.931 \\ 0.099 \end{cases}$$

1.5.6 A Lower Confidence Limit for the Multiple Correlation Coefficient. In most cases the multiple correlation coefficient is significantly different from zero and interest may center in putting a lower confidence bound on it. For regressions containing as many as 5 independent variables, a lower 95 percent confidence limit can be obtained from Table Z of Appendix II. This is a table of the upper 5 percent points of the distribution of the sample multiple correlation coefficient from a multivariate normal universe. It is a selection from a larger table published in *Biometrika* 59 (1972), pp. 178–88 by Yoong-sin Lee.[7] In the table, ρ stands for the universe multiple correlation coefficient, $n_1 = p - 1$, where p is the number of variables, and $\nu = 60/\sqrt{n_2}$ where $n_2 = n - p$.

[7] For lower percentage points of the distribution of the sample multiple correlation coefficient, see K. H. Kramer's table in *Biometrika Tables for Statisticians,* Vol. 2 (Cambridge, England: University Press, 1972).

Let us illustrate the use of Table Z for a trivariate regression which is the case in which $n_1 = 2$. For the cotton yarn problem, we had a sample multiple correlation coefficient equal to

$$R_{1.23} = 0.8037$$

The sample size was 20, $n_1 = 2$ and $n_2 = 20 - 3 = 17$. For this case $\nu = 60/\sqrt{n_2} = 14.55$. Linear interpolation on ν and then on ρ will be sufficiently accurate for our purposes.[8]

We have from the table

ν	$\rho = 0.5$	$\rho = 0.6$
14	0.7599	0.8133
16	0.7895	0.8367

Interpolation on ν thus gives

ν	$\rho = 0.5$	$\rho = 0.6$
14.55	0.7762	0.8262

and further interpolation on ρ yields 0.5550 as the lower 0.95 confidence limit for $R'_{1.23}$.

1.6. Solving the Least-Squares Equations and Computing Standard Errors for a Trivariate Regression[9]

The following is a method for solving the least-squares equations and computing standard errors that is systematic and compact. It also has the advantages that (1) it may readily be extended to more than two independent variables, and (2) if another dependent variable, say X_4, is also to be correlated with the same values of the same independent variables, this method avoids duplication of calculations. The steps are as follows:

1. Set up two sets of equations, each of which is a modification of the original least-squares equations. Thus, from the equations [Cf. equations (33.5) above.]

(33.19)

$$b_{12.3}\sum x_2^2 + b_{13.2}\sum x_2x_3 = \sum x_1x_2$$

$$b_{12.3}\sum x_2x_3 + b_{13.2}\sum x_3^2 = \sum x_1x_3$$

we form the two sets of auxiliary equations

(33.20)

$$c_{22}\sum x_2^2 + c_{23}\sum x_2x_3 = 1$$

$$c_{22}\sum x_2x_3 + c_{23}\sum x_3^2 = 0$$

---

[8] Lee indicates that linear interpolation on ν generally gives good results and linear interpolation is sufficient for larger values of ρ.

[9] For the theory underlying this section see Appendix I (39) and (40).

and

(33.21)
$$c_{23}\sum x_2^2 + c_{33}\sum x_2x_3 = 0$$
$$c_{23}\sum x_2x_3 + c_{33}\sum x_3^2 = 1$$

2. We solve the auxiliary equations for c_{22}, c_{33}, and c_{23} (the last being the same, whether derived from the first or the second auxiliary set).

(33.22)
$$b_{12.3} = c_{22}\sum x_1x_2 + c_{23}\sum x_1x_3$$
$$b_{13.2} = c_{23}\sum x_1x_2 + c_{33}\sum x_1x_3$$

If we wanted also to compute $b_{42.3}$ and $b_{43.2}$ for the regression equation in which X_4 is the dependent variable and X_2 and X_3 are the independent variables, then, for the same values of X_2 and X_3, we would have

(33.23)
$$b_{42.3} = c_{22}\sum x_4x_2 + c_{23}\sum x_4x_3$$
$$b_{43.2} = c_{23}\sum x_4x_2 + c_{33}\sum x_4x_3$$

In other words, the only additional calculations required would be the computation of Σx_4x_2 and Σx_4x_3.

4. The standard errors of the b's may also be computed very easily.

(33.24)
$$s_{b_{12.3}} = s_{1.23}\sqrt{c_{22}}$$
$$s_{b_{13.2}} = s_{1.23}\sqrt{c_{33}}$$

Table 33.4 presents a compact procedure for solving the two sets of auxiliary equations in one set of operations. The data pertain to the least-squares equations for the cotton yarn data of Section 1.3. The steps are as follows:

1. Write down the coefficients of the auxiliary equations, omitting the coefficient of the second equation to the left of the diagonal. Enter in the sum column the total of *all* the coefficients in each equation (including the omitted coefficient in the second equation), together with the constant terms (now preceded by a negative sign).
2. In line 1 write auxiliary equation I in abbreviated form.
3. Compute the reciprocal of the first coefficient in this equation, give it a minus sign, and multiply line 1 by it. Note that the sum of items in line 2, from column (1) through column (4), should equal (within the error caused by decimals) the figure in the sum column in line 2. Compute this cross-sum and enter it in the check column.
4. Write down equation II in abbreviated form. Multiply line 1 by the figure in line 2, column 2, and enter the results in line 4, omitting the figures in column 1. Add line 4 to line 3 to get line 5. Make cross-check.
5. Compute the reciprocal of the first figure in line 5 and multiply line 5 by the negative of this reciprocal. Make cross-check.

TABLE 33.4
Illustration of the Solution of Auxiliary Equations (three-variable problem)

Line	*Reciprocal*	*(1)*	*(2)*	*(3)*	*(4)*	Σ	*Check*	*Comments*
I	. . . .	1,216.95	24.90	−1	0	1,240.85	. . . .	. . . .
II	. . . .	. . . .	303.80	0	−1	327.70	. . . .	. . . .
1	0.00082172644	1,216.95	24.90	−1	0	1,240.85	. . . .	Eq. I
2	. . . .	−1	−0.0204610	0.0008217	0	−1.019639	−1.019639	1 ÷ −1,216.95
3	. . . .	. . . .	303.80	0	−1	327.70	. . . .	Eq. II
4	. . . .	. . . .	−0.509479	0.0204610	0	−25.389032	. . . .	1 × (−0.020461)
5	. . . .	. . . .	303.290521	0.0204610	−1	302.310968	302.310982	3 + 4
6	0.0032971686	. . . .	−1	−0.00006746	0.003297	−0.996770	0.996770	5 ÷ 303.290521

$c_{22} = -0.00006746$
$c_{22} = -0.020461\ (-0.00006746) + 0.0008217 = 0.0008231$
$c_{33} = 0.003297$

6. The figure in line 6, column (3), is c_{23}.
7. For the given data line 2 is to be read

$$c_{22} = -0.020461 c_{23} + 0.0008217$$

Substitute the value of c_{23} in this and derive the value of c_{22}.

8. The figure in line 6, column (4), is c_{33}.

The values of c_{22}, c_{33}, and c_{23} are now applied to finding $b_{12.3}$ and $b_{13.2}$ as follows. For the given data we have

$$\begin{aligned} b_{12.3} &= 0.0008231(1{,}332.60) - 0.00006746(335.20) \\ &= 1.074 \\ b_{13.2} &= -0.00006746(1{,}332.60) + 0.003297(335.20) \\ &= 1.015 \end{aligned}$$

This is how the values for $b_{12.3}$ and $b_{13.2}$ given in Section 1.3 were derived.

The standard errors of $b_{12.3}$ and $b_{13.2}$ are given by

$$s_{b\,12.3} = 7.56\sqrt{0.0008203} = 0.217$$

$$s_{b\,13.2} = 7.56\sqrt{0.003297} = 0.434$$

The explanation of the use of equations (33.20) and (33.21) to solve the simultaneous equations (33.19) will be readily understood by those who are acquainted with elementary matrix algebra. The matrix of the coefficients of equations (33.19) is

(33.25)
$$S = \begin{pmatrix} \sum x_2^2 & \sum x_2 x_3 \\ \sum x_2 x_3 & \sum x_3^2 \end{pmatrix}$$

Let the column vector B be defined as

(33.26)
$$B = \begin{pmatrix} b_{12.3} \\ b_{13.2} \end{pmatrix}$$

and let the column vector K be defined as

(33.27)
$$K = \begin{pmatrix} \sum x_1 x_2 \\ \sum x_1 x_3 \end{pmatrix}$$

Then equation (33.19) may be written

(33.19a)
$$SB = K$$

To solve equation (33.19a) we multiply by the inverse of S, i.e.,

$$S^{-1}SB = S^{-1}K$$

yielding

$$B = S^{-1}K$$

Call S^{-1} the C matrix. Then

(33.28) $$B = CK$$

Let the elements of the C matrix be

(33.29) $$C = \begin{pmatrix} c_{22} & c_{23} \\ c_{23} & c_{33} \end{pmatrix}$$

Then (33.28) may be written

(33.28a) $$\begin{pmatrix} b_{12.3} \\ b_{13.2} \end{pmatrix} = \begin{pmatrix} c_{22} & c_{23} \\ c_{23} & c_{33} \end{pmatrix} \begin{pmatrix} \sum x_1 x_2 \\ \sum x_1 x_3 \end{pmatrix}$$

which when multiplied out yields equations (33.22).

To find the elements of the C matrix (i.e., the inverse of S) we note that by definition of an inverse

$$SC = I$$

or

(33.20–21a) $$\begin{pmatrix} \sum x^2 & \sum x_2 x_3 \\ \sum x_2 x_3 & \sum x_3^2 \end{pmatrix} \begin{pmatrix} c_{22} & c_{23} \\ c_{23} & c_{33} \end{pmatrix} = \begin{pmatrix} 1 & 0 \\ 0 & 1 \end{pmatrix}$$

When multiplied out this matrix equation yields the two sets of equations (33.20) and (33.21). Solving these two sets of equations is thus a method of finding the inverse of the matrix S and that is what is done in Table 33.4. Any other method of getting the inverse of a matrix could be used.

1.7. Confidence Limits for a Trivariate Regression and Prediction Limits for Individual Values

We found in Chapter 32 that the confidence limits for regression values were a pair of hyperbolic loci on each side of the sample line. It also can be shown that the confidence limits for multiple regression values are a pair of hyperbolic surfaces on either side of the sample plane of regression. The equations for these limiting surfaces are

(33.30) $$X_{1r} = \bar{X}_1 + b_{12.3}x_2 + b_{13.2}x_3$$
$$\pm t_{0.025} \text{ (for } \nu = n - 3) s_{1.23}\sqrt{1/n + c_{22}x_2^2 + c_{33}x_3^2 + 2c_{23}x_2x_3}$$

where the origin is at $X_2 = \bar{X}_2$, $X_3 = \bar{X}_3$, and $X_1 = 0$. The chances are 0.95 that these limiting loci include the universe regression value at given values of X_2 and X_3. The chances are also 0.95 that a single value picked at random in the future will lie within the limits

(33.31) $$\bar{X}_1 + b_{12.3}x_2 + b_{13.2}x_3 \pm t_{0.025}\, s_{1.23}\sqrt{1/n + c_{22}x_2^2 + c_{33}x_3^2 + 2c_{23}x_2x_3 + 1}$$

It will be noticed again that the most reliable estimates are given by the sample regression equation when the values of the independent variables are near their means.[10]

2. REGRESSIONS INVOLVING MORE THAN THREE VARIABLES

2.1. Multivariate Distributions

When more than three variables are involved in a regression analysis, we cannot make use of diagrams and models to illustrate a plane of regression. Just as with three variables the density function became a surface of higher dimensions, so now with more than three variables not only do the density functions become "hypersurfaces," but the planes of regression become "hyperplanes." We shall continue to talk about them, however, by analogy with their three-dimensional counterparts.

If we have n sets of p measurements each, e.g., 20 sets of measurements of 4 characteristics of cotton yarn, we can figuratively plot each set of measurements as a point in p-dimensional space. If the p scales are divided into class intervals and "planes" run through the limits of each class interval, these planes will form "cells" in the p-dimensional space. If we record the number of points in each of these cells, the result will be a multivariate distribution of the n sets of measurements. If the frequency of each cell is divided by n, we get relative frequencies that can be taken as measures of the densities of the points in the various cells. Then, if we indefinitely increase the number of sets of measurements and continuously reduce the size of our class intervals, the variations in density from cell to cell will become less and less abrupt, and the limit will be a continuous density function. This density function becomes the mathematical description of the distribution of the infinite universe of measurements. Subsequently, when we speak of a universe of cases, we shall have in mind an infinite universe of this kind or at least a very large set of values that can be approximated by a density function, just as a histogram of a large number of cases can be approximated by a frequency curve.

With reference to the multivariate distribution or density function, there is a mean value of X_1, say, for any given set of values of $X_2, X_3, \ldots, X_p$. The locus of this mean value as we change the values for $X_2, X_3, \ldots, X_p$ will describe the regression of X_1 on $X_2, X_3, \ldots, X_p$. If the locus is a "plane," the regression will be linear. Otherwise, it will be curved. There are also regressions for each of the other $p - 1$ variables. For any point on a regression it is possible to compute a standard deviation of the deviations

[10] Compare Chapter 32, Section 7.5.

from the regression. If this standard deviation is constant for all points on the regression locus, the variation is said to be "homoscedastic." In this text we shall refer only to homoscedastic distributions. In such cases each regression will have associated with it a standard error of estimate represented by the symbol $\sigma'_{i.jk...p}$. With each linear regression there will also be associated a multiple correlation coefficient $R'_{i.jk...p}$ that will be defined by the equation

(33.32) $$R'^2_{i.jk...p} = 1 - \frac{\sigma'^2_{i.jk...p}}{\sigma'^2_i}$$

2.2. Estimation of a Linear Regression of a Multivariate Universe

As in previous cases, the practical problem is to estimate a universe regression from a set of sample data. This section will describe the use of the method of least squares in estimating a "plane" of regression for a universe of p variables. We shall consider only the regression of X_1 on $X_2, X_3, X_4, \ldots,$ X_p. The other planes of regression can be obtained in the same way by simply interchanging subscripts.

2.2.1. Summary of General Procedure. Let the universe plane of regression of X_1 on $X_2, X_3, \ldots, X_p$ be represented by the equation

(33.33) $$X'_{1r} = a'_{1.23...p} + b'_{12.3...p}X_2 + b'_{13.2...p}X_3 + \ldots + b'_{1p.23...p-1}X_p$$

and let the estimate of this derived from a sample be

(33.34) $$X_{1r} = a_{1.23...p} + b_{12.3...p}X_2 + b_{13.2...p}X_3 + \ldots + b_{1p.23...p-1}X_p$$

The method of least squares will determine equation (33.34) such that $\Sigma(X_1 - X_{1r})^2$ is a minimum. The equations yielded by the minimizing process are[11]

(33.35)

$$b_{12.3...p}\sum x_2^2 + b_{13.2...p}\sum x_2x_3 + \ldots + b_{1p.23...p-1}\sum x_2x_p = \sum x_1x_2$$

$$b_{12.3...p}\sum x_2x_3 + b_{13.2...p}\sum x_3^2 + \ldots + b_{1p.23...p-1}\sum x_3x_p = \sum x_1x_3$$

$$\cdots\cdots\cdots\cdots\cdots\cdots\cdots\cdots\cdots\cdots\cdots\cdots$$

$$b_{12.3...p}\sum x_2x_p + b_{13.2...p}\sum x_3x_p + \ldots + b_{1p.23...p-1}\sum x_p^2 = \sum x_1x_p$$

The simultaneous solution of these least-squares equations will give $b_{12.3...p}$, $b_{13.2...p}$, etc., and $a_{1.23...p}$ may be derived from the equation

(33.36) $$a_{1.23...p} = \bar{X}_1 - b_{12.3...p}\bar{X}_2 - \ldots - b_{1p.23...p-1}\bar{X}_p$$

[11] See Appendix I (39).

To solve the above equations, we first solve the following $p - 1$ sets of auxiliary equations:[12]

$$
\begin{array}{lll}
 & & i = (2)(3) \ldots (p) \\
 & c_{i2}\sum x_2^2 + c_{i3}\sum x_2x_3 + \ldots + c_{ip}\sum x_2x_p = & 1 \; 0 \; \ldots \; 0 \\
(33.37) & c_{i2}\sum x_2x_3 + c_{i3}\sum x_3^2 + \ldots + c_{ip}\sum x_3x_p = & 0 \; 1 \; \ldots \; 0 \\
 & \cdots\cdots\cdots\cdots\cdots\cdots\cdots\cdots\cdots & \\
 & c_{i2}\sum x_2x_p + c_{i3}\sum x_3x_p + \ldots + c_{ip}\sum x_p^2 = & 0 \; 0 \; \ldots \; 1
\end{array}
$$

The procedure is illustrated for four variables in the next section. Having found the c's, we compute

$$
\begin{array}{ll}
 & b_{12.3\ldots p} = c_{22}\sum x_1x_2 + c_{23}\sum x_1x_3 + \ldots c_{2p}\sum x_1x_p \\
(33.38) & b_{13.2\ldots p} = c_{23}\sum x_1x_2 + c_{33}\sum x_1x_3 + \ldots + c_{3p}\sum x_1x_p \\
 & \cdots\cdots\cdots\cdots\cdots\cdots\cdots\cdots\cdots \\
 & b_{1p.23\ldots p-1} = c_{2p}\sum x_1x_2 + c_{3p}\sum x_1x_3 + \ldots + c_{pp}\sum x_1x_p
\end{array}
$$

(The mathematically trained reader will note that this solution of equations (33.35) is again simply a solution of the matrix equation $SB = K$, where S, B, and K are generalizations of the quantities defined by equations (33.25) to (33.27) for $p - 1$ dimensions.)

The sum of the squares of the deviations from the regression equation is

$$
(33.39) \quad \sum u_{1.23\ldots p}^2 = \sum x_1^2 - b_{12.3\ldots p}\sum x_1x_2 - b_{13.2\ldots p}\sum x_1x_3 \ldots - b_{1p.23\ldots p-1}\sum x_1x_p
$$

and an unbiased estimate of the square of the universe standard error of estimate ($\sigma'^2_{1.23\ldots p}$) is

$$
(33.40) \quad s_{1.23\ldots p}^2 = \frac{\Sigma u_{1.23\ldots p}^2}{n - p}
$$

From this we estimate the standard errors of the regression parameters [origin at $X_2 = \bar{X}_2$, $X_3 = \bar{X}_3$, . . . , $X_p = \bar{X}_p$, $X_1 = 0$] as follows:

$s_{\bar{X}_1} = \dfrac{s_{1.23\ldots p}}{\sqrt{n}}$ [note that with origin at $\bar{X}_2$, $\bar{X}_3$ . . . , $\bar{X}_p$, 0 the constant term equals $\bar{X}_1$].

$$
\begin{array}{ll}
 & s_{b_{12.3\ldots p}} = s_{1.23\ldots p}\sqrt{c_{22}} \\
 & s_{b_{13.2\ldots p}} = s_{1.23\ldots p}\sqrt{c_{33}} \\
(33.41) & \cdots\cdots\cdots\cdots \\
 & s_{b_{1p.2\ldots p-1}} = s_{1.23\ldots p}\sqrt{c_{pp}}
\end{array}
$$

---

[12] Note that $c_{23} = c_{32}$ and, in general, $c_{ij} = c_{ji}$.

TABLE 33.5
Values of k for Use in Estimating $R'^2_{i.jk...p}$

n	p = 4	p = 5	p = 6
10	1.167	1.400	1.750
15	1.091	1.200	1.333
20	1.063	1.133	1.214
25	1.048	1.100	1.158
30	1.038	1.080	1.125
40	1.028	1.057	1.088
50	1.022	1.044	1.068
60	1.018	1.036	1.056
80	1.013	1.027	1.041
100	1.010	1.021	1.032
200	1.005	1.010	1.015

Ninety-five percent confidence limits for a universe regression coefficient are given by the sample coefficient plus and minus $t_{0.025}$ (for $\nu = n - p$) times the estimated standard error of the coefficient, and the equations for 0.95 confidence limits for the regression values are

(33.42)
$$X_{1r} = \bar{X}_1 + b_{12.3...p}x_2 + . . . + b_{1p.23...p-1}x_p$$
$$\pm t_{0.025}s_{1.23...p}\sqrt{1/n + c_{22}x_2^2 + . . . + c_{pp}x_p^2 + 2c_{23}x_2x_3 + . . . + 2c_{2p}x_2x_p + . . . + 2c_{3p}x_3x_p + . . .}$$

where $t_{0.025}$ is that for $\nu = n - p$. If we add 1 to the expression under the radical, we shall get 0.95 "prediction limits" for an individual value.

The sample multiple correlation coefficient associated with the "plane" of regression of X_1 on $X_2, X_3, . . . , X_p$ is defined by

(33.43)
$$R_{1.23...p} = \sqrt{1 - \frac{\Sigma u^2_{1.23...p}}{\Sigma x_1^2}}$$

and an unbiased estimate of $R'^2_{1.23...p}$ is given approximately by[13]

(33.44)
$$I(R^2_{1.23...p}) = kR^2_{1.23...p} - \frac{2k(1 - R^2_{1.23...p})^2}{n - p + 2} + 1 - k$$

where $k = (n - 3)/(n - p)$ and p is the number of variables. Values of k for $p = 4$, 5, and 6 are given in Table 33.5 for $n = 10$ to 200. Another estimate frequently used is the "adjusted" sample multiple correction coefficient

(33.45)
$$\tilde{R}^2_{1.23...p} = 1 - \frac{(n - 1)\Sigma u^2_{1.23...p}}{(n - p)\,\Sigma x_1^2}$$

[13] Derived from a formula given by Ingram Olkin and John W. Pratt, "Unbiased Estimation of Certain Correlation Coefficients," *Annals of Mathematical Statistics* 29 (1958), p. 211.

For a regression involving up to 5 variables, Table Z of Appendix II can be used to get a lower 0.95 confidence limit for $R'_{i.jk...p}$. This was described briefly in Section 1.5.6. above. For a table running up to 41 variables and for values to be used to get a lower 99 percent confidence limit, the reader is referred to the original article by Yoong-Sin Lee.[14] Another table by K. H. Kramer that can be used to get both upper and lower confidence limits will be found in *Biometrika Tables for Statisticians,* Vol. 2 (1972).

2.2.2. An Example of a Four-Variable Regression Analysis. To illustrate a four-variable regression problem let us refer back to the data of Table 33.1 and let us presume that we are interested in the plane of regression of skein strength (X_1) on fiber length (X_2), fiber tensile strength (X_3), and fiber fineness (X_4). Our example will thus be the extension of the example of Section 1 to include the variable fiber fineness (X_4).

The computational work involved in this four-variable problem will not be discussed here[15] since electronic computers are generally available now to run a regression analysis for a large number of variables. For the data of Table 33.1, let us assume that a computer program gives the following results:

1. The means and standard deviations of all the variables:

$$\bar{X}_1 = 95.40 \qquad s_1 = 12.01$$

$$\bar{X}_2 = 77.05 \qquad s_2 = 8.00$$

$$\bar{X}_3 = 75.10 \qquad s_3 = 4.00$$

$$\bar{X}_4 = 41.25 \qquad s_4 = 5.47$$

2. The matrix of sums of squares and cross-products. (This is the S-matrix. Only the upper half of S is given since it is symmetric.)

$\sum x_1^2 = 2742.80$	$\sum x_1x_2 = 1332.60$	$\sum x_1x_3 = 335.20$	$\sum x_1x_4 = -608.00$
	$\sum x_2^2 = 1216.95$	$\sum x_2x_3 = 24.90$	$\sum x_2x_4 = -29.25$
		$\sum x_3^2 = 303.80$	$\sum x_3x_4 = -276.50$
			$\sum x_4^2 = 567.75$

3. The matrix of sample correlation coefficients. This is obtained from S by dividing the respective elements by $n\sqrt{\Sigma x_i^2 \, \Sigma x_j^2}$.

[14] See *Biometrika* 59 (1972), pp. 175–89.

[15] Those who are interested are referred to the 3d edition of this book, Chapter 33, Section 2.2.2.

1	$r_{12} = 0.7295$	$r_{13} = 0.3672$	$r_{14} = -0.4874$
	1	$r_{23} = 0.0410$	$r_{24} = -0.0352$
		1	$r_{34} = -0.660$
			1

Only the upper half is given since again the matrix is symmetric and $r_{12} = r_{21}$.

4. The inverse of the S-matrix (= the C-matrix). C is symmetric and only the upper half is given.

$c_{11} = 0.05$	$c_{12} = 0$	$c_{23} = -0.00005186$	$c_{24} = 0.00001715$
	$c_{22} = 0.0008232$	$c_{33} = 0.005915$	$c_{34} = 0.002878$
			$c_{44} = 0.003164$

5. The estimates of the constant term, regression coefficients and their standard errors, together with the ratio of an estimated parameter to its estimated standard error designated as t.

$$\bar{X}_1 = 95.40 \qquad s_{\bar{X}_1} = 1.47 \qquad t = 64.90$$

$$b_{12.34} = 1.069 \qquad s_{b_{12.34}} = 0.189 \qquad t = 5.66$$

$$b_{13.24} = 0.164 \qquad s_{b_{13.24}} = 0.507 \qquad t = 0.32$$

$$b_{14.23} = -0.936 \qquad b_{b_{14.23}} = 0.371 \qquad t = 2.52$$

$$t_{0.025}(\nu = 16) = 2.12 \text{ and } t_{0.01}(\nu = 16) = 2.583$$

6. The sample sum of squares of residuals.

$$\sum u^2_{1.234} = 694.19$$

7. The sample standard error of estimate

$$s_{1.234} = \sqrt{\frac{\Sigma u^2_{1.234}}{n-4}} = 6.59$$

8. The sample multiple correlation coefficient,

$$R_{1.234} = \sqrt{1 - \frac{\Sigma u^2_{1.234}}{\Sigma x^2_1}}$$

$$= \sqrt{1 - \frac{694.19}{2742.80}} = 0.8642$$

9. A stepwise analysis of the total sum of squares:

	Sum of Squares	Percent
Total sum of squares $(= \Sigma x_1^2)$	= 2,742.8	100.00
Part accounted for by X_2 $= \Sigma x_1^2 - \Sigma u_{1.2}^2 = 2{,}742.8 - 1{,}283.6$	= 1.459.2	53.20
Part accounted for by adding X_4 $= \Sigma u_{1.2}^2 - \Sigma u_{1.24}^2 = 1{,}283.6 - 698.5$	= 585.1	21.33
Additional part accounted for by adding X_3 $= \Sigma u_{1.24}^2 - \Sigma u_{1.234}^2 = 698.5 - 694.2$	= 4.3	0.16
Total accounted for by X_2, X_3, and X_4	= 2,048.6	74.69
Part unaccounted for	= 694.2	25.31

With such a computer output we would conclude the following:

1. The sample multiple correlation coefficient of 0.8642 suggests that about 75 percent $(= 0.8642)^2$ of the total variation in skein strength is accounted for by the three fiber characteristics (cf. (9) above). This figure is not adjusted, however, for the fact that the sample multiple correlation coefficient is not an unbiased estimate of the universe multiple correlation coefficient. An unbiased estimate of the percentage would be given by

$$I(R_{1.234}^2) = kR_{1.234}^2 - \frac{2k(1 - R_{1.234}^2)^2}{n - p + 2} + 1 - k$$

$$= 1.063(0.7469) - \frac{2(1.063)(0.2531)^2}{20 - 4 + 2} + 1 - 1.063$$

$$= 0.7940 - 0.0076 - 0.063 = 0.7940 - 0.0706 = 0.7244$$

or roughly 72 percent.

(Compare equation 33.45 above.) This is equivalent to an estimate of $R'_{1.234} = 0.8511$.

We might want to make some allowance, however, for the possibility that the universe multiple correlation coefficient is actually less than the sample result obtained. We could then turn to Table Z in Appendix II and calculate the lower 0.95 confidence limit for $R'_{1.234}$. We would have $n_1 = 3$ and $\nu = 60/\sqrt{16} = 15$. With $R_{1.234} = 0.8642$, this yields by interpolation on both ν and ρ, a lower 0.95 confidence limit on $R'_{1.234}$ of 0.6638. This is equivalent to $(0.6638)^2 = 44$ percent determination of skein strength. We thus conclude that although 72 percent is a good estimate of the percent of the variation in skein strength that is accounted for by the three fiber characteristics, there is a reasonable possibility that it is actually as low as 44 percent. A much closer confidence bound could have been obtained with a larger sample.

2. From the t-values we note that $\bar{X}_1$ and the regression coefficient $b_{12.34}$ are highly significant. The regression coefficient $b_{13.24}$ is not significant. In

other words, it might be concluded that fiber tensile strength (X_3) has no effect on skein strength (X_1). This conclusion is supported by the analysis of the sums of squares given in (9) above where X_3 is shown to account for less than 1 percent of the variation in X_1 after the effects of X_2 and X_4 are taken out.

The above conclusion about the effects of X_3 must be adopted with great caution, however, for X_3 and X_4 are correlated ($r_{34} = -0.67$). If we calculate $\Sigma u_{1.23}^2 = 975.4$, we find that the part that would have been accounted for by X_3 after X_2 is taken out but before X_4 is taken out would have been $\Sigma u_{1.2}^2 - \Sigma u_{1.23}^2 = 1283.6 - 975.4 = 308.2$ or 11.24 percent. Then when X_4 is added, an additional part $= \Sigma u_{1.23}^2 - \Sigma u_{1.234}^2 = 975.4 - 694.2 = 281.2$ or 10.25 percent would have been accounted for instead of the 21.33 percent that X_4 accounted for when it was added prior to X_3.

Thus fiber tensile strength (X_3) and fiber fineness (X_4) *together* account is some physical theory as to how the various fiber characteristics affect skein strength. In other words, statistical analysis alone cannot give us the answer. A physical model is necessary for valid conclusions.

3. NONLINEAR REGRESSIONS

3.1. General Discussion

If any of the variables appears in the regression equation to a higher degree than 1, the regression equation becomes nonlinear. The regression is also nonlinear if there are cross-product or "interaction" terms in the equation. If the nonlinear terms enter into the equation in an additive manner, however, this complication need not affect the method of estimation. For if we take a higher order term or an interaction term as simply an extra variable, the equation can be derived in exactly the same way as described in Section 2. For example, the equation

$$X_{1r} = a + bX_2 + cX_3 + dX_2^2$$

may be written

$$X_{1r} = a + bX_2 + cX_3 + dX_4$$

where

$$X_4 = X_2^2$$

Likewise

$$X_{1r} = a + bX_2 + cX_3 + dX_2X_3$$

may be written

$$X_{1r} = a + bX_2 + cX_3 + dX_5$$

where $X_5 = X_2X_3$.

If the function is nonlinear in the parameters, it may be possible to convert the relationship to a linear form by a general transformation of the variables. For example, an equation of the form $X_{1r} = AX_2^b X_3^c X_4^d$ can be converted to a linear form by taking logs. Thus we have

$$\log X_{1r} = \log A + b \log X_2 + c \log X_3 + d \log X_4$$

which may be written

$$Z_{1r} = a + bZ_2 + cZ_3 + dZ_4$$

where $Z_{1r} = \log X_{1r}$, $Z_2 = \log X_2$, $Z_3 = \log X_3$, *and* $Z_4 = \log X_4$. It is this last equation, then, that is fitted by the method of least squares. Other transformations, such as taking reciprocals, may help to "straighten out" the regression. The essential thing is that the new expression be *linear in the parameters.* As we have seen, it makes no difference whether the variables themselves enter in a linear manner. The possibility of using the method of least squares, however, depends on whether the expression to be fitted is linear in the parameters themselves.

If no transformation can be found that will make the regression linear in the parameters, but first derivatives of the regression function can be obtained with respect to the various variables, and outside information suggests that the regression function may not be far from linear in the neighborhood of a given point of interest, then the regression function can be expanded in a Taylor series around that point and the hyperplane given by the first order terms of the expansion can be taken as an approximation to the regression function in that area that may be fitted by least squares. Thus if the regression function is $F(X_2, X_3, X_4, a, b, c)$, the approximating hyperplane can be taken as

(33.46)

$$\begin{aligned} X_{1r} &= F(X_{20}, X_{30}, X_{40}, a, b, c) \\ &\quad + \frac{\partial F}{\partial X_2}(X_{20}, X_{30}, X_{40}, a, b, c)(X_2 - X_{20}) \\ &\quad + \frac{\partial F}{\partial X_3}(X_{20}, X_{30}, X_{40}, a, b, c)(X_3 - X_{30}) \\ &\quad + \frac{\partial F}{\partial X_4}(X_{20}, X_{30}, X_{40}, a, b, c)(X_4 - X_{40}) \end{aligned}$$

where $X_2 = X_{20}$, $X_3 = X_{30}$, $X_4 = X_{40}$ is the point of special interest.

Given a sample set of data in the neighborhood of the point $(X_2 = X_{20}, X_3 = X_{30}, X_4 = X_{40})$, the constant term and coefficients of the hyperplane (33.46) can be estimated by least squares as explained in Section 2.2.1 above.

3.2. Fitting an Approximating Second-Degree Polynomial

3.2.1. Theory. If Taylor's expansion for several variables is carried out to include squared and cross-product terms, we will get a second-degree polynomial. Often a second-degree polynomial is locally a good approximation to a given multivariate surface and will yield some understanding of the nature of the surface in the specified locality. We will conclude the chapter, therefore, with a discussion of the fitting of a multivariate second-degree polynomial. It will be assumed that values of the dependent variable are determined experimentally and that we can select the values of the independent variables to be used.

For the purposes of exposition we will consider a multivariate surface which expresses a dependent variable as a function of two independent variables. This will have the general form $Y = F(X_1, X_2)$. The second-degree polynomial that we will fit to it as an approximation will be

$$(33.47) \qquad Y_r = a' + b'_1X_1 + b'_2X_2 + b'_{11}X_1^2 + b'_{12}X_1X_2 + b'_{22}X_2^2$$

This may be put in standard regression form by setting $Z_{1r} = Y_r$, $Z_2 = X_1$, $Z_3 = X_2$, $Z_4 = X_1^2$, $Z_5 = X_1X_2$, and $Z_6 = X_2^2$. We will then have

$$(33.48) \qquad Z_{1r} = a' + b'_1Z_2 + b'_2Z_3 + b'_{11}Z_4 + b'_{12}Z_5 + b'_{22}Z_6$$

This is the regression equation we will fit to a selected set of points.

In observing the surface at a selected point we will assume that the reported value consists of the true surface value, $F(X_1, X_2)$, plus an observational error. These errors will be assumed to be random in nature and uncorrelated. They will also be assumed to be homoscedastic in that their variance will be assumed to have the same value, σ'^2, at every point of observation. The deviations of the fitted polynomial from the given surface will thus be a function of two elements, the degree of approximation of the polynomial and the effect of the random errors on the fitting process.

In fitting equation (33.48) we must have at least 6 observations since there are 6 constants, a', b'_1, b'_2, and so on, to be determined. Further observations will be needed for measurement of the random variance σ'^2. The optimal layout of the observations for fitting a second-degree polynomial has been the subject of much discussion. See for example the Box and Hunter references at the end of the chapter. It will be sufficient here to note that a hexagonal layout with several observations in the middle will give good results. They will be good in the sense that confidence limits for the approximating polynomial that would have been derived if there had been no errors of observation are equally wide at equal distances from the center of this layout. This hexagonal design belongs to the class of what are called rotable designs. In our example we will take four additional observations in the middle.

If we shift the origin for measuring the factors X_1 and X_2 to the center of the layout, a hexagonal design (plus four points[16] at the center) is obtained by adopting the following factor values measured in terms of spacing units. (Small x's are used since their means are all zero.)

---

[16] If we run six cases in the center, we shall get a completely "orthogonal" design. This means that in Table 33.6 *all* the z_iz_j cross-product terms will sum to zero. The amount of extra computation resulting from the lack of complete orthogonality is not very great, however.

Test Number	x_1	x_2
1	1	0
2	0.5	0.866*
3	−0.5	0.866*
4	−1	0
5	−0.5	−0.866*
6	0.5	−0.866*
7	0	0
8	0	0
9	0	0
10	0	0

* Exactly the $\sqrt{0.75}$.

Spacing units will be distances on the X_1 and X_2 axes that are related to the size of the area chosen for appoximating the given surface. The analysis will be in terms of these spacing units. This design is shown in Figure 33.2.

Now set $z_1 = Y - \bar{Y}$, where the Y's are actual observed values and $\bar{Y}$ their mean. Let $z_2 = x_1$, $z_3 = x_2$, $z_4 = x_1^2 -$ mean x_1^2, $z_5 = x_1x_2$, and $z_6 = x_2^2 -$ mean x_2^2. Since mean $x_1^2 = 0.3$ and mean $x_2^2 = 0.3$, we have $z_4 =$

FIGURE 33.2
Hexagonal Design for Second-Order Model

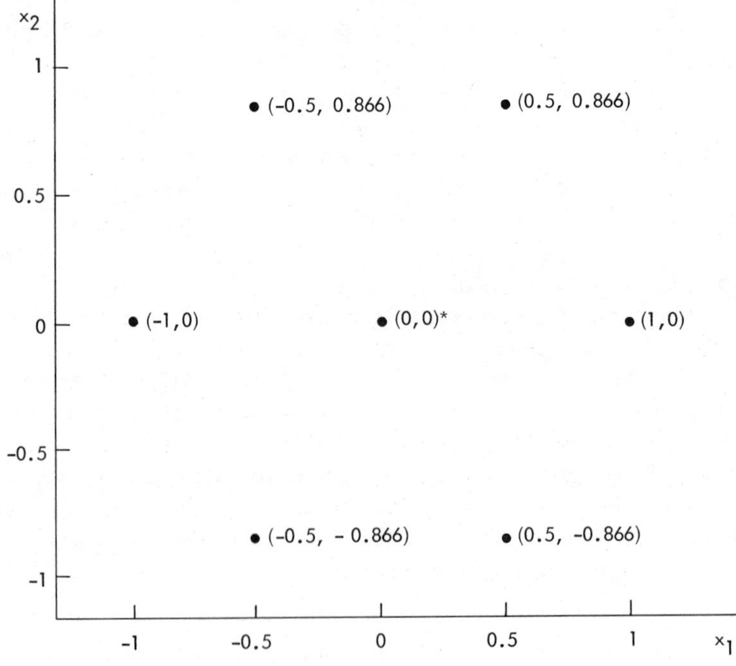

* Four observations taken at (0, 0).

$x_1^2 - 0.3$ and $z_6 = x_2^2 - 0.3$. With these definitions the c elements that we shall use in fitting the second-degree polynomial will be given by solution of the sets of simultaneous equations[17] (33.49).

(33.49) Constant terms for $i = 2\ 3\ 4\ 5\ 6$

$$c_{i2}\sum z_2^2 + c_{i3}\sum z_2 z_3 + c_{i4}\sum z_2 z_4 + c_{i5}\sum z_2 z_5 + c_{i6}\sum z_2 z_6 = 1\ 0\ 0\ 0\ 0$$

$$c_{i2}\sum z_2 z_3 + c_{i3}\sum z_3^2 + c_{i4}\sum z_3 z_4 + c_{i5}\sum z_3 z_5 + c_{i6}\sum z_3 z_6 = 0\ 1\ 0\ 0\ 0$$

$$c_{i2}\sum z_2 z_4 + c_{i3}\sum z_3 z_4 + c_{i4}\sum z_4^2 + c_{i5}\sum z_4 z_5 + c_{i6}\sum z_4 z_6 = 0\ 0\ 1\ 0\ 0$$

$$c_{i2}\sum z_2 z_5 + c_{i3}\sum z_3 z_5 + c_{i4}\sum z_4 z_5 + c_{i5}\sum z_5^2 + c_{i6}\sum z_5 z_6 = 0\ 0\ 0\ 1\ 0$$

$$c_{i2}\sum z_2 z_6 + c_{i3}\sum z_3 z_6 + c_{i4}\sum z_4 z_6 + c_{i5}\sum z_5 z_6 + c_{i6}\sum z_6^2 = 0\ 0\ 0\ 0\ 1$$

(Note $c_{ij} = c_{ji}$)

The values of the z_i's for the hexagonal design of Figure 33.2 are given in Table 33.6. From this table it is seen that equations (33.49) reduce to the following simple set.

For $i = 2$, we have

(33.49a)

$$\begin{aligned} 3c_{22} &= 1 & & c_{22} = \tfrac{1}{3} \\ 3c_{23} &= 0 & \text{which yield}\quad & c_{23} = 0 \\ 1.35c_{24} - 0.15c_{26} &= 0 & & c_{24} = 0 \\ 0.75c_{25} &= 0 & & c_{25} = 0 \\ -0.15c_{24} + 1.35c_{26} &= 0 & & c_{26} = 0 \end{aligned}$$

For $i = 3$, we have

(33.49b)

$$\begin{aligned} 3c_{23} &= 0 & & c_{23} = 0 \\ 3c_{33} &= 1 & \text{which yield}\quad & c_{33} = \tfrac{1}{3} \\ 1.35c_{34} - 0.15c_{36} &= 0 & & c_{34} = 0 \\ 0.75c_{35} &= 0 & & c_{35} = 0 \\ -0.15c_{34} + 1.35c_{36} &= 0 & & c_{36} = 0 \end{aligned}$$

---

[17] See Section 1.6 above.

TABLE 33.6
Squares and Cross-Products of the z_i for a Hexagonal Design Plus Four Points in the Center

Test Number	z_2	z_3	z_4	z_5	z_6	z_2z_3	z_2z_4	z_2z_5	z_2z_6
1	1	0	0.70	0	−0.30	0	0.70	0	−0.30
2	0.5	0.866	−0.05	0.433	0.45	0.433	−0.025	0.2165	0.225
3	−0.5	0.866	−0.05	−0.433	0.45	−0.433	0.025	0.2165	−0.225
4	−1	0	0.70	0	−0.30	0	−0.70	0	0.30
5	−0.5	−0.866	−0.05	0.433	0.45	0.433	0.025	−0.2165	−0.225
6	0.5	−0.866	−0.05	−0.433	0.45	−0.433	−0.025	−0.2165	0.225
7	0	0	−0.30	0	−0.30	0	0	0	0
8	0	0	−0.30	0	−0.30	0	0	0	0
9	0	0	−0.30	0	−0.30	0	0	0	0
10	0	0	−0.30	0	−0.30	0	0	0	0
Σ	0	0	0	0	0	0	0	0	0

TABLE 33.7
Computations for Fitting Second-Degree Polynomial: Hexagonal Design Plus Four Points at the Center

Test Number	z_2	z_3	z_4	z_5
1	1.00	0	0.70	0
2	0.50	0.866	−0.05	0.433
3	−0.50	0.866	−0.05	−0.433
4	−1.00	0	0.70	0
5	−0.50	−0.866	−0.05	0.433
6	0.50	−0.866	−0.05	−0.433
7	0	0	−0.30	0
8	0	0	−0.30	0
9	0	0	−0.30	0
10	0	0	−0.30	0
Total	0	0	0	0

For $i = 4$, we have

$$3c_{24} = 0 \qquad\qquad c_{24} = 0$$

$$3c_{34} = 0 \qquad \text{which yield} \qquad c_{34} = 0$$

(33.49c) $$1.35c_{44} - 0.15c_{46} = 1 \qquad\qquad c_{44} = 3/4$$

$$0.75c_{45} = 0 \qquad\qquad c_{45} = 0$$

$$-0.15c_{44} + 1.35c_{46} = 0 \qquad\qquad c_{46} = 1/12$$

z_2^2	z_3z_4	z_3z_5	z_3z_6	z_3^2	z_4z_5	z_4z_6	z_4^2	z_5z_6	z_5^2	z_6^2
1	0	0	0	0	0	−0.21	0.49	0	0	0.09
0.25	−0.0433	0.375	0.375	0.75	−0.02165	−0.0225	0.0025	0.19485	0.1875	0.2025
0.25	−0.0433	−0.375	0.375	0.75	0.02165	−0.0225	0.0025	−0.19485	0.1875	0.2025
1	0	0	0	0	0	−0.21	0.49	0	0	0.09
0.25	0.0433	−0.375	−0.375	0.75	−0.02165	−0.0225	0.0025	0.19485	0.1875	0.2025
0.25	0.0433	0.375	−0.375	0.75	0.02165	−0.0225	0.0025	−0.19485	0.1875	0.2025
0	0	0	0	0	0	0.09	0.09	0	0	0.09
0	0	0	0	0	0	0.09	0.09	0	0	0.09
0	0	0	0	0	0	0.09	0.09	0	0	0.09
0	0	0	0	0	0	0.09	0.09	0	0	0.09
3	0	0	0	3	0	−0.15	1.35	0	0.75	1.35

z_6	Y	z_2Y	z_3Y	z_4Y	z_5Y	z_6Y	Y^2
−0.30	2.00	2.00	0	1.40	0	−0.60	4.00
0.45	1.29	0.64	1.12	−0.06	0.56	0.58	1.66
0.45	−0.74	0.37	−0.64	0.04	0.32	−0.33	0.55
−0.30	0.03	−0.03	0	0.02	0	−0.01	0.00
0.45	−0.27	0.14	0.23	0.01	−0.12	−0.12	0.07
0.45	1.90	0.95	−1.65	−0.10	−0.82	0.85	3.61
−0.30	1.53	0	0	−0.46	0	−0.46	2.34
−0.30	2.23	0	0	−0.67	0	−0.67	4.97
−0.30	0.93	0	0	−0.28	0	−0.28	0.86
−0.30	1.03	0	0	−0.31	0	−0.31	1.06
0	9.93	4.07	−0.94	−0.41	−0.06	−1.35	19.12

For $i = 5$, we have

$$
\begin{aligned}
&3c_{25} = 0 & & c_{25} = 0 \\
&3c_{35} = 0 & \text{which yield} \quad & c_{35} = 0 \\
\textbf{(33.49d)} \quad &1.35c_{45} - 0.15c_{56} = 0 & & c_{45} = 0 \\
&0.75c_{55} = 1 & & c_{55} = 4/3 \\
&-0.15c_{45} + 1.35c_{56} = 0 & & c_{56} = 0
\end{aligned}
$$

For $i = 6$, we have

$$\begin{aligned} &3c_{26} = 0 & & c_{26} = 0 \\ &3c_{36} = 0 & \text{which yield} \quad & c_{36} = 0 \\ \text{(33.49e)} \quad &1.35c_{46} - 0.15c_{56} = 0 & & c_{46} = 1/12 \\ &0.75c_{56} = 0 & & c_{56} = 0 \\ &-0.15c_{46} + 1.35c_{66} = 1 & & c_{66} = 3/4 \end{aligned}$$

These results give us[18] the following formulas for estimating the parameters of the second-degree equation:

$$\text{(33.50)} \quad \begin{aligned} a &= \Sigma Y/n - b_{11}\bar{Z}_4^2 - b_{22}\bar{Z}_6^2 & &[\text{Note: } \textstyle\sum z_1 z_2 \\ b_1 &= \tfrac{1}{3} \textstyle\sum z_2 Y & &= \textstyle\sum z_2(Y - \bar{Y}) \\ b_2 &= \tfrac{1}{3} \textstyle\sum z_3 Y & &= \textstyle\sum z_2 Y \\ b_{11} &= \tfrac{3}{4} \textstyle\sum z_4 Y + \tfrac{1}{12} \textstyle\sum z_6 Y & &\text{since } \textstyle\sum z_2 \bar{Y} \\ b_{12} &= \tfrac{4}{3} \textstyle\sum Z_5 Y & &= \bar{Y} \textstyle\sum z_2 = 0 \\ b_{22} &= \tfrac{1}{12} \textstyle\sum z_4 Y + \tfrac{3}{4} \textstyle\sum z_6 Y & &\text{etc.}] \end{aligned}$$

We also have[19]

$$\text{(33.51)} \quad \begin{aligned} \sum u^2 = \sum Y^2 - \frac{(\Sigma Y)^2}{n} \\ - b_1 \sum z_2 Y - b_2 \sum z_3 Y - b_{11} \sum z_4 Y - b_{12} \sum z_5 Y - b_{22} \sum z_6 Y \end{aligned}$$

and the error sum of squares at the center point will be given by the formula

$$\text{(33.52)} \quad Y_7^2 + Y_8^2 + Y_9^2 + Y_{10}^2 - \frac{(Y_7 + Y_8 + Y_9 + Y_{10})^2}{4}$$

The quantities $\Sigma z_i Y$ can be obtained readily by setting up a work sheet in which the z_i of Table 33.6 are multiplied respectively by Y and summed. See, for example, Table 33.7.

3.2.2. An Example. To make our ideas more concrete let us consider a surface representing the yield of a given chemical as a function of temperature and concentration. Suppose for our experiment that we have the following combinations of the independent variables, expressed first in original units and second as deviations from the center of the design measured in spacing units of 2.5 and 5, respectively.

[18] Cf. formulas (33.22).

[19] Cf. formula (33.8).

Test Number	*Temperature* (T)	*Concentration* (C)	$x_2 = \frac{T-62.5}{2.5}$	$x_2 = \frac{C-59.4}{5}$
1	65.00°	59.4%	1	0
2	63.75	63.7	0.5	0.866
3	61.25	63.7	−0.5	0.866
4	60.00	59.4	−1	0
5	61.25	55.1	−0.5	−0.866
6	63.75	55.1	0.5	−0.866
7	62.50	59.4	0	0
8	62.50	59.4	0	0
9	62.50	59.4	0	0
10	62.50	59.4	0	0

Suppose that we run the above experiments and obtain the following responses:

Test Number	*Response*	*Coded Response* (Y = *Response* − 97)
1	99.00	2.00
2	98.29	1.29
3	96.26	−0.74
4	97.03	0.03
5	96.73	−0.27
6	98.90	1.90
7	98.53	1.53
8	99.23	2.23
9	97.93	0.93
10	98.03	1.03

The computations for fitting a second-degree polynomial to these results are shown in Table 33.7. From this table and from formulas (33.50) and (33.51) we have:

$$a = \frac{9.93}{10} - (-0.42)(0.3) - (-1.05)(0.3) = 1.43$$

$$b_1 = \tfrac{1}{3}(4.07) = 1.36$$

$$b_2 = \tfrac{1}{3}(-0.94) = -0.31$$

$$b_{11} = \tfrac{3}{4}(-0.41) + \tfrac{1}{12}(-1.35) = -0.42$$

$$b_{12} = \tfrac{4}{3}(-0.06) = -0.08$$

$$b_{22} = \tfrac{1}{12}(-0.41) + \tfrac{3}{4}(-1.35) = -1.05$$

The estimated second-degreee polynomial thus becomes

$$Y_r = 98.43 + 1.36x_1 - 0.31x_2 - 0.42x_1^2 - 0.08x_1x_2 - 1.05x_2^2$$

where the x's, it will be recalled, are in spacing units of 2.5 and 5, respectively. In terms of T and C we have

$$Y_r = -366 + 9.3T + 5.3C - 0.067T^2 - 0.0064TC - 0.042C^2$$

We also have

$$\sum u^2 = 19.12 \frac{(9.93)^2}{10} - (1.36)(4.07) - (-0.31)(-0.94) - (-0.42)(-0.41)$$

$$-(-0.08)(-0.06) - (-1.05)(-1.35)$$

$$= 19.12 - 9.86 - 5.54 - 0.29 - 0.17 - 0.00 - 1.42 = 1.84$$

which yields an estimated overall residual variance of 1.84/4 = 0.46.

For the points at the center we have

$$Y_7^2 + Y_8^2 + Y_9^2 + Y_{10}^2 - \frac{(Y_7 + Y_8 + Y_9 + Y_{10})^2}{4} = 9.24 - \frac{32.72}{4} = 1.06$$

which yields an estimated observational variance of 1.06/3 = 0.35. From this we may estimate σ' to be equal to $\sqrt{0.35} = 0.59$.

It can thus be concluded that of the total residual sum of squares (= 1.84), 57.6 percent (= 1.06/1.84) is due to errors of measurement and 42.4 percent to lack of fit of the second degree polynomial.

4. PROBLEMS

33.1. A research and development plan was carried out to determine the effect of thickness and Brinell Hardness on the "ballistic limit" of armor plate placed at an angle of 40° obliquity to the line of fire. ("Ballistic limit" is defined as the average of two velocities within 50 feet per second of each other—one, the lowest velocity of projectiles fired at the plate that give complete penetrations and the other, the highest velocity of incomplete or partial penetrations.) In this program, armor-piercing bullets were fired against armor plate of the following thicknesses and Brinell Hardness numbers with the resulting ballistic limits as indicated:[20]

X_1 *Ballistic Limit in Feet/Sec.*	X_2 *Thickness in 0.001 Inches*	X_3 *Brinell Hardness No.*
927	253	317
978	258	321
1,028	259	341
906	247	350
1,159	256	352

[20] Cf. Leslie E. Simon, " 'Quality Control' in Research," *First Annual Regional Quality Control Clinic Proceedings, September 17, 1948, Philadelphia and Delaware Section,* American Society for Quality Control and Community College Center of Temple University, Philadelphia, Pa., p. 81.

X_1 *Ballistic Limit in Feet/Sec.*	X_2 *Thickness in 0.001 Inches*	X_3 *Brinell Hardness No.*
1,055	246	363
1,335	257	365
1,392	262	375
1,362	255	373
1,374	258	391
1,393	253	407
1,401	252	426
1,436	246	432
1,327	250	469
950	242	275
998	243	302
1,144	239	331
1,080	242	355
1,276	244	385
1,062	234	426

a. Estimate the universe plane of regression of X_1 on X_2 and X_3 by the method of least squares. (It might be helpful to code each series by referring it to an arbitrary origin.)

b. Estimate the universe standard error of estimate.

c. If a plate is 0.250 inches thick and if its Brinell Hardness number is 400, what estimate would you make of its ballistic limit?

d. Suppose the estimates of the universe values in *a* and *b* are close to the true values and let the deviations from the universe plane of regression be normally distributed. Given these conditions, determine 0.95 limits for the ballistic limit when the plate is 0.250 inches thick and its Brinell Hardness number is 400.

e. Compute standard errors for $a_{1.23}$, $b_{12.3}$, and $b_{13.2}$.

f. Determine confidence limits for the regression coefficients.

g. Assuming that deviations from the universe plane of regression are normally distributed and allowing for sampling errors, determine 0.95 prediction limits for the ballistic limit of armor plate that is 0.250 inches thick and has a Brinell Hardness number of 400. Compare your answer with that given to part *d.*

h. Estimate the universe multiple correlation coefficient for the universe plane of regression of X_1 on X_2 and X_3. What percent of the total variation in ballistic limit is apparently accounted for by plate thickness and hardness?

i. How much more of the variation in ballistic limit is accounted for by plate thickness and hardness jointly than is accounted for by thickness alone? By hardness alone?

j. How much correlation, if any, is there between plate thickness and hardness?

33.2. Zircon, acid, and asbestos are mixed, molded, and baked to form an arc chute for an electric circuit breaker. Data on green strength (i.e., flexural strength before baking), hydraulic oil pressure used in the molding process, and acid concentration on 20 chutes are as follows:[21]

Chute	*Green Strength* (X_1) *Psi*	*Hydraulic Pressure* (X_2) *Psi*	*Acid Concentration* (X_3) *Percent of Nominal Rate*
1	6,650	1,100	116
2	6,180	1,190	104
3	6,200	1,380	94
4	5,780	1,300	86
5	6,820	1,430	110
6	5,940	1,330	87
7	7,220	1,470	114
8	7,000	1,420	106
9	6,810	1,250	107
10	6,950	1,350	106
11	6,640	1,520	98
12	5,480	1,180	86
13	6,200	1,550	87
14	5,950	1,280	96
15	7,400	1,460	120
16	6,700	1,320	108
17	6,400	1,300	104
18	5,900	1,120	91
19	5,700	1,130	92
20	6,400	1,200	100

a. Estimate the universe plane of regression of X_1 on X_2 and X_3 by the method of least squares. (It might be helpful to code each series by referring it to an arbitrary origin.)

b. Estimate the universe standard error of estimate.

c. If hydraulic pressure is 1,400 psi and acid concentration is 105 percent of the nominal rate, what estimate would you make of the green strength?

d. Suppose that the estimates of the universe values in *a* and *b* are close to the true values and let the deviations from the universe plane of regression be normally distributed. Given these conditions, determine 0.95 limits for the green strength when hydraulic pressure is 1,400 psi and acid concentration is 105 percent.

e. Compute standard errors for $a_{1.23}$, $b_{12.3}$, and $b_{13.2}$.

f. Determine 0.95 confidence limits for the regression coefficients.

g. Assuming that deviations from the universe plane of regression are normally distributed and allowing for sampling errors, determine 0.95 "prediction

[21] These are fictitious data based on a chart derived from actual data. See William Masser, "Trouble Shooting with Statistical Quality Control," *Second Annual Quality Control Conference Papers, Philadelphia and Delaware Sections,* American Society for Quality Control and the Community College and Technical Institute of Temple University, February 10 and 11, 1950, p. 164.

limits" for green strength when hydraulic pressure is 1,400 psi and acid concentration is 105 percent of normal. Compare your answer with that given to *d.*

h. Estimate the universe multiple correlation coefficient for the universe plane of regression of X_1 on X_2 and X_3. What percent of the total variation in green strength is accounted for by hydraulic pressure and acid concentration?

i. How much more of the variation in green strength is accounted for by hydraulic pressure and acid concentration jointly than is accounted for by hydraulic pressure alone? By acid concentration alone?

33.3. The following are 20 more sets of the same data as those given in Table 33.1. They pertain to 22s cotton yarn.

Set number	*Yarn Strength (X_1) (Pounds)*	*Fiber Length (X_2) (0.01 Inch)*	*Fiber Tensile Strength (X_3) (1,000 Psi)*	*Fiber Fineness (X_4) (0.1 Micrograms)*
21	96	72	83	40
22	102	70	76	32
23	106	72	78	34
24	89	74	71	44
25	91	72	74	50
26	90	72	72	37
27	88	71	69	42
28	83	70	77	49
29	92	72	78	39
30	91	73	71	44
31	90	73	78	49
32	92	71	77	41
33	90	72	76	44
34	108	75	81	38
35	96	73	76	51
36	91	72	78	48
37	102	74	82	49
38	98	78	85	52
39	108	76	83	43
40	100	79	77	45

a. Estimate the universe plane of regression of X_1 on X_2, X_3, and X_4 by the method of least squares. Compare your results with those obtained in the text.

b. Estimate the universe standard error of estimate.

c. Assume that the values obtained in *a* and *b* are close to the universe values and that the deviations from the universe plane of regression are normally distributed. If fiber length is 0.75 inches, fiber tensile strength is 80,000 pounds per square inch, and fiber fineness is 4.5 micrograms, what would you estimate the skein strength to be? Determine 0.95 "prediction limits" for the actual skein strength of a set at these levels of fiber length, fiber tensile strength, and fiber fineness.

d. Estimate the standard errors of the regression coefficients and determine confidence limits for the universe values. Do these limits overlap with those obtained from the data of Table 33.1 (see Section 1.5.5.)?

e. Assume that deviations from the plane of regression are normally distributed and, allowing for sampling errors, determine 0.95 prediction limits for skein strength when fiber length is 0.75 inches, fiber tensile strength is 80,000 pounds per square inch, and fiber fineness is 4.5 micrograms. Compare with the results obtained in *(c)*.

33.4. Observations on the detergency of a given detergent at various times, temperatures, and concentrations are as follows:[22]

Detergency (X_1)	*Time* (X_2)	*Concentration* (X_3)	*Temperature* (X_4)
37	5	10	100
42	5	20	100
46	5	30	100
48	5	40	100
53	5	10	120
62	5	20	120
79	5	30	120
84	5	40	120
59	5	10	140
74	5	20	140
88	5	30	140
102	5	40	140
39	10	10	100
45	10	20	100
53	10	30	100
56	10	40	100
61	10	10	120
74	10	20	120
84	10	30	120
95	10	40	120
68	10	10	140
82	10	20	140
98	10	30	140
123	10	40	140
43	15	10	100
47	15	20	100
55	15	30	100
63	15	40	100
68	15	10	120
79	15	20	120
92	15	30	120
104	15	40	120
71	15	10	140
89	15	20	140
115	15	30	140
122	15	40	140

[22] Cf. Casper Goffman, "The Effect of a Number of Variables on a Quality Characteristic," *Industrial Quality Control,* January 1946, p. 4.

a. Estimate the universe plane of regression of X_1 on X_2, X_3, and X_4 by the method of least squares. (Note the simplicity that results from taking time, concentration, and temperature at uniformly spaced intervals. Further simplification can be obtained by coding the levels of X_2 and X_4 as -1, 0, and 1, and coding the levels of X_3 as -3, -1, 1, and 3. These arrangements, it will be noted, greatly simplify the auxiliary equations.)

b. Estimate the universe standard error of estimate.

c. Assume that the values obtained in *a* and *b* are close to the universe values and that deviations from the universe plane of regression are normally distributed. If the temperature is 100°, the concentration 20 percent, and the time 15 minutes, estimate the detergency of the detergent. Lay off 0.95 "prediction limits" for the actual detergency at these levels of temperature, concentration, and time. How much of an increase in the detergency would you expect to result from increasing the time from 10 to 15 minutes? What is the standard error of this estimate?

d. Determine standard errors and 0.95 confidence limits for your regression coefficients.

5. SELECTED REFERENCES*

(In addition to those of Chapter 32)

Box and Hunter (P '57), Brownless (B '48 and B '60), Brunt (B '17. A classical work on least squares), Davies (B '54), Deming (P '37 and B '43), Duncan and Kenney (B '46), Dwyer (P '45), Hahn and Shapiro (P '66), Hocking and Leslie (P '67), Kramer (P '63), Lee (P '72), Pearson and Hartley (B '72), Plackett (B '60), Schultz (B '38), Toothill (P '63), Williams (B '59), and Winter (P '61).

---

* B and P refer to the Book and Periodical section, respectively, of the Cumulative List of References in Appendix V.

34

Analysis of Covariance

1. USES OF ANALYSIS OF COVARIANCE

R. A. Fisher points out that the analysis of covariance combines the advantages and reconciles the requirements of the two widely applicable procedures, regression and analysis of variance.[1] Consider the following problem. Suppose an experiment is to be run on the effect of concentration of an inert solvent on the yield of a chemical. It is known, however, that another variable, say temperature, affects the yield, but it is not possible under the conditions under which the experiment has to be run to control this variable sufficiently well to attain predetermined fixed levels. Temperature cannot therefore be entirely "balanced out" in a two-way analysis, as was done with such factors in the problems of Chapters 29 and 30. If the deviations in temperature from target levels are randomly distributed, a two-way temperature-concentration analysis of variance could be run, but these temperature deviations would increase the "error" variance and make the experiment less able to detect real differences due to the inert solvent. Analysis of covariance offers a means of adjusting for the effects of a "nuisance" variable of this kind, without increasing the "experimental" error variance.

Analysis of covariance also is a means, as its name implies, of analyzing covariance. Thus it is of use in testing for differences between slopes of lines of regression. For example, in Table 32.1 there are data on the shear strength and weld diameter of 10 spot welds of 0.064″–0.064″ steel. The regression of the first of these variables on the second is shown graphically in Figure 32.6 and in the accompanying text is determined numerically. Similar data might be derived for steel of other gauges, and other regressions might be

[1] *Statistical Methods for Research Workers,* 13th rev. ed. (Edinburgh: Oliver and Boyd, Ltd., 1958), p. 281.

TABLE 34.1
Results of an Experiment on Methods of Applying a Rust Arrestor (X is initial rust on the coupon; Y is amount of rust added during exposure to a salt spray)

Brushing			*Spraying*			*Dipping*		
Coupon Number	*Y*	*X*	*Coupon Number*	*Y*	*X*	*Coupon Number*	*Y*	*X*
1	63	16	6	81	48	11	72	40
2	77	45	7	73	40	12	54	31
3	81	50	8	59	24	13	57	40
4	60	19	9	74	33	14	59	33
5	63	18	10	77	41	15	52	20
Totals	344	148		364	186		294	164
Means	68.8	29.6		72.8	37.2		58.8	32.8
						Grand Total	1,002	498
						Grand Mean	66.8	33.2

SOURCE: See B. B. Day, F. R. Del Priore, and Edward Sax, "The Technique of Regression Analysis," *Quality Control Conference Papers, 1953,* American Society for Quality Control (1953), p. 417.

computed. The identity or dissimilarity of the slopes of the various lines of regression might then become a matter of interest.[2]

There are other uses of analysis of covariance. For an excellent review the reader is referred to the article by William G. Cochran in *Biometrics,* September 1957, pp. 261–81.

2. ADJUSTING FOR NUISANCE EFFECTS BY ANALYSIS OF COVARIANCE

2.1. One-Way Classification

Let us consider first the use of analysis of covariance in making simple adjustments for "nuisance" effects.

Table 34.1 contains the results of an experiment on a rust arrestor which are graphed in Figure 34.1. In the experiment, steel coupons, already in a measurable state of rust, were coated with a rust arrestor compound by one of three methods: brushing, spraying, and dipping. Five coupons coated by each method were then exposed in a salt spray cabinet, and the additional rust due to this exposure was measured. The information desired is whether the three methods of coating show significantly different results after adjustment has been made for variation in the initial state of rust. We shall answer this question through an analysis of covariance.

[2] This problem is studied in Section 3.1.

FIGURE 34.1
Relationship between Initial Rust and Rust Added, Classified by Method of Application of a Rust Arrestor (data from Table 34.1)

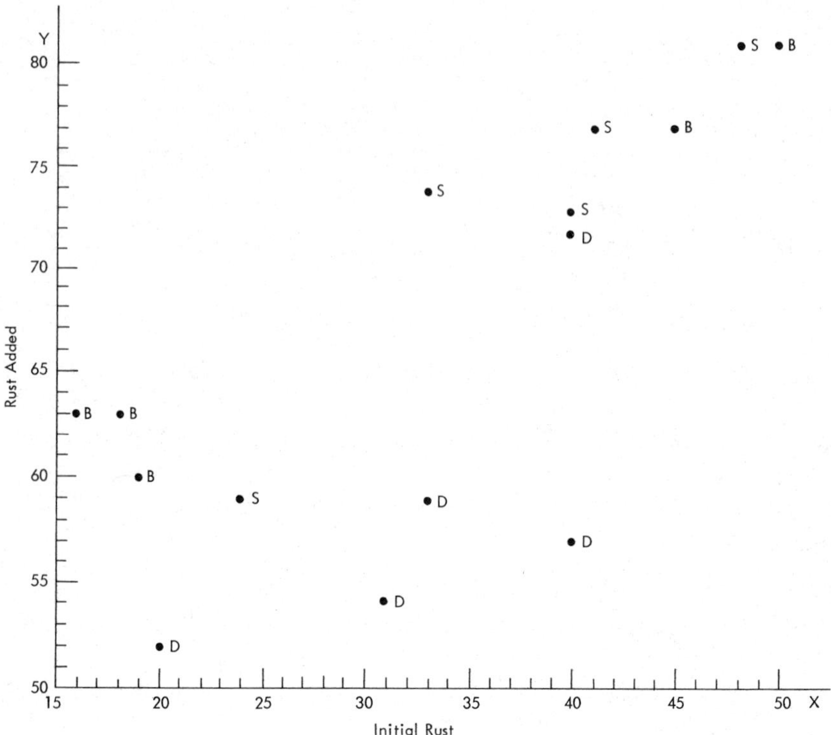

2.1.1. Theory. We shall attack the above problem by fitting by the method of least squares a combined analysis of variance and regression equation of the type[3]

$$Y_{ij} = \mu + \theta_j + \beta(X_{ij} - \bar{\bar{X}}) + \epsilon_{ij} \quad \begin{cases} i = 1, 2, \ldots, r \\ j = 1, 2, \ldots, c \end{cases} \tag{34.1}$$

Here the symbols Y and X are found more convenient to use than the X_1 and X_2 employed in Chapter 32. The quality Y_{ij} is the dependent variable (amount of rust added); μ is an overall constant; θ_j represents a "differential" that is independent of or "adjusted" for variation in X_{ij} and may be zero in any or all of the c classes (brushing, spraying, or dipping); X_{ij} is a nuisance variable (initial rust) for which adjustment is to be made; and ϵ_{ij} is a random

[3] We shall follow here the analysis of variance practice of using Greek letters to represent universe parameters. Sample estimates of these universe parameters will be indicated by putting "hats" on the symbols. $\bar{\bar{X}}$ and $\bar{\bar{Y}}$ are used to represent grand means.

variable that is normally distributed with zero mean and the same variance σ'^2 for all c classes and all values of X_{ij}. Since the θ_j's are differentials, it will be assumed that $\Sigma\theta_j = 0$. The $\bar{\bar{X}}$ in equation (34.1) will be the grand mean of the "fixed set" of X_{ij}'s that are given in Table 34.1. The sampling variation and associated significance tests will be with respect to a "conditional" type of sampling that requires each sample to always have this particular fixed set of X_{ij}'s.

Based on a random sample of r items from c classes, the least-squares estimates of μ, θ_j, and β are as follows:[4]

$$\hat{\mu} = \frac{\sum_i \sum_j Y_{ij}}{n} = \bar{\bar{Y}}$$

(34.2)
$$\hat{\theta}_j = \frac{\sum_i Y_{ij}}{r} - \hat{\mu} - \hat{\beta}(\bar{X}_j - \bar{\bar{X}}) = \bar{Y}_j - \hat{\mu} - \hat{\beta}(\bar{X}_j - \bar{\bar{X}})$$

$$\hat{\beta} = \frac{{}_CE_{xy}}{{}_CE_{xx}}$$

where ${}_CE_{xy}$ and ${}_CE_{xx}$ are defined in row 4 of Table 34.2.

To test for the significance of column effects adjusted for regression of Y on X we shall need to compute two residual sums of squares. These are

1. Σu_C^2 which is the residual sum of squares obtained when equation (34.1) is fitted, and
2. Σu_0^2 which is the residual sum of squares obtained when equation (34.1) is fitted *omitting* the θ_j terms.

It can be shown that[5]

(34.3) $$\sum u_C^2 = S_{yy} - C_{yy} - {}_CE_{xy}^2/{}_CE_{xx} = {}_CE_{yy} - {}_CE_{xy}^2/{}_CE_{xx}$$

which is the ordinary residual sum of squares for a one-way analysis of variance (i.e., $S_{yy} - C_{yy} = {}_CE_{yy}$) minus the contribution to the total sum of squares attributed to the regression of Y on X when Y and X are adjusted for column effects (i.e., ${}_CE_{xy}^2/{}_CE_{xx}$). It also can be shown that[6]

(34.1) $$\sum u_0^2 = S_{yy} - S_{xy}^2/S_{xx}$$

which is the total sum of squares minus the contribution attributed to the regression of the unadjusted Y on the unadjusted X. (For definitions of the symbols, see Table 34.2).

The additional reduction in the total sum of squares that may be attributed to including the θ_j in equation (34.1) is the difference between Σu_0^2 and

[4] See Appendix I (42).
[5] See Appendix I (42).
[6] Ibid.

TABLE 34.2
Sums of Squares and Cross-Products for One-Way and Two-Way Analysis of Covariance, Common Slopes*

Source of Variation	*Sums of Squares for Y*	*Sums of Cross-Products for X and Y*	*Sums of Squares for X*
1. Row differences	$R_{yy} = c \sum_i (\bar{Y}_i - \bar{\bar{Y}})^2$ $= \sum_i \frac{R_{yi}^2}{c} - K_y$	$R_{xy} = c \sum_i (\bar{X}_i - \bar{\bar{X}})(\bar{Y}_i - \bar{\bar{Y}})$ $= \sum_i \frac{R_{xi}R_{yi}}{c} - K_{xy}$	$R_{xx} = c \sum_i (\bar{X}_i - \bar{\bar{X}})^2$ $= \sum_i \frac{R_{xi}^2}{c} - K_x$
2. Column differences	$C_{yy} = r \sum_j (\bar{Y}_j - \bar{\bar{Y}})^2$ $= \sum_j \frac{C_{yj}^2}{r} - K_y$	$C_{xy} = r \sum_j (\bar{X}_j - \bar{\bar{X}})(\bar{Y}_j - \bar{\bar{Y}})$ $= \sum_j \frac{C_{xj}C_{yj}}{r} - K_{xy}$	$C_{xx} = r \sum_j (\bar{X}_j - \bar{\bar{X}})$ $= \sum_j \frac{C_{xj}^2}{r} - K_x$
3. Residual or error (when model contains only row effects)	${}_RE_{yy} = \sum_i \sum_j (Y_{ij} - \bar{Y}_i)^2$ $= S_{yy} - R_{yy}$	${}_RE_{xy} = \sum_i \sum_j (X_{ij} - \bar{X}_i)(Y_{ij} - \bar{Y}_i)$ $= S_{xy} - R_{xy}$	${}_RE_{xx} = \sum_i \sum_j (X_{ij} - \bar{X}_i)^2$ $= S_{xx} - R_{xx}$
4. Residual or error (when model contains only column effects)	${}_CE_{yy} = \sum_i \sum_j (Y_{ij} - \bar{Y}_j)^2$ $= S_{yy} - C_{yy}$	${}_CE_{xy} = \sum_i \sum_j (X_{ij} - \bar{X}_j)(Y_{ij} - \bar{Y}_j)$ $= S_{xy} - C_{xy}$	${}_CE_{xx} = \sum_i \sum_j (X_{ij} - \bar{X}_j)^2$ $= S_{xx} - C_{xx}$
5. Residual or error (when model contains both row and column effects)	${}_{RC}E_{yy} = \sum_i \sum_j (Y_{ij} - \bar{Y}_i - \bar{Y}_j + \bar{\bar{X}})^2$ $= S_{yy} - R_{yy} - C_{yy}$	${}_{RC}E_{xy} = \sum_i \sum_j (X_{ij} - \bar{X}_i - \bar{X}_j + \bar{\bar{X}})$ $(Y_{ij} - \bar{Y}_i - \bar{Y}_j + \bar{\bar{Y}})$ $= S_{xy} - R_{xy} - C_{xy}$	${}_{RC}E_{xx} = \sum_i \sum_j (X_{ij} - \bar{X}_i - \bar{X}_j - \bar{\bar{X}})^2$ $= S_{xx} - R_{xx} - C_{xx}$
6. Total	$S_{yy} = \sum_i \sum_j (Y_{ij} - \bar{\bar{Y}})^2$ $= \sum_i \sum_j Y_{ij}^2 - K_y$	$S_{xy} = \sum_i \sum_j (X_{ij} - \bar{\bar{X}})(Y_{ij} - \bar{\bar{Y}})$ $= \sum_i \sum_j X_{ij}Y_{ij} - K_{xy}$	$S_{xx} = \sum_i \sum_j (X_{ij} - \bar{\bar{X}})^2$ $= \sum_i \sum_j X_{ij}^2 - K_x$

where $R_{xi} = \sum_j X_{ij}$, $C_{ij} = \sum_i X_{ij}$, $R_{yi} = \sum_j Y_{ij}$, $C_{yi} = \sum_i Y_{ij}$, $K_x = \frac{(\Sigma\Sigma X_{ij})^2}{n}$, $K_y = \frac{(\Sigma\Sigma Y_{ij})^2}{n}$, and $K_{xy} = \frac{(\Sigma\Sigma X_{ij})(\Sigma\Sigma Y_{ij})}{n}$.

* The first formula in each cell is the definition of the quantity listed and the second formula the one useful for computation. The method of deriving the computational formulas is indicated in Appendix I (34).

Σu_C^2. This is usually called the "reduced sum of squares" due to adjusted column effects.[7] We shall represent it by RSS_θ. We thus have

$$\text{(34.5)} \qquad RSS_\theta = \sum u_{0C}^2 - \sum u_C^2$$

Now it can be shown that in repeated random sampling with the fixed set of X_{ij}'s the quantity $\Sigma u_C^2/\sigma'^2$ will have a χ^2 distribution with degrees of freedom equal to $c(r - 1) - 1$ or $n - c - 1$ (since $rc = n$) and $s_e^2 = \Sigma u_C^2/(n - c - 1)$ will be an unbiased estimate of σ'^2. Similarly, if the θ_j are all individually zero, then independently of Σu_C^2, RSS_θ/σ'^2 will have a χ^2 distribution with $c - 1$ degrees of freedom and $s_\theta^2 = RSS_\theta/(c - 1)$ will be an unbiased estimate of σ'^2. The ratio of these two independent estimates of σ'^2 will have an F distribution.[8] Hence to test the hypothesis that $\theta_j = 0$ for all j, we compute s_θ^2/s_e^2, and compare it with F for $\nu_1 = c - 1$, $\nu_2 = n - c - 1$.

The above calculations are usually summarized in an analysis of covariance table. The general form of this is indicated in Table 34.3 and illustrated numerically in Table 34.4. The various quantities are defined in Table 34.2, and formulas are given there for their computation.

Charts I and II can be used, as in analysis of variance, to determine the 0.50 and 0.10 points of the OC curves of the above tests.[9] In using them, we take $\lambda = \dfrac{\Sigma\theta_j^2}{c\sigma'^2}$ and $\phi = \lambda\sqrt{r}$.

Besides testing for significant differences between the classes we can also make estimates of the adjusted class means.

An estimate of a class mean adjusted for the nuisance variable will be

$$\text{(34.6)} \qquad \text{Estimate of } \mu + \theta_j = \hat{\mu} + \hat{\theta}_j = \bar{Y}_j - \hat{\beta}(\bar{X}_j - \bar{\bar{X}})$$

and 0.95 confidence limits for this estimate will be

$$\text{(34.7)} \qquad \bar{Y}_j - \hat{\beta}(\bar{X}_j - \bar{\bar{X}}) \pm t_{0.025} s_e \sqrt{\frac{1}{r} + \frac{(\bar{X}_j - \bar{\bar{X}})^2}{cE_{xx}}}$$

where

$$s_e^2 = \sum u_C^2/(n - c - 1)$$

and $t_{0.025}$ is taken for $n - c - 1$ degree of freedom. Ninety-five percent confidence limits for the difference between any two adjusted class means with an error rate of 0.05 per comparison will be given by

$$\text{(34.8)} \qquad (\bar{Y}_1 - \bar{Y}_2) - \hat{\beta}(\bar{X}_1 - \bar{X}_2) \pm t_{0.025} s_e \sqrt{\frac{2}{r} + \frac{(\bar{X}_1 - \bar{X}_2)}{cE_{xx}}}$$

---

[7] Cf. Section 1.5.4 of Chapter 33.

[8] Compare Chapter 26, Section 3.1.

[9] See Chapter 29, Section 1.4.

TABLE 34.3
An Analysis of Covariance Table, One-Way Classification

Source of Variation (1)	*Sum of Squares for Y* (2)	*Sum of Cross-Products for X and Y* (3)	*Sum of Squares for X* (4)	*Regression Coefficient* (5)	*Amount Explained by Regression* (6)	*Adjusted Sum of Squares for Y* (7)	*df* (8)	*Mean Square* (9)	F *Ratio* (10)
Between columns	C_{yy}	C_{xy}	C_{xx}			$\Sigma u_0^2 - \Sigma u_C^2 = C_{yy} - \left(\frac{S_{xy}^2}{S_{xx}} - \frac{{}_CE_{xy}^2}{{}_CE_{xx}}\right)$	$c - 1$	$s_\theta^2 = \frac{(7)}{(8)}$	s_θ^2/s_e^2
Residual or error	${}_CE_{yy}$	${}_CE_{xy}$	${}_CE_{xx}$	$\hat{\beta}_C = \frac{{}_CE_{xy}}{{}_CE_{xx}}$	$\frac{{}_CE_{xy}^2}{{}_CE_{xx}}$	$\Sigma u_C^2 = {}_CE_{yy} = \frac{{}_CE_{xy}^2}{{}_CE_{xx}}$	$n - c - 1$	$s_e^2 = \frac{(7)}{(8)}$	
Total	S_{yy}	S_{xy}	S_{xx}	$\hat{\beta}_0 = \frac{S_{xy}}{S_{xx}}$	$\frac{S_{xy}^2}{S_{xx}}$	$\Sigma u_0^2 = S_{yy} - \frac{S_{xy}^2}{S_{xx}}$	$n - 2$		

TABLE 34.4
Analysis of Covariance Table for Data of Table 34.1

Source of Variation (1)	*Sum of Squares for* Y *(2)*	*Sum of Cross-Products for* X *and* Y *(3)*	*Sum of Squares for* X *(4)*	*Regression Coefficient (5)*	*Amount Explained by Regression (6)*	*Adjusted Sum of Squares for* Y *(7)*	*df (8)*	*Mean Square (9)*	F *Ratio (10)*
Between columns	520.0	100.0	145.6			651.5 − 199.8 = 451.7	2	225.9	12.4
Residual or error	1,404.4 − 520.0 = 884.4	1,174.6 − 100 = 1,074.6	1,832.4 − 145.6 = 1,686.8	$\frac{1,074.6}{1,686.8}$ = 0.6371	(0.6371)(1,074.6) = 684.6	884.4 − 684.6 = 199.8	11	18.16	
Total	1,404.4	1,174.6	1,832.4	$\frac{1,174.6}{1,832.4}$ = 0.6410	(0.6410)(1,174.6) = 752.9	1,404.4 − 752.9 = 651.5	13		

2.1.2. Application to the Rust Arrestor Problem. Let us now apply the previous theoretical discussion to the rust arrestor data of Table 34.1. For these data we have

$$K_y = \frac{(\Sigma\Sigma Y_{ij})^2}{n} = \frac{(1{,}002)^2}{15} = 66{,}933.6$$

$$K_x = \frac{(\Sigma\Sigma X_{ij})^2}{n} = \frac{(498)^2}{15} = 16{,}533.6$$

$$K_{xy} = \frac{(\Sigma\Sigma X_{ij})(\Sigma\Sigma Y_{ij})}{n} = \frac{(498)(1{,}002)}{15} = 33.266.4$$

$$C_{yy} = \frac{\sum_j C_{yj}^2}{r} - K_y = \frac{344^2 + 364^2 + 294^2}{5} - 66.933.6 = 520.0$$

$$C_{xx} = \frac{\sum_j C_{xj}^2}{r} - K_x = \frac{148^2 + 186^2 + 164^2}{5} - 16{,}533.6 = 145.6$$

$$C_{xy} = \frac{\sum_j C_{xj}C_{yj}}{r} - K_{xy} = \frac{(344)(148) + (364)(186) + (294)(164)}{5}$$

$$-33{,}266.4 = 100.0$$

$$S_{yy} = 63^2 + 77^2 + 81^2 + 60^2 + 63^2 + 81^2 + 73^2 + 59^2 + 74^2$$
$$+ 77^2 + 72^2 + 54^2 + 57^2 + 59^2 + 52^2 - 66{,}933.6 = 1{,}404.4$$

$$S_{xx} = 16^2 + 45^2 + 50^2 + 19^2 + 18^2 + 48^2 + 40^2 + 24^2 + 33^2 + 41^2$$
$$+ 40^2 + 31^2 + 40^2 + 33^2 + 20^2 - 16.533.6 = 1{,}832.4$$

$$S_{xy} = (63)(16) + (77)(45) + (81)(50) + (60)(19) + (63)(18) + (81)(48)$$
$$+ (73)(40) + (59)(24) + (74)(33) + (77)(41) + (72)(40) + (54)(31)$$
$$+ (57)(40) + (59)(33) + (52)(20) - 33{,}266.4 = 1{,}174.6$$

These results are entered in an analysis of covariance table (Table 34.4) and the remaining quantities $_CE_{yy}$, $_CE_{xy}$, $_CE_{xx}$, etc., are computed. The final result is an F ratio of 12.4 which is far beyond the $F_{0.05}$ value for $\nu_1 = 2$, $\nu_2 = 11$, viz approximately 4.00. We therefore conclude that there is a significant difference between the methods of applying the rust arrestor as to the amount of rust added during the exposure to a salt spray after adjustment has been made for initial rust.

We also have

$$\hat{\mu} = \frac{\Sigma\Sigma Y_{ij}}{n} = 66.80$$

$$\hat{\beta} = \frac{{}_C E_{xy}}{{}_C E_{xx}} = 0.6371$$

$$\hat{\sigma} = s_e = \sqrt{18.16} = 4.261$$

Estimated adjusted mean rust added by *brushing* =

$$\hat{\mu} + \hat{\theta}_B = \bar{Y}_B - \hat{\beta}(\bar{X}_B - \bar{\bar{X}})$$

$$= 68.80 - 0.6371(29.6 - 33.2) = 71.09$$

with 0.95 confidence limits for $\mu + \theta_B$ equal to

$$71.09 \pm 2.201(4.261)\sqrt{\frac{1}{5} + \frac{(-3.6)^2}{1{,}686.8}} = 71.09 \pm 4.27 = \begin{cases} 75.36 \\ 66.83 \end{cases}$$

Estimated adjusted mean rust added by *spraying* =

$$\hat{\mu} + \hat{\theta}_S = \bar{Y}_S - \hat{\beta}(\bar{X}_S - \bar{\bar{X}})$$

$$= 72.8 - \ 0.6371(37.2 - 33.2) = 70.25$$

with 0.95 confidence limits for $\mu + \theta_S$ equal to

$$70.25 \pm 2.201(4.261)\sqrt{\frac{1}{5} + \frac{(4)^2}{1{,}686.8}} = 72.25 \pm 4.29 = \begin{cases} 77.5 \\ 68.0 \end{cases}$$

Estimated adjusted mean rust added by *dipping* =

$$\mu + \hat{\theta}_D = \bar{Y}_D - \hat{\beta}(\bar{X}_D - \bar{X})$$

$$= 58.8 - 0.6371(32.8 - 33.2) = 59.05$$

with 0.95 confidence limits for $\mu + \theta_D$ equal to

$$59.05 \pm 2.201(4.261)\sqrt{\frac{1}{5} + \frac{(-0.4)^2}{1{,}686.8}} = 59.05 \pm 4.19 = \begin{cases} 63.24 \\ 54.86 \end{cases}$$

2.2. Two-Way Classification

Let us consider now adjustment of data that are grouped into a two-way classification. Consider the data of Table 34.5. This lists the skein strength of carded 22*s* cotton yarn and its fiber length classified by variety of cotton and state of origin. Suppose that we would like to know whether the character of this cotton varies significantly from variety to variety and state to state with respect to the qualities that affect skein strength *other than the lengths of the fibers.* We know that skein strength is affected by the length of the fibers making up the cotton yarn[10] and that this varies greatly from one piece of yarn to another made from the same variety of cotton from the same state. In our study fiber length will thus be a "nuisance variable."

[10] Cf. Chapter 32, Section 5.

TABLE 34.5

Data on Skein Strength and Fiber Length of Five Different Varieties of Cotton from Different States, Crop of 1944* (Y = skein strength in pounds of carded 22*s* cotton yarn, X = mean fiber length in 0.01 inches)

Variety	Area									Totals		Means	
	Georgia			Mississippi			Texas						
	Test No.	*Y*	*X*	*Test No.*	*Y*	*X*	*Test No.*	*Y*	*X*	*Y*	*X*	*Y*	*X*
Coker 4 in 1–6	1	100.	74	6	103	92	11	102	83	305	249	101.7	83.0
Empire A	2	94	77	7	95	81	12	99	68	288	226	96.0	75.3
Deltapine 14 (833)	3	93	80	8	103	82	13	94	62	290	224	96.7	74.7
Stoneville 2B (8275)	4	106	72	9	105	79	14	93	65	304	216	101.3	72.0
Coker 100–8	5	88	70	10	92	82	15	106	75	286	227	95.3	75.7
Totals		481	373		498	416		494	365	1,473	1,142		
Means		96.2	74.6		99.6	83.2		98.8	70.6			98.2	76.1

* See footnotes to Table 32.3.

The covariance analysis will test for differences between varieties and between states when such differences have been adjusted for differences in fiber length.

2.2.1. Theory. The covariance adjustment model for a two-way classification is

$$Y_{ij} = \mu + \tau_i + \theta_j + \beta(X_{ij} - \bar{\bar{X}}) + \epsilon_{ij} \begin{cases} i = 1, 2, \ldots, r \\ j = 1, 2, \ldots, c \end{cases} \tag{34.9}$$

where μ is a constant; τ_i is an adjusted "row differential" that sums to zero over the rows, i.e., $\sum_i \tau_i = 0$; θ_j is an adjusted "column differential" that sums to zero over the columns, i.e., $\sum_j \theta_j = 0$; β is the common slope of the regression of Y on X; and ϵ_{ij} is a random variable that is normally distributed with zero mean and variance σ'^2.

If this equation is fitted to a set of sample data by the method of least squares, we get[11]

$$\begin{aligned} \hat{\mu} &= \sum_i \sum_j Y_{ij}/n = \bar{\bar{Y}} \\ \hat{\tau}_i &= \frac{\sum_j Y_{ij}}{c} - \hat{\mu} - \hat{\beta}(\bar{X}_i - \bar{\bar{X}}) = \bar{Y}_i - \hat{\mu} - \hat{\beta}(\bar{X}_i - \bar{\bar{X}}) \\ \hat{\theta}_j &= \frac{\sum_i Y_{ij}}{r} - \hat{\mu} - \hat{\beta}(\bar{X}_j - \bar{\bar{X}}) = \bar{Y}_j - \hat{\mu} - \hat{\beta}(\bar{X}_j - \bar{\bar{X}}) \\ \hat{\beta} &= \frac{{}_{RC}E_{xy}}{{}_{RC}E_{xx}} \end{aligned} \tag{34.10}$$

where ${}_{RC}E_{xx}$, ${}_{RC}E_{yy}$, and ${}_{RC}E_{xy}$ are derived as indicated in Table 34.2.

If we assume in our theory that sampling is with respect to a fixed set of X_{ij}'s, viz. those listed in Table 34.5, and that an equation like (34.9) is fitted to each sample by the method of least squares, we can proceed to test for the significance of row and column effects in a manner similar to that followed in Section 2.1. We proceed to compute three residual sums of squares. These are

1. Σu_{RC}^2, which is the residual sum of squares obtained when the full equation (34.9) is fitted;
2. Σu_R^2, which is the residual sum of squares that would be obtained if an equation like (34.9) were fitted *omitting* the θ_j terms; and
3. Σu_C^2, which is the residual sum of squares that would be obtained if an equation like (34.9) were fitted *omitting* the τ_i terms.

It can be shown that these qualities are given by the following:[12]

---

[11] Cf. Appendix I (42).

[12] Cf. Appendix I (42).

$$\sum u_{RC}^2 = {}_{RC}E_{yy} - {}_{RC}E_{xy}^2/{}_{RC}E_{xx}$$

(34.11)
$$\sum u_R^2 = {}_{R}E_{yy} - {}_{R}E_{xy}^2/{}_{R}E_{xx}$$

$$\sum u_C^2 = {}_{C}E_{yy} - {}_{C}E_{xy}^2/{}_{C}E_{xx}$$

(The Σu_C^2 will be the same as that computed in Section 2.1.)

If now we take the difference between the third and first of the above residual sums of squares and the difference between the second and the first, we shall get, respectively, the "reduced sum of squares for adjusted row effects" and the "reduced sum of squares for adjusted column effects." Thus we have

$$\text{RSS for adjusted row effects} = \sum u_C^2 - \sum u_{RC}^2$$

(34.12)
$$\text{RSS for adjusted column effects} = \sum u_R^2 - \sum u_{RC}^2$$

These are the contributions of the adjusted row effects (τ_i) and adjusted column effects (θ_j) to the total variation in Y_{ij}; in other words, they represent the amount of the variation in Y_{ij} that is "explained,"[13] statistically speaking, by τ_i and θ_j.

We can, then, calculate three mean squares defined as follows;

$$s_\tau^2 = \text{mean square for adjusted row effects} = \frac{\Sigma u_C^2 - \Sigma u_{RC}^2}{r-1}$$

(34.13)
$$s_\theta^2 = \text{mean square for adjusted column effects} = \frac{\Sigma u_R^2 - \Sigma u_{RC}^2}{c-1}$$

$$s_e^2 = \text{residual or error mean square} = \sum u_{RC}^2/(n-r-c)$$

To test the hypothesis that the adjusted row effects (the τ_i) are all individually zero, we compute s_τ^2/s_e^2. If this ratio is greater than $F_{0.05}$ for $\nu_1 = r - 1$, $\nu_2 = n - r - c$, we reject; otherwise we accept. To test the hypothesis that the adjusted column effects (the θ_j) are all individually zero, we compute s_θ^2/s_e^2. If this ratio is greater than $F_{0.05}$ for $\nu_1 = c - 1$, $\nu_2 = n - r - c$, we reject; otherwise we accept. The whole procedure is summarized in Table 34.6. The 0.50 and 0.10 points of the OC curves for the tests will be given by Charts I and II of Appendix II if we take $\lambda = \sqrt{\frac{\Sigma \tau_i^2}{r\sigma'^2}}$ or $\lambda = \sqrt{\frac{\Sigma \theta_j^2}{c\sigma'^2}}$ and $\phi = \lambda\sqrt{r}$ or $\phi = \lambda\sqrt{c}$, respectively.[14]

We can estimate the various class means as follows. An unbiased estimate of an adjusted row mean is

(34.14a)
$$\text{Estimate of } \mu + \tau_i = \bar{Y}_i - \hat{\beta}(\bar{X}_i - \bar{\bar{X}})$$

---

[13] Cf. Section 2.1.1 above and Section 1.5.4 of Chapter 33.

[14] Cf. Chapter 29, Section 2.5.

TABLE 34.6
Analysis of Covariance Table, Two-Way Classification

Source of Variation (1)	Sums of Squares for Y (2)	Sums of Cross-Products for X and Y (3)	Sums of Squares for X (4)	Regression Coefficient (5)	Amount Explained by Regression (6)	Adjusted Sum of Squares for Y (7)	df (8)	Mean Square (9)	F Ratio (10)
1. Between rows	R_{yy}	R_{xy}	R_{xx}			$RSS_R = \Sigma u_C^2 - \Sigma u_{RC}^2$	$r-1$	$s_\tau^2 = \frac{(7)}{(8)}$	s_τ^2/s_e^2
2. Between columns	C_{yy}	C_{xy}	C_{xx}			$RSS_C = \Sigma u_R^2 - \Sigma u_{RC}^2$	$c-1$	$s_\theta^2 = \frac{(7)}{(8)}$	s_θ^2/s_e^2
3. Residual or error (when model contains only row effects)	${}_RE_{yy}$	${}_RE_{xy}$	${}_RE_{xx}$	$\hat{\beta}_r = \frac{{}_RE_{xy}}{{}_RE_{xx}}$	$\frac{{}_RE_{xy}^2}{{}_RE_{xx}}$	$\Sigma u_R^2 = {}_RE_{yy} - \frac{{}_RE_{xy}^2}{{}_RE_{xx}}$			
4. Residual or error (when model contains only column effects)	${}_CE_{yy}$	${}_CE_{xy}$	${}_CE_{xx}$	$\hat{\beta}_C = \frac{{}_CE_{xy}}{{}_CE_{xx}}$	$\frac{{}_CE_{xy}^2}{{}_CE_{xx}}$	$\Sigma u_C^2 = {}_CE_{yy} - \frac{{}_CE_{xy}^2}{{}_CE_{xx}}$			
5. Residual or error (when model contains both row and column effects)	${}_{RC}E_{yy}$	${}_{RC}E_{xy}$	${}_{RC}E_{xx}$	$\hat{\beta}_{RC} = \frac{{}_{RC}E_{xy}}{{}_{RC}E_{xx}}$	$\frac{{}_{RC}E_{xy}^2}{{}_{RC}E_{xx}}$	$\Sigma u_{RC}^2 = {}_{RC}E_{yy} - \frac{{}_{RC}E_{xy}^2}{{}_{RC}E_{xx}}$	$n-r-c$	$s_e^2 = \frac{(7)}{(8)}$	
6. Total	S_{yy}	S_{xy}	S_{xx}	$\hat{\beta}_0 = \frac{S_{xy}}{S_{xx}}$	$\frac{S_{xy}^2}{S_{xx}}$	$\Sigma u_0^2 = S_{yy} - \frac{S_{xy}^2}{S_{xx}}$	$n-2$		

and an unbiased estimate of an adjusted column mean is

(34.14b) $$\text{Estimate of } \mu + \theta_j = \bar{Y}_j - \hat{\beta}(\bar{X}_j - \bar{\bar{X}})$$

Ninety-five percent confidence limits for an adjusted row mean will be given by

(34.15a) $$\bar{Y}_i - \hat{\beta}(\bar{X}_i - \bar{\bar{X}}) \pm t_{0.025} s_e \sqrt{\frac{1}{c} + \frac{(\bar{X}_i - \bar{\bar{X}})^2}{{}_{RC}E_{xx}}}$$

and 0.95 confidence limits for an adjusted column mean will be given by

(34.15b) $$\bar{Y}_j - \hat{\beta}(\bar{X}_j - \bar{\bar{X}}) \pm t_{0.025} s_e \sqrt{\frac{1}{r} + \frac{(\bar{X}_j - \bar{\bar{X}})^2}{{}_{RC}E_{xx}}}$$

where $t_{0.025}$ in both cases is taken for $n - r - c$ degrees of freedom.

For the difference between any two-row or any two-column means, 0.95 confidence limits (on a per comparison basis)[15] will be given by

(34.16) $$(\bar{Y}_1 - \bar{Y}_2) - \hat{\beta}(\bar{X}_1 - \bar{X}_2) \pm t_{0.025} s_e \sqrt{\frac{2}{m} + \frac{(\bar{X}_1 - \bar{X}_2)^2}{{}_{RC}E_{xx}}}$$

where m is the number of elements in the row or column, depending on which is involved.

Ninety-five percent confidence limits for β will be given by

(34.17) $$\hat{\beta} \pm t_{0.025} \frac{s_e}{\sqrt{{}_{RC}E_{xx}}}$$

where $t_{0.025}$ is taken for $n - r - c$ degrees of freedom.

2.2.2. Application to a Cotton Quality Problem. Let us illustrate the analysis by reference to the data of Table 34.5.

We put to these data the following questions: (1) If we adjust for differences in fiber length, is there a significant difference between the skein strengths of the cotton grown in the three states? (2) If we adjust for differences in fiber length, is there a significant difference in the skein strengths of the five varieties of cotton? The answers are worked out in Table 34.7, which fiber length, is there a significant difference in the skein strengths of the five varieties of cotton? The answers are worked out in Table 34.7, which is the application of Table 34.6 to the data of Table 34.5.

The procedure is to work from the left of the table to the right, filling in rows 3, 4, 5 of column (7). The formulas for computing the R_{yy}, R_{xy}, and so on, are given in Table 34.2. After we have calculated Σu^2_{RC}, Σu^2_R and Σu^2_C, we compute rows (1) and (2) of column (7) by subtraction. Finally we compute the three mean squares and the two F ratios desired. Since the F ratios, s^2_τ/s^2_e and s^2_θ/s^2_e, are both less than one, we conclude immediately that when adjustment has been made for fiber length, neither differences in

[15] Cf. Chapter 31, Section 1.1.3.

TABLE 34.7
Analysis of Covariance Table for Data of Table 34.5

Source of Variation (1)	Sum of Squares for Y (2)	Sum of Cross-Products for X and Y (3)	Sum of Squares for X (4)	Regression Coefficient (5)	Amount Explained by Regression (6)	Adjusted Sum of Squares for Y (7)	df (8)	Mean Square (9)	F Ratio (10)
1. Between rows	$R_{yy} =$ 111.7	$R_{xy} =$ 48.6	$R_{xx} =$ 201.7			$RSS_r = 391.3 - 287.9 = 103.4$	4	$s_\tau^2 =$ 25.85	0.63
2. Between columns	$C_{yy} =$ 31.6	$C_{xy} =$ 50.2	$C_{xx} =$ 414.5			$RRS_\theta = 327.0 - 287.9 = 39.1$	2	$s_\theta^2 =$ 19.55	0.48
3. Residual or error (when model contains only row effects)	${}_RE_{yy} =$ 362.7	${}_RE_{xy} =$ 154.0	${}_RE_{xx} =$ 664.0	$\hat{\beta}_R = \frac{154.0}{664.0} = 0.2319$	$(0.2319)(154.0) = 35.71$	$\Sigma u_R^2 = 362.7 - 35.71 = 327.0$			
4. Residual or error (when model contains only column effects	${}_CE_{yy} =$ 442.8	${}_CE_{xy} =$ 152.4	${}_CE_{xx} =$ 451.2	$\hat{\beta}_C = \frac{152.4}{451.2} = 0.3378$	$(0.3378)(152.4) = 51.48$	$\Sigma u_C^2 = 442.8 - 51.48 = 391.3$			
5. Residual or error (when model contains both row and column effects)	${}_{RC}E_{yy} =$ 331.1	${}_{RC}E_{xy} =$ 103.8	${}_{RC}E_{xx} =$ 249.5	$\hat{\beta}_{RC} = \frac{103.8}{249.5} = 0.4160$	$(0.4160)(103.8) = 43.18$	$\Sigma u_{RC}^2 = 331.1 - 43.18 = 287.9$	7	$s_e^2 =$ 41.13	
6. Total	$S_{yy} =$ 474.4	$S_{xy} =$ 202.6	$S_{xx} =$ 865.7	$\hat{\beta}_0 = \frac{202.6}{856.7} = 0.2340$	$(0.2340)(202.6) = 47.41$	$\Sigma u_0^2 = 474.4 - 47.41$ 427.0	13		

variety nor differences in geographical origin lead to differences in skein strength of the cotton.

We also have the following:

The slope of the regression of skein strength on fiber length is estimated at

$$\hat{\beta}_{RC} = \frac{103.8}{249.5} = 0.4161$$

and 0.95 confidence limits for β are

$$\hat{\beta}_{RC} \pm t_{0.025} \frac{s_e}{\sqrt{_{RC}E_{xx}}} = 0.4161 \pm 2.365 \frac{(6.413)}{\sqrt{249.5}}$$

$$= 0.4161 \pm 0.9599 = \begin{cases} 1.3760 \\ -0.5438 \end{cases}$$

The latter suggests that when allowance is made for possible differences between varieties and states the regression of skein strength on fiber length may after all be zero.

The adjusted mean skein strength of Deltapine cotton is estimated at

$$\hat{\bar{Y}}_{\text{Deltapine}} = \bar{Y}_{\text{Del.}} - \hat{\beta}(\bar{X}_{\text{Del.}} - \bar{\bar{X}}) = 96.7 - 0.4161(74.7 - 76.1) = 97.3$$

and 0.95 confidence limits for the universe mean skein stength of Deltapine cotton, adjusted for fiber length, are

$$97.3 \pm 2.365(6.413)\sqrt{\frac{1}{3} + \frac{(74.7 - 76.1)^2}{249.47}} = 97.3 \pm 8.9 = \begin{cases} 106.2 \\ 88.4 \end{cases}$$

The adjusted mean skein strength of other varieties of cotton and of cotton from any of the states can be computed in the same way. Since the cotton showed no significant difference in skein strength between varieties or between states, after adjustment was made for differences in fiber length, these other adjusted means were not computed.

3. COMPARISON OF SLOPES OF LINES OF REGRESSION

As indicated above, a common problem that is solved by an analysis of covariance is the comparison of the slopes of lines of regression. The simplest of these problems is the comparison of the slopes of two lines of regression. This will be discussed in Section 3.1. The comparison of the slopes of several lines of regression will be discussed in Section 3.2.

3.1. Comparison of Two Slopes

Although the comparison of the slopes of two lines of regression is a special case of the comparison of the slopes of several lines of regression, it can also be handled by methods of analysis developed in earlier chapters. This special way of carrying out the analysis is employed here for the sake of those who are primarily interested in the two-slope problem and who do not particularly want to delve into the broader aspects of analysis of covariance. For the student it may serve as a stepping stone to the more general analysis.

3.1.1. Theory. Suppose that we have two separate bivariate universes. In each, the mean value of Y is a linear function of X and the deviations from this mean line are normally distributed with a mean of zero and a standard deviation ($\sigma'_{y \cdot x}$) that is called the standard error of estimate. Thus our model is

Universe I	*Universe II*
$Y_1 = a'_1 + b'_1X_1 + \epsilon_1$ where ϵ_1 has a normal distribution with zero mean and standard deviation $\sigma'_{y_1 \cdot x_1}$	$Y_2 = a'_2 + b'_2X_2 + \epsilon_2$ where ϵ_2 has a normal distribution with zero mean and standard deviation $\sigma'_{y_2 \cdot x_2}$

(Note that it is more convenient to use Y and X here instead of the X_1 and X_2 that were employed in Chapter 32.)

Let us take a random sample of n_1 from Universe I and another independent random sample of n_2 from Universe II and let us fit least-squares lines of regression to each sample. Then let us undertake to test the hypothesis that $b'_1 = b'_2$ (we make no assumptions regarding a'_1 and a'_2). To do this we need to note the following.

When dealing with a single universe in Chapter 32, it was indicated[16] that if we restrict outselves to a conditional type of sampling in which the X's are fixed at the same values for all samples and the only variation is in the ϵ's, the variance from sample to sample in the regression coefficients fitted to these samples by least squares would be given by $\sigma'^2_{y \cdot x}/\Sigma x^2$. With only a single sample available, $\sigma'^2_{y \cdot x}/\Sigma x^2$ is estimated as equal to $s^2_{y \cdot x}/\Sigma x^2$. A common way of putting this last is to say that under sampling with fixed X's, the estimated "standard error" of b is $s_{y \cdot x}/\sqrt{\Sigma x^2}$. In Chapter 32, Section 7.1, $s_{y \cdot x}$ is defined as equal to

$$s^2_{y \cdot x} = \frac{\Sigma u^2}{n - 2} \tag{34.18}$$

where Σu^2 is the sum of the squared deviations (in the Y direction) from the sample line of regression.

Now if we wish to compare two independently derived sample regression coefficients, we can make use of the relationship that the sampling variance (square of the standard error) of the difference between two regression coefficients is the sum of their two separate sampling variances.[17] In other words,

---

[16] Chapter 32, Section 7.1. Note that $\Sigma x^2 = (n - 1)s^2_x$.

[17] Compare Chapter 26, Section 2.1.

if b_1 is the slope of the first sample regression and b_2 is the slope of the second sample regression, the sampling variance of the difference between two such sample regression coefficients is

(34.19)
$$\sigma'^2_{b_1-b_2} = \frac{\sigma'^2_{y_1 \cdot x_1}}{\Sigma x_1^2} + \frac{\sigma'^2_{y_2 \cdot x_2}}{\Sigma x_2^2}$$

and the standard error of the difference is the square root of this.

If the standard errors of estimate[18] for the two universes are the same, i.e., $\sigma'_{y_1 \cdot x_1} = \sigma'_{y_2 \cdot x_2} = \sigma'_{y \cdot x}$, the above formula reduces to

(34.20)
$$\sigma'^2_{b_1-b_2} = \sigma'^2_{y \cdot x}\left[\frac{1}{\Sigma x_1^2} + \frac{1}{\Sigma x_2^2}\right]$$

In this case, we can estimate the common standard error of estimate by pooling the sums of the standard deviations from the two sample lines of regression. Thus we can take

(34.21)
$$s^2_{y \cdot x} = \frac{\Sigma(Y_1 - Y_{1r})^2 + \Sigma(Y_2 - Y_{2r})^2}{n_1 - 2 + n_2 - 2} = \frac{\Sigma u_1^2 + \Sigma u_2^2}{n_1 - 2 + n_2 - 2}$$

where in each case Σu^2 is computed from the formula

(34.22)
$$\sum u^2 = \sum Y^2 - na - b \sum XY = \sum y^2 - b \sum xy = \sum y^2 - \frac{(\Sigma xy)^2}{\Sigma x^2}$$

in which $x = X - \bar{X}$ and $y = Y - \bar{Y}$.

In the event then that we know that the standard errors of estimate are the same for the universe being compared, we can test the hypothesis $b'_1 = b'_2$ by computing the statistic:

(34.23)
$$t = \frac{b_1 - b_2}{s_{b_1-b_2}} = \frac{b_1 - b_2}{s_{y \cdot x}\sqrt{\dfrac{1}{\Sigma x_1^2} + \dfrac{1}{\Sigma x_2^2}}}$$

and noting whether it exceeds $\pm t_{0.025}$ for $\nu = n_1 - 2 + n_2 - 2$. (This assumes a two-tail test with $\alpha = 0.05$.) OC curves for the test with a $\alpha = 0.05$ will be given by Figure 15.9 (one-tail) and Figure 25.6 (two-tail) if we take

[18] The reader should note that "standard error" alone refers to a sampling error, but "standard error of estimate" pertains to the variation in the dependent variable not accounted for by the independent variable.

(34.24)
$$\lambda = \frac{b_1' - b_2'}{\sigma_{y\cdot x}' \sqrt{n\left(\frac{1}{\Sigma x_1^2} + \frac{1}{\Sigma x_2^2}\right)}}$$

where $n = n_1 + n_2 - 3$.

Ninety-five percent confidence limits for the difference between two regression coefficients will be given by

(34.25)
$$b_1 - b_2 \pm t_{0.025} s_{y\cdot x} \sqrt{\frac{1}{\Sigma x_1^2} + \frac{1}{\Sigma x_2^2}}$$

It will be noted that the procedure described above is similar to that used in comparing two sample means when universe variances were unknown, but were known to be equal. If the universe standard errors of estimate were known, we would have used, as in the comparison of two means, a normal distributed statistic, viz

(34.23a)
$$z = \frac{b_1 - b_2}{\sqrt{\frac{\sigma_{y_1\cdot x_1}'^2}{\Sigma x_1^2} + \frac{\sigma_{y_2\cdot x_2}'^2}{\Sigma x_2^2}}}$$

If the universe standard errors of estimate are unknown and not known to be equal, we can use procedures similar to the Aspin-Welch procedures for comparing two means with unknown universe standard deviations.[19] In this case the statistic we would use would be

(34.23b)
$$t = \frac{b_1 - b_2}{\sqrt{\frac{s_{y_1\cdot x_1}^2}{\Sigma x_1^2} + \frac{s_{y_2\cdot x_2}^2}{\Sigma x_2^2}}}$$

and this would be compared with $\pm t_{0.025}$ (for a two-tail test) with degrees of freedom given by

(34.26)
$$\text{df} = \frac{1}{\frac{c^2}{n_1 - 2} + \frac{(1-c)^2}{n_2 - 2}}$$

where

(34.27)
$$c = \frac{s_{y_1\cdot x_1}^2 / \Sigma x_1^2}{\frac{s_{y_1\cdot x_1}^2}{\Sigma x_1^2} + \frac{s_{y_2\cdot x_2}^2}{\Sigma x_2^2}}$$

3.1.2. Application to a Welding Problem. Let us consider the data of Table 34.8 pertaining to the relationship between shear strength of a spot

[19] Cf. Chapter 26, Section 2.3.

TABLE 34.8
Shear Strength and Weld Diameter of Spot Welds of Two Gauges of Steel

0.040"—0.040"			0.064"—0.064"		
Test No.	*Shear Strength (Y_1) (Lbs.)*	*Weld Diameter (X_1) (0.001 In.)*	*Test No.*	*Shear Strength (Y_2) (Lbs.)*	*Weld Diameter (X_2) (0.001 In.)*
1	350	140	11	680	190
2	380	155	12	800	200
3	385	160	13	780	209
4	450	165	14	885	215
5	465	175	15	975	215
6	485	165	16	1,025	215
7	535	195	17	1,100	230
8	555	185	18	1,030	250
9	590	195	19	1,175	265
10	605	210	20	1,300	250

FIGURE 34.2
Regression of Shear Strength on Weld Diameter for Two Gauges of Steel Welds (data from Table 34.8)

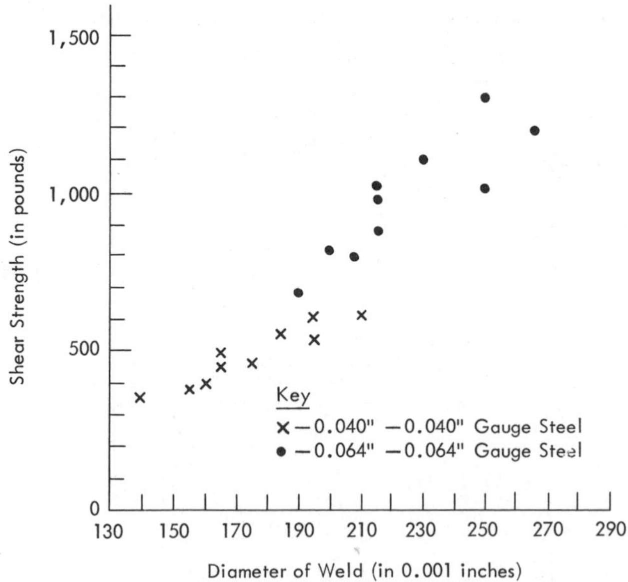

weld and its diameter for steel of two different gauges. The data are shown graphically in Figure 34.2.

Taking Y_1 to be the shear strength of the 0.040"–0.040" steel weld and X_1 its diameter and taking Y_2 to be the shear strength of the 0.064"–0.064" steel weld and X_2 its diameter, we have the following:

$$n_1 = 10 \qquad n_2 = 10$$

$$\sum Y_1 = 4{,}800 \qquad \sum Y_2 = 9{,}750$$

$$\sum X_1 = 1{,}745 \qquad \sum X_2 = 2{,}239$$

$$\sum X_1 Y_1 = 854{,}250 \qquad \sum X_2 Y_2 = 2{,}219{,}370$$

$$\sum Y_1^2 = 2{,}377{,}450 \qquad \sum Y_2^2 = 9{,}836{,}860$$

$$\sum X_1^2 = 308{,}675 \qquad \sum X_2^2 = 506{,}581$$

From these we compute:

$$\sum x_1 y_1 = \sum X_1 Y_1 - \frac{(\Sigma X_1)(\Sigma Y_1)}{n_1}$$

$$= 854{,}250 - \frac{(1{,}745)(4{,}800)}{10} = 16{,}650$$

$$\sum x_1^2 = \sum X_1^2 - \frac{(\Sigma X_1)^2}{n_1}$$

$$= 308{,}675 - \frac{(1{,}745)^2}{10} = 4{,}172.5$$

$$\sum y_1^2 = \sum Y_1^2 - \frac{(\Sigma Y_1)^2}{n_1}$$

$$= 2{,}377{,}450 - \frac{(4{,}800)^2}{10} = 73{,}450$$

$$b_1 = \frac{\Sigma x_1 y_1}{\Sigma x_1^2} = \frac{16{,}650}{4{,}172.5} = 3.9904$$

$$\sum u_1^2 = \sum y_1^2 - b_1 \sum x_1 y_1$$

$$= 73{,}450 - (3.9904)(16{,}650) = 7{,}009.8$$

$$\sum x_2 y_2 = \sum X_2 Y_2 - \frac{(\Sigma X_2)(\Sigma Y_2)}{n_2}$$

$$= 2{,}219{,}370 - \frac{(2{,}239)(9{,}750)}{10} = 36{,}345$$

$$\sum x_2^2 = \sum X_2^2 - \frac{(\Sigma X_2)^2}{n_2}$$

$$= 506{,}581 - \frac{(2{,}239)^2}{10} = 5{,}268.9$$

$$\sum y_2^2 = \sum Y_2^2 - \frac{(\Sigma Y)^2}{n_2}$$

$$= 9{,}936{,}800 - \frac{(9{,}750)^2}{10} = 330{,}550$$

$$b_2 = \frac{\Sigma x_2 y_2}{\Sigma x_2^2} = \frac{36{,}345}{5{,}268.9} = 6.8980$$

$$\sum u_2^2 = \sum y_2^2 - b_2 \sum x_2 y_2$$

$$= 330{,}550 - (6.8980)(36{,}345) = 79{,}842$$

$$s_{y \cdot x}^2 = \frac{\Sigma u_1^2 + \Sigma u_2^2}{n_1 + n_2 - 4}$$

$$= \frac{7{,}009.8 + 79{,}842}{10 + 10 - 4} = 5{,}428.2$$

$$s_{y \cdot x} = 73.676$$

$$t = \frac{b_1 - b_2}{s_{y \cdot x}\sqrt{\dfrac{1}{\Sigma x_1^2} + \dfrac{1}{\Sigma x_2^2}}}$$

$$= \frac{3.9904 - 6.8980}{73.676\sqrt{\dfrac{1}{4{,}172.5} + \dfrac{1}{5{,}268.9}}} = -1.90$$

With $\nu = n_1 + n_2 - 4 = 10 + 10 - 4 = 16$ the critical t's for $\alpha = 0.05$ and a two-tail test are ± 2.120. Since the sample t lies between these critical values, the hypothesis of equal slopes is accepted.

In view of the appearance of Figure 34.2, it seems difficult to agree with the conclusion that the slopes do not differ significantly. The explanation would appear to lie in the small size of the samples employed and the relatively large $\sigma'_{y \cdot x}$. Suppose, for example, that $\sigma'_{y \cdot x}$ actually equals 70 and we have, as in the given problem, $\Sigma x_1^2 = 4{,}173$ and $\Sigma x_2^2 = 5{,}269$. Suppose further that we are interested in detecting a difference between b'_1 and b'_2 of 3 points. Then with $n_1 = n_2 = 10$, we shall have [from equation (34.24)]

$$\lambda = \frac{3}{70\sqrt{17\left(\dfrac{1}{4{,}173} + \dfrac{1}{5{,}269}\right)}} = 0.502,$$

and Figure 25.6 shows that with $n = n_1 + n_2 - 3 = 17$ there is a chance of only 0.50 of detecting the difference.

It is also to be noted that $s_{y_2 \cdot x_2}^2$ is over 11 times $s_{y_1 \cdot x_1}^2$ which is far beyong the $F_{0.05}$ value marking a significant difference. This suggests that we should not have assumed $\sigma'_{y_1 \cdot x_1} = \sigma'_{y_2 \cdot x_2}$, but should have used Aspin-

Welch procedures. This, however, would have reduced the degrees of freedom even further, and the critical t would have been even greater than 2.120.

3.2. Comparison of Several Slopes

The comparison of the slopes of several lines of regression can best be handled as part of a general analysis of covariance. Consider the data of Table 34.9. Here are listed measurements of printability of newsprint and associated measures of roughness of five rolls of newsprint from each of four newsprint suppliers. The data are graphed in Figure 34.3. In general there seems to be a negative relationship between roughness of the newsprint and its printability. Let us consider the question of whether the slopes of these individual regressions can reasonably be considered to be the same for all suppliers.

3.2.1. Theory. The general model to be used in comparing several slopes is the following:

$$Y_{ij} = \mu + \theta_j + \beta_m(\bar{X}_j - \bar{\bar{X}}) + \beta_a(X_{ij} - \bar{X}_j) + \delta_j(X_{ij} - \bar{X}_j) + \epsilon_{ij} \tag{34.28}$$

Here μ is a constant; θ_j is a differential pertinent to the jth class (newsprint supplier) adjusted for values of X (roughness); β_m is the slope of the regression of Y class means on X class means (mean supplier printability on mean supplier roughness); β_a is the slope of the regression of Y on X *within* classes that is common to all classes (the regression of printability on roughness

TABLE 34.9
Data on Printability (Y) and Roughness (X) of Five Rolls of Newsprint from Each of Four Suppliers*

Supplier I			*Supplier II*			*Supplier III*			*Supplier IV*		
Test No.	*Y*	*X*	*Test No.*	*Y*	*X*	*Test No.*	*Y*	*X*	*Test No.*	*Y*	*X*
1	66.1	123.5	6	62.7	153.3	11	75.9	46.5	16	65.9	86.7
2	71.7	83.8	7	62.7	110.9	12	73.1	67.0	17	70.5	78.8
3	69.4	88.6	8	67.7	132.5	13	72.6	76.0	18	68.3	93.4
4	69.0	98.7	9	63.4	131.1	14	71.2	74.6	19	67.4	72.1
5	66.0	112.5	10	65.9	139.0	15	73.0	66.2	20	67.4	72.6
Σ	342.2	507.1	Σ	322.4	666.8	Σ	365.8	330.3	Σ	339.5	403.6
Mean	68.44	101.42		64.48	133.36		73.16	66.06		67.90	80.72
						Grand Total			$\Sigma\Sigma Y_{ij} =$ 1,369.9		$\Sigma\Sigma X_{ij} =$ 1,907.8
						Grand Mean			$\bar{\bar{Y}} = 68.50$		$\bar{\bar{X}} = 95.39$

* Printability is measured by the La Rocque printability test. This is described as follows: "A precision 6" × 7" plate is inked with a controlled quantity of ink and printed on a sample of newsprint using a printing pressure of 60 lb. per inch. The resulting print is expressed numerically through the use of the Hunter reflectometer by comparing printed and unprinted reflectance." (See *ANPA Research Bulletin,* No. 145, December 2, 1957, p. 5). Printability is measured in percentage. In the table, Y is the mean measure of printability for five different sheets from the same roll each of which is itself a mean of the printability of five different sections of the sheet.

Roughness (called smoothness in the *ANPA Research Bulletin)* is measured by the Bendtsen Smoothness and Porosity Tester which works on the principle of measuring the quantity of regulated air flow escaping between the head and the surface of the sample. Roughness is measured in these "Bendtsen units." In the table, X is the mean measure of roughness for ten sheets of newsprint from the same roll.

Both X and Y are measured on the "felt side" of the newsprint.

SOURCE: Kindly supplied to the author by the American Newspaper Publishers Association Research Institute, Inc.

FIGURE 34.3
Regression of Printability (*Y*) on Roughness (*X*) of Newsprint, Classified by Suppliers
(data from Table 34.9)

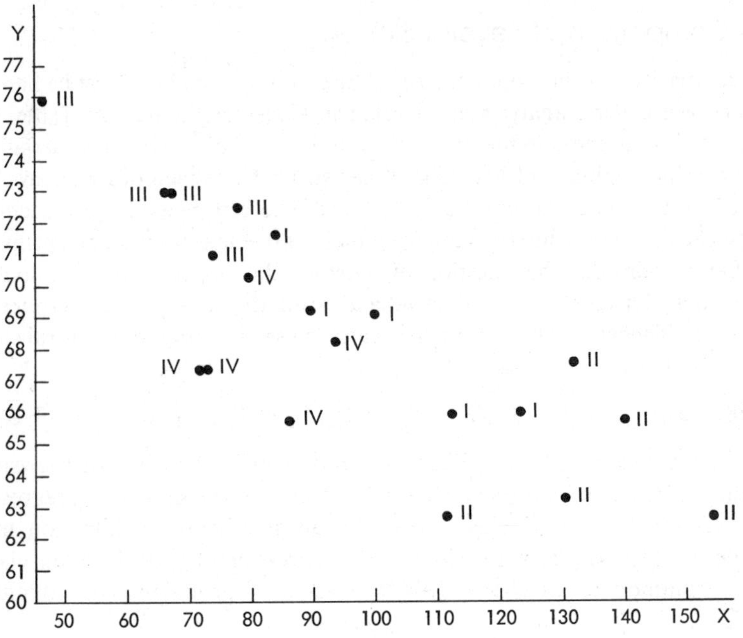

within supplier that is common to all suppliers); δ_j is a slope differential peculiar to class j (supplier j); and ϵ_{ij} is a random, normally distributed variable with zero mean and variance σ'^2, this variance being the same for all classes and all values of X. If equation (34.28) is fitted to a set of sample data by the method of least squares, unbiased estimates of the various parameters are given as follows:[20]

(34.29)

$$\hat{\mu} = \sum_i \sum_j Y_{ij}/n = \bar{\bar{Y}}$$

$$\hat{\theta}_j = \bar{Y}_j - \hat{\mu} - \hat{\beta}_m(\bar{X}_j - \bar{\bar{X}})$$

$$\hat{\beta}_m = \frac{C_{xy}}{C_{xx}}$$

$$\hat{\beta}_a = \frac{E_{xy}}{C_{xx}}$$

$$\hat{\delta}_j = \frac{E_{xyj}}{E_{xxj}} - \hat{\beta}_a = \frac{E_{xyj}}{E_{xxj}} - \frac{E_{xy}}{E_{xx}}$$

where C_{xx}, C_{xy}, E_{xx}, E_{xy}, E_{xxj}, and E_{xyj} are defined in Table 34.10.

[20] See Appendix I (34).

TABLE 34.10
Sums of Squares and Cross-Products for One-Way Analysis of Covariance, Different Slopes*

Source of Variation	*Sums of Squares for Y*	*Sums of Cross-Products for X and Y*	*Sums of Squares for X*
1. Differences between columns	$C_{yy} = r \sum_j (\bar{Y}_j - \bar{\bar{Y}})^2$ $= \sum_j \frac{C_{yj}^2}{r} - K_y$	$C_{xy} = r \sum_j (\bar{X}_j - \bar{\bar{X}})(\bar{Y}_j - \bar{\bar{Y}})$ $= \sum_j \frac{C_{xj} C_{yj}}{r} - K_{xy}$	$C_{xx} = r \sum_j (\bar{X}_j - \bar{\bar{X}})^2$ $= \sum_j \frac{C_{xj}^2}{r} - K_x$
2. Differences within column j	$E_{yyj} = \sum_i (Y_{ij} - \bar{Y}_j)^2$ $= \sum_i Y_{ij}^2 - K_{yj}$	$E_{xyj} = \sum_i (X_{ij} - \bar{X}_j)(Y_{ij} - \bar{Y}_j)$ $= \sum_i X_{ij} Y_{ij} - K_{xyj}$	$E_{xxj} = \sum_i (X_{ij} - \bar{X}_j)^2$ $= \sum_i X_{ij}^2 - K_{xj}$
3. Sum of differences within columns	$E_{yy} = \sum_j \sum_i (Y_{ij} - \bar{Y}_j)^2$ $= \sum_j E_{yyj}$	$E_{xy} = \sum_j \sum_i (X_{ij} - \bar{X}_j)(Y_{ij} - \bar{Y}_j)$ $= \sum_j E_{xyj}$	$E_{xx} = \sum_j \sum_i (X_{ij} - \bar{X}_j)^2$ $= \sum_j E_{xxj}$
4. Total	$S_{yy} = \sum_i \sum_j (Y_{ij} - \bar{\bar{Y}})^2$ $= \sum_i \sum_j Y_{ij}^2 - K_y$	$S_{xy} = \sum_i \sum_j (X_{ij} - \bar{\bar{X}})(Y_{ij} - \bar{\bar{Y}})$ $= \sum_i \sum_j X_{ij} Y_{ij} - K_{xy}$	$S_{xx} = \sum_i \sum_j (X_{ij} - \bar{\bar{X}})^2$ $= \sum_i \sum_j X_{ij}^2 - K_x$

where $K_x = \left(\sum_i \sum_j X_{ij}\right)^2/n$; $K_y = \left(\sum_i \sum_j Y_{ij}\right)^2/n$; $K_{xj} = \left(\sum_i X_{ij}\right)^2/r$; $K_{yj} = \left(\sum_i Y_{ij}\right)^2/r$; $K_{xy} = \left(\sum_i \sum_j X_{ij}\right)\left(\sum_i \sum_j Y_{ij}\right)/n$; $K_{xyj} = \left(\sum_i X_{ij}\right)\left(\sum_i Y_{ij}\right)/r$; $C_{xj} = \sum_i X_{ij}$; and $C_{yj} = \sum_i Y_{ij}$

* The first formula in each cell is the definition; the second a computational formula.

To test the hypothesis that the slopes of regressions *within* the various classes are all the same, i.e., that the δ_j are all individually zero, we need to compute two residual sums of squares. One is the residual sum of squares for equation (34.28). This is given by

(34.30) $$\sum u_e^2 = E_{yy} - \sum_j \frac{E_{xyj}^2}{E_{xxj}}$$

The other is the residual sum of squares for an equation that is the same as equation (34.28) *except that it does not contain the* δ_j *terms.* This is given by

(34.31) $$\sum u_{(\delta)}^2 = E_{yy} - \frac{E_{xy}^2}{E_{xx}}$$

The difference between these two residual sums of squares is the "reduced sum of squares" for the δ_j differentials. Thus the reduced sum of squares for δ_j is given by

(34.32) $$RSS_\delta = \sum u_{(\delta)}^2 - \sum u_e^2 = \sum_j \frac{E_{xyj}^2}{E_{xxj}} - \frac{E_{xy}^2}{E_{xx}}$$

Now if the δ_j are actually all zero, the quantity RSS_δ/σ'^2 will have a χ^2 distribution with $c - 1$ degrees of freedom and

(34.33) $$s_\delta^2 = \frac{RSS_\delta}{c - 1}$$

will yield an unbiased estimate of σ'^2. At the same time, independently of s_δ^2, the quantity $\Sigma u_e^2/\sigma'^2$ will have a χ^2 distribution with $c(r - 2)$ degrees of freedom and the quantity

(34.34) $$s_e^2 = \frac{\Sigma u_e^2}{c(r-2)}$$

will be an unbiased estimate of σ'^2. Hence, the ratio s_δ^2/s_e^2 will have an F distribution with $\nu_1 = c - 1$, and $\nu_2 = c(r - 2)$, so that the hypothesis that $\delta_j = 0$ for all j can be tested by computing s_δ^2/s_e^2 and comparing it with $F_{0.05}$ for $\nu_1 = c - 1$, $\nu_2 = c(r - 2)$.

With the comprehensive model represented by equation (34.28) we can do more, however, than merely test for a difference between the slopes of the j classes. It is shown in Appendix I (42) that the total sum of squares can be divided as follows:

(34.35) $$\sum_i \sum_j (Y_{ij} - \bar{Y})^2 = Q_e + Q_\delta + Q_\theta + Q_{m-a} + Q_0$$

where the Q's are components of the total sum of squares that represent variations the significance of which we might be interested in testing. Formulas for computing the Q's are given in Table 34.11, together with the nature

TABLE 34.11
Analysis of Covariance

Source of Variation	*Component of Total Sum of Squares*	*df*	*MS*
Deviations from regressions within groups	$Q_e = E_{yy} - \sum_j \frac{E_{xyj}^2}{E_{xxj}}$	$n - 2c$	$s_e^2 = Q_e/(n - 2c)$
Differences between regressions within groups	$Q_\delta = \sum_j \frac{E_{xyj}^2}{E_{xxj}} - \frac{E_{xy}^2}{E_{xx}}$	$c - 1$	$s_\delta^2 = Q_\delta/(c - 1)$
Deviations within classes from common regression β_a	$Q_a = Q_e + Q_\delta$	$n - c - 1$	$s_a^2 = Q_a/(n - c - 1)$
Deviations between groups from linear regressions β_m	$Q_\theta = C_{yy} - \frac{C_{xy}^2}{C_{xx}}$	$c - 2$	$s_\theta^2 = Q_\theta/(c - 2)$
Difference between β_a and β_m	$Q_{m-a} = \frac{C_{xy}^2}{C_{xx}} + \frac{E_{xy}^2}{E_{xx}} - \frac{S_{xy}^2}{S_{xx}}$	1	$s_{m-a}^2 = Q_{m-a}$
Common over-all regression β_0	$Q_0 = \frac{S_{xy}^2}{S_{xx}}$	1	$s_0^2 = Q_0$
Total	S_{yy}	$n - 1$	

of the variation to which each pertains. Thus, in terms of the printability problem, Q_e pertains to "residual" variations in printability other than those related to differences in suppliers and variations in roughness. Q_δ pertains to possible differences in the slopes of the regressions of printability on roughness within each supplier as they vary from supplier to supplier. Q_θ pertains to possible differences in the average level of printability from supplier to supplier other than those that can be explained (linearly) by differences in the roughness of the newsprint. Q_{m-a} pertains to a possible difference between the *common part* of the slope of the regression of printability on roughness that occurs *within* each supplier and the slope of the regression of printability on roughness as these qualities vary *between* suppliers. (In the newsprint problem it is to be expected that this difference between β_m and β_a would presumably be zero, but it need not be and hence should be tested.[21]) Finally Q_0 pertains to the slope of the overall regression of printability on roughness.

Each component Q divided by its degrees of freedom (see Table 34.11) will yield a "mean square" (s^2). The test that we ran above of the hypothesis that the regression slopes within supplier did not vary from supplier to supplier consisted of computing the ratio s_δ^2/s_e^2 and comparing it with $F_{0.05}$ $[\nu_1 = c - 1, \nu_2 = c(r - 2)]$.

---

[21] Compare the results actually obtained, however, and the discussion thereof.

TABLE 34.12
Sums of Squares and Cross-Products for Data of Table 34.9

Source of Variation	*Sums of Squares for Y*	*Sums of Cross-Products for X and Y*	*Sums of Squares for X*
1. Differences between suppliers	$C_{yy} = 191.20$	$C_{xy} = -1{,}404.37$	$C_{xx} = 12{,}767.70$
2.1. Differences within supplier I	$E_{yy\,1} = 23.29$	$E_{xy\,1} = 149.97$	$E_{xx\,1} = 1{,}092.51$
2.2. Differences within supplier II	$E_{yy\,2} = 19.89$	$E_{xy\,2} = 12.17$	$E_{xx\,2} = 939.71$
2.3. Differences within supplier III	$E_{yy\,3} = 11.69$	$E_{xy\,3} = -75.98$	$E_{xx\,3} = 555.23$
2.4. Differences within supplier IV	$E_{yy\,4} = 11.42$	$E_{xy\,4} = -3.51$	$E_{xx\,4} = 340.47$
3. Pooled differences within suppliers	$E_{yy} = 66.29$	$E_{xy} = -217.29$	$E_{xx} = 2{,}927.92$
4. Total	$S_{yy} = 257.49$	$S_{xy} = -1{,}621.66$	$S_{xx} = 15{,}695.62$

We can, however, run other tests. To test whether there is any difference in printability between suppliers that is not (linearly) explained by differences in roughness (the problem of Section 2) we compute the ratio s_δ^2/s_e^2 and compare it with $F_{0.05}\,[\nu_1 = c - 2,\ \nu_2 = c(r - 2)]$. If the difference between slopes within suppliers is not significant, we can increase the degrees of freedom of this second test by pooling[22] Q_e and Q_δ, computing s_a^2 and s_θ^2/s_a^2 and comparing the latter with $F_{0.05}\,[\nu_1 = c - 2,\ \nu_2 = n - c - 1]$.

We can also test whether the (common) slope of the regression of printability on roughness within supplier is the same as the slope of the regression of printability on roughness between suppliers by computing s_{m-a}^2/s_e^2 or $s_{m-a}^2/s_{a'}^2$ depending on whether or not we decide to pool, and comparing the ratio with $F_{0.05}\,[\nu_1 = 1,\ \nu_2 = c(r - 2)]$ or $F_{0.05}\,[\nu_1 = 1,\ \nu_2 = n - c - 1]$, respectively.

A single test of whether there is any departure from one overall regression can be run by computing the ratio

$$\frac{(Q_\delta + Q_\theta + Q_{m-a})/(2c - 2)}{s_e^2}$$

and comparing it with $F_{0.05}$ for $\nu_1 = 2c - 2$, $\nu_2 = c(r - 2)$. In fact, it is well to run this test before we test the individual Q components.

3.2.2. Application to a Problem. Let us apply the theory of the previous section to the data of Table 34.9. To answer the various questions of interest

[22] This involves a risk of bias, however. See discussion of pooling, Chapter 29, Section 2.4.

it is convenient to set up the numerical counterpart of Table 34.10. This is done in Table 34.12. From Tables 34.9 and 34.12 we compute

$$\hat{\mu} = \sum_i \sum_j Y_{ij}/n = \frac{1{,}369.9}{20} = 68.50$$

$$\hat{\beta}_m = \frac{C_{xy}}{C_{xx}} = \frac{-1{,}404.37}{12{,}767.70} = -0.1100$$

$$\hat{\beta}_a = \frac{E_{xy}}{E_{xx}} = \frac{-217.29}{2{,}927.92} = -0.07421$$

$$\hat{\theta}_{\mathrm{I}} = \bar{Y}_{\mathrm{I}} - \hat{\mu} - \hat{\beta}_m(\bar{X}_{\mathrm{I}} - \bar{\bar{X}})$$

$$= 68.44 - 68.50 + 0.1100(101.42 - 95.39) = 0.60$$

$$\hat{\theta}_{\mathrm{II}} = \bar{Y}_{\mathrm{II}} - \hat{\mu} - \hat{\beta}_m(\bar{X}_{\mathrm{II}} - \bar{\bar{X}})$$

$$= 64.48 - 68.50 + 0.1100(133.36 - 95.39) = 0.16$$

$$\hat{\theta}_{III} = \bar{Y}_{\mathrm{III}} - \hat{\mu} - \hat{\beta}_m(\bar{X}_{\mathrm{III}} - \bar{\bar{X}})$$

$$= 73.16 - 68.50 + 0.1100(66.06 - 95.39) = 1.43$$

$$\hat{\theta}_{\mathrm{IV}} = \bar{Y}_{\mathrm{IV}} - \hat{\mu} - \hat{\beta}_m(\bar{X}_{\mathrm{IV}} - \bar{\bar{X}})$$

$$= 67.90 - 68.50 + 0.1100(80.72 - 95.39) = -2.21$$

$$\hat{\beta}_a + \hat{\delta}_{\mathrm{I}} = \frac{E_{xy\,1}}{E_{xx\,1}} = \frac{-149.97}{1{,}092.51} = -0.1372$$

$$\hat{\beta}_a + \hat{\delta}_{\mathrm{II}} = \frac{E_{xy\,2}}{E_{xx\,2}} = \frac{12.17}{939.71} = 0.01295$$

$$\hat{\beta}_a + \hat{\delta}_{\mathrm{III}} = \frac{E_{xy\,3}}{E_{xx\,3}} = \frac{-75.98}{555.23} = -0.1376$$

$$\hat{\beta}_a + \hat{\delta}_{\mathrm{IV}} = \frac{E_{xy\,4}}{E_{xx\,4}} = \frac{-3.51}{340.47} = -0.01031$$

The various Q's and s^2's that we will want in running different tests are computed in the analysis of covariance table, Table 34.13.

Let us begin by running a general test for departure from a common overall regression. We do this by computing

$$s^2_{\text{overall}} = \frac{Q_\delta + Q_\theta + Q_{m-a}}{3+2+1} = \frac{15.05 + 36.73 + 3.05}{6} = 9.14$$

The error variance (s_e^2) based upon the deviations from the within supplier regressions is computed in Table 34.13 as equal to 2.926. To test the hypothesis of no departure from a single overall regression we compute

TABLE 34.13
Analysis of Covariance for Data of Table 34.9

Source of Variation	*Component of Total Sum of Squares*	*df*	*MS*
Deviation from regressions within groups	$Q_e = 66.29 - 31.18 = 35.11$	12	$s_e^2 = \frac{35.11}{12} = 2.926$
Differences between regressions within groups	$Q_\delta = 31.18 - 16.13 = 15.05$	3	$s_\delta^2 = \frac{15.05}{3} = 5.017$
Deviations within classes from common regression β_a	$Q_a = 35.11 + 15.05 = 50.16$	15	$s_a^2 = \frac{50.16}{15} = 3.344$
Deviations between groups from linear regression β_m	$Q_\theta = 191.20 - 154.47 = 36.73$	2	$s_\theta^2 = \frac{36.73}{2} = 18.36$
Difference between β_a and β_m	$Q_{m-a} = 15.47 + 16.13 - 167.55 = 3.05$	1	$s_{m-a}^2 = 3.05$
Common overall regression β_0	$Q_0 = 167.55$	1	$s_0^2 = 167.55$
Total	$S_{yy} = 257.49$	19	

$$\frac{s_{\text{overall}}^2}{s_e^2} = \frac{9.14}{2.926} = 3.12$$

The $F_{0.05}$ for $\nu_1 = 6$, $\nu_2 = 12$ is 3.00, so we reject the null hypothesis and conclude that there is a departure from a single overall regression of printability on roughness.

It is next of interest to note whether the slopes of the regressions of printability on roughness within suppliers differ significantly. To test the hypothesis that there is no difference (the null hypothesis) we compute from Table 34.13

$$\frac{s_\delta^2}{s_e^2} = \frac{5.017}{2.926} = 1.71$$

The $F_{0.05}$ for $\nu_1 = 3$, $\nu_2 = 12$ is 3.49, which is greater than our sample F ratio, 1.71. In this case we do not reject the null hypothesis. We cannot conclude that the slope of the regression of printability on roughness within supplier differs from supplier to supplier.

As a third step let us determine whether the common slope of the regressions of printability on roughness *within* suppliers is significantly different from the slope of the regression *between* suppliers. To do this we compute from Table 34.13

$$\frac{s_{m-a}^2}{s_e^2} = \frac{3.05}{2.926} = 1.04$$

Since $F_{0.05}$ for $\nu_1 = 1$, $\nu_2 = 12$ is 4.75, we do not reject the null hypothesis and conclude that there is no significant difference between the slope of the regression of printability on roughness *within* suppliers and that *between* suppliers.

Finally, let us test whether supplier differentials in printability are significant when adjusted for differences in roughness. For this we compute from Table 34.13

$$\frac{s_\theta^2}{s_e^2} = \frac{18.36}{2.926} = 6.27$$

The $F_{0.05}$ for $\nu_1 = 2$ and $\nu_2 = 12$ is 3.89, so again we conclude that there is a significant difference. This suggests that the suppliers' newsprint must also differ in other qualities than roughness in such a way as to affect printability.

4. PROBLEMS

34.1. The following data[23] pertain to breaking strength and thickness of two kinds of glue. Could you infer from these data that there is a significant difference between the breaking strength of the two glues after adjustment has been made for differences in thickness?

Glue A			*Glue C*		
Specimen Number	*Breaking Strength (Lbs.)*	*Thickness (0.001 In.)*	*Specimen Number*	*Breaking Strength (Lbs.)*	*Thickness (0.001 In.)*
1	52.1	270	9	53.2	244
2	52.0	289	10	54.9	209
3	55.2	250	11	54.3	223
4	54.3	246	12	52.6	254
5	53.0	273	13	56.1	225
6	52.4	294	14	56.2	220
7	51.7	284	15	52.1	235
8	51.2	267	16	54.6	215

34.2. The following are data[24] on carded 36*s* cotton yarn similar to the data given in Table 34.5 on carded 22*s* cotton yarn. Determine whether there is a significant

[23] From a mimeographed handout by D. B. DeLury, "Elements of the Analysis of Covariance" (Department of Mathematical Statistics. Ontario Research Foundation, Toronto 5, Canada).

[24] For source and exact meaning of measurements, see footnotes to Table 34.5.

difference in skein strength (adjusted for fiber length) of the carded 36*s* yarn (1) between states and (2) between varieties. Compare your results in each case with those obtained in the text for 22*s* yarn (*Y* in pounds, *X* in 0.01 inches).

	Area								
	Georgia			Mississippi			Texas		
Variety	Test No.	Y	X	Test No.	Y	X	Test No.	Y	X
Coker 100 Wilt 4	1	45	82	5	52	88	9	51	79
Deltapine 14 (833)	2	49	80	6	54	82	10	49	65
Empire A	3	50	77	7	51	81	11	56	69
Stoneville 2B (8275)	4	56	72	8	55	79	12	50	65

34.3. The following are data[25] on printability and roughness of newsprint similar to the data of Table 34.9. Determine whether the slope of the regression of printability on roughness is the same for the "wire" side of the paper as for the "felt" side.

Felt Side			Wire Side		
Test Number	Printability	Roughness	Test Number	Printability	Roughness
1	68.7	103.6	16	76.4	97.9
2	72.8	87.3	17	73.7	91.9
3	70.9	100.1	18	71.7	117.4
4	66.1	123.5	19	67.5	159.0
5	73.0	59.4	20	68.3	119.6
6	74.3	67.7	21	75.1	52.5
7	71.9	56.7	22	62.4	141.2
8	67.1	104.8	23	61.5	177.5
9	65.4	131.4	24	72.3	82.8
10	72.6	76.0	25	66.2	108.3
11	71.2	58.0	26	65.8	108.6
12	67.5	132.6	27	70.8	97.5
13	66.6	121.4	28	74.3	83.3
14	71.7	83.8	29	71.2	94.5
15	69.5	128.6	30	67.2	167.8

34.4. The following are data on printability (*Y*) and hardness[26] (*X*) of the "felt" side of newsprint. These are similar to the data of Table 34.9.

[25] For nature of the measurements see footnote to Table 34.9. Data are supplied by courtesy of the American Newspaper Publishers Association Research Institute, Inc.

[26] Hardness is measured in the same manner as roughness except that a 5kg. weight is added to the head of the Bendtsen tester. Hardness is taken as the ratio of the measure of roughness with the weight in the head to the measure of roughness without it, multiplied by 100.

Supplier I			Supplier II			Supplier III			Supplier IV		
Test No.	Y	X	Test No.	Y	X	Test No.	Y	X	Test No.	Y	X
1	66.1	35.4	6	62.7	38.5	11	75.9	36.3	16	65.9	38.8
2	71.7	36.4	7	62.7	37.7	12	73.1	36.8	17	70.5	40.2
3	69.4	33.0	8	67.7	40.9	13	72.6	37.3	18	68.3	41.0
4	69.0	37.4	9	63.4	38.3	14	71.2	37.6	19	67.4	43.1
5	66.0	40.6	10	65.9	40.9	15	73.0	39.1	20	67.4	41.7

a. Determine whether there is a significant departure in these data from a single regression of printability on hardness (i.e., in terms of equation (34.35) whether $Q_\delta + Q_\theta + Q_{m-a}$ as a combination is significant).

b. If you find a significant difference in *a,* then determine (1) whether there is a significant difference between suppliers in the slopes of the regression of printability on hardness within suppliers, (2) whether there is a significant difference in printability of newsprint from the various suppliers when this has been adjusted for its correlation with hardness, and (3) whether the slope of the regression of printability on hardness from supplier to supplier is significantly different from the common slope of the regression of printability on hardness within suppliers.

34.5. The following are data on printability *(Y)* and roughness *(X)* of the "wire side" of newsprint. They are similar to the data of Table 34.9.

Supplier I			Supplier II			Supplier III			Supplier IV		
Test No.	Y	X	Test No.	Y	X	Test No.	Y	X	Test No.	Y	X
1	66.1	135.5	6	61.5	177.5	11	75.1	52.5	16	65.8	108.6
2	71.4	102.0	7	62.7	121.7	12	71.8	88.9	17	68.4	102.6
3	70.2	103.3	8	67.2	167.8	13	73.3	93.0	18	68.2	111.2
4	70.8	117.8	9	64.5	179.9	14	70.2	111.9	19	66.9	73.6
5	65.6	112.5	10	66.4	141.2	15	72.0	89.0	20	67.7	83.1

a. Determine whether there is a significant departure in these data from a single regression of printability on roughness (i.e., in terms of equation (34.35) whether $Q_\delta + Q_\theta + Q_{m-a}$ as a combination is significant).

b. If you find a significant difference in *a,* then determine (1) whether there is a significant difference between suppliers in the slopes of the regression of printability on roughness, (2) whether there is a significant difference in printability of newsprint from the various suppliers when this has been adjusted for its correlation with roughness, and (3) whether the slope of the regression of printability on roughness from supplier to supplier is significantly different from the common slope of the regression of printability on roughness within suppliers.

5. SELECTED REFERENCES*

Anderson and Bancroft (B '52), Bennett and Franklin (B '54), *Biometrics* '82, Brownlee (B '60), Cochran (P '57. This whole issue of *Biometrics* is devoted to analysis of covariance.), Cowden (B '57), Davies (B '57), Day and Fisher (P '37), Day, Del Priore, and Sax (P '53), Dixon and Massey (B '69), Fisher (B '58), Johnson (B '49), Kendall (B '48), Robson and Atkinson (P '60), Walker and Lev (B '53), and Wilsdon (P '34 with appendixes by E. S. Pearson and F. E. James.).

---

* B and P refer to the Book and Periodical sections, respectively, of the Cumulative List of References in Appendix V.

35

The Design of Experiments

The previous chapters of Part 5 have been concerned with statistical tools for evaluating the results of industrial experiments. In this chapter and the next we shall consider the question of experimental design. This is an important and rather extensive subject. We cannot do more here than break ground for the reader.

In every experiment there is an experimental error that arises from two sources: lack of uniformity of the material and the inherent variability in the experimental technique. This is the basic unit error that applies to a single elementary experiment. There is nothing one can do statistically to change this error. It can be reduced only by improving the material or the experimental technique.

A complete experiment, however, usually consists of a series or combination of elementary experiments, and the results take the form of a mean value or a set of mean values. The purpose of the complete experiment is to test hypotheses pertaining to the universe mean values, and possibly to estimate the means and the various components of variance. Often one experimental design will lead to greater efficiency than will other designs in testing hypothesis and making estimates. It is the purpose of this chapter to consider the relative efficiency of some of the more elementary experimental designs. In the next chapter we shall discuss special techniques, such as confounding and fractional replication, that have been found particularly useful by the industrial experimenter.

---

[1] In this chapter the author has drawn heavily on W. G. Cochran and Gertrude M. Cox, *Experimental Designs* (2d ed.; New York: John Wiley & Sons, 1957).

1. DESIGNING AN EXPERIMENT FOR ANALYZING THE EFFECTS OF A SINGLE FACTOR

Let us begin by considering how we would design an experiment for testing the differential effect of two lots of raw material. Suppose that one lot of raw material is obtained from supplier A and another lot from supplier B, and suppose that we are interested in whether the two raw materials are different insofar as they affect the quality of the output. We shall distinguish the raw materials as "raw material A" and "raw material B."

1.1. Complete Confounding

Let us suppose that the manufacturing operation consists of processing the raw material on certain machines and that, in carrying out our tests, we need to consider the machines as well as the raw material. For example, unless the machines are known to differ but little, it is obviously a faulty experimental design to use raw material A on one machine only and raw material B on another and then compare the two results. In such an instance, we would not know whether the difference is due to the difference in raw materials or to the difference in machines. The machine effect, in other words, would be completely "confounded" with the raw material effect, and the two could not be separated.

1.2. A Completely Randomized Experiment

A much better procedure, the results of which can be evaluated by standard statistical techniques, is to assign the raw materials to a number of machines at random. Say we select n machines where n is an even number. Raw material A is to be used on half of these machines; raw material B on the other half. Number the machines from 1 to n. To assign the machines at random, subject to the restriction that half of the machines are to be used with A and half with B, we select the first $n/2$ numbers from a randomly chosen section of Table DD in Appendix II, that are less than or equal to n and use raw material A on the machines with these numbers. Raw material B is used on the remaining machines.

This second experimental design is called a completely randomized experiment. The principal disadvantage of this type of experiment is that it throws machine variability together with experimental errors and may thus decrease the efficiency of the experiment.[2] For example, if we ran an analysis of variance

[2] Cochran and Cox (ibid., p. 95) list the following advantages of this experimental design: (1) It allows for complete flexibility. Any number of factor classes and replications may be used. (2) The statistical analysis is relatively simple, even if we do not have the same number of replicates for each factor class or if the experimental errors are not the same from class to class of this factor. (3) The method of analysis remains simple when data are missing or rejected, and the loss of information due to missing data is smaller than with any design.

on the results of the experiment described above, we would have an analysis of variance table like the following:[3]

Source of Variation	*Sum of Squares*	*Degrees of Freedom*	*Mean Square*	*Unbiased Estimate of** —
Column (raw materials)	—	1	$s^2_{Columns}$	$\sigma'^2_m + \sigma'^2_e + \left(\frac{n}{2}\right)\sigma'^2_c$
Remainder (machines plus experimental error)	—	$n-2$	$s^2_{Remainder}$	$\sigma'^2_m + \sigma'^2_e$
Total	—	$n-1$	—	—

* σ'^2_e = error variance; σ'^2_m = machine variance; σ'^2_c = raw material or column variance.

If machine variance (σ^2_m) is relatively large, this design makes for an unnecessarily large remainder mean square and may thus reduce the sensitivity of the F test for detecting raw material differences.[4]

1.3. Single Grouping to Reduce Errors: Randomized Blocks

1.3.1. Pairing when There Are Only Two Factor Classes. A third design that may have greater efficiency than a completely randomized experiment is to make "paired" tests and take differences. Following this plan, we would select a set of machines at random; try both raw materials on each machine (preferably in random order); take the differences in results, machine by machine; and test whether this average net difference is significantly different from zero.

The equivalent of this procedure would be to run a two-factor analysis of variance, using the machines as the second factor and raw materials, of course, as the first. The analysis of variance table would run as follows:

---

[3] In Chapters 29 and 30, the analysis of variance models that were discussed were all "measurement-error" models in the sense that the ϵ residuals were assumed to be independently normally distributed. In the experimental designs discussed in this chapter and the next, the residuals may consist, in part at least, of a limited number of "fixed effects" to which the various experimental combinations are assigned at random. The distribution of F is thus over all possible randomizations. E. J. G. Pitman has shown that with some restrictions the ordinary analysis of variance F tests work out quite well when applied to these "randomization" models. (See E. J. G. Pitman, "Significance Tests that May Be Applied to Samples from Any Population, III The Analysis of Variance Test," *Biometrika* 29 (1937), pp. 322–35. Also see book by Kempthorne in Cumulative List of References.) For this reason the analysis of variance procedures of Chapters 29 and 30 are applied directly to the problems discussed in Chapters 35 and 36.

[4] For comparison of the efficiency of this design with that of a randomized block design, see Section 1.3.3 below.

It is not necessary for a completely randomized experiment to have equal numbers in each class; e.g., the number of machines used with raw material A may be different from the number used with raw material B. It is better, however, to have equal numbers in each class if possible.

Source of Variation	Sum of Squares	Degrees of Freedom	Mean Square	Unbiased Estimate of* —
Columns (raw materials)	—	1	$s^2_{Columns}$	$\sigma'^2_e + \left(\frac{n}{2}\right)\sigma'^2_c$
Rows (machines)	—	$\frac{n}{2} - 1$	s^2_{Rows}	$\sigma'^2_e + 2\sigma'^2_m$
Remainder (experimental error plus any interaction)	—	$\frac{n}{2} - 1$	$s^2_{Remainder}$	σ'^2_e

* σ'^2_e = the unit error variance plus any interaction variance that might exist.

In this instance we have separated σ'^2_m from σ'^2_e and may thus obtain a more sensitive test of a difference than in the case of a completely randomized experiment. This need not necessarily follow, because the number of degrees of freedom upon which the estimate of the remainder variance is based are now $\frac{n-2}{2}$ instead of $n - 2$. If the machine variance (σ'^2_m) is not very large, its separation from the error variance (σ'^2_e) may not be sufficient to offset the reduction in the degrees of freedom. In this instance, the advantage may be with the completely randomized experiment. A formula for comparing the efficiency of both designs will be found in Section 1.3.3. below.

The experimental design just described is an example of a "randomized block" design. The name comes from agricultural research, in which a field may be divided into several blocks and various "treatments" assigned in a random fashion to plots in each block. In our case the machines are the "blocks," the raw materials the "treatments," and the order of using the raw materials the "plots." Thus we would have the following experimental design:

	Raw Materials (Treatments) and Order of Use (Plots)	
Machine 1 (block 1)	A	B
Machine 2 (block 2)	A	B
Machine 3 (block 3)	A	B
⋮	⋮	

1.3.2. Grouping when There Are Several Factor Classes. If there were several raw materials, used in random order on each machine, the following would be the randomized block design employed:

Machines	*Raw Materials in Order of Use*
1	A | B | C | D
2	B | A | C | D
3	C | A | D | B
⋮	⋮
	(Order of A, B, C, D selected at random)

The analysis of variance table for a randomized block design will, in general, have the following form:

Source of Variation	*Sum of Squares*	*Degrees of Freedom*	*Mean Squares*	*Unbiased Estimate of*—*
Treatments	—	$c-1$	s_t^2	$\sigma_e'^2 + r\sigma_t'^2$
Blocks	—	$r-1$	s_b^2	$\sigma_e'^2 + c\sigma_b'^2$
Remainder	—	$(r-1)(c-1)$	s_e^2	$\sigma_e'^2$

* $\sigma_e'^2$ = experimental error variance plus any interaction.

By comparing the treatment mean square with the remainder mean square, we can decide by an F test whether the treatments have any effect, regardless of whether there is a significant variation from block to block.

1.3.3. Relative Efficiency of the Randomized Block Design as Compared with a Completely Randomized Experiment. If $s_{e.rb}^2$ is the mean square for error of the randomized block design and s_b^2 is the mean square for blocks, and if $\nu_{e.rb}$, ν_b, and ν_t are the degrees of freedom for error, blocks, and treatments, Cochran and Cox[5] have shown that

(35.1)
$$s_{e.cr}^2 = \frac{\nu_b s_b^2 + (\nu_t + \nu_{e.rb})s_{e.rb}^2}{\nu_b + \nu_t + \nu_{e.rb}}$$

is an estimate of the error variance of a completely randomized design with the same experimental material.

To compare the relative efficiency of the two designs, we make use of the formula:[6]

(35.2)
$$\text{Relative efficiency of randomized block design} = \frac{(\nu_1 + 1)(\nu_2 + 3)s_{e.cr}^2}{(\nu_2 + 1)(\nu_1 + 3)s_{e.rb}^2}$$

[5] Cochran and Cox, *Experimental Designs,* p. 112.

[6] Cochran and Cox, ibid., p. 34. The formula is attributable to R. A. Fisher (see his *Design of Experiments* [3d ed.; London: Oliver & Boyd, Ltd., 1942], Section 74).

where ν_1 is the number of degrees of freedom associated with the mean square for error of the randomized block design and ν_2 the number of degrees of freedom associated with the mean square for error of the completely randomized design for the same experimental material. The use of this formula will indicate whether the randomized block design is more efficient for the problem in hand than the completely randomized design.

It is to be noted that any gain of the randomized block design arises from the grouping that is employed in the design. In our example, the product was grouped by machines, since it was logical to believe that the output of the same machine might be less variable than the output of different machines (i.e., that σ'^2_m was not negligible). Possibly, grouping on some other basis would yield even better results. The principle is to use groupings within which the product is likely to be less variable than the product as a whole.

It is also to be noted that designs can be compared on other bases than their relative "efficiency," as defined by formula (35.2). They can be compared on the basis of the confidence intervals derived from the residual error, or they can be compared with respect to their sensitivity in detecting a given difference (i.e., with respect to their OC curves). Cochran and Cox follow R. A. Fisher in using formula (35.2), but they point out[7] that this gives results that closely approximate comparisons on the basis of OC curves if the degrees of freedom are not less than 5.

1.4. Double Grouping to Reduce Errors: Latin Squares

1.4.1. The Latin Square Design. It is possible that the error variance may be reduced more by a double grouping than by a single grouping. For example, in testing differences of raw materials, a more efficient design might be obtained if the product is grouped both by machines and by operators. A special double grouping design is the Latin square.

To use a Latin square it is essential that the number of machines and the number of operators both be the same as the number of raw materials being tested. With this arrangement the raw materials are assigned at random, subject to the restriction that each raw material is used once on each machine and once by each operator. The following is an example of a 4 × 4 Latin square.[8]

---

[7] Cochran and Cox, *Experimental Designs,* Section 2.3.1.

[8] This was derived as follows. One of the four Latin squares given in Fisher and Yates's *Statistical Tables for Biological, Agricultural and Medical Research* (London: Oliver & Boyd, Ltd., 1953) was chosen at random. The columns of the Fisher and Yates square were numbered from 1 to 4 and then rearranged in accordance with the order taken by the numbers 1, 2, 3, 4 in a randomly selected section of a table of random numbers (such as Table DD of Appendix II). After the columns were permuted in this way, the last three rows were permuted in a similar fashion (compare Cochran and Cox, *Experimental Designs,* p. 121).

Machines	Operators I	II	III	IV	
1	A	C	B	D	Raw materials represented by A, B, C, and D
2	C	A	D	B	
3	B	D	C	A	
4	D	B	A	C	

The results of an experiment might be entered in a Latin square as follows:

Machines	Operators I	II	III	IV	Totals
1	A, + 2.0	C, + 0.5	B, + 1.0	D, + 1.5	5.0
2	C, + 1.5	A, 0.0	D, − 0.5	B, − 1.0	0.0
3	B, + 1.0	D, + 1.0	C, + 1.0	A, + 0.5	3.5
4	D, + 1.0	B, + 1.5	A, + 1.0	C, + 1.0	4.5
Totals ...	5.5	3.0	2.5	2.0	13.0

Totals for Raw Materials

A	B	C	D
3.5	2.5	4.0	3.0

1.4.2. Analysis of Variance for a Latin Square Design. Computations for the analysis of variance of a Latin square follow much the same pattern as that of any other analysis of variance. The steps to be taken in the analysis of the data of the previous section are as follows:

1. Compute the "correction factor" *(K)* by squaring the grand total and dividing by the number of observations. Thus,

$$K = \frac{(13.0)^2}{16} = \frac{169}{16} = 10.56$$

2. Compute the total sum of squres by summing the squares of the individual observations and subtracting the correction factor. Thus,

$$\begin{aligned}\text{Total sum of squares} &= 2.0^2 + 1.5^2 + 1.0^2 + 1.0^2 + 0.5^2 + 1.0^2 \\ &\quad + 1.5^2 + 1.0^2 + (-0.5)^2 + 1.0^2 + 1.0^2 \\ &\quad + 1.5^2 + (-1.0)^2 + 0.5^2 + 1.0^2 - 10.56 \\ &= 19.50 - 10.56 = 8.94\end{aligned}$$

3. Compute the row sum of squares by summing the squares of the row sums, dividing by the number of items in a row, and subtracting the correction factor. Thus,

$$\text{Row sum of squares} = \tfrac{1}{4}(5.0^2 + 0.0^2 + 3.5^2 + 4.5^2) - 10.56$$
$$= 14.375 - 10.56 = 3.815$$

4. Compute the column sum of squares by summing the squares of the column sums, dividing by the number of items in a column,[9] and subtracting the correction factor. Thus,

$$\text{Column sum of squares} = \tfrac{1}{4}(5.5^2 + 3.0^2 + 2.5^2 + 2.0^2) - 10.56$$
$$= 12.375 - 10.56 = 1.815$$

5. Compute the "treatment" (in our problem the raw material) sum of squares by summing the squares of the treatment (raw material) sums, dividing by the number of treatments (raw materials), and subtracting the correction factor. Thus,

$$\text{Treatment (raw material) sum of squares} = \tfrac{1}{4}(3.5^2 + 2.5^2 + 4.0^2 + 3.0^2) - 10.56$$
$$= 10.875 - 10.56 = 0.315$$

6. Compute the remainder sum of squares by subtracting the sum of 3, 4, and 5 from 2. Thus,

$$\text{Remainder sum of squares} = 8.94 - 3.815 - 1.815 - 0.315 = 2.995$$

7. Enter these sums of squares in an analysis of variance table and compute the various mean squares. This is done in Table 35.1.

8. The last step is to run an F test by comparing the treatment mean square with the remainder mean square. In the given problem, however, the former is smaller than the latter, so that it is immediately concluded that the raw material variation is not significant.[10]

1.4.3. Efficiency of a Latin Square Relative to a Randomized Block Design. The effectiveness of a row or column grouping in a Latin square may be tested by the formula[11]

---

[9] The number of items in a column is, of course, the same as the number of items in a row, which, again, is the same as the number of raw materials.

[10] The Latin square model assumes that interactions between treatments and row and column groupings are nonexistent. Since each treatment occurs only once in each row or column, if interactions are present it is possible for them to cause an apparently significant difference between treatments. This is one of the reasons why it is important to choose rows and columns of a particular Latin square in a random way (see footnote 8 above). Interactions if present can then be viewed as random elements that are part of the error treatment. They blow up the error variance and make the test less efficient but their randomization still allows for a valid theoretical test.

[11] Cochran and Cox, *Experimental Designs,* p. 127.

TABLE 35.1
Analysis of Variance Table for an $r \times r$ Latin Square

Source of Variation	*Sum of Squares*	*Degrees of Freedom*	*Mean Square*
Rows (machines)	3.815	$r-1=3$	$s_r^2=1.272$
Columns (operators)	1.815	$r-1=3$	$s_c^2=0.605$
Treatments (raw materials)	0.315	$r-1=3$	$s_t^2=0.105$
Remainder (experimental error plus interactions)	2.995	$(r-1)(r-2)=3\times 2=6$	$s_e^2=0.499$
Total	8.94	....	....

(35.3)
$$s^2_{e.rb.col.} = \frac{\nu_r s_r^2 + (\nu_t + \nu_{e.ls})s^2_{e.ls}}{\nu_r + \nu_t + \nu_{e.ls}}$$

In this expression $s^2_{e.rb.col.}$ is an estimate of the error mean square that would have been obtained if the row grouping had not been used, i.e., if we had used a randomized block design with the columns as blocks. Also, s_r^2 and $s^2_{e.ls}$ are the row mean square and the remainder mean square obtained by the use of the Latin square, and ν_r, ν_t, and $\nu_{e.ls}$ are the degrees of freedom for rows, treatments, and remainder.

To compare the efficiency of the Latin square with that of a randomized block design with the columns serving as blocks, we use the formula

(35.2a)
$$\text{Realtive efficienty of the Latin square} = \frac{(\nu_1+1)(\nu_2+3)s^2_{e.rb.col.}}{(\nu_2+2)(\nu_1+3)s^2_{e.ls}}$$

where ν_1 is the number of degrees of freedom associated with the mean square for error of the Latin square ($s^2_{e.ls}$) and ν_2 is the number of degrees of freedom associated with mean square for error of randomized block design ($s^2_{e.rb.col.}$). Similar formulas may be used to compare the efficiency of the Latin square with a randomized block design in which the rows are taken as blocks.

1.5. Triple Grouping to Reduce Errors: Graeco-Latin Squares

Triple grouping may increase the efficiency of an experiment still more than double or single grouping. For example, in testing raw materials, it might be desirable to group the product not only by machines and operators but also by days. In this case, we could arrange it so that the treatments and days are assigned at random, with the restriction that each raw material be used once on each day, once by each operator, and once on each machine. The result would be a Graeco-Latin square. An example of a Graeco-Latin square is the following:

	Operators				
Machines	*I*	*II*	*III*	*IV*	
					Raw materials
1	Aa	Bc	Cd	Db	represented by
2	Bb	Ad	Dc	Ca	A, B, C, and D.
3	Cc	Da	Ab	Bd	
4	Dd	Cb	Ba	Ac	Days represented
					by a, b, c, and d.

The analysis of variance of a Graeco-Latin square proceeds in a manner similar to that of a Latin square, except that there is now a third grouping (days). Its sum of squares would be computed by summing the squares of the sums of the a's, b's, c's, and d's, dividing by r (in our case 4), and subtracting the correction factor (i.e., the squared sum of all the observations divided by r^2). This sum of squares is then divided by the degrees of freedom for the fourth factor, which is the same as that for all the others, viz $r - 1$. The degrees of freedom of the remainder sum of squares for the Graeco-Latin square is $(r - 1)(r - 3)$.

1.6. Further Discussion

Another experimental design that might be used to test whether the two raw materials have different effects on the quality of output is the "crossover" design. There is also the problem of what to do when some of the data are lost or fail to be squared. For discussion of these topics the reader is referred to William G. Cochran and Gertrude M. Cox, *Experimental Designs.* [12]

2. DESIGNING AN EXPERIMENT FOR ANALYZING THE EFFECTS OF SEVERAL FACTORS

In some experiments we are primarily interested in a single factor and are concerned only with other factors insofar as they become confounded with the factor we are studying or tend to increase the overall experimental error. This was the case in the previous section. There we sought ways of treating these other factors so as conveniently to eliminate their effects. In other experiments, however, we are interested in the effects of several factors and their interrelations. Tools for analyzing the results of such experiments have already been developed in the earlier chapters on analysis of variance and covariance and multiple regression. The purpose here is to consider designs of experiments that will allow the most effective use of these tools.

---

[12] Chapter iv.

2.1. The Traditional Procedure

The traditional method of studying the effects of several factors is the one-at-a-time method. Suppose that we wish to study the effect of two factors X_2 and X_3 on the quality of output X_1. The traditional method would be to hold X_3 constant at some prescribed level and then note how X_1 varies with X_2. This might be repeated for several other levels of X_3. Then X_2 would be held constant at several levels and the effects of X_3 studied. Replication of each elementary experiment might be made to get an estimate of the experimental error. If we let X_2 take four different values, say, for each of four different levels of X_3 with a single replication of each elementary experiment and if we let X_3 take four different values for each of four levels of X_2 with replication, the total number of elementary experiments would be[13] $4 \times 4 \times 2 + 4 \times 4 \times 2$ or 64.

2.2. Factorial Experiments

The modern theory of experimental design suggests that the number of elementary experiments called for by the traditional one-at-a-time method of analysis may be unnecessarily large when (1) we have somewhat limited objectives and (2) it appears justified to make special assumptions regarding the factors involved in the experiment and the nature of the experimental errors. The first recommendation of modern theory is thus to state clearly the objective of your experiment. The second is to note any a priori knowledge that you may have regarding the factors and the technique of experimentation. For example, it might be helpful if there were theoretical grounds for believing that the factors are independent. Again, it might be helpful to know that, in past experiments of the type proposed, the experimental errors appeared to be about the same for all levels of the factors.

If we have reason to believe that the factors are independent and that experimental precision is the same for all factor combinations, we can plan the experiment so as to apply a simple two-factor analysis of variance. Thus we can set up the experiment, for example, so that we make measurements of X_1 at four levels of X_2 for each of four levels of X_3, the experimental design being as follows:

Levels of Factor X_3	*Levels of Factor* X_2			
	1	2	3	4
1	$X_1(1, 1)$	$X_1(1, 2)$	$X_1(1, 3)$	$X_1(1, 4)$
2	$X_1(2, 1)$	$X_1(2, 2)$	$X_1(2, 3)$	$X_1(2, 4)$
3	$X_1(3, 1)$	$X_1(3, 2)$	$X_1(3, 3)$	$X_1(3, 4)$
4	$X_1(4, 1)$	$X_1(4, 2)$	$X_1(4, 3)$	$X_1(4, 4)$

[13] Cf. Leonard A. Seder, "The Technique of Experimenting in the Factory," *Industrial Quality Control,* March 1948, pp. 6–14.

An experiment such as this, that calls for combining the levels of each factor with the levels of all the other factors, is technically known as a factorial experiment.

The above experimental design requires only 16 elementary experiments as compared with 64 undertaken by the traditional procedure. If we can assume that the factors are independent and the experimental precision the same for all factor combinations, the above factorial experiment will enable us to

1. Test whether X_2 has any effect on X_1 over the range of the experiment.
2. Estimate the extent of that effect, if it exists, for each level of X_2 at which the experiment was run.
3. Determine confidence limits for the estimates of 2, these limits being based on an estimate of experimental error derived from the whole 16 elementary experiments.
4. Test whether X_3 has any effect on X_1 over the range of the experiment.
5. Estimate the extent of that effect, if it exists, for each level of X_3 at which the experiment was run.
6. Determine confidence limits for the estimates of 5, these limits being based on an estimate of experimental error derived from the whole 16 elementary experiments.

Thus, if we are simply interested in noting how X_2 affects X_1 and how X_3 affects X_1, we can gain as much information about these relationships from this set of 16 elementary experiments as we could if we made separate studies of the effect of X_2 on X_1 alone, making 4 replications of each experiment at each of 4 levels for X_2, and the effect of X_3 on X_1 alone, making 4 replications of each experiment at each of 4 levels for X_3.[14]

It is possible that we cannot make the assumption of independence of the factors. In that case, in order to test for possible interactions and to estimate the experimental error, it is necessary either to replicate the elementary experiments or to introduce a third factor. Replication will at least double the number of elementary experiments, raising the total to a minimum of 32. If we use the above experimental design, however, our results will be in a form that will permit the use of analysis of variance techniques and may thus yield more information than might be obtained from a less well-organized procedure.

When we employ a factorial experiment with replications and apply the ordinary analysis of variance, we can carry out all the tests and make all the estimates listed above and, in addition—

7. Test for the existence of interaction between X_2 and X_3 in their effect on X_1 over the range of that experiment.

---

[14] The assumption is that in the separate studies we make separate estimates of the experimental error.

8. Estimate for each X_2, X_3 combination the extent of the interaction effect if it exists, as distinct from the main effects of X_2 and X_3.
9. Determine confidence limits for the estimates of 8, these limits being based on an estimate of experimental error derived from 16 degrees of freedom.

The introduction of a third factor instead of replicating the elementary experiments of a two-factor design is a more costly way of getting the interactions between the original two factors, since the number of total elementary experiments is now raised to r^3 (in our problem 64). We obtain knowledge of the third factor, however, which may more than make up for the additional expense. With a third factor we can get all the second-order interactions between the factors and also, if there is no third-order interaction, an estimate of the unit experimental error. The introduction of a third factor may thus easily pay for itself.

The general advantage of a factorial experiment thus lies in its efficiency in extracting certain kinds of information. It is possible that other information might better be extracted by the traditional one-at-a-time method. It may be, for example, that the latter method is more efficient in finding the combination of the factors that produces the maximum result. It is well, therefore, for the experimenter to have clearly in his mind what he wishes to learn from the experiment and then find from the theory of experimental design what type of experiment will give him this information with the least cost.

It is to be noted that the experimental designs described in Section 1 can also be applied in undertaking factorial experiments. We can thus have factorial experiments that employ randomized block designs, Latin squares, and Graeco-Latin squares. As the number of factors or number of levels of a factor is increased, however, these designs may become impracticable. For this reason, "incomplete block" designs, "split-block" designs, and "lattice" designs have been developed. For discussion of these special designs the reader is referred to the books by Brownlee, Cochran and Cox, Federer, Finney, and Kempthorne which are given in the Cumulative List of References in Appendix V.

2.3. Planning an Experiment for a Multiple Regression Analysis

If our interest is in estimating the functional form of the relationship between output X_1 and factors X_2 and X_3, the results of our experiment would be analyzed by the multiple regression technique described in Chapter 33. In planning an experiment the results of which are to be analyzed by multiple regression, the rather lengthy calculations can be simplified by spacing the levels of the independent factors at equal intervals. We can in this instance code the levels of X_2 and X_3, for example, as -2, -1, 0, 1, and 2 if there are five levels, or -3, -1, 1, and 3 if there are four levels (see, for example, Problem 33.4 of Chapter 33).

If the "factors" under study are simply powers of a single factor, in other words, if the functional form to be estimated is a nonlinear function of a single factor, equal spacing of factor levels will permit the use of orthogonal polynomials, which helps greatly in curtailing the amount of calculation and any subsequent adjustment that might arise from discarding of old terms (because of their nonsignificance) or adding of new terms.[15] For the use of orthogonal polynomials the reader is referred[16] to R. A. Fisher, *Statistical Methods for Research Workers*[17] and R. A. Fisher and F. Yates, *Statistical Tables for Biological, Agricultural, and Medical Research.*[18]

If the principal purpose of an experiment is to estimate the slope of a linear regression or to extrapolate the regression, certain gains can be obtained through proper selection of the values of the independent variable. For example, if the distribution is homoscedastic, a slope of a linear regression is estimated most precisely if the values of the independent variables are chosen in equal quantitites at the extreme ends of the range of measurable variation. For further discussion of this aspect of the design of experiments, the reader is referred to Cuthbert Daniel and Nicholas Heerema, "Design of Experiments for Most Precise Slope Estimation in Linear Extrapolation," *Journal of the American Statistical Association* 45 (1950), pp. 546–56.

3. PROBLEMS

35.1 The following table gives the (coded) results of nine different determinations of the viscosity of silicone gum rubber. The three factors considered are: (1) drums of the material that make up a batch, (2) the operator who makes the test, and (3) the various types *(A, B, C)* of viscosity measuring jars.

	Operator		
Drum	*I*	*II*	*III*
1	9*(A)*	8*(B)*	−3*(C)*
2	17*(B)*	−2*(C)*	7*(A)*
3	−2*(C)*	41*(A)*	2*(B)*

Analyze these results for significant effects. Write the equation for the analysis of variance model and state clearly the assumptions underlying your analysis.

35.2 In an experiment to test for relative biases of three instruments employed in sampling bags of fertilizer, three men (1, 2, and 3) used each of the three instruments (*X, Y,* and *Z*) in three different orders (I, II, and III). The experiment

---

[15] For the use of a hexagonal, rotatable design in fitting a second degree polynomial in two independent variables, see Chapter 33, Section 3.2.

[16] Also see J. G. Smith and A. J. Duncan, *Elementary Statistics and Applications* (New York: McGraw-Hill, 1944), chap. xxii.

[17] Section 27 ff.

[18] See Table xxiii and discussion in Introduction.

was replicated on two different bags, with the following results. The data are deviations (in tenths of a percent) of the actual percentage total phosphorus found in the samples from the nominal 10 percent stamped on the bag.[19]

First Bag

	Order		
Men	*I*	*II*	*III*
1	10*(X)*	9*(Z)*	9*(Y)*
2	10*(Y)*	11*(X)*	9*(Z)*
3	11*(Z)*	10*(Y)*	10*(X)*

Second Bag

	Order		
Men	*I*	*II*	*III*
1	14*(Z)*	13*(X)*	12*(Y)*
2	13*(X)*	12*(Y)*	13*(Z)*
3	12*(Y)*	13*(Z)*	15*(X)*

Do you find any evidence of significant instrumental biases? Justify your answer. State the assumptions on which your analysis rests. (*Hint:* Analyze the total set of data by individual degrees of freedom. See Chapter 29, Section 1.7.)

4. SELECTED REFERENCES*

(For additional references, see Chapter 36)

Brownlee (B '48 and B '60), Cochran and Cox (B '57), Daniel (P '54), Daniel and Heerema (P '50), Day (P '49), Federer (B '55), Fenech (P '79), Finney (B '55), Fisher (B '42), Gore (P '47), Hahn (P '77), Kempthorne (B '52), Pitman (P '37), Purcell (P '51), Seder (P '48), Snedecor and Cochran (B '56), Steinberg and Hunter (P '83), Tippett (B '50), Tukey (P '51), Wooding (P '73), and Yates (B '37).

---

[19] Data provided by the courtesy of the National Plant Food Institute.

* B and P refer to the Book and Periodical sections, repsectively, of the Cumulative List of References in Appendix V.

36

The Design of Experiments (continued)

1. CONFOUNDING IN FACTORIAL EXPERIMENTS

When the number of combinations of the factors in a factorial experiment is large, two difficulties may arise. First, it may be difficult to test all factor combinations under homogeneous conditions. For example, it may take several days to run all the combinations called for and the different conditions on the different days may make it difficult to make comparisons. Second, to run all the combinations may unduly increase the expense and the total time of the experiment. To meet both of these difficulties statisticians have introduced the device of "confounding." Before we consider this device, however, we need to develop a special notation.

1.1. A Special Notation for a 2^k Experiment

Consider an experiment in which each of k factors is run at 2 levels.[1] This is called a 2^k experiment. If $k = 3$, we have a 2^3 experiment which would require 8 different combinations of the factors. The degrees of freedom for a 2^3 experiment would be distributed as follows:

Main Effects	*df*
A	1
B	1
C	1

[1] When a factor, like temperature, can be varied quantitatively, "two levels" mean simply two different quantitative values. When a factor is qualitative, however, e.g., a new method of operation as against an old, "two levels" mean simply two different kinds. In a subsequent example, we shall be concerned with three factors: (1) concentration of a solvent, (2) temperature, and (3) catalyst. The first two will have two different quantitative levels. The two "levels" of the last will be simply two different catalysts.

Two-Factor Interactions	*df*
AB	1
AC	1
BC	1

Three-Factor Interaction or Error	*df*
ABC	1

In discussing a 2^k experiment of this type it is helpful to employ a special notation. This consists of representing the level at which a given factor occurs in a particular combination of factors by the presence or absence of a lowercase letter. For example, the combination in which A occurs at the low level, B at the high level, and C at the high level would be indicated by bc, the a not appearing because A is at the low level. When A is at the high level and both B and C are at the low level, we write simply a. When all factors are at the low level, the symbol used is (1). Thus, for a 2^3 experiment the various combinations of factors are as follows:

		C *Low*		C *High*	
		B		B	
		Low	*High*	*Low*	*High*
A	*Low*	(1)	*b*	*c*	*bc*
	High	*a*	*ab*	*ac*	*abc*

1.2. Symbolic Representation of Various Comparisons

The special notation noted above is useful in indicating the various comparisons that we test in analysis of variance. Thus in a 2^3 experiment the main effect of A is given by ¼ of the difference between all combinations with A at the high level and all combinations with A at the low level. Symbolically, the A comparison is represented by

(36.1a) $$A = \tfrac{1}{4}[a + ab + ac + abc - (1) - b - c - bc]$$

The comparisons representing other main effects would be

(36.1b) $$B = \tfrac{1}{4}[b + ab + bc + abc - (1) - a - c - ac]$$

and

(36.1c) $$C = \tfrac{1}{4}[c + ac + bc + abc - (1) - a - b - ab]$$

Suppose, for example, that we are studying the effects of temperature, type of catalyst, and concentration of a solvent on the yield of a given chemical. Let the results of this experiment be the following:[2]

	Catalyst (C)			
	I		*II*	
	Temperature (B)		*Temperature* (B)	
Concentration of Solvent (A)	*Low*	*High*	*Low*	*High*
Low	45.1 [(1)]	44.8 [*b*]	33.0 [*c*]	32.6 [*bc*]
High	44.9 [*a*]	44.7 [*ab*]	53.8 [*ac*]	54.2 [*abc*]

If A represents the concentration of solvent, B the temperature, and C the catalyst, the above results mean that—

For factor combination (1) the yield was 45.1
For factor combination a the yield was 44.9
For factor combination b the yield was 44.8
For factor combination c the yield was 33.0
For factor combination ab the yield was 44.7
For factor combination ac the yield was 53.8
For factor combination bc the yield was 32.6
For factor combination abc the yield was 54.2

Hence, the main effect of A (concentration of solvent) would be measured by

$$\tfrac{1}{4}[(44.9 + 44.7 + 53.8 + 54.2) - (45.1 + 44.8 + 33.0 + 32.6)]$$

$$= \frac{42.10}{4} = 10.525$$

The main effect of B (temperature) would be measured by

$$\tfrac{1}{4}[(44.8 + 44.7 + 32.6 + 54.2) - (45.1 + 44.9 + 33.0 + 53.8)]$$

$$= -\frac{0.50}{4} = -0.125$$

and the main effect of C (catalyst) would be measured by

$$\tfrac{1}{4}[(33.0 + 53.8 + 32.6 + 54.2) - (45.1 + 44.9 + 44.8 + 44.7)]$$

$$= -\frac{5.90}{4} = -1.475$$

[2] See footnote 1.

The interaction terms may be represented by similar comparisons. For two factors, A and B at each of the two levels, the interaction between A and B is defined as ½ the difference between *(i)* the difference between A at the high level and A at the low level when B is at a high level and *(ii)* the difference between A at the high level and A at the low level when B is at a low level. In other words, it is a difference of differences. Symbolically it would be written

$$\tfrac{1}{2}[(ab - b) - (a - (1))]$$

When there are more than two factors, the AB interaction is still the same as defined above except that the difference is for the averages over the other factors. Thus, when there are three factors, the AB interaction may be written

$$\begin{aligned} AB &= \tfrac{1}{2}\{\tfrac{1}{2}[(ab - b) + (abc - bc)] - \tfrac{1}{2}[(a - (1)) + (ac - c)]\} \\ &= \tfrac{1}{4}[(1) + c + ab + abc - a - b - ac - bc] \end{aligned} \tag{36.2a}$$

Other two-factor interactions would be written in the same way. Thus, with three factors,

$$\begin{aligned} AC &= \tfrac{1}{2}\{\tfrac{1}{2}[(ac - c) + (abc - bc)] - \tfrac{1}{2}[(a - (1)) + (ab - b)]\} \\ &= \tfrac{1}{4}[(1) + b + ac + abc - a - c - ab - bc] \end{aligned} \tag{36.2b}$$

and

$$\begin{aligned} BC &= \tfrac{1}{2}\{\tfrac{1}{2}[(bc - c) + (abc - ac)] - \tfrac{1}{2}[(b - (1)) + (ab - a)]\} \\ &= \tfrac{1}{4}[(1) + a + bc + abc - b - c - ab - ac] \end{aligned} \tag{36.2c}$$

In terms of the numerical example given above, the AB interaction would be measured by

$$\begin{aligned} AB &= \tfrac{1}{2}\{\tfrac{1}{2}[(44.7 - 44.8) + (54.2 - 32.6] - \tfrac{1}{2}[(44.9 - 45.1) + (53.8 - 33.0)]\} \\ &= \tfrac{1}{4}[(45.1 + 33.0 + 44.7 + 54.2) - (44.9 + 44.8 + 53.8 + 32.6)] \\ &= \frac{0.90}{4} = 0.225 \end{aligned}$$

Similarly, the AC interaction would be measured by

$$\begin{aligned} AC &= \tfrac{1}{2}\{\tfrac{1}{2}[(53.8 - 33.0) + (54.2 - 32.6)] - \tfrac{1}{2}[(44.9 - 45.1) + (44.7 - 44.8)]\} \\ &= \tfrac{1}{2}[45.1 + 44.8 + 53.8 + 54.2 - 44.9 - 33.0 - 44.7 - 32.6] \\ &= \frac{42.7}{4} = 10.675 \end{aligned}$$

and the BC interaction would be measured by

$$BC = \tfrac{1}{2}\{\tfrac{1}{2}[32.6 - 33.0) + (54.2 - 53.8)] - \tfrac{1}{2}(44.8 - 45.1) + (44.7 - 44.9)]\}$$
$$= \tfrac{1}{4}[45.1 + 44.9 + 32.6 + 54.2 - 44.8 - 33.0 - 44.7 - 53.8]$$
$$= \frac{0.50}{4} = 0.125$$

When there are three factors, the three-factor interaction ABC is ½ the difference between the two-factor interaction AB when C is at a high level and the two-factor interaction AB when C is at a low level. Thus

$$ABC = \tfrac{1}{2}\{\tfrac{1}{2}[(abc - bc) - (ac - c)] - \tfrac{1}{2}[(ab - b) - (a - [1))]\}$$

(36.3) $$= \tfrac{1}{4}[a + b + c + abc - (1) - ab - bc - ac]$$

For the given numerical example, the ABC interaction would be measured by

$$ABC = \tfrac{1}{2}\{\tfrac{1}{2}[(54.2 - 32.6) - (53.8 - 33.0)] - \tfrac{1}{2}[(44.7 - 44.8) - (44.9 - 45.1)]\}$$
$$= \tfrac{1}{4}[(44.9 + 44.8 + 33.0 + 54.2) - (45.1 + 44.7 + 32.6 + 52.8)]$$
$$= \frac{0.70}{4} = 0.175$$

It will be noted that with a single replication all comparisons in a 2^3 experiment involve a multiplier of ¼. For a 2^k experiment with a single replication, this multiplier will be $\frac{1}{2^{k-1}}$.

A simple way of remembering the various main effects and interaction terms for a three-factor experiment is to note that they are the algebraic expansions of the product of certain simple binomials. Thus,

(36.4)

$$A = \tfrac{1}{4}(a - 1)(b + 1)(c + 1) = \tfrac{1}{4}[a + ab + ac + abc - (1) - b - c - bc]$$

$$B = \tfrac{1}{4}(b - 1)(a + 1)(c + 1) = \tfrac{1}{4}[b + ab + bc + abc - (1) - a - c - ac]$$

$$C = \tfrac{1}{4}(c - 1)(a + 1)(b + 1) = \tfrac{1}{4}[c + ac + bc + abc - (1) - a - b - ab]$$

$$AB = \tfrac{1}{4}(a - 1)(b - 1)(c + 1) = \tfrac{1}{4}[(1) + c + ab + abc - a - b - ac - bc]$$

$$AC = \tfrac{1}{4}(a - 1)(c - 1)(b + 1) = \tfrac{1}{4}[(1) + b + ac + abc - a - c - ab - bc]$$

$$BC = \tfrac{1}{4}(b - 1)(c - 1)(a + 1) = \tfrac{1}{4}[(1) + a + bc + abc - b - c - ab - ac]$$

$$ABC = \tfrac{1}{4}(a - 1)(b - 1)(c - 1) = \tfrac{1}{4}[a + b + c + abc - (1) - ab - bc - ac]$$

For a 2^k experiment the various comparisons would be as follows:

$$A = \frac{1}{2^{k-1}}(a-1)(b+1)(c+1)(d+1)(e+1)\ldots$$

$$B = \frac{1}{2^{k-1}}(a+1)(b-1)(c+1)(d+1)(e+1)\ldots$$

$$\vdots$$

(36.5)

$$AB = \frac{1}{2^{k-1}}(a-1)(b-1)(c+1)(d+1)(e+1)\ldots$$

$$\vdots$$

$$ABC = \frac{1}{2^{k-1}}(a-1)(b-1)(c-1)(d-1)(e+1)\ldots$$

Etc.

1.3. Orthogonal Character of the Various Comparisons

The various comparisons noted above have the interesting and very useful property of being orthogonal. If we write down in tabular form the sums of the factor combinations that make up the various main effects and interactions, we have the following:

	(1)	*a*	*b*	*c*	*ab*	*ac*	*bc*	*abc*	*Multiplier*
Sum	+1	+1	+1	+1	+1	+1	+1	+1	$1/\sqrt{8}$
A main effect	−¼	+¼	−¼	−¼	+¼	+¼	−¼	+¼	$\sqrt{2}$
B main effect	−¼	−¼	+¼	−¼	+¼	−¼	+¼	+¼	$\sqrt{2}$
C main effect	−¼	−¼	−¼	+¼	−¼	+¼	+¼	+¼	$\sqrt{2}$
AB interaction	+¼	−¼	−¼	+¼	+¼	−¼	−¼	+¼	$\sqrt{2}$
AC interaction	+¼	−¼	+¼	−¼	−¼	+¼	−¼	+¼	$\sqrt{2}$
BC interaction	+¼	+¼	−¼	−¼	−¼	−¼	+¼	+¼	$\sqrt{2}$
ABC interaction	−¼	+¼	+¼	+¼	−¼	−¼	−¼	+¼	$\sqrt{2}$

It will be noted that the matrix of coefficients as it stands has the orthogonal property that the product of every row by any other row sums to zero. Furthermore, by multiplying each row by the "multiplier" on the right, we shall get a matrix that will be completely orthogonal in that the squares of the elements of each row sum to one.

The result is of considerable interest to us for two reasons. First, the orthogonality will enable us to compute the sum of squares for a particular comparison required in the analysis of variance directly from the value of the comparison concerned. The reader is referred back to Chapter 29, Section 1.7. Second, a very important result is that if we find it desirable to "confound" a comparison with some extraneous factor, the orthogonality of the other comparisons assures us that they will not also be confounded with this extraneous factor.

1.4. Confounding to Get Error Control

Suppose that in running the 2^3 experiment we cannot run more than four factor combinations in any one day. This may raise a problem because a comparison involving factor combinations tested on different days will contain not only experimental errors but also the differences that exist between the effects of the two days, unless the factor combinations are assigned to the two days in such a way that the day differentials cancel out for the given comparison. This last can be accomplished for seven comparisons if we are willing to make the sacrifice of confounding the eighth comparison with the day-to-day comparison.

For example, suppose that we consider the *ABC* interaction as being negligible (it usually would be identified as an error term, not an interaction). Then we would arrange the eight factor combinations as follows:

First Day				Second Day			
(1)	*ab*	*ac*	*bc*	*a*	*b*	*c*	*abc*

This arrangement "confounds" the day effect with the *ABC* interaction in that ¼ of the difference between the combinations run the first day and those run the second day is the measure we have adopted for the *ABC* comparison. Any other comparison, however, will be orthogonal to this, and the day effect will cancel out.

Thus, consider the main effect *A*. This will be

$$\tfrac{1}{4}[(ab + ac - bc - (1)) + (a + abc - b - c)]$$

where the combinations are grouped according to the days in which they are run. It will be noted that the effect of the first day appears twice in the *A* comparison with a plus sign and twice with a minus sign and similarly for the second day's effect. Hence, the day effect cancels out. Since all other comparisons are also orthogonal to the *ABC* comparison, the day effect will cancel out in each case.

The analysis of variance in this case will be as follows:

Source of SS	SS	df	MS	
A	$2A^2$	1	$2A^2$	
B	$2B^2$	1	$2B^2$	
C	$2C^2$	1	$2C^2$	
AB	$2(AB)^2$	1	$2(AB)^2$	Take some or all as error to test *A*, *B*, and *C*
AC	$2(AC)^2$	1	$2(AC)^2$	
BC	$2(BC)^2$	1	$2(BC)^2$	
Days (and *ABC*)	$2(ABC)^2$	1	$2(ABC)^2$	

In running such a limited experiment, if we do not have an outside estimate of error, we would have to assume that some or all of the two-factor interactions did not exist and use these as an error term to test the main effects.

To illustrate let us suppose that it was necessary to run the chemical experiments discussed in Section 1.2 on two different days. Let the *ABC* interaction be confounded with days and let the effect of day 1 be the addition of 1.5 to all yields and the effect of day 2 be the subtraction of 1.5 from all yields. Then all other comparisons would be the same as before, since the day effect will cancel out, but the *ABC* interaction, when confounded with days, will now be 3.175 instead of 0.175 and the analysis of variance will be as follows:

Factor	*SS*	*df*	*MS*
A	221.55125	1	221.55125
B	0.03125	1	0.03125
C	4.35125	1	4.35125
AB	0.10125	1	0.10125
AC	227.91125	1	227.91125
BC	0.03125	1	0.03125
ABC + days . . .	20.16125	1	20.16125

If, a priori, we have reason to think that the *AB* and *BC* interactions are nonexistent, we can pool these two SS and take their average MS as an estimate of error. Thus, for the above we would have the error MS = $\frac{0.10125 + 0.03125}{2} = 0.06625$, and if we used this to test other terms, we would find a significant *AC* interaction and a significant *A* main effect. This would indicate that the effect of concentration of solvent varied significantly from one catalyst to the other. The fact that the average over both catalysts (the "main effect") was also significant would probably have little meaning.

1.5. Partial Confounding

If we are willing to replicate a complete experiment, it is possible to get some measure of a confounded comparison by confounding different comparisons in different replications. Suppose, for example, that we replicated the chemical experiment of the previous section. Then on the first pair of days we could confound the *ABC* interaction as before, and on the second pair of days, we could confound the *BC* interaction.

The analysis of variance in this case would be the following:

Source of SS	*SS**	*df*	*MS*
A	$4A^2$	1	$4A^2$
B	$4B^2$	1	$4B^2$
C	$4C^2$	1	$4C^2$
AB	$4(AB)^2$	1	$4(AB)^2$
AC	$4(AC)^2$	1	$4(AC)^2$
BC†	$2(BC)^2$	1	$2(BC)^2$ [½ precision]
ABC†	$2(ABC)^2$	1	$2(ABC)^2$ [½ precision]
Days, BC, ABC . . .	$4\Sigma(\bar{X}_d - \bar{\bar{X}})^2$	3	$4\Sigma(\bar{X}_d - \bar{\bar{X}})^2/3$
Residual	(= Error)	5	

* Here A stands for the average A effect over the two replications. In this average each of the 16 results will be divided by 8 and the multiplier must now be 2 to yield a strictly orthogonal comparison. The same remarks apply to B, C, AB, and so on.

† The sum is for those parts of the BC and ABC comparisons unconfounded with days.

Here, "½ precision" means that the number of results over which these particular interaction terms are averaged is ½ of those over which the main effects and other interaction terms are averaged.

Partial confounding may be illustrated by a numerical example. Suppose that a replicate of the chemical experiment that yielded the results on page 926 above turned out as follows:

	Catalyst (C)			
	I		*II*	
	Temperature (B)		*Temperature* (B)	
Concentration of Solvent (A)	*Low*	*High*	*Low*	*High*
Low	44.6 [(1)]	46.7 [b]	35.0 [c]	33.7 [bc]
High	45.3 [a]	44.8 [ab]	51.7 [ac]	53.2 [abc]

Suppose further that the experimental design employed was as follows:

Replicate I		*Replicate II*	
Day 1	*Day 2*	*Day 3*	*Day 4*
(1)	a	(1)	b
ab	b	a	c
ac	c	bc	ab
bc	abc	abc	ac
[ABC confounded with days]		[BC confounded with days]	

For this design, we would have

$$A = \tfrac{1}{8}[(44.9 + 45.3 + 44.7 + 44.8 + 53.8 + 51.7 + 54.2 + 53.2) - (45.1 + 44.6 + 44.8 + 46.7 + 33.0 + 35.0 + 32.6 + 33.7)] = \frac{77.1}{8} = 9.6375$$

$$B = \tfrac{1}{8}[(44.8 + 46.7 + 44.7 + 44.8 + 32.6 + 33.7 + 54.2 + 53.2) - (45.1 + 44.6 + 44.9 + 45.3 + 33.0 + 35.0 + 53.8 + 51.7)] = \frac{1.3}{8} = 0.1625$$

$$C = \tfrac{1}{8}[(33.0 + 35.0 + 53.8 + 51.7 + 32.6 + 33.7 + 54.2 + 53.2) - (45.1 + 44.6 + 44.9 + 45.3 + 44.8 + 46.7 + 44.7 + 44.8)] = \frac{-13.7}{8} = -1.7125$$

$$AB = \tfrac{1}{8}[(45.1 + 44.6 + 35.0 + 33.0 + 44.7 + 44.8 + 54.2 + 53.2) - (44.9 + 45.3 + 44.8 + 46.7 + 53.8 + 51.7 + 32.6 + 33.7)] = \frac{1.10}{8} = 0.1375$$

$$AC = \tfrac{1}{8}[(45.1 + 44.6 + 44.8 + 46.7 + 53.8 + 51.7 + 54.2 + 53.2) - (44.9 + 45.3 + 33.0 + 35.0 + 44.7 + 44.8 + 32.6 + 33.7)] = \frac{80.1}{8} = 10.0125$$

$$BC = \tfrac{1}{4}[(45.1 + 44.9 + 32.6 + 54.2) - (44.8 + 33.0 + 44.7 + 53.8)] = \frac{0.50}{4} = 0.125$$

$$ABC = \tfrac{1}{4}[(45.3 + 46.7 + 35.0 + 53.2) - (44.6 + 44.8 + 33.7 + 51.7)] = \frac{5.4}{4} = 1.35$$

The analysis of variance would be as shown on the top of page 934:

Source of Variation		*SS*	*df*	*MS*
A	$4A^2$ =	371.5256	1	371.5256
B	$4B^2$ =	0.1056	1	0.1056
C	$4C^2$ =	11.7306	1	11.7306
AB	$4(AB)^2$ =	0.0756	1	0.0756
AC	$4(AC)^2$ =	401.0006	1	401.0006
BC	$2(BC)^2$ =	0.0312	1	0.0312 (½ precision)
ABC	$2(ABC)^2$ =	3.6450	1	3.6450 (½ precision)
Days, *BC*, *ABC** ...		22.9925	3	7.6642
Error†		853.1808	5	170.6362
Total‡		1,664.2875	15	

* Computed from formula, $SS = \frac{T_1^2 + T_2^2 + T_3^2 + T_4^2}{4} - \frac{T^2}{16}$, where T_1, T_2, T_3, T_4 are the totals for each day, $T = T_1 + T_2 + T_3 + T_4$. Four is the number of results per day, and 16 is the total number of results. Cf. above, Chapter 29, Section 1.3.

† Computed as a residual.

‡ Computed from the formula $\Sigma X^2 - T^2/16$.

2. FRACTIONAL REPLICATION

Sometimes in industrial experiments we have reasons to believe that higher order interactions do not exist or are negligible. In such a case, we do not need to run all factor combinations to test for the main effects and interaction terms that would appear to be important. For example, if we run a 2^7 experiment, there are altogether 128 different factor combinations. To test each of these may involve more money than is available or more time than can be allowed. If higher order interactions are believed to be negligible, however, we can run a half replicate or even a quarter replicate, which will cut down the total number of factor combinations to be tested from 128 to 64 or from 128 to 32. In a half replicate, the procedure is to divide (on paper) the total number of factor combinations into two blocks, so assigned that one of the higher order (usually the highest order) interaction is confounded with the blocks. Then we run only one block of the tests and ignore the other.

To see what happens in this case, consider again a 2^3 experiment. (Fractional replication would not be applied to such a small experiment, but it is easier to talk about for purposes of exposition.) If we confound the *ABC* interaction with the two blocks, we will have (as above) the following:

Block I				*Block II*			
(1)	*ab*	*ac*	*bc*	*a*	*b*	*c*	*abc*

Now, if we test only the first group of factor combinations, what can we do with the results?

To understand what is involved, let us consider the general model for the experiment. This runs

(36.6) $$Y_{ijk} = \mu + a_i^A + a_j^B + a_k^C + a_{ij}^{AB} + a_{ik}^{AC} + a_{jk}^{BC} + \epsilon_{ijk}$$

where i, j, and k take on the values 1, 2, and where $\sum_i a_i^A = \sum_j a_j^B = \sum_k a_k^C = 0$ and $\sum_i a_{ij}^{AB} = \sum_j a_{ij}^{AB} = \sum_i a_{ik}^{AC} = \sum_k a_{ik}^{AC} = \sum_j a_{jk}^{BC} = \sum_k a_{jk}^{BC} = 0$.

The factor combinations (1), *ab*, *ac*, and *bc* will give the following results:

$$(1) \text{ will give } Y_{111} = \mu + a_1^A + a_1^B + a_1^C + a_{11}^{AB} + a_{11}^{AC} + a_{11}^{BC} + \epsilon_{111}$$

$$(ab) \text{ will give } Y_{221} = \mu + a_2^A + a_2^B + a_1^C + a_{22}^{AB} + a_{21}^{AC} + a_{21}^{BC} + \epsilon_{221}$$

$$(ac) \text{ will give } Y_{212} = \mu + a_2^A + a_1^B + a_2^C + a_{21}^{AB} + a_{22}^{AC} + a_{12}^{BC} + \epsilon_{212}$$

$$(bc) \text{ will give } Y_{122} = \mu + a_1^A + a_2^B + a_2^C + a_{12}^{AB} + a_{12}^{AC} + a_{22}^{BC} + \epsilon_{122}$$

If we try to measure the A effect by taking

$$A = \tfrac{1}{2}[(ab + ac) - ((1) + bc)]$$

we get

(36.7) $$A = \tfrac{1}{2}[2(a_2^A - a_1^A) - (a_{11}^{BC} - a_{12}^{BC} - a_{21}^{BC} + a_{22}^{BC}) + \text{terms in } \epsilon]$$

all other elements summing to zero. Hence, aside from an error term, we get the A effect minus the BC interaction effect.[3] If however, we know that the BC interaction is negligible, then we can take the above to represent the A effect.

In general, if the ABC interaction is used as the basis for forming the half replicate, we can estimate the B effect from the half replicate if there is no AC interaction and the C effect if there is no AB interaction. This can be proved as above. A shorter procedure is the following: Take the ABC interaction; multiply it by A and call A^2 unity. Then we have

$$A(ABC) = BC$$

This relationship is described by saying that with respect to "the defining contrast"[4] ABC, A and BC are "aliases." For the half replicate of the 2^3 experiment, based on ABC as the defining contrast, we also have

$$B(ABC) = AC$$

$$C(ABC) = AB$$

so that B and AC are aliases and C and AB are aliases. Thus, in a half replicate of a 2^3 experiment, we can estimate all the main effects if the interactions are all negligible. For such a small experiment, we could not test the

[3] That $(a_{11}^{BC} - a_{12}^{BC} - a_{21}^{BC} + a_{22}^{BC})$ is twice the BC "interaction effect" can be seen directly from the meaning of the term. It can be "verified" by substituting the model (36.6) in ¼ [(1) + a + bc + abc − b − c − ab − ac].

[4] A "defining contrast" is a contrast or comparison, in most cases an interaction, that is selected to be confounded with the blocks defining the experimental fractions.

significance of these effects unless we had an outside estimate of the experimental error.

Consider now a half replicate of a 2^5 experiment. If we take the *ABCDE* interaction as the defining contrast, this interaction will be measured by

$$\tfrac{1}{16}(a-1)(b-1)(c-1)(d-1)(e-1)$$

which equals[5]

(36.8)

$$\tfrac{1}{16}[(abcde + abc + abd + abe + acd + ace + ade + bcd + bce + bde + cde + a + b + c + d + e) - (abcd + abce + abde + acde + bcde + ab + ac + ad + ae + bc + bd + be + cd + ce + de + (1))]$$

Let the half replicate of factor combinations for which experiments are actually run be the block containing the factor combination (1), called the *principal block.*

With *ABCDE* as the defining contrast, the various aliases will be as follows:

Term	*Alias*
A	*BCDE*
B	*ACDE*
C	*ABDE*
D	*ABCE*
E	*ABCD*
AB	*CDE*
AC	*BDE*
AD	*BCE*
AE	*BCD*
BC	*ADE*
BD	*ACE*
BE	*ACD*
CD	*ABE*
CE	*ABD*
DE	*ABC*

Hence, if the interactions on the right can be considered negligible, we can use a half replicate of a 2^5 experiment to measure all the main effects and the two-factor interactions.

The analysis of variance will run as follows:

[5] The positive block is made up of all the terms involving an odd number of letters, and the negative block all the terms involving an even number of letters.

Source of Variation	SS	df	MS
A	$4A^2$	1	$4A^2$
B	$4B^2$	1	$4B^2$
C	$4C^2$	1	$4C^2$
D	$4D^2$	1	$4D^2$
E	$4E^2$	1	$4E^2$
AB	$4(AB)^2$	1	$4(AB)^2$
AC	$4(AC)^2$	1	$4(AC)^2$
AD	$4(AD)^2$	1	$4(AD)^2$
AE	$4(AE)^2$	1	$4(AE)^2$
BC	$4(BC)^2$	1	$4(BC)^2$
BD	$4(BD)^2$	1	$4(BD)^2$
BE	$4(BE)^2$	1	$4(BE)^2$
CD	$4(CD)^2$	1	$4(CD)^2$
CE	$4(CE)^2$	1	$4(CE)^2$
DE	$4(DE)^2$	1	$4(DE)^2$

The estimates of the various main effects and interactions will be as follows:

$$A = \tfrac{1}{8}\,[ab + ac + ad + abcd + ae + abce + abde + acde - (1) - bc - bd - cd - be - ce - de - bcde]$$

$$B = \tfrac{1}{8}\,[ab + bd + abcd + bc + abce + abde + bcde + be - (1) - ac - ad - cd - ae - ce - de - acde]$$

$$C = \tfrac{1}{8}\,[ac + bc + cd + abcd + ce + abce + acde + bcde - (1) - ab - ad - bd - ae - be - de - abde]$$

$$D = \tfrac{1}{8}\,[ad + bd + cd + abcd + de + abde + acde + bcde - (1) - ab - ac - bc - ae - be - ce - abce]$$

$$E = \tfrac{1}{8}\,[ae + be + ce + abce + de + abde + acde + bcde - (1) - ab - ac - bc - ad - bd - cd - abcd]$$

$$AB = \tfrac{1}{8}\,[(1) + ab + cd + abcd + ce + abce + de + abde - ac - bc - ad - bd - ae - be - acde - bcde]$$

$$AC = \tfrac{1}{8}\,[(1) + ac + bd + abcd + be + abce + de + acde - ab - bc - ad - cd - ae - ce - abde - bcde]$$

$$BC = \tfrac{1}{8}\,[(1) + bc + ad + abcd + ae + abce + de + bcde - ab - ac - bd - cd - be - ce - abde - acde]$$

$$AD = \tfrac{1}{8}\,[(1) + bc + ad + abcd + be + ce + abde + acde - ab - ac - bd - cd - ae - abce - de - bcde]$$

$$BD = \tfrac{1}{8}\,[(1) + ac + bd + abcd + ae + ce + abde + bcde - ab - bc - ad - cd - be - abce - de - acde]$$

$$CD = \tfrac{1}{8}\,[(1) + ab + cd + abcd + ae + be + acde + bcde - ac - bc - ad - bd - ce - abce - de - abde]$$

$$AE = \tfrac{1}{8}\,[(1) + bc + bd + cd + ae + abce + abde + acde - ab - ac - ad - abcd - be - ce - de - bcde]$$

$$BE = \tfrac{1}{8}\,[(1) + ac + ad + cd + be + abce + abde + bcde - ab - bc - bd - abcd - ae - ce - de - acde]$$

$$CE = \tfrac{1}{8}\,[(1) + ab + ad + bd + ce + abce + acde + bcde - ac - bc - cd - abcd - ae - be - de - abde]$$

$$DE = \tfrac{1}{8}\,[(1) + ab + ac + bc + de + abde + acde + bcde - ad - bd - cd - abcd - ae - be - ce - abce]$$

The above equations will be referred to as equations (36.9).

If some or all of the interaction terms can be viewed as error terms, we can use them to test the significance of the main effects and other interaction terms. If we had taken a half replicate of a 2^6 experiment, the analysis of variance would have been

Source of Variation	*Total df*
Main effects (*A*, etc.)	6
Two-factor interactions (*AB*, *BC*, etc.)	15
Error	10

In other words, with a half replicate, we need a 2^6 or an experiment of higher order to give us an independent error term to test all the two-factor interactions.

With larger experiments such as a 2^7, for example, we might be interested in ¼ replicates or even ⅛ replicates. For the details of how to set up a quarter replicate, the reader is referred to one of the more advanced texts listed at the end of the chapter. The principles involved, however, may be illustrated with reference to a 2^5 experiment.

When we took the *ABCDE* interaction as a basis for forming the half replicate, the experiments that were contained in the same block as (1) were those that were run. We can, if we wish, take another interaction term to split this half replicate. Let this be the *ABC* interaction. This will yield the same results as if we used its alias the *DE* interaction. Let us match both of these with the *ABCDE* interaction, as follows:[6]

The Half Replicate Containing (1)				*The Other Half Replicate*			
Combination	*ABCDE*	*ABC*	*DE*	*Combination*	*ABCDE*	*ABC*	*DE*
(1)	−	−	+	*a*	+	+	+
ab	−	−	+	*b*	+	+	+
ac	−	−	+	*c*	+	+	+
ad	−	+	−	*d*	+	−	−
ae	−	+	−	*e*	+	−	−
bc	−	−	+	*abc*	+	+	+
bd	−	+	−	*abd*	+	−	−
be	−	+	−	*abe*	+	−	−
cd	−	+	−	*acd*	+	−	−
ce	−	+	−	*ace*	+	−	−
de	−	−	+	*ade*	+	+	+
abcd	−	+	−	*bcd*	+	−	−
abce	−	+	−	*bce*	+	−	−
abde	−	−	+	*bde*	+	+	+
acde	−	−	+	*cde*	+	+	+
bcde	−	−	+	*abcde*	+	+	+

To get the quarter replicate, we can take all combinations which have minus signs for both *ABCDE* and *ABCD* or all those that have minus signs for *ABCDE* and plus signs for *DE.* In either case, these are (1), *ab, ac, bc, de, abde, acde,* and *bcde.* These will be the eight combinations actually run.

With *ABCDE, ABC,* and *DE* confounded with the quarter replicate, we have the following group of aliases (obtained by multiplying the given factor by one of the above and ignoring all squared terms):

Term	*Aliases*		
A	*BCDE*	*BC*	*ADE*
B	*ACDE*	*AC*	*BDE*
C	*ABDE*	*AB*	*CDE*
D	*ABCE*	*ABCD*	*E*
CD	*ABE*	*ABD*	*CE*
AD	*BCE*	*BCD*	*AE*
BD	*ACE*	*ACD*	*BE*

It will be noted immediately that not all main effects can be obtained in a quarter replicate of a 2^5 experiment. One main effect must be considered negligible if the others are to be obtained. The analysis of variance for the above quarter replicate would be

[6] *ABC* can be found by expanding $\frac{1}{16}(a - 1)(b - 1)(c - 1)(d + 1)(e + 1)$ and *DE* by expanding $\frac{1}{16}(a + 1)(b + 1)(c + 1)(d - 1)(e - 1)$.

Main effects . . . . Total of 4 df
Second-order interactions used as error . . . Total of 3 df

With a larger experiment, all main effects could be tested in a quarter replicate and possibly some or all of the two-factor interactions.

For a more extended discussion of fractional replication the reader is referred to Owen L. Davies (ed.), *The Design and Analysis of Industrial Experiments.*

3. PROBLEMS

36.1. In running experiments on a detergency, it is decided to consider three factors: temperature, concentration, and time. Each factor is to be run only at two levels, since it is decided that linear effects are predominant. Because of the introduction of the time factor, however, it is possible to run only four experiments a day. Lay out an experimental program, assigning the different "treatments" (factor combinations) to various days, such that all of the eight different treatments will be replicated twice. Interactions between temperature and concentration, and temperature and time are to be determined with the same precision as main effects. The interaction between concentration and time and the three-factor interaction—time × temperature × concentration—may be determined with less precision, but the precision should be equal for both.

36.2. You wish to explore the effect of each of eight factors on the yield of a process. If a complete factorial is run with each factor at two levels, this will require $2^8 = 256$ experiments. As an initial step you decide to run only 1/16*th* of this total, or 16 tests. You want this block of tests to contain the test at which all factors are kept at their low level. The results of the experiment will be used to test the main effects. Interactions will be deemed to be of minor importance compared with main effects, and for the purposes of this initial experiment will be treated as error terms.

State the various combinations of the factors that will be tested and give the aliases of your main effects. Note the degrees of freedom for the "error" term to be used in testing a main effect.

36.3. In a 2^6 experiment you run only the following 16 experiments with the results indicated:

Combination	*Yield*	*Combination*	*Yield*
(1)	740	*df*	780
ae	1,108	*adef*	1,184
bef	888	*bde*	864
abf	1,008	*abd*	1,312
cef	784	*cde*	1,000
acf	1,156	*acd*	1,376
bc	692	*bcdf*	860
abce	984	*abcdef*	1,292

State the specific conclusions, if any, that you can draw from these results and indicate the assumptions on which your analysis rests. (First assume all interactions are nonexistent and, using these as error terms, test for existence of main effects. What are the "aliases" of these main effects? Would you have any error mean square for testing the two-factor interactions? If so, what is this error mean square and what are its degrees of freedom?)

4. SELECTED REFERENCES*

Box and Hunter (P '61), Daniel (P '54 and P '59), Davies (B '54), Davies and Hay (P '50), U.S. Dept. of Commerce, National Bureau of Standards (B '57 and B '59).

---

* B and P refer to the Book and Periodical sections, respectively, of the Cumulative List of References in Appendix V.

Appendixes

APPENDIX I
Selected Mathematical Proofs and Technical Material

(1) Symbolic Summation

Suppose that a variable X takes on the values $X_1 = 2$, $X_2 = 5$, $X_3 = 7$, and $X_4 = 8$. Then the sum of the X values is

$$X_1 + X_2 + X_3 + X_4 = 2 + 5 + 7 + 8 = 22$$

If we wish we can use the operator Σ to mean "sum of" and write

$$\sum_{i=1}^{4} X_i = 22$$

The subscript i distinguishes the various values of X, and, by writing $i = 1$ at the bottom of the summation sign and 4 at the top, we mean the variable X is to be summed from X_1 to X_4, inclusive. Sometimes the subscripts are left off and we write simply

$$\sum X_i = 22 \quad \text{or} \quad \sum X = 22$$

in which case it is implied that X is summed for all values.

Sometimes the variable that is being summed is one of several variables in a function. For example, consider the simple function X^c. If we write

$$\sum_{c=0}^{n} X^c$$

we mean the series

$$X^0 + X^1 + X^2 + \ldots + X^n$$

Here, so far as the summation is concerned, the quantity X is treated as a constant. The variable for which summation is carried out is c.

In more complicated problems we have double and triple summation. For example, suppose that values of X are arranged in a square, such as the following:

$$\begin{matrix} 4 & 2 & 1 \\ 5 & 9 & 3 \\ 8 & 7 & 6 \end{matrix}$$

Then we can designate a particular X by reference to its position in the square. If we let an i subscript indicate which row it is in and a j subscript indicate the column, then any X can be indicated as X_{ij}. Thus $X_{21} = 5$, $X_{12} = 2$, $X_{33} = 6$, and so on. If we write

$$\sum_i \sum_j X_{ij}$$

we mean the complete summation of X over both rows and columns. Thus

$$\sum_i \sum_j X_{ij} = 4+5+8$$
$$+2+9+7$$
$$+1+3+6=45$$

If we write

$$\sum_i X_{ij}$$

we mean the sum of the elements in the j th column. Thus

$$\sum_i X_{i\,2} = 2+9+7=18$$

Likewise,

$$\sum_j X_{3j} = 8+7+6=21$$

The rules of summation are simple.

(1)
$$\sum_i (X_i + Y_i) = \sum_i X_i + \sum_i Y_i$$

For

$$\sum_i (X_i + Y_i) = (X_1 + Y_1) + (X_2 + Y_2) + \ldots$$
$$= X_1 + X_2 + \ldots + Y_1 + Y_2 + \ldots = \sum_i X_i + \sum_i Y_i$$

(2)
$$\sum_i (X_i - Y_i) = \sum_i X_i - \sum_i Y_i$$

(3)
$$\sum_i aX_i = a \sum_i X_i$$

For

$$aX_1 + aX_2 + \ldots = a(X_1 + X_2 + \ldots)$$

(4) $\sum_{i=1}^{n} a = na$, if a is a constant; for a added n times is na

(5)
$$\sum_{i=1}^{n} (X_i - a)^2 = \sum_{i=1}^{n} (X_i^2 - 2aX_i + a^2)$$
$$= \sum_{i=1}^{n} X_i^2 - 2a \sum_{c=1}^{n} X_i + na^2$$

(6)
$$\sum_i \sum_j X_i Y_j = \left(\sum_i X_i\right)\left(\sum_j Y_j\right)$$

(2) Proof of the Short Method for Computing the Arithmetic Mean

By definition

$$\bar{X} = \frac{\Sigma F_j X_j}{n}$$

But $X_j = A + i\xi_j$ (see Figure 3.9). Hence

$$\bar{X} = \frac{\Sigma F_j (A + i\xi_j)}{n}$$

Removing the parentheses yields

$$\bar{X} = \frac{\Sigma F_j A}{n} + \frac{\Sigma F_j i \xi_j}{n} = A\frac{\Sigma F_j}{n} + \frac{\Sigma F_j \xi_j}{n} i$$

Since $n = \Sigma F_j$, we have

$$\bar{X} = A + \frac{\Sigma F_j \xi_j}{n} i$$

If we define

$$\mu_{a\,1} = \frac{\Sigma F_j \xi_j}{n} i$$

then

$$\bar{X} = A + \mu_{a\,1}$$

(3) Proof of the Short Method for Computing the Standard Deviation

By definition,

$$\mu_2 = \frac{\Sigma F_j (X_j - \bar{X})^2}{n}$$

But

$$X_j - \bar{X} = (i\xi_j + A) - (A + \mu_{a\,1}) = i\xi_j - \mu_{a\,1}$$

Hence

$$\mu_2 = \frac{\Sigma F_j (i\xi_j - \mu_{a\,1})^2}{n}$$

$$= \frac{\Sigma F_j [(i\xi_j)^2 - 2\mu_{a\,1} i\xi_j + \mu_{a\,1}^2]}{n}$$

$$= \frac{\Sigma F_j (i\xi_j)^2 - 2\mu_{a\,1}\Sigma F_j i\xi_j + \mu_{a\,1}^2 \Sigma F_j}{n}$$

$$= \frac{\Sigma F_j (i\xi_j)^2}{n} - 2\mu_{a\,1}\frac{\Sigma F_j i\xi_j}{n} + \mu_{a\,1}^2 \frac{\Sigma F_j}{n}$$

$$= \frac{\Sigma F_j (i\xi_j)^2}{n} - 2\mu_{a\,1}^2 + \mu_{a\,1}^2$$

If we set

$$\mu_{a\,2} = \frac{\Sigma F_j \xi_j^2}{n} i^2$$

then

$$\mu_2 = \mu_{a\,2} - 2\mu_{a\,1}^2 + \mu_{a\,1}^2$$
$$= \mu_{a\,2} - \mu_{a\,1}^2$$

By definition

$$s = \sqrt{\left(\frac{n}{n-1}\right)\mu_2}$$

Hence

$$s = \sqrt{\left(\frac{n}{n-1}\right)(\mu_{a\,2} - \mu_{a\,1}^2)}$$

(4) Derivation of a Formula for Computing μ_3

By definition

$$\mu_3 = \frac{\sum_j F_j (X_j - \bar{X})^3}{n}$$

But $X_j - \bar{X} = (A + i\xi_j) - (A + \mu_{a\,1}) = i\xi_j - \mu_{a\,1}$

Hence

$$\mu_3 = \frac{\sum_j F_j (i\xi_j - \mu_{a\,1})^3}{n}$$

$$= \frac{\sum_j F_j [(i\xi_j)^3 - 3(i\xi_j)^2 \mu_{a\,1} + 3(i\xi_j)\mu_{a\,1}^2 - \mu_{a\,1}^3]}{n}$$

$$= \frac{\sum_j F_j \xi_j^3 (i)^3}{n} - \frac{3\sum_j F_j \xi_j^2 (i)^2 \mu_{a\,1}}{n} + \frac{3\sum_j F_j \xi_j (i) \mu_{a\,1}^2}{n} - \frac{n\mu_{a\,1}^3}{n}$$

Hence, if we define $\mu_{a\,3}$ as equal to $\dfrac{\Sigma F_j \xi_j^3}{n} i^3$, then

$$\mu_3 = \mu_{a\,3} - 3\mu_{a\,2}\mu_{a\,1} + 3\mu_{a\,1}^3 - \mu_{a\,1}^3$$
$$= \mu_{a\,3} - 3\mu_{a\,2}\mu_{a\,1} + 2\mu_{a\,1}^3$$

(5) Derivation of a Formula for Computing μ_4

By definition

$$\mu_4 = \frac{\sum_j F_j(X_j - \bar{X})^4}{n}$$

But, as in (4), we have $X_j = \bar{X} = (i\xi_j - \mu_{a\,1})$, so

$$\mu_4 = \frac{\sum_j F_j(i\xi_j - \mu_{a\,1})^4}{n}$$
$$= \frac{\sum_j F_j[(i\xi_j)^4 - 4(i\xi_j)^3\mu_{a\,1} + 6(i\xi_j)^2\mu_{a\,1}^2 - 4i\xi_j\mu_{a\,1}^3 + \mu_{a\,1}^4]}{n}$$
$$= \frac{\sum_j F_j\xi_j^4(i)^4}{n} - \frac{4\sum_j F_j\xi_j^3(i)^3\mu_{a\,1}}{n} + \frac{6\sum_j F_j\xi_j^2(i)^2\mu_{a\,1}^2}{n} - \frac{4\sum_j F_j\xi_j(i)\mu_{a\,1}^3}{n} + \frac{n\mu_{a\,1}^4}{n}$$

Hence, if we define $\mu_{a\,4}$ as equal to $\dfrac{\Sigma F_j \xi_j^4}{n} i^4$, then

$$\mu_4 = \mu_{a\,4} - 4\mu_{a\,3}\mu_{a\,1} + 6\mu_{a\,2}\mu_{a\,1}^2 - 4\mu_{a\,1}^4 + \mu_{a\,1}^4$$
$$= \mu_{a\,4} - 4\mu_{a\,3}\mu_{a\,1} + 6\mu_{a\,2}\mu_{a\,1}^2 - 3\mu_{a\,1}^4$$

(6) Proof that the Mean of a Binomial Distribution Equals p' if X/n Is the Variable

We have[1]

$$E\left(\frac{X}{n}\right) = \sum_{X=0}^{X=n} \frac{n!}{X!(n-X)!} p'^X(1-p')^{n-X}\left(\frac{X}{n}\right)$$
$$= \sum_{X=1}^{X=n} \frac{(n-1)!}{(X-1)!(n-X)!} p'^X(1-p')^{n-X}$$

---

[1] It will be recalled that E is used as an operator to signify "mean of." Thus $E(X)$ means "mean of X" (see Chapter 4, Section 4.1.)

$$=p' \sum_{X-1=0}^{X-1=n-1} \frac{(n-1)!}{(X-1)![(n-1)-(X-1)]!} p'^{X-1}(1-p')^{(n-1)-(X-1)}$$

$$=p' \sum_{y=0}^{y=m} \frac{m!}{y!(m-y)!} P'^{y}(1-p')^{m-y}$$

where $y = X - 1$ and $m = n - 1$. But the sum term is the sum of *all* the ordinates of a binominal distribution, which equals 1. Hence,

$$E\left(\frac{X}{n}\right)=p'$$

(7) Proof that the Variance of a Binomial Distribution Equals $\frac{p'(1-p')}{n}$ if X/n Is the Variable

By definition we have

$$\sigma_p'^2 = E\left(\frac{X}{n}-p'\right)^2 = E\left(\frac{X}{n}\right)^2 - 2p'E\left(\frac{X}{n}\right) + (p')^2 = E\left(\frac{X}{n}\right)^2 - 2p'^2 + p'^2$$

$$= E\left(\frac{X}{n}\right)^2 - p'^2$$

But

$$E\left(\frac{X}{n}\right)^2 = \sum \frac{n!}{X!(n-X)!} p'^X(1-p')^{n-X}\left(\frac{X}{n}\right)^2$$

$$= \sum \frac{n!}{X!(n-X)!} p'^X(1-p')^{n-X}\left(\frac{X(X-1)+X}{n^2}\right)$$

$$= \frac{(n-1)}{\text{n}} p'^2 \sum \frac{(n-2)!p'^{X-2}(1-p')^{(n-2)-(X-2)}}{(X-2)![(n-2)-(X-2)]!}$$

$$+\frac{1}{n} \sum \frac{n!}{X!(n-X)!} p'^X(1-p')^{n-X}\left(\frac{X}{n}\right)$$

But the first sum $= 1$ and the second $= p'$. Hence

$$\sigma_p'^2 = E\left(\frac{X}{n}-p'\right)^2 = \frac{n-1}{n}p'^2 + \frac{p'}{n} - p'^2 = \frac{p'-p'^2}{n} = \frac{p'(1-p')}{n}$$

(8) Proof that the Mean of a Poisson Distribution Equals μ'

By definition

$$E(X) \quad \text{or} \quad \bar{X} = \sum_{X=0}^{X=\infty} \frac{(\mu')^X e^{-\mu'}}{X!} X$$

$$= \mu' \sum_{X-1=0}^{X-1=\infty} \frac{(\mu')^{X-1}}{(X-1)!} e^{-\mu'}$$

$$= \mu'$$

since the second factor is the sum of the ordinates of the Poisson distribution over the whole range of the variable and consequently equals 1.

(9) Proof that the Variance of a Poisson Distribution Equals μ'

By definition

$$\sigma'^2_X = \sum_{X=0}^{X=\infty} P(X)(X-\bar{X})^2 = \sum_{X=0}^{X=\infty} P(X)X^2 - 2\bar{X}\sum_{X=0}^{X=\infty} P(X)X + \bar{X}^2 \sum_{X=0}^{X=\infty} P(X)$$

$$= \sum_{X=0}^{X=\infty} P(X)X^2 - \bar{X}^2$$

$$\text{since } \sum_{X=0}^{X=\infty} P(X)X = \bar{X} \quad \text{and} \quad \sum_{X=0}^{X=\infty} P(X) = 1$$

Now

$$\sum_{X=0}^{X=\infty} P(X)X^2 = \sum_{X=0}^{X=\infty} \frac{(\mu')^X e^{-\mu'} X^2}{X!}$$

$$= \sum_{X=0}^{X=\infty} \frac{(\mu')^X e^{-\mu'}}{X!}(X)(X-1) + \sum_{X=0}^{X=\infty} \frac{(\mu')^X e^{-\mu'}}{X!} X$$

$$= (\mu')^2 \sum_{X-2=0}^{X-2=\infty} \frac{(\mu')^{X-2} e^{-\mu'}}{(X-2)!} + \mu' \sum_{X-1=10}^{X-1=\infty} \frac{(\mu')^{X-1} e^{-\mu'}}{(X-1)!}$$

$$= (\mu')^2 + \mu'$$

But $\bar{X}^2 = (\mu')^2$, as shown in Section (8). Hence

$$\sigma'^2_X = (\mu')^2 + \mu' - (\mu')^2 = \mu'$$

(10) Derivation of Tchebychev's Inequality

By definition the variance of a variable X equals[2]

$$\sigma'^2 = \sum_{i=1}^{N} P_i(X_i - \bar{X}')^2$$

$$= P_1(X_1 - \bar{X}')^2 + P_2(X_2 - \bar{X}')^2 + \ldots + P_N(X - \bar{X}')^3$$

[2] The proof here is for the discrete case. It is equally valid for the continuous case.

Let all values of X for which $X_i - \bar{X}'$ exceeds $k\sigma'$ be represented by X_s. Then we have

$$\sigma'^2 \geq \sum_s P_s(X_s - \bar{X}')^2$$

and

$$\sigma'^2 \geq \sum_s P_s(k\sigma')^2$$

But $\sum_s P_s$ = probability that $X - \bar{X}'$ exceeds $k\sigma'$. Hence,

$$P(X - \bar{X}' \geq k\sigma') \leq 1/k^2$$

(11) Proof that $E(X \pm Y) = E(X) \pm E(Y)$

Let the joint distribution of X and Y be represented by the following diagram:

	Y_1	Y_2	Y_3	$\cdot$	Y_n	*Total*
X_1	P_{11}	P_{12}	P_{13}	$\cdot$	P_{1n}	$P_{1.}$
X_2	P_{21}	P_{22}	P_{23}	$\cdot$	P_{2n}	$P_{2.}$
X_3	P_{31}	P_{32}	P_{33}	$\cdot$	P_{3n}	$P_{3.}$
$\cdot$	$\cdot$	$\cdot$	$\cdot$	$\cdot$	$\cdot$	$\cdot$
X_n	P_{n1}	P_{n2}	P_{n3}	$\cdot$	P_{nn}	$P_{n.}$
Total	$P_{.1}$	$P_{.2}$	$P_{.3}$	$\cdot$	$P_{.n}$	1.00

in which $X_1, X_2, X_3, \ldots, X_n$ represent the various categories of X; $Y_1, Y_2, Y_3, \ldots, Y_n$ represent the various categories of Y; P_{ij} represents the joint probability of X_i and Y_j; and $P_{i.}$ represents the probability of X_i and $P_{.j}$ represents the probability of Y_j.

By definition, the mean value of the distribution of $X \pm Y$ equals $\sum_i \sum_j P_{ij}(X_i \pm Y_j)$. Written in tabular form, this sum becomes

$$\begin{aligned} &P_{11}(X_1 \pm Y_1) + P_{12}(X_1 \pm Y_2) \ldots + P_{1n}(X_1 \pm Y_n) \\ &+ P_{21}(X_2 \pm Y_1) + P_{22}(X_2 \pm Y_2) + \ldots + P_{2n}(X_2 \pm Y_n) \\ &\quad \cdot \quad \cdot \quad \cdot \quad \cdot \quad \cdot \quad \cdot \\ &+ P_{n1}(X_n \pm Y_1) + P_{n2}(X_n \pm Y_2) + \ldots + P_{nn}(X_n \pm Y_n) \end{aligned}$$

If in this tabular array we remove the parentheses and sum the X terms by rows and the Y terms by columns, we get

$$\begin{aligned} \sum_i \sum_j P_{ij}(X_i \pm Y_j) \\ &= P_{1.}X_1 + P_{2.}X_2 + \ldots + P_{n.}X_n \\ &\pm P_{.1}Y_1 \pm P_{.2}Y_2 \pm \ldots \pm P_{.n}Y_n \end{aligned}$$

which equals

$$\sum_i P_{i.}X_i \pm \sum_j P_{.j}Y_j$$

$$= E(X) \pm E(Y)$$

(12) Proof that $E(XY) = E(X)E(Y)$ if X and Y Are Independent

By definition

$$E(XY) = \sum_i \sum_j P_{ij}(X_i Y_j)$$

But, on the assumption of independence,

$$\sum_i \sum_j P_{ij}(X_i Y_j) = \sum_i \sum_j P_{i.}P_{.j}(X_i Y_j) = \sum_i \sum_j P_{i.}(X_i)P_{.j}(Y_j)$$

If we refer to the diagram used in (11), we see that

$$\sum_i \sum_j P_{i.}(X_i)P_{.j}(Y_j)$$

$$= P_{1.}(X_1)P_{.1}(Y_1) + P_{1.}(X_1)P_{.2}(Y_2) + \ldots + P_{1.}(X_1)P_{.n}(Y_n)$$

$$+ P_{2.}(X_2)P_{.1}(Y_1) + P_{2.}(X_2)P_{.2}(Y_2) + \ldots + P_{2.}(X_2)P_{.n}(Y_n)$$

$$\cdot \quad \cdot \quad \cdot \quad \cdot \quad \cdot \quad \cdot$$

$$+ P_{n.}(X_n)P_{.1}(Y_1) + P_{n.}(X_n)P_{.2}(Y_2) + \ldots + P_{n.}(X_n)P_{.n}(Y_n)$$

$$= \sum_i P_{i.}(X_i)[P_{.1}(Y_1) + P_{.2}(Y_2) + \ldots + P_{.n}(Y_n)]$$

$$= \sum_i P_{i.}(X_i) \sum_j P_{.j}(Y_j)$$

$$= E(X)E(Y)$$

(13) Proof that $\sigma'^2_{X\pm Y} = \sigma'^2_X + \sigma'^2_Y$ if X and Y Are Independent

Refer once again to the diagram used in (11). Let us now assume that the variables are measured from their means. This we can do without loss of generality, since the scales on which we measure X and Y are purely arbitrary.

By definition,

$$\sigma'^2_{X\pm Y} = E[(X \pm Y) - E(X \pm Y)]^2$$

But by Section (11) $E(X \pm Y) = E(X) \pm E(Y)$. Since, by assumption, X and Y are deviations from their means, we have $E(X) = E(Y) = 0$. Hence

$$\sigma'^2_{X\pm Y} = E(X \pm Y)^2$$

On expansion we have

$$\sigma'^2_{X\pm Y} = E(X^2 + Y^2 \pm 2XY)$$

or from Section (11)

$$\sigma'^2_{X \pm Y} = E(X^2) + E(Y^2) \pm 2E(XY)$$
$$= \sigma'^2_X + \sigma'^2_Y \pm 2E(XY)$$

But, by Section (12),

$$E(XY) = E(X)E(Y) \quad \text{if } X \text{ and } Y \text{ are independent}$$

In our case $E(X)E(Y)$ equals 0 because we are measuring our variables from their means. Therefore,

$$\sigma'^2_{X \pm Y} = \sigma'^2_X + \sigma'^2_Y$$

(14) Proof that $\sigma'^2_{XY} = \bar{Y}'^2\sigma'^2_X + \bar{X}'^2\sigma'^2_Y + \sigma'^2_X\sigma'^2_Y$ if X and Y Are Independent

Set $X = \bar{X}' + x$ and $Y = \bar{Y}' + y$, then $(XY) = (\bar{X}' + x)(\bar{Y}' + y)$. By definition

$$\sigma'^2_{XY} = E(XY - E(XY))^2 = E(X^2Y^2) - [E(XY)]^2 = E(X^2Y^2) - \bar{X}'^2\bar{Y}'^2$$

But

$$E(X^2Y^2) = E[(\bar{X}' + x)^2(\bar{Y}' + y)^2] = E[(\bar{X}'^2 + 2x\bar{X}' + x^2)(\bar{Y}'^2 + 2y\bar{Y}' + y^2)]$$
$$= E[(\bar{X}'^2\bar{Y}'^2 + 2y\bar{X}'^2\bar{Y}' + y^2\bar{X}'^2 + 2x\bar{X}'\bar{Y}'^2 + 4xy\bar{X}'\bar{Y}' + 2xy^2\bar{X}'$$
$$+ x^2\bar{Y}'^2 + 2x^2y\bar{Y}' + x^2y^2]$$

Hence

$$E(X^2Y^2) - \bar{X}'^2\bar{Y}'^2 = \bar{X}'^2\sigma'^2_y + \bar{Y}'^2\sigma'^2_x + \sigma'^2_x\sigma'^2_y$$

since $E(2y\bar{X}'\bar{Y}')$ and similar terms are all zero, $E(x^2\bar{Y}'^2) = \bar{Y}'^2\sigma'^2_x$ and $E(y^2\bar{X}'^2) = \bar{X}'^2\sigma'^2_y$, and $E(x^2y^2) = E(x^2)E(y^2) = \sigma'^2_x\sigma'^2_y$ because x and y are independent and $E(x^2) = \sigma'^2_x$ and $E(y^2) = \sigma'^2_y$ by definition. We thus have

$$\sigma'^2_{XY} = \bar{Y}'^2\sigma'^2_X + \bar{X}'^2\sigma'^2_Y + \sigma'^2_X\sigma'^2_Y \text{ since } \sigma'^2_X = \sigma'^2_x \text{ and } \sigma'^2_Y = \sigma'^2_y$$

(15) Outline of Proof that the Binomial Distribution May Be Approximated by the Normal Distribution[3]

The formula for the binomial distribution is

$$\text{(1)} \qquad P(X) = \frac{n!}{X!(n-X)!}p'^X(1-p')^{n-X}$$

[3] For a more complete proof see J. G. Smith and A. J. Duncan, *Sampling Statistics and Applications* (New York: McGraw-Hill, 1945), pp. 68 ff.

First set $x = X - np'$. A sufficiently good approximation is obtained if we assume that x is an integer. Substitution of $X = x + np'$ in the formula for the binomial distribution yields

(2) $$P(x) = P(0)\frac{np'!nq'!}{(np'+x)!(nq'-x)!}p'^{x}q'^{-x}$$

where $q' = 1 - p'$ and

(3) $$P(0) = \frac{n!}{np'!nq'!}p'^{np'}q'^{nq'}$$

The second step is to use Stirling's approximation for factorials, viz

(4) $$a! \doteq a^a e^{-a}\sqrt{2\pi a}$$

Evaluation of the factorials in (2) by this formula yields (after considerable reduction)

(5) $$\frac{P(x)}{P(0)} \doteq \frac{1}{\left(1+\frac{x}{np'}\right)^{np'+x+1/2}\left(1-\frac{x}{nq'}\right)^{nq'-x+1/2}}$$

The third step is to take logs. This yields

(6) $$\log_e \frac{P(x)}{P(0)} \doteq -\left(np' + x + \frac{1}{2}\right)\log_e\left(1+\frac{x}{np'}\right) - \left(nq' - x + \frac{1}{2}\right)\log_e\left(1-\frac{x}{nq'}\right)$$

Finally we expand the logs in power series [note that $\log_e(1 + a) = a - \frac{a^2}{2} + \frac{a^3}{3} + \frac{a^4}{4} + \ldots$] and multiply out various terms, arranging the results in rising powers of $1/n$ and neglecting all terms of order equal to or higher than $1/\sqrt{n}$. [Note that $\sigma'_{x/n} = \sqrt{\frac{p'(1-p')}{n}}$, so that x/n is of the order $1/\sqrt{n}$. Hence, we keep only terms such as x or $x/\sqrt{n}$. We drop terms such as x/n, $x/n^{3/2}$, x/n^2, and so on.] The result of the final step is

$$\log_e \frac{P(x)}{P(0)} \doteq -\frac{x^2}{2\sigma_x'^2} \quad \text{where } \sigma_x'^2 = np'q'$$

Hence

$$\frac{P(x)}{P(0)} \doteq e^{-x^2/2\sigma_x'^2}$$

Evaluating $P(0)$ by Stirling's formula yields

$$P(x) \doteq \frac{1}{\sigma'_x\sqrt{2\pi}} e^{-x^2/2\sigma'^2_x}$$

which is the formula for the normal curve.

(16) Proof that the Poisson Distribution Approximates the Binomial Distribution if p' Is Small and n Is Large

The formula for the binomial distribution is

$$P(X) = \frac{n!}{X!(n-X)!} P'^X(1-p')^{n-X}$$

Let $p'n = u'$ and substitute u'/n for p', giving

$$P(X) = \frac{n!}{X!(n-X)!}\left(\frac{u'}{n}\right)^X\left(1-\frac{u'}{n}\right)^{n-X}$$

This may be rewritten

$$P(X) = \frac{u'^X}{X!}\left(1-\frac{1}{n}\right)\left(1-\frac{2}{n}\right)\ldots\left(1-\frac{X-1}{n}\right)\left(1-\frac{u'}{n}\right)^n\left(1-\frac{u'}{n}\right)^{-X}$$

Now, if n is increased, the expression approaches

$$P(X) = \frac{u'^X e^{-u'}}{X!}$$

since $1/n$, $2/n$, and so forth, all approach zero and the limit of $\left(1-\frac{u'}{n}\right)^n$ as n approaches infinity is $e^{-u'}$. The Poisson distribution is thus the limit of the binomial distribution as n approaches infinity but $p'n$ remains constant.

(17) Binomial Approximations to the Hypergeometric Distribution

If N, m, and $N-m$ are large relative to n and X, the hypergeometric distribution can be approximated by a binomial distribution with $p' = m/N$ and n (binomial) $= n$ (hypergeometric). This can be seen by noting that the formula for the hypergeometric distribution may be written

(1)
$$P(X/n) = \frac{C_{n-X}^{N-m}C_X^m}{C_m^N} = \frac{\dfrac{(N-m)!}{(n-X)!(N-m-n+X)!}\cdot\dfrac{m!}{X!(m-X)!}}{\dfrac{N!}{n!(N-n)!}}$$

or

(1a)
$$P(X/n) = \frac{n!(N-n)!m!(N-m)!}{(n-X)!(m-X)!X!N!(N-m-n+X)!}$$

which can be written

(1b) $P(X/n) =$

$$\frac{\left(\frac{(N-m)(N-m-1)\ldots(N-m-n+X+1)}{(n-X)!}\right)\left(\frac{m(m-1)\ldots(m-X+1)}{X!}\right)}{\frac{N(N-1)\ldots(N-n+1)}{n!}}$$

In (1b) factor $(N - m)^{n-X}$ out of the first expression in the numerator, m^X out of the second expression and N^n out of the denominator. Set $p' = m/N$, $(1 - p') = (N - m)/N$ and note that terms like $(n - X + 1)/(N - m)$, $(X + 1)/m$ and $(n - 1)/N$ can be neglected if N, $N - m$, and m are large relative to n and X. Then we shall have

(1c) $$P(X/n) \doteq \frac{n!}{X!(n-X)!}(p')^X(1-p')^{a-X}$$

This approximation is called Binomial I.

If N and n are large, but m or $N - m$ is relatively small, the hypergeometric distribution can be approximated by a binomial distribution with $p' = n/N$ and n(binomial) $= m$(hypergeometric). This follows from (1a) in exactly the same way that (1c) did because of the symmetry of (1a) in n and m. We thus have

(1d) $$P(X/n) \doteq \frac{m!}{X!(m-X)!}(p')^X(1-p')^{m-X}$$

where $p' = n/N$. This approximation is called Binomial II.

The formulas (1c) and (1d) are more often used in cumulative form. Thus,

(1e) $$\sum_{X=0}^{c} P(X/n) \doteq \sum_{X=0}^{c} \frac{n!}{X!(n-X)!}\left(\frac{m}{N}\right)^X\left(1-\frac{m}{N}\right)^{n-X}$$

and

(1f) $$\sum_{X=0}^{c} P(X/n) \doteq \sum_{X=0}^{c} \frac{m!}{X!(m-X)!}\left(\frac{n}{N}\right)^X\left(1-\frac{n}{N}\right)^{m-X}$$

(18) Leslie E. Simon's I_Q Charts

General Leslie E. Simon[4] has prepared certain charts for evaluating binomial probabilities. These are his $I_Q = 0.005$, $I_Q = 0.1$, $I_Q = 0.5$, $I_Q = 0.9$, and

[4] See enclosures in his book *An Engineers' Manual of Statistical Methods* (New York: John Wiley & Sons, 1941).

$I_Q = 0.995$ charts. For values of the binomial p' (Simon's Q) from 0.00001 to 0.50000 and for values of n up to 500, Simon's 0.005 chart gives the values of X (called c by Simon) which bracket the upper 0.005 point of the binomial distribution. The c's are to be read as 1 more than those given on the chart. More concretely, if p' or $Q = 0.01$, say, and $n = 200$, Simon's 0.005 chart shows that the probability that the number of nonconforming units will equal 7 or more is somewhat less than 0.005 and the probability that the number of nonconforming units will equal 6 or more is somewhat greater than 0.005. Thus, $c = 7$ and $c = 6$ may be said to bracket the 0.005 point.[5] His 0.1 chart gives the c's that bracket the upper 0.1 point. (Again take actual c's as 1 more than those read from the chart.) His 0.995 chart and 0.9 chart give the c's which bracket the lower 0.005 and lower 0.1 points. In the case of the 0.9 and 0.995 charts, however, it should be noted that the actual c's are to be taken equal to those given directly by the chart.

Simon's charts are based on the Incomplete Beta Function and are therefore subject only to the graphical errors involved in reading from the chart.[6] Their usefulness is accordingly limited only by their restriction to selected probability points and the limitation of sample sizes to 500 or less. Simon's charts are especially suited for finding confidence limits for p' when n and c are given. Their use in this connection is discussed in Section (30) below.

(19) Karl Pearson's Type III Distribution

Sometimes a formula for a continuous skewed distribution is found very useful. Such a formula is provided by Karl Pearson's Type III distribution. The density function for Karl Pearson's Type III distribution may be written

$$y = y_0\left(1 + \frac{\gamma_1'}{2} z\right)^{\frac{4}{\gamma_1'^2} - 1} e^{-\frac{2}{\gamma_1'} z}$$

where y_0 is the ordinate at the origin and the origin is at the mean of the distribution. Tables of ordinates and areas under the standardized Type III curve[7] have been prepared by L. R. Salvosa and may be found in the *Annals*

---

[5] Since the binomial distribution is discrete, there is not likely to be an exact 0.005 point.

[6] Of these charts Simon writes: "Except for very high and very low ranges of Q [the binomial p'], the data for the charts were calculated to three significant figures. The plots were made to a large scale, the curves carefully smoothed, and the charts finally photostated down to reduced size. The grid of the charts is especially designed to spread the curves over the surface and to keep the precision in significant figures as nearly uniform as possible. Actually, the calculations of the points were carried to four figures and then rounded off; and this precision, together with that lent by graphical smoothing on a large scale, tends to give the charts considerably greater precision than that to which they can be read. It should be noted, therefore, that for practical purposes the accuracy of results obtained from the chart is, in the case of a priori probabilities, a function of the accuracy of the data with which they are entered." (Ibid., pp. 186–87.)

[7] "Standardized" in the sense that the variable is $z = \dfrac{X - \bar{X}'}{\sigma'}$.

of Mathematical Statistics, Volume I (1930).[8] These tables give the ordinates and areas from $-\infty$ to z for values of z from -4.99 to $+4.99$ and values of γ_1' from 0 (the normal curve case) to 1.1. If it is deemed that a distribution can be sufficiently well approximated by a Type III curve, these tables will be found very useful in plotting the curve or in computing probabilities for specified intervals.

(20) Proof that $E(\bar{X}) = \bar{X}'$

In Section (11) it was proved that the mean value of a sum $X_1 + X_2$ equals the mean value of X_1 plus the mean value of X_2. By definition, the mean of a sample of n items is

$$\bar{X} = \frac{X_1 + X_2 + \ldots + X_n}{n}$$

Hence

$$E(\bar{X}) = \frac{1}{n} E(X_1 + X_2 + \ldots + X_n) = \frac{1}{n} [E(X_1) + E(X_2) + \ldots + E(X_n)]$$

But $X_1, X_2 \ldots X_n$ are all from the same universe. Hence

$$E(\bar{X}) = \frac{1}{n} [nE(X)] = E(X) = \bar{X}'$$

(21) Proof that the Variance of $\bar{X}$ Equals σ'^2/n

This can be proved with the help of the corollary to Theorem 4.2 in Chapter 4. This stated that the variance of the distribution of $aX_1 + bX_2$ equals a^2 times the variance of X_1 plus b^2 times the variance of X_2, if X_1 and X_2 are independent. In the present instance we are concerned with random sampling, and in random sampling there is no correlation between the items picked (otherwise the sampling would not be random). Hence we can immediately apply the above corollary.

Proof:

By definition

$$\bar{X} = \frac{X_1 + X_2 + \ldots + X_n}{n} = \frac{X_1}{n} + \frac{X_2}{n} + \ldots + \frac{X_n}{n}$$

Let $\sigma^2(X)$ mean the variance of X, i.e., let $\sigma^2(\ )$ be an "operator," which means finding the variance of what is in the parentheses. Then, by the corollary to Theorem 4.2,

[8] Salvosa's tables have also been published separately by the Institute of Mathematical Statistics, the publisher of the *Annals.*

$$\sigma^2(\bar{X}) = \sigma^2\left(\frac{X_1}{n} + \frac{X_2}{n} + \ldots + \frac{X_n}{n}\right)$$

$$= \frac{1}{n^2}[\sigma^2(X_1) + \sigma^2(X_2) + \ldots + \sigma^2(X_n)]$$

But $\sigma^2(X_1)$ is the same as $\sigma^2(X_2)$ and the same as $\sigma^2(X_n)$, and so on, since the X_i all come from the same universe. Hence we have

$$\sigma^2(\bar{X}) = \frac{n\,\sigma^2(X)}{n^2} = \frac{\sigma^2(X)}{n}$$

or, using the normal notation in which σ^2 is the variance, we have

$$\sigma_{\bar{X}}'^2 = \sigma_X'^2/n$$

(22) Proof that the Mean of s^2 Equals σ'^2

$$E(s^2) = E\left[\frac{\sum_i (X_i - \bar{X})^2}{n-1}\right] = E\left[\frac{\sum_i [(X_i - \bar{X}') - (\bar{X} - \bar{X}')]^2}{n-1}\right]$$

$$= E\left[\frac{\sum_i (X_i - \bar{X}')^2 + n(\bar{X} - \bar{X}')^2 - 2(\bar{X} - \bar{X}') \sum_i (X_i - \bar{X}')}{n-1}\right]$$

$$= E\left[\frac{\sum (X_i - \bar{X}')^2 - n(\bar{X} - \bar{X}')^2}{n-1}\right]$$

since $\Sigma(X_i - \bar{X}') = n(\bar{X} - \bar{X}')$.

But

$$E(\bar{X} - \bar{X}')^2 = E\left[\frac{(X_1 - \bar{X}') + (X_2 - \bar{X}') + \ldots + (X_n - \bar{X}')}{n}\right]^2$$

$$= E\left[\frac{(X_1 - \bar{X}')^2 + (X_2 - \bar{X}')^2 + \ldots + (X_n - \bar{X}')^2}{n^2}\right]$$

since $E(X_i - \bar{X}')(X_j - \bar{X}') = 0$ for $i \neq j$.

Hence

$$E(s^2) = E\left[\frac{\sum (X_i - \bar{X}')^2 - \sum (X_i - \bar{X}')^2/n}{n-1}\right]$$

$$= \frac{n\sigma'^2 - n\sigma'^2/n}{n-1} = \sigma'^2$$

since by definition $E[(X_i - \bar{X}')^2] = \sigma'^2$.

(23) Proof that the Variance of s^2 Equals $\frac{1}{n}\left[\mu'_4 - \frac{n-3}{n-1}\mu'^2_2\right]$

In (22) we saw that

$$s^2 = \frac{\sum (X_i - \bar{X})^2}{n-1} = \frac{\sum (X_i - \bar{X}')^2 - n(\bar{X} - \bar{X}')^2}{n-1}$$

Likewise we have by definition

$$\textit{Variance of } s^2 = E[s^2 - E(s^2)]^2 = E[s^2 - \sigma'^2]^2$$
$$= E(s^2)^2 - 2\sigma'^2 E(s^2) + \sigma'^4 = E(s^2)^2 - \sigma'^4$$

We then have

$$E(s^2)^2 = E\left[\frac{\sum (X_i - \bar{X})^2}{n-1}\right]^2 = \frac{E\left[\sum (X_i - \bar{X}')^2 - n(\bar{X} - \bar{X}')^2\right]^2}{(n-1)^2}$$
$$= \frac{E\left[\left(\sum (X_i - \bar{X})^2\right)^2 - 2n(\bar{X} - \bar{X}')^2 \sum (X_i - \bar{X}')^2 + n^2(\bar{X} - \bar{X}')^4\right]}{(n-1)^2}$$

Treating each term in the numerator separately, we have:

(A)

$$E\left[\sum (X_i - \bar{X})^2\right]^2 = E\,[X_1 - \bar{X}')^2 + (X_2 - \bar{X}')^2 + \ldots + (X_n - \bar{X}')^2]^2$$
$$= nE(X_i - \bar{X}')^4 + n(n-1)E[(X_i - \bar{X}')^2(X_j - \bar{X}')^2]$$

when $i \neq j$

$$= n\mu'_4 + n(n-1)\mu'^2_2$$

since X_i and X_j are independent.

(B)

$$E\left[2n(\bar{X} - \bar{X}')^2 \sum (X_i - \bar{X}')^2\right]$$
$$= 2nE\left\{\left[\frac{(X_1 - \bar{X}')^2 + (X_2 - \bar{X}')^2 + \ldots + (X_n - \bar{X}')^2 + 2(X_1 - \bar{X}')(X_2 - \bar{X}') + 2(X_1 - \bar{X}')(X_3 - \bar{X}') + \ldots}{n^2}\right]\left[(X_1 - \bar{X}')^2 + (X_2 - \bar{X}')^2 + \ldots + (X_n - \bar{X}')^2\right]\right\}$$
$$= 2n\left[\frac{nE(X_i - \bar{X}')^4 + n(n-1)E(X_i - \bar{X}')^2(X_j - \bar{X}')^2}{n^2}\right]$$

since $E(X_i - \bar{X}')^3(X_j - \bar{X}') = 0$

and $E(X_i - \bar{X}')^2(X_j - \bar{X}')(X_k - \bar{X}') = 0$

all for $i \neq j \neq k$.

$$= 2[\mu'_4 + (n-1)\mu'^2_2]$$

(C) $$E[n^2(\bar{X} - \bar{X}')^4] = n^2 E\left[\frac{(X_1 - \bar{X}') + (X_2 + \bar{X}') + \ldots + (X_n - \bar{X}')}{n}\right]^4$$

$$= \frac{nE(X_i - \bar{X}')^4 + 3n(n-1)E(X_i - \bar{X}')^2(X_j - \bar{X}')^2}{n^2}$$

$$= \frac{\mu'_4 + 3(n-1)\mu'^2_2}{n}$$

Hence

$$E(s^2) = \frac{A - B + C}{(n-1)^2}$$

$$= \frac{[n\mu'_4 + n(n-1)\mu'^2_2] - [2(\mu'_4 + (n-1)\mu'^2_2)] + \left[\frac{\mu'_4 + 3(n-1)\mu'^2_2}{n}\right]}{(n-1)^2}$$

$$= \frac{\mu'_4}{n} + \frac{(n-1)^2 + 2}{n(n-1)}\mu'^2_2$$

Finally we have

$$\text{Variance } s^2 = E(s^2) - \sigma'^4$$

$$= \frac{\mu'_4}{n} + \frac{(n-1)^2 + 2}{n(n-1)}\mu'^2_2 - \mu'^2_2$$

$$= \frac{1}{n}\left[\mu'_4 - \frac{n-3}{n-1}\mu'^2_2\right]$$

(24) Proof that the Standard Deviation of s Equals Approximately $\frac{\sigma'}{\sqrt{2n}}\sqrt{\frac{\gamma'_2 + 1}{2}}$

By an elementary theorem in algebra,

$$(s^2 - \sigma'^2) = (s + \sigma')(s - \sigma')$$

Hence

$$s - \sigma' = \frac{(s^2 - \sigma'^2)}{s + \sigma'}$$

If n is large, we shall have $s + \sigma'$ close to $2\sigma'$. Hence we can write the above as

$$(s - \sigma') \doteq \frac{(s^2 - \sigma'^2)}{2\sigma'}$$

Since, from Theorem 6.6 of Chapter 6, $E(s - \sigma') \doteq 0$ if n is large, the variance of $(s - \sigma')$ will be given approximately by

$$E(s - \sigma')^2 \doteq \frac{E(s^2 - \sigma'^2)^2}{4\sigma'^2}$$

But, from Section (23), $E[(s^2 - \sigma'^2)]^2 = \frac{1}{n}\left[\mu_4' - \frac{n-3}{n-1}\mu_2'^2\right]$ which for large n equals approximately $\frac{\mu_4' - \mu_2'^2}{n}$.

Hence

$$E(s - \sigma')^2 \doteq \frac{E(s^2 - \sigma'^2)^2}{4\sigma'^2} \doteq \frac{\mu_4' - \mu_2'^2}{4n\sigma'^2} = \frac{\mu_2'^2(\gamma_2' + 2)}{4n\mu_2'}$$

where $\gamma_2' = \frac{\mu_4'}{\mu_2'^2} - 3$. Accordingly, for the standard deviation of sample standard deviations we have

$$\sigma_s' \doteq \sigma'\sqrt{\frac{\gamma_2' + 2}{4n}} = \frac{\sigma'}{\sqrt{2n}}\sqrt{\frac{\gamma_2'}{2} + 1}$$

(25) Derivation of a Unit Sequential-Sampling Plan with a Given p_1', p_2', α, and β

Let the prescribed characteristics for the sampling plan be p_1', p_2', α, and β. To find the plan desired, use is made of the "sequential probability ratio." This is the ratio of the probability of a cumulated sample result computed on the assumption that p_2' is the true fraction nonconforming to the probability of the cumulated sample result computed on the assumption that p_1' is the true fraction nonconforming. For example, suppose that p_1' is set at 0.01 and p_2' at 0.08, and consider a point that lies slightly beyond the reject limit of Figure 8.3. Such a point would be given by $n = 45$, $c = 3$. This point could be reached, for example, alone a path such as that outlined in the diagram. If the true fraction nonconforming was 0.08, the probability of such a series of cases would be $(0.08)^3(1 - 0.08)^{42}$. Let there be K such ways of reaching the point (45,3) without going outside the reject or accept limits until the last case.[9] Then, on the assumption that $p' = p_2' = 0.08$,

[9] K does not equal the binominal coefficient $\frac{n!}{X!(n-X)!}$, since this includes several paths of reaching the point (45,3) that are outside the limits. For example, the binomial coefficient includes the occurrence of 3 nonconforming items followed by 42 conforming items. But under the sequential-sampling plan this would lead to rejection on the second item inspected and hence is not an "eligible" path for reaching the point (45,3).

the probability of reaching the point (45,3) is $K(0.08)^3(1-0.08)^{42}$. Similarly, the probability of reaching the point (45,3) on the assumption that $p' = p_1' = 0.01$, is $K(0.01)^3(1-0.01)^{42}$. Hence the sequential probability ratio for that point is

$$\text{SPR} = \frac{K(0.08)^3(1-0.08)^{42}}{K(0.01)^3(1-0.01)^{42}} = \frac{(0.08)^3(1-0.08)^{42}}{(0.01)^3(1-0.01)^{42}}$$

In general, the sequential probability ratio is given by

$$\text{(1)} \qquad \text{SPR} = \frac{p_2'^X(1-p_2')^{n-X}}{p_1'^X(1-p_1')^{n-X}} = \left(\frac{p_2'}{p_1'}\right)^X \left(\frac{1-p_2'}{1-p_1'}\right)^{n-X}$$

where n is the size of the cumulated sample and X is the cumulated number of nonconforming units. If we take logs, we get

$$\text{(2)} \qquad \log \text{SPR} = X \log\left[\frac{p_2'}{p_1'}\right] + (n-X) \log\left[\frac{1-p_2'}{1-p_1'}\right]$$

$$= X \log\left[\frac{p_2'(1-p_1')}{p_1'(1-p_2')}\right] + n \log\left[\frac{1-p_2'}{1-p_1'}\right]$$

Since $p_2' > p_1'$, the coefficient of X in the above equation will always be positive and the coefficient of n always negative. Hence the sequential probability ratio increases with X and decreases with n.

Since the sequential probability ratio varies directly with n and X, a sequential-sampling plan may be set up as follows. If the sequential probability ratio for a cumulated sample of n units is found to be less than or equal to a designated quantity B, the lot is to be accepted. If the ratio is found to be equal to or greater than another designated quantity A, A being greater than B, the lot is to be rejected. If the ratio is found to lie between A and B, another unit is added to the sample and the analysis repeated. The problem then becomes that of finding an A and a B that will satisfy the initial requirements of the sampling plan, viz that, if the lot fraction nonconforming is p_1', the probability of acceptance will be $1-\alpha$; if it is p_2', the probability of acceptance will be β.

With reference to Figure A1, the set of points that will lead to acceptance are the heavy dots lying on the acceptance line. The set of points that will lead to rejection is the set of points falling on the heavy lines to the left of the rejection line. For the heavy point i on the acceptance line we must have

$$\frac{K_i p_2'^{X_i}(1-p_2')^{n_i-X_i}}{K_i p_1'^{X_i}(1-p_1')^{n_i-X_i}} \leq B$$

or

$$K_i p_2'^{x_i}(1-p_2')^{n_i-X_i} \leq B\, K_i p_1'^{X_i}(1-p_1')^{n_i-X_i}$$

FIGURE A1
An Illustration Pertaining to the Derivation of a Sequential-Sampling Plan

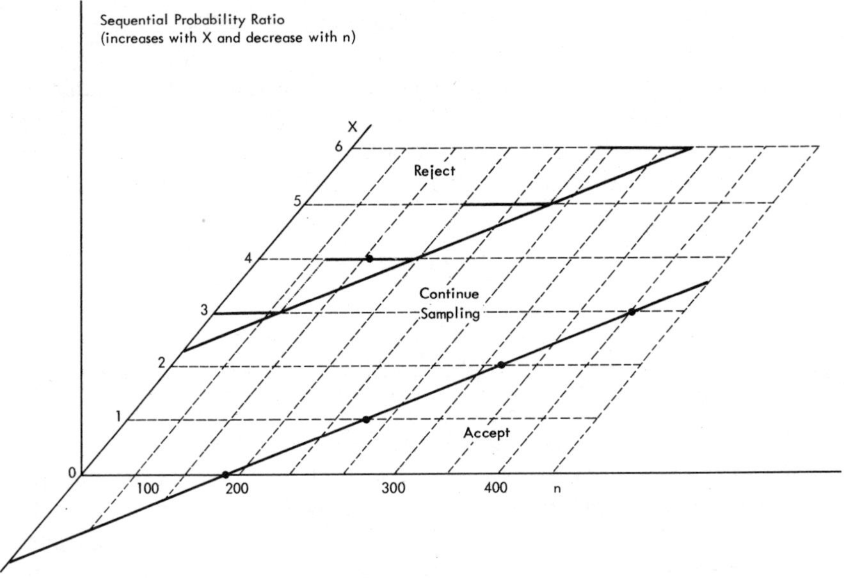

Similar relationships exist for all the other heavy points on the acceptance line. Hence

$$\sum_i K_i p_2'^{X_i}(1-p_2')^{n_i-X_i} \leq B \sum_i K_i p_1'^{X_i}(1-p_1')^{n_i-X_i}$$

But the sum on the left is the probability of acceptance on the assumption that $p' = p_2'$, which, according to the design of our plan, is to equal β, while the sum on the right is the probability of acceptance on the assumption that $p' = p_1'$, which, according to design, is to equal $1 - \alpha$. Hence the specifications for our sampling plan will be met if

$$\beta \leq B(1-\alpha)$$

or if

$$B \geq \frac{\beta}{1-\alpha} \tag{3}$$

In like manner, for the point j in Figure A1, we must have

$$\frac{K_j p_2'^{X_j}(1-p_2')^{n_j-X_j}}{K_j p_1'^{X_j}(1-p_1')^{n_j-X_j}} \geq A'$$

or

$$K_j p_2'^{X_j}(1-p_2')^{n_j-X_j} \geq A\, K_j p_1'^{X_j}(1-p_1')^{n_j-X_j}$$

This will be true for any of the points on the heavy lines to the left of the rejection line. Hence

$$\sum_j K_j p_2'^{X_j}(1-p_2')^{n_j-X_j} \geq A \sum_j K_j p_1'^{X_j}(1-p_1')^{n_j-X_j}$$

But the sum on the left is the probability of rejection on the assumption that $p' = p_2'$, which, according to the design of the plan, is to equal $1-\beta$, while the sum on the right is the probability of rejection on the assumption that $p' = p_1'$, which, according to design, is to equal α. Hence the specifications will be met if

$$1-\beta \geq A\alpha$$

or if

(4) $$A \leq \frac{1-\beta}{\alpha}$$

The second of the above inequalities gives $(1-\beta)/\alpha$ as an upper limit for A and the first gives $\beta/(1-\alpha)$ as a lower limit for B. In practice, it is convenient to ignore the inequality signs and to put $B = \beta/(1-\alpha)$ and $A = (1-\beta)/\alpha$. This may result in giving an operating characteristic curve which has values of α and β different from those initially prescribed, but the differences will be slight. If $1-\alpha'$ and β' are the probabilities of acceptance of p_1' and p_2' quality given by the operating characteristic curve that results from putting $B = \beta/(1-\alpha)$ and $A = (1-\beta)/\alpha$, then it can be shown[10] that $\alpha' + \beta' \leq \alpha + \beta$ and $\alpha' \leq \frac{\alpha}{1-\beta}$ and $\beta' \leq \frac{\beta}{(1-\alpha)}$. Thus if α and β are small, α' and β' will not differ much from them.

From here on, the solution of the problem is straight algebra. The limits to the sequential probability ratio are given by the two equations.

(5)
$$\frac{p_2'^X(1-p_2')^{n-X}}{p_1'^X(1-p_1')^{n-X}} = A = \frac{1-\beta}{\alpha}$$

$$\frac{p_2'^X(1-p_2')^{n-X}}{p_1'^X(1-p_1')^{n-X}} = B = \frac{\beta}{1-\alpha}$$

Following the analysis suggested by equation (2), equations (5) can be put in a form that expresses X as a function of n, viz

(6)
$$X = -h_1 + sn$$
$$X = h_2 + sn$$

where

[10] A. Wald, *Sequential Analysis* (New York: John Wiley & Sons, 1947), pp. 45–46.

(7)
$$h_1 = \frac{\log\left(\frac{1-\alpha}{\beta}\right)}{\log\left[\frac{p_2'(1-p_1')}{p_1'(1-p_2')}\right]}$$

$$h_2 = \frac{\log\left(\frac{1-\beta}{\alpha}\right)}{\log\left[\frac{p_2'(1-p_1')}{p_1'(1-p_2')}\right]}$$

$$s = \frac{\log\left[\frac{1-p_1')}{(1-p_2')}\right]}{\log\left[\frac{p_2'(1-p_1')}{p_1'(1-p_2')}\right]}$$

(26) Proof that s Lies between p_1' and p_2'[11]

Let $g_1 = \log p_2'/p_1'$ and $g_2 = \log \frac{1-p_1'}{1-p_2'}$ and note that p_1' is, by the conditions of the problem, less than p_2'.

By definition, $s = \frac{g_2}{g_1+g_2}$. Then $sg_1 + sg_2 = g_2$. Therefore, $g_2(1-s) = sg_1$ and $g_2 = \frac{sg_1}{1-s}$.

By the theorem of the mean of the differential calculus, we have

$$g_1 = \log p_2' - \log p_1' = (p_2' - p_1')\frac{1}{e_1} \quad \text{where} \quad p_1' < e_1 < p_2'$$

$$g_2 = \log(1-p_1') - \log(1-p_2') = (p_2' - p_1')\frac{1}{e_2}$$

$$\text{where } (1-p_2') < e_2 < (1-p_1')$$

Hence

$$(p_2' - p_1')\frac{1}{e_2} = \frac{s}{1-s}(p_2' - p_1')\frac{1}{e_1}$$

Therefore,

$$\frac{s}{1-s} = e_1/e_2$$

Hence

[11] The author is indebted to Dr. Samuel Bourne for this proof.

$$\frac{p_1'}{1-p_1'} < \frac{s}{1-s} < \frac{p_2'}{1-p_2'}$$

and

$$1+\frac{p_1'}{1-p_1'} < 1+\frac{s}{1-s} < 1+\frac{p_2'}{1-p_2'}$$

$$\frac{1}{1-p_1'} < \frac{1}{1-s} < \frac{1}{1-p_2'}$$

Therefore, $1 - s < 1 - p_1'$ and $p_1' < s$. Also $1 - p_2' < 1 - s$ and $s < p_2'$. Hence $p_1' < s < p_2'$, as was to be proved.

(27) Derivation of a Sampling Plan Based on $\bar{X}$ in which the Fraction Nonconforming of the Lot Is the Criterion of Acceptability. Given the p_1', p_2', α, and β for the Plan; σ' Known and Constant, Universe of Type III, and a Single Engineering Specification Limit

When the distribution of product is nonnormal, the procedure is similar to that outlined in Section 4.1 of Chapter 11, except that the relationship between $\bar{X}'$ and p' will be different. To get that relationship, we must know the form of the distribution. To show how this is done when the distribution of product is nonnormal, we shall assume that the distribution can be represented by Karl Pearson's Type III distribution.[12] We shall consider only a single lower specification limit.

To illustrate the procedure, let us take up again the problem of Section 4.1. As before, let the desired p_1' for our plan be 0.01 and the p_2', 0.08. Let the specification limit be 17,000 psi and $\sigma' = 800$, but let γ_1 now equal 0.5 instead of 0. Our first task is to find a z such that the probability of falling short of 17,000 psi is 0.01 in one case and 0.08 in the other case. For $\gamma_1 = 0.5$, Salvosa's tables of Type III distribution show that the lower 0.01 point of the distribution comes at $z = -1.955$ and the lower 0.08 point comes at -1.312. Hence, the $\bar{X}'$'s that are the equivalent of $p' = 0.01$ and 0.08 are

$$\frac{17{,}000-\bar{X}'}{\sigma'} = -1.955$$

or

$$\bar{X}' = 18{,}564$$

and

$$\frac{17{,}000-\bar{X}'}{\sigma'} = -1.312$$

[12] See Section (19), above.

or

$$\bar{X}' = 18{,}050$$

These are the $\bar{X}_1'$ and $\bar{X}_2'$ for our variables plan.

In this nonnormal case it is easiest to operate the plan directly in terms of an acceptance $\bar{X}$, designated as $\bar{X}_a$. Thus we solve the following equations for $\bar{X}_a$ and n; viz

$$\frac{\bar{X}_a - 18{,}564}{800/\sqrt{n}} = -1.645$$

and

$$\frac{\bar{X}_a - 18{,}050}{800/\sqrt{n}} = 1.282$$

which yield $n \doteq 21$ and $\bar{X}_a = 18{,}275$.[13] (Note that the distribution of $\bar{X}$ is taken as normal even though X is not normally distributed.) The desired plan is to accept if $\bar{X} \geq \bar{X}_a$.

The OC curve for the plan is given by the probability that $\frac{\bar{X} - \bar{X}'}{\sigma'/\sqrt{n}} \geq \frac{\bar{X}_a - \bar{X}'}{\sigma'/\sqrt{n}}$ or the probability that a standard normal deviate (z) is greater than or equal to $\frac{\bar{X}_a - \bar{X}'}{\sigma'/\sqrt{n}}$, which is given directly by Table A2, Appendix II. Such calculations will give a curve with reference to abscissa values of $\bar{X}'$. To derive the equivalent p' scale, however, we use Salvosa's tables of the Type III distribution instead of the normal distribution. For example, for $\bar{X}' = 18{,}300$, we find from the *normal* table that the probability of a sample $\bar{X}$'s falling above 18,275 is the probability of a z equal to or greater than

$$\frac{18{,}275 - 18{,}300}{800/\sqrt{21}} = -0.14$$

which is 0.56. This is the ordinate of the OC curve at the abscissa point $\bar{X}' = 18{,}300$. Now, to convert 18,300 into a p' value, we use Salvosa's tables. Thus we have

$$\frac{17{,}000 - 18{,}300}{800} = -\frac{1{,}300}{800} = -1.62$$

For $z = -1.62$ with $\gamma_1 = 0.5$, Salvosa's tables give the probability of an equal or larger negative value as 0.034. Hence, corresponding to the point $\bar{X}' = 18{,}300$ on the $\bar{X}'$ scale, we have the point $p' = 0.034$ on the p' scale.

The foregoing results are based on the assumption that the universe has a Type III distribution with $\gamma_1 = 0.5$. It is interesting to note the results

[13] One equation gives 18,273.8, the other 18,276.8.

obtained if $\gamma_1 = -0.5$, i.e., if the universe is skewed negatively. This means that the specification limit is on the long tail of the distribution and not the short tail.

When $\gamma_1 = -0.5$, a lot whose fraction nonconforming is 0.01 will, according to Salvosa's tables, have a mean given by the following equation:

$$\frac{17{,}000 - \bar{X}'}{800} = -2.686$$

and a lot whose fraction nonconforming is 0.08 will have a mean given by the equation:

$$\frac{17{,}000 - \bar{X}'}{800} = -1.474$$

These yield the values of $\bar{X}' = 19{,}149$ and $\bar{X}' = 18{,}179$.

Hence, $\bar{X}_a$ and n are given by the equations

$$\frac{\bar{X}_a - 19{,}149}{800/\sqrt{n}} = -1.645$$

and

$$\frac{\bar{X}_a - 18{,}179}{800/\sqrt{n}} = 1.282$$

These yield $n = 5.81$, or approximately 6, and $\bar{X}_a = 18{,}605$.

It will be noticed that, when the specification limit is on the long tail of the universe distribution, the size of the sample needed to meet the requirements is smaller than when the specification limit is on the short tail. This is because on the $\bar{X}'$ scale the distance between the 0.01 and 0.08 points is greater on the long tail than on the short tail. Hence, the discriminating power of the variables plan, and therefore the sample size, need not be as great to satisfy the desired criteria.

(28) Derivation of the OC Curve for a Sampling Plan Based on s

In Chapter 15, Section 3, it was pointed out that a sampling plan based on s might run as follows. A random sample of n items is taken from a lot, and the s of the sample computed. If this exceeds a given critical value s_a, the lot is rejected. If it equals or falls short of s_a, the lot is accepted. In Chapter 15, OC curves for such a plan were given for $\alpha = 0.05$ and a variety of sample sizes. It is the purpose here to show how the OC curve for a particular plan is derived.

The ordinate of the OC curve is the probability (P_a) that we shall accept the lot when the lot standard deviation (σ') equals a given value, say, σ'_g. If s_a is the acceptance limit, then P_a is the probability of getting a sample

s less than or equal to s_a when the lot standard deviation is σ'_g. Symbolically, we can write this as follows:

$$P_a = \text{Prob. } (s \le s_a \,|\, \sigma' = \sigma'_g)$$

But the equality or inequality is not affected if we square both sizes and multiply by $(n - 1)/\sigma'^2_g$. Thus we have

$$P_a = \text{Prob. } \left(\frac{(n-1)s^2}{\sigma'^2_g} \le \frac{(n-1)s_a^2}{\sigma'^2_g} \,|\, \sigma'^2 = \sigma'^2_g \right)$$

But

$$\frac{(n-1)s_a^2}{\sigma'^2_g} = \frac{(n-1)s_a^2}{\sigma'^2_g}\,\frac{\sigma'^2_1}{\sigma'^2_1} = \frac{(n-1)s_a^2}{\sigma'^2_1}\,\frac{\sigma'^2_1}{\sigma'^2_g} = \frac{(n-1)s_a^2}{\lambda^2 \sigma'^2_1}$$

where $\lambda = \sigma'_g/\sigma'_1$. Also note that $\dfrac{(n-1)s_a^2}{\sigma'^2_1} = \chi_a^2$, since s_a^2 was found by setting $\dfrac{(n-1)s_a^2}{\sigma'^2_1} = \chi_a^2$. Hence,

$$P_a = \text{Prob. } \left(\frac{(n-1)s_a^2}{\sigma'^2_g} \le \frac{\chi_a^2}{\lambda^2} \,|\, \sigma'^2 = \sigma'^2_g \right)$$

But if $\sigma'^2 = \sigma'^2_g$, sample values of $(n = 1)s_a^1/\sigma'^2_g$ will form a χ^2 distribution with $\nu = n - 1$ (cf. Chapter 6, Section 2.3). Hence P_a is the probability that a variable distributed as χ^2 with $\nu = n - 1$ will be equal to or less than χ_a^2/λ^2, i.e., $P_a = \text{Prob. } (\chi^2 \le \chi_a^2/\lambda^2)$ for $\nu = n - 1$, or, reversing the inequality, $P_a = 1 - \text{Prob. } (\chi^2 > \chi_a^2/\lambda^2)$ for $\nu = n - 1$. Conversely, if we specify P_a, we can compute λ by taking $P = 1 - P_a$, locating χ^2 for $\nu = n - 1$ under this P, and then taking $\lambda^2 = \chi_a^2/\chi_P^2$.

(29) Confidence Limits for σ'^2 and σ' when the Sample Is Large and Universe Nonnormal

When the universe is nonnormal, we cannot set exact confidence limits for the variance or standard deviation of the population. We have two courses open. We can either use the normal theory, with the hope that the results are not too far off, or we can use Theorem 6.5 of Chapter 6 and estimate our probabilities from the Camp-Meidel modification of Tchebychev's inequality. In the latter event, we should have several hundred cases so we can get a reasonably good estimate of γ'_2 of the universe. For, from Appendix (23), we have

$$\sigma'^2_{s^2} = \frac{1}{n}\left[\mu'_4 - \frac{n-3}{n-1}\mu'^2_2\right] = \mu'^2_2\left[\frac{\gamma'_2 + 3}{n} - \frac{n-3}{n(n-1)}\right]$$

$$= \sigma'^4\left[\frac{\gamma'_2}{n} + \frac{2}{n-1}\right]$$

Hence, we need to estimate γ_2' to use this formula. If we have enough data to do this,[14] we can write, after Tchebychev, Camp, and Meidel,

$$\text{Prob.}\left(\sigma'^2 - k\sigma'^2\sqrt{\frac{\gamma_2'}{n}+\frac{2}{n-1}} \le s^2 \le \sigma'^2 + k\sigma'^2\sqrt{\frac{\gamma_2'}{n}+\frac{2}{n-1}}\right) \ge 1 - \frac{1}{2.25k^2}$$

or

$$\text{Prob.}\left(\sigma'^2 \le \frac{s^2}{1-k\sqrt{\dfrac{\gamma_2'}{n}+\dfrac{2}{n-1}}} \text{ and } \frac{s^2}{1+k\sqrt{\dfrac{\gamma_2'}{n}+\dfrac{2}{n-1}}} \ge \sigma'^2\right) \ge 1 - \frac{1}{2.25k^2}$$

or

$$\text{Prob.}\left(\frac{s^2}{1+k\sqrt{\dfrac{\gamma_2'}{n}+\dfrac{2}{n-1}}} \le \sigma'^2 \le \frac{s^2}{1-k\sqrt{\dfrac{\gamma_2'}{n}+\dfrac{2}{n-1}}}\right) \ge 1 - \frac{1}{2.25k^2}$$

To determine 0.95 confidence limits, we can determine k by setting

$$1 - \frac{1}{2.25k^2} = 0.95 \quad \text{or} \quad k^2 = \frac{1}{2.25(0.05)} = 8.9$$

which yields $k = 2.98$. Hence, 0.95 confidence limits will be approximately

$$\text{Lower limit} = \frac{s^2}{1 + 2.98\sqrt{\dfrac{\gamma_2'}{n}+\dfrac{2}{n-1}}}$$

and

$$\text{Upper limit} = \frac{s^2}{1 - 2.98\sqrt{\dfrac{\gamma_2'}{n}+\dfrac{2}{n-1}}}$$

The confidence coefficient in this case will be 0.95 *or better.*

For example, if we test 400 pieces of cotton yarn and find that the sample variance is $(11)^2 = 121$ and if γ_2 for the sample is 1.5, then 0.95 confidence limits for σ'^2 will be given approximately by

$$\text{Lower limit} = \frac{(11)^2}{1 + 2.98\sqrt{\dfrac{1.5}{400}+\dfrac{2}{399}}} = \frac{121}{1.279} = 98.4$$

$$\text{Upper limit} = \frac{(11)^2}{1 - 2.98\sqrt{\dfrac{1.5}{400}+\dfrac{2}{399}}} = \frac{121}{0.721} = 168$$

and the corresponding limits for σ' will be 9.75 and 13.0. In contrast, if we had been able to use straight normal theory with $n = 400$, the limits

[14] See Chapter 6, Section 2.2, for a discussion of the large sampling errors of μ_4.

would have been 10.3 and 11.8. There is, therefore, a definite gain in using normal theory when it can be applied. There is also a distinct danger, however, that our confidence interval will be too small if we apply the normal theory to cases in which the kurtosis of the universe is distinctly in excess of that of a normal distribution.

(30) Confidence Limits for p' Derived from Leslie E. Simon's I_Q Charts

If 0.99 confidence limits are desired for a universe proportion p', use may be made of special charts designed by Leslie E. Simon and attached to his book *An Engineers' Manual of Statistical Methods.* These charts are his 0.005 and 0.995 I_Q charts and may be used for sample sizes up to 500. To find the upper 0.99 confidence limit for p', given c nonconforming units out of a sample of n, locate on the 0.995 I_Q chart the intersection of the vertical line through n and the curve $c + 1$, and read off the ordinate of this point. To find the lower confidence limit, locate on the 0.005 I_Q chart the intersection of the vertical line through n and the curve $c - 1$, and read off the ordinate. For example, if a sample of 200 contains 5 nonconforming items, 0.99 confidence limits for the lot fraction nonconforming are read from the charts as 0.069 and 0.0055.

(31) A Method for Testing the Difference between Two Proportions[15] when the Proportions Are Relatively Small

The methods described in the text for testing the difference between two proportions are approximate methods that are satisfactory if the sample sizes are large and the probabilities not very small. The following method is an exact method that is valid for any size samples and any probabilities, provided that the universes sampled are "infinitely" large. The results are reasonably accurate when the universes are at least 10 times the samples. The method is especially useful when sample sizes are small or moderate and probabilities are small. For larger sample sizes and probabilities it may not be practical.

Suppose that the results of inspecting two random samples from two large lots are as follows:

	Nonconforming	*Conforming*	*Total*
Lot I	2	98	100
Lot II	7	193	200
Total	9	291	300

[15] See R. A. Fisher, *Statistical Methods for Research Workers* (13th rev. ed.; Edinburgh: Oliver & Boyd, Ltd., 1958), Section 21.02; and Statistical Research Group, Columbia University, *Techniques of Statistical Analysis* (New York: McGraw-Hill, 1947), pp. 250–53.

The question to which an answer is sought is whether the fractions nonconforming of the two lots are different. The procedure of analysis is to set up the hypothesis that the lot fractions nonconforming are the same ($p_1' = p_2'$) and to compute the probability of getting the given sample result or a result that deviates further from that expected on the basis of the given hypothesis. If this is less than the level of significance (α) adopted for the test, the hypothesis is rejected; otherwise, it is accepted.

To carry out this procedure we take the following steps:

1. We determine the probability of getting 2 nonconforming items out of 100 on the assumption that the common lot fraction nonconforming is p'. Thus, using the binominal formula, we have

$$\text{Prob. (2 out of 100)} = \frac{100!}{2!98!} p'^2(1 - p')^{98}$$

2. We determine the probability of getting 7 nonconforming items out of 200 on the assumption that the common lot fraction nonconforming is p'. Thus we have

$$\text{Prob. (7 out of 200)} = \frac{200!}{7!193!} p'^7(1 - p')^{193}$$

3. We determine the joint probability of the two events (1) and (2). Thus we have

$$\text{Prob. (2 out of 100 and 7 out of 200)} = \frac{100!200!}{2!7!98!193!} p'^9(1 - p')^{291}$$

This is the probability that, if we took two samples from a lot with fraction nonconforming p', the first sample containing 100 items and the second 200 items, the result would be 2 nonconforming items in the first sample and 7 nonconforming items in the second. But this may be looked at from another angle. It is also the joint probability of two events: one, that out of a sample of 300 items we would get 9 nonconforming items; the other, that 2 of the 9 nonconforming items would fall in the first 100 and 7 in the last 200 items.

Now it is the probability of the second of these last two events that we are interested in to solve our problem. So far as the immediate problem is concerned, we are not interested in the probability of getting 9 nonconforming items out of a combined sample of 300 items. What we are interested in is the way in which the 9 nonconforming items are divided between the first and second samples. In other words, we are interested in the conditional probability of the total nonconforming items being divided in a 2 to 7 ratio, given a total of 9 nonconforming items out of 300. The next two steps are thus to determine the probability of getting 9 nonconforming items out of 300 and then divide this into the joint probability determined above. [For it will be recalled that $P(AB) = P(A)P(B|A)$. *Hence* $P(B|A) = P(AB)/P(A)$.]

4. In accordance with the above reasoning, we determine the probability of getting 9 nonconforming items in a sample of 300 from a lot the fraction nonconforming of which is p'. Thus we have

$$\text{Prob. (9 out of 300)} = \frac{300!}{9!291!} p'^9 (1 - p')^{291}$$

5. We compute the desired conditional probability by dividing the probability determined in step 3 by the probability determined in step 4. Thus we have

$$\text{Prob. (2 out of 9, given 9 out of 300)} = \frac{100!200!9!291!}{2!7!98!193!300!} = 0.2353$$

the calculations being carried out with the help of a table of logs of factorials (see Table CC of Appendix II). If α has been set at the usual value of 0.05, the above result immediately gives the answer to our problem. Since the computed probability is greater than 0.05, the hypothesis that the two lots have equal fractions nonconforming is accepted.

6. If the result of step 5 is a probability that is less than α, we continue on to compute the conditional probability of the 9 nonconforming items being divided in the ratio 1 to 8. This would be given by

$$\frac{100!200!9!291!}{1!8!99!192!300!}$$

If this, together with the probability computed in step 5, is equal to or greater than α, we accept the null hypothesis; if not, we continue on.

7. If the probabilities computed in steps 5 and 6 do not add up to α, we compute the conditional probability of the 9 nonconforming items being divided in the ratio 0 to 9. This would be given by[16]

$$\frac{100!200!9!291!}{0!9!100!191!300!}$$

If the total of the three probabilities computed in steps 5 to 7 equals or exceeds α, we accept the null hypothesis.

We have now computed the conditional probability of each possible division of the 9 nonconforming items that deviates at least as much as the given sample from the ratio that would have been expected on the basis of the null hypothesis. Hence, if the total of the probabilities in steps 5 to 7 is less than α, we reject the hypothesis that $p_1' = p_2'$ and conclude that the fractions nonconforming of the two lots are different.

---

[16] The reader will note that $0! = 1$.

(32) The Component Parts of the Total Sum of Squares in a Single-Factor Analysis of Variance

Let there be c sets of data distinguished as follows:

$$\begin{matrix} X_{11} & X_{12} & \cdots & X_{1c} \\ X_{21} & X_{22} & \cdots & X_{2c} \\ \cdot & \cdot & \cdots & \cdot \\ \cdot & \cdot & \cdots & \cdot \\ \cdot & \cdot & \cdots & \cdot \\ X_{r1} & X_{r2} & \cdots & X_{rc} \end{matrix}$$

These might be c samples of r each from c different lots or from c different processes or from the same process at c different times. In the discussion the samples are assumed to be all of the same size, but this is not necessary for the present analysis.

Regardless of the assumptions made in the subsequent analysis, the following is valid for the above set of data, viz

(1) $$\sum_i \sum_j (X_{ij} - \bar{X})^2 \equiv r \sum_j (\bar{X}_j - \bar{X})^2 + \sum_i \sum_j (X_{ij} - \bar{X}_j)^2$$

This says that the sum of the squares of the deviations of the individual items from the grand mean $(\bar{X})$ is identically equal to the weighted sum of the squared deviations of the column means from the grand mean plus the sum of the squared deviations of the individual items in each column from the mean of the column summed for all columns. The r in the formula is the number of items in each column.

Identity (1) can be proved as follows:

$$\sum \sum (X_{ij} - \bar{X})^2 \equiv \sum \sum (X_{ij} - \bar{X}_j + \bar{X}_j - \bar{X})^2$$

$$\equiv \sum \sum (X_{ij} - \bar{X}_j)^2 + 2 \sum \sum (X_{ij} - \bar{X}_j)(\bar{X}_j - \bar{X})$$

$$+ \sum \sum (\bar{X}_j - \bar{X})^2$$

But

$$\sum \sum (X_{ij} - \bar{X}_j)(\bar{X}_j - \bar{X}) = \sum \sum X_{ij}\bar{X}_j - \sum \sum X_{ij}\bar{X} - \sum \sum \bar{X}_j^2$$

$$+ \sum \sum \bar{X}_j\bar{X}$$

$$= r \sum \bar{X}_j\bar{X}_j - n\bar{X}\bar{X} - r \sum \bar{X}_j^2 + r \sum \bar{X}_j\bar{X} = r \sum \bar{X}_j^2 - n\bar{X}^2 - r \sum \bar{X}_j^2$$

$$+ n\bar{X}^2 = 0$$

since $\sum_j X_{ij}/r = \bar{X}_j$, $\sum \sum X_{ij}/n = \bar{X}$ and $\sum_j \bar{X}_j/c = \bar{X}$ taking $n = rc$. Also $\sum_i \sum_j (\bar{X}_j - \bar{X})^2 = r \sum_j (\bar{X}_j - \bar{X})^2$. Hence

$$\sum \sum (X_{ij} - \bar{X})^2 \equiv r \sum (\bar{X}_j - \bar{X})^2 + \sum \sum (X_{ij} - \bar{X}_j)^2$$

(33) Estimates of Components of Variance in a Single-Factor Analysis of Variance

Suppose for the moment that the variable X in Section (32) is made up of a constant A plus a column factor θ_j that varies from column to column plus a random variable ϵ_{ij} that has a mean of zero and a variance σ'^2 for each column.

Thus

$$X_{ij} = A + \theta_j + \epsilon_{ij} \tag{1}$$

It will be noted that no assumptions are made regarding the form of the distribution of ϵ and that initially the θ_j are taken as a fixed set of deviations.

Let the mean of the jth column be $\bar{X}_j$ and the mean of all the data $\bar{X}$. Then we have

$$\bar{X}_j = A + \theta_j + \bar{\epsilon}_j \tag{2}$$

$$\bar{X} = A + \bar{\theta} + \bar{\epsilon} \tag{3}$$

$$X_{ij} - \bar{X}_j = A + \theta_j + \epsilon_{ij} - A - \theta_j - \bar{\epsilon}_j = \epsilon_{ij} - \bar{\epsilon}_j \tag{4}$$

$$\bar{X}_j - \bar{X} = A + \theta_j + \bar{\epsilon}_j - A - \bar{\theta} - \bar{\epsilon} = (\theta_j - \bar{\theta}) + (\bar{\epsilon}_j - \bar{\epsilon}) \tag{5}$$

Applying the operator E to the random variable ϵ and using the theorems of Chapter 6, we have

$$E\left[\frac{r \sum_j (\bar{X}_j - \bar{X})^2}{c-1}\right] = r\left\{E\frac{\sum_j [(\theta_j - \bar{\theta}) + (\bar{\epsilon}_j - \bar{\epsilon})]^2}{c-1}\right\}$$

$$= rE\frac{\sum_j (\theta_j - \bar{\theta})^2}{c-1} + 2rE\frac{\sum_j (\theta_j - \bar{\theta})(\bar{\epsilon}_j - \bar{\epsilon})}{c-1} + rE\frac{\sum_j (\bar{\epsilon}_j - \bar{\epsilon})^2}{c-1}$$

But

$$E\sum_j (\theta_j - \bar{\theta})(\bar{\epsilon}_j - \bar{\epsilon})$$

$$= (\theta_1 - \bar{\theta})E(\bar{\epsilon}_1 - \bar{\epsilon}) + (\theta_2 - \bar{\theta})E(\bar{\epsilon}_2 - \bar{\epsilon}) + \ldots = 0$$

since

$$E(\bar{\epsilon}_j - \bar{\epsilon}) = E(\bar{\epsilon}_j) - E(\bar{\epsilon}) = \bar{\epsilon}' - \bar{\epsilon}' = 0$$

Hence,

$$E\left[\frac{r \sum_j (\bar{X}_j - \bar{X})^2}{c-1}\right] = rE\frac{\sum_j (\theta_j - \bar{\theta})^2}{c-1} + rE\frac{\sum_j (\bar{\epsilon}_j - \bar{\epsilon})^2}{c-1} \tag{6}$$

$$= r\frac{\sum_j (\theta_j - \bar{\theta})^2}{c-1} + r\frac{\sigma'^2}{r} = r\frac{\sum_j (\theta_j - \bar{\theta})^2}{c-1} + \sigma'^2 \tag{6a}$$

by Theorems 6.4 and 6.2 of Chapter 6. Likewise

$$E\frac{\sum_i \sum_j (X_{ij}-\bar{X}_j)^2}{c(r-1)} = E\frac{\sum_i \sum_j (\epsilon_{ij}-\bar{\epsilon}_j)^2}{c(r-1)}$$

$$= \frac{1}{c(r-1)} E \sum_j \sum_i (\epsilon_{ij}^2 - 2\epsilon_{ij}\bar{\epsilon}_j + \bar{\epsilon}_j^2)$$

$$= \frac{1}{c(r-1)} E \sum_j \left(\sum_i \epsilon_{ij}^2 - 2r\bar{\epsilon}_j^2 + r\bar{\epsilon}_j^2\right)$$

$$= \frac{1}{c(r-1)} E \sum_j \left(\sum_i \epsilon_{ij}^2 - r\bar{\epsilon}_j^2\right)$$

$$= \frac{1}{c(r-1)} \left[E \sum_j \sum_i \epsilon_{ij}^2 - rE \sum_j \bar{\epsilon}_j^2\right]$$

$$= \frac{1}{c(r-1)} \left[E \sum_j \sum_i \epsilon_{ij}^2 - rE \sum_j \frac{(\epsilon_{1j}+\epsilon_{2j}+\ldots+\epsilon_{rj})^2}{r^2}\right]$$

$$= \frac{1}{c(r-1)} \left[E \sum_j \sum_i \epsilon_{ij}^2 - \frac{1}{r}\sum_j \left\{E(\epsilon_{1j}^2)+E(\epsilon_{2j}^2)+\ldots+E(\epsilon_{rj}^2) + 2E(\epsilon_{1j}\epsilon_{2j})+\ldots\right\}\right]$$

But $E(\epsilon_{ij})(\epsilon_{kj}) = 0$, when $i \neq k$, since ϵ is a random variable. Hence, the above equals

$$\frac{1}{c(r-1)}\left[cr\sigma'^2 - \frac{cr}{r}\sigma'^2\right] = \frac{c(r-1)}{c(r-1)}\sigma'^2 = \sigma'^2$$

Hence

$$E\frac{\sum_i \sum_j (X_{ij}-\bar{X}_j)^2}{c(r-1)} = \sigma'^2 \tag{7}$$

It therefore follows that if the null hypothesis is true, viz

$$H_0: \theta_1 = \theta_2 = \ldots = \theta_c = 0 \tag{8}$$

then both

$$s_1^2 = \frac{r\sum_j (\bar{X}_j - \bar{X})^2}{c-1} \quad \text{and } s_2^2 = \frac{\sum_j \sum_i (X_{ij}-\bar{X}_j)^2}{c(r-1)}$$

are unbiased estimates of σ'^2.

If the θ_j's are not all 0, then the bias in using s_1^2 as an estimate of σ'^2 is indicated in equation (6a). If θ itself is viewed as a random variable and the θ_j's as a sample of c cases from the θ universe, then from (6) we have

(9)
$$E\left[\frac{r\sum_{j}(\bar{X}_j-\bar{X})^2}{c-1}\right]=r\sigma_{\theta}'^2+\sigma'^2$$

where $\sigma_{\theta}'^2$ is the variance of the θ universe. This explains the entry in the first row, last column, of Table 31.3.

(34) Derivation of Practical Formulas for Use in a Single-Factor Analysis of Variance

(1)
$$\begin{aligned}\sum_i\sum_j(X_{ij}-\bar{X})^2&=\sum_i\sum_j X_{ij}^2-2\bar{X}\sum_i\sum_j X_{ij}+n\bar{X}^2\\&=\sum_i\sum_j X_{ij}^2-2n\bar{X}^2+n\bar{X}^2\\&=\sum_i\sum_j X_{ij}^2-n\bar{X}^2\\&=\sum_i\sum_j X_{ij}^2-\frac{\left(\sum_i\sum_j X_{ij}\right)^2}{n}\end{aligned}$$

(2)
$$\begin{aligned}r\sum_j(\bar{X}_j-\bar{X})^2&=r\sum_j\bar{X}_j^2-2r\bar{X}\sum_j\bar{X}_j+r\sum_j\bar{X}^2\\&=r\sum_j\bar{X}_j^2-2rc\bar{X}^2+rc\bar{X}^2\\&=r\sum_j\left(\sum_i X_{ij}/r\right)^2-n\bar{X}^2\\&=\sum_j\frac{\left(\sum_i X_{ij}\right)^2}{r}-\frac{\left(\sum_i\sum_j X_{ij}\right)^2}{n}\end{aligned}$$

Note that from Section (32) we have

$$\sum_i\sum_j(X_{ij}-\bar{X}_j)^2=\sum_i\sum_j(X_{ij}-\bar{X})^2-r\sum_j(X_j-\bar{X})^2$$

Hence, if we compute

$$(a)\quad\sum_j\frac{\left(\sum_i X_{ij}\right)^2}{r}$$

$$(b)\quad\sum_i\sum_j X_{ij}^2$$

$$(c)\quad\left(\sum_i\sum_j X_{ij}\right)^2\Big/n$$

then

$$r \sum_j (\bar{X}_j - \bar{X})^2 = (a) - (c)$$

$$\sum_i \sum_j (X_{ij} - \bar{X}_j)^2 = (b) - (a)$$

(35) The Hotelling T^2 Test for a Mixed-Effects Analysis of Variance Model[17]

The Hotelling T^2 test, as applied to a mixed-effects analysis of variance model, is best explained by an example. Let us apply it to the data of Table 30.1A.[18]

HYPOTHESIS: *Analysts have no biases.*

Step 1. In Table 30.1A subtract the last row of the cell totals from the others, yielding the results:

	A	B	C	D	Total	Mean $(\bar{d}_{i.3})$
I—III $= d_{1.3} =$	1.0	0	−2.0	1.5	0.5	0.125
II—III $= d_{2.3} =$	−0.5	−1.5	−3.0	−2.5	−7.5	−1.875

Step 2. Compute:

$$s_{11} = \frac{1^2 + 0^2 + (-2)^2 + 1.5^2 - (0.5)^2/4}{4-1} = 2.396$$

$$s_{22} = \frac{(-0.5)^2 + (-1.5)^2 + (-3)^2 + (-2.5)^2 - (-7.5)^2/4}{4-1} = 1.229$$

$$s_{12} = \frac{(1)(-0.5) + (0)(-1.5) + (-2)(-3) + (1.5)(-2.5) - (0.5)(-7.5)/4}{4-1}$$

$$= 0.896$$

Step 3. Compute:

$$D = \begin{vmatrix} s_{11} & s_{12} \\ s_{12} & s_{22} \end{vmatrix} = s_{11}s_{22} - s_{12}^2$$

$$= (2.396)(1.229) - (0.896)^2 = 2.1419$$

If the original table had had four analysts, then D would have been defined as follows:

---

[17] Based on the article by Henry Scheffé, "A 'Mixed Model' for the Analysis of Variance," *Annals of Mathematical Statistics* 17 (1956), pp. 23–36. See particularly p. 32. This test is also often spoken of as the Hotelling T test. The test belongs to the class of multivariate tests and has many more uses than that described here. For interesting industrial applications see C. H. Hicks, "Some Applications of Hotelling's T," *Industrial Quality Control* 11, No. 9 (June 1955), p. 23. Hotelling's original article appeared in *Annals of Mathematical Statistics* 2 (1931), pp. 360–78.

[18] See Chapter 30, Section 1.1.

$$D = \begin{vmatrix} s_{11} & s_{12} & s_{13} \\ s_{12} & s_{22} & s_{23} \\ s_{13} & s_{23} & s_{33} \end{vmatrix}$$

where $s_{ij} = \dfrac{\Sigma(d_{i.k})(d_{j.k}) - (\Sigma d_{i.k})(\Sigma d_{j.k})/4}{c-1}$ and k is the row subtracted. And so forth for $r > 4$. (Note that r is the number of rows in the original table, here the number of analysts.)

Step 4. Compute:

$$T^2 = c\left[\frac{D_{11}}{D}(\bar{d}_{1.3})^2 + \frac{D_{22}}{D}(\bar{d}_{2.3})^2 + 2\frac{D_{12}}{D}(\bar{d}_{1.3})(\bar{d}_{2.3})\right]$$

$$= 4\left[\frac{1.229(0.125)^2}{2.1419} + \frac{(2.396)(-1.875)^2}{2.1419} - \frac{2(0.896)(0.125)(-1.875)}{2.1419}\right]$$

$$= 15.872$$

In general $T^2 = \dfrac{c\,\Sigma D_{ij}(\bar{d}_{i.k})(\bar{d}_{j.k})}{D}$, where D_{ij} is the cofactor of the ij th element in D and c is the number of columns in the original table.

Step 5. Compute:

$$F = \frac{(c - r + 1)T^2}{(r-1)} = 15.872$$

Step 6. Note:

The critical value for comparison with the sample F is $F_{0.05}$ for $\nu_1 = r - 1$, $\nu_2 = c - 2$. For the given values of r and c this is 19.0. Since $F < F_{0.05}$, the hypothesis is not rejected.

For a discussion of the Hotelling T^2 test as applied to analysis of variance of mixed-effects models, the reader is referred to Chapter 30, Section 2.2.3.

(36) Derivation of a Line of Regression by the Method of Least Squares

Let $a_{1.2}$ and b_{12} be estimates of $a'_{1.2}$ and b'_{12} in the regression equation $X'_{1r} = a'_{1.2} + b'_{12}X_2$. To be "least-squares" estimates they must be such as to minimize $\Sigma(X_1 - a_{1.2} - b_{12}X_2)^2$. This requires that the partial derivatives with respect to $a_{1.2}$ and b_{12} must both be zero. Thus

(1)
$$\frac{\partial\Sigma(X_1 - a_{1.2} - b_{12}X_2)^2}{\partial a_{1.2}} = 2\sum(X_1 - a_{1.2} - b_{12}X_2)(-1) = 0$$

$$\frac{\partial\Sigma(X_1 - a_{1.2} - b_{12}X_2)^2}{\partial b_{12}} = 2\sum(X_1 - a_{1.2} - b_{12}X_2)(-X_2) = 0$$

These are the least-squares "normal equations" and may be written

$$\sum X_1 - na_{1.2} - b_{12} \sum X_2 = 0$$

(2) (origin at $X_1 = 0, X_2 = 0$)

$$\sum X_1X_2 - a_{1.2} \sum X_2 - b_{12} \sum X_2^2 = 0$$

If the origin is shifted to $X_1 = \bar{X}_1$, $X_2 = \bar{X}_2$, so that the equation for the fitted line becomes $x_{1r} = A + b_{12}x_2$, then the least-squares equations become

$$nA + b_{12} \sum x_2 = \sum x_1$$

(origin at $X_1 = \bar{X}_1$, $X_2 = \bar{X}_2$)

$$A \sum x_2 + b_{12} \sum x_2^2 = \sum x_1x_2$$

Since $\Sigma x_1 = \Sigma x_2 = 0$, the first gives $A = 0$ when the origin is at $X_1 = \bar{X}_1$, $X_2 = \bar{X}_2$, and the second gives

(3) $$b_{12} = \frac{\Sigma x_1x_2}{\Sigma x_2^2}$$

Hence with reference to the origin $(\bar{X}_2, \bar{X}_1)$ the regression equation is

$$x_{1r} = b_{12}x_2$$

If we shift back to the original origin, we get

$$(X_{1r} - \bar{X}_1) = b_{12}(X_2 - \bar{X}_2)$$

or

$$X_{1r} = \bar{X}_1 - b_{12}\bar{X}_2 + b_{12}X_2 \quad \text{(origin at } X_1 = 0, X_2 = 0\text{)}$$

From this it follows that

(4) $$a_{1.2} = \bar{X}_1 - b_{12}\bar{X}_2 \quad \text{(origin at } X_1 = 0, X_2 = 0\text{)}$$

If we set $u_{1.2} = X_1 - a_{1.2} - b_{12}X_2$, we have from equations (1)

(5) $$\sum u_{1.2} = 0 \qquad \sum u_{1.2}X_2 = 0$$

These are useful for theoretical purposes. In practice we derive $a_{1.2}$ and b_{12} by computing $\bar{X}_1$, $\bar{X}_2$, Σx_1x_2, and Σx_2^2 for the sample set of data, and substituting these values in equations (3) and (4).

(37) Derivation of a Formula for $\Sigma u_{1.2}^2$

If $u_{1.2} = X_1 - a_{1.2} - b_{12}X_2$, then

$$\sum u_{1.2}^2 = \sum u_{1.2}u_{1.2} = \sum u_{1.2}(X_1 - a_{1.2} - b_{12}X_2)$$

$$= \sum u_{1.2}X_1 - a_{1.2} \sum u_{1.2} - b_{12} \sum u_{1.2}X_2$$

But from equations (5) of Section 36 we have $\Sigma u_{1.2} = 0$ and $\Sigma u_{1.2}X_2 = 0$. Hence

$$\textbf{(1)}\quad \begin{aligned}\sum u_{1.2}^2 = \sum u_{1.2}X_1 &= \sum (X_1 - a_{1.2} - b_{12}X_2)X_1 \\ &= \sum X_1^2 - a_{1.2}\sum X_1 - b_{12}\sum X_1X_2\end{aligned}$$

If X_1 and X_2 are measured from their means, we have

$$\textbf{(2)}\quad \sum u_{1.2}^2 = \sum x_1^2 - b_{12}\sum x_1x_2$$

(38) For a Bivariate Normal Universe the Slope of the Line of Regression of X_1 or X_2 is $r'_{12}\sigma'_1/\sigma'_2$

The density function for the bivariate normal distribution is

$$f(X_1, X_2) = \frac{1}{2\pi\sigma'_1\sigma'_2\sqrt{1-r'^2_{12}}}\, e^{-\frac{1}{2(1-r'^2_{12})}\left[\left(\frac{X_1-\bar{X}'_1}{\sigma'_1}\right)^2 - 2r'_{12}\left(\frac{X_1-\bar{X}'_1}{\sigma'_1}\right)\left(\frac{X_2-\bar{X}'_2}{\sigma'_2}\right) + \left(\frac{X_2-\bar{X}'_2}{\sigma'_2}\right)^2\right]}$$

The marginal distributions are given by

$$f_1(X_1) = \int_{-\infty}^{\infty} f(X_1, X_2)dX_2 \quad\text{and}\quad f_2(X_2) = \int_{\infty}^{\infty} f(X_1, X_2)dX_1$$

which yield

$$f_1(X_1) = \frac{1}{\sigma'_1\sqrt{2\pi}}\, e^{-(X_1-\bar{X}'_1)^2/2\sigma_1^2}$$

and

$$f_2(X_2) = \frac{1}{\sigma_2\sqrt{2\pi}}\, e^{-(X_2-\bar{X}_2)^2/2\sigma_2^2}$$

showing that each marginal distribution is also normal.

The conditional distribution of X_1 given X_2 has the density function

$$f(X_1|X_2) = f(X_1, X_2)/f_2(X_2)$$

which on making the proper substitution can be reduced to

$$f(X_1|X_2) = \frac{1}{\sqrt{2\pi}\,\sigma_1\sqrt{1-r'^2_{12}}}\, e^{-\frac{1}{2\sigma'^2_1(1-r'^2_{12})}\left[(X_1-\bar{X}_1) - \frac{r'_{12}\sigma'_1}{\sigma'_2}(X_2-\bar{X}'_2)\right]^2}$$

This is of the form of a normal distribution with a mean equal to

$$\frac{r'_{12}\sigma'_1}{\sigma'_2}(X_2 - \bar{X}'_2)$$

and variance equal to

$$\sigma'^2_1(1 - r'^2_{12})$$

The mean of X_1 for a given X_2 thus varies linearly with X_2 and has the slope $r'_{12}\sigma'_1/\sigma'_2$.[19]

(39) Estimation of a Universe Plane of Regression by the Method of Least Squares, p Variables

Let the universe plane of regression be

(1) $$X'_{1r} = a'_{1.23\ldots p} + b'_{12.3\ldots p}X_2 + \ldots + b_{1p.23\ldots p-1}X_p$$

and let sample estimates of the coefficients be $a_{1.2\ldots p}$, $b_{12.3\ldots p}$, and so on. Then these will be "least-squares" estimates if

$$\sum u^2_{1.23\ldots p} = \sum (X_1 - a_{1.23\ldots p} - b_{12.3\ldots p}X_2 - \ldots - b_{1p.23\ldots 1-p}X_p)^2$$

equals a minimum. For this to be true, it is necessary that

(2)
$$\frac{\partial \Sigma u^2_{1.23\ldots p}}{\partial a_{1.23\ldots p}} = 0$$
$$\frac{\partial \Sigma u^2_{1.23\ldots p}}{\partial b_{12.3\ldots p}} = 0$$
$$\cdot \quad \cdot \quad \cdot$$
$$\frac{\partial \Sigma u^2_{1.23\ldots p}}{\partial b_{1p.2\ldots p-1}} = 0$$

These yield the equations

$$na_{1.23\ldots p} + b_{12.3\ldots p}\sum X_2 + \ldots + b_{1p.23\ldots p-1}\sum X_p = \sum X_1$$

$$a_{1.23\ldots p}\sum X_2 + b_{12.3\ldots p}\sum X_2^2 + \ldots + b_{1p.23\ldots p-1}\sum X_2X_p = \sum X_1X_2$$

$$\cdot \quad \cdot \quad \cdot \quad \cdot \quad \cdot$$

$$a_{1.23\ldots p}\sum X_p + b_{12.3\ldots p}\sum X_2X_p + \ldots + b_{1p.23\ldots p-1}\sum X_p^2 = \sum X_1X_p$$

(origin at $0, 0, \ldots, 0$)

If we shift to $X_1 = \bar{X}_1$, $X_2 = \bar{X}_2$, $\ldots$, $X_p = \bar{X}_p$ as the origin, these equations become

$nA = 0$ where A is the new constant term

$$b_{12.3\ldots p}\sum x_2^2 + \ldots + b_{1p.23\ldots p-1}\sum x_2x_p = \sum x_1x_2$$

(3) $$b_{12.3\ldots p}\sum x_2x_p + \ldots + b_{1p.23\ldots p-1}\sum x_p^2 = \sum x_1x_p$$

(origin at $\bar{X}_1, \bar{X}_2, \ldots, \bar{X}_p$)

---

[19] See A. M. Mood and F. A. Graybill, *Introduction to the Theory of Statistics* (2d ed., New York: McGraw-Hill, 1963), p. 202.

Set

$$D = \begin{vmatrix} \sum x_2^2 & \sum x_2x_3 & \cdots & \sum x_2x_p \\ \sum x_2x_3 & \sum x_3^2 & \cdots & \sum x_3x_p \\ \cdot & \cdot & \cdots & \cdot \\ \cdot & \cdot & \cdots & \cdot \\ \cdot & \cdot & \cdots & \cdot \\ \sum x_2x_p & \sum x_3x_p & \cdots & \sum x_p^2 \end{vmatrix}$$

Then

$$\begin{aligned} b_{12.3\ldots p} &= \frac{1}{D}\left[D_{11} \sum x_1x_2 + D_{21} \sum x_1x_3 + \ldots + D_{m1} \sum x_1x_p\right] \\ b_{13.2\ldots p} &= \frac{1}{D}\left[D_{12} \sum x_1x_2 + D_{22} \sum x_1x_3 + \ldots + D_{m2} \sum x_1x_p\right] \\ &\cdot \quad \cdot \quad \cdot \quad \cdot \quad \cdot \\ b_{1p.2\ldots p-1} &= \frac{1}{D}\left[D_{1m} \sum x_1x_2 + D_{2m} \sum x_1x_3 + \ldots + D_{mm} \sum x_1x_p\right] \end{aligned} \tag{4}$$

where D_{ij} is the ijth cofactor of D and $m = p - 1$. Note, however, that, if we set up the auxiliary equations,

$$\begin{aligned} c_{22} \sum x_2^2 + c_{23} \sum x_2x_3 + \ldots + c_{2p} \sum x_2x_p &= 1 \\ c_{22} \sum x_2x_3 + c_{23} \sum x_3^2 + \ldots + c_{2p} \sum x_3x_p &= 0 \\ \cdot \quad \cdot \quad \cdot \quad \cdot \quad \cdot & \\ c_{22} \sum x_2x_p + c_{23} \sum x_3x_p + \ldots + c_{2p} \sum x_p^2 &= 0 \end{aligned} \tag{5}$$

then

$$\begin{aligned} c_{22} &= \frac{D_{11}}{D} \\ c_{23} &= \frac{D_{12}}{D} \\ &\cdot \\ &\cdot \\ &\cdot \\ c_{2p} &= \frac{D_{1m}}{D} \end{aligned}$$

But, because of the symmetry of D, we have

$$c_{22} = \frac{D_{11}}{D}$$

$$c_{23} = \frac{D_{21}}{D}$$

$$\cdot$$

$$\cdot$$

$$\cdot$$

$$c_{2p} = \frac{D_{m1}}{D}$$

Hence, from equations (4),

(6) $$b_{12.3\ldots p} = c_{22} \sum x_1x_2 + c_{23} \sum x_1x_3 + \ldots + c_{2p} \sum x_1x_p$$

If we set up the auxiliary equations

(7)
$$\begin{aligned}
c_{23} \sum x_2^2 + c_{33} \sum x_1x_3 + \ldots + c_{3p} \sum x_2x_p &= 0 \\
c_{23} \sum x_2x_3 + c_{33} \sum x_3^2 + \ldots + c_{3p} \sum x_3x_p &= 1 \\
\cdot \quad \cdot \quad \cdot \quad \cdot \quad \cdot & \\
c_{23} \sum x_2x_p + c_{33} \sum x_3x_p + \ldots + c_{3p} \sum x_p^2 &= 0
\end{aligned}$$

then

$$c_{23} = \frac{D_{21}}{D} = \frac{D_{12}}{D}$$

$$c_{33} = \frac{D_{22}}{D} = \frac{D_{22}}{D}$$

$$\cdot \quad \cdot \quad \cdot \quad \cdot \quad \cdot$$

$$c_{3p} = \frac{D_{2m}}{D} = \frac{D_{m2}}{D}$$

Hence

(8) $$b_{13.2\ldots p} = c_{23} \sum x_1x_2 + c_{33} \sum x_1x_3 + \ldots + c_{3p} \sum x_1x_p,$$

and so forth for the other b's.

Since the constant term is zero when the origin is at $X_1 = \bar{X}_1$, $X_2 = \bar{X}_2, \ldots, X_p = \bar{X}_p$, we have

$$(X_{1r} - \bar{X}_1) = b_{12.3...p}(X_2 - \bar{X}_2) + \ldots + b_{1p.2...p-1}(X_p - \bar{X}_p)$$

or

$$X_{1r} = \bar{X}_1 - b_{12.3...p}\bar{X}_2 - \ldots - b_{1p.2...1-p}\bar{X}_p + b_{12.3...p}X_2 + \ldots + b_{1p.2...p-1}X_p$$

Hence

(9)
$$a_{1.23...p} = \bar{X}_1 - b_{12.3...p}\bar{X}_2 - \ldots - b_{1p.2...p-1}\bar{X}_p$$

(origin at 0, 0, . . . , 0)

(40) Standard Errors of Estimates of Regression Parameters

Note from equation (8) of Section 39 that

$$b_{12.3...p} = c_{22} \sum x_1x_2 + c_{23} \sum x_1x_3 + \ldots + c_{2p} \sum x_1x_p$$

In repeated sampling let the same values of X_2, X_3, . . . , X_p be selected for each sample so that the only variations from sample to sample are the chance variations in the values of X_1 which we shall designate as the $u'_{1.23...p}$'s. Under these conditions, the error in $b_{12.3...p}$ derived from a single sample of n sets of p values each equals

(1)
$$c_{22} \sum u'_{1.23...p}x_2 + c_{23} \sum u'_{1.23...p}x_3 + \ldots + c_{2p} \sum u'_{1.23...p}x_p$$

where $u'_{1.23...}$ is a deviation from the universe line of regression and summation is over the n sets. If we designate by $x_2^{[1]}$ the value of $X_2 - \bar{X}_2$ for the first set of values, by $x_2^{[2]}$ the value of $X_2 - \bar{X}_2$ for the second set of values, by $u'^{[2]}_{1.23...p}$ the error in X_1 associated with the second set of values, and so on, then the error in $b_{12.3...p}$ is equal to

(2)
$$\begin{aligned}
&[c_{22}x_2^{[1]} + c_{23}x_3^{[1]} + \ldots + c_{2p}x_p^{[1]}]u'^{[1]}_{1.23...p} \\
&+ [c_{22}x_2^{[2]} + c_{23}x_3^{[2]} + \ldots + c_{2p}x_p^{[2]}]u'^{[2]}_{1.23...p} \\
&\quad \cdot \quad \cdot \quad \cdot \quad \cdot \quad \cdot \\
&+ [c_{22}x_2^{[n]} + c_{23}x_3^{[n]} + \ldots + c_{2p}x_p^{[n]}]u'^{[n]}_{1.23...p}
\end{aligned}$$

Set

(3)
$$\begin{aligned}
a_1 &= c_{22}x_2^{[1]} + c_{23}x_3^{[1]} + \ldots + c_{2p}x_p^{[1]} \\
a_2 &= c_{22}x_2^{[2]} + c_{23}x_3^{[2]} + \ldots + c_{2p}x_p^{[2]} \\
&\quad \cdot \quad \cdot \quad \cdot \quad \cdot \quad \cdot \\
a_n &= c_{22}x_2^{[n]} + c_{23}x_3^{[n]} + \ldots + c_{2p}x_p^{[n]}
\end{aligned}$$

Then, by the corollary of Theorem 4.2,

$$\sigma'^2_{b12.3..p} = \sigma'^2_{1.23..p} \sum_i a_i^2$$

where $\sigma'^2_{1.23..p}$ is the variance of $u'_{1.23..p}$. To find $\sum_i a_i^2$, multiply equations (3) by $a_1, a_2, \ldots, a_n$, respectively, and add. Thus

$$a_1^2 = c_{22}x_2^{[1]}a_1 + c_{23}x_3^{[1]}a_1 + \ldots + c_{2p}x_p^{[1]}a_1$$
$$a_2^2 = c_{22}x_2^{[2]}a_2 + c_{23}x_3^{[2]}a_2 + \ldots + c_{2p}x_p^{[2]}a_2$$
$$\cdot \quad \cdot \quad \cdot \quad \cdot \quad \cdot$$
$$a_n^2 = c_{22}x_2^{[n]}a_n + c_{23}x_3^{[n]}a_n + \ldots + c_{2p}x_p^{[n]}a_n$$

and

$$\sum_i a_i^2 = c_{22} \sum_i a_i x_2^{[i]} + c_{23} \sum_i a_i x_3^{[i]} + \ldots + c_{2p} \sum_i a_i x_p^{[i]}$$

But from equations (3) above[20]

$$\sum_i a_i x_2^{[i]} = c_{22} \sum x_2^2 + c_{23} \sum x_2 x_3 + \ldots + c_{2p} \sum x_2 x_p$$
$$= \frac{D_{11}}{D} \sum x_2^2 + \frac{D_{12}}{D} \sum x_2 x_3 + \ldots + \frac{D_{1m}}{D} \sum x_2 x_p$$
$$= \frac{D}{D} = 1$$

and

$$\sum a_i x_3^{[i]} = c_{22} \sum x_2 x_3 + c_{23} \sum x_3^2 + \ldots + c_{2p} \sum x_3 x_p$$
$$= \frac{D_{11}}{D} \sum x_2 x_3 + \frac{D_{12}}{D} \sum x_3^2 + \ldots + \frac{D_{1m}}{D} \sum x_3 x_p = 0$$

since the sum of the cofactors of one row of a determinant times the elements of another row is zero. Likewise

$$\sum a_i x_4^{[i]} = \ldots = \sum a_i x_p^{[i]} = 0$$

Hence

$$\sigma'^2_{b12.3..p} = c_{22}\sigma'^2_{1.23..p}$$

and

$$\sigma'_{b12.3..p} = \sqrt{c_{22}}\, \sigma'_{1.23..p}$$

[20] The superscripts are implied but not shown on the right-hand side of the equations, since they might become confused with the exponents.

In the same manner it can be shown that

$$\sigma'_{b\,13.2..p} = \sqrt{c_{33}}\,\sigma'_{1.23..p}$$

and

$$\sigma'_{b\,1p.2..p-1} = \sqrt{c_{pp}}\,\sigma'_{1.23..p}$$

Since

$$a_{1.23..p} = \bar{X}_1 - b_{12.3..p}\bar{X}_2 - \ldots - b_{1p\,2...1-p}\bar{X}_p$$

then,
when the origin is chosen at $X_2 = \bar{X}_2$, $X_3 = \bar{X}_3$, . . . , $X_p = \bar{X}_p$, $X_1 = 0$,

$$\sigma'_{a\,1.23..p} = \sigma'_{\bar{X}_1} = \sigma'_{1.23..p}/\sqrt{n}$$

(41) Proof that $R^2_{1.23} = r^2_{12} + r^2_{13}$ if $r_{23} = 0$

By definition,

$$R^2_{1.23} = 1 - \frac{\Sigma u^2_{1.23}}{\Sigma x^2_1}$$

By an argument similar to that used in (37) above, it can be shown that

$$\Sigma u^2_{1.23} = \Sigma x^2_1 - b_{12.3}\Sigma x_1 x_2 - b_{13.2}\Sigma x_2 x_3$$

Hence

$$R^2_{1.23} = \frac{\Sigma x^2_1 - \Sigma x^2_1 + b_{12.3}\Sigma x_1 x_2 + b_{13.2}\Sigma x_1 x_3}{\Sigma x^2_1}$$

$$= \frac{b_{12.3}\Sigma x_1 x_2 + b_{13.2}\Sigma x_1 x_3}{\Sigma x^2_1}$$

But, if r_{23} is 0, then $b_{12.3} = b_{12}$ and $b_{13.2} = b_{13}$, since holding X_3 constant does not affect X_2 and holding X_2 constant does not affect X_3. Also from equations (32.16) of Chapter 32 and from (36) above we have $\Sigma x^2_1 = n\hat{\sigma}^2_1$, $r_{ij} = \dfrac{\Sigma x_i x_j}{n\hat{\sigma}_i\hat{\sigma}_j}$ and $b_{ij} = \dfrac{\Sigma x_i x_j}{\Sigma x^2_j} = \dfrac{\Sigma x_i x_j}{n\hat{\sigma}^2_j}$ so that $\Sigma x_1 x_2 = n\hat{\sigma}_1\hat{\sigma}_2 r_{12}$, $\Sigma x_1 x_3 = n\hat{\sigma}_1\hat{\sigma}_3 r_{13}$, $b_{12} = r_{12}\hat{\sigma}_1/\hat{\sigma}_2$ and $b_{13} = r_{13}\hat{\sigma}_1/\hat{\sigma}_3$ where $\hat{\sigma}_i = RMSD_i$. Thus if $r_{23} = 0$,

$$R^2_{1.23} = \frac{nr^2_{12}\hat{\sigma}^2_1 + nr^2_{13}\hat{\sigma}^2_1}{n\hat{\sigma}^2_1} = r^2_{12} + r^2_{13}$$

The above proof is for a sample plane of regression fitted by the method of least squares. The relationship is also valid for a universe plane of regression of a multivariate normal distribution.

(42) The Regression Approach to Analysis of Variance and Covariance

a. One-Way Analysis of Variance Model. The understanding of analysis of variance and covariance will be greatly facilitated by viewing them as special cases of regression analysis. Let us consider, for example, the fitting of the simple one-way analysis of variance model

(1) $$Y_{ij} = \mu + \theta_j + \epsilon_{ij}$$

by the method of least squares. Let there be c groups of r items each, which means that i will run from 1 to r and j from 1 to c. It will be assumed, as usual, that ϵ_{ij} is a normally distributed random variable with zero mean and standard deviation σ'.

The least-squares estimate of μ and the θ_j will be given by minimizing

(2) $$\sum u^2 = \sum_i \sum_j (Y_{ij} - \hat{\mu} - \hat{\theta}_j)^2$$

or, to spell it out, by minimizing

$$\begin{aligned} &(Y_{11} - \hat{\mu} - \hat{\theta}_1)^2 + (Y_{21} - \hat{\mu} - \hat{\theta}_1)^2 + \ldots + (Y_{r1} - \hat{\mu} - \hat{\theta}_1)^2 \\ &+ (Y_{12} - \hat{\mu} - \hat{\theta}_2)^2 + (Y_{22} - \hat{\mu} - \hat{\theta}_2)^2 + \ldots + (Y_{r2} - \hat{\mu} - \hat{\theta}_2)^2 \\ &\quad . \quad . \quad . \quad . \quad . \\ &+ (Y_{1c} - \hat{\mu} - \hat{\theta}_c)^2 + (Y_{2c} - \hat{\mu} - \hat{\theta}_c)^2 + \ldots + (Y_{rc} - \hat{\mu} - \hat{\theta}_c)^2 \end{aligned}$$

with respect to $\hat{\mu}$ and the $\hat{\theta}_j$. This is accomplished by setting equal to zero the partial derivatives of Σu^2 with respect to $\hat{\mu}$ and the $\hat{\theta}_j$ and getting the common solution of these "least-squares equations." We have

$$\begin{aligned} \frac{\partial \Sigma u^2}{\partial \hat{\mu}} &= -2(Y_{11} - \hat{\mu} - \hat{\theta}_1) - 2(Y_{21} - \hat{\mu} - \hat{\theta}_1) - \ldots - 2(Y_{r1} - \hat{\mu} - \hat{\theta}_1) \\ &\quad -2(Y_{12} - \hat{\mu} - \hat{\theta}_2) - 2(Y_{22} - \hat{\mu} - \hat{\theta}_2) - \ldots - 2(Y_{r2} - \hat{\mu} - \hat{\theta}_2) \\ &\quad . \quad . \quad . \quad . \quad . \\ &\quad -2(Y_{1c} - \hat{\mu} - \hat{\theta}_c) - 2(Y_{2c} - \hat{\mu} - \hat{\theta}_c) - \ldots - 2(Y_{rc} - \hat{\mu} - \hat{\theta}_c) \\ &= \sum_i \sum_j Y_{ij} - n\hat{\mu} - r\hat{\theta}_1 - r\hat{\theta}_2 - \ldots - r\hat{\theta}_c = 0 \end{aligned}$$

since the −2's cancel out. (Note that $n = rc$.) Likewise we have

$$\begin{aligned} \frac{\partial \Sigma u^2}{\partial \hat{\theta}_1} &= -2(Y_{11} - \hat{\mu} - \hat{\theta}_1) - 2(Y_{21} - \hat{\mu} - \hat{\theta}_1) - \ldots - 2(Y_{r1} - \hat{\mu} - \hat{\theta}_1) \\ &= \sum_i Y_{ij} - r\hat{\mu} - r\hat{\theta}_1 = 0 \end{aligned}$$

and there will be similar results for $\dfrac{\partial \Sigma u^2}{\partial \hat{\theta}_2}, \dfrac{\partial \Sigma u^2}{\partial \hat{\theta}_3}$, and so on.

In summary, then, the least-squares equations will be as follows:

$$\text{1 equation: } \sum_i \sum_j Y_{ij} - n\hat{\mu} - r\hat{\theta}_1 - r\hat{\theta}_2 - \ldots - r\hat{\theta}_c = 0$$

(3) $$c \text{ equations of the form: } \sum_i Y_{ij} - r\hat{\mu} - r\hat{\theta}_j = 0$$

Since there are $c + 1$ unknowns, $\hat{\mu}$ and the c $\hat{\theta}_j$'s, and $c + 1$ equations, it appears at first glance as if our problem is solved. On more careful scrutiny, however, it will be noted that the first equation is merely the sum of the last c equations, and in reality we have only c independent equations. This means that there is an infinite set of solutions. We get a unique solution, however, if we add the additional equation $\sum_j \hat{\theta}_j = 0$. This means that we will view the θ_j as differentials, which seems a sensible way out of the difficulty. With this extra assumption we now have

(3a)
$$\hat{\mu} = \sum_i \sum_j Y_{ij}/n = \bar{Y}$$

$$\hat{\theta}_j = \frac{\sum_i Y_{ij}}{r} - \hat{\mu} = \bar{Y}_j - \hat{\mu}$$

In line with the usual regression analysis[21] we will have

$$\begin{aligned}\sum \sum u^2 = \sum \sum uY = \sum \sum (Y_{ij} - \hat{\mu} - \hat{\theta}_j)Y_{ij} \\ &= \sum \sum Y_{ij}^2 - \hat{\mu} \sum \sum Y_{ij} - \sum_j \hat{\theta}_j \sum_i Y_{ij} \\ &= \sum \sum Y_{ij}^2 - \frac{(\Sigma\Sigma Y_{ij})^2}{n} - \sum_j \left(\frac{\sum_i Y_{ij}}{r} - \hat{\mu} \right) \sum_i Y_{ij} \\ &= \sum \sum Y_{ij}^2 - K_y - \left[\frac{\sum_j \left(\sum_i Y_{ij} \right)^2}{r} - K_y \right]\end{aligned}$$

where $K_y = \left(\sum_i \sum_j Y_{ij} \right)^2 \Big/ n$ and is called the correction factor.

In analysis of variance language, we have thus proved that if $Y_{ij} = \mu + \theta_j + \epsilon_{ij}$ is fitted by least squares, then the

Total sum of squares = Column sum of squares + Residual sum of squares

(4) $$\left[\sum \sum Y_{ij}^2 - K_y \right] = \left[\frac{\sum_j \left(\sum_i Y_{ij} \right)^2}{r} - K_y \right] + \sum \sum u^2$$

---

[21] Cf. Appendix I (37).

b. Two-Way Analysis of Variance with a Single Case in Each Class. With an $r \times c$ two-way arrangement with a single case in each class, the analysis of variance model is

$$Y_{ij} = \mu + \tau_i + \theta_j + \epsilon_{ij} \tag{5}$$

When this is fitted by minimizing $\Sigma\Sigma u_{ij}^2$, we get the following least-squares equations:

(6)

$$\text{1 equation: } \frac{\partial \Sigma\Sigma u_{ij}^2}{\partial \hat{\mu}} = \Sigma \, \Sigma \, Y_{ij} - n\hat{\mu} - c\hat{\tau}_1 - \ldots - c\hat{\tau}_r - r\hat{\theta}_1 - \ldots - r\hat{\theta}_c = 0$$

$$r \text{ equations: } \frac{\partial \Sigma\Sigma u_{ij}^2}{\partial \hat{\tau}_i} = \sum_j Y_{ij} - c\hat{\mu} - c\hat{\tau}_i - \hat{\theta}_1 - \ldots - \hat{\theta}_c = 0$$

$$c \text{ equations: } \frac{\partial \Sigma\Sigma u_{ij}^2}{\partial \hat{\theta}_j} = \sum_i Y_{ij} - r\hat{\mu} - \hat{\tau}_1 - \ldots - \hat{\tau}_r - r\hat{\theta}_j = 0$$

These equations are not all independent, however, since both the sum of the last c equations and the sum of the next r equations each yield the first equation. To get a solution we assume that the τ_i and the θ_j are differentials and add to the least-squares equations the conditions $\Sigma\hat{\tau}_i = 0$ and $\Sigma\hat{\theta}_j = 0$. We then get the following

(7)

$$\hat{\mu} = \frac{\sum_i \sum_j Y_{ij}}{n} = \bar{Y}$$

$$\hat{\tau}_i = \frac{\sum_j Y_{ij}}{c} - \hat{\mu} = \bar{Y}_i - \hat{\mu}$$

$$\hat{\theta}_j = \frac{\sum_i Y_{ij}}{r} - \hat{\mu} = \bar{Y}_j - \hat{\mu}$$

As before we have

$$\begin{aligned}
\sum_i \sum_j u_{ij}^2 &= \Sigma \, \Sigma \, (Y_{ij} - \hat{\mu} - \hat{\tau}_i - \hat{\theta}_j) Y_{ij} \\
&= \Sigma \, \Sigma \, Y_{ij}^2 - \hat{\mu} \, \Sigma \, \Sigma \, Y_{ij} - \sum_i \hat{\tau}_i \sum_j Y_{ij} - \sum_j \hat{\theta}_j \sum_i Y_{ij} \\
&= \Sigma \, \Sigma \, Y_{ij}^2 - \frac{(\Sigma\Sigma Y_{ij})^2}{n} - \left[\frac{\sum_i \left(\sum_j Y_{ij} \right)^2}{c} - \hat{\mu} \sum_i \sum_j Y_{ij} \right] \\
&\quad - \left[\frac{\sum_j \left(\sum_i Y_{ij} \right)^2}{r} - \hat{\mu} \sum_i \sum_j Y_{ij} \right]
\end{aligned}$$

$$= \sum \sum Y_{ij}^2 - K_y - \left[\frac{\sum_i \left(\sum_j Y_{ij} \right)^2}{c} - K_y \right] - \left[\frac{\sum_j \left(\sum_i Y_{ij} \right)^2}{r} - K_y \right]$$

Or again we have

(8) Total SS = Row SS + Column SS + Residual SS

$$\sum_i \sum_j Y_{ij}^2 - K_y = \left[\frac{\sum_i \left(\sum_j Y_{ij} \right)^2}{c} - K_y \right] + \left[\frac{\sum_j \left(\sum_i Y_{ij} \right)^2}{r} - K_y \right] + \sum_i \sum_j u_{ij}^2$$

c. One-Way Analysis of Covariance. A simple one-way covariance model is

(9) $$Y_{ij} = \mu + \theta_j + \beta x_{ij} + \epsilon_{ij}$$

where $x_{ij} = X_{ij} - \bar{\bar{X}}$ and the other quantities have the same meaning as before. [Note that we are using $\bar{\bar{X}}$ here to represent the grand mean.]

When we fit this by least squares, we get the following equations (note $\Sigma\Sigma x_{ij} = 0$):

$$1 \text{ equation: } \frac{\partial \Sigma\Sigma u_{ij}^2}{\partial \hat{\mu}} = \sum_i \sum_j Y_{ij} - n\hat{\mu} - r\hat{\theta}_1 - r\hat{\theta}_2 - \ldots - r\hat{\theta}_c = 0$$

$$c \text{ equations: } \frac{\partial \Sigma\Sigma u_{ij}^2}{\partial \hat{\theta}_j} = \sum_i Y_{ij} - r\hat{\mu} - r\hat{\theta}_j - \hat{\beta} \sum_i x_{ij} = 0$$

$$1 \text{ equation: } \frac{\partial \Sigma\Sigma u_{ij}^2}{\partial \hat{\beta}} = \sum_i \sum_j x_{ij} Y_{ij} - \sum_j \hat{\theta}_j \sum_i x_{ij} - \hat{\beta} \sum_i \sum_j x_{ij}^2 = 0$$

If we add the second c equations, we shall get the first and to make up the deficiency we add the condition $\Sigma\hat{\theta}_j = 0$. This gives us then

$$\hat{\mu} = \frac{\sum_i \sum_j Y_{ij}}{n} = \bar{\bar{Y}}$$

$$\hat{\theta}_j = \frac{\sum_i Y_{ij}}{r} - \hat{\mu} - \hat{\beta} \frac{\sum_i x_{ij}}{r} = \bar{Y}_j - \hat{\mu} - \hat{\beta}\bar{x}_j$$

$$\hat{\beta} = \frac{\sum_i \sum_j x_{ij} Y_{ij} - \dfrac{\sum_j \left(\sum_i x_{ij} \right) \left(\sum_i Y_{ij} \right)}{r}}{\sum_i \sum_j x_{ij}^2 - \sum_j \left(\sum_i x_{ij} \right)^2 / r}$$

$$= \frac{S_{xy} + K_{xy} - C_{xy} - K_{xy}}{S_{xx} + K_x - C_{xx} - K_x} = \frac{{}_cE_{xy}}{{}_cE_{xx}}$$

the ${}_cE_{xy}$, ${}_cE_{xx}$, K_{xy}, and K_y being defined in Table 34.2.

We also have

$$\sum_i \sum_j u_{ij}^2 = \sum_i \sum_j (Y_{ij} - \hat{\mu} - \hat{\theta}_j - \hat{\beta} x_{ij}) Y_{ij}$$

$$= \sum_i \sum_j Y_{ij}^2 - \hat{\mu} \sum_i \sum_j Y_{ij} - \sum_j \hat{\theta}_j \sum_i Y_{ij} - \hat{\beta} \sum_i \sum_j x_{ij} Y_{ij}$$

$$= \sum_i \sum_j Y_{ij}^2 - \frac{\left(\sum_i \sum_j Y_{ij}\right)^2}{n} - \sum_j (\bar{Y}_j - \hat{\mu} - \hat{\beta}\bar{x}_j) \sum_i Y_{ij} - \hat{\beta} \sum_i \sum_j x_{ij} Y_{ij}$$

$$= \sum_i \sum_j Y_{ij}^2 - Ky - \left[\sum_j \frac{\left(\sum_i Y_{ij}\right)^2}{r} - K_y \right] - \hat{\beta} \left[\sum_i \sum_j x_{ij} Y_{ij} - \sum_j \frac{\left(\sum_i x_{ij}\right)\left(\sum_i Y_{ij}\right)}{r} \right]$$

This means that

Total SS	=	Column SS	+	SS due to regression	+	Residual SS
$\left[\sum_i \sum_j Y_{ij}^2 - K_y \right]$	=	$\left[\sum_j \frac{\left(\sum_i Y_{ij}\right)^2}{r} - K_y \right]$	+	$\frac{{}_cE_{xy}^2}{{}_cE_{xx}}$	+	$\sum u^2$

or in the notation of Table 34.2 and Chapter 34

$$S_{yy} = C_{yy} + \frac{{}_cE_{xy}^2}{{}_cE_{xx}} + \sum u_C^2$$

Compare equation (34.3).

d. A Basic Theorem on Testing Regression Coefficients. Anderson and Bancroft[22] prove the following basic theorem pertaining to the testing of regression coefficients. It will merely be stated here.

Consider the regression model

$$X_1 = a'_{1.23\ldots} + b'_{12.34\ldots} X_2 + b'_{13.24\ldots} X_3 + \ldots + b'_{1k.23\ldots} X_k + \ldots + b'_{1p.23\ldots} X_p + \epsilon$$

where ϵ is normally distributed with zero mean and variance σ'^2.

[22] *Statistical Theory in Research* (New York: McGraw-Hill, 1952) chap. xiv.

If the full equation is fitted by least squares, then in an infinite set of random samples with fixed independent variables X_2 to X_p, the quantity

$$\frac{(n-p)s_e^2}{\sigma'^2} = \frac{\Sigma u^2_{\text{Full Equation}}}{\sigma'^2}$$

will have a χ^2 distribution with $n - p$ degrees of freedom. (Note p is the number of variables, both independent and dependent.)

If terms X_{k+1} to X_p are dropped from the equation and this diminished equation is fitted by least squares, we can compute $\Sigma u^2_{\text{Diminished Equation}}$. If the coefficients of X_{k+1} to X_p are all individually zero, it can be shown that the quantity

$$\frac{(p-k)s_R^2}{\sigma'^2} = \frac{\Sigma u^2_{\text{Diminished Equation}} - \Sigma u^2_{\text{Full Equation}}}{\sigma'^2}$$

will have a χ^2 distribution with $p - k$ degrees of freedom. The quantity $\Sigma u^2_{\text{Diminished Equation}} - \Sigma u^2_{\text{Full Equation}}$ is called the Reduced Sum of Squares, and the mean square is given a subscript R.

The mean squares s_e^2 and s_R^2 are independently distributed. Hence if the regression coefficients $b'_{1(k+1).23\ldots}$ to $b'_{1p.23\ldots}$ are all individually zero, the quantity s_R^2/s_e^2 will have an F distribution with $\nu_1 = p - k$ and $\nu_2 = n - p$. Therefore to test the hypothesis that coefficients $b_{1(k+1).23\ldots}$ to $b'_{1p.23\ldots}$ are all individually zero, we compute s_R^2/s_e^2 and compare it with F_α for $\nu_1 = p - k$ and $\nu_2 = n - p$.

e. A General One-Way Classification Covariance Model. In conclusion let us consider a general one-way classification model in which there may be class differentials in slopes as well as in levels and also a difference in slopes between and within classes. Let the mathematical model be

(1) $$Y = \mu + \theta_j + \beta_m \bar{x}_j + \beta_a(x_{ij} - \bar{x}_j) + \delta_j(x_{ij} - \bar{x}_j) + \epsilon_{ij}$$

where the grand mean of x_{ij} is zero and ϵ_{ij} is a normally distributed random variable with zero mean and variance σ'^2. The θ_j's are "adjusted" class differentials, β_m is the slope of the regression of class means of Y on class means of x, β_a is the slope of that part of the regression of Y on x within classes that is common to all classes, and the δ_j's are class slope differentials.

If we fit the above equation to a random set of data consisting of r items in each of c classes, we get the following least-squares equations:

1 equation given by partial differentiation of $\Sigma\Sigma u^2$ by $\hat{\mu}$, viz

$$\sum_i \sum_j Y_{ij} - n\hat{\mu} - r\hat{\theta}_1 - \ldots - r\hat{\theta}_c - r\hat{\beta}_m \bar{x}_1 - \ldots - r\hat{\beta}_m \bar{x}_c$$

$$-\hat{\beta}_a \sum_i \sum_j (x_{ij} - \bar{x}_j) - \hat{\delta}_1 \sum_i (x_{i1} - \bar{x}_1) - \ldots - \hat{\delta}_c \sum_i (x_{ic} - \bar{x}_c) = 0$$

which reduces to

(2) $$\sum_i \sum_j Y_{ij} - n\hat{\mu} - r \sum_j \hat{\theta}_j = 0$$

since $\sum_j \bar{x}_j = 0$, $\sum_i \sum_j (x_{ij} - \bar{x}_j) = 0$ and $\sum_i (x_{ij} - \bar{x}_j) = 0$ for all j.

c equations given by partial differentiation of $\Sigma\Sigma u^2$ by $\hat{\theta}_j$, viz

(3) $$\sum_i Y_{ij} - r\hat{\mu} - r\hat{\theta}_j - r\hat{\beta}_m \bar{x}_j = 0$$

1 equation given by partial differential of $\Sigma\Sigma u^2$ by $\hat{\beta}_m$, viz

(4) $$\sum_i \sum_j \bar{x}_j Y_{ij} - r \sum_j \hat{\theta}_j \bar{x}_j - r\hat{\beta}_m \sum \bar{x}_j^2 = 0$$

1 equation given by partial differentiation of $\Sigma\Sigma u^2$ by $\hat{\beta}_a$, viz

(5) $$\sum_i \sum_j (x_{ij} - \bar{x}_j) Y_{ij} - \hat{\beta}_a \sum_i \sum_j (x_{ij} - \bar{x}_j)^2 - \sum_j \hat{\delta}_j \sum_i (x_{ij} - \bar{x}_j)^2 = 0$$

c equations given by partial differentiation of $\Sigma\Sigma u^2$ by $\hat{\delta}_j$, viz

(6) $$\sum_i (x_{ij} - \bar{x}_j) Y_{ij} - \hat{\beta}_a \sum_i (x_{ij} - \bar{x}_{ij})^2 - \hat{\delta}_j \sum_i (x_{ij} - \bar{x}_j)^2 = 0$$

Not all these equations are independent. If we sum the c equations (3), we get (2), since $\Sigma\bar{x}_j = 0$. If we sum the c equations (6), we get (5). Finally, if we multiply each equation of (3) by the corresponding $\bar{x}_j$ and sum, we get (4). Hence to get a unique solution we need to add three other equations. It is mathematically convenient and generally reasonable to add the following:

$\sum_j \hat{\theta}_j = 0$ (the $\hat{\theta}_j$ differentials sum to zero)

(7) $\sum_j \hat{\theta}_j \bar{x}_j = 0$ (the covariance between $\hat{\theta}_j$ and $\bar{x}_j$ is zero)

$\sum_j \hat{\delta}_j \sum_i (x_{ij} - \bar{x}_j)^2 = 0$ (the weighted sum of the slope differentials $\hat{\delta}_j$ is zero)

Our least-squares equations thus reduce to the following:

(8)

$$\hat{\mu} = \sum_i \sum_j Y_{ij}/n = \bar{\bar{Y}}$$

$$\hat{\theta}_j = \sum_i \frac{Y_{ij}}{r} - \hat{\mu} - \hat{\beta}_m \bar{x}_j = \bar{Y}_j - \hat{\mu} - \beta_m \bar{x}_j$$

$$\hat{\beta}_m = \frac{\sum_j \bar{x}_j \bar{Y}_j}{\sum_j \bar{x}_j^2} = \frac{\sum_j (\bar{x}_j - \bar{\bar{x}})(\bar{Y}_j - \bar{\bar{Y}})}{\sum_j (\bar{x}_j - \bar{\bar{x}})^2} = \frac{C_{xy}}{C_{xx}} \quad \text{[Note: } \bar{\bar{x}} = 0.\text{]}$$

$$\hat{\beta}_a = \frac{\sum_i \sum_j (x_{ij} - \bar{x}_j) Y_{ij}}{\sum_i \sum_j (x_{ij} - \bar{x}_j)^2} = \frac{\sum_i \sum_j x_{ij} - \bar{x}_j)(Y_{ij} - \bar{Y}_j)}{\sum_i \sum_j (x_{ij} - \bar{x}_j)^2} = \frac{E_{xy}}{E_{xx}}$$

$$\hat{\delta}_j = \frac{\sum_i (x_{ij} - \bar{x}_j)(Y_{ij})}{\sum_i (x_{ij} - \bar{x}_j)^2} - \hat{\beta}_a$$

$$= \frac{\sum_i (x_{ij} - \bar{x}_j)(Y_{ij} - \bar{Y}_j)}{\sum_i (x_{ij} - \bar{x}_j)^2} - \hat{\beta}_a = \frac{E_{xyj}}{E_{xxj}} - \hat{\beta}_a$$

where the quantities C_{xy}, C_{xx}, E_{xy}, E_{xx}, E_{xyj}, and E_{xxj} are defined in Table 34.10.

The residual sum of squares is derived as follows:

From Section 37 of this Appendix we have

$$\sum_i \sum_j u_{ij}^2 = \sum_i \sum_j u_{ij} Y_{ij}$$

$$= \sum_i \sum_j \left[Y_{ij} - \hat{\mu} - \hat{\theta}_j - \hat{\beta}_m \bar{x}_j - \hat{\beta}_a (x_{ij} - \bar{x}_j) - \hat{\delta}_j (x_{ij} - \bar{x}_j) \right] Y_{ij}$$

But this equals

$$\sum_i \sum_j Y_{ij}^2 - \hat{\mu} \sum_i \sum_j Y_{ij} - \sum_j \hat{\theta}_j \sum_i Y_{ij} - \hat{\beta}_m \sum_j \bar{x}_j \sum_i Y_{ij}$$

$$- \hat{\beta}_a \sum_i \sum_j (x_{ij} - \bar{x}_j) Y_{ij} - \sum_j \hat{\delta}_j \sum_i (x_{ij} - \bar{x}_j) Y_{ij}$$

which from equations (8) equals

$$\left[\sum_i \sum_j Y_{ij}^2 - \frac{\left(\sum_i \sum_j Y_{ij} \right)^2}{n} \right] - \left[\sum_j \frac{\left(\sum_i Y_{ij} \right)^2}{r} - \frac{\left(\sum_i \sum_j Y_{ij} \right)^2}{n} - r\hat{\beta}_m \sum_j \bar{x}_j \bar{Y}_j \right]$$

$$- r\hat{\beta}_m \sum_j \bar{x}_j \bar{Y}_j - \hat{\beta}_a \sum_i \sum_j (x_{ij} - \bar{x}_j)(Y_{ij} - \bar{Y}_j) - \sum_j \hat{\delta}_j \sum_i (x_{ij} - \bar{x}_j)(Y_{ij} - \bar{Y}_j)$$

or in terms of the notation of Table 34.10

$$\sum_i \sum_i u_{ij}^2 = S_{yy} - \left[C_{yy} - \frac{C_{xy}^2}{C_{xx}} \right] - \frac{C_{xy}^2}{C_{xx}} - \frac{E_{xy}^2}{E_{xx}} - \left[\sum_j \frac{E_{xyj}^2}{E_{xxj}} - \frac{E_{xy}^2}{E_{xx}} \right]$$

$$= S_{yy} - C_{yy} - \sum_j \frac{E_{xyj}^2}{E_{xxj}}$$

or

(9) $$\sum_i \sum_j u_{ij}^2 = E_{yy} - \sum_j \frac{E_{xyj}^2}{E_{xxj}}$$

By similar reasoning it can be shown that if equation (1) without the δ_j terms is fitted by least squares the residual sum of squares will equal

(10) $$\sum_i \sum_j u^2_{\text{without } \delta_j \text{ terms}} = E_{yy} - \frac{E_{xy}^2}{E_{xx}}$$

Hence the reduced sum of squares for δ is

(11) $$\text{RSS}_\delta = \sum_j \frac{E^2_{xyj}}{E_{xxj}} - \frac{E^2_{xy}}{E_{xx}}$$

which is equation (34.32) of Chapter 34.

It also follows directly from the above that if we let

$$Q_e = \sum_i \sum_j u^2_{ij} \text{ (for full equation (1))}$$

$$Q_\delta = \sum_j \frac{E^2_{xyj}}{E_{xxj}} - \frac{E^2_{xy}}{E_{xx}}$$

(12) $$Q_\theta = C_{yy} - \frac{C^2_{xy}}{C_{xx}}$$

$$Q_{m-a} = \frac{C^2_{xy}}{C_{xx}} + \frac{E^2_{xy}}{E_{xx}} - \frac{S^2_{xy}}{S_{xx}}$$

$$Q_0 = \frac{S^2_{xy}}{S_{xx}}$$

then since $\sum_i \sum_j (Y_{ij} - \bar{\bar{Y}})^2 = S_y$ by definition.

(13) $$\sum_i \sum_j (Y_{ij} - \bar{\bar{Y}})^2 = Q_c + Q_\delta + Q_\theta + Q_{m-a} + Q_0$$

which is equation (34.35) of Chapter 34.

APPENDIX II
Special Tables and Charts

NOTES ON TABLES

Tables A1 and A2. In these tables z represents a standardized variable, i.e., a variable measured from its mean and expressed in standard deviation units. The normal curve is symmetrical. Consequently, the ordinates of Table A1 listed for z are also valid for $-z$. To fit these ordinates to any given histogram they must be divided by σ'.

Table A2 gives the area under the normal curve from $-\infty$ to $+z$. To find the area from $-\infty$ to $-z$, subtract the tabular value from 1.

For use of the tables, see Chapter 4, Section 6.1.

Table B. The probabilities of Table B are those of equaling or exceeding the *absolute* values of t listed in the table.

For values of ν between 30 and 120, it is better to interpolate linearly with respect to the reciprocals of ν rather than with respect to ν itself. Thus, if we were given the $t_{0.025}$ values for $\nu = 30$ and $\nu = 60$ and we wished to find the $t_{0.025}$ point for $\nu = 40$, we would proceed as follows:

$$t_{0.025}(\nu = 40)$$

$$= t_{0.025}(\nu = 30) - \frac{1/30 - 1/40}{1/30 - 1/60}[t_{0.025}(\nu = 30) - t_{0.025}(\nu = 60)]$$

$$= 2.042 - \frac{0.25}{0.50}[2.042 - 2.000]$$

$= 2.021$ which is the value tabulated in Table B.

By straight-line interpolation with reference to ν, we would have obtained

$$t_{0.025}\ (\nu = 40) = 2.028$$

Table C. This gives the area under the χ^2 curve from χ^2 to ∞. For larger values of ν set $z = \sqrt{2\chi^2} - \sqrt{2\nu - 1}$ and use Table A2.

Table D3. This is an extension of a table originally compiled by P. B. Patnaik and later extended by H. A. David. It is a basic table that may be used whenever the average range is used in lieu of s.

Table F. For large samples from a normal universe, the standard error of γ_1 is $\sqrt{6/n}$. Approximate probabilities can be obtained by setting $z = \gamma_1/\sqrt{6/n}$ and using Table A2. Thus, for $n = 150$, we can estimate the upper 0.05 point for γ_1 by setting $\gamma_1/\sqrt{6/n} = 1.645$, which yields $\gamma_1 = 0.329$. The table value is 0.321.

Table G. For large samples from a normal universe, the standard error of γ_2 is $\sqrt{24/n}$. Rough estimates of probabilities can be obtained for large samples by setting $z = \gamma_2/\sqrt{24/n}$ and using Table A2. Thus, for $n = 500$, we can estimate the 0.05 points for γ_2 by setting $\gamma_2/\sqrt{24/n} = \pm 1.645$, which yields $\gamma_2 = \pm 0.361$. The tabular values are -0.33 and 0.37.

Table J. Table J gives the area under the F curve from F to ∞. The $F_{1-\alpha}$ point of the F distribution for ν_1 and ν_2 degrees of freedom is equal to the reciprocal of the F_α point for ν_2 and ν_1 degrees of freedom. Interpolation for ν_1 and ν_2 is more accurate if we interpolate linearly on the reciprocals of ν_1 and ν_2 instead of on ν_1 and ν_2 directly. See notes on Table B.

Table L. Table L will be found useful in making a rough sketch of the OC curve for a test of the ratio of two variances when $n_1 \neq n_2$. By the nature of the test it automatically follows that $\beta = 0.95$ when $\phi = 1$. The values of ϕ for $\beta = 0.10$ can be obtained easily from Table L by taking the square root of the tabular entry. Other values can be computed by taking ϕ equal to the square root of the product $F_{0.05}(\nu_1, \nu_2)\ F_\beta(\nu_2, \nu_1)$.

Table R. Table R provides quantities that are useful in setting up a unit sequential-sampling plan. For $\alpha = 0.05$ and $\beta = 0.10$, it gives, for selected values of p_1' and p_2', values of (1) the X intercept of the rejection line *(h_2)*; (2) the X intercept of the acceptance line *(h_1)*; (3) the slope of the limit lines *(s)*; and (4) the ordinates of the ASN curve *($\bar{n}$)* for $p' = 0$, $p' = 1$, $p' = p_1'$, $p' = s$, and $p' = p_2'$.

Table S. This is a conversion by Cowden of a table originally computed by Mahalanobis in *Sankhyā,* Vol. I, p. 121.

Table Y. This is a condensation of Bechhofer's original table.

Charts I and II. These are special condensations of much more extensive charts by Pearson and Hartley. See *Biometrika* 38 Parts 1 and 2 (1951), pp. 112–30.

Table Z. Table Z is from a larger table published in *Biometrika* 59 (1972), pp. 178–88 by Yoong-Sin Lee. For a table of lower percentage points see K. H. Kramer's table in *Biometrika Tables for Statisticians* 2 (1972). (Also see book by Ezekiel and article by Kramer in Cumulative List of References) For an example of the use of Table Z, see Chapter 33, Section 1.5.6.

Table DD. Table DD is a set of random sampling numbers. They are helpful in random sampling when the items of the universe can be associated with a succession of numbers. In that instance a selection of a group of numbers from the table will yield a random sample from the universe. One may start at any point in the table, move in any preassigned direction, and the numbers so obtained will be a random set. For an illustration of the use of the table, see Chapter 2, Section 4.

TABLE A1
Ordinates of the Normal Probability Distribution

z	.00	.01	.02	.03	.04	.05	.06	.07	.08	.09
.0	.3989	.3989	.3989	.3988	.3986	.3984	.3982	.3980	.3977	.3973
.1	.3970	.3965	.3961	.3956	.3951	.3945	.3939	.3932	.3925	.3918
.2	.3910	.3902	.3894	.3885	.3876	.3867	.3857	.3847	.3836	.3825
.3	.3814	.3802	.3790	.3778	.3765	.3752	3739	.3725	.3712	.3697
.4	.3683	.3668	.3653	.3637	.3621	.3605	.3589	.3572	.3555	.3538
.5	.3521	.3503	.3485	.3467	.3448	.3429	.3410	.3391	.3372	.3352
.6	.3332	.3312	.3292	.3271	.3251	.3230	.3209	.3187	.3166	.3144
.7	.3123	.3101	.3079	.3056	.3034	.3011	.2989	.2966	.2943	.2920
.8	.2897	.2874	.2850	.2827	.2803	2780	.2756	.2732	.2709	.2685
.9	.2661	.2637	.2613	.2589	.2565	.2541	.2516	.2492	.2468	.2444
1.0	.2420	.2396	.2371	.2347	.2323	.2299	.2275	.2251	.2227	.2203
1.1	.2179	.2155	.2131	.2107	.2083	2059	.2036	.2012	.1989	.1965
1.2	.1942	.1919	.1895	.1872	.1849	.1826	.1804	.1781	1758	.1736
1.3	.1714	.1691	.1669	.1647	.1626	1604	.1582	.1561	1539	.1518
1.4	.1497	.1476	.1456	.1435	.1415	.1394	.1374	.1354	.1334	.1315
1.5	.1295	.1276	.1257	.1238	.1219	.1200	.1182	.1163	.1145	.1127
1.6	.1109	.1092	.1074	.1057	.1040	.1023	1006	.0989	.0973	.0957
1.7	.0940	.0925	.0909	.0893	.0878	.0863	.0848	.0833	.0818	.0804
1.8	.0790	.0775	.0761	.0748	.0734	.0721	.0707	.0694	.0681	.0669
1.9	.0656	.0644	.0632	.0620	.0608	.0596	.0584	.0573	.0562	.0551
2.0	.0540	.0529	.0519	.0508	.0498	.0488	.0478	.0468	.0459	.0449
2.1	.0440	.0431	.0422	.0413	.0404	.0396	.0387	.0379	.0371	.0363
2.2	.0355	.0347	.0339	.0332	.0325	.0317	.0310	.0303	.0297	.0290
2.3	.0283	.0277	.0270	.0264	.0258	.0252	.0246	.0241	.0235	.0229
2.4	.0224	.0219	.0213	.0208	.0203	.0198	.0194	.0189	.0184	.0180
2.5	.0175	.0171	.0167	.0163	.0158	.0154	.0151	.0147	.0143	.0139
2.6	.0136	.0132	.0129	.0126	.0122	.0119	.0116	.0113	.0110	.0107
2.7	.0104	.0101	0099	.0096	.0093	.0091	.0088	.0086	.0084	.0081
2.8	.0079	.0077	.0075	.0073	.0071	.0069	.0067	.0065	.0063	.0061
2.9	.0060	.0058	.0056	.0055	.0053	.0051	.0050	.0048	.0047	.0046
3.0	.0044	.0043	.0042	.0040	.0039	.0038	.0037	.0036	.0035	.0034
3.1	.0033	.0032	.0031	.0030	.0029	.0028	.0027	.0026	.0025	.0025
3.2	.0024	.0023	.0022	.0022	.0021	.0020	.0020	.0019	.0018	.0018
3.3	.0017	.0017	.0016	.0016	.0015	.0015	.0014	.0014	.0013	.0013
3.4	.0012	.0012	.0012	.0011	.0011	.0010	.0010	.0010	.0009	.0009
3.5	.0009	0008	.0008	.0008	.0008	.0007	.0007	0007	.0007	.0006
3.6	.0006	.0006	.0006	.0005	.0005	.0005	.0005	.0005	.0005	.0004
3.7	.0004	.0004	.0004	.0004	.0004	.0004	.0003	.0003	.0003	.0003
3.8	.0003	.0003	.0003	.0003	.0003	.0002	.0002	.0002	.0002	.0002
3.9	.0002	.0002	.0002	.0002	.0002	.0002	.0002	.0002	.0001	.0001

SOURCE: Table A1 is reprinted with permission from A. M. Mood, *Introduction to the Theory of Statistics* (New York: McGraw-Hill, 1950), p. 422.

Table A2

Cumulative Probabilities of the Normal Probability Distribution (areas under the normal curve from $-\infty$ to z)

z	.00	.01	.02	.03	.04	.05	.06	.07	.08	.09
.0	.5000	.5040	.5080	.5120	.5160	.5199	.5239	.5279	.5319	.5359
.1	.5398	.5438	.5478	.5517	.5557	.5596	.5636	.5675	.5714	.5753
.2	.5793	.5832	.5871	.5910	.5948	.5987	.6026	.6064	.6103	.6141
.3	.6179	.6217	.6255	.6293	.6331	.6368	.6406	.6443	.6480	.6517
.4	.6554	.6591	.6628	.6664	.6700	.6736	.6772	.6808	.6844	.6879
.5	.6915	.6950	.6985	.7019	.7054	.7088	.7123	.7157	.7190	.7224
.6	.7257	.7291	.7324	.7357	.7389	.7422	.7454	.7486	.7517	.7549
.7	.7580	.7611	.7642	.7673	.7704	.7734	.7764	.7794	.7823	.7852
.8	.7881	.7910	.7939	.7967	.7995	.8023	.8051	.8078	.8106	.8133
.9	.8159	.8186	.8212	.8238	.8264	.8289	.8315	.8340	.8365	.8389
1.0	.8413	.8438	.8461	.8485	.8508	.8531	.8554	.8577	.8599	.8621
1.1	.8643	.8665	.8686	.8708	.8729	.8749	.8770	.8790	.8810	.8830
1.2	.8849	.8869	.8888	.8907	.8925	.8944	.8962	.8980	.8997	.9015
1.3	.9032	.9049	.9066	.9082	.9099	.9115	.9131	.9147	.9162	.9177
1.4	.9192	.9207	.9222	.9236	.9251	.9265	.9279	.9292	.9306	.9319
1.5	.9332	.9345	.9357	.9370	.9382	.9394	.9406	.9418	.9429	.9441
1.6	.9452	.9463	.9474	.9484	.9495	.9505	.9515	.9525	.9535	.9545
1.7	.9554	.9564	.9573	.9582	.9591	.9599	.9608	.9616	.9625	.9633
1.8	.9641	.9649	.9656	.9664	.9671	.9678	.9686	.9693	.9699	.9706
1.9	.9713	.9719	.9726	.9732	.9738	.9744	.9750	.9756	.9761	.9767
2.0	.9772	.9778	.9783	.9788	.9793	.9798	.9803	.9808	.9812	.9817
2.1	.9821	.9826	.9830	.9834	.9838	.9842	.9846	.9850	.9854	.9857
2.2	.9861	.9864	.9868	.9871	.9875	.9878	.9881	.9884	.9887	.9890
2.3	.9893	.9896	.9898	.9901	.9904	.9906	.9909	.9911	.9913	.9916
2.4	.9918	.9920	.9922	.9925	.9927	.9929	.9931	.9932	.9934	.9936
2.5	.9938	.9940	.9941	.9943	.9945	.9946	.9948	.9949	.9951	.9952
2.6	.9953	.9955	.9956	.9957	.9959	.9960	.9961	.9962	.9963	.9964
2.7	.9965	.9966	.9967	.9968	.9969	.9970	.9971	.9972	.9973	.9974
2.8	.9974	.9975	.9976	.9977	.9977	.9978	.9979	.9979	.9980	.9981
2.9	.9981	.9982	.9982	.9983	.9984	.9984	.9985	.9985	.9986	.9986
3.0	.9987	.9987	.9987	.9988	.9988	.9989	.9989	.9989	.9990	.9990
3.1	.9990	.9991	.9991	.9991	.9992	.9992	.9992	.9992	.9993	.9993
3.2	.9993	.9993	.9994	.9994	.9994	.9994	.9994	.9995	.9995	.9995
3.3	.9995	.9995	.9995	.9996	.9996	.9996	.9996	.9996	.9996	.9997
3.4	.9997	.9997	.9997	.9997	.9997	.9997	.9997	.9997	.9997	.9998

z	1.282	1.645	1.960	2.326	2.576	3.090	3.291	3.891	4:417
$F(z)$	.90	.95	.975	.99	.995	.999	.9995	.99995	.999995
$2[1 - F(z)]$	.20	.10	.05	.02	.01	.002	.001	.0001	.00001

SOURCE: Table A2 is reprinted with permission from A. M. Mood, *Introduction to the Theory of Statistics* (New York: McGraw-Hill, 1950), p. 423.

Table B

Percentage Points of the *t* Distribution (Probabilities refer to the sum of the two-tail areas. For a single tail divide the probability by 2.)

Probability (P).

ν	·9	·8	·7	·6	·5	·4	·3	·2	·1	·05	·02	·01	·001
1	·158	·325	·510	·727	1·000	1·376	1·963	3·078	6·314	12·706	31·821	63·657	636·619
2	·142	·289	·445	·617	·816	1·061	1·386	1·886	2·920	4·303	6·965	9·925	31·598
3	·137	·277	·424	·584	·765	·978	1·250	1·638	2·353	3·182	4·541	5·841	12·941
4	·134	·271	·414	·569	·741	·941	1·190	1·533	2·132	2·776	3·747	4·604	8·610
5	·132	·267	·408	·559	·727	·920	1·156	1·476	2·015	2·571	3·365	4·032	6·859
6	·131	·265	·404	·553	·718	·906	1·134	1·440	1·943	2·447	3·143	3·707	5·959
7	·130	·263	·402	·549	·711	·896	1·119	1·415	1·895	2·365	2·998	3·499	5·405
8	·130	·262	·399	·546	·706	·889	1·108	1·397	1·860	2·306	2·896	3·355	5·041
9	·129	·261	·398	·543	·703	·883	1·100	1·383	1·833	2·262	2·821	3·250	4·781
10	·129	·260	·397	·542	·700	·879	1·093	1·372	1·812	2·228	2·764	3·169	4·587
11	·129	·260	·396	·540	·697	·876	1·088	1·363	1·796	2·201	2·718	3·106	4·437
12	·128	·259	·395	·539	·695	·873	1·083	1·356	1·782	2·179	2·681	3·055	4·318
13	·128	·259	·394	·538	·694	·870	1·079	1·350	1·771	2·160	2·650	3·012	4·221
14	·128	·258	·393	·537	·692	·868	1·076	1·345	1·761	2·145	2·624	2·977	4·140
15	·128	·258	·393	·536	·691	·866	1·074	1·341	1·753	2·131	2·602	2·947	4·073
16	·128	·258	·392	·535	·690	·865	1·071	1·337	1·746	2·120	2·583	2·921	4·015
17	·128	·257	·392	·534	·689	·863	1·069	1·333	1·740	2·110	2·567	2·898	3·965
18	·127	·257	·392	·534	·688	·862	1·067	1·330	1·734	2·101	2·552	2·878	3·922
19	·127	·257	·391	·533	·688	·861	1·066	1·328	1·729	2·093	2·539	2·861	3·883
20	·127	·257	·391	·533	·687	·860	1·064	1·325	1·725	2·086	2·528	2·845	3·850
21	·127	·257	·391	·532	·686	·859	1·063	1·323	1·721	2·080	2·518	2·831	3·819
22	·127	·256	·390	·532	·686	·858	1·061	1·321	1·717	2·074	2·508	2·819	3·792
23	·127	·256	·390	·532	·685	·858	1·060	1·319	1·714	2·069	2·500	2·807	3·767
24	·127	·256	·390	·531	·685	·857	1·059	1·318	1·711	2·064	2·492	2·797	3·745
25	·127	·256	·390	·531	·684	·856	1·058	1·316	1·708	2·060	2·485	2·787	3·725
26	·127	·256	·390	·531	·684	·856	1·058	1·315	1·706	2·056	2·479	2·779	3·707
27	·127	·256	·389	·531	·684	·855	1·057	1·314	1·703	2·052	2·473	2·771	3·690
28	·127	·256	·389	·530	·683	·855	1·056	1·313	1·701	2·048	2·467	2·763	3·674
29	·127	·256	·389	·530	·683	·854	1·055	1·311	1·699	2·045	2·462	2·756	3·659
30	·127	·256	·389	·530	·683	·854	1·055	1·310	1·697	2·042	2·457	2·750	3·646
40	·126	·255	·388	·529	·681	·851	1·050	1·303	1·684	2·021	2·423	2·704	3·551
60	·126	·254	·387	·527	·679	·848	1·046	1·296	1·671	2·000	2·390	2·660	3·460
120	·126	·254	·386	·526	·677	·845	1·041	1·289	1·658	1·980	2·358	2·617	3·373
∞	·126	·253	·385	·524	·674	·842	1·036	1·282	1·645	1·960	2·326	2·576	3·291

SOURCE: Table B is reprinted with permission from Table III of R. A. Fisher and F. Yates, *Statistical Tables for Biological, Agricultural and Medical Research* (Edinburgh: Oliver & Boyd, Ltd, 1953).

TABLE C
Percentage Points of the χ^2 Distribution

ν/P	0.995	0.990	0.975	0.950	0.900	0.80	0.70	0.50	0.30	0.20	0.10	0.05	0.025	0.010	0.005	0.001
1	0.0^4393	0.0^3157	0.0^3982	0.00393	0.0158	0.0642	0.148	0.455	1.07	1.64	2.71	3.84	5.02	6.63	7.88	10.8
2	0.0100	0.0201	0.0506	0.103	0.211	0.446	0.713	1.39	2.41	3.22	4.61	5.99	7.38	9.21	10.6	13.8
3	0.0717	0.115	0.216	0.352	0.584	1.01	1.42	2.37	3.66	4.64	6.25	7.82	9.35	11.3	12.8	16.3
4	0.207	0.297	0.484	0.711	1.06	1.65	2.19	3.36	4.88	5.99	7.78	9.49	11.1	13.3	14.9	18.5
5	0.412	0.554	0.831	1.15	1.61	2.34	3.00	4.35	6.06	7.29	9.24	11.1	12.8	15.1	16.7	20.5
6	0.676	0.872	1.24	1.64	2.20	3.07	3.83	5.35	7.23	8.56	10.6	12.6	14.4	16.8	18.5	22.5
7	0.989	1.24	1.69	2.17	2.83	3.82	4.67	6.35	8.38	9.80	12.0	14.1	16.0	18.5	20.3	24.3
8	1.34	1.65	2.18	2.73	3.49	4.59	5.53	7.34	9.52	11.0	13.4	15.5	17.5	20.1	22.0	26.1
9	1.73	2.09	2.70	3.33	4.17	5.38	6.39	8.34	10.7	12.2	14.7	16.9	19.0	21.7	23.6	27.9
10	2.16	2.56	3.25	3.94	4.87	6.18	7.27	9.34	11.8	13.4	16.0	18.3	20.5	23.2	25.2	29.6
11	2.60	3.05	3.82	4.57	5.58	6.99	8.15	10.3	12.9	14.6	17.3	19.7	21.9	24.7	26.8	31.3
12	3.07	3.57	4.40	5.23	6.30	7.81	9.03	11.3	14.0	15.8	18.5	21.0	23.3	26.2	28.3	32.9
13	3.57	4.11	5.01	5.89	7.04	8.63	9.93	12.3	15.1	17.0	19.8	22.4	24.7	27.7	29.8	34.5
14	4.07	4.66	5.63	6.57	7.79	9.47	10.8	13.3	16.2	18.2	21.1	23.7	26.1	29.1	31.3	36.1
15	4.60	5.23	6.26	7.26	8.55	10.3	11.7	14.3	17.3	19.3	22.3	25.0	27.5	30.6	32.8	37.7
16	5.14	5.81	6.91	7.96	9.31	11.2	12.6	15.3	18.4	20.5	23.5	26.3	28.8	32.0	34.3	39.3
17	5.70	6.41	7.56	8.67	10.1	12.0	13.5	16.3	19.5	21.6	24.8	27.6	30.2	33.4	35.7	40.8
18	6.26	7.01	8.23	9.39	10.9	12.9	14.4	17.3	20.6	22.8	26.0	28.9	31.5	34.8	37.2	42.3
19	6.84	7.63	8.91	10.1	11.7	13.7	15.4	18.3	21.7	23.9	27.2	30.1	32.9	36.2	38.6	43.8
20	7.43	8.26	9.59	10.9	12.4	14.6	16.3	19.3	22.8	25.0	28.4	31.4	34.2	37.6	40.0	45.3
21	8.03	8.90	10.3	11.6	13.2	15.4	17.2	20.3	23.9	26.2	29.6	32.7	35.5	38.9	41.4	46.8
22	8.64	9.54	11.0	12.3	14.0	16.3	18.1	21.3	24.9	27.3	30.8	33.9	36.8	40.3	42.8	48.3
23	9.26	10.2	11.7	13.1	14.8	17.2	19.0	22.3	26.0	28.4	32.0	35.2	38.1	41.6	44.2	49.7
24	9.89	10.9	12.4	13.8	15.7	18.1	19.9	23.3	27.1	29.6	33.2	36.4	39.4	43.0	45.6	51.2
25	10.5	11.5	13.1	14.6	16.5	18.9	20.9	24.3	28.2	30.7	34.4	37.7	40.6	44.3	46.9	52.6
26	11.2	12.2	13.8	15.4	17.3	19.8	21.8	25.3	29.2	31.8	35.6	38.9	41.9	45.6	48.3	54.1
27	11.8	12.9	14.6	16.2	18.1	20.7	22.7	26.3	30.3	32.9	36.7	40.1	43.2	47.0	49.6	55.5
28	12.5	13.6	15.3	16.9	18.9	21.6	23.6	27.3	31.4	34.0	37.9	41.3	44.5	48.3	51.0	56.9
29	13.1	14.3	16.0	17.7	19.8	22.5	24.6	28.3	32.5	35.1	39.1	42.6	45.7	49.6	52.3	58.3
30	13.8	15.0	16.8	18.5	20.6	23.4	25.5	29.3	33.5	36.3	40.3	43.8	47.0	50.9	53.7	59.7

SOURCE: Columns 0.995, 0.975, 0.025, and 0.005 are abridged with permission from Catherine M. Thompson, "Table of Percentage Points of the χ^2 Distribution," *Biometrika* 32 Part II (1941), pp. 188–89. The remainder of the table is abridged with permission from Table IV of R. A. Fisher and F. Yates, *Statistical Tables for Biological, Agricultural and Medical Research* (Edinburgh: Oliver & Boyd, Ltd., 1953). For a more comprehensive table, see H. Leon Harter, "A New Table of Percentage Points of the Chi-Square Distribution," *Biometrika* 51 (1964), pp. 231–39.

TABLE D1

Percentage Points of the Distribution of the Relative Range $w = R/\sigma'$, Normal Universe

n	Mean w or d_2	σ'_w or d_3	Probability That w Is Less than or Equal to Tabular Entry†									
			0.001	0.005	0.010	0.025	0.050	0.950	0.975	0.990	0.995	0.999
2	1.128	0.8525	0.00	0.01	0.02	0.04	0.09	2.77	3.17	3.64	3.97	4.65
3	1.693	0.8884	0.06	0.13	0.19	0.30	0.43	3.31	3.68	4.12	4.42	5.06
4	2.059	0.8798	0.20	0.34	0.43	0.59	0.76	3.63	3.98	4.40	4.69	5.31
5	2.326	0.8641	0.37	0.55	0.66	0.85	1.03	3.86	4.20	4.60	4.89	5.48
6	2.534	0.8480	0.54	0.75	0.87	1.06	1.25	4.03	4.36	4.76	5.03	5.62
7	2.704	0.833	0.69	0.92	1.05	1.25	1.44	4.17	4.49	4.88	5.15	5.73
8	2.847	0.820	0.83	1.08	1.20	1.41	1.60	4.29	4.61	4.99	5.26	5.82
9	2.970	0.808	0.96	1.21	1.34	1.55	1.74	4.39	4.70	5.08	5.34	5.90
10	3.078	0.797	1.08	1.33	1.47	1.67	1.86	4.47	4.79	5.16	5.42	5.97
11	3.173	0.787	1.20	1.45	1.58	1.78	1.97	4.55	4.86	5.23	5.49	6.04
12	3.258	0.778	1.30	1.55	1.68	1.88	2.07	4.62	4.92	5.29	5.54	6.09

† Note that, in contrast to Table C, these probabilities are cumulated from the lower end of the distribution.

SOURCE: Probabilities reproduced with permission from E. S. Pearson, "The Probability Integral of the Range in Samples of n Observations from a Normal Population," *Biometrika* 32 (1941–42), pp. 301–8. Mean and σ' reproduced with permission from E. S. Pearson, "The Percentage Limits for the Distribution of Range in Samples from a Normal Population," *Biometrika* 24 (1932), pp. 404–17.

TABLE D2

Upper 0.05 Points of the Distribution of the Studentized Range, $q = \dfrac{(X_U - X_L)}{s_v}$ (n = number of items from which range is obtained, and ν is the degrees of freedom of s_ν.)

ν \ n	2	3	4	5	6	7	8	9	10	11	12	13	14	15	16	17	18	19	20
1	17·97	26·98	32·82	37·08	40·41	43·12	45·40	47·36	49·07	50·59	51·96	53·20	54·33	55·36	56·32	57·22	58·04	58·83	59·56
2	6·08	8·33	9·80	10·88	11·74	12·44	13·03	13·54	13·99	14·39	14·75	15·08	15·38	15·65	15·91	16·14	16·37	16·57	16·77
3	4·50	5·91	6·82	7·50	8·04	8·48	8·85	9·18	9·46	9·72	9·95	10·15	10·35	10·52	10·69	10·84	10·98	11·11	11·24
4	3·93	5·04	5·76	6·29	6·71	7·05	7·35	7·60	7·83	8·03	8·21	8·37	8·52	8·66	8·79	8·91	9·03	9·13	9·23
5	3·64	4·60	5·22	5·67	6·03	6·33	6·58	6·80	6·99	7·17	7·32	7·47	7·60	7·72	7·83	7·93	8·03	8·12	8·21
6	3·46	4·34	4·90	5·30	5·63	5·90	6·12	6·32	6·49	6·65	6·79	6·92	7·03	7·14	7·24	7·34	7·43	7·51	7·59
7	3·34	4·16	4·68	5·06	5·36	5·61	5·82	6·00	6·16	6·30	6·43	6·55	6·66	6·76	6·85	6·94	7·02	7·10	7·17
8	3·26	4·04	4·53	4·89	5·17	5·40	5·60	5·77	5·92	6·05	6·18	6·29	6·39	6·48	6·57	6·65	6·73	6·80	6·87
9	3·20	3·95	4·41	4·76	5·02	5·24	5·43	5·59	5·74	5·87	5·98	6·09	6·19	6·28	6·36	6·44	6·51	6·58	6·64
10	3·15	3·88	4·33	4·65	4·91	5·12	5·30	5·46	5·60	5·72	5·83	5·93	6·03	6·11	6·19	6·27	6·34	6·40	6·47
11	3·11	3·82	4·26	4·57	4·82	5·03	5·20	5·35	5·49	5·61	5·71	5·81	5·90	5·98	6·06	6·13	6·20	6·27	6·33
12	3·08	3·77	4·20	4·51	4·75	4·95	5·12	5·27	5·39	5·51	5·61	5·71	5·80	5·88	5·95	6·02	6·09	6·15	6·21
13	3·06	3·73	4·15	4·45	4·69	4·88	5·05	5·19	5·32	5·43	5·53	5·63	5·71	5·79	5·86	5·93	5·99	6·05	6·11
14	3·03	3·70	4·11	4·41	4·64	4·83	4·99	5·13	5·25	5·36	5·46	5·55	5·64	5·71	5·79	5·85	5·91	5·97	6·03
15	3·01	3·67	4·08	4·37	4·59	4·78	4·94	5·08	5·20	5·31	5·40	5·49	5·57	5·65	5·72	5·78	5·85	5·90	5·96
16	3·00	3·65	4·05	4·33	4·56	4·74	4·90	5·03	5·15	5·26	5·35	5·44	5·52	5·59	5·66	5·73	5·79	5·84	5·90
17	2·98	3·63	4·02	4·30	4·52	4·70	4·86	4·99	5·11	5·21	5·31	5·39	5·47	5·54	5·61	5·67	5·73	5·79	5·84
18	2·97	3·61	4·00	4·28	4·49	4·67	4·82	4·96	5·07	5·17	5·27	5·35	5·43	5·50	5·57	5·63	5·69	5·74	5·79
19	2·96	3·59	3·98	4·25	4·47	4·65	4·79	4·92	5·04	5·14	5·23	5·31	5·39	5·46	5·53	5·59	5·65	5·70	5·75
20	2·95	3·58	3·96	4·23	4·45	4·62	4·77	4·90	5·01	5·11	5·20	5·28	5·36	5·43	5·49	5·55	5·61	5·66	5·71
24	2·92	3·53	3·90	4·17	4·37	4·54	4·68	4·81	4·92	5·01	5·10	5·18	5·25	5·32	5·38	5·44	5·49	5·55	5·59
30	2·89	3·49	3·85	4·10	4·30	4·46	4·60	4·72	4·82	4·92	5·00	5·08	5·15	5·21	5·27	5·33	5·38	5·43	5·47
40	2·86	3·44	3·79	4·04	4·23	4·39	4·52	4·63	4·73	4·82	4·90	4·98	5·04	5·11	5·16	5·22	5·27	5·31	5·36
60	2·83	3·40	3·74	3·98	4·16	4·31	4·44	4·55	4·65	4·73	4·81	4·88	4·94	5·00	5·06	5·11	5·15	5·20	5·24
120	2·80	3·36	3·68	3·92	4·10	4·24	4·36	4·47	4·56	4·64	4·71	4·78	4·84	4·90	4·95	5·00	5·04	5·09	5·13
∞	2·77	3·31	3·63	3·86	4·03	4·17	4·29	4·39	4·47	4·55	4·62	4·68	4·74	4·80	4·85	4·89	4·93	4·97	5·01

SOURCE: Table D2 is reproduced with permission from J. Pachares, "Table of the Upper 10% Points of the Studentized Range," *Biometrika* 46 (1959), pp. 461–66.

TABLE D3

Values Associated with the Distribution of the Average Range* [$\nu(\bar{R}/d_2^*)^2/\sigma'^2$ is distributed approximately as χ^2 with ν degrees of freedom; $\bar{R}$ is the average range of g subgroups, each of size (m)]

Number of Samples (g)	Size of Samples (m) 2: ν	2: d_2^*	3: ν	3: d_2^*	4: ν	4: d_2^*	5: ν	5: d_2^*	6: ν	6: d_2^*	7: ν	7: d_2^*	8: ν	8: d_2^*
1	1.0	1.41	2.0	1.91	2.9	2.24	3.8	2.48	4.7	2.67	5.5	2.83	6.3	2.96
2	1.9	1.28	3.8	1.81	5.7	2.15	7.5	2.40	9.2	2.60	10.8	2.77	12.3	2.91
3	2.8	1.23	5.7	1.77	8.4	2.12	11.1	2.38	13.6	2.58	16.0	2.75	18.3	2.89
4	3.7	1.21	7.5	1.75	11.2	2.11	14.7	2.37	18.1	2.57	21.3	2.74	24.4	2.88
5	4.6	1.19	9.3	1.74	13.9	2.10	18.4	2.36	22.6	2.56	26.6	2.73	30.4	2.87
6	5.5	1.18	11.1	1.73	16.6	2.09	22.0	2.35	27.1	2.56	31.8	2.73	36.4	2.87
7	6.4	1.17	12.9	1.73	19.4	2.09	25.6	2.35	31.5	2.55	37.1	2.72	42.5	2.87
8	7.2	1.17	14.8	1.72	22.1	2.08	29.3	2.35	36.0	2.55	42.4	2.72	48.5	2.87
9	8.1	1.16	16.6	1.72	24.8	2.08	32.9	2.34	40.5	2.55	47.7	2.72	54.5	2.86
10	9.0	1.16	18.4	1.72	27.6	2.08	36.5	2.34	44.9	2.55	52.9	2.72	60.6	2.86
11	9.9	1.16	20.2	1.71	30.3	2.08	40.1	2.34	49.4	2.55	58.2	2.72	66.6	2.86
12	10.8	1.15	22.0	1.71	33.0	2.07	43.7	2.34	53.9	2.55	63.5	2.72	72.7	2.85
13	11.6	1.15	23.9	1.71	35.7	2.07	47.4	2.34	58.4	2.55	68.8	2.71	78.7	2.85
14	12.5	1.15	25.7	1.71	38.5	2.07	51.0	2.34	62.8	2.54	74.0	2.71	84.7	2.85
15	13.4	1.15	27.5	1.71	41.2	2.07	54.6	2.34	67.3	2.54	79.3	2.71	90.8	2.85
d_2		1.13		1.69		2.06		2.33		2.53		2.70		2.85
c.d.	0.88		1.82		2.74		3.62		4.47		5.27		6.03	

Number of Samples (g)	Size of Samples (m) 9: ν	9: d_2^*	10: ν	10: d_2^*	11: ν	11: d_2^*	12: ν	12: d_2^*	13: ν	13: d_2^*	14: ν	14: d_2^*	15: ν	15: d_2^*
1	7.0	3.08	7.7	3.18	8.4	3.27	9.0	3.35	9.6	3.42	10.2	3.49	10.8	3.55
2	13.8	3.02	15.1	3.13	16.5	3.22	17.8	3.30	19.0	3.38	20.2	3.45	21.3	3.51
3	20.5	3.01	22.6	3.11	24.6	3.21	26.5	3.29	28.4	3.37	30.2	3.43	31.9	3.50
4	27.3	3.00	30.1	3.10	32.7	3.20	35.3	3.28	37.8	3.36	40.1	3.43	42.4	3.49
5	34.0	2.99	37.5	3.10	40.9	3.19	44.1	3.28	47.1	3.35	50.1	3.42	52.9	3.49
6	40.8	2.99	45.0	3.10	49.0	3.19	52.8	3.27	56.5	3.35	60.1	3.42	63.5	3.49
7	47.5	2.99	52.4	3.10	57.1	3.19	61.6	3.27	65.9	3.35	70.0	3.42	74.0	3.48
8	54.3	2.98	59.9	3.09	65.2	3.19	70.3	3.27	75.3	3.35	80.0	3.42	84.6	3.48
9	61.0	2.98	67.3	3.09	73.3	3.18	79.1	3.27	84.6	3.35	90.0	3.42	95.1	3.48
10	67.8	2.98	74.8	3.09	81.5	3.18	87.9	3.27	94.0	3.34	99.9	3.42	105.6	3.48
11	74.6	2.98	82.3	3.09	89.6	3.18	96.6	3.27	103.4	3.34	109.9	3.41	116.2	3.48
12	81.3	2.98	89.7	3.09	97.7	3.18	105.4	3.27	112.8	3.34	119.9	3.41	126.7	3.48
13	88.1	2.98	97.2	3.09	105.8	3.18	114.1	3.27	122.2	3.34	129.9	3.41	137.3	3.48
14	94.8	2.98	104.6	3.08	113.9	3.18	122.9	3.27	131.5	3.34	139.8	3.41	147.8	3.48
15	101.6	2.98	112.1	3.08	122.1	3.18	131.7	3.26	140.9	3.34	149.8	3.41	158.3	3.48
d_2		2.97		3.08		3.17		3.26		3.34		3.41		3.47
c.d.	6.76		7.45		8.12		8.76		9.38		9.97		10.54	

[a] In general, the degrees of freedom will be given approximately by the reciprocal of $(-2 + 2\sqrt{1 + 2(\text{c.v.})^2/g})$ where c.v. is the coefficient of variation (d_3/d_2) of the range and g is the number of subgroups. Also, d_2^* is given approximately by d_2 (i.e., the infinity value of d_2^*) times $(1 + 1/4\nu)$. Values of ν are also very readily built up from the constant differences. Table D 3 is a basic table that may be used whenever the average range is used in lieu of s.

Note: c.d. = constant difference.

SOURCE: Table D3 is reproduced, with permission, in part from Table 30A of *Biometrika Tables for Statisticians,* Vol. 1, and in part from Acheson J. Duncan, "The Use of Ranges in Comparing Variabilities," *Industrial Quality Control* 9 No. 5 (February 1955). Part of Table D3 has been computed by the author. It first appeared as a whole in the *Journal of the American Statistical Association* 53 (1958), p. 548.

TABLE E
Summation of Terms of the Poisson Distribution
Entries in body of table give the probability (decimal point omitted) of X or less nonconformities (or nonconforming units), when the expected number is that given in the left margin of the table.

u' or $p'n$ \ X	0	1	2	3	4	5	6	7	8	9
0.02	980	1,000								
0.04	961	999	1,000							
0.06	942	998	1,000							
0.08	923	997	1,000							
0.10	905	995	1,000							
0.15	861	990	999	1,000						
0.20	819	982	999	1,000						
0.25	779	974	998	1,000						
0.30	741	963	996	1,000						
0.35	705	951	994	1,000						
0.40	670	938	992	999	1,000					
0.45	638	925	989	999	1,000					
0.50	607	910	986	998	1,000					
0.55	577	894	982	998	1,000					
0.60	549	878	977	997	1,000					
0.65	522	861	972	996	999	1,000				
0.70	497	844	966	994	999	1,000				
0.75	472	827	959	993	999	1,000				
0.80	449	809	953	991	999	1,000				
0.85	427	791	945	989	998	1,000				
0.90	407	772	937	987	998	1,000				
0.95	387	754	929	984	997	1,000				
1.00	368	736	920	981	996	999	1,000			
1.1	333	699	900	974	995	999	1,000			
1.2	301	663	879	966	992	998	1,000			
1.3	273	627	857	957	989	998	1,000			
1.4	247	592	833	946	986	997	999	1,000		
1.5	223	558	809	934	981	996	999	1,000		
1.6	202	525	783	921	976	994	999	1,000		
1.7	183	493	757	907	970	992	998	1,000		
1.8	165	463	731	891	964	990	997	999	1,000	
1.9	150	434	704	875	956	987	997	999	1,000	
2.0	135	406	677	857	947	983	995	999	1,000	

SOURCE: Reprinted with permission from E. C. Molina, *Poisson's Exponential Binomial Limit* (New York: D. Van Nostrand Co. Inc., 1947).

TABLE E ***(continued)***
Summation of Terms of the Poisson Distribution

X / u' or $p'n$	0	1	2	3	4	5	6	7	8	9
2.2	111	355	623	819	928	975	993	998	1,000	
2.4	091	308	570	779	904	964	988	997	999	1,000
2.6	074	267	518	736	877	951	983	995	999	1,000
2.8	061	231	469	692	848	935	976	992	998	999
3.0	050	199	423	647	815	916	966	988	996	999
3.2	041	171	380	603	781	895	955	983	994	998
3.4	033	147	340	558	744	871	942	977	992	997
3.6	027	126	303	515	706	844	927	969	988	996
3.8	022	107	269	473	668	816	909	960	984	994
4.0	018	092	238	433	629	785	889	949	979	992
4.2	015	078	210	395	590	753	867	936	972	989
4.4	012	066	185	359	551	720	844	921	964	985
4.6	010	056	163	326	513	686	818	905	955	980
4.8	008	048	143	294	476	651	791	887	944	975
5.0	007	040	125	265	440	616	762	867	932	968
5.2	006	034	109	238	406	581	732	845	918	960
5.4	005	029	095	213	373	546	702	822	903	951
5.6	004	024	082	191	342	512	670	797	886	941
5.8	003	021	072	170	313	478	638	771	867	929
6.0	002	017	062	151	285	446	606	744	847	916

	10	11	12	13	14	15	16
2.8	1,000						
3.0	1,000						
3.2	1,000						
3.4	999	1,000					
3.6	999	1,000					
3.8	998	999	1,000				
4.0	997	999	1,000				
4.2	996	999	1,000				
4.4	994	998	999	1,000			
4.6	992	997	999	1,000			
4.8	990	996	999	1,000			
5.0	986	995	998	999	1,000		
5.2	982	993	997	999	1,000		
5.4	977	990	996	999	1,000		
5.6	972	988	995	998	999	1,000	
5.8	965	984	993	997	999	1,000	
6.0	957	980	991	996	999	999	1,000

TABLE E *(continued)*
Summation of Terms of the Poisson Distribution

u' or $p'n$ \ X	0	1	2	3	4	5	6	7	8	9
6.2	002	015	054	134	259	414	574	716	826	902
6.4	002	012	046	119	235	384	542	687	803	886
6.6	001	010	040	105	213	355	511	658	780	869
6.8	001	009	034	093	192	327	480	628	755	850
7.0	001	007	030	082	173	301	450	599	729	830
7.2	001	006	025	072	156	276	420	569	703	810
7.4	001	005	022	063	140	253	392	539	676	788
7.6	001	004	019	055	125	231	365	510	648	765
7.8	000	004	016	048	112	210	338	481	620	741
8.0	000	003	014	042	100	191	313	453	593	717
8.5	000	002	009	030	074	150	256	386	523	653
9.0	000	001	006	021	055	116	207	324	456	587
9.5	000	001	004	015	040	089	165	269	392	522
10.0	000	000	003	010	029	067	130	220	333	458
	10	11	12	13	14	15	16	17	18	19
6.2	949	975	989	995	998	999	1,000			
6.4	939	969	986	994	997	999	1,000			
6.6	927	963	982	992	997	999	999	1,000		
6.8	915	955	978	990	996	998	999	1,000		
7.0	901	947	973	987	994	998	999	1,000		
7.2	887	937	967	984	993	997	999	999	1,000	
7.4	871	926	961	980	991	996	998	999	1,000	
7.6	854	915	954	976	989	995	998	999	1,000	
7.8	835	902	945	971	986	993	997	999	1,000	
8.0	816	888	936	966	983	992	996	998	999	1,000
8.5	763	849	909	949	973	986	993	997	999	999
9.0	706	803	876	926	959	978	989	995	998	999
9.5	645	752	836	898	940	967	982	991	996	998
10.0	583	697	792	864	917	951	973	986	993	997
	20	21	22							
8.5	1,000									
9.0	1,000									
9.5	999	1,000								
10.0	998	999	1,000							

TABLE E ***(continued)***
Summation of Terms of the Poisson Distribution

X / u' or $p'n$	0	1	2	3	4	5	6	7	8	9
10.5	000	000	002	007	021	050	102	179	279	397
11.0	000	000	001	005	015	038	079	143	232	341
11.5	000	000	001	003	011	028	060	114	191	289
12.0	000	000	001	002	008	020	046	090	155	242
12.5	000	000	000	002	005	015	035	070	125	201
13.0	000	000	000	001	004	011	026	054	100	166
13.5	000	000	000	001	003	008	019	041	079	135
14.0	000	000	000	000	002	006	014	032	062	109
14.5	000	000	000	000	001	004	010	024	048	088
15.0	000	000	000	000	001	003	008	018	037	070
	10	11	12	13	14	15	16	17	18	19
10.5	521	639	742	825	888	932	960	978	988	994
11.0	460	579	689	781	854	907	944	968	982	991
11.5	402	520	633	733	815	878	924	954	974	986
12.0	347	462	576	682	772	844	899	937	963	979
12.5	297	406	519	628	725	806	869	916	948	969
13.0	252	353	463	573	675	764	835	890	930	957
13.5	211	304	409	518	623	718	798	861	908	942
14.0	176	260	358	464	570	669	756	827	883	923
14.5	145	220	311	413	518	619	711	790	853	901
15.0	118	185	268	363	466	568	664	749	819	875
	20	21	22	23	24	25	26	27	28	29
10.5	997	999	999	1,000						
11.0	995	998	999	1,000						
11.5	992	996	998	999	1,000					
12.0	988	994	997	999	999	1,000				
12.5	983	991	995	998	999	999	1,000			
13.0	975	986	992	996	998	999	1,000			
13.5	965	980	989	994	997	998	999	1,000		
14.0	952	971	983	991	995	997	999	999	1,000	
14.5	936	960	976	986	992	996	998	999	999	1,000
15.0	917	947	967	981	989	994	997	998	999	1,000

TABLE E *(concluded)*
Summation of Terms of the Poisson Distribution

u' or $p'n$ \ X	4	5	6	7	8	9	10	11	12	13
16	000	001	004	010	022	043	077	127	193	275
17	000	001	002	005	013	026	049	085	135	201
18	000	000	001	003	007	015	030	055	092	143
19	000	000	001	002	004	009	018	035	061	098
20	000	000	000	001	002	005	011	021	039	066
21	000	000	000	000	001	003	006	013	025	043
22	000	000	000	000	001	002	004	008	015	028
23	000	000	000	000	000	001	002	004	009	017
24	000	000	000	000	000	000	001	003	005	011
25	000	000	000	000	000	000	001	001	003	006
	14	15	16	17	18	19	20	21	22	23
16	368	467	566	659	742	812	868	911	942	963
17	281	371	468	564	655	736	805	861	905	937
18	208	287	375	469	562	651	731	799	855	899
19	150	215	292	378	469	561	647	725	793	849
20	105	157	221	297	381	470	559	644	721	787
21	072	111	163	227	302	384	471	558	640	716
22	048	077	117	169	232	306	387	472	556	637
23	031	052	082	123	175	238	310	389	472	555
24	020	034	056	087	128	180	243	314	392	473
25	012	022	038	060	092	134	185	247	318	394
	24	25	26	27	28	29	30	31	32	33
16	978	987	993	996	998	999	999	1,000		
17	959	975	985	991	995	997	999	999	1,000	
18	932	955	972	983	990	994	997	998	999	1,000
19	893	927	951	969	980	988	993	996	998	999
20	843	888	922	948	966	978	987	992	995	997
21	782	838	883	917	944	963	976	985	991	994
22	712	777	832	877	913	940	959	973	983	989
23	635	708	772	827	873	908	936	956	971	981
24	554	632	704	768	823	868	904	932	953	969
25	473	553	629	700	763	818	863	900	929	950
	34	35	36	37	38	39	40	41	42	43
19	999	1,000								
20	999	999	1,000							
21	997	998	999	999	1,000					
22	994	996	998	999	999	1,000				
23	988	993	996	997	999	999	1,000			
24	979	987	992	995	997	998	999	999	1,000	
25	966	978	985	991	994	997	998	999	999	1,000

TABLE F
0.05 and 0.01 Points of the Distribution of γ_1, Normal Universe*
(approximate values)

	Probability that γ_1 Will Exceed Listed Value in Positive Direction Is	
n	*0.05*	*0.01*
25	0.714	1.073
30	0.664	0.985
35	0.624	0.932
40	0.587	0.869
45	0.558	0.825
50	0.533	0.787
60	0.492	0.723
70	0.459	0.673
80	0.432	0.631
90	0.409	0.596
100	0.389	0.567
125	0.350	0.508
150	0.321	0.464
175	0.298	0.430
200	0.280	0.403
250	0.251	0.360
300	0.230	0.329
400	0.200	0.285
500	0.179	0.255
750	0.146	0.208
1,000	0.127	0.180

* The points listed are on the positive tail of the distribution. With a minus sign attached, they are equally valid for the negative tail. Also see Ralph D'Agostino and E. S. Pearson, "Tests for Departure from Normality. Empirical Results for the Distribution of b_2 and $\sqrt{b_1}$," *Biometrika* 60 (1973), pp. 613–22. Since the publication of this paper, Prof. F. J. Anscombe has pointed out that b_2 and $\sqrt{b_1}$ are *not* independent in samples from a normal universe. Hence the joint tests suggested in Section 6 of the paper are not correct, though they might be nearly so. [Letters from E. S. Pearson, April 4, 1974.] For a discussion of the operating characteristics of a γ_1 test for normality, see K. O. Bowman and L. R. Shenton, "Notes on the Distribution of $\sqrt{b_1}$ in Sampling from Pearson Distributions," *Biometrika* 60 (1973), pp. 155–67.

SOURCE: Table F is abridged with permission from R. C. Geary and E. S. Pearson, *Tests of Normality* (London: Biometrika Office, University College, 1938) and from Ralph B. Agostino and Gary L. Tietjen, "Approaches to the Null Distribution of $\sqrt{b_1}$," *Biometrika* 60 (1973), pp. 169–73.

TABLE G
Percentage Points of the Distribution of γ_2, Normal Universe*
(approximate values)

	Probability that γ_2 Falls below Listed Value Is:		*Probability that γ_2 Falls above Listed Value Is:*	
n	*0.01*	*0.05*	*0.05*	*0.01*
25	−1.28	−1.09	1.16	2.30
30	−1.21	−1.02	1.11	2.21
40	−1.11	−0.93	1.06	2.04
50	−1.05	−0.85	0.99	1.88
75	−0.92	−0.73	0.87	1.59
100	−0.82	−0.65	0.77	1.39
125	−0.76	−0.60	0.71	1.24
150	−0.71	−0.55	0.65	1.13
200	−0.63	−0.49	0.57	0.98
250	−0.58	−0.45	0.52	0.87
300	−0.54	−0.41	0.47	0.79
500	−0.43	−0.33	0.37	0.60
1,000	−0.32	−0.24	0.26	0.41
2,000	−0.23	−0.17	0.18	0.28
5,000	−0.15	−0.11	0.12	0.17

* The last reference contains data for values of *n* from 7 to 50 additional to those listed in Table G. Also see Ralph D'Agostino and E. S. Pearson, "Tests for Departure from Normality. Empirical Results for the Distributions of b_2 and $\sqrt{b_1}$," *Biometrika,* Vol. 60 (1973), pp. 613–22, for charts of the percentage points of the distribution of $b_2 = \gamma_2 + 3$. Since the publication of this paper, Prof. F. J. Anscombe has pointed out that b_2 and $\sqrt{b_1}$ are *not* independent in samples from a normal universe. Hence the joint tests suggested in Section 6 of the paper are not correct, though they might be nearly so. [Letters from E. S. Pearson, April 4, 1974.] For a discussion of the operating characters of a γ_2 test for normality, see K. O. Bowman, "Power of the Kurtosis Statistic, b_2 in Tests of Departure from Normality." *Biometrika* 60 (1973), pp. 623–28.

SOURCE: Table G is adapted with permission from R. C. Geary and E. S. Pearson, *Tests of Normality* (London: Biometrika Office, University College, 1938), E. S. Pearson, "Tables of Percentage Points of $\sqrt{b_1}$ and b_2 in Normal Samples; A Rounding Off." *Biometrika,* Vol. 52 (1965), pp. 282–85 and Ralph B. D'Agostino and Gary L. Tietjen, "Simulation Probability Points of b_2 for Small Samples," *Biometrika* 58 (1971), pp. 669–72.

TABLE H
Percentage Points of the Distribution of a = AD/RMSD Normal Universe

	Probability that a *Falls below Listed Value Is:*			*Probability that* a *Falls above Listed Value Is:*		
ν	*0.01*	*0.05*	*0.10*	*0.10*	*0.05*	*0.01*
10	0.6675	0.7153	0.7409	0.8899	0.9073	0.9359
15	0.6829	0.7236	0.7452	0.8733	0.8884	0.9137
20	0.6950	0.7304	0.7495	0.8631	0.8768	0.9001
25	0.7040	0.7360	0.7530	0.8570	0.8686	0.8901
30	0.7110	0.7404	0.7559	0.8511	0.8625	0.8827
40	0.7216	0.7470	0.7604	0.8436	0.8540	0.8722
50	0.7291	0.7518	0.7636	0.8385	0.8481	0.8648
60	0.7347	0.7554	0.7662	0.8349	0.8434	0.8592
70	0.7393	0.7583	0.7683	0.8321	0.8403	0.8549
80	0.7430	0.7607	0.7700	0.8298	0.8376	0.8515
100	0.7487	0.7644	0.7726	0.8264	0.8344	0.8460
200	0.7629	0.7738	0.7796	0.8178	0.8229	0.8322
300	0.7693	0.7781	0.7828	0.8140	0.8183	0.8260
400	0.7731	0.7807	0.7847	0.8118	0.8155	0.8223
500	0.7757	0.7825	0.7861	0.8103	0.8136	0.8198
700	0.7791	0.7848	0.7878	0.8084	0.8112	0.8164
1,000	0.7822	0.7869	0.7894	0.8066	0.8090	0.8134

* The size of the sample is n, and $\nu = n - 1$.

SOURCE: Table H is abridged with permission from R. C. Geary and E. S. Pearson, *Tests of Normality* (London: Biometrika Office, University College, 1938).

TABLE J
Percentage Points of the F Distribution

ν_2	P	$\nu_1 = 1$	2	3	4	5	6	7	8
	0.500	1.00	1.50	1.71	1.82	1.89	1.94	1.98	2.00
	0.100	39.9	49.5	53.6	55.8	57.2	58.2	58.9	59.4
	0.050	161	200	216	225	230	234	237	239
1	0.025	648	800	864	900	922	937	948	957
	0.010	4,050	5,000	5,400	5,620	5,760	5,860	5,930	5,980
	0.005	16,200	20,000	21,600	22,500	23,100	23,400	23,700	23,900
	0.001	405,284	500,000	540,379	562,500	576,405	585,937	...	598,144
	0.500	0.667	1.00	1.13	1.21	1.25	1.28	1.30	1.32
	0.100	8.53	9.00	9.16	9.24	9.29	9.33	9.35	9.37
	0.050	18.5	19.0	19.2	19.2	19.3	19.3	19.4	19.4
2	0.025	38.5	39.0	39.2	39.2	39.3	39.3	39.4	39.4
	0.010	98.5	99.0	99.2	99.2	99.3	99.3	99.4	99.4
	0.005	199	199	199	199	199	199	199	199
	0.001	998.5	999.0	999.2	999.2	999.3	999.3	...	999.4
	0.500	0.585	0.881	1.00	1.06	1.10	1.13	1.15	1.16
	0.100	5.54	5.46	5.39	5.34	5.31	5.28	5.27	5.25
	0.050	10.1	9.55	9.28	9.12	9.01	8.94	8.89	8.85
3	0.025	17.4	16.0	15.4	15.1	14.9	14.7	14.6	14.5
	0.010	34.1	30.8	29.5	28.7	28.2	27.9	27.7	27.5
	0.005	55.6	49.8	47.5	46.2	45.4	44.8	44.4	44.1
	0.001	167.5	148.5	141.1	137.1	134.6	132.8	...	130.6
	0.500	0.549	0.828	0.941	1.00	1.04	1.06	1.08	1.09
	0.100	4.54	4.32	4.19	4.11	4.05	4.01	3.98	3.95
	0.050	7.71	6.94	6.59	6.39	6.26	6.16	6.09	6.04
4	0.025	12.2	10.6	9.98	9.60	9.36	9.20	9.07	8.98
	0.010	21.2	18.0	16.7	16.0	15.5	15.2	15.0	14.8
	0.005	31.3	26.3	24.3	23.2	22.5	22.0	21.6	21.4
	0.001	74.1	61.3	56.2	53.4	51.7	50.5	...	49.0
	0.500	0.528	0.799	0.907	0.965	1.00	1.02	1.04	1.05
	0.100	4.06	3.78	3.62	3.52	3.45	3.40	3.37	3.34
	0.050	6.61	5.79	5.41	5.19	5.05	4.95	4.88	4.82
5	0.025	10.0	8.43	7.76	7.39	7.15	6.98	6.85	6.76
	0.010	16.3	13.3	12.1	11.4	11.0	10.7	10.5	10.3
	0.005	22.8	18.3	16.5	15.6	14.9	14.5	14.2	14.0
	0.001	47.0	36.6	33.2	31.1	29.8	28.8	...	27.6
	0.500	0.515	0.780	0.886	0.942	0.977	1.00	1.02	1.03
	0.100	3.78	3.46	3.29	3.18	3.11	3.05	3.01	2.98
	0.050	5.99	5.14	4.76	4.53	4.39	4.28	4.21	4.15
6	0.025	8.81	7.26	6.60	6.23	5.99	5.82	5.70	5.60
	0.010	13.7	10.9	9.78	9.15	8.75	8.47	8.26	8.10
	0.005	18.6	14.5	12.9	12.0	11.5	11.1	10.8	10.6
	0.001	35.5	27.0	23.7	21.9	20.8	20.0	...	19.0
	0.500	0.506	0.767	0.871	0.926	0.960	0.983	1.00	1.01
	0.100	3.59	3.26	3.07	2.96	2.88	2.83	2.78	2.75
	0.050	5.59	4.74	4.35	4.12	3.97	3.87	3.79	3.73
7	0.025	8.07	6.54	5.89	5.52	5.29	5.12	4.99	4.90
	0.010	12.2	9.55	8.45	7.85	7.46	7.19	6.99	6.84
	0.005	16.2	12.4	10.9	10.1	9.52	9.16	8.89	8.68
	0.001	29.2	21.7	18.8	17.2	16.2	15.5		14.6

SOURCE: The 0.001 points of Table J are abridged with permission from Table V of R. A. Fisher and F. Yates, *Statistical Tables for Biological, Agricultural and Medical Research* (Edinburgh; Oliver & Boyd Ltd., 1953). The rest of the table is abridged with permission from Maxine Merrington and Catherine M. Thompson, "Tables of Percentage Points of the Inverted Beta (F) Distribution," *Biometrika* 33 (1943–46), pp. 73–88.

TABLE J *(continued)*

9	10	12	15	20	24	30	60	120	∞
2.03	2.04	2.07	2.09	2.12	2.13	2.15	2.17	2.18	2.20
59.9	60.2	60.7	61.2	61.7	62.0	62.3	62.8	63.1	63.3
241	242	244	246	248	249	250	252	253	254
963	969	977	985	993	997	1,001	1,010	1,010	1,020
6,020	6,060	6,110	6,160	6,210	6,235	6,260	6,310	6,340	3,370
24,100	24,200	24,400	24,600	24,800	24,900	25,000	25,300	25,400	25,500
...	...	610,667	...	...	623,497	...	...	...	636,619
1.33	1.34	1.36	1.38	1.39	1.40	1.41	1.43	1.43	1.44
9.38	9.39	9.41	9.42	9.44	9.45	9.46	9.47	9.48	9.49
19.4	19.4	19.4	19.4	19.4	19.5	19.5	19.5	19.5	19.5
39.4	39.4	39.4	39.4	39.4	39.5	39.5	39.5	39.5	39.5
99.4	99.4	99.4	99.4	99.4	99.5	99.5	99.5	99.5	99.5
199	199	199	199	199	199	199	199	199	200
...	...	999.4	..	...	999.5	...	...	...	999.5
1.17	1.18	1.20	1.21	1.23	1.23	1.24	1.25	1.26	1.27
5.24	5.23	5.22	5.20	5.18	5.18	5.17	5.15	5.14	5.13
8.81	8.79	8.74	8.70	8.66	8.64	8.62	8.57	8.55	8.53
14.5	14.4	14.3	14.3	14.2	14.1	14.1	14.0	13.9	13.9
27.3	27.2	27.1	26.9	26.7	26.6	26.5	26.3	26.2	26.1
43.9	43.7	43.4	43.1	42.8	42.6	42.5	42.1	42.0	41.8
...	...	128.3	...	...	125.9	...	...	...	123.5
1.10	1.11	1.13	1.14	1.15	1.16	1.16	1.18	1.18	1.19
3.94	3.92	3.90	3.87	3.84	3.83	3.82	3.79	3.78	3.76
6.00	5.96	5.91	5.86	5.80	5.77	5.75	5.69	5.66	5.63
8.90	8.84	8.75	8.66	8.56	8.51	8.46	8.36	8.31	8.26
14.7	14.5	14.4	14.2	14.0	13.9	13.8	13.7	13.6	13.5
21.1	21.0	20.7	20.4	20.2	20.0	19.9	19.6	19.5	19.3
...	...	47.4	...	...	45.8	...	...	...	44.1
1.06	1.07	1.09	1.10	1.11	1.12	1.12	1.14	1.14	1.15
3.32	3.30	3.27	3.24	3.21	3.19	3.17	3.14	3.12	3.11
4.77	4.47	4.68	4.62	4.56	4.56	4.53	4.43	4.40	4.37
6.68	6.62	6.52	6.43	6.33	6.28	6.23	6.12	6.07	6.02
10.2	10.1	9.89	9.72	9.55	9.47	9.38	9.20	9.11	9.02
13.8	13.6	13.4	13.1	12.9	12.8	12.7	12.4	12.3	12.1
...	...	26.4	...	...	25.1	...	...	...	23.8
1.04	1.05	1.06	1.07	1.08	1.09	1.10	1.11	1.12	1.12
2.96	2.94	2.90	2.87	2.84	2.82	2.80	2.76	2.74	2.72
4.10	4.06	4.00	3.94	3.87	3.84	3.81	3.74	3.70	3.67
5.52	5.46	5.37	5.27	5.17	5.12	5.07	4.96	4.90	4.85
7.98	7.87	7.72	7.56	7.40	7.31	7.23	7.06	6.97	6.88
10.4	10.2	10.0	9.81	9.59	9.47	9.36	9.12	9.00	9.88
...	...	18.0	...	...	16.9	...	...	...	15.8
1.02	1.03	1.04	1.05	1.07	1.07	1.08	1.09	1.10	1.10
2.72	2.70	2.67	2.63	2.59	2.58	2.56	2.51	2.49	2.47
3.68	3.64	3.57	3.51	3.44	3.41	3.38	3.30	3.27	3.23
4.82	4.76	4.67	4.57	4.47	4.42	4.36	4.25	4.20	4.14
6.72	6.62	6.47	6.31	6.16	6.07	5.99	5.82	5.74	5.65
8.51	8.38	8.18	7.97	7.75	7.65	7.53	7.31	7.19	7.08
...	...	13.7	...	...	12.7	...	...	...	11.7

TABLE J *(continued)*

ν_2	P	$\nu_1 = 1$	2	3	4	5	6	7	8
	0.500	0.499	0.757	0.860	0.915	0.948	0.971	0.988	1.00
	0.100	3.46	3.11	2.92	2.81	2.73	2.67	2.62	2.59
	0.050	5.32	4.46	4.07	3.84	3.69	3.58	3.50	3.44
8	0.025	7.57	6.06	5.42	5.05	4.82	4.65	4.53	4.43
	0.010	11.3	8.65	7.59	7.01	6.63	6.37	6.18	6.03
	0.005	14.7	11.0	9.60	8.81	8.30	7.95	7.69	7.50
	0.001	25.4	18.5	15.8	14.4	13.5	12.9	. . .	12.0
	0.500	0.494	0.749	0.852	0.906	0.939	0.962	0.978	0.990
	0.100	3.36	3.01	2.81	2.69	2.61	2.55	2.51	2.47
	0.050	5.12	4.26	3.86	3.63	3.48	3.37	3.29	3.23
9	0.025	7.21	5.71	5.08	4.72	4.48	4.32	4.20	4.10
	0.010	10.6	8.02	6.99	6.42	6.06	5.80	5.61	5.47
	0.005	13.6	10.1	8.72	7.96	7.47	7.13	6.88	6.69
	0.001	22.9	16.4	13.9	12.6	11.7	11.1	. . .	10.4
	0.500	0.490	0.743	0.845	0.899	0.932	0.954	0.971	0.983
	0.100	3.29	2.92	2.73	2.61	2.52	2.46	2.41	2.38
	0.050	4.96	4.10	3.71	3.48	3.33	3.22	3.14	3.07
10	0.025	6.94	5.46	4.83	4.47	4.24	4.07	3.95	3.85
	0.010	10.0	7.56	6.55	5.99	5.64	5.39	5.20	5.06
	0.005	12.8	9.43	8.08	7.34	6.87	6.54	6.30	6.12
	0.001	21.0	14.9	12.6	11.3	10.5	9.92	. . .	9.20
	0.500	0.484	0.735	0.835	0.888	0.921	0.943	0.959	0.972
	0.100	3.18	2.81	2.61	2.48	2.39	2.33	2.28	2.24
	0.050	4.75	3.89	3.49	3.26	3.11	3.00	2.91	2.85
12	0.025	6.55	5.10	4.47	4.12	3.89	3.73	3.61	3.51
	0.010	9.33	6.93	5.95	5.41	5.06	4.82	4.64	4.50
	0.005	11.8	8.51	7.23	6.52	6.07	5.76	5.52	5.35
	0.001	18.6	13.0	10.8	9.63	8.89	8.38	. . .	7.71
	0.500	0.478	0.726	0.826	0.878	0.911	0.933	0.949	0.960
	0.100	3.07	2.70	2.49	2.36	2.27	2.21	2.16	2.12
	0.050	4.54	3.68	3.29	3.06	2.90	2.79	2.71	2.64
15	0.025	6.20	4.77	4.15	3.80	3.58	3.41	3.29	3.20
	0.010	8.68	6.36	5.42	4.89	4.56	4.32	4.14	4.00
	0.005	10.8	7.70	6.48	5.80	5.37	5.07	4.85	4.67
	0.001	16.6	11.34	9.34	8.25	7.57	7.09	. . .	6.47
	0.500	0.472	0.718	0.816	0.868	0.900	0.922	0.938	0.950
	0.100	2.97	2.59	2.38	2.25	2.16	2.09	2.04	2.00
	0.050	4.35	3.49	3.10	2.87	2.71	2.60	2.51	2.45
20	0.025	5.87	4.46	3.86	3.51	3.29	3.13	3.01	2.91
	0.010	8.10	5.85	4.94	4.43	4.10	3.87	3.70	3.56
	0.005	9.94	6.99	5.82	5.17	4.76	4.47	4.26	4.09
	0.001	14.8	9.95	8.10	7.10	6.46	6.02	. . .	5.44
	0.500	0.469	0.714	0.812	0.863	0.895	0.917	0.932	0.944
	0.100	2.93	2.54	2.33	2.19	2.10	2.04	1.98	1.94
	0.050	4.26	3.40	3.01	2.78	2.62	2.51	2.42	2.36
24	0.025	5.72	4.32	3.72	3.38	3.15	2.99	2.87	2.78
	0.010	7.82	5.61	4.72	4.22	3.90	3.67	3.50	3.36
	0.005	9.55	6.66	5.52	4.89	4.49	4.20	3.99	3.83
	0.001	14.0	9.34	7.55	6.59	5.98	5.55	. . .	4.99

TABLE J *(continued)*

9	10	12	15	20	24	30	60	120	∞
1.01	1.02	1.03	1.04	1.05	1.06	1.07	1.08	1.08	1.09
2.56	2.54	2.50	2.46	2.42	2.40	2.38	2.34	2.32	2.29
3.39	3.35	3.28	3.22	3.15	3.12	3.08	3.01	2.97	2.93
4.36	4.30	4.20	4.10	4.00	3.95	3.89	3.78	3.73	3.67
5.91	5.81	5.67	5.52	5.36	5.28	5.20	5.03	4.95	4.86
7.34	7.21	7.01	6.81	6.61	6.50	6.40	6.18	6.06	5.95
...	...	11.2	...	...	10.3	...	...	...	9.34
1.00	1.01	1.02	1.03	1.04	1.05	1.05	1.07	1.07	1.08
2.44	2.42	2.38	2.34	2.30	2.28	2.25	2.21	2.18	2.16
3.18	3.14	3.07	3.01	2.94	2.90	2.86	2.79	2.75	2.71
4.03	3.96	3.87	3.77	3.67	3.61	3.56	3.45	3.39	3.33
5.35	5.26	5.11	4.96	4.81	4.73	4.65	4.48	4.40	4.31
6.54	6.42	6.23	6.03	5.83	5.73	5.62	5.41	5.30	5.19
...	...	9.57	...	...	8.72	...	...	...	7.81
0.992	1.00	1.01	1.02	1.03	1.04	1.05	1.06	1.06	1.07
2.35	2.32	2.28	2.24	2.20	2.18	2.16	2.11	2.08	2.06
3.02	2.98	2.91	2.84	2.77	2.74	2.70	2.62	2.58	2.54
3.78	3.72	3.62	3.52	3.42	3.37	3.31	3.20	3.14	3.08
4.94	4.85	4.71	4.56	4.41	4.33	4.25	4.08	4.00	3.91
5.97	5.85	5.66	5.47	5.27	5.17	5.07	4.86	4.75	4.64
...	...	8.45	...	...	7.64	...	...	...	6.76
0.981	0.989	1.00	1.01	1.02	1.03	1.03	1.05	1.05	1.06
2.21	2.19	2.15	2.10	2.06	2.04	2.01	1.96	1.93	1.90
2.80	2.75	2.69	2.62	2.54	2.51	2.47	2.38	2.34	2.30
3.44	3.37	3.28	3.18	3.07	3.02	2.96	2.85	2.79	2.72
4.39	4.30	4.16	4.01	3.86	3.78	3.70	3.54	3.45	3.36
5.20	5.09	4.91	4.72	4.53	4.43	4.33	4.12	4.01	3.90
...	...	7.00	...	...	6.25	...	...	...	5.42
0.970	0.977	0.989	1.00	1.01	1.02	1.02	1.03	1.04	1.05
2.09	2.06	2.02	1.97	1.92	1.90	1.87	1.82	1.79	1.76
2.59	2.54	2.48	2.40	2.33	2.29	2.25	2.16	2.11	2.07
3.12	3.06	2.96	2.86	2.76	2.70	2.64	2.52	2.46	2.40
3.89	3.80	3.67	3.52	3.37	3.29	3.21	3.05	2.96	2.87
4.54	4.42	4.25	4.07	3.88	3.79	3.69	3.48	3.37	3.26
...	...	5.81	...	...	5.10	...	...	...	4.31
0.959	0.966	0.977	0.989	1.00	1.01	1.01	1.02	1.03	1.03
1.96	1.94	1.89	1.84	1.79	1.77	1.74	1.68	1.64	1.61
2.39	2.35	2.28	2.20	2.12	2.08	2.04	1.95	1.90	1.84
2.84	2.77	2.68	2.57	2.46	2.41	2.35	2.22	2.16	2.09
3.46	3.37	3.23	3.09	2.94	2.86	2.78	2.61	2.52	2.42
3.96	3.85	3.68	3.50	3.32	3.22	3.12	2.92	2.81	2.69
...	...	4.82	...	...	4.15	...	...	...	3.38
0.953	0.961	0.972	0.983	0.994	1.00	1.01	1.02	1.02	1.03
1.91	1.88	1.83	1.78	1.73	1.70	1.67	1.61	1.57	1.53
2.30	2.25	2.18	2.11	2.03	1.98	1.94	1.84	1.79	1.73
2.70	2.64	2.54	2.44	2.33	2.27	2.21	2.08	2.01	1.94
3.26	3.17	3.03	2.89	2.74	2.66	2.58	2.40	2.31	2.21
3.69	3.59	3.42	3.25	3.06	2.97	2.87	2.66	2.55	2.43
...	...	4.39	...	...	3.74	...	...	...	2.97

TABLE J *(continued)*

ν_2	P	$\nu_1 = 1$	2	3	4	5	6	7	8
	0.500	0.466	0.709	0.807	0.858	0.890	0.912	0.927	0.939
	0.100	2.88	2.49	2.28	2.14	2.05	1.98	1.93	1.88
	0.050	4.17	3.32	2.92	2.69	2.53	2.42	2.33	2.27
30	0.025	5.57	4.18	3.59	3.25	3.03	2.87	2.75	2.65
	0.010	7.56	5.39	4.51	4.02	3.70	3.47	3.30	3.17
	0.005	9.18	6.35	5.24	4.62	4.23	3.95	3.74	3.58
	0.001	13.29	8.77	7.05	6.12	5.53	5.12	. . .	4.58
	0.500	0.461	0.701	0.798	0.849	0.880	0.901	0.917	0.928
	0.100	2.79	2.39	2.18	2.04	1.95	1.87	1.82	1.77
	0.050	4.00	3.15	2.76	2.53	2.37	2.25	2.17	2.10
60	0.025	5.29	3.93	3.34	3.01	2.79	2.63	2.51	2.41
	0.010	7.08	4.98	4.13	3.65	3.34	3.12	2.95	2.82
	0.005	8.49	5.80	4.73	4.14	3.76	3.49	3.29	3.13
	0.001	11.97	7.76	6.17	5.31	4.76	4.37	. . .	3.87
	0.500	0.458	0.697	0.793	0.844	0.875	0.896	0.912	0.923
	0.100	2.75	2.35	2.13	1.99	1.90	1.82	1.77	1.72
	0.050	3.92	3.07	2.68	2.45	2.29	2.18	2.09	2.02
120	0.025	5.15	3.80	3.23	2.89	2.67	2.52	2.39	2.30
	0.010	6.85	4.79	3.95	3.48	3.17	2.96	2.79	2.66
	0.005	8.18	5.54	4.50	3.92	3.55	3.28	3.09	2.93
	0.001	11.38	7.31	5.79	4.95	4.42	4.04	. . .	3.55
	0.500	0.455	0.693	0.789	0.839	0.870	0.891	0.907	0.918
	0.100	2.71	2.30	2.08	1.94	1.85	1.77	1.72	1.67
	0.050	3.84	3.00	2.60	2.37	2.21	2.10	2.01	1.94
∞	0.025	5.02	3.69	3.12	2.79	2.57	2.41	2.29	2.19
	0.010	6.63	4.61	3.78	3.32	3.02	2.80	2.64	2.51
	0.005	7.88	5.30	4.28	3.72	3.35	3.09	2.90	2.74
	0.001	10.8	6.91	5.42	4.62	4.10	3.74	. . .	3.27

TABLE J *(concluded)*

9	10	12	15	20	24	30	60	120	∞
0.948	0.955	0.966	0.978	0.989	0.994	1.00	1.01	1.02	1.02
1.85	1.82	1.77	1.72	1.67	1.64	1.61	1.54	1.50	1.46
2.21	2.16	2.09	2.01	1.93	1.89	1.84	1.74	1.68	1.62
2.57	2.51	2.41	2.31	2.20	2.14	2.07	1.94	1.87	1.79
3.07	2.98	2.84	2.70	2.55	2.47	2.39	2.21	2.11	2.01
3.45	3.34	3.18	3.01	2.82	2.73	2.63	2.42	2.30	2.18
...	...	4.00	...	...	3.36	...	...	...	2.59
0.937	0.945	0.956	0.967	0.978	0.983	0.989	1.00	1.01	1.01
1.74	1.71	1.66	1.60	1.54	1.51	1.48	1.40	1.35	1.29
2.04	1.99	1.92	1.84	1.75	1.70	1.65	1.53	1.47	1.39
2.33	2.27	2.17	2.06	1.94	1.88	1.82	1.67	1.58	1.48
2.72	2.63	2.50	2.35	2.20	2.12	2.03	1.84	1.73	1.60
3.01	2.90	2.74	2.57	2.39	2.29	2.19	1.96	1.83	1.69
...	...	3.31	...	...	2.69	...	...	...	1.90
0.932	0.939	0.950	0.961	0.972	0.978	0.983	0.994	1.00	1.01
1.68	1.65	1.60	1.55	1.48	1.45	1.41	1.32	1.26	1.19
1.96	1.91	1.83	1.75	1.66	1.61	1.55	1.43	1.35	1.25
2.22	2.16	2.05	1.95	1.82	1.76	1.69	1.53	1.43	1.31
2.56	2.47	2.34	2.19	2.03	1.95	1.86	1.66	1.53	1.38
2.81	2.71	2.54	2.37	2.19	2.09	1.98	1.75	1.61	1.43
...	...	3.02	...	...	2.40	...	...	...	1.56
0.927	0.934	0.945	0.956	0.967	0.972	0.978	0.989	0.994	1.00
1.63	1.60	1.55	1.49	1.42	1.38	1.34	1.24	1.17	1.00
1.88	1.83	1.75	1.67	1.57	1.52	1.46	1.32	1.22	1.00
2.11	2.05	1.94	1.83	1.71	1.64	1.57	1.39	1.27	1.00
2.41	2.32	2.18	2.04	1.88	1.79	1.70	1.47	1.32	1.00
2.62	2.52	2.36	2.19	2.00	1.90	1.79	1.53	1.36	1.00
...	...	2.74	...	...	2.13	...	...	...	1.00

TABLE K
Values for Testing Fixed Main-Effects in a Mixed Model (upper 5 percent-point of $F_{h, mh}$)

h \ m	2	3	4	5	6	7	8
1.00	18.513	10.128	7.709	6.608	5.987	5.591	5.318
1.05	16.936	9.554	7.366	6.358	5.785	5.418	5.163
1.10	15.613	9.053	7.062	6.133	5.603	5.261	5.023
1.15	14.488	8.613	6.790	5.931	5.436	5.117	4.893
1.20	13.523	8.223	6.545	5.746	5.284	4.984	4.774
1.25	12.687	7.875	6.323	5.578	5.145	4.862	4.664
1.30	11.957	7.563	6.121	5.424	5.016	4.749	4.562
1.35	11.314	7.281	5.937	5.282	4.897	4.644	4.466
1.40	10.744	7.025	5.768	5.150	4.786	4.547	4.377
1.45	10.236	6.792	5.612	5.029	4.683	4.455	4.294
1.50	9.780	6.579	5.468	4.915	4.587	4.370	4.216
1.60	8.997	6.202	5.210	4.711	4.413	4.214	4.073
1.70	8.350	5.880	4.986	4.532	4.258	4.076	3.946
1.80	7.806	5.602	4.789	4.373	4.121	3.952	3.831
1.90	7.343	5.358	4.615	4.231	3.997	3.840	3.728
2.00	6.944	5.143	4.459	4.103	3.885	3.739	3.634
2.10	6.597	4.952	4.319	3.987	3.784	3.646	3.548
2.20	6.293	4.781	4.192	3.882	3.691	3.562	3.469
2.30	6.024	4.627	4.077	3.786	3.606	3.484	3.396
2.40	5.784	4.487	3.972	3.697	3.527	3.412	3.328
2.50	5.568	4.360	3.875	3.616	3.455	3.345	3.266
2.60	5.374	4.244	3.786	3.541	3.387	3.283	3.207
2.70	5.198	4.137	3.704	3.471	3.325	3.225	3.153
2.80	5.038	4.038	3.628	3.405	3.266	3.171	3.102
2.90	4.892	3.947	3.557	3.344	3.211	3.120	3.054
3.00	4.757	3.863	3.490	3.287	3.160	3.073	3.009
3.20	4.518	3.710	3.370	3.183	3.066	2.985	2.926
3.40	4.313	3.577	3.264	3.091	2.982	2.907	2.852
3.60	4.134	3.459	3.169	3.008	2.907	2.836	2.785
3.80	3.977	3.354	3.084	2.934	2.838	2.772	2.724
4.00	3.838	3.259	3.007	2.866	2.776	2.714	2.668
4.20	3.714	3.174	2.937	2.804	2.720	2.661	2.618
4.40	3.602	3.096	2.873	2.748	2.667	2.612	2.571
4.60	3.501	3.026	2.815	2.696	2.619	2.566	2.527
4.80	3.410	2.961	2.761	2.648	2.575	2.524	2.487
5.00	3.326	2.901	2.711	2.603	2.534	2.485	2.450
5.20	3.249	2.846	2.665	2.561	2.495	2.449	2.414
5.40	3.178	2.795	2.621	2.523	2.459	2.415	2.382
5.60	3.113	2.747	2.581	2.487	2.425	2.382	2.351
5.80	3.052	2.703	2.544	2.452	2.394	2.352	2.322

SOURCE: Reproduced with permission from J. P. Imhoff, "Testing the Hypothesis of No Fixed Main-Effects in Scheffe's Mixed Model," *Annals of Mathematical Statistics* 33 (1962), pp. 1085–94. For explanation of use, see Chapter 30, Sections 1.2.3. and 1.3.4.

TABLE K *(concluded)*
(upper 1 percent-point of $F_{h, mh}$)

h \ m	2	3	4	5	6	7	8
1.00	98.503	34.116	21.198	16.258	13.745	12.246	11.259
1.05	83.954	30.786	19.606	15.236	12.985	11.631	10.733
1.10	72.597	28.025	18.247	14.349	12.317	11.086	10.265
1.15	63.570	25.706	17.075	13.571	11.726	10.600	9.846
1.20	56.278	23.736	16.054	12.884	11.199	10.164	9.467
1.25	50.303	22.045	15.159	12.274	10.726	9.770	9.124
1.30	45.345	20.581	14.367	11.727	10.299	9.413	8.812
1.35	41.183	19.302	13.662	11.236	9.913	9.088	8.526
1.40	37.655	18.178	13.032	10.791	9.561	8.790	8.264
1.45	34.635	17.183	12.464	10.387	9.239	8.516	8.022
1.50	32.029	16.296	11.950	10.017	8.943	8.264	7.798
1.60	27.779	14.788	11.058	9.368	8.419	7.814	7.397
1.70	24.481	13.554	10.310	8.815	7.968	7.425	7.048
1.80	21.862	12.528	9.673	8.339	7.576	7.084	6.741
1.90	19.743	11.663	9.125	7.924	7.231	6.783	6.469
2.00	18.000	10.925	8.649	7.559	6.927	6.515	6.226
2.10	16.545	10.287	8.231	7.236	6.655	6.275	6.008
2.20	15.314	9.732	7.861	6.948	6.411	6.059	5.810
2.30	14.262	9.243	7.532	6.689	6.191	5.863	5.631
2.40	13.354	8.811	7.237	6.455	5.991	5.684	5.467
2.50	12.563	8.425	6.970	6.242	5.808	5.521	5.317
2.60	11.869	8.079	6.729	6.049	5.641	5.371	5.178
2.70	11.255	7.768	6.509	5.871	5.487	5.232	5.050
2.80	10.709	7.485	6.308	5.708	5.345	5.104	4.932
2.90	10.220	7.227	6.123	5.557	5.214	4.985	4.821
3.00	9.780	6.992	5.953	5.417	5.092	4.874	4.718
3.20	9.020	6.577	5.649	5.166	4.872	4.674	4.531
3.40	8.389	6.222	5.386	4.948	4.679	4.497	4.367
3.60	7.856	5.915	5.156	4.755	4.508	4.340	4.220
3.80	7.400	5.648	4.953	4.584	4.355	4.200	4.088
4.00	7.006	5.412	4.773	4.431	4.218	4.074	3.969
4.20	6.662	5.203	4.611	4.293	4.095	3.960	3.862
4.40	6.360	5.015	4.465	4.168	3.982	3.856	3.763
4.60	6.091	4.847	4.333	4.054	3.880	3.760	3.673
4.80	5.852	4.694	4.213	3.950	3.786	3.673	3.590
5.00	5.636	4.556	4.103	3.855	3.699	3.592	3.514
5.20	5.442	4.429	4.001	3.767	3.619	3.517	3.443
5.40	5.265	4.313	3.908	3.685	3.545	3.448	3.377
5.60	5.104	4.205	3.822	3.610	3.476	3.383	3.315
5.80	4.956	4.106	3.741	3.539	3.411	3.322	3.258

TABLE L

0.10 Points of the OC Curves for One-Tail Tests of the Hypothesis $\sigma_1' = \sigma_2'$, Normal Universe, $\alpha = 0.05$ (values in the table are those of $\phi^2 = \sigma_1'^2/\sigma_2'^2$, for $\nu = n_1 - 1$, and $\nu_2 = n_2 - 1$)

ν_2 \ ν_1	1	2	3	4	5	6	7	8	9	10	12	15	20	24	30	40	60	120	∞
1	6436.	1701.	1195.	1021.	934.5	883.5	849.9	826.0	808.3	794.6	774.8	755.9	737.8	729.0	720.4	712.1	704.0	695.9	688.1
2	916.4	171.0	104.7	83.24	72.93	66.95	63.04	60.30	58.28	56.72	54.49	52.37	50.35	49.38	48.44	47.52	46.62	45.74	44.89
3	542.8	87.51	50.01	38.21	32.62	29.40	27.32	25.86	24.79	23.96	22.78	21.67	20.61	20.11	19.61	19.13	18.66	18.21	17.77
4	430.4	64.19	35.22	26.24	22.02	19.60	18.04	16.95	16.15	15.54	14.66	13.83	13.05	12.67	12.31	11.95	11.61	11.27	10.95
5	378.2	53.77	28.72	21.03	17.44	15.38	14.06	13.14	12.46	11.94	11.20	10.50	9.837	9.521	9.213	8.913	8.622	8.339	8.063
6	348.5	47.96	25.14	18.18	14.94	13.09	11.89	11.06	10.46	9.990	9.324	8.696	8.102	7.818	7.541	7.273	7.011	6.757	6.509
7	329.4	44.29	22.89	16.39	13.38	11.65	10.55	9.777	9.211	8.779	8.160	7.577	7.026	6.762	6.505	6.255	6.012	5.775	5.545
8	316.1	41.77	21.35	15.18	12.31	10.68	9.632	8.902	8.367	7.957	7.371	6.818	6.296	6.046	5.802	5.565	5.334	5.109	4.890
9	306.3	39.93	20.24	14.30	11.55	9.979	8.972	8.272	7.757	7.364	6.802	6.271	5.770	5.529	5.295	5.067	4.844	4.628	4.416
10	298.8	38.53	19.40	13.63	10.97	9.449	8.474	7.796	7.298	6.917	6.373	5.859	5.372	5.139	4.912	4.690	4.474	4.263	4.057
12	288.2	36.55	18.20	12.70	10.15	8.703	7.773	7.127	6.652	6.289	5.769	5.279	4.814	4.590	4.372	4.159	3.952	3.748	3.549
15	278.1	34.70	17.10	11.82	9.394	8.012	7.124	6.507	6.054	5.707	5.210	4.740	4.294	4.079	3.870	3.664	3.464	3.266	3.072
20	268.6	32.98	16.06	11.02	8.693	7.372	6.523	5.933	5.499	5.167	4.691	4.240	3.810	3.603	3.400	3.200	3.004	2.811	2.618
24	264.1	32.16	15.57	10.64	8.361	7.069	6.239	5.662	5.237	4.912	4.445	4.003	3.581	3.376	3.176	2.978	2.783	2.590	2.397
30	259.7	31.36	15.10	10.27	8.042	6.777	5.965	5.400	4.984	4.666	4.208	3.773	3.358	3.156	2.957	2.761	2.567	2.373	2.177
40	255.4	30.59	14.65	9.912	7.734	6.497	5.701	5.148	4.741	4.428	3.979	3.551	3.141	2.942	2.744	2.549	2.353	2.156	1.954
60	251.3	29.85	14.21	9.569	7.437	6.226	5.447	4.905	4.506	4.199	3.758	3.336	2.931	2.733	2.536	2.339	2.141	1.937	1.723
120	247.2	29.13	13.78	9.239	7.151	5.965	5.202	4.670	4.279	3.977	3.543	3.128	2.726	2.528	2.330	2.130	1.926	1.710	1.465
∞	243.3	28.43	13.37	8.920	6.875	5.713	4.965	4.444	4.059	3.763	3.335	2.925	2.524	2.326	2.125	1.919	1.702	1.457	1.000

SOURCE: Table L is reprinted with permission from Table 8.3 of Statistical Research Group, Columbia University, *Techniques of Statistical Analysis* (New York: McGraw-Hill, 1947), p. 288. *Note carefully that ϕ^2 of Table L equals ϕ in the Techniques' Table 8.3.*

TABLE M
Factors Useful in the Construction of Control Charts

Number of Observations in Sample, n	Chart for Averages: Factors for Control Limits			Chart for Standard Deviations — Factors for: Central Line	Control Limits				Chart for Ranges: Factors for Central Line		Factors for Control Limits					
	A	A_2	A_3	c_4	B_3	B_4	B_5	B_6	d_2	$1/d_2$	d_3	D_1	D_2	D_3	D_4
2	2.121	1.880	2.659	0.7979	0	3.267	0	2.606	1.128	0.8865	0.853	0	3.686	0	3.267
3	1.732	1.023	1.954	0.8862	0	2.568	0	2.276	1.693	0.5907	0.888	0	4.358	0	2.575
4	1.500	0.729	1.628	0.9213	0	2.266	0	2.088	2.059	0.4857	0.880	0	4.698	0	2.282
5	1.342	0.577	1.427	0.9400	0	2.089	0	1.964	2.326	0.4299	0.864	0	4.918	0	2.115
6	1.225	0.483	1.287	0.9515	0.030	1.970	0.029	1.874	2.534	0.3946	0.848	0	5.078	0	2.004
7	1.134	0.419	1.182	0.9594	0.118	1.882	0.113	1.806	2.704	0.3698	0.833	0.205	5.203	0.076	1.924
8	1.061	0.373	1.099	0.9650	0.185	1.815	0.179	1.751	2.847	0.3512	0.820	0.387	5.307	0.136	1.864
9	1.000	0.337	1.032	0.9693	0.239	1.761	0.232	1.707	2.970	0.3367	0.808	0.546	5.394	0.184	1.816
10	0.949	0.308	0.975	0.9727	0.284	1.716	0.276	1.669	3.078	0.3249	0.797	0.687	5.469	0.223	1.777
11	0.905	0.285	0.927	0.9754	0.321	1.679	0.313	1.637	3.173	0.3152	0.787	0.812	5.534	0.256	1.744
12	0.866	0.266	0.886	0.9776	0.354	1.646	0.346	1.610	3.258	0.3069	0.778	0.924	5.592	0.284	1.716
13	0.832	0.249	0.850	0.9794	0.382	1.618	0.374	1.585	3.336	0.2998	0.770	1.026	5.646	0.308	1.692
14	0.802	0.235	0.817	0.9810	0.406	1.594	0.399	1.563	3.407	0.2935	0.762	1.121	5.693	0.329	1.671
15	0.775	0.223	0.789	0.9823	0.428	1.572	0.421	1.544	3.472	0.2880	0.755	1.207	5.737	0.348	1.652
16	0.750	0.212	0.763	0.9835	0.448	1.552	0.440	1.526	3.532	0.2831	0.749	1.285	5.779	0.364	1.636
17	0.728	0.203	0.739	0.9845	0.466	1.534	0.458	1.511	3.588	0.2787	0.743	1.359	5.817	0.379	1.621
18	0.707	0.194	0.718	0.9854	0.482	1.518	0.475	1.496	3.640	0.2747	0.738	1.426	5.854	0.392	1.608
19	0.688	0.187	0.698	0.9862	0.497	1.503	0.490	1.483	3.689	0.2711	0.733	1.490	5.888	0.404	1.596
20	0.671	0.180	0.680	0.9869	0.510	1.490	0.504	1.470	3.735	0.2677	0.729	1.548	5.922	0.414	1.586
21	0.655	0.173	0.663	0.9876	0.523	1.477	0.516	1.459	3.778	0.2647	0.724	1.606	5.950	0.425	1.575
22	0.640	0.167	0.647	0.9882	0.534	1.466	0.528	1.448	3.819	0.2618	0.720	1.659	5.979	0.434	1.566
23	0.626	0.162	0.633	0.9887	0.545	1.455	0.539	1.438	3.858	0.2592	0.716	1.710	6.006	0.443	1.557
24	0.612	0.157	0.619	0.9892	0.555	1.445	0.549	1.429	3.895	0.2567	0.712	1.759	6.031	0.452	1.548
25	0.600	0.153	0.606	0.9896	0.565	1.435	0.559	1.420	3.931	0.2544	0.709	1.804	6.058	0.459	1.541
Over 25	$\frac{3}{\sqrt{n}}$	$\frac{3}{\sqrt{n}}$	.....	.....	*	**	.....	.....	.....	.....	.....	.....	.....	.....	.....

$* \; 1 - \frac{3}{\sqrt{2n}}$ $\quad ** \; 1 + \frac{3}{\sqrt{2n}}$

TABLE M *(concluded)*

Chart	*Central Line*	3σ *Control Limits*
$\bar{X}$	$\bar{\bar{X}}$	$\bar{\bar{X}} \pm A_3\bar{s}$
		$\bar{\bar{X}} \pm A_2\bar{R}$
	$\bar{X}''$	$\bar{X}'' \pm A\sigma''$
R	$\bar{R}$	$D_3\bar{R}$ and $D_4\bar{R}$
	$d_2\sigma''$	$D_1\sigma''$ and $D_2\sigma''$
s	$\bar{s}$	$B_3\bar{s}$ and $B_4\bar{s}$
	$c_4\sigma''$	$B_5\sigma''$ and $B_6\sigma''$

Definitions: $A = 3/\sqrt{n}$, $A_2 = \dfrac{3}{d_2\sqrt{n}}$, $A_3 = 3/(c_4\sqrt{n})$, $B_3 = 1 - \dfrac{K}{c_4}$, $B_4 = 1 + \dfrac{K}{c_4}$,

$B_5 = c_4 - 3\sqrt{1 - c_4^2}$, $B_6 = c_4 + 3\sqrt{1 - c_4^2}$. $D_1 = d_2 - 3d_3$, $D_2 = d_2 + 3d_3$,

$$D_3 = 1 - 3\frac{d_3}{d_2}, \text{ and } D_4 = 1 + 3\frac{d_3}{d_2}, \text{ where}$$

$$c_4 = \sqrt{\frac{2}{n-1}}\left[\frac{\Gamma(n/2)}{\Gamma[(n-1)/2]}\right], \text{ and } K = 3\sqrt{1 - c_4^2}$$

Note that d_2 and d_3 are the same as mean w and σ_w' appearing in Table D and have the same original source.

Warning: The fourth significant figures for D_1, D_2, D_3, and D_4 are in doubt for n greater than 5.

SOURCE: A, A_2, B_3, B_4, d_2, $1/d_2$, d_3 and $D_1 - D_4$ reproduced with permission from Table B2 of the *A.S.T.M. Manual on Quality Control of Materials,* p. 115. The quantities A_3, B_5, B_6, and c_4 reproduced with permission from ASQC Standard A1, Table 1. For nonnormal variations, see Irving W. Burr, "The Effect of Non-Normality on Constants for $\bar{X}$ and R Charts," *Industrial Quality Control,* May 1967, pp. 563–69.

TABLE N1

Table for Testing Randomness of Grouping in a Sequence of Alternatives (probability of an equal or smaller number of runs than that listed is $P = 0.005$)

s = cases on one side of average; r = cases on other side of average } r always taken as the smaller number of cases; s the larger

s \ r	6	7	8	9	10	11	12	13	14	15	16	17	18	19	20
6	2														
7	2	3													
8	3	3	3												
9	3	3	3	4											
10	3	3	4	4	5										
11	3	4	4	5	5	5									
12	3	4	4	5	5	6	6								
13	3	4	5	5	5	6	6	7							
14	4	4	5	5	6	6	7	7	7						
15	4	4	5	6	6	7	7	7	8	8					
16	4	5	5	6	6	7	7	8	8	9	9				
17	4	5	5	6	7	7	8	8	8	9	9	10			
18	4	5	6	6	7	7	8	8	9	9	10	10	11		
19	4	5	6	6	7	8	8	9	9	10	10	10	11	11	
20	4	5	6	7	7	8	8	9	9	10	10	11	11	12	12

SOURCE: Freda S. Swed and C. Eisenhart, "Tables for Testing Randomness of Grouping in a Sequence of Alternatives," *Annals of Mathematical Statistics* 14 (1943), pp. 66–87.

TABLE N2

Table for Testing Randomness of Grouping in a Sequence of Alternatives (probability of an equal or smaller number of runs than that listed is $P = 0.05$)

s = cases on one side of average; r = cases on other side of average } r always taken as the smaller number of cases; s the larger

s \ r	6	7	8	9	10	11	12	13	14	15	16	17	18	19	20
6	3														
7	4	4													
8	4	4	5												
9	4	5	5	6											
10	5	5	6	6	6										
11	5	5	6	6	7	7									
12	5	6	6	7	7	8	8								
13	5	6	6	7	8	8	9	9							
14	5	6	7	7	8	8	9	9	10						
15	6	6	7	8	8	9	9	10	10	11					
16	6	6	7	8	8	9	10	10	11	11	11				
17	6	7	7	8	9	9	10	10	11	11	12	12			
18	6	7	8	8	9	10	10	11	11	12	12	13	13		
19	6	7	8	8	9	10	10	11	12	12	13	13	14	14	
20	6	7	8	9	9	10	11	11	12	12	13	13	14	14	15

SOURCE: The same as that of Table N1.

TABLE N3
Limiting Values for the Total Number of Runs above and below the Median of a Set of Values

	Probability of an Equal or Smaller Value			*Probability of an Equal or Smaller Value*	
r = *s*	*0.005*	*0.05*	*r* = *s*	*0.005*	*0.05*
10	4	6	55	42	46
11	5	7	56	42	47
12	6	8	57	43	48
13	7	9	58	44	49
14	7	10	59	45	50
15	8	11			
16	9	11	60	46	51
17	10	12	61	47	52
18	10	13	62	48	53
19	11	14	63	49	54
			64	49	55
20	12	15	65	50	56
21	13	16	66	51	57
22	14	17	67	52	58
23	14	17	68	53	58
24	15	18	69	54	59
25	16	19			
26	17	20	70	55	60
27	18	21	71	56	61
28	18	22	72	57	62
29	19	23	73	57	63
			74	58	64
30	20	24	75	59	65
31	21	25	76	60	66
32	22	25	77	61	67
33	23	26	78	62	68
34	23	27	79	63	69
35	24	28			
36	25	29	80	64	70
37	26	30	81	65	71
38	27	31	82	66	71
39	28	32	83	66	72
			84	67	73
40	29	33	85	68	74
41	29	34	86	69	75
42	30	35	87	70	76
43	31	35	88	71	77
44	32	36	89	72	78
45	33	37			
46	34	38	90	73	79
47	35	39	91	74	80
48	35	40	92	75	81
49	36	41	93	75	82
			94	76	83
50	37	42	95	77	84
51	38	43	96	78	85
52	39	44	97	79	86
53	40	45	98	80	87
54	41	45	99	81	87
			100	82	88

SOURCE: The same as that of Table N1.

TABLE P
Limiting Values for Lengths of Runs on Either Side of the Median of *n* Cases (probability of getting at least one run of specified size or more)

n	*0.05*	*0.01*	*0.001*
10	5	—	—
20	7	8	9
30	8	9	—
40	9	10	12
50	10	11	—

(Larger runs than these suggest existence of nonrandom influences.)
SOURCE: F. Mosteller, "Note on Application of Runs to Quality Control Charts," *Annals of Mathematical Statistics* 12 (1941), p. 232.

TABLE Q
Limiting Values for Lengths of Runs up and down in a Series of *n* Numbers*

	Probability Equal to or Less than 0.0032		*Probability Equal to or Less than 0.0567*	
n	*Run*	*Probability of an Equal or Greater Run*	*Run*	*Probability of an Equal or Greater Run*
4	4	0.0028	4	0.0028
5	5	0.0004	4	0.0165
6	5	0.0028	4	0.0301
7	6	0.0004	4	0.0435
8	6	0.0007	4	0.0567
9	6	0.0011	5	0.0099
10	6	0.0014	5	0.0122
11	6	0.0018	5	0.0146
12	6	0.0021	5	0.0169
13	6	0.0025	5	0.0193
14	6	0.0028	5	0.0216
15	6	0.0032	5	0.0239
20	7	0.0006	5	0.0355
40	7	0.0015	6	0.0118
60	7	0.0023	6	0.0186
80	7	0.0032	6	0.0254
100	8	0.0005	6	0.0322
200	8	0.0010	7	0.0085
500	8	0.0024	7	0.0215
1,000	9	0.0005	7	0.0428
5,000	9	0.0025	8	0.0245

* Probabilities based on approximation of exact distribution by the Poisson exponential. See P. Olmstead, "Distribution of Sample Arrangement for Runs Up and Down," *Annals of Mathematical Statistics* 17 (1946), p. 29.

TABLE R

Characteristic Qualities of Sequential Tests of the Binomial Distribution Computed for Various Combinations of p'_1, p'_2, $\alpha = 0.05$, and $\beta = 0.10$

p'_1	p'_2	h_2	h_1	s	$\bar{n}_0$	$\bar{n}_1$	$\bar{n}_{p'_1}$	$\bar{n}_s$	$\bar{n}_{p'_2}$
0.005	0.01	4.1398	3.2245	0.007216	447	5	1,289	1,863	1,222
	0.02	2.0624	1.6064	0.01084	149	3	244	309	185
	0.03	1.5906	1.2389	0.01400	89	2	122	143	82
	0.04	1.3664	1.0643	0.01693	63	2	79	87	49
	0.05	1.2305	0.9585	0.01970	49	2	58	61	33
	0.06	1.1371	0.8857	0.02237	40	2	45	46	25
	0.07	1.0679	0.8318	0.02496	34	2	37	36	19
0.010	0.03	2.5829	2.0118	0.01824	111	3	216	290	181
	0.04	2.0397	1.5887	0.02172	74	3	120	153	92
	0.05	1.7510	1.3639	0.02499	55	2	81	98	58
	0.06	1.5678	1.2211	0.02811	44	2	60	70	40
	0.07	1.4391	1.1209	0.03113	37	2	47	53	30
	0.08	1.3426	1.0458	0.03406	31	2	38	43	24
0.015	0.03	4.0796	3.1776	0.02166	147	5	423	612	402
	0.04	2.8716	2.2367	0.02554	88	3	188	258	163
	0.05	2.3307	1.8153	0.02917	63	3	113	149	92
	0.06	2.0169	1.5710	0.03263	49	3	79	100	61
	0.07	1.8089	1.4089	0.03596	40	2	60	74	44
0.02	0.03	6.9527	5.4154	0.02467	220	8	1,027	1,565	1,073
	0.04	4.0495	3.1541	0.02889	110	5	314	455	300
	0.05	3.0509	2.3763	0.03282	73	4	164	228	146
	0.06	2.5348	1.9743	0.03655	55	3	106	142	89
	0.07	2.2146	1.7250	0.04012	43	3	76	99	61
	0.08	1.9941	1.5532	0.04359	36	3	58	74	45
	0.09	1.8315	1.4265	0.04696	31	2	47	58	35
	0.10	1.7056	1.3285	0.05025	27	2	39	47	28

SOURCE: Table R is abridged with permission from Table 2.23 of Statistical Research Group, Columbia University, *Sequential Analysis of Statistical Data: Applications* (New York: Columbia University Press, 1945), pp. 2.39–2.42.

TABLE S
The Upper 0.05 Points of the Distribution of Λ_0 (g = number of groups, m = number per group)

g	m											
	2	3	4	5	6	7	8	9	10	12	15	20
2	52.6	5.82	3.05	2.26	1.99	1.78	1.62	1.51	1.41	1.33	1.25	1.17
3	46.1	5.73	3.12	2.33	2.06	1.84	1.68	1.54	1.43	1.35	1.26	1.18
4	38.8	5.42	3.00	2.26	2.01	1.81	1.66	1.54	1.43	1.35	1.26	1.18
5	33.2	5.14	2.92	2.22	1.97	1.78	1.62	1.50	1.42	1.34	1.25	1.18
10	21.3	4.36	2.64	2.06	1.84	1.66	1.54	1.44	1.38	1.31	1.23	1.16
20	15.2	3.79	2.41	1.93	1.74	1.58	1.44	1.38	1.32	1.26	1.20	1.14
25	13.9	3.64	2.38	1.87	1.68	1.54	1.42	1.36	1.31	1.25	1.19	1.13
50	11.4	3.29	2.15	1.74	1.59	1.48	1.38	1.32	1.27	1.21	1.17	1.12

SOURCE: Reproduced with permission from Dudley J. Cowden, *Statistical Methods in Quality Control* (New York: Prentice-Hall, Inc., 1957), p. 687.

TABLE T

Upper 0.05 Point of the Ratio, $s^2_{max.}/s^2_{min.}$ ($s^2_{max.}$ is the largest and $s^2_{min.}$ is the smallest in a set of g independent mean squares, each based on ν degrees of freedom)

ν \ g	2	3	4	5	6	7	8	9	10	11	12
2	39.0	87.5	142	202	266	333	403	475	550	626	704
3	15.4	27.8	39.2	50.7	62.0	72.9	83.5	93.9	104	114	124
4	9.60	15.5	20.6	25.2	29.5	33.6	37.5	41.1	44.6	48.0	51.4
5	7.15	10.8	13.7	16.3	18.7	20.8	22.9	24.7	26.5	28.2	29.9
6	5.82	8.38	10.4	12.1	13.7	15.0	16.3	17.5	18.6	19.7	20.7
7	4.99	6.94	8.44	9.70	10.8	11.8	12.7	13.5	14.3	15.1	15.8
8	4.43	6.00	7.18	8.12	9.03	9.78	10.5	11.1	11.7	12.2	12.7
9	4.03	5.34	6.31	7.11	7.80	8.41	8.95	9.45	9.91	10.3	10.7
10	3.72	4.85	5.67	6.34	6.92	7.42	7.87	8.28	8.66	9.01	9.34
12	3.28	4.16	4.79	5.30	5.72	6.09	6.42	6.72	7.00	7.25	7.48
15	2.86	3.54	4.01	4.37	4.68	4.95	5.19	5.40	5.59	5.77	5.93
20	2.46	2.95	3.29	3.54	3.76	3.94	4.10	4.24	4.37	4.49	4.59
30	2.07	2.40	2.61	2.78	2.91	3.02	3.12	3.21	3.29	3.36	3.39
60	1.67	1.85	1.96	2.04	2.11	2.17	2.22	2.26	2.30	2.33	2.36
∞	1.00	1.00	1.00	1.00	1.00	1.00	1.00	1.00	1.00	1.00	1.00

SOURCE: Reproduced with permission from Table 31 of *Biometrika Tables for Statisticians,* Vol. I.

TABLE U

Upper Percentage Points of the Extreme Studentized Deviate from the Sample Mean, $(X_H - \bar{X})/s_v$ or $(\bar{X} - X_L)/s_v$: Normal Universe (n = size of sample from which X_H or X_L comes, s_v is an estimate of σ'_X which is based on ν degrees of freedom and is independent of the numerator.)

0.05 Points

ν \ n	3	4	5	6	7	8	9	10	12
10	2.01	2.27	2.46	2.60	2.72	2.81	2.89	2.96	3.08
11	1.98	2.24	2.42	2.56	2.67	2.76	2.84	2.91	3.03
12	1.96	2.21	2.39	2.52	2.63	2.72	2.80	2.87	2.98
13	1.94	2.19	2.36	2.50	2.60	2.69	2.76	2.83	2.94
14	1.93	2.17	2.34	2.47	2.57	2.66	2.74	2.80	2.91
15	1.91	2.15	2.32	2.45	2.55	2.64	2.71	2.77	2.88
16	1.90	2.14	2.31	2.43	2.53	2.62	2.69	2.75	2.86
17	1.89	2.13	2.29	2.42	2.52	2.60	2.67	2.73	2.84
18	1.88	2.11	2.28	2.40	2.50	2.58	2.65	2.71	2.82
19	1.87	2.11	2.27	2.39	2.49	2.57	2.64	2.70	2.80
20	1.87	2.10	2.26	2.38	2.47	2.56	2.63	2.68	2.78
24	1.84	2.07	2.23	2.34	2.44	2.52	2.58	2.64	2.74
30	1.82	2.04	2.20	2.31	2.40	2.48	2.54	2.60	2.69
40	1.80	2.02	2.17	2.28	2.37	2.44	2.50	2.56	2.65
60	1.78	1.99	2.14	2.25	2.33	2.41	2.47	2.52	2.61
120	1.76	1.96	2.11	2.22	2.30	2.37	2.43	2.48	2.57
∞	1.74	1.94	2.08	2.18	2.27	2.33	2.39	2.44	2.52

0.001 Points

ν \ n	3	4	5	6	7	8	9	10	12
10	4.0	4.3	4.6	4.8	5.0	5.2	5.3	5.4	5.6
11	3.8	4.2	4.5	4.7	4.8	5.0	5.1	5.2	5.3
12	3.7	4.1	4.3	4.5	4.7	4.8	4.9	5.0	5.1
13	3.6	4.0	4.2	4.4	4.5	4.6	4.7	4.8	5.0
14	3.5	3.9	4.1	4.3	4.4	4.5	4.6	4.7	4.9
15	3.5	3.8	4.0	4.2	4.3	4.4	4.5	4.6	4.8
16	3.4	3.7	4.0	4.1	4.3	4.4	4.5	4.5	4.7
17	3.4	3.7	3.9	4.1	4.2	4.3	4.4	4.5	4.6
18	3.3	3.6	3.9	4.0	4.1	4.2	4.3	4.4	4.5
19	3.3	3.6	3.8	4.0	4.1	4.2	4.3	4.4	4.5
20	3.3	3.6	3.8	3.9	4.0	4.1	4.2	4.3	4.4
24	3.2	3.5	3.7	3.8	3.9	4.0	4.1	4.2	4.3
30	3.1	3.4	3.6	3.7	3.8	3.9	4.0	4.0	4.1
40	3.0	3.3	3.5	3.6	3.7	3.7	3.8	3.9	4.0
60	2.9	3.2	3.4	3.5	3.6	3.6	3.7	3.8	3.8
120	2.9	3.1	3.3	3.4	3.5	3.5	3.6	3.6	3.7
∞	2.8	3.0	3.2	3.3	3.4	3.4	3.5	3.5	3.6

SOURCE: Reproduced with permission from *Biometrika* 43 Parts 3 and 4 (1956), note by H. A. David, pp. 449–51. For 0.05 points for $\nu = 1 - 9$, see *Biometrika* 46 1959, p. 473, note by K. S. C. Pillar.

TABLE W
Criteria and Critical Values for Testing an Extreme Value

Statistic†	Number of Obs., n	Critical Values	
		$\alpha = 0.05$	$\alpha = 0.01$
$r_{10} = \dfrac{X_2 - X_1}{X_n - X_1}$	3	0.941	0.988
	4	0.765	0.889
	5	0.642	0.780
	6	0.560	0.698
	7	0.507	0.637
$r_{11} = \dfrac{X_2 - X_1}{X_{n-1} - X_1}$	8	0.554	0.683
	9	0.512	0.635
	10	0.477	0.597
$r_{21} = \dfrac{X_3 - X_1}{X_{n-1} - X_1}$	11	0.576	0.679
	12	0.546	0.642
	13	0.521	0.615
$r_{22} = \dfrac{X_3 - X_1}{X_{n-2} - X_1}$	14	0.546	0.641
	15	0.525	0.616
	16	0.507	0.595
	17	0.490	0.577
	18	0.475	0.561
	19	0.462	0.547
	20	0.450	0.535
	21	0.440	0.524
	22	0.430	0.514
	23	0.421	0.505
	24	0.413	0.497
	25	0.406	0.489

† For alternative forms, see Chapter 31, Section 1.4.

SOURCE: Reproduced with permission from W. J. Dixon, "Processing Data for Outliers," *Biometrics* 9 (1953), pp. 74–89.

TABLE Y

Table of $\lambda\sqrt{n}$ Corresponding to Various Probabilities (to be used for designing experiments involving k normal universes to decide which t have the largest [or smallest] universe means)

	Probability of Correct Ranking		
k, t	0.90	0.95	0.99
2, 1	1.8124	2.3262	3.2900
3, 1	2.2302	2.7101	3.6173
4, 1	2.4516	2.9162	3.7970
4, 2	2.6353	3.0808	3.9323
5, 1	2.5997	3.0552	3.9196
5, 2	2.8505	3.2805	4.1058
6, 1	2.7100	3.1591	4.0121
6, 2	2.9948	3.4154	4.2244
7, 1	2.7972	3.2417	4.0861
7, 2	3.1024	3.5164	4.3140
8, 1	2.8691	3.3099	4.1475
8, 2	3.1876	3.5968	4.3858
9, 1	2.9301	3.3679	4.1999
9, 2	3.2579	3.6633	4.4455
10, 1	2.9829	3.4182	4.2456
10, 2	3.3176	3.7198	4.4964
11, 2	3.3693	3.7689	4.5408
11, 3	3.5239	3.9099	4.6602
12, 3	3.5751	3.9584	4.7039
13, 4	3.7134	4.0867	4.8158
14, 5	3.8166	4.1831	4.9005

SOURCE: Robert E. Bechhofer, "A Single-Sample Multiple Decision Procedure for Ranking Means of Normal Populations with Known Variances," *Annals of Mathematical Statistics* 25 (1954), pp. 16–39.

CHART I
The 10 Percent and 50 Percent Points of the Operating Characteristic Curves of Fixed Effects Analysis of Variance *F* Tests, $\alpha = 0.05$

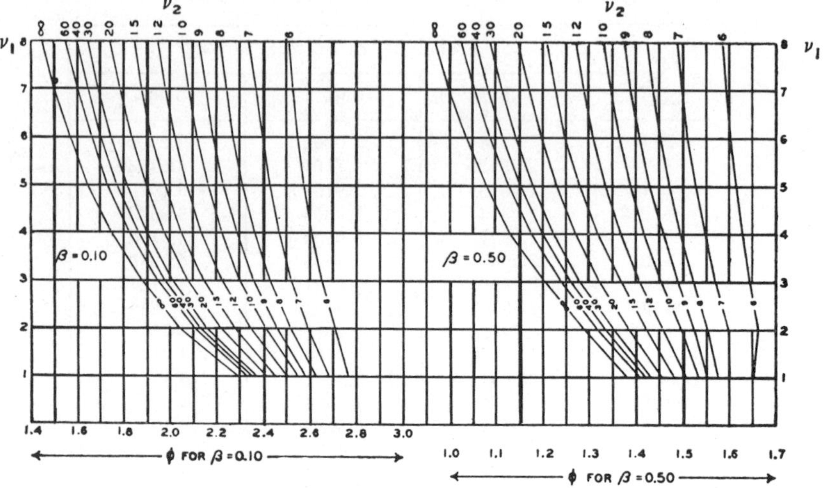

CHART II
The 10 Percent and 50 Percent Points of the Operating Characteristic Curves of Fixed Effects Analysis of Variance *F* Tests, $\alpha = 0.01$

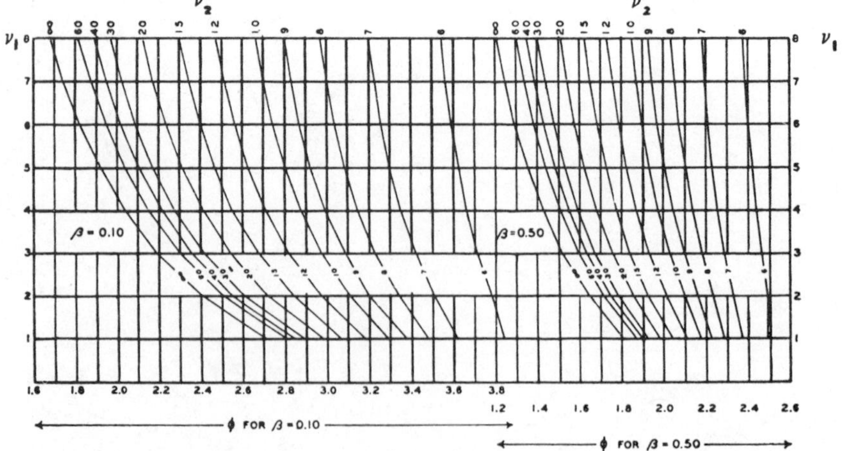

SOURCE: The above charts are reproduced from Acheson J. Duncan, "Charts of the 10% and 50% Points of the Operating Characteristic Curves for Fixed Effects Analysis of Variance F Tests, $\alpha = 0.01$ and 0.05," *Journal of the American Statistical Association* 52 (1957), pp. 345–49.

TABLE Z

Upper 5 Percent Points of the Sample Multiple Correlation Coefficient (normal universe)

ν \ ρ	0·0	0·1	0·2	0·3	0·4	0·5	0·6	0·7	0·8	0·9
					$n_1 = 2$					
1	0·0408	0·1283	0·2268	0·3252	0·4231	0·5205	0·6175	0·7139	0·8098	0·9051
2	0·0815	0·1583	0·2544	0·3507	0·4463	0·5409	0·6347	0·7275	0·8193	0·9101
3	0·1219	0·1896	0·2826	0·3764	0·4694	0·5612	0·6517	0·7408	0·8285	0·9149
4	0·1621	0·2216	0·3112	0·4023	0·4925	0·5812	0·6683	0·7538	0·8375	0·9196
5	0·2019	0·2543	0·3400	0·4281	0·5153	0·6009	0·6847	0·7664	0·8462	0·9241
6	0·2412	0·2874	0·3689	0·4538	0·5380	0·6204	0·7007	0·7788	0·8547	0·9284
8	0·3179	0·3540	0·4264	0·5046	0·5822	0·6580	0·7315	0·8024	0·8707	0·9366
10	0·3916	0·4202	0·4831	0·5540	0·6249	0·6940	0·7606	0·8245	0·8856	0·9441
12	0·4616	0·4845	0·5382	0·6015	0·6656	0·7280	0·7879	0·8450	0·8994	0·9510
14	0·5276	0·5459	0·5912	0·6469	0·7041	0·7599	0·8133	0·8640	0·9120	0·9573
16	0·5890	0·6038	0·6415	0·6897	0·7401	0·7895	0·8367	0·8814	0·9234	0·9629
18	0·6456	0·6575	0·6886	0·7297	0·7736	0·8169	0·8582	0·8972	0·9338	0·9680
20	0·6972	0·7068	0·7323	0·7668	0·8045	0·8419	0·8778	0·9116	0·9432	0·9726
					$n_1 = 3$					
1	0·0466	0·1295	0·2274	0·3256	0·4234	0·5207	0·6176	0·7140	0·8098	0·9052
2	0·0930	0·1625	0·2567	0·3522	0·4473	0·5417	0·6352	0·7278	0·8195	0·9102
3	0·1390	0·1978	0·2873	0·3796	0·4716	0·5628	0·6528	0·7415	0·8290	0·9151
4	0·1846	0·2346	0·3189	0·4074	0·4961	0·5838	0·6702	0·7550	0·8383	0·9199
5	0·2294	0·2723	0·3511	0·4356	0·5207	0·6048	0·6874	0·7683	0·8474	0·9246
6	0·2735	0·3105	0·3835	0·4639	0·5452	0·6255	0·7043	0·7813	0·8562	0·9291
8	0·3587	0·3866	0·4485	0·5200	0·5933	0·6660	0·7371	0·8062	0·8730	0·9376
10	0·4391	0·4606	0·5120	0·5746	0·6398	0·7047	0·7681	0·8295	0·8887	0·9455
12	0·5139	0·5307	0·5729	0·6267	0·6839	0·7412	0·7971	0·8512	0·9031	0·9527
14	0·5827	0·5959	0·6302	0·6757	0·7251	0·7751	0·8239	0·8711	0·9162	0·9592
16	0·6450	0·6554	0·6832	0·7211	0·7632	0·8062	0·8484	0·8891	0·9280	0·9650
18	0·7009	0·7090	0·7313	0·7625	0·7979	0·8344	0·8705	0·9054	0·9387	0·9702
20	0·7502	0·7566	0·7744	0·7998	0·8291	0·8598	0·8903	0·9199	0·9481	0·9748
					$n_1 = 4$					
1	0·0513	0·1307	0·2280	0·3260	0·4236	0·5209	0·6177	0·7140	0·8099	0·9052
2	0·1023	0·1665	0·2589	0·3537	0·4483	0·5424	0·6357	0·7282	0·8197	0·9103
3	0·1529	0·2056	0·2919	0·3826	0·4738	0·5643	0·5639	0·7423	0·8294	0·9154
4	0·2028	0·2465	0·3263	0·4125	0·4997	0·5864	0·6720	0·7563	0·8390	0·9203
5	0·2517	0·2885	0·3615	0·4429	0·5259	0·6086	0·6901	0·7701	0·8485	0·9251
6	0·2995	0·3307	0·3972	0·4735	0·5521	0·6306	0·7079	0·7837	0·8577	0·9298
8	0·3909	0·4139	0·4682	0·5343	0·6038	0·6736	0·7425	0·8098	0·8752	0·9386
10	0·4758	0·4932	0·5370	0·5931	0·6534	0·7147	0·7752	0·8343	0·8916	0·9468
12	0·5534	0·5668	0·6018	0·6486	0·7002	0·7531	0·8056	0·8569	0·9065	0·9542
14	0·6232	0·6335	0·6613	0·6998	0·7433	0·7884	0·8334	0·8775	0·9200	0·9609
16	0·6850	0·6930	0·7150	0·7463	0·7824	0·8204	0·8585	0·8959	0·9321	0·9669
18	0·7391	0·7453	0·7626	0·7877	0·8173	0·8489	0·8809	0·9123	0·9428	0·9721
20	0·7858	0·7906	0·8042	0·8242	0·8481	0·8741	0·9006	0·9268	0·9522	0·9767

n = number of cases, p = number of variables, $n_1 = p - 1$, $n_2 = n - p$, $\nu = 60/\sqrt{n_2}$ and $\rho = R_{i.j...p}$

SOURCE: Table Z is reproduced with permission from *Biometrika* 59 (1972), pp. 179–80.

TABLE Z *(concluded)*

$n_1 = 5$

$\nu \backslash \rho$	0·0	0·1	0·2	0·3	0·4	0·5	0·6	0·7	0·8	0·9
1	0·0554	0·1318	0·2287	0·3264	0·4239	0·5211	0·6179	0·7141	0·8099	0·9052
2	0·1105	0·1704	0·2612	0·3551	0·4494	0·5432	0·6363	0·7285	0·8199	0·9104
3	0·1649	0·2130	0·2964	0·3857	0·4759	0·5659	0·6550	0·7430	0·8299	0·9156
4	0·2184	0·2577	0·3334	0·4174	0·5032	0·5890	0·6738	0·7575	0·8398	0·9206
5	0·2707	0·3032	0·3715	0·4499	0·5310	0·6122	0·6927	0·7719	0·8496	0·9256
6	0·3215	0·3488	0·4100	0·4827	0·5588	0·6354	0·7114	0·7860	0·8591	0·9304
8	0·4178	0·4376	0·4862	0·5476	0·6136	0·6808	0·7477	0·8133	0·8774	0·9396
10	0·5059	0·5207	0·5590	0·6098	0·6660	0·7240	0·7819	0·8388	0·8943	0·9481
12	0·5851	0·5963	0·6263	0·6677	0·7148	0·7640	0·8134	0·8622	0·9097	0·9557
14	0·6550	0·6635	0·6869	0·7203	0·7591	0·8002	0·8420	0·8833	0·9236	0·9625
16	0·7158	0·7223	0·7405	0·7671	0·7986	0·8326	0·8674	0·9020	0·9358	0·9685
18	0·7678	0·7728	0·7869	0·8079	0·8333	0·8610	0·8897	0·9184	0·9465	0·9738
20	0·8119	0·8157	0·8266	0·8431	0·8633	0·8857	0·9090	0·9326	0·9558	0·9783

$n_1 = 6$

$\nu \backslash \rho$	0·0	0·1	0·2	0·3	0·4	0·5	0·6	0·7	0·8	0·9
1	0·0591	0·1330	0·2293	0·3268	0·4242	0·5213	0·6180	0·7142	0·8100	0·9052
2	0·1177	0·1742	0·2633	0·3566	0·4504	0·5439	0·6368	0·7289	0·8201	0·9105
3	0·1756	0·2200	0·3008	0·3886	0·4781	0·5674	0·6561	0·7438	0·8304	0·9158
4	0·2323	0·2681	0·3403	0·4222	0·5067	0·5915	0·6756	0·7587	0·8406	0·9210
5	0·2875	0·3168	0·3810	0·4567	0·5359	0·6158	0·6953	0·7736	0·8506	0·9261
6	0·3408	0·3653	0·4221	0·4915	0·5653	0·6402	0·7148	0·7883	0·8605	0·9311
8	0·4411	0·4586	0·5027	0·5601	0·6230	0·6878	0·7526	0·8167	0·8794	0·9406
10	0·5316	0·5445	0·5786	0·6251	0·6777	0·7327	0·7882	0·8431	0·8969	0·9493
12	0·6116	0·6213	0·6475	0·6848	0·7280	0·7739	0·8206	0·8671	0·9127	0·9571
14	0·6811	0·6884	0·7086	0·7380	0·7730	0·8108	0·8497	0·8885	0·9268	0·9640
16	0·7404	0·7459	0·7614	0·7845	0·8126	0·8433	0·8752	0·9073	0·9391	0·9700
18	0·7904	0·7946	0·8064	0·8244	0·8467	0·8714	0·8973	0·9236	0·9497	0·9752
20	0·8320	0·8352	0·8443	0·8582	0·8756	0·8953	0·9162	0·9375	0·9588	0·9797

$n_1 = 7$

$\nu \backslash \rho$	0·0	0·1	0·2	0·3	0·4	0·5	0·6	0·7	0·8	0·9
1	0·0624	0·1341	0·2299	0·3271	0·4245	0·5215	0·6181	0·7143	0·8100	0·9053
2	0·1244	0·1779	0·2655	0·3580	0·4514	0·5446	0·6373	0·7292	0·8204	0·9106
3	0·1853	0·2268	0·3051	0·3916	0·4802	0·5689	0·6572	0·7445	0·8308	0·9160
4	0·2449	0·2779	0·3470	0·4270	0·5101	0·5940	0·6774	0·7599	0·8413	0·9213
5	0·3026	0·3294	0·3901	0·4633	0·5408	0·6194	0·6978	0·7754	0·8517	0·9266
6	0·3582	0·3803	0·4335	0·4999	0·5715	0·6448	0·7181	0·7906	0·8619	0·9317
8	0·4617	0·4774	0·5179	0·5718	0·6319	0·6944	0·7574	0·8199	0·8814	0·9415
10	0·5540	0·5654	0·5962	0·6391	0·6885	0·7408	0·7941	0·8471	0·8994	0·9504
12	0·6343	0·6428	0·6662	0·7000	0·7400	0·7830	0·8273	0·8717	0·9155	0·9584
14	0·7030	0·7093	0·7271	0·7535	0·7854	0·8203	0·8567	0·8934	0·9297	0·9654
16	0·7608	0·7655	0·7790	0·7995	0·8246	0·8527	0·8822	0·9122	0·9420	0·9714
18	0·8087	0·8123	0·8225	0·8383	0·8580	0·8802	0·9039	0·9282	0·9526	0·9766
20	0·8480	0·8507	0·8585	0·8706	0·8859	0·9034	0·9223	0·9418	0·9614	0·9809

TABLE AA
Four-Place Common Logarithms of Numbers

N	0	1	2	3	4	5	6	7	8	9
10	0000	0043	0086	0128	0170	0212	0253	0294	0334	0374
11	0414	0453	0492	0531	0569	0607	0645	0682	0719	0755
12	0792	0828	0864	0899	0934	0969	1004	1038	1072	1106
13	1139	1173	1206	1239	1271	1303	1335	1367	1399	1430
14	1461	1492	1523	1553	1584	1614	1644	1673	1703	1732
15	1761	1790	1818	1847	1875	1903	1931	1959	1987	2014
16	2041	2068	2095	2122	2148	2175	2201	2227	2253	2279
17	2304	2330	2355	2380	2405	2430	2455	2480	2504	2529
18	2553	2577	2601	2625	2648	2672	2695	2718	2742	2765
19	2788	2810	2833	2856	2878	2900	2923	2945	2967	2989
20	3010	3032	3054	3075	3096	3118	3139	3160	3181	3201
21	3222	3243	3263	3284	3304	3324	3345	3365	3385	3404
22	3424	3444	3464	3483	3502	3522	3541	3560	3579	3598
23	3617	3636	3655	3674	3692	3711	3729	3747	3766	3784
24	3802	3820	3838	3856	3874	3892	3909	3927	3945	3962
25	3979	3997	4014	4031	4048	4065	4082	4099	4116	4133
26	4150	4166	4183	4200	4216	4232	4249	4265	4281	4298
27	4314	4330	4346	4362	4378	4393	4409	4425	4440	4456
28	4472	4487	4502	4518	4533	4548	4564	4579	4594	4609
29	4624	4639	4654	4669	4683	4698	4713	4728	4742	4757
30	4771	4786	4800	4814	4829	4843	4857	4871	4886	4900
31	4914	4928	4942	4955	4969	4983	4997	5011	5024	5038
32	5051	5065	5079	5092	5105	5119	5132	5145	5159	5172
33	5185	5198	5211	5224	5237	5250	5263	5276	5289	5302
34	5315	5328	5340	5353	5366	5378	5391	5403	5416	5428
35	5441	5453	5465	5478	5490	5502	5514	5527	5539	5551
36	5563	5575	5587	5599	5611	5623	5635	5647	5658	5670
37	5682	5694	5705	5717	5729	5740	5752	5763	5775	5786
38	5798	5809	5821	5832	5843	5855	5866	5877	5888	5899
39	5911	5922	5933	5944	5955	5966	5977	5988	5999	6010
40	6021	6031	6042	6053	6064	6075	6085	6096	6107	6117
41	6128	6138	6149	6160	6170	6180	6191	6201	6212	6222
42	6232	6243	6253	6263	6274	6284	6294	6304	6314	6325
43	6336	6345	6355	6365	6375	6385	6395	6405	6415	6425
44	6435	6444	6454	6464	6474	6484	6493	6503	6513	6522
45	6532	6542	6551	6561	6571	6580	6590	6599	6609	6618
46	6628	6637	6646	6656	6665	6675	6684	6693	6702	6712
47	6721	6730	6739	6749	6758	6767	6776	6785	6794	6803
48	6812	6821	6830	6839	6848	6857	6866	6875	6884	6893
49	6902	6911	6920	6928	6937	6946	6955	6964	6972	6981
50	6990	6998	7007	7016	7024	7033	7042	7050	7059	7067
51	7076	7084	7093	7101	7110	7118	7126	7135	7143	7152
52	7160	7168	7177	7185	7193	7202	7210	7218	7226	7235
53	7243	7251	7259	7267	7275	7284	7292	7300	7308	7316
54	7324	7332	7340	7348	7356	7364	7372	7380	7388	7396

TABLE AA *(concluded)*

N	0	1	2	3	4	5	6	7	8	9
55	7404	7412	7419	7427	7435	7443	7451	7459	7466	7474
56	7482	7490	7497	7505	7513	7520	7528	7536	7543	7551
57	7559	7566	7574	7582	7589	7597	7604	7612	7619	7627
58	7634	7642	7649	7657	7664	7672	7679	7686	7694	7701
59	7709	7716	7723	7731	7738	7745	7752	7760	7767	7774
60	7782	7789	7796	7803	7810	7818	7825	7832	7839	7846
61	7853	7860	7868	7875	7882	7889	7896	7903	7910	7917
62	7924	7931	7938	7945	7952	7959	7966	7973	7980	7987
63	7993	8000	8007	8014	8021	8028	8035	8041	8048	8055
64	8062	8069	8075	8082	8089	8096	8102	8109	8116	8122
65	8129	8136	8142	8149	8156	8162	8169	8176	8182	8189
66	8195	8202	8209	8215	8222	8228	8235	8241	8248	8254
67	8261	8267	8274	8280	8287	8293	8299	8306	8312	8319
68	8325	8331	8338	8344	8351	8357	8363	8370	8376	8382
69	8388	8395	8401	8407	8414	8420	8426	8432	8439	8445
70	8451	8457	8463	8470	8476	8482	8488	8494	8500	8506
71	8513	8519	8525	8531	8537	8543	8549	8555	8561	8567
72	8573	8579	8585	8591	8597	8603	8609	8615	8621	8627
73	8633	8639	8645	8651	8657	8663	8669	8675	8681	8686
74	8692	8698	8704	8710	8716	8722	8727	8733	8739	8745
75	8751	8756	8762	8768	8774	8779	8785	8791	8797	8802
76	8808	8814	8820	8825	8831	8837	8842	8848	8854	8859
77	8865	8871	8876	8882	8887	8893	8899	8904	8910	8915
78	8921	8927	8932	8938	8943	8949	8954	8960	8965	8971
79	8976	8982	8987	8993	8998	9004	9009	9015	9020	9025
80	9031	9036	9042	9047	9053	9058	9063	9069	9074	9079
81	9085	9090	9096	9101	9106	9112	9117	9122	9128	9133
82	9138	9143	9149	9154	9159	9165	9170	9175	9180	9186
83	9191	9196	9201	9206	9212	9217	9222	9227	9232	9238
84	9243	9248	9253	9258	9263	9269	9274	9279	9284	9289
85	9294	9299	9304	9309	9315	9320	9325	9330	9335	9340
86	9345	9350	9355	9360	9365	9370	9375	9380	9385	9390
87	9395	9400	9405	9410	9415	9420	9425	9430	9435	9440
88	9445	9450	9455	9460	9465	9469	9474	9479	9484	9489
89	9494	9499	9504	9509	9513	9518	9523	9528	9533	9538
90	9542	9547	9552	9557	9562	9566	9571	9576	9581	9586
91	9590	9595	9600	9605	9609	9614	9619	9624	9628	9633
92	9638	9643	9647	9652	9657	9661	9666	9671	9675	9680
93	9685	9689	9694	9699	9703	9708	9713	9717	9722	9727
94	9731	9736	9741	9745	9750	9754	9759	9763	9768	9773
95	9777	9782	9786	9791	9795	9800	9805	9809	9814	9818
96	9823	9827	9832	9836	9841	9845	9850	9854	9859	9863
97	9868	9872	9877	9881	9886	9890	9894	9899	9903	9908
98	9912	9917	9921	9926	9930	9934	9939	9943	9948	9952
99	9956	9961	9965	9969	9974	9978	9983	9987	9991	9996

TABLE BB
Squares of Numbers

N	0	1	2	3	4	5	6	7	8	9
100	10000	10201	10404	10609	10816	11025	11236	11449	11664	11881
110	12100	12321	12544	12769	12996	13225	13456	13689	13924	14161
120	14400	14641	14884	15129	15376	15625	15876	16129	16384	16641
130	16900	17161	17424	17689	17956	18225	18496	18769	19044	19321
140	19600	19881	20164	20449	20736	21025	21316	21609	21904	22201
150	22500	22801	23104	23409	23716	24025	24336	24649	24964	25281
160	25600	25921	26244	26569	26896	27225	27556	27889	28224	28561
170	28900	29241	29584	29929	30276	30625	30976	31329	31684	32041
180	32400	32761	33124	33489	33856	34225	34596	34969	35344	35721
190	36100	36481	36864	37249	37636	38025	38416	38809	39204	39601
200	40000	40401	40804	41209	41616	42025	42436	42849	43264	43681
210	44100	44521	44944	45369	45796	46225	46656	47089	47524	47961
220	48400	48841	49284	49729	50176	50625	51076	51529	51984	52441
230	52900	53361	53824	54289	54756	55225	55696	56169	56644	57121
240	57600	58081	58564	59049	59536	60025	60516	61009	61504	62001
250	62500	63001	63504	64009	64516	65025	65536	66049	66564	67081
260	67600	68121	68644	69169	69696	70225	70756	71289	71824	72361
270	72900	73441	73984	74529	75076	75625	76176	76729	77284	77841
280	78400	78961	79524	80089	80656	81225	81796	82369	82944	83521
290	84100	84681	85262	85849	86436	87025	87616	88209	88804	89401
300	90000	90601	91204	91809	92416	93025	93636	94249	94864	95481
310	96100	96721	97344	97969	98596	99225	99856	100489	101124	101761
320	102400	103041	103684	104329	104976	105625	106276	106929	107584	108241
330	108900	109561	110224	110889	111556	112225	112896	113569	114244	114921
340	115600	116281	116964	117649	118336	119025	119716	120409	121104	121801
350	122500	123201	123904	124609	125316	126025	126736	127449	128164	128881
360	129600	130321	131044	131769	132496	133225	133956	134689	135424	136161
370	136900	137641	138384	139129	139876	140625	141376	142129	142884	143641
380	144400	145161	145924	146689	147456	148225	148996	149769	150544	151321
390	152100	152881	153664	154449	155236	156025	156816	157609	158404	159201
400	160000	160801	161604	162409	163216	164025	164836	165649	166464	167281
410	168100	168921	169744	170569	171396	172225	173056	173889	174724	175561
420	176400	177241	178084	178929	179776	180625	181476	182329	183184	184041
430	184900	185761	186624	187489	188356	189225	190096	190969	191844	192721
440	193600	194481	195364	196249	197136	198025	198916	199809	200704	201601
450	202500	203401	204304	205209	206116	207025	207936	208849	209764	210681
460	211600	212521	213444	214369	215296	216225	217156	218089	219024	219961
470	220900	221841	222784	223729	224676	225625	226576	227529	228484	229441
480	230400	231361	232324	233289	234256	235225	236196	237169	238144	239121
490	240100	241081	242064	243049	244036	245025	246016	247009	248004	249001
500	250000	251001	252004	253009	254016	255025	256036	257049	258064	259081
510	260100	261121	262144	263169	264196	265225	266256	267289	268324	269361
520	270400	271441	272484	273529	274576	275625	276676	277729	278784	279841
530	280900	281961	283024	284089	285156	286225	287296	288369	289444	290521
540	291600	292681	293764	294849	295936	297025	298116	299209	300304	301401

TABLE BB *(concluded)*

N	0	1	2	3	4	5	6	7	8	9
550	302500	303601	304704	305809	306916	308025	309136	310249	311364	312481
560	313600	314721	315844	316969	318096	319225	320356	321489	322624	323761
570	324900	326041	327184	328329	329476	330625	331776	332929	334084	335241
580	336400	337561	338724	339889	341056	342225	343396	344569	345744	346921
590	348100	349281	350464	351649	352836	354025	355216	356409	357604	358801
600	360000	361201	362404	363609	364816	366025	367236	368449	369664	370881
610	372100	373321	374544	375769	376996	378225	379456	380689	381924	383161
620	384400	385641	386884	388129	389376	390625	391876	393129	394384	395641
630	396900	398161	399424	400689	401956	403225	404496	405769	407044	408321
640	409600	410881	412164	413449	414736	416025	417316	418609	419904	421201
650	422500	423801	425104	426409	427716	429025	430336	431649	432964	434281
660	435600	436921	438244	439569	440896	442225	443556	444889	446224	447561
670	448900	450241	451584	452929	454276	455625	456976	458329	459684	461041
680	462400	463761	465124	466489	467856	469225	470596	471969	473344	474721
690	476100	477481	478864	480249	481636	483025	484416	485809	487204	488601
700	490000	491401	492804	494209	495616	497025	498436	498849	501264	502681
710	504100	505521	506944	508369	509796	511225	512656	514089	515524	516961
720	518400	519841	521284	522729	524176	525625	527076	528529	529984	531441
730	532900	534361	535824	537289	538756	540225	541696	543169	544644	546121
740	547600	549081	550564	552049	553536	555025	556516	558009	559504	561001
750	562500	564001	565504	567009	568516	570025	571536	573049	574564	576081
760	577600	579121	580644	582169	583696	585225	586756	588289	589824	591361
770	592900	594441	595984	597529	599076	600625	602176	603729	605284	606841
780	608400	609961	611524	613089	614656	616225	617796	619369	620944	622521
790	624100	625681	627264	628849	630436	632025	633616	635209	636804	638401
800	640000	641601	643204	644809	646416	648025	649636	651249	652864	654481
810	656100	657721	659344	660969	662596	664225	665856	667489	669124	670761
820	672400	674041	675684	677329	678976	680625	682276	683929	685584	687241
830	688900	690561	692224	693889	695556	697225	698896	700569	702244	703921
840	705600	707281	708964	710649	712336	714025	715716	717409	719104	720801
850	722500	724201	725904	727609	729316	731025	732736	734449	736164	737881
860	739600	741321	743044	744769	746496	748225	749956	751689	753424	755161
870	756900	758641	760384	762129	763876	765625	767376	769129	770884	772641
880	774400	776161	777924	779689	781456	783225	784996	786769	788544	790321
890	792100	793881	795664	797449	799236	801025	802816	804609	806404	808201
900	810000	811801	813604	815409	817216	819025	820836	822649	824464	826281
910	828100	829921	831744	833569	835396	837225	839056	840889	842724	844561
920	846400	848241	850084	851929	853776	855625	857476	859329	861184	863041
930	864900	866761	868624	870489	872356	874225	876096	877969	879844	881721
940	883600	885481	887364	889249	891136	893025	894916	896809	898704	900601
950	902500	904401	906304	908209	910116	912025	913936	915849	917764	919681
960	921600	923521	925444	927369	929296	931225	933156	935089	937024	938961
970	940900	942841	944784	946729	948676	950625	952576	954529	956484	958441
980	960400	962361	964324	966289	968256	970225	972196	974169	976144	978121
990	980100	982081	984064	986049	988036	990025	992016	994009	996004	998001

TABLE CC
Logarithms of Factorials

	0	1	2	3	4	5	6	7	8	9
00	0.0000	0.0000	0.3010	0.7782	1.3802	2.0792	2.8573	3.7024	4.6055	5.5598
10	6.5598	7.6012	8.6803	9.7943	10.9404	12.1165	13.3206	14.5511	15.8063	17.0851
20	18.3861	19.7083	21.0508	22.4125	23.7927	25.1906	26.6056	28.0370	29.4841	30.9465
30	32.4237	33.9150	35.4202	36.9387	38.4702	40.0142	41.5705	43.1387	44.7185	46.3096
40	47.9116	49.5244	51.1477	52.7811	54.4246	56.0778	57.7406	59.4127	61.0939	62.7841
50	64.4831	66.1906	67.9066	69.6309	71.3633	73.1037	74.8519	76.6077	78.3712	80.1420
60	81.9202	83.7055	85.4979	87.2972	89.1034	90.9163	92.7359	94.5619	96.3945	98.2333
70	100.0784	101.9297	103.7870	105.6503	107.5196	109.3946	111.2754	113.1619	115.0540	116.9516
80	118.8547	120.7632	122.6770	124.5961	126.5204	128.4498	130.3843	132.3238	134.2683	136.2177
90	138.1719	140.1310	142.0948	144.0632	146.0364	148.0141	149.9964	151.9831	153.9744	155.9700
100	157.9700	159.9743	161.9829	163.9958	166.0128	168.0340	170.0593	172.0887	174.1221	176.1595
110	178.2009	180.2462	182.2955	184.3485	186.4054	188.4661	190.5306	192.5988	194.6707	196.7462
120	198.8254	200.9082	202.9945	205.0844	207.1779	209.2748	211.3751	213.4790	215.5862	217.6967
130	219.8107	221.9280	224.0485	226.1724	228.2995	230.4298	232.5634	234.7001	236.8400	238.9830
140	241.1291	243.2783	245.4306	247.5860	249.7443	251.9057	254.0700	256.2374	258.4076	260.5808
150	262.7569	264.9359	267.1177	269.3024	271.4899	273.6803	275.8734	278.0693	280.2679	282.4693
160	284.6735	286.8803	289.0898	291.3020	293.5168	295.7343	297.9544	300.1771	302.4024	304.6303
170	306.8608	309.0938	311.3293	313.5674	315.8079	318.0509	320.2965	322.5444	324.7948	327.0477
180	329.3030	331.5606	333.8207	336.0832	338.3480	340.6152	342.8847	345.1565	347.4307	349.7071
190	351.9859	354.2669	356.5502	358.8358	361.1236	363.4136	365.7059	368.0003	370.2970	372.5959

TABLE CC *(concluded)*

	0	1	2	3	4	5	6	7	8	9
200	374.8969	377.2001	379.5054	381.8129	384.1226	386.4343	388.7482	391.0642	393.3822	395.7024
210	398.0246	400.3489	402.6752	405.0036	407.3340	409.6664	412.0009	414.3373	416.6758	419.0162
220	421.3587	423.7031	426.0494	428.3977	430.7480	433.1002	435.4543	437.8103	440.1682	442.5281
230	444.8898	447.2534	449.6189	451.9862	454.3555	456.7265	459.0994	461.4742	463.8508	466.2292
240	468.6094	470.9914	473.3752	475.7608	478.1482	480.5374	482.9283	485.3210	487.7154	490.1116
250	492.5096	494.9093	497.3107	499.7138	502.1186	504.5252	506.9334	509.3433	511.7549	514.1682
260	516.5832	518.9999	521.4182	523.8381	526.2597	528.6830	531.1078	533.5344	535.9625	538.3922
270	540.8236	543.2566	545.6912	548.1273	550.5651	553.0044	555.4453	557.8878	560.3318	562.7774
280	565.2246	567.6733	570.1235	572.5753	575.0287	577.4835	579.9399	582.3977	584.8571	587.3180
290	589.7804	592.2443	594.7097	597.1766	599.6449	602.1147	604.5860	607.0588	609.5330	612.0087
300	614.4858	616.9644	619.4444	621.9258	624.4087	626.8930	629.3787	631.8659	634.3544	636.8444
310	639.3357	641.8285	644.3226	646.8182	649.3151	651.8134	654.3131	656.8142	659.3166	661.8204
320	664.3255	666.8320	669.3399	671.8491	674.3596	676.8715	679.3847	681.8993	684.4152	686.9324
330	689.4509	691.9707	694.4918	697.0143	699.5380	702.0631	704.5894	707.1170	709.6460	712.1762
340	714.7076	717.2404	719.7744	722.3097	724.8463	727.3841	729.9232	732.4635	735.0051	737.5479
350	740.0920	742.6373	745.1838	747.7316	750.2806	752.8308	755.3823	757.9349	760.4888	763.0439
360	765.6002	768.1577	770.7164	773.2764	775.8375	778.3997	780.9632	783.5279	786.0937	788.6608
370	791.2290	793.7983	796.3689	798.9406	801.5135	804.0875	806.6627	809.2390	811.8165	814.3952
380	816.9749	819.5559	822.1379	824.7211	827.3055	829.8909	832.4775	835.0652	837.6540	840.2440
390	842.8351	845.4272	848.0205	850.6149	853.2104	855.8070	858.4047	861.0035	863.6034	866.2044

TABLE DD
Random Numbers III

22 17 68 65 84	68 95 23 92 35	87 02 22 57 51	61 09 43 95 06	58 24 82 03 47
19 36 27 59 46	13 79 93 37 55	39 77 32 77 09	85 52 05 30 62	47 83 51 62 74
16 77 23 02 77	09 61 87 25 21	28 06 24 25 93	16 71 13 59 78	23 05 47 47 25
78 43 76 71 61	20 44 90 32 64	97 67 63 99 61	46 38 03 93 22	69 81 21 99 21
03 28 28 26 08	73 37 32 04 05	69 30 16 09 05	88 69 58 28 99	35 07 44 75 47
93 22 53 64 39	07 10 63 76 35	87 03 04 79 88	08 13 13 85 51	55 34 57 72 69
78 76 58 54 74	92 38 70 96 92	52 06 79 79 45	82 63 18 27 44	69 66 92 19 09
23 68 35 26 00	99 53 93 61 28	52 70 05 48 34	56 65 05 61 86	90 92 10 70 80
15 39 25 70 99	93 86 52 77 65	15 33 59 05 28	22 87 26 07 47	86 96 98 29 06
58 71 96 30 24	18 46 23 34 27	85 13 99 24 44	49 18 09 79 49	74 16 32 23 02
57 35 27 33 72	24 53 63 94 09	41 10 76 47 91	44 04 95 49 66	39 60 04 59 81
48 50 86 54 48	22 06 34 72 52	82 21 15 65 20	33 29 94 71 11	15 91 29 12 03
61 96 48 95 03	07 16 39 33 66	98 56 10 56 79	77 21 30 27 12	90 49 22 23 62
36 93 89 41 26	29 70 83 63 51	99 74 20 52 36	87 09 41 15 09	98 60 16 03 03
18 87 00 42 31	57 90 12 02 07	23 47 37 17 31	54 08 01 88 63	39 41 88 92 10
88 56 53 27 59	33 35 72 67 47	77 34 55 45 70	08 18 27 38 90	16 95 86 70 75
09 72 95 84 29	49 41 31 06 70	42 38 06 45 18	64 84 73 31 65	52 53 37 97 15
12 96 88 17 31	65 19 69 02 83	60 75 86 90 68	24 64 19 35 51	56 61 87 39 12
85 94 57 24 16	92 09 84 38 76	22 00 27 69 85	29 81 94 78 70	21 94 47 90 12
38 64 43 59 98	98 77 87 68 07	91 51 67 62 44	40 98 05 93 78	23 32 65 41 18
53 44 09 42 72	00 41 86 79 79	68 47 22 00 20	35 55 31 51 51	00 83 63 22 55
40 76 66 26 84	57 99 99 90 37	36 63 32 08 58	37 40 13 68 97	87 64 81 07 83
02 17 79 18 05	12 59 52 57 02	22 07 90 47 03	28 14 11 30 79	20 69 22 40 98
95 17 82 06 53	31 51 10 96 46	92 06 88 07 77	56 11 50 81 69	40 23 72 51 39
35 76 22 42 92	96 11 83 44 80	34 68 35 48 77	33 42 40 90 60	73 96 53 97 86
26 29 13 56 41	85 47 04 66 08	34 72 57 59 13	82 43 80 46 15	38 26 61 70 04
77 80 20 75 82	72 82 32 99 90	63 95 73 76 63	89 73 44 99 05	48 67 26 43 18
46 40 66 44 52	91 36 74 43 53	30 82 13 54 00	78 45 63 98 35	55 03 36 67 68
37 56 08 18 09	77 53 84 46 47	31 91 18 95 58	24 16 74 11 53	44 10 13 85 57
61 65 61 68 66	37 27 47 39 19	84 83 70 07 48	53 21 40 06 71	95 06 79 88 54
93 43 69 64 07	34 18 04 52 35	56 27 09 24 86	61 85 53 83 45	19 90 70 99 00
21 96 60 12 99	11 20 99 45 18	48 13 93 55 34	18 37 79 49 90	65 97 38 20 46
95 20 47 97 97	27 37 83 28 71	00 06 41 41 74	45 89 09 39 84	51 67 11 52 49
97 86 21 78 73	10 65 81 92 59	58 76 17 14 97	04 76 62 16 17	17 95 70 45 80
69 92 06 34 13	59 71 74 17 32	27 55 10 24 19	23 71 82 13 74	63 52 52 01 41
04 31 17 21 56	33 73 99 19 87	26 72 39 27 67	53 77 57 68 93	60 61 97 22 61
61 06 98 03 91	87 14 77 43 96	43 00 65 98 50	45 60 33 01 07	98 99 46 50 47
85 93 85 86 88	72 87 08 62 40	16 06 10 89 20	23 21 34 74 97	76 38 03 29 63
21 74 32 47 45	73 96 07 94 52	09 65 90 77 47	25 76 16 19 33	53 05 70 53 30
15 69 53 82 80	79 96 23 53 10	65 39 07 16 29	45 33 02 43 70	02 87 40 41 45
02 89 08 04 49	20 21 14 68 86	87 63 93 95 17	11 29 01 95 80	35 14 97 35 33
87 18 15 89 79	85 43 01 72 73	08 61 74 51 69	89 74 39 82 15	94 51 33 41 67
98 83 71 94 22	59 97 50 99 52	08 52 85 08 40	87 80 61 65 31	91 51 80 32 44
10 08 58 21 66	72 68 49 29 31	89 85 84 46 06	59 73 19 85 23	65 09 29 75 63
47 90 56 10 08	88 02 84 27 83	42 29 72 23 19	66 56 45 65 79	20 71 53 20 25
22 85 61 68 90	49 64 92 85 44	16 40 12 89 88	50 14 49 81 06	01 82 77 45 12
67 80 43 79 33	12 83 11 41 16	25 58 19 68 70	77 02 54 00 52	53 43 37 15 26
27 62 50 96 72	79 44 61 40 15	14 53 40 65 39	27 31 58 50 28	11 39 03 34 25
33 78 80 87 15	38 30 06 38 21	14 47 47 07 26	54 96 87 53 32	40 36 40 96 76
13 13 92 66 99	47 24 49 57 74	32 25 43 62 17	10 97 11 69 84	99 63 22 32 98

SOURCE: Table DD is reprinted with permission from Random Numbers III and IV of Table XXXIII of R. A. Fisher and F. Yates, *Statistical Tables for Biological, Agricultural and Medical Research* (Edinburgh: Oliver & Boyd, Ltd., 1953).

TABLE DD *(concluded)*

Random Numbers IV

10 27 53 96 23	71 50 54 36 23	54 31 04 82 98	04 14 12 15 09	26 78 25 47 47
28 41 50 61 88	64 85 27 20 18	83 36 36 05 56	39 71 65 09 62	94 76 62 11 89
34 21 42 57 02	59 19 18 97 48	80 30 03 30 98	05 24 67 70 07	84 97 50 87 46
61 81 77 23 23	82 82 11 54 08	53 28 70 58 96	44 07 39 55 43	42 34 43 39 28
61 15 18 13 54	16 86 20 26 88	90 74 80 55 09	14 53 90 51 17	52 01 63 01 59
91 76 21 64 64	44 91 13 32 97	75 31 62 66 54	84 80 32 75 77	56 08 25 70 29
00 97 79 08 06	37 30 28 59 85	53 56 68 53 40	01 74 39 59 73	30 19 99 85 48
36 46 18 34 94	75 20 80 27 77	78 91 69 16 00	08 43 18 73 68	67 69 61 34 25
88 98 99 60 50	65 95 79 42 94	93 62 40 89 96	43 56 47 71 66	46 76 29 67 02
04 37 59 87 21	05 02 03 24 17	47 97 81 56 51	92 34 86 01 82	55 51 33 12 91
63 62 06 34 41	94 21 78 55 09	72 76 45 16 94	29 95 81 83 83	79 88 01 97 30
78 47 23 53 90	34 41 92 45 71	09 23 70 70 07	12 38 92 79 43	14 85 11 47 23
87 68 62 15 43	53 14 36 59 25	54 47 33 70 15	59 24 48 40 35	50 03 42 99 36
47 60 92 10 77	88 59 53 11 52	66 25 69 07 04	48 68 64 71 06	61 65 70 22 12
56 88 87 59 41	65 28 04 67 53	95 79 88 37 31	50 41 06 94 76	81 83 17 16 33
02 57 45 86 67	73 43 07 34 48	44 26 87 93 29	77 09 61 67 84	06 69 44 77 75
31 54 14 13 17	48 62 11 90 60	68 12 93 64 28	46 24 79 16 76	14 60 25 51 01
28 50 16 43 36	28 97 85 58 99	67 22 52 76 23	24 70 36 54 54	59 28 61 71 96
63 29 62 66 50	02 63 45 52 38	67 63 47 54 75	83 24 78 43 20	92 63 13 47 48
45 65 58 26 51	76 96 59 38 72	86 57 45 71 46	44 67 76 14 55	44 88 01 62 12
39 65 36 63 70	77 45 85 50 51	74 13 39 35 22	30 53 36 02 95	49 34 88 73 61
73 71 98 16 04	29 18 94 51 23	76 51 94 84 86	79 93 96 38 63	08 58 25 58 94
72 20 56 20 11	72 65 71 08 86	79 57 95 13 91	97 48 72 66 48	09 71 17 24 89
75 17 26 99 76	89 37 20 70 01	77 31 61 95 46	26 97 05 73 51	53 33 18 72 87
37 48 60 82 29	81 30 15 39 14	48 38 75 93 29	06 87 37 78 48	45 56 00 84 47
68 08 02 80 72	83 71 46 30 49	89 17 95 88 29	02 39 56 03 46	97 74 06 56 17
14 23 98 61 67	70 52 85 01 50	01 84 02 78 43	10 62 98 19 41	18 83 99 47 99
49 08 96 21 44	25 27 99 41 28	07 41 08 34 66	19 42 74 39 91	41 96 53 78 72
78 37 06 08 43	63 61 62 42 29	39 68 95 10 96	09 24 23 00 62	56 12 80 73 16
37 21 34 17 68	68 96 83 23 56	32 84 60 15 31	44 73 67 34 77	91 15 79 74 58
14 29 09 34 04	87 83 07 55 07	76 58 30 83 64	87 29 25 58 84	86 50 60 00 25
58 43 28 06 36	49 52 83 51 14	47 56 91 29 34	05 87 31 06 95	12 45 57 09 09
10 43 67 29 70	80 62 80 03 42	10 80 21 38 84	90 56 35 03 09	43 12 74 49 14
44 38 88 39 54	86 97 37 44 22	00 95 01 31 76	17 16 29 56 63	38 78 94 49 81
90 69 59 19 51	85 39 52 85 13	07 28 37 07 61	11 16 36 27 03	78 86 72 04 95
41 47 10 25 62	97 05 31 03 61	20 26 36 31 62	68 69 86 95 44	84 95 48 46 45
91 94 14 63 19	75 89 11 47 11	31 56 34 19 09	79 57 92 36 59	14 93 87 81 40
80 06 54 18 66	09 18 94 06 19	98 40 07 17 81	22 45 44 84 11	24 62 20 42 31
67 72 77 63 48	84 08 31 55 58	24 33 45 77 58	80 45 67 93 82	75 70 16 08 24
59 40 24 13 27	79 26 88 86 30	01 31 60 10 39	53 58 47 70 93	85 81 56 39 38
05 90 35 89 95	01 61 16 96 94	50 78 13 69 36	37 68 53 37 31	71 26 35 03 71
44 43 80 69 98	46 68 05 14 82	90 78 50 05 62	77 79 13 57 44	59 60 10 39 66
61 81 31 96 82	00 57 25 60 59	46 72 60 18 77	55 66 12 62 11	08 99 55 64 57
42 88 07 10 05	24 98 65 63 21	47 21 61 88 32	27 80 30 21 60	10 92 35 36 12
77 94 30 05 39	28 10 99 00 27	12 73 73 99 12	49 99 57 94 82	96 88 57 17 91
78 83 19 76 16	94 11 68 84 26	23 54 20 86 85	23 86 66 99 07	36 37 34 92 09
87 76 59 61 81	43 63 64 61 61	65 76 36 95 90	18 48 27 45 68	27 23 65 30 72
91 43 05 96 47	55 78 99 95 24	37 55 85 78 78	01 48 41 19 10	35 19 54 07 73
84 97 77 72 73	09 62 06 65 72	87 12 49 03 60	41 15 20 76 27	50 47 02 29 16
87 41 60 76 83	44 88 96 07 80	83 05 83 38 96	73 70 66 81 90	30 56 10 48 59

APPENDIX III
Glossary of Symbols[1]

Note: *No attempt is made to list sample statistics and universe parameters separately. In the text, universe parameters are generally distinguished by attaching a prime to a symbol; sample statistics are unprimed. In analysis of variance, Greek letters are used for the most part as universe values; and sample estimates of these may be distinguished by "hats" on the Greek letters.*

A = arbitrary origin to which a variable is referred to simplify computations.
A = in the construction of an $\bar{X}$-chart, A is the factor $3/\sqrt{n}$.
A_1 = in the construction of an $\bar{X}$-chart, A_1 is the factor $3/c_2\sqrt{n}$.
A_2 = in the construction of an $\bar{X}$-chart, A_2 is the factor $3/d_2\sqrt{n}$.
A_i = the ith class of the A classification.
A, B, C, etc. = main effects in the analysis of variance; also used as coefficients of p_1, p_2, p_3, etc., in a linear equation.
A, B, C, K = matrices involved in solution of least-squares equations.
a = AD/RMSD; it is a measure of kurtosis.
a = combination of factors $A, B, C, \ldots, K$ in a 2^k experiment in which only A occurs at the high level; similarly for $b, c, \ldots, k$.
$a_i^A, a_j^B, a_{ij}^{AB}$, etc. = terms in an analysis of variance model.
$a_{1.2}, a_{1.23}, \ldots, a_{i.jk\ldots p}$ = constant terms in regression equations in which X_1 (in general, X_i) is the dependent variable.
α = probability of rejection; called producer's risk, level of significance, and risk of an error of the first kind.
AD = average deviation.
AOQ = average outgoing quality.
AOQL = average outgoing quality limit.
APL = acceptable process level.
AQL = acceptable quality level.
ARL = average run length.
ASN = average sample number.
ATI = average total inspection.
B_j = the jth class of the B classification.
$b_{12}, b_{12.3}, \ldots, b_{ij.k\ldots p}$ = coefficients of X_2 (in general, X_j) in regression equations in which X_1 (in general, X_i) is the dependent variable.
$B_{12}, B_{12.3}, \ldots, B_{ij.k\ldots p}$ = hypothetical values for $b'_{12}, b'_{12.3}, \ldots, b'_{ij.k\ldots p}$.
$b_1, b_2, b_{11}, b_{12}, b_{22}$ = coefficients of the variables and their powers in a second-degree polynomial.
β = probability of acceptance; also called risk of an error of the second kind; frequently an alternative to the symbol P_a; when $p' = p'_t$, β is the "consumers risk." The "power of a test" equals $1 - \beta$.
C_r^n = number of combinations of n things, r at a time.
CSP = continuous sampling plan.

[1] Some symbols with a limited range of usage in the text are not listed here.

c = number of nonconformities or nonconforming units; often used as the acceptance number in a single-sampling plan.
c_1 = the acceptance number for the first sample of a double-sampling plan.
c_2 = the final acceptance number of a double-sampling plan.
c_1 = unit cost of tests of sample I.
c_2 = unit cost of tests of sample II.
c_2 = an adjustment factor used to estimate the standard deviation of a universe; estimate of $\sigma' = \mathbf{RMSD}/c_2$.
$\bar{c}$ = average number of nonconforming units or average number of nonconformities in a sample of several inspection units.
c_{ij} = special quantities used in solving least-squares equations. These are the elements in the inverse matrix.
γ_1 and γ_2 = coefficients of skewness and kurtosis, respectively.
D_1 = in the construction of an R-chart, D_1 is the factor $(d_2 - 3\sigma'_w)$.
D_2 = in the construction of an R-chart, D_2 is the factor $(d_2 + 3\sigma'_w)$.
D_3 = in the construction of an R-chart, D_3 is the factor $(1 - 3\sigma'_w/d_2)$.
D_4 = in the construction of an R-chart, D_4 is the factor $(1 + 3\sigma'_w/d_2)$.
$D_{0.001}$ = in the construction of an R-chart, $D_{0.001}$ is the factor $w_{0.001}/d_2$.
$D_{0.999}$ = in the construction of an R-chart, $D_{0.999}$ is the factor $w_{0.999}/d_2$.
D_i = divisor of Z_i^2 to get contribution of an individual comparison to the total sum of squares.
d = in tests of differences between means, d equal $\bar{X}_1 - \bar{X}_2$.
d = lead distance on a **V**-mask.
d'_0 = a hypothetical value for d'.
d_2 = an adjustment factor used to estimate the standard deviation of a universe; estimate of $\sigma' = \bar{R}/d_2$. By definition, d_2 equals $\bar{w}$.
d_2^* = an adjustment factor for use with average range of g groups of m each. Approaches d_2 as g approaches infinity.
d_3 = reciprocal of σ'_w.
df = degrees of freedom.
δ = a difference that is deemed important to detect.
δ = a slope differential in analysis of covariance.
E = an operator signifying "mean value of."
E_{xx} and E_{yy} = error sum of squares in X and Y. Sometimes distinguished for special situations by ${}_CE_{xx}$, ${}_CE_{yy}$, ${}_{RC}E_{xx}$, and ${}_{RC}E_{yy}$.
E_{xy} = error cross-product sum. ${}_CE_{xy}$, etc., pertain to special situations.
e = the constant 2.718+; it is the limit of $\left(1 + \frac{1}{X}\right)^X$ as X approaches infinity.
ϵ = a random variable, usually assumed to be normally distributed with zero mean and unit standard deviation.
F = frequency.
F_j = frequency of the jth class.
$f(X)$ = density function for the random variable X.
$F(X)$ = cumulative probability function for the random variable X.

F = in sampling, the symbol F is often identified with the ratio s_1^2/s_2^2 or $s_1^2\sigma_2'^2/s_2^2\sigma_1'^2$, because this statistic is distributed in the form of an F distribution.

F_α = the α point of an F distribution.

f = fraction of product sampled in a continuous sampling plan.

f_1, f_2 = fractions of the lot sampled in a sequential rectifying inspection.

g = number of groups or sets or, in a three-way analysis of variance, the number in a group.

$G(r_{12})$ = an unbiased estimate of r'_{12}. See Chapter 32, Section 8.2.

h = decision interval for a cusum chart.

h = a constant to be used in the analysis of a mixed model in analysis of variance.

h_1 = the lower intercept of the line of acceptance on a sequential-sampling chart.

h_2 = the upper intercept of the line of rejection on a sequential-sampling chart.

θ = a column differential in an analysis of variance.

θ = half the angle of a **V**-mask.

i = the class interval of a frequency distribution.

i = used as a subscript to differentiate a variable.

i = length of run of conforming items required in a continuous sampling plan.

$I'_{i.j}$ = an index of nonlinear correlation.

$I'_{i.jk}$ = an index of multiple (nonlinear) correlation.

$I(R^2_{i.jk...p})$ = unbiased estimate of $R'^2_{i.jk...p}$

K_x, K_y, K_{xy} = correction factors in analysis of variance and covariance.

k = a critical value for a normal variable; e.g., a critical value in a variables sampling plan.

k = used as a subscript to differentiate a quantity.

k = in a c-chart or nonconformities-per-unit sampling plan, the number of inspection units.

k = the number of levels in a multilevel continuous sampling plan.

k = reference value for a cusum chart.

k_1 = per unit cost of inspection.

k_2 = cost of passing a nonconforming unit. It may be the cost, for example, to dismantle, repair, reassemble, and test an assembly that fails because a nonconforming part was put into the production line.

L = lower specification limit.

LCL = lower control limit on a control chart.

LQ = limiting quality.

$L(m)$ = the ARL at m.

L_a = the ARL at m_a.

L_r = the ARL at m_r.

$l(\,)$ = loss function.

λ = the abscissa for OC curves.

λ = the process average number of nonconformities per unit area of opportunity, used when that average is subject to random variation.

λ_r = the lot break-even number of nonconformities per unit area of opportunity.

Λ_0 = a statistic used in testing for general homogeneity of normal data.

M, H, L, etc. = letters used in a capital-letter display in exploratory data analysis. The *M* here refers to the median and is the same as the *Mi* listed below.

M = a critical value for the sample percent nonconforming estimated from $\bar{X}$ and *s*.

M_1 = a hypothetical value for $\bar{X}'_1$.

MSD = maximum value for *s* in a variables sampling plan.

MAR = maximum value for the average sample range in a variables sampling plan.

Mi = median; the value of the middle case or the average of the two middle cases when arranged in order of size.

Mo = mode or most typical value.

m = number of items in a group when there are *g* groups.

m_a = an acceptable quality level for a process mean.

m_r = a barely tolerable (or rejectable) quality level for a process mean.

m_i = expected value of *i*th cell in a χ^2 test.

m'_i = alternative expected value.

$m_x(t)$ = the moment generating function of *X*.

μ = constant term in an analysis of variance model.

μ_k = the *k*th moment around the mean.

μ_{ak} = the *k*th moment around an arbitrating origin *A*.

μ_{ok} = the *k*th moment around zero.

n = number of cases; size of a sample.

N = size of lot or finite universe.

N! = *N* factorial; means $N(N-1)(N-2)\ldots 1$.

ν = degrees of freedom. (This is a Greek nu and is to be carefully distinguished from an English italic lower case *v*.)

ξ = deviation from an arbitrary origin ($= X - A$). If data are grouped, it is measured in class interval units $\left(=\frac{X-A}{i}\right)$.

0! = zero factorial; equals 1.

OC = operating characteristic.

p = proportion or fraction. p' is lot fraction or process average.

p = the number of variables in a regression equation.

p_a = in tests of hypotheses, the acceptance limit for *p*.

p_r = the lot break-even fraction of nonconforming units, used when discussing a stream of lots from a process.

p_{La} = in tests of hypotheses, the lower acceptance limit for *p*.

p_{Ua} = in tests of hypotheses, the upper acceptance limit for *p*.

p_i = one of a set of orthogonal polynomials.
p'_0 = the hypothetical value of p' that is being tested.
p'_t = lot tolerance fraction nonconforming.
p'_1 = a value of p' for which the probability of acceptance is high.
p'_2 = a value of p' for which the probability of acceptance is low.
$\bar{p}$ = the weighted average of several sample fractions nonconforming.
$\hat{p}$ = an estimate of p'.
p'' = a standard value for p' to be used in setting up a p-chart.
P_a = probability of acceptance.
P_r^n = number of permutations of n things, r at a time.
$P(A)$ or Prob. (A) = probability of A.
$P(A|B)$ = probability of A, given B; a conditional probability.
$P(AB)$ = probability of A and B; a joint probability.
Q_1 and Q_3 = the lower and upper quartiles.
Q = the semiquartile range [$= (Q_3 - Q_1)/2$].
Q = a component of a sum of squares.
Q_U and Q_L = quality indexes for upper and lower specification limits in a variables sampling plan.
q = a variable that has a studentized range distribution.
$q = 1 - p$.
R = the range; equals $X_H - X_L$.
R_a = the acceptance grade in a sampling-inspection plan based on R.
$\bar{R}$ = mean of several sample ranges.
$\bar{\bar{R}}$ = mean of $\bar{R}$'s for several sets of samples.
RMSD = the root-mean-square deviation.
RPL = rejectable process level.
RSS = reduced sum of squares in regression analysis and analysis of covariance.
r = number of rows in an analysis of variance.
r = the (smaller) number of cases above or below an average.
r_{ij} = the coefficient of correlation between X_i and X_j; measures linear correlation.
$\tilde{r}_{12}$, $\tilde{r}_{ij}$ = special estimates of r'_{12} and r'_{ij}. See Chapter 32, Section 8.2.
$R_{1.23} \ldots R_{i.jk...p}$ = the multiple correlation coefficient between X_1 (or, in general, X_i), on the one hand, and a linear function of X_2 and X_3 (in general, $X_j, X_k \ldots X_p$), on the other hand.
P = equals $\bar{p}/p_r$, the ratio of the average fraction nonconforming for the process to the lot break-even quality level. It also may equal λ/λ_r, the ratio of the average number of nonconformities per unit area of opportunity for the process to the lot break-even quality level.
s = the (larger) number of cases above or below an average.
s = the slope of the lines of acceptance and rejection on a sequential-sampling chart.
s = the sample standard deviation.
s_1 = the standard deviation of set 1 of data.

s = a parameter of the beta distribution and beta-binomial distribution; also a parameter of the gamma-Poisson distribution.
s_a = in a sampling plan, the acceptance limit for s.
s^2 = the sample variance.
$s^2_{1.2}, s^2_{1.23}, \ldots, s^2_{i.jk\ldots p}$ = unbiased estimates of $\sigma'^2_{1.2}, \sigma'^2_{1.23}, \ldots, \sigma'^2_{i.jk\ldots p}$.
s^2_C = column mean square in an analysis of variance.
s^2_R = row mean square in an analysis of variance.
s^2_{RC} = interaction mean square in an analysis of variance.
s^2_e = error mean square in analysis of variance.
$s^2_{\text{res.}}$ = residual mean square in an analysis of variance.
$s_{y.x}$ = estimate of standard error of estimate for regression of Y on X.
SPR = sequential probability ratio.
Σ = an operator signifying "sum of."
σ' = the universe standard deviation.
σ'^2 = the universe variance; $\sigma'^2 = \int(X - \bar{X}')^2 f(X)dX$.
$\hat{\sigma}$ = estimate of σ'.
σ'_0 = the hypothetical value of σ' that is being tested.
σ'_p = standard error of p.
$\hat{\sigma}_p$ = an estimate of σ'_p.
σ'_c = standard error of c.
$\hat{\sigma}_c$ = an estimate of σ'_c.
$\sigma'_{\bar{X}}$ = standard error of $\bar{X}$.
$\hat{\sigma}_X$ = an estimate of σ'_X.
σ'_w = standard deviation of w.
σ'_R = the standard deviation of R.
$\hat{\sigma}_R$ = an estimate of σ'_R.
σ'_s = standard error of the standard deviation.
$\sigma'_{p_1-p_2}$ = standard error of $p_1 - p_2$.
$\hat{\sigma}_{p_1-p_2}$ = an estimate of $\sigma'_{p_1-p_2}$.
$\sigma'_{\bar{X}_1}$ = standard error of $\bar{X}_1$.
$\sigma'_{\bar{X}_1-\bar{X}_2}$ = standard error of the difference between two means.
σ'_d = same as $\sigma'_{\bar{X}_1-\bar{X}_2}$.
$\hat{\sigma}_d$ = an estimate of σ'_d.
$\sigma'^2_A, \sigma'^2_{AB}$, etc., = component of variance due to main effect A, to interaction AB, etc.
$\sigma'_{b_{12}}, \sigma'_{b_{12.3}}, \ldots, \sigma'_{b_{ij.k\ldots p}}$ = standard errors of $b_{12}, b_{12.3}, \ldots, b_{ij.k\ldots p}$.
$\sigma'_{1.2}, \sigma'_{1.23}, \ldots, \sigma'_{i.jk\ldots p}$ = standard deviations of the deviations from the universe regressions X_1 on X_2, X_1 on X_2 and X_3, ... X_i on X_j, X_k ... X_p; standard errors of estimate.
$\sigma'_{y.x}$ = standard error of estimate for regression of Y on X.
σ'_1 = a value of the variability of a quality characteristic for which the probability of acceptance is high. Also the standard deviation of universe I.
σ'_2 = a value of the variability of a quality characteristic for which the

probability of acceptance is low. Also the standard deviation of universe II.

σ'' = standard value for σ' to be used in setting up variables control charts.

σ''_R = standard value for σ'_R based on the standard for σ'.

t = usually used to represent the deviation of a variable from its mean divided by an estimate of its true or universe standard error based on sample data.

t = a parameter of the beta distribution and beta-binomial distribution.

t_α = α point of a t distribution; area beyond t_α in one direction is α.

τ = a differential row element in analysis of variance.

U = upper specification limit.

$\bar{u}$ = the sample average number of nonconformities per unit when samples vary in size.

u = average number of pieces inspected under screening in a continuous sampling plan.

$u_{1.2}$, $u_{1.23}$, . . . , $u_{i.jk...p}$ = deviations from the regression of X_1 on X_2, X_1 on X_2 and X_3, . . . , X_i on X_j, X_k . . . X_p.

v = in a continuous sampling plan the average number of pieces inspected under sampling before a nonconformity is found.

W_i = $Z_i/\sqrt{D_i}$ = an "orthogonalized" comparison in analysis of variance.

w = relative range; equals R/σ'.

w = the scale factor for a cusum chart.

w_r = the fraction of the process prior distribution of lots for which the lot fraction nonconforming exceeds the lot break-even quality level.

X = a variable.

X_i = the ith variable or ith value of X.

X_H = highest value in a set.

X_L = lowest value in a set.

$(X/N)_r$ = lot break-even fraction nonconforming.

$\bar{X}$ = mean of X; also used as the grand mean.

$\bar{\bar{X}}$ = grand mean; weighted mean of several means.

$\bar{X}_j$ = mean of the jth column.

$\bar{X}'_0$ = the hypothetical value of $\bar{X}'$ that is being tested.

$\bar{X}_a$ = in a sampling plan, the acceptance limit for $\bar{X}$.

$\bar{X}_{La}$ = in a sampling plan, the lower acceptance limit for $\bar{X}$.

$\bar{X}_{Ua}$ = in a sampling plan, the upper acceptance limit for $\bar{X}$.

$\bar{X}'_1$ = a value of the average of a quality characteristic for which the probability of acceptance is high. Also the mean of universe I.

$\bar{X}'_2$ = a value of the average of a quality characteristic for which the probability of acceptance is low. Also the mean of universe II.

$\bar{X}'_{U2}$ = an upper value of $\bar{X}'_2$.

$\bar{X}'_{L2}$ = a lower value of $\bar{X}'_2$.

$\bar{X}''$ = a standard value for $\bar{X}'$ to be used in setting up an $\bar{X}$-chart.

X_{1r} = the regression value for X_1.

x = deviation of X from its mean; equals $X - \bar{X}$.
Y = another variable like X.
y = a value used to compute the AOQL of a sampling plan.
z = a variable that has a normal distribution with zero mean and unit standard deviation.
z_1 = a value of z for which the probability of acceptance is high.
z_2 = a value of z for which the probability of acceptance is low.
Z_i = a variable in a multiple regression equation.
Z_i = a particular comparison or contrast in an analysis of variance. Has one degree of freedom.
z_i = a variable in a multiple regression equation measured from its own mean.
$z_{p'}$ = the normal deviate that is exceeded by $100p'$ percent of the deviates in a specified direction. (Similarly for z_α, z_β, etc.)
z_L = the normal deviate for the lower limit L.
Z_U = the normal deviate for the upper limit U.
φ = a quantity related to λ in the determination of risks of errors of the second kind in an analysis of variance and comparison of two variances. See Chapter 29, Section 1.4.
χ^2 = usually used to represent a sample sum of squares divided by a universe variance.
χ^2_α = the α point of a χ^2 distribution.
ψ = an interaction element in an analysis of variance.

APPENDIX IV
Glossary of Special Technical Terms

acceptance control chart. a chart used for acceptance of a process.

accuracy. refers to the deviation of an estimate from the true value. Note that this is not the same as precision. May also mean bias.

alias. an element which is confounded with another.

attribute sampling. sampling in which the characteristic determined is simply a quality or attribute.

average run length (ARL) curve. for a control chart that is used to monitor a process, the ARL curve shows the average number of samples that will be taken before a specified change in the process is detected.

Bayesian sampling plan. a sampling plan that is based on a prior distribution of lot quality.

beta distribution. a particular distribution often used as a prior distribution in Bayesian procedures. (For formula, see p. 322.)

beta-binomial distribution. If a process is in control at a constant proportion of nonconforming units p, the proportion of nonconforming units in random samples from the process will follow a binomial distribution. If p varies, however, in accordance with a beta distribution, the proportion of nonconforming units in random samples from the process will follow a beta-binomial distribution. (For formulas, see p. 323.)

bias. difference between mean value of a sample statistic and the universe parameter it is used to estimate.

BIPP. binomial probability paper. (See Chapter 23, Section 4.)

block. a group of tests or experiments, often a complete replication of a set of unit experiments.

box plot. a chart of the median, upper, and lower hinges is represented by a box, the width of which is arbitrary and has no significance. By extending lines from the hinges to the extremes we obtain a "box and whiskers plot."

coding. means referring a variable to an arbitrary origin and possibly expressing it in some other unit.

column mean square. the column sum of squares divided by its degrees of freedom.

column sum of squares. the sum of the squares of the deviations of the column means from their grand mean weighted by the number of cases in a column.

completely randomized experiment. an experiment in which other factors than that being studied are assigned in a random manner.

composite design. a design used in mapping a response surface.

compressed limit gauging. gauging to artificially narrow limits to get a more efficient sampling plan.

conditional probability. $P(A|B)$.

confidence coefficient. the chance that a confidence interval has of including the universe value.

confidence interval. an interval that has a designated chance (the confidence coefficient) of including the universe value.

confidence limits. the end points of a confidence interval.

confidence loci. the loci that give confidence limits for X'_{ir} as a function of the independent variable or variables.

confounding. confusing two or more factors so their effects cannot be separated.

consistent estimate. When the sample size is continuously increased, a consistent estimate of a universe parameter has a decreasing probability of deviating from the universe value by any preassigned amount.

consumer's risk. the chance of lot tolerance quality being accepted.

continuous sampling plan. a plan applied to continuous production.

controlled process. a process the control chart for which shows no points outside the limits and no nonrandom variation within the limits.

correction factor. an adjustment factor usually associated with the measurement of cases from an origin other than their mean.

covariance. variation together; more technically
$\iint (X - \bar{X})(Y - \bar{Y}) f(X, Y) dX\, dY$.

cusum chart. a chart that shows cumulative sums of sample data.

defining contrast(s). the comparison(s) that are selected as the basis for confounding in fractional replication.

degrees of freedom. the number of "free" elements associated with a sum or some other function. (See Chapter 27, Section 2; Chapter 28, Section 1; and Chapter 29, Section 1.3.)

density function. a function giving the rate at which the cumulative probability is increasing at a given point.

depth. in a stem and leaf display used in exploratory data analysis, the cumulative number of leaves on a given stem and the number of leaves on stems closer to the nearer end of the batch. The depth of the median is $(n + 1)/2$.

duplicates. results of unit experiments produced under the same conditions; used whether these number two or more.

efficiency of a statistic. If a statistic (X) is approximately normally distributed in large samples with a mean equal approximately to the parameter being estimated and a sampling variance equal to σ'^2_X, its efficiency is the ratio that the sampling variance of the maximum likelihood estimate bears to σ'^2_X in large samples. Maximum likelihood estimates have 100 percent efficiency in large samples.

efficiency of an experimental design. If the error mean square of a given experimental design is less than that of another design applied to the same experimental material, the efficiency of the first is greater than that of the second.

error mean square. equals residual mean square when residual variation is associated with experimental error.

error of the first kind. the error of rejecting a hypothesis when it is true. Sometimes called Type I error.

error of the second kind. the error of accepting a hypothesis when it is not true. Sometimes called Type II error.

error rate. rate at which errors of the first kind will be allowed.

expected value. the mean value of a variable.

***F* distribution.** a particular type of sampling distribution.

factorial experiment. an experiment in which the levels of each factor are combined with all levels of every other factor.

fences. In exploratory data analysis inner fences are defined as the lower hinge − (1.5 × H-spread) and the upper hinge + (1.5 × H-spread). Outer fences are defined as the lower hinge − (3 × H-spread) and the upper hinge + (3 × H-spread).

fixed factors, fixed variables, and fixed effects. factors and variables and their effects that are the same for all samples.

fractional factorial experiment. a factorial experiment in which not all combinations of the factors are employed.

gamma distribution. a particular distribution used in this text as a prior distribution for the process number of nonconformities per unit area of opportunity. (For formulas, see footnote 16, p. 323.)

gamma-Poisson distribution. If a process is in control at a constant number of nonconformities per unit area of opportunity, the number of nonconformities in random samples of n unit areas from the process will follow a Poisson distribution. If the Poisson parameter λ varies, however, in accordance with a gamma distribution, the number of nonconformities in random samples of n unit areas from the process will follow a gamma-Poisson distribution. (For formulas, see p. 323.)

Gauss-Doolittle method. a special procedure for solving least-squares equations.

Graeco-Latin square. similar to a Latin square, except that results are grouped in three ways.

heteroscedastic. having nonuniform scatter.

hierarchal model. *see* **nested model.**

hinges. in exploratory data analysis, values that show the location of the upper and lower quarters of the data. The depth of a hinge equals (Depth of median + 1)/2 so that the hinges may be somewhat closer to the median than the quartiles of the data.

H-spread. spread between the upper and lower hinges of a set of data.

histogram. a bar diagram representing a frequency distribution.

homoscedastic. having uniform scatter.

independence. If A and B are independent, then $P(AB) = P(A)P(B)$. Also for density functions, $f(x,y) = f_1(x)f_2(y)$.

individual degrees of freedom. components that make up a total sum of squares such that each has only one degree of freedom.

interaction. means dependence; in analysis of variance it means the tendency for the combination of factors *A* and *B* to produce a result that is different from the mere sum of their two individual contributions.

interaction mean square. equals the interaction sum of squares divided by its degrees of freedom.

interaction sum of squares. equals the sum of the squares of the deviations of the means of cells from some expected value made up of a constant plus an independent contribution by factor *A* plus an independent contribution by factor *B*.

joint probability. *P(AB)*.

kurtosis. degree of peakedness.

Latin square. an experimental design in which the effects of one factor are grouped according to levels of two other factors, the levels of the first factor being assigned at random, with the restriction that no one level of the first factor will appear more than once with any given level of either of the other two factors.

least-squares equations. the equations used to derive the line or curve estimated by the method of least squares. Also called normal equations.

least-squares method. *see* **method of least squares.**

letter value display. in exploratory data analysis a display of the median, "hinges" (quarters), eighths, etc., of the data, represented by the capital letters M, H, E, D, etc., together with the midpoints of the upper and lower values for the same capital letter and the spreads between these upper and lower values.

linear approximation. a first-degree equation representing a line, a plane, or hyperplane.

lot break-even quality level. for attribute inspection, the quality of a lot for which the acceptance of the lot without inspection costs the same as 100 percent inspection. (See pp. 318.)

lot plot method. a scheme of acceptance sampling based on a sample frequency distribution.

marginal probability. the probability of an item in a row or column as contrasted with the probability of an item in a single cell in the row or column.

maximum likelihood estimate. an estimate of a universe parameter that, if it were the true value, would maximize the probability of the sample on which the estimate is based.

mean square for rows. the row sum of squares divided by its degrees of freedom.

method of least squares. a method of estimating a function (a point, a line, a curve) that minimizes the sum of the squares of the deviations from the estimate.

mixed model. an analysis of variance model in which one or more factors are fixed and one or more are random.

mixed variables-attributes sampling plan. If a plot is not accepted under a given variables plan, it is submitted to an attributes sampling plan.

modified control limits. control limits based on specification limits. See Chapter 23, Section 1.

moment. The kth moment around origin A is the expected value of the k-powered deviations from A. "Moment coefficient" would be a more precise term.

multiple comparison. comparison between members of a set.

nested model. an analysis of variance model in which the effects of certain factors are pyramided. "Levels" of "lower" factors are "nested" in levels of factors higher up the pyramid but do not cut across the levels of the higher factor.

normal. unless otherwise stated, refers to the property of being distributed in the form of a normal frequency distribution.

normal equations. same as least-squares equations.

nuisance variable. a variable that is correlated with the dependent variable in an analysis of variance and for the effects of which the dependent variable must be adjusted.

one-tail test. a test of a hypothesis that has only one critical value, either an upper or lower acceptance limit.

operating characteristic curve or OC curve. the curve that gives the probability of acceptance as a function of some universe value.

orthogonal. literally, perpendicular or at right angles. An experimental design is "orthogonal" if the effects of one factor balance out (sum to zero) across the effects of other factors.

orthogonal polynomials. polynomials that are orthogonal to each other in the sense that for the values of the variable for which they are evaluated they individually sum to zero and their cross-products also sum to zero.

orthogonal transformation. a transformation that rotates axes but maintains their perpendicular relationship. To be strictly orthogonal the coefficients of the transformation (i.e., elements in the matrix of the transformation) must be such that when the coefficients (elements) in any row are squared they sum to one, and when they are cross-multiplied row by row they sum to zero.

outlier. an extreme individual or extreme mean.

outside value. a data value that lies beyond an inner fence. If a data value lies beyond an outer fence, it is referred to as a "far outside value."

parameter. a constant or coefficient of a universe that describes some characteristic of its distribution.

partial confounding. an experimental design in which a comparison (main effect or interaction) is confounded with block effects in one but not all replications of the experiment.

permutation. an arrangement.

pooling. usually means combining sums or sums of squares to get a combined mean or combined mean square.

posterior probability. a conditional probability for the existence of a specified universe given a certain result.

power function. equals 1 − ordinate of OC curve.

power of a test. equals $1 - \beta$.
precision. refers to the standard deviation of a variable; the smaller the standard deviation, the higher the precision.
prediction limits. limits for the actual value of X_i associated with given values of the independent variables.
prior probability. probability of existence of a given universe.
probability. the relative frequency of objects or things in a given class of a probability set.
probability distribution. a distribution of the relative frequencies of a random variable.
probability set. the set of objects or things for which probabilities are computed, divided in a specified way into a mutually exclusive and exhaustive group of classes.
process average. the average fraction nonconforming, or mean value of a quality characteristic, of the output of a process.
process capabilities. actual or potential capabilities of the process with respect to meeting specifications. Interest may center on the standard deviation of the process, ultimately on the percent nonconforming.
producer's risk. the chance of a lot being rejected. Usually pertains to lots of AQL quality.
random factors. factors in an analysis of variance the values of which are randomly selected in each sample from a universe of such values.
random numbers. usually refer to numbers from some table of random sampling numbers.
randomized block experiment. an experiment in which the treatments to be compared are assigned at random to positions in each of several blocks of replications. In a complete block experiment each block contains all treatments.
randomness. an intuitive concept referring to a condition of disorder and unpredictability of individual results.
rectifying inspection. an inspection, 100 percent or otherwise, that aims to improve the quality of the lots or process submitted for inspection.
regression. the tendency for the mean value of one variable for given values of one or more other variables to vary with the other variables.
regression equation. the function that indicates how the average value of one variable for given values of one or more other variables varies with these other variables.
replication. a whole experiment. If repeated k times, there are k replications. A complete "block" is a replication. Sometimes called a replicate.
residual mean square. equals the residual sum of squares divided by its degrees of freedom.
residual sum of squares. the residual after the column, row, and other special sums of squares are subtracted from the total sum of squares.
response function or surface. the functional relationship that exists between some physical output or economic outcome and the productive factors.

robust test. usually refers to a sample test that is not seriously affected by moderate departures of the universe from the normal form.

rotable design. a design used in the mapping of response surfaces. A pentagonal or hexagonal layout in two dimensions is an example.

row sum of squares. the sum of the squares of the deviations of the row means from their grand mean weighted by the number of cases in a row.

sample. a set of objects or things from a larger set called a universe. *Unless otherwise specified, all samples are assumed to be random samples.*

sample space. the totality of all possible results. Used in discussions of probability.

sampling distribution. distribution of a given statistic in the set of all possible samples from a given universe.

sampling plan. a plan that designates sample size(s) and acceptance criteria.

sampling scheme. a coordinated set of sampling plans that may be applied to attain a given objective.

sensitivity. refers to the power of a test or experiment to detect deviations from some hypothetical value.

significant. means that a result deviates from some hypothetical value by more than can reasonably be attributed to the chance errors of sampling.

skewed. asymmetrical.

standard error. the standard deviation of a sampling distribution.

standard error of estimate. the standard deviation of the deviations from a line, curve, plane, or surface of regression; not to be confused with an ordinary standard error. (See Chapter 32, Section 2.3.)

state of chaos. a term used to describe the character of a process. (See p. 327.)

statistical tolerances. engineering tolerances based on statistical analysis of product variation. Also can mean limits that have a given chance of including a specified percent of the universe. (See Chapter 4, Section 10.2 and Chapter 6, Section 4.)

statistic. a function of sample observations used to estimate a universe parameter.

stem and leaf display. in exploratory data analysis a way of showing the distribution of an ordered set of data.

studentized range. the difference between the highest and lowest values of a sample of X divided by a mean square estimate of the standard deviation of X that is independent of the numerator.

Student's distribution. another name for the t distribution.

subgrouping. dividing a group of items into subgroups. To be of analytical value, the segregation of the items should be such that if assignable causes are present they will be common to all elements of one or more subgroups.

sufficient estimate. an estimate that contains all the information obtainable in a sample about the universe parameter it is used to estimate.

***t* distribution.** a particular type of sampling distribution.

total sum of squares. the sum of the squares of the deviations of the individual items from the mean of all the data.

2^k experiment. a factorial experiment in which k factors each occur at two levels.

two-tail test. a test of a hypothesis that has two critical values, both an upper and a lower acceptance limit.

Type A OC curve. a curve that gives the chance of a lot being accepted in a long series of lots of identical quality.

Type B OC curve. a curve that gives the chance of a lot being accepted in a long series of binomially distributed lots.

unbiased estimator. a statistic the mean of which in a universe of samples is equal to the corresponding parameter of the universe of individual values.

under control. *see* **controlled process.**

universe. the larger set of objects or things from which samples are drawn. Usually assumed to be infinitely large or at least very large relative to the sample.

variables sampling. a sampling in which the characteristic of interest is measured rather than merely classified qualitatively as in sampling by attributes.

variance. the square of the standard deviation.

V-mask. a **V**-shaped mask used with a cusum chart. (See Chapter 22, Section 3.2.)

warning limits. limits on a control chart, usually 2 sigma limits. If these are exceeded, it indicates that more data should be taken immediately for study of the process.

χ^2 distribution. a particular type of sampling distribution.

APPENDIX V
Cumulative List of References

BOOKS, GOVERNMENT STANDARDS, AND THE LIKE*

ABC WORKING GROUP, *Minutes of Meeting.* Mimeographed, 1960–61.

Acceptance Sampling—A Symposium. Washington, D.C.: American Statistical Association, 1950.

ACTON, FORMAN S. *Analysis of Straight-Line Data.* New York: John Wiley & Sons, 1959.

AMERICAN SOCIETY FOR QUALITY CONTROL. *ANSI/ASQC Standard A1–1978—Definitions, Symbols, Formulas, and Tables for Control Charts.*

———. *ANSI/ASQC Standard A2–1978—Terms, Symbols and Definitions for Acceptance Sampling.*

———. *ANSI/ASQC Standard A3–1978—Quality Systems Terminology.*

———. *ANSI/ASQC Standard Z1.1–1985—Guide for Quality Control Charts.*

———. *ANSI/ASQC Standard Z1.2–1985—Control Chart Method of Analyzing Data.*

———. *ANSI/ASQC Standard Z1.3–1985—Control Chart Method of Controlling Quality during Production.*

———. *ANSI/ASQC Standard Z1.4–1981—Sampling Procedures and Tables for Inspection by Attributes.*

———. *ANSI/ASQC Standard Z1.9–1980—Sampling Procedures and Tables for Inspection by Variables for Percent Nonconforming.*

———. *ANSI/ASQC Standard C1–1968 (ANSI Z1.8–1971)—General Requirements for a Quality Program.*

———. *ANSI/ASQC Standard Z1.15–1979—Generic Guide Lines for Quality Systems.*

———. "How To" booklets on technical subjects related to Statistical Quality Control. An updated list is obtainable from ASQC, 230 W. Wells St., Milwaukee, WI 53203.

AMERICAN SOCIETY FOR TESTING AND MATERIALS. *ASTM Manual on Presentation of Data and Control Chart Analysis.* STP No. 15–D. Philadelphia, 1976.

———. *ASTM Manual on Fitting Straight Lines.* STP No. 313, Philadelphia, 1962.

———. *Recommended Practice for Dealing with Outlying Observations* (E178–74), Philadelphia, 1974.

ANDERSON, R. L., AND BANCROFT, T. A. *Statistical Theory in Research.* New York: McGraw-Hill, 1952.

ARMITAGE, P. *Sequential Medical Tests.* Springfield, Ill.: Charles C Thomas, 1961.

AROIAN, LEO A. *Tables and Percentage Points of the Distribution Function of a Product* (for normal variables). Culver City, Calif.: Hughes Aircraft Co., Systems Development Laboratories. (Offset.)

* No attempt has been made to determine whether a possible later edition of a listed book provides as good or better coverage of the material for which the book is offered as a reference. The reader is advised to see if his questions are adequately answered by whatever edition is readily available.

BAZOVSKY, IGOR. *Reliability Theory and Practice.* Englewood Cliffs, N.J.: Prentice-Hall, 1961.

BENNETT, CARL A., AND FRANKLIN, NORMAN L. *Statistical Analysis in Chemistry and the Chemical Industry.* New York: John Wiley & Sons, 1954.

BOWKER, A. H., AND GOODE, H. P. *Sampling Inspection by Variables.* New York: McGraw-Hill, 1952.

______, AND LIEBERMAN, G. L. *Handbook of Industrial Statistics.* Englewood Cliffs, N.J.: Prentice-Hall, 1955.

______, AND ______. *Engineering Statistics.* 2d ed. Englewood Cliffs, N.J.: Prentice-Hall, 1972.

BOWLEY, ARTHUR L. *Elements of Statistics.* London: P. S. King and Son, Ltd., 1926.

BOX, GEORGE E. P., AND JENKINS, GWILYM M., *Time Series Analysis, Forecasting and Control.* San Francisco: Holden-Day, 1970.

BRITISH STANDARDS INSTITUTION, BS 6001, Supplement 1: 1984, *Sampling Plans Indexed by Limiting Quality (LQ).*

BROOKS, S. "Comparison of Methods for Estimating the Optimal Factor Combination" (ScD. thesis). Baltimore: The Johns Hopkins Press, 1955.

BROWNLEE, K. A. *Industrial Experimentation.* 2d rev. American ed. Brooklyn, N.Y.: Chemical Publishing Co., 1948.

______. *Statistical Theory and Methodology in Science and Engineering.* New York: John Wiley and Sons, 1960.

BRUNT, DAVID. *The Combination of Observations.* Cambridge, England: Cambridge University Press, 1917. New York: Macmillan.

BURR, IRVING W. *Statistical Quality Control Methods,* New York: Marcel Dekker, 1976.

______. *Applied Statistical Methods.* New York: Academic Press, 1974.

CHEW, VICTOR (ed.). *Experimental Designs in Industry.* New York: John Wiley & Sons, 1958.

COCHRAN, W. G., AND COX, GERTRUDE M. *Experimental Designs.* 2d ed. New York: John Wiley & Sons, 1957.

COWDEN, DUDLEY J. *Statistical Methods in Quality Control.* Englewood Cliffs, N.J.: Prentice-Hall, 1957.

CRAMER, HAROLD. *Mathematical Methods of Statistics.* Princeton, N.J. Princeton University Press, 1946.

DAVID, F. N. *Tables of the Ordinate and Probability Integral of the Distribution of the Correlation Coefficient in Small Samples.* Cambridge, England: Cambridge University Press, 1938.

DAVIES, OWEN L. (ed.). *The Design and Analysis of Industrial Experiments.* London: Oliver & Boyd, Ltd., 1954.

______. *Statistical Methods in Research and Production with Special Reference to the Chemical Industry.* 3d ed. London: Oliver & Boyd, Ltd., 1957.

DEMING, W. EDWARDS. *Statistical Adjustment of Data.* New York: John Wiley & Sons, 1943.

———. *Some Theory of Sampling.* New York: John Wiley & Sons, 1950.

———. *Out of the Crisis.* Cambridge, Mass: Massachusetts Institute of Technology Center for Advanced Engineering Study, 1986.

DIXON, WILFRED J., AND MASSEY, FRANK J., JR. *Introduction to Statistical Methods.* 3d ed. New York: McGraw-Hill, 1969.

DODGE, HAROLD F. *A General Procedure for Sampling Inspection by Attributes—Based on the AQL Concept.* Technical Report No. 10. Statistics Center, Rutgers: The State University, December 15, 1959.

DODGE, H. F., AND ROMIG, HARRY G. *Sampling Inspection Tables, Single and Double Sampling.* 2d ed. New York: John Wiley & Sons, 1959.

DRAPER, N. R., AND SMITH, H. *Applied Regression Analysis.* New York: John Wiley and Sons, 1967.

DUDDING, B. P., AND JENNETT, W. J. *Quality Control Charts* (British Standard 600 R: 1942). London: British Standards Institution, 1942.

———. *Quality Control Chart Technique When Manufacturing to a Specification,* August 1944. (Distributed in U.S.A. by Gryphon Press, Arlington, Va.)

DUNCAN, DAVID B., AND KENNEY, JOHN F. *On the Solution of Normal Equations and Related Topics.* Ann Arbor, Mich.: Edwards Bros., 1946. (Lithoprinted.)

ENRICK, N. L. *Quality Control and Reliability.* 6th ed. New York: Industrial Press, 1972.

EZEKIEL, MORDECAI AND FOX, KARL A. *Methods of Correlation and Regression Analysis, Linear and Curvilinear,* 3d ed. New York: John Wiley and Sons, 1959.

FEDERER, WALTER T. *Experimental Design, Theory and Application.* New York: Macmillan, 1955.

FELLER, W. *An Introduction to Probability Theory and Its Applications.* 3d ed. New York: John Wiley & Sons, 1968.

FINNEY, D. J. *Experimental Design and Its Statistical Basis.* Chicago: University of Chicago Press, 1955.

FISHER, R. A. *The Design of Experiments.* 3d ed. Edinburgh: Oliver & Boyd, Ltd., 1942.

———. *Statistical Methods and Scientific Inference.* New York: Hafner Publishing Co., 1956.

———. *Statistical Methods for Research Workers.* 13th rev. ed. Edinburgh: Oliver & Boyd, Ltd., 1958.

———, AND YATES, FRANK. *Statistical Tables for Biological, Agricultural and Medical Research.* Edinburgh: Oliver & Boyd, Ltd., 1953.

FREEMAN, H. A. *Industrial Statistics, Statistical Technique Applied to Problems in Industrial Research and Quality Control.* New York: John Wiley and Sons, 1942.

FREUND, JOHN E. *Mathematical Statistics.* Englewood Cliffs, N.J.: Prentice-Hall, 1962.

———. *Modern Elementary Statistics.* 3d ed. Englewood Cliffs, N.J.: Prentice-Hall, 1967.

FRY, T. C. *Probability and Its Engineering Uses.* New York: D. Van Nostrand Co., 1928.

GEARY, R. C., AND PEARSON, E. S. *Tests of Normality.* London: Biometrika Office, 1938.

GENERAL ELECTRIC CO. *Tables of the Individual and Cumulative Terms of Poisson Distribution.* New York: D. Van Nostrand Co., 1962.

GIRSCHICK, M. A. *A Sequential Inspection Plan for Quality Control.* Technical Report No. 16. Stanford University, Calif.: Applied Mathematics and Statistics Laboratory, July 1954.

GOEL, A. L. "A Comparative and Economic Investigation of $\bar{X}$ and Cumulative Sum Control Charts" (Ph.D. thesis). University of Wisconsin, Madison, Wisconsin, 1968, available on microfilm.

GOULDEN, C. H. *Methods of Statistical Analysis.* New York: John Wiley & Sons, 1939.

GRANT, EUGENE L., AND LEAVENWORTH, RICHARD S. *Statistical Quality Control.* 4th ed. New York: McGraw-Hill, 1972.

HALD, A. *Statistical Theory with Engineering Applications.* New York: John Wiley & Sons, 1952.

______, AND THYREGOD, P. *Bayesian Single Sampling Plans Based on Linear Costs and the Poisson Distribution.* The Royal Danish Academy of Sciences and Letters, Matematisk-fysishe Shrifter 3,7. Kobenhavn, Denmark: Munksgaard, 1971.

______. *Statistical Theory of Sampling Inspection by Attributes.* New York: Academic Press, 1981.

HANSEN, BERTRAND L. *Quality Control: Theory and Applications.* Englewood Cliffs, N.J.: Prentice-Hall, 1963.

HARTER, H. LEON, AND LUM, MARY D. *Partially Hierachal Models in the Analysis of Variance.* WADC Technical Report 55–33. Dayton, Ohio: Wright-Patterson Air Force Base, 1955.

HARVARD UNIVERSITY COMPUTING LABORATORY. *Tables of the Cumulative Binomial Probability Distribution.* Cambridge, Mass.: Harvard University Press, 1955.

HASSAN, M. F. "Some Multilevel Continuous Sampling Plans" (Thesis in Central Library), Illinois Institute of Technology, Chicago, Illinois, 1965.

HOEL, P. C. *Introduction to Mathematical Statistics.* 4th ed. New York: John Wiley & Sons, 1971.

HUBER, PETER J. *Robust Statistics.* New York: John Wiley & Sons, 1981.

HUNTER, J. S. "The Box Method of Experimentation" (1957). Dittoed notes for a course.

ISHIKAWA, KAORU, *Guide to Quality Control.* Tokyo: Asian Productivity Organization, 1980. Obtainable in the U.S. from Unipub, 345 Park Avenue South, New York, N.Y. 10016.

INTERNATIONAL ORGANIZATION FOR STANDARDIZATION. *ISO International Standard 2859–1974—Sampling procedures and tables for inspection by attributes.* Geneva, Switzerland. Also obtainable from American National Standards Institute, 1430 Broadway, New York, N.Y. 10018. Revised standard expected in 1986.

______. *ISO International Standard 3951–1981—Sampling procedures and charts for inspection by variables for percent defective.* Geneva, Switzerland. Also obtainable from ANSI; see note on ISO 2859 above.

JOHNSON, NORMAN L., AND LEONE, FRED C. *Statistics and Experimental Design in Engineering and the Physical Sciences.* 2 vols. New York: John Wiley & Sons, 1964.

JOHNSON, PALMER O. *Statistical Methods in Research.* New York: Prentice-Hall, 1949.

JURAN INSTITUTE, INC. 866 United Nations Plaza, New York, N.Y. 10017.

———. JURAN, J. M. (ed.). *Quality Control Handbook.* 3d ed. New York: McGraw-Hill, 1974.

———. The Juran Report. Six numbers have been issued as of the winter of 1986.

———, AND GRYNA, FRANK M., JR. *Quality Planning and Analysis.* 2d ed. New York: McGraw-Hill, 1980.

KEMPTHORNE, OSCAR. *The Design and Analysis of Experiments.* New York: John Wiley & Sons, 1952.

KENDALL, M. G. *The Advanced Theory of Statistics.* 2 vols. London: Charles Griffin & Co., Ltd., 1948. (This has appeared in various revised forms since 1948. Inquire Hafner Publishing Co., New York, N.Y., for latest information.)

KENNEDY, C. W. *Quality Control Methods.* Englewood Cliffs, N.J.: Prentice-Hall, 1948.

KOLMOGOROV, A. N. *Foundations of the Theory of Probability.* New York: Chelsea Publishing Co., 1950.

LIEBERMAN, GERALD J., AND OWENS, DONALD B. *Tables of the Hypergeometric Probability Distribution.* Stanford, Calif.: Stanford University Press, 1961.

LORENZEN, THOMAS J. AND VANCE, LONNIE C. *The Economic Design of Control Charts: A Unified Approach.* (GMR–4792, August 1984) Mathematics Dept., General Motors Research Laboratories: Warren, Mich., 48090.

MAINLAND, DONALD; HERRERA, LEE; AND SUTCLIFFE, MARION I. *Tables for Use with Binomial Samples.* New York: New York University College of Medicine, 1956.

MANDEL, JOHN. *The Statistical Analysis of Experimental Data.* New York: Interscience Publishers, 1964.

MANN, H. B. *Analysis and Design of Experiments.* New York: Dover Publications, 1949.

MEYER, HERBERT A. (ed.). *Symposium on Monte Carlo Methods.* New York: John Wiley and Sons, 1956.

MOLINA, E. C. *Poisson's Exponential Binomial Limit.* New York: D. Van Nostrand Co., 1949.

MOOD, A. M., AND GRAYBILL, F. A. *Introduction to the Theory of Statistics,* 2d ed. New York: McGraw-Hill, 1963.

MOSTELLER, F.; ROURKE, R. E. K.; AND THOMAS, G. B., JR. *Probability with Statistical Applications.* Reading, Mass.: Addison-Wesley Publishing, 1961.

MUNROE, M. E. *The Theory of Probability.* New York: McGraw-Hill, 1951.

Natrella, Mary G. *Experimental Statistics.* National Bureau of Standards Handbook 91, 1963. Obtainable from the Superintendent of Documents, U.S. Government Printing Office, Washington, D. C. Also obtainable from John Wiley and Sons, New York.

Neyman, J. *Lectures and Conferences on Mathematical Statistics.* Washington, D.C.: Graduate School of the United States Department of Agriculture, 1938.

______. *First Course in Probability and Statistics.* New York: Henry Holt & Co., 1950.

Olds, Edwin G. *The Power to Detect a Single Slippage and the Probability of a Type 1 Error for the Upper Three-Sigma Limit Control Chart for Fraction Defective. No Standard Given.* Carnegie Institute of Technology, Department of Mathematics, Technical Report No. 3, 1956. Published by the American Society for Quality Control as *General Publication No. 4.*

Operations Research Center, Massachusetts Institute of Technology. *Notes on Operations Research, 1959.* Cambridge: Technology Press.

OPRD. *Quality Control Reports,* Nos. 1, 3, 8, and 9. See United States War Production Board.

Orsini, Joyce M. "Simple Rule to Reduce the Total Cost of Inspection and Correction of Product in State of Chaos" (Ph.D. thesis). New York University, 1982.

Ostle, Bernard. *Statistics in Research.* Ames, Iowa: Iowa State College Press, 1954.

Ott, Ellis, R. *Process Quality Control, Trouble Shooting and Interpretation of Data.* New York: McGraw-Hill, 1975.

Peach, Paul. *Industrial Statistics and Quality Control.* 2d ed. Raleigh, N.C.: Edwards & Broughton Co., 1947.

Pearson, E. S., and Hartley, H. O. (eds.). *Biometrika Tables for Statisticans,* Vol. I. 2d ed. (1958) and Vol. II (1972). Cambridge, England: Cambridge University Press.

Plackett, R. L. *Regression Analysis.* Oxford: University Press, 1960.

Proschan, Frank and Serfling, R. J. (eds.) *Reliability and Biometry—Statistical Analysis of Lifelength.* Philadelphia: Society for Industrial and Applied Mathematics, 1974.

Rand Corporation. *A Million Random Digits with 100,000 Normal Deviates.* Glencoe, Ill.: The Free Press, 1955.

Resnikoff, George L. *A New Two-Sided Acceptance Region for Sampling by Variables.* Applied Mathematics and Statistics Laboratory, Stanford University, 1952.

______, and Lieberman, G. J. *Tables of the Non-Central t-Distribution.* Stanford, Calif.: Stanford University Press, 1957.

Rice, William B. *Control Charts in Factory Management.* New York: John Wiley & Sons, 1947.

Rickmers, Albert D., and Todd, Hollis, N. *Statistics. An Introduction.* New York: McGraw-Hill, 1967.

Rietz, H. L. *Mathematical Statistics.* Carus Mathematical Monographs, No. 3. Chicago: Open Court Publishing Co., 1927.

ROMIG, H. C. "Allowable Average in Sampling Inspection" (Ph.D. thesis). Columbia University, New York, 1939. Can probably be borrowed from Columbia University Library.

______. *50–100 Binomial Tables.* New York: John Wiley and Sons, 1947.

RUTHERFORD, JOHN G. *Quality Control in Industry—Methods and Systems.* New York: Pitman Publishing Corp., 1948.

SAVAGE, I. RICHARD. *Bibliography of Nonparametric Statistics.* Cambridge, Mass.: Harvard University Press, 1962.

SAVAGE, LEONARD J. *Foundations of Statistics.* New York: John Wiley & Sons, 1954.

SCHEFFÉ, HENRY. *The Analysis of Variance.* New York: John Wiley & Sons, 1959.

SCHILLING, EDWARD G. *Acceptance Sampling in Quality Control.* New York: Marcel Dekker, 1982.

SCHROCK, EDWARD M. *Quality Control and Statistical Methods.* New York: Reinhold Publishing Corp., 1950.

SCHULTZ, HENRY. *The Theory and Measurement of Demand.* Chicago: The University of Chicago Press, 1938.

SEDER, LEONARD A., AND COWAN, DAVID. *The Span Plan Method Process Capability Analysis.* General Publication No. 3. Milwaukee: American Society for Quality Control, 1956.

SHEWHART, W. A. *Economic Control of Quality of Manufactured Product.* New York: D. Van Nostrand Co., 1931.

______. *Statistical Method from the Viewpoint of Quality Control.* Washington, D.C.: Graduate School, Department of Agriculture, 1939.

SIMON, LESLIE E. *An Engineers' Manual of Statistical Methods.* New York: John Wiley & Sons, 1941.

SMITH, EDWARD S. *Control Charts.* New York: McGraw-Hill, 1947.

SMITH, J. G., AND DUNCAN, A. J. *Elementary Statistics and Applications.* New York: McGraw-Hill, 1944.

______, AND ______. *Sampling Statistics and Applications.* New York: McGraw-Hill, 1945.

SNEDECOR, GEORGE W., AND COCHRAN, W. G. *Statistical Methods Applied to Experiments in Agriculture and Biology.* 6th ed. Ames, Iowa: Iowa State College Press, 1967.

STATISTICAL RESEARCH GROUP, COLUMBIA UNIVERSITY. *Sequential Analysis of Statistical Data: Applications.* New York: Columbia University Press, 1945.

______. *Techniques of Statistical Analysis.* New York: McGraw-Hill Book Co., 1947.

______. *Sampling Inspection.* New York: McGraw-Hill, 1948.

TIPPETT, L. H. C. *Technological Applications of Statistics.* New York: John Wiley & Sons, 1950.

TUKEY, JOHN W. "The Problem of Multiple Comparisons." Unpublished dittoed notes. Princeton University, 1953.

______. *Propagation of Errors.* Statistical Research Group, Princeton University, *Technical Reports,* Nos. 10, 11 and 12.

______, *Exploratory Data Analysis.* Reading, Mass.: Addison-Wesley, 1977.

UNITED KINGDOM, QUALITY ASSURANCE DIRECTORATE. Defence Standard 05–30/1 (1973), *Sampling Procedures and Charts for Inspection by Variables.*

UNITED STATES DEPARTMENT OF THE ARMY, CHEMICAL CORPS ENGINEERING AGENCY. ENASR No. PR-7 (1953). *A Method of Discrimination for Single and Double Sampling OC Curves Utilizing The Tangent of the Point of Inflection.*

______. *Master Sampling Plans for Single, Duplicate, Double and Multiple Sampling.* Manual No. 2. Army Chemical Center, Md., 1953.

______. ENASR No. PR-12 (1954). *Method of Fitting and Comparing Variables and Attributes Operating Characteristic Curves Using the Inflection Tangent.*

______. ENASR No. ES-1 (1954). *Inflection Tangents for Fitting and Comparing Sequential Sampling Power Curves with Other Sampling Power Curves.*

UNITED STATES DEPARTMENT OF THE ARMY. Military Standard, *Single and Multilevel Continuous Sampling Procedures and Tables for Inspection by Attributes* (Mil. Std. 1235 [ORD]). Washington, D.C.: U.S. Government Printing Office, 1962.

UNITED STATES DEPARTMENT OF COMMERCE, NATIONAL BUREAU OF STANDARDS. *Fractional Factorial Experiment Designs for Factors at Two Levels.* Applied Mathematics Series No. 48, 1957.

______. *Fractional Factorial Experiment Designs for Factors at Three Levels.* Applied Mathematics Series No. 54, 1959.

______. *Handbook of Mathematical Functions with Formulas, Graphs, and Mathematical Tables.* Applied Mathematics Series 59, 1964.

UNITED STATES DEPARTMENT OF DEFENSE. *Administration of Sampling Procedures for Acceptance Inspection (H 105).* Supply and Logistics Handbook. Washington, D.C.: Office of the Assistant Secretary of Defense (Supply and Logistics), 1954.

______. Military Standard, *Sampling Procedures and Tables for Inspection by Variables for Percent Defective* (Mil. Std. 414). Washington, D.C.: U.S. Government Printing Office, 1957.

______. *Multi-Level Continuous Sampling Procedures and Tables for Inspection by Attributes (H 106).* Inspection and Quality Control Handbook (Interim). Washington, D.C.: Office of the Assistant Secretary of Defense (Supply and Logistics), 1958.

______. *Single Level Continuous Sampling Procedures and Tables for Inspection by Attributes (H 107).* Inspection and Quality Control Handbook (Interim). Washington, D.C. Office of the Assistant Secretary of Defense (Supply and Logistics), 1959.

______. *Evaluation of Contract or Quality Control Systems (H 110).* Quality Control and Reliability Handbook (Interim). Washington, D.C.: Office of the Assistant Secretary of Defense (Supply and Logistics), 1960.

______. Military Standard, *Sampling Procedures and Tables for Inspection by Attributes* (Mil. Std. 105D). Washington, D.C.: U.S. Government Printing Office, 1963. Sometimes referred to as Mil. Std. 105D (ABC).

______. Military Specification, *Quality Control System Requirements* (Mil-Q-9858). Prepared by U.S. Air Force, Headquarters, Air Material Command, Wright-Patterson Air Force Base, Ohio. A copy may be obtained from any military procurement activity.

UNITED STATES DEPARTMENT OF HEALTH AND HUMAN SERVICES, NATIONAL INSTITUTE FOR OCCUPATIONAL SAFETY AND HEALTH. *Dealing with Outlying Observations 556.* Text for course given by the Division of Training and Manpower Development, 1983.

UNITED STATES DEPARTMENT OF THE NAVY, BUREAU OF ORDNANCE. NAVORD OSTD 81 (1952). *Sampling Procedures and Tables for Inspection by Attributes on a Moving Line.*

______. *Mil. Std. 414 Technical Memorandum.*

______. Naval Electronic Systems Command. *Mil. Std. 781C, Reliability Design Qualification and Production Acceptance Tests: Exponential Distribution,* 1977.

VELLEMAN, PAUL F., AND HOAGLIN, DAVID C. *Applications, Basics, and Computing of Exploratory Data Analysis.* Boston: Duxbury Press, 1981.

VON MISES, R. *Probability, Statistics, and Truth.* New York: Macmillan, 1939.

WALD, A. *Sequential Analysis.* New York: John Wiley & Sons, 1947.

WALKER, HELEN M. AND LEV, JOSEPH. *Statistical Inference.* New York: Henry Holt and Co., 1953.

WALLIS, W. A., AND MOORE, G. H. *A Significance Test for Time Series.* National Bureau of Economic Research, Technical Papers, No. 1, 1941.

WEISS, LIONEL. *Statistical Decision Theory.* New York: McGraw-Hill, 1961.

WETHERILL, G. BARRIE. *Sampling Inspection and Quality Control.* London: Methuen; New York: Barnes and Noble, 1969.

WILDE, DOUGLASS J. *Optimum Seeking Methods.* Englewood Cliffs, N.J.: Prentice-Hall, 1964.

WILKS, S. S. *Mathematical Statistics.* Princeton, N.J.: Princeton University Press, 1946. Paper cover, offset.

______. *Elementary Statistical Analysis.* Princeton, N.J.: Princeton University Press, 1948.

______. *Mathematical Statistics.* New York: John Wiley & Sons, 1962.

WILLIAMS, E. J. *Regression Analysis.* New York: John Wiley & Sons, 1959.

WORKING, HOLBROOK. *A Guide to Utilization of the Binomial and Poisson Distributions in Industrial Quality Control.* Stanford, Calif.: Stanford University Press, 1943.

WORKING, HOLBROOK, AND OLDS, EDWIN G. *Organizations Concerned with Statistical Quality Control.* OPRD Quality Control Reports, No. 2.

YATES, F. *The Design and Analysis of Factorial Experiments.* Imperial Bureau of Soil Science, Technical Communications, No. 35. Harpenden, England, 1937.

YULE, G. UDNEY, AND KENDALL, MAURICE G. *An Introduction to the Theory of Statistics.* London: Charles Griffin & Co., Ltd., 1949.

UNITED STATES WAR PRODUCTION BOARD, OFFICE OF PRODUCTION RESEARCH AND DEVELOPMENT (OPRD). *Quality Control* Reports, 1945–.

PERIODICAL ARTICLES, CONFERENCE PAPERS, AND THE LIKE

ABRAHAM, BOVAS, AND BOX, GEORGE E. P. "Sampling Interval and Feedback Control." *Technometrics* 21 (1979), pp. 1–8.

ACTON, F. S., AND OLDS E. G. "Tolerances—Additive or Pythagorean?" *Industrial Quality Control.*" November 1948, pp. 6–12.

ALBRECHT, A., GULDE, H., MACLEAN, A., AND THOMPSON, P. "Continuous Sampling at Minneapolis-Honeywell." *Industrial Quality Control,* September 1955, pp. 4–9.

AMERICAN SOCIETY FOR QUALITY CONTROL. Conference Papers and Convention Transactions, 1946–

ANDREWS, DAVID F. AND PREGIBDON, DARYL. "Finding the Outliers that Matter." *Journal of the Royal Statistical Society* B 40 (1978), pp. 85–93.

ANSCOMBE, F. J. "Tables of Sequential Inspection Schemes to Control Fraction Defective." *Journal of the Royal Statistical Society* A 112 Part II (1949), pp. 180–206.

———. "Rectifying Inspection of a Continuous Output." *Journal of the American Statistical Association* 53 (1958), pp. 702–19.

———. "Rejection of Outliers." *Technometrics* 2 (1960), pp. 123–47.

———. "Rectifying Inspection of Lots." *Journal of the American Statistical Association* 56 (1961), pp. 807–33.

———. "Tests of Goodness of Fit." *Journal of the Royal Statistical Society* B 25 (1963), pp. 81 ff.

———, AND GLYNN, WILLIAM J. "Distribution of the Kurtosis Statistic b, for Normal Samples." *Biometrika* 70 (1983), pp. 227–234.

ARMSTRONG, G. R., AND CLARKE, P. C. "Frequency Distribution vs. Acceptance Table." *Industrial Quality Control,* September 1946, pp. 22–27.

ARMSTRONG, WILLIAM M. "Foundry Application of Quality Control." *Industrial Quality Control,* May 1946, pp. 12–16.

AROIAN, LEO A., AND LEVENE, HOWARD. "The Effectiveness of Quality Control Charts." *Journal of the American Statistical Association* 45 (1950), pp. 520–29.

ASPIN, ALICE A. "Tables for Use in Comparisons Whose Accuracy Involves Two Variances, Separately Estimated." (with Appendix by B. L. Welch) *Biometrika* 36 (1949), pp. 290–96.

BACHE, NIELS. "Approximate Percentage Points of the Distribution of a Product of Independent Positive Random Variables." *Applied Statistics* 28 (1979), pp. 158–62.

BAGSHAW, M., AND JOHNSON, R. A. "The Influence of Reference Values and Estimated Variance on the ARL of Cusum Tests." *Journal of the Royal Statistical Society* B 37 (1975), pp. 413–20.

BAINBRIDGE, T. R. "Staggered, Nested Designs for Estimating Variance Components." *Industrial Quality Control,* July 1965, pp. 12–20.

BANCROFT, T. A., AND HAN, CHIEN-PAI. "A Note on Pooling Variances." *Journal of the American Statistical Association* 76 (1983), pp. 981–3.

BANZHAF, R. A., AND BRUGGER, R. M. "Mil. Std. 1235 (ORD), Single- and Multi-Level Continuous Sampling Procedures and Tables for Inspection by Attributes." *Journal of Quality Technology* 2 (1970), pp. 40–53.

BARKER, THOMAS B. "Quality Engineering by Design—The Taguchi Approach," American Society for Quality Control, *39th Annual Quality Congress Transactions* (1985), pp. 675–88. May be obtained from ASQC, 230 W. Wells St., Milwaukee, WI 53203.

BARNARD, G. A. "Sequential Tests in Industrial Statistics." *Journal of the Royal Statistical Society* B 8 (1946), pp. 1–21.

______. "Control Charts and Stochastic Processes." *Journal of the Royal Statistical Society* B 21 (1959), pp. 239 ff.

BARNETT, V. "Economic Choice of Sample Size for Sampling Inspection Plans." *Applied Statistics* 23 (1974), pp. 149–57.

______. "The Study of Outliers: Purpose and Model." *Applied Statistics* 27 (1978), pp. 242–50. Has sizable list of references.

BARRACLOUGH, ELIZABETH D., AND PAGE, E. S. "Tables for Wald Tests for the Mean of a Normal Distribution." *Biometrika* 46 (1959), pp. 169–77.

BARTLETT, M. S. "Properties of Sufficiency and Statistical Tests." *Proceedings of the Royal Society of London* A 160 (1937) pp. 268–82.

BARTKY, W. "Multiple Sampling with Constant Probability." *Annals of Mathematical Statistics* 14 (1943), pp. 363–77.

BARTON, D. E. "Answer to Query 25—Completed Runs of Length *k* Above and Below Median." *Technometrics* 9 (1967), pp. 682–94.

BASU, D. "Randomization Analysis of Experimental Data: The Fisher Randomization Test," with 5 discussants and a rejoinder. *Journal of the American Statistical Association* 75 (1980), pp. 575–95.

BAUER, PETER, AND HACKL, PETER. "The Use of MOSUMS for Quality Control." *Technometrics* 20 (1978), pp. 431–36.

______, AND ______. "An Extension of the MOSUM Technique for Quality Control." *Technometrics* 22 (1980), pp. 1–7.

BAYES, THOMAS. "An Essay Towards Solving a Problem in the Doctrine of Chances." Reprinted in *Biometrika* 45 (1958), pp. 296–315, from *Philosophical Transactions of the Royal Society of London* (1763).

BEATTIE, D. W. "A Continuous Acceptance Sampling Procedure Based upon a Cumulative Sum Chart for the Number of Defectives." *Applied Statistics* 11 (1962), pp. 137–47.

BECHHOFER, ROBERT E. "A Single-Sample Multiple Decision Procedure for Ranking Means of Normal Populations with Known Variances." *Annals of Mathematical Statistics* 25 (1954), pp. 16–39.

______, DUNNETT, C. W.; AND SOBEL, M. "A Two-Sample Multiple Decision Procedure for Ranking Means of Normal Populations with a Common Unknown Variance." *Biometrika* 41 (1954), pp. 170–76.

BECKMAN, R. J., AND COOK, R. D. "Outlier . . . s," with Discussion and Response. *Technometrics* 25 (1983), pp. 119–63.

BENDER, ARTHUR, JR. "Sampling by Variables to Control the Fraction Defective: Part II." *Journal of Quality Technology* 7 (1975), pp. 139–43.

BENNETT, B. M., AND HSU, P. "On the Power Function of the Exact Test for the 2 × 2 Contingency Table." *Biometrika* 47 (1960), pp. 393–94.

BENJAMINI, YOAV. "Is the t-Test Really Conservative When the Parent Distribution is Long-Tailed?" *Journal of the American Statistical Association* 78 (1983), pp. 645–54.

BERKSON, JOSEPH. "Estimation of a Linear Function for a Calibration Line: Consideration of a Recent Proposal." *Technometrics* 2 (1969), pp. 649–60.

BIDLACK, W.; CLOSE, E. R.; AND WARREN, J. C. "Applications of Quality Control at Bausch and Lomb Optical Co." *Industrial Quality Control,* July 1945, pp. 3–9.

BISHOP, T. A., AND DUDEWICZ, E. J. "Exact Analysis of Variance with Unequal Variances: Test Procedures and Tables." *Technometrics* 20 (1978), pp. 419–30.

———, AND ———. "Heteroscedastic ANOVA." *Sankhyā* 43 (1981), pp. 40–57. Gives mathematics for their *Technometrics* paper.

BISSELL, A. F. "The Performance of Control Charts and Cusums under Linear Trend." *Applied Statistics* 33 (1984), pp. 145–151.

BLACKWELL, M. T. R. "The Effect of Short Production Runs on CSP-1." *Technometrics* 19 (1977), pp. 259–63.

BLISS, C. I.; COCHRAN, W. G.; AND TUKEY, J. W. "A Rejection Criterion Based upon the Range." *Biometrika* 43 (1956), pp. 418–22.

BOWDEN, DAVID C. "Query 26—Tolerance Interval in Regression." *Technometrics* 10 (1968), pp. 207–9.

BOWKER, ALBERT H. "Continuous Sampling Plans." *Proceedings of the Third Berkeley Symposium on Mathematical Statistics and Probability,* 1954–55, pp. 75–85.

BOX, G. E. P. "Multi-Factor Designs of First Order." *Biometrika* 39 (1952), pp. 49–57.

———. "The Exploration and Exploitation of Response Surfaces: Some General Considerations and Examples." *Biometrics* 10 (1954), pp. 16–60.

———. "Evolutionary Operation: A Method for Increasing Industrial Productivity." *Applied Statistics* 6 (1957), pp. 81–101.

———. "Use and Abuse of Regression." *Technometrics* 8 (1966), pp. 625–29.

———, AND DRAPER, NORMAN R. "A Basis for the Selection of a Response Surface Design." *Journal of the American Statistical Association* 54 (1959), pp. 622–54.

———, AND HUNTER, J. S. "A Confidence Region for the Solution of a Set of Simultaneous Equations with an Application to Experimental Design." *Biometrika* 41 (1954), pp. 190–99.

———, AND ———. "Multi-Factor Experimental Designs for Exploring Response Surfaces." *Annals of Mathematical Statistics* 28 (1957), pp. 195–241.

———, AND ———. "The 2^{k-p} Fractional Factorial Designs." *Technometrics* 3 (1961), pp. 311–51 and 449–58.

———, AND JENKINS, G. M. "Some Statistical Aspects of Adaptive Optimization and Control." *Journal of the Royal Statistical Society* B 24 (1962), pp. 297 ff.

———, AND MEYER, DANIEL. "Analysis of Unreplicated Factorials Allowing for Possibly Faulty Observations." MRC Technical Summary Report #2799 (1985), University of Wisconsin, Madison, Wisc.

———, AND ———. "Studies in Quality Improvement: I Dispersion Effects From Fractional Designs and II An Analysis for Unreplicated Fractional Factorals." MRC Technical Summary Reports #2796 and #2797 (1985), University of Wisconsin, Madison, Wisc.

———, AND TIDWELL, P. W. "Transformation of the Independent Variables." *Technometrics* 4 (1962), pp. 531–50.

———, AND WATSON, G. S. "Robustness to Non-Normality of Regression Tests." *Biometrics* 49 (1962), pp. 93–106.

———, AND WILSON, K. B. "On the Experimental Attainment of Optimum Conditions." *Journal of the Royal Statistical Society* B 13 (1951), pp. 1–38.

BRADLEY, RALPH ALLEN. "Determination of Optimum Operating Conditions by Experimental Methods: Part I—Mathematics and Statistics Fundamental to the Fitting of Response Surfaces." *Industrial Quality Control,* July 1958, pp. 16–20.

BROFFITT, JAMES D., AND RANDLES, RONALD H. "A Power Approximation for the Chi-Square Goodness of Fit Test: Simple Hypothesis Case." *Journal of the American Statistical Association* 72 (1977), pp. 604–07.

BROOKS, S., AND MICKEY, M. RAY. "Optimum Estimation of Gradient Direction in Steepest Ascent Experiments." *Biometrics* 27 (1961), pp. 48–56.

BROWN, B. M. "A Short-Cut Test for Outliers Using Residuals." *Biometrika* 62 (1975), pp. 623–29.

BROWN, MORTON B. "Identification of the Sources of Significance in Two-Way Contingency Tables." *Applied Statistics* 23 (1974), pp. 405–413.

———. "Exploring Interaction Effects in the ANOVA." *Applied Statistics* 24 (1975), pp. 288–298.

BRUNBAUGH, M. A. "The Analysis of Sampling Plans in Receiving Inspection." *Industrial Quality Control,* January 1945, pp. 6–15.

———. "A Report on Pre-War Quality." *Industrial Quality Control,* September 1945, pp. 11–12.

BUDNE, T. A. "The Application of Random Balance Designs." *Technometrics* 1 (1959), pp. 139–55.

BURDICK, RICHARD K., AND SULKEN, ROBERT L., JR. "Exact Confidence Intervals for Linear Combinations of Variance Components in Nested Classifications." *Journal of the American Statistical Association* 73 (1978), pp. 632–35.

BURGESS, A. R. "A Graphical Method of Determining a Single Sampling Plan." *Industrial Quality Control,* May 1948, pp. 25–27.

BURR, IRVING W. "A New Method for Approving a Machine or Process Setting." *Industrial Quality Control,* January 1949, pp. 12–18; September 1949, pp. 15–19; and November 1949, pp. 13–16.

———. "Average Sample Number under Curtailed or Truncated Sampling." *Industrial Quality Control,* February 1957, pp. 5–7. Also see "Correction Notice." *Industrial Quality Control,* July 1957, p. 17.

______. "Some Theoretical and Practical Aspects of Tolerances for Mating Parts." *Industrial Quality Control* 15 (September 1958), pp. 18–22.

______. "The Effect of Non-Normality on Constants for $\bar{X}$ and R Charts." *Industrial Quality Control* 23, May 1967, pp. 563–69.

CASE, KENNETH E.; BENNETT, G. KEMBLE; AND SCHMIDT, J. W. "The Effect of Inspection Error on Average Outgoing Quality." *Journal of Quality Technology* 7 (1975), pp. 28–33.

CHAMPERNOWNE, D. G. "The Economics of Sequential Sampling Procedures for Defectives." *Applied Statistics* 2 (1953), pp. 118–30.

CHANDRA, RAMESH, AND HAHN, GERALD. "Confidence Bounds on the Probability of Meeting a Poisson or Binomial Distribution Specification Limit." *Journal of Quality Technology* 13 (1981), pp. 241–8.

CHAO, MIN-TE, AND GLASER, RONALD E. "The Exact Distribution of Bartlett's Test Statistic for Homogeneity of Variances with Unequal Sample Sizes." *Journal of the American Statistical Association* 73 (1978), pp. 422–26.

CHENG, R. C. H., AND ILES, T. C. "Confidence Bands for Cumulative Distribution Functions of Continuous Random Variables." *Technometrics* 25 (1983), pp. 77–86.

CHERNOFF, H., AND LIEBERMAN, G. J. "Use of Normal Probability Paper." *Journal of the Americal Statistical Association* 49 (1954), pp. 778–85.

______, AND ______. "Sampling Inspection by Variables with No Calculations." *Industrial Quality Control,* January 1957, pp. 5–7.

CHIU, W. K., AND WETHERILL, G. B. "A Simplified Scheme for the Economic Design of $\bar{X}$-Charts." *Journal of Quality Technology* 6 (1974), pp. 63–69.

______, AND CHEUNG, K. C. "An Economic Study of $\bar{X}$-Charts with Warning Limits." *Journal of Quality Technology* 9 (1977), pp. 166–71.

CHOU, YOUN-MIN, AND OWEN, D. B. "One-Sided Confidence Regions on the Upper and Lower Tail Areas of the Normal Distribution." *Journal of Quality Technology* 16 (1984), pp. 150–58. Has tables.

CLIFFORD, PAUL C. "Acceptance Sampling Inspection by Variables." *Industrial Quality Control,* March 1947, pp. 12–15.

CLUNIES-ROSS, C. W. "Interval Estimation of the Parameter of a Binomial Distribution." *Biometrika* 44 (1958), pp. 275–79.

COCHRAN, W. G. "The χ^2-Correction for Continuity." *Iowa State College Journal of Science* 16 (1942), pp. 421–36.

______. "Some Consequences When the Assumptions for the Analysis of Variance are Not Satisfied." *Biometrics* 3, no. 1 (1947), pp. 22–38.

______. "Testing a Linear Relationship among Variances." *Biometrics* 7 (1951), pp. 17–32.

______. "The χ^2 Test of Goodness of Fit." *Annals of Mathematical Statistics* 23 (1952), pp. 315–45.

______. "A Test of a Linear Function of the Deviations between Observed and Expected Numbers." *Journal of the American Statistical Association* 50 (1955), pp. 377–97.

______. "Analysis of Covariance: Its Nature and Uses." *Biometrics* 13 (1957), pp. 261–81.

______. "The Design of Experiments." In *Operations Research and Systems Engineering,* ed. Charles D. Flagle; William H. Huggins; and Robert H. Roy. Baltimore: The Johns Hopkins Press, 1960.

COHEN, A. CLIFFORD, JR., AND SIMON, LESLIE E. "Quality Control at Picatinny Arsenal, 1934–1945." American Society for Quality Control, *Annual Technical Conference Transactions,* 1971, pp. 273–79.

COLLETT, D., AND LEWIS, T. "The Subjective Nature of Outlier Rejection Procedures." *Applied Statistics* 25 (1976), pp. 228–37.

CONOVER, W. J.; JOHNSON, MARK E.; AND JOHNSON, MYRLE M. "A Comparative Study of Tests for Homogeneity of Variance, with Applications to the Outer Continental Shelf Bidding Data." *Technometrics* 23 (1981), pp. 351–61.

CORNFIELD, JEROME. "Some Finite Sampling Concepts in Experimental Statistics" (Abstract), *Journal of the American Statistical Association,* 49 (1954), pp. 369–70.

CRADDOCK, J. M., AND FLOOD, C. R. "The Distribution of the χ^2 Statistic in Small Contingency Tables." *Applied Statistics* 19 (1970), pp. 171–81.

CRAIG, CECIL C. "Some Remarks Concerning the Lot Plot Plan." *Industrial Quality Control,* September 1953, pp. 41–48.

CROW, EDWIN L., AND GARDNER, ROBERT S. "Confidence Intervals for the Expectation of a Poisson Variable." *Biometrika* 46 (1959), pp. 441–53.

CURTISS, J. H. "Acceptance Sampling by Variables, with Special Reference to the Case in Which Quality Is Measured by Average or Dispersion." Research Paper, RP 1827, *Journal of Research of the National Bureau of Standards* 39 (September 1947), pp. 271–90.

D'AGOSTINO, R. B. "Transformation to Normality of the Null Distribution of g_1" *Biometrika* 57 (1970), pp. 679–81.

______, AND TIETJEN, GARY L. "Simulation Probability Points of b_2 for Small Samples." *Biometrika* 58 (1971), pp. 669–72.

______, AND ROSEMAN, BERNARD. "The Power of Geary's Test of Normality." *Biometrika* 61 (1974), pp. 181–84.

DANIEL, CUTHBERT. "Fractional Replication in Industrial Research." *Proceedings of the Third Berkeley Symposium on Mathematical Statistics and Probability,* 1954, pp. 87–98.

______, AND HEEREMA, N. "Design of Experiments for Most Precise Slope Estimation in Linear Extrapolation." *Journal of the American Statistical Association* 45 (1950), pp. 546–56.

______. "On Varying One Factor at a Time." *Biometrics* 14 (1958), pp. 430–31.

______. "Use of Half-Normal Plots in Interpreting Factorial Two-Level Experiments." *Technometrics* 1 (1959), pp. 311–41.

DAVID, H. A. "Revised Upper Percentage Points of the Extreme Studentized Deviate from the Sample Mean." *Biometrika* 43 (1956), pp. 449–51.

______, AND PAULSON, A. S. "The Performance of Several Tests for Outliers." *Biometrika* 52 (1965), pp. 429–36.

DAVIES, O. L. "Some Statistical Aspects of the Economics of Analytical Testing." *Technometrics* 1 (1959), pp. 49–61.

———, AND HAY, W. A. "The Construction and Uses of Fractional Factorial Designs in Industrial Research." *Biometrics* 6 (1950), pp. 233–49.

DAY, BESSE B. "Application of Statistical Methods to Research and Development in Engineering." *Review of the International Statistical Institute,* nos. 3, 4, 1949, pp. 129–55.

———, AND FISHER, R. A. "The Comparison of Variability in Populations Having Unequal Means. An Example of the Analysis of Covariance with Multiple Dependent and Independent Variates." *Annals of Eugenics* 7 (1937), pp. 332–48.

———; DEL PRIORE, F. R.; AND SAX, EDWARD. "The Technique of Regression Analysis." *Quality Control Conference Papers,* American Society for Quality Control, Seventh Annual Convention, 1953, pp. 399–418.

DEBANN, R. "Response Surface Designs for Three Factors at Three Levels." *Technometrics* 1 (1959), pp.1–8.

DEBUSH, R. E. "Experience in Evolutionary Operations at Tennessee Eastman Co." *Industrial Quality Control,* October 1962, pp. 15–21.

DEELY, J. J., AND LINDLEY, D. V. "Bayes Empirical Bayes." *Journal of the American Statistical Association* 76 (1981), pp. 833–41.

DEGROOT, M. H., AND NADLER, JACK. "Some Aspects of the Use of the Sequential Probability Ratio Tests." *Journal of the American Statistical Association* 53 (1958), pp. 187–99.

DELURY, D. B. "Designing Experiments to Isolate Sources of Variation." *Industrial Quality Control,* September 1954, pp. 22–24.

DEMING, W. EDWARDS. "On the Significant Figures of Least Squares and Correlations." *Science* 85 (1937), pp. 451–54.

DERMAN, C.; LITTAUER, S.; AND SOLOMON, H. "Tightened Multi-Level Continuous Sampling Plans." *Annals of Mathematical Statistics* 28 (1957), pp. 395–404.

———; JOHNS M. V., JR.; AND LIEBERMAN, G. J. "Continuous Sampling Procedures without Control." *Annals of Mathematical Statistics* 30 (1959), pp. 1175–91.

DIXON, W. J. "Processing Data for Outliers." *Biometrics* 9 (1953), pp. 74–89.

DODGE, HAROLD F. "A Method of Rating Manufactured Products." *Bell System Technical Journal* 7 (1928), pp. 350–68. (Reprinted by the Bell Telephone Laboratories as Reprint B-315.)

———. "A Sampling Inspection Plan for Continuous Production." *Annals of Mathematical Statistics* 14 (1943), pp. 264–79.

———. "Chain Sampling Inspection Plans." *Industrial Quality Control,* January 1955, pp. 10–13.

———. "Notes on the Evolution of Acceptance Sampling Plans." Part I, *Journal of Quality Technology,* April 1969, pp. 77–88; Part II, *Journal of Quality Technology,* July 1969, pp. 155–62; Part III, *Journal of Quality Technology,* October 1969, pp. 225–32; Part IV, *Journal of Quality Technology,* January 1970, pp. 1–8.

———, AND TORREY, M. N. "Additional Continuous Sampling Inspection Plans." *Industrial Quality Control,* November 1951, pp. 5–9.

DRAPER, NORMAN R. "Second Order Rotatable Designs in Four or More Dimensions." *Annals of Mathematical Statistics* 31 (1960), pp. 23–33.

______. "Third Order Rotatable Designs in Three Dimensions." *Annals of Mathematical Statistics* 31 (1960), pp. 865–74.

______. "A Third Order Rotatable Design in Four Dimensions." *Annals of Mathematical Statistics* 31 (1960), pp. 875–77.

______. " 'Ridge Analysis' of Response Surfaces," *Technometrics* 5 (1963), pp. 469–79.

DUDDING, BERNARD P. "The Introduction of Statistical Methods to Industry." *Applied Statistics* 1 (1952), pp. 1–20.

DUNCAN, ACHESON J. "The Use of Ranges in Comparing Variabilities." *Industrial Quality Control,* February 1955, pp. 18 ff. (Also see May 1955, issue p. 70.)

______. "The Economic Design of $\bar{X}$ Charts Used to Maintain Current Control of a Process." *Journal of the American Statistical Association* 51 (1956), pp. 228–42.

______. "Design and Operation of a Double-Limit Variables Sampling Plan." *Journal of the American Statistical Association* 53 (1958), pp. 543–50.

______. "Process Capability Studies." *Proceedings, Middle Atlantic Conference,* American Society for Quality Control, February 1958, pp. 163–76.

______. "Bulk Sampling: Problems and Lines of Attack." *Technometrics* 4 (1962), pp. 319–44.

______. "The Economic Design of $\bar{X}$ Charts When There Is a Multiplicity of Assignable Causes." *Journal of the American Statistical Association* 66 (1971), pp. 107–21. (See W. K. Chiu, *J.A.S.A.* 68, pp. 919–21.)

______. "Sampling by Variables to Control the Fraction Defective: Part I." *Journal of Quality Technology* 7 (1975), pp. 34–42.

______. "The Economic Design of p-Charts to Maintain Current Control of a Process: Some Numerical Results." *Technometrics* 20 (1978), pp. 235–43.

DUNCAN, DAVID B. "Multiple Range and Multiple F Tests." *Biometrics* 11 (1955), pp. 1–42.

______. "Multiple Range Tests for Correlated and Heteroscedastic Means." *Biometrics* 13 (1957), pp. 164–76.

DWYER, P. S. "The Square Root Method and Its Use in Correlation and Regression." *Journal of the American Statistical Association* 40 (1945), pp. 493–503.

EATON, H. C. "Engineered Assembly Tolerance." *Industrial Quality Control,* September 1947, pp. 16–17.

EISENHART, CHURCHILL. "The Assumptions Underlying the Analysis of Variance." *Biometrics* 3 (1947), pp. 1–21.

ELDER, ROBERT S., AND MUSE, H. DAVID. "An Approximate Method for Evaluating Mixed Sampling Plans." *Technometrics* 24 (1982), pp. 207–11.

______; PROVOST, LLOYD P.; AND ECKER, OWEN M. "U.S. Department of Agriculture Cusum Acceptance Sampling Procedures." *Journal of Quality Technology* 13 (1981), pp. 59–64.

ENELL, JOHN W. "Which Sampling Plan Should I Choose?" *Journal of Quality Technology* 16 (1984), pp. 168–71.

ENRICK, NORBERT L. "Operating Characteristics of Reject Limits for Measurements." *Industrial Quality Control,* September 1945, pp. 9–10.

ENTERS, J. H., AND HAMAKER, H. C. "Multiple Sampling in Theory and Practice." A paper from Royal Statistical Society, *Statistical Method in Industrial Production.* Papers given at a Conference held in Sheffield in 1950. (Printed for private circulation, 1951.)

EPSTEIN, B. "Elements of the Theory of Extreme Values." *Technometrics* 2 (1960), pp. 27–41.

EVANS, DAVID H. "Statistical Tolerancing: The State of the Art." Part I "Background." *Journal of Quality Technology* 6 (1974), pp. 188–95; Part II "Methods for Estimating Moments." *Journal of Quality Technology* 7 (1975), pp. 11–12; Part III "Shifts and Drifts." *Journal of Quality Technology* 7 (1975), pp. 72–76.

EWAN, W. D. "When and How to Use Cu-Sum Charts." *Technometrics* 5 (1963), pp. 1–32.

———, AND KEMP, K. W. "Sampling Inspection of Continuous Processes with No Autocorrelation between Successive Results." *Biometrika* 47 (1960), pp. 363 ff.

Factory Management and Maintenance, February 1953, p. 93. This gives details for construction of blocks to demonstrate statistical determination of tolerances.

FEDER, PAUL I. "Some Differences Between Fixed, Mixed, and Random Effects Analysis of Variance Models." *Journal of Quality Technology* 6 (1974), pp. 98–106.

FENECH, ALAN PAUL. "Tukey's Method of Multiple Comparisons in Randomized Blocks Model." *Journal of the American Statistical Association* 74 (1979), pp. 881–84.

FERRELL, ENOCH B. "Control Charts Using Midranges and Medians." *Industrial Quality Control,* March 1953, pp. 30–34.

FERRIS, C. D.; GRUBBS, F. E.; AND WEAVER, C. L. "Operating Characteristics of the Common Statistical Tests of Significance." *Annals of Mathematical Statistics* 17 (1946), pp. 178–97.

FERTIG, KENNETH V., AND MANN, NANCY R. "A Prediction Interval Approach to Obtaining Variables Sampling Plans for Small Lots: Single Sampling from Gaussian Processes." *Journal of the American Statistical Association* 72 (1977), pp. 585–92.

———, AND ———. "Sample Sizes for Variables Sampling Plans Based on a Decision Theoretic Approach." *Journal of Quality Technology* 15 (1983), pp. 63–67.

FISHBURN, PETER C. "Preference-Based Definitions of Subjective Probability." *Annals of Mathematical Statistics* 38 (1967), pp. 1605–17.

———. "A General Theory of Subjective Probabilities and Expected Utilities." *Annals of Mathematical Statistics* 40 (1969), pp. 1419–29.

FIX, EVELYN. "Tables of Non-Central χ^2." *University of California Publications in Statistics* 1, no. 2 (1949), pp. 15–19.

FOLKS, JOHN LEROY, AND BLANKENSHIP, JOHN H. "A Note on Probability Plotting." *Industrial Quality Control,* April 1967, pp. 495–96.

FOLKS, JOHN LEROY; PIERCE, DONALD A.; AND STEWART, CHARLES. "Estimating the Fraction of Acceptable Product." *Technometrics* 7 (1965), pp. 43–50.

FOSTER, JULIE A. "Kolmogorov-Smirnov Test for Goodness of Fit: What It Is and How to Apply It." *Industrial Quality Control,* January 1962, pp. 4–8.

FREEMAN, H. A. "Statistical Methods for Quality Control." *Mechanical Engineering,* April 1937, p. 261.

FREUND, RICHARD A. "Acceptance Control Charts." *Industrial Quality Control,* October 1957, pp. 13–23.

______. "Graphical Process Control." *Industrial Quality Control,* January 1962, pp. 15–22.

______. "Contrast Analysis of Experiments." *Journal of Quality Technology,* 6 (1974), pp. 2–21.

FRIEDMAN, M., AND SAVAGE, L. J. "Planning Experiments Seeking Maxima." In *Techniques of Statistical Analysis.* New York: McGraw-Hill, 1947.

GABRIEL, K. R. "Analysis of Variance of Proportions with Unequal Frequencies." *Journal of the American Statistical Association* 58 (1963), pp. 1133–57.

GASTWIRTH, JOSEPH L., AND OWENS, M. E. B. "On Classical Tests of Normality." *Biometrika* 64 (1977), pp. 135–139.

GAYLORD, D. W., AND HOPPER, F. N. "Estimating the Degrees of Freedom for Linear Combinations of Mean Squares by Satterthwaite's Formula." *Technometrics* 11 (1969), pp. 691–706.

GENTLEMAN, J. F., AND WILK, M. B. "Detecting Outliers in a Two-Way Table: I Statistical Behavior of Residuals." *Technometrics* 17 (1975), pp. 1–17.

______, AND ______. "Detection of Outliers II. Supplementing the Direct Analysis of Residuals." *Biometrics* 31 (1975), pp. 387–410.

GHOSH, B. K. "A Comparison of Some Approximate Confidence Intervals for the Binomial Parameter." *Journal of the American Statistical Association* 74 (1979), pp. 894–900.

GIBRA, ISAAC N. "Economically Optimal Determination of the Parameters of np-Charts." *Journal of Quality Technology* 10 (1978), pp. 12–19.

______. "Economic Design of Attribute Control Charts for Multiple Assignable Causes." *Journal of Quality Technology* 13 (1981), pp. 93–99.

GLASSER, RONALD E. "Exact Critical Values for Bartlett's Test for Homogeneity of Variances." *Journal of the American Statistical Association* 71 (1976), pp. 488–490.

GODFREY, A. BLANTON, AND MUNDEL, AUGUST B. "Guide for Selection of an Acceptence Sampling Plan." *Journal of Quality Technology* 16 (1984) pp. 50–55. A review of an International Organization for Standardization (ISO) draft proposal.

GOLDMAN, A. "Sample Size for a Specified Width Confidence Interval on the Ratio of Variances from Two Independent Normal Populations." *Biometrics* 19 (1963), pp. 465–77.

GOLDSMITH, C. H., AND GAYLOR, D. W. "Three Stage Nested Designs for Estimating Variance Components." *Technometrics* 12 (1970), pp. 487–98.

GOLDSMITH, P. L., AND WHITFIELD, H. W. "Average Run Lengths in Cumulative Chart Quality Control Schemes." *Technometrics* 3 (1961), pp. 11–20.

GOODMAN, LEO A. "The Variance of a Product of K Random Variables." *Journal of the American Statistical Association* 57 (1962), pp. 54–60.

GORE, W. L. "Statistical Techniques for Research and Development." *Conference Papers, First Annual Convention American Society for Quality Control and Second Midwest Quality Control Conference, June 5 and 6, 1947,* pp. 145–51. Chicago: John L. Swift Co., 1947.

GRANDAGE, A. "Orthogonal Coefficients for Unequal Intervals." *Biometrics* 14 (1958), pp. 287–89.

GRANT, EUGENE L. "Industrialists and Professors in Quality Control—A Look Back and a Look Forward." *Industrial Quality Control,* February 1953, pp. 31–35.

GRUBBS, F. E. "The Difference Control Chart with an Example of Its Use." *Industrial Quality Control,* July 1946, pp. 22–25.

———. "On Designing Single Sampling Inspection Plans." *Annals of Mathematical Statistics* 20 (1949), pp. 242–56.

———. "On Estimating Precision of Measuring Instruments and Product Variability." *Journal of the American Statistical Association* 43 (1948), pp. 243–64.

———. "Procedures for Detecting Outlying Observations in Samples." *Technometrics* 11 (1969), pp. 1–21.

———, AND BECK, GLENN. "Extension of Sample Sizes and Percentage Points for Significance Tests of Outlying Observations." *Technometrics* 14 (1972), pp. 847–54.

———, COON, HELEN J., AND PEARSON, E. S. "On the Use of Patnaik Type Chi Approximations to the Range in Significance Tests." *Biometrika* 53 (1966), pp. 248–52.

———, AND WEAVER, C. L. "The Best Unbiased Estimate of the Population Standard Deviation Based on Group Ranges." *Journal of the American Statistical Association* 42 (1947), pp. 224–41.

———. "An Optimum Procedure for Setting Machines or Adjusting Processes." *Journal of Quality Technology* 15 (1983), pp. 186–89.

GUENTHER, WILLIAM C. "On the Determination of Single Sampling Attribute Plans Based upon a Linear Cost Model and a Prior Distribution." *Technometrics* 13 (1971), pp. 483–98.

———. "A Sample Size Formula for the Hypergeometric." *Journal of Quality Technology* 5 (1973), pp. 167–70.

GUILD, RICHARD D., AND RALEA, IDA I. D. G. "Effective Sampling Plans Based on a Prior Distribution." *Journal of Quality Technology* 12 (1980), pp. 88–93.

GURLAND, J., AND MCCOLLOUGH, R. S. "Testing Equality of Means after a Preliminary Test of Equality of Variances." *Biometrika* 49 (1962), pp. 403–17.

GUTHRIE, D., JR., AND JOHNS, M. V. JR. "Bayes Acceptance Sampling Procedures for Large Lots." *Annals of Mathematical Statistics* 30 (1959), pp. 896–925.

HABER, MICHAEL. "A Comparison of Some Continuity Corrections for the Chi-Square Test on 2 × 2 Tables." *Journal of the American Statistical Association* 75 (1980), pp. 510–515.

HAHN, GERALD J. "Statistical Intervals for a Normal Population. Part I. Tables, Examples and Applications." *Journal of Quality Technology,* July 1970, pp. 115–25; "Part II. Formulas, Assumptions, Some Derivations." *Journal of Quality Technology,* October 1970, pp. 195–206.

———, AND SHAPIRO, S. S. "The Use and Misuse of Multiple Regression." *Industrial Quality Control,* October 1966, pp. 184–89.

———. "Minimum Size Sampling Plans." *Journal of Quality Technology* 6 (1974), pp. 121–127.

———. "Some Things Engineers Should Know About Experimental Design." *Journal of Quality Technology* 9 (1977), pp. 13–20.

———. "Fitting Regression Models with No Intercept Term." *Journal of Quality Technology* 9 (1977), pp. 56–61.

———, AND BOARDMAN, THOMAS J. "The Statistician's Role in Quality Improvement." *Amstat News,* March 1985 (No. 113), pp. 5–8.

HALD, A. "The Compound Hypergeometric Distribution and a System of Single Sampling Inspection Plans Based on Prior Distributions and Costs." *Technometrics* 2 (1960), pp. 275–340.

———, AND THYREGOD, P. "The Composite Operating Characteristic under Normal and Tightened Sampling Inspection by Attributes." International Statistical Institute, *Proceedings of the 35th Session,* 1965, Book 1, pp. 517–29. (Included in Vol. 41 of the *Bulletin* of the International Statistical Institute.)

HALL, IRVING J. "Approximate One-Sided Tolerance Limits for the Difference or Sum of Two Independent Normal Variates." *Journal of Quality Technology* 16 (1984), pp. 15–19.

———, AND SAMPSON, C. B. "Tolerance Limits for the Product and Quotient of Normal Variates." *Biometrics* 29 (1973), pp. 109–19.

HALL, JAMES E. "Minimum Variance and VOQL Sampling Plans." *Technometrics* 21 (1979), pp. 555–65.

HALL, PETER. "Improving the Normal Approximation When Constructing One-Sided Confidence Intervals for Binomial or Poisson Parameters." *Biometrika* 69 (1982), pp. 647–652.

HALPERIN, MAX. "Fitting of Straight Lines and Prediction When Both Variables Are Subject to Error." *Journal of the American Statistical Association* 56 (1961), pp. 657–69.

HAMAKER, H. C. "Lot Inspection by Sampling." *Philips Technical Review* 11 (1949), pp. 176–82.

———. "The Theory of Sampling Inspection Plans." *Philips Technical Review* 11 (1950), pp. 260–70.

———. "The Efficiency of Sequential Sampling for Attributes." *Philips Research Reports* 8 (1953), pp. 35–46 and 427–33.

———. "Some Basic Principles of Sampling Inspection by Attributes." *Applied Statistics* 7 (1958), pp. 149–59.

———. "Acceptance Sampling for Percent Defective by Variables and by Attributes." *Journal of Quality Technology* 11 (1979), pp. 139–48.

———, AND VAN STRIK, R. "The Efficiency of Double Sampling for Attributes." *Journal of the American Statistical Association* 50 (1955), pp. 830–49.

HARRISON, M. J., AND MCCABE, B. P. M. "A Test for Heteroscedasticity Based on Ordinary Least Squares Residuals." *Journal of the American Statistical Association* 74 (1979), pp. 494–99.

HARTER, H. LEON. "Error Rates and Sample Sizes for Range Tests in Multiple Comparisons." *Biometrics* 13 (1957), pp. 511–36.

HARTLEY, H. O. "Smallest Composite Designs for Quadratic Response Surfaces." *Biometrics* 15 (1959), pp. 611–24.

HAWKES, CUYLER J. "Curves for Sample Size Determination in Lot Sensitive Sampling Plans." *Journal of Quality Technology* 11 (1979), pp. 205–10.

HEALY, W. C., JR. "Limits for a Variance Component with an Exact Confidence Coefficient." *Annals of Mathematical Statistics* 32 (1961), pp. 466–76.

HICKS, CHARLES R. "Fundamentals of Analysis of Variance," published in three parts in *Industrial Quality Control,* August 1956, pp. 17–20; September 1956, pp. 5–8; and October 1956, pp. 13–16. This condensed review was subsequently put out in reprint form by the American Society for Quality Control, 230 West Wells St., Milwaukee, Wisconsin, 53203.

———. "Have You Tried *BIPP?*" *Industrial Quality Control,* November 1956, pp. 15–17.

HILL, DAVID. "Modified Control Limits." *Applied Statistics* 5 (1956), pp. 12–19.

HILL, I. D.; HORSNELL, G.; AND WARNER, B. T. "Deferred Sentencing Schemes." *Applied Statistics* 8 (1959), pp. 78–91.

———. "The Economic Incentive Provided by Sampling Inspection." *Applied Statistics* 9 (1960), pp. 69–81.

———. "The Design of the Mil. Std. 105D Sampling Tables." *Journal of Quality Technology* 5 (1973), pp. 80–83.

HILLIARD, J. E., AND LASATER, H. ALAN. "Type One Risks When Several Tests Are Used Together on Control Charts for Means and Ranges." *Industrial Quality Control,* August 1966, pp. 56–61.

HILLIER, FREDERICK S. "New Criteria for Selecting Continuous Sampling Plans." *Technometrics* 6 (1964), pp. 161–78.

HINKLEY, D. V. "Ratio of Correlated Normal Random Variables." *Biometrika* 56 (1969), 635–39.

HOADLEY, B. "The Quality Measurement Plan (QMP)." *The Bell System Technical Journal* 60 (1981), pp. 215–74.

HOCKING, R. R., AND LESLIE, R. N. "Selection of the Best Subset in Regression Analysis." *Technometrics* 9 (1967), pp. 531–40.

———. "Developments in Linear Regression Methodology: 1959–1982," with Discussion and Response. *Technometrics* 25 (1983), pp. 219–49.

HOMILING, MORRIS. "Bayesian Decision Theory and Statistical Quality Control." *Industrial Quality Control,* December 1962, pp. 10–14.

HOMMEL, GERHARD. "Tail Probabilities for Contingency Tables with Small Expectations." *Journal of the American Statistical Association* 73 (1978), pp. 764–66.

HORSNEL, GARETH. "The Determination of Single-Sample Schemes for Percentage Defectives." *Applied Statistics* 3 (1954), pp. 150–58.

HOTELLING, HAROLD. "The Generalization of Student's Ratio." *Annals of Mathematical Statistics* 2 (1931), pp. 360–78.

———. "Experimental Determination of the Maximum of a Function." *Annals of Mathematical Statistics* 12 (1941), pp. 20–45.

HOWELL, JOHN M. "Control Chart for Largest and Smallest Values." *Annals of Mathematical Statistics* 20 (1949), pp. 305–9.

———, AND JOHNSON, LEE. "Statistical Control of Accidents." *Industrial Standardization,* May 1947, pp. 117–18.

HSU, P. L. "Notes on Hotelling's Generalized T." *Annals of Mathematical Statistics* 9 (1938), pp. 231–43.

HUDSON, J. D., JR., AND KRUTCHKOFF, R. C. "A Monte Carlo Investigation of the Size and Power of Tests Employing Satterthwaite's Synthetic Mean Squares." *Biometrika* 55 (1968), pp. 431–33.

HUNTSBERGER, D. V. "A Generalization of a Preliminary Testing Procedure for Pooling Data." *Annals of Mathematical Statistics* 26 (1955), pp. 734–43.

HWANG, F. K. "Robust Group Testing." *Journal of Quality Technology* 16 (1984), pp. 189–95.

IMHOFF, J. P., "Testing the Hypothesis of No Fixed Main-Effects in Scheffe's Mixed Model." *Annals of Mathematical Statistics* 33 (1962), pp. 1085–94.

INDIANAPOLIS SECTION, AMERICAN SOCIETY FOR QUALITY CONTROL, ALLISON DIVISION G.M.C. "How Effective Is Lot Plot." *Industrial Quality Control,* March 1952, pp. 33–34.

Industrial Quality Control, 1944.

JACKSON, J. EDWARD. "Bibliography on Sequential Analysis." *Journal of the American Statistical Association* 55 (1960), pp. 561–80.

JAECH, JOHN L. "Determination of Acceptance Numbers and Sample Sizes for Attribute Sampling Plans." *Journal of Quality Technology* 12 (1980), pp. 187–90.

JENNETT, W. J., AND WELCH, B. L. "The Control of Proportion Defective as Judged by a Single Quality Characteristic Varying on a Continuous Scale." *Journal of the Royal Statistical Society* B, 6 (1939), pp. 80–88.

JENNISON, CHRISTOPHER, AND TURNBULL, BRUCE W. "Confidence Intervals for a Binomial Parameter Following a Multistage Test with Application to Mil. Std. 105D and Medical Trials." *Technometrics* 25 (1983), pp. 49–58.

JOHN, J. A. "Outliers in Factorial Experiments." *Applied Statistics* 27 (1978), pp. 111–19.

JOHNSON, ERIC E., AND COUNTS, RICHARD W. "Cyclic Data Control Charts." *Journal of Quality Technology* 11 (1979), pp. 28–36.

JOHNSON, N. L. "A Simple Theoretical Approach to Cumulative Sum Control Charts." *Journal of the American Statistical Association* 56 (1961), pp. 83 ff.

———, KOTZ, SAMUEL, AND RODRIGUEZ, ROBERT N. "Statistical Effects of Imperfect Inspection Sampling I. Some Basic Distributions." *Journal of Quality Technology* 17 (1985), pp. 1–31.

———, AND LEONE, F. C. "Cumulative Sum Control Charts—Mathematical Principles Applied to Their Construction and Use." *Industrial Quality Control,* June 1962, pp. 15–21; July 1962, pp. 29–36; and August 1962, pp. 22–28.

JOHNSON, N. S. "C_α Method for Testing for Significance in the r × c Contingency Table." *Journal of the American Statistical Association* 70 (1975), pp. 942–45.

KACKAR, HAGHU N. "Off-Line Quality Control, Parameter Design, and the Taguchi Method," with discussion, *Journal of Quality Technology* 17 (1985), pp. 176–209.

KAO, JOHN H. K. "Mil. Std. 414 Sampling Procedures and Tables for Inspection by Variables for Percent Defective," *Journal of Quality Technology* 3 (1971), pp. 28–37.

KEEFE, GORDON J. "Attribute Sampling—Mil. Std. 105." *Industrial Quality Control,* April 1963, pp. 7–12.

KEEN, JOAN, AND PAGE, DENYS J. "Estimating Variability from the Differences between Successive Readings" (with appendix: "Some Theoretical Aspects," by H. O. Hartley). *Applied Statistics* 2 (1953), pp. 13–23.

KEMP, KENNETH W. "Formulas for Calculating the Operating Characteristic and the Average Sample Number of Some Sequential Tests." *Journal of the Royal Statistical Society* B 20 (1958), pp. 379–86.

———. "The Average Run Length of the Cumulative Sum Chart When a V-Mask Is Used." *Journal of the Royal Statistical Society* B 23 (1961), pp. 149–53.

———. "The Use of Cumulative Sums for Sampling Inspection Schemes." *Applied Statistics* 11 (1962), pp. 16–31.

KENDALL, M. G. "Ronald Aylmer Fisher, 1890–1962." *Biometrika* 50 (1963), pp. 1–16.

KEULS, M. "The Use of 'Studentized Range' in Connection with an Analysis of Variance." *Euphytica* 1 (1952), pp. 112–22.

KHAN, RASUL A. "A Sequential Detection Procedure and the Related Cusum Procedure." *Sankhyā* B 40 (1979), pp. 146–62.

———. "A Note on Page's Two-Sided Cumulative Sum Procedure." *Biometrika* 68 (1981), pp. 717–19.

KIMBALL, BRADFORD F. "On the Choice of Plotting Positions on Probability Paper." *Journal of the American Statistical Association* 55 (1960), pp. 546–60.

KITAGAWA, GENSHIRO. "On the Use of AIC for the Detection of Outliers." *Technometrics* 21 (1979), pp. 193–99.

KNOWLER, LLOYD A. "Fundamentals of Quality Control." *Industrial Quality Control,* July 1946, pp. 7–18.

KOEHLER, TRUMAN L. "Evolutionary Operation: A Program for Optimizing Plant Operation, through the Application of Statistics to Scale-Up Problems." *Chemical Engineering Progress,* October 1959, pp. 76–79.

KOGURE, M. "On the New Random Dice and Their Applications," *Reports of Statistical Application Research, Union of Japanese Scientists and Engineers* 5, no. 2 (1958), pp. 35–48.

KOYAMA, TAKESHI. "Switching Characteristics under Mil. Std. 105D." *Technometrics* 21 (1979), pp. 9–19.

KRAMER, K. H. "Tables for Constructing Confidence Limits for the Multiple Correlation Coefficient." *Journal of the American Statistical Association* 58 (1963), pp. 1082–85.

KRUSKAL, W. H. "Some Remarks on Wild Observations." *Technometrics* 2 (1960), pp. 1–3.

KURTZ, T. E.; LINK, R. L.; TUKEY, J. W.; AND WALLACE, D. L. "Correlation of Ranges of Correlated Deviates." *Biometrika* 53 (1966), pp. 191–97.

LADANY, SHAUL P. "Graphical Determination of Single-Sample Attribute Plans for Industrial Small Lots." *Journal of Quality Technology,* July 1971, pp. 115–19.

LAM, K., AND CHIU, W. K. "On the Probability of Correctly Selecting the Best of Several Normal Populations." *Biometrika* 63 (1976), pp. 410–11.

LAUER, G. NICHOLAS. "Acceptance Probabilities for Sampling Plans Where the Proportion Defective Has a Beta Distribution." *Journal of Quality Technology* 10 (1978), pp. 52–55.

LAWAL, H. B. "Tables of Percentage Points of Pearson's Goodness-of-Fit Statistic for Use with Small Expectations." *Applied Statistics* 29 (1980), pp. 292–98. Gives tables.

LEE, AUSTIN F. S., AND GARLAND, JOHN. "Size and Power of Tests for Equality of Means of Two Normal Populations with Unequal Variances." *Journal of the American Statistical Association* 70 (1975), pp. 933–41.

LEE, YOONG-SIN. "Tables of Upper Percentage Points of the Multiple Correlation Coefficient." *Biometrika* 59 (1972), pp. 175–89.

LEONE, FRED C., AND NELSON, LLOYD S. "Sampling Distributions of Variance Components. I. Empirical Studies of Balanced Nested Design." *Technometrics* 8 (1966), pp. 457–68.

———; NELSON, L. S.; JOHNSON, N. L.; AND EISENSTAT, STAN. "Sampling Distributions of Variance Components. II. Empirical Studies of Unbalanced Nested Designs." *Technometrics* 10 (1968), pp. 719–37.

LIEBERMAN, G. J., AND RESNIKOFF, G. J. "Sampling Plans for Inspection by Variables." *Journal of the American Statistical Association* 50 (1955), pp. 457–516 and p. 1333.

———, AND SOLOMON, H. "Multi-Level Continuous Sampling Plans." *Annals of Mathematical Statistics* 26 (1955), pp. 686–704.

LIEBESMAN, B. S. "The Use of Mil. Std. 105D to Control Average Outgoing Quality." *Journal of Quality Technology* 11 (1979), pp. 36–43.

———, AND HAWLEY, F. C. "Small Acceptance Number Plans for Use in Military Standard 105D." *Journal of Quality Technology* 16 (1984), pp. 219–31.

———, AND SAPERSTEIN, B. "A Proposed Attribute Skip-Lot Sampling Program." *Journal of Quality Technology* 15 (1983), pp. 130–40.

LINDLEY, D. V. "Professor Hogben's 'Crisis'—A Survey of the Foundations of Statistics." *Applied Statistics* 7 (1958), pp. 186–98.

______. "The Choice of Variables in Multiple Regression." *Journal of the Royal Statistical Society* B 30 (1968), pp. 31–66.

LING, ROBERT F. "A Study of the Accuracy of Some Approximations for t, χ^2 and F Tail Probabilities." *Journal of the American Statistical Association* 73 (1978), pp. 274–83.

LITTAUER, S. B. "The Development of Statistical Quality Control in the United States." *American Statistician,* October 1950, pp. 14–20.

LORD, E. "The Use of Range in Place of Standard Deviation in the *t*-Test." *Biometrika* 34 (1947), pp. 41–67.

LUCAS, JAMES M. "The Design and Use of V-Mask Control Schemes." *Journal of Quality Technology* 8 (1976), pp. 1–12.

______. "Combined Shewhart-CUSUM Quality Control Schemes." *Journal of Quality Technology* 14 (1982), pp. 51–59.

______. "Counted Data CUSUMS." *Technometrics* 27 (1985), pp. 129–44.

______, AND CROSIER, RONALD B. "Fast Initial Response for CUSUM Quality Control Schemes: Give Your CUSUM a Head Start." *Technometrics* 24 (1982), pp. 199–205.

______, AND ______. "Robust CUSUM." *Communications in Statistics, Part A—Theory and Methods* 11 (1982), pp. 2669–87.

MCARTHUR, D. S., AND HEIGL, J. J. "Strategy in Research," *National Convention Transactions,* American Society for Quality Control, 1957, pp. 1–18.

MCELRATH, GAYLE, W., AND BEARMAN, JACOB E. "Sampling by Variables III: Protection on Variability." *National Convention Transactions,* American Society for Quality Control, 1957, pp. 447–62.

MAGHSOODLOO, S., AND BUSH, B. K. "The Effects of Inspection Error on Double Sampling by Attributes." *Journal of Quality Technology* 17 (1985), pp. 32–39.

MAHALANOBIS, P. C. "Tables for L-Tests." *Sankhyā: The Indian Journal of Statistics* 1 (1933), pp. 109–22.

MALLIOS, WILLIAM S. "A Generalized Application of Instrumental Variable Estimation of Straight-line Relations When Both Variables Are Subject to Error." *Technometrics* 11 (1969), pp. 255–63.

MANDEL, JOHN. "Fitting Straight Lines When Both Variables Are Subject to Error." *Journal of Quality Technology* 16 (1984), pp. 1–14.

MANUELE, J. "Control Chart for Determining Tool Wear," *Industrial Quality Control,* May 1945, pp. 7–10.

MARITZ, J. S., AND JARRETT, R. G. "A Note on Estimating the Variance of the Sample Median." *Journal of the American Statistical Association* 73 (1978), pp. 194–96.

MARQUARDT, DONALD W. "New Technical and Educational Directions for Managing Product Quality." *The American Statistician* 38 (1984), pp. 8–14.

MARTZ, H. F., AND LIAN, M. G. "Empirical Bayes Estimation of the Binomial Parameter." *Biometrika* 61 (1974), pp. 517–23.

MASSEY, FRANK J., JR. "The Kolmogorov-Smirnov Test for Goodness of Fit." *Journal of the American Statistical Association* 46 (1951), pp. 68–78.

MATHAI, A. M., AND KATIYAR, R. S. "Exact Percentage Points for Testing Independence." *Biometrika* 66 (1979), pp. 353–56. Gives tables.

MEHTA, CYRUS R., AND PATEL, NITTIN R. "A Network Algorithm for Performing Fisher's Exact Test in r × c Contingency Tables." *Journal of the American Statistical Association* 78 (1983), pp. 427–34.

MEMON, M. V. "Estimation of the Shape and Scale Parameters of the Weibull Distribution." *Technometrics* 5 (1963), pp. 175–82.

MERRINGTON, MAXINE, AND THOMPSON, CATHERINE, M., "Tables of Percentage Points of the Inverted Beta (F) Distribution." *Biometrika* 33 (1943–46), pp. 73–88.

MITTEN, L. G., AND SANOH, A. "The $\bar{X}$ Warning Limit Chart." *Industrial Quality Control,* August 1961, pp. 15–19.

MOLENNAR, W. "Simple Approximations to the Poisson, Binomial, and Hypergeometric Distributions." *Biometrics* 29 (1973), pp. 403–7.

MOOD, A. M. "On the Dependence of Sampling Inspection Plans on Population Distributions." *Annals of Mathematical Statistics* 14 (1943), pp. 415–25.

MONTGOMERY, DOUGLAS C. "The Economic Design of Control Charts: A Renewal and Literature Survey." *Journal of Quality Technology* 12 (1980), pp. 75–87.

MOORE, P. G. "Some Properties of Runs in Quality Control Procedures." *Biometrika* 45 (1958), pp. 89–95.

MORRIS, CARL N. "Parametric Empirical Bayes Inference: Theory and Applications," with Comments and Rejoinder. *Journal of the American Statistical Association* 78 (1983), pp. 47–65.

MOSES, LINCOLN E. "Some Theoretical Aspects of the Lot Plot Sampling Inspection Plan." *Journal of the American Statistical Association* 51 (1956), pp. 84–107.

MOSTELLER, F. "Note on Application of Runs to Control Charts." *Annals of Mathematical Statistics* 12 (1941), p. 232.

______, AND TUKEY, J. W. "Practical Applications of New Theory: A Review. I. Location and Scale: Tables." *Industrial Quality Control,* September 1949, pp. 5–8.

______, AND ______. "The Uses and Usefulness of Binomial Probability Paper." *Journal of the American Statistical Association* 44 (1949), pp. 174–212.

MULHOLLAND, H. P. "On the Null Distribution of $\sqrt{b_1}$ for Samples of Size at Most 25, with Tables." *Biometrika* 64 (1977), pp. 401–09.

MULLET, GARY M. "Process Analysis Using the Runs Test." *Journal of Quality Technology* 9 (1977), pp. 1–5.

MUMFORD, A. G. "A Control Chart Based on Cumulative Scores." *Applied Statistics* 29 (1980), pp. 252–58.

MUNDEL, AUGUST B. "Group Testing." *Journal of Quality Technology* 16 (1984), pp. 181–88.

______. "Group Testing and the Effects of Cost Ratios." *American Society for Quality Control, 39th Annual Quality Congress Transactions,* pp. 563–68.

MURPHY, R. B., "Stopping Rules with CSP-1 Sampling Inspection Plans." *Industrial Quality Control* 16 (November 1959), pp. 10–16.

NAGEL, E. "Principles of the Theory of Probability." *International Encyclopaedia of Unified Science, Foundation of the Unity of Science* 1, no. 6 (1939).

NCUBE, M. M., AND WOODALL, W. H. "A Combined Shewhart-Cumulative Score Quality Control Chart." *Applied Statistics* 33 (1984), pp. 259–65.

NELSON, LLOYD S. "Nomograph for Two-Sided Distribution Free Tolerance Intervals." *Industrial Quality Control,* June 1963, pp. 11–12.

———. "Factors for the Analysis of Means." *Journal of Quality Technology* 6 (1974), pp. 175–81.

———. "Nomograph for Samples Having Zero Defectives." *Journal of Quality Technology* 10 (1978), pp. 42–43.

———. "A Simple Test for Normality." *Journal of Quality Technology* 13 (1981), pp. 76–77.

———. "Nomograph for Determining Variables Sampling Plans." *Journal of Quality Technology* 13 (1981), pp. 270–1.

———. "Control Charts for Medians." *Journal of Quality Technology* 14 (1982), pp. 226–27.

———. "An Early-Warning Test for Use with the Shewhart p Control Chart." Journal of Quality Technology 15 (1983), pp. 68–71.

NEWMAN, D. "The Distribution of the Range in Samples from a Normal Population, Expressed in Terms of an Independent Estimate of Standard Deviation." *Biometrika* 31 (1939), pp. 20–30.

OLDS, EDWIN G. "The Nature of the Standard Control Chart, and Some of Its Competitors." *Industrial Quality Control,* October 1956, pp. 3–8.

OLKIN, INGRAM, AND PRATT, JOHN W. "Unbiased Estimation of Certain Correlation Coefficients." *Annals of Mathematical Statistics* 29 (1958), pp. 201–11.

OLMSTEAD, P. S. "Distribution of Sample Arrangements for Runs Up and Down." *Annals of Mathematical Statistics* 17 (1946), pp. 24–33.

OTT, E. R., AND MUNDEL, A. B. "Narrow Limit Gaging." *Industrial Quality Control,* March 1954, pp. 21–28.

OWEN, DONALD B. "One-sided Variables Sampling Plans." *Industrial Quality Control* 22 (1966), pp. 450–56.

———. "Variables Sampling Plans Based on the Normal Distribution." *Technometrics* 9 (1967), pp. 417–23.

———. "Summary of Recent Work on Variables Acceptance Sampling with Emphasis on Non-Normality." *Technometrics* 11 (1969), pp. 631–37.

———, AND FRAWLEY, W. H. "Factors for Tolerance Limits Which Control Both Tails of the Normal Distribution." *Journal of Quality Technology,* April 1971, pp. 69–79.

———, AND CHOU, YOUN-MIN. "Effect of Measurement Error and Instrument Bias on Operating Characteristics for Variables Sampling Plans." *Journal of Quality Technology* 15 (1983), pp. 107–17.

PABST, W. R., JR. "Mil. Std. 105D." *Industrial Quality Control* 20 (November 1963), pp. 4–9.

——— (ed.). "Reviews of Standards and Specifications." *Journal of Quality Technol-*

ogy, April 1971, pp. 87–94. (Discussion of Switching Rules of JIS Z9015 and Mil. Std. 105D.)

PACHORES, JAMES. "Tables of Confidence Limits for the Binomial Distribution." *Journal of the American Statistical Association* 55 (1960), pp. 521–23.

PAGANO, MARCELLO, AND HALVORSEN, KATHERINE TAYLOR. "An Algorithm for Finding the Exact Significance Levels of r × c Contingency Tables." *Journal of the American Statistical Association* 76 (1981), pp. 931–41.

PAGE, E. S. "Continuous Inspection Schemes." *Biometrika* 41 (1954), pp. 100–14.

______. "Control Charts with Warning Lines." *Biometrika* 42 (1955), pp. 243 ff.

______. "On Problems in Which a Change in a Parameter Occurs at an Unknown Point." *Biometrika* 44 (1957), pp. 248–52.

______. "Pseudo-Random Elements for Computers." *Applied Statistics* 8 (1959), pp. 124–31.

______. "Cumulative Sum Charts." *Technometrics* 3 (1961), pp. 1–9.

______. "Controlling the Standard Deviation by Cusums and Warning Lines." *Technometrics* 5 (1963), pp. 307–15.

PALUMBO, FRANK A., AND STRUGALA, EDWARD S. "Fraction Defective of Battery Adapter Used in Handie-Talkie." *Industrial Quality Control,* November 1945, pp. 6–8.

PAPADAKIS, EMMANUAL P. "The Deming Inspection Criterion for Choosing Zero or 100 Percent Inspection." *Journal of Quality Technology* 17 (1985), pp. 121–27.

"Papers on the Analysis of Covariance." *Biometrics* 38 (1982), pp. 540–747.

PATNAIK, P. B. "The Non-Central χ^2 and F-Distributions and Their Applications." *Biometrika* 36 (1949), pp. 202–32.

______. "The Use of Mean Range as an Estimator of Variance in Statistical Tests," *Biometrika* 37 (1950), pp. 78–87.

PAULL, A. E. "On a Preliminary Test for Pooling Mean Squares in the Analysis of Variance." *Annals of Mathematical Statistics* 21 (1950), pp. 539–56.

PEACH, PAUL. "A Comparison of Acceptance Inspection Plans." *Industrial Quality Control,* May 1945, pp. 11–14.

PEARSON, E. S. "A Survey of the Uses of Statistical Method in the Control and Standardization of the Quality of Manufactured Product." *Journal of the Royal Statistical Society* A 96 (1933), pp. 21–60.

______. "Charts of the Power Function for Analysis of Variance Tests, Derived from the Non-Central F-Distribution." *Biometrika* 38 (1951), pp. 112–30.

______. "Tables of Percentage Points of $\sqrt{b_1}$ and b_2 in Normal Samples; a Rounding Off." *Biometrika* 52 (1965), pp. 282–85.

______. "Some Historical Reflections on the Introduction of Statistical Methods in Industry." *The Statistician* 22 (1973), pp. 165–79.

______; D'AGOSTINO, R. B.; AND BOWMAN, K. O. "Tests for Departure from Normality: Comparison of Powers." *Biometrika* 64 (1977), pp. 231–46.

PETTITT, A. N., AND STEPHENS, M. A. "The Kolmogorov-Smirnov Goodness-of-

Fit Statistic with Discrete and Grouped Data." *Technometrics* 19 (1977), pp. 207–10.

PFANZAL, J. "Sampling Procedures Based on Prior Distribution and Cost." *Technometrics* 5 (1963), pp. 47–61.

PITMAN, E. J. G. "Significance Tests That May Be Applied to Samples from Any Populations. III. The Analysis of Variance Tests." *Biometrika* 29 (1937), pp. 322–35.

PROSCHAN, FRANK. "Rejection of Outlying Observations." *American Journal of Physics* 21 (1953), pp. 520–25.

______, AND SAVAGE, I. R. "Starting a Control Chart. The Effect of Number and Size of Samples on the Level of Significance at the Start of a Control Chart for Sample Means." *Industrial Quality Control,* September 1960, pp. 12–13.

PURCELL, WARREN R. "Balancing and Randomizing in Experiments." *Industrial Quality Control,* January 1951, pp. 7–14.

RAO, PODURI S. R. S.; KAPLAN, JACK; AND COCHRAN, WILLIAM G. "Estimators for the One-Way Random Effects Model with Unequal Error Variances." *Journal of the American Statistical Association* 76 (1981), pp. 89–97.

READ, D. R. AND BEATTIE, D. W. "The Variable Lot-Size Acceptance Sampling Plan for Continuous Production." *Applied Statistics* 10 (1961), pp. 147–56.

RESNIKOFF, GEORGE L. *A New Two-Sided Acceptance Region for Sampling by Variables.* App. Math. and Statistics Lab., Stanford Univ. (1952).

REYNOLDS, J. H. "THE RUN SUM CONTROL CHART PROCEDURE." *Journal of Quality Technology,* January 1971, pp. 23–27.

ROBERTS, S. W. "Properties of Control Chart Zone Tests." *The Bell System Technical Journal* 37 (1958), pp. 83–114.

______. "Control Chart Tests Based on Geometric Moving Averages." *Technometrics* 1 (1959), pp. 239–50.

ROBINSON, P. B., AND HO, T. Y. "Average Run Lengths of Geometric Moving Average Charts by Numerical Methods." *Technometrics* 20 (1978), pp. 85–93.

ROBSON, D. S., AND ATKINSON, G. F. "Individual Degrees of Freedom for Testing Homogeneity of Regression Coefficients in a One-Way Analysis of Covariance." *Biometrics* 16 (1960), pp. 593–605.

ROMIG, H. G. "New Statistical Approaches in Aviation and Allied Fields." *Proceedings, Aircraft Quality Control Conference, November 1953,* Aircraft Technical Committee, American Society for Quality Control.

ROSS, ELIZABETH G. "Response-Surface Techniques as a Statistical Approach to Research and Development in Ultrasonic Welding." *Annual Convention Transactions,* American Society for Quality Control, 1961, pp. 445–56.

ROY, S. N. "Some Further Results in Simultaneous Confidence Interval Estimation." *Annals of Mathematical Statistics* 25 (1954), pp. 752–61.

______, AND BOSE, R. C. "Simultaneous Confidence Interval Estimation." *Annals of Mathematical Statistics* 24 (1953), pp. 513–36.

ROYSTON, J. P. "An Extension of Shapiro and Wilk's W Test for Normality to Large Samples." *Applied Statistics* 31 (1982), pp. 115–24.

RUSSELL, T. S., AND BRADLEY, R. A. "One-Way Variance in a Two-Way Classification." *Biometrika* 45 (1958), pp. 111–29.

SACKROWITZ, HAROLD. "A Note on Unrestricted AOQLs." *Journal of Quality Technology* 7 (1975), pp. 77–80.

SAMIUDDIN, M., AND ATIQUILLAH, M. "A Test for Equality of Variances." *Biometrika* 63 (1976), pp. 206–08.

———. "Bayesian Test of Homogeneity of Variance." *Journal of the American Statistical Association* 71 (1976), pp. 515–17.

———; HANIF, M.; AND ASAD, H. "Some Comparisons of the Bartlett and Cube Root Tests of Homogeneity of Variance." *Biometrika* 65 (1978), pp. 218–21.

SATTERTHWAITE, F. E. "Synthesis of Variance." *Psychometrika* 6 (1941), pp. 309–16.

———. "An Approximate Distribution of Estimates of Variance Components." *Biometrics Bulletin* 2 (1946), pp. 110–14.

———. "Comparison of Two Fractions Defective." *Industrial Quality Control,* November 1956, pp. 17–18.

———. "New Developments in Experimental Design." *Proceedings of the All Day Conference on Quality Control,* American Society for Quality Control and Rutgers University, 1956, pp. 55–57.

———. "Binomial and Poisson Confidence Limits." *Industrial Quality Control,* May 1957, pp. 56–59.

———. "Random Balance Experimentation." *Technometrics* 1 (1959), pp. 111–37.

SAVAGE, I. RICHARD. "A Production Model and Continuous Sampling Plan." *Journal of the American Statistical Association* 54 (1959), pp. 231–47.

SCHEFFÉ, HENRY. "Operating Characteristics of Average and Range Charts." *Industrial Quality Control,* May 1949, pp. 13–18.

———. "A Method for Judging All Contrasts in the Analysis of Variance." *Biometrika* 40 (1953), pp. 87–104.

———. "A 'Mixed Model' for the Analysis of Variance." *Annals of Mathematical Statistics* 27 (1956), pp. 23–36.

———. "Alternative Models for the Analysis of Variance." *Annals of Mathematical Statistics* 27 (1956), pp. 251–71.

SCHILLING, EDWARD G. "A Systematic Approach to the Analysis of Means: Part I. Analysis of Treatment Effects." *Journal of Quality Technology* 5 (1973), pp. 93–108.

———. "A Systematic Approach to the Analysis of Means, Part II. Analysis of Contrasts; Part III. Analysis of Non-Normal Data." *Journal of Quality Technology* 5 (1973), pp. 147–59.

———. "Revised Attributes Acceptance Sampling Standard—ANSI/ASQC Z1.4 (1981)." *Journal of Quality Technology* 14 (1982), pp. 215–19.

———. "An Overview of Acceptance Control." American Society for Quality Control, *37th Annual Quality Congress Transactions* (1983), pp. 64–69.

———, AND DODGE, H. F. "Procedures and Tables for Evaluating Dependent Mixed Acceptance Sampling Plans." *Technometrics* 11 (1969), pp. 341–72.

———, AND JOHNSON, LUCILLE L. "Tables for the Construction of Matched Single, Double and Multiple Sampling Plans With Application to Mil. Std. 105D." *Journal of Quality Technology* 12 (1980), pp. 220–29.

SCHWARTZ, DAVID H., AND KAUFMAN, PAUL. "Determining Conformance with Requirements for Lot Average." *Proceedings, Annual Middle Atlantic Regional Conference,* American Society for Quality Control, February 16 and 17, 1951, pp. 114–25. (This article is reprinted in *Industrial Quality Control,* March 1951, pp. 18–21. The original charts were in error, and corrected forms appear in *Industrial Quality Control,* July 1951, pp. 30–31.)

SCHWARZ, GIDEON. "A Class of Factorial Designs with Unequal Cell-Frequencies." *Annals of Mathematical Statistics* 31 (1960), pp. 749–55.

SEDER, LEONARD. "The Technique of Experimenting in the Factory." *Industrial Quality Control,* March 1948, pp. 6–14.

———. "Diagnosis with Diagrams." Part I and II. *Industrial Quality Control,* January 1950, pp. 11–19; and March 1950, pp. 7–11.

SHAHANI, A. K. "Wald-Wolfowitz Type Sampling Plans for Continuous Production." *Technometrics* 21 (1979), pp. 21–31.

SHAININ, DORIAN. "The Hamilton Standard Lot Plot Method of Acceptance Sampling by Variables." *Industrial Quality Control,* July 1950, pp. 15–34.

———. "Recent Lot Plot Experiences Around the Country." *Industrial Quality Control,* March 1952, pp. 20–29.

———. "The Hamilton Standard Reset-Run Card." *Industrial Quality Control,* September 1952, pp. 12–16.

SHAFER, J. L. "Sampling Incoming Material." *Industrial Quality Control,* September 1946, pp. 16–20.

SHAPIRO, S. S. AND WILK, M. B. "An Analysis of Variance Test of Normality (Complete Samples)." *Biometrika* 52 (1965), pp. 591–611.

———, ———, AND CHEN, H. J. "A Comparative Study of Various Tests of Normality." *Journal of the American Statistical Association* 63 (1968), pp. 1343–72.

SHEWHART, W. A. "Application of Statistical Method in Mass Production." *Proceedings of the Industrial Statistics Conference Held at Massachusetts Institute of Technology,* September 8–9, 1938. New York: Pitman Publishing, 1939.

SHUKLA, G. K. "Testing the Homogeneity of Variances in a Two-Way Classification." *Biometrika* 69 (1982), pp. 411–16.

———. "Some Exact Tests of Hypotheses about Grubbs Estimators." *Biometrics* 29 (1973), pp. 373–77.

SIMON, LESLIE E. "On the Relation of Instrumentation to Quality Control." *Instruments* 19 (1946), pp. 654–56.

———, AND COHEN, A. CLIFFORD, JR. "Quality Control at Picatinny Arsenal, 1934–1945." *American Society for Quality Control, Annual Technical Conference Transactions, 1971,* pp. 173–79.

SINGH, G. "On the Distribution of Range of Samples from Non-Normal Populations." *Biometrika* 57 (1970), pp. 451–56.

SNEE, RONALD D. "Graphical Analysis of Process Variation Studies." *Journal of Quality Technology* 15 (1983), pp. 76–88.

SOUNDARARAJAN, V. "Procedures and Tables for Construction and Selection of Chain Sampling Plans (Ch SP-1) Part 1." *Journal of Quality Technology* 10 (1978), pp. 56–60; and Part 2: "Tables for Selection of Chain Sampling Plans." *Journal of Quality Technology* 10 (1978), pp. 99–103.

———, AND GOVINDARAJU, K. "Construction and Selection of Chain Sampling Plans ChSP (0,1)." *Journal of Quality Technology* 15 (1983), pp. 180–85.

SPIEGELHALTER, D. J. "A Test for Normality Against Symmetric Alternatives." *Biometrika* 64 (1977), pp. 415–18.

SRIVASTASA, A. B. L. "Effect of Non-Normality on the Power Function of t-Tests." *Biometrika* 45 (1958), pp. 421 ff.

———. "Effect of Non-Normality on the Power of the Analysis of Variance Test." *Biometrika* 46 (1959), pp. 114–22.

STECK, G. P., AND OWEN, D. B. "Percentage Points for the Distribution of Outgoing Quality." *Journal of the American Statistical Association* 54 (1959), pp. 689–94.

STEINBERG, DAVID M., AND HUNTER, WILLIAM G. "Experimental Design: Review and Comment," with Discussion and Response. *Technometrics* 36 (1983), pp. 71–130.

STEPHENS, KENNETH S. "CSP-1 for Consumer Protection." *Journal of Quality Technology* 13 (1981), pp. 249–53.

———, AND LARSON, K. E. "An Evaluation of the Mil. Std. 105D System of Sampling Plans." *Industrial Quality Control,* January 1967, pp. 310–19.

———, AND DODGE, H. F. "Comparison of Chain Sampling Plans with Single and Double Sampling Plans." *Journal of Quality Technology* 8 (1976), pp. 24–33.

STEPHENS, LARRY JOE. "A Closed Form Solution for Single Sample Acceptance Sampling Plans." *Journal of Quality Technology* 10 (1978), pp. 159–63.

STEVENS, W. L. "Distribution of Groups in a Sequence of Alternatives." *Annals of Eugenics* 9, part 7 (1939), pp. 10–17.

———. "Control by Gauging." *Journal of the Royal Statistical Society* B 10 (1948), pp. 54–108.

STEWART, D. C. "Calibrating the Accident Barometer." *National Safety News,* November 1948, pp. 36 ff.

SUBRAHMANIAM, KATHLEEN, AND SUBRAHMANIAM, KOCHERLAKOTA. "Some Extensions to Miss F. N. David's Tables of Sample Correlation Coefficient: Distribution Function and Percentiles." *Sankhyā* B 45 (1983), pp. 75–147.

SWAN, A. W. "The χ^2 Significance Test—Expected vs. Observed Results." *The Engineer,* December 31, 1948, pp. 679 ff.

SWED, S., AND EISENHART, C. "Tables for Testing Randomness of Sampling in a Sequence of Alternatives." *Annals of Mathematical Statistics* 14 (1943), pp. 66–87.

TANG, P. C. "The Power Function of the Analysis of Variance Tests with Tables and Illustrations of Their Use." University of London, Department of Statistics. *Statistical Research Memoirs* 2 (1938), pp. 126 ff.

Technometrics. May 1965. (Whole issue devoted to multiple comparisons.)

THOMPSON, W. A., JR. "The Problem of Negative Estimates of Variance." *Annals of Mathematical Statistics* 33 (1962), pp. 273–89.

______. "Precision of Simultaneous Management Procedures," *Journal of the American Statistical Association* 58 (1963), pp. 474–79.

TIETJEN, G. L. AND MOORE, R. H. "Some Grubbs-Type Statistics for Detection of Several Outliers." *Technometrics* 14 (1972), pp. 583–97.

TIPPETT, L. H. C. "A View of Quality Control in the United Kingdom." *Industrial Quality Control,* September 1962, pp. 15–17.

TOMSKY, J. L.; NAKANO, K.; AND IWASHIKA, M. "Confidence Limits for the Number of Defectives in a Lot." *Journal of Quality Technology* 11 (1979), pp. 199–204.

TOOTILL, J. P. R. "Routine Least Squares Estimations from Models Containing a Single Non-Linearity." *Biometrics* 19 (1963), pp. 118–43.

TRUAX, H. MACK. "Cumulative Sum Charts and Their Application to the Chemical Industry." *Industrial Quality Control,* December 1961, pp. 18–25.

______. "An Introduction to Cumulative Sum Charts." *Annual Convention Transactions,* American Society for Quality Control, 1961, pp. 103–12.

TUKEY, JOHN W. "Quick and Dirty Methods in Statistics, II. Simple Analyses of Standard Designs." *Quality Control Conference Papers 1951, Fifth Annual Convention,* American Society for Quality Control, May 23–24, 1951, pp. 189–97.

______. "Conclusions vs. Decisions." *Technometrics* 2 (1960), pp. 423–33.

VAGHOLHAR, M. K., AND WETHERILL, G. R. "The Most Economical Binomial Sequential Probability Ratio Test." *Biometrika* 47 (1960), pp. 103–9.

VANCE, LONNIE C., AND MCDONALD, GARY C. "A Class of Multiple Run Sampling Plans." *Technometrics* 21 (1979), pp. 141–46.

______. "A Bibliography of Statistical Quality Control Chart Techniques, 1970–1980." *Journal of Quality Technology* 15 (1983), pp. 59–62.

VAN DER WAERDEN, B. L. "Sampling Inspection as a Minimum Loss Problem." *Annals of Mathematical Statistics* 31 (1960), pp. 369–84.

VAN REST, EDWARD D. "Quality Control in the U.S.A." *Applied Statistics* 2 (1953), pp. 141–51.

WALD, A., AND WOLFOWITZ, J. "Sampling Inspection Plans for Continuous Production Which Insure a Prescribed Limit on the Outgoing Quality." *Annals of Mathematical Statistics* 16 (1945), pp. 30–49.

______, AND ______. "Optimum Character of the Sequential Probability Ratio Test." *Annals of Mathematical Statistics* 19 (1948), pp. 326–39.

______, AND ______. "On a Test Whether Two Samples Are from the Same Population." *Annals of Mathematical Statistics* 11 (1949), pp. 147–62.

WATSON, G. S. "On Chi-Square Goodness-of-Fit Tests for Continuous Distributions," with Discussion. *Journal of the Royal Statistical Society* B 20 (1958), pp. 44–72.

______. "The χ^2 Goodness-of-Fit Test for Normal Distributions," *Biometrika* 44 (1957), pp. 336–48.

______. "Some Recent Results in Chi-Square Goodness-of-Fit Tests." *Biometrics* 15 (1959), pp. 440–68.

WEAVER, WADE R. "Quality Control's March of Progress." *Proceedings, Annual Middle Atlantic Regional Conference,* American Society for Quality Control, February 16 and 17, 1951, pp. 8–13.

WEAVER, WARREN. "Probability, Rarity, Interest and Surprise." *The Scientific Monthly* 67 (1948), pp. 390–92.

WEISS, LIONEL. "On Sequential Tests Which Minimize the Maximum Expected Sample Size." *Journal of the American Statistical Association* 57 (1962), pp. 551–66.

WELCH, B. L. "A Generalization of 'Student's' Problem When Several Different Population Variances Are Involved." *Biometrika* 34 (1947), pp. 28–35.

WESCOTT, MASON E. "Attribute Charts in Quality Control." *Conference Papers, First Annual Convention, American Society for Quality Control and Second Midwest Quality Control Conference, June 5 and 6, 1947,* pp. 49–80. Chicago: John S. Swift Co., 1947.

WETHERILL, G. B. "Some Remarks on the Bayesian Solutions of the Single Sampling Inspection Scheme." *Technometrics* 2 (1960), pp. 341 ff.

———. "Bayesian Sequential Analysis." *Biometrika* 48 (1961), pp. 281–92.

———, AND CAMPLING, G. E. G. "The Decision Theory Approach to Sampling Inspection." *Journal of the Royal Statistical Society* B 28 (1966), pp. 381–406.

———, AND CHIU, W. K. "A Simplified Attribute Sampling Scheme." *Applied Statistics* 23 (1974), pp. 143–48.

———, AND OFOSU, J. B. "Selection of the Best of k Normal Populations." *Applied Statistics* 23 (1974), pp. 253–77. (Contains an extensive bibliography.)

WHEELER, DONALD F. "Detecting a Shift in Process Average: Tables of the Power Function for $\bar{X}$ Charts." *Journal of Quality Technology* 15 (1983), pp. 155–70.

WHITE, JOHN S. "A New Graph for Determining CSP-1 Sampling Plans." *Industrial Quality Control* 17 (May 1961), pp. 18–19.

WILLIAMS, E. J. "Regression Methods in Calibrations." *Technometrics* 11 (1969), pp. 189–92.

WILLIAMS, WILLIAM W.; LOONEY, STEPHEN W.; AND PETERS, MICHAEL H. "Use of Curtailed Sampling Plans in the Economic Design of np-Control Charts." *Technometrics* 27 (1985), pp. 57–63.

WILSDON, B. H. "Discrimination by Specification Statistically Considered and Illustrated by the Standard Specification for Portland Cement." *Journal of the Royal Statistical Society* B 1 (1934), pp. 152–92, with appendixes by E. S. Pearson and F. E. James.

WINTER, R. F. "A Simplified Method for Solving Normal Equations." *Industrial Quality Control,* October 1961, pp. 14–16.

WINTERHALTER, ALFRED J. "Development of Reject Limits for Measurements." *Industrial Quality Control,* January 1945, pp. 12–15; and March 1945, pp. 12–13.

WISHART, JOHN, AND METAKIDES, THEOCHARIS. "Orthogonal Polynomial Fitting." *Biometrika* 40 (1953), pp. 361–69.

WOODING, WILLIAM M. "The Split-Plot Design." *Journal of Quality Technology* 5 (1973), pp. 16–33.

WOODS, RAYMOND F. "Effective, Economic Quality Through the Use of Acceptance Control Charts." *Journal of Quality Technology* 8 (1976), pp. 81–85.

WORTHAN, A. W., AND MOGY, J. W. "A Technical Note on Average Outgoing Quality." *Journal of Quality Technology* 2 (1970), pp. 30–31.

YANG, GRACE L. "A Renewal-Process Approach to Continuous Sampling Plans." *Technometrics* 25 (1983), pp. 59–67.

YATES, F. "The Analysis of Multiple Classifications with Unequal Numbers in the Different Classes." *Journal of the American Statistical Association* 29 (1934), pp. 51–66.

YOUDEN, W. J. "Graphical Analysis of Interlaboratory Test Results." *Industrial Quality Control,* May 1959, pp. 24–28.

______; KEMPTHORNE, O.; TUKEY, J. W.; BOX, G. E. P.; AND HUNTER, J. S. "Discussion of the Papers of Messrs. Satterthwaite and Budne." *Technometrics* 1 (1959), pp. 157 ff.

ZIMMER, WILLIAM J., AND BURR, IRVING W. "Variables Sampling Plans Based on Non-Normal Populations." *Industrial Quality Control,* July 1963, pp. 18–36.

APPENDIX VI

A List of Computer Programs Selected from Those Reviewed in the *Journal of Quality Technology**

1. *Sample Frequency Distribution Analysis*

   Summarization of Data (Pertains to analysis of frequency distribution sample data), Vol. 1 (1969), p. 68.

2. *Probability Distributions*

   Machine Plotting of Probability Charts, Vol. 5 (1973), p. 135.

   Cumulative Probabilities for the Standard Normal, χ^2 , F and t Distributions, Vol. 16 (1984), p. 232.

3. *Sampling Plans*

   Single Sampling Plans Given an AQL, LTPD, Producer, and Consumer Risks, Vol. 4 (1972), p. 168.

   Minimum Sample Size Single Sampling Plans, Vol. 12 (1980), p. 230.

   Double Sampling Plans (Derivation and reporting of operating characteristics), Vol. 4 (1972), p. 205.

   An Algorithm for Determining Double Attribute Sampling Plans, Vol. 14 (1982), p. 166.

   Solution of Multiple Sampling Plans, Vol. 5 (1973), p. 39.

   Wald Sequential Sampling for Attribute Inspection, Vol. 16 (1984), p. 172.

   ASN of Curtailed Attributes Sampling Plans, Vol. 17 (1985), p. 108.

   Mil. Std. 414, Vol. 9 (1977), p. 82.

   Determination of Rectifying Inspections Plans for Single Sampling by Attributes, Vol. 16 (1984) p. 56.

   Dodge's Continuous Sampling Plans, Vol. 7 (1975), p. 43.

   GRASP: A General Routine for Attribute Sampling Plan Evaluation, Vol. 10 (1978), p. 125.

4. *Control Charts*

   Plotting p and np charts, Vol. 1 (1969), p. 217.

   Plotting c- and u-charts, Vol. 1 (1969), p. 285.

   Plotting $\bar{X}$- and R-Charts, Vol. 1 (1969), p. 149.

   Economic Design of $\bar{X}$-Control Charts, Vol. 14 (1982), p. 40.

   Plotting Cumulative Sum Charts, Vol. 2 (1970), p. 54.

   Exponentially Smoothed Average Control Charts (A program for plotting), Vol. 5 (1973), p. 84.

---

* The author has not personally checked the computer programs listed in Appendix VI. It is a good rule in general to check a program that is to be used by initially applying it in the solution of a problem for which the answer is already known.

It should also be noted that the March 1986 issue of *Quality Progress* contains a directory of Quality Assurance and Quality Control Software commercially available as of that date.

5. *Estimation and Testing Hypotheses*

Randomization Tests for K Sample Binomial Data, Vol. 14 (1982), p. 220.

Student's t-Test (For a single mean or difference between two means), Vol. 2 (1970), p. 243.

Computation of a Two-Tailed Fisher's Test, Vol. 11 (1979), p. 44.

A Small Sample Test for Non-Normality, Vol. 11 (1979), p. 95.

Testing for Normality, Vol. 15 (1983), p. 141.

6. *Analysis of Variance, Analysis of Means, and Design of Experiments*

Analysis of Variance and Analysis of Means, Vol. 12 (1980), p. 106.

Analysis of Means (Comparison of K Samples Involving Variables or Attributes Data), Vol. 12 (1980), p. 47.

Factoral Experiments (Program for Analysis of Two-Level Factorial Experiments), Vol. 14 (1982), p. 95.

Analysis of Means for Balanced Experimental Designs, Vol. 15 (1983), p. 45.

Simultaneous Pairwise Comparison Tests among Treatment Means, Vol. 13 (1981), p. 65.

Randomized Complete Block Design (Analysis of data), Vol. 10 (1978), p. 40.

Replicated Randomized Complete Block Design (Program for analysis of data), Vol. 10 (1978), p. 84.

Variance Components for Unbalanced N-Level Hierarchic Designs, Vol. 7 (1973), p. 144.

Variance Estimation Using Staggered, Nested Design, Vol. 15 (1983), p. 195.

Latin Square (Analysis of Variance of an n × n L.S., with a Subroutine for Duncan's New Multiple Range Test), Vol. 7 (1975), p. 90.

A Run Test for Sample Non-Randomness, Vol. 7 (1975), p. 196.

The Two-Sample Kolmogorov-Smirnov Test, Vol. 13 (1982), p. 139.

7. *Correlation and Regression*

Scatter Plots, Vol. 3 (1971), p. 38.

The Coefficient of Correlation, Vol. 3 (1971), p. 95.

Simple Linear Regression, Vol. 3 (1971), p. 138.

Multiple Linear Regression, Vol. 3 (1971), p. 184.

Doolittle Technique, Vol. 6 (1974), p. 160.

Interior Analysis of the Observations in Multiple Linear Regression, Vol. 12 (1980), p. 165.

Constructing Orthogonal Polynomials when the Independent Variable is Unequally Spaced, Vol. 6 (1974), p. 113.

Orthogonal Polynomial Regression for Unequal Spacing and Frequencies, Vol. 10 (1978), p. 170.

8. *Special Computational Procedures*

Lagrange Interpolation, Vol. 2 (1970), p. 174.

A Sequential Simplex Program for Solving Minimization Problems, Vol. 6 (1974), p. 53.

INDEX

This book has been set VideoComp in 10 and 9 point Times Roman, leaded 2 points. Part and chapter numbers are set in 30 point Avant Garde Book; part and chapter titles are set in 18 point Avante Garde Book. The size of the type page is 27 by 47 picas.